Nonvertical line, slope m

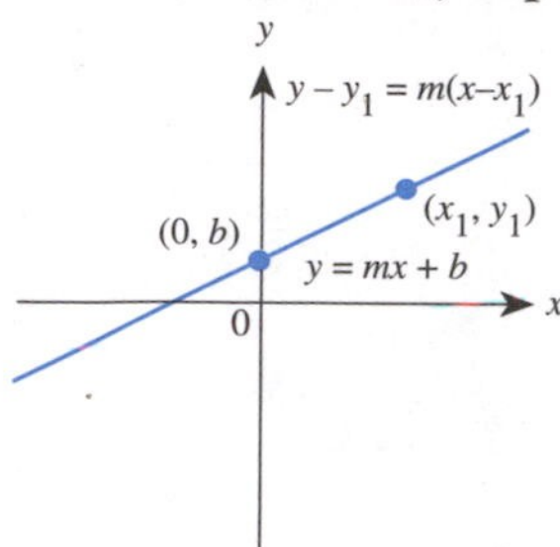

Horizontal line, vertical line

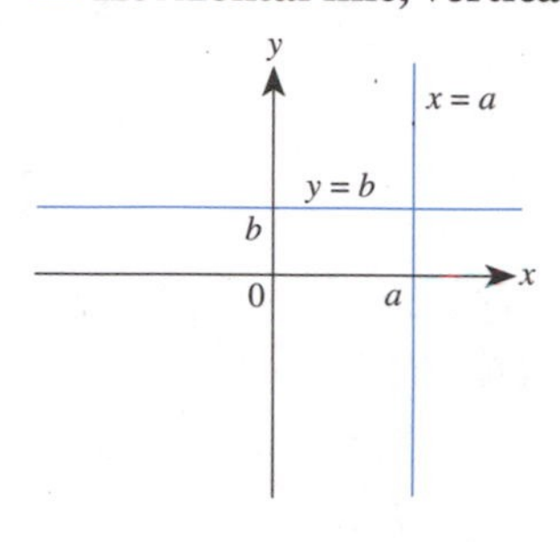

Distance formula, slope m of a nonvertical line

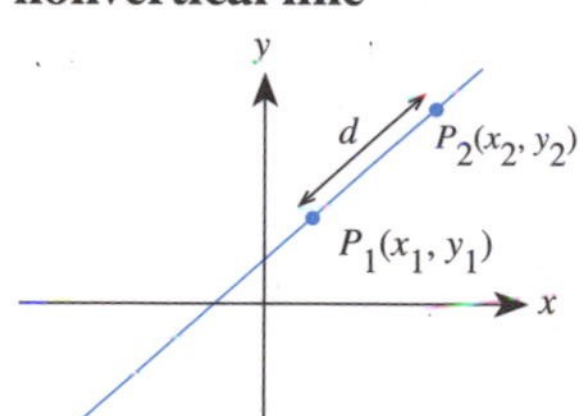

$$d = \sqrt{(x_2 - x_1)^2 + (y_2 - y_1)^2}$$

$$m = \frac{y_2 - y_1}{x_2 - x_1}$$

$y = x^2$

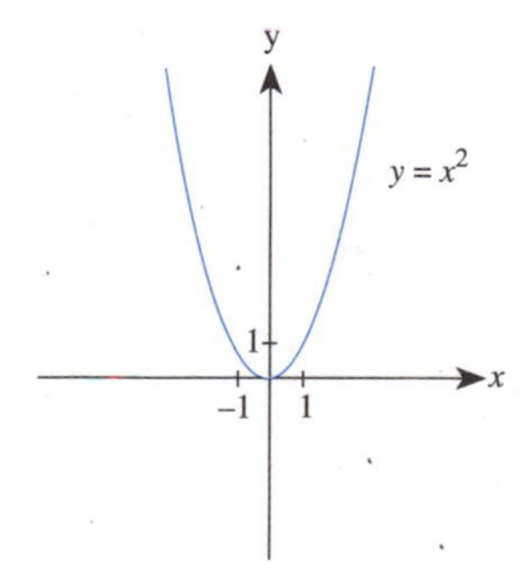

$y = x^3$

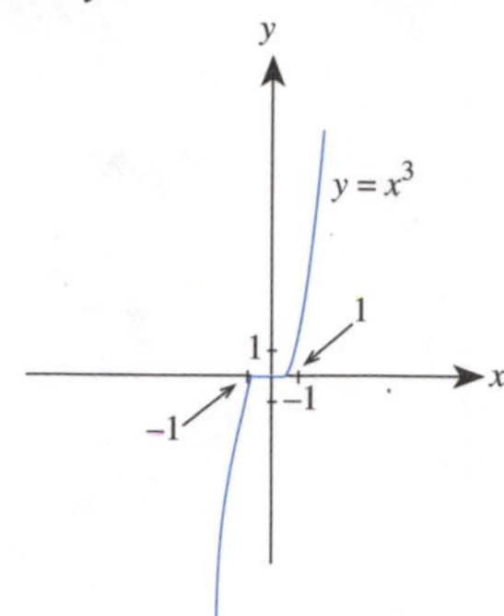

$y = \sqrt{x}$

y

$y = \sqrt{x}$

1

1 x

$y = b^x, b > 0, b \neq 1$

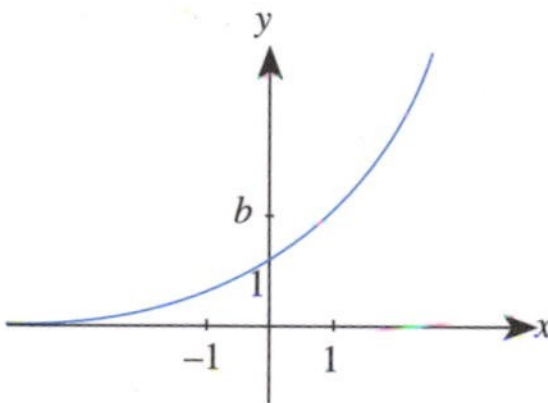

$y = \log_b x, b > 1$

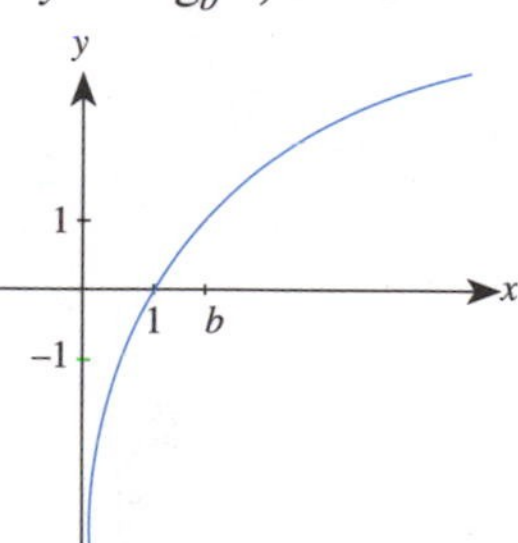

$(x-h)^2 + (y-k)^2 = r^2$

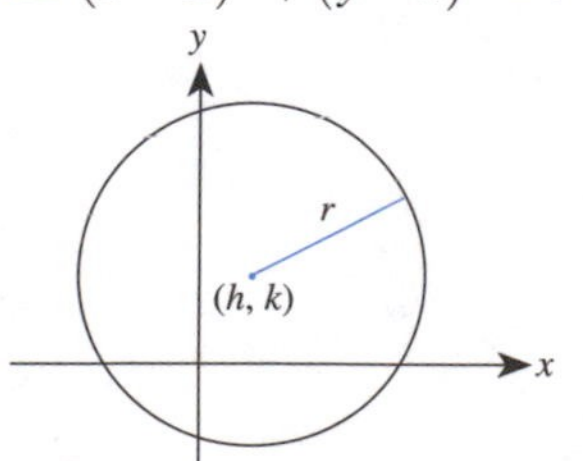

$4p(y-k) = (x-h)^2, p > 0$

$\dfrac{(x-h)^2}{a^2} + \dfrac{(y-k)^2}{b^2} = 1, a > b$

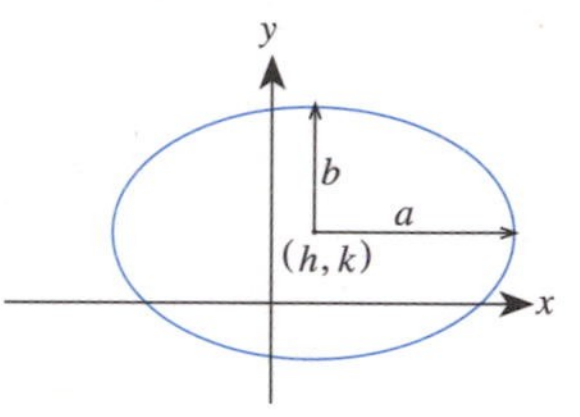

$\dfrac{(x-h)^2}{a^2} - \dfrac{(y-k)^2}{b^2} = 1$

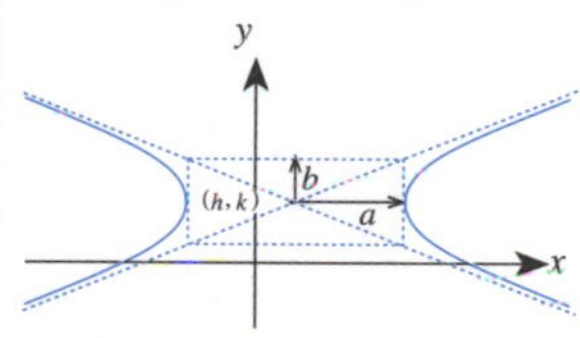

$y = \sin x$

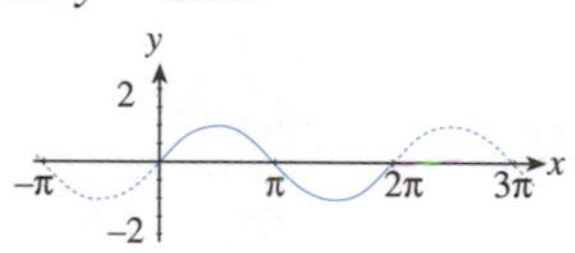

$y = \cos x$

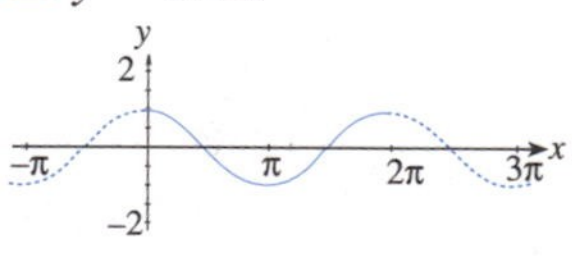

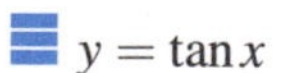

$y = \tan x$

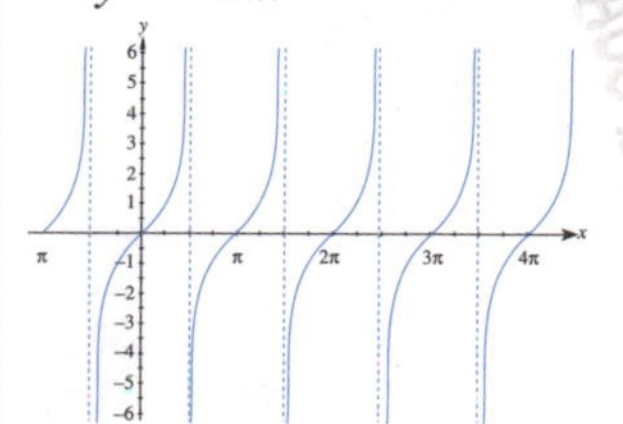

Technical Calculus

WITH ANALYTIC GEOMETRY

Technical Calculus

WITH ANALYTIC GEOMETRY

John C. Peterson

Chattanooga State Technical Community College

I(T)P™ An International Thomson Publishing Company

Albany • Bonn • Boston • Cincinnati • Detroit • London • Madrid • Melbourne
Mexico City • New York • Pacific Grove • Paris • San Francisco • Singapore • Tokyo
Toronto • Washington

NOTICE TO THE READER

Publisher does not warrant or guarantee any of the products described herein or perform any independent analysis in connection with any of the product information contained herein. Publisher does not assume, and expressly disclaims, any obligation to obtain and include information other than that provided to it by the manufacturer.

The reader is expressly warned to consider and adopt all safety precautions that might be indicated by the activities described herein and to avoid all potential hazards. By following the instructions contained herein, the reader willingly assumes all risks in connection with such instructions.

The publisher makes no representations or warranties of any kind, including but not limited to, the warranties of fitness for particular purpose or merchantability, nor are any such representations implied with respect to the material set forth herein, and the publisher takes no responsibility with respect to such material. The publisher shall not be liable for any special, consequential, or exemplary damages resulting, in whole or in part, from the readers' use of, or reliance upon, this material.

Cover photo: Gary Conner

Delmar Staff
Publisher: Robert D. Lynch
Acquisitions Editor: Paul Shepardson
Production Manager: Larry Main
Art and Design Coordinator: Nicole Reamer

Printed in the United States of America

For more information, contact:

Delmar Publishers Inc.
3 Columbia Circle, Box 15015
Albany, NY 12212-5015

International Thomson Publishing Europe
Berkshire House 168-173
High Holborn
London, WC1V 7AA
England

Thomas Nelson Australia
102 Dodds Street
South Melbourne, 3205
Victoria, Australia

Nelson Canada
1120 Birchmont Road
Scarborough, Ontario
Canada, M1K 5G4

International Thomson Editores
Campos Eliseos 385, Piso 7
Col Polanco
11560 Mexico D F Mexico

International Thomson Publishing GmbH
Konigswinterer Strasse 418
53227 Bonn
Germany

International Thomson Publishing Asia
221 Henderson Road
#05-10 Henderson Building
Singapore 0315

International Thomson Publishing—Japan
Hirakawacho Kyowa Building, 3F
2-2-1 Hirakawacho
Chiyoda-ku, Tokyo 102
Japan

1 2 3 4 5 6 7 8 9 10 XXX 02 01 00 99 98 97

Library of Congress Cataloging-in-Publication Data

Peterson, John C. (John Charles), 1939-
Technical calculus with analytic geometry / John C. Peterson.
p. cm.
Includes indexes.
ISBN 0-8273-7415-1
1. Mathematics. I. Title.
QA37.2.P48 1996
510--dc20
95-13516
CIP

Contents

Preface

Technical Calculus with Analytic Geometry is designed to be easily read and understood by students who are preparing for technical or scientific careers. Mathematics can be a difficult subject to read or study, especially when the math terms become tangled with the technical terms used in the applications. This text offers clear, readable development of the math concepts. But learning mathematics takes more than reading—it requires extensive practice. Therefore, *Technical Calculus with Analytic Geometry* provides abundant opportunities to practice problem solving. Together, these two essential elements have formed our constant focus in the development of this text: thorough, uncomplicated discussions followed by examples and problem sets that draw on real-world applications of mathematics.

Organization and Approach

This text retains the same organization and approach as the highly successful *Technical Mathematics with Calculus*. The first two chapters of this text provide a brief review of graphing in the real number system, using graphing calculators, and plane alalytic geometry. Much of the material in these chapters may be familiar to students and may be covered quickly or skipped entirely. Chapters 3–15 are devoted to calculus and differential equations.

Practical applications of mathematics in the technical and scientific areas are provided throughout the text. Students also receive ample opportunities to solve problems using scientific calculators, graphing calculators, and microcomputers.

This textbook strives to make the reading and comprehension of its contents easier for students. The uncomplicated writing style makes difficult discussions easier to grasp. Each important concept is supported by numerous examples, applications, and practice problems

Features

The following major themes have guided the development of *Technical Calculus with Analytic Geometry*:

Standards

Throughout this book, we have attempted to embrace the curriculum standards adopted by the National Council of Teachers of Mathematics (NCTM) and the American Mathematical Association for Two-Year Colleges (AMATYC). In particular, in the text and its supplements, more emphasis has been placed on mathematics as problems solving, modeling, communications, reasoning, and connections, and on teaching mathematics through the use of appropriate technology.

Rule of Four

Most sections reflect an increased emphasis on the graphical and written approaches as well as the numerical and symbolic approaches. This is a reflection of the increased emphasis that is being placed in mathematics on the rule of four: every topic should be presented (and studied) geometrically, numerically, symbolically, and verbally. Because this is a text on technical mathematics, we might add to the list that each topic should be presented using appropriate technology. The rule of four is prevalent in much of the calculus reform materials, and is very consistent with the NCTM and AMATYC standards.

Problem Solving and Modeling

Students often have trouble assessing a real situation and sorting out the pertinent information. Our approach to problem solving encourages students to visualize problems, organize information, and develop their intuitive abilities. The applications, examples, and exercises are intended to teach students how to interpret real-world situations. Many examples and exercises are accompanied by illustrations and photos designed to guide students in analyzing information.

Problem solving is developed throughout the book. Hints, notes, and cautions are offered to hone problem-solving techniques, and many boxed features include guidelines for solving certain types of problems. Modeling activities are part of problem solving, but are especially prevalent in the calculator, computer, and Internet supplements described later.

Communications

Each nonreview exercise set concludes with at least two writing exercises. These can be used to provide students with opportunities to organize, synthesize, enhance, and express their understanding of mathematical ideas in their own words. These writing exercises also help students realize that writing is an important part of the communications skills that are essential for success in many fields. There are 200 of these writing exercises in the text.

For students to develop good problem-solving skills, they must learn to visualize mathematical problems. This text retains a feature that helps students see the important information in a problem. Each chapter opens with a photograph that demonstrates mathematics at work in the real world. Later in the chapter, this photograph is analyzed in an example and used to solve a problem based on the picture. A sketch placed next to the photo extracts the important mathematical elements. Where possible, this sketch is imposed over the photograph to show students how to cull the important elements from a real situation, and how to create a sketch that will help to solve a problem.

Reasoning

Ample opportunity is provided to help students apply mathematical reasoning. The text explains why something is true mathematically. Students are asked to explain mathematically why they can or cannot perform a certain operation or when are the appropriate times to use a certain math procedure.

Connections

The book contains frequent, realistic examples that show how to apply mathematical concepts. Examples cover both routine mathematical manipulations and applications. Over 20 percent of these examples have been selected from trade, vocational, technical, and industrial areas to demonstrate how mathematics is applied to all fields of technology.

One of the primary purposes of a technical mathematics course is to prepare students to solve applied problems in their chosen technical fields. For this reason, we have taken care to include exercises that show how to apply math in a variety of technical fields. The text also offers an index of the applications by field.

Technology

This text emphasizes the wise use of technologies—scientific calculators, graphing calculators, and microcomputers—because they can take the drudgery out of repetitive computations and can help students learn faster. Reducing the drudgery can help students focus on the important aspects of the operations they are performing. However, students must be aware that not all solutions are aided by technology. They must also learn how to determine whether a calculator or computer has given a correct answer, a reasonable answer, or a wrong answer.

The graphing calculator is used extensively throughout the text. In fact, while it is not necessary for the understanding of most of the material in the text, it is assumed that students are using a graphing calculator (or computer graphing software). Most graphing calculator figures have been generated by using a Texas Instruments TI-82. However, the text does not assume that students are using any particular calculator brand or model. In most instances, specific keystrokes have not been given. (Supplements are available which contain activities and specific keystrokes for each of the most popular graphing calculators.) Some examples are specifically indicated for showing how a graphing calculator can be used to help do the work of that section. Some exercises have been marked as "graphing calculator exercises" because they require the use of a graphing calculator. These instances are indicated by the placement of a graphing calculator icon . Rather than indicate that only certain problems should be solved with a calculator or computer, the author has assumed that any problem can be solved with the help of one of these tools.

Help for Teaching and Learning

Well written, carefully illustrated discussions form the foundation of any textbook, but students will form their mathematical skills with *practice*. We have developed a package of learning features that will inspire students to practice hard, and this will make their efforts, and the instructor's, effective.

Exercises

Ample practice is provided in more than 3,800 exercises. Over 17% of the exercises show the great variety of technical and scientific applications for mathematics. Each chapter ends with a set of review exercises and a chapter test. Both of these should encourage students to review concepts studied in the chapter.

Methods for solving most of the exercises can be found by reviewing the worked examples. Answers to the odd-numbered questions appear in the back of the book. A separate solutions manual for all problems is available to instructors, with a manual of selected solutions available to students.

Pedagogical Features

Throughout the book, you will find rules, formulas, guidelines, and hints that are boxed for easy identification. The boxes make the material easy to locate, and also make the book a valuable reference after the course is finished.

Six icons appear frequently in the margins:

The graphing calculator icon indicates that a certain example or exercise may be used with a graphing calculator. However, students may prefer to use a computer graphing program.

The computer icon indicates that a certain problem requires the student to do some computer programming.

Many years of teaching experience have shown the author where students make common errors. These places are indicated by the "caution" icon. This icon not only points out the potential errors, but how to avoid them.

The "hint" icon points out a valuable technique or learning hint that students can use to solve problems.

The "note" icon refreshes students' memories about a concept, or points out interesting or unusual ideas. These notes highlight ideas that might easily be overlooked, alternative ways to solve a problem, or a possible shortcut.

The "In Your Words" icon is used to indicate that the following exercises are specifically designed to improve your students' written communication skills. This icon appears just prior to the last set of exercises of each nonreview exercise set.

Chapter Review

Each chapter concludes with a list of the important terms covered in the chapter, a generous set of review exercises, and a chapter test. The review exercises include both routine computations and applications. Many of these exercises draw on more than one section of the chapter, requiring students to think harder and synthesize their learning. The chapter test gives students an idea of the types of questions they might expect for a 50-minute exam.

Supplements

We have carried our developmental themes of straightforward discussions and plenty of practice into our supplements package. This flexible package will allow emphasis on the teaching strategies a course demands, and gives students even more problem-solving experience.

The *Solutions Manual* contains fully worked solutions for every text problem. The *Student Solutions Manual* contains fully worked solutions for every odd exercise. All of these solutions have been thoroughly examined for accuracy.

A *Computerized Test Bank* is available on 3 1/2-inch disks. Illustrations are included on the disks for maximum flexibility. A printed version is also offered.

Computer software is available in two forms. A disk is available containing all of the debugged computer programs that appear in the Computer Programs section. In addition, a tutorial program is available to help teach problem-solving skills.

A package of two-color transparencies and transparency masters is available.

Graphing calculator activity manuals are available for the following brands and models of calculators: Casio *fx-7700* and *fx-9700*; Texas Instruments *TI-82* and *TI-85*. Each manual contains a series of activities with specific keystrokes for each calculator.

Acknowledgments

I first want to thank my editors, Mary Clyne and Paul Shepardson, for their patience and for skill in coordinating all the many aspects of producing a book. I would also like to thank T. Gerald Brooks for overseeing the art manuscript, performing much of the copy editing, and coordinating the work with the LaTeX expert, Ed Sznyter.

Anyone who writes a book enjoys the direct and indirect assistance of many people. First let me thank the reviewers of the manuscript for this edition. Their many suggestions have made this a better text. These reviewers include:

Haya Adner, Queensborough Community College, Bayside, N.Y.
Doris Bratcher, Northwest Iowa Technical College, Iowa
Granville Brown, Grand Rapids Community College, Grand Rapids, Mich.
Ronald Bukowski, Johnson Technical Institute, Scranton, Pa.
Ed Champy, Northern Essex Community College
Bill Ferguson, Columbus State Community College, Columbus, Ohio
Maggie Flint, Northeast State Technical Community College, Blountville, Tenn.
Marion Graziano, Montgomery County Community College, Blue Bell, Pa.
Henry Hosek, Purdue University, Calumet, Ind.
David Hutchinson, Mohawk College of Technical and Applied Arts, Hamilton, Ont.
Allan Hymers, Centennial College, Scarborough, Ont.
Stanley Koper, Triangle Tech, Belle Vernon, Pa.
Larry Lance, Columbus State Community College, Columbus, Ohio
Tom McCollow, DeVry Institute of Technology, Phoenix, Ariz.
Peter Moreno, ITT Technical Institute, San Bernardino, Calif.

Robert Opel, Waukesha County Technical College, Waukesha, Wis.
Steven B. Ottman, Southeast Community College, Nebraska
Gordon Schlafmann, Western Wisconsin Technical College
Rita Shillabeer, University at Alberta, Canada
Cathy Vollstedt, North Central Technical College, Schofield, Wis.
Anita Walker, Asheville, N.C.
Phyliss Wiegman, Ivy Tech, Fort Wayne, Ind.
Nancy Woods, Des Moines Area Community College, Des Moines, Iowa

During several rigorous accuracy checks, the following professionals examined the text and solutions in detail:

C. Lea Campbell, Lamar University, Port Arthur, Tex.
Gerry East, Tulane University, New Orleans, La.
Diane Ferris, Portland Community College, Portland, Oreg.
Alan Herweyer, Chattanooga State Technical College, Chattanooga, Tenn.
Barbara Karp, Newcastle, Ind.
Amy Relyea, Miami University, Oxford, Ohio
Gordon Schlafmann, Western Wisconsin Technical College
Randall Sowell, Central Virginia Community College, Lynchburg, Va.
John Wisthoff, Anne Arundel Community College, Md.
George Wyatt, Rensselaer Polytechnic Institute, Troy, N.Y.

Several enthusiastic instructors generously agreed to test the final pages in their classrooms:

C. Lea Campbell, Lamar University, Port Arthur, Tex.
Bill Ferguson, Columbus State Community College, Columbus, Ohio
Marion Graziano, Montgomery County Community College, Blue Bell, Pa.
Henry Hosek, Purdue University, Calumet, Ind.
Robert Opel, Waukesha County Technical College, Waukesha, Wis.
Arthur J. Varie, Trumbull County JVS, Warren, Ohio

I would especially like to thank George Glass, Chattanooga State Technical Community College, and John Darby for their many suggestions for applications in the forestry and medical technology areas, respectively. Special thanks go to Alan Herweyer for preparing the *Solutions Manual*. Al gave many helpful and insightful suggestions that made the text a much better finished product.

No one could complete an effort such as this without the support, encouragement, and tolerance of an understanding family. I want to thank my wife, Dr. Marla Peterson, our son, Matt, and daughter-in-law, Jennifer, for giving me the time and understanding to complete this work.

John C. Peterson

CHAPTER

1

Rectangular Graphing in the Plane

Some technicians have to take water samples at a given distance from a source of pollution. In Section 1.2, we will see how to use functions and their graphs to help analyze these samples and predict the effects of pollution.

Courtesy of Janet Essman, New York State Department of Environmental Conservation

In previous mathematics courses you have studied algebraic equations and functions. You learned ways to solve equations and how to take information and write an equation or a function that showed how the information was related.

A function shows an important relationship between two variables. Functions are often expressed in the form of an equation. They can also be expressed as ordered pairs and used to create a table of values. In this chapter, we will see how we can better understand functions by graphing them.

1.1 RECTANGULAR COORDINATES

You should know how to graph points on a number line. In this section we will expand our idea of graphing to a plane.

Why is a graph so important? A graph gives us a picture of an equation. By looking at a graph, we can often get a better idea of what we can expect an equation to do. The graphs will also help us find solutions to some of our problems.

The numbers can be represented by points on a line. To get a graph in a plane, we need two number lines. The two number lines are usually drawn perpendicular to each other and intersect at the number zero as shown in Figure 1.1. The point of intersection is called the **origin**. If one of the lines is horizontal; then the other is vertical. The horizontal number line is called the **x-axis** and the vertical number line the **y-axis**. This is called the **rectangular coordinate system** or the **Cartesian coordinate system** in honor of the man who invented it, René Descartes.

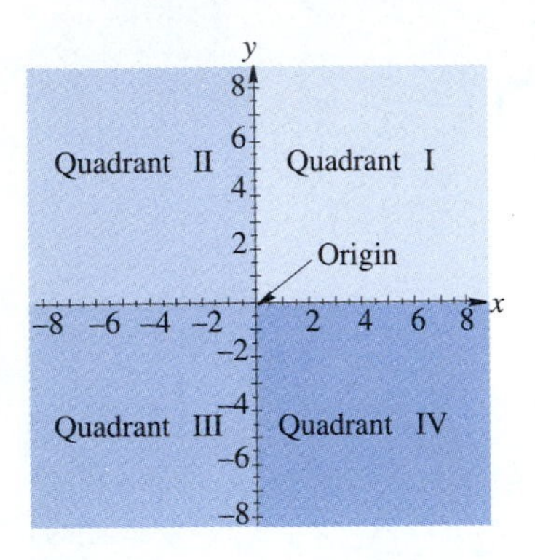

FIGURE 1.1

On the x-axis, positive numbers are to the right of the origin and negative numbers to the left. On the y-axis, positive numbers are above the origin and negative values are below it. The two axes divide the plane into four regions called **quadrants**, with the first quadrant in the upper right section of the plane and the others numbered in a counterclockwise rotation around the origin. (See Figure 1.1.)

Suppose that P is a point in the plane. The coordinates of P can be determined by drawing a perpendicular line segment from P to the x-axis. If this perpendicular segment meets the x-axis at the value a then the x-coordinate of the point P is a. Now draw a perpendicular segment from P to the y-axis. It meets the y-axis at b, so the y-coordinate of P is b. The coordinates of P are the ordered pair (a, b).

Note

The x-coordinate is always listed first. The order in which the coordinates are written is very important. In most cases, the point (a, b) is different than the point (b, a). (See Figure 1.2.)

EXAMPLE 1.1

The positions of $A(2, 4)$, $B(4, 2)$, $C(-4, 5)$, $D(7, 0)$, $E(-4, -3)$, $F(0, -3)$, and $G(5, -4)$ are shown in Figure 1.3. Note that $A(2, 4)$ and $B(4, 2)$ are different points as are $C(-4, 5)$ and $G(5, -4)$. ▪

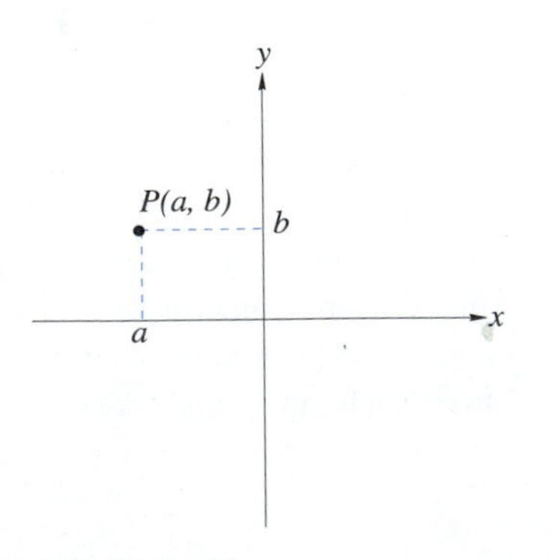

FIGURE 1.2

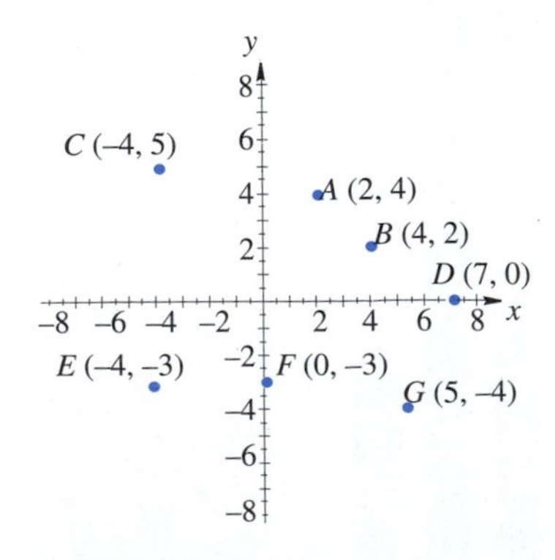

FIGURE 1.3

EXAMPLE 1.2

$A(2,7)$, $B(-3,7)$ and $C(-3,-1)$ are three vertices of a rectangle as shown in Figure 1.4. What are the coordinates of the fourth vertex, D?

Solution If we plot points A, B, and C, we can see that the missing vertex D will have the same x-coordinate as A, 2, and the same y-coordinate as C, -1. So, D has the coordinates $(2, -1)$.

EXAMPLE 1.3

Some ordered pairs for the equation $y = 2x + 7$ are $(-2,3)$, $(-1,5)$, $(0,7)$, $(\frac{1}{2},8)$, $(1,9)$, and $(2,11)$. Plot these ordered pairs on a rectangular coordinate system.

Solution The ordered pairs are plotted on the graph in Figure 1.5.

Application

EXAMPLE 1.4

The following table shows the results of a series of drillings used to determine the depth of the bedrock at a building site. These drillings were taken along a straight line down the middle of the lot, from the front to the back, where the building will be placed. In the table, x is the distance from the front of the parking lot and y is the corresponding depth. Both x and y are given in feet.

x	0	20	40	60	80	100	120	140	160
y	33	35	40	45	42	38	46	40	48

Plot these ordered pairs on a rectangular coordinate system. Connect the points in order to get an estimate of the profile of the bedrock.

Solution The ordered pairs are plotted on the graph in Figure 1.6. The points have been connected in order. Note that this graph is "upside down." That is, the top of the dirt is along the x-axis, while the bedrock, which is below ground, is above the x-axis.

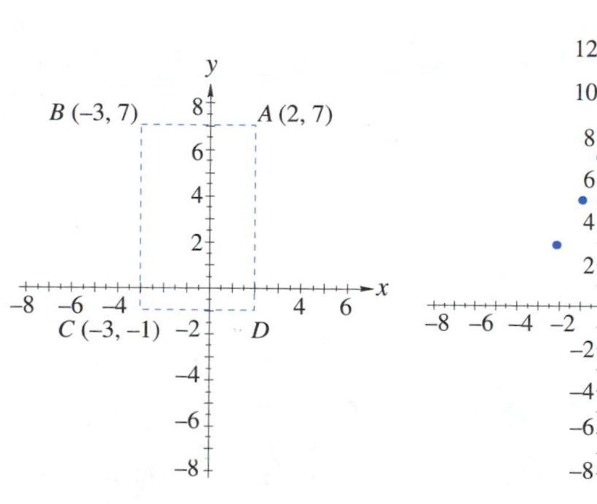

FIGURE 1.4

FIGURE 1.5

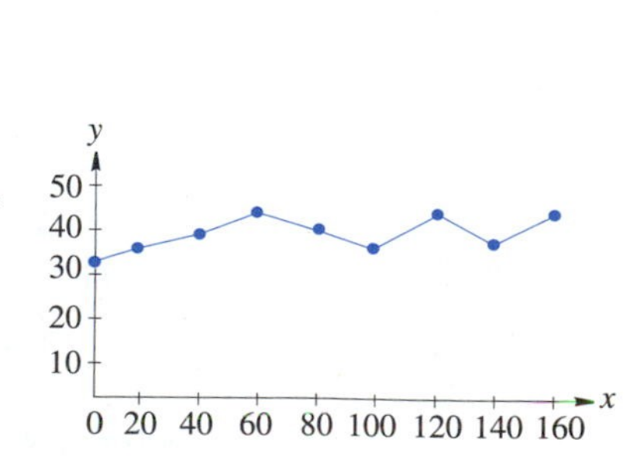

FIGURE 1.6

Exercise Set 1.1

In Exercises 1–10, plot the points on a rectangular coordinate system.

1. $(4, 5)$
2. $(1, 7)$
3. $(7, 1)$
4. $(-2, 4)$
5. $(-3, -5)$
6. $(6, -1)$
7. $(0, 0)$
8. $(-\frac{5}{2}, 3)$
9. $(2, -\frac{3}{2})$
10. $(4.5, -1.5)$

Solve Exercises 11–28.

11. If $A(2, 5)$, $B(-1, 5)$, and $C(2, -4)$ are three vertices of a rectangle, what are the coordinates of the fourth vertex?

12. Some of the ordered pairs of the equation $y = x^2 - 3$ are $(-3, 6)$, $(-2, 1)$, $(-1, -2)$, $(0, -3)$, $(1, -2)$, $(2, 1)$, and $(3, 6)$. Plot these ordered pairs on the same rectangular coordinate system.

13. Plot the ordered pairs $(-5, 3)$, $(-3, 3)$, $(0, 3)$, $(1, 3)$, $(4, 3)$, and $(6, 3)$. What do all of these points have in common?

14. Plot the ordered pairs $(-2, 6)$, $(-2, 4)$, $(-2, 1)$, $(-2, -1)$, $(-2, -3)$, and $(-2, -4)$. What do all of these points have in common?

15. *Automotive technology* Graph these ordered pairs from the conversion formula 6.9 kPa = 1 psi: $(138, 20)$, $(172.5, 25)$, $(207, 30)$, $(241.5, 35)$, $(276, 40)$, $(345, 50)$, $(414, 60)$. Use a ruler and draw the segment connecting $(138, 20)$ and $(414, 60)$. What seems to happen?

16. Where are all the points whose x-coordinates are 0?

17. Where are all the points whose y-coordinates are -2?

18. Where are all the points whose y-coordinates are 7?

19. Where are all the points whose x-coordinates are -5?

20. Where are all the points whose y-coordinates are 0?

21. Where are all the points (x, y) for which $x > -3$?

22. Where are all the points (x, y) for which $x > 0$ and $y < 0$?

23. Where are all the points (x, y) for which $x > 1$ and $y < -2$?

24. *Automotive technology* If the antifreeze content of the coolant is increased, the boiling point of the coolant is also increased, as shown in the table below.

Percent antifreeze in coolant	0	10	20	30	40	50	60	70	80	90	100
Boiling temperature °F	210	212	214	218	222	228	236	246	258	271	330

(a) Plot these ordered pairs on a rectangular coordinate system.
(b) What will be the boiling point of the coolant if the system is filled with a recommended solution of 50% water and 50% antifreeze?
(c) What will be the boiling point of the coolant if the system is filled with a recommended solution of 40% water and 60% antifreeze?

25. *Automotive technology* The temperature-pressure relationship of the refrigerant R-12 is very important to maintain proper operation and for diagnosis. The table below indicates the pressure of R-12 at various temperatures.

Temperature °F	−20	−10	0	10	20	30	40	50	60	70	80
Atmospheric pressure (psi)	2.4	4.5	9.2	14.7	21.1	28.5	37.0	46.8	57.7	70.1	84.1

(a) Plot these ordered pairs on a rectangular coordinate system. Connect the ordered pairs in order to help you answer **(b)** and **(c)**.
(b) One day the early morning temperature was 45°F. What was the pressure, in psi, when the temperature was 45°F?
(c) If a technician connected a pressure gauge to an air-conditioning system filled with R-12 on a 90°F summer day, what pressure, in psi, would the gauge indicate?

26. *Automotive technology* Recent automotive air conditioning systems use R-134a rather than R-12. This action is based on evidence which indicates that R-12 is causing depletion of the earth's protective ozone layer. As with R-12, the temperature-pressure relationship of refrigerant R-134a is very important to maintain proper operation and diagnosis. The table below indicates the pressure of R-134a at various temperatures.

Temperature (°C)	−20	−10	0	10	20	30
R-134a pressure (kPa)	31.4	99.4	191.4	312.9	469.5	666.7

(a) Plot these ordered pairs on a rectangular coordinate system. Connect the ordered pairs in order to help you answer parts (b), (c), and (d).
(b) One day the early morning temperature was 5°C. What was the pressure, in kPa, when the temperature was 5°C?
(c) If a technician connected a pressure gauge to an air conditioning system filled with R-134a on a 35°C summer day, what pressure, in kPa, would the gauge indicate?
(d) Suppose a technician connected a pressure gauge to an air conditioning system filled with R-134a and got a reading of 500 kPa. What is the temperature in °C?

27. *Machine technology* The following table gives actual recording times and counter values as obtained on one particular cassette tape deck.

Time (minutes)	1	2	3	4	5	10	15	20	25	30
Counter reading	30	60	88	115	141	262	369	466	556	640

(a) Plot these ordered pairs on a rectangular coordinate system. Connect the ordered pairs in order to help you answer parts (b) and (c).
(b) What would you expect the counter to read after 12.5 min?
(c) How much time has passed if the counter has a value of 500?

28. *Forestry* The following table gives the girth of a pine tree measured in feet at shoulder height and the amount of lumber in board feet that was finally obtained.

Girth (ft)	15	18	20	22	25	28	33	35	40	43
Lumber (bd ft)	9	27	42	60	90	126	198	231	324	387

(a) Plot these ordered pairs on a rectangular coordinate system. Connect the ordered pairs in order to help you answer parts (b) and (c).
(b) How many board feet of lumber would you expect to obtain from a tree with a girth of 30 ft?
(c) What was the girth of a tree that produced 275 board feet of lumber?

In Your Words

29. Look again at Example 1.4. Describe how you would have changed the graph, or the way the data was recorded, so that the graph would appear rightside up instead of upside down.

30. Gather some data from your field of interest. Write a problem that will require that data to be graphed. Then write some questions that are answered by reading the graph. Put your answers on the back of the paper. Give your problem to a classmate and ask him or her to solve the problem. Rewrite your exercise or solution to clarify places where your classmate had difficulty.

1.2 GRAPHS

In the last section, we learned how to graph a point on a plane. Earlier, we said that a function could be represented by a graph. In this section, we will see how we can use graphs to represent functions and we will use the computer to help us draw graphs.

> **Graph of an Equation**
>
> The graph of an equation in two variables x and y is formed by all the points $P(x, y)$ whose coordinates (x, y) satisfy the given equation.

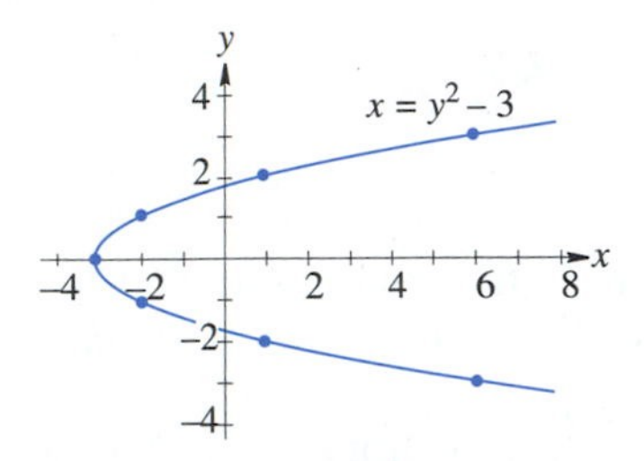

FIGURE 1.7

To graph an equation such as $x = y^2 - 3$, you can set up a table of values, plot the points in the table, and then connect the points smoothly. To fill the table, we will select values for y and use the equation to solve for the corresponding values for x. Because the ordered pairs have x listed first, we will list x as the top value in the table.

x	6	1	−2	−3	−2	1	6
y	−3	−2	−1	0	1	2	3

These points are plotted in Figure 1.7 and a smooth curve has been used to connect the points. Remember, this equation has an infinite number of possible ordered pairs. We have selected only seven of them. We must plot enough points to be confident that the points give an outline of the curve that is complete enough for us to tell what the actual curve looks like. Later, you will learn some ways to use mathematics to help determine when you have a sufficient number of points to sketch a curve.

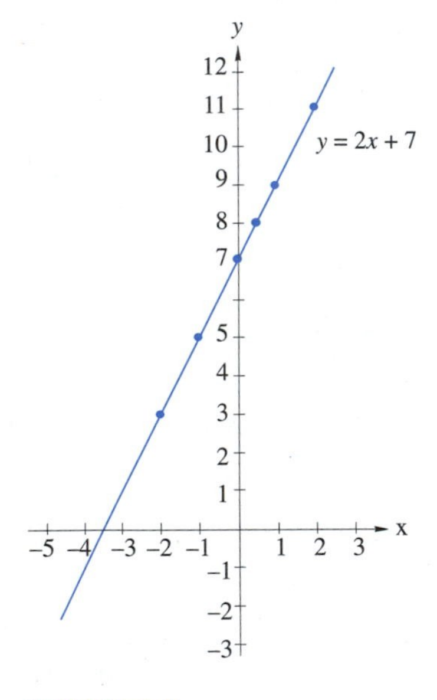

FIGURE 1.8

In Figure 1.5 we plotted six ordered pairs for the equation $y = 2x + 7$: $(-2, 3)$, $(-1, 5)$, $(0, 7)$, $(\frac{1}{2}, 8)$, $(1, 9)$, and $(2, 11)$. If we connect these points, we see that we get the straight line in Figure 1.8. In fact, any equation of the form $Ax + By + C = 0$ where A, B, and C are constants and both A and B are not 0 is a **linear equation** whose graph is a straight line. The equation $y = 2x + 7$ is also a function and can be written as $f(x) = 2x + 7$.

Intercepts

The graph of the equation $y = 2x + 7$ crosses both the x-axis and the y-axis. These points are called the **intercepts**. The graph crosses the y-axis at $(0, 7)$, so we say that the ***y*-intercept** is 7. It crosses the x-axis at $\left(-\frac{7}{2}, 0\right)$, so the ***x*-intercept** is $-\frac{7}{2}$.

Any time you have a linear equation, you will only need to plot two points in order to sketch the graph of that equation. The intercepts are often the easiest two points to determine and use when graphing a linear equation.

EXAMPLE 1.5

Plot the graph of $2x + 3y = 12$.

Solution This is a linear equation, so we will use two points to determine the line. If we find the intercepts, we get the points $(6, 0)$ and $(0, 4)$. These points and the line they determine are shown in Figure 1.9.

Slope

One important property of the graph of a straight line is its slope. The slope refers to the steepness or inclination of the line. The idea of the slope is so important that we will keep coming back to it. In fact, it is a major topic in calculus.

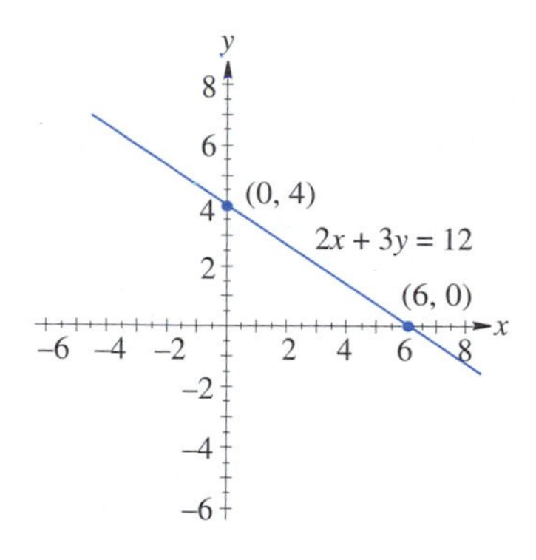

FIGURE 1.9

Slope

The **slope**, m, of a straight line is a measure of its steepness with respect to the x-axis. If (x_1, y_1) and (x_2, y_2) are two different points on a line, then the slope of the line is defined as

$$m = \frac{y_2 - y_1}{x_2 - x_1}$$

provided $x_2 - x_1 \neq 0$.

Some people like to remember the slope as the "rise over the run" or $\frac{\text{rise}}{\text{run}}$. Here "rise" means the vertical change and "run" the horizontal change.

EXAMPLE 1.6

What is the slope of the line $2x+3y=12$ in Example 1.5?

Solution We know that $(6,0)$ and $(0,4)$ are two points on this line. If we let $x_1 = 6$, $y_1 = 0$, $x_2 = 0$, and $y_2 = 4$, then

$$\begin{aligned} m &= \frac{y_2 - y_1}{x_2 - x_1} \\ &= \frac{4-0}{0-6} = \frac{4}{-6} = -\frac{2}{3} \end{aligned}$$

The slope of this line is $m = -\frac{2}{3}$.

EXAMPLE 1.7

Graph $4y - 6x = 12$ and find the intercepts and the slope.

Solution This is a straight line. The intercepts are $(0,3)$ and $(-2,0)$. Plotting these two points and the line through them gives the graph in Figure 1.10. Letting $x_1 = 0$, $y_1 = 3$, $x_2 = -2$, and $y_2 = 0$, we determine that the slope is

$$m = \frac{0-3}{-2-0} = \frac{-3}{-2} = \frac{3}{2}$$

EXAMPLE 1.8

Graph $8x = 2 + y$, and find the intercepts and the slope.

Solution Again, this is a straight line. If we let $x = 0$, then we get $0 = 2 + y$ or $y = -2$, and so one intercept is $(0, -2)$. If we let $y = 0$, then $8x = 2$ and $x = \frac{1}{4}$.

EXAMPLE 1.8 (Cont.)

Thus, $\left(\frac{1}{4}, 0\right)$ is the other intercept. The slope is

$$m = \frac{-2-0}{0-\frac{1}{4}} = \frac{-2}{-\frac{1}{4}} = 8$$

Plotting the two intercepts and drawing the line through them produces the line in Figure 1.11. ▪

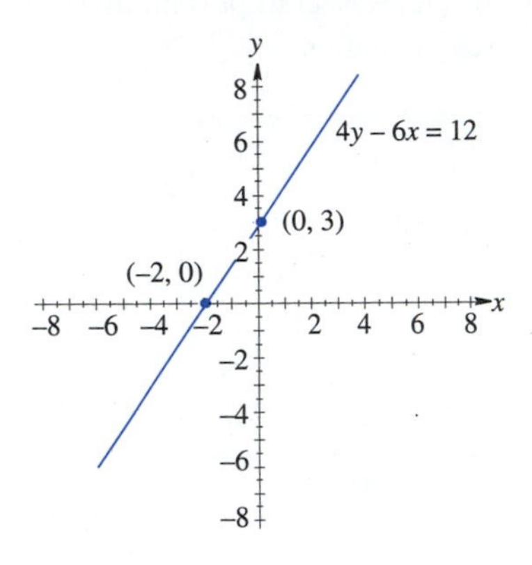

FIGURE 1.10

Look again at the last three examples. In Example 1.6, the slope of the line was $-\frac{2}{3}$. Notice that this line falls as the x-values get larger. In Example 1.7, the slope of the line was $\frac{3}{2}$. In Example 1.8, the slope of the line was 8. Both of these lines rose as the values of x got larger. Notice that the line in Example 1.8 was steeper than the line in Example 1.7.

In general, as the values of x increase, a straight line rises if it has a positive slope and falls if it has a negative slope. This is demonstrated in Figures 1.12 and 1.13.

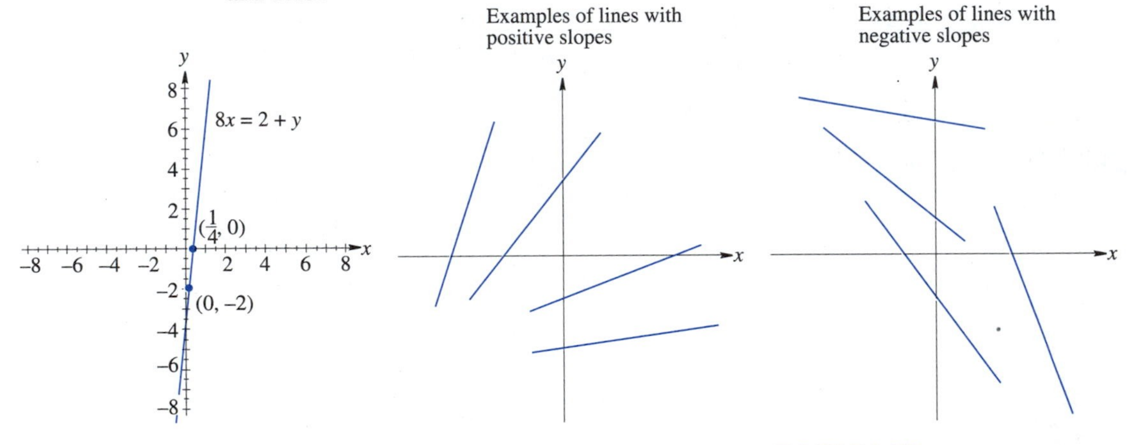

FIGURE 1.11 **FIGURE 1.12** **FIGURE 1.13**

Graphs That Are Not Lines

Not every graph is a straight line. The next four examples show some graphs of functions that are not lines. This is followed by a test that tells how to use a graph to determine if it is a graph of a function.

EXAMPLE 1.9

Graph the function $y = x^2 - 3$.

Solution Some ordered pairs that satisfy $y = x^2 - 3$ are shown in the following table.

x	−3	−2	−1	0	1	2	3
y	6	1	−2	−3	−2	1	6

EXAMPLE 1.9 (Cont.)

If we plot these seven points and connect them with a smooth curve, we get the graph in Figure 1.14. Notice that this graph has a y-intercept at $y = -3$ and that it has two x-intercepts, when $x = \sqrt{3}$ and when $x = -\sqrt{3}$.

EXAMPLE 1.10

Graph the function $f(x) = x^2 - 2x$.

Solution Here is a table of some of the values for this function.

x	−3	−2	−1	0	1	2	3	4	5
$f(x)$	15	8	3	0	−1	0	3	8	15

These points and the curve connecting them are shown in Figure 1.15.

EXAMPLE 1.11

Graph the function $g(x) = \frac{x-1}{x}$.

Solution Note that this function is not defined when $x = 0$. A partial table of values for this function follows.

x	−5	−4	−3	−2	−1	−0.4	−0.1
$g(x)$	1.2	1.25	$1.3\overline{3}$	1.5	2	3.5	11

x	0	0.1	0.4	1	2	3	4	5
$g(x)$	DNE*	−9	−1.5	0	0.5	$1.6\overline{6}$	0.75	0.80

*DNE means Does Not Exist.

The graph is shown in Figure 1.16.

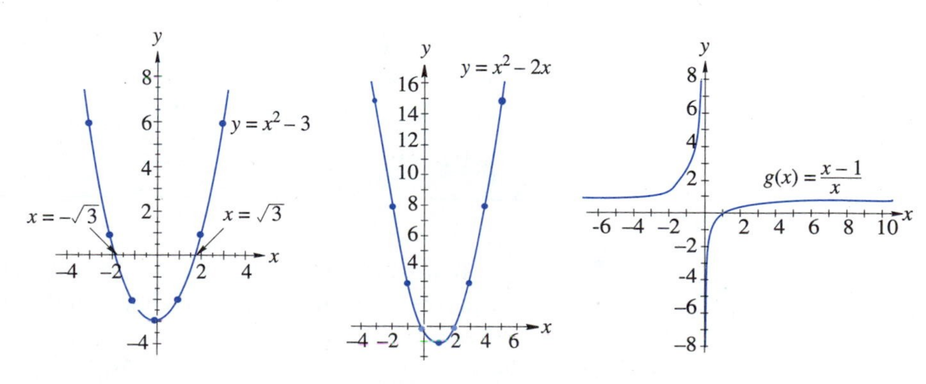

FIGURE 1.14 **FIGURE 1.15** **FIGURE 1.16**

Another interesting graph is shown by Example 1.12.

EXAMPLE 1.12

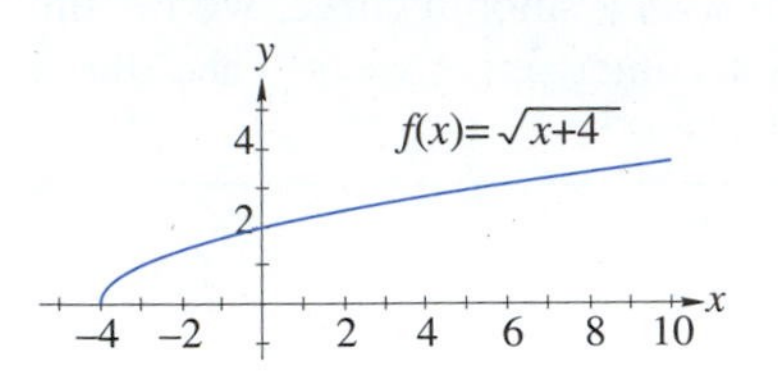

FIGURE 1.17

Graph the function $f(x) = \sqrt{x+4}$.

Solution You can only take the square root of a nonnegative number. Thus, the function is only defined for $x \geq -4$, and the domain of the function is $\{x : x \geq -4\}$. A partial table of values for this function follows. Its graph is shown in Figure 1.17.

x	-4	-3	-2	-1	0	1	2	3	4	5
y	0	1	$\sqrt{2} \approx$ 1.414	$\sqrt{3} \approx$ 1.732	2	$\sqrt{5} \approx$ 2.236	$\sqrt{6} \approx$ 2.449	$\sqrt{7} \approx$ 2.646	$\sqrt{8} \approx$ 2.828	3

Application

EXAMPLE 1.13

The total cost to manufacture x items of a certain product is given by $C(x) = \dfrac{8}{x} + 0.5x$, where $x > 0$. Set up a partial table of values and sketch the graph of this function.

Solution This function is only defined for $x > 0$. A partial table of values is shown. Its graph is shown in Figure 1.18.

x	1	2	3	4	5	6	7	8	9	10
y	8.5	5	$4\frac{1}{6}$	4	4.1	$4\frac{1}{3}$	$4\frac{9}{14}$	5	$5\frac{7}{18}$	5.8

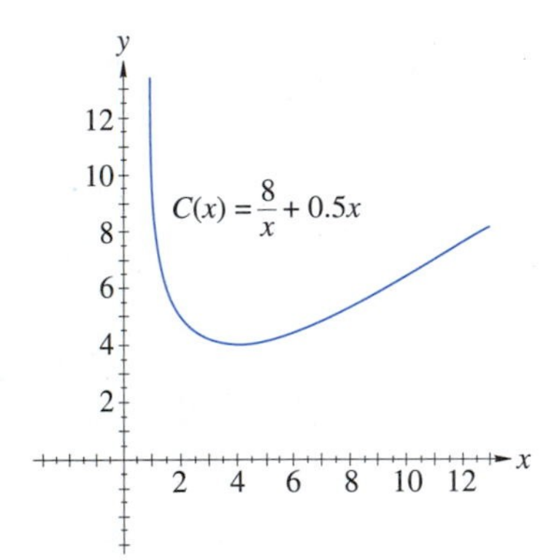

FIGURE 1.18

Vertical Line Test

A graph can provide us with an easy test to see if the equation that has been graphed is a function. This test is called the **vertical line test**.

Vertical Line Test

The vertical line test states that a graph is the graph of a function if no vertical line intersects the graph more than once.

To use the vertical line test, you look at a graph and determine if there are any places where a vertical line would intersect the graph more than once. If you cannot find any such places, then this is the graph of a function. If you can, then this is not the graph of a function. Figures 1.19a and 1.19b give some examples. The graph in Figure 1.19a is not the graph of a function, because the vertical line intersects the graph three times. On the other hand, the graph in Figure 1.19b is the graph of a function because no vertical line can intersect the graph more than once.

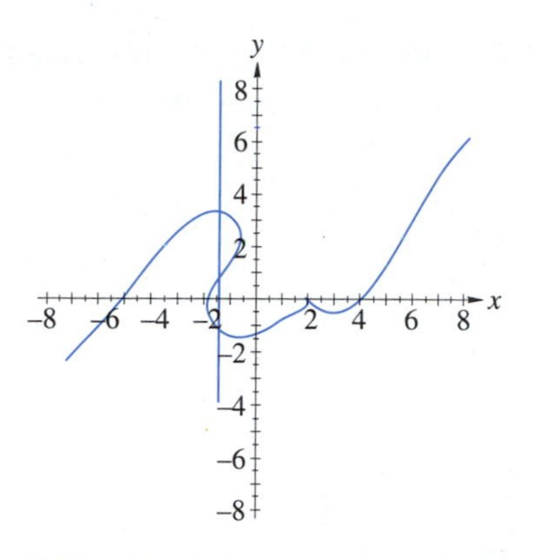

FIGURE 1.19a

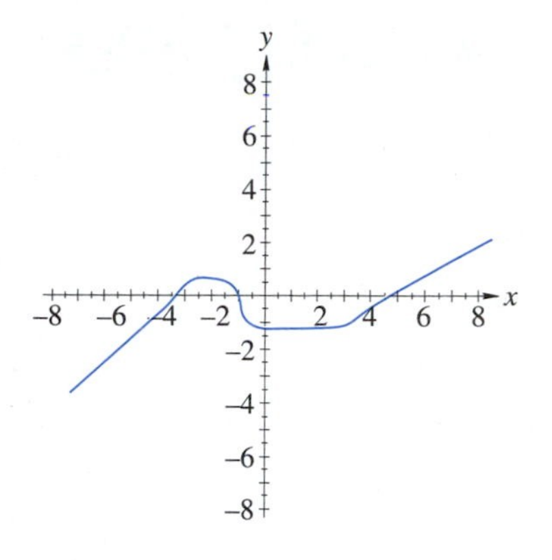

FIGURE 1.19b

Application

EXAMPLE 1.14

The concentration in parts per million (ppm) of a certain pollutant m mi from a certain factory is given by

$$C(m) = \frac{50}{m^2+4}, \quad m \geq 0$$

Set up a partial table of values and sketch the graph of this function.

Solution This function is defined for values of $m \geq 0$. A partial table of values follows. The graph of the function is shown in Figure 1.20. Notice that this graph passes the vertical line test.

m	0	1	2	3	4	5	6
$C(m)$	12.5	10	6.25	3.85	2.5	1.72	1.25

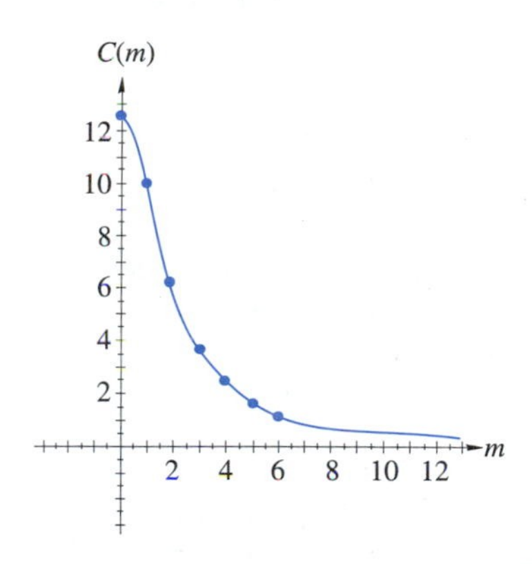

FIGURE 1.20

Using a Graph to Determine the Domain and Range of a Function

We just used the vertical line test to determine if a graph is the graph of a function. We can use a variation of that test to help us determine the domain of the function. To determine the domain of a function, think of vertical lines drawn from each point on the x-axis. If a vertical line intersects the graph of a function, then the x-coordinate determined by that line is in the domain. If the vertical line does not intersect the graph, then the x-coordinate is not in the domain.

We can also use a graph to help us determine the range of a function. To determine the range of a function, think of horizontal lines drawn from each point on the y-axis. If a horizontal line intersects the graph, then the y-coordinate determined by that line is in the range. If it does not intersect the graph, then the y-coordinate is not in the range.

EXAMPLE 1.15

Determine the domain and range of the function shown by the graph in Figure 1.21.

Solution Because the graph extends from $x = -6$ to, but not including, $x = 3$, the domain is $\{x: -6 \leq x < 3\}$. Since the graph extends from $y = 5$ to $y = -2$, the range is $\{y: -2 \leq y \leq 5\}$. Notice that the range was not determined by the y-values when $x = -6$ and when $x = 3$.

EXAMPLE 1.16

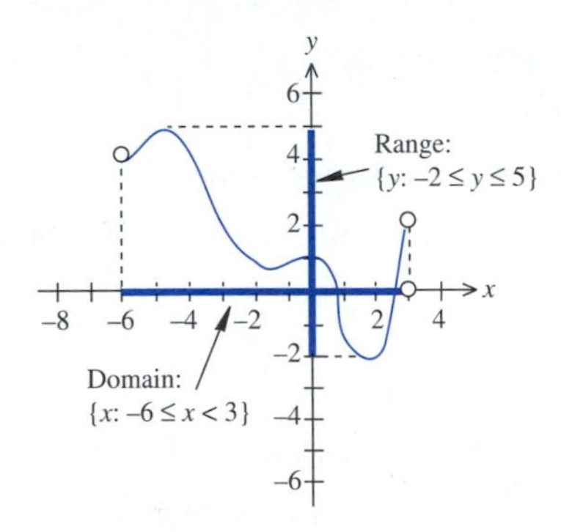

FIGURE 1.21

Show how to use the graph of a function to determine the domain and range of the functions $y = x + 2$ and $y = \dfrac{x^2 - 4}{x - 2}$.

Solution The graph of $y = x + 2$ is shown in Figure 1.22a and the graph of $y = \dfrac{x^2 - 4}{x - 2}$ is shown in Figure 1.22b.

The graph of the equation $y = x + 2$ is the line shown in Figure 1.22a. While we cannot see the entire graph, we know that this is a line and are led to believe that any vertical line will intersect this graph exactly once. Similarly, it appears as if any horizontal line will intersect the graph. Hence, the domain and range are both the set of real numbers.

Look at the graph of $y = \frac{x^2 - 4}{x - 2}$ in Figure 1.22b. There appears to be a "hole" in the graph at $(2, 4)$. Because the denominator is 0 when $x = 2$, the domain of this function is $\{x: x \neq 2\}$. If you draw a vertical line through the "hole," it will hit the x-axis at $x = 2$. This graphically shows that $x = 2$ is not in the domain of this function. If you draw a horizontal line through the "hole" at $(2, 4)$, it hits the y-axis at $y = 4$. This leads us to conclude that the range is $\{y: y \neq 4\}$.

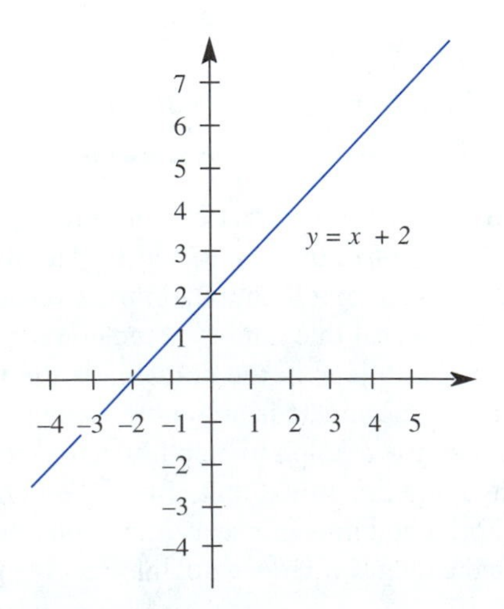

FIGURE 1.22a

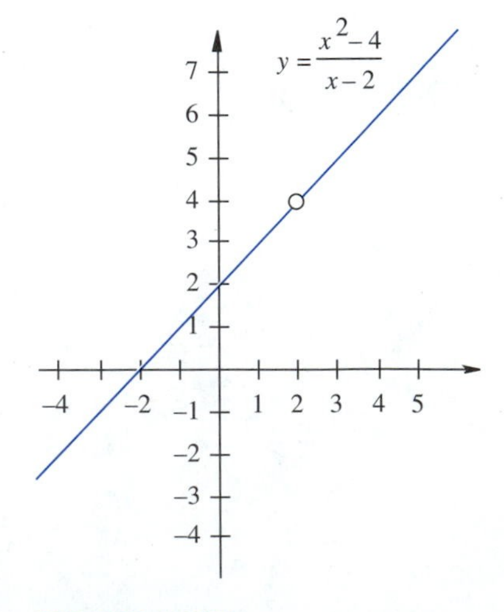

FIGURE 1.22b

You must be careful when you use the graph of a function to determine the domain and the range. Most graphs that you will draw or see show only part of the graph. You must use your algebra skills, your experience, *and* the graph to fully determine the range.

The next section will further explore how you can determine the domain and range with the aid of a graphing calculator.

Exercise Set 1.2

For each of Exercises 1–10, (a) draw the graphs of the linear equations, (b) determine the intercepts, and (c) calculate the slope.

1. $y = x$
2. $y = 2x$
3. $y = -3x$
4. $y = \frac{1}{2}x - 3$
5. $y = -4x + 2$
6. $x + y = 2$
7. $x - y = 2$
8. $2x + y = 1$
9. $3x - 6y = 9$
10. $3x - 6y = -6$

In Exercises 11–20, (a) set up a table of values, (b) graph the given functions, and (c) determine the domain and range of each function.

11. $y = x^2$
12. $y = x^2 + 3$
13. $y = x^2 - 2$
14. $y = x^2 + 2x + 1$
15. $y = x^2 - 6x + 9$
16. $y = -x^2 + 2$
17. $y = \dfrac{1}{x}$
18. $y = \dfrac{1}{x+3}$
19. $y = x^3$
20. $y = \sqrt[3]{x}$

In Exercises 21–26, set up a table of values and graph each of the equations.

21. $y^2 + x^2 = 25$
22. $y = \sqrt{25 - x^2}$
23. $y = |x + 2|$
24. $\dfrac{x^2}{4} + \dfrac{y^2}{9} = 1$
25. $\dfrac{x^2}{4} - \dfrac{y^2}{9} = 1$
26. $y = x^3 - x^2$

Which of the graphs in Exercises 27–30 are functions?

27.

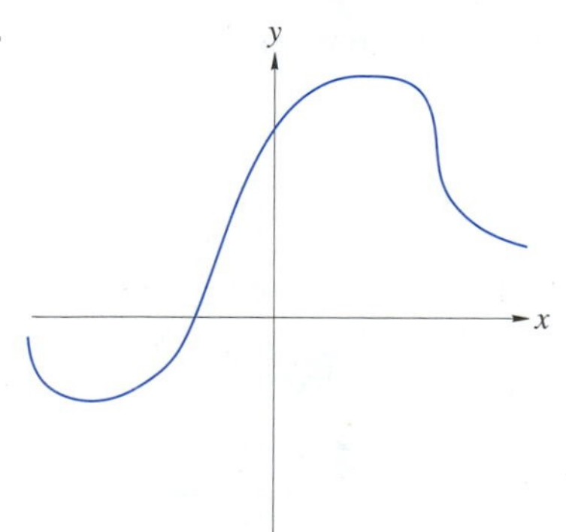

28.

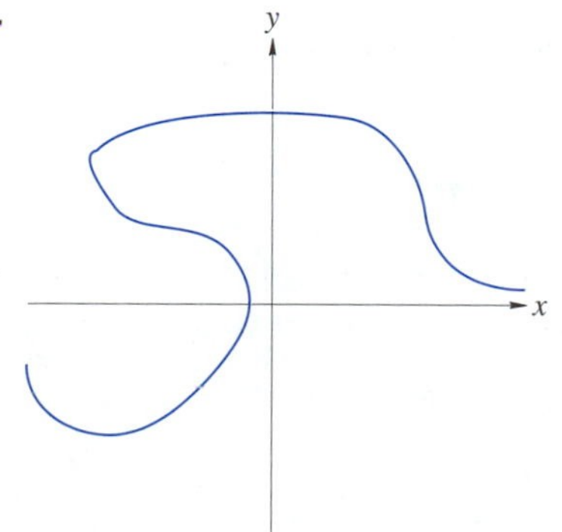

29.

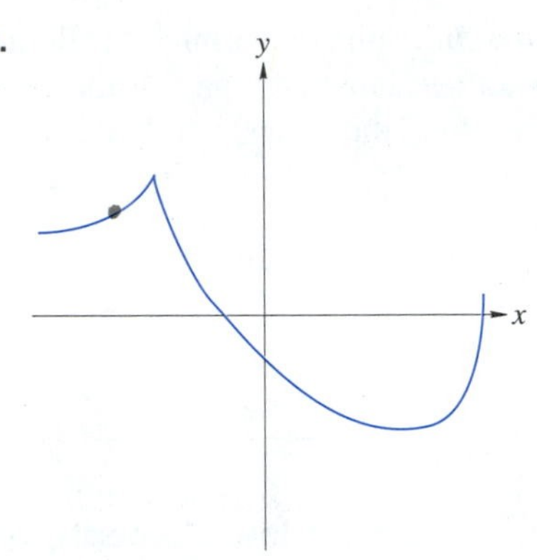

30.

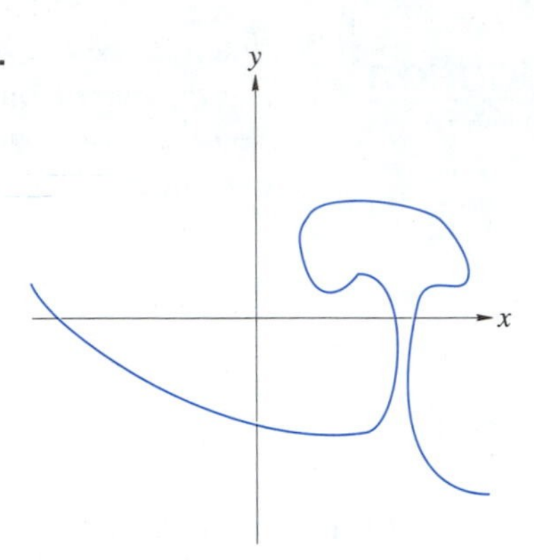

Solve Exercises 31–34.

31. *Ecology* An ecologist is investigating the effects of air pollution from an industrial city in the plants surrounding the city. She estimates that the percentage of diseased plants is given by

$$p(k) = \frac{25}{2k+1}$$

(a) Set up a table of values and graph the equation for $0 \le k \le 9$.
(b) Does your graph pass the vertical line test?

32. *Ecology* A second ecologist thinks that the function in the previous problem does not give the correct percentages. She thinks that the percentage of diseased plants k km from the city is given by

$$p(k) = \frac{25}{\sqrt{k+1}}$$

Make a table of values and graph the function for $0 \le k \le 9$

33. *Business* Based on past experience, a company decides that the weekly demand (in thousands) for a new microwaveable food product is $d(p) = -p^2 + 2.5p + 10$, where p is the price (in dollars) of the product.
(a) What is the domain of this function?
(b) Make a table of values for each \$0.25 change in price of $0 < p \le \$2.00$.
(c) Sketch a graph of this function.
(d) What appears to be the price that will result in the most sales?

34. *Automotive technology* In a chrome-electroplating process, the mass m in grams of the chrome plating increases according to the formula $m = 1 + 2^{t/2}$, where t is the time in minutes.
(a) Set up a table of values and graph the equation for $1 \le t \le 15$.
(b) How long does it take to form 100 g of plating?

In Your Words

35. Describe how to use the graph of a function to help determine the range of the function.

36. Explain what the slope of a line tells you.

1.3 GRAPHING CALCULATORS AND COMPUTER-AIDED GRAPHING

The introduction of microcomputers has allowed people to use inexpensive computers in their work for writing and for performing calculations quickly and accurately; but people discovered that words and numbers were not enough, and so computer graphics were introduced.

Some of these graphing capabilities are now available on calculators. To some extent, the use of a graphing calculator or a computer removes some of the drudgery of plotting equations by hand. We will show you how you can use computer and calculator graphing capabilities to demonstrate some mathematical concepts and to relieve you of some work. However, we will also show you how to take advantage of these technological devices and show how to interpret the information they give you.

The graphics capabilities of computers and calculators vary widely. Examples in this section, and other graphics examples in this text, were run on either a Casio fx-7700G or a Texas Instruments TI-82 graphing calculator or on a Macintosh computer.

Rather than teach you how to write a graphing program for a computer, we will assume you have access to a program that can be used to graph functions. Some, such as *Master Grapher,* are specialized graphing programs; some, such as *Excel* and *Lotus 1-2-3,* are computer "spreadsheets" that have graphing capabilities. Other more specialized programs include *Mathematica, Derive, Microcalc, MathCad*, and *Maple,* all of which might be classified as computer algebra systems.

The purpose of this book is not to explain how to use the graphing capabilities of these programs, but to help you use these computer programs to graph curves and to interpret their graphs.

Using a Graphing Calculator

We begin with a discussion on using graphing calculators. The examples and the instructions in this book are meant to supplement the user's guide for your calculator, not to replace it. Additional details on using a graphing calculator are in Appendix A and the *Student's Handbook.*

The first and third examples (Examples 1.17 and 1.19) use the Casio fx-7700G. The second example (Example 1.18) uses a Texas Instruments TI-82 graphing calculator. Later examples will be more generic and, unless specified, will be generated using a TI-82 graphing calculator.

The first equation will be of a straight line, $y = 2x + 7$. Before we begin, we must determine the domain of this function. The range of this function is determined by the allowable values of y that can appear on the calculator's screen.

EXAMPLE 1.17

Use a Casio fx-7700G graphing calculator to sketch the graph of $f(x) = 2x + 7$.

Solution First, enter [SHIFT] [F5] [EXE] to clear the graphics screen of any previous graphs that might be on the screen or in the calculator's memory.

We begin by setting the domain and range we want displayed on the screen. Press the [Range] key. The screen should be similar to Figure 1.23a. Now, press the [SHIFT] [F1] [Range] [Range] keys. This clears the memory and sets the domain and range to the default values in the machine.

EXAMPLE 1.17 (Cont.)

Range
Xmin:-4.7
max:4.7
scl:1.
Ymin:-3.1
max:3.1
scl:1.
INIT

FIGURE 1.23a

The part of the coordinate system that will be shown on the screen is indicated by Xmin (the smallest value of x displayed on the screen), Xmax (the corresponding largest x-value), and Xscl (the x-axis' scale or, more precisely, the distance between "tick marks" on the x-axis). The range is similarly restricted by Ymin, Ymax, and Yscl. As explained in Appendix A, these values can be changed and we will do that in later graphing calculator activities. However, for this example, let's use these preset values.

You are now ready to input a formula describing the function you want graphed, in this example $f(x) = 2x + 7$. To input $f(x) = 2x + 7$, press the following sequence of keys:

[Graph] 2 [X,θ,T] [+] 7 [EXE]

Be careful! To input the variable x, you press the [X,θ,T] key or the [ALPHA] [X]. Do not confuse the [X] with the multiplication [×] key.

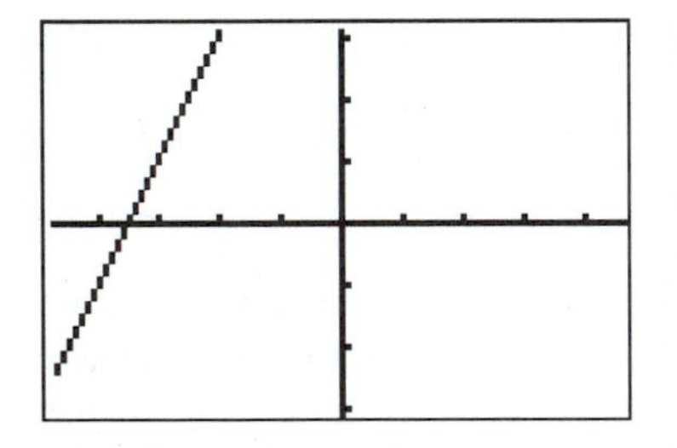

FIGURE 1.23b

The result is the graph shown in Figure 1.23b. This graph is fine, but it does not show us where the graph crosses the y-axis. Let's see one way that we can change the domain and range of the display screen. Press the [SHIFT] [F2] [F4] keys. This causes the calculator to "zoom out" from the middle of the screen. The result is the graph in Figure 1.23d. Press [Range] to see how the domain and range values have changed. As you can see, they have doubled.

You still cannot see where the graph crosses the y-axis, so press the [F2] [F4] keys again. The calculator "zooms out" once again and produces the graph in Figure 1.23c.

You can now see where the line crosses the x- and y-axes. It is hard to determine the coordinates of these intercepts by looking at the calculator screen. In the next section, we will see how to determine the approximate coordinates of the points.

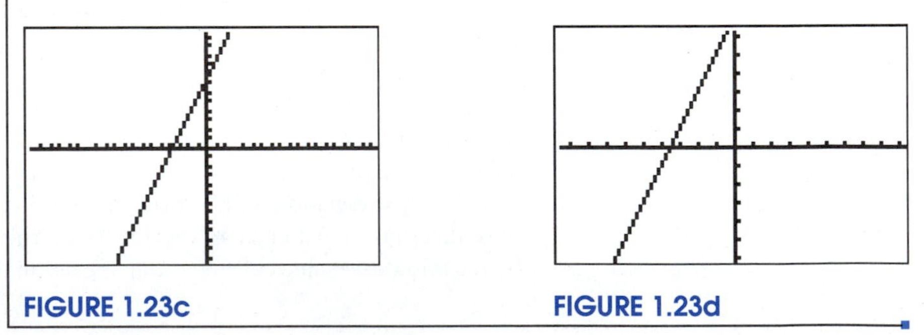

FIGURE 1.23c FIGURE 1.23d

EXAMPLE 1.18

Use a Texas Instruments TI-82 graphing calculator to sketch the graph of $f(x) = 2x + 15$.

Solution First press [ZOOM] 6. You may have to wait for the screen to clear. This clears the memory and sets the domain and range to the default values in the

EXAMPLE 1.18 (Cont.)

machine. Wait until the xy-axes appear on the screen, and then press WINDOW. The screen should show an image something like that shown in Figure 1.24a.

```
WINDOW FORMAT
Xmin=-10
Xmax=10
Xscl=1
Ymin=-10
Ymax=10
Yscl=1
```

FIGURE 1.24a

The part of the coordinate system that will be shown on the screen is indicated by Xmin (the smallest value of x displayed on the screen), Xmax (the corresponding largest x-value), and Xscl (the x-axis scale or, more precisely, the distance between tick marks on the x-axis). The range is similarly restricted by Ymin, Ymax, and Yscl. As explained in Appendix A, these values can be changed, and we will do that in later graphing calculator activities. However, for this example, let's use these preset values.

You are now ready to input a formula describing the function you want graphed, in this example $f(x) = 2x + 15$. To input $f(x) = 2x + 15$, press the following sequence of keys:

Y= CLEAR 2 X,T,θ + 15 GRAPH

Be careful! To input the variable x, you press the X,T,θ key or the ALPHA X keys. Do not confuse the X key with the multiplication, or ×, key.

The result is the graph shown in Figure 1.24b.

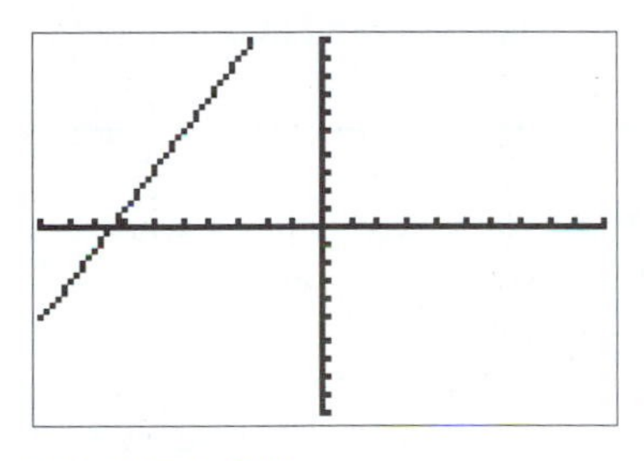

FIGURE 1.24b

This graph is fine, but you will notice that it does not show us where the graph crosses the y-axis. In this case we will need to zoom out to see more of the graph. Let's see how using the calculator's "zoom-out" ability changes the domain and range of the display screen. Press the ZOOM 3 ENTER keys. This causes the calculator to "zoom out" from the middle of the screen. The result may look like the graph in Figure 1.24c. Press WINDOW to see how the domain and range values have changed. In Figure 1.24c, they are Xmin= −40, Xmax= 40, Xscl= 1, Ymin= −40, Ymax= 40, and Yscl= 1.

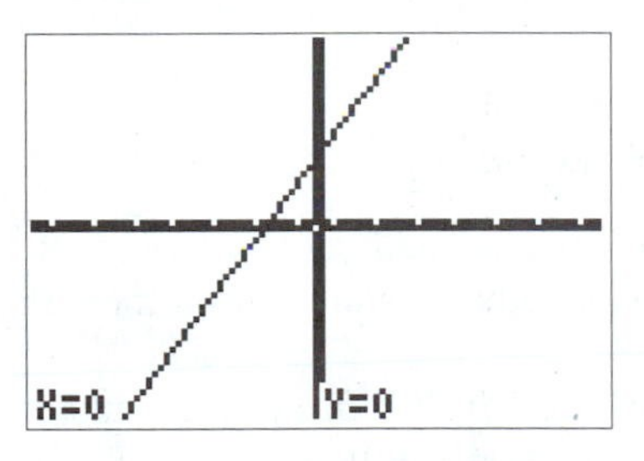

[−40, 40, 1] x [−40, 40, 1]

FIGURE 1.24c

It is hard to determine the coordinates of these intercepts by looking at the calculator screen. In the next section, we will see how to determine the approximate coordinates of the points.

Note

Whenever we are not using the default window settings on a calculator, we will adopt the notation [Xmin, Xmax, Xscl] × [Ymin, Ymax, Yscl]. Thus, as indicated in Figure 1.24c, the viewing window is $[-40, 40, 1] \times [-40, 40, 1]$. If Xscl and Yscl are both 1 we may shorten this to $[-40, 40] \times [-40, 40]$.

Let's make the graph of a curve, which is not a straight line. In Section 1.2, we graphed $y = x^2 - 3$ with the result shown in Figure 1.14. Let's see what that looks like on a calculator.

EXAMPLE 1.19

Use a graphing calculator to sketch the graph of $y = x^2 - 3$.

Solution We will show how to do this on a Casio fx-7700G. You will have to make a few changes if you are using a different machine.

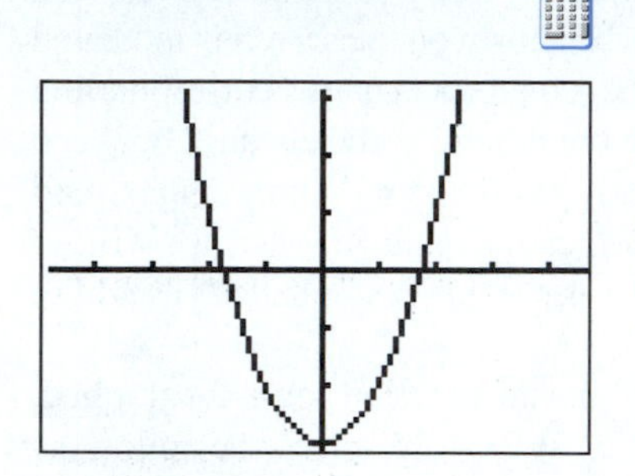

FIGURE 1.25

First, enter $\boxed{\text{Shift}}\,\boxed{\text{F5}}\,\boxed{\text{EXE}}$ to clear the graphics screen of the previous graphs from the screen. As before, press the $\boxed{\text{Range}}\,\boxed{\text{Shift}}\,\boxed{\text{F1}}\,\boxed{\text{Range}}\,\boxed{\text{Range}}$ keys to clear the memory and set the domain and range to the default values in the machine.

You are now ready to input a formula describing the function you want graphed. To input $y = x^2 - 3$, press the following sequence of keys:

$$\boxed{\text{GRAPH}}\,\boxed{\text{X},\theta,\text{T}}\,\boxed{\text{Shift}}\,\boxed{x^2}\,\boxed{-}\,3\,\boxed{\text{EXE}}$$

The result is the graph shown in Figure 1.25. Compare this to the earlier graph of this same function in Figure 1.14.

From this graph we can see that the curve crosses the y-axis near $y = -3$ and the x-axis around $x = \pm 1.8$. In the next section, we shall see how to obtain better approximations of these values.

By now you have probably noticed that in order to tell the calculator what you want it to graph on a Casio fx-7700G, you press the $\boxed{\text{GRAPH}}$ key and on a TI-82, you press the $\boxed{\text{Y=}}$ key. In each case you obtain a phrase like Casio's "Graph Y=" and TI-82's "Y1=" on the calculator's screen. This should remind you that you can only graph a function. It may be necessary to solve an equation for y before you can graph it.

EXAMPLE 1.20

Use a graphing calculator to sketch the graph of $f(x) = \dfrac{x^2+1}{x+2}$.

Solution Graphing this is very similar to the previous graphing you have done. However, you must be careful to insert enough parentheses and to insert them in the right spots.

You might want to write this equation on one line. First, since your calculator will be graphing something in the $y =$ form, you should replace the $f(x) =$ with $y =$. Then write the fraction on one line. Remember, everything over the fraction bar is to be grouped, and everything under the fraction bar is another group. Parentheses are used to group terms. So, we end up writing the equation as

$$y = (x^2+1)/(x+2)$$

You are now ready to input this equation into your graphing calculator. On a TI-82, you would press the following sequence of keys:

$$\boxed{\text{Y=}}\,\boxed{(}\,\boxed{\text{X,T},\theta}\,\boxed{x^2}\,\boxed{+}\,1\,\boxed{)}\,\boxed{/}\,\boxed{(}\,\boxed{\text{X,T},\theta}\,\boxed{+}\,2\,\boxed{)}\,\boxed{\text{GRAPH}}$$

The result is shown in Figure 1.26a.

Sometimes, as in Figure 1.26a, the calculator draws in some points (usually shown as vertical lines) that are really not part of the graph. This may also happen with some computer graphing software. Sometimes it is possible to adjust the viewing window, as in Figure 1.26b, to eliminate these extraneous points. Consult the users' manual for your calculator or software program for suggestions.

FIGURE 1.26a

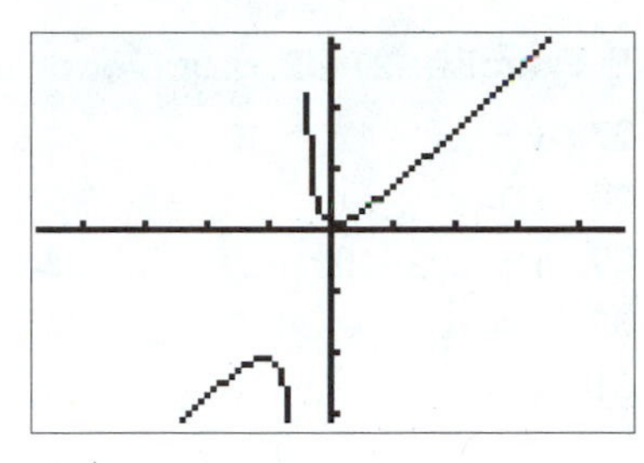

[−18.8, 18.8, 4] x [−12.4, 12.4, 4]

FIGURE 1.26b

EXAMPLE 1.21

Use a graphing calculator to sketch the graph of $y+5=\sqrt{x+2}$.

Solution First, solve this equation for y, obtaining $y=\sqrt{x+2}-5$.

You will want to put parentheses around the $x+2$. If you do not, the same as $y=\sqrt{x}-3$. On a TI-82 you press the following sequence of keys:

[Y=] [2nd] [√] [(] [X,T,θ] [+] 2 [)] [−] 5 [GRAPH]

The result is shown in Figure 1.27. ▪

FIGURE 1.27

Exercise Set 1.3

Graph each of these functions with a calculator or on a computer. Compare the machine's graph with the graph you made of the same functions in Section 1.2.

1. $y=x$
2. $y=2x$
3. $y=-3x$
4. $y=\frac{1}{2}x-3$
5. $y=-4x+2$
6. $x+y=2$
7. $x-y=2$
8. $2x+y=1$
9. $3x-6y=9$
10. $3x-6y=-6$
11. $y=x^2$
12. $y=x^2+3$
13. $y=x-2$
14. $y=x^2+2x+1$
15. $y=x^2-6x+9$
16. $y=-x^2+2$
17. $y=\dfrac{1}{x}$
18. $y=\dfrac{1}{x+3}$
19. $y=x^3$
20. $y=\sqrt[3]{x}$
21. $y^2+x^2=25$ (HINT: Solve for y, then graph two equations.)
22. $y=\sqrt{25-x^2}$
23. $y=|x+2|$ (HINT: Use the [Abs] key. Read your manual to see how to access this on your calculator.)
24. $\dfrac{x^2}{4}+\dfrac{y^2}{9}=1$ (HINT: Solve for y, then graph two equations.)
25. $\dfrac{x^2}{4}-\dfrac{y^2}{9}=1$
26. $y=x^3-x^2$

In Exercises 27–42, graph each of the given functions with a calculator or on a computer.

27. $y = x^2 - 4x$
28. $y = x^2 + 6x$
29. $y = x^2 - 10x + 37$
30. $y = 0.1x^2 - 20$
31. $y = -x^2 - 4x$
32. $y = -x^2 + 6x$
33. $y = \sqrt{x+5}$
34. $y = \sqrt{x-5}$
35. $y = \sqrt{9-x}$
36. $y = -\sqrt{5-x}$
37. $y = \dfrac{x+2}{x}$
38. $y = \dfrac{x}{x+2}$
39. $y = \dfrac{x^2+3}{x+2}$
40. $y = \dfrac{x+2}{x^2+3}$
41. $y = \dfrac{x^2-4}{x+2}$
42. $y = \dfrac{x+2}{x^2-4}$

In Your Words

43. Explain the notation $[-15, 15, 2] \times [-10, 20, 4]$.
44. Write a description that tells how to zoom in and zoom out on your calculator.

1.4 USING GRAPHS TO SOLVE EQUATIONS

A graph not only gives you a picture of what an equation looks like, but it can be used to help solve the equation. In Section 1.2, we discussed the points where the graph crosses the axes. The y-intercepts were the points where the graph crossed or touched the y-axis; the x-intercepts were the points where the graph crossed or touched the x-axis. The x-intercepts are also known as the **roots** or **solutions** of the equation, because these are the values of x for which y will equal zero. If the equation is a function of x, these are called the roots (or **zeros**) of the function because they are the values of x when $f(x) = 0$.

EXAMPLE 1.22

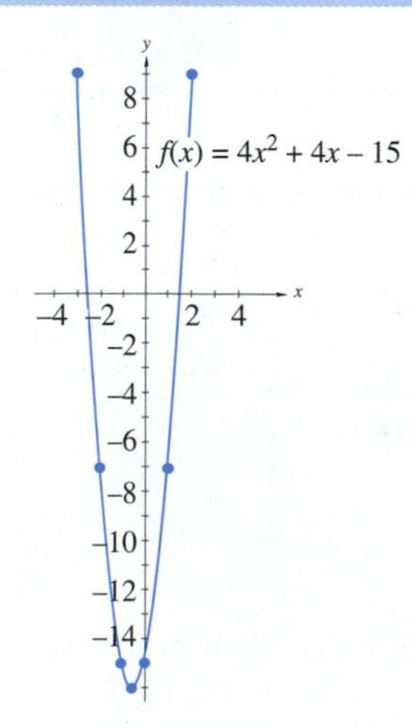

FIGURE 1.28

Find graphically the approximate roots of $4x^2 + 4x - 15 = 0$.

Solution First, set $f(x) = 4x^2 + 4x - 15$. Then set up a partial table of values for the function f.

x	-3	-2	-1	0	1	2	3
$y = f(x)$	9	-7	-15	-15	-7	9	33

These points are plotted and connected to form the curve in Figure 1.28. From the graph we can see that the x-intercepts look as if they are at $x = -2\frac{1}{2}$ and $x = 1\frac{1}{2}$. A check of the function confirms that $f\left(-2\frac{1}{2}\right) = 0$ and $f\left(1\frac{1}{2}\right) = 0$. These are the roots of this equation.

EXAMPLE 1.23

Find graphically the approximate roots of $f(x) = x^2 - 3x - 1$.

Solution Again, if we set up a table of values we get

x	−2	−1	0	1	2	3	4	5
$f(x)$	9	3	−1	−3	−3	−1	3	9

Graphing the curve determined by these points, we get the curve in Figure 1.29. From the graph we can see that there appear to be roots at $x = -0.25$ and $x = 3.25$. If we evaluate the function at these two values, we get $f(-0.25) = -0.1875$ and $f(3.25) = -0.1875$. This shows two things. First, it shows that -0.25 and 3.25 are *not* roots of this function, since $f(x) \neq 0$ at either of these two points. Second, it does show that the roots of this function are "close" to -0.25 and 3.25. ▪

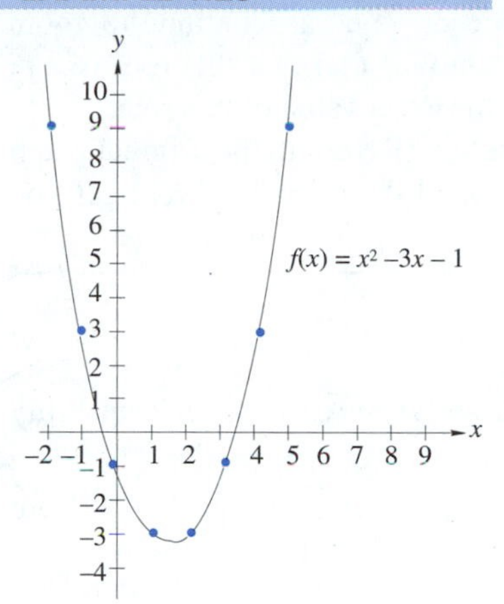

FIGURE 1.29

If we wanted more accurate approximations of the roots of the function in Example 1.23, we could substitute different values for x until we got an approximation that was as accurate as we wanted. In a later chapter, we will find out how to determine the exact roots to this function. In the next example, we will show how to use a graphing calculator to approximate this function's roots.

EXAMPLE 1.24

Use a graphing calculator or computer graphing software with "zoom" capability to find the approximate roots of $f(x) = x^2 - 3x - 1$.

Solution This is the same function we graphed in Example 1.23. We already know that the roots are near $x = -0.25$ and $x = 3.25$. Details on how to use the "zoom" features of a Casio fx-7700G and Texas Instrument's TI-81 graphing calculator are in Appendix A.

When you graph the function using the calculator's default settings for the range, you obtain the graph in Figure 1.30a on an fx-7700G or the graph in Figure 1.30b on a TI-82.

As you can see from these figures, the graph crosses the x-axis at two points; these are the x-intercepts or roots. Use the "trace" function of your calculator until the cursor is positioned near the x-intercept on the left. Figure 1.30c shows the fx-7700G screen and indicates that the cursor is at $(-0.3, -0.01)$, and Figure 1.30d shows that the TI-82 places the cursor at $(-0.212766, -0.3164328)$.

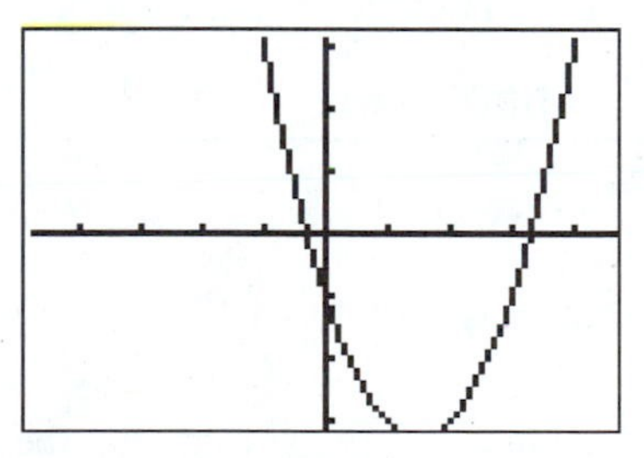

FIGURE 1.30a

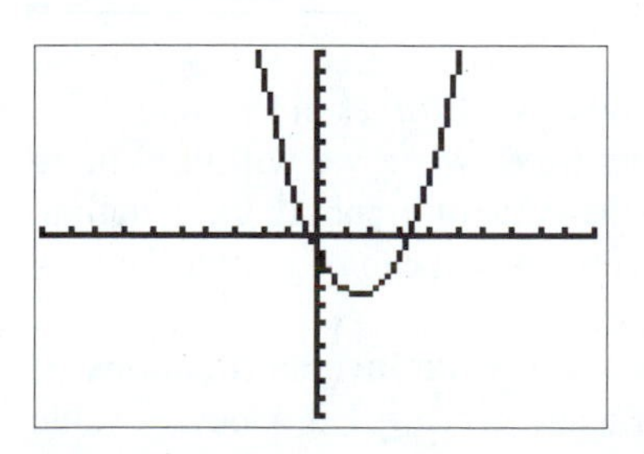

FIGURE 1.30b

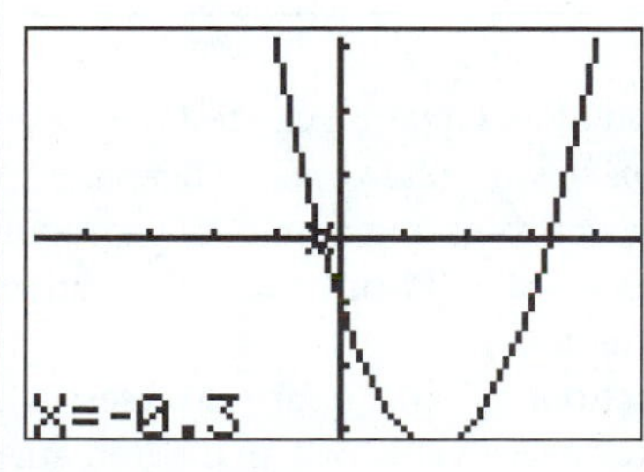

FIGURE 1.30c

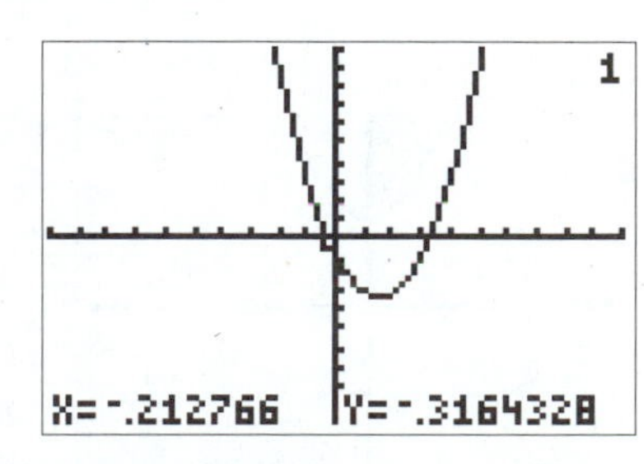

FIGURE 1.30d

EXAMPLE 1.24 (Cont.)

For more accurate approximations of this root, you can continue to zoom in around this point on the graph. If you do, you will find that this root is $x \approx -0.302775638$. Later we will show how to find the exact value of this root.

Now, approximate the other root of this function. Return to the original graph in either Figure 1.30a or 1.30b, and zoom in around the right-hand root. If you zoom in enough, you will obtain $x \approx 3.302775638$.

EXAMPLE 1.25

Use a graphing calculator or computer graphing software with "zoom" capability to find the approximate roots of $y = x^4 - 2x^3 - 1$.

Solution The graph of this function on a TI-82 is shown in Figure 1.31a. Here Xmin = −5, Xmax = 5, Ymin = −3, and Ymax = 3. Zoom in, as in Figure 1.31b, press [Trace] to move the cursor, and see that $x \approx -0.7180851$.

Notice that the second root is not on the screen. Returning to the original graph, we move the cursor over near the second root. Zooming in around this point and using the trace function, we see that the other root is approximately $x \approx 2.105$.

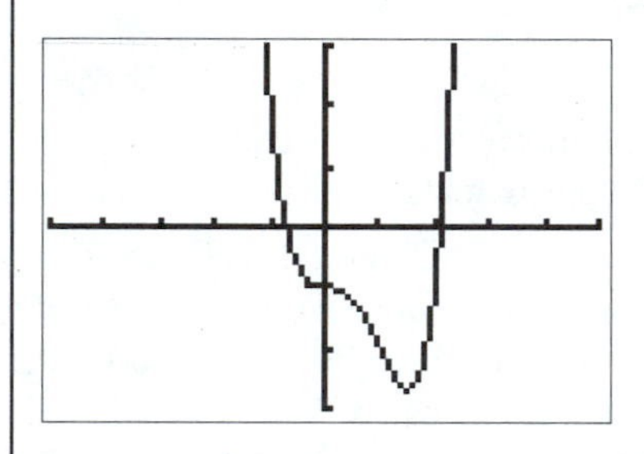

FIGURE 1.31a

1

X=-.7180851 Y=.00644676

FIGURE 1.31b

EXAMPLE 1.26

Use a graphing calculator or computer graphing software with "zoom" capability to determine any x-intercepts of $y = x^2 + 1.5$

Solution The graph of this function is shown in Figure 1.32. As you can see, the graph does not cross the x-axis. Thus, this function does not have any real roots.

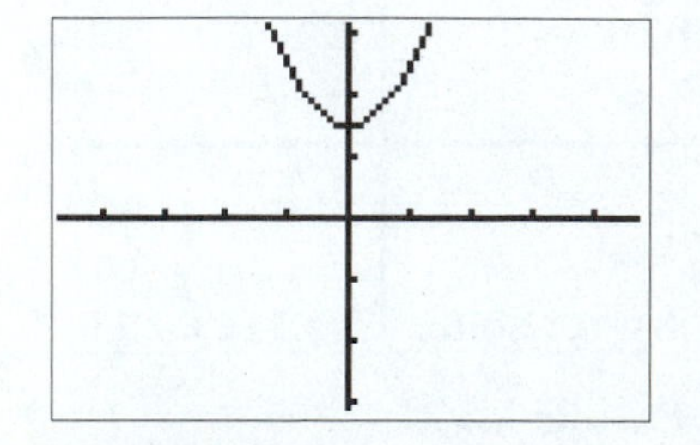

FIGURE 1.32

The graph of an equation can tell the approximate roots or, as in Example 1.26, if the equation has no real roots. There are many times when we will need more accurate values of roots and a quicker procedure than that of graphing the equation. In later chapters, we will learn some alternative methods that can be used to find the roots of an equation.

Two functions, f and g, are inverses of each other, **or inverse functions**, if $f[g(x)] = x$ for every value of x in the domain of g and $g[f(x)] = x$ for every value of x in the domain of f. We also said that if g was the inverse of f, we would write $g(x) = f^{-1}(x)$.

The graph of a function can be used to determine if the function has an inverse and to sketch the graph of the inverse function. You may remember that the vertical line test indicates that if no vertical line intersected a graph more than once, then the graph represents a function. A similar test can be used to determine if a function has an inverse.

Horizontal Line Test

The horizontal line test states that a function has an inverse function if no horizontal line intersects its graph at more than one point.

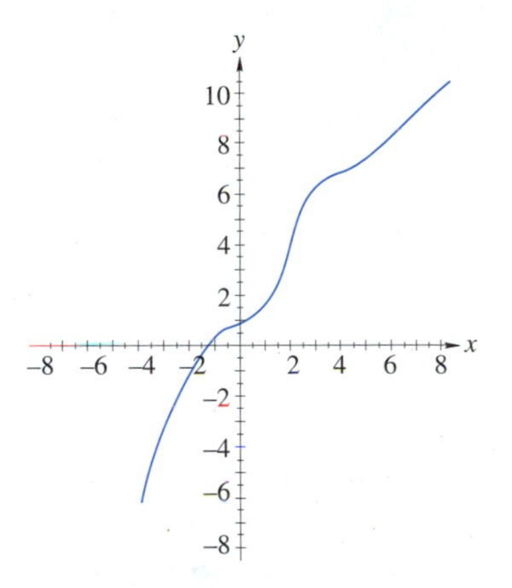

FIGURE 1.33a

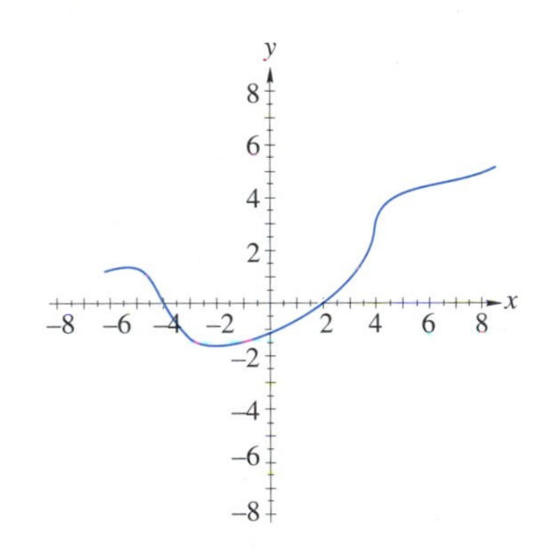

FIGURE 1.33b

For example, Figures 1.33a and 1.33b show the graphs of two functions. The function in Figure 1.33a has an inverse function, because it is not possible to draw a horizontal line that will intersect the graph at more than one point. The function in Figure 1.33b does not have an inverse because it is possible to draw a horizontal line that will intersect the graph twice. In fact, the x-axis intersects the graph in Figure 1.33b in at least two places.

Now let's use our graphing ability to graph the inverse of a function. The graph of $f(x) = \sqrt{x-1}$ is shown in Figure 1.34. Using the horizontal line test, we can see that this function has an inverse. Now, draw the line $y = x$. (In Figure 1.34 this is the dashed line.) Suppose you placed a mirror on the line $y = x$. The image you would see is indicated by the colored curve in Figure 1.34.

What this essentially does is take any point (a, b) on the function and convert it to its mirror image (b, a). Remember that if (a, b) is a point on a function f, then $f(a) = b$; and if (b, a) is a point on some function g, then $g(b) = a$. But, since $f[f^{-1}(a)] = f(b) = a$, we can see that g must be f^{-1}.

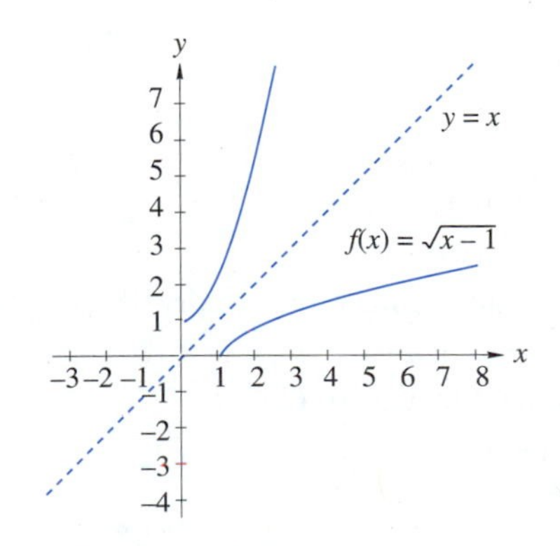

FIGURE 1.34

EXAMPLE 1.27

Graph the inverse function of $y = x^3$.

Solution The graph of $y = x^3$ is given in Figure 1.35a and the line $y = x$ is shown by a dashed line. The reflection of $y = x^3$ in the line $y = x$ is shown in Figure 1.35b and is the graph of $y = \sqrt[3]{x}$. If $f(x) = x^3$, then $f^{-1}(x) = \sqrt[3]{x}$.

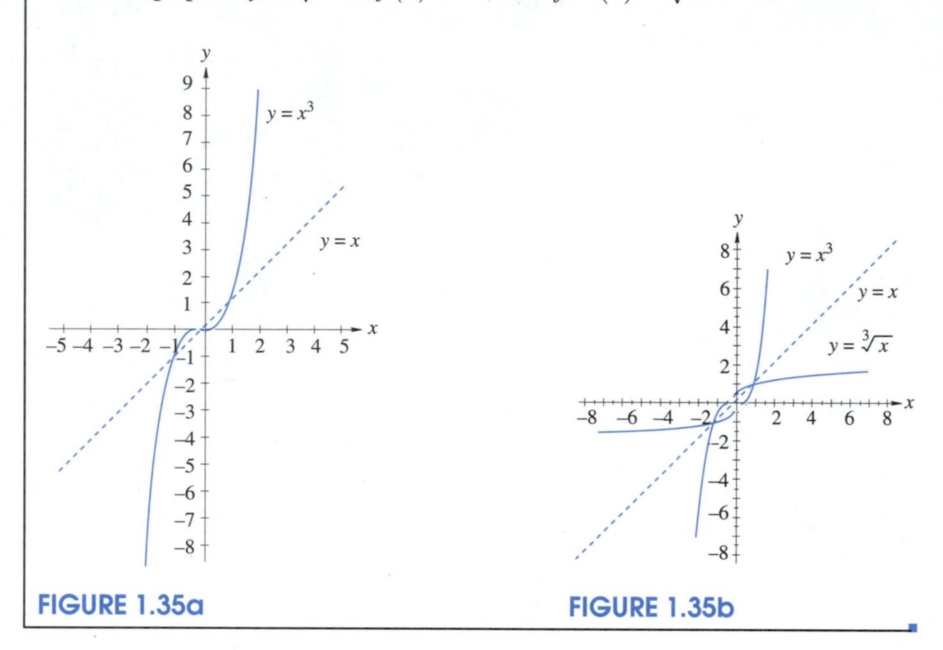

FIGURE 1.35a

FIGURE 1.35b

Application

EXAMPLE 1.28

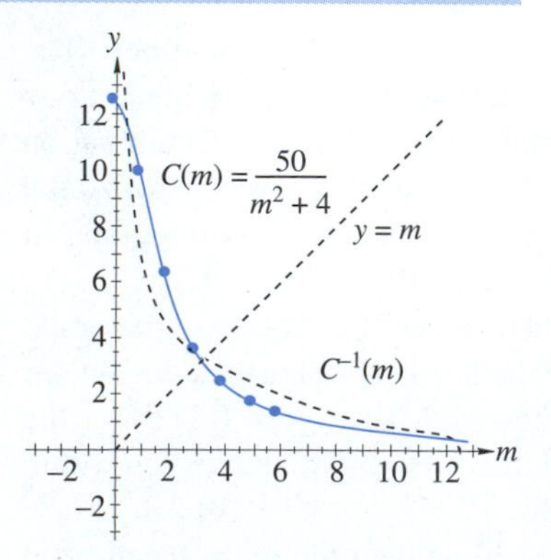

FIGURE 1.36a

In Example 1.14, we made the following partial table for the concentration in parts per million (ppm) of a certain pollutant m mi from a certain factory as given by the function

$$C(m) = \frac{50}{m^2+4}, \qquad m \geq 0$$

m	0	1	2	3	4	5	6
$C(m)$	12.5	10	6.25	3.85	2.5	1.72	1.25

Next, we used our table to sketch the graph of this function.

(a) Use the table to sketch the graph of the inverse of this function.

(b) Use the table or the graph to estimate how far you are from the factory, if the concentration of the pollutant is 3.08 ppm.

(c) Determine the equation that describes the inverse function of C.

(d) Use your equation for C^{-1} to determine $C^{-1}(3.08)$.

Solutions

(a) The graph of the function $y = C(m)$ is shown in Figure 1.36b. The reflection of the graph in the line $y = m$ produces the inverse function shown by the dotted curve in Figure 1.36a.

EXAMPLE 1.28 (Cont.)

(b) We will use the graph of the original function to determine $C^{-1}(3.08)$. Move up the vertical axis for the graph of $C(m)$ until you reach the point that is approximately at 3.08 on the vertical axis. Draw a horizontal line from this point until it reaches the graph of C. Draw a vertical line until it crosses the horizontal axis. This should be near 3.5 as shown in Figure 1.36b. Thus, $C^{-1}(3.08) \approx 3.5$ and we conclude that the pollutant was collected about 3.5 mi from the factory.

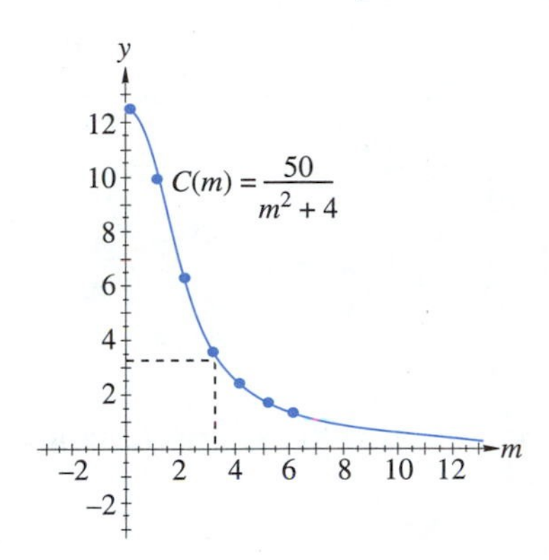

FIGURE 1.36b

(c) As described in Section 1.2, we will let $y = \dfrac{50}{m^2+4}$. Solving for m, we obtain

$$m = \sqrt{\frac{50}{y} - 4}, \text{ for } 0 < y \le 12.5. \text{ Thus, } C^{-1}(m) = \sqrt{\frac{50}{m} - 4}, \text{ for } 0 < m \le 12.5.$$

(d) Substituting 3.08 in the formula for C^{-1}, we obtain

$$C^{-1}(3.08) = \sqrt{\frac{50}{3.08} - 4} \approx 3.50$$

Thus, the formula and the graph give us approximately the same answers. ■

Application

EXAMPLE 1.29

Use a graphing calculator to rework parts of Example 1.28.

(a) Sketch the graph of $C(m) = \dfrac{50}{m^2+4}$, $m \ge 0$.

(b) Use the graph and your calculator to estimate how far you are from the factory if the concentration of the pollutant is 3.08 ppm.

Solutions

(a) You must make two changes before we begin graphing. First, you must replace $C(m)$ with y. Second, your calculator will only allow you to use certain variables when you graph a function. At present the only variable you can use is x. So, rewrite the given function as $y = \dfrac{50}{x^2+4}$, $x \ge 0$. Don't forget the correct use of parentheses. You will want to graph this as $y = \dfrac{50}{(x^2+4)}$. Also, to get the graph to look like the one in Figure 1.36b, change the viewing window to $[0, 14, 1] \times [0, 14, 1]$. The result is the graph shown in Figure 1.37.

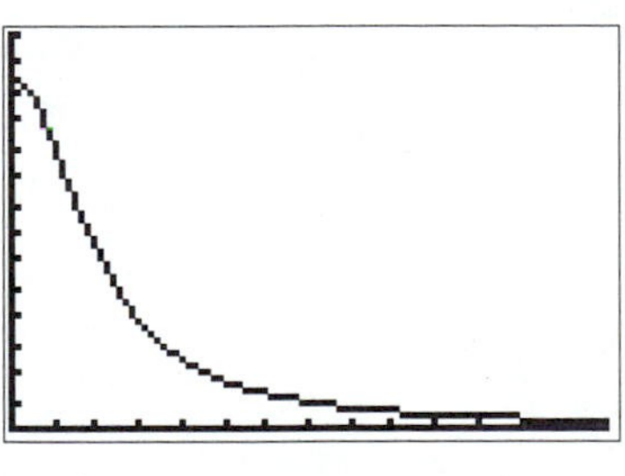

[0, 14, 1] x [0, 14, 1]

FIGURE 1.37

(b) We want to know when the concentration of the pollutant is 3.08 ppm. This means we want to know the value of x when $y = 3.08$. Press TRACE and move the cursor until you have a y-value near 3.08. On this particular calculator, the y-values change from around 2.98 to 3.17. Zooming in around one of these points and using the trace function, we see that when $y \approx 3.08$, then $x \approx 3.4976$. Once again, we conclude that the pollutant was collected about 3.5 mi from the factory. ■

Exercise Set 1.4

The equations in Exercises 1–18 all have at least one root between −10 and 10. Write each equation in the form *y* = *f*(*x*). Graph each equation to find the approximate value of the roots. If possible, use a graphing calculator or a computer graphing program to help you.

1. $2x+5=0$
2. $5x-9=0$
3. $x^2-9=0$
4. $4x^2-10=0$
5. $x^2=5x$
6. $x^2=4x-3$
7. $x^2+5x-3=0$
8. $3x^2+5x-3=0$
9. $20x^2+21x=54$
10. $x^4-4x^3-25x^2+x+6=0$
11. $\sqrt{x+1}=0$
12. $\sqrt{x-2.5}=0$
13. $\sqrt{x+1}=3$
14. $\sqrt{x-3}=2$
15. $\dfrac{x}{x+1}=-3$
16. $\dfrac{x}{x-2}=5$
17. $\dfrac{x}{x-2}=2-x^2$
18. $\dfrac{x^2+5}{x-3}=x^2-2$

In Exercises 19–34, graph each of the functions with a calculator or on a computer. Use the graph to help determine the domain and range of each function.

19. $y=x^2-4x$
20. $y=x^2+6x$
21. $y=x^2-10x+37$
22. $y=0.1x^2-20$
23. $y=-x^2-4x$
24. $y=-x^2+6x$
25. $y=\sqrt{x+5}$
26. $y=\sqrt{x-5}$
27. $y=\sqrt{9-x}$
28. $y=-\sqrt{5-x}$
29. $y=\dfrac{x+2}{x}$
30. $y=\dfrac{x}{x+2}$
31. $y=\dfrac{x^2-1}{x+2}$
32. $y=\dfrac{x+2}{x^2+3}$
33. $y=\dfrac{x^2-4}{x+2}$
34. $y=\dfrac{x+2}{x^2-4}$

Use the horizontal line test to determine whether or not the functions graphed in Exercises 35–38 have inverse functions.

35.

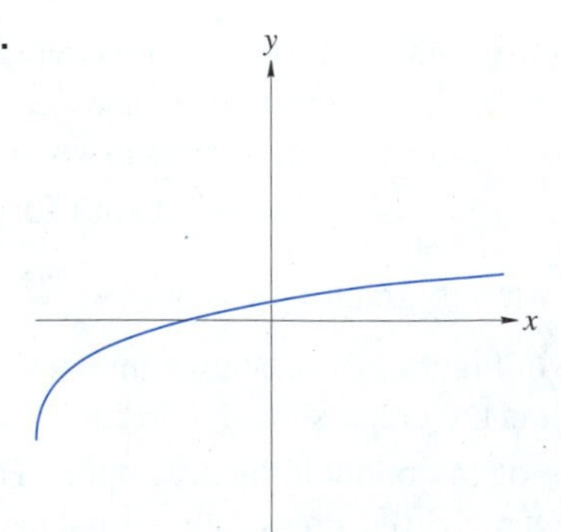

36.

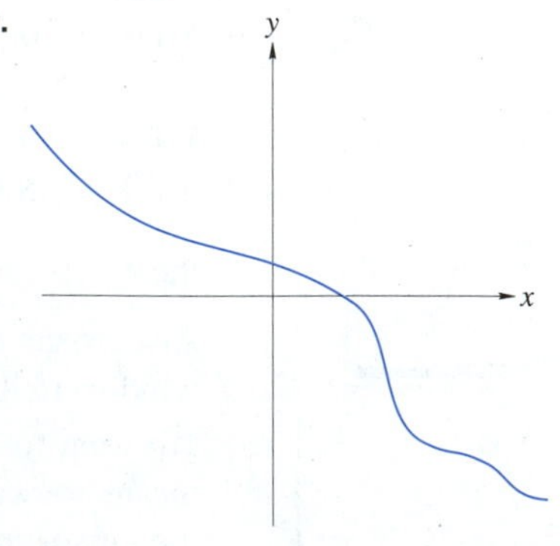

37.

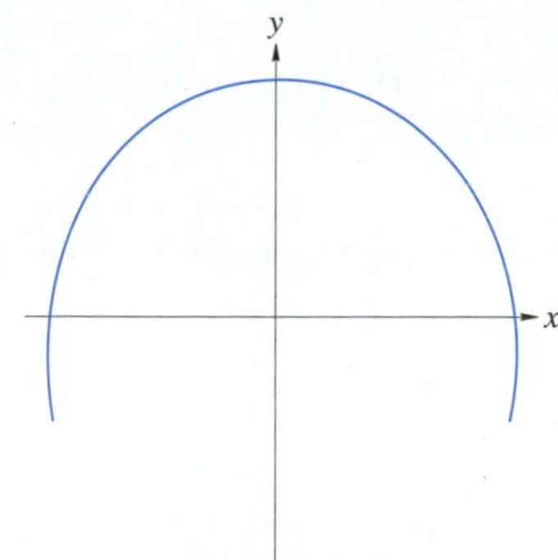

38.

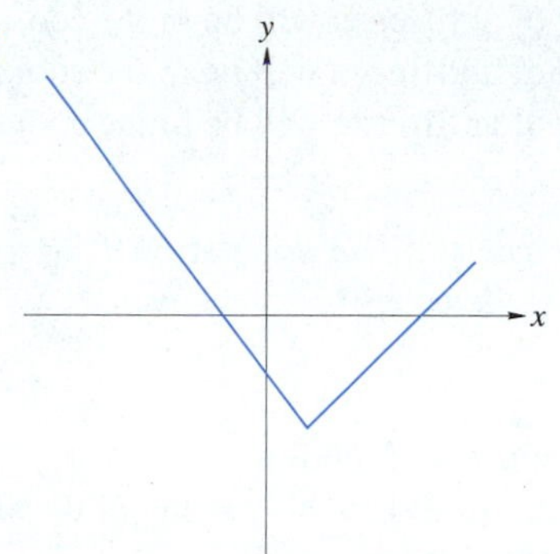

Each of the functions in Exercises 39–42 have inverses. Sketch the graph of the inverse of each function.

39.

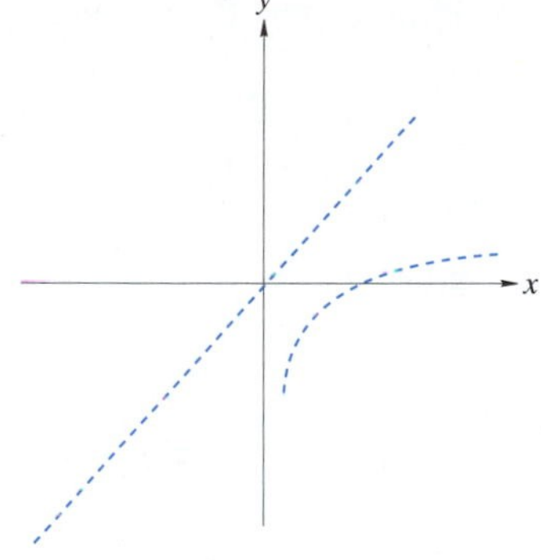

40.

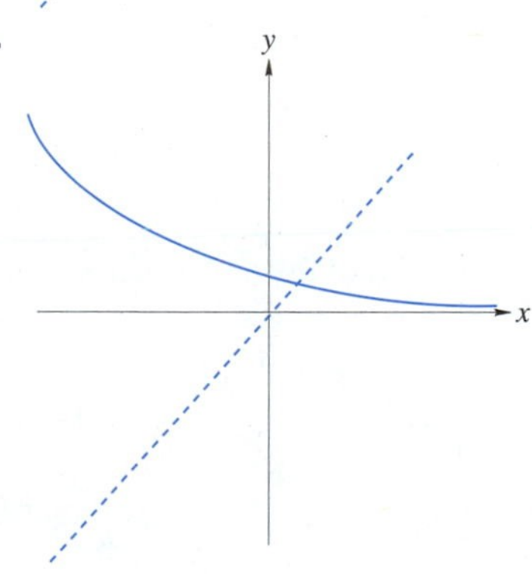

41.

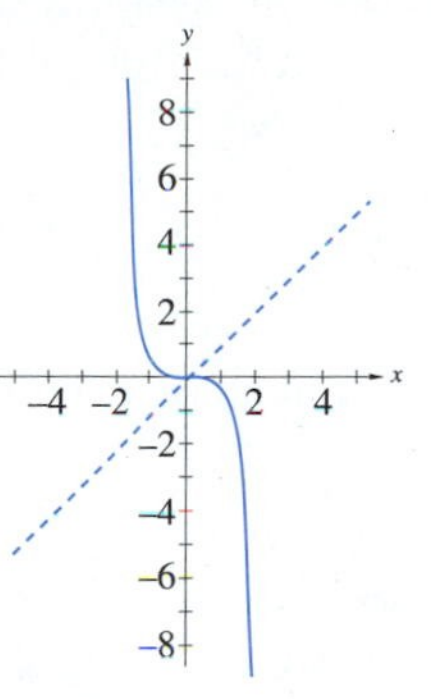

42.

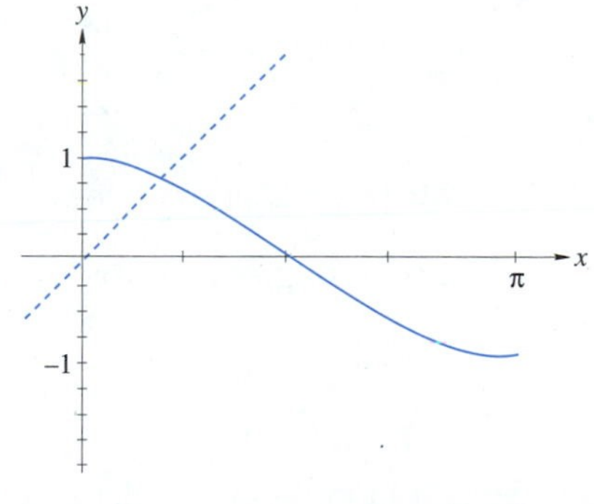

Solve Exercises 43–46.

43. *Automotive technology* Study the following table. It shows that if the antifreeze content of a coolant is increased, the boiling point of the coolant is also increased.

Percent antifreeze in coolant	0	10	20	30	40	50	60	70	80	90	100
Boiling temperature °F	210	212	214	218	222	228	236	246	258	271	330

(a) What percent of antifreeze will be in the coolant if the boiling point of the system is 218°F?
(b) What percent of antifreeze will be in the coolant if the boiling point of the system is 236°F?
(c) What percent of antifreeze will be in the coolant if the boiling point of the system is 250°F?

44. *Environmental science* The population P of a certain species of animal depends on the number n of a smaller animal on which it feeds, with

$$P(n) = 5\sqrt{n} - 10$$

(a) Determine the inverse function for P.
(b) If the population of P is 5, how many of the small animals are there?
(c) Graph the inverse function of P.

45. *Construction* This table contains the results of a series of drillings used to determine the depth of the bedrock at a building site. Drillings were taken along a straight line down the middle of the lot, where the building will be placed. In the table, x is the distance from the front of the parking lot and y is the corresponding depth. Both x and y are given in feet.

x	0	20	40	60	80	100	120	140	160
y	33	35	40	45	42	38	46	40	48

These ordered pairs have been plotted on the following graph. Does this graph have an inverse function?

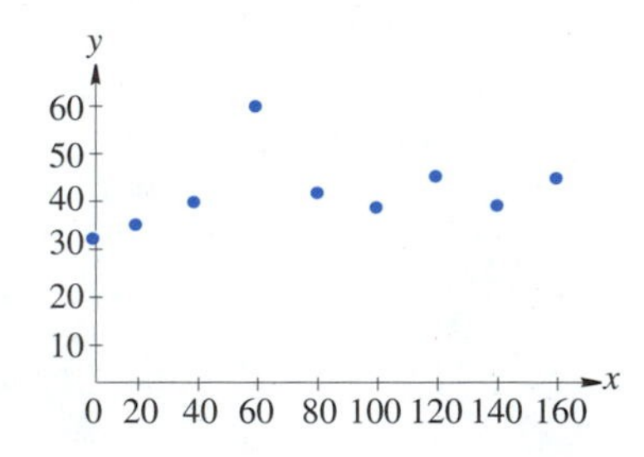

46. *Environmental science* Suppose a cost-benefit model is given by

$$C(x) = \frac{6.4x}{100 - x}$$

where $C(x)$ is the cost in millions of dollars for removing x percent of a given pollutant.
(a) What is the cost of removing 30% of the pollutant?
(b) What is the cost of removing 60% of the pollutant?
(c) What is the inverse function of C?
(d) If a community can only spend $12,000,000, what percent of the pollutant can be removed?
(e) Graph the given function and its inverse function.

In Your Words

47. Explain how to use your calculator to solve an equation.

48. On a sheet of paper, explain the horizontal line test. Do not look at the definition in the book.

CHAPTER 1 REVIEW

Important Terms and Concepts

Cartesian coordinate system
Dependent variable
Domain
Horizontal line test
Intercepts
Inverse function
Linear equation
Origin
Quadrants
Range
Rectangular coordinate system
Relation
Roots
Slope
Solutions
Vertical line test
x-axis
x-intercept
y-axis
y-intercept
Zeros

Review Exercises

For Exercises 1–6 (a) graph each of the relations, (b) determine the domain, range, *x*-intercept, and *y*-intercept of each relation, (c) use the vertical line test to determine if each relation is a function, (d) use the horizontal line test to determine if each function has an inverse function, and (e) graph each inverse function that exists.

1. $y = 8x - 7$
2. $y = 2x^2 - 4$
3. $y = \sqrt{x} - 3$
4. $y = \frac{1}{2}x^3 + 2$
5. $y = x^2 - 2x$
6. $y = x^2 + 4$

In Exercises 7–12, given the function $f(x) = 4x - 12$, determine the following.

7. $f(0)$
8. $f(-2)$
9. $f(3)$
10. $f(a)$
11. $f(a-2)$
12. $f(x+h)$

In Exercises 13–19, given the function $g(x) = \dfrac{x^2-9}{x^2+9}$, determine the following.

13. $g(0)$
14. $g(3)$
15. $g(-3)$
16. $g(-2)$
17. $g(4)$
18. $g(-5)$
19. Use your values from Exercises 13–18 to graph $g(x)$. From your graph, determine the zeros of g.

In Exercises 20–26, graph each of these functions or relations with a calculator or computer.

20. $y = \dfrac{5}{x-4}$
21. $y = -\dfrac{7}{x+3}$
22. $y = \sqrt{2x-3}$
23. $y = \sqrt{5-4x}$
24. $y = \dfrac{x}{\sqrt{x+2}}$
25. $x^2 + y^2 = 9$
26. $\frac{1}{2}x^2 + y^2 = 16$

Each of the equations in Exercises 27–30 has a root between −10 and 10. Write each equation in the form $y = f(x)$. Graph each equation to find the approximate value of the roots.

27. $4x + 7y = 0$

28. $x^2 - 20 = 0$

29. $2x^2 + 10x + 4 = 0$

30. $8x^3 - 20x^2 - 34x + 21 = 0$

Solve Exercises 31–34.

31. *Business* The manager of a videotape store has found that n videotapes can be sold if the price is $P(n) = 35 - \frac{n}{20}$ dollars.

(a) What price should be charged in order to sell 101 videotapes? 350 videotapes? 400 videotapes?

(b) Find an expression for the revenue from the sale of n vidoetapes, where revenue = demand × price.

(c) How much revenue can be expected if 101 videotapes are sold? if 350 are sold? if 400 are sold?

32. *Business* A videotape store has learned that the function $R(n) = 30n - \frac{n^2}{20}$ is a good predictor of its revenue, in dollars, from the sale of n tapes. The cost of operating the store is given by $C(n) = 550 + 10n$.

(a) If the profit P is given by $P(n) = R(n) - C(n)$, what is the profit function?

(b) How much profit will the store make if it sells 30 videotapes? 100 videotapes? 150 videotapes? 300 videotapes? 400 videotapes?

33. *Medical technology* A measure of cardiac output can be determined by injecting a dye into a vein near the heart and measuring the concentration of the dye. In a normal heart, the concentration of the dye is given by the function

$$h(t) = -0.02t^4 + 0.2t^3 - 0.3t^2 + 3.2t$$

where t is the number of seconds since the dye was injected. Set up a partial table of values for $0 \leq t \leq 10$, and sketch the graph of this function.

34. *Automotive technology* The distance s, in feet, needed to stop a car traveling v mph is given by

$$s(v) = 0.04v^2 + v$$

(a) Set up a partial table of values for $s(v)$ with $0 \leq v \leq 70$.

(b) What was the velocity of a car that took 265 ft. to stop?

(c) Sketch the graph of s and s^{-1} on the same set of axes.

CHAPTER 1 TEST

Use the function $f(x) = 7x - 5$ in Exercises 1 and 2, and determine the indicated value.

1. $f(-2)$

2. $f(3 - a)$

In Exercises 3 and 4, let $g(x) = \frac{x^2 - 2x - 15}{x + 3}$ and determine the indicated value.

3. $g(0)$

4. $g(5)$

Solve Exercise 5.

5. **(a)** Graph $h(x) = \frac{1}{2}\sqrt{x + 4} - 3$.

(b) What is the domain of h?

(c) What is the range of h?

(d) What is the x-intercept of h?

(e) What is the y-intercept of h?

In Exercises 6–11, let $f(x) = 3x - 15$ and $g(x) = \frac{x-5}{x+5}$, and determine the indicated value.

6. $(f+g)(x)$
7. $(f-g)(x)$
8. $(f \cdot g)(x)$
9. $(f/g)(x)$
10. $(f \circ g)(x)$
11. $(g \circ f)(x)$

In Exercises 12–16, sketch the graph of each of the following functions or relations.

12. $y = 3x - 4$
13. $f(x) = x^2 - 2x + 1$
14. $g(x) = \frac{x+5}{x-1}$
15. $\frac{x^2}{x+1} = 2x - 1$
16. $\sqrt{x^2 - 1} = 3 - x$

Solve Exercises 17 and 18.

17. A researcher in physiology has decided that the function $r(s) = -s^2 + 12s - 20$ is a good mathematical model for the number of impulses fired after a nerve has been stimulated. Here, r is the number of responses per millisecond (ms) and s is the number of milliseconds since the nerve was stimulated.
   (a) Graph this function.
   (b) How many responses can be expected after 3 ms?
   (c) If there are 16 responses, how many ms have elapsed since the nerve was stimulated?
   (d) If there are 12 responses, how many ms have elapsed since the nerve was stimulated?

18. The cost of removing a certain pollutant is given by

$$C(x) = \frac{5x}{100 - x}$$

   where $C(x)$ is the cost in thousands of dollars of removing x percent of the pollutant.
   (a) Graph this function.
   (b) How much will it cost to remove 50% of the pollutant?
   (c) How much will it cost to remove 90% of the pollutant?
   (d) What is the inverse function of C?
   (e) If you had only $15,000, how much of the pollutant could you remove?

CHAPTER

2

An Introduction to Plane Analytic Geometry

A lithotripter is a medical instrument. Its design is based on an ellipse. In Section 2.4, we will see how the ellipse makes this instrument effective.

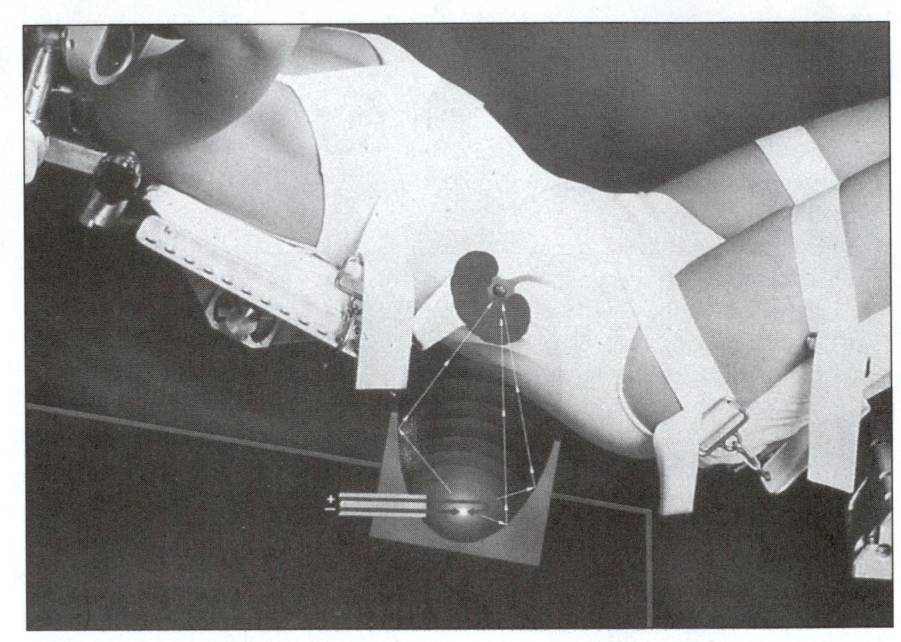

Courtesy of Dornier Medical Systems Inc.

Many scientific and technical applications make use of four special curves: the circle, parabola, ellipse, and hyperbola. These are four of the seven conic sections that can be formed by a plane intersecting a cone. (The others are a point, a line, and two intersecting lines.)

One medical instrument that uses an ellipse is a lithotripter, an instrument that uses sound waves to break up kidney stones. An ellipse has two foci. (Foci is the plural of focus.) A sound wave transmitter is placed at one focus and the patient's kidney is placed at the other. The elliptical shape of the lithotripter causes sound waves to be reflected off the ellipse and focused on the kidney stones.

In Chapter 1, the rectangular coordinate system was introduced. We began by graphing equations on the Cartesian coordinate system. The process of using algebra to describe geometry is called analytic geometry. In this chapter, we first review the line and then look at circles, parabolas, ellipses, and hyperbolas.

2.1 BASIC DEFINITIONS AND STRAIGHT LINES

We introduced graphing in Chapter 1. In this section, we will introduce some new ideas and review some of those that were introduced earlier.

If we have two points in a plane, we often want to know some things about them. Besides the location of the points, some of the things we might want to know include the distance between them, the point halfway between them, and the equation of the line through those two points, including its slope and intercepts. Let's look at these and a few other ideas, one at a time.

The Distance Formula

For most of this discussion we will be using two points, P_1 and P_2. Point P_1 has coordinates (x_1, y_1) and P_2 has coordinates (x_2, y_2). Using the Pythagorean theorem, we can find the distance d between these points.

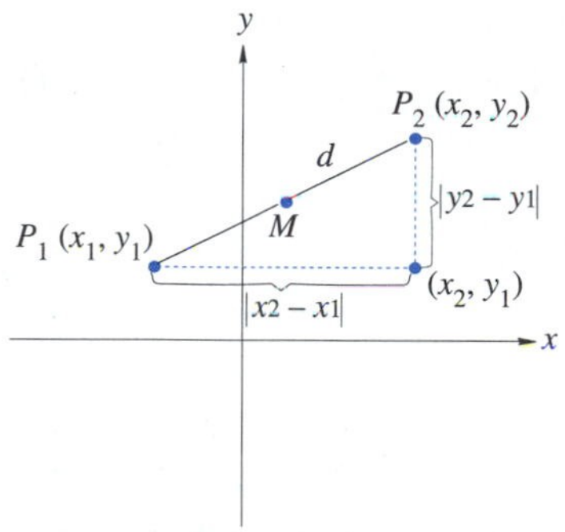

FIGURE 2.1

The Distance Formula

The distance d between any two points $P_1\,(x_1, y_1)$ and $P_2\,(x_2, y_2)$ in a plane is given by the **distance formula**

$$d = \sqrt{(x_1 - x_2)^2 + (y_1 - y_2)^2}$$

If there are several points, the symbol $d\,(P_1, P_2)$ would then represent the two points that are being used. (See Figure 2.1.)

EXAMPLE 2.1

Find the lengths of the sides of the triangle with vertices $A(-1, -3)$, $B(6, 1)$, and $C(2, -7)$.

Solutions

$$d\,(C, A) = \sqrt{(-1-2)^2 + (-3+7)^2} = \sqrt{(-3)^2 + 4^2}$$
$$= \sqrt{9+16} = \sqrt{25} = 5$$

$$d\,(B, A) = \sqrt{(-1-6)^2 + (-3-1)^2} = \sqrt{(-7)^2 + (-4)^2}$$
$$= \sqrt{49+16} = \sqrt{65}$$

$$d\,(C, B) = \sqrt{(6-2)^2 + (1+7)^2} = \sqrt{4^2 + 8^2}$$
$$= \sqrt{16+64} = \sqrt{80} = 4\sqrt{5}$$

The Midpoint Formula

If you want to find the coordinates of the point M halfway between two points P_1 and P_2, then you use the **midpoint formula**.

> **The Midpoint Formula**
>
> The coordinates of the midpoint M between any two points $P_1(x_1, y_1)$ and $P_2(x_2, y_2)$ are given by the **midpoint formula**
>
> $$M = \left(\frac{x_1 + x_2}{2}, \frac{y_1 + y_2}{2}\right)$$

The midpoint formula gives you the coordinates of the point on the segment $\overline{P_1 P_2}$ halfway between P_1 and P_2.

EXAMPLE 2.2

Find the midpoints of the sides of the triangle with vertices $A(-1, -3)$, $B(6, 1)$, and $C(2, -7)$. Note that this is the same triangle we used in Example 2.1.

Solutions

$$M(A, B) = \left(\frac{-1+6}{2}, \frac{-3+1}{2}\right) = \left(\frac{5}{2}, \frac{-2}{2}\right) = \left(\frac{5}{2}, -1\right)$$

$$M(A, C) = \left(\frac{-1+2}{2}, \frac{-3-7}{2}\right) = \left(\frac{1}{2}, \frac{-10}{2}\right) = \left(\frac{1}{2}, -5\right)$$

$$M(B, C) = \left(\frac{6+2}{2}, \frac{1-7}{2}\right) = \left(\frac{8}{2}, \frac{-6}{2}\right) = (4, -3)$$

Slope

As we have discussed before, the slope of a line measures the steepness of the line. The slope also indicates whether, as values of x increase, a line is rising (positive slope), falling (negative slope), or constant (slope of 0). A formula for determining the slope of the line through two points follows.

> **Slope of a Line**
>
> If a line through two points $P_1(x_1, y_1)$ and $P_2(x_2, y_2)$ is not vertical, then the **slope**, m, of this line is
>
> $$m = \frac{y_2 - y_1}{x_2 - x_1}$$
>
> All lines except vertical lines have a slope. The slope of a vertical line is undefined. The slope of a horizontal line is 0.

EXAMPLE 2.3

What is the slope of the line through the points $A(2,-5)$ and $B(-4,7)$?

Solution

$$m=\frac{7-(-5)}{-4-2}=\frac{7+5}{-4-2}=\frac{12}{-6}=-2$$

Any line that is not parallel to the x-axis must eventually cross the x-axis. The angle measure in a positive direction from the x-axis to a line is called the **inclination** of the line. The inclination of a line parallel to the x-axis is defined as 0. The inclination provides us with an alternative definition for the slope. If α is the inclination that a line makes with the x-axis, as shown in Figures 2.2a and 2.2b, then the slope is

$$m=\tan\alpha, \qquad 0^\circ \le \alpha < 180^\circ \qquad \text{or} \qquad 0 \le \alpha < \pi$$

Of course, since $\tan 90^\circ$ and $\tan \frac{\pi}{2}$ are undefined, the inclination of a vertical line is undefined.

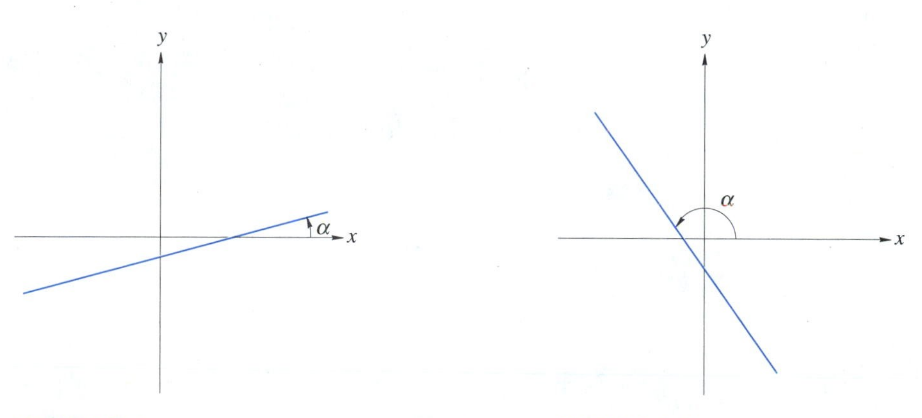

FIGURE 2.2a

FIGURE 2.2b

Many technical areas have their own terms that refer to the slope of a line or surface. For example, construction workers talk about the **pitch** of a roof, truck drivers refer to the **grade** of a hill, and, in some fields, **declination** refers to a line with negative slope.

EXAMPLE 2.4

What is the slope of a line with an inclination of **(a)** 30°, **(b)** 0.85 rad, and **(c)** 115°?

Solutions

(a) $m=\tan 30^\circ = 0.5773503$

(b) $m=\tan 0.85\text{ rad} = 1.1383327$

(c) $m=\tan 115^\circ = -2.1445069$

EXAMPLE 2.5

The line in Example 2.3 has a slope of −2. What is the inclination of this line?

Solution

$$m = \tan\alpha = -2 \text{ so } \alpha = \tan^{-1}(-2) \approx 116.56505°$$

[If you used a calculator, $\tan^{-1}(-2) \approx -63.43498°$. To obtain the inclination, you need to add 180°, since the inclination must be between 0° and 180°, with the result $180° + (-63.43498°) = 116.56505°$.]

If two lines are parallel, they have the same inclination, and so parallel lines have the same slope. If two lines are perpendicular, they intersect at a 90° angle. This means that their inclinations must differ by 90°. It also means that the slope of one of these lines is the negative reciprocal of the other.

Slopes of Perpendicular Lines

If m_1 is the slope of a line and m_2 is the slope of a line perpendicular to the first line, then

$$m_1 = -\frac{1}{m_2}$$

This can also be written as $m_1 m_2 = -1$.

The above rule does not apply if one of the lines is horizontal because the other line would then be vertical and not have a slope.

EXAMPLE 2.6

What is the slope of a line that is perpendicular to a line with a slope of −2?

Solution If $m_1 = -2$, then the slope of the perpendicular line is $m_2 = -\frac{1}{-2} = \frac{1}{2}$

There are several formulas for the equation of a line. Three of them are given below. If b represents the y-intercept of a line, then the line crosses the y-axis at the point $(0, b)$. If the slope of this line is m, then the **slope-intercept form for the equation of a line** is

$$y = mx + b$$

EXAMPLE 2.7

Write an equation for a line that has a slope of $\frac{1}{4}$ and a y-intercept of 5.

Solution We have $m = \frac{1}{4}$ and $b = 5$, so $y = \frac{1}{4}x + 5$. Writing this without fractions, we get $4y = x + 20$.

EXAMPLE 2.8

What are the slope and y-intercept of the line $4x+6y-3=0$?

Solution We need to rewrite this equation in the form $y=mx+b$.

$$\begin{aligned}4x+6y-3&=0\\6y&=-4x+3\\y&=-\frac{4}{6}x+\frac{3}{6}\\&=-\frac{2}{3}x+\frac{1}{2}\end{aligned}$$

The slope is $-\frac{2}{3}$ and the y-intercept is $\frac{1}{2}$.

If you know the slope of a line is m and $P(x_1, y_1)$ is a point on the line, then the equation for the line can be written in the **point-slope form for the equation of a line**.

$$y-y_1=m(x-x_1)$$

EXAMPLE 2.9

What is the equation of the line through the point $(2, -3)$ and with a slope of 5?

Solution In this example, $x_1=2$, $y_1=-3$, and $m=5$. Substitute these values into the equation $y-y_1=m(x-x_1)$.

$$\begin{aligned}y-(-3)&=5(x-2)\\y+3&=5x-10\\y&=5x-13\end{aligned}$$

Every straight line can be written in the **general form for the equation of a line**. This form is represented by the equation

$$Ax+By+C=0$$

where A, B, and C represent constants and A and B are not both 0.

EXAMPLE 2.10

What are the slope and intercepts of the line $4x+3y-24=0$?

Solution We know that the intercepts are the points where the line crosses the axes. The line crosses the x-axis at $(a, 0)$ and the y-axis at $(0, b)$.

If we let $y=0$, we then get $x=6$, and if $x=0$, then $y=8$; so the x-intercept is $(6, 0)$ and the y-intercept is $(0, 8)$. We will use these two points to determine the following slope.

$$m=\frac{8-0}{0-6}=\frac{8}{-6}=-\frac{4}{3}$$

The three forms for the equation of a line are summarized in the following box.

Different Forms for the Equation of a Line

If a line has a slope of m, a y-intercept of $(0, b)$, and $P(x_1, y_1)$ is a point on the line, then the equation for the line can be written in any of these three forms:

Type of Equation	Equation for Line
Point-slope form	$y - y_1 = m(x - x_1)$
Slope-intercept form	$y = mx + b$
General form	$Ax + By + C = 0$

Application

EXAMPLE 2.11

A mechanic charges \$96 for a job that takes 2 h to finish and \$135 if the job is completed in 5 h. **(a)** Find a linear equation that describes how much the mechanic should charge for a job of x hours. **(b)** Use your equation to determine how much should be charged for a job that takes 7 h 15 min.

Solutions **(a)** We begin by thinking of the given information as points on a line. We give each point as an ordered pair of the form (x, C), where x represents the number of hours a job takes to complete and C stands for the amount the mechanic charges for this job.

A 2-h job with a charge of \$96 can be thought of as the point $(2, 96)$, and the 5-h job as the point $(5, 135)$. From this information we determine that the slope of this line is

$$m = \frac{135 - 96}{5 - 2} = \frac{39}{3} = 13$$

Using this value for the slope and selecting the point $(2, 96)$ we can use the point-slope form for the equation of a line to determine the needed equation.

$$C - 96 = 13(x - 2)$$
$$C - 96 = 13x - 26$$
$$\text{or} \qquad C = 13x + 70$$

Thus the desired equation is $C = 13x + 70$. **(b)** A 7 h 15 min job can be thought of as a 7.25-h job. Substituting 7.25 for x in the equation $C = 13x + 70$, we obtain $C = 13(7.25) + 70 = 94.25 + 70 = 164.25$. The mechanic should charge \$164.25 for a job that will take 7 h and 15 min.

Application

EXAMPLE 2.12

The resistance of a circuit element varies directly with its temperature. The resistance at 0°C is found to be 5.00 Ω. Further tests show that the resistance increases by 1.5 Ω for every 1°C increase in temperature. **(a)** Determine the equation for the resistance R in terms of the temperature T. **(b)** Use the equation from part (a) to find the resistance at 27°C.

Solutions **(a)** In order to find R in terms of T, we should think of R as the dependent variable. Since resistance varies directly with temperature, the two quantities have a linear relationship. From the given information, we notice that since $R = 5.00$ when $T = 0$, we have the R-intercept. As a result, we shall use the slope-intercept form for the equation of a line.

$$R = mT + b$$
$$R = mT + 5.00$$

To find the slope m, we will use

$$m = \frac{\text{Change in } R}{\text{Change in } T} = \frac{1.5\,\Omega}{1°\text{C}} = 1.5\,\Omega/°\text{C}$$

Substituting 1.5 for m in the slope-intercept equation, we are able to complete the equation.

$$R = 1.5T + 5.00$$

(b) To find the resistance at 27°C, we will substitute 27°C for T in the equation from part (a).

$$R = 1.5(27) + 5.00 = 45.5\,\Omega$$

Exercise Set 2.1

In Exercises 1–8, find the distance between the given pairs of points.

1. $(2, 4)$ and $(7, -9)$
2. $(-3, 5)$ and $(0, 1)$
3. $(5, -6)$ and $(-3, -5)$
4. $(-8, -4)$ and $(6, 10)$
5. $(12, 1)$ and $(3, -13)$
6. $(11, -2)$ and $(4, -8)$
7. $(-2, 5)$ and $(5, 5)$
8. $(-4, -4)$ and $(-4, 7)$

In Exercises 9–16, find the midpoints of the given pairs of points in Exercises 1–8.

In Exercises 17–24, find the slopes of the lines through the points in Exercises 1–8.

In Exercises 25–32, use the point-slope form of the equation for the line to write the equation of the lines through the points in Exercises 1–8.

In Exercises 33–36, find the slopes of the lines with the given inclinations.

33. 45°

34. 175°

35. 0.15 rad

36. 1.65 rad

In Exercises 37–40, find the inclinations of the lines with the given slopes.

37. 2.5

38. 0.30

39. −0.50

40. −1.475

In Exercises 41–44, determine the slope of a line that is perpendicular to a line of the given slope.

41. 3

42. $\frac{2}{5}$

43. $-\frac{1}{2}$

44. −5

In Exercises 45–56, find the equation of each line with the given properties.

45. Passes through $(2, -5)$ with a slope of 6

46. Passes through $(6, -2)$ with a slope of $-\frac{1}{2}$

47. Passes through $(2, 3)$ and $(4, -2)$

48. Passes through $(-5, 2)$ and $(-3, -4)$

49. Passes through $(-2, -4)$ with an inclination of 60°

50. Passes through $(5, 1)$ with an inclination of 0.75 rad

51. Has an inclination of 20° and a y-intercept of $(0, 3)$

52. Has an inclination of 2.5 rad and a y-intercept of $(0, -2)$

53. Passes through $(2, 5)$ and is parallel to $2x - 3y + 4 = 0$

54. Passes through $(-3, 2)$ and is parallel to $3x + 4y = 12$

55. Passes through $(4, -1)$ and is perpendicular to $2x + 5y = 20$

56. Passes through $(-1, -6)$ and is perpendicular to $8x - 3y = 24$

In Exercises 57–60, determine the slope and intercepts of each line.

57. $3x + 2y = 12$

58. $5x - 3y = 15$

59. $x - 3y = 9$

60. $6x + y = 9$

Solve Exercises 61–68.

61. *Physics* The instantaneous velocity v of an object under constant acceleration a during an elapsed time t is given by $v = v_0 + at$, where v_0 is its initial velocity. If an object has an initial velocity of 2.6 m/s and a velocity of 5.8 m/s after 8 s of constant acceleration, write the equation relating velocity to time.

62. *Physics* The **coefficient of linear expansion** α is the change in length of a solid due to the change in temperature.

(a) Determine the coefficient of linear expansion of a copper rod that is 1.000 000 cm at 10°C and expands to 1.000 084 cm at 15°C.

(b) Determine the coefficient of linear expansion of a copper rod that is 4.000 000 cm at 10°C and is 4.000 336 cm at 15°C.

(c) The coefficient of linear expansion for any specific solid is a constant. This means that the answers for parts (a) and (b) should have been the same. Assume that the answer in part (a) is correct. What changes need to be made in the way you determine the coefficient of linear expansion?

(d) Determine the coefficient of linear expansion of a glass rod that expands from 72.000 024 cm at 10°C to 72.005 208 cm at 18°C.

63. *Electronics* In a dc circuit, when the internal resistance of the voltage source is taken into account, the voltage E is a linear function of the current I and is given by $E = IR + Ir$, where R is the circuit resistance and r is the internal resistance. If the resistance of the circuit $R = 4.0\,\Omega$ and $I = 2.5\,A$ when $E = 12.0$ V, find r.

64. *Electronics* The resistance of a circuit element is found to increase by $0.006\,\Omega$ for every 1°C increase in temperature over a wide range of temperatures. If the resistance is $7.000\,\Omega$ at 0°C, **(a)** write the equation relating the resistance R to the temperature T, and **(b)** find R, when $T = 17$°C.

65. *Industrial management* The production of a certain computer component has fixed costs of \$1,225 and additional costs of \$1.25 per component manufactured.

(a) Write an equation relating the total cost, C, to the number of components produced, n.

(b) What is the cost of producing 20,000 components?

66. *Accounting* The linear method for determining the depreciated value, V, of a deductible item is given by

$$V = C\left(1 - \frac{n}{N}\right)$$

where C is the original cost, n is the number of years since the depreciation began, and N is the usable life of the item. Graph V as a function of C when $n = 5$ and $N = 8$.

67. *Environmental science* One estimate of the concentration, C, of carbon dioxide (CO_2) in parts per million (ppm) in year t is given by

$$C = 1.44t + 282.88$$

where t is the number of years after 1940.

(a) What was the CO_2 concentration in 1995?

(b) Predict the CO_2 concentration for the year 2050.

(c) Predict the CO_2 concentration for the current year.

68. *Energy technology* The width, W, and height, H, of a roof overhang and the latitude, L, of a north-south facing room determine whether the room will get any summer sunshine. The overhang dimensions that give no summer sunshine are $W = H \tan B$, where $B = L - 23.5°$, is the angle of inclination of the sun at summer solstice. (For houses at either latitude 37° N or 37° S, let $L = 37$.)

(a) Chattanooga, Tennessee is at latitude $31°2'$ N. What overhang width on an $8'$-high room would ensure that it gets no summer sunshine?

(b) The latitude of Vancouver, British Columbia, is $49°16'$ N. What overhang width on a 2.7 m-high room would ensure that it gets no summer sunshine?

(c) The latitude of Anchorage, Alaska, is $61°13'$ N. Will a room with an overhang that is $8'6''$ high and $6'3''$ wide get any summer sunshine?

In Your Words

69. Describe how to determine if two lines are perpendicular.

70. Compare and contrast the point-slope, slope-intercept, and general forms for the equation of a line. How are they alike and how are they different?

2.2 THE CIRCLE

In Section 2.1, we developed a general equation for a straight line. In this section, we will work with the circle.

A **circle** is the set of all points in a plane that are at some fixed distance from a fixed point in the plane. The fixed point is called the **center** of the circle and the fixed distance is called the **radius**. In general, suppose we have a circle with radius r and center (h, k). An seen in Figure 2.3, if (x, y) is any point on the circle, then the distance from (x, y) to (h, k) must be r. Using the distance formula, we have the equation

$$\sqrt{(x-h)^2 + (y-k)^2} = r$$

Squaring both sides, we have

$$(x-h)^2+(y-k)^2=r^2$$

Thus, we have established the following standard equation of a circle.

Standard Equation of a Circle

The standard equation of a circle with center at (h,k) and radius r is

$$(x-h)^2+(y-k)^2=r^2$$

EXAMPLE 2.13

The equation $(x-5)^2+(y+3)^2=36$ is a circle with center at $(5,-3)$ and a radius of 6. Remember, since the equation states that $y+3=y-k$, you obtain $k=-3$. Note very carefully the signs of the numbers in the equation and the signs of the coordinates of the center. This circle is shown in Figure 2.4.

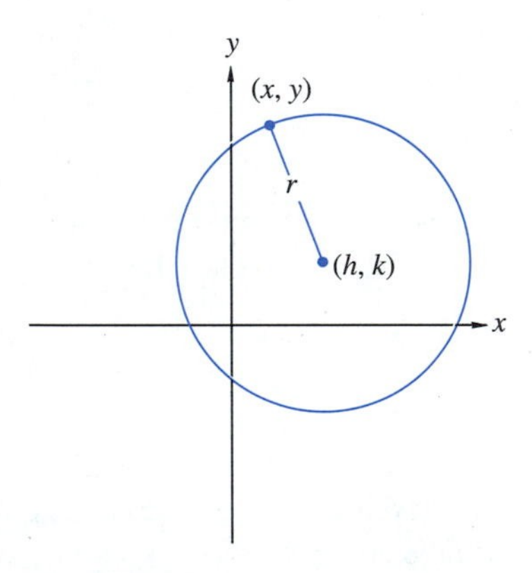

FIGURE 2.3

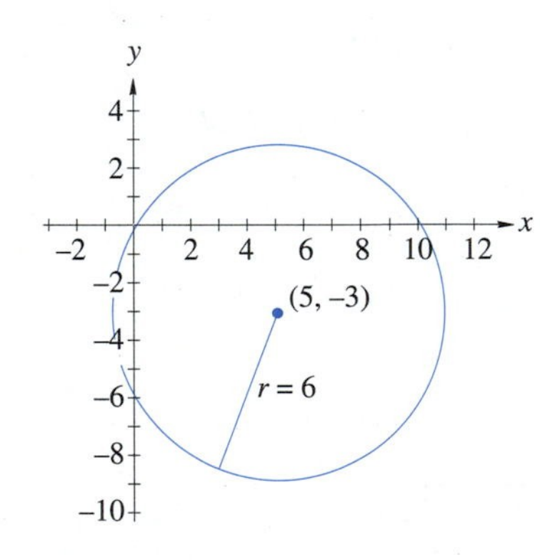

FIGURE 2.4

EXAMPLE 2.14

Write the equation for the circle with center at $(-2,4)$ and radius 7.

Solution The circle is shown in Figure 2.5. Since the center is $(-2,4)$, we have $h=-2$ and $k=4$, and since the radius is 7, then $r=7$. So,

$$(x-h)^2+(y-k)^2=r^2$$
$$[x-(-2)]^2+(y-4)^2=7^2$$
$$(x+2)^2+(y-4)^2=49$$

Notice that if the center is at the origin, then $h = 0$ and $k = 0$. The equation becomes

$$x^2 + y^2 = r^2$$

If we expand the general equation for a circle, we get

$$(x-h)^2 + (y-k)^2 = r^2$$
$$x^2 - 2hx + h^2 + y^2 - 2ky + k^2 = r^2$$

Since h^2, k^2, and r^2 are all constants, we will let $h^2 + k^2 - r^2 = F$. If we let the constants $-2h = D$ and $-2k = E$, then the equation for a circle can be written as the following general equation.

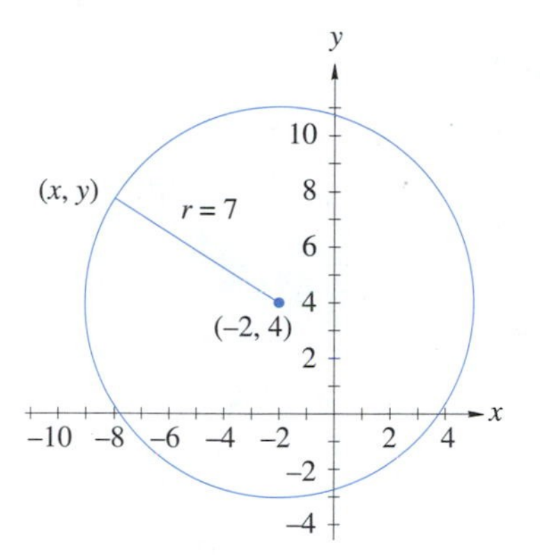

FIGURE 2.5

General Equation of a Circle

If A, D, E, and F are any real number constants, $A \neq 0$, then the general equation for a circle is

$$Ax^2 + Ay^2 + Dx + Ey + F = 0$$

EXAMPLE 2.15

Write $x^2 + y^2 + 4x - 8y + 4 = 0$ in the standard form for the equation of a circle and determine the center and radius.

Solution To convert the equation to the standard form, we need to complete the square. We first write the constant on the right-hand side and group the terms containing x together and then group those containing y. The ○ and □ show the missing constants that we must determine in order to complete the square.

$$x^2 + y^2 + 4x - 8y + 4 = 0$$
$$(x^2 + 4x + \bigcirc) + (y^2 - 8y + \square) = -4 + \bigcirc + \square$$

The coefficient of the x-term is 4. If we take half of that and square it, we can add the result, 4, to both sides of the equation. Similarly, half of -8 is -4, and we also add $(-4)^2 = 16$ to both sides. The equation then becomes

$$(x^2 + 4x + \textcircled{4}) + \left(y^2 - 8y + \boxed{16}\right) = -4 + 4 + 16$$
$$(x+2)^2 + (y-4)^2 = 16$$

Since $16 = 4^2$, the radius is 4, and the center is $(-2, 4)$.

EXAMPLE 2.16

Write the equation $x^2 + y^2 + x - 5y + 2 = 0$ in the standard form for the equation of a circle and determine the center and radius.

EXAMPLE 2.16 (Cont.)

Solution Again we will complete the square and use a ◯ and a □ to show where the constants need to be placed.

$$x^2+y^2+x-5y+2=0$$
$$(x^2+x+\bigcirc)+(y^2-5y+\square)=-2+\bigcirc+\square$$
$$\left(x^2+x+\frac{1}{4}\right)+\left(y^2-5y+\frac{25}{4}\right)=-2+\frac{1}{4}+\frac{25}{4}$$
$$\left(x+\frac{1}{2}\right)^2+\left(y-\frac{5}{2}\right)^2=\frac{18}{4}$$

The center is $\left(-\frac{1}{2},\frac{5}{2}\right)$ and the radius is $\sqrt{\frac{18}{4}}=\frac{3\sqrt{2}}{2}$.

EXAMPLE 2.17

Write the equation $4x^2+4y^2-8x-24y-9=0$ in the standard form for the equation of a circle and determine the center and radius.

Solution Again, we will complete the square. First, move the constant to the right-hand side and group the x-terms and the y-terms.

$$4x^2-8x+4y^2-24y=9$$

Next, factor the 4 out of the x-terms and the y-terms.

$$4(x^2-2x)+4(y^2-6y)=9$$

We now use a ◯ and a □ to indicate the missing constants. Notice that a 4◯ and a 4□ are added on the right-hand side of the equation to show where the constants need to be placed.

$$4(x^2-2x+\bigcirc)+4(y^2-6y+\square)=9+4\bigcirc+4\square$$
$$4(x^2-2x+1)+4(y^2-6y+9)=9+4(1)+4(9)$$
$$4(x-1)^2+4(y-3)^2=49$$
$$(x-1)^2+(y-3)^2=\frac{49}{4}$$

The center is $(1,3)$ and the radius $r=\sqrt{\frac{49}{4}}=\frac{7}{2}$.

While we could have started by dividing this entire equation by 4, we waited until the end in order to delay working with fractions. This example uses a process that we will need in the sections on ellipses and hyperbolas.

A circle with an equation of the form $(x-h)^2+y^2=r^2$ has its center on the x-axis [at $(h,0)$] and thus is symmetrical with the x-axis. In the same way, a circle with an equation of the form $x^2+(y-k)^2=r^2$ has its center on the y-axis [at $(0,k)$] and is symmetrical with the y-axis.

If $(x-h)^2+(y-k)^2=0$, the circle has a radius of 0. This is sometimes called a **point circle**. If the radius is 1, the circle is often referred to as a **unit circle**.

Finally, if $r^2<0$, then the equation does not define a circle and this equation would have no graph on the Cartesian plane.

Application

EXAMPLE 2.18

A square concrete post is reinforced axially with eight rods arranged symmetrically around a circle, much like those shown in the photo in Figure 2.6a. The cross section in Figure 2.6b provides a schematic drawing of the post and rods. **(a)** Determine the equation of the circle using the upper left-hand corner of the post as the origin. **(b)** Find the coordinates of each rod's location.

Solution **(a)** While it is not specifically stated, we will assume that the eight rods are placed around a circle that shares the center of the square post. This would place the center at the point $(15,-15)$. The radius of the circle is $2.925+7.075\text{ cm}=10\text{ cm}$. Thus, the equation for this circle is $(x-15)^2+(y+15)^2=100$. **(b)** The coordinates of rods 1, 3, 5, and 7 should be fairly easy to determine from the given information. Determining the coordinates of the four remaining rods is a little more difficult. We will show how to determine the coordinates for Rod 2 and then use the symmetry of the rod's placement to determine the coordinates of the other three rods.

The radius of the circle on which the rods are placed is 10 cm. Thus, each rod is located 10 cm from the center of the post. Rod 2 is located directly above a point that is 7.075 cm from the center of the post. Using the Pythagorean theorem, we find that this rod is placed $\sqrt{10^2-7.075^2}=\sqrt{100-50.055\,625}=\sqrt{49.944\,375}\approx 7.067\text{ cm}$ above the line through Rods 7 and 3. Then the y-coordinate of Rod 2 is $-15+7.067=-7.933$. This places Rod 2 at the coordinates $(22.075,-7.933)$. The eight rods have the coordinates $(15,-5)$, $(22.075,-7.933)$, $(25,-15)$, $(22.075,-22.067)$, $(15,-25)$, $(7.925,-22.067)$, $(5,-15)$, and $(7.925,-7.933)$.

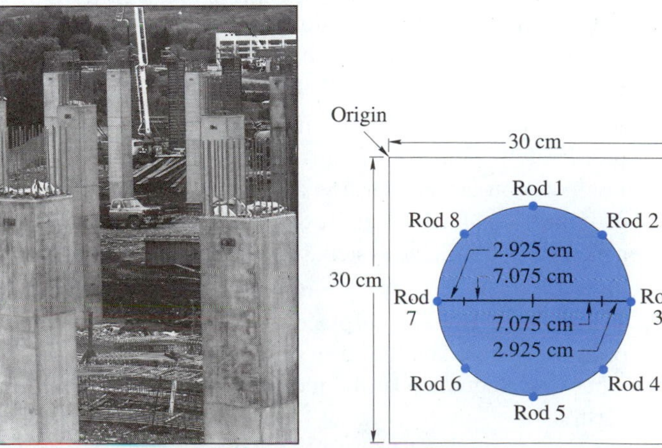

Courtesy of Barry, Bette, and Led Duke Inc.

FIGURE 2.6a **FIGURE 2.6b**

Exercise Set 2.2

In Exercises 1–8, find the standard and general form of an equation for the circle with the given center C and radius r.

1. $C=(2,5), r=3$
2. $C=(4,1), r=8$
3. $C=(-2,0), r=4$
4. $C=(0,-7), r=2$
5. $C=(-5,-1), r=\frac{5}{2}$
6. $C=(-4,-5), r=\frac{5}{4}$
7. $C=(2,-4), r=1$
8. $C=(5,-2), r=\sqrt{3}$

In Exercises 9–14, give the center and radius of the circle described by each equation. Sketch the circle.

9. $(x-3)^2+(y-4)^2=9$
10. $(x-7)^2+(y+5)^2=25$
11. $\left(x+\frac{1}{2}\right)^2+\left(y+\frac{13}{4}\right)^2=7$
12. $\left(x+\frac{7}{2}\right)^2+\left(y-\frac{7}{3}\right)^2=\frac{11}{6}$
13. $x^2+\left(y-\frac{7}{3}\right)^2=6$
14. $(x+3)^2+y^2=1.21$

In Exercises 15–26, describe the graph of each equation. If it is a circle, give the center and radius.

15. $x^2+y^2+4x-6y+4=0$
16. $x^2+y^2-10x+2y+22=0$
17. $x^2+y^2+10x-6y-47=0$
18. $x^2+y^2+2x-12y-27=0$
19. $x^2+y^2-2x+2y+3=0$
20. $x^2+y^2+2x+2y-3=0$
21. $x^2+y^2+6x-16=0$
22. $x^2+y^2-8y-9=0$
23. $x^2+y^2+5x-9y=9.5$
24. $x^2+y^2-7x+3y+2.5=0$
25. $9x^2+9y^2+18x-15y+27=0$
26. $25x^2+25y^2-10x+30y+1=0$

Solve Exercises 27–34.

27. *Physics* When a particle with mass m and charge q enters a magnetic field of induction β with a velocity v at right angle to β, the path of the particle is a circle. The radius of the path is $\sigma=\dfrac{mv}{\beta q}$. If a proton of mass 1.673×10^{-27} kg and charge 1.5×10^{-19} C enters a magnetic field with induction 4×10^{-4} T (tesla $= \text{V}\cdot\text{s}\cdot\text{m}^2$) with a velocity of 1.186×10^{7} m/s, find the equation of the path of the electron.

28. *Astronomy* The earth's orbit around the sun is approximately a circle of radius 1.495×10^{8} km. The moon's orbit around earth is approximately a circle of radius 3.844×10^{5} km. If the sun is placed at the center of a coordinate system, what is the equation of the earth's orbit?

29. *Astronomy* If the sun is placed at the center of a coordinate system and the earth on the positive x-axis, what is the equation of the moon's orbit around the earth?

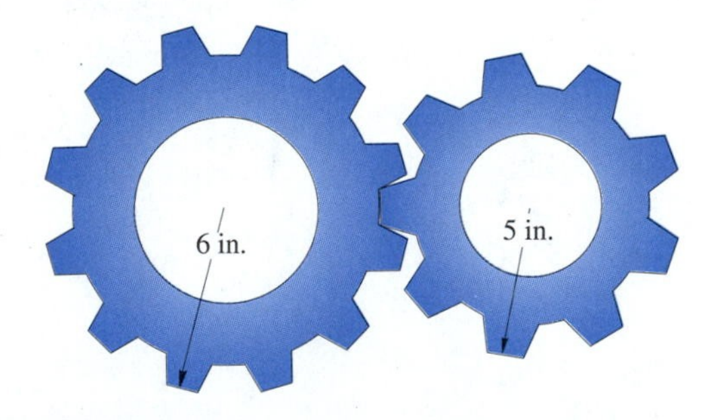

FIGURE 2.7

30. *Industrial design* A drafter is drawing two gears that intermesh. They are represented by two intersecting circles, as shown in Figure 2.7. The first circle has a radius of 6 in. and the second has a radius of 5 in. The section of intersection has a maximum depth of 1 in. What is the equation of each circle, if the center of the first circle is at the origin and the positive x-axis passes through the center of the second circle?

31. *Industrial design* In designing a tool, an engineer calls for a hole to be drilled with its center 0.9 cm directly above a specified origin. If the hole must have a diameter of 1.1 cm, find the equation of the hole.

32. *Industrial engineering* In designing dies for blanking a sheet-metal piece, the configuration shown in Figure 2.8 is used. Write the equation of the small circle with its center at B, using the point marked A as the origin. Use the following values: $r = 3$ in., $a = 3.25$ in., $b = 1.0$ in., and $c = 0.25$ in.

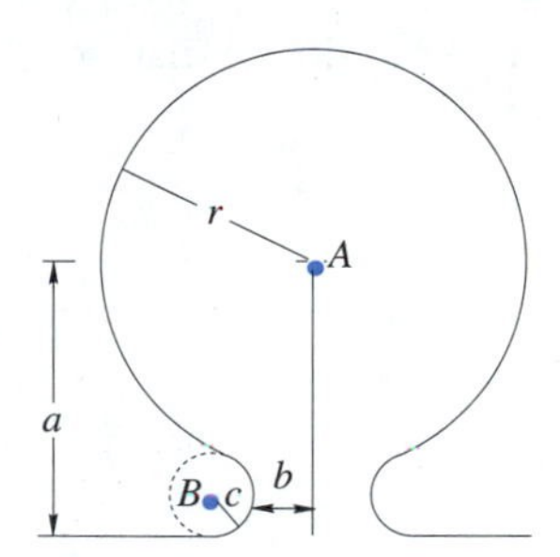

FIGURE 2.8

33. *Industrial engineering* A flywheel that is 26 cm in diameter is to be mounted so that its shaft is 5 cm above the floor, as shown in Figure 2.9. **(a)** Write an equation for the path followed by a point on the rim. Use the surface of the floor as the horizontal axis and the perpendicular line through the center as

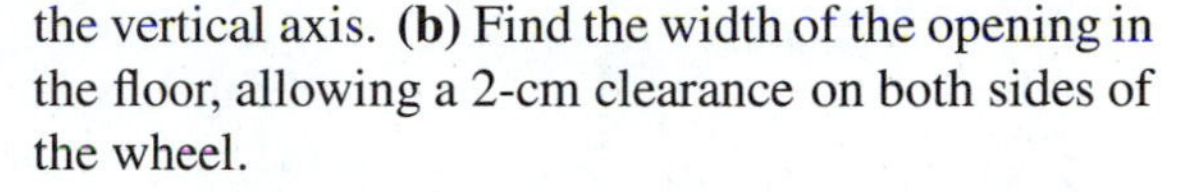

the vertical axis. **(b)** Find the width of the opening in the floor, allowing a 2-cm clearance on both sides of the wheel.

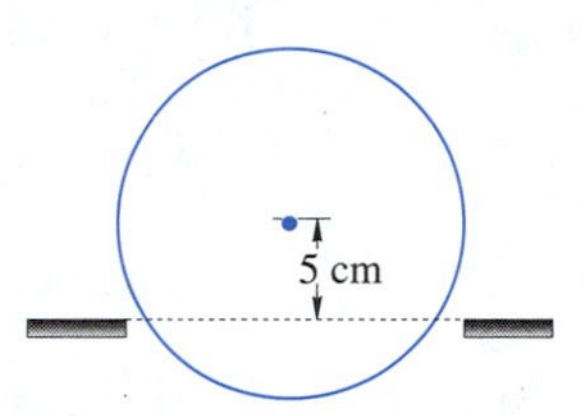

FIGURE 2.9

34. *Electricity* The impedance, Z, in an ac circuit is given by

$$Z = \sqrt{R^2 + (X_L - X_C)^2}$$

where R is the resistance, X_L the inductive reactance, and X_C the inductive capacitance.

(a) If $R = 6\,\Omega$, $X_L = 10\,\Omega$, and $X_C = 12\,\Omega$, determine Z.

(b) If $R = 8\,\Omega$ and $X_C = 12.5\,\Omega$, what are the possible values for Z?

(c) If $Z = 40\,\Omega$ and $X_C = 50\,\Omega$, what are the possible values for X_L?

(d) Use your result from **(c)** to graph the relationship between resistance and inductive reactance when $Z = 40\,\Omega$ and $X_C = 50\,\Omega$. (Put R on the horizontal axis.)

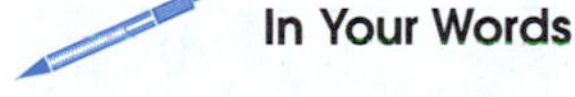

In Your Words

35. Without looking in the text, write the standard equation of a circle. Explain what each constant represents.

36. Describe the procedures you would use to change the general equation of a circle to the standard equation, and vice versa.

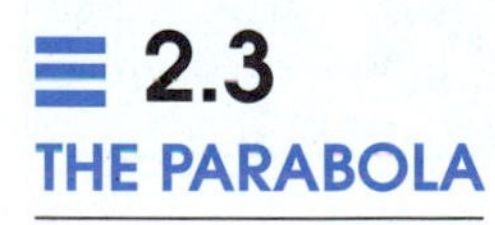

2.3 THE PARABOLA

The second curve we will study in this chapter is the parabola. A **parabola** is the set of points in a plane for which the distance from a fixed line is equal to its distance from a fixed point not on the line. The given line is called the **directrix** and the given point is the **focus**.

In Figure 2.10, we have indicated the line ℓ is the directrix, and F is the focus. The line through F perpendicular to the directrix is called the **axis** of the parabola. Examination of the parabola in Figure 2.10 indicates that the axis of the parabola is the only axis of symmetry for the parabola. The point V on the axis that is halfway between the focus and the directrix is the **vertex**.

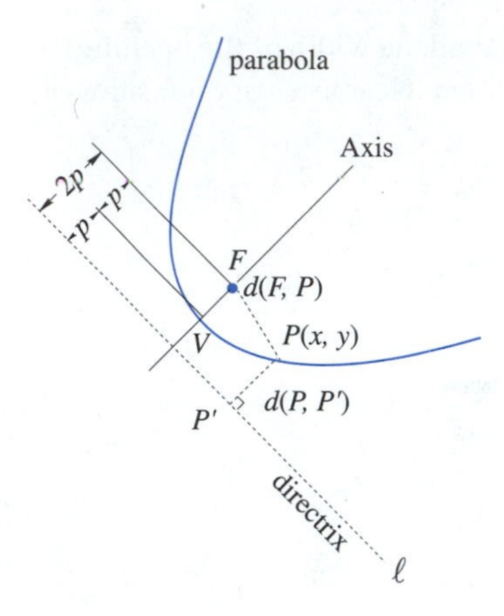

FIGURE 2.10

Let P be any point on the parabola. Draw a line through P that is perpendicular to the directrix at a point P'. According to the definition, the distance from P to P' is the same as the distance from P to F.

Let's set up a similar drawing on a coordinate system so that we can develop a formula for a parabola. We will let the vertex be at the origin and the focus F at $(0, p)$. Since the vertex is midway between the focus and the directrix, the directrix is the line $y = -p$, as shown in Figure 2.11. If $P(x, y)$ is any point on the parabola, then the distance from F to P is, by the distance formula,

$$\sqrt{(x-0)^2 + (y-p)^2}$$

Since P' is on the directrix, its coordinates are $(x, -p)$. So the distance from P to P' is

$$\sqrt{(x-x)^2 + (y+p)^2}$$

Since these two distances are equal, we have

$$\sqrt{(x-0)^2 + (y-p)^2} = \sqrt{(x-x)^2 + (y+p)^2}$$

Squaring both sides and simplifying, we obtain

$$\begin{aligned} x^2 + (y-p)^2 &= (y+p)^2 \\ x^2 + y^2 - 2py + p^2 &= y^2 + 2py + p^2 \\ x^2 &= 4py \end{aligned}$$

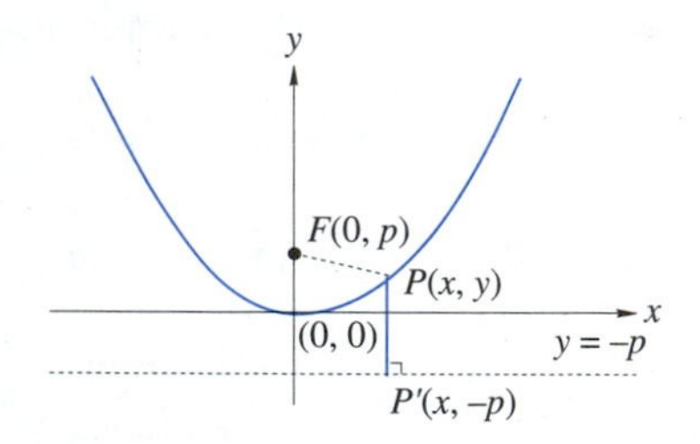

FIGURE 2.11

We have established the following standard equation of a vertical parabola. The term vertical parabola is used because the axis is a vertical line. Notice that $|p|$ is the distance from the focus to the vertex.

Standard Equation of a Vertical Parabola

The standard equation of a vertical parabola with focus F at $(0, p)$ and directrix $y = -p$ is

$$x^2 = 4py$$

If $p > 0$, the parabola opens upward, as shown in Figure 2.11.

If $p < 0$, the parabola opens downward.

EXAMPLE 2.19

Find the focus and directrix of the parabola with the equation $x^2 = -10y$. Sketch its graph.

Solution Since $x^2 = 4py$, then $4p = -10$ and $p = -\frac{5}{2}$. The focus is at $\left(0, -\frac{5}{2}\right)$. The equation of the directrix is $y = -p$ or $y = \frac{5}{2}$. The graph is sketched in Figure 2.12. Notice that $p < 0$ and that this parabola opens downward.

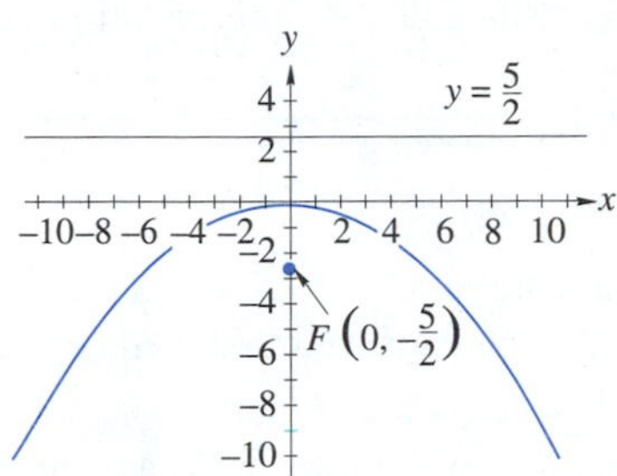

FIGURE 2.12

EXAMPLE 2.20

Find the equation of the parabola that has its vertex at the origin, opens upward, and passes through $(-5, 9)$.

Solution The general form of the equation is $x^2 = 4py$. Since it opens upward, $p > 0$, and since $(-5, 9)$ is on the parabola, then $(-5)^2 = 4p(9)$. Solving for p, we get $p = \frac{25}{36}$. Substituting this value for p in the equation, we get

$$x^2 = 4\left(\frac{25}{36}\right)y$$

$$x^2 = \frac{25}{9}y$$

$$\text{or} \qquad 9x^2 = 25y$$

If we had used the x-axis as the axis of the parabola, and the vertex was still at the origin, the focus would then have been at $(p, 0)$, and the directrix would have had the equation $x = -p$. Using the same method that we used earlier, we would get the following standard equation for a horizontal parabola. It is called a horizontal parabola because the axis is horizontal.

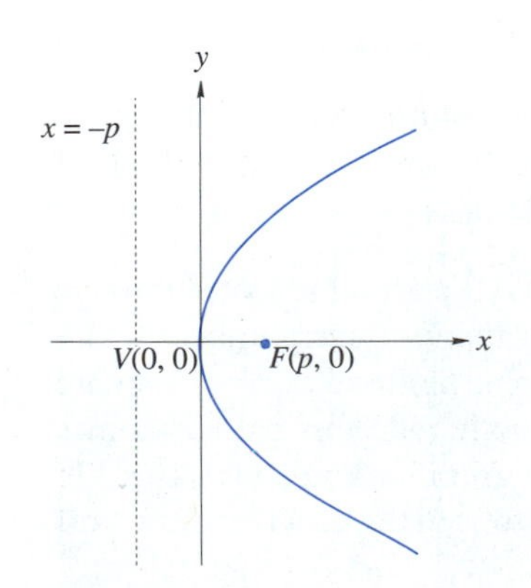

FIGURE 2.13

Standard Equation of a Horizontal Parabola

The standard equation of a horizontal parabola with focus F at $(p, 0)$ and directrix $x = -p$ is

$$y^2 = 4px$$

If $p > 0$, the parabola opens to the right, as shown in Figure 2.13.

If $p < 0$, the parabola opens to the left.

EXAMPLE 2.21

Find the equation of the parabola with its vertex at the origin and focus at $(5,0)$.

Solution Since the focus is at $(5,0)$, $p=5$. A parabola with focus on the x-axis and vertex at the origin is of the form $y^2=4px$. Since $p=5$, the equation is $y^2=20x$.

EXAMPLE 2.22

Sketch the graphs of one horizontal and one vertical parabola, each of which has its vertex at the origin and passes through the point $(3,-6)$. Determine the equation for each parabola.

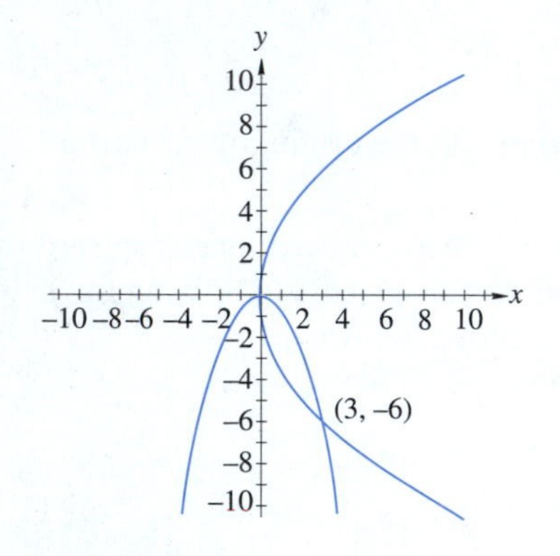

FIGURE 2.14

Solution The graphs are in Figure 2.14. One graph is of the form $x^2=4py$ and the other is of the form $y^2=4px$. Substituting 3 for x and -6 for y in each equation, we get

$$\begin{aligned} x^2 &= 4py & y^2 &= 4px \\ 3^2 &= 4p(-6) & (-6)^2 &= 4p(3) \\ 9 &= -24p & 36 &= 12p \\ p &= -\frac{9}{24} = -\frac{3}{8} & p &= 3 \\ x^2 &= 4\left(-\frac{3}{8}\right)y & y^2 &= 4(3)x \\ x^2 &= -\frac{3}{2}y & y^2 &= 12x \\ \text{or}\quad 2x^2 &= -3y & & \end{aligned}$$

The vertical parabola has the equation $x^2=-\frac{3}{2}y$ and the horizontal parabola has the equation $y^2=12x$.

Reflective Properties of Parabolas

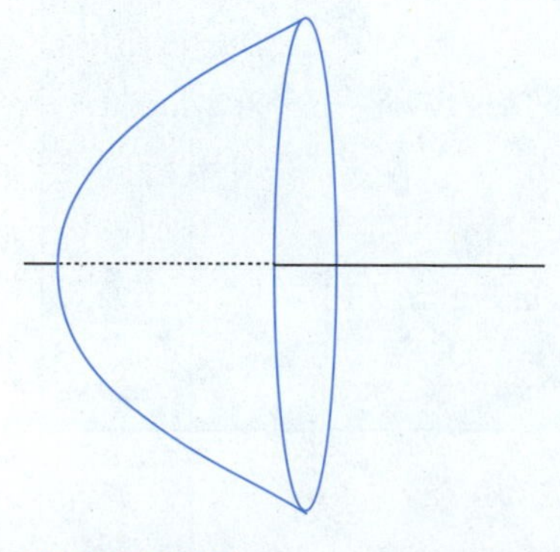

FIGURE 2.15

A three-dimensional object called a **paraboloid of revolution** is formed when a parabola is revolved around its axis of symmetry. (See Figure 2.15.) Paraboloids of revolution have many uses based on the following two facts.

1. Rays entering a parabola along lines parallel to its axis are all reflected through its focus. Many examples exist for the different types of energy rays. Radio telescopes, radar antennae, and satellite television dishes used as downlinks are all examples of paraboloids of revolution. Parabolic reflectors are sometimes used at sports events to pick out specific sounds from background noises. In each of these cases, a ray or wave that is directed toward the dish is reflected off the sides of the paraboloid through its focus.
2. Rays drawn through a parabola's focus are all reflected along lines parallel to the parabola's axis. This uses rays in the reverse of the first property. Applications include flashlights, automobile headlights, search lights, and satellite dishes used as uplinks.

Paraboloids are sometimes used to both send and receive information. For example, a radar transmitter alternately sends and receives waves.

Application

EXAMPLE 2.23

The television satellite dish shown in Figure 2.16a is in the shape of a paraboloid of revolution measuring 0.4 m deep and 2.5 m across at its opening. Where should the receiver be placed in order to pick up the incoming signals?

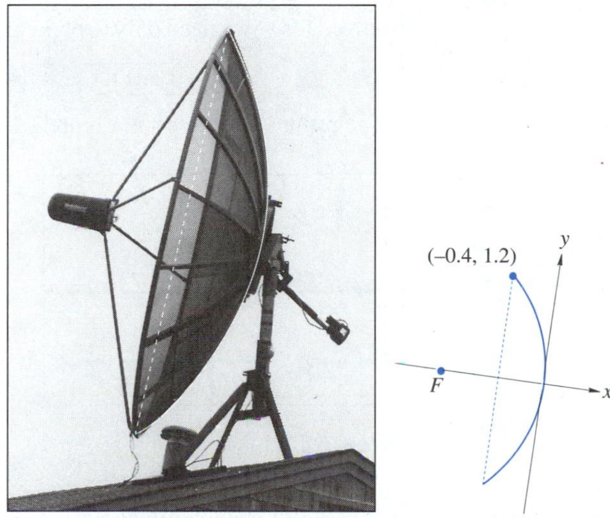

Courtesy of Michael A. Gallitelli, Metroland Photo Inc.

FIGURE 2.16a **FIGURE 2.16b**

Solution A cross-section through the axis of this satellite dish has been drawn over the photograph of the dish in Figure 2.16a. Because the cross-section is a parabola, the receiver should be placed at the parabola's focus. We will next determine the location of that focus.

If we place the drawing of this cross-section on a coordinate system so that the vertex is at the origin and the focus is placed on the x-axis, we get Figure 2.16b. We know that the parabola in Figure 2.16b has an equation of the form $y^2 = 4px$. From the given data, the point $(-0.4, 1.2)$ is on the parabola. This means that $p = \dfrac{y^2}{4x} = \dfrac{1.2^2}{-1.6} = -0.9$. Since the focus is at the point $(0, -p)$, this means that we can place the focus 0.9 m from the vertex.

Using a Graphing Calculator

Graphing calculators can be used to sketch the graph of any of the curves in this chapter. You must remember, however, that these calculators will only sketch the graphs of functions. This means that you will first need to solve the equation algebraically for y and then graph the one or two equations that result. The next two examples will demonstrate how this is done.

EXAMPLE 2.24

Use a graphing calculator to sketch the graphs of the parabolas **(a)** $y = \frac{3}{4}x^2$ and **(b)** $y = \frac{3}{4}x^2 - 3$.

Solution As usual, after you turn on the calculator, clear the screen of any previous graphs.

Now you are ready to graph $y = \frac{3}{4}x^2$. Since this equation has already been solved for y, we need only to press the following key sequence on a Casio fx-7700G.

[Graph] [(] 3 [÷] 4 [)] [X,θ,T] [SHIFT] [x^2] [EXE]

The result is the graph in Figure 2.17a.

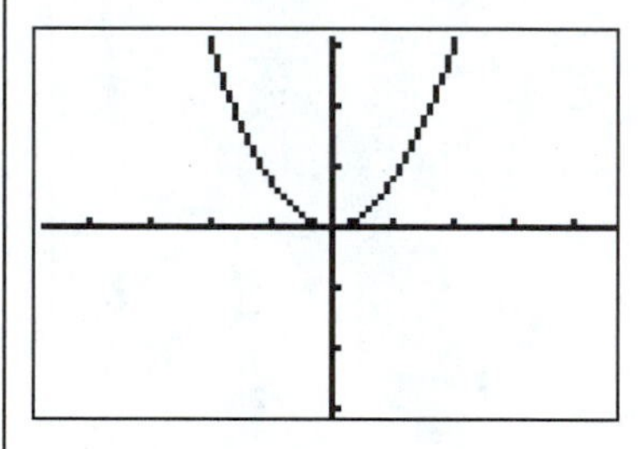

FIGURE 2.17a

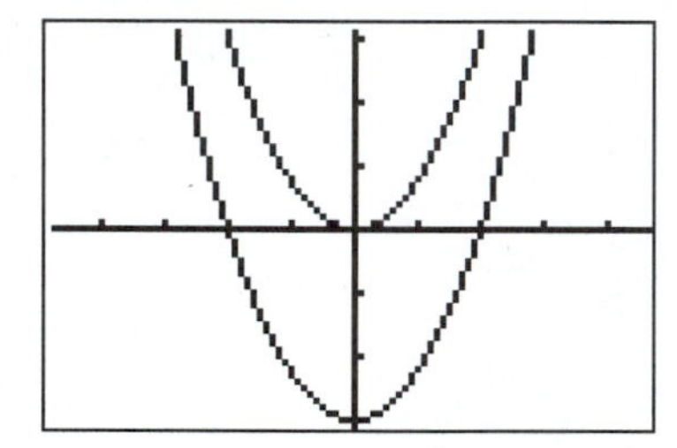

FIGURE 2.17b

If you want to graph $y = \frac{3}{4}x^2 - 3$, enter the key sequence

[Graph] [(] 3 [÷] 4 [)] [X,θ,T] [SHIFT] [x^2] [−] 3 [EXE]

The result is the graph in Figure 2.17b. (If you did not clear the screen after you graphed $y = \frac{3}{4}x^2$, you will see both graphs on your screen.) Notice how the graph of $y = \frac{3}{4}x^2 - 3$ is simply the graph of $y = \frac{3}{4}x^2$ shifted down 3 units.

EXAMPLE 2.25

Use a TI-82 graphing calculator to sketch the graph of the circle $x^2 + y^2 + 4x - 8y + 4 = 0$.

Solution This is the same equation we used in Example 2.15. This equation can be simplified to

$$(x+2)^2 + (y-4)^2 = 16$$

and represents the equation of a circle with center $(-2, 4)$ and radius $r = 4$.

In order to use the graphing calculator, we need to solve the equation for y.

$$\begin{aligned}(x+2)^2 + (y-4)^2 &= 16 \\ (y-4)^2 &= 16 - (x+2)^2\end{aligned}$$

EXAMPLE 2.25 (Cont.)

$$y-4 = \pm\sqrt{16-(x+2)^2}$$
$$y = 4\pm\sqrt{16-(x+2)^2}$$

Since this result has two solutions, we will have to graph the two equations $y=4+\sqrt{16-(x+2)^2}$ and $y=4-\sqrt{16-(x+2)^2}$.

Now, we know that since this circle has its center at $(-2,4)$ and a radius of 4, the x-values will be from $x=-2-4=-6$ to $x=-2+4=2$ and the y-values will be from $y=4-4=0$ to $y=4+4=8$. We will want to set the calculator screen's "range" values so that both of these appear on the screen. We will let x be over the interval $[-8,4]$ and y be over the interval $[-2,10]$.

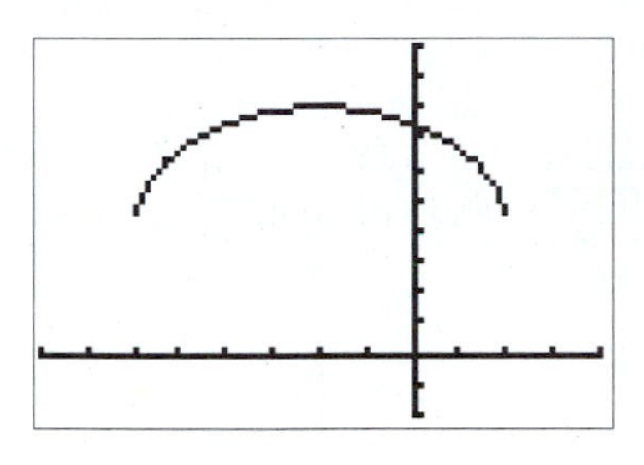

[–8, 4] x [–2, 10]

FIGURE 2.18a

The graph of $y=4+\sqrt{16-(x+2)^2}$ is obtained by first pressing [Y=] and then pressing the key sequence

4 [+] [√] [(] 16 [−] [(] [X,T,θ] [+] 2 [)] [SHIFT] [x^2] [)] [GRAPH]

The result is the graph in Figure 2.18a. Now graph $y=4-\sqrt{16-(x+2)^2}$ by pressing [Y=], moving the cursor to Y2, and pressing the sequence

4 [−] [√] [(] 16 [−] [(] [X,T,θ] [+] 2 [)] [SHIFT] [x^2] [)] [GRAPH]

This graph is displayed over the earlier graph as shown in Figure 2.18b.

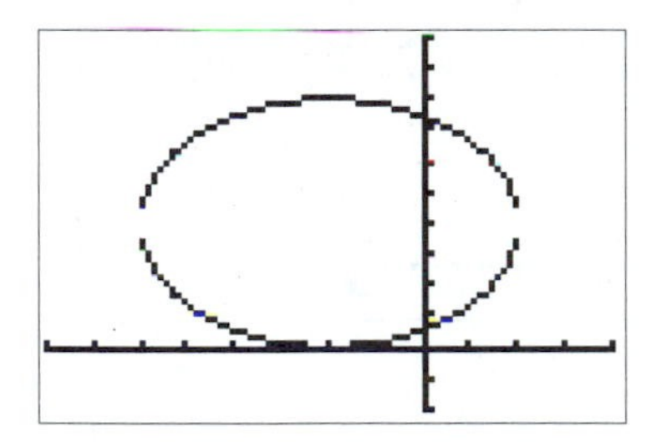

[–8, 4] x [–2, 10]

FIGURE 2.18b

This does not look like a circle. First, the top and bottom parts do not touch, and it does not have a circular shape. The fact that parts do not touch is a common error of graphing calculators and computer graphing programs. The noncircular shape is caused by the fact that the calculator screen's ranges for both the x- and y-values are 12 units long. But, the calculator does not have a square screen. In fact, the ratio of the screen's horizontal to vertical size on calculators such as the Casio fx-7700, TI-81, and TI-82 is 3 : 2; on the TI-85 the ratio is approximately 7 : 4. So, if we want this graph of a circle to *look* like a circle, we need to have the ratio of the TI-82's calculator screen's range of x-values to the range of y-values be 3 : 2.

Let's leave the y-values at $[-2,10]$. This is 12 units, so the x-values should cover 18 units. We will change them to $[-10,8]$. Now press [GRAPH] and the graphs should be redrawn as shown in Figure 2.18c.

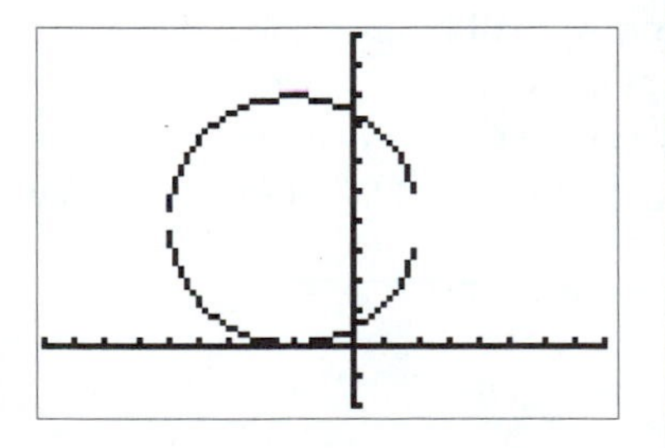

[–10, 8] x [–2, 10]

FIGURE 2.18c

If you use a Texas Instruments TI-82 graphics calculator, the following procedure makes the units on the axes the same length. After you have graphed the two functions and get a graph like the one in Figure 2.18b, press the [ZOOM] button on the top row of the calculator. You should see a screen similar to the one in shown in Figure 2.18d. This zoom menu presents you with nine options. Pressing a [5] selects option 5: ZSquare. This option adjusts the current range values so the width of the dots on the x- and y-axes are equalized. ▪

Some calculators will draw a circle if you enter the coordinates of the center and the radius. The next example illustrates this on a TI-82.

```
ZOOM MEMORY
1:ZBox
2:Zoom In
3:Zoom Out
4:ZDecimal
5:ZSquare
6:ZStandard
7↓ZTrig
```

FIGURE 2.18d

EXAMPLE 2.26

Use a TI-82 to draw the circle

$$(x+2)^2+(y-4)^2=25$$

Solution Here $h=-2$, $k=4$, and $r=5$. The "width" of this circle will stretch from $h-r=-7$ to $h+r=3$ and the "height" from $k-r=-1$ to $k+r=9$. A reasonable viewing window that will show this entire circle and will show a circular-looking circle is $[-9.4, 9.4]\times[-2.2, 10.2]$. (Notice that the ratio of width to height is $18.8:12.4\approx 3:2$. Press [2nd] [DRAW] 9. Your screen should look like that shown in Figure 2.19a. Next, enter the values of h, k, and r separated by commas, by pressing [(−)] 2 [,] 4 [,] 5 [)]. Press [ENTER] and you should get a result similar to the one in Figure 2.19b.

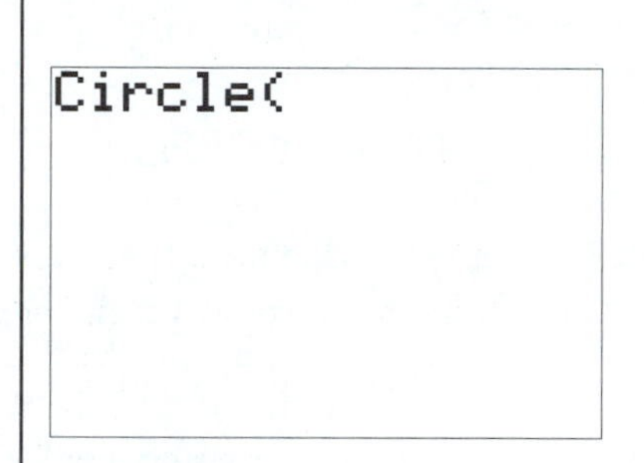

FIGURE 2.19a

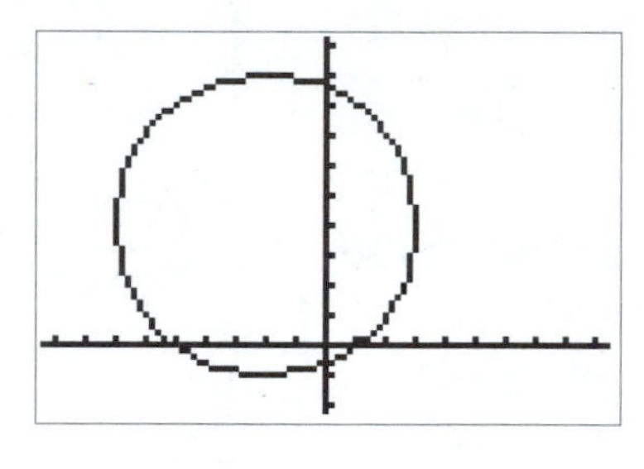

[−9.4, 9.4] x [−2.2, 10.2]

FIGURE 2.19b

EXAMPLE 2.27

Use a graphing calculator to draw the parabola

$$y^2=-12x$$

Solution In order to put this in our calculator, we will have to solve this for y, getting $y=\pm\sqrt{-12x}$. This will have to be graphed as two separate functions, $y_1=\sqrt{-12x}$ and $y_2=-\sqrt{-12x}$. Make sure you graph both y_1 and y_2. The graph of y_1 is the "top half" of the parabola, and the graph of y_2 is the "bottom half." The result is shown in Figure 2.20.

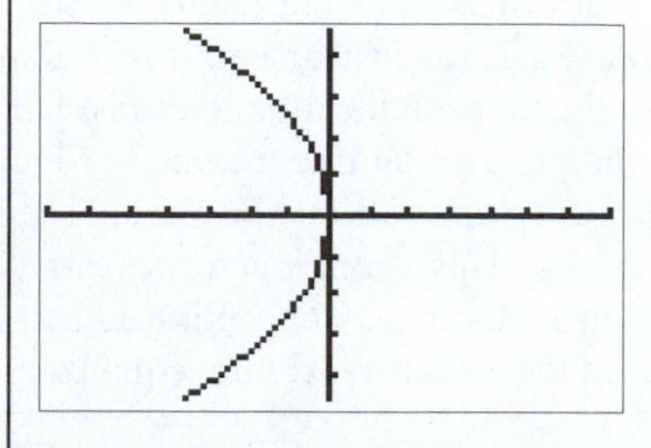

[−14.1, 14.1, 2] x [−9.3, 9.3, 2]

FIGURE 2.20

Exercise Set 2.3

In Exercises 1–12, determine the coordinates of the focus and the equation of the directrix of each parabola. State the direction in which each parabola opens and sketch each curve.

1. $x^2 = 4y$
2. $x^2 = 12y$
3. $y^2 = -4x$
4. $y^2 = -20x$
5. $x^2 = -8y$
6. $x^2 = -16y$
7. $y^2 = 10x$
8. $y^2 = -14x$
9. $x^2 = 2y$
10. $x^2 = -17y$
11. $y^2 = -21x$
12. $y^2 = 5x$

In Exercises 13–20, determine the equation of the parabola satisfying the given conditions.

13. Focus $(0, 4)$, directrix $y = -4$
14. Focus $(0, -5)$, directrix $y = 5$
15. Focus $(-6, 0)$, directrix $x = 6$
16. Focus $(3, 0)$, directrix $x = -3$
17. Focus $(2, 0)$, vertex $(0, 0)$
18. Focus $(0, -7)$, vertex $(0, 0)$
19. Vertex $(0, 0)$, directrix $x = -\frac{3}{2}$
20. Vertex $(0, 0)$, directrix $y = \frac{3}{4}$

Solve Exercises 21–28.

21. *Civil engineering* In a suspension bridge, the main cables are in a parabolic shape. This is because a parabola is the only shape that will bear the total weight load evenly. The twin towers of a certain bridge extend 90 m above the road surface and are 360 m apart, as shown in Figure 2.21. The cables are suspended from the tops of the towers and are tangent to the road surface at a point midway between the towers. What is the height of the cable above the road surface at a point 100 m from the center of the bridge?

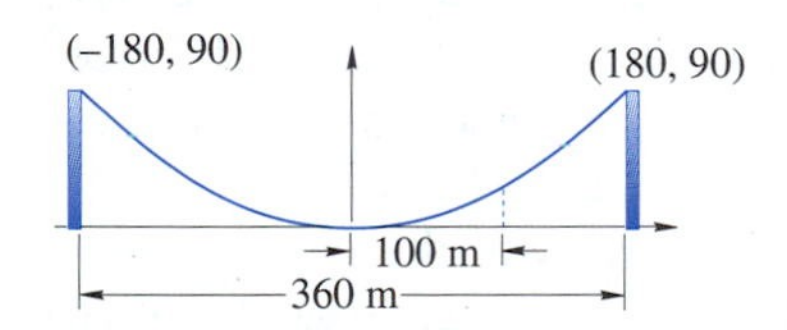

FIGURE 2.21

22. *Optics* A parabolic reflector is a mirror formed by rotating a parabola around its axis. If a light is placed at the focus of the mirror, all the light rays starting from the focus will be reflected off the mirror in lines parallel to the axis. A radar antenna is constructed in the shape of a parabolic reflector. The receiver is placed at the focus. If the reflector has a diameter of 2 m and a depth of 0.4 m, what is the location of the receiver?

23. *Electronics* When rays from a distant source strike a parabolic reflector, they will be reflected to a single point—the focus. A parabolic antenna is used to catch the television signals from a satellite. The antenna is 5 m across and 1.5 m deep. If the receiver is located at the focus, what is its location?

24. *Energy* The rate at which heat is dissipated in an electric current is referred to as the power loss. This loss, P, is given by the relationship $P = I^2R$, where I is the current and R is the resistance. If the resistance is $10\,\Omega$, sketch a graph of the power loss as a function of the current.

25. *Electronics* For a simple linear resistance R in Ω, the power, P, in W dissipated in the circuit depends on the current, I, in A according to the equation $P = RI^2$. Graph P as a function of I for $0 \leq I \leq 6.0$ when $R = 0.25\,\Omega$.

26. *Thermodynamics* The heat, H, in joules (J) produced by a voltage, V, across a heating coil is given by $H = \frac{V^2}{R}t$, where t is the time and R the resistance. If $t = 2$ h and $R = 15\,\Omega$, sketch the graph of H as a function of V from $V = 0$ to $V = 120$ V.

27. *Broadcasting* A cable television "dish" is 2 ft deep and measures 10 ft across at its opening. A technician must place the receiver at the focus. How far is this from the vertex?

28. *Astronomy* Astronomers and optical experts have been exploring the possibility of making mirrors from liquids. When a liquid is spun in a container, the surface of the liquid takes on a parabolic shape. If a liquid metal, such as mercury, is used, then the optical quality is about as good as that of more expensive ground-glass mirrors. The parabolic shape is kept as long as the mirror is kept spinning. The focal length of the mirror is proportional to the square of the rotation period, p. (The rotation period is the reciprocal of the rotational frequency.) If a mirror spinning at 45 rpm has a focal length of 22.1 cm, then
(a) determine the focal length of a mirror spinning at $33\frac{1}{3}$ rpm
(b) determine the focal length of a mirror spinning at 15 rpm

In Your Words

29. Describe how you can tell by looking at a standard equation for a parabola whether it is a horizontal or a vertical parabola.

30. Explain how to tell from the equation if a parabola opens upward, downward, left, or right.

2.4 THE ELLIPSE

The third curve we will study in this chapter is the ellipse. An **ellipse** is the set of points in a plane that have the sum of their distances from two fixed points a constant. Each of the fixed points is called a **focus**. The **major axis** of an ellipse is the line segment through the two foci with endpoints on the ellipse. The endpoints of the major axis are called the **vertices**. In Figure 2.22, the foci are labeled F and F' and the vertices are V and V'. The **center** C of the ellipse is the midpoint of the segment joining the foci. The segment through the center and perpendicular to the major axis is called the **minor axis**. The endpoints of the minor axis are on the ellipse. In Figure 2.22, the endpoints of the minor axis are M and M'. The major axis is always longer than the minor axis.

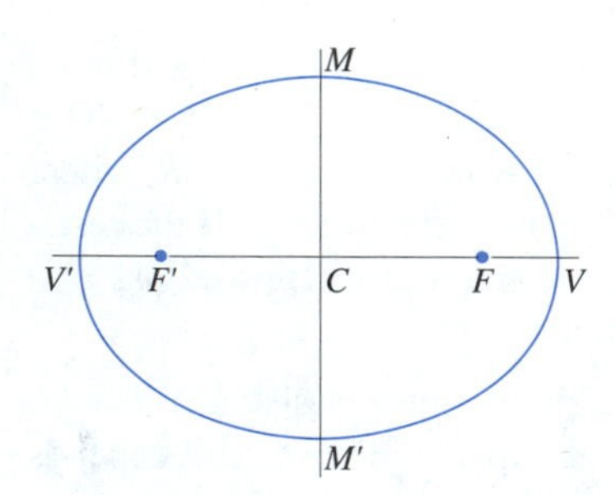

FIGURE 2.22

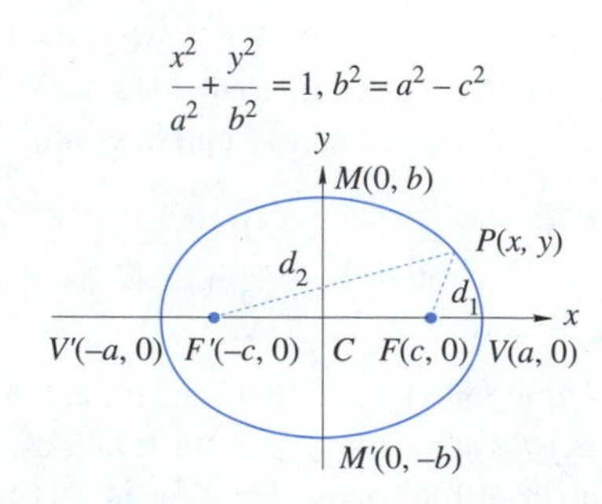

FIGURE 2.23

If we were to draw an ellipse with its center at the origin and its major axis along the x-axis, we would get a figure like the one in Figure 2.23. We will let the vertices have the coordinates $V(a,0)$ and $V'(-a,0)$. The foci will have the coordinates $F(c,0)$ and $F'(-c,0)$. If $P(x,y)$ is any point on the ellipse, then the distance from P to F, d_1, plus the distance from P to F', d_2, is a constant, which we will call $2a$. The sum of these distances, $d_1+d_2=(a-c)+(a+c)=2a$.

The points F, F', and M form an isosceles triangle with the lengths of two equal sides, $\overline{FM}$ and $\overline{F'M}$, being a. Thus, F, M, and C form a right triangle with legs of lengths c and b and hypotenuse of length a. As a result, we see that $a^2=b^2+c^2$.

Applying the distance formula, we see that $d_1=\sqrt{(x-c)^2+y^2}$ and $d_2=\sqrt{(x+c)^2+y^2}$. Since $d_1+d_2=2a$, we have $\sqrt{(x-c)^2+y^2}+\sqrt{(x+c)^2+y^2}=2a$. Using techniques for solving radical equations, we get

$$\sqrt{(x-c)^2+y^2}=2a-\sqrt{(x+c)^2+y^2}$$

$$\left[\sqrt{(x-c)^2+y^2}\right]^2=\left[2a-\sqrt{(x+c)^2+y^2}\right]^2 \qquad \text{Square both sides.}$$

$$(x-c)^2+y^2=(2a)^2-4a\sqrt{(x+c)^2+y^2}+(x+c)^2+y^2$$

$$4a\sqrt{(x+c)^2+y^2}=4a^2+4cx$$

$$a\sqrt{(x+c)^2+y^2}=a^2+cx$$

$$\left[a\sqrt{(x+c)^2+y^2}\right]^2=\left[a^2+cx\right]^2 \qquad \text{Square again.}$$

$$a^2\left[(x+c)^2+y^2\right]=a^4+2a^2cx+c^2x^2$$

$$a^2x^2+2a^2cx+a^2c^2+a^2y^2=a^4+2a^2cx+c^2x^2$$

$$(a^2-c^2)x^2+a^2y^2=a^2(a^2-c^2)$$

But, $a^2=b^2+c^2$, so $a^2-c^2=b^2$. The equation becomes

$$b^2x^2+a^2y^2=a^2b^2$$

Dividing both sides by a^2b^2 and simplifying results in one of the standard equations for an ellipse:

$$\frac{x^2}{a^2}+\frac{y^2}{b^2}=1$$

Standard Equation: Ellipse with a Horizontal Major Axis

The standard equation of an ellipse with center at $(0,0)$ and the major axis on the x-axis is

$$\frac{x^2}{a^2}+\frac{y^2}{b^2}=1$$

where $a>b$.

The vertices are $(-a,0)$ and $(a,0)$.

The endpoints of the minor axis are $(0,-b)$ and $(0,b)$.

The foci are at $(-c,0)$ and $(c,0)$, where $c^2=a^2-b^2$.

EXAMPLE 2.28

Sketch the ellipse $4x^2+25y^2=100$. Give the coordinates of the vertices, foci, and endpoints of the minor axis.

Solution This equation is not in the standard form, since the right-hand side is not 1. To get the equation into the standard form, we must divide both sides of the equation by 100. This produces the equation

$$\frac{x^2}{25}+\frac{y^2}{4}=1$$

From this, we see that $a^2=25$, so $a=5$ and $b^2=4$ or $b=2$. Since $c^2=a^2-b^2=25-4=21$, we see that $c=\sqrt{21}$. So, the vertices are $V(5,0)$ and $V'(-5,0)$, and the foci are $F(\sqrt{21},0)$ and $F'(-\sqrt{21},0)$. The endpoints of the minor axis are $M(0,2)$ and $M'(0,-2)$. A sketch of this ellipse is shown in Figure 2.24.

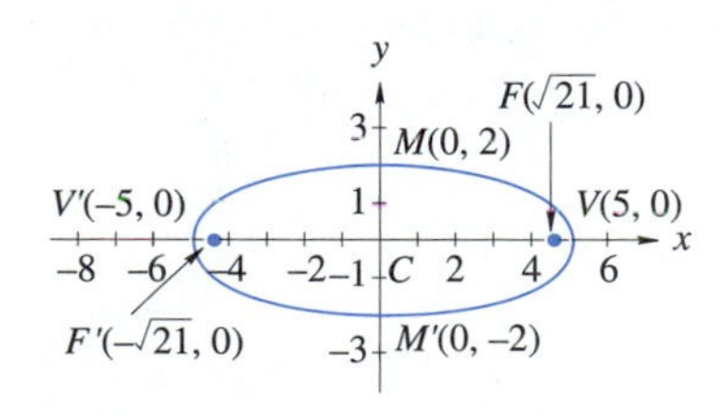

FIGURE 2.24

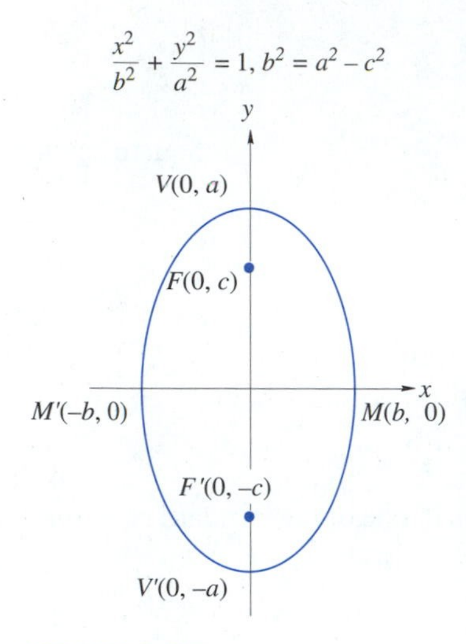

FIGURE 2.25

Not all ellipses have their major axis along the x-axis. For some situations, it is more convenient if the center is at the origin and the major axis is along the y-axis. When that is the case, the vertices are $V(0, a)$ and $V'(0, -a)$, the foci are $F(0, c)$ and $F'(0, -c)$, and the endpoints of the minor axis are $M(b, 0)$ and $M'(-b, 0)$. In this case, we have the following standard form of the equation for an ellipse. A typical graph of an ellipse with center at the origin and a vertical major axis is shown in Figure 2.25.

Standard Equation: Ellipse with a Vertical Major Axis

The standard equation of an ellipse with center at $(0, 0)$ and the major axis on the y-axis is

$$\frac{x^2}{a^2} + \frac{y^2}{b^2} = 1$$

where $a < b$.

The vertices are $(0, -b)$ and $(0, b)$.

The endpoints of the minor axis are $(-a, 0)$ and $(a, 0)$.

The foci are at $(0, -c)$ and $(0, c)$, where $c^2 = b^2 - a^2$.

If we put these two types of ellipses together, we see that the equation of an ellipse with center at the origin and foci on a coordinate axis can always be written in the form

$$\frac{x^2}{p^2} + \frac{y^2}{q^2} = 1 \qquad \text{or} \qquad q^2x^2 + p^2y^2 = p^2q^2$$

where p and q are both positive. If $p^2 > q^2$, then the major axis is along the x-axis. If $p^2 < q^2$, then the major axis is along the y-axis.

Note The x-axis is the major, or longer, axis if the denominator of x^2 is the larger denominator, and the y-axis is the major axis if the denominator of y^2 is the larger denominator.

Remember, c^2 is found by subtracting the smaller denominator from the larger. You may want to remember this as $c^2 = |a^2 - b^2|$.

EXAMPLE 2.29

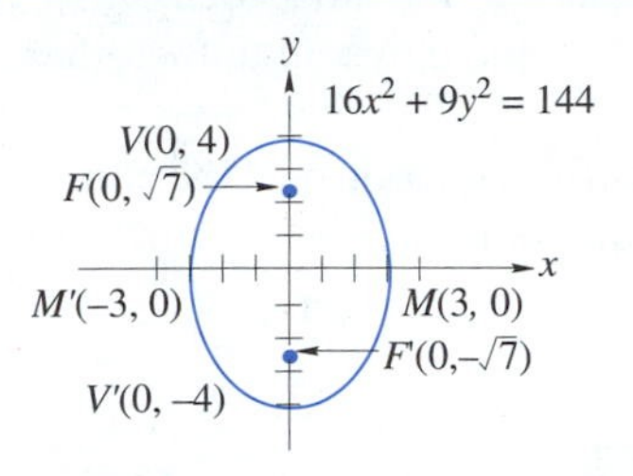

FIGURE 2.26

Discuss and graph the equation $16x^2+9y^2=144$.

Solution Dividing both sides of the equation by 144, we get the standard equation

$$\frac{16x^2}{144}+\frac{9y^2}{144}=\frac{144}{144}$$
$$\frac{x^2}{9}+\frac{y^2}{16}=1$$

So, $a^2=9, b^2=16$, and $c^2=b^2-a^2=7$. Since $a=3$ and $b=4$, we see that $a<b$, so the major axis is along the y-axis. The vertices are $V(0,4)$ and $V'(0,-4)$, the foci are $F(0,\sqrt{7})$ and $F'(0,-\sqrt{7})$, and the endpoints of the minor axis are $M(3,0)$ and $M'(-3,0)$. The sketch of this ellipse is shown in Figure 2.26.

EXAMPLE 2.30

Find the equation of the ellipse with center $(0,0)$, one focus at $(5,0)$, and a vertex at $(-8,0)$.

Solution The second focus is at $(-5,0)$ and the other vertex is at $(8,0)$. Since the major axis is horizontal (along the x-axis), the equation is of the form

$$\frac{x^2}{a^2}+\frac{y^2}{b^2}=1$$

where $c^2=a^2-b^2, a>b$.

Since one focus is at $(5,0)$, $c=5$ and the fact that one vertex is at $(8,0)$ tells us that $a=8$. So, $5^2=8^2-b^2$ or $b^2=64-25=39$, and $b=\sqrt{39}$. The equation of this ellipse is $\frac{x^2}{64}+\frac{y^2}{39}=1$.

Reflective Properties of Ellipses

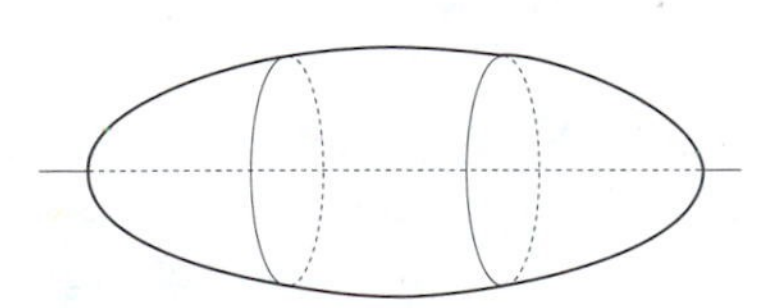

FIGURE 2.27

A three-dimensional object called a **ellipsoid of revolution** is formed when an ellipse is revolved around one of its axes of symmetry. In Figure 2.27, the ellipse has been revolved around its longer axis.

These ellipsoids of revolution have uses based on the fact that rays leaving or passing through one focus of an ellipse are all reflected by the ellipse so that they pass through the other focus of the ellipse. One architectural application of these ellipsoids of revolution is in the construction of whispering galleries, such as the one in Statuary Hall of the Capitol in Washington, D.C. or St. Paul's Cathedral in London, England. If you position yourself at one focus in a whispering gallery, you will be able to hear anything said at the other focus, regardless of the direction in which the speakers address their remarks.

Application

EXAMPLE 2.31

A lithotripter is a medical device that is used to break up kidney stones with shock waves through water. An ellipsoid is cut in half and careful measurements are used to place the patient's kidney stones at one focus of the ellipse. A source

EXAMPLE 2.31 (Cont.)

that produces ultra–high-frequency shock waves is placed at the other focus. (See Figures 2.28a and b.) The shock waves are reflected by the ellipsoid to the other focus, where they break up the kidney stones.

If the endpoints of the major axis are located 6 in. from the center of the ellipse that is rotated to make a lithotripter and one end of the minor axis is 2.5 in. from the center, what are the locations of the lithotripter's foci?

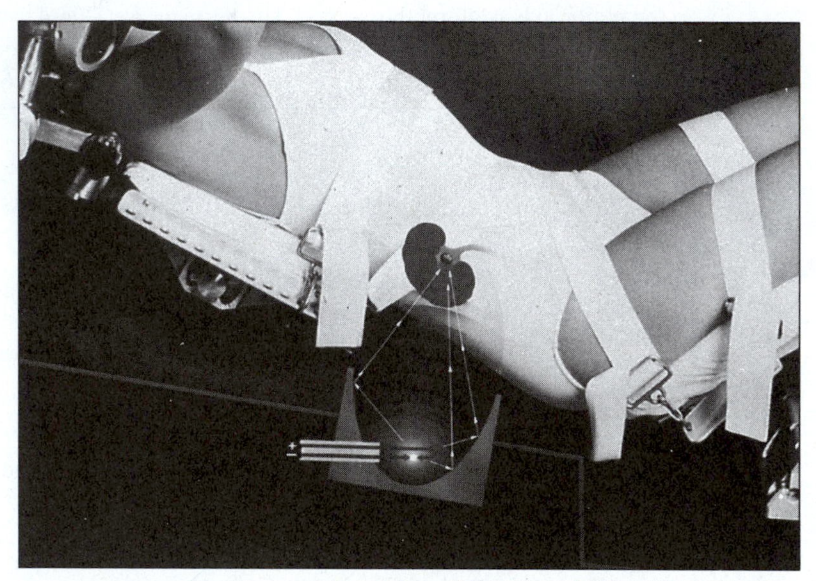

Courtesy of Dornier Medical Systems Inc.

FIGURE 2.28a

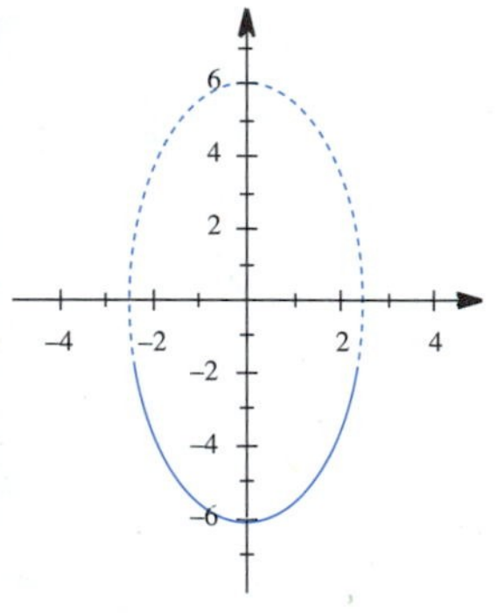

FIGURE 2.28b

Solution This appears to be a vertically oriented ellipse. If we place the drawing of the cross-section of the ellipse in Figure 2.28a on a coordinate system so that the center is at the origin and the major axis is on the y-axis, we get Figure 2.28b. The solid curve in Figure 2.28b represents the lithotripter. The dotted curve represents the portion of the ellipse that has been removed. From the given information, we know that the ellipse in Figure 2.28b has endpoints on the major axis at $(0, -6)$ and $(0, 6)$, and that the endpoints of the minor axis are $(-2.5, 0)$ and $(2.5, 0)$. From the given data, we know that the desired ellipse has the equation $\frac{x^2}{a^2} + \frac{y^2}{b^2} = \frac{x^2}{2.5^2} + \frac{y^2}{6^2} = 1$. Thus, $a = 2.5$ and $b = 6$ and, since $c^2 = b^2 - a^2$, we have $c^2 = 6^2 - 2.5^2 = 29.75$ and so $c = \sqrt{29.75} \approx 5.45$. Thus, the foci are at $(0, -5.45)$ and $(0, 5.45)$.

Using a Graphing Calculator

When using a graphing calculator, choose a viewing window that places the endpoints of the major and minor axes are in the window. Again, since the calculator will graph only a function, you will need to solve the equation algebraically for y and then graph the two equations that result.

EXAMPLE 2.32

Use a graphing calculator to graph $\frac{x^2}{9} + \frac{y^2}{64} = 1$.

Solution Since this is in standard form, we see that this is a vertically oriented ellipse. Here $a^2 = 9$, so $a = 3$ and $b^2 = 64$, so $b = 8$. A viewing window that will include the entire ellipse would set Xmin $= -3$, Xmax $= 3$, Ymin $= -8$, and Ymax $= 8$. As we will see, this does not graph the ellipse so that it appears to be vertically oriented.

Now, solve the equation for y.

$$\begin{aligned} \frac{x^2}{9} + \frac{y^2}{64} &= 1 \\ \frac{y^2}{64} &= 1 - \frac{x^2}{9} \\ y^2 &= 64\left(1 - \frac{x^2}{9}\right) \\ y &= \pm\sqrt{64\left(1 - \frac{x^2}{9}\right)} \\ &= \pm 8\sqrt{1 - \frac{x^2}{9}} \end{aligned}$$

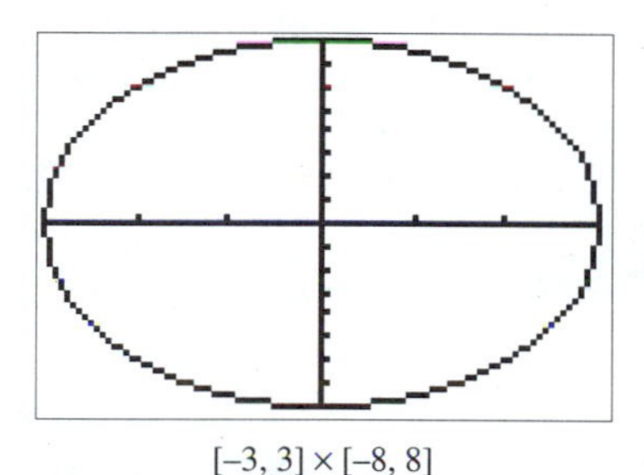

[−3, 3] × [−8, 8]

FIGURE 2.29a

We will graph $y_1 = 8\sqrt{1 - \frac{x^2}{9}}$ and $y_2 = -8\sqrt{1 - \frac{x^2}{9}}$. This result is shown in Figure 2.29a.

This ellipse does not look as if it is a vertical ellipse. If we set the ratio (Xmax − Xmin) : (Ymax − Ymin) = 3 : 2 and leave the Ymin and Tmax settings alone, we get

$$\begin{aligned} \frac{3}{2} &= \frac{\text{Xmax} - \text{Xmin}}{16} \\ 24 &= \text{Xmax} - \text{Xmin} \end{aligned}$$

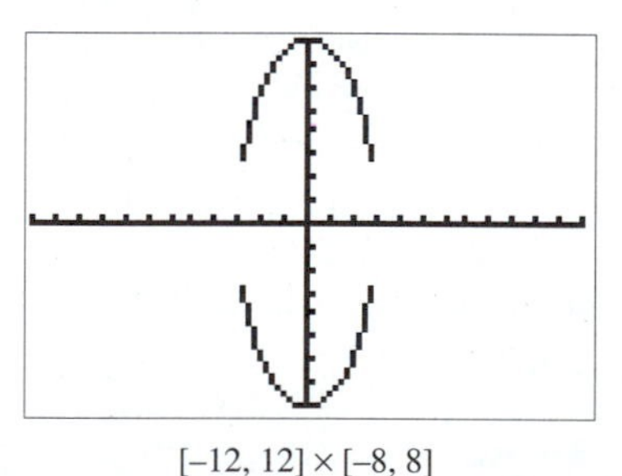

[−12, 12] × [−8, 8]

FIGURE 2.29b

This ellipse is centered at the origin, so Xmin and Xmax should be the same distance from the center. Let Xmin $= -12$ and Xmax $= 12$. The resulting graph is shown in Figure 2.29b.

Exercise Set 2.4

In Exercises 1–8, find the equation of the ellipse with the stated properties. Each ellipse has its center at (0,0).

1. Focus at $(4,0)$, vertex at $(6,0)$
2. Focus at $(9,0)$, vertex at $(12,0)$
3. Focus at $(0,2)$, vertex at $(0,-4)$
4. Focus at $(0,-3)$, vertex at $(0,5)$
5. Focus at $(3,0)$, vertex at $(5,0)$
6. Focus at $(0,-2)$, vertex at $(0,-3)$
7. Length of the minor axis 6, vertex at $(4,0)$
8. Length of the minor axis 10, vertex at $(0,-6)$

In Exercises 9–16, give the coordinates of the vertices, foci, and endpoints of the minor axis, and sketch each curve.

9. $\frac{x^2}{4}+\frac{y^2}{9}=1$
10. $\frac{x^2}{9}+\frac{y^2}{4}=1$
11. $\frac{x^2}{4}+\frac{y^2}{1}=1$
12. $\frac{x^2}{16}+\frac{y^2}{36}=1$
13. $4x^2+y^2=4$
14. $9x^2+4y^2=36$
15. $25x^2+36y^2=900$
16. $9x^2+y^2=18$

Solve Exercises 17–24.

17. *Astronomy* The orbit of the earth is an ellipse with the sun at one focus. The major axis has a length of 2.992×10^8 km and the ratio of $\frac{c}{a}=\frac{1}{60}$. What is the length of the minor axis? $\left(\text{The ratio } \frac{c}{a} \text{ is called the}\right.$ **eccentricity**.$\left.\right)$

18. *Civil engineering* In order to support a bridge, an arch in the shape of the upper half of an ellipse is built. As shown in Figure 2.30, the bridge is to span a river 80 ft wide and the center of the arch is to be 24 ft above the water. If the water level is used as the x-axis and the y-axis passes through the center of the bridge, write an equation for the ellipse that forms the arch of the bridge.

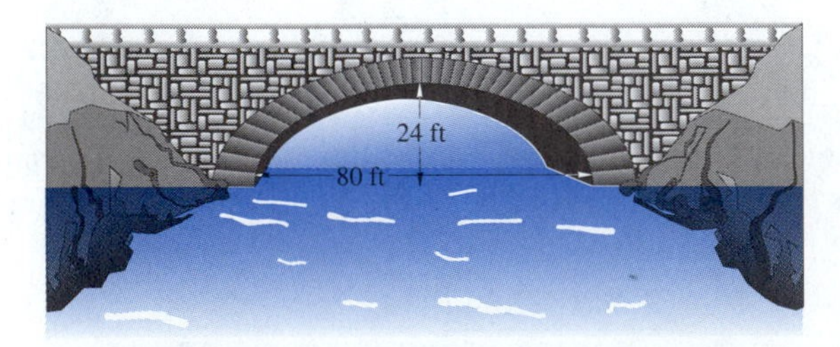

FIGURE 2.30

19. *Mechanical engineering* A circular pipe has an inside diameter of 10 in. One end of the pipe is cut at an angle of 45°, as shown in Figure 2.31. What are the lengths of the major and minor axes of the elliptical opening?

FIGURE 2.31

20. *Mechanical engineering* Elliptical gears are used in some machinery to provide a quick-return mechanism or a slow power stroke (for heavy cutting) in each revolution. Figure 2.32 shows two congruent gears that remain in contact as they rotate around the indicated foci. If the distance between the foci used at the centers of rotation is 6 cm and the shortest distance from a focus to the edge of a gear is 1 cm, what is the length of the minor axis?

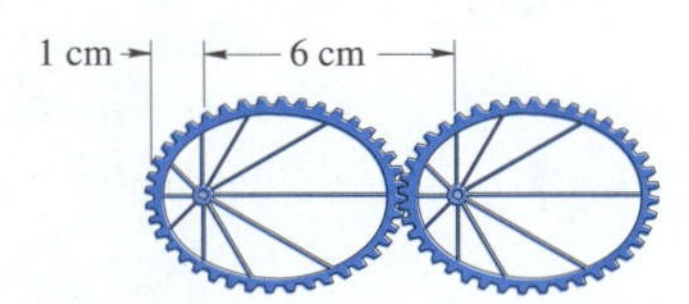

FIGURE 2.32

21. *Astronomy* The comet Kahoutek has major and minor axes of lengths 3 600 and 44 astronomical units. One astronomical unit is about 1.495×10^8 km. What is the eccentricity of the comet's orbit?
22. *Space technology* A satellite orbits the earth in an elliptical path with focus at the center of the earth. The altitude of the satellite ranges from 764.4 km to 3 017.5 km. If the radius of the earth is approximately 6 373 km, what is the equation of the path of the orbit?
23. *Mechanical engineering* Two circular pipes that are each 8.0 in. in diameter are joined at a 45° angle as shown in Figure 2.33. Find the length of the major axis for the elliptical intersection.

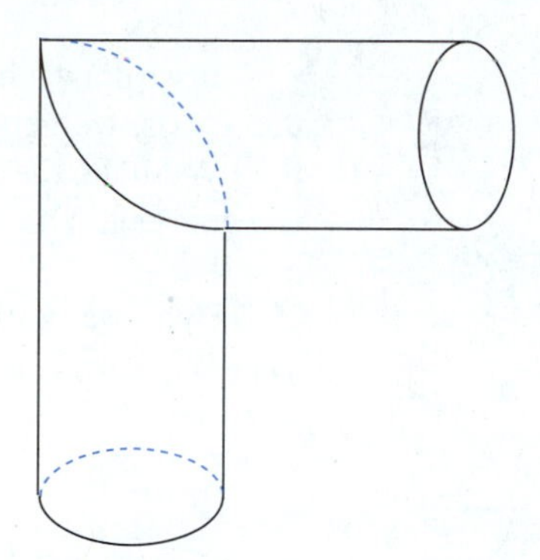

FIGURE 2.33

24. *Transportation engineering* A road passes through a tunnel whose cross-section is a semiellipse 52 ft wide and 12.5 ft high at the center. How tall is the tallest vehicle that can pass under the tunnel at a point 14 ft from the center?

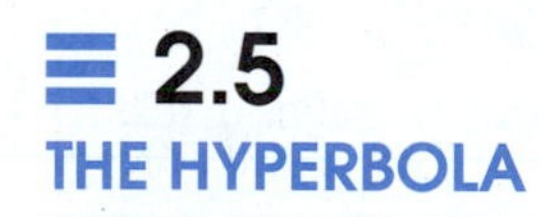

In Your Words

25. Describe how you can tell by inspecting the standard equation of an ellipse whether it is a horizontal or a vertical ellipse.
26. Explain how to use the endpoints of the axes to locate the foci of an ellipse.

2.5 THE HYPERBOLA

The fourth and last conic section that we will study is the hyperbola. A **hyperbola** is the set of points in a plane for which the difference of the distances of the points from two fixed points is a constant. As with the ellipse, each of the fixed points is called a **focus**. The **transverse axis** is the line segment through the two foci with its endpoints on the hyperbola. The endpoints of the transverse axis are called the **vertices**.

A hyberbola is shown by the blue curve in Figure 2.34. In Figure 2.34, the foci are labeled F and F' and the vertices V and V'. The **center** C of the hyperbola is the midpoint of the foci. The line segment through the center that is perpendicular to the transverse axis is called the **conjugate axis**. The endpoints of the conjugate axis, labeled W and W' in Figure 2.34, are not on the hyperbola. We will learn how to determine the length of the conjugate axis later in this section.

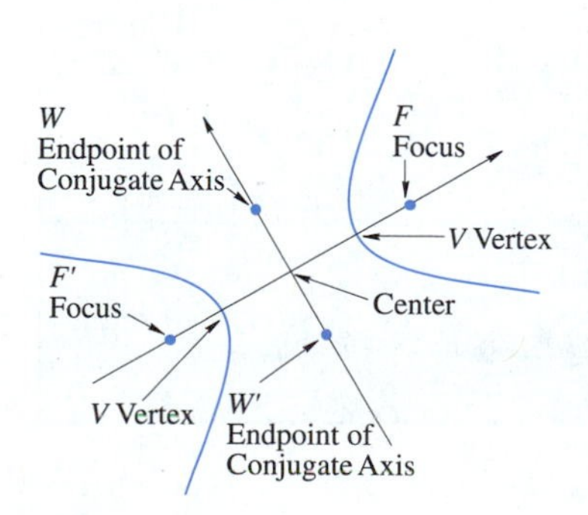

FIGURE 2.34

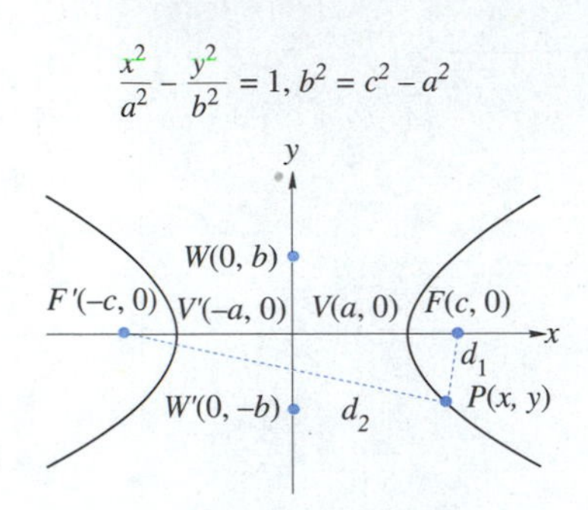

FIGURE 2.35

If we draw a hyperbola with its center at the origin and its transverse axis along the x-axis, we get a figure like the one in Figure 2.35. If $P(x, y)$ is any point on the hyperbola, the distance from P to F, d_1, minus the distance from P to F', d_2, is a constant. The difference of the distances is $|d_1 - d_2| = |(c-a)-(c+a)| = 2a$, $a > 0$.

Here we will use the distance formulas $d_1 = \sqrt{(x-c)^2+y^2}$ and $d_2 = \sqrt{(x+c)^2+y^2}$. As a result, we compute the difference of the distances as

$$\left|\sqrt{(x-c)^2+y^2} - \sqrt{(x+c)^2+y^2}\right| = 2a$$

or

$$\sqrt{(x-c)^2+y^2} - \sqrt{(x+c)^2+y^2} = \pm 2a$$

Using the same technique we used on the ellipse, we get

$$c^2x^2 - a^2x^2 - a^2y^2 = a^2c^2 - a^4$$

or

$$x^2(c^2-a^2) - a^2y^2 = a^2(c^2-a^2)$$

If we let $b^2 = c^2 - a^2$, this can be written as $b^2x^2 - a^2y^2 = a^2b^2$. Dividing both sides by a^2b^2, the equation becomes

$$\frac{x^2}{a^2} - \frac{y^2}{b^2} = 1$$

which is one of the standard equations for a hyperbola. It can be shown that the lines $y = \frac{b}{a}x$ and $y = -\frac{b}{a}x$ are the hyperbola's asymptotes.

Standard Equation: Hyperbola with a Horizontal Major Axis

The standard equation of a hyperbola with center at $(0, 0)$ and the major axis on the x-axis is

$$\frac{x^2}{a^2} - \frac{y^2}{b^2} = 1$$

The vertices are $(-a, 0)$ and $(a, 0)$.

The endpoints of the conjugate axis are $(0, -b)$ and $(0, b)$.

The foci are at $(-c, 0)$ and $(c, 0)$, where $c^2 = a^2 + b^2$.

The lines $y = \frac{b}{a}x$ and $y = -\frac{b}{a}x$ are **asymptotes** of this hyperbola.

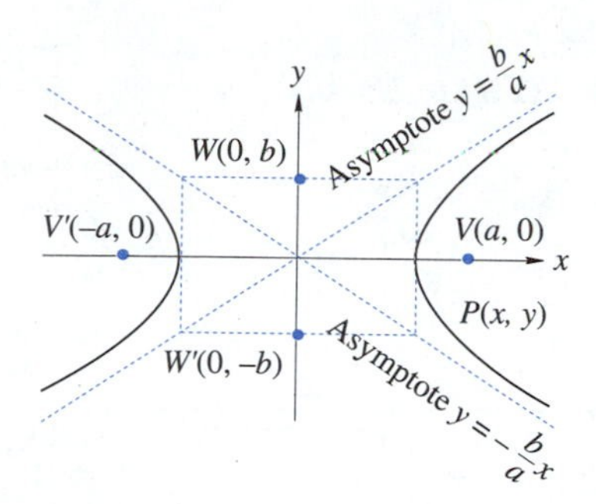

FIGURE 2.36

The asymptotes provide convenient guidelines for drawing hyperbolas. Asymptotes are easily drawn as the diagonals of a rectangle. The midpoints of the sides of the rectangles are the vertices and the endpoints of the conjugate axis, as shown in Figure 2.36.

If the transverse axis is along the y-axis and the conjugate axis is along the x-axis, the hyperbola has a second standard form.

Standard Equation: Hyperbola with a Vertical Major Axis

The standard equation of a hyperbola with center at $(0, 0)$ and the major axis on the y-axis is

$$\frac{y^2}{a^2} - \frac{x^2}{b^2} = 1$$

The vertices are $(0, -a)$ and $(0, a)$.

The endpoints of the conjugate axis are $(-b, 0)$ and $(b, 0)$.

The foci are at $(0, -c)$ and $(0, c)$, where $c^2 = a^2 + b^2$.

The lines $y = \frac{a}{b}x$ and $y = -\frac{a}{b}x$ are **asymptotes** of this hyperbola.

Caution

The formulas relating a, b, and c for the hyperbola are different from the corresponding formulas for the ellipse.

For the ellipse

$c^2 = |a^2 - b^2|$ (See Figure 2.37a.) $a > c$ (See Figure 2.37b.)

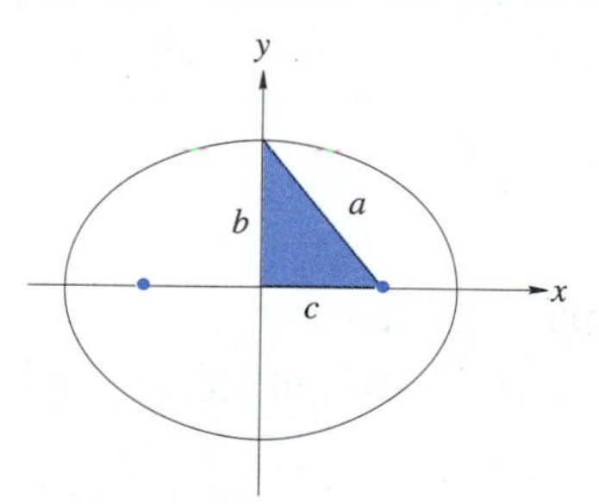

FIGURE 2.37a

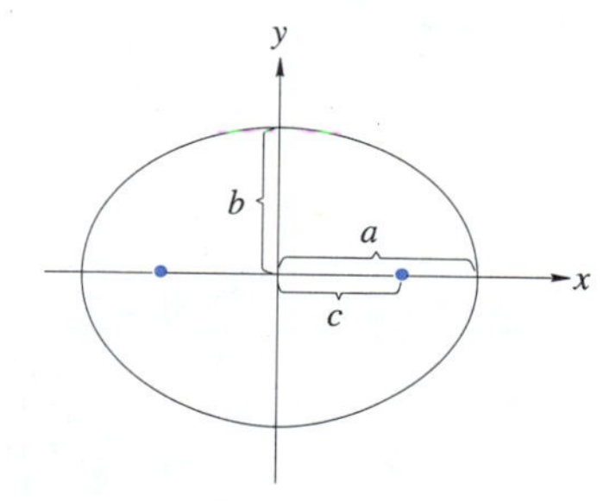

FIGURE 2.37b

For the hyperbola

$c^2 = a^2 + b^2$ (See Figure 2.38a.) $\quad c > a$ (Figure 2.38b.)

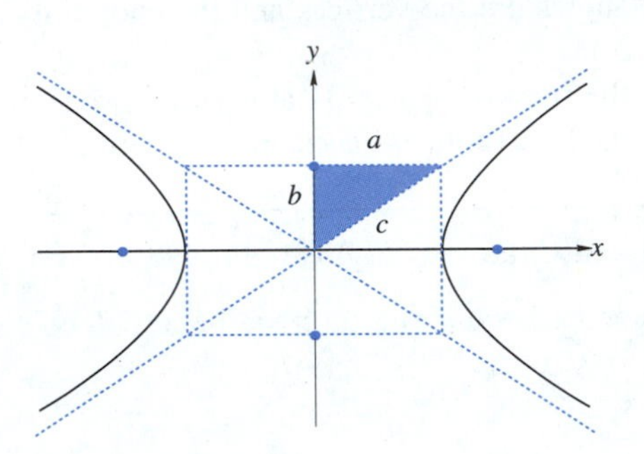

FIGURE 2.38a

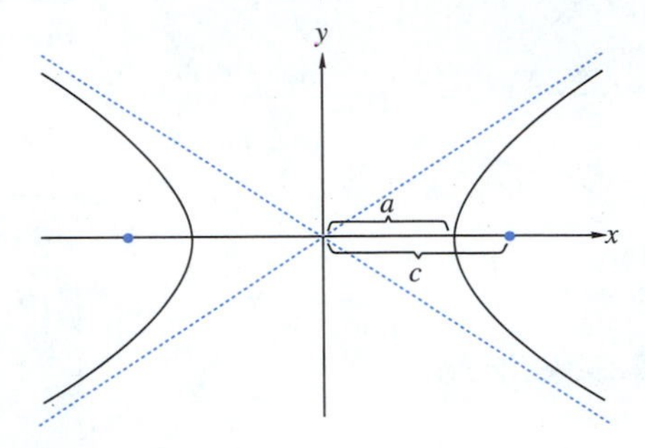

FIGURE 2.38b

EXAMPLE 2.33

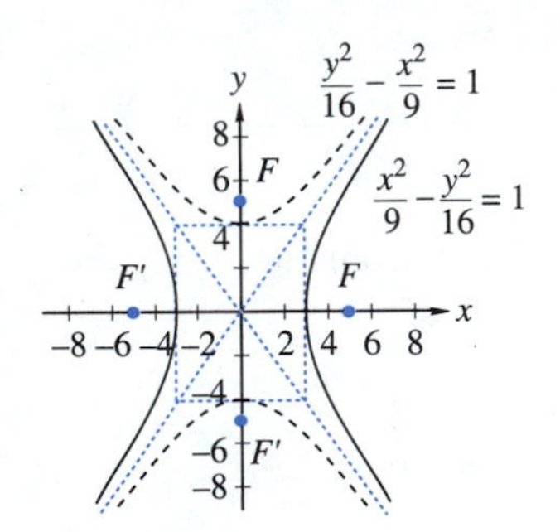

FIGURE 2.39

Discuss and graph the equation $16x^2 - 9y^2 = 144$.

Solution To get the equation in standard form, we divide both sides by 144 and obtain the equation

$$\frac{x^2}{9} - \frac{y^2}{16} = 1$$

This is the equation of a hyperbola with center $(0,0)$. The transverse axis is the x-axis. Also, $a^2 = 9$ and $b^2 = 16$, so $c^2 = a^2 + b^2 = 9 + 16 = 25$. The foci are at $F(5,0)$ and $F'(-5,0)$, the vertices at $V(3,0)$ and $V'(-3,0)$, and the endpoints of the conjugate axis at $W(0,4)$ and $W'(0,-4)$. If we form the rectangle with midpoints of its sides V, V', W, and W' and draw its diagonals, we get the asymptotes $y = \frac{4}{3}x$ and $y = -\frac{4}{3}x$ shown by the dotted lines in Figure 2.39. The sketch of the graph is shown by the solid curve in Figure 2.39.

EXAMPLE 2.34

Discuss and graph the equation $9y^2 - 16x^2 = 144$.

Solution Again, we divide both sides by 144 to get

$$\frac{y^2}{16} - \frac{x^2}{9} = 1$$

This is the equation of a hyperbola with center $(0,0)$. The transverse axis is the y-axis. Also, $a^2 = 16$ and $b^2 = 9$, so $c^2 = 25$. The foci are at $F(0,5)$ and $F'(0,-5)$, the vertices at $V(0,4)$ and $V'(0,-4)$, and the endpoints of the conjugate axis at $W(3,0)$ and $W'(-3,0)$. This hyperbola has the same asymptotes as the hyperbola in Example 2.33. The sketch of this hyperbola is shown by the dotted curves in Figure 2.39. The hyperbolas in Examples 2.33 and 2.34 are called **conjugate hyperbolas**, because they have the same set of asymptotes.

EXAMPLE 2.35

Find the equation of a hyperbola with center $(0,0)$, a focus at $(6,0)$, and a vertex at $(-3,0)$.

Solution The second focus is at $(-6,0)$, and the second vertex at $(3,0)$. Since the foci and vertices are on the x-axis, it is the transverse axis. The equation is

$$\frac{x^2}{a^2}-\frac{y^2}{b^2}=1$$

Since $a=3$ and $c=6$, then $b^2=c^2-a^2=36-9=27$. The equation of this hyperbola is

$$\frac{x^2}{9}-\frac{y^2}{27}=1$$

or $27x^2-9y^2=243$.

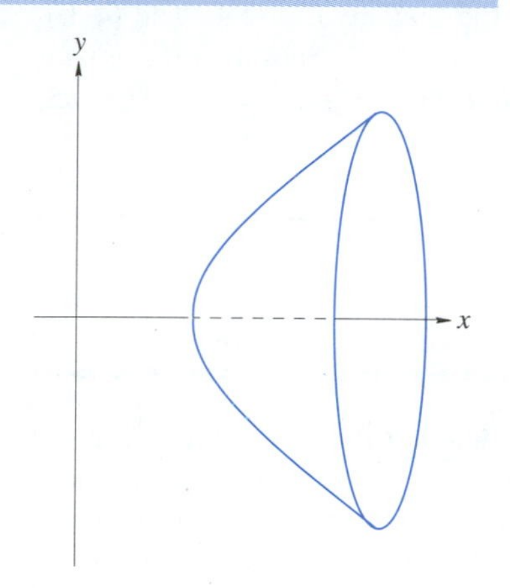

FIGURE 2.40

Reflective Properties of Hyperbolas

Like the other conic sections, hyperbolas are used in many of today's technological applications. A three-dimensional object called a **hyperboloid of revolution** is formed when a hyperbola is revolved around its transverse axis of symmetry. (See Figure 2.40.) Notice that a hyperbola of revolution normally uses only one branch of the hyperbola. Uses of these hyperboloids of revolution are based on the fact that rays aimed at one focus of a hyperbolic reflector are reflected so that they pass through the other focus.

Hyperboloids are frequently used to locate the source of a sound or radio signal. In Exercise Set 2.5, Exercise 18 describes how a system called LORAN (for LOng RAnge Navigation) uses hyperbolas as aids for navigation.

A telescope can use both a hyperbolic lens and a parabolic lens. The main lens is parabolic with its focus at F_1 and vertex at F_2. A second lens is hyperbolic with its foci at F_1 and F_2, as shown in Figure 2.41. The eye is positioned at F_2. The key to this arrangement is that the same point, F_1, is both the focus of the parabola and a focus of the hyperbola.

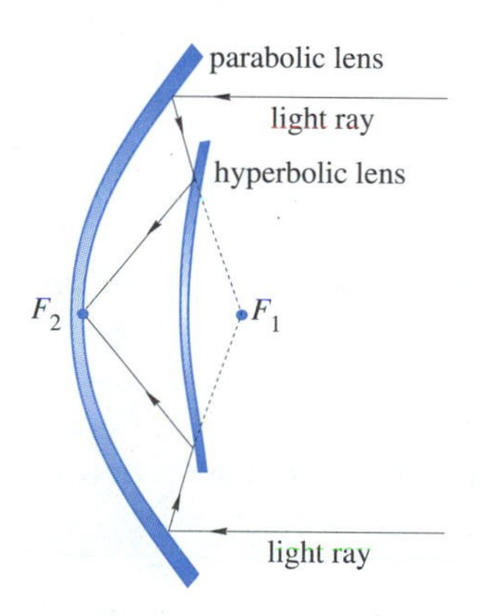

FIGURE 2.41

Application

EXAMPLE 2.36

As we will see in Section 2.6, many telescopes use both a parabolic reflector and a hyperbolic reflector. In this example, we shall examine a simple hyperbolic reflector. In Section 2.6, we shall consider an application that involves both parabolic and hyperbolic reflectors.

A hyperbolic reflector is designed to fit in a confined area. It is possible to get the foci so they are 8 ft apart and one vertex is 3 ft from a focus. What is the equation for this hyperbolic reflector?

Solution We shall arrange this reflector on a coordinate system, so that the transverse axis is on the x-axis and the center is at the origin. A cross-section through the axis of this hyperbolic reflector has been drawn on the desired coordinate system in Figure 2.42.

EXAMPLE 2.36 (Cont.)

Since the foci are 8 ft apart, this means that the foci are at $(-4,0)$ and $(4,0)$. The given vertex is 3 ft from one of the foci (and hence, 5 ft from the other), so it is at $(-1,0)$ or $(1,0)$. We have $a=1$ and $c=4$, so $b^2=c^2-a^2=4^2-1^2=15$. Thus, the desired equation is $\dfrac{x^2}{1}-\dfrac{y^2}{15}=1$.

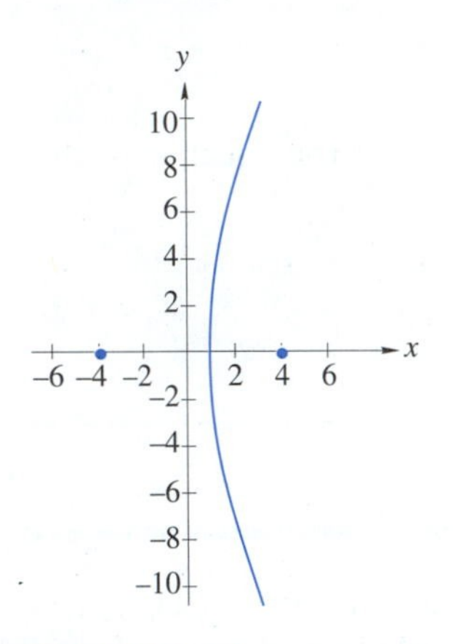

FIGURE 2.42

Using a Graphing Calculator

When using a graphing calculator, choose a viewing window that places the endpoints of the transverse and conjugate axes in substantially from the edge of the window. The next example will give a way to estimate where to put the viewing window. Again, since the calculator will graph only a function, you will need to solve the equation algebraically for y and then graph the two equations that result.

EXAMPLE 2.37

Use a graphing calculator to graph $\dfrac{x^2}{25}-\dfrac{y^2}{81}=1$.

Solution Since this is in standard form, we see that this is a horizontally oriented hyperbola. Here $a^2=25$, so $a=5$ and $b^2=81$, so $b=9$. A viewing window that sets Xmin $=-5$, Xmax $=5$, Ymin $=-9$, and Ymax $=9$ would only show the vertices of the hyperbola. Thus, we need a wider window. As rough rule, we will use $2a$ and $2b$ to set the window settings. For this hyperbola, we will use a viewing window that sets Xmin $=-10$, Xmax $=10$, Ymin $=-18$, and Ymax $=18$.

Next, solve the equation for y.

$$\frac{x^2}{25}-\frac{y^2}{81}=1$$
$$\frac{y^2}{81}=\frac{x^2}{25}-1$$
$$y^2=81\left(\frac{x^2}{25}-1\right)$$
$$y=\pm\sqrt{81\left(\frac{x^2}{25}-1\right)}$$
$$y=\pm 9\sqrt{\frac{x^2}{25}-1}$$

We will graph $y_1=9\sqrt{\dfrac{x^2}{25}-1}$ and $y_2=-9\sqrt{\dfrac{x^2}{25}-1}$. This result is shown in Figure 2.43.

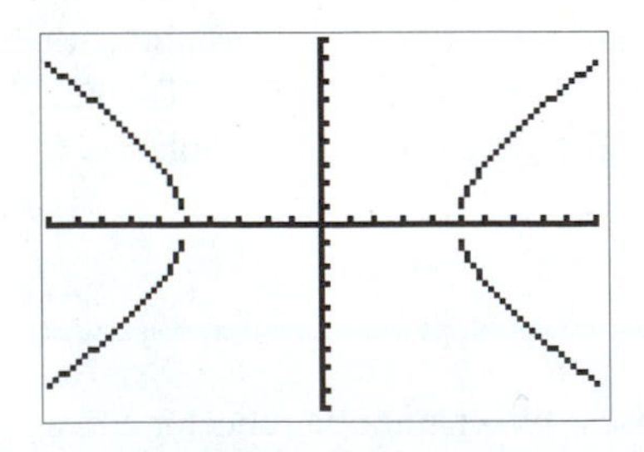

[−10, 10, 1] x [−18, 18, 2]

FIGURE 2.43

Exercise Set 2.5

In Exercises 1–8, find the equation of the hyperbola with the stated properties. Each hyperbola has its center at (0,0).

1. Focus at $(6,0)$, vertex at $(4,0)$
2. Focus at $(12,0)$, vertex at $(9,0)$
3. Focus at $(0,5)$, vertex at $(0,-3)$
4. Focus at $(0,-4)$, vertex at $(0,2)$
5. Focus at $(5,0)$, vertex at $(3,0)$
6. Focus at $(0,-3)$, vertex at $(0,-2)$
7. Focus at $(4,0)$, length of conjugate axis, 6
8. Focus at $(0,-6)$, length of conjugate axis, 10

In Exercises 9–16, give the coordinates of the vertices, foci, and endpoints of the conjugate axis and sketch each curve.

9. $\frac{x^2}{4} - \frac{y^2}{9} = 1$
10. $\frac{x^2}{9} - \frac{y^2}{4} = 1$
11. $\frac{y^2}{4} - \frac{x^2}{1} = 1$
12. $\frac{y^2}{36} - \frac{x^2}{16} = 1$
13. $4x^2 - y^2 = 4$
14. $9x^2 - 4y^2 = 36$
15. $36y^2 - 25x^2 = 900$
16. $y^2 - 9x^2 = 18$

Solve Exercises 17–26.

17. The **eccentricity** e of the hyperbola is defined as $\frac{c}{a}$. Since $c > a$, then $e > 1$. (Remember, for the ellipse, $0 < e < 1$.) What is the equation of a hyperbola with center at the origin, vertex at $(7,0)$, and $e = 1.5$?
18. *Navigation* Hyperbolas are used in long-range navigation as part of the LORAN system of navigation. A transmitter is located at each focus and radio signals are sent to the navigator simultaneously from each station. The difference in time at which the signals are received allows the navigator to determine the position. Suppose the transmitting towers are 1 000 km apart on an east-west line. A ship on the line between two towers determines its position to be 250 km from the west tower.
    (a) Sketch the signal hyperbola on which the ship is located. Place the center at $(0,0)$ and let the x-axis be the transverse axis.
    (b) Find the equation of the hyperbola.
19. A hyperbola for which $a = b$ is called an **equilateral hyperbola**. Find the eccentricity of an equilateral hyperbola.
20. *Civil engineering* A proposed design for a cooling tower at a nuclear power plant is a branch of a hyperbola rotated about a conjugate axis. Find the equation of the hyperbola passing through the point $(238, -616)$ with center $(0,0)$ and one vertex at $(153,0)$.
21. *Electronics* For any given voltage the product of current I and resistance R in a simple circuit is constant. If the current in a circuit is 3.2 A when the resistance is 18 Ω, sketch the graph of current as a function of resistance for this value of the voltage.
22. *Civil engineering* The silhouette of the cooling tower of a nuclear reactor forms a hyperbola, like the one in Figure 2.44. If the asymptotes are the lines given by $y = 1.25x$ and $y = -1.25x$, find the equation of the hyperbola.

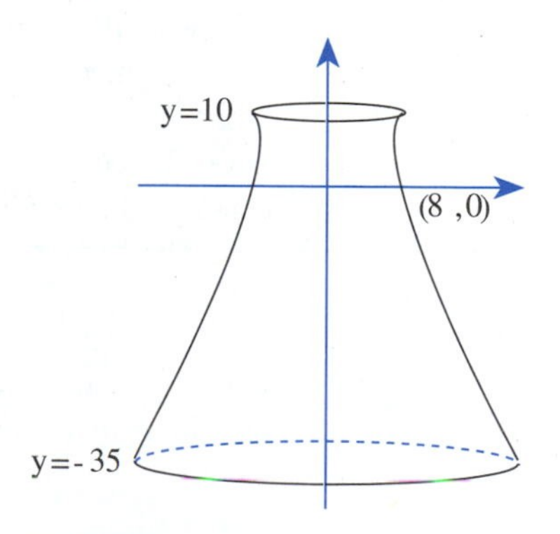

FIGURE 2.44

23. *Aeronautical engineering* A supersonic jet plane flying at a constant speed produces a shock wave in the shape of a cone. If the plane's path is parallel

to the ground, the intersection of the cone with the ground is a hyperbola. Find the equation of the hyperbola if the center is at the origin of a coordinate system, one vertex is at $(-42, 0)$, and the hyperbola passes through the point $(-126, 30\sqrt{3})$.

24. *Electronics* Ohm's law states that voltage V, current I, and resistance R are related by the formula $V = IR$. If $V = 14.0$ V, sketch the graph of I as a function of positive values of R.

25. *Wastewater technology* The formula $A = 8.34FC$ is used to determine the amount, A, of chlorine (in lb) to add to a basin with a flow rate of F million gallons per day to achieve a chlorine concentration of C parts per million (ppm). If 2250 lb of chlorine is added to a certain reservoir, sketch the graph of flow rate as a function of chlorine concentration for this amount of added chlorine.

26. *Wastewater technology* The formula $A = 8.34FC$ is used to determine the amount, A, of chlorine in kg to add to a basin with a flow rate of F million liters per day to achieve a chlorine concentration of C parts per million (ppm). If 750 kg of chlorine is added to a certain reservoir, sketch the graph of flow rate as a function of chlorine concentration for this amount of added chlorine.

In Your Words

27. Describe how you can tell by inspecting the standard equation of a hyperbola whether it is a horizontal or a vertical hyperbola.

28. Explain how to use the endpoints of the axes to locate the foci of an hyperbola.

29. In Exercise 19, it was stated that an equilateral hyperbola is one in which $a = b$. Why is this called an equilateral hyperbola?

2.6 TRANSLATION OF AXES

So far, the equations that we have used for the ellipse and the hyperbola have all been centered at the origin. You may remember from our study of circles that some circles had their center at the origin and some did not. In this section, we will look at cases where one of the axes of an ellipse is parallel to one of the coordinate axes. To do this, we will use translation axes.

The method we will use to translate axes will be very similar to the one we used when we were solving the equation for a circle. We took each equation that was not centered at the origin and completed the square. This told us the location of the center and the radius.

The equations that we will be working with are all in the form

$$Ax^2 + Cy^2 + Dx + Ey + F = 0$$

where A and C cannot both be zero at the same time. The process can best be explained by following the next example.

EXAMPLE 2.38

Discuss and sketch the graph of $9x^2 + 4y^2 - 36x + 40y + 100 = 0$.

Solution As we just mentioned, we will complete the square. First we will group the terms, and use ◯ and □ to indicate the missing constants that we must determine

EXAMPLE 2.38 (Cont.)

in order to complete the square.

$$(9x^2 - 36x) + (4y^2 + 40y) + 100 = 0$$
$$9(x^2 - 4x + \bigcirc) + 4(y^2 + 10y + \square) = -100 + 9\bigcirc + 4\square$$

Next, we complete the square of each expression within parentheses by adding 4 to the x terms and 25 to the y terms. Remember that on the right-hand side of the equation, each of these numbers (4 and 25) must be multiplied by the number outside the parentheses (9 and 4):

$$9(x^2 - 4x + 4) + 4(y^2 + 10y + 25) = -100 + 9(4) + 4(25)$$

or

$$9(x-2)^2 + 4(y+5)^2 = 36$$

Dividing both sides by 36, we get

$$\frac{(x-2)^2}{4} + \frac{(y+5)^2}{9} = 1$$

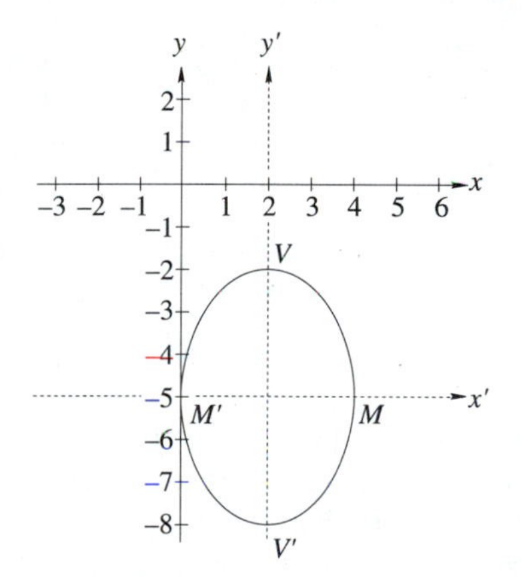

FIGURE 2.45

This looks like the equation for an ellipse. If we let $x' = x - 2$ and $y' = y + 5$, then the equation can be written as

$$\frac{(x')^2}{4} + \frac{(y')^2}{9} = 1$$

This is an ellipse with center at $(x', y') = (0', 0')$. If we draw a new coordinate system, the $x'y'$-system, with its origin at $(2, -5)$, we can draw our ellipse on this new coordinate system. Thus, this ellipse is centered at $(0', 0')$, the vertices are at $V(0', 3')$ and $V'(0', -3')$, the endpoints of the minor axis at $M(2', 0')$ and $M'(-2', 0')$, and the foci are at $F(0', \sqrt{5}')$ and $F'(0', -\sqrt{5}')$. Notice that we have used $3', 0', 2'$, and so on, to show that these are points on the $x'y'$-axes. A sketch of this graph is shown in Figure 2.45. The $x'y'$-axes have been traced over the xy-system.

We can use a table to show the coordinates in both systems. Since we know the coordinates in the $x'y'$-system and we know that $x' = x - 2$ and $y' = y + 5$, to find the coordinates in the xy-system, we solve these equations for x and y: $x = x' + 2$ and $y = y' - 5$. The table follows.

	$x'y'$-system	xy-system
center	$(0', 0')$	$(2, -5)$
vertices	$(0', \pm 3')$	$(2, -2), (2, -8)$
foci	$(0', \pm\sqrt{5}')$	$(2, \sqrt{5} - 5), (2, -\sqrt{5} - 5)$
endpoints of minor axis	$(\pm 2', 0')$	$(4, -5), (0, -5)$

In general, if we want the point (h, k) in the xy-coordinate system to become identified as the origin of the $x'y'$-coordinate system, we use the substitution

$$x' = x - h \qquad \text{and} \qquad y' = y - k$$

This procedure is called a **translation of axes**. As a result of a translation of axes, new axes are formed that are parallel to the old ones.

EXAMPLE 2.39

Discuss and sketch the graph of $x^2 - 18y^2 + 6x - 36y - 45 = 0$.

Solution Again, we will complete the square.

$$(x^2 + 6x) + (-18y^2 - 36y) = 45$$
$$(x^2 + 6x + \bigcirc) - 18(y^2 + 2y + \square) = 45 + \bigcirc - 18\square$$

We add 9 to complete the square of $x^2 + 6x$ and 1 to complete $y^2 + 2y$.

$$(x^2 + 6x + 9) - 18(y^2 + 2y + 1) = 45 + 9 - 18(1)$$
$$(x+3)^2 - 18(y+1)^2 = 36$$

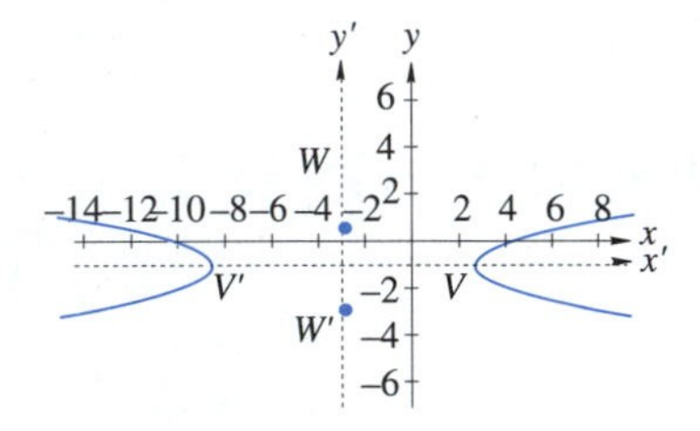

FIGURE 2.46

Dividing both sides by 36, we get

$$\frac{(x+3)^2}{36} - \frac{(y+1)^2}{2} = 1$$

If $x' = x + 3$ and $y' = y + 1$, then the equation becomes

$$\frac{x'^2}{36} - \frac{y'^2}{2} = 1$$

which is a hyperbola with center at $(0', 0')$ and vertices at $(\pm 6', 0')$ of the $x'y'$-system. The origin of the $x'y'$-system is found at $(-3, -1)$ of the xy-system. The sketch of this hyperbola on both coordinate systems is shown in Figure 2.46. Corresponding coordinates in the two systems are shown in the following table.

	$x'y'$-system	xy-system
center	$(0', 0')$	$(-3, -1)$
vertices	$(\pm 6', 0')$	$(-9, -1), (3, -1)$
foci	$(\pm\sqrt{38}', 0')$	$(-3+\sqrt{38}, -1)$, $(-3-\sqrt{38}, -1)$
endpoints of conjugate axis	$(0', \pm\sqrt{2}')$	$(-3, -1+\sqrt{2}), (-3, -1-\sqrt{2})$

EXAMPLE 2.40

Discuss and sketch the graph of $y^2 + 12x + 2y - 23 = 0$.

Solution There is no x^2-term, so this is the equation for a parabola. We will complete the square on just the y-terms. We will also put the y-terms on one side of the equation and all the other terms on the other side.

$$y^2 + 2y = -12x + 23$$
$$y^2 + 2y + 1 = -12x + 23 + 1$$
$$(y+1)^2 = -12(x-2)$$

EXAMPLE 2.40 (Cont.)

If $y' = y + 1$ and $x' = x - 2$, the equation becomes

$$y'^2 = -12x'$$

which is a parabola with vertex at $(0', 0')$, a horizontal axis, and that opens to the left. Since $4p = -12$, $p = -3$ and the directrix is the line $x = 3'$. The focus is at $(-3', 0')$. The sketch of this curve is shown in Figure 2.47.

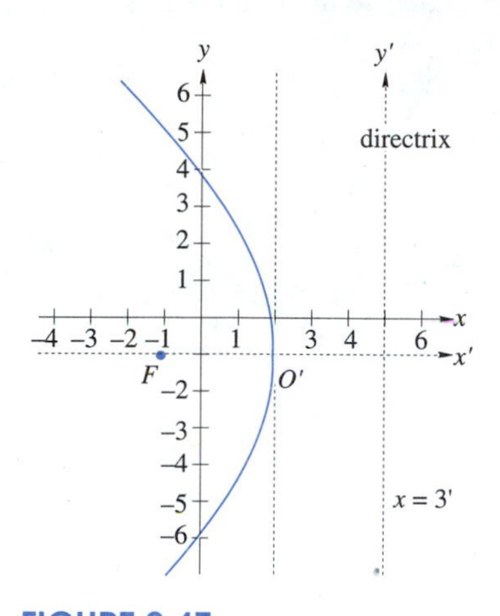

FIGURE 2.47

EXAMPLE 2.41

Find the equation of the hyperbola with vertices $(3, 1)$ and $(-5, 1)$, and focus $(5, 1)$.

Solution The center of a hyperbola is at the midpoint between the vertices. Using the midpoint formula, we can determine that the center is at $\left(\frac{3+(-5)}{2}, \frac{1+1}{2}\right) = (-1, 1)$. The distance from the vertex to the center is 4, so $a = 4$. The distance from the focus to the center is 6, so $c = 6$. This is a hyperbola, so $b^2 = c^2 - a^2 = 6^2 - 4^2 = 20$. Finally, this is a hyperbola with foci on the x'-axis and the equation is

$$\frac{x'^2}{16} - \frac{y'^2}{20} = 1$$

Since $x' = x - h$ and $y' = y - k$, we have $x' = x + 1$ and $y' = y - 1$, so the equation becomes

$$\frac{(x+1)^2}{16} - \frac{(y-1)^2}{20} = 1$$

Application

EXAMPLE 2.42

A telescope is shown in Figure 2.48a. The focus of one branch of the hyperbolic lens is 20.4 cm from the vertex of the other branch of this same lens. The parabolic lens is 63 cm deep and measures 168 cm across. Determine **(a)** the location of the focus for the parabolic lens **(b)** an equation for the parabola, and **(c)** an equation for the hyperbola.

Solution A cross-section through the axis of this telescope has been drawn on a coordinate system in Figure 2.48b. Because the focus of one branch of the hyper-

EXAMPLE 2.42 (Cont.)

bolic lens is also the vertex of the parabolic lens, we have placed the given focus at the origin. The given vertex of the other branch has been located at $(20.4, 0)$. The parabolic lens is 63 cm deep and 168 cm across and so the point $(63, 84)$ is on this lens.

FIGURE 2.48a

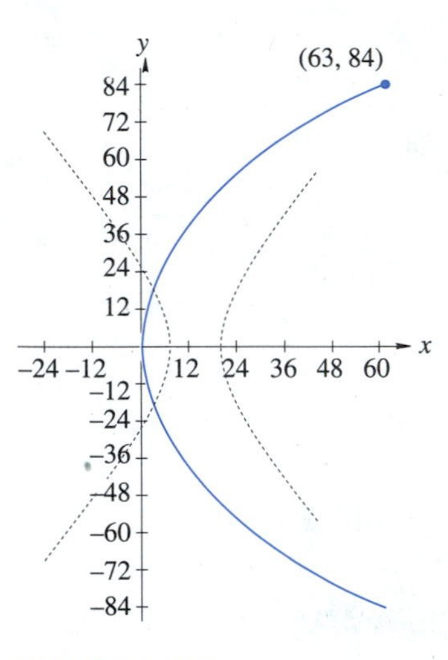

FIGURE 2.48b

This is a parabola. Its equation is of the form $y^2 = 4px$. We know that $(63, 84)$ is a point on the parabola, hence we have $p = \dfrac{y^2}{4x} = \dfrac{84^2}{4(63)} = 28$. Thus, the parabola's focus is at $(28, 0)$ or 28 cm from the vertex and the equation of the parabola is $y^2 = 4(28)x = 112x$.

The foci of the hyperbola are at $(0, 0)$ and $(28, 0)$. Thus, the center of the hyperbola is at $(14, 0)$, and so $c = 28 - 14 = 14$. Since one of the hyperbola's vertices is at $(20.4, 0)$, $a = 20.4 - 14 = 6.4$. Using these values of a and c, we can determine that $b^2 = c^2 - a^2 = 14^2 - 6.4^2 = 155.04$. Thus, the standard equation of the hyperbola for this lens is

$$\frac{(x-14)^2}{6.4^2} - \frac{y^2}{155.04} = 1$$

Using a Graphing Calculator

When using a graphing calculator, choose a viewing window that puts the center of the conic section at the center of the window. Then use the guidelines presented earlier for determining the viewing window. The next example shows how to do this with a translated ellipse. Again, since the calculator will graph only a function, you will need to solve the equation algebraically for y and then graph the two equations that result.

EXAMPLE 2.43

Use a graphing calculator to graph $9x^2+4y^2-36x+40y+100=0$.

Solution This is the same curve we graphed in Example 2.42. From that example, we know that the conic is an ellipse with standard equation $\frac{(x-2)^2}{4}+\frac{(y+5)^2}{9}=1$, center at $(2,-5)$, $a=2$, and $b=3$. For the viewing window we will use the following settings, where c_x and c_y represent the x- and y-coordinates of the center.

$$\begin{aligned}
\text{Xmin} = \quad c_x-a &= 2-2=0\\
\text{Xmax} = \quad c_x+a &= 2+2=4\\
\text{Ymin} = \quad c_y-b &= -5-3=-8\\
\text{Ymax} = \quad c_y+b &= -5+3=-2
\end{aligned}$$

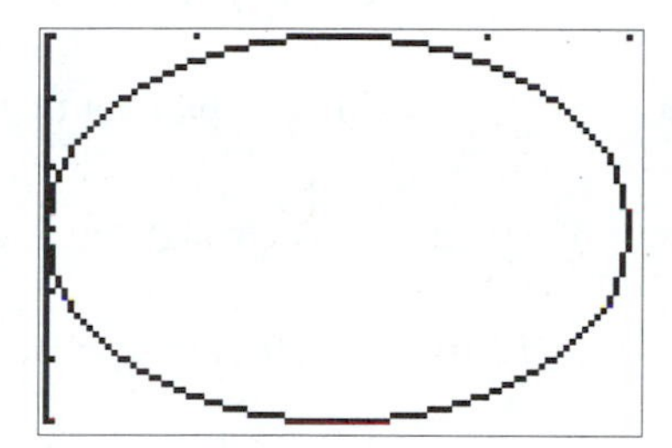

FIGURE 2.49a

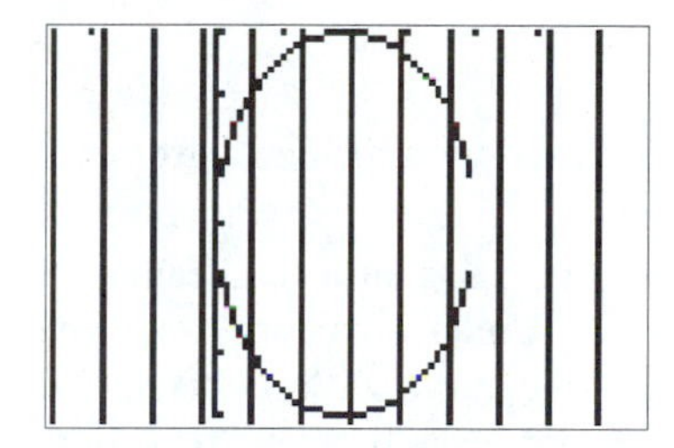

FIGURE 2.49b

Next, solve the equation for y.

$$\begin{aligned}
\frac{(x-2)^2}{4}+\frac{(y+5)^2}{9}&=1\\
\frac{(y+5)^2}{9}&=1-\frac{(x-2)^2}{4}\\
(y+5)^2&=9\left(1-\frac{(x-2)^2}{4}\right)\\
y+5&=\pm\sqrt{9\left(1-\frac{(x-2)^2}{4}\right)}\\
y&=-5\pm\sqrt{9\left(1-\frac{(x-2)^2}{4}\right)}\\
y&=-5\pm3\sqrt{1-\frac{(x-2)^2}{4}}
\end{aligned}$$

We will graph $y_1=-5+3\sqrt{1-\frac{(x-2)^2}{4}}$ and $y_2=-5-3\sqrt{1-\frac{(x-2)^2}{4}}$. This result is shown in Figure 2.49a. Pressing ZOOM 5, we obtain the proportionally correct graph shown in Figure 2.49b. ▪

Exercise Set 2.6

Sketch the graphs in Exercises 1–12, after making suitable translations of axes.

1. $\frac{(x-4)^2}{9}+\frac{(y+3)^2}{4}=1$
2. $\frac{(x+5)^2}{16}-\frac{(y+3)^2}{25}=1$
3. $\frac{(y-3)^2}{36}-(x+4)^2=1$
4. $(y+5)^2=12(x+1)$
5. $100(x+5)^2-4y^2=400$
6. $(y+3)^2=8(x-2)$
7. $16x^2+4y^2+64x-12y+57=0$

8. $x^2+y^2-2x+2y-2=0$

9. $25x^2+4y^2-250x-16y+541=0$

10. $100x^2-180x-100y+81=0$

11. $2x^2-y^2-16x+4y+24=0$

12. $9x^2-36x+16y^2-32y-524=0$

In Exercises 13–22, determine the equation of each of the curves described by the given information.

13. Parabola, vertex at $(2,-3)$, $p=8$, axis parallel to y-axis

14. Hyperbola, vertex at $(2,1)$, foci at $(-6,1)$ and $(8,1)$

15. Ellipse, center at $(4,-3)$, focus at $(8,-3)$, vertex at $(10,-3)$

16. Ellipse, center at $(-2,0)$, focus at $(6,0)$, vertex at $(9,0)$

17. Hyperbola, center at $(-3,2)$, focus at $(-3,7)$, transverse axis 6 units

18. Ellipse, center at $(3,5)$, focus at $(3,8)$, minor axis 2 units

19. Parabola, vertex at $(-5,1)$, $p=-4$, axis parallel to x-axis

20. Parabola, vertex at $(-3,-6)$, $p=-12$, axis parallel to y-axis

21. Ellipse, center at $(-4,1)$, focus at $(-4,9)$, minor axis 12 units

22. Hyperbola, center at $(2,8)$, vertex at $(6,8)$, conjugate axis 10 units

Solve Exercises 23–34.

23. *Physics* The height s of a ball thrown vertically upward is given by the equation $s=29.4t-4.9t^2$, where s is in meters and t is the elapsed time in seconds. Graph this curve. Discuss the curve. Determine the maximum height of the ball.

24. *Astronomy* Satellites often orbit earth in an elliptical path with the center of the earth at one of the foci. For a certain satellite, the maximum altitude is 140 mi above the surface of the earth and the minimum is 90 mi. If the radius of earth is approximately 4,000 mi, find the equation of the orbit.

25. *Navigation* Two navigational transmitting towers A and B are 1 000 km apart along an east-west line. Radio signals are sent (traveling at 300 m/μs) simultaneously from both towers. An airplane is located somewhere north of the line joining the two towers. The signal from A arrives at the plane 600 μs after the signal from B. The signal sent from B and reflected by the plane takes a total of 8 000 μs to reach A. What is the location of the plane?

26. *Astronomy* The orbit of Halley's comet is an ellipse with the sun at one focus. The major and minor semi-axes of this orbit are 18.09 A.U. and 4.56 A.U., respectively. (A **semi-axis** is half an axis. One A.U. is one astronomical unit or about 1.496×10^8 km.) What is the equation of this orbit, if the sun is at the origin and the major axis is along the x-axis? What are the maximum and minimum distances from the sun to Halley's comet?

27. *Mechanical engineering* A cantilever beam is a beam fixed at one end. Under a uniform load, the beam assumes a parabolic curve with the fixed end as the vertex. For a cantilever beam 2.00 m long, the equation of the load parabola is approximately $x^2+200y-4.00=0$, where x is the distance in meters from the fixed end and y is the displacement. See Figure 2.50.

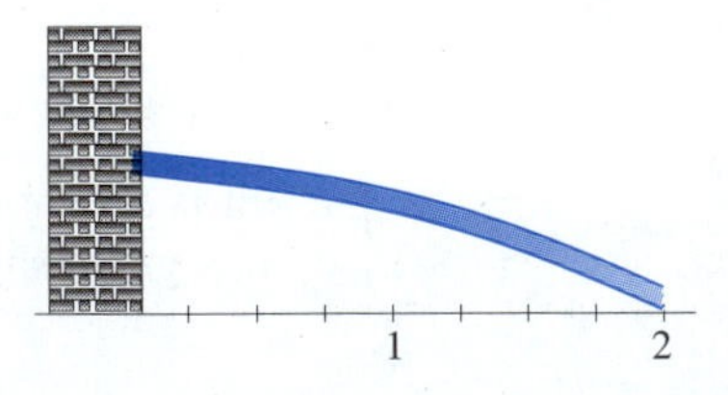

FIGURE 2.50

(a) Change this equation to the standard equation for a parabola.

(b) How far is the free end of the beam displaced from its no-load position?

28. *Mechanical engineering* A cantilever beam is a beam fixed at one end. Under a uniform load the beam assumes a parabolic curve, with the fixed end as the vertex. For a cantilever beam 8.0 ft long, the equation of the load parabola is approximately $x^2 + 100y - 4.00 = 0$, where x is the distance in feet from the fixed end and y is the displacement.

(a) Change this equation to the standard equation.

(b) How far is the free end of the beam displaced from its no-load position?

29. *Air traffic control* An airplane starts at the origin and flies in a straight line northeast, that is, up and to the right along the line $y = x$. Transmitters at coordinates (in miles) $(0, 0)$ and $(0, 250)$ send out synchronized radio signals, and instruments in the plane measure the difference in their arrival times. If the speed of the radio signal is known, then it is possible to locate the set of points where a plane is 50 mi farther from $(0, 0)$ than from $(0, 250)$ is one branch of a hyperbola.

(a) Find the equation of the hyperbola and identify the correct branch.

(b) Find the location of the airplane when it is 50 mi farther from $(0, 0)$ than from $(0, 250)$.

30. *Air traffic control* In the same setting as Exercise 29, a second plane flies along the line $y = 2x$. Comparison of radio signals shows that it is 50 mi farther from $(0, 0)$ than from $(0, 250)$. Find the location of the airplane.

31. *Automotive technology* The design of an automobile cam consists of the graph of the circle $0.25x^2 + 0.25y^2 = 0.0625$ and the ellipse $0.4x^2 + 0.16y^2 = 0.1$.

(a) Solve each of these equations so their graphs, when drawn on the same pair of axes, will produce this cam.

(b) Use your equations from (a) to sketch this cam.

32. *Police science* The stopping distance, d, in ft of a car moving at v mph is given approximately by the equation $d = v + \frac{1}{20}v^2$. Plot d as a function of v for $0 \leq v \leq 80$.

33. *Industrial management* A factory normally packages items in lots of 24. It costs \$0.90/item to package a lot of exactly 24 items. However, the cost per item is reduced by \$0.01 for each item over 24.

(a) Determine a function for the cost in terms of the number of items produced.

(b) Graph your function from (a).

(c) Using your graph in (b), determine the lot size that will produce the highest packaging cost.

(d) What is the highest packaging cost?

34. *Electronics* For a certain temperature-sensitive electronic device, the voltage, V, is given by $V = 5.0T - 0.025T^2$, where T is the ambient temperature in °C.

(a) Plot V as a function of t for values of $V \geq 0$.

(b) At what temperature is $V = 0$?

In Your Words

35. Explain how you can tell by looking at the equation $Ax^2 + Cy^2 + Dx + Ey + F = 0$ whether the graph of the equation is a circle, parabola, ellipse, or hyperbola.

36. Describe how to change an equation of the form $Ax^2 + Cy^2 + Dx + Ey + F = 0$ into the standard form for its conic section.

2.7 ROTATION OF AXES; THE GENERAL SECOND-DEGREE EQUATION

The conic sections that we have considered so far have all had their axes on a coordinate axis or parallel to a coordinate axis. All of these could be written as second-degree equations of the form

$$Ax^2 + Cy^2 + Dx + Ey + F = 0$$

Notice that this equation does not contain an xy-term.

In this section, we will work with curves in which an axis is not parallel to a coordinate axis. However, the axes are still perpendicular to each other. These equations will all have an xy-term. In order to recognize its graph, we will change to a new coordinate system. In Section 2.6, we changed to a new coordinate system through a translation of axes. In this section, we will use a rotation of axes and we will work with the general second-degree equation.

General Form of a Second-Degree Equation

The **general form of a second-degree equation** is

$$Ax^2 + Bxy + Cy^2 + Dx + Ey + F = 0$$

where A, B, and C are not all 0.

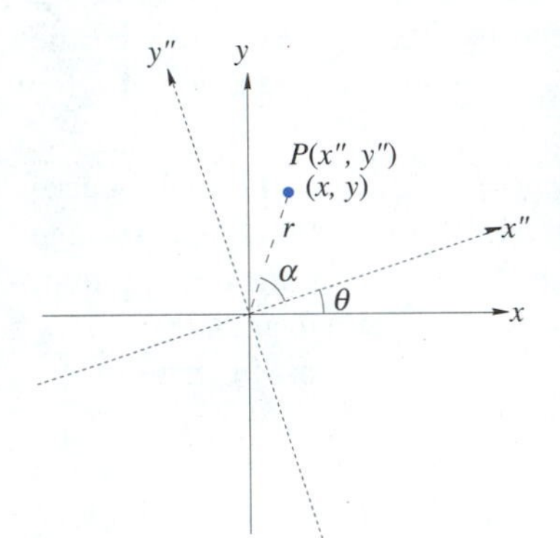

FIGURE 2.51

In a rotation, the origin remains fixed, while the x-axis and y-axis are rotated through a positive acute angle θ. These rotated axes will be labeled the x''-axis and the y''-axis. (See Figure 2.51.) If $P(x, y)$ is any general point in the xy-coordinate system, then that same point would have the coordinates $P(x'', y'')$ in the $x''y''$-coordinate system.

We can convert from one coordinate system to the other with the help of a pair of equations that express x and y in terms of x'' and y''.

$$x = x'' \cos\theta - y'' \sin\theta$$
$$y = y'' \cos\theta + x'' \sin\theta$$

These appear to be terrible equations to work with, but the following examples should show that they are not as difficult to use as they seem.

EXAMPLE 2.44

Transform the equation $xy = 1$ by rotating the axes through an angle of 45°.

Solution Since $\theta = 45°$, the conversion equations become

$$x = x'' \cos 45° - y'' \sin 45° = 0.7071x'' - 0.7071y''$$
$$y = y'' \cos 45° + x'' \sin 45° = 0.7071y'' + 0.7071x''$$

Substituting these into the original equation, $xy = 1$, we get

$$\cdot[0.7071(y'' + x'')] = 1$$
$$0.5(x''^2 - y''^2) = 1$$
$$\frac{x''^2}{2} - \frac{y''^2}{2} = 1$$

This is an equation of a hyperbola with center at the origin of the $x''y''$-coordinate system and transverse axis along the x''-axis. Here $a = b = \sqrt{2}$ and $c = 2$. The asymptotes of this hyperbola happen to be the original xy-axes. A graph of this curve is shown in Figure 2.52.

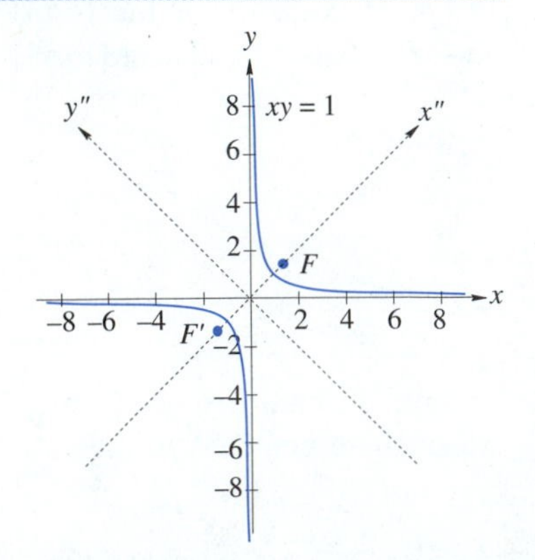

FIGURE 2.52

Any equation of the type in Example 2.44 is called an **equilateral hyperbola** or a **rectangular hyperbola** because the asymptotes are perpendicular. Equilateral hyperbolas are of the form $xy = k$, where k is a constant.

Now, suppose we have an equation that is a general second-degree equation and $B \neq 0$. What we want to do is rotate the axes so that we will get an equation of the form

$$A''(x^2) + B''xy + C''(y^2) + D''x + E''y + F'' = 0$$

where $B'' \neq 0$. Then we will be able to put the equation in the $x''y''$-coordinate system in a recognizable form of one of the conic sections.

We can cause B'' to be 0 if we let θ be the unique acute angle where

$$\cot 2\theta = \frac{A - C}{B} \qquad 0 < \theta < 90^\circ \qquad 0 < \theta < \frac{\pi}{2}$$

If you use a calculator, you have to be careful. First there is no COT key on your calculator, so you will have to take the reciprocal of the value. Since $\frac{1}{\cot\theta} = \tan\theta$, you can then use the INV tan keys. But, the arctan function will give answers between $-\frac{\pi}{2}$ and $\frac{\pi}{2}$. If you get a negative angle, you will have to add π rad or 180°. Let's see how it works in the next example.

EXAMPLE 2.45

Determine the graph of the equation $29x^2 + 24xy + 36y^2 - 54x - 72y - 135 = 0$.

Solution We will first determine the angle of rotation using the formula $\cot 2\theta = \frac{A - C}{B}$. Here $A = 29$, $B = 24$, and $C = 36$. The following description shows how to use a graphing calculator to determine θ.

PRESS	DISPLAY	
(29 − 36) ÷ 24 ENTER	−0.2916666667	
x^{-1} ENTER	−3.428571429	Changes from $\cot 2\theta$ to $\tan 2\theta$
2nd TAN 2nd ANS ENTER	−73.73979529	2θ in degrees
+ 180	106.2602047	Add 180° because 2θ is negative
÷ 2	53.13010235	This is θ

So, using our conversion formulas and the fact that $\sin\theta = 0.8$ and $\cos\theta = 0.6$, we get

$$x = x''\cos\theta - y''\sin\theta = 0.6x'' - 0.8y''$$
$$y = y''\cos\theta + x''\sin\theta = 0.6y'' + 0.8x''$$

Substituting these values for x and y in the given equation, $29x^2 + 24xy + 36y^2 - 54x - 72y - 135 = 0$, we obtain

$$29(0.6x'' - 0.8y'')^2 + 24(0.6x'' - 0.8y'')(0.6y'' + 0.8x'')$$
$$+ 36(0.6y'' + 0.8x'')^2 - 54(0.6x'' - 0.8y'')$$

EXAMPLE 2.45 (Cont.)

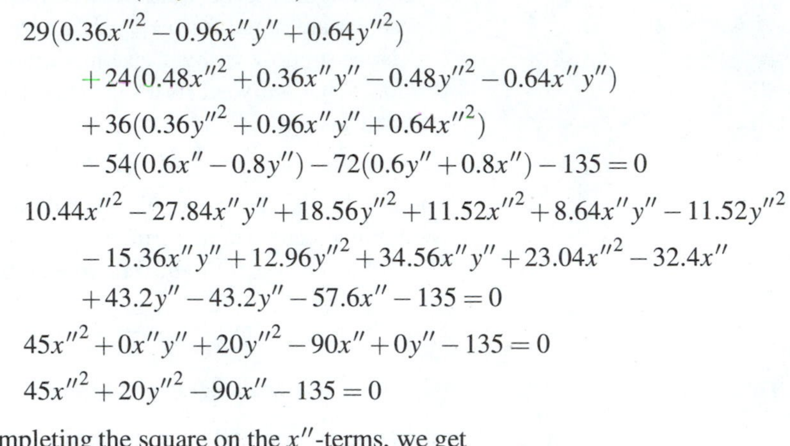

$$
\begin{aligned}
&- 72(0.6y'' + 0.8x'') - 135 = 0\\
&29(0.36x''^2 - 0.96x''y'' + 0.64y''^2)\\
&\quad + 24(0.48x''^2 + 0.36x''y'' - 0.48y''^2 - 0.64x''y'')\\
&\quad + 36(0.36y''^2 + 0.96x''y'' + 0.64x''^2)\\
&\quad - 54(0.6x'' - 0.8y'') - 72(0.6y'' + 0.8x'') - 135 = 0\\
&10.44x''^2 - 27.84x''y'' + 18.56y''^2 + 11.52x''^2 + 8.64x''y'' - 11.52y''^2\\
&\quad - 15.36x''y'' + 12.96y''^2 + 34.56x''y'' + 23.04x''^2 - 32.4x''\\
&\quad + 43.2y'' - 43.2y'' - 57.6x'' - 135 = 0\\
&45x''^2 + 0x''y'' + 20y''^2 - 90x'' + 0y'' - 135 = 0\\
&45x''^2 + 20y''^2 - 90x'' - 135 = 0
\end{aligned}
$$

Completing the square on the x''-terms, we get

$$
\begin{aligned}
45(x''^2 - 2x'' + 1) + 20y''^2 &= 135 + 45\\
45(x'' - 1)^2 + 20y''^2 &= 180
\end{aligned}
$$

Dividing both sides by 180 we obtain

$$\frac{(x''-1)^2}{4} + \frac{y''^2}{9} = 1$$

This is the equation of an ellipse with a major axis of 6 ($a = 3$) and a minor axis of 4 ($b = 2$) with the major axis along the vertical axis. The center of this ellipse is at $(1'', 0'')$. The graph of this ellipse is shown in Figure 2.53.

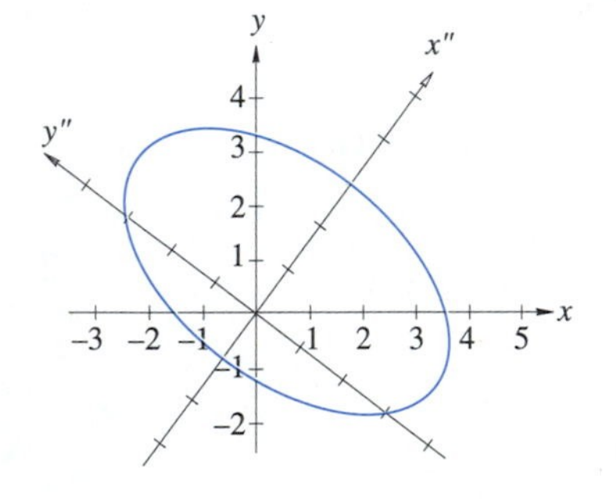

FIGURE 2.53

The Discriminant

There is an easy method for determining the nature of the curve described by a general second-degree equation. It turns out that the **discriminant**, $B^2 - 4AC$, will tell the type of curve described by the equation.

> **Classifying the Graph of a General Quadratic Equation**
>
> The graph of a general quadratic equation
>
> $$Ax^2 + Bxy + Cy^2 + Dx + Ey + F = 0$$
>
> is either a conic or a degenerate conic. If it is a conic, then the discriminant, $B^2 - 4AC$, can be used to classify the graph of the equation by using the following:
>
> If $B^2 - 4AC > 0$, the curve is a hyperbola.
> If $B^2 - 4AC = 0$, the curve is a parabola.
> If $B^2 - 4AC < 0$, the curve is an ellipse.

The majority of the problems we will work are not as long as the last example. We are more likely to get problems similar to the one in the next example.

Application

EXAMPLE 2.46

When the power P in an electric circuit is constant, the voltage V is inversely proportional to the current I, as indicated by the equation $P = IV$. If the power is 120 W, sketch the relationship of I vs V.

Solution The equation is $IV = 120$. A table of values is

I	1	2	4	6	8	10	12	15	20	30	60	120
V	120	60	30	20	15	12	10	8	6	4	2	1

Negative values do not apply. The graph is one branch of an equilateral hyperbola, as shown in Figure 2.54.

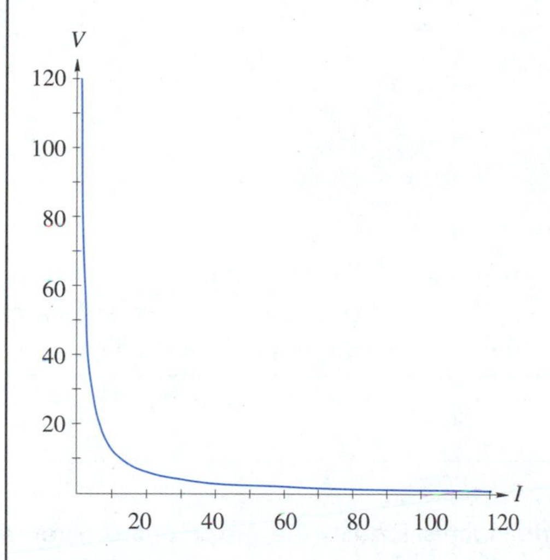

FIGURE 2.54

Using a Graphing Calculator

The most difficult part about using a graphing calculator to graph a rotated conic is getting the equation for the conic in a form that the calculator can use. This requires completing the square in a way you have probably not seen.

EXAMPLE 2.47

Use a graphing calculator to graph $29x^2 + 24xy + 36y^2 - 54x - 72y - 135 = 0$.

Solution This is the same curve we graphed in Example 2.45. From that example, we know that the conic is an ellipse rotated about 53.13° and with standard equation on the rotated axes $\dfrac{(x'' - 1)^2}{4} + \dfrac{y''^2}{9} = 1$ on the rotated axes.

EXAMPLE 2.47 (Cont.)

We will use the original equation and complete the square in order to solve the equation for y.

$$29x^2+24xy+36y^2-54x-72y-135=0$$
$$24xy+36y^2-72y=135-29x^2+54x$$
$$36y^2+24xy-72y=135+54x-29x^2$$
$$36y^2+24y(x-2)=135+54x-29x^2$$
$$36\left[y^2+\frac{2}{3}y(x-2)\right]=135+54x-29x^2$$
$$y^2+\frac{2}{3}y(x-2)=\frac{1}{36}(135+54x-29x^2)$$
$$y^2+\frac{2}{3}y(x-2)+\left[\frac{1}{3}(x-2)\right]^2=\frac{1}{36}(135+54x-29x^2)+\left[\frac{1}{3}(x-2)\right]^2$$
$$\left[y+\frac{1}{3}(x-2)\right]^2=\frac{1}{36}(151+38x-25x^2)$$
$$y+\frac{1}{3}(x-2)=\pm\frac{1}{6}\sqrt{(151+38x-25x^2)}$$
$$y=-\frac{1}{3}(x-2)\pm\frac{1}{6}\sqrt{(151+38x-25x^2)}$$

We will graph the equations $y_1=-\frac{1}{3}(x-2)+\frac{1}{6}\sqrt{(151+38x-25x^2)}$ and $y_2=-\frac{1}{3}(x-2)-\frac{1}{6}\sqrt{(151+38x-25x^2)}$. This result is shown in Figure 2.55. Compare this to the earlier result in Figure 2.53.

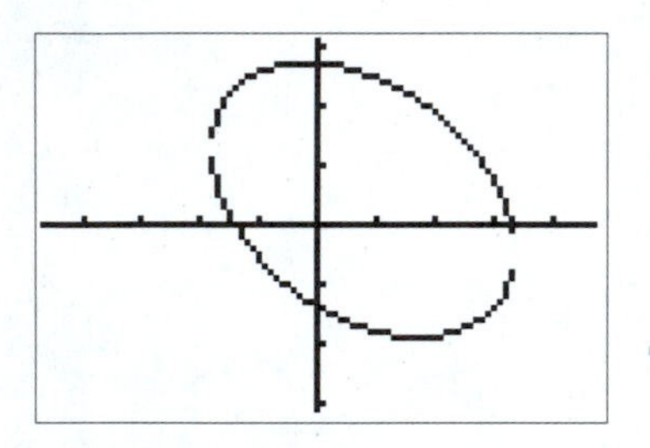

FIGURE 2.55

Exercise Set 2.7

In Exercises 1–12, (a) use the discriminant to identify the graph of the given equation, (b) determine the angle θ needed to rotate the coordinate axes to remove the xy-term, (c) rotate the axes through the angle θ, (d) determine the equation of the conic in the $x''y''$-coordinate system, and (e) sketch the graph.

1. $xy=-9$
2. $xy=9$
3. $x^2-6xy+y^2-8=0$
4. $x^2+4xy-2y^2-6=0$
5. $52x^2-72xy+73y^2-100=0$
6. $11x^2+10\sqrt{3}xy+y^2-4=0$
7. $x^2-2xy+y^2+x+y=0$
8. $3x^2+2\sqrt{3}xy+y^2-2x+2\sqrt{3}y=0$
9. $2x^2+12xy-3y^2-42=0$
10. $7x^2-20xy-8y^2+52=0$
11. $6x^2-5xy+6y^2-26=0$
12. $9x^2-6xy+y^2-12\sqrt{10}x-36\sqrt{10}y=0$

Solve Exercises 13–16.

13. *Physics* Boyle's law states that the volume of a gas is inversely proportional to its pressure, provided that the mass and temperature are constant. Thus, if P is the pressure and V the volume, $PV=k$, where k is a constant. Draw the graph of P vs V, if 15 mm^3 of gas is under a pressure of 400 kPa at a constant temperature.

14. *Electricity* For a given alternating current circuit, the capacitance C and the capacitive reactance X_C are related by the equation $X_C = \frac{1}{\omega C}$, where ω is the angular frequency. What kind of curve is this? Sketch a graph of the equation if $\omega = 280$ rad/s.

15. *Thermodynamics* When a cross-section of a pipe is suddenly enlarged, as shown in Figure 2.56, the loss of heat of the fluid through the pipe is related by the formula $19.62h_L = (\overline{v_1} - \overline{v_2})^2$, where h_L is the heat loss and $\overline{v_1}$ and $\overline{v_2}$ are the average velocities in the two pipes. If h_L is to be held to less that 5°C, describe this curve. Sketch a graph of the curve with $(\overline{v_1} - \overline{v_2})$ on one axis and h_L on the other.

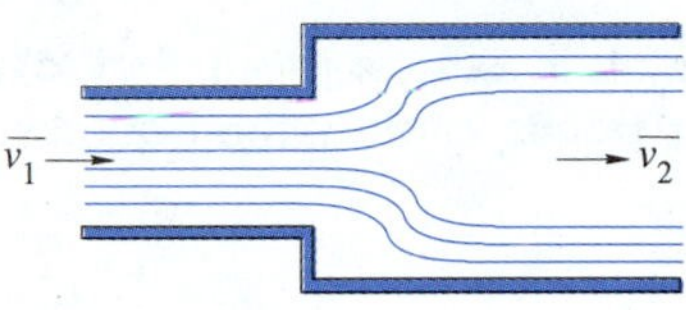
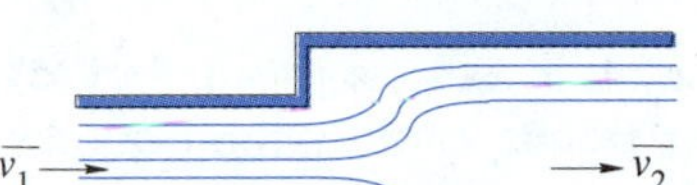

FIGURE 2.56

16. The sides of a rectangle are $3x$ and y and the diagonal is $x + 10$. What kind of curve is represented by the equation relating x and y? Sketch the curve.

In Your Words

17. (a) What is the discriminant of the general quadratic equation

$$Ax^2 + Bxy + Cy^2 + Dx + Ey + F = 0$$

(b) Explain how to use the discriminant to classify the graph of a general quadratic equation.

18. Describe how you can graph an equation of the form $Ax^2 + Bxy + Cy^2 + Dx + Ey + F = 0$ on your graphing calculator.

CHAPTER 2 REVIEW

Important Terms and Concepts

- Angle of inclination
- Circle
  - Center
  - Radius
- Degenerate conics
- Distance
- Ellipse
  - Center
  - Foci
  - Major axis
  - Minor axis
  - Vertices
- Hyperbola
  - Asymptotes
  - Center
  - Conjugate axis
  - Foci
  - Transverse axis
  - Vertices
- Midpoint
- Parabola
  - Axis
  - Directrix
  - Foci
  - Vertices
- Rotation of axes
- Slope
- Translation of axes
- x-intercept
- y-intercept

Review Exercises

For each pair of points in Exercises 1–8, find (a) the distance between them, (b) their midpoint, (c) the slope of the line through the points, and (d) the equation of the line through the points.

1. $(2,5)$ and $(-1,9)$
2. $(-2,-5)$ and $(10,-10)$
3. $(1,-4)$ and $(3,6)$
4. $(2,-5)$ and $(-6,3)$
5. For each line in Exercises 1–8,find the slope of one of its perpendiculars.
6. For each pair of points in Exercises 1–8, write the equation for the line passing through the midpoint of each pair and perpendicular to the line through them.
7. What is the equation of the line that passes through the point $(-3,5)$ and is parallel to $2y+4x=9$?
8. What is the equation of the line through $(2, -7)$ with a slope of 4?

Sketch each of the conic sections in Exercises 9–24.

9. $x^2+y^2=16$
10. $x^2-y^2=16$
11. $x^2+4y^2=16$
12. $y^2=16x$
13. $(x-2)^2+(y+4)^2=16$
14. $(x-2)^2-(y+4)^2=16$
15. $(x-2)^2+4(y+4)^2=16$
16. $(y+4)^2=16(x-2)$
17. $x^2+y^2+6x-10y+18=0$
18. $x^2-4y^2+6x+40y-107=0$
19. $x^2+4y^2+6x-40y+93=0$
20. $x^2+6x-4y+29=0$
21. $2x^2+12xy-3y^2-42=0$
22. $5x^2-4xy+8y^2-36=0$
23. $3x^2+2\sqrt{3}xy+y^2+8x-8\sqrt{3}y=32$
24. $2x^2-4xy-y^2=6$

Solve Exercises 25 and 26.

25. **(a)** Graph $y^2+6y=16x+13$.
    (b) Identify the graph.
    (c) Specify all the "important" points, such as foci, vertices, etc.
26. **(a)** Graph $16x^2-9y^2=144$.
    (b) Identify the graph.
    (c) Specify all the "important" points, such as foci, vertices, etc.

Solve Exercises 27–30.

27. *Civil engineering* A concrete bridge for a highway overpass is constructed in the shape of half an elliptic arch, as shown in Figure 2.57. The arch is 100 m long. In order to have enough clearance for tall vehicles, the arch must be 6.1 m high at a point 5 m from the end of the arch. What is the equation of the ellipse for this arch, if the origin is at the midpoint of the major axis?

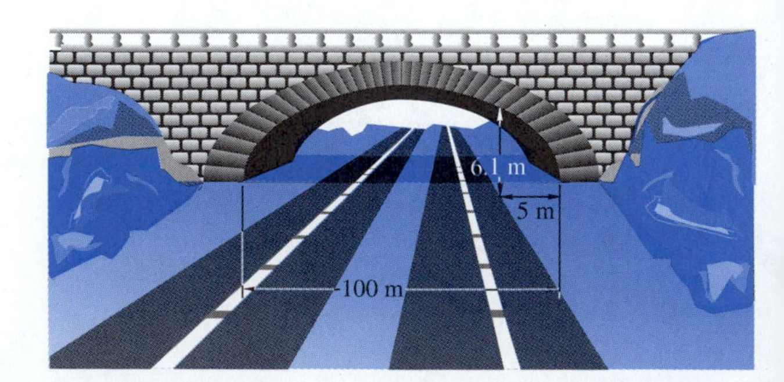

FIGURE 2.57

28. *Civil engineering* The main cables on a suspension bridge approximate a parabolic shape. The twin towers of a suspension bridge are to be 120 m above the road surface and are 400 m apart. If the cable's lowest point is 10 m above the road surface, what is the equation of the parabola for the main cable?

29. *Navigation* An airplane sends out an impulse that travels at the speed of sound (320 m/μs). The plane is 50 km south of the line connecting two receiving stations. The stations are on an east-west line with station A 400 km west of station B. Station A receives the signal from the plane 500 μs after station B. What is the location of the plane?

30. *Astronomy* A map of the solar system is drawn so that the surface of earth is represented by the equation $x^2 + y^2 - 2x + 4y - 6{,}361 = 0$. A satellite orbits earth in a circular orbit 0.8 units above earth. What is the equation of the satellite's orbit on this map?

CHAPTER 2 TEST

1. Find the focus and the directrix of the parabola $y^2 = -18x$, and sketch the graph.

2. Graph the hyperbola $9x^2 - 4y^2 = 36$. Specify the foci, vertices, and endpoints of the conjugate axis.

3. **(a)** Graph $16x^2 + 9y^2 = 144$.
(b) Identify the graph.
(c) Specify all the "important" points, such as foci, vertices, etc.

4. Determine the distance between the points $(-2, 4)$ and $(5, -6)$.

5. What is the equation of the line that is perpendicular to the line through the points $(4, -5)$ and $(2, -8)$ and passes through their midpoint?

6. The x-intercept of a line is 3, and its angle of inclination is 30°. Write the equation of the line in slope-intercept form.

7. **(a)** Graph $14(x+1)^2 - 9(y-2)^2 = 36$.
(b) Identify the graph.
(c) Specify all the "important" points, such as foci, vertices, etc.

8. **(a)** Write the equation $3x^2 + 4y^2 - 6x + 16y + 7 = 0$ in the appropriate standard form. **(b)** Determine the type of conic section described by this equation. **(c)** Determine the coordinates of all "significant" points, such as vertices and foci. **(d)** Graph the equation.

9. **(a)** Determine the angle needed to rotate the coordinate axes so that the transformed equation of $x^2 + xy + y^2 = 4$ has no xy-term. **(b)** Identify the graph.

10. Identify and sketch the graph of the polar equation $r = \dfrac{2}{1 + 4\cos\theta}$.

11. A cable hangs in a parabolic curve between two vertical supports that are 120 ft apart. At a distance 48 ft in from each support, the cable is 3.0 ft above its lowest point. How high up is the cable attached on each support?

CHAPTER

3

An Introduction to Calculus

In order to pave this parking lot, we must figure out how much asphalt is needed. In this chapter, we will see how calculus can be used to find this answer.

Courtesy of Michael A. Gallitelli, Metroland Photo Inc.

In the middle of the seventeenth century, scientists began to study speed, motion, and rates of change. From this study evolved a new branch of mathematics called calculus. For the next 250 years, most of the important developments in mathematics and science were connected with calculus. Two men, Isaac Newton (1642–1727) and Gottfried Wilhelm Leibniz (1646–1716) are given credit for its discovery.

The development of calculus was a result of attempts to answer several questions in geometry. One question dealt with the slope of the tangent line to a curve, which led to differential calculus. A second question, concerning the area enclosed by the graph of a function and the x-axis, led to integral calculus. The solutions to these questions, the tangent question, and the area question will be explored in Sections 3.1 and 3.2. In Chapters 4 and 6 we will begin a more detailed look at ways to solve these questions.

This chapter forms a foundation for your study of calculus. As you proceed through these next few chapters, you will be given opportunities to use calculus in such areas as heat, light, sound, electricity, and magnetism. Computer programs and calculators will help your study of the ideas behind this powerful field of mathematics.

3.1 THE TANGENT QUESTION

In Chapter 2, we studied the slope of a straight line. In particular, we found that if $P(x_1, y_1)$ and $Q(x_2, y_2)$ are two points on a line, then the slope of the line is given by the formula

$$m = \frac{y_2 - y_1}{x_2 - x_1}$$

But, what about curves that are not straight lines?

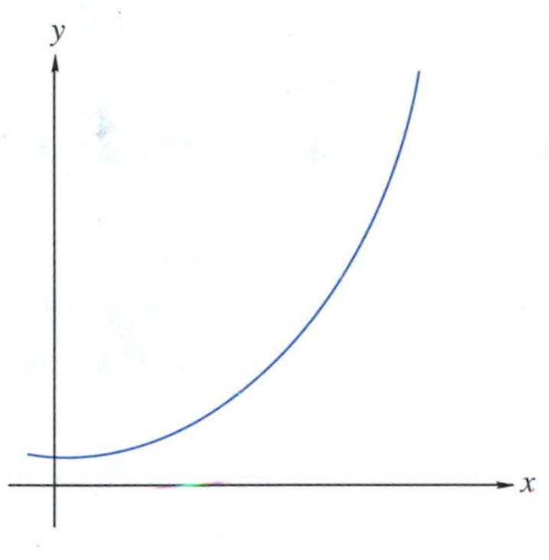

FIGURE 3.1

Consider the curve in Figure 3.1. As you can see, this is not a straight line, but it is the curve of a function. Slope is an indication of the "steepness" of a line. The slope of a straight line is the same at every point on the line. The curve in Figure 3.1 gives the impression that it gets steeper as the values of x increase. We might expect that the slope of a nonlinear curve would be different at different points on the curve. We would like a way to measure the steepness, or slope, of a nonlinear curve at any particular point on that curve.

One way to think of the slope of a curve at some point is to draw the tangent line to the curve at that point. Look at the curve in Figure 3.2. We have drawn the tangent to this curve at point P. We have also drawn the tangent to the curve at point Q. We can tell at a glance that the slope of the tangent at Q is greater than the slope of the tangent at P. Thus, we can let the slope of the tangent to a curve at some point be used for the slope of the curve at that point.

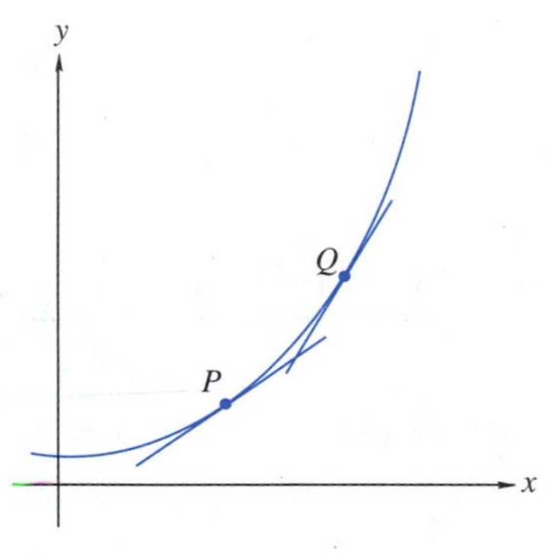

FIGURE 3.2

This is a helpful idea. Now, all we have to do is figure out some way to determine the slope of the tangent to a curve at any point. At present we do not have the background to determine the slope of a tangent line to a curve. We will substitute another idea until we have the necessary background.

Look at the graph of the function in Figure 3.3. This is the same curve we used in the previous example. As in Figure 3.2, P and Q are different points on the curve. P has the coordinates (x_1, y_1) and Q has the coordinates (x_2, y_2). The line that passes through points P and Q is called a **secant line**. The slope of the secant line through points P and Q is given by

$$m_{PQ} = \frac{y_2 - y_1}{x_2 - x_1}$$

Average Slope

This is an important idea. We have found that the slope of the secant line is a way to approximate the slope of the tangent line. What the secant line really gives is the **average rate of change** or **average slope** of the curve over the interval from (x_1, y_1) to (x_2, y_2). We will use a bar to indicate an average. Thus, the symbol for the average slope will be $\bar{m}$, the symbol for the average velocity $\bar{v}$, and so on. The average rate of change through any two points such as $P(x_1, y_1)$ and $Q(x_2, y_2)$ is the same as the

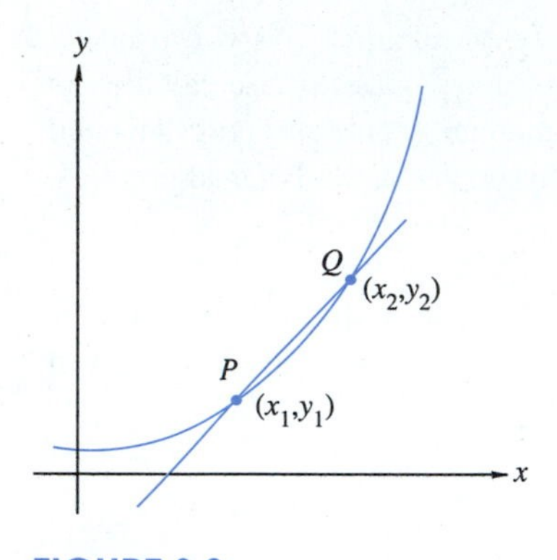

FIGURE 3.3

slope of the line through those two points, and is given by

$$\bar{m} = \frac{y_2 - y_1}{x_2 - x_1}$$

or, since $y_1 = f(x_1)$ and $y_2 = f(x_2)$,

$$\bar{m} = \frac{f(x_2) - f(x_1)}{x_2 - x_1}$$

EXAMPLE 3.1

Consider the function $f(x) = x^2 + 3x - 4$. Find the average slope $\bar{m}$ of the curve over the interval from $x = 3$ to $x = 6$.

Solution If we let $x_1 = 3$ and $x_2 = 6$, then, since $f(x) = x^2 + 3x - 4$, we have $f(3) = 14$ and $f(6) = 50$. Figure 3.4 shows a portion of the graph of $f(x) = x^2 + 3x - 4$ and the secant line through the points (3, 14) and (6, 50). Thus, the average slope of $\bar{m}$ is derived as follows:

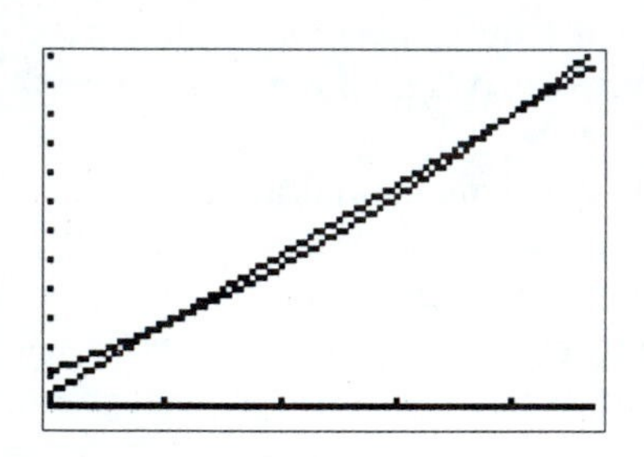

[2, 6.7, 1] x [–2, 60, 5]

FIGURE 3.4

$$\begin{aligned}\bar{m} &= \frac{f(x_2) - f(x_1)}{x_2 - x_1} \\ &= \frac{f(6) - f(3)}{6 - 3} \\ &= \frac{50 - 14}{6 - 3} \\ &= \frac{36}{3} \\ &= 12\end{aligned}$$

The average slope of $f(x) = x^2 + 3x - 4$ from $x = 3$ to $x = 6$ is 12.

EXAMPLE 3.2

Find the average slope of the function $g(x) = x^3 - 4x + 1$ from $x = -1$ to $x = 2$.

Solution We begin by letting $x_1 = -1$ and $x_2 = 2$, and evaluating $g(-1)$ and $g(2)$.

$$\begin{aligned}g(-1) &= (-1)^3 - 4(-1) + 1 \\ &= 4 \\ \text{and } g(2) &= 2^3 - 4(2) + 1 \\ &= 1\end{aligned}$$

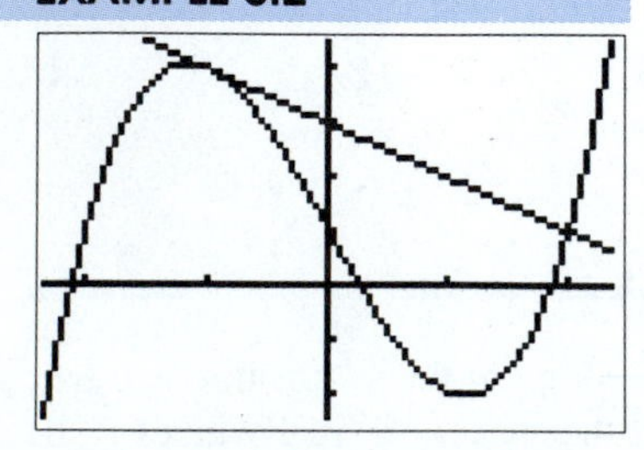

[–2.35, 2.35, 1] x [–2.5, 4.5, 1]

FIGURE 3.5

Figure 3.5 shows the graph of $g(x) = x^3 - 4x + 1$ over the interval $[-2.35, 2.35]$ and the secant line through the points $(-1, 4)$ and (2, 1). Thus,

$$\bar{m} = \frac{g(2) - g(-1)}{2 - (-1)}$$

EXAMPLE 3.2 (Cont.)

$$= \frac{1-4}{2+1}$$
$$= \frac{-3}{3}$$
$$= -1$$

The average slope of $g(x) = x^3 - 4x + 1$ from $x = -1$ to $x = 2$ is -1.

EXAMPLE 3.3

Find the average slope of the curve $h(x) = 2x^2 + 5x - 6$ from $x = -4$ to $x = x_1$, where $x_1 \neq -4$.

Solution This is a more general version of the average slope problem. When $x = -4$, $h(x) = h(-4) = 6$ and when $x = x_1$, $h(x_1) = 2x_1^2 + 5x_1 - 6$. Figure 3.6 shows the graph of h and a line through the points $(-4, 6)$ and $(x_1, h(x_1))$. The average slope is

$$\bar{m} = \frac{h(x_1) - h(-4)}{x_1 - (-4)}$$
$$= \frac{(2x_1^2 + 5x_1 - 6) - 6}{x_1 + 4}$$
$$= \frac{2x_1^2 + 5x_1 - 12}{x_1 + 4}$$
$$= \frac{(2x_1 - 3)(x_1 + 4)}{x_1 + 4}$$
$$= 2x_1 - 3$$

since $x_1 \neq -4$.

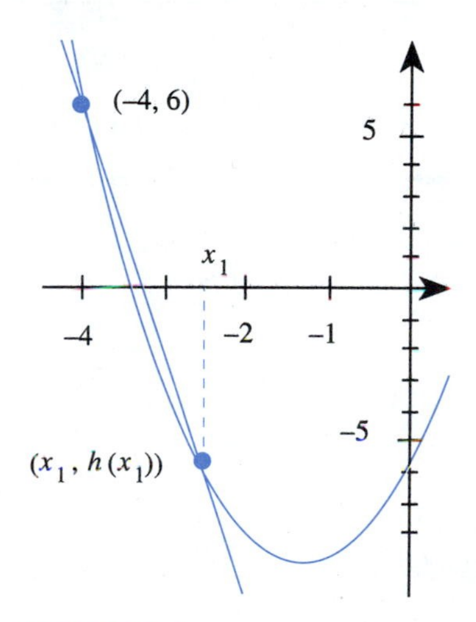

FIGURE 3.6

This last example provides a general formula for finding the average slope from $x = -4$ to any other point on the curve. For example, if we want the average slope from -4 to -5, we can use this formula with $x_1 = -5$ and get $\bar{m} = 2x_1 - 3 = 2(-5) - 3 = -13$.

The idea for finding the average slope can be applied to finding any type of average change. This can best be shown by the following example.

Application

EXAMPLE 3.4

A tank is filled with water by opening a valve on an inlet pipe. The volume V in liters of water in the tank t min after the valve is opened is given by the formula $V(t) = 5t^2 + 4t$. **(a)** What is the average rate of increase in the volume from the first minute to the second minute? **(b)** What is the average rate of increase in the volume during the next 30 seconds?

EXAMPLE 3.4 (Cont.)

Solution We want to find the average rate of change in the volume.

(a) When $t = 2$, $V(t) = V(2) = 28$ and when $t = 1$, $V(t) = V(1) = 9$. If R_{Vol} is the rate of change in the volume, then

$$\bar{R}_{\text{Vol}} = \frac{V(2) - V(1)}{2 - 1} = \frac{28 - 9}{2 - 1} = 19$$

The volume changed at the rate of 19 L/min from the first to the second minute.

(b) The next 30 seconds will be from $t = 2.0$ min to $t = 2.5$ min. When $t = 2.5$, we determine that $V(t) = V(2.5) = 41.25$. Thus,

$$\bar{R}_{\text{Vol}} = \frac{V(2.5) - V(2)}{2.5 - 2.0} = \frac{41.25 - 28}{2.5 - 2.0} = \frac{13.25}{0.5} = 26.5$$

Thus, the volume changed at the rate of 26.5 L/min from minutes 2 to 2.5.

Using a Graphing Calculator

Many graphing calculators can be used to approximate the slope of a tangent line at a particular point on a curve. Example 3.5 will show how this can be done on the function used in Example 3.3.

EXAMPLE 3.5

Use a graphing calculator to find the slope of the tangent line to $h(x) = 2x^2 + 5x - 6$ at $x = -4$.

Solution Begin by graphing h. The result of graphing h on a TI-82 as $y_1 = 2x^2 + 5x - 6$ is shown in Figure 3.7a .

Next, press TRACE and move the cursor to $x = -4$ and then press 2nd DRAW 5 ENTER[1]. (Note that pressing 2nd DRAW 5 accesses the "`Draw-Tangent`" option on the calculator.) The result is shown in Figure 3.7b. In the lower left-hand corner of the calculator screen shown in Figure 3.7b, you should see $dy/dx = -11$. The notation $dy/dx = -11$ indicates that the slope of this tangent line is -11. In Example 3.3 , we found that the average slope of the tangent line to $h(x) = 2x^2 + 5x - 6$ near $x = -4$ was $\overline{m} = 2x_1 - 3$. If you let $x_1 = -4$, then $\overline{m} = 2(-4) - 3 = -11$, the value of dy/dx.

---

[1]On a TI-85, press GRAPH MORE MATH MORE MORE TANLN and then move the cursor to (or near) $x = -4$ and press ENTER. Setting `xMin=-6.3` and `xMax=6.3` will allow the cursor to stop at $x = -4$.

EXAMPLE 3.5 (Cont.)

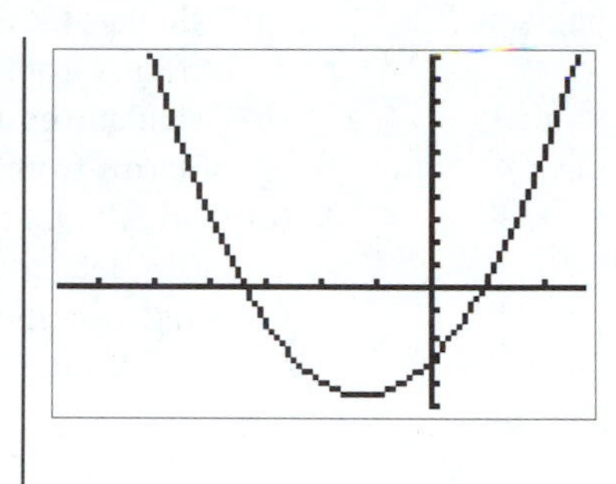

[−6.7, 2.7, 1] x [−10, 20, 2]

FIGURE 3.7a

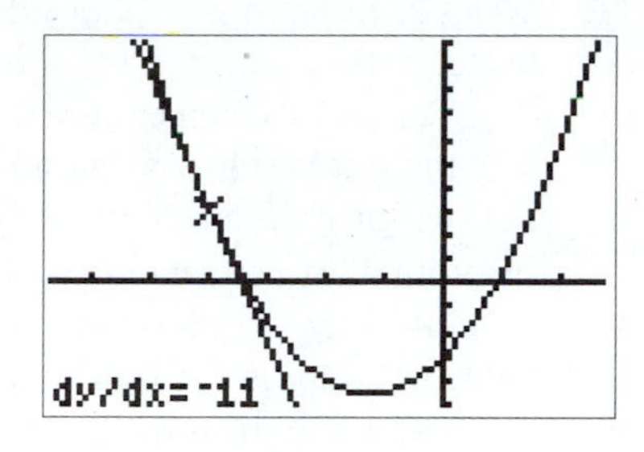

[−6.7, 2.7, 1] x [−10, 20, 2]

FIGURE 3.7b

Exercise Set 3.1

Find the average slope over the indicated intervals for the functions in Exercises 1–14.

1. $f(x) = x^2 + 5x$ from $x = 1$ to $x = 3$
2. $g(x) = 5x - 3$ from $x = 4$ to $x = 7$
3. $h(x) = 7x + 1$ from $x = -3$ to $x = -1$
4. $j(x) = 4x^2 - 5$ from $x = -2$ to $x = 0$
5. $k(x) = 3x^3 + 7x - 1$ from $x = -2$ to $x = -1$
6. $m(x) = 2x^4 - 5x^2 + x - 1$ from $x = 0$ to $x = 1$
7. $f(x) = \frac{5}{x}$ from $x = 2$ to $x = 5$
8. $g(x) = \frac{8}{x+2}$ from $x = -6$ to $x = -3$
9. $h(x) = x^2$ from $x = 1$ to $x = 0$
10. $j(x) = x^2 + 3x + 5$ from $x = -2$ to $x = b$, $b \neq -2$
11. $k(x) = x^2 + x - 7$ from $x = 1$ to $x = x_1$, $x_1 \neq 1$
12. $m(x) = x^3 + 4$ from $x = 1$ to $x = x_1$, $x_1 \neq 1$
13. $f(x) = x^2 + 1$ from $x = x_1$ to $x = x_1 + h$, $h \neq 0$
14. $g(x) = x^2 + 4x - 1$ from $x = x_1$ to $x = x_1 + h$, $h \neq 0$

Solve Exercises 15–22.

15. *Machine technology* What was the average rate of change in the volume for the tank in Example 3.4 **(a)** from the second to the third minute, **(b)** from the third to the fourth minute, and **(c)** from the second to the fourth minute?

16. *Physics* A stone is dropped into a pool of water and causes a ripple that travels in the shape of a circle out from the point of impact at a rate of 2 m/s. What is the average change in the area within this circle from the third to the fourth second?

17. Let $f(x) = 3x^2 - 5$. Find the average slope of the curve from $x_1 = 4$ to **(a)** $x_2 = 6$, **(b)** $x_3 = 5$, **(c)** $x_4 = 4.5$, **(d)** $x_5 = 4.25$, **(e)** $x_6 = 4.1$, and **(f)** $x_7 = 4.05$.

18. For the same function in Exercise 17, $f(x) = 3x^2 - 5$, find the average slope of the curve from $x_1 = 4$ to **(a)** $x_2 = 2$, **(b)** $x_3 = 3$, **(c)** $x_4 = 3.5$, **(d)** $x_5 = 3.75$, **(e)** $x_6 = 3.9$, and **(f)** $x_7 = 3.95$.

19. Write a computer program to determine the average slope between any two points on a curve. Use your program to check your results from Exercises 17 and 18.

20. *Medical technology* Consider the following sentence:

The child's temperature has been rising for the last two hours, but not as rapidly since we gave the antibiotic an hour ago.

(a) With this statement in mind, sketch a graph of the child's temperature as a function of time.

(b) How did you (or how can you) use tangents to your graph to show that your graph is consistent with the statement?

21. *Transportation* On a 50-minute trip, a car travels for 15 min with an average velocity of 20 mph and then 35 min with an average velocity of 54 mph. Determine **(a)** the total distance traveled and **(b)** the average velocity for the entire trip.

22. *Agriculture* The graph in Figure 3.8 gives the number of $f(t)$ farms in a certain Iowa county t years after 1980.

(a) Calculate the average rate of change in the number of farms from 1980 to 1985.

(b) Calculate the average rate of change in the number of farms from 1990 to 1995.

(c) During what two-year period was the number of farms decreasing most rapidly?

(d) What was the rate of change in the number of farms for the two years you answered in (c)?

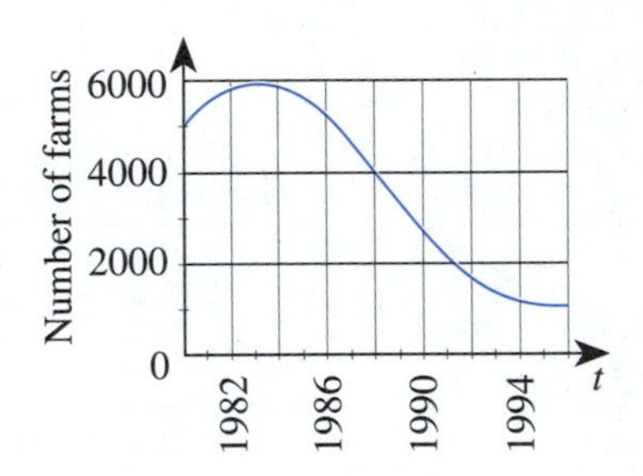

FIGURE 3.8

In Your Words

23. You are in an automobile that is traveling from your home to school. Explain how you can determine the average rate that the car is traveling.

24. What is the difference between a secant line and a tangent line?

3.2 THE AREA QUESTION

In Section 3.1, we examined one of the basic questions that led to the development of calculus. In this section, we will look at the other basic question—the area question.

The area question concerns the area between the graph of a function and the x-axis. There are some restrictions on this problem that can best be described by examining Figure 3.9. The function $f(x)$ should be nonnegative over a closed interval; that is, it should not fall below the x-axis in some closed interval. If the closed interval is $[a, b]$, then the desired area lies between the curve of $f(x)$, the x-axis, and the vertical lines $x = a$ and $x = b$. This is indicated by the shading in Figure 3.9.

We have used several different formulas for the area. We will use two of them to help determine the shaded area in Figure 3.9.

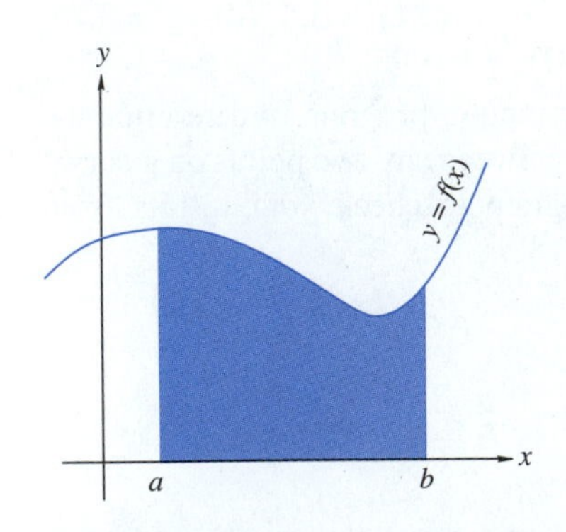

FIGURE 3.9

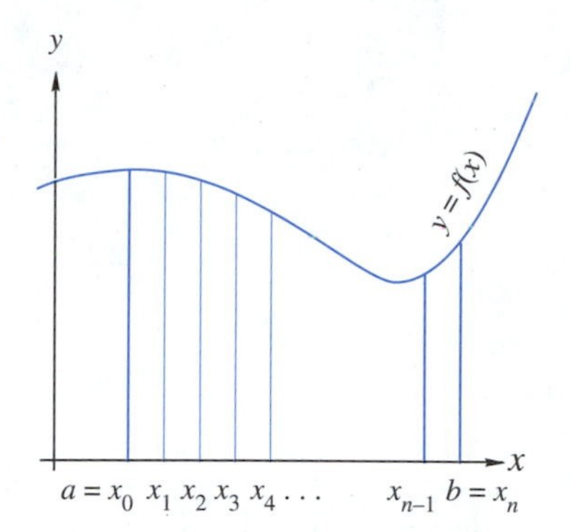

FIGURE 3.10

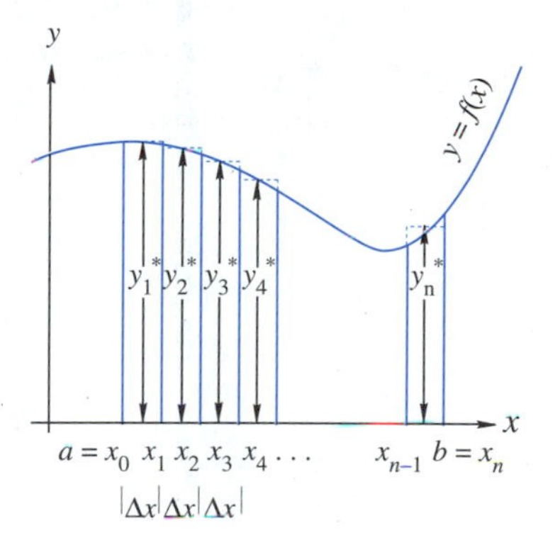

FIGURE 3.11

Rectangular Approximation

The first method we will use will approximate the area by adding the areas of several rectangles. To begin, we will divide the interval $[a, b]$ into n segments. While each segment can be a different length, we will make them all the same length. If we let $a = x_0$ and $b = x_n$, the first segment is $[x_0, x_1]$; the second segment $[x_1, x_2]$; the third segment $[x_2, x_3]$; and so on. The last segment is $[x_{n-1}, x_n]$. In general, the ith segment would be $[x_{i-1}, x_i]$. At each of these points, $x_0, x_1, x_2, \ldots, x_n$, we will erect a line perpendicular to the x-axis. This creates a group of n strips or bands, as shown in Figure 3.10.

We want to draw a line segment parallel to the x-axis across the top of each of these strips. Where we draw these segments is up to us. We would like to draw them so that we can get a close approximation to the area under the curve. What we will do is use an approach that should give a reasonable approximation.

Look at Figure 3.11. We have selected a point in the middle of each strip and drawn the line segment so that it went through the y-coordinate at that point. For example, in the first strip, the midpoint between x_0 and x_1 is the point we will call x_1^*. The height of the curve at x_1^* is $f(x_1^*)$, which we will call y_1^*. Now, since each segment is the same width, the width of each rectangle is the same. This width, which we will call Δx, is equal to $\frac{x_n - x_0}{n}$. This means that the area of the first rectangle is $(\Delta x)(y_1^*)$. In the same way, the area of the second rectangle would be $(\Delta x)(y_2^*)$ and the area of the last rectangle would be $(\Delta x)(y_n^*)$.

We can now get an idea of the area under this curve by adding all the areas of these rectangles. Thus, the area under the curve, A, can be written as

$$\begin{aligned} A &\approx (y_1^*)(\Delta x) + (y_2^*)(\Delta x) + (y_3^*)(\Delta x) + \cdots + (y_n^*)(\Delta) \\ &= (y_1^* + y_2^* + y_3^* + \cdots + y_n^*)(\Delta x) \\ &= \left(\sum_{i=1}^{n} y_i^*\right)\Delta x \end{aligned}$$

EXAMPLE 3.6

Suppose $f(x) = x^2 + 3$. Find an approximation of the area between this curve and the x-axis from $x = 1$ to $x = 6$.

Solution We will divide the interval $[1, 6]$ into 5 equal segments. (The number of segments is purely arbitrary and we selected 5 because it means that each segment has a length of 1, or $\Delta x = 1$.) The midpoints of the segments are at 1.5, 2.5, 3.5, 4.5, and 5.5 as can be seen in Figure 3.12. The value of the function at each of these points can be seen in the following table.

x^*	1.50	2.50	3.50	4.50	5.50
$f(x^*) = y^*$	5.25	9.25	15.25	23.25	33.25

EXAMPLE 3.6 (Cont.)

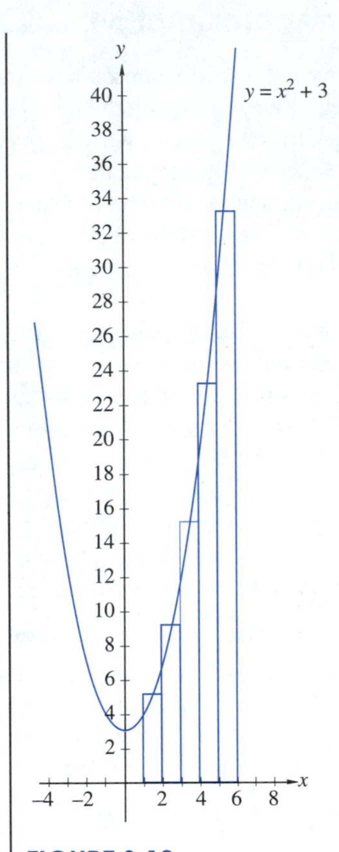

FIGURE 3.12

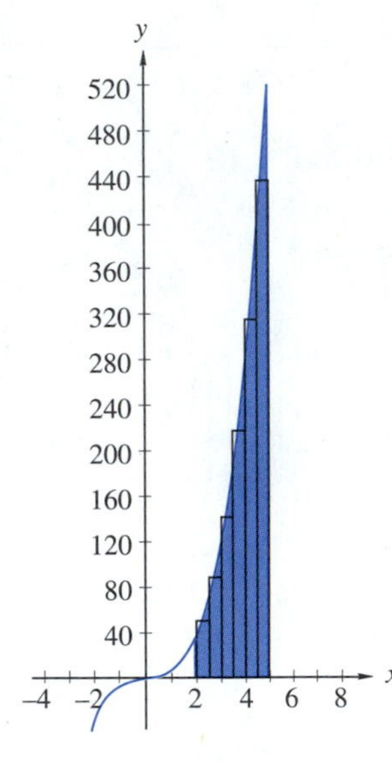

FIGURE 3.13

The area then is found using our new formula for area:

$$\begin{aligned} A &\approx (y_1^*)(\Delta x) + (y_2^*)(\Delta x) + (y_3^*)(\Delta x) + \cdots + (y_n^*)(\Delta x) \\ &\approx (y_1^* + y_2^* + y_3^* + \cdots + y_n^*)(\Delta x) \\ &= (5.25 + 9.25 + 15.25 + 23.25 + 33.25)(1) \\ &= 86.25 \end{aligned}$$

The area is approximately 86.25 square units.

We will work another example, only this time we will make the lengths of the segments something other than 1.

EXAMPLE 3.7

Approximate the area under the curve $g(x) = 4x^3 + 2x - 1$ from $x = 2$ to $x = 5$.

Solution The curve is shown in Figure 3.13 with the shaded region indicating the area we want to approximate. The total length of the interval is 3 units. We will divide this into 6 segments, so each segment will be 0.5 units long or $\Delta x = 0.5$. The following table gives the values of the function at the midpoints of each of these intervals.

EXAMPLE 3.7 (Cont.)

x^*	2.25	2.7500	3.2500	3.7500	4.2500	4.7500
y^*	49.0625	87.6875	142.8125	217.4375	314.5625	437.1875

Again, using our formula for approximating the area we get

$$\begin{aligned} A &\approx (y_1^* + y_2^* + y_3^* + \cdots + y_n^*)(\Delta x) \\ &= (49.0625 + 87.6875 + 142.8125 + 217.4375 \\ &\quad + 314.5625 + 437.1875)(0.5) \\ &= (1{,}248.75)(0.5) \\ &= 624.375 \end{aligned}$$

The approximate area under this curve from $x = 2$ to $x = 5$ is 624.375 square units.

Application

EXAMPLE 3.8

The following table gives the results of a series of drillings to determine the depth of the bedrock at a building site. These drillings were taken along a straight line down the middle of the lot where the building will be placed. In the table, x is the distance from the front of the parking lot and y is the corresponding depth. Both x and y are given in feet.

x	0	20	40	60	80	100	120	140	160
y	33	35	40	45	42	38	46	40	48

Approximate the area of this cross-section.

Solution The total length of the interval is 160 ft. This has been divided into 8 segments, with each segment 20 ft long, so $\Delta x = 20$. Since we do not know the function, we will have to approximate the depth at the midpoints of the intervals. We will do this by taking the average of the values at each end of an interval. Thus, the length at the midpoint of the first interval is $y_1^* = \frac{y_1 + y_2}{2} = \frac{33+35}{2} = 34$. The next table gives the approximate values at the midpoints of these intervals.

x^*	10	30	50	70	90	110	130	150
y^*	34	37.5	42.5	43.5	40	42	43	44

Once again, using our formula for approximating the area, we get

$$\begin{aligned} A &\approx (y_1^* + y_2^* + y_3^* + \cdots + y_8^*)\Delta x \\ &= (34 + 37.5 + 42.5 + 43.5 + 40 + 42 + 43 + 44)(20) \\ &= (326.5)(20) \\ &= 6{,}530\,\text{ft}^2 \end{aligned}$$

In Section 6.5, we will find another method to approximate the area of this cross-section, at which point we will get a more accurate approximation of 6,460 ft^2.

Application

EXAMPLE 3.9

Before the parking lot for the building in Example 3.8 is completed it will have to be paved. A photograph of the completed lot is shown in Figure 3.14a. In order to pave this parking lot, its area had to be determined. This area was then multiplied by the thickness of the asphalt to obtain the total volume of the asphalt that was needed.

In order to find the area, the contractor drew a base line through the "middle" of the parking lot and then drew lines perpendicular to the base line every 50 feet. Each of these lines were drawn until they reached the other side of the parking lot, as shown in Figure 3.14b. The following table shows the lengths labeled y (in feet) of each of these lines.

x	0	50	100	150	200	250	300	350	400	450	500	550	600
y	72	104	160	200	200	200	190	190	190	180	180	180	180

Use the data in the table to approximate the area of the parking lot.

Courtesy of Michael A. Gallitelli, Metroland Photo Inc.

FIGURE 3.14a

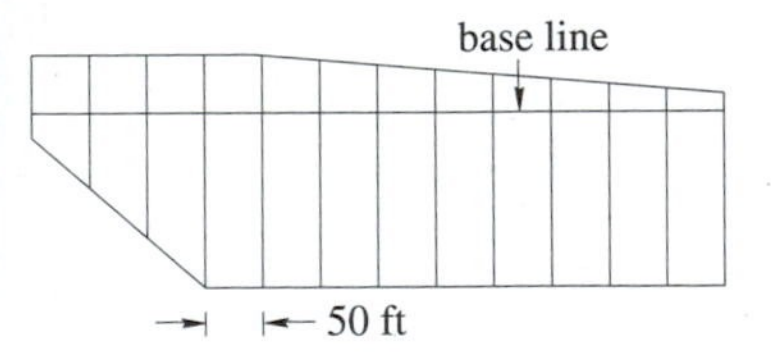

FIGURE 3.14b

Solution We are given that the length of each interval is 50 ft, and so $\Delta x = 50$. As in Example 3.8, we will approximate the lengths at the midpoints by taking the average of the values at each end of the intervals. The next table gives the approximate values at the midpoints of these intervals.

x^*	25	75	125	175	225	275	325	375	425	475	525	575
y^*	88	132	180	200	200	195	190	190	185	180	180	180

EXAMPLE 3.9 (Cont.)

As before, we use our formula for approximating the area and obtain

$$\begin{aligned}A &\approx (y_1^* + y_2^* + y_3^* + \cdots + y_{12}^*)\Delta x \\ &= (88 + 132 + 180 + 200 + 200 + 195 + 190 \\ &\quad + 190 + 185 + 180 + 180 + 180)(50) \\ &= (2{,}100)(50) \\ &= 105{,}000\end{aligned}$$

Thus, we see that the area of this parking lot is approximately 105,000 ft^2.

Exercise Set 3.2

In Exercises 1–10, find the area under the graph of the function from *a* to *b* using the method described in this section. The value of *n* in each problem indicates the number of segments into which you should divide the interval.

1. $f(x) = 3x + 2;\ a = 1,\ b = 7,\ n = 6$
2. $g(x) = 7 - 4x;\ a = -4,\ b = 1,\ n = 5$
3. $h(x) = x^2 + 1;\ a = 0,\ b = 3,\ n = 6$
4. $k(x) = 3x^2 - 2;\ a = 1,\ b = 3,\ n = 4$
5. $j(x) = 4x^2 + 3x - 5;\ a = 1,\ b = 5,\ n = 8$
6. $m(x) = x^3 + 2;\ a = -1,\ b = 2,\ n = 6$
7. $f(x) = 16 - x^2;\ a = -2,\ b = 4,\ n = 8$
8. $g(x) = 4x - x^3;\ a = 0,\ b = 2,\ n = 10$
9. $h(x) = 3x^2 - 2x + 7;\ a = -5,\ b = -1,\ n = 8$
10. $k(x) = x^4 - 3;\ a = 2,\ b = 4,\ n = 8$

In Exercises 11–12, find the approximate area under the curve defined by graphing the sets of experimental data.

11.

x	5	6	7	8	9	10	11
y	4.2	3.9	3.8	4.0	3.5	3.4	3.9

x	12	13
y	4.1	4.3

12.

x	1.00	1.25	1.50	1.75	2.00
y	16.32	16.48	16.73	16.42	16.38

x	2.25	2.50
y	16.29	16.25

Solve Exercises 13–18.

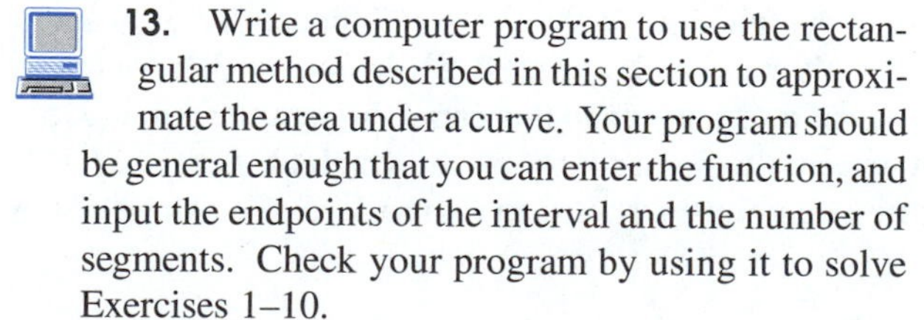

13. Write a computer program to use the rectangular method described in this section to approximate the area under a curve. Your program should be general enough that you can enter the function, and input the endpoints of the interval and the number of segments. Check your program by using it to solve Exercises 1–10.

14. Use the program you wrote in Exercise 13 to get more accurate approximations of the area under the curve in Exercises 9–10. Make the value of n **(a)** 15, **(b)** 24, **(c)** 48, and **(d)** 96.

15. *Business* The marginal profit for a certain type of coat is given by $P'(x) = 78 - 0.8x$ dollars per coat, where x is the number of coats produced and sold weekly. The profit for the first n coats that are produced and sold is determined by finding the area under the graph of P' from $a = 0$ to $b = n$. What is the profit for the first 40 coats produced and sold?

16. *Environmental science* The level of pollution in San Juan Pedro Bay, due to a sewage spill, is estimated to be $f(t) = \dfrac{1200t}{\sqrt{t^2+10}}$ parts per million, where t is the time in days since the spill occurred. Find the total amount of pollution during the first 5 days of the oil spill by using the method described in this section and 10 divisions.

17. *Electronics* The charge on a capacitor in millicoulombs can be estimated by finding the area under the graph of $f(t) = 0.6t^2 - 0.2t^3$ from $t = 1$ to $t = 3$. Estimate the charge on this capacitor by using the method described in this section and 8 divisions.

18. *Ecology* A town wants to drain and fill the swamp shown in Figure 3.15.
    (a) What is the surface area of the swamp?
    (b) If the swamp has an average depth of 6 ft, how many cubic yards of dirt will it take to fill the "hole" that is left after the swamp is drained?

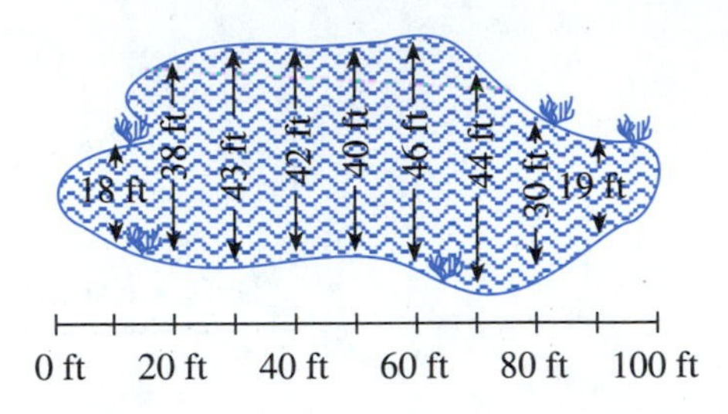

FIGURE 3.15

In Your Words

19. Describe how to use the procedures of this chapter to find the area of an irregular-shaped region.

20. What changes would you have to make in the procedures of this chapter if the rectangles were not all the same width?

21. Use the procedures described in Example 3.19 on an irregular-shaped region such as a parking lot or a lake to determine its area.

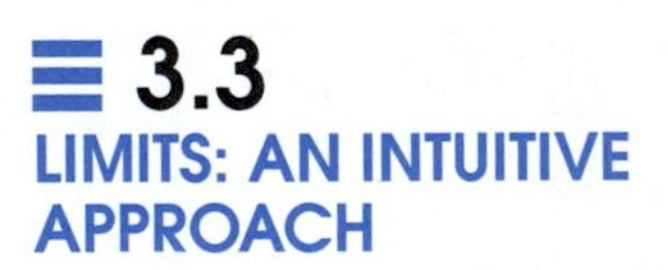

3.3 LIMITS: AN INTUITIVE APPROACH

In Sections 3.1 and 3.2, we have looked at ways in which we could attempt to solve two of the questions that led to the development of calculus. These two questions are known as the tangent question and the area question. So far we have laid a foundation for answering them. In this section, we will learn about a third part of the foundation—the limit.

To begin our introduction to the limit, we will return to our look at the tangent question. Consider the curve of the function $f(x) = x^2 - 4$. Suppose that we want to find the slope of the tangent to this curve at the point $(3, 5)$. In Section 3.1, we found out how to determine the slope of a secant line to this curve through the point $(3, 5)$. Let $(x, f(x))$ be any point, except the point $(3, 5)$, on the graph of $f(x) = x^2 - 4$. The secant line through these two points is shown in Figure 3.16. The slope of the secant line is given by

$$\vec{m} = \frac{f(x) - 5}{x - 3} = \frac{(x^2 - 4) - 5}{x - 3} = \frac{x^2 - 9}{x - 3}$$

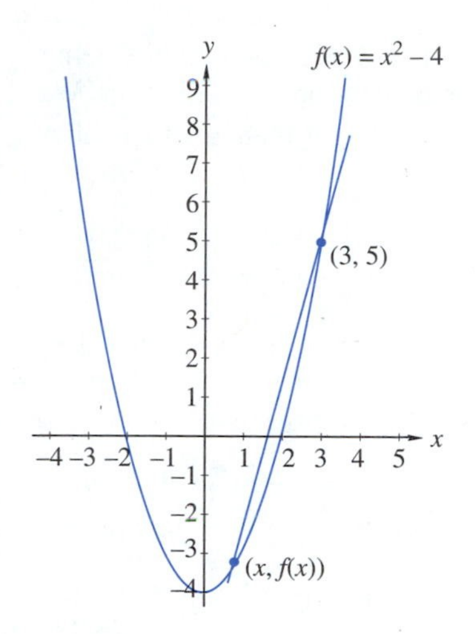

FIGURE 3.16

Since the point $(x, f(x))$ is different from $(3, 5)$, we know that $x \neq 3$ and so we can simplify our equation to

$$\bar{m} = \frac{(x+3)(x-3)}{x-3} = x+3$$

So far so good! Everything that we have done is just like what we did in Section 3.1. But now, we want to look at the problem a little closer. What happens as x gets closer to 3? The values of $\bar{m} = x+3$ get closer to 6. We put this idea into symbols by writing

$$\bar{m}_{\tan} = \lim_{x \to 3} \frac{x^2-9}{x-3} = \lim_{x \to 3}(x+3) = 6$$

Since we know that the slope of the tangent line at the point $(3, 5)$ is 6, we are able to determine that the equation of the tangent is

$$\begin{aligned} y-5 &= 6(x-3) \\ y &= 6x-13 \end{aligned}$$

This is the basic idea behind a limit. We will now begin to look at a more detailed explanation.

Limit of a Function

If a function $y = f(x)$ approaches the real number L as x approaches a particular value a, then we say that L is the **limit** of f as x approaches a. The notation for this concept is

$$\lim_{x \to a} f(x) = L.$$

The $\lim_{x \to a} f(x)$ may or may not exist for any function at any particular a. The following examples use numerical calculations to locate limits.

The Numerical Approach to a Limit

To get a better idea of a limit, we will return to Exercises 17–18 from Exercise Set 3.1.

In these two exercises $f(x) = 3x^2 - 5$. We will set up two tables of values. Each table will contain values for x and $\bar{m} = \dfrac{f(x)-f(4)}{x-4} = \dfrac{f(x)-43}{x-4}$. In the first, we will put the values for Exercise 17, and in the second, the values for Exercise 18.

In Exercise 18, all of the values of x were larger than 4 with the first one being the largest, 6, and each one getting smaller. The last value was the smallest, 4.05. We will continue this process by selecting additional numbers larger than 4 (and smaller than 4.05), each one getting closer to 4.

x	6	5	4.5	4.25	4.10	4.05	4.01	4.001	4.0001
$\overline{m}$	30	27	25.5	24.75	24.30	24.15	24.03	24.003	24.0003

The table of values for Exercise 18 contains values of x that are all smaller than 4 with the largest being 3.95. Again, we will continue the process. This time the numbers will all be smaller than 4, each getting increasingly closer to 4.

x	2	3	3.5	3.75	3.9	3.95	3.99	3.999	3.9999
$\overline{m}$	18	21	22.5	23.75	23.70	23.85	23.97	23.997	23.9997

These tables give us a numerical method for approximating the $\lim_{x\to 4}(3x^2-5)$. As you can see from the two tables, when x is 4.0001, $m \approx 24.0004$ and when x is 3.9999, $m \approx 23.9995$. It appears that if we were to pick values of x that were closer to 4, we could then get a value of $f(x) = 3x^2 - 5$ that was closer to 24.

EXAMPLE 3.10

Use a numerical approach to find $\lim_{x\to 1}\frac{x^3-1}{x-1}$.

Solution As in the previous explanation, we will construct a table (using the program from Exercise 19, Exercise Set 3.1). This time, since we want to find the limits as x gets closer to 1, we will select values of x that keep getting closer to 1. Some of the values selected are less than 1, and some are larger than 1. The results are shown in the following table.

x	0.500	0.900	0.990	0.999	1.001	1.010	1.100	1.500
$\frac{x^3-1}{x-1}$	1.75	2.71	2.9701	2.997	3.003	3.0301	3.31	4.75

It appears from the table that $\lim_{x\to 1}\frac{x^3-1}{x-1} = 3$.

This result is confirmed by looking at the graph of $y = \frac{x^3-1}{x-1}$, as shown in Figure 3.17. The "hole" in the graph appears to be at $(1, 3)$.

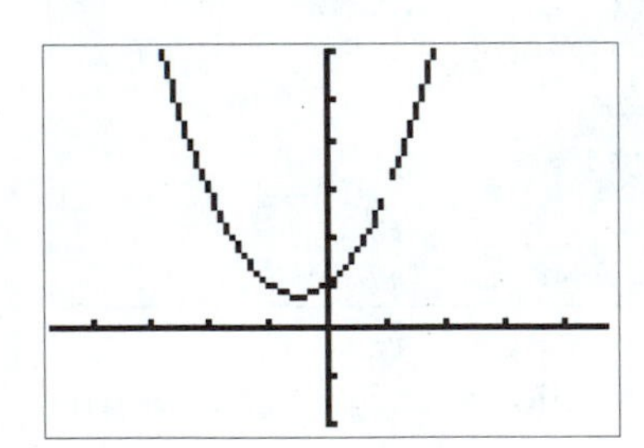

[-4.7, 4.7, 1] x [-2, 6, 1]

FIGURE 3.17

You may remember that this factors as $x^3 - 1 = (x-1)(x^2+x+1)$. We can use this knowledge to see that as long as $x \neq 1$, $g(x) = \frac{x^3-1}{x-1} = x^2+x+1$. This makes it a lot easier to see that as x gets closer to 1, the value of the function gets closer to 3.

This next example will show that this type of simplification does not work for all limits.

EXAMPLE 3.11

Use a numerical approach to find $\lim_{x\to 0} \frac{\sin x}{x}$.

Solution Since we cannot factor an x out of $\sin x$, the simplification method we have been using will not work. But, if we carefully construct a table of values for x and $g(x) = \frac{\sin x}{x}$ for values of x near 0, we can get an idea of what this limit will be. The following table shows this. Notice that the values of x are in radians.

x (radians)	0.200	0.100	0.010	−0.010	−0.100	−0.200
$\sin x$	0.1987	0.0998	0.0099	−0.0099	−0.0998	−0.1987
$\frac{\sin x}{x}$	0.9933	0.9983	0.9999	0.9999	0.9983	0.9933

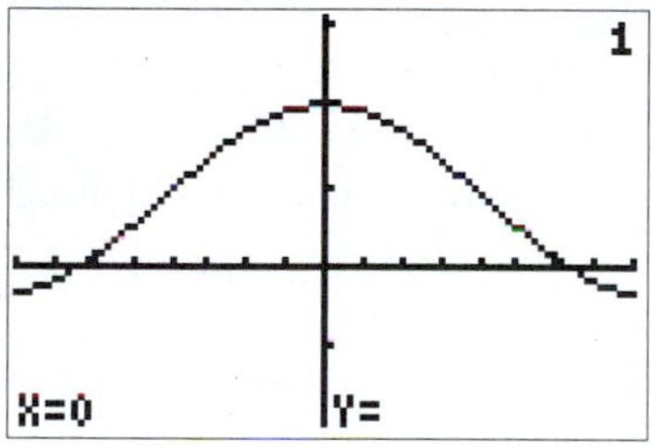

From this table of values it appears that

$$\lim_{x\to 0} \frac{\sin x}{x} = 1$$

The graph in Figure 3.18 seems to reinforce this numerical method that shows $\lim_{x\to 0} \frac{\sin x}{x} = 1$. Notice that if you TRACE this curve, there is no y-value when $x = 0$. That is because 0 is not in the domain of the function.

1

X=0 Y=

[−4, 4, 0.5] x [−1, 1.5, 1]

FIGURE 3.18

The Graphical Approach to a Limit

The numerical approach is often a good way to approximate a limit. There are some times, however, when a graphical approach will be easier. We will use the graphical approach in the next two examples.

EXAMPLE 3.12

Use a graphical approach to find $\lim_{x\to 1} h(x)$, where

$$h(x) = \begin{cases} 2x - 4 & \text{if } x \neq 1 \\ 5 & \text{if } x = 1 \end{cases}$$

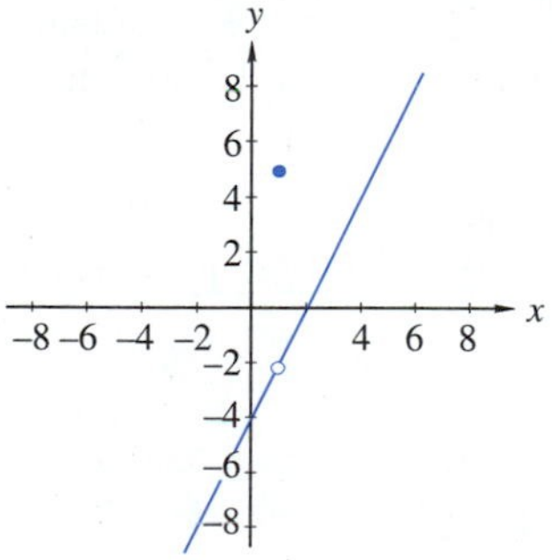

Solution The graph of this function is shown in Figure 3.19. As you can see, as the values of x get close to 1 the values of $h(x)$ get very close to −2. In fact, if we select x close enough to 1, we can get $h(x)$ as close to −2 as we want. We decide, based on our observation, that $\lim_{x\to 1} h(x) = -2$. Notice, however, that when $x = 1$, $h(x) = 5$. Thus, $\lim_{x\to 1} h(x) \neq h(1)$.

y
8
6
4
2
−8 −6 −4 −2
−2
4 6 8
x
−4
−6
−8

FIGURE 3.19

Note While Example 3.12 is somewhat artificial, you can see that the limit of a function at a given point is not always the same as the value of the function at that point.

The limit of a function at a certain point may not always exist. Until now, all of the functions we have examined have approached a certain number as the values of x got close to a particular value. The next example will show that this does not always happen.

EXAMPLE 3.13

Use a graphical approach to find $\lim_{x \to 0} k(x)$, where

$$k(x) = \begin{cases} 2 & \text{if } x < 0 \\ -4 & \text{if } x = 0 \\ 5 & \text{if } x > 0 \end{cases}$$

Solution The graph of this function is shown in Figure 3.20. As you can see, if the values of x are close to 0 and negative, the value of $k(x)$ is 2. On the other hand, for values of x that are close to 0 and positive, the value of $k(x)$ is 5. There is no specific number that the values of $k(x)$ are near when x is close to 0. We must conclude that the $\lim_{x \to 0} k(x)$ does not exist. The fact that the function is defined when $x = 0$, $k(0) = -4$, has no effect on whether the limit of the function exists at that point.

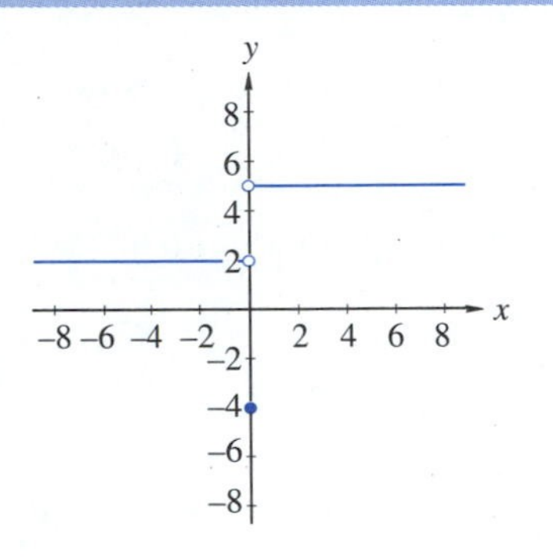

FIGURE 3.20

We will use the graphical approach to show two basic rules for limits:

Rules for Limits

Rule 1: $\lim_{x \to c} A = A$, where A and c are real numbers

Rule 2: $\lim_{x \to c} x = c$, where c is a real number

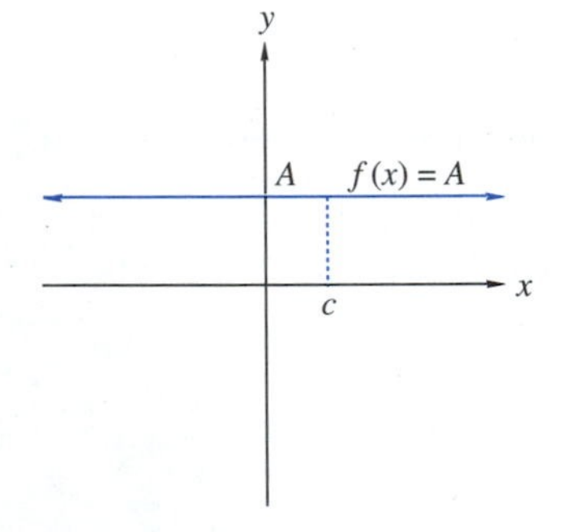

FIGURE 3.21

The graph for the first rule is shown in Figure 3.21. As you can see, the value of $f(x)$ is always A and so, as the values of x approach c, $f(x) = A$, thus we have $\lim_{x \to c} f(x) = A$.

The second rule is demonstrated in Figure 3.22. The graph of $f(x) = x$ is a straight line. For any value of c, as x gets close to c, the values of $f(x)$ are just as close to c, since $f(x) = x$. This shows that $\lim_{x \to c} f(x) = c$.

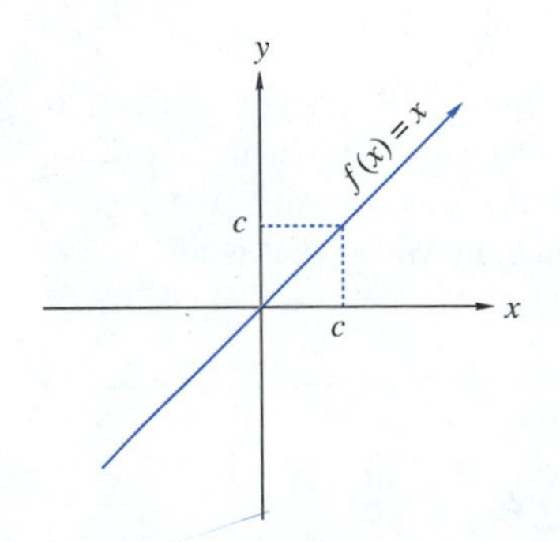

FIGURE 3.22

EXAMPLE 3.14

Evaluate each of the following limits: **(a)** $\lim_{x\to 3} 6$, **(b)** $\lim_{x\to -2} 7$, **(c)** $\lim_{x\to -5} x$, and **(d)** $\lim_{x\to 7} x$.

Solution The first two, (a) and (b), follow rule 1, $\lim_{x\to c} A = A$. In (a), $A = 6$ so, $\lim_{x\to 3} 6 = 6$ and in (b), $A = 7$ and $\lim_{x\to -2} 7 = 7$.

The last two parts, (c) and (d), use rule 2, $\lim_{x\to c} x = c$. In (c) we have $\lim_{x\to -5} x = -5$ and in (d) $\lim_{x\to 7} x = 7$.

In this section, we have tried to give you a foundation for the idea of a limit. The following exercises are designed to further develop this foundation. In Section 3.4 we will take a more detailed look at the limit concept.

Exercise Set 3.3

In Exercises 1–8, complete each table and use it to determine the indicated limit.

1.

x	0.9	0.99	0.999	0.9999	1.0001	1.001	1.01	1.1
$f(x) = 3x$								

$\lim_{x\to 1} 3x = ?$

2.

x	2.9	2.99	2.999	2.9999	3.0001	3.001	3.01	3.1
$g(x) = x - 4$								

$\lim_{x\to 3} (x-4) = ?$

3.

x	−1.1	−1.01	−1.001	−1.0001	−0.9999	−0.999	−0.99	−0.9
$h(x) = x^2 + 2$								

$\lim_{x\to -1} (x^2+2) = ?$

4.

x	−2.1	−2.01	−2.001	−2.0001	−1.9999	−1.999	−1.99	−1.9
$k(x) = \frac{x^2-4}{x+2}$								

$\lim_{x\to -2} \frac{x^2-4}{x+2} = ?$

5.

x	−0.1	−0.01	−0.001	−0.0001	0.0001	0.001	0.01	0.1
$f(x) = \frac{\tan x}{x}$								

$\lim_{x\to 0} \frac{\tan x}{x} = ?$

6.

x	−0.1	−0.01	−0.001	−0.0001	0.0001	0.001	0.01	0.1
$g(x) = \frac{1-\cos x}{x^2}$								

$\lim_{x\to 0} \frac{1-\cos x}{x^2} = ?$

7.

x	0.9	0.99	0.999	0.9999	1.0001	1.001	1.01	1.1
$h(x) = \frac{x}{x-1}$								

$\lim_{x\to 1} \frac{x}{x-1} = ?$

8.

x	−0.1	−0.01	−0.001	−0.0001	0.0001	0.001	0.01	0.1
$k(x) = \frac{1}{x^2}$								

$\lim_{x\to 0} \frac{1}{x^2} = ?$

Solve Exercises 9 and 10.

9. For the function f in the given figure, find the following limits or state that the limit does not exist. Give a reason for your answers. **(a)** $\lim_{x\to 0} f(x)$ **(b)** $\lim_{x\to 2} f(x)$ **(c)** $\lim_{x\to 3} f(x)$

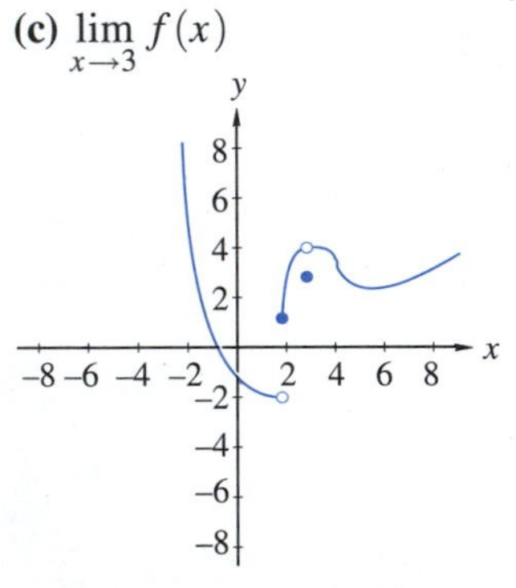

10. For the function g in the given figure, find the following limits or state that the limit does not exist. Give reasons for your answers. **(a)** $\lim_{x\to -2} g(x)$ **(b)** $\lim_{x\to 0} g(x)$ **(c)** $\lim_{x\to 3} g(x)$

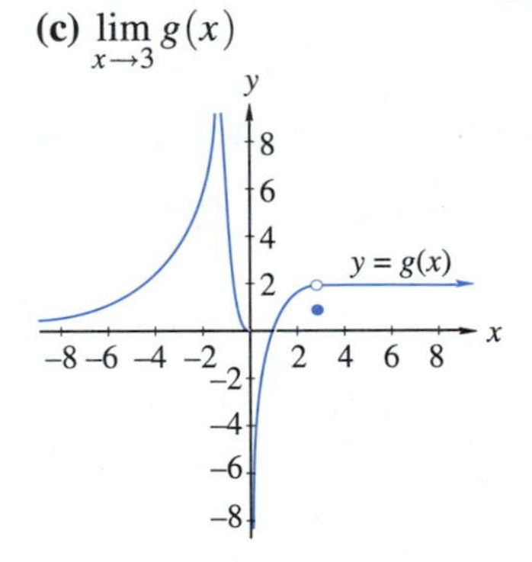

In Exercises 11–18, sketch the graphs of the functions and find out if $\lim_{x\to c}$ exists at the given value of c or state that the limit does not exist.

11. $f(x) = \begin{cases} 3x-2 & \text{if } x \neq 1 \\ 0 & \text{if } x = 1 \end{cases}$ $\quad c = 1$

12. $g(x) = \begin{cases} x+1 & \text{if } x < 0 \\ x^2+1 & \text{if } x \geq 0 \end{cases}$ $\quad c = 0$

13. $h(x) = \begin{cases} 5-2x & \text{if } x < 0 \\ 0 & \text{if } x = 0 \\ 4x+5 & \text{if } x > 0 \end{cases}$ $\quad c = 0$

14. $k(x) = \begin{cases} x+1 & \text{if } x < 1 \\ \text{not defined} & \text{if } x = 1 \\ 2x & \text{if } x > 1 \end{cases}$ $\quad c = 1$

15. $f(x) = \begin{cases} 1-3x & \text{if } x < 0 \\ 3x+1 & \text{if } x \geq 0 \end{cases}$ $\quad c = 0$

16. $g(x) = \begin{cases} 3x+1 & \text{if } x \leq -1 \\ 1-3x & \text{if } x > -1 \end{cases}$ $\quad c = -1$

17. $h(x) = \begin{cases} 3x+1 & \text{if } x < 0 \\ 4 & \text{if } x = 0 \\ 1-3x & \text{if } x > 0 \end{cases}$ $\quad c = 0$

18. $k(x) = \begin{cases} 3x+1 & \text{if } x < 0 \\ 2 & \text{if } x = 0 \\ x+1 & \text{if } x > 0 \end{cases}$ $\quad c = 0$

Evaluate the limits in Exercises 19–24.

19. $\lim_{x \to 2}(-15)$

20. $\lim_{x \to 0} x$

21. $\lim_{x \to -4} x$

22. $\lim_{x \to -7} 19$

23. $\lim_{x \to 8} x$

24. $\lim_{x \to -5}(-9)$

In Exercises 25–28, use a graphing calculator or graphing software to guess whether each of the following limits exist. If the limit does exist, estimate its value.

25. $\lim_{x \to 0} \frac{2x^2}{x^2 + \sin x}$

26. $\lim_{x \to 0} \frac{5x^2}{2x^2 + \sin^2 x}$

27. $\lim_{x \to 0} \frac{2x^2}{x^2 + \cos x - 1}$

28. $\lim_{x \to 0} \frac{3x^2}{x^2 + \tan x}$

In Your Words

29. Explain the notation $\lim_{x \to a} f(x) = L$.

30. Suppose that $y = g(x)$ is a function and $\lim_{x \to 2} g(x) = 4$ and $g(2) = -3$.
(a) Sketch a possible graph for g.
(b) Use your graph to explain what is happening to g for values of x close to $x = 2$.

31. Suppose that the function $y = h(x)$ is not defined at $x = 4$. Is it possible for $\lim_{x \to 4} h(x)$ to exist? Explain your answer.

3.4 ONE-SIDED LIMITS

In using both the numerical and the graphical approaches to finding $\lim_{x \to c} f(x)$, we studied the behavior of the function f on both sides of c. There are times when it is necessary to investigate the limit on just one side of c. When this is done, we are investigating one-sided limits.

There are two types of **one-sided limits**. One, the **left-hand limit**, is written as $\lim_{x \to c^-} f(x)$, and the other, the **right-hand limit**, is written as $\lim_{x \to c^+} f(x)$.

To get a better idea of one-sided limits, we will return to an example we worked earlier. In Example 3.13, we wanted to find $\lim_{x \to 0} k(x)$, where

$$k(x) = \begin{cases} 2 & \text{if } x < 0 \\ -4 & \text{if } x = 0 \\ 5 & \text{if } x > 0 \end{cases}$$

The graph of the function is shown in Figure 3.23.

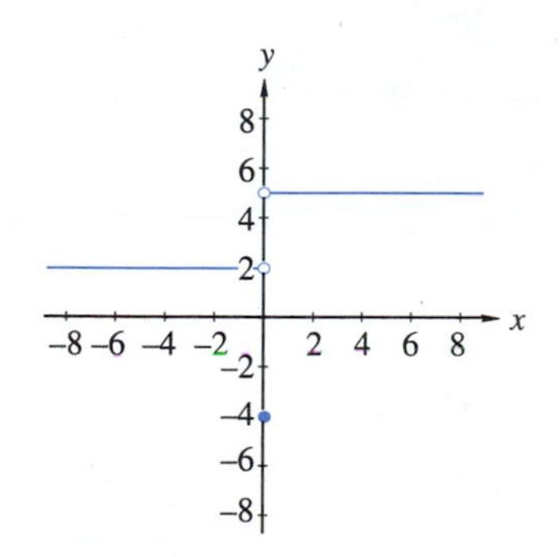

FIGURE 3.23

We know that $\lim_{x \to 0} k(x)$ does not exist, because when x is positive, the value of $k(x)$ is 5, and when x is negative, the value of $k(x)$ is 2. What we have really said is that on the right-hand side of 0 the limit of $k(x)$ is 5 and on the left-hand side, the limit is 2. These are examples of one-sided limits.

In general, the **one-sided limits** of a function $f(x)$ at a point c are the **left-hand limit**, $\lim_{x \to c^-} f(x)$, and the **right-hand limit**, $\lim_{x \to c^+} f(x)$. In the function in Example 3.13, $\lim_{x \to 0^-} k(x) = 2$ and $\lim_{x \to 0^+} k(x) = 5$.

EXAMPLE 3.15

Determine the one-sided limits of $f(x)$ at $x = 3$, if

$$f(x) = \begin{cases} 2x+1 & \text{if } x < 1 \\ 4-x & \text{if } 1 \le x \le 3 \\ x & \text{if } x > 3 \end{cases}$$

Solution The way in which $f(x)$ behaves depends on whether $x < 3$ or $x > 3$, as shown in Figure 3.24. We have different definitions for $f(x)$ for each one-sided limit. When $x < 3$ (but close to 3), f is defined by $f(x) = 4 - x$, and so the left-hand limit is

$$\lim_{x\to 3^-} f(x) = \lim_{x\to 3^-} (4-x) = 1$$

But, when $x > 3$, f is defined by $f(x) = x$, and so the right-hand limit is

$$\lim_{x\to 3+} f(x) = \lim_{x\to 3+} x = 3$$

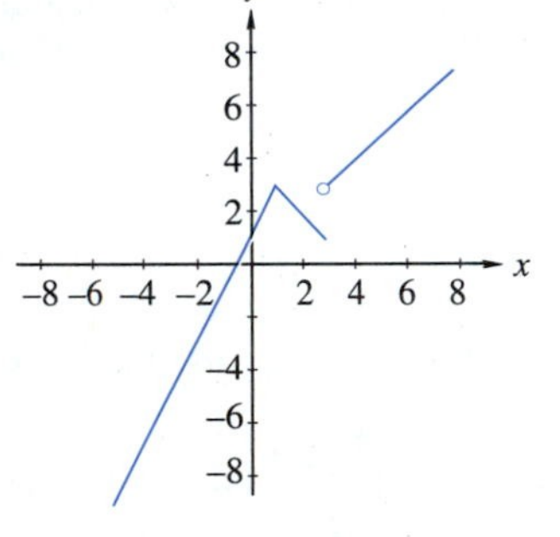

FIGURE 3.24

One-sided limits provide a way to determine whether or not the limit of a function exists at a point. The limit of a function $f(x)$ will exist at a point c, if and only if the right-hand limit of $f(x)$ at c and the left-hand limit of $f(x)$ at c both equal the same real number. Symbolically, this relationship is written as follows.

Relationship Between One-Sided and Two-Sided Limits

$\lim_{x\to c} f(x) = L$, if and only if $\lim_{x\to c^-} = L$ and $\lim_{x\to c^+} f(x) = L$, where L is a real number.

EXAMPLE 3.16

Determine the one-sided limits of $f(x)$ at $x = 1$, where

$$f(x) = \begin{cases} 2x+1 & \text{if } x < 1 \\ 4-x & \text{if } 1 \le x \le 3 \\ x & \text{if } x > 3 \end{cases}$$

Solution This is the same function we studied in Example 3.15. We will take the one-sided limits at $x = 1$. Here, when $x < 1$, f behaves as if $f(x) = 2x + 1$, and we get

$$\lim_{x\to 1^-} f(x) = \lim_{x\to 1^-} (2x+1) = 3$$

When $x > 1$ (but close to 1), we use $f(x) = 4 - x$, with the result that

$$\lim_{x\to 1+} f(x) = \lim_{x\to 1+} (4-x) = 3$$

Since $\lim_{x\to 1^-} f(x) = \lim_{x\to 1+} f(x) = 3$, we can say that $\lim_{x\to 1} f(x) = 3$.

We just specified that in order for the limit to exist, both of the one-sided limits must equal the same real number. The next example will show why we must make sure that it is a real number.

EXAMPLE 3.17

Find the one-sided limits of $g(x) = \dfrac{1}{x^2}$, where $x = 0$.

Solution We will use a table of values and a graph to give us an indication of these limits. The following table and the graph in Figure 3.25 seem to indicate that as $x \to 0$, from both the left and the right, $g(x)$ increases without bound or without limit.

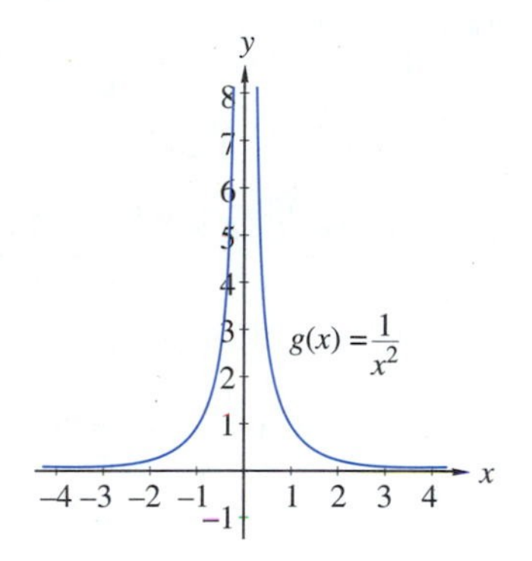

FIGURE 3.25

x	± 1	± 0.5	± 0.1	± 0.01	± 0.001	± 0.0001
$g(x)$	1	4	100	10,000	1,000,000	100,000,000

So, there is no limit for $g(x)$ at $x = 0$. But, because both one-sided limits increase without bound, we say that $g(x)$ becomes positively infinite as x approaches 0. Symbolically, we write

$$\lim_{x \to 0} \frac{1}{x^2} = \infty$$

Caution

Be careful! The fact that we used an equal sign in this last limit does not mean that the limit exists. In fact, just the opposite is true. The infinity symbol ∞ here means that the limit does not exist and shows this because the one-sided limits increase without bound.

A somewhat different result is shown in the next example, where we look at $\dfrac{1}{x}$ as it gets close to 0.

EXAMPLE 3.18

Find the one-sided limits of $h(x) = \dfrac{1}{x}$ as $x \to 0$.

Solution Again, we will use a table of values and a graph to demonstrate our answer. The table follows and the graph of $h(x)$ is in Figure 3.26.

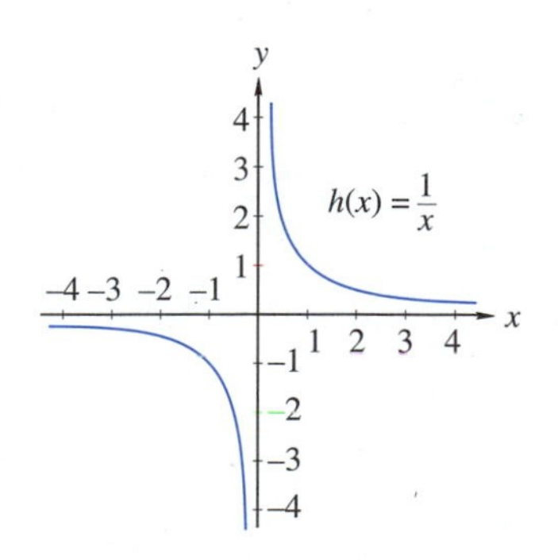

FIGURE 3.26

x	-1	-0.1	-0.01	-0.001	0.001	0.01	0.1	1
$h(x)$	-1	-10	-100	$-1{,}000$	1,000	100	10	1

As we can see, $\lim\limits_{x \to 0^+} \dfrac{1}{x} = \infty$. But, what happens when x is negative? In this case, the values of $h(x)$ become negatively infinite, so

$$\lim_{x \to 0^-} \frac{1}{x} = -\infty$$

Thus we see two reasons that $\lim\limits_{x \to 0} \frac{1}{x}$ does not exist. One reason is that the left-hand and right-hand limits are not equal; the other is that infinity, either positive or negative, is not a real number.

Limits at Infinity

What happens to $h(x)$ as x gets increasingly larger? The following table gives some indication of what happens when the values of x get larger.

x	100	1,000	10,000	100,000	1,000,000
$h(x) = \frac{1}{x}$	0.01	0.001	0.0001	0.00001	0.000001

This table and the graph in Figure 3.26 indicate that

$$\lim_{x\to\infty} \frac{1}{x} = 0$$

That is, as x gets larger without bound, the value of $\frac{1}{x}$ approaches 0. Similarly, as x gets smaller without bound, the value of $\frac{1}{x}$ approaches 0. This is written as

$$\lim_{x\to-\infty} \frac{1}{x} = 0$$

These two results are summarized in the following box.

Limits at Infinity

$$\lim_{x\to\infty} \frac{1}{x} = 0$$

and

$$\lim_{x\to-\infty} \frac{1}{x} = 0$$

Exercise Set 3.4

Solve Exercises 1 and 2.

1. For the function f given in the figure below, find the following limits or state that the limit does not exist.

   (a) $\lim_{x\to2^-} f(x)$ **(c)** $\lim_{x\to3^-} f(x)$
   (b) $\lim_{x\to2^+} f(x)$ **(d)** $\lim_{x\to3^+} f(x)$

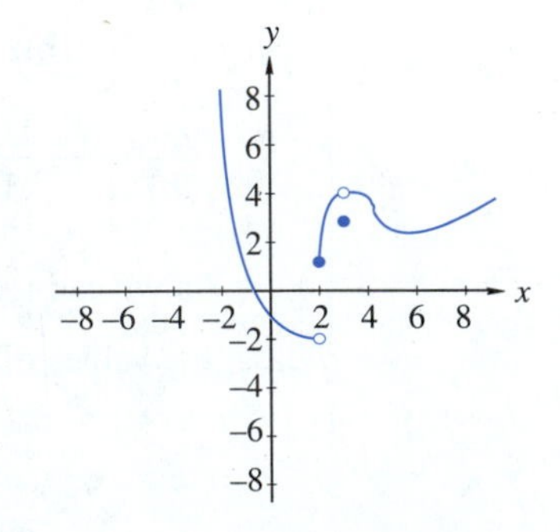

2. For the function g given in the figure below, find the following limits or state that the limit does not exist.

(a) $\lim_{x\to-1^-} g(x)$ (c) $\lim_{x\to 0^-} g(x)$ (e) $\lim_{x\to\infty} g(x)$

(b) $\lim_{x\to-1^+} g(x)$ (d) $\lim_{x\to 0^+} g(x)$ (f) $\lim_{x\to-\infty} g(x)$

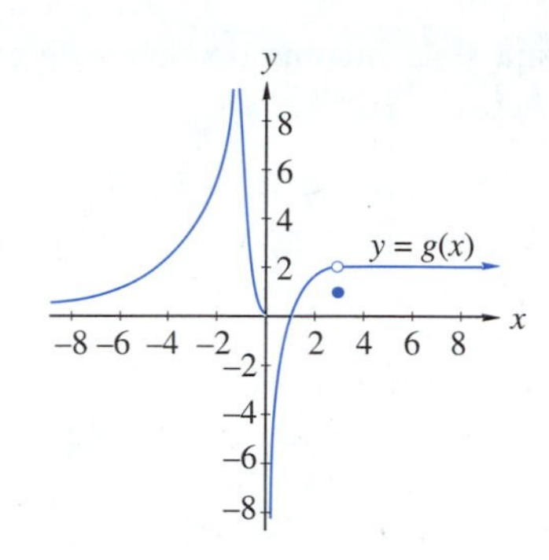

In Exercises 3–18, use a table or a graph to determine the one-sided limits.

3. $\lim_{x\to 5^+} (x-2)$

4. $\lim_{x\to 2^-} (x^2-1)$

5. $\lim_{x\to 4^+} \dfrac{x}{x-4}$

6. $\lim_{x\to 3^-} \dfrac{2x}{x-3}$

7. $\lim_{x\to -2^+} \dfrac{x^2-4}{x+2}$

8. $\lim_{x\to -5^-} \dfrac{x^2+25}{x+5}$

9. $\lim_{x\to 1^-} \dfrac{|x-1|}{x-1}$

10. $\lim_{x\to 1^+} \dfrac{|x-1|}{x-1}$

11. $\lim_{x\to 2^-} \dfrac{|x-2|}{x-2}$

12. $\lim_{x\to 2^+} \dfrac{|x-2|}{x-2}$

13. $\lim_{x\to 2^+} \sqrt{x-2}$

14. $\lim_{x\to -4^+} \sqrt{x+4}$

15. $\lim_{x\to\infty} \dfrac{9}{3x}$

16. $\lim_{x\to-\infty} \dfrac{-8}{4x}$

17. $\lim_{x\to-\infty} \dfrac{1}{2x+1}$

18. $\lim_{x\to\infty} \dfrac{-5}{9x-4}$

In Exercises 19–24, use a graphing calculator or graphing software to guess whether each of the following limits exists. If the limit does exist, estimate its value.

19. $\lim_{x\to\infty} \dfrac{2x^2}{x^2+\sin x}$

20. $\lim_{x\to\infty} \dfrac{5x^2}{2x^2+\sin^2 x}$

21. $\lim_{x\to 0^-} \dfrac{2x^2}{x^2-\cos x}$

22. $\lim_{x\to 0^+} \dfrac{2x^2}{x^2-\tan x}$

23. $\lim_{x\to\infty} \dfrac{5^x-5}{2^x-2}$

24. $\lim_{x\to-\infty} \dfrac{5^x-5}{2^x-2}$

Solve Exercises 25–28.

25. *Medical technology* The concentration of a drug in a patient's bloodstream t hours after it was injected is given by

$$C(t) = \frac{0.25t}{t^2+5}$$

Find each of the following. **(a)** $C(0.5)$, **(b)** $C(2)$, **(c)** $\lim_{t\to\infty} C(t)$

26. *Business* It has been determined that after t weeks of training, a certain student in a word processing class can keyboard at the rate of $W(t) = 65 + \dfrac{70t^2}{t^2+15}$ words per minute. What are the most words per minute that we can expect this student to keyboard, that is, what is $\lim_{t\to\infty}\left(65 + \dfrac{70t^2}{t^2+15}\right)$?

27. *Business* The cost, in dollars, of an overseas telephone call is given by the function

$$C(t) = \begin{cases} 4.75 & \text{if } 0 < t \le 3 \\ 0.65t + 2.80 & \text{if } t > 3 \end{cases}$$

where t is the length of the call in minutes. Determine each of the following:

(a) $\lim_{t\to 3^-} C(t)$

(b) $\lim_{x\to 3^+} C(t)$

(c) $\lim_{x\to 3} C(t)$

28. *Finance* A certain state income tax schedule can be given by the function

$$T(x) = \begin{cases} 0.04x & \text{if } 0 < x \leq 12{,}500 \\ 0.03x + 125 & \text{if } 12{,}500 < x \leq 35{,}000 \\ 0.02x + 465 & \text{if } x > 35{,}000 \end{cases}$$

where x is the taxable income in dollars, $0 < x$, and $T(x)$ is in dollars. Determine each of the following:

(a) $\lim_{x \to 12{,}500-} T(x)$
(b) $\lim_{x \to 12{,}500+} T(x)$
(c) $\lim_{x \to 12{,}500} T(x)$
(d) $\lim_{x \to 35{,}000-} T(x)$
(e) $\lim_{x \to 35{,}000+} T(x)$
(f) $\lim_{x \to 35{,}000} T(x)$

In Your Words

29. **(a)** Describe the differences among $\lim_{x \to a^-} f(x)$, $\lim_{x \to a^+} f(x)$, and $\lim_{x \to a} f(x)$.
(b) Are all three of the limits in (a) ever the same? If so, what does that mean?
(c) Can two of the limits in (a) ever be the same (and the third one different)? If so, what does that mean?
(d) Can all three of the limits in (a) be different? If so, what does that mean?

30. Explain the difference between $\lim_{x \to c} f(x) = \infty$ and $\lim_{x \to \infty} f(x) = c$.

3.5 ALGEBRAIC TECHNIQUES FOR FINDING LIMITS

Until now, we have found limits by using graphs or using tables of values. In many cases it would be a lot easier to use our knowledge of algebra to determine a limit. The following rules of limits will help us find a limit with much less difficulty.

In Section 3.3, we had the following two rules for limits:

Rules 1 and 2 for Limits

Rule 1: $\lim_{x \to c} A = A$, where A and c are real numbers

Rule 2: $\lim_{x \to c} x = c$, where c is a real number

For all of the following rules, we will assume that f and g are two functions and that $\lim_{x \to c} f(x) = L$ and $\lim_{x \to c} g(x) = M$, where L and M are both real numbers.

Rule 3: Limit of a Sum or Difference $\lim_{x \to c}[f(x) \pm g(x)] = \lim_{x \to c} f(x) \pm \lim_{x \to c} g(x)$
$= L \pm M$

Thus, in order to find the limit of the sum (or difference) of two functions, you can find the sum (or difference) of their limits.

EXAMPLE 3.19

Evaluate $\lim_{x\to 9}(x+6)$.

Solution

$$\begin{aligned}\lim_{x\to 9}(x+6) &= \lim_{x\to 9} x + \lim_{x\to 9} 6 && \text{(rule 3)}\\ &= 9+6 && \text{(rules 1 and 2)}\\ &= 15\end{aligned}$$

EXAMPLE 3.20

Evaluate $\lim_{x\to 2}(3-x)$.

Solution

$$\begin{aligned}\lim_{x\to 2}(3-x) &= \lim_{x\to 2} 3 - \lim_{x\to 2} x && \text{(rule 3)}\\ &= 3-2 && \text{(rules 1 and 2)}\\ &= 1\end{aligned}$$

Rule 4: Limit of a Product

$$\begin{aligned}\lim_{x\to c}[f(x)g(x)] &= \left[\lim_{x\to c} f(x)\right]\left[\lim_{x\to c} g(x)\right]\\ &= LM\end{aligned}$$

This rule states that the limit of the product of two functions is the product of their limits.

EXAMPLE 3.21

Evaluate $\lim_{x\to -3} 7x$.

Solution

$$\begin{aligned}\lim_{x\to -3} 7x &= \left(\lim_{x\to -3} 7\right)\left(\lim_{x\to -3} x\right) && \text{(rule 4)}\\ &= (7)(-3) && \text{(rules 1 and 2)}\\ &= -21\end{aligned}$$

EXAMPLE 3.22

Evaluate $\lim_{x\to 4} 6x^2$.

Solution

$$\lim_{x\to 4} 6x^2 = \left(\lim_{x\to 4} 6x\right)\left(\lim_{x\to 4} x\right) \qquad \text{(rule 4)}$$

EXAMPLE 3.22 (Cont.)

$$= 6\left(\lim_{x \to 4} x\right)\left(\lim_{x \to 4} x\right) \qquad \text{(rules 1 and 4)}$$
$$= 6 \cdot 4 \cdot 4 \qquad \text{(rule 2)}$$
$$= 96$$

Rule 5: Limit of a Quotient

$$\lim_{x \to c} \frac{f(x)}{g(x)} = \frac{\lim_{x \to c} f(x)}{\lim_{x \to c} g(x)} = \frac{L}{M}, \quad \text{if } M \neq 0$$

If the limit of the denominator is not 0, then the limit of the quotient of two functions is the quotient of their limits.

EXAMPLE 3.23

Evaluate $\lim_{x \to 5} \frac{2x+4}{x-1}$.

Solution

$$\lim_{x \to 5} \frac{2x+4}{x-1} = \frac{\lim_{x \to 5}(2x+4)}{\lim_{x \to 5}(x-1)} \qquad \text{(rule 5)}$$
$$= \frac{14}{4} = \frac{7}{2}$$

Rule 6: Limit of $[f(x)]^n$ or $\sqrt[n]{f(x)}$.

If n is a positive integer,

$$\lim_{x \to c} [f(x)]^n = \left[\lim_{x \to c} f(x)\right]^n = L^n$$

and
$$\lim_{x \to c} \sqrt[n]{f(x)} = \sqrt[n]{\lim_{x \to c} f(x)} = \sqrt[n]{L}$$

EXAMPLE 3.24

Evaluate $\lim_{x \to -2} (2x-1)^3$.

Solution

$$\lim_{x \to -2} (2x-1)^3 = \left[\lim_{x \to -2} (2x-1)\right]^3 \qquad \text{(rule 6)}$$
$$= (-5)^3 = -125$$

EXAMPLE 3.25

Evaluate $\lim_{x\to 3} \sqrt[6]{(3x-1)^2}$.

Solution

$$\begin{aligned}\lim_{x\to 3} \sqrt[6]{(3x-1)^2} &= \sqrt[6]{\lim_{x\to 3}(3x-1)^2} && \text{(rule 6)}\\ &= \sqrt[6]{[\lim_{x\to 3}(3x-1)]^2} && \text{(rule 6)}\\ &= \sqrt[6]{(8)^2} = \sqrt[6]{64} = 2\end{aligned}$$

EXAMPLE 3.26

Evaluate $\lim_{x\to 2} \frac{x^2-5x+6}{x^2-4}$.

Solution Checking the denominator, we see that $\lim_{x\to 2}(x^2-4) = 0$. This means that we cannot use rule 5. But, this does not mean the limit does not exist. If we factor both the numerator and denominator, we see that

$$\frac{x^2-5x+6}{x^2-4} = \frac{(x-2)(x-3)}{(x-2)(x+2)}$$

We are interested in the limits as x approaches 2 and not when x has the value of 2. Thus, the factor $x-2$ is not 0 for these values and we can cancel this common factor. We then get

$$\begin{aligned}\lim_{x\to 2} \frac{x^2-5x+6}{x^2-4} &= \lim_{x\to 2} \frac{(x-2)(x-3)}{(x-2)(x+2)}\\ &= \lim_{x\to 2} \frac{x-3}{x+2}\\ &= \frac{2-3}{2+2} = \frac{-1}{4}\end{aligned}$$

In Example 3.26, substituting 2 for x in the expression $\frac{x^2-5x+6}{x^2-4}$ resulted in $\frac{0}{0}$. Whenever substitution results in $\frac{0}{0}$, we must do more work to determine whether a limit exists.

Until now, we have applied the six rules for limits only for the limit of a function as x approaches a specific point. As the next examples show, we often have to be more careful when we find limits at infinity.

EXAMPLE 3.27

Evaluate $\lim_{x\to\infty} \frac{1}{x^n}, n > 0$.

Solution

$$\lim_{x\to\infty} \frac{1}{x^n} = \left(\lim_{x\to\infty} \frac{1}{x}\right)^n = 0^n = 0$$

Notice from this example that as $x \to \infty$, $\frac{1}{x^2}, \frac{1}{x^3}, \frac{1}{x^4}$, etc., all have limits of 0. Using this with rule 3, we see that

$$\lim_{x\to\infty} \frac{c}{x^n} = \left(\lim_{x\to\infty} c\right)\left(\lim_{x\to\infty} \frac{1}{x^n}\right) = c \cdot 0 = 0$$

Thus, we have the following.

Limits at Infinity

If c is a constant, then

$$\lim_{x\to\infty} \frac{c}{x^n} = 0$$

We will now show some techniques for finding limits at infinity.

EXAMPLE 3.28

Evaluate $\lim_{x\to\infty} \frac{4x^3+2x^2+5}{5x^3+x}$.

Solution We will first divide both the numerator and denominator by the largest power of x in the denominator, in this case x^3. This will make each term into a constant or a term with a variable in the denominator and allow us to use the properties for limits at infinity.

$$\begin{aligned}\lim_{x\to\infty} \frac{4x^3+2x^2+5}{5x^3+x} &= \lim_{x\to\infty} \frac{\frac{4x^3+2x^2+5}{x^3}}{\frac{5x^3+x}{x^3}} \\ &= \lim_{x\to\infty} \frac{\frac{4x^3}{x^3}+\frac{2x^2}{x^3}+\frac{5}{x^3}}{\frac{5x^3}{x^3}+\frac{x}{x^3}} \\ &= \lim_{c\to\infty} \frac{4+\frac{2}{x}+\frac{5}{x^3}}{5+\frac{1}{x^2}}\end{aligned}$$

EXAMPLE 3.28 (Cont.)

Now, according to the properties for limits at infinity, all the terms with an x in the denominator have a limit of 0. So, we now have

$$\lim_{x\to\infty} \frac{4+\frac{2}{x}+\frac{5}{x^3}}{5+\frac{1}{x^2}} = \frac{4+0+0}{5+0} = \frac{4}{5}$$

When you want to find a limit at infinity, divide both the numerator and denominator by the largest power of the variable in the denominator.

EXAMPLE 3.29

Evaluate $\lim\limits_{x\to\infty} \dfrac{7x^5-2x^2+1}{4x^6+5x^3+x-5}$.

Solution As in Example 3.28, we will divide both the numerator and denominator by the largest power of x in the denominator, in this case x^6.

$$\lim_{x\to\infty} \frac{7x^5-2x^2+1}{4x^6+5x^3+x-5} = \lim_{x\to\infty} \frac{\frac{7x^5-2x^2+1}{x^6}}{\frac{4x^6+5x^3+x-5}{x^6}}$$

$$= \lim_{x\to\infty} \frac{\frac{7x^5}{x^6}-\frac{2x^2}{x^6}+\frac{1}{x^6}}{\frac{4x^6}{x^6}+\frac{5x^3}{x^6}+\frac{x}{x^6}-\frac{5}{x^6}}$$

$$= \lim_{x\to\infty} \frac{\frac{7}{x}-\frac{2}{x^4}+\frac{1}{x^6}}{4+\frac{5}{x^3}+\frac{1}{x^5}-\frac{5}{x^6}}$$

$$= \frac{0-0+0}{4+0+0-0} = \frac{0}{4} = 0$$

Thus, $\lim\limits_{x\to\infty} \dfrac{7x^5-2x^2+1}{4x^6+5x^3+x-5} = 0$.

Exercise Set 3.5

In Exercises 1–34, evaluate the given limit using algebraic techniques.

1. $\lim_{x\to 5} 8$
2. $\lim_{x\to -4} x$
3. $\lim_{x\to 8}(x-3)$
4. $\lim_{x\to -7}(x+4)$
5. $\lim_{x\to -6}(x+7)$
6. $\lim_{x\to 6}(9-x)$
7. $\lim_{x\to 2} 3x$
8. $\lim_{t\to -3} 5t$
9. $\lim_{x\to 6} \frac{2}{3}x+5$
10. $\lim_{x\to 4}\left(\frac{3}{2}x-2\right)$
11. $\lim_{p\to 3}\frac{p^2+6}{p}$
12. $\lim_{x\to -2}\frac{x^2-6}{x}$
13. $\lim_{s\to 5}\frac{3s^2+5}{2s}$
14. $\lim_{x\to 6}\frac{2x^2-3}{3-x}$
15. $\lim_{x\to 2}(x+1)^3$
16. $\lim_{y\to 5}(y-1)^3$
17. $\lim_{x\to -1}(2x^2-1)^2$
18. $\lim_{t\to 4}(3t^2+2)^2$
19. $\lim_{x\to 3}\sqrt{x+3}$
20. $\lim_{x\to -6}\sqrt{6-7x}$
21. $\lim_{x\to 2}\frac{x^2-2x}{x-2}$
22. $\lim_{w\to 3}\frac{w^2-1}{w+1}$
23. $\lim_{x\to -1}\frac{x^2-1}{x+1}$
24. $\lim_{y\to 3}\frac{y^2-7y+12}{y-3}$
25. $\lim_{x\to -4}\frac{x^2+2x-8}{x^2+5x+4}$
26. $\lim_{x\to \frac{1}{2}}\frac{2x^2+5x-3}{4x^2-2x}$
27. $\lim_{t\to -7}\frac{2t^2+15t+7}{t^2+5t-14}$
28. $\lim_{x\to 0}\frac{(x+3)^2-9}{x}$
29. $\lim_{h\to 0}\frac{(4+h)^2-4^2}{h}$
30. $\lim_{x\to c}\frac{x^3-c^3}{x-c}$
31. $\lim_{x\to\infty}\frac{2x^2+4}{5x^2+3x+2}$
32. $\lim_{x\to\infty}\frac{6x^3+9x+1}{2x^3+4x+8}$
33. $\lim_{x\to\infty}\frac{7x^4+5x^2}{x^5+2}$
34. $\lim_{x\to\infty}\frac{4x^5+3x}{2x^4+1}$

In Your Words

35. Suppose that $\lim_{x\to 1} f(x)=0$ and $\lim_{x\to 1} g(x)=\infty$.
 (a) Give an example of two functions f and g so that $\lim_{x\to 1}[f(x)+g(x)]=\infty$.
 (b) Give an example of two functions f and g so that $\lim_{x\to 1}[f(x)\cdot g(x)]=0$.
 (c) Give an example of two functions f and g so that $\lim_{x\to 1}[f(x)\cdot g(x)]=5$.
36. Suppose that neither $\lim_{x\to 2} f(x)$ and $\lim_{x\to 2} g(x)$ exists. Give an example of two functions f and g so that $\lim_{x\to 2}[f(x)+g(x)]$ does exist or explain why it is not possible.

3.6 CONTINUITY

In Sections 3.4 and 3.5, we have talked about two of the important ideas leading up to calculus. In this section, we will look at the last new idea we need before we begin calculus—**continuity at a point**.

Let's consider the two functions

$$f(x)=x$$

and

$$g(x)=\begin{cases} x & \text{if } x\neq 2 \\ 3 & \text{if } x=2 \end{cases}$$

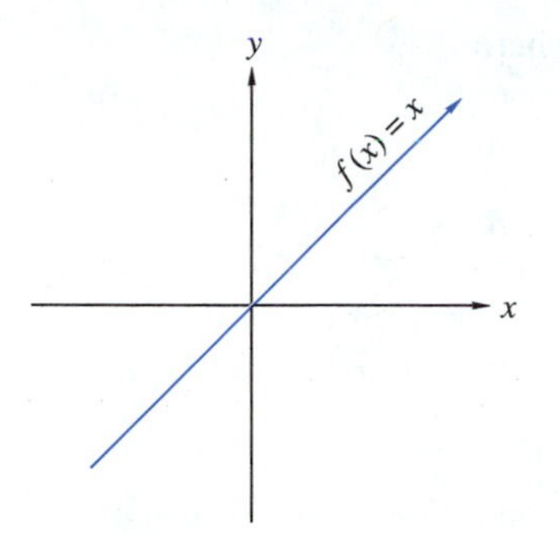

FIGURE 3.27

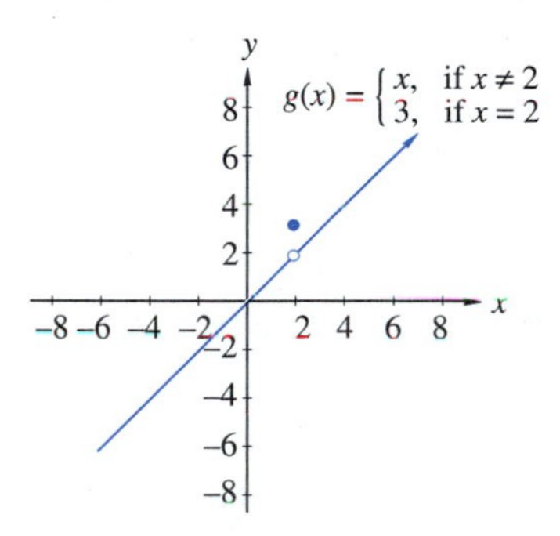

FIGURE 3.28

The graphs of these functions appear in Figures 3.27 and 3.28, respectively. Looking at either the graphs or the algebraic definitions, you can see that the two functions are exactly the same, except at one point. At that point, when $x = 2$, there is a "break" in the graph of g, but no "break" in the graph of f. The function g is an example of a function that is not continuous, while f demonstrates a continuous function.

The break in g is used to determine that the function is not continuous. One way of thinking of a continuous function is that it is a function you can draw without lifting your pencil from the paper. Look again at the graphs of f and g. You could draw f without lifting your pencil. You could not draw g without lifting your pencil.

Continuity at a Point

As we saw above, we can describe a function as either continuous or noncontinuous by looking at how it is drawn. We can use the idea of limits to provide an algebraic definition of continuity. Look again at the definitions of f and g. What is $\lim_{x \to 2} f(x)$? What are $f(2)$ and $g(2)$?

As you can see, $\lim_{x \to 2} f(x) = f(2) = 2$. We know that f is a continuous function at $x = 2$. But $\lim_{x \to 2} g(x) = 2 \neq g(2) = 3$, thus g is not a continuous function, or is **discontinuous** at $x = 2$. This provides us with a more formal definition:

Continuity at a Point

A function f is **continuous** at $x = c$, if and only if the following three conditions are satisfied:

(a) $f(x)$ is defined at $x = c$,

(b) $\lim_{x \to c} f(x)$ exists, and

(c) $\lim_{x \to c} f(x) = f(c)$.

A function that is not continuous at $x = c$ is **discontinuous** at that point.

Note

A function is discontinuous at $x = c$ if any one of these three conditions are not satisfied.

EXAMPLE 3.30

Discuss the continuity of the function f at $x = 4$, where

$$f(x) = \begin{cases} \dfrac{x^2 - 16}{x - 4} & \text{if } x \neq 4 \\ 8 & \text{if } x = 4 \end{cases}$$

Solution In order to determine if f is continuous, we need to check the three conditions in the definition: **(a)** The function is defined at $x = 4$, since $f(4) = 8$, **(b)** $\lim_{x \to 4} \dfrac{x^2 - 16}{x - 4} = \lim_{x \to 4} \dfrac{(x+4)(x-4)}{x - 4} = 8$, and **(c)** $\lim_{x \to 4} \dfrac{x^2 - 16}{x - 4} = 8 = f(4)$.
The three conditions are satisfied, so this function is continuous at $x = 4$.

EXAMPLE 3.31

Discuss the continuity of each of the following functions at the indicated points:

(a) $g(x) = \dfrac{1}{x}$ at $x = 0$

(b) $h(x) = \begin{cases} x - 1 & \text{if } x < 2 \\ x + 1 & \text{if } x \geq 2 \end{cases}$ at $x = 2$

(c) $j(x) = \begin{cases} x^2 + 2 & \text{if } x \neq 0 \\ 0 & \text{if } x = 0 \end{cases}$ at $x = 0$

Solutions None of these are continuous at the designated points.

(a) $g(x)$ is not defined when $x = 0$, so it does not satisfy the first condition.

(b) $h(x)$ is defined at $x = 2$, but we can see that $\lim_{x \to 2^-} h(x) = 1$ and $\lim_{x \to 2^+} h(x) = 3$, so the $\lim_{x \to 2} h(x)$ does not exist and the second condition is not satisfied.

(c) $j(x)$ is defined at 0, $j(0) = 0$. $\lim_{x \to 0} j(x) = 2$. The third condition is not satisfied, since $\lim_{x \to 0} j(x) = 2 \neq 0 = j(0)$.

The graphs in these three functions are shown in Figures 3.29a, 3.29b, and 3.29c, respectively.

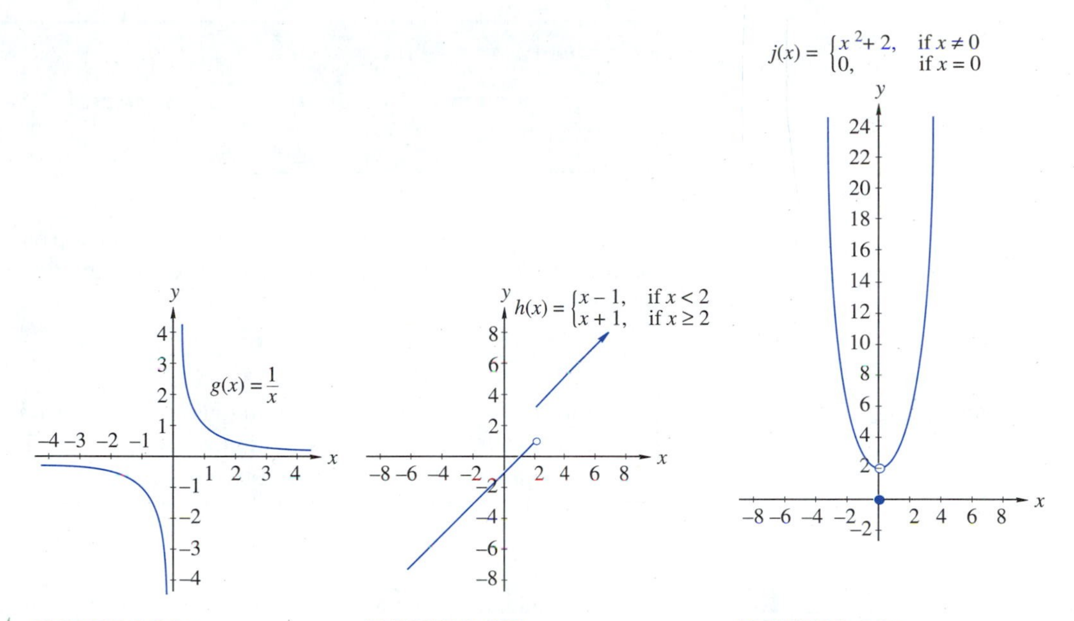

FIGURE 3.29a FIGURE 3.29b FIGURE 3.29c

Continuity on Intervals

When we talk about functions, we usually consider them over an interval. An **interval** is a set of all points on the number line between a and b. Because the endpoints are not always included in the intervals, notation has been developed to make these distinctions. We will use the following notations and terms:

Interval Notation

$(a, b) : a < x < b$

$[a, b) : a \leq x < b$

$(a, b] : a < x \leq b$

$[a, b] : a \leq x \leq b$

The interval (a, b) is called an **open interval** and does not include either endpoint. The interval $[a, b]$ is a **closed interval** and includes both endpoints. The other two intervals, $[a, b)$ and $(a, b]$, are **half-open intervals**. Because $+\infty$ and $-\infty$ are not numbers, any interval that includes either of these is open on that end. For example $a < x < \infty$, or simply $a < x$, is the interval, (a, ∞) and $-\infty < x \leq b$, or simply $x \leq b$, is the interval $(-\infty, b]$.

The definition of continuity we just gave is for a function continuous at a point. We are usually more interested in the continuity of a function over an interval.

Continuity Over an Interval

We say that a function is **continuous on an interval** if it is continuous at each point in the interval.

If you are concerned with the continuity of a function on an interval that includes an endpoint of the interval, then you need to examine the one-sided limits at that endpoint.

EXAMPLE 3.32

The function $k(x) = \sqrt{x-2}$ is continuous on the interval $[2, \infty)$.
The function $m(x) = \sqrt{3-x}$ is continuous on the interval $(-\infty, 3]$.
The function $f(x) = \sqrt[3]{x}$ is continuous on $(-\infty, \infty)$.

Note All polynomial functions are continuous at every point or on the interval $(-\infty, \infty)$.

Functions that are continuous at all points, such as a polynomial, are said to be **continuous everywhere**, or continuous, and are referred to as **continuous functions**. Rational functions are continuous over their domains.

EXAMPLE 3.33

Find any points of discontinuity for each of the following functions:

(a) $f(x) = \dfrac{x^3 - 3x^2 + 17x - 5}{x^2 - 1}$

(b) $g(x) = \begin{cases} x^2 & \text{if } x < 2 \\ x^2 + 5 & \text{if } x \geq 2 \end{cases}$

(c) $h(x) = \begin{cases} 6 & \text{if } x < 3 \\ x + 3 & \text{if } x > 3 \end{cases}$

Solutions

(a) Possible points of discontinuity occur when the denominator is 0. The denominator is 0 when x is ± 1, so f is not defined at these two points. By condition (a), if the function is not defined at ± 1, it is not continuous at those two points. These are the only points of discontinuity for f.

(b) Possible points of discontinuity occur when $x = 2$, since this is the only place where a break may occur. When $x < 2$, $\lim\limits_{x \to 2^-} g(x) = \lim\limits_{x \to 2^-} x^2 = 4$ and when $x > 2$, $\lim\limits_{x \to 2^+} g(x) = \lim\limits_{x \to 2^+} (x^2 + 5) = 9$. The limit does not exist at $x = 2$, so this is a point of discontinuity.

(c) The function h is not defined at $x = 3$, so the function is not continuous at this point. It is continuous for all other values of x.

Properties of Continuous Functions

The formal definition that we gave for a continuous function was based on a limit. The rules that apply to limits help us develop three properties that apply to continuous functions.

> **Properties of Continuous Functions**
>
> Property 1: If f and g are both continuous at a point c, then so are $f+g$, $f-g$, and $f \cdot g$.
>
> Property 2: If f and g are both continuous at c and $g(c) \neq 0$, then $f/g = \dfrac{f}{g}$ is also continuous at c.
>
> Property 3: If f is continuous at $g(c)$ and g is continuous at c, then $f \circ g = f(g(x))$ is continuous at c.

We used property 2 in the solution of Example 3.33a. There, we let f be the polynomial $x^3-3x^2+17x-5$ and g the polynomial x^2-1. Since f and g are both polynomials they are continuous everywhere, so f/g is continuous where $g \neq 0$, namely when $x \neq \pm 1$.

This concludes this section on continuity and completes your foundation for calculus. Limits and continuity will both provide necessary tools as we continue to develop our knowledge of calculus.

Exercise Set 3.6

In Exercises 1–10, determine where the function is continuous at c. If the function is not continuous, give a reason.

1. $f(x) = \begin{cases} 3x+1 & \text{if } x \leq 4 \\ x^2-3 & \text{if } x > 4 \end{cases} \qquad c = 4$

2. $g(x) = \begin{cases} 2x-1 & \text{if } x \leq 1 \\ 2x & \text{if } x > 1 \end{cases} \qquad c = 1$

3. $h(x) = \begin{cases} 3x+1 & \text{if } x < 1 \\ 3 & \text{if } x = 1 \\ 4x & \text{if } x > 1 \end{cases} \qquad c = 1$

4. $j(x) = \begin{cases} 5x-1 & \text{if } x < 1 \\ 4 & \text{if } x = 1 \\ 4x & \text{if } x > 1 \end{cases} \qquad c = 1$

5. $k(x) = \begin{cases} x^2+2 & \text{if } x > 2 \\ 3x & \text{if } x < 2 \end{cases} \qquad c = 2$

6. $f(x) = \begin{cases} x^3+1 & \text{if } x > 2 \\ 0 & \text{if } x = 2 \\ x^2+5 & \text{if } x < 2 \end{cases} \qquad c = 2$

7. $g(x) = \begin{cases} x^2 & \text{if } x < -2 \\ 3 & \text{if } x = -2 \\ -4x+2 & \text{if } x > -2 \end{cases} \qquad c = -2$

8. $h(x) = \begin{cases} x^2 & \text{if } x \leq -2 \\ 2x & \text{if } x > -2 \end{cases} \qquad c = -2$

9. $j(x) = \begin{cases} x^3-4 & \text{if } x < 2 \\ 5 & \text{if } x = 2 \\ x^2 & \text{if } x > 2 \end{cases} \qquad c = 2$

10. $k(x) = \dfrac{x^3 - 1}{x - 1}$ $\quad c = 1$

In Exercises 11–20, find all points of discontinuity.

11. $f(x) = 2x^2 - 5$
12. $g(x) = \dfrac{4}{x - 5}$
13. $h(x) = \dfrac{x^2 + 3x + 2}{x + 1}$
14. $j(x) = 7t - 4$
15. $k(x) = \dfrac{x + 1}{x^2 + x - 6}$
16. $f(x) = \dfrac{x^2 + x - 6}{x^2 + 5x + 6}$
17. $g(x) = \dfrac{x}{x^2 + 1}$
18. $h(x) = \dfrac{x}{x}$
19. $j(x) = \begin{cases} x^2 + 2 & \text{if } x < 2 \\ \sqrt{x - 2} & \text{if } x \geq 2 \end{cases}$
20. $k(x) = \begin{cases} \dfrac{1}{x} & \text{if } x \neq 5 \\ 3 & \text{if } x = 5 \end{cases}$

In Exercises 21–30, determine the intervals over which the given function is continuous.

21. $f(x) = \dfrac{x}{x + 3}$
22. $g(x) = \dfrac{x}{x^2 - 1}$
23. $h(x) = \dfrac{x + 2}{x^2 - 4}$
24. $j(x) = \dfrac{x^2 + 3x - 4}{x + 4}$
25. $k(x) = \sqrt{x + 3}$
26. $f(x) = \sqrt{5 - x}$
27. $g(x) = \sqrt{x^2 - 9}$
28. $h(x) = \sqrt{x^2 + 4x + 4}$
29. $j(x) = \dfrac{1}{x}$
30. $k(x) = \dfrac{1}{\sqrt{x}}$

Solve Exercises 31–34.

31. *Physics* The acceleration due to gravity, g, varies with the height above the surface of the Earth. The acceleration varies in a different way below the Earth's surface. It has been determined that, as a function of r, the distance from the center of the Earth, g is described by

$$g(r) = \begin{cases} \dfrac{GMr}{R^3} & \text{for } r < R \\ \dfrac{GM}{r^2} & \text{for } r \geq R \end{cases}$$

where $R \approx 6.371 \times 10^3$ km is the radius of the Earth, $M \approx 5.975 \times 10^{24}]$ kg is the mass of the Earth, and $G \approx 6.672 \times 10^{-11}$ N·m²/kg² is the gravitational constant.

(a) Is g a continuous function of r? Explain your answer.

(b) Sketch a graph of g near $r = R$.

32. *Business* The cost, in dollars, of an overseas telephone call is given by the function

$$C(t) = \begin{cases} 8.50 & \text{if } 0 < t \leq 3 \\ 0.85t + 5.95 & \text{if } t > 3 \end{cases}$$

where t is the length of the call in minutes. Is C a continuous function at $t = 3$?

33. *Finance* A recent federal income tax schedule can be given by the function

$$T(x) = \begin{cases} 0.15x & \text{if } 0 < x \leq 23{,}900 \\ 0.28x - 3{,}107 & \text{if } 23{,}900 < x \leq 61{,}650 \\ 0.33x - 6{,}189.50 & \text{if } 61{,}650 < x \leq 123{,}790 \end{cases}$$

where x is the taxable income in dollars, $0 < x \leq 123{,}790$, and $T(x)$ is in dollars. Determine each of the following:

(a) $\lim_{x \to 23{,}900-} T(x)$

(b) $\lim_{x \to 23{,}900+} T(x)$

(c) $\lim_{x \to 23{,}900} T(x)$

(d) Is T continuous at $x = 23{,}900$?

(e) $\lim_{x \to 61{,}650-} T(x)$

(f) $\lim_{x \to 61{,}650+} T(x)$

(g) $\lim_{x \to 61{,}650} T(x)$

(h) Is T continuous at $x = 61{,}650$?

34. *Business* The local gas company uses the following function for computing their customers' monthly gas bills:

$$C(t) = \begin{cases} 0.47x + 2.95 & \text{if } 0 < x \leq 24 \\ 0.89x - 7.13 & \text{if } x > 24 \end{cases}$$

where x is the number of thermal units (therms) used by the customer and $C(x)$ is the cost in dollars. Determine each of the following:

(a) $\lim_{x \to 24-} C(x)$

(b) $\lim_{x \to 24+} C(x)$

(c) $\lim_{x \to 24} C(x)$

(d) Is C continuous at $x = 24$?

In Your Words

35. The definition of continuity at a point has three conditions that must be satisfied if the function is to be continuous at a specific point.

(a) List each of these conditions.

(b) For each condition, give an example of a function which is not continuous because it does not satisfy that condition but does satisfy any previous conditions.

36. Suppose that f and g are functions and that neither is continuous as $x = 5$. Explain how $f + g$ can be continuous at $x = 5$.

CHAPTER 3 REVIEW

Important Terms and Concepts

Area under a graph
Average slope
Continuity
  At a point
  Over an interval
Continuous function
Interval notation
Limits
  At infinity
  Left-hand
  One-sided
  Right-hand
Properties of continuous functions
Rules for limits

Review Exercises

Find the average slope over the indicated intervals for the given functions in Exercises 1–6.

1. $f(x) = x^2 - 7$ from $x = 0$ to $x = 2$
2. $g(x) = 4x + 5$ from $x = -3$ to $x = -1$
3. $h(x) = 2x^2 + 1$ from $x = -3$ to $x = -2$
4. $j(x) = \dfrac{8}{x+4}$ from $x = 0$ to $x = \frac{1}{2}$
5. $k(x) = x^2 + 2x$ from $x = -1$ to $x = b$
6. $m(x) = x^2 - 5x$ from $x = 6$ to $x = x_1$

In Exercises 7–10, find the areas under the graph of the given function over the indicated interval. The value of *n* indicates the number of segments into which each interval should be divided.

7. $f(x) = -(4x+7)$ over $[-5, -2], n = 6$
8. $g(x) = 2x^2 + 5$ over $[0, 4], n = 8$
9. $h(x) = 3x^2 + 2x - 1$ over $[2, 4], n = 8$
10. $j(x) = 5 - 2x^2$ over $[-1, 1], n = 8$

In Exercise 11–20, use either the graphical or algebraic approach to determine the limit of the given function at the indicated point, or state that the limit does not exist.

11. $\lim\limits_{x \to 1} (3x^2 + 2x - 5)$
12. $\lim\limits_{x \to -2} \dfrac{3x^2 - 12}{x + 2}$
13. $\lim\limits_{x \to 0} \dfrac{3x^2 - 12}{x - 2}$
14. $\lim\limits_{x \to -3} \dfrac{x^2 - 9}{3x + 9}$
15. $\lim\limits_{x \to \infty} \dfrac{6x^2 + 3}{2x^2}$
16. $\lim\limits_{x \to \infty} \dfrac{9x^3 + 2x - 1}{3x^2 + x + 1}$
17. $\lim\limits_{x \to 5^+} \sqrt{x - 5}$
18. $\lim\limits_{x \to 4^-} \dfrac{x + 4}{x^2 - 16}$
19. $\lim\limits_{x \to \infty} \dfrac{x + 4}{2x^2 + 1}$
20. $\lim\limits_{x \to 1} f(x)$

    if $f(x) = \begin{cases} x^2 - 1 & \text{if } x < 1 \\ 1 - x & \text{if } x > 1 \end{cases}$

In Exercises 21–24, determine the points of discontinuity for the given function.

21. $f(x) = \dfrac{x}{x - 5}$
22. $g(x) = \dfrac{3}{x^3}$
23. $h(x) = \begin{cases} x + 5 & \text{if } x < -3 \\ 5 - x & \text{if } x \geq -3 \end{cases}$
24. $g(x) = \begin{cases} \dfrac{x}{x+1} & \text{if } x < 1 \\ \dfrac{1}{3 - x} & \text{if } x \geq 1 \end{cases}$

In Exercises 25–28, determine the intervals where the given function is continuous.

25. $f(x) = \dfrac{5}{x + 7}$
26. $g(x) = \dfrac{x^2 - 4}{x^2 + 5x + 6}$
27. $h(x) = \dfrac{x + 5}{\sqrt{x^2 - 25}}$
28. $j(x) = \sqrt{\dfrac{3 - x}{3 + x}}$

CHAPTER 3 TEST

Solve Exercises 1 and 2.

1. Find the average slope of $f(x) = 2x^3 - 3$ from $x = 0$ to $x = 1$.
2. Find the area under the graph of $g(x) = 3x^2 - 2x$ over the interval $[1, 5]$, when it is divided into $n = 4$ subintervals.

In Exercises 3–6, determine the limit of the given function or state that the limit does not exist.

3. $\lim_{x \to 2} (5x^2 - 7)$

4. $\lim_{x \to 2} \left(\dfrac{x^2 - 4}{3x - 6} \right)$

5. $\lim_{x \to 1+} f(x)$ where $f(x) = \begin{cases} x^3 - 1 & \text{if } x < 1 \\ 4 & \text{if } x = 1 \\ 1 + x & \text{if } x > 1 \end{cases}$

6. $\lim_{x \to \infty} \dfrac{5x^3 - 4x + 1}{7x^3 - 2}$

Solve Exercises 7 and 8.

7. Determine the points of discontinuity for $h(x) = \begin{cases} x + 7 & \text{if } x < 2 \\ \dfrac{1}{x^2 - 9} & \text{if } x \geq 2. \end{cases}$

8. Determine the intervals where $j(x) = \dfrac{x^2 - 9}{x^2 - x - 12}$ is continuous.

CHAPTER

4

The Derivative

When working with lenses, optical technicians need to be able to determine the line perpendicular to the surface of the lens at the point where the light enters the lens. Section 4.5 explains how to determine the equations of these normal lines.

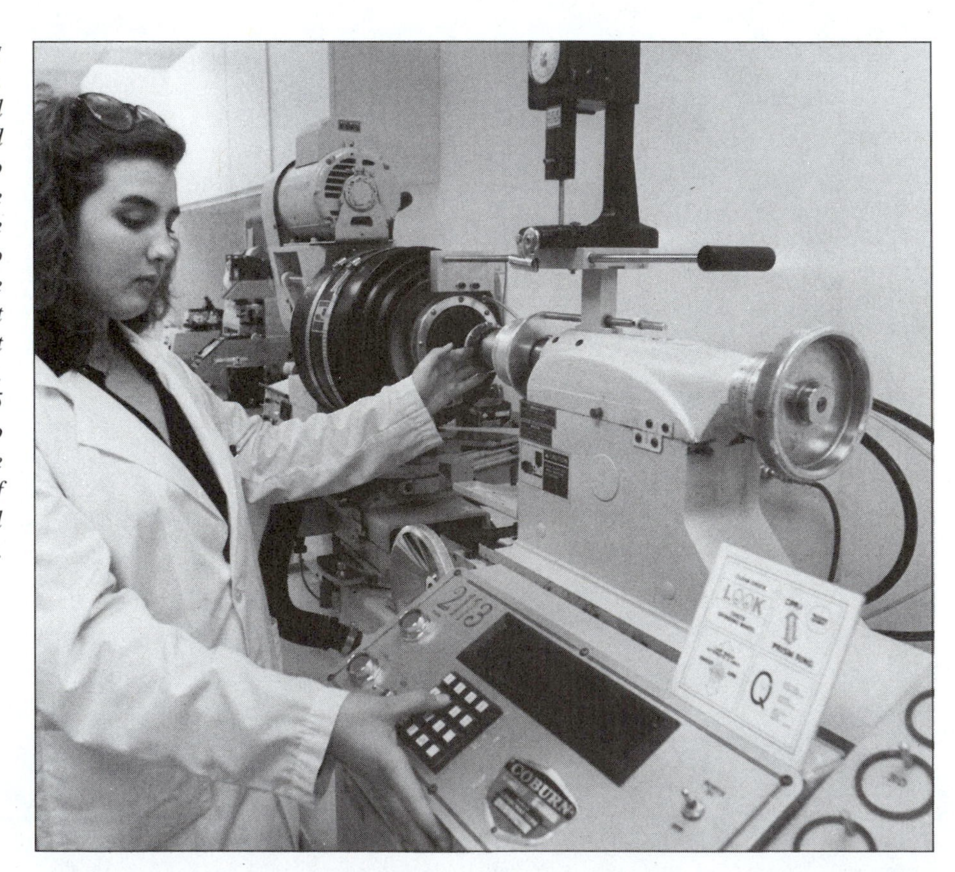

In Chapter 3, we looked at one of the questions that led to the development of calculus—the tangent question. The answer to that question depends on the use of limits. As you pursue the answer to the tangent question, you will develop an understanding of a part of calculus called the derivative. In this chapter, you will not only develop an understanding of a derivative but you will develop techniques for finding derivatives of functions.

≡ 4.1 THE TANGENT QUESTION AND THE DERIVATIVE

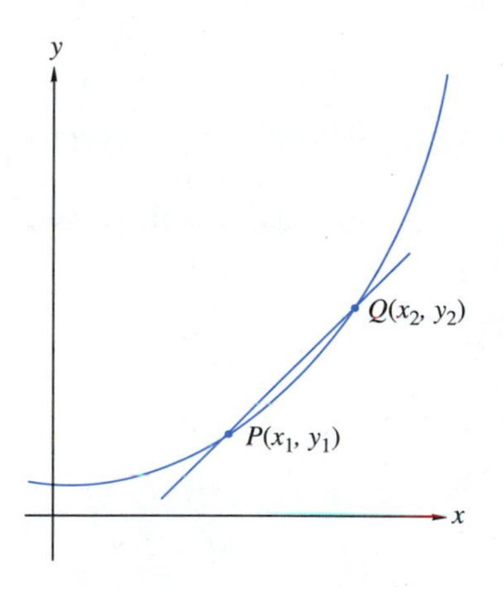

FIGURE 4.1

We will begin with the graph of the function we examined in Section 3.1. The graph is shown again in Figure 4.1. The curve has two points, P and Q. P has the coordinates (x_1, y_1) and Q the coordinates (x_2, y_2). The slope of the secant line through P and Q is

$$m_{PQ} = \frac{y_2 - y_1}{x_2 - x_1}$$

But, we are not really interested in the slope of a secant line. What we want to determine is the slope of the line that is tangent to this curve at the point P. To do this we are going to look at the various secant lines that we get by leaving P fixed. Figure 4.2 contains the same curve as in Figure 4.1, again with the points P and Q. A series of points Q_1, Q_2, Q_3, and Q_4 have been selected with each of these points lying between P and Q, each closer to P than the previous point. As each point gets closer to P, the slope of the secant line through that point and P approaches the slope of the tangent to the curve at P.

A similar thing happens if points Q^{I}, Q^{II}, Q^{III}, and Q^{IV} are on the curve but on the opposite side of P, as shown in Figure 4.3. As each point gets closer to P, the slope of the secant line through that point and P approaches the slope of the tangent line through P. Both of these examples lead us to the idea that the slope of a curve at a point P is the slope of the tangent to the curve at P.

It looks as if both of these examples involve the idea of a limit. Let's look at a more general case for the slope of these secant lines. Again, we will examine a function $y = f(x)$ as shown by the curve in Figure 4.4. We want to find an expression for the slope of the curve, or the slope of the tangent to the curve, at the point P.

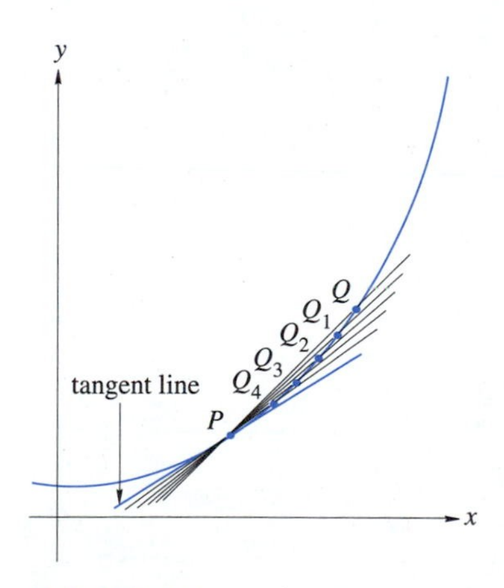

FIGURE 4.2

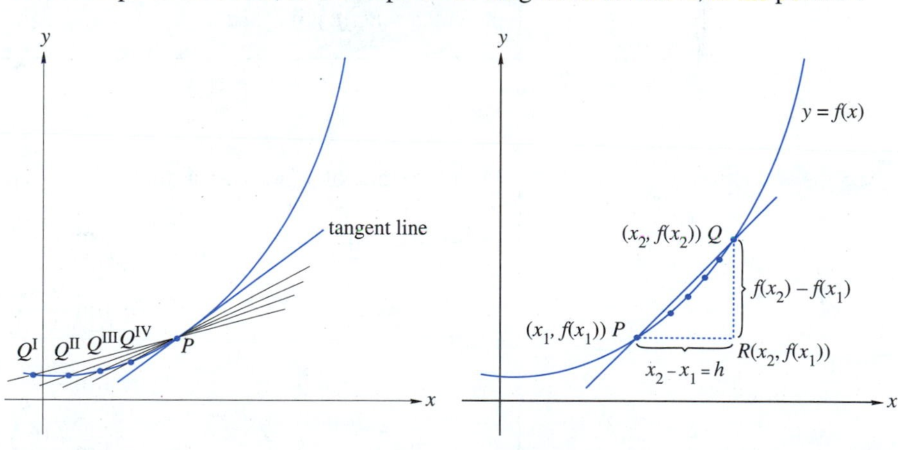

FIGURE 4.3

FIGURE 4.4

The point P has the coordinates $(x_1, y_1) = (x_1, f(x_1))$. We will let Q be any other point on the curve and let its coordinates be $(x_2, f(x_2))$. The slope of the secant line through P and Q is

$$m_{PQ} = \frac{f(x_2) - f(x_1)}{x_2 - x_1}$$

Now draw a line through Q parallel to the y-axis and a line through P parallel to the x-axis. These two lines intersect at the point R, which has the coordinates $(x_2, f(x_1))$. The distance from P to R is $x_2 - x_1$. If we let $h = x_2 - x_1$, then $x_2 = x_1 + h$. This allows us to write the slope of the secant line through P and Q as

$$\begin{aligned} m_{PQ} &= \frac{f(x_1+h) - f(x_1)}{(x_1+h) - x_1} \\ &= \frac{f(x_1+h) - f(x_1)}{h} \end{aligned}$$

We know that h is not 0 because if $x_1 = x_2$, either we would not have two distinct points or we would not have a function.

As Q is picked closer to P, x_2 approaches x_1. This means that h is getting closer to 0. The slope of the tangent line will be the limit

$$\lim_{h \to 0} \frac{f(x_1+h) - f(x_1)}{h}$$

This formula gives us the slope of the curve at the point $(x_1, f(x_1))$.

Slope of a Curve

The slope of a curve $y = f(x)$ at the point $(x_1, f(x_1))$ is

$$\lim_{h \to 0} \frac{f(x_1+h) - f(x_1)}{h}$$

provided that this limit exists.

EXAMPLE 4.1

Find the slope of $f(x) = x^2$ at the point $(3, 9)$.

Solution

$$\begin{aligned} \lim_{h \to 0} \frac{f(3+h) - f(3)}{h} &= \lim_{h \to 0} \frac{(3+h)^2 - 3^2}{h} \\ &= \lim_{h \to 0} \frac{9 + 6h + h^2 - 9}{h} \\ &= \lim_{h \to 0} \frac{6h + h^2}{h} \\ &= \lim_{h \to 0} \frac{h(6+h)}{h} \\ &= \lim_{h \to 0} (6+h) = 6 \end{aligned}$$

So, the slope of the tangent line to $y = x^2$ at $(3, 9)$ is 6.

The process that we have been describing will determine the slope of a tangent to a curve or the slope of the curve at a point. This process can be used to describe how any function is changing at a particular point. In other words, we have been using the method for finding the average change in a function and then the limit of that average change as the denominator approaches 0. This gives the instantaneous change of the function at that point. We will see this by taking a different look at Example 3.4.

Application

EXAMPLE 4.2

A tank is filled with water by opening an inlet pipe. The volume V in liters of water in the tank t minutes after the valve is opened is given by $V(t) = 5t^2 + 4t$. What is the instantaneous rate of increase in the volume at the first minute?

Solution We want to find the instantaneous rate of change R in the volume when $t = 1$. We can do this by using the formula for the rate of change $R_{\text{Vol}} = \dfrac{V(t_2) - V(t_1)}{t_2 - t_1}$ and letting $t_1 = 1$ and $t_2 = t_1 + h = 1 + h$, then finding the limit when h approaches 0.

$$\begin{aligned}
R_{\text{Vol}} &= \lim_{h\to 0} \frac{V(1+h)^2 - V(1)}{h} \\
&= \lim_{h\to 0} \frac{[5(1+h)^2 + 4(1+h)] - [5(1)^2 + 4(1)]}{(1+h) - 1} \\
&= \lim_{h\to 0} \frac{[5(1+2h+h^2) + 4(1+h)] - [5+4]}{h} \\
&= \lim_{h\to 0} \frac{(5+10h+5h^2+4+4h) - (9)}{h} \\
&= \lim_{h\to 0} \frac{9+14h+5h^2-9}{h} \\
&= \lim_{h\to 0} \frac{14h+5h^2}{h} \\
&= \lim_{h\to 0}(14+5h) = 14
\end{aligned}$$

When $t = 1$, the volume of the tank is increasing at the rate of 14 L/min. ▪

The Derivative

All of this discussion about instantaneous change of a function leads us to the following definition of the derivative.

Derivative of a Function f at a Number c

If $y = f(x)$ defines a function f and if the point c is in the domain of f, then the **derivative of f at c**, written as $f'(c)$, is defined as

$$f'(c) = \lim_{h\to 0} \frac{f(c+h) - f(c)}{h}$$

provided that this limit exists.

We used the notation $f'(c)$ to indicate the derivative of the function at the point c. In general, the derivative of the function is the following.

Derivative of a Function

$$f'(x) = \lim_{h \to 0} \frac{f(x+h) - f(x)}{h}$$

Note

The notation $f'(x)$ is not a unique notation for the derivative. Other notations include $D_x y$, $\frac{dy}{dx}$, $\frac{d}{dx}[f(x)]$, y', $Df(x)$, $\frac{df}{dx}$, and $D_x[f(x)]$. There are times when a meaning is clearer if one of these notations is used instead of $f'(x)$.

Caution

$\frac{dy}{dx}$, $\frac{df}{dx}$, and $\frac{d}{dx}$ are not fractions but are symbols for a derivative.

Four-Step Method

We can calculate the derivative of a function f at the point c by using the following four steps.

Four Steps for Calculating $f'(c)$

Step 1: Find $f(c+h)$.

Step 2: Subtract $f(c)$ from $f(c+h)$.

Step 3: Divide the result in Step 2 by h to get the difference quotient.

Step 4: Find the limit, if it exists, of the difference quotient in Step 3, as h approaches 0.

EXAMPLE 4.3

Find the derivative of $f(x) = x^3$ at $x = 3$ by using the four-step method.

Solution

Step 1: $f(3+h) = (3+h)^3 = 27 + 27h + 9h^2 + h^3$

Step 2: Since $f(3) = 3^3 = 27$, $f(3+h) - f(3) = [27 + 27h + 9h^2 + h^3] - 27$

$$= 27h + 9h^2 + h^3$$

Step 3: $\frac{f(3+h) - f(3)}{h} = \frac{27h + 9h^2 + h^3}{h} = 27 + 9h + h^2$

Step 4: $\lim_{h \to 0}(27 + 9h + h^2) = 27$.

So, the derivative of $f(x) = x^3$ at $x = 3$ is $f'(3) = 27$.

We can also use the four-step method to find the derivative of a function at any point x, rather than at a particular point.

EXAMPLE 4.4

Find the derivative of $g(x) = 3x^2 + x$ using the four-step method.

Solution

Step 1: $$\begin{aligned} g(x+h) &= 3(x+h)^2 + (x+h) \\ &= 3(x^2 + 2xh + h^2) + (x+h) \\ &= 3x^2 + 6xh + 3h^2 + x + h \end{aligned}$$

Step 2: $$\begin{aligned} g(x+h) - g(x) &= (3x^2 + 6xh + 3h^2 + x + h) - (3x^2 + x) \\ &= 6xh + 3h^2 + h \end{aligned}$$

Step 3: $$\frac{g(x+h) - g(x)}{h} = \frac{6xh + 3h^2 + h}{h} = 6x + 3h + 1\text{u}$$

Step 4: $$\lim_{h \to 0} \frac{g(x+h) - g(x)}{h} = \lim_{h \to 0}(6x + 3h + 1) = 6x + 1$$

So, the derivative of $g(x) = 3x^2 + x$ is $g'(x) = 6x + 1$.

Once you have the derivative of a function, you can evaluate it at any particular point.

EXAMPLE 4.5

If $g(x) = 3x^2 + x$, what are **(a)** $g'(2)$, **(b)** $g'(0)$, and **(c)** $g'(-3)$?

Solutions In Example 4.4, we learned that the derivative of g is $g'(x) = 6x + 1$. As a result, we have

(a) $g'(2) = 6(2) + 1 = 13$

(b) $g'(0) = 6(0) + 1 = 1$

(c) $g'(-3) = 6(-3) + 1 = -17$

EXAMPLE 4.6

Use the four-step method to find the derivative of $y = \dfrac{1}{x+2}$.

Solution

Step 1: $y = f(x)$, so $f(x+h) = \dfrac{1}{x+h+2}$

Step 2: $f(x+h) - f(x) = \dfrac{1}{x+h+2} - \dfrac{1}{x+2}$.

In order to conduct the subtraction, we need to first find a common denominator and then subtract. The common denominator is $(x+2)(x+h+2)$.

$$\frac{1}{x+h+2} - \frac{1}{x+2} = \frac{1}{x+h+2}\left(\frac{x+2}{x+2}\right) - \frac{1}{x+2}\left(\frac{x+h+2}{x+h+2}\right)$$

EXAMPLE 4.6 (Cont.)

$$= \frac{x+2}{(x+h+2)(x+2)} - \frac{x+h+2}{(x+h+2)(x+2)}$$
$$= \frac{(x+2)-(x+h+2)}{(x+h+2)(x+2)}$$
$$= \frac{-h}{(x+h+2)(x+2)}$$

Step 3: $$\frac{f(x+h)-f(x)}{h} = \frac{\frac{-h}{(x+h+2)(x+2)}}{h}$$
$$= \frac{-h}{h(x+h+2)(x+2)}$$
$$= \frac{-1}{(x+h+2)(x+2)}$$

Step 4: $$\lim_{h\to 0} \frac{-1}{(x+h+2)(x+2)} = \frac{-1}{(x+2)^2}$$

So, $y' = \dfrac{-1}{(x+2)^2}$.

EXAMPLE 4.7

Use the four-step method to find the derivative of $f(x) = \sqrt{x-3}$ at $x = 4$.

Solution We will first find the general expression for $f'(x)$ and then determine $f'(4)$.

Step 1: $f(x+h) = \sqrt{x+h-3}$

Step 2: $f(x+h) - f(x) = \sqrt{x+h-3} - \sqrt{x-3}$

Step 3: $$\frac{f(x+h)-f(x)}{h} = \frac{\sqrt{x+h-3}-\sqrt{x-3}}{h}$$

In order to simplify this expression, we will rationalize the numerator by multiplying both numerator and denominator by $\sqrt{x+h-3}+\sqrt{x-3}$.

$$\frac{\sqrt{x+h-3}-\sqrt{x-3}}{h} = \frac{(\sqrt{x+h-3}-\sqrt{x-3})(\sqrt{x+h-3}+\sqrt{x-3})}{h(\sqrt{x+h-3}+\sqrt{x-3})}$$
$$= \frac{(x+h-3)-(x-3)}{h(\sqrt{x+h-3}+\sqrt{x-3})}$$
$$= \frac{h}{h(\sqrt{x+h-3}+\sqrt{x-3})}$$
$$= \frac{1}{\sqrt{x+h-3}+\sqrt{x-3}}$$

EXAMPLE 4.7 (Cont.)

Step 4: $\lim_{h \to 0} \frac{1}{\sqrt{x+h-3}+\sqrt{x-3}} = \frac{1}{\sqrt{x-3}+\sqrt{x-3}}$

$= \frac{1}{2\sqrt{x-3}}$

Now that we know $f'(x) = \frac{1}{2\sqrt{x-3}}$, we can find $f'(4) = \frac{1}{2\sqrt{4-3}} = \frac{1}{2}$.

Thus, the derivative of $f(x) = \sqrt{x-3}$ at $x = 4$ is $\frac{1}{2}$.

Exercise Set 4.1

In Exercises 1–24, use the definition of the derivative (that is, use the four-step method) to differentiate the given function.

1. $f(x) = 3x + 2$
2. $g(x) = 5x - 7$
3. $f(x) = 4 - 3x$
4. $g(t) = 6 - 4t$
5. $y = 3x^2$
6. $y = -5t^2$
7. $y = 2x^2 + 5$
8. $j(x) = 3x^2 - x + 2$
9. $k(x) = 3 - 4x^2$
10. $g(x) = 2x - 7x^2$
11. $y = -4x^3$
12. $y = x^2 - 2x^3$
13. $y = t - 3t^3$
14. $f(x) = \frac{2x+1}{3}$
15. $f(x) = \frac{1}{x+1}$
16. $g(x) = \frac{5}{2x-3}$
17. $j(x) = \frac{4}{1-x^2}$
18. $f(x) = \frac{9}{x^2+2x}$
19. $s = 16t^2 - 6t + 3$
20. $q = t^3 - 4t^2 + 5t - 2$
21. $y = \sqrt{x^2+4}$
22. $y = \sqrt{5-x}$
23. $y = \sqrt{2x-x^2}$
24. $y = \sqrt{5x-4x^3}$

In Exercises 25–36, evaluate the derivative at each of the indicated points.

25. $f(x) = 3x + 7$; $f'(2)$, $f'(9)$
26. $g(x) = x^2 - 2$; $g'(2)$, $g'(9)$
27. $s(t) = 16t^2 + 2t$; $s'(3)$, $s'(2)$, $s'(0)$
28. $s(t) = 2t^2 - 1$; $s'(1)$, $s'(5)$
29. $q(t) = t^3 - 4t^2$; $q'(0)$, $q'(3)$
30. $j(x) = \frac{1}{\sqrt{x}}$; $j'(4)$, $j'(9)$
31. $k(x) = \frac{1}{x^2-4}$; $k'(5)$, $k'(-5)$
32. $f(x) = \frac{4}{x^2+1}$; $f'(0)$, $f'(3)$, $f'(-5)$
33. $j(t) = \sqrt{5t-t^2}$; $j'(1)$, $j'(4)$
34. $g(x) = \frac{1}{\sqrt{x+4}}$; $g'(0)$, $g'(5)$, $g'(-2)$
35. $f(x) = \frac{5}{\sqrt{25-x^2}}$; $f'(0)$, $f'(3)$, $f'(-4)$
36. $h(x) = \sqrt{x^2+4x+9}$; $h'(-4)$, $h'(0)$, $h'(4)$

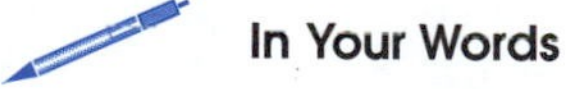

In Your Words

37. Without looking in the book, write the four steps used to calculate $f'(c)$.
38. Explain the relationship between the slope of a curve at some point x_0 and the derivative of the function that graphs that curve at x_0.

4.2 DERIVATIVES OF POLYNOMIALS

At this point, the definition by the four-step method is our only way to take derivatives. The four-step method is a very important process and a method that we will use again and again. But because of its length, it is time consuming and there are many opportunities to make mistakes. In this section, we are going to use the definition to develop some formulas for taking the derivative of any polynomial. In the remainder of the chapter, we will develop some rules for taking more involved derivatives.

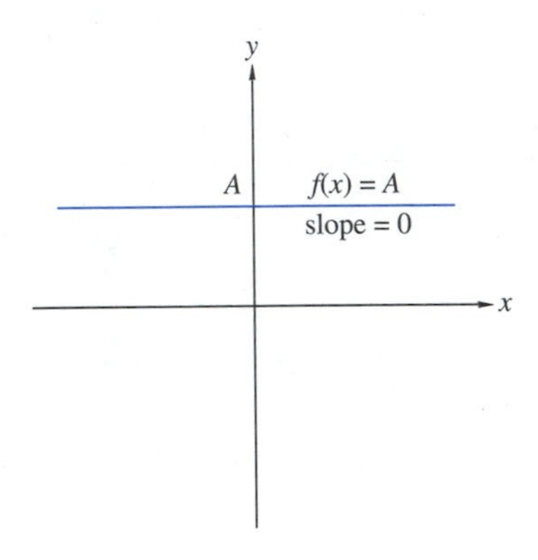

FIGURE 4.5

Constant Function

This first polynomial we will study is the constant function, $f(x) = k$. A look at the graph of this function in Figure 4.5 will show us what to expect. The slope of a constant function is 0, therefore, since the derivative is a measure of the slope of the tangent to the curve, we should expect the derivative to be 0. Let's use the four-step method to confirm our theory.

Step 1: $f(x+h) = k$

Step 2: $f(x+h) - f(x) = k - k = 0$

Step 3: $\dfrac{f(x+h) - f(x)}{h} = \dfrac{0}{h} = 0$

Step 4: $\lim\limits_{h\to 0} \dfrac{f(x+h) - f(x)}{h} = \lim\limits_{h\to 0} 0 = 0$

Thus, we have determined the following derivative of a constant function.

> **Rule 1: Derivative of $f(x) = k$**
>
> For the function $f(x) = k$, where k is a constant, the derivative is $f'(x) = 0$.

Linear Function

The next polynomial is the linear function $f(x) = mx + b$. Notice that we have written this function in the slope-intercept form of the equation. We know that the slope of this line is m. The derivative is supposed to give the slope of a tangent to the curve. Since this curve is a line, and its tangent is the same line, we expect that its derivative is also m. We use the four-step method:

Step 1: $f(x+h) = m(x+h) + b = mx + mh + b$

Step 2: $f(x+h) - f(x) = (mx + mh + b) - (mx + b) = mh$

Step 3: $\dfrac{f(x+h) - f(x)}{h} = \dfrac{mh}{h} = m$

Step 4: $\lim\limits_{h\to 0} \dfrac{f(x+h) - f(x)}{h} = \lim\limits_{h\to 0} m = m$

Again, we confirm our expectation and have shown the following for the derivative of a straight line.

Rule 2: Derivative of $f(x) = mx + b$

For the linear function $f(x) = mx + b$, the derivative is $f'(x) = m$.

EXAMPLE 4.8

Find the derivative of **(a)** $f(x) = -8$ and **(b)** $g(x) = 9x - 4$.

Solution

(a) $f(x) = -8$ is a constant function, so $f'(x) = 0$.
(b) $g(x) = 9x - 4$ is a linear function, so $g'(x) = 9$.

Polynomials of the Form $f(x) = x^n$

We will now consider higher-degree polynomials. This will be done in three steps. The first step will consider two specific forms of x^n, where n is 2 and 3. The second step will generalize this for any real number n. The third step will consider the sums and differences of these terms.

In order to develop the rule for the derivative of x^n, we will determine derivatives for x^2 and x^3. We begin with $f(x) = x^2$.

Step 1: $f(x+h) = (x+h)^2 = x^2 + 2xh + h^2$

Step 2: $f(x+h) - f(x) = (x^2 + 2xh + h^2) - x^2$
$= 2xh + h^2$

Step 3: $\dfrac{f(x+h) - f(x)}{h} = \dfrac{2xh + h^2}{h} = 2x + h$

Step 4: $\lim\limits_{h \to 0} \dfrac{f(x+h) - f(x)}{h} = \lim\limits_{h \to 0} (2x + h) = 2x$

So, the derivative of $f(x) = x^2$ **is** $f'(x) = 2x$.

Next, we find the derivative of $g(x) = x^3$.

Step 1: $g(x+h) = (x+h)^3 = x^3 + 3x^2h + 3xh^2 + h^3$

Step 2: $g(x+h) - g(x) = (x^3 + 3x^2h + 3xh^2 + h^3) - x^3$
$= 3x^2h + 3xh^2 + h^3$

Step 3: $\dfrac{g(x+h) - g(x)}{h} = \dfrac{3x^2h + 3xh^2 + h^3}{h} = 3x^2 + 3xh + h^2$

Step 4: $\lim\limits_{h \to 0} \dfrac{g(x+h) - g(x)}{h} = \lim\limits_{h \to 0} (3x^2 + 3xh + h^2) = 3x^2$

Thus, the derivative of $g(x) = x^3$ **is** $g'(x) = 3x^2$.

If we continued this process, we could show that the derivative of x^4 is $4x^3$, the derivative of x^5 is $5x^4$, the derivative of x^6 is $6x^5$, and so on. We could also show that the derivative of x^{-7} is $-7x^{-8}$, the derivative of $x^{5/2}$ is $\frac{5}{2}x^{3/2}$, and the derivative of $x^{-7/3}$ is $-\frac{7}{3}x^{-10/3}$.

What we have demonstrated is how to find the derivative of x^n, when n is a rational number. Thus, we now have

Rule 3: Derivative of $f(x) = x^n$

The derivative of any function $f(x) = x^n$, n a rational number, is $f'(x) = nx^{n-1}$.

EXAMPLE 4.9

Find the derivative of **(a)** $f(x) = x^7$, **(b)** $g(x) = x^{-3}$, **(c)** $h(x) = x^{2/3}$, and **(d)** $j(x) = x^{-5/2}$.

Solution

(a) $f(x) = x^7$ and so $f'(x) = 7x^{7-1} = 7x^6$

(b) $g(x) = x^{-3}$ and $g'(x) = -3x^{-3-1} = -3x^{-4}$

(c) $h(x) = x^{2/3}$ and $h'(x) = \frac{2}{3}x^{\frac{2}{3}-1} = \frac{2}{3}x^{-\frac{1}{3}}$

(d) $j(x) = x^{-5/2}$ and $j'(x) = -\frac{5}{2}x^{-\frac{5}{2}-1} = -\frac{5}{2}x^{-7/2}$

The next example demonstrates the usefulness of fractional exponents—both positive and negative.

EXAMPLE 4.10

Find the derivatives of **(a)** $y = x^2\sqrt{x}$ and **(b)** $f(x) = \dfrac{1}{x^{5/2}}$.

Solution

(a) $y = x^2\sqrt{x}$. To find the derivative of y, or $D_x y$, we must first rewrite $x^2\sqrt{x}$ as $x^2x^{1/2} = x^{5/2}$. So, $y = x^{5/2}$ and

$$D_x y = \frac{5}{2}x^{3/2} = \frac{5}{2}x\sqrt{x}$$

(b) $f(x) = \dfrac{1}{x^{5/2}}$. To apply rule 3, first write $f(x)$ as $f(x) = x^{-5/2}$. In Example 4.9(d), we showed that $f'(x) = -\frac{5}{2}x^{-7/2}$.

Be careful. Do not just differentiate the denominator. If you do, you get

$$D_x\left(\frac{1}{x^{5/2}}\right) \neq \frac{1}{\frac{5}{2}x^{3/2}} = \frac{2}{5}x^{-3/2}$$

which is not the correct answer.

Three General Formulas

Suppose we have a function $F(x) = kf(x)$, where k is a constant and $f'(x)$ exists. [This might be a function such as $G(x) = 9x^3$, where $k = 9$ and $g(x) = x^3$.] We will use the four-step method to find the derivative of $F(x)$.

Step 1: $F(x+h) = kf(x+h)$

Step 2: $F(x+h) - F(x) = kf(x+h) - kf(x) = k[f(x+h) - f(x)]$

Step 3:
$$\frac{F(x+h) - F(x)}{h} = \frac{k[f(x+h) - f(x)]}{h} = k\left[\frac{f(x+h) - f(x)}{h}\right]$$

Step 4:
$$\lim_{h\to 0} \frac{F(x+h) - F(h)}{h} = \lim_{h\to 0} k\left[\frac{f(x+h) - f(x)}{h}\right] = k \lim_{h\to 0} \frac{f(x+h) - f(x)}{h} = kf'(x)$$

Rule 4: Derivative of $F(x) = kf(x)$

If $F(x) = kf(x)$, k is a constant and $f'(x)$ exists, then

$$F'(x) = kf'(x)$$

EXAMPLE 4.11

Find the derivative of **(a)** $f(x) = 9x^3$, **(b)** $g(x) = 7x^{-5}$, and **(c)** $h(x) = \frac{2}{3}x^{5/2}$.

Solutions

(a) $f(x) = 9x^3$ so $f'(x) = 9(3x^2) = 27x^2$

(b) $g(x) = 7x^{-5}$ and $g'(x) = 7(-5x^{-6}) = -35x^{-6}$

(c) $h(x) = \frac{2}{3}x^{5/2}$ and $h'(x) = \frac{2}{3}(\frac{5}{2}x^{3/2}) = \frac{5}{3}x^{3/2}$

The next two formulas have to do with the sum and difference of two functions. These last two rules will allow us to find derivatives of higher-degree polynomials.

If f and g are two functions that can be differentiated, then what is the derivative of their sum or difference? It turns out that the derivative of a sum (or difference) is the sum (or difference) of the derivative of each term. These are stated formally in the following box, which contains the last two rules for this section.

Rules 5 and 6: Derivative of $F(x) = f(x) \pm g(x)$

Rule 5: If $F(x) = f(x) + g(x)$ and $f'(x)$ and $g'(x)$ exist, then

$$F'(x) = f'(x) + g'(x)$$

Rule 6: If $F(x) = f(x) - g(x)$ and $f'(x)$ and $g'(x)$ exist, then

$$F'(x) = f'(x) - g'(x)$$

EXAMPLE 4.12

Find the derivative of $f(x) = 7x^5 - 4x^3 + 2x - 8$.

Solution According to the last two rules, the derivative of a sum (or difference) is the sum (or difference) of the individual derivatives. We will differentiate this term by term:

$$\begin{aligned} f'(x) &= 7(5x^4) - 4(3x^2) + 2(1) - 0 \\ &= 35x^4 - 12x^2 + 2 \end{aligned}$$

Exercise Set 4.2

In Exercises 1–37, find the derivative of the function by using the six rules given in this section.

1. $f(x) = 19$
2. $g(x) = -4$
3. $h(x) = 7x - 5$
4. $j(x) = -3x + 7$
5. $k(x) = 9x^2$
6. $m(x) = 15x^3$
7. $f(x) = \frac{1}{3}x^{15}$
8. $g(x) = \frac{2}{5}x^{10}$
9. $h(x) = 5x^{-2}$
10. $j(x) = -3x^{-4}$
11. $k(x) = -\frac{2}{3}x^{3/2}$
12. $j(x) = \dfrac{4\sqrt{x}}{3}$
13. $h(x) = 4\sqrt{3x}$
14. $m(x) = \frac{3}{4}x^{-5/3}$
15. $g(x) = \dfrac{3}{\sqrt[3]{x}}$
16. $f(x) = 4x^3 - 5x^2 + 2$
17. $g(x) = 9x^2 + 3x - 4$
18. $n(x) = 7x^7 - 5x^5$
19. $h(x) = \frac{1}{3}x^3 + \frac{1}{2}x^2 - 5x + \frac{2}{3}x^{-2} + 5$
20. $j(x) = \frac{3}{4}x^4 - \frac{2}{3}x^3 + \frac{5}{4}x^2 - \frac{1}{2}x + 2x^{-3} + x^{-4} + 7$
21. $f(x) = \frac{4}{5}x^5 - \frac{3}{2}x^3 - 7x^0 + \frac{1}{2}x^{-2} - \frac{2}{3}x^{-3}$
22. $k(x) = \sqrt{2}x^5 - \sqrt{3}x^3 + 3x + \pi$
23. $\ell(x) = 3\sqrt{3}x^4 + 3\sqrt{5x} + \sqrt{4x^4}$
24. $m(x) = \sqrt{7}x^6 - \sqrt{4}x^3 + 5\sqrt{3}x^2 - \sqrt{7}$
25. $s(t) = 16t^2 - 32t + 5$
26. $q(t) = 4.3t^3 - 2.7t + 3.0$
27. $\alpha(t) = 30 - 4.0t^2 + 2t^{1/2}$
28. $s(t) = \frac{5}{3}t^4 + 7\sqrt{2}t^3 - 8t^{2/3} + 7t^{-1/2}$
29. $q(t) = \sqrt{6t} - 4t\sqrt{3t} + \sqrt[3]{t^2}$
30. $f(x) = 3x\sqrt{x}$
31. $g(x) = 5x^2\sqrt[3]{x}$
32. $j(x) = \dfrac{3x^3}{\sqrt[3]{x}}$
33. $h(x) = \dfrac{4}{x^2}$

34. $j(x) = 7x^2 + 2x\sqrt{x} + \dfrac{5}{x^3}$

35. $K(x) = \dfrac{4\sqrt[3]{x}}{\sqrt{x}}$

36. $M(x) = \dfrac{3\sqrt{x^3}}{\sqrt{3x}}$

37. $f(x) = \dfrac{4\sqrt{3x} + 3x^2 - 7\sqrt{x^3} + 2x^{-1}}{\sqrt{x}}$

Solve Exercises 38–44.

38. Derive rule 6.

39. Find all values of x where $f'(x) = 0$, if $f(x) = x^3 + 3x^2$.

40. Find all values of t where $\theta'(t) = 0$, if $\theta(t) = 3t^3 - t$.

41. Find all values of x where $g'(x) = 0$, if $g(x) = 2x^3 + x^2 - 4x$.

42. If $\omega(t) = 4.0t^2 - 2.0\sqrt{t}$, find the values of t for which $\omega'(t) = 0$.

43. If $j(t) = 4t^3 + \dfrac{1}{t}$, for what values of t is $j'(t) = 0$?

44. If $h(x) = 4 + \dfrac{1}{x} - \dfrac{8}{x^3}$, for what values of x is $h'(x) = 0$?

Solve Exercises 45–56.

45. *Demography* The population of a certain city t months from now can be estimated by the formula $P(t) = 146{,}000 + 64t - 0.04t^2$. Find

(a) The population of the city six months from now

(b) The rate of change in the city's population six months from now

46. *Business* A manufacturer estimates that the revenue in dollars from the sale of c cellular telephone is $R(c) = 120c - c^2$. Find

(a) The revenue from the sale of 30 telephones

(b) The rate of change in the revenue when $c = 30$

47. *Physics* A ball is rolled down an incline. The distance in feet of the ball from the starting point after t seconds is given by $s(t) = 15t + 7t^2$. Find

(a) The distance the ball has traveled when $t = 4$ s

(b) The velocity (rate of change) of the ball when $t = 4$ s

48. *Petroleum engineering* When an oil tank is drained for cleaning, the volume, V, of oil in gallons left in the tank t minutes after the drain valve is opened is given by the formula $V(t) = 75{,}000 - 3{,}000t + 30t^2$.

(a) How much oil is left in the tank 20 min after the drain is opened?

(b) What is the average rate that the oil drains during the first 20 min?

(c) What is the rate of change that the oil is draining when $t = 20$?

49. *Demography* A certain city's population t years from now can be estimated by using the function $P(t) = 125{,}000 + 1{,}500t - 36\sqrt{t}$. Find

(a) The city's population six years from now

(b) The rate at which the city's population is growing six years from now

(c) The city's population ten years from now

(d) The rate at which the city's population is growing ten years from now

50. *Business* The total cost, C, in dollars, of producing n units of a certain product is given by $C(n) = 5{,}000 + 25n - 0.4n^2$, where $0 \le n \le 60$.

(a) Determine the rate of change in the cost when $n = 10$.

(b) Determine the rate of change in the cost when $n = 30$.

(c) Determine the rate of change in the cost when $n = 50$.

51. *Electronics* The power, P, in watts, delivered to a $25\,\Omega$ resistor is $P = 25i^2$. Find the instantaneous rate of change of P with respect of i when $i = 2.4$ A.

52. *Material science* For certain temperatures, the tensile strength, S, in pounds, of a piece of material as a function of its temperature is described by $S(T) = 520 - 0.000085T^2$. Find the rate of change of S at the instant that $T = 135°$F.

53. *Business* A manufacturer has determined that the revenue in dollars from the sale of n cellular telephones is given by $R(n) = 78n - 0.025n^2$. The cost of producing n telephones is $C(n) = 9{,}800 + 22.5n$.
(a) If the profit, P, is the difference between the revenue and the cost, what is the profit function?
(b) Find the marginal profit function, P'.
(c) Find $P'(50)$.

54. *Electronics* The output voltage, V, in volts, of a circuit increases gradually with time t, in seconds, according to the function $V(t) = 30.0 + 0.04t^2$. Find an expression for the instantaneous rate of change of the voltage with time.

55. *Physics* The altitude, h, of a rocket fired from an airplane flying over the ocean can be described by the function $h(t) = -t^4 + 12t^3 - 16t^2 + 605$, where t is in seconds and the altitude is in feet above sea level. The velocity of the rocket at time t is $v(t) = h'(t)$.
(a) From what altitude was the rocket launched?
(b) At what time does the rocket fall into the ocean?
(c) What is the rocket's altitude and velocity at $t = 2$ s?
(d) What is the rocket's altitude and velocity at $t = 4$ s?
(e) What is the rocket's altitude and velocity at $t = 6$ s?
(f) What is the rocket's altitude and velocity at $t = 10$ s?
(g) What is the rocket's velocity when it falls into the ocean?

56. *Navigation* Upon reversing the engines, the distance traveled in meters by a supertanker in a straight line before coming to a stop is given by $s(t) = 84.5t - \frac{1}{6}t^3$, where t is in minutes.
(a) If the velocity of the supertanker at time t is $v(t) = s'(t)$, find an equation for the velocity.
(b) What was the velocity of the tanker when the engines were put in reverse?
(c) How long does it take for the supertanker to stop? That is, when is $v(t) = 0$?
(d) How far does the supertanker travel from the time the engines are reversed until it comes to a stop?

In Your Words

57. Describe how to find the derivative of $f(x) = kx^n$.

58. Describe how to find the derivative of the sum of two functions.

4.3 DERIVATIVES OF PRODUCT AND QUOTIENTS

In this section, we will learn general formulas for finding the derivatives of functions that are products or quotients.

The Product Rule

As before, we will assume that we have two functions, $f(x)$ and $g(x)$, and that their derivatives $f'(x)$ and $g'(x)$ exist. If we let F be the product of f and g, then $F(x) = f(x)g(x)$. We want to find $F'(x)$. Again, we will use the four-step method. The first two steps will be quite straightforward. In the third step, we will have to employ some more complicated algebra, which will be explained at that time.

Step 1: $F(x+h) = f(x+h)g(x+h)$

Step 2: $F(x+h) - F(x) = f(x+h)g(x+h) - f(x)g(x)$

Step 3: $\dfrac{F(x+h) - F(x)}{h} = \dfrac{f(x+h)g(x+h) - f(x)g(x)}{h}$

This does not look too promising. But, what we will do is add a fancy version of 0 to the numerator. We will add $-f(x+h)g(x)+f(x+h)g(x)$ to the numerator. Since this is equal to 0, the value of the numerator is not changed and we get

$$\frac{f(x+h)g(x+h)-f(x+h)g(x)+f(x+h)g(x)-f(x)g(x)}{h}$$
$$=f(x+h)\frac{g(x+h)-g(x)}{h}+\frac{f(x+h)-f(x)}{h}g(x)$$

Now we will go to Step 4 and take the limit of the quantity. Remember that the limit of a product is the product of the limits.

Step 4: $$\lim_{h\to 0}\frac{F(x+h)-F(x)}{h}=\lim_{h\to 0}f(x+h)\lim_{h\to 0}\frac{g(x+h)-g(x)}{h}+\lim_{h\to 0}\frac{f(x+h)-f(x)}{h}\lim_{h\to 0}g(x)$$
$$=f(x)g'(x)+f'(x)g(x)$$

This gives us the product rule for derivatives.

Product Rule for Derivatives

If $F(x)=f(x)g(x)$ and $f'(x)$ and $g'(x)$ exist, then

$$F'(x)=f(x)g'(x)+f'(x)g(x)$$

In words, this can be stated as "the derivative of the product of two functions, whose derivatives exist, is equal to the first function times the derivative of the second plus the second function times the derivative of the first."

Note

Note carefully that the derivative of a product does not behave as nicely as the limit of a product. **The derivative of a product is not the product of the derivatives.**

EXAMPLE 4.13

Find the derivative of $F(x)=(x^2+x)(x^3-1)$.

Solution Here the function $F(x)$ is the product of $f(x)=x^2+x$ and $g(x)=x^3-1$. We calculate the derivatives of each of these two functions as $f'(x)=2x+1$ and $g'(x)=3x^2$. According to the new rule

$$\begin{aligned}F'(x)&=f(x)g'(x)+f'(x)g(x)\\&=(x^2+x)(3x^2)+(2x+1)(x^3-1)\\&=3x^4+3x^3+2x^4+x^3-2x-1\\&=5x^4+4x^3-2x-1\end{aligned}$$

Caution

Once again we stress that $F'(x) \neq f'(x)g'(x)$. In Example 4.13 we found that if $F(x) = (x^2 + x)(x^3 - 1)$, then $F'(x) = 5x^4 + 4x^3 - 2x - 1$. But, $f'(x)g'(x) = (2x+1)(3x^2) = 6x^3 + 3x^2 \neq F'(x)$.

EXAMPLE 4.14

Find $G'(x)$, if $G(x) = (x^3 + 2x^2 + x)(x^2 - 5x)$.

Solution We could multiply these two polynomials and then take the derivative of that product, but we will use the product rule instead. The time will come when we will have to use the product rule, so we might as well get some practice. Also, the product rule makes the task somewhat easier. We will let

$$f(x) = x^3 + 2x^2 + x \text{ and } g(x) = x^2 - 5x$$

so

$$f'(x) = 3x^2 + 4x + 1 \text{ and } g'(x) = 2x - 5$$

$$\begin{aligned} G'(x) &= f(x)g'(x) + f'(x)g(x) \\ &= (x^3 + 2x^2 + x)(2x - 5) + (3x^2 + 4x + 1)(x^2 - 5x) \\ &= 5x^4 - 12x^3 - 27x^2 - 10x \end{aligned}$$

The Quotient Rule

Suppose now that $F(x) = \dfrac{f(x)}{g(x)}$, where $g(x) \neq 0$, and $f'(x)$ and $g'(x)$ both exist. What is $F'(x)$? There are a couple of ways to develop this formula. One way uses the four-step method. We are going to use a slightly different approach that depends on the assumption that $F'(x)$ exists.

If $F(x) = \dfrac{f(x)}{g(x)}$, then $F(x)g(x) = f(x)$. Now, using the product rule on the second equation, we see that

$$F(x)g'(x) + F'(x)g(x) = f'(x)$$

Solving this for $F'(x)$ we get

$$F'(x) = \frac{f'(x) - F(x)g'(x)}{g(x)}$$

But, $F(x) = \dfrac{f(x)}{g(x)}$, and so

$$F'(x) = \frac{f'(x) - \dfrac{f(x)}{g(x)}g'(x)}{g(x)} = \frac{g(x)f'(x) - f(x)g'(x)}{[g(x)]^2}$$

And so we have the second, and last, rule of this section, the quotient rule.

Quotient Rule for Derivatives

If $F(x) = \frac{f(x)}{g(x)}$, $g(x) \neq 0$ and both $f'(x)$ and $g'(x)$ exist, then

$$F(x) = \frac{g(x)f'(x) - f(x)g'(x)}{[g(x)]^2}$$

EXAMPLE 4.15

If $F(x) = \frac{x^2+1}{3x-2}$, find $F'(x)$.

Solution According to the quotient rule,

$$F'(x) = \frac{g(x)f'(x) - f(x)g'(x)}{[g(x)]^2}$$

We will let $f(x) = x^2 + 1$ and $g(x) = 3x - 2$, so $f'(x) = 2x$ and $g'(x) = 3$. Thus,

$$\begin{aligned} F'(x) &= \frac{(3x-2)(2x) - (x^2+1)(3)}{(3x-2)^2} \\ &= \frac{(6x^2-4x) - (3x^2+3)}{(3x-2)^2} \\ &= \frac{3x^2-4x-3}{(3x-2)^2} \end{aligned}$$

Note

Note that the derivative of the quotient of two functions is **not** the quotient of the derivatives. Thus, if $F(x) = \frac{f(x)}{g(x)}$, then $F'(x) \neq \frac{f'(x)}{g'(x)}$. In Example 4.15, $\frac{f'(x)}{g'(x)} = \frac{2x}{3} \neq F'(x)$.

EXAMPLE 4.16

If $y = \frac{1}{x^3}$, find y'.

Solution Here y can be considered a quotient with $f(x) = 1$ and $g(x) = x^3$. This means that $f'(x) = 0$ and $g'(x) = 3x^2$. Using the quotient rule, we get

$$\begin{aligned} y' &= \frac{x^3(0) - (1)(3x^2)}{(x^3)^2} \\ &= \frac{0 - 3x^2}{x^6} = \frac{-3}{x^4} \end{aligned}$$

EXAMPLE 4.16 (Cont.)

We could have used rule 3 to solve this by rewriting $y = \frac{1}{x^3} = x^{-3}$. Then we would have gotten $y' = -3x^{-4} = \frac{-3}{x^4}$.

EXAMPLE 4.17

If $f(x) = \frac{4x^2 - 8\sqrt{x}}{x^3 + \sqrt{x^5}}$, find $f'(x)$.

Solution We begin by writing the radicals with fractional exponents. Thus, we get

$$f(x) = \frac{4x^2 - 8x^{1/2}}{x^3 + x^{5/2}}$$

We now use the quotient rule:

$$\begin{aligned} f'(x) &= \frac{(x^3 + x^{5/2})(8x - 4x^{-1/2}) - (4x^2 - 8x^{1/2})(3x^2 + \frac{5}{2}x^{3/2})}{(x^3 + x^{5/2})^2} \\ &= \frac{(8x^4 - 4x^{5/2} + 8x^{7/2} - 4x^2) - (12x^4 + 10x^{7/2} - 24x^{5/2} - 20x^2)}{(x^3 + x^{5/2})^2} \\ &= \frac{-4x^4 - 2x^{7/2} + 20x^{5/2} + 16x^2}{(x^3 + x^{5/2})^2} \end{aligned}$$

Normal Lines

Until now we have concentrated on finding the slope of a line that is tangent to a curve at a particular point. We are also interested in the line that is perpendicular to the curve at a given point. This is called the **normal line**. If the slope of the tangent line is m_1 and the slope of the normal line is m_2, then, since the lines are perpendicular, we know that $m_1 = \frac{-1}{m_2}$.

EXAMPLE 4.18

Find an equation of the tangent and normal lines to $y = \frac{1}{x^3}$ at the point $\left(2, \frac{1}{8}\right)$.

Solution In Example 4.16, we found that the derivative of this function is $y' = \frac{-3}{x^4}$. The derivative is used to find the slope of a line tangent to y. Evaluating y' at the point $\left(2, \frac{1}{8}\right)$, we find that the line tangent to y at this point will have slope $m = \frac{-3}{2^4} = \frac{-3}{16}$. We now know that the slope is $-\frac{3}{16}$ at $\left(2, \frac{1}{8}\right)$, so the equation of the tangent line is

$$\begin{aligned} y - y_0 &= m(x - x_0) \\ y - \frac{1}{8} &= -\frac{3}{16}(x - 2) \end{aligned}$$

EXAMPLE 4.18 (Cont.)

$$\text{or} \qquad y = -\frac{3}{16}x + \frac{1}{2}$$

The slope of the normal line is $-\dfrac{1}{-3/16} = \frac{16}{3}$. The equation of the normal line to $y = \dfrac{1}{x^3}$ at the point $\left(2, \frac{1}{8}\right)$ is

$$y - \frac{1}{8} = \frac{16}{3}(x-2)$$

$$\text{or} \qquad y = \frac{16}{3}x - \frac{253}{24}$$

The next example will require the use of the quotient rule to find the slope of a normal line to a curve.

EXAMPLE 4.19

Determine the equation for the normal line to $f(x) = \dfrac{x^2+3x-1}{x+4}$ at $(3,2)$.

Solution We begin by finding the slope of the tangent line at this point. To do this, we find the derivative of f. Using the quotient rule,

$$f'(x) = \frac{(x+4)(2x+3)-(x^2+3x-1)(1)}{(x+4)^2}$$

To find the slope of f at $(3,2)$, we evaluate f' at $x = 3$. Notice that it is not necessary to simplify the derivative in order to evaluate it.

$$f'(3) = \frac{(7)(9)-(17)(1)}{7^2} = \frac{46}{49}$$

Since the slope of the tangent line is $\frac{46}{49}$, the slope of the normal line is $-\frac{1}{46/49} = -\frac{49}{46}$. We now know the slope of the normal line to f at $(3,2)$, and so the equation of that normal line is

$$y - 2 = -\frac{49}{46}(x-3)$$

$$\text{or} \qquad y = -\frac{49}{46}x + \frac{239}{46}$$

EXAMPLE 4.20

Find the equation of the tangent and normal lines to the curve

$$y = \frac{(x+1)(x^2+2x+5)}{3-x}$$

at the point $(1,8)$.

Solution We have one point on the line, $(1,8)$. What we need to find is the slope. To do this, we will find the derivative y' and evaluate it at the point $(1,8)$. While y

EXAMPLE 4.20 (Cont.)

is in the form $\frac{f(x)}{g(x)}$, the function f is a product of two functions. We will use the quotient rule, but indicate the derivatives that need to be taken by using the notation D_x. This way we get

$$y' = \frac{(3-x)D_x[(x+1)(x^2+2x+5)] - (x+1)(x^2+2x+5)D_x(3-x)}{(3-x)^2}$$

We use the product rule to find

$$D_x[(x+1)(x^2+2x+5)] = (x+1)(2x+2) + (1)(x^2+2x+5)$$

We also have $D_x(3-x) = -1$, so

$$y' = \frac{(3-x)[(x+1)(2x+2)+(x^2+2x+5)] - (x+1)(x^2+2x+5)(-1)}{(3-x)^2}$$

To find the slope of y at $(1,8)$, we need to evaluate y' when $x=1$. It is not necessary to simplify the derivative to do this. When $x=1$,

$$y' = \frac{(2)[(2)(4)+8] - (2)(8)(-1)}{2^2} = \frac{2(16)+16}{4} = \frac{48}{4} = 12$$

We now know that the slope of the tangent line is 12 and a point on the line is $(1,8)$, so an equation of the line is

$$y - 8 = 12(x-1)$$

The slope of the normal line is $\frac{-1}{12}$, so an equation of that line is $y - 8 = \frac{-1}{12}(x-1)$.

Exercise Set 4.3

In Exercises 1–38, differentiate the function.

1. $f(x) = (3x+1)(2x-7)$
2. $g(x) = (6x-2)(5-4x)$
3. $h(x) = (2x^2+x-1)(3x-5)$
4. $k(x) = (4x^3-1)(x^2-7x)$
5. $j(x) = (4-3x)(6x^2+6x-4)$
6. $s(t) = (6-4t)(3t^2-7t)$
7. $q(t) = 4(t^2-3t)^2$
8. $r(p) = (4p-3p^2)(2p-4)$
9. $f(w) = (3w^3-4w^2+2w-5)(w^2-w^{-1})$
10. $g(r) = (2r^4-4r^2+2r)(r^2+2r-1)$
11. $f(x) = \frac{4x-1}{2x+3}$
12. $h(y) = (y^3-1)(y^2-1)(y-1)$
13. $g(x) = \frac{9x^2+2}{4x-1}$
14. $h(x) = 4x^2 - \frac{4}{x^2}$
15. $j(s) = 6s^2 - \frac{1}{6s^2}$
16. $K(s) = \frac{3-4s}{5s-2s^2}$
17. $f(t) = 4t^3 - \frac{2t}{t-2}$
18. $j(x) = \frac{1+2x}{1-2x}$
19. $H(x) = \frac{x^3-1}{x-1}$

20. $k(x) = \dfrac{x^4+4}{3x}$

21. $f(\phi) = \dfrac{\phi^2}{3\phi^2-1}$

22. $L(t) = \dfrac{t^2+3t+2}{t^2-4t-4}$

23. $h(x) = \dfrac{3x^3-x+1}{x^3-3x-1}$

24. $m(x) = \dfrac{x^2+5x+6}{x^2-5x+6}$

25. $f(t) = \dfrac{3t^2-t-1}{\sqrt[3]{t}}$

26. $F(x) = \dfrac{2x^{3/2}-4x^{1/2}}{4x^{1/2}-2}$

27. $g(w) = 12w^2 + \dfrac{w-1}{w+1}$

28. $j(t) = \dfrac{t^2+1}{2t+1} + \dfrac{t-1}{2t+1}$

29. $h(x) = \dfrac{1}{x^2+1}$

30. $k(x) = \dfrac{5}{x^3-1}$

31. $H(x) = \dfrac{4-x^3}{(2-x^2)(3x-x^3)}$

32. $L(x) = \dfrac{(2x+1)(3x^2-4x)}{7x-1}$

33. $n(s) = \dfrac{2s^3}{(s^2-1)(s-1)}$

34. $m(z) = \dfrac{(z-5)(z^2+7)}{z^2+3z}$

35. $y = \dfrac{(2x-3)(x^2-4x+1)}{3x^3+1}$

36. $y = \dfrac{5x}{5-x}(25-x^2)$

37. $y = \dfrac{(t^2-1)}{(2t+1)} \cdot \dfrac{(t-1)}{(2t+1)}$

38. $y = \left(\dfrac{1-x}{x}\right)(1-x^2)$

Solve Exercises 39 and 40.

39. Find the slope of the curve $y = (3x^2+2x-1) \times (x^3-x+1)$ at $(1,4)$.

40. Find the slope of the curve $y = \dfrac{x^3}{x^2+1}$ at $(-1, -\frac{1}{2})$.

In Exercises 41–46, find the equation of the tangent and normal lines to the curves at the given points.

41. $y = \dfrac{8}{x+1}$ at $(3,2)$

42. $y = \dfrac{4x+1}{5x}$ at $(1,1)$

43. $y = \dfrac{5x+4}{2x+1}$ at $(1,3)$

44. $y = (x^2-3x-6)(2x-6)$ at $(-1,16)$

45. $y = \dfrac{9x+2}{x^2(x+4)}$ at $(-2,-2)$

46. $y = (x^3+5x^2-1)(x^2-2x)$ at $(1,-5)$

Solve Exercises 47–52.

47. *Demography* It is estimated that t years from now the population of a certain city, in thousands of people, will be $P(t) = (0.8t-6)(0.5t+9)+87$. How fast will the population be growing in 5 years?

48. *Business* The profit from the sale of n items of a certain product is given by $P(n) = (4-0.2n)(2.6n+7)$, where $P(n)$ is in hundreds of dollars and $1 \leq n \leq 20$. If the marginal profit is $P'(n)$, find the marginal profit when $n = 7$.

49. *Electronics* In a voltage divider circuit the output, V, in volts, depends on the value of a variable resistor, R, in ohms, according to the function $V = \dfrac{110R}{R+60}$. Find $V' = \dfrac{dV}{dR}$.

50. *Electronics* The impedance, Z, in ohms, of a Wein bridge depends on the resistance, R, in ohms, of one arm of the bridge so that $Z = \dfrac{(1+8R)^2+8R}{8(1+8R)}$. Find $Z' = \dfrac{dZ}{dR}$.

51. *Medical technology* It is estimated by a medical research team that the population of a bacterial culture after t hours is approximately $N(t) = \dfrac{t^2-2t+1}{3\sqrt{t}+2}$, where $N(t)$ is in thousands and $0 \le t \le 12$. **(a)** Find $N'(t)$. **(b)** Find the rate of growth after 5 hours.

52. *Optics* For thin lenses the object distance, s, is given by $s = \dfrac{s'f}{s'-f}$, where f is the focal length and s' is the image distance. If f is a constant, find the rate of change of s with respect to s', that is, find $\dfrac{ds}{ds'}$. (Note that in this standard optics formula, s' does not indicate the derivative of s.)

In Your Words

53. Without looking in the text, explain how to determine the derivative of the product of two functions.

54. Without looking in the text, explain how to determine the derivative of the quotient of two functions.

4.4 DERIVATIVES OF COMPOSITE FUNCTIONS

None of the functions that we have differentiated have involved composite functions. In this section, we will learn how to find the derivative of a function that is the composite of two differentiable functions. The rule for finding these derivatives is called the chain rule.

We will begin by briefly reviewing composite functions. We will then look at a special case of the chain rule, called the power rule.

Suppose that f and g are two functions. The composite function of f and g is $(f \circ g)(x) = f(g(x))$, where the domain of f is the range of g.

EXAMPLE 4.21

(a) If $f(x) = x^2$ and $g(x) = x+1$, then $(f \circ g)(x) = (x+1)^2$

(b) If $f(x) = 2x^3 + 5x$ and $g(x) = x^4$, then

$$\begin{aligned}(f \circ g)(x) &= 2(x^4)^3 + 5(x^4)\\ &= 2x^{12} + 5x^4\end{aligned}$$

(c) If $g(w) = \sin 2w$ and $h(w) = 3w^2 + 1$, then

$$\begin{aligned}(g \circ h)(w) &= \sin[2(3w^2+1)]\\ &= \sin(6w^2+2)\end{aligned}$$

It is just as important to be able to recognize a composite function and to decompose it into its individual functions. In this case, you begin with $y = f \circ g$ and determine f and g.

EXAMPLE 4.22

(a) $y = (f \circ g)(x) = (x^2 - 2x + 1)^3$, where $f(x) = x^3$ and $g(x) = x^2 - 2x + 1$
(b) $y = (f \circ g)(x) = (x^5 - 4)^{-2}$, where $f(x) = x^{-2}$ and $g(x) = x^5 - 4$
(c) $y = (f \circ g)(x) = \sqrt{x^3 - 2x}$, where $f(x) = \sqrt{x} = x^{1/2}$ and $g(x) = x^3 - 2x$
(d) $y = (f \circ g)(x) = \ln x^5$, where $f(x) = \ln x$ and $g(x) = x^5$
(e) $y = (f \circ g)(x) = \tan 2x$, where $f(x) = \tan x$ and $g(x) = 2x$

The Power Rule

We will begin our exploration of the chain rule by looking at some special cases. In each of these cases, $f(x) = x^n$ and so $(f \circ g)(x) = [g(x)]^n$. We will use the product rule to look at the derivatives when $n = 2$, 3, and 4. Because it is easier, we will use the $\frac{d}{dx}$ notation to indicate what is being differentiated.

$n = 2$

$$\begin{aligned}\frac{d}{dx}[g(x)]^2 &= \frac{d}{dx}[g(x)g(x)] \\ &= g(x)g'(x) + g'(x)g(x) \\ &= 2g(x)g'(x)\end{aligned}$$

$n = 3$

$$\begin{aligned}\frac{d}{dx}[g(x)]^3 &= \frac{d}{dx}\{[g(x)]^2 g(x)\} \\ &= [g(x)]^2 g'(x) + \frac{d}{dx}[g(x)]^2 g(x)\end{aligned}$$

From the example when $n = 2$, we know that

$$\frac{d}{dx}[g(x)]^2 = 2g(x)g'(x)$$

and so we get

$$\begin{aligned}\frac{d}{dx}[g(x)]^3 &= [g(x)]^2 g'(x) + 2g(x)g'(x)g(x) \\ &= 3[g(x)]^2 g'(x)\end{aligned}$$

$n = 4$

$$\begin{aligned}\frac{d}{dx}[g(x)]^4 &= \frac{d}{dx}\{[g(x)]^3 g(x)\} = [g(x)]^3 g'(x) + \left\{\frac{d}{dx}[g(x)]^3\right\} g(x) \\ &= [g(x)]^3 g'(x) + 3[g(x)]^2 g'(x)g(x) \\ &= 4[g(x)]^3 g'(x)\end{aligned}$$

Look at these results:

$$\begin{aligned}\frac{d}{dx}[g(x)]^2 &= 2g(x)g'(x) \\ \frac{d}{dx}[g(x)]^3 &= 3[g(x)]^2 g'(x) \\ \frac{d}{dx}[g(x)]^4 &= 4[g(x)]^3 g'(x)\end{aligned}$$

These results seem to indicate a pattern. It is possible to show that the pattern results in the following rule, called the **general power rule**.

Rule 9: General Power Rule for Derivatives

If g is a function and its derivative g' exists, then

$$\frac{d}{dx}[g(x)]^n = n[g(x)]^{n-1}g'(x).$$

Notice that this is very similar to rule 3, which states that if $f(x) = x^n$, then $f'(x) = nx^{n-1}$. The main difference is the last factor, $g'(x)$. If $g(x) = x$, then $g'(x) = 1$, and rule 9 is the same as rule 3.

EXAMPLE 4.23

Find the derivatives of **(a)** $(x^2+1)^3$, **(b)** $(4-x^3)^5$, and **(c)** $(x^3-2x^2+1)^{-4}$.

Solutions

(a) If $y = (x^2+1)^3$, then this is $[g(x)]^3$, where $g(x) = x^2+1$. Since $g'(x) = 2x$ we have

$$y' = 3[g(x)]^2 g'(x) = 3(x^2+1)^2(2x)$$

(b) Here $y = (4-x^3)^5 = [g(x)]^5$, where $g(x) = 4-x^3$ and $g'(x) = -3x^2$.

$$\begin{aligned} y' &= 5[g(x)]^4 g'(x) \\ &= 5(4-x^3)^4(-3x^2) \\ &= -15x^2(4-x^3)^4 \end{aligned}$$

(c) $y = (x^3-2x^2+1)^{-4} = [g(x)]^{-4}$, where $g(x) = x^3-2x^2+1$ and $g'(x) = 3x^2-4x$.

$$\begin{aligned} y' &= -4[g(x)]^{-5} g'(x) \\ &= -4(x^3-2x^2+1)^{-5}(3x^2-4x) \end{aligned}$$

The general power rule is not always used alone. It can be used with other rules, such as the product rule or the quotient rule.

EXAMPLE 4.24

Find the derivatives of **(a)** $f(x) = 3x(x^3-1)^4$ and **(b)** $f(x) = \left(\frac{4x+1}{x^2-2}\right)^3$.

Solutions

(a) Here we have a product of $3x$ and $(x^3-1)^4$. We will first use the product rule.

$$f'(x) = 3x\left[\frac{d}{dx}(x^3-1)^4\right] + (x^3-1)^4(3)$$

EXAMPLE 4.24 (Cont.)

Now we use the general power rule to determine that

$$\frac{d}{dx}(x^3-1)^4 = 4(x^3-1)^3(3x^2) = 12x^2(x^3-1)^3$$

Putting this value into the formula for $f'(x)$, we get

$$\begin{aligned} f'(x) &= 3x[12x^2(x^3-1)^3] + (x^3-1)^4(3) \\ &= 36x^3(x^3-1)^3 + 3(x^3-1)^4 \\ &= [36x^3 + 3(x^3-1)](x^3-1)^3 \\ &= (39x^3-3)(x^3-1)^3 \end{aligned}$$

(b) Here $g(x) = \dfrac{4x+1}{x^2-2}$ and $f(x) = [g(x)]^3$

Using the general power rule, we see that

$$\begin{aligned} f'(x) &= 3[g(x)]^2 g'(x) \\ &= 3\left(\frac{4x+1}{x^2-2}\right)^2 g'(x) \end{aligned}$$

We will use the quotient rule to get

$$\begin{aligned} g'(x) &= \frac{(x^2-2)(4)-(4x+1)(2x)}{(x^2-2)^2} \\ &= \frac{(4x^2-8)-(8x^2+2x)}{(x^2-2)^2} \\ &= \frac{-4x^2-2x-8}{(x^2-2)^2} \end{aligned}$$

Putting this value for $g'(x)$ into the answer for $f'(x)$, we get

$$\begin{aligned} f'(x) &= 3\left(\frac{4x+1}{x^2-2}\right)^2\left(\frac{-4x^2-2x-8}{(x^2-2)^2}\right) \\ &= -3\frac{(4x+1)^2(4x^2+2x+8)}{(x^2-2)^4} \end{aligned}$$

Note In Example 4.24(a), after we found the derivative as $f'(x) = 3x[12x^2(x^3-1)^3] + (x^3-1)^4(3)$, we continued until it was factored as $f'(x) = (39x^3-3)(x^3-1)^3$ because the factored form is more useful in applications.

The Chain Rule

The power rule is a special case of the chain rule. Another way of writing the composition function $y = (f \circ g)(x) = f(g(x))$ is

$$y = f(u) \quad \text{where } u = g(x)$$

The major change here is that y is now a function of u and u is a function of x. By substituting, we can get y as a function of x.

Rule 10: Chain Rule

If $y = f(u)$ and $u = g(x)$, and both $f'(u)$ and $g'(x)$ exist, then

$$y' = f'(u)g'(x)$$

or

$$y' = f'[g(x)] \cdot g'(x)$$

Note The chain rule is often written as

$$\frac{dy}{dx} = \frac{dy}{du} \cdot \frac{du}{dx}$$

EXAMPLE 4.25

Find y', if $y = (x^3 - 2x^2 + 1)^{-4}$.

Solution If $y = (x^3 - 2x^2 + 1)^{-4}$, then we will let $y = f(u) = u^{-4}$ and $u = g(x) = x^3 - 2x^2 + 1$. $f'(u) = -4u^{-5}$ and $g'(x) = 3x^2 - 4x$, so

$$\begin{aligned} y' &= f'(u)g'(u) \\ &= -4u^{-5}(3x^2 - 4x) \end{aligned}$$

Substituting for u, we get

$$y' = -4(x^3 - 2x^2 + 1)^{-5}(3x^2 - 4x)$$

This is the same answer we got when we worked this problem in Example 4.23(c).

EXAMPLE 4.26

Find y', if $y = \sqrt[3]{x^5 - 4x}$.

Solution $y = f(u) = \sqrt[3]{u} = u^{1/3}$ and $u = g(x) = x^5 - 4x$. $f'(u) = \frac{1}{3}u^{-2/3}$ and $g'(x) = 5x^4 - 4$, and so

$$\begin{aligned} y &= f'(u)g'(x) \\ &= \frac{1}{3}u^{-2/3}(5x^4 - 4) \\ &= \frac{1}{3}(x^5 - 4x)^{-2/3}(5x^4 - 4) \qquad \text{Substitute for } u. \end{aligned}$$

It is possible to extend the chain rule. For example, suppose $y = f(u), u = g(v)$, and $v = h(x)$. Then the composite function

$$y = (f \circ g \circ h)(x) = f(g(h(x)))$$

has a derivative

$$y' = f'(u)g'(v)h'(x)$$

or

$$= \frac{dy}{du}\frac{du}{dv}\frac{dv}{dx}$$

EXAMPLE 4.27

If $y = \left[6\left(\frac{5}{x^2}\right)^4 - 7\right]^{10}$, find y'.

Solution This looks more difficult than it is. First, if we let $v = h(x) = \frac{5}{x^2} = 5x^{-2}$, then we have $y = (6v^4 - 7)^{10}$. Now let $u = g(v) = 6v^4 - 7$ and we have $y = f(u) = u^{10}$. So, if $f(u) = u^{10}$, $g(v) = 6v^4 - 7$, and $h(x) = 5x^{-2}$, then

$$\begin{aligned}
y' &= f'(u)g'(v)h'(x) \\
&= 10u^9(24v^3)(-10x^{-3}) \\
&= 10(6v^4 - 7)^9(24v^3)(-10x^{-3}) && \text{Substitute for } u. \\
&= 10[6(5x^{-2})^4 - 7]^9[24(5x^{-2})^3](-10x^{-3}) && \text{Substitute for } v. \\
&= 10\left[6\left(\frac{625}{x^8}\right) - 7\right]^9\left(24 \cdot \frac{125}{x^6}\right)\left(\frac{-10}{x^3}\right) \\
&= -\frac{300,000}{x^9}\left[6\left(\frac{625}{x^8}\right) - 7\right]^9
\end{aligned}$$

■

Application

EXAMPLE 4.28

After a sewage spill, the level of pollution in Herd Weyer Bay is estimated by $P(t) = \frac{100t + 25}{\sqrt{5t^2 + 10}}$ parts per million (ppm), where t is the time in days since the spill occurred.

(a) What is the rate of change in the level of pollution?

(b) What is the rate of change in the level of pollution after 1.5 days?

(c) What is the rate of change in the level of pollution after 2.5 days?

EXAMPLE 4.28 (Cont.)

Solution

(a) We find the rate of change in the level of pollution by finding the derivative of the level of pollution function $P(t) = \dfrac{100t+25}{\sqrt{5t^2+10}} = \dfrac{100t+25}{(5t^2+10)^{1/2}}$.

$$P'(t) = \frac{(5t^2+10)^{1/2}(100) - (100t+25)\frac{1}{2}(5t^2+10)^{-1/2}}{\left((5t^2+10)^{1/2}\right)^2}$$

$$= \frac{(5t^2+10)^{1/2}(100) - \left(50t+\frac{25}{2}\right)(5t^2+10)^{-1/2}}{5t^2+10}$$

$$= \frac{(5t^2+10)^{1/2}(100) - \left(50t+\frac{25}{2}\right)(5t^2+10)^{-1/2}}{5t^2+10} \cdot \frac{(5t^2+10)^{1/2}}{(5t^2+10)^{1/2}}$$

$$= \frac{(5t^2+10)(100) - \left(50t+\frac{25}{2}\right)}{(5t^2+10)^{3/2}}$$

$$= \frac{500t^2+1000-50t+\frac{25}{2}}{(5t^2+10)^{3/2}} = \frac{500t^2-50t+1012.5}{(5t^2+10)^{3/2}}$$

Thus, the pollution is changing at the rate of $P'(t) = \dfrac{500t^2-50t+1012.5}{(5t^2+10)^{3/2}}$ ppm/day.

(b) The rate of change in the level of pollution after 1.5 days is

$$P'(1.5) = \frac{500(1.5)^2-50(1.5)+1012.5}{(5(1.5)^2+10)^{3/2}} \approx 21.055 \text{ ppm/day.}$$

(c) The rate of change in the level of pollution after 2.5 days is

$$P'(2.5) = \frac{500(2.5)^2-50(2.5)+1012.5}{(5(2.5)^2+10)^{3/2}} \approx 15.145 \text{ ppm/day.}$$

So far, we have had ten rules for derivatives. These are summarized in the following box.

Ten Rules for Derivatives

1. If $f(x) = k$, k a constant, then $f'(x) = 0$.
2. If $f(x) = mx + b$, then $f'(x) = m$.
3. If $f(x) = x^n$, n a rational number, then $f'(x) = nx^{n-1}$.
4. If $F(x) = kf(x)$, k a constant, and $f'(x)$ exists, then $F'(x) = kf'(x)$.
5. If $F(x) = f(x) + g(x)$ and $f'(x)$ and $g'(x)$ exist, then $F'(x) = f'(x) + g'(x)$.
6. If $F(x) = f(x) - g(x)$ and both $f'(x)$ and $g'(x)$ exist, then $F'(x) = f'(x) - g'(x)$.
7. Product rule: If $F(x) = f(x)g(x)$, and both $f'(x)$ and $g'(x)$ exist, then
$$F'(x) = f(x)g'(x) + f'(x)g(x)$$
8. Quotient rule: If $F(x) = f(x)/g(x)$, $g(x) \neq 0$, and both $f'(x)$ and $g'(x)$ exist, then
$$F'(x) = \frac{g(x)f'(x) - f(x)g'(x)}{[g(x)]^2}$$
9. General power rule: If $g'(x)$ exists, then
$$\frac{d}{dx}[g(x)]^n = n[g(x)]^{n-1}g'(x)$$
10. Chain rule: If $y = f(u)$, $u = g(x)$, and both $f'(u)$ and $g'(x)$ exist, then $y' = f'(u)g'(x)$ or
$$\frac{dy}{dx} = \frac{dy}{du} \cdot \frac{du}{dx}$$

Exercise Set 4.4

In Exercises 1–22, find the derivative of the given function by using the general power rule.

1. $f(x) = (3x - 6)^4$
2. $g(x) = (7 - 2x)^3$
3. $h(x) = (5x - 7)^{-5}$
4. $k(x) = (9x + 5)^{-3}$
5. $f(x) = (x^2 + 3x)^4$
6. $h(x) = \sqrt{4x^2 + 7}$
7. $H(x) = \dfrac{1}{\sqrt{4x^2 + 7}}$
8. $g(x) = (x^7 - 9)^6$
9. $s(t) = (t^4 - t^3 + 2)^3$
10. $g(t) = (3t^5 + 2t^3 - t)^{10}$
11. $f(u) = (u^3 + 2u^{-4})^3$
12. $f(u) = (4u^2 - 3u^{-1})^2$
13. $g(x) = \left[(x - 2)(3x^2 - x)\right]^3$
14. $j(t) = 4t^3\sqrt{t^2 - 4}$
15. $h(v) = (v^2 + 1)^2(2v - 5)^3$
16. $g(x) = \left[\left(x^2 - 4x\right)\left(2x^3 - 7\right)\right]^3$
17. $h(x) = \left[\left(x^3 - 6x^2\right)\left(5x - 6x^2 + x^3\right)\right]^4$
18. $j(w) = (w^3 + 2w)^4(3w - 5)^{-2}$
19. $k(x) = (3x^2 + 2)^3\sqrt{x^3 - 7x}$
20. $h(x) = \left(\dfrac{3x + 4}{2x^2 - 1}\right)^4$

21. $g(v) = \dfrac{(v^2-4v)^3}{(v^3-9)^2}$

22. $f(u) = \left(\dfrac{4u^2+5}{6u^3-3u}\right)^5$

In Exercises 23–42, find the derivative of the given function by using the chain rule.

23. $y = f(u) = u^6, u = g(x) = x^5+4$
24. $y = f(u) = u^3, u = g(x) = 2x^3-5x$
25. $y = f(u) = \sqrt{u}, u = g(x) = 4x^2-5$
26. $y = f(u) = 4u^5, u = g(x) = \sqrt{3x^2+5x}$
27. $y = f(u) = \sqrt[3]{u^2}, u = g(x) = 7x^3-9x$
28. $y = f(u) = (u+1)^2, u = g(x) = \dfrac{2}{x}$
29. $y = f(u) = (u^2+1)^3, u = g(x) = \dfrac{1}{x+3}$
30. $y = f(u) = u^3-4u, u = x^4+5$
31. $y = f(u) = \sqrt[3]{u^2-2u}, u = x^3+4$
32. $y = f(u) = 2u^3-8u, u = 6x^2-5x-1$
33. $y = g(x) = 4x^4-3x, x = 3t^2-4t$
34. $y = f(u) = 5u^3-7u^2-5u+1, u = 6t^2-8t$
35. $y = (9x^2+4x)^6$
36. $y = 3(2x^2-5x+1)^{5/3}$
37. $y = (11x^5-2x+1)^{10}$
38. $y = \dfrac{8}{(x^3-5x+2)^4}$
39. $y = \dfrac{7}{(9x^2-4)^8}$
40. $y = \left(\sqrt[3]{2x^5-x^3}\right)^4$
41. $y = \sqrt[4]{(2x^2-5)^3}$
42. $y = \dfrac{1}{\sqrt[5]{(7x-4x^3)^2}}$

Solve Exercises 43–56.

43. Find $\dfrac{dy}{dx}$, if $y = u^4, u = 2v^3-1$, and $v = \dfrac{4}{x^2}$.
44. Find $\dfrac{dy}{dt}$, if $y = 4u^2-u, u = x^3-8$, and $x = 6t+4$.
45. Find $\dfrac{dy}{dx}$, if $y = 3u^2, u = \dfrac{4}{v}$, and $v = x^5$.
46. Find an equation of the tangent line to y at $(4, 1)$, if $y = \sqrt[3]{(x-5)^2}$.
47. Find an equation of the tangent line to y at $(2, -8)$, if $y = (4x^2-18)^3$.
48. Find an equation for the normal line to y at $(-3, 1)$, if $y = \dfrac{4}{\sqrt{25-x^2}}$.
49. *Energy technology* The energy output of an electrical heater varies with time t according to the equation $E = 6\left(1+4t^2\right)^3$.
    (a) Find the power, P, in watts, generated by the heater if $P = E' = \dfrac{dE}{dt}$.
    (b) The answer in (a) is in watts. Rewrite your answer so that it is in kilowatts.
    (c) Find the power, in kilowatts, at $t = 2.5$ s
50. *Demography* It is estimated that t years from now the population of a certain city will be $P(t) = 10(45+3.5t)^2-1750t$.
    (a) What is the rate of change in the population?
    (b) How fast will the population be growing in 5 years?
51. *Environmental science* After a sewage spill, the level of pollution in San Juan Pedro Bay is estimated by $P(t) = \dfrac{250t^2}{\sqrt{t^2+15}}$ parts per million (ppm), where t is the time in days since the spill occurred.
    (a) What is the rate of change in the level of pollution?
    (b) What is the rate of change in the level of pollution after 5.0 days?
52. *Environmental science* It has been estimated that t years from now the level of pollution in the air will be $P(t) = \dfrac{0.6\sqrt{8t^2+11t+60}}{(t+1)^2}$ ppm.
    (a) What is the rate of change in the level of pollution?
    (b) Find the rate of change in the level of pollution in 10.0 years?
53. *Optics* When light passes from air into a transparent medium, the ratio of reflected light to incident light is given approximately by $R = \left(\dfrac{n-1}{n+1}\right)^2$, where n is the index of reflection for the medium. Find the derivative of R with respect to n.

54. *Electronics* The charge on a capacitor over a short interval of time, t, in seconds, is given by $q = 5t^2 + \sqrt{2 - 5t}$. The formula for the current is the derivative of the formula for the charge, that is, $I = q'$.
(a) What is the formula for the current?
(b) Find the value of I when $t = 0.25$ s?

55. *Electronics* In an RC circuit the current, I, is given by $I = \dfrac{V}{\sqrt{R^2 + X_C{}^2}}$, where V is the voltage in volts, R is the resistance in ohms, and X_C is the capacitative resistance in ohms. For this exercise, assume that $V = 120$ V and $R = 30\,\Omega$.
(a) What is the rate of change in the current with respect to the capacitative resistance?
(b) Find the value of I' when $X_C = 25\,\Omega$?

56. *Environmental technology* Some studies have shown that the average level of certain pollutants in the air is given by $L = 1 + 0.25x + 0.001x^2$ ppm when the population is x thousand people. It is estimated that t years from now the population will be $x = \dfrac{200}{\sqrt{9 - 0.4t}}$, in thousands of people.
(a) Find the rate of change in the level of pollutants with respect to the number of years from now.
(b) Find the rate of change in the level of pollutants when $t = 5$ years.
(c) Find the rate of change in the level of pollutants when $t = 10$ years.

In Your Words

57. Explain how to use the chain rule to differentiate the composition of two functions.

58. The general power rule for differentiation is a special case of the chain rule. Describe a function where you could use the chain rule but not the general power rule to find its derivative.

≡ 4.5 IMPLICIT DIFFERENTIATION

So far, we have considered only functions in the explicit form $y = f(x)$. If the functional relationship between the independent variable x and the dependent variable y is not of this form, they we say that x and y are related implicitly. Examples of x and y being implicitly related are

$$4x - 8y + 6 = 0 \qquad 4x^2 + 9y^2 - 36 = 0 \qquad y \le 0 \qquad xy + 9 = 0$$

Sometimes it is possible to form an explicit function from an expression where x and y are related implicitly. When this can be done, the relationship is solved for y, and a function $y = f(x)$ results. For example, the implicit equation $4x - 8y + 6 = 0$ results in the explicit function $y = \frac{1}{2}x + \frac{3}{4}$. In this section, we will learn how to take derivatives of implicit functions. The result of differentiating an implicit function is called an **implicit derivative** and the process is called **implicit differentiation.**

For many equations it is very difficult, or even impossible, to solve for y in terms of x. In these cases, implicit differentiation is a much easier method of finding the derivative. We will begin by taking the derivative in both the explicit and the implicit form.

We will begin with the implicit function $4x - 8y + 6 = 0$, which we write in the explicit form $y = \frac{1}{2}x + \frac{3}{4}$. The derivative of the explicit form is $y' = \frac{1}{2}$.

In taking an implicit derivative, it is not necessary to solve for y in terms of x. Instead, we will use the general power rule or the chain rule. One important realization is the different way to take the derivative of x^2 and of y^2. At present we know that $\frac{dx^2}{dx} = 2x$. But $\frac{dy^2}{dx} = 2y\frac{dy}{dx}$ because we are differentiating with respect to x. This is an application of the general power rule with $g(x)$ replaced by y. In general, if $y = f(x)$, then

$$\begin{aligned}\frac{d}{dx}[f(x)]^n &= n[f(x)]^{n-1}\frac{d}{dx}[f(x)]\\ &= ny^{n-1}\frac{dy}{dx}\\ &= ny^{n-1}y'\end{aligned}$$

Let's return to our example, $4x - 8y + 6 = 0$. We will differentiate each term with respect to x and get

$$\begin{aligned}\frac{d}{dx}(4x) - \frac{d}{dx}(8y) + \frac{d}{dx}(6) &= \frac{d}{dx}(0)\\ 4 - 8\frac{dy}{dx} + 0 &= 0\end{aligned}$$

Solving this for $\frac{dy}{dx}$ we get

$$\frac{dy}{dx} = y' = \frac{-4}{-8} = \frac{1}{2}$$

This is the same answer we got before. Notice that we used the chain rule to differentiate $8y$. Thus $\frac{d}{dx}(8y) = \frac{d}{dy}(8y)\frac{dy}{dx} = 8\frac{dy}{dx}$.

In general, use the following four steps in order to differentiate implicitly.

Four Steps for Implicit Differentiation

In the list below, we assume that x is the independent variable and that we are trying to determine dy/dx.

1. Differentiate both sides of the equation with respect to the independent variable, x.
2. Collect the terms with dy/dx on one side of the equation and place the remaining terms on the other side of the equals sign.
3. Factor out dy/dx.
4. Solve for dy/dx.

In the next example, we will again find the derivative using both the explicit and implicit methods.

EXAMPLE 4.29

Find $\frac{dy}{dx}$ of $4x^2+9y^2-36=0$.

Solution

Explicit form

This is the equation of an ellipse. If we solve this for y in terms of x, we get

$$y=\pm\sqrt{\frac{36-4x^2}{9}}=\pm\frac{1}{3}\sqrt{36-4x^2}$$

This can be written as two expressions:

$$y=\frac{1}{3}\sqrt{36-4x^2} \text{ and } y=-\frac{1}{3}\sqrt{36-4x^2}$$

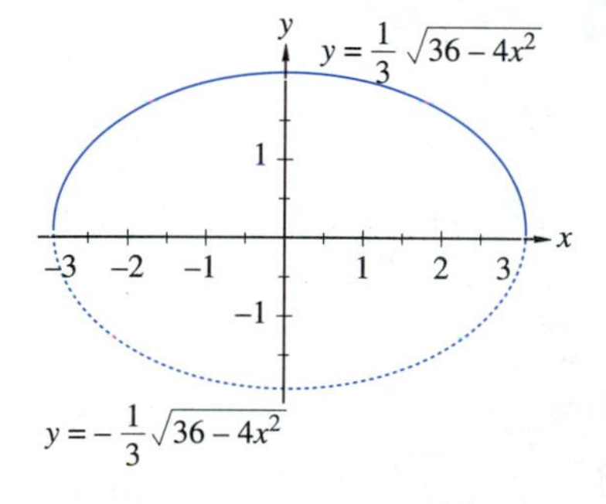

FIGURE 4.6

Each separate expression is now a function. As you can see in Figure 4.6, $y=\frac{1}{3}\sqrt{36-4x^2}$ is the upper half of the ellipse and $y=-\frac{1}{3}\sqrt{36-4x^2}$ is the lower half.

If you want to find the slope of the tangent at a given point, you need to first determine which half of the ellipse the point is on and then differentiate the appropriate equation. We will differentiate the equation for the upper half:

$$\begin{aligned} y &= \frac{1}{3}\sqrt{36-4x^2}=\frac{1}{3}(36-4x^2)^{1/2} \\ y' &= \frac{1}{6}(36-4x^2)^{-1/2}(-8x) \\ &= \frac{-4x}{3\sqrt{36-4x^2}} \end{aligned}$$

Let's compare this process to the implicit method.

Implicit form

The original equation is $4x^2+9y^2-36=0$. We will differentiate both sides with respect to x.

$$\begin{aligned} \frac{d}{dx}(4x^2+9y^2-36) &= \frac{d}{dx}(0) \\ 8x+18y\frac{dy}{dx} &= 0 \\ 8x+18yy' &= 0 \end{aligned}$$

Solving for y', we get

$$y'=\frac{-8x}{18y}=\frac{-4x}{9y}$$

This is certainly a much simpler equation. But, do we get the same results?

The easiest way to find out is to substitute one of the other equations for y. You will see that when $y=\frac{1}{3}\sqrt{36-4x^2}$ is substituted, you get

$$y=\frac{-4x}{3\sqrt{36-4x^2}}$$

It works! The results are not only the same, but by using implicit differentiation, we got an easier result to use. ▪

EXAMPLE 4.30

Use implicit differentiation to find the derivative of $9x^2 - 4y^2 = 2x$.

Solution We differentiate both sides of the equation with respect to x.

$$\begin{aligned}
\frac{d}{dx}(9x^2 - 4y^2) &= \frac{d}{dx}(2x) \\
18x - 8y\frac{dy}{dx} &= 2 \\
\frac{dy}{dx} &= \frac{2 - 18x}{-8y} \\
\frac{dy}{dx} &= \frac{18x - 2}{8y} = \frac{9x - 1}{4y}, \text{ if } y \neq 0
\end{aligned}$$

EXAMPLE 4.31

Find $\frac{dy}{dx}$, if $xy^2 - x^2y + 4x - 5y = 16$.

Solution

$$\begin{aligned}
\frac{d}{dx}(xy^2 - x^2y + 4x - 5y) &= \frac{d}{dx}(16) \\
\frac{d}{dx}(xy^2) - \frac{d}{dx}(x^2y) + \frac{d}{dx}(4x) - \frac{d}{dx}(5y) &= \frac{d}{dx}(16)
\end{aligned}$$

We must use the product rule on the first two terms.

$$\begin{aligned}
x\frac{d}{dx}(y^2) + y^2\frac{d}{dx}(x) - x^2\frac{d}{dx}y - y\frac{d}{dx}(x^2) + \frac{d}{dx}(4x) - \frac{d}{dx}(5y) &= \frac{d}{dx}(16) \\
x\left(2y\frac{dy}{dx}\right) + y^2(1) - x^2\frac{dy}{dx} - y(2x) + 4 - 5\frac{dy}{dx} &= 0
\end{aligned}$$

If $\frac{dy}{dx} = y'$, then we get

$$\begin{aligned}
2xyy' + y^2 - x^2y' - 2xy + 4 - 5y' &= 0 \\
(2xy - x^2 - 5)y' &= 2xy - y^2 - 4 \\
y' &= \frac{2xy - y^2 - 4}{2xy - x^2 - 5}
\end{aligned}$$

if the denominator is not 0.

EXAMPLE 4.32

Find y', if $\sqrt[3]{x^2 - y} = x$.

Solution

$$\begin{aligned}
\frac{d}{dx}(x^2 - y)^{1/3} &= \frac{d}{dx}(x) \\
\frac{1}{3}(x^2 - y)^{-2/3}\left[\frac{d}{dx}(x^2 - y)\right] &= \frac{d}{dx}(x)
\end{aligned}$$

EXAMPLE 4.32 (Cont.)

$$\frac{1}{3}(x^2-y)^{-2/3}\left[2x-\frac{dy}{dx}\right]=1$$

And if $(x^2-y)^{-2/3}\neq 0$,

$$\begin{aligned} 2x-y' &= \frac{1}{\frac{1}{3}(x^2-y)^{-2/3}} \\ &= 3(x^2-y)^{2/3} \\ y' &= 2x-3(x^2-y)^{2/3} \end{aligned}$$

EXAMPLE 4.33

Find the slope of the tangent line to the graph of $x^3+3xy+12y^2=10$ at the point $(-2,-1)$.

Solution We first find the derivative, y'.

$$\begin{aligned} \frac{d}{dx}(x^3+3xy+12y^2) &= \frac{d}{dx}(10) \\ 3x^2+\frac{d}{dx}(3xy)+12\frac{d}{dx}y^2 &= 0 \\ 3x^2+(3xy'+3y)+24yy' &= 0 \\ (3x+24y)y' &= -3x^2-3y \\ y' &= \frac{-3x^2-3y}{3x+24y} \\ &= -\frac{x^2+y}{x+8y} \end{aligned}$$

Now, at the point $(-2,-1)$, we see that

$$y'=-\frac{(-2)^2+(-1)}{-2+8(-1)}=-\frac{4-1}{-2-8}=\frac{3}{10}$$

Thus, the slope of the tangent line to the graph of $x^3+3xy+12y^2=10$ at the point $(-2,-1)$ is $\frac{3}{10}$.

In Section 4.3, we learned how to use derivatives to calculate the equation of tangent and normal lines to a curve. When light enters a lens, the light forms an angle with the line perpendicular to the surface of the lens at the light's entry point.

The perpendicular line is **normal** to the surface of the lens. The next example shows how to use implicit differentiation to determine the equation of this normal line.

Application

EXAMPLE 4.34

A profile of a lens is shown in Figure 4.7a. This particular lens is described by the quadratic equation $x^2-xy+y^2=13$. Find the equation of the normal line to this lens at the point $(-1,3)$.

EXAMPLE 4.34 (Cont.)

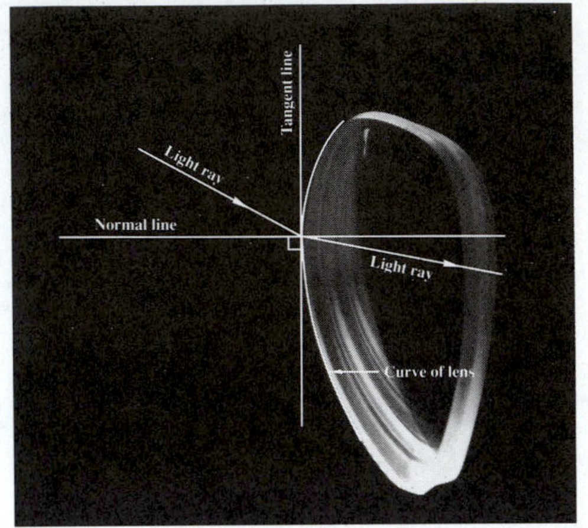

Courtesy of Ruby Gold

FIGURE 4.7a

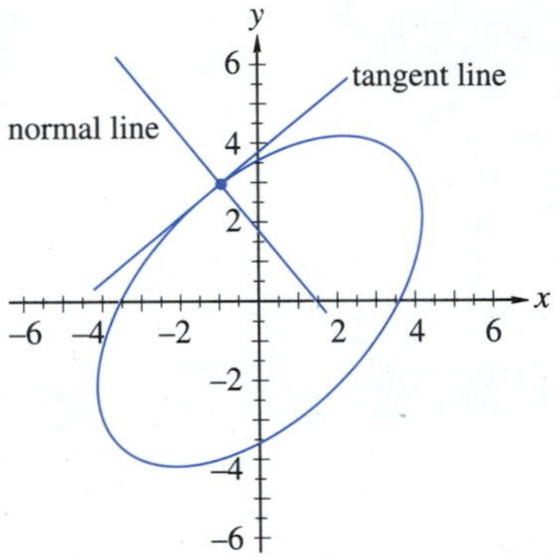

FIGURE 4.7b

Solution The sketch in Figure 4.7a shows a lens, the tangent and normal lines, where the normal line is the path of a "typical" light ray. In Figure 4.7b, we have sketched the lens on a coordinate system and shown the tangent and normal lines at the point $(-1,3)$. We begin by implicitly differentiating the equation that describes the lens.

$$\frac{d}{dx}(x^2-xy+y^2)=\frac{d}{dx}(13)$$

$$\frac{d}{dx}(x^2)-\frac{d}{dx}(xy)+\frac{d}{dx}y^2=\frac{d}{dx}(13)$$

Using the product rule on the middle term, we obtain

$$\frac{d}{dx}(x^2)-\left[x\frac{d}{dx}(y)+y\frac{d}{dx}(x)\right]+\frac{d}{dx}y^2=\frac{d}{dx}(13)$$

$$2x-x\frac{dy}{dx}-y+2y\frac{dy}{dx}=0$$

Letting $\dfrac{dy}{dx}=y'$, produces

$$2x-xy'-y+2yy'=0$$

or

$$(2y-x)y'=y-2x$$

and so,

$$y'=\frac{y-2x}{2y-x}$$

Now, at the point $(-1,3)$, we see that

$$y'=\frac{3-2(-1)}{2(3)-(-1)}=\frac{3+2}{6+1}=\frac{5}{7}$$

EXAMPLE 4.34 (Cont.)

Thus, the slope of the tangent line to the lens at $(-1,3)$ is $\frac{5}{7}$, so the slope of the normal line at this point is $-\frac{7}{5}$. Using the point-slope form for the line, we see that the equation of the normal line is

$$y-3=-\frac{7}{5}(x+1)$$

$$\text{or} \quad 7x+5y=8$$

Exercise Set 4.5

In Exercises 1–34, use implicit differentiation to find $\frac{dy}{dx}$.

1. $4x+5y=0$
2. $6x-7y=3$
3. $x-y^2=4$
4. $x^2+y^2=9$
5. $x^2-y^2=16$
6. $x^2-2y^2=2x$
7. $9x^2+16y^2=144$
8. $16x^2-9y^2=144$
9. $4x^2-y^2=4y$
10. $x^2y+xy^2=x$
11. $xy+5xy^2=x^2$
12. $x^2y-3xy^2=y^3$
13. $x^2+4xy+y^2=y$
14. $x^3y-3xy^2=2x$
15. $x+5x^2-10y^2=3y$
16. $x^2-2xy+y^2=x$
17. $4x^2+y^3=9$
18. $y^4-9x^2=9$
19. $6y^3+x^3=xy$
20. $x^2y=7y$
21. $x^2y=x+1$
22. $x^2+\frac{1}{y}=2x$
23. $\frac{1}{x}+\frac{1}{y}=16$
24. $\frac{1}{x^2}-\frac{1}{y^2}=xy$
25. $\frac{3}{x+1}+x^2y=5$
26. $y^2=\frac{x}{y+1}$
27. $xy+\frac{y}{x}=x$
28. $x^3-8x^2y^2+y=9x$
29. $x^4-6x^2y^2+y^2=10x$
30. $(x+2y)^2=4x$
31. $\sqrt{x^2+y^2}=\frac{2y}{x}$
32. $\sqrt{x+xy}=x^2y^2$
33. $(x^2+y^2)^3=y$
34. $(x^2y+4)^2=3x$

Solve Exercises 35–44.

35. Find the equation of the line tangent to the circle $x^2+y^2=25$ at the point $(3,-4)$.
36. Find the equation of the line normal to the circle $x^2+y^2-2x-4y=20$ at the point $(-2,6)$.
37. Find the equation of the line tangent to the ellipse $9x^2+16y^2=100$ at the point $(-2,2)$.
38. Find the slopes of the lines tangent and normal to the parabola $y^2=-16x$ at the point $(-1,4)$.
39. Find the slopes of the lines tangent and normal to the hyperbola $12x^2-16y^2=192$ at the point $(-8,-6)$.
40. *Physics* The position of a particle at time t is described by the relation $s^3-4st+2t^3-5t=0$. Find the velocity, ds/dt, of the particle.
41. *Automotive technology* An automobile's position, s, in miles, on a track at time t, in minutes, is described by the relation $2s^2+\sqrt{st}-3t=0$.
    (a) Find $\frac{ds}{dt}$, the velocity of the car.
    (b) What is the position of the car when $t=2$?
    (c) What is the velocity of the car when $t=2$?
42. *Economics* The **Cobb-Douglas production formula**, $P=Cx^ay^{1-a}$, is frequently used by economists to relate the cost of a production process, P, to labor and capital, where C is a constant. Suppose that a firm's level of production is given by $P=20x^{1/5}y^{4/5}$, where x represents the units of labor and y the units of capital. Currently, the company is using 32 units of labor and 243 units of capital. If labor is increasing by 25% per month, what change in units of capital are needed to maintain the same level of production?

43. *Economics* The manager of an electronics store has determined that the number of digital satellites and the number of television sets sold weekly are related by the equation $0.7y^2 = 12x + xy$, where x is the number of digital satellites and y is the number of television sets.

(a) Find $\dfrac{dy}{dx}$ when $x = 10$ and $y = 25$.

(b) Interpret your answer.

44. *Business* The number of pairs of trousers, x, and the number of shirts, y, sold at a clothing store are related by the equation $48x = 12y + 0.01x^2y$.

(a) Find $\dfrac{dy}{dx}$ when $x = 10$ and $y = 40$.

(b) Interpret your answer.

In Your Words

45. Describe how to differentiate an implicit function.

46. Explain how an implicit function differs from an explicit function.

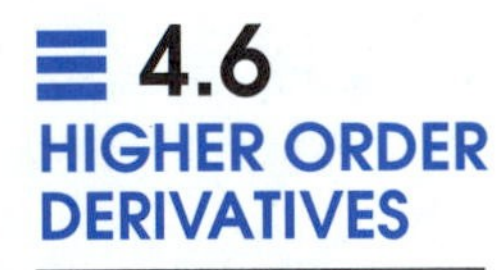

4.6 HIGHER ORDER DERIVATIVES

The derivative of a function, such as $f(x)$, is also a function, $f'(x)$, called the **derivative function** f'. At present, the only application of the derivative that we have examined is that of the slope of the curve or of a tangent to the curve. In Chapter 5, we will look at more applications of the derivative. As we work these applications there will be times when we need to take the derivative of a derivative.

When it exists, the derivative of the derivative f' is called the **second derivative** and is denoted by f''. It is possible to continue this process: The derivative of the second derivative is the **third derivative**, f''', and its derivative is the **fourth derivative**, $f^{(4)}$. The process can be continued indefinitely, as long as the resulting function has a derivative.

In general, if n is a positive integer, then $f^{(n)}$ represents the nth derivative of f and is found by starting with f and differentiating successively, n times. The integer n is called the **order** of the derivative $f^{(n)}$ and, as a group, these are known as **higher order derivatives.**

As with the derivative, there are many different notations. If $y = f(x)$, then

$$y'' = f''(x) = D_x^2 y = \frac{d^2y}{dx^2} = \frac{d^2}{dx^2}f(x)$$

all represent second derivatives,

$$y''' = f'''(x) = D_x^3 y = \frac{d^3y}{dx^3} = \frac{d^3}{dx^3}f(x)$$

all represent third derivatives,

$$y^{(4)} = f^{(4)}(x) = D_x^4 y = \frac{d^4y}{dx^4} = \frac{d^4}{dx^4}f(x)$$

all represent fourth derivatives, and in general,

$$y^{(n)} = f^{(n)}(x) = D_x^n y = \frac{d^ny}{dx^n} = \frac{d^n}{dx^n}f(x)$$

all represent the nth derivative.

EXAMPLE 4.35

Find the first four derivatives of $f(x) = 2x^5 - 3x$.

Solution

$$\begin{aligned} f'(x) &= 10x^4 - 3 \\ f''(x) &= 40x^3 \\ f'''(x) &= 120x^2 \\ f^{(4)}(x) &= 240x \end{aligned}$$

EXAMPLE 4.36

Find the first three derivatives of $f(x) = 4x^3 + \frac{2}{x}$.

Solution We will first rewrite $f(x)$ as $4x^3 + 2x^{-1}$.

$$\begin{aligned} f'(x) &= 12x^2 - 2x^{-2} \\ f''(x) &= 24x + 4x^{-3} \\ f'''(x) &= 24 - 12x^{-4} \end{aligned}$$

EXAMPLE 4.37

Use implicit differentiation to find y' and y'' of $x^2 + 2xy - y^2 = 25$.

Solution Using implicit differentiation, we get

$$\begin{aligned} \frac{d}{dx}(x^2 + 2xy - y^2) &= \frac{d}{dx}(25) \\ \frac{d}{dx}(x^2) + \frac{d}{dx}(2xy) - \frac{d}{dx}(y^2) &= \frac{d}{dx}(25) \\ 2x + 2y + 2xy' - 2yy' &= 0 \end{aligned}$$

$$y' = -\frac{2x + 2y}{2x - 2y} = \frac{x + y}{y - x}, \quad \text{if } y \neq x$$

To find y'', we will use the quotient rule:

$$\begin{aligned} y'' &= \frac{d}{dx}\left(\frac{x + y}{y - x}\right) \\ &= \frac{(y - x)\frac{d}{dx}(x + y) - (x + y)\frac{d}{dx}(y - x)}{(y - x)^2} \\ &= \frac{(y - x)(1 + y') - (x + y)(y' - 1)}{(y - x)^2} \\ &= \frac{2y - 2xy'}{(y - x)^2} \end{aligned}$$

EXAMPLE 4.37 (Cont.)

Substituting $\frac{x+y}{y-x}$ for y', we get

$$\begin{aligned} y'' &= \frac{2y-2x\left(\frac{x+y}{y-x}\right)}{(y-x)^2} \\ &= \frac{\frac{2y(y-x)-2x(x+y)}{y-x}}{(y-x)^2} \\ &= \frac{2y^2-2yx-2x^2-2xy}{(y-x)^3} \\ &= \frac{2y^2-4xy-2x^2}{(y-x)^3} \\ &= \frac{-2(x^2+2xy-y^2)}{(y-x)^3} \\ &= \frac{-2(25)}{(y-x)^3} = \frac{-50}{(y-x)^3} \end{aligned}$$

This concludes our introduction to derivatives. In Chapter 5, we will look at many of the applications of derivatives.

Exercise Set 4.6

In Exercises 1–19, find the indicated higher derivatives of the given function.

1. $y = 4x^3 - 6x^2 + 3x - 10$; find y''.
2. $y = 9x^4 + x^3 - 10$; find y'''.
3. $y = 7x^5 - 3x^3 + x$; find $y^{(4)}$.
4. $f(x) = 10x - 1$; find $f''(x)$.
5. $f(x) = x + \frac{1}{x}$; find $f''(x)$.
6. $f(t) = t^3 - \frac{1}{t^2}$; find $f'''(t)$.
7. $f(t) = t^2 + \frac{1}{t+1}$; find $f''(t)$.
8. $h(x) = (x^2 + 2)(x - 1)$; find $h''(x)$.
9. $g(u) = \frac{u-1}{u^2}$; find $g'''(u)$.
10. $y = 3\sqrt[3]{x^2+1}$; find $\frac{d^2y}{dx^2}$.
11. $y = \sqrt{x+1}$; find $\frac{d^2y}{dx^2}$.
12. $y = (4x - x^2)^2$; find $\frac{d^2y}{dx^2}$.
13. $y = (2x + 1)^3$; find $\frac{d^3y}{dx^3}$.
14. $y = x^4 - 3x^2 + 2$; find $\frac{d^3y}{dx^3}$.
15. $y = \frac{1}{\sqrt[3]{x^2}}$; find $\frac{d^2y}{dx^2}$.
16. $y = \frac{x^3}{\sqrt[3]{x}}$; find $\frac{d^3y}{dx^3}$.
17. $f(x) = \frac{x}{x+1}$; find $D_x^4 f(x)$.
18. $f(t) = \frac{t+1}{t+2}$; find $D_t{}^3 f(t)$.
19. $f(w) = (3w^2 + 1)^{-2}$; find $D_w^2 f(w)$.

In Exercises 20–27, find the indicated derivatives.

20. $\frac{d^5}{dx^5}(x^4-3x^2+1)$

21. $\frac{d^4}{dx^4}(x^7-4x^5-x^3)$

22. $\frac{d^8}{dx^8}(x^8-3x^2+2)$

23. $\frac{d^3}{dx^3}\left(x-\frac{1}{x^2}\right)$

24. $\frac{d^2}{dx^2}\left(\frac{x^2}{2x+3}\right)$

25. $\frac{d^2}{dx^2}\left(\frac{x^3}{3x-1}\right)$

26. $\frac{d^2}{dx^2}\left[(x^2+2x)(x-1)\right]$

27. $\frac{d^2}{dx^2}\left(\frac{x}{\sqrt{x+1}}\right)$

In Exercises 28–34, find y' and y'' in terms of x and y.

28. $x^2+y^2=9$

29. $x^2-4y^2=x$

30. $x^2-y^2=16$

31. $xy^2+4x=y$

32. $xy^2-5x=4y$

33. $x^2y=4$

34. $xy+y^2=2x^2$

Solve Exercises 35–38.

35. *Police science* Due to the rapid increase in major crimes, the mayor of a large city plans to organize a major crime task force. It is estimated that for each 1,000 people in the city, the number of major crimes will be $N(t)=62+5t^2-0.08t^{5/2}$, where t is the number of months after the task force has been organized and $0 \le t \le 12$. **(a)** Find $N'(t)$. **(b)** Find $N''(t)$.

36. *Meteorology* Meteorological results for a certain city indicate that for the month of April, the daily temperature in °F between midnight and 6:00 P.M. can be approximated by $T(t)=-0.04t^3+1.16t^2-9.3t+56$, where t is the number of hours after midnight and $0 \le t \le 18$. **(a)** Find $T'(t)$. **(b)** Find $T''(t)$.

37. *Meteorology* A weather balloon is released from the ground and rises in the atmosphere. The balloon's height above the ground, in meters, from the moment it is released is given by $h(t)=120t-12t^2$, where the time, t, varies from 0 to 6 min. **(a)** Find $h'(t)$. **(b)** Find $h''(t)$.

38. *Automotive* A car accelerates from rest to a maximum velocity at $t=45$ s and then decelerates to a stop at $t=120$ s. The distance traveled in meters is given by $s(t)=3t^2-\frac{1}{60}t^3$. **(a)** Find $s'(t)$. **(b)** Find $s''(t)$. (Note: $s'(t)$ is the velocity and $s''(t)$ is the acceleration.)

In Your Words

39. If the derivative of a function describes the rate of change of the function, what information does the second derivative give?

40. Explain the difference in the meanings of the notations $f^{(4)}(x)$ and $f^4(x)$.

CHAPTER 4 REVIEW

Important Terms and Concepts

Chain rule
Derivative
  At a point
  Four-step method
  Of a curve
  Of polynomials
Higher order derivatives
Implicit differentiation
Normal line
Power rule
Product rule
Quotient rule
Rules for derivatives
  Chain rule
  Constant function
  Of a composite function

Linear function
Of a sum or difference
Power rule
Product rule
Quotient rule
Second derivative
Third derivative

Review Exercises

In Exercises 1–6, use the four-step method to differentiate the given function.

1. $f(x) = 2x^2 - x$
2. $g(x) = x^3 - 1$
3. $f(x) = x - 2x^3$
4. $k(x) = \dfrac{4}{x^2}$
5. $m(x) = \sqrt{x}$
6. $j(t) = \sqrt{t+4}$

In Exercises 7–32, find the first derivative of the given function.

7. $f(x) = 3x^2 + 2x - 1$
8. $g(x) = 4x^{-5} + 2x$
9. $H(x) = 3x^5 + \dfrac{4}{x^3} + 3\sqrt{x}$
10. $y = \dfrac{1}{x^2+1}$
11. $y = \dfrac{3x}{x+3}$
12. $y = (2x+1)(x^2+3)$
13. $y = (x-4)(x^3-3x)$
14. $F(x) = \dfrac{\sqrt{x+1}}{x}$
15. $g(x) = \dfrac{x+1}{\sqrt{x+1}}$
16. $G(x) = \dfrac{x^2}{\sqrt{x+1}}$
17. $h(x) = (x^2+4)(x-2)(x^3-2)$
18. $f(x) = (3x^2-4x)$ $(5x^3+2x-1)$
19. $F(x) = \sqrt{(x^3+7)(x+5)}$
20. $g(x) = (x+5)^3$
21. $y = (x^2+5x)^2(x^3+1)^{-2}$
22. $h(x) = (x^2+7x-1)^4$
23. $y = (x+1)^2(x-1)^3$
24. $y = (x+4)^3\sqrt{x-1}$
25. $y = \dfrac{5}{(x^2+1)^3}$
26. $y = \dfrac{4x}{(3x+4)^2}$
27. $y = \dfrac{x^2}{\sqrt{(x^2-2)^3}}$
28. $f(t) = \dfrac{4t^3-3}{t^2-2}$
29. $g(u) = (u^2 + \frac{1}{u} - \frac{4}{u^2})^{-4}$
30. $R(t) = \sqrt{t^2-2t+1}\sqrt[3]{t^2+1}$
31. If $y = u^3 - u$, $u = x^2+1$, find $\dfrac{dy}{dx}$.
32. If $y = (u+4)^3$, $u = 2t-5$, find $\dfrac{dy}{dt}$.

In Exercises 33–38, find the implicit derivatives.

33. $x^2 + 4xy + 4y^2 = 16$
34. $x^3 - 2xy = y^2$
35. $2xy - x^2 = y^2x$
36. $x^2y + yx^3 = 10$
37. $x^2y^3 + \dfrac{1}{y} = x$
38. $4x^2 + xy^2 + y^{-1} = 2x$

In Exercises 39–49, find the indicated derivative.

39. $f(x) = x^7 + 2x^4 + 3x$; find $f''(x)$.
40. $h(x) = \sqrt{x} - \dfrac{1}{\sqrt{x}}$; find $h'''(x)$.
41. $g(x) = 4x^3 - 5x + \dfrac{2}{x}$; find $g'''(x)$.
42. $y = 4x^5 - 2x^{-3}$; find $\dfrac{d^2y}{dx^2}$.
43. $y = 7x^3 + \dfrac{7}{x^3}$; find $\dfrac{d^3y}{dx^3}$.
44. $y = x^{-3} + 4x^{-1}$; find $\dfrac{d^3y}{dx^3}$.

45. $f(x) = 5x^3 + \sqrt{x}$; find $D_x^3 f(x)$.

46. $g(x) = (x^2+1)^3$; find $\dfrac{d^3}{dx^3} g(x)$.

47. $x^2 y = y + 2x$; find y''.

48. $y^2 = xy^2 + x^3$; find y'.

49. $xy = x^2 + y^2$; find y' and y''.

Solve Exercises 50–60.

50. Find the slopes of the tangent and the normal to the curve $y = 4x^2 - 9x$ at $(3, 9)$.

51. Find the slope of the tangent to the curve $f(x) = 7x^3 - 3x + 6$ at $(-1, 2)$

52. Find the slope of the tangent to the ellipse $4x^2 + 9y^2 = 40$ at $(-1, 2)$

53. Find the slopes of the tangent and the normal to the curve $4x^2 + 3y^2 - 5xy - 4x + 10y - 11 = 0$ at $(2, -1)$.

54. If $f(x) = x^3 - 3x - 1$, find all values of x for which $f'(x) = 0$.

55. If $g(x) = \dfrac{x^2}{x+1}$, find all values of x for which $g'(x) = 0$.

56. Find the equation of the tangent to the curve $y = x^2 + 4$ at $(-1, 5)$.

57. Find the equation of the normal to the ellipse $4x^2 + 5y^2 = 36$ at $(-2, 2)$.

58. If $y = \sqrt{x+4}$, find all values of x for which $y' = 0$.

59. Find all values of x where $f''(x) = 0$ if $f(x) = 3x^4 - 6x^2 + 2$.

60. If $y = x^3 - 9x^2 - 4$, find the value(s) of x for which $y'' = 0$.

CHAPTER 4 TEST

In Exercises 1–8, find the first derivative of the given functions.

1. $f(x) = 5x^3 - 4$
2. $g(x) = 3x^{-4} + 5x^2$
3. $h(x) = \sqrt[3]{x^6 - 2}$
4. $j(x) = (2x+1)(3x^2 - x)$
5. $k(x) = \dfrac{2x+1}{\sqrt{4x^2+1}}$

Solve Exercises 6–10.

6. Find the second and third derivatives of $f(x) = 5x^4 - 4x^{-3}$.

7. Find y', if $2xy + y^3 x = x^3$.

8. Find the equation of the tangent to the curve $g(x) = 4x^3 - 2x^2 + 4$ at $(1, 6)$.

9. A particle moving in a straight line is at a distance of $s(t) = 4.5t^2 + 27t$ ft from its starting point after t s. If the velocity $v(t)$ of the particle at time, t, is $v(t) = s'(t)$, find the velocity of this particle.

10. The population of a certain bacterial culture after t hours is approximated by $P(t) = \dfrac{t^2 - 5t}{2\sqrt{t} + 7}$, where $P(t)$ is in thousands of bacteria. **(a)** Find an expression for the rate of growth of these bacteria. **(b)** Determine the rate of growth after 5 hours.

CHAPTER

5

Applications of Derivatives

A telephone company is asked to provide service to the house shown here. In Section 5.4, you will learn where to connect this line so that it will cost the least.

Courtesy of Michael A. Gallitelli, Metroland Photo Inc.

In Chapter 4, we learned how to take derivatives of many functions. In this chapter, we will look at some of the applications of derivatives. We will begin by looking at motion applications. We will then look at ways to determine maxima and minima of functions. Finally, we will learn two tests that use derivatives.

5.1 RATES OF CHANGE

One idea that is important to the study of mathematics is **change**. As a technician, one of your responsibilities may be to observe something and measure it at certain time intervals. When these observations are completed, it will be necessary to see if any changes resulted and how those changes occurred. One of the powerful aspects of calculus is that it can be used to determine the rate at which something changes.

Velocity, Speed, and Acceleration

Suppose the distance x of an object p from the origin at any time t is given by $s(t)$. It is often convenient to consider **rectilinear motion** where the object or particle moves along a straight line. Usually we use a horizontal axis to represent rectilinear motion and select the origin as the initial position of the particle. Thus, $x = s(t)$ represents the distance of the object p from the origin 0 at t seconds. (See Figure 5.1.) Because $x = s(t)$ gives the position of p at time t, it is often referred to as a **position function**.

FIGURE 5.1

In order to measure the velocity of p at some given time c, we look at the average velocity of p from c to $c+h$. Now, the average velocity is the distance traveled divided by the time it took to go that distance. In this case, the average velocity.

$$\bar{v} = \frac{s(c+h)-s(c)}{h}$$

As you can see, if we take the limit as h approaches 0, we will get the **instantaneous velocity** of p at $t = c$ or $v(c)$. Thus,

$$v(c) = \lim_{h\to 0}\frac{s(c+h)-s(c)}{h}$$

But, the expression on the right-hand side is the derivative of s and so

$$v(c) = s'(c)$$

Thus, the **velocity function** for p is the derivative of the position function s. In other words,

$$v(t) = s'(t) = D_t s = \frac{ds}{dt}$$

The units of s and t determine the units for the velocity. If s is in miles and t in hours, the velocity is in miles per hour (mph). If s is in meters or feet and t is in seconds, the velocity is in meters per second (m/s) or feet per second (ft/s).

We should probably point out that the velocity of an object is not the same as the speed of an object.

Note **Speed** indicates the magnitude of the velocity and is defined as the absolute value of velocity, $|v(t)|$. Thus, speed is always nonnegative. Velocity can be positive, negative, or zero.

If the velocity is positive, the object is moving in a positive direction. If it is negative, the object is moving in the opposite direction. Thus, speed tells us how fast the object is moving while velocity tells us both the speed and the direction.

Now, if we differentiate the velocity, we get $v'(t) = s''(t)$. This measures the instantaneous rate of change of the velocity and is known as the **acceleration**. If we let $a(t)$ denote the acceleration, then

$$a(t) = v'(t) = s''(t) = \frac{d^2s}{dt^2}$$

The units for the acceleration are cm/s^2 (centimeters per second per second), m/s^2 (meters per second per second), mi/h^2 (miles per hour per hour), and so on.

EXAMPLE 5.1

The position function s of a particle p on a coordinate line is given by

$$s(t) = t^3 - 9t^2 + 24t - 10$$

where t is in seconds and $s(t)$ in millimeters. Where is the particle when $t = 3$ and what is its velocity and acceleration at that time?

Solution The position is given by

$$\begin{aligned} s(3) &= 3^3 - 9(3^2) + 24(3) - 10 \\ &= 27 - 81 + 72 - 10 \\ &= 8 \text{ mm or } 8 \text{ mm to the right of the origin} \end{aligned}$$

The first derivative of s, $s'(t)$, is the velocity function

$$v(t) = s'(t) = 3t^2 - 18t + 24$$

and when $t = 3$,

$$\begin{aligned} v(t) &= 3(3^2) - 18(3) + 24 \\ &= 27 - 54 + 24 = -3 \text{ mm/s} \end{aligned}$$

The object is moving with a speed of 3 mm/s in a negative direction. The acceleration function is the second derivative of the position function, and so

$$a(t) = s''(t) = 6t - 18$$

and when $t = 3$,

$$a(3) = 6(3) - 18 = 0 \text{ mm/s}^2$$

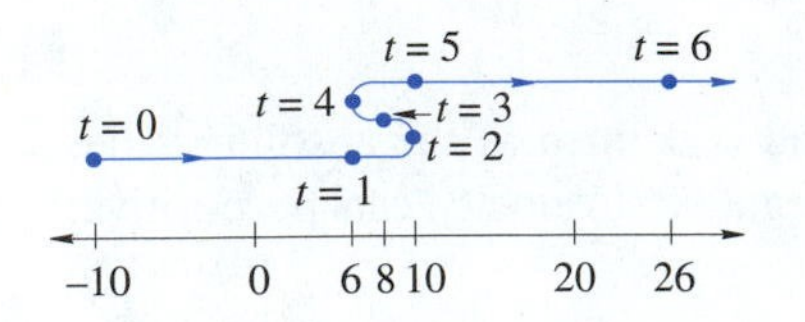

FIGURE 5.2

The following table shows the values for the position, velocity, and acceleration of the particle in Example 5.1, each second from $t = 0$ through $t = 6$.

t	0	1	2	3	4	5	6
s(t)	−10	6	10	8	6	10	26
v(t)	24	9	0	−3	0	9	24
a(t)	−18	−12	−6	0	6	12	18

The motion of an object such as the particle in Example 5.1 can often be represented by a drawing like that shown in Figure 5.2. This curve does not show the path of the particle, but only shows how the particle moves. The direction of the motion is indicated by the arrows.

Not all motion is horizontal. The motion of a projectile fired straight upward is another example of rectilinear motion. In this instance, the axis is vertical rather than horizontal, the origin is at ground level, and the positive direction is upward.

Application

EXAMPLE 5.2

A projectile is fired straight upward with a velocity of 192 ft/s. Its distance above the ground at t seconds is given by $s(t) = -16t^2 + 192t$. Find the time when the projectile hits the ground. What is its velocity and acceleration at the time of impact?

Solution We are given the distance function as $s(t) = -16t^2 + 192t$. When $s(t) = 0$, the projectile will be at ground level, so we need to solve $s(t) = -16t^2 + 192t = 0$. We can factor t out of the expression to obtain

$$t(-16t + 192) = 0$$

and so $s(t) = 0$ when $t = 0$ or when $t = 12$. Since $t = 0$ when the projectile is fired, it must strike the ground when $t = 12$.

The velocity is represented by $v(t) = s'(t) = -32t + 192$ and the acceleration by $a(t) = s''(t) = v'(t) = -32$. When $t = 12$, $v(12) = -192$ and $a(t) = -32$. So, the velocity is 192 ft/s downward and the acceleration is 32 ft/s^2 downward.

Electricity

Time is not the only variable that can be used to study rates of change. In the next example, we will look at the rate at which the current changes with respect to the resistance.

Application

EXAMPLE 5.3

The current I in a certain electrical circuit is given by the formula $I = \dfrac{120}{R}$, where R represents the resistance. What is the rate of change of I with respect to R when the resistance is 60 Ω?

Solution We are given $I = \dfrac{120}{R}$ and want to find $\dfrac{dI}{dR}$ when $R = 60$.

$$\begin{aligned} I &= \frac{120}{R} = 120R^{-1} \\ \frac{dI}{dR} &= -120R^{-2} \\ &= \frac{-120}{R^2} \end{aligned}$$

So, when $R = 60$, $\dfrac{dI}{dR} = \dfrac{-120}{60^2} = \frac{-1}{30}$. Thus, when $R = 60\,\Omega$, the current is decreasing at a rate of $\frac{1}{30}$ A/Ω (amperes per ohm).

Exercise Set 5.1

In Exercises 1–8, position functions of points moving rectilinearly are defined. Find the functions for the velocity and acceleration at time t. Make a table of values for each of the three functions at each second. (You might want to write a computer program to help you.) Illustrate the motion by means of a diagram like the one shown in Figure 5.2.

1. $s(t) = 3t^2 - 12t + 5$ from $t = 0$ to $t = 5$
2. $s(t) = 3t^2 - 18t + 10$ from $t = 0$ to $t = 5$
3. $s(t) = t^3 - 12t + 2$ from $t = -4$ to $t = 4$
4. $s(t) = t^3 - 27t + 20$ from $t = -4$ to $t = 4$
5. $s(t) = t + 4/t$ from $t = 1$ to $t = 4$
6. $s(t) = t + 8/t$ from $t = 1$ to $t = 4$
7. $s(t) = 2\sqrt{t} + \dfrac{1}{\sqrt{t}}$ from $t = 1$ to $t = 4$
8. $s(t) = t^3 - 8$ from $t = 0$ to $t = 4$

In Exercises 9–12, an object is fired straight upward. Its height is represented by the given formula. (a) When will it hit the ground? (b) What are its velocity and acceleration when it hits the ground? (c) What was its height when it was fired?

9. $s(t) = 144t - 16t^2$ ft
10. $s(t) = 250 + 256t - 16t^2$ ft
11. $s(t) = 29.4t - 4.9t^2$ m
12. $s(t) = 180 + 98t - 4.9t^2$ m

Solve Exercises 13–28.

13. *Electronics* The repulsion F, in dynes, between a certain pair of electrical charges varies with the distance apart, s, in centimeters.
    (a) If $F(s) = \dfrac{50}{s^2}$, what is the rate of change of F with respect to s?
    (b) What is the rate of change when $s = 3$ cm?
14. *Electronics* The current at a given point in a circuit i, in amperes (A), is defined as the instantaneous rate of change of the charge q, in coulombs (C), at any given time t, in seconds (s). The expression for the charge is given by
    $$q = 25 + 10t - 2.0t^2$$
    (a) What is the general expression for the current?
    (b) What is the initial charge; that is, what is the charge when $t = 0.0$ s?
    (c) What is the current at $t = 0.0$ s and $t = 4.0$ s?
15. *Physics* If the relationship between F, the temperature in degrees Fahrenheit (°F), and C, the temperature in degrees Celsius (°C), is given by $F = \frac{9}{5}C + 32$, find **(a)** $\dfrac{dF}{dC}$, **(b)** $\dfrac{dC}{dF}$, and **(c)** $\dfrac{dC}{dF}$ at $C = 20$°C.
16. *Thermodynamics* When metal is heated it expands. The area of a circle is related to its radius.
    (a) If a circular metal plate is heated, what is the rate of change of the area with respect to the radius?
    (b) What is the rate when $r = 1.5$ cm?
    (c) What is the rate when $r = 3$ cm?
17. *Hydrology* As water leaks out of a tank, the quantity Q of water in gallons remaining in the tank after t min is given by the formula $Q(t) = 900 - 50t + 0.5t^2$.
    (a) What is the rate of change in the quantity of water?
    (b) What is the rate of change when $t = 4$ min?
    (c) What is the rate of change when $t = 8$ min?
18. *Physics* An object rolls down a ramp. The distance in millimeters it rolls in t seconds is given by the formula $s(t) = 5t^2 + 2t$.
    (a) What is the formula for its velocity?
    (b) What is the formula for its acceleration?
    (c) What distance has it rolled after 3 s, and what are its velocity and acceleration at that time?
    (d) At what time is its velocity 46 mm/s?
19. *Physics* The position of a ball that is projected upward on a frictionless inclined plane is given by $s(t) = 8.0t - 2.0t^2$, where $s(t)$ is in meters (m) and t is time in seconds (s).
    (a) What is the initial velocity?

(b) How far will the ball move up the inclined plane before it stops and starts to roll back down the plane?
(c) How long will it take for the ball to roll up the plane and return to the initial position?

20. Find the rate of change of the area A of a circle with respect to its circumference C.

21. *Electronics* The induced voltage in a coil v, in volts (V), is given by $v = -N\frac{d\phi}{dt}$, where N is the number of turns in the coil, ϕ is the magnetic flux in webers (Wb), and t is time in seconds (s). If a coil of 50 turns is connected by a magnetic flux of $\phi = 2.4t^{1/2} - 0.4t^2$, find
(a) the general expression for the induced voltage and
(b) the induced voltage at $t = 1.0$ s and $t = 9.0$ s.

22. The height of a cone h is twice the radius r of the base of the cone.
(a) Find the instantaneous rate of change of the volume V with respect to the radius.
(b) What is $\frac{dV}{dr}$ when $r = 24.3$ mm?

23. *Hydrology* A stone dropped in a pond makes a circular ripple that travels out from the point of impact at 1.7 m/s. At what rate is the area of the circle increasing when $t = 8$ s?

24. *Electricity* The current i, in amperes (A), in a resistance of $r\,\Omega$ when the voltage V, in volts (V), is given by $i = \frac{V}{r}$. The resistance changes with time t in seconds (s) according to $r = 3.0t^{1/2}$ and the voltage changes according to $V = t^2 - 1.2$.
(a) Give an expression for i in terms of t.
(b) Find $\frac{di}{dt}$.
(c) Find $\frac{di}{dt}$ at $t = 2.4$ s.

25. *Physics* The pressure of a gas P in kilopascals (kPa) varies inversely with the volume V in cubic meters. If the pressure of the gas is $P = 5.21$ kPa when the volume is 1.23 m^3,
(a) state the equation for P, and
(b) find the instantaneous rate of change of the pressure with respect to the volume.

26. *Physics* The angular displacement of a rotating wheel θ in radian measure (rad) is given by $\theta(t) = 3.4t - 1.2t^2$, where t is time in seconds (s).
(a) Find the angular velocity, $\omega(t)$.
(b) Find the angular acceleration, $\alpha(t)$.
(c) What is the initial angular velocity and acceleration?

27. *Automotive* A car accelerates from rest to a maximum velocity at $t = 45$ s and then decelerates to a stop at $t = 120$ s. The distance traveled in meters is given by $s(t) = 3t^2 - \frac{1}{60}t^3$.
(a) What is the velocity as a function of time t of the car during this period?
(b) What is the acceleration as a function of time t of the car during this period?
(c) What is the speed of the car during this period?

28. *Meteorology* A weather balloon is released from the ground and rises in the atmosphere. The balloon's height above the ground, in meters, for the first six minutes after it is released is given by $h(t) = 120t - 12t^2$.
(a) What is the velocity of the balloon during this period?
(b) What is the acceleration of the balloon during this period?

In Your Words

29. Describe how the position, velocity, and acceleration of an object are related.

30. What is the relationship between the speed and the velocity of an object?

5.2 EXTREMA AND THE FIRST DERIVATIVE TEST

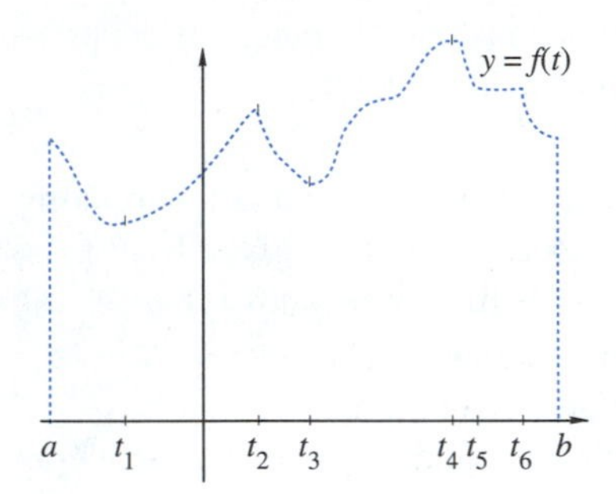

FIGURE 5.3

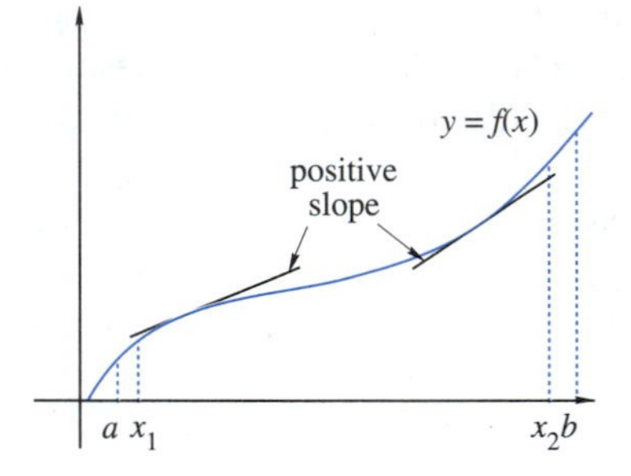

FIGURE 5.4

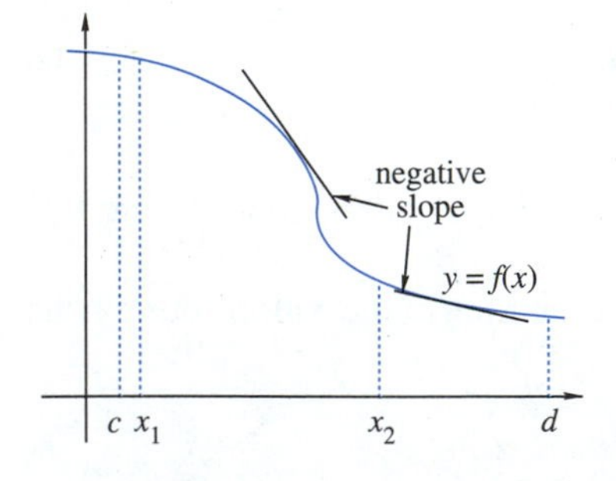

FIGURE 5.5

In Section 5.1, we looked at several applications of derivatives, particularly as they apply to the distance function. We were unable to determine when the distance, velocity, or acceleration is the greatest or least, or when the distance or speed stops increasing and begins decreasing. In this section, we will look at methods for using calculus to determine this type of information.

We will begin by examining the graph of a function f shown in Figure 5.3. This graph might have resulted when a recording instrument was used to measure something over a period of time. Several times have been marked because they are of interest. For example, the graph decreased in the time interval $[a, t_1]$ and increased in the interval $[t_1, t_2]$, decreased in $[t_2, t_3]$, and so on.

You can also see that, in the time period immediately surrounding t_1, the function is at a low point at t_1. Similarly, you can see that the function reaches a high point at t_2, another low point at t_3, and another high point at t_4. Because t_1 represents a place where the graph is at its lowest in an interval around the point, it is referred to as a **local minimum** or **relative minimum**. Another local minimum is at t_3. Similarly, t_2 and t_4 each represent a **local maximum**, because each one is a place where the graph is at its highest over some interval. Now let's look at all of these ideas more closely.

Increasing and Decreasing Functions

The first thing we noticed in Figure 5.3 was that over some intervals the function increased, over some it decreased, and on one interval it neither increased nor decreased.

Look at the graph of $y = f(x)$ in Figure 5.4. As the values of x increase (going from left to right) on the interval $[a, b]$, the corresponding values of $f(x)$ also increase and the curve rises. More formally we would state the following.

Increasing Function

If x_1 and x_2 are any two points in the interval $[a, b]$ with $x_1 < x_2$ and if $f(x_1) < f(x_2)$, then f is an **increasing function** on the interval.

As you can see, the curve has a positive slope at every point in this interval and so $f'(x) > 0$ for all values of x in $[a, b]$.

In a similar manner, we should look at the function in Figure 5.5. The graph shows a decreasing function. In particular, as the values of x increase on the interval $[c, d]$, the corresponding values of f decrease and the curve falls.

Decreasing Function

If x_1 and x_2 are any two points in the interval $[a, b]$ with $x_1 < x_2$ and if $f(x_1) > f(x_2)$, then f is a **decreasing function** on the interval.

The tangent lines of a decreasing function all have a negative slope and so $f'(x) < 0$ for all points in the interval $[c, d]$.

This leads us to an important rule dealing with increasing and decreasing functions.

> **Rule 1: Test to See if a Function is Increasing or Decreasing**
>
> If $f'(x) < 0$ on an interval, then f is a decreasing function on that interval.
>
> If $f'(x) > 0$ on an interval, then f is an increasing function on that interval.
>
> If $f'(x) = 0$ on an interval, then f is a constant function on that interval.

Determining the intervals when a derivative is positive or negative will often require your ability to solve inequalities. The next two examples will show how to determine when a function is increasing and when it is decreasing.

EXAMPLE 5.4

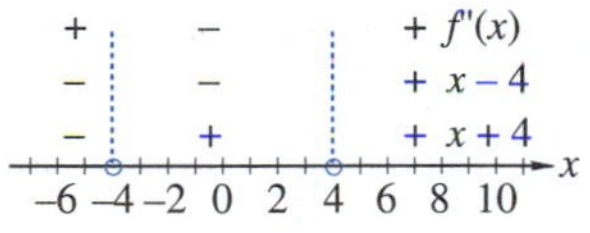

+ − + f'(x)
− − + x − 4
− + + x + 4
−6 −4 −2 0 2 4 6 8 10 x

FIGURE 5.6

Find the intervals on which $f(x) = \frac{4}{3}x^3 - 64x$ is increasing or decreasing.

Solution Rule 1 says that we can determine when f is decreasing or increasing by looking at f'. The derivative of f is

$$\begin{aligned} f'(x) &= 4x^2 - 64 \\ &= 4(x^2 - 16) \\ &= 4(x+4)(x-4) \end{aligned}$$

The sign chart in Figure 5.6 shows where $f'(x)$ is positive and where it is negative. Since $f'(x)$ is negative on the interval $(-4, 4)$, we can say the f decreases on $(-4, 4)$. Similarly, f increases over the intervals $(-\infty, -4)$ and $(4, \infty)$.

EXAMPLE 5.5

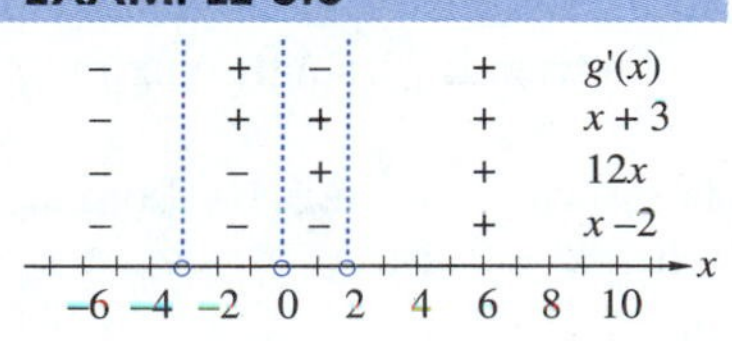

FIGURE 5.7

Find the intervals on which $g(x) = 3x^4 + 4x^3 - 36x^2$ is increasing or decreasing.

Solution The derivative of $g(x)$ is

$$\begin{aligned} g'(x) &= 12x^3 + 12x^2 - 72x \\ &= 12(x^3 + x^2 - 6x) \\ &= 12x(x^2 + x - 6) \\ &= 12x(x+3)(x-2) \end{aligned}$$

From Figure 5.7 we can see that $g'(x) < 0$ on the intervals $(-\infty, -3)$ and $(0, 2)$, so $g(x)$ is decreasing on those intervals. Similarly, since $g'(x) > 0$ on $(-3, 0)$ and $(2, \infty)$, $g(x)$ is increasing on these two intervals.

Extrema and Critical Points

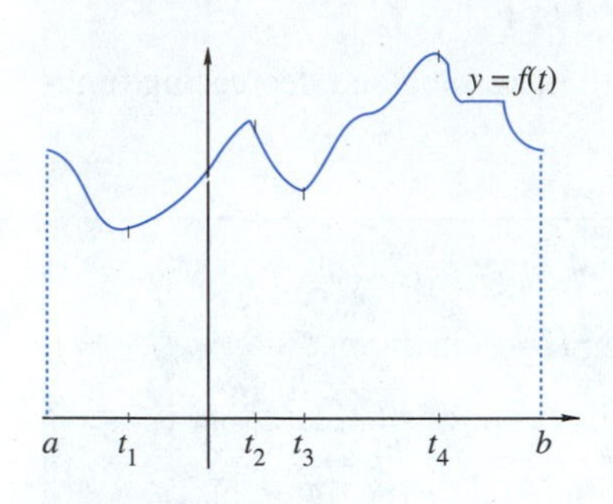

FIGURE 5.8

Let's look again at the graph that was used to begin this section. The graph is repeated in Figure 5.8. At t_2 and t_4 we see that the function is higher than any other "nearby" points on the curve. The function is lower than any other nearby points at t_1 and t_3. Since these may not be the highest (or lowest) points on the entire curve, we say that f has a local maximum or a relative maximum at t_2 and t_4. If we can show that the curve does have a highest point then it is called the **absolute maximum**. Just as we described a function as increasing or decreasing over an interval, we say that a function has a local maximum at a point if the value of the function at that point is the highest value in the interval. When finding the maxima or minima over an interval you must always check the endpoints of the interval.

As you look at the local maxima and minima in Figure 5.8, there is something else you should notice. To the left of each relative minimum the function decreases; to the right, the function increases. Just the opposite is true for relative maxima.

EXAMPLE 5.6

Determine the relative maxima and minima of

$$g(x) = 3x^4 + 4x^3 - 36x^2$$

Solution This is the same function we examined in Example 5.5. We found that on

$$(-\infty, -3), \ g \text{ is decreasing.}$$
$$(-3, 0), \ g \text{ is increasing.}$$
$$(0, 2), \ g \text{ is decreasing.}$$
$$(2, \infty), \ g \text{ is increasing.}$$

Thus, we have a minimum when $x = -3$, a maximum when $x = 0$, and a minimum when $x = 2$.

The graph of g, as shown in Figure 5.9, supports the conclusions we reached using calculus.

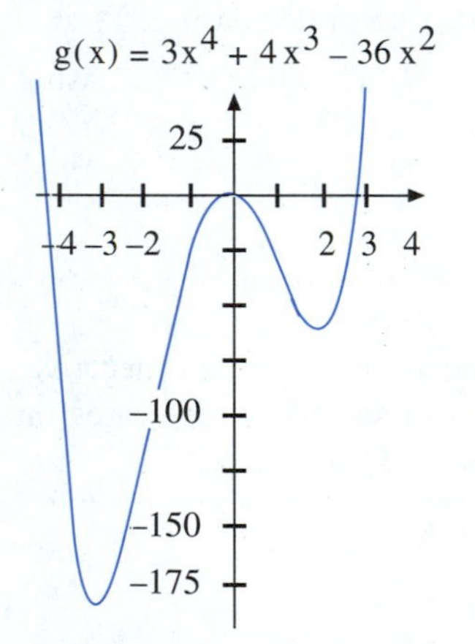

FIGURE 5.9

In the previous example, we began a practice we will follow from now on. We will let maximum refer to either an absolute or a relative maximum. We will also use minimum for either an absolute or relative minimum. Collectively, maximum and minimum values are referred to as **extreme values** or **extrema**. (A single maximum or minimum is called an **extremum**.)

Consider the function $h(x) = x^{2/3}$. Its graph is shown in Figure 5.10. Looking at the graph, we can see that $h(x)$ has a minimum when $x = 0$. But,

$$h'(x) = \frac{2}{3}x^{-1/3} = \frac{2}{3x^{1/3}}$$

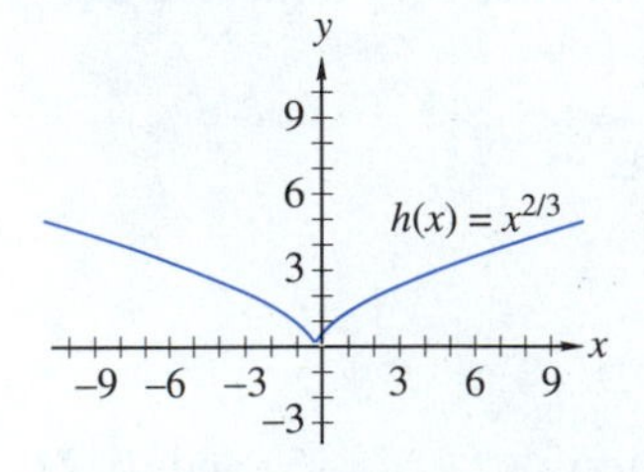

FIGURE 5.10

When $x = 0$, $h'(x)$ is not defined.

The example above leads to a definition of critical values (and critical points) for a function.

Critical Values and Critical Points

Extrema may occur at values of x in the domain of a function $y = f(x)$ for which $f'(x) = 0$ or $f'(x)$ is not defined. These values of x are referred to as **critical values**. If x is a critical value, then the point $(x, f(x))$ is called a **critical point**.

First Derivative Test

We are now ready to talk about a method for testing critical values to see if they are extrema. This method is called the **first derivative test** for $f(x)$ and it has four steps.

First Derivative Test

1. Find $f'(x)$.
2. Determine the critical values.
3. Use the critical values to determine intervals when f is decreasing $(f'(x) < 0)$ or increasing $(f'(x) > 0)$.
4. At each critical value c, determine if $f'(x)$ changes sign as x increases through c. To do this, select a value of x less than c and one greater than c. If, as x increases,
   (a) $f'(x)$ changes from $+$ to $-$, then $f(c)$ is a maximum;
   (b) $f'(x)$ changes from $-$ to $+$, then $f(c)$ is a minimum;
   (c) the sign of $f'(x)$ does not change, then $f(c)$ is neither a maximum nor a minimum.

EXAMPLE 5.7

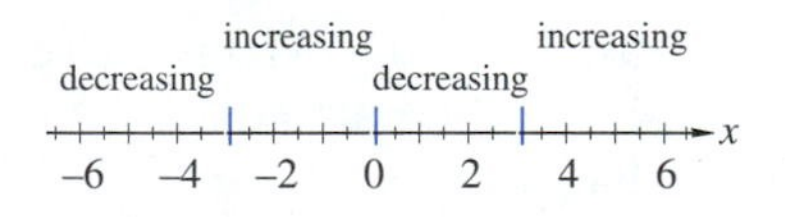

FIGURE 5.11

If $f(x) = x^4 - 18x^2$, use the first derivative test to find the intervals where f is increasing or decreasing and to locate all extrema.

Solution

1. $f'(x) = 4x^3 - 36x = 4x(x+3)(x-3)$
2. Letting $f'(x) = 0$ gives the critical values when $x = 0, -3$, and 3.
3. There are four intervals to consider. Using the same procedures we used in Examples 5.4 and 5.5, we find that on

$$(-\infty, -3),\ f'(x) < 0, \text{ so } f \text{ is decreasing}$$
$$(-3, 0),\ f'(x) > 0, \text{ so } f \text{ is increasing}$$
$$(0, 3),\ f'(x) < 0, \text{ so } f \text{ is decreasing}$$
$$(3, \infty),\ f'(x) > 0, \text{ so } f \text{ is increasing}$$

   This is summarized in Figure 5.11.

EXAMPLE 5.7 (Cont.)

4. At $x = -3$, there is a minimum since $f'(x)$ changes from $-$ to $+$. When $x = 0$, there is a maximum since $f'(x)$ changes from $+$ to $-$. At $x = 3$, there is a minimum since $f'(x)$ changes from $-$ to $+$.

   The graph of this function, $f(x) = x^4 - 18x^2$, is shown in Figure 5.12.

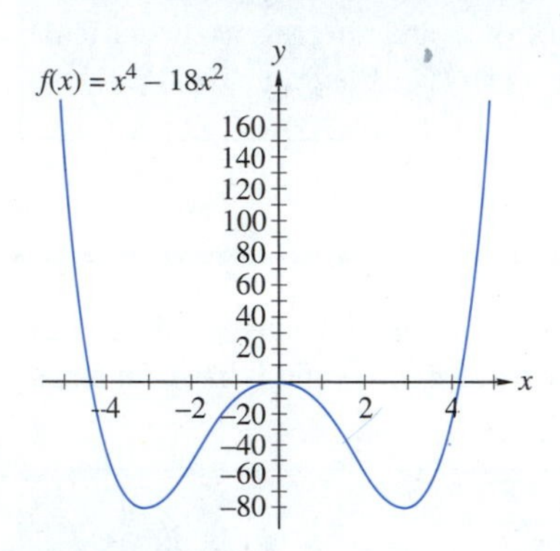

FIGURE 5.12

EXAMPLE 5.8

If $g(x) = x + \dfrac{9}{x+2}$, use the first derivative test to find all extrema.

Solution Notice that -2 is not in the domain of g and that the graph of g has a vertical asymptote at $x = -2$.

1. $$\begin{aligned} g'(x) &= 1 - \frac{9}{(x+2)^2} \\ &= \frac{(x+2)^2}{(x+2)^2} - \frac{9}{(x+2)^2} \\ &= \frac{(x+2)^2 - 9}{(x+2)^2} \\ &= \frac{x^2+4x+4-9}{(x+2)^2} \\ &= \frac{x^2+4x-5}{(x+2)^2} \\ &= \frac{(x+5)(x-1)}{(x+2)^2} \end{aligned}$$
2. $g'(x) = 0$ when $(x+5)(x-1) = 0$ or when $x = -5$ or $x = 1$. $g'(x)$ is undefined when $(x+2)^2 = 0$ or when $x = -2$. However, since -2 is not in the domain of g, it is not a critical value. The only critical values are $x = -5$ and $x = 1$.
3. There are four intervals to consider. Remember that $(x+2)^2$ is never negative. Solving we get the following results.

   $$\begin{aligned} &(-\infty, -5),\ g'(x) > 0, \text{ so } g \text{ is increasing} \\ &(-5, -2),\ g'(x) < 0, \text{ so } g \text{ is decreasing} \\ &(-2, 1),\ g'(x) < 0, \text{ so } g \text{ is decreasing} \\ &(1, \infty),\ g'(x) > 0, \text{ so } g \text{ is increasing} \end{aligned}$$

   This is summarized in Figure 5.13.

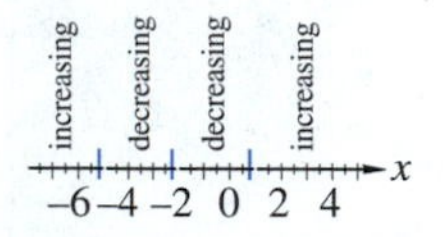

FIGURE 5.13

EXAMPLE 5.8 (Cont.)

4. At $x = -5$, there is a maximum since $g'(x)$ changes from $+$ to $-$. At $x = 1$, there is a minimum since $g'(x)$ changes from $-$ to $+$. The value $x = -2$ is not in the domain of g. There is no extrema at that point. The graph of g is shown in Figure 5.14.

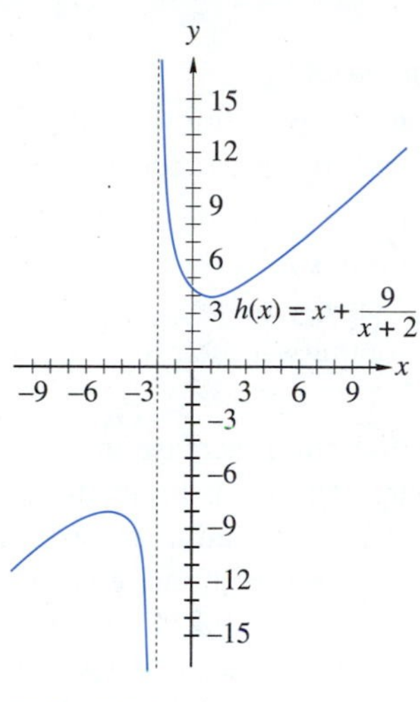

FIGURE 5.14

Exercise Set 5.2

In Exercises 1–26, determine when the function is increasing or decreasing, locate all critical values, and determine all extrema. Do not sketch the graph of the function.

1. $f(x) = x^2 - 8x$
2. $g(x) = 10x - x^2$
3. $m(x) = 16 - x^2$
4. $h(x) = 3x^2 - 12x + 10$
5. $j(x) = 15 + 16x - 4x^2$
6. $k(x) = x^2 - 4x + 11$
7. $f(x) = 4x^2 - 16x + 1$
8. $s(t) = -t^2 + 7t + 4$
9. $q(t) = t^3 - 3t$
10. $v(t) = 21t + 9t^2 - t^3$
11. $y = \frac{1}{4}x^4 - 18x^2 + 16$
12. $t = 3.2 - 12.5w^2 + 0.25w^4$
13. $f(z) = (z+4)^3$
14. $g(x) = \left(2x^2 - 8\right)^2$
15. $g(v) = 3v^4 - 4v^3 - 12v^2 + 12$
16. $y = \frac{5}{x}$
17. $s(t) = \frac{6}{t+1}$
18. $h(x) = x + \frac{1}{x}$
19. $g(x) = x^2 - \frac{2}{x}$
20. $j(x) = x + \frac{1}{x+3}$
21. $k(x) = \frac{x^2}{1-x}$
22. $m(b) = 4\sqrt{b}$
23. $f(t) = t\sqrt{1-t^2}$
24. $g(s) = (s+2)^2(s-1)^{2/3}$
25. $y = x(x-1)^{2/3}$
26. $f(x) = 9\sqrt[3]{x} - 4x$

Solve Exercises 27–36.

27. *Physics* An object is fired straight upward with its height, given in feet, represented by $s(t) = 250 + 256t - 16t^2$. **(a)** At what time is the altitude a maximum? **(b)** What is the maximum altitude?

28. *Physics* An object is fired straight upward with its height, given in meters, represented by $s(t) = 180 + 98t - 4.9t^2$. **(a)** At what time is its height a maximum? **(b)** What is the maximum height?

29. *Electricity* The power P in watts (W) in a resistor is given by $P = 4.7(t-2.0)^2$, where t is time in seconds (s).
(a) What is the lowest power in the resistor?
(b) When does this occur?

30. *Electricity* The current i in amperes (A) at a given point in a circuit is given by $i = 4.8t - 1.2t^2$ where t is time in seconds (s).
(a) At what time is the current maximum?
(b) What is the maximum current?

31. *Electricity* When two resistors R_1 and R_2 are connected in parallel, their total resistance R is

$$R = \frac{R_1 R_2}{R_1 + R_2}$$

If $R_1 = 9\,\Omega$ and R_2 is a variable resistor, then

$$R = \frac{9R_2}{9 + R_2}$$

(a) What values of R_2 provide the minimum and maximum values of R?
(b) What are the minimum and maximum values of R?

32. *Physics* A particle moves along the x-axis according to the position function $x(t) = 8t^3 - t^4$. Find the velocity at the instant the acceleration is a maximum.

33. *Business* The profit from the sale of n items of a certain product is given by $P(n) = (4-0.2n)(2.5n+7)$, where $P(n)$ is in hundreds of dollars and $1 \le n \le 20$.
(a) If the marginal profit is $P'(n)$, find the marginal profit function.
(b) How many items must be sold to make the most profit?
(c) What is the maximum profit?

34. *Energy technology* The energy output in joules of an electrical heater varies with time, t, according to the equation $E = 0.01\left(6t - t^2\right)^3$, where $0 \le t \le 6$.
(a) When is the energy output a maximum?
(b) What is the maximum energy output?
(c) When are the energy outputs at their lowest levels?

35. *Police science* Due to the rapid increase in major crimes, the mayor of a large city plans to organize a major crime task force. It is estimated that for each 1,000 people in the city, the number of major crimes will be $N(t) = 132 + 4\sqrt{t} + 2t^{3/2} - 0.8t^{5/2}$, where t is the number of months after the task force has been organized and $0 \le t \le 6$. **(a)** Find $N'(t)$. **(b)** When is the number of crimes at its peak?

36. *Meteorology* Meteorological results for a certain city indicate that for the month of April, the daily temperature in °F between midnight and 6:00 P.M. can be approximated by $T(t) = -0.04t^3 + 1.26t^2 - 9.3t + 56$, where t is the number of hours after midnight and $0 \le t \le 18$.
(a) Find $T'(t)$.
(b) When is the temperature the highest?
(c) When is the temperature the lowest?

In Your Words

37. Explain the first derivative test and what information it gives about a function.

38. **(a)** What determines whether a function is increasing or decreasing?
(b) How can you use calculus to determine whether a function is increasing or decreasing?

5.3 CONCAVITY AND THE SECOND DERIVATIVE TEST

In Section 5.2, we used the first derivative to determine when a function is increasing or decreasing and to locate its extrema. In this section, we will learn some uses for the second derivative. In order to be able to use the second derivative, we need to take a closer look at a curve and its shape.

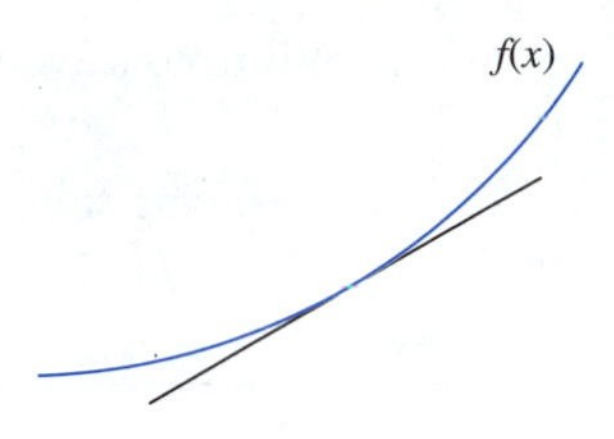

FIGURE 5.15a

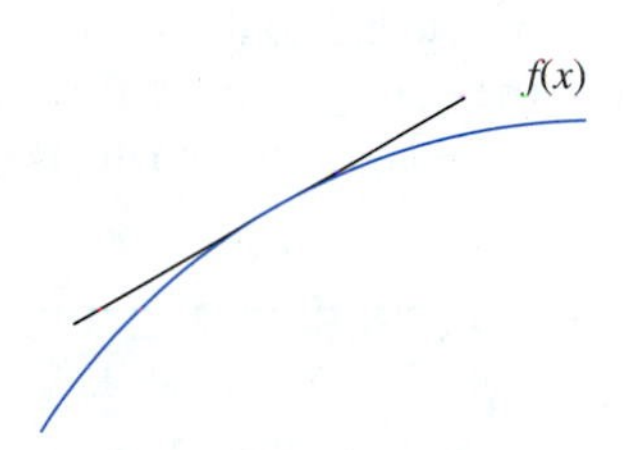

FIGURE 5.15b

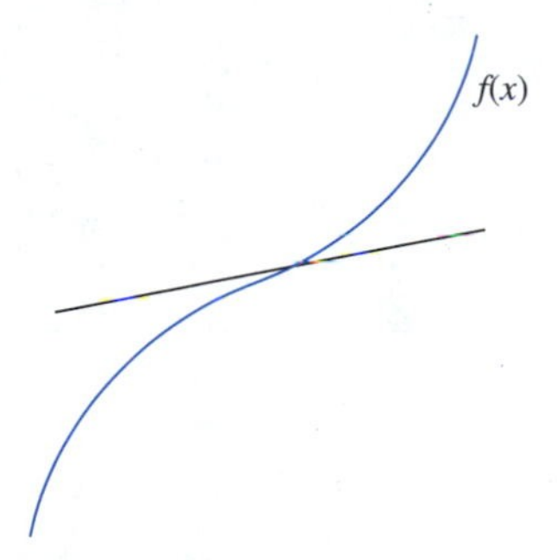

FIGURE 5.15c

Concavity

We will begin by looking at the three curves $f(x)$ in Figures 5.15a–c. In all three cases a tangent line has been drawn. Looking at the tangents, we can see that they all have positive slopes and so $f'(x) > 0$. These figures illustrate the three possible situations. In Figure 5.15a, the curve lies above the tangent line; in Figure 5.15b, the curve lies below the tangent line; and in Figure 5.15c, the curve crosses the tangent line. This example is not unique. We could have the same three situations if the first derivative were negative or if it were zero.

The three situations demonstrated in Figures 5.15a–c give us the chance to introduce some new terminology. If there is an open interval around a point where the graph of a function f is above the tangent lines for all points in that interval, then the graph of f is **concave up** at that point. This is demonstrated in Figure 5.15a and in Figure 5.16. There is an open interval (a, b) that contains the point c. The graph of $y = f(x)$ passes through $(c, f(c))$ and is above the tangent line through that point. Pick any other point in (a, b) and draw the tangent lines through that point. If the curve is above that tangent line, then the curve is concave up. In Figure 5.16, the points x_1 and x_2 represent any two points in (a, b), except c. The tangent lines to the curve at each of the points $(x_1, f(x_1))$ and $(x_2, f(x_2))$ are drawn. In both cases the curve is above the tangent line and so the curve is concave up.

We have a similar situation if the curve is below the tangent lines as in Figure 5.15b. In this case, if there is an open interval around a point where the graph of a function is below the tangent lines for all points in that interval, then the graph of the function is **concave down** at that point.

Test for Concavity

It would be rather cumbersome to have to examine the tangent lines to a graph every time we wanted to determine if the graph was concave up or down. Fortunately, a test has been developed that allows us to determine the concavity of a curve. The test of concavity uses the second derivative and states the following.

Test for Concavity

The graph of a function f is

(a) concave up on any interval where $f''(x) > 0$

(b) concave down on any interval where $f''(x) < 0$

You might want to use these happy and sad faces to help remember how to use the second derivative to check the concavity of a function.

If the second derivative is positive, the graph is concave upward.

If the second derivative is negative, the graph is concave downward.

EXAMPLE 5.9

Determine when $f(x) = x^3 - 6x^2 + 3x + 5$ is concave up and when it is concave down.

Solution We find the second derivative of f is $f''(x) = 6x - 12$. Solving this, we see that $f''(x) < 0$ when $x < 2$ and $f''(x) > 0$ when $x > 2$. Thus, $f(x)$ is concave down on the interval $(-\infty, 2)$ and concave up on $(2, \infty)$.

We have two items that have not been examined. What happens when a curve crosses a tangent line as in Figure 5.15c? What happens when $f''(x) = 0$?

It turns out that we can use Figure 5.15c to help answer both of these questions. To the left of the tangent line the curve is concave down and so the second derivative is negative. To the right of the tangent line the curve is concave up and so the second derivative is positive. At the tangent line in Figure 5.15c the curve changes from concave down to concave up. A point on the graph at which the concavity changes is called an **inflection point**.

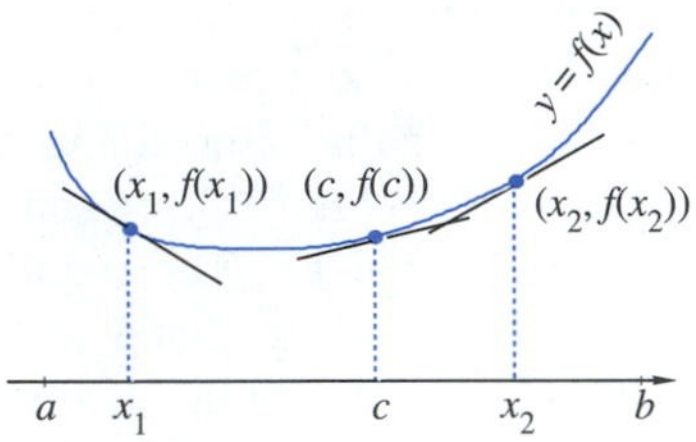

FIGURE 5.16

Locating Inflection Points

There are two ways to locate possible inflection points.

- Determine when the second derivative is zero.
- Determine when the second derivative is undefined.

Once you have found all of these points, look at the sign of the second derivative on both sides of each point. If the signs are different, then the concavity changes and it is a point of inflection.

EXAMPLE 5.10

Locate all inflection points of $f(x) = x^3 - 6x^2 + 3x + 5$.

Solution The second derivative is $f''(x) = 6x - 12$. Solving this, we see that $f''(x) = 0$ when $x = 2$. In Example 5.9, we showed that $f''(x) < 0$ when $x < 2$ and $f''(x) > 0$ when $x > 2$. Since the sign of f'' is negative to the left of 2 and positive to the right of 2, the point $(2, f(2)) = (2, -5)$ is an inflection point. The graph of f is shown in Figure 5.17.

The fact that the second derivative is zero at some point does not mean that it is an inflection point. Consider the graph of $f(x) = x^4$. The second derivative is $f''(x) = 12x^2$ and $f''(x) = 0$ when $x = 0$. But $f''(x) > 0$ for all $x \neq 0$. The sign of the second derivative is the same on both sides of 0. Thus, there is no inflection point when $x = 0$, as can be seen in the graph of $f(x) = x^4$ in Figure 5.18.

Second Derivative Test

The second derivative can sometimes tell us more than the concavity of a curve and the location of possible inflection points. Suppose a curve is concave up on some interval. If there is a point c in this interval with a first derivative of zero, then it must

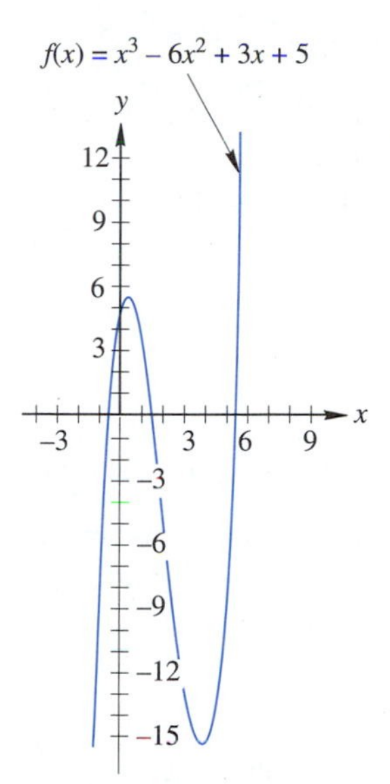

FIGURE 5.17

have a horizontal tangent line. Since the curve is concave up, the tangent line is below the curve. Thus the point is minimum. This forms the basis for what is known as the **second derivative test**.

Second Derivative Test

If f is a function whose second derivative exits, and if there is a point c where $f'(c) = 0$:

(a) if $f''(c) < 0$, then $f(c)$ is a maximum,
(b) if $f''(c) > 0$, then $f(c)$ is a minimum,
(c) if $f''(c) = 0$, then the second derivative will not tell if $f(c)$ is a maximum or minimum.

EXAMPLE 5.11

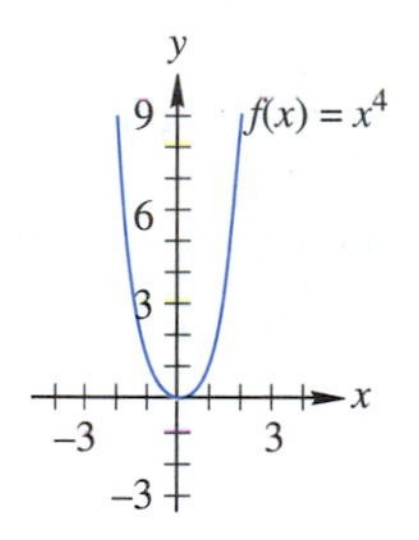

FIGURE 5.18

Use the second derivative test to find all extrema of $g(x) = 2x^3 - 15x^2 + 36x - 25$.

Solution Using the first derivative $g'(x) = 6x^2 - 30x + 36$, we find that $x = 2$ and $x = 3$ are critical values. The second derivative is $g''(x) = 12x - 30$. Evaluating the second derivative at each critical value, we get $g''(2) = -6$ and $g''(3) = 6$. Since $g''(2) < 0$, there is a maximum at 2 and since $g''(3) > 0$, there is a minimum at 3. Remember that the minimum occurs when $x = 3$ so the location of the minimum is at $(3,\ g(3))$ or $(3, 2)$. Similarly, the maximum is at $(2, g(2))$ or $(2, 3)$.

In tabular form, we get the following results from the first derivative:

Interval	$g'(x)$	$g(x)$
$(-\infty, 2)$	$+$	increasing
$(2, 3)$	$-$	decreasing
$(3, \infty)$	$+$	increasing

Setting $g''(x) = 0$, we see that the only possible inflection point is when $x = \frac{30}{12} = \frac{5}{2} = 2.5$. Using the concavity test, we get the following results:

Interval	$g''(x)$	$g(x)$
$(-\infty, 2.5)$	$-$	concave down
$(2.5, \infty)$	$+$	concave up

The graph of g is shown in Figure 5.19.

EXAMPLE 5.12

Sketch the graph of $h(x) = 3 + x + 3x^{2/3}$.

Solution $h'(x) = 1 + \dfrac{2}{x^{1/3}} = \dfrac{x^{1/3} + 2}{x^{1/3}}$. The critical values are when $h'(x) = 0$ and when $h'(x)$ is not defined. Now, since a fractional expression is 0 only when the

EXAMPLE 5.12 (Cont.)

numerator is 0, then $h'(x) = \frac{x^{1/3}+2}{x^{1/3}} = 0$ is equivalent to $x^{1/3}+2=0$ or $x^{1/3}=-2$ and $x=-8$. Now, $h'(x)$ is not defined whenever the denominator is 0, or when $x^{1/3}=0$, which is when $x=0$. Since $x=0$ is in the domain of h, the critical values are $x=-8$ where $h'(x)=0$ and $x=0$ where $h'(x)$ is not defined.

The second derivative is $h''(x) = -\frac{2}{3x^{4/3}}$. When $x=-8$, then $h''(-8) = -\frac{2}{3\left(-8^{4/3}\right)} = -\frac{2}{48} < 0$, and so $(-8,7)$ is a maximum. When $x=0$, then $h''(0) = -\frac{2}{3\left(0^{4/3}\right)}$, which is not defined and the second derivative test fails. Using the first derivative test, we see that when $x>0$, $h'(x)>0$ and so h is increasing. When x is in the interval $(-8,0)$, $h'(x)<0$ and so h is decreasing. Thus, $x=0$ produces the minimum $(0,h(0))$ or $(0,3)$. Summarizing our findings and adding those of the test for concavity, we get the following results.

Interval	$h'(x)$	h	Interval	$h''(x)$	h
$(-\infty,-8)$	+	increasing	$(-\infty,0)$	−	concave down
$(-8,0)$	−	decreasing	$(0,\infty)$	−	concave down
$(0,\infty)$	+	increasing			

The graph of this function is shown in Figure 5.20. ▪

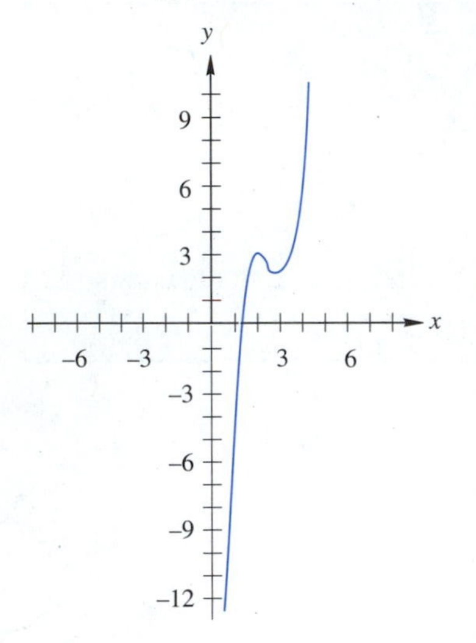

FIGURE 5.19

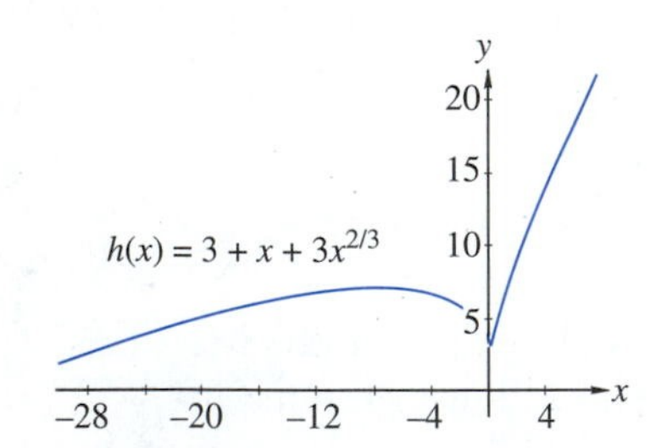

FIGURE 5.20

In Example 5.12, the domain of $h(x) = 3 + x + 3x^{2/3}$ was all real numbers. It is important to notice the domain. For example, graphing `h(x) = 3 + x + 3x^(2/3)` on a graphing calculator or a computer graphing program may result in something like the graph in Figure 5.21a. However, graphing h as `h(x) = 3 + x + 3(x^2)^(1/3)` produces the correct graph shown in Figure 5.21b.

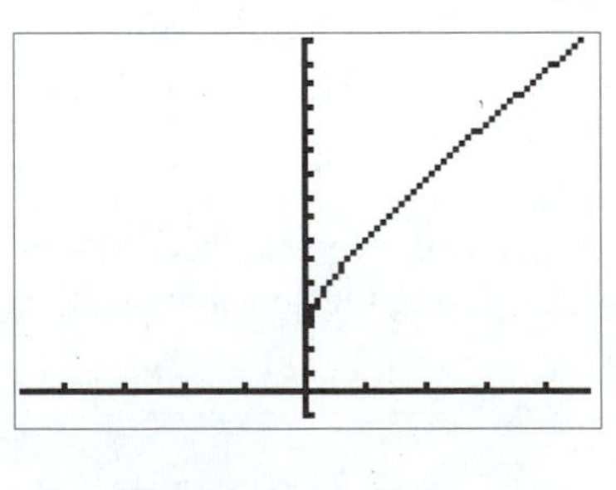

[-3.1, 3.1, 1] x [-1, 16, 1]

FIGURE 5.21a

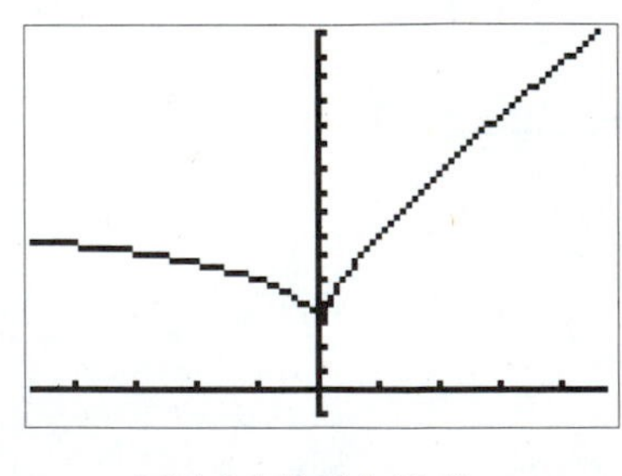

[-3.1, 3.1, 1] x [-1, 16, 1]

FIGURE 5.21b

In an earlier course, you learned how to find some horizontal and vertical asymptotes. Basically, in order to test for horizontal asymptotes of $f(x)$, we evaluate $\lim_{x\to\infty} f(x)$ and $\lim_{x\to-\infty} f(x)$. Vertical asymptotes occur when the function $f(x)$ is not defined. Consider the next example.

EXAMPLE 5.13

Sketch the graph of $k(x) = \dfrac{x}{x^2+4}$.

Solution We will first check for asymptotes.

$$\lim_{x\to\infty} \frac{x}{x^2+4} = \lim_{x\to\infty} \frac{1/x}{1+4/x^2} = \frac{0}{1} = 0$$

and $\displaystyle\lim_{x\to-\infty} \frac{x}{x^2+4} = 0$. So, the x-axis, or $y=0$, is a horizontal asymptote.

Finding the derivatives of $k(x)$, we get

$$k'(x) = \frac{4-x^2}{(x^2+4)^2} \text{ and } k''(x) = \frac{2x^3-24x}{(x^2+4)^3} = \frac{2x(x^2-12)}{(x^2+4)^3}.$$

From the first derivative we see that the critical values are ± 2. In tabular form, we get the following results from the first derivative.

Interval	$k'(x)$	$k(x)$
$(-\infty,-2)$	$-$	decreasing
$(-2,2)$	$+$	increasing
$(2,\infty)$	$-$	decreasing

Thus, when $x=-2$, we have a minimum at $(-2,-0.25)$, and at $x=2$, we have a maximum at $(2, 0.25)$.

Setting $k''(x)=0$, we see that the possible inflection points are when $x=0$ and $x=\pm\sqrt{12}=\pm 2\sqrt{3}$. Using the concavity test, we get the following results.

Interval	$k''(x)$	$k(x)$
$(-\infty,-2\sqrt{3})$	$-$	concave down
$(-2\sqrt{3},0)$	$+$	concave up
$(0,2\sqrt{3})$	$-$	concave down
$(2\sqrt{3},\infty)$	$+$	concave up

Thus, we have points of inflection at all three points. The sketch of the graph is shown in Figure 5.22.

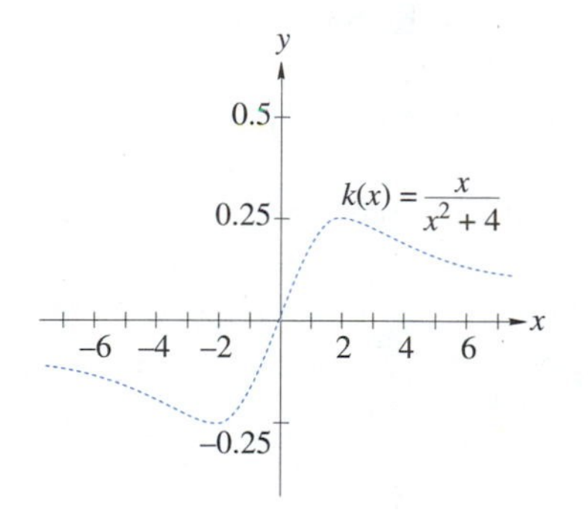

FIGURE 5.22

Exercise Set 5.3

In Exercises 1–28, find the extrema, determine the concavity, locate the inflection points, and sketch the graph.

1. $f(x) = -3x^2+12x$
2. $g(x) = 3x^2-6x+5$
3. $g(x) = x^3-12x+12$
4. $f(x) = x^3+3x^2-5$
5. $j(x) = x^3-6x^2+12x-4$
6. $k(x) = x^3-6x^2-20$
7. $h(x) = -2x^3+3x^2-12x+5$
8. $m(x) = x^4-4x^3+15$
9. $j(x) = 4x^2-\frac{4}{3}x^3-x^4$
10. $f(x) = 3x^4-4x^3+8$
11. $h(x) = (1-x)^3$
12. $g(x) = 3x^5-5x^3$
13. $h(x) = \sqrt[3]{x}+2$
14. $j(x) = x-x^{2/3}$
15. $k(x) = x+\dfrac{4}{x}$
16. $m(x) = x-\dfrac{9}{x}$
17. $n(x) = x^2-\dfrac{2}{x}$
18. $f(x) = \dfrac{1}{x-2}$
19. $g(x) = \dfrac{x^3}{3}-\dfrac{x^2}{2}-6x$
20. $h(x) = x^4-32x+48$

21. $j(x)=2\sqrt{x}-x$
22. $k(x)=-\sqrt{x^2-4}$
23. $g(x)=2+x^{2/3}$
24. $h(x)=2x-3x^{2/3}$
25. $f(x)=x^{2/3}(x-10)$
26. $g(x)=\dfrac{2}{x^2-4}$
27. $h(x)=\dfrac{x}{x^2+1}$
28. $j(x)=\dfrac{x^{2/3}}{x-4}$

Solve Exercises 29–32.

29. *Agriculture* The catfish population of a commercial catfish farm can be approximated by $F(t)=30+0.621t^2-0.009t^3$, where t is the number of weeks since this pond was started, $0\le t\le 52$. To produce the maximum yield of catfish, the best time to harvest the catfish is at the point of the fastest growth in the fish population. This point is an inflection point. When is the best time to harvest this fish population?

30. *Economics* When a company starts a successful advertising campaign, sales and the rate at which sales grow both increase. At the **point of diminishing returns**, even though more money is spent on advertising, sales continue to grow but at a slower rate. The point of diminishing returns occurs at an inflection point where the sales curve changes from concave upward to concave downward. Find the point of diminishing returns for the sales function $S(x)=97+2.1x^2-0.1x^3$, where x represents thousands of dollars on advertising, $0\le x\le 10$, and S is sales in thousands of dollars.

31. *Police science* Due to the rapid increase in major crimes, the mayor of a large city plans to organize a major crime task force. It is estimated that for each 1,000 people in the city, the number of major crimes will be $N(t)=67.4+3t^2-0.8t^{5/2}$, where t is the number of months after the task force has been organized and $0\le t\le 12$.
(a) Find the maximum of $N(t)$.
(b) When is the maximum rate of increase in $N(t)$?

32. *Meteorology* Meteorological results for a certain city indicate that for the month of April, the daily temperature in °F between midnight and 6:00 P.M. can be approximated by $T(t)=-0.04t^3+1.26t^2-9.3t+56$, where t is the number of hours after midnight and $0\le t\le 18$.
(a) What are the maximum and minimum temperatures?
(b) What is the maximum rate of increase in the temperature?

In Your Words

33. Explain the second derivative test and what information it gives about a function.

34. What is an inflection point?

5.4 APPLIED EXTREMA PROBLEMS

Up to this point, most of the applications for extrema have been made to curve sketching. We have used derivatives to help determine the high and low points on the curve, when it is increasing or decreasing, and other useful information. A curve is a picture of a function and functions are often used to represent real situations. In this section, we will look at some ways to solve problems that involve using derivatives.

As each example is worked, you should make sure that you understand what was done. The most important part will be in understanding how to set up the function; then, being aware of how derivatives are used to find possible solutions and to determine the correct answer should be understood.

Application

EXAMPLE 5.14

A company needs to erect a fence around a rectangular storage yard next to a warehouse. They have 400 m of available fencing. If they do not fence in the side next to the warehouse, what is the largest area that can be enclosed?

Solution The clue to the desired answer is "largest area." We need to develop a function for the area and then determine when it is a maximum.

Look at Figure 5.23. The fenced-in rectangular yard has dimensions that have been labeled x and y. The area of the fenced-in region is $A = xy$. We need to write A in terms of a single variable. The total amount of fencing is $400\,\text{m} = y + x + y$. Thus, we have

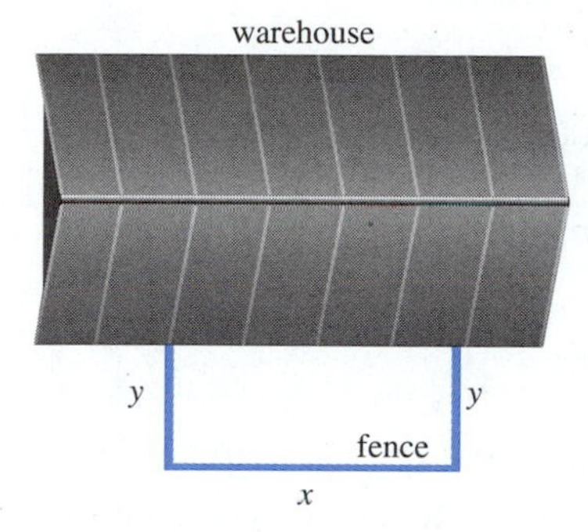

FIGURE 5.23

$$2y + x = 400$$

or

$$x = 400 - 2y$$

If we substitute this in the area formula, we obtain

$$\begin{aligned} A &= xy \\ &= (400 - 2y)y \\ &= 400y - 2y^2 \end{aligned}$$

The area is now expressed in terms of one variable, y, and so we have

$$A(y) = 400y - 2y^2$$

We want to know when $A(y)$ is a maximum. Taking the first derivative, we get

$$A'(y) = 400 - 4y$$

The only critical value occurs when $A'(y) = 0$, which is when $y = 100$.

Is the area a maximum when $y = 100$? Using the second derivative test, we see that $A''(y) = -4$. Since $A''(100) < 0$, the area is a maximum when $y = 100$.

We have one other item to check—is this an acceptable solution? We know that $y > 0$. We can also determine that the largest possible value for y is 200 m. Thus, the answer must be in the interval $[0, 200]$.

Now, when $y = 100$, $x = 200$, and so the maximum area is $A = (200)(100) = 20\,000\,\text{m}^2$. ▪

Application

EXAMPLE 5.15

When a battery or generator of voltage V and an internal resistance r is connected to an external resistance R, the total resistance in the circuit is $R + r$ and the current flow is $I = \dfrac{V}{R + r}$. The power P dissipated in the load R is $P = I^2 R$. If the voltage is 100 V and $r = 5\,\Omega$, what value of R will produce the maximum load power? (See Figure 5.24.)

EXAMPLE 5.15 (Cont.)

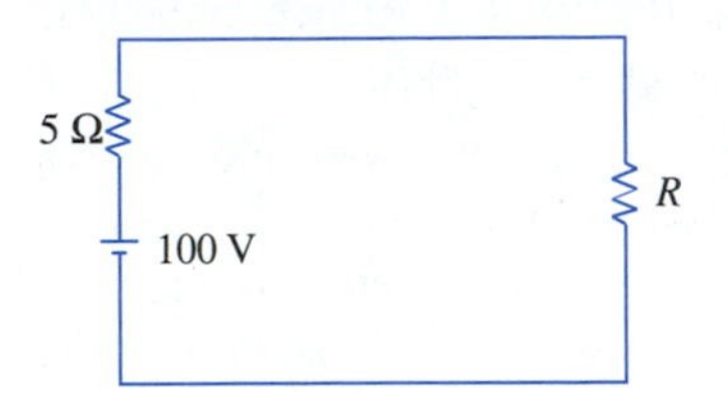

FIGURE 5.24

Solution We need to develop a formula in terms of the external resistance R. Since $I = \dfrac{V}{R+r}$, we have $I = \dfrac{100}{R+5}$. Substituting this into the power formula, we get

$$P = I^2R = \left(\frac{100}{R+5}\right)^2 R = \frac{10\,000R}{(R+5)^2}$$

Differentiating produces

$$\begin{aligned} P' &= \frac{(R+5)^2(10\,000) - 10\,000R(2)(R+5)}{(R+5)^4} \\ &= \frac{(R+5)10\,000 - 20\,000R}{(R+5)^3} \\ &= \frac{50\,000 - 10\,000R}{(R+5)^3} \end{aligned}$$

Since $R > 0$, the only critical value occurs when $P' = 0$ or when $R = 5$. Using the first derivative test, we see that this is a maximum and so the maximum load power occurs when $R = 5\,\Omega$.

We could generalize this last example to show that the maximum load power occurs when the external load is equal to the internal resistance.

Application

EXAMPLE 5.16

A container is to be made in the form of a right circular cylinder and is to contain 1 L. What dimensions of the container will require the least amount of material?

Solution We know that the volume of a right circular cylinder is given by the formula $V = \pi r^2 h$ and the total surface area is $A = 2\pi rh + 2\pi r^2$. We will assume that the thickness of the material is negligible. We want to minimize the area. The formula for the area is given in terms of two variables, h and r. We know that $V = 1\text{ L} = 1\,000\text{ cm}^3$. We can solve the equation $1\,000 = \pi r^2 h$ for either r or h and substitute the result into the area formula. If we solve for h we get

$$h = \frac{1\,000}{\pi r^2}$$

Substituting, we get the area as a function of r:

$$\begin{aligned} A(r) &= 2\pi r\left(\frac{1\,000}{\pi r^2}\right) + 2\pi r^2 \\ &= \frac{2\,000}{r} + 2\pi r^2 \end{aligned}$$

Differentiating with respect to r, we have

$$A'(r) = -2\,000r^{-2} + 4\pi r$$

EXAMPLE 5.16 (Cont.)

Critical values occur when $r=0$, because $A'(r)$ is not defined when $r=0$, and when $A'(r)=0$. If we set the derivative equal to zero, we obtain

$$\frac{2\,000}{r^2}=4\pi r$$

or $$r^3=\frac{2\,000}{4\pi}=\frac{500}{\pi}$$

and so $$r=\sqrt[3]{\frac{500}{\pi}}$$

Using the second derivative test, we see that

$$A''(r)=4\,000r^{-3}+4\pi$$

When $r=0$, $A''(r)$ is not defined, but we know that $r>0$. (Why? What happens if $r=0$?)

When $r=\sqrt[3]{\frac{500}{\pi}}$, $A''(r)>0$, so the area is a minimum.

When $r=\sqrt[3]{\frac{500}{\pi}}\approx 5.4193\,\text{cm}$, $h=\frac{1\,000}{\pi r^2}\approx 10.8385\,\text{cm}$ and the area is $A\approx 553.5810\,\text{cm}^2$.

Application

EXAMPLE 5.17

A telephone company is asked to provide telephone service to a customer whose house is 0.5 km away from the road along which the telephone lines run. The nearest telephone box is 5 km down the road. The cost to connect the telephone line is \$80 per kilometer along the road and \$100 per kilometer away from the road. At which point along the road should the company connect the line in order to minimize the cost?

Solution Examine the sketch of this problem in Figure 5.25. We let $5-x$ represent the distance from the box B to the place where the connection is made, A. The distance x is from the connection A to the point on the road closest to the house, H. The telephone line will go from B to A and then from A to H. The length AH is $\sqrt{x^2+0.5^2}=\sqrt{x^2+0.25}$. The total cost C is $C(x)=80(5-x)+100\sqrt{x^2+0.25}$. We want $C(x)$ to be a minimum.

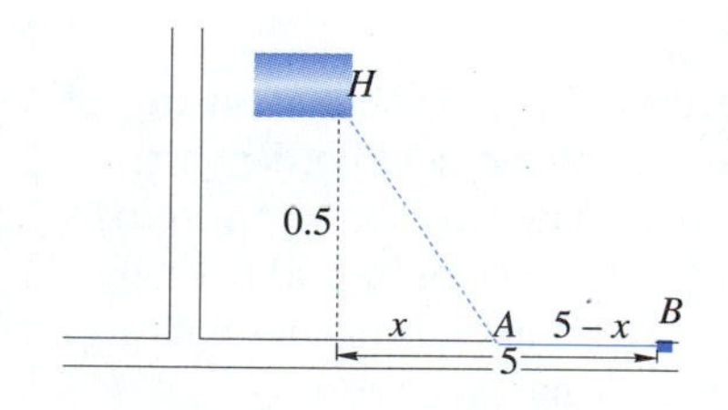

FIGURE 5.25

$$C'(x)=-80+\frac{100x}{\sqrt{x^2+0.25}}$$

$$=\frac{-80\sqrt{x^2+0.25}+100x}{\sqrt{x^2+0.25}}$$

EXAMPLE 5.17 (Cont.)

The critical value occurs when the numerator is 0; that is, when

$$80\sqrt{x^2+0.25} = 100x$$
$$6\,400(x^2+0.25) = 10\,000x^2$$
$$1\,600 = 3\,600x^2$$
$$\frac{16}{36} = x^2$$

or when

$$x = \pm\frac{2}{3}$$

When $x = \frac{2}{3}$ km, the cost will be a minimum. The minimum cost is \$430. ▪

When solving extrema problems, there are some helpful procedures to use. If you go back and look at the examples in this section, you will see that we used each of them.

Hint

Hints for Solving Extrema Problems

1. When possible, draw a figure to illustrate the problem. Label all the parts that are important to the problem.
2. Write an equation for the quantity that is going to be a maximum or a minimum.
3. If more than one variable is involved, eliminate all but one of them.
4. Take the derivative and locate the critical values.
5. Use the first and/or second derivative test on each critical value to see if it provides a maximum or a minimum.
6. If the function is defined for a limited range of values, examine the endpoints for possible extrema values.

Exercise Set 5.4

Solve Exercises 1–36.

1. *Agriculture* A company has 10,000 ft of available fencing. It wants to use the fencing to enclose a rectangular field. One side of the field is bordered by a river. If no fencing is placed on the side next to the river, what is the largest area that can be enclosed?

2. *Agriculture* If the company in Exercise 1 decides to fence the side bordering the river, what is the largest area that can be enclosed?

3. *Agriculture* A rectangular field contains an area of 6,000,000 ft^2. One side of the field borders a river. Because of difficulties caused by the river, it costs \$2.00/ft to fence along the river and \$1.00/ft to fence the other three sides. **(a)** What dimensions will give the minimum cost? **(b)** What is the cost?

4. *Agriculture* A 200-m piece of link fencing is to be cut so that one piece encloses a square section of garden and the other piece encloses a circular section. Where should the fencing be cut if the enclosed areas are to be a maximum?

5. *Industrial design* An open box is to be formed by cutting a square from each corner of a square piece of cardboard 24 cm on a side. After the squares are removed from each corner, the sides are folded up as shown in the following figure. How large a square must be removed from each corner in order for the box to have the largest volume?

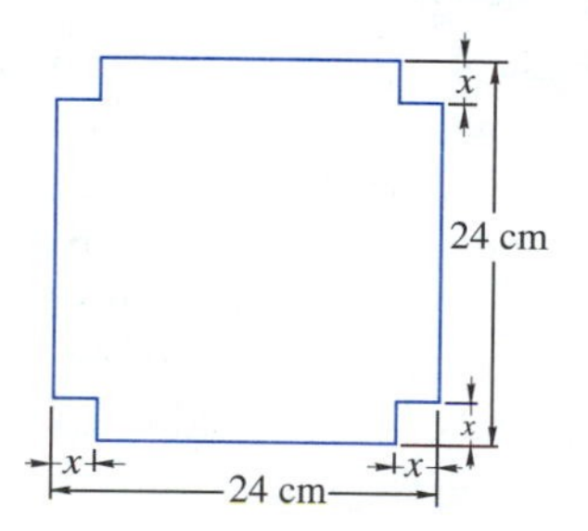

6. *Business* A rectangular sign is to be painted on the end of a semicircular-shaped building that has a diameter of 20.00 m. What dimensions will produce a sign with the largest area?

7. What is the largest rectangle that can be inscribed in the first quadrant of the ellipse $9x^2 + 16y^2 = 144$?

8. *Industrial design* An open rectangular box is to be made from a piece of cardboard 8 in. wide and 15 in. long by cutting a square from each corner and bending up the sides. How large a square should be removed from each corner in order for the box to have the largest volume?

9. *Industrial design* A box with a square base and an open top is to hold 32 in.^{3} What dimensions will require the least amount of material?

10. *Industrial design* If the box in Exercise 9 is to be closed on top, what dimensions will require the least amount of material?

11. *Industrial design* The total surface area of a right-circular cylindrical tin is 480 cm^2. What are the dimensions of the cylinder that has the largest volume?

12. *Mechanical engineering* The cutting speed s (in meters per minute) of a saw used to cut a certain type of material is given by $s(t) = \sqrt{t - 9t^2}$, where t is the time in seconds. What is the maximum speed of this product?

13. *Printing* A page of a book is to have 580 cm^2 with a 3-cm margin at the bottom and sides and a 2-cm margin at the tope. Find the dimensions of the page that will allow the largest printed area.

14. Find the point on the curve $y = 2x^2$ that is closest to the point $(5, 1)$.

15. *Construction* The strength of a rectangular beam is directly proportional to the width of the beam and the square of its depth. What are the dimensions of the strongest beam that can be cut from a log whose cross-section is a circle of radius 15 in.? What are the dimensions of the strongest beam cut from a circular log whose cross-section has a radius of r?

16. *Sheet metal technology* A metal trough is to be made by bending a rectangular sheet of metal in the middle to form a "V." If the piece of metal is 300 mm wide, what angle between the sides will give the maximum capacity?

17. *Sheet metal technology* A gutter is to be made from a long sheet of metal 320 mm wide by bending up equal widths along the edges into vertical positions. What dimensions will give the largest capacity?

18. *Electronics* The current I in amperes (A) through a resistor of $R\,\Omega$ with a constant voltage E in volts (V), is given by

$$I = \frac{E}{\sqrt{R^2 + X^2}}$$

(a) Assuming that R is constant, what value of X will make the current in the resistor a maximum?
(b) What is this maximum current?

19. *Electricity* The output of a battery is given by the formula $P = VI - RI^2$, where V is the voltage, I the current, and R the resistance. Find the current for which the output is a maximum if $V = 10$ V and $R = 4.0\,\Omega$.

20. *Chemistry* The rate of a certain autocatalytic reaction is given by $v(x) = 5x(a - x)$ for x in $[0, a]$, where a is the original amount of the substance produced by the reaction. Determine when the reaction is a maximum?

21. *Construction* A power company wishes to install a temporary power line. The purchase price of the wire is proportional to the cross-sectional area A of the wire. The total cost C of the wire is given by the formula $C(A) = bA + \frac{c}{A}$, where b and c are constants. What cross-sectional area will produce the least cost?

22. *Architecture* A window above the entrance to a room that has a cathedral ceiling is to be designed in the shape of a rectangle with a semicircle on top. If the area of the rectangular part is to be $2.00\,\text{m}^2$, what is the diameter of the semicircle that will result in a window with the smallest perimeter?

23. *Architecture* What would be the diameter of the semicircular part of the window in Exercise 22 if the perimeter is 8.00 m and the total area of the window is to be a maximum?

24. *Business* The cost of operating a truck on an interstate highway is $0.35 + \frac{s}{300}$ dollars per mile, where s is the speed of the truck in miles per hour. The driver's salary of $12 per hour is extra. **(a)** At what speed should the truck driver operate the truck to make an 800-mi trip for the least cost? **(b)** What is the cost of this 800-mi trip?

25. *Electricity* The power P in watts (W) in a circuit is given by $P = I^2R$, where I is current in amperes (A) and R is resistance in ohms (Ω). If $R = 25.4\,\Omega$ and $I = 4.5 - t$, where t in time is seconds (s), find
(a) the time when there is minimum power, and
(b) the minimum power.

26. *Electricity* The current in a circuit is given by $\frac{dq}{dt}$, where q is the charge in coulombs (C) and t is the time in seconds (s). If $q = -4.36t^2 + 2.14t^3$, find the minimum current in the circuit and the time at which this occurs.

27. *Lighting technology* The intensity of illumination at a point varies inversely as the square of the distance between the point and the light source. Two lights are 8 m apart. One of the lights has an intensity four times that of the other. At which point between the lights is the illumination the least?

28. *Aerodynamics* The drag D on an airplane traveling at velocity v is $D = av^2 + \frac{b}{v^2}$, where a and b are positive constants. At what speeds does the airplane have the least drag?

29. *Lighting technology* The deflection Y of a beam of length L at a horizontal distance x from one end is given by $Y = k(2x^4 - 5Lx^3 + 3L^2x^2)$, where k is a positive constant. For what value of x does the maximum deflection occur?

30. *Mechanical engineering* The efficiency E of a screw is given by $E = \frac{T(1 - T\mu)}{T + \mu}$ where μ is the coefficient of friction and T is the tangent of the pitch angle of the screw. For what value of T is the efficiency the greatest?

31. *Civil engineering* A proposed tunnel has a fixed cross-sectional area. The tunnel will have a horizontal floor, vertical walls of equal height, and the ceiling will be a semicircle. The ceiling cost four times as much per square meter as the floor and walls. What ratio of the diameter of the semicircular ceiling to the height of the walls will produce the least cost?

32. A person in a rowboat is 4 km offshore, where the nearest point P is on a straight shoreline of the lake. A town is 20 km down the shore road from P. This person can row 2.5 km/h and can walk 4.5 km/h. At which point should the boat be landed in order to arrive in town in the shortest time?

33. *Petroleum engineering* An oil company wants to lay a pipeline from its offshore drilling rig to a storage tank on shore. The rig is 5 mi offshore, and the storage tank is 12 mi down the straight shoreline. The cost of laying pipe underwater is $1,200 per mi and along the shoreline the cost is $600 per mi. How far from the storage tank should they locate the connection of the underwater pipe to the shoreline pipe?

34. *Package design* A manufacturer is designing a closed rectangular shipping crate with a square base. The volume of the crate is to be $36\,\text{ft}^3$. Material for the top costs $1 per square foot, material for the sides costs $\$0.75/\text{ft}^2$, and material for the bottom costs $\$1.50/\text{ft}^2$.
(a) Find the dimensions of the box that will minimize the total cost of material.
(b) Determine the minimum cost.

35. *Construction* A rectangular warehouse is being constructed with a total floor area of $4{,}200\,\text{ft}^2$. A single interior wall will partition the area into storage space and office space. It costs \$175/ft to construct an exterior wall and \$115/ft to erect the interior wall.

(a) What are the dimensions of the warehouse that will minimize the cost?

(b) Determine the minimum cost.

36. *Package design* The U.S. Postal Service has a limit of 108 in. on the combined length and girth of a rectangular package that can be mailed. Find the dimensions of the package of maximum volume that has a square cross-section. (The girth is defined as the smallest perimeter of a rectangular cross-section of the box.)

In Your Words

37. Without looking in the text, describe how you solve a problem that asks you to find a maximum or a minimum.

38. Write an application in your technology area of interest that requires you to find a maximum or a minimum. Give your problem to a classmate and see if he or she understands and can solve your problem using the techniques of this and earlier chapters. Rewrite the problem as necessary to remove any difficulties encountered by your classmate.

5.5 RELATED RATES

It is not unusual to have two or more quantities that vary with time and an equation that expresses some relationship between them. Normally, the values of these quantities at some time are given. We are also given all but one of the rates of change. The problem is to find the rate of change that is not given. We do this by taking the derivative of the expression (with respect to time) that relates the variables. The following examples show the basic method for solving these problems.

EXAMPLE 5.18

A ladder 30 ft long leans against a vertical wall. The foot of the ladder is slipping away from the wall at the rate of 4 ft/s. How fast is the top of the ladder sliding down the wall at the instant when the foot of the ladder is 18 ft from the wall?

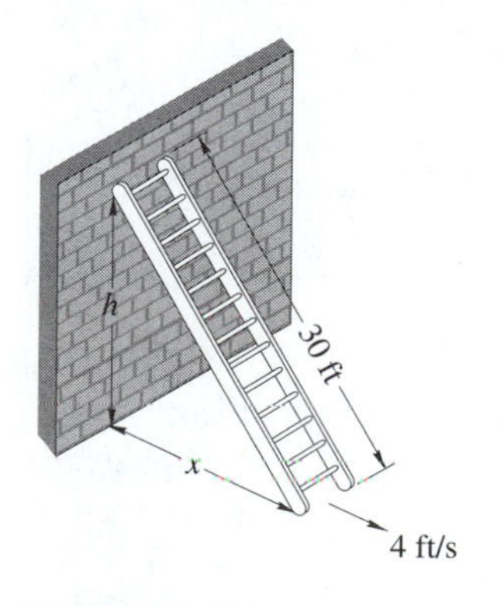

FIGURE 5.26

Solution Figure 5.26 shows the ladder against the wall. The constants are the length of the ladder, 30 ft, and the rate at which the foot of the ladder moves, 4 ft/s. The quantities that change are the height of the top of the ladder, h, and the distance of the foot of the ladder from the wall, x.

We want to find a rate. If h denotes the height of the top of the ladder above the ground in feet and t denotes the time in seconds, then we want to find dh/dt: the rate at which the height of the ladder is changing (dh) with respect to time (dt). In particular, we want to find dh/dt when $x = 18$ ft.

We have a right triangle with the right angle where the wall meets the ground. The ladder is the hypotenuse of the triangle. Thus, $x^2 + h^2 = 30^2$. Using implicit differentiation with respect to time t, we get

$$\frac{d(x^2)}{dt} + \frac{d(h^2)}{dt} = 0$$

or

$$2x\frac{dx}{dt} + 2h\frac{dh}{dt} = 0$$

EXAMPLE 5.18 (Cont.)

Thus, the rate we want to find is

$$\frac{dh}{dt} = \left(\frac{-x}{h}\right)\left(\frac{dx}{dt}\right)$$

When $x = 18, h = 24$, and $\frac{dx}{dt} = 4\,\text{ft/s}$, we obtain

$$\frac{dh}{dt} = -\frac{18}{24}(4) = -3\,\text{ft/s}$$

Thus, the top of the ladder is falling at the rate of 3 ft/s.

We solved Example 5.18 for the general case when the foot of the ladder was x-feet from the bottom of the wall. Once we had this answer, we solved it for the specific case when $x = 18$. Thus, by first solving for x we are then able to determine the rate for any value of x.

As with the extrema problems, there are some guidelines to help us solve related rate problems. Some of the guidelines are virtually the same as those for extrema, but are restated here to keep the guidelines as a group.

Hints for Solving Related Rates Problems

1. Make a diagram of the problem. Which quantities change; which remain constant?
2. Write down what you are to find. (In Example 5.18, it was a rate.) Express it in terms of a variable (in Example 5.18, dh/dt). Show the variable in the sketch (h in Figure 5.26).
3. Identify the other variables in the problem. Show them in the sketch. Write down any numerical information that is given about them.
4. Write equations that relate the variables.
5. Use substitution and differentiation to establish additional relationships among the variables and their derivatives.
6. Substitute numerical values for the variables and the derivatives. Solve for the unknown rate.

A common error is to introduce specific values for the rates and variable quantities too early in the solution. Remember to first get a general formula that involves the rates of change at any time t. After you have the general formula, specific values can be substituted to get the desired solution.

Application

EXAMPLE 5.19

A weather balloon in the shape of a sphere is being inflated at the rate of 12 m^3/min. Find the rate at which the surface area of the balloon is changing when the radius is 6 m.

Solution The variables in this problem are:

$$\begin{aligned} t &= \text{time in minutes} \\ r &= \text{radius in meters of the balloon at time } t \\ V &= \text{volume in cubic meters at time } t \\ S &= \text{surface area in square meters at time } t \end{aligned}$$

We have the following rates of changes:

$$\begin{aligned} \frac{dr}{dt} &= \text{rate of change of radius with respect to time} \\ \frac{dV}{dt} &= \text{rate of change of volume with respect to time} \\ \frac{dS}{dt} &= \text{rate of change of surface area with respect to time} \end{aligned}$$

We are told that $\frac{dV}{dt} = 12\,\text{m}^3/\text{min}$. We want to find $\frac{dS}{dt}$ when $r = 6$ m.

The volume of a sphere is $V = \frac{4}{3}\pi r^3$ and the surface area is $S = 4\pi r^2$. Differentiating each of these with respect to t, we get

$$\frac{dV}{dt} = 4\pi r^2 \frac{dr}{dt}$$

$$\text{and} \qquad \frac{dS}{dt} = 8\pi r \frac{dr}{dt}$$

Since we want to find $\frac{dS}{dt}$ and we know the value of $\frac{dV}{dt}$, we will solve the $\frac{dV}{dt}$ equation for $\frac{dr}{dt}$ and substitute this into the $\frac{dS}{dt}$ equation. Solving for $\frac{dr}{dt}$ we get

$$\frac{dr}{dt} = \frac{1}{4\pi r^2} \cdot \frac{dV}{dt}$$

Substituting, this gives

$$\frac{dS}{dt} = 8\pi r \left(\frac{1}{4\pi r^2}\right) \frac{dV}{dt} = \frac{2}{r} \frac{dV}{dt}$$

When $r = 6$ m and $\frac{dV}{dt} = 12\,\text{m}^3/\text{min}$, we get

$$\frac{dS}{dt} = \frac{2}{6} \cdot 12 = 4\,\text{m}^2/\text{min}$$

The surface area is increasing at a rate of 4 m^2/min when the radius is 6 m. ▪

Application

EXAMPLE 5.20

Water is running out of a conical funnel at a rate of $5\,\text{cm}^3/\text{s}$. If the radius of the base of the funnel is 10 cm and the altitude is 20 cm, find the rate at which the water level is dropping when it is 5 cm from the top.

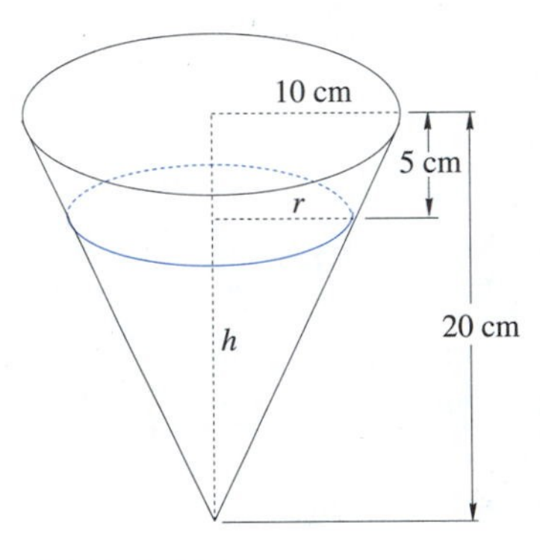

FIGURE 5.27

Solution A sketch of this problem is given in Figure 5.27. The variables are:

$$t = \text{time in seconds}$$
$$r = \text{radius in cm of water level at time } t$$
$$h = \text{height in cm of surface of water at time } t$$
$$V = \text{volume in cm}^3 \text{ of water in cone at time } t$$

We have the following rates of change:

$$\frac{dh}{dt} = \text{rate of change of height of water with respect to time}$$
$$\frac{dV}{dt} = \text{rate of change of volume of water with respect to time}$$

By similar triangles, we have $\frac{r}{10} = \frac{h}{20}$ or $r = \frac{1}{2}h$. The volume of a cone is $V = \frac{1}{3}\pi r^3$, and substituting for r, we get

$$V = \frac{1}{3}\pi\left(\frac{1}{2}h\right)^3 = \frac{1}{24}\pi h^3$$

Differentiating with respect to t we get

$$\frac{dV}{dt} = \frac{dV}{dh}\cdot\frac{dh}{dt} = \frac{1}{8}\pi h^2\frac{dh}{dt}$$

We are given $\frac{dV}{dt} = -5$. This is negative because the volume is decreasing. When the water level is 5 cm from the top, $h = 20 - 5 = 15$ cm and so we have

$$-5 = \frac{1}{8}\pi(15)^2\frac{dh}{dt}$$

or
$$\frac{dh}{dt} = -\frac{40}{225\pi} = \frac{-8}{45\pi}\ \text{cm/s}$$

Application

EXAMPLE 5.21

When air changes volume adiabatically (without the addition of heat), the pressure P and volume V satisfy the formula $PV^{1.4} = k$, where k is a constant. At a certain instant, the pressure is $75\,\text{lb/in.}^2$ and the volume is $25\,\text{in.}^3$ and is decreasing at the rate of $3\,\text{in.}^3/\text{s}$. How rapidly is the pressure changing at that time?

EXAMPLE 5.21 (Cont.)

Solution Since the equation is given, a drawing is not needed for this problem. The variables are:

$$t = \text{time in seconds}$$
$$P = \text{pressure in lb/in.}^2$$
$$V = \text{volume in in.}^3$$

We have the following rates of change:

$$\frac{dV}{dt} = \text{rate of change of volume with respect to time}$$
$$\frac{dP}{dt} = \text{rate of change of pressure with respect to time}$$

From the formula $PV^{1.4} = k$ and the fact that when $P = 75$, $V = 25$, we can determine that $k \approx 6{,}794.81$.

We are given $\frac{dV}{dt} = -3$ when $V = 25$ and need to find $\frac{dP}{dt}$. We know that $P = kV^{-1.4}$ or

$$\frac{dP}{dt} = \frac{dP}{dV} \cdot \frac{dV}{dt} = -1.4kV^{-2.4}\frac{dV}{dt}$$
$$= \frac{-1.4(6{,}794.81)(-3)}{(25)^{2.4}}$$
$$= 12.6$$

The pressure is increasing at the rate of 12.6 lb/in.^{2} ▪

Exercise Set 5.5

Solve Exercises 1–44.

1. Gas is escaping from a balloon at the rate of 20 L/min. How fast is the surface area changing when the radius is 300 cm? (1 liter = 1000 cm^3)
2. *Construction* Sand is emptied down a chute at the rate of 10 ft^3/s forms a conical pile whose height is always twice the radius. At what rate is the radius changing when the height is 4 ft?
3. Water is emptied from a spherical tank of radius 24 ft. The water in the tank is 8 ft deep and decreasing at the rate of 2 ft/min. At this time, what is the rate of change of the radius of the top of the water?
4. *Environmental technology* A circular oil slick of uniform thickness is caused by a spill of 10 m^3 of oil. The thickness of the oil slick is decreasing at the rate of 1 mm/h. At what rate is the radius of the oil slick changing when the radius is 12 m?
5. *Transportation* A helicopter flies parallel to the ground at an altitude of 0.75 km and at a speed of 3 km/min. If the helicopter flies in a straight line that will pass directly over you, at what rate is the distance between you and the helicopter changing 1 min after it passes over you?
6. *Recreation* A rectangular swimming pool 15 m long and 10 m wide is 4 m deep at one end and 1 m deep at the other, with a constant drop from the shallow end to the deep end. Water is pumped into the pool at the rate of 50 L/min. At what rate is the water rising when it is 2 m deep at the deep end? (1 000 L = 1 m^3)

7. *Landscaping* A fish pond is in the shape of the bottom half of a sphere with a radius of 5 ft. Water is pumped into the pond at the rate of $10\,\text{ft}^3/\text{min}$. **(a)** At what rate is the water rising when it is 2 ft deep? **(b)** What is the rate when the water is 3 ft deep? **(c)** What is the rate when the water is 4 ft deep? [Note: $V_{\text{liquid}} = \frac{1}{3}\pi h^2(3R-h) = \frac{1}{6}\pi h(3r^2+h^2)$.]

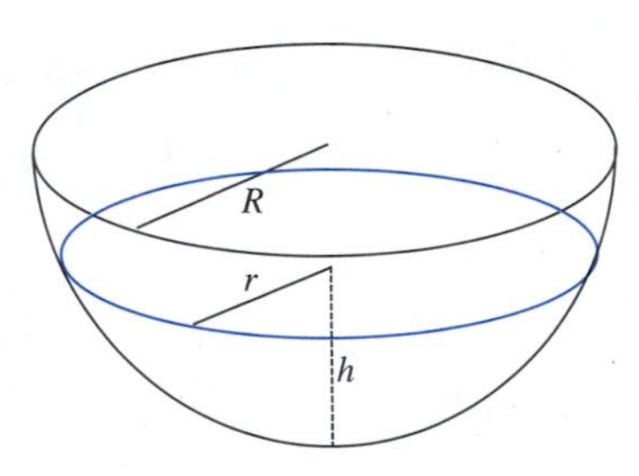

8. *Landscaping* When the pond in Exercise 7 is drained, the water drains at the rate of $45\,\text{ft}^3/\text{min}$. **(a)** At what rate does the water drop when it is 4 ft deep? **(b)** What is the rate when it is 2 ft deep?
9. A horizontal trough 10.00 m long has a cross-section in the shape of an equilateral triangle with a top width of 0.500 m. If the trough is filled with water at the rate of 25.0 L/min, how fast is the surface of the water rising when the depth is 0.175 m?
10. The base of a triangle is increasing at the rate of 5.00 cm/s and the height is decreasing at the rate of 4.00 cm/s. At what rate is the area of the triangle changing when the base is 2.625 m and the height is 1.375 m?
11. *Environmental science* Air is pumped into a spherical weather balloon at the rate of $60\,\text{cm}^3/\text{s}$. How fast is the surface area changing when the radius is 8 cm?
12. The height of a right-circular cone with radius of 2.43 m is increasing at the rate of 0.425 m/h. How fast is the lateral area of the cone changing when the height is 4.27 m?
13. A balloon is in the shape of a cylinder with a hemisphere on each end. The height of the cylinder is four times its radius. Air is pumped into the balloon at the rate of $60\,\text{cm}^3/\text{s}$. How fast is the surface area changing when the radius is 8 cm?
14. *Thermodynamics* When a metal plate is heated it expands. If the shape of the plate is circular and the radius is increasing at the rate of 0.2 mm/s, at what rate is the area of the top surface changing when the radius is 20 mm?
15. *Thermodynamics* If a metal plate in the shape of a square is heated and the length of one side is increasing at the rate of 0.3 mm/s, at what rate is the area of the top surface changing when the length of one side is 40 mm?
16. At a certain instant when air is changing adiabatically, the pressure is $60\,\text{dynes/cm}^2$ and is increasing at a rate of $4\,\text{dynes/cm}^2/\text{s}$. If the volume at that time is $100\,\text{cm}^3$, what is the rate at which the volume is changing?
17. *Physics* The pressure P in kilopascals (kPa) and volume V in cubic meters of a gas is given by $PV^{1.2} = 400$. If the pressure is increasing at a rate of 10% of itself per hour, find the rate at which the volume is changing expressed as a percent of itself.
18. *Meteorology* A weather balloon is rising vertically at the rate of 5 ft/s. An observer is 500 ft from a point on the ground directly below the balloon. At what rate is the distance between the balloon and the observer changing when the altitude of the balloon is 2,000 ft?
19. *Space technology* A radar station is following the path of a rocket which has been launched vertically from a site 2.430 km away. How fast is the rocket rising when it is 3.750 km high, if the distance between the rocket and the radar site is increasing at the rate of 325.0 m/s?
20. *Physics* **Boyle's law** for gases states that $PV = k$, where P is the pressure and V the volume, and k is a constant. At a certain instant the volume is $50\,\text{in.}^3$, the pressure $20\,\text{lb/in.}^2$, and the pressure is increasing at the rate of $1.5\,\text{lb/in.}^2$ every minute. At what rate is the volume changing at this instant?
21. *Physics* The displacement s in meters and the velocity v of an object in meters per second are related according to $v^2 = 1200 - 36.0s$. **(a)** Find the acceleration of the object. **(b)** What is the acceleration when $s = 2.45$ m?

22. A 10-m ladder is leaning against a vertical wall. If the base of the ladder slips away from the wall at the rate of 0.5 m/s, how fast is the top of the ladder going down the wall when the base of the ladder is **(a)** 2 m from the wall? **(b)** 3 m from the wall? **(c)** 4 m from the wall? **(d)** 6 m from the wall?

23. *Navigation* Two radar stations are at points A and B. Point B is 8 mi east of A. The two stations are tracking a ship. At a certain instant, the ship is 8 mi from A and this distance is increasing at the rate of 14 mph. At the same instant, the ship is also 6 mi from B and this distance is increasing at the rate of 2 mph. **(a)** Where is the ship? **(b)** How fast is it moving? **(c)** In what direction is it moving?

24. *Transportation* A car leaves an intersection and travels due north with a speed of 80.0 km/h. Another car leaves the same intersection 0.5 h later going due west at a speed of 100.0 km/h. How fast are the two cars separating

(a) 1.00 h after the second car started; and

(b) 2.00 h after the second car started?

25. A person 6 ft tall walks at the rate of 5 ft/s away from a street light that is 20 ft above the ground. **(a)** At what rate is the top of the person's shadow moving? **(b)** At what rate is the length of the shadow changing when the person is 10 ft from the base of the light? **(c)** What is the rate of change of the shadow when the person is 15 ft from the base of the light?

26. *Physics* A 75-lb weight is attached to a rope 50 ft long, which passes over a pulley at P, 20 ft above the ground. The other end of the rope is attached to the bumper of a truck at a point 2 ft above the ground. If the truck moves at the rate of 8 ft/s, how fast is the weight rising when it is 10 ft above the ground?

27. *Electricity* The electric resistance R of a certain resistor is a function of the temperature T, as shown by $R = 6 + 0.008T^2$, where R is in ohms and T in degrees Celsius. If the temperature is increasing at the rate of 0.01°C/s, how fast is the resistance changing when $T = 40$°C?

28. *Transportation* Two cars leave the same point, traveling on straight roads that are perpendicular. One car is traveling east at the rate of 30 mph and the other car is traveling north at the rate of 40 mph. How fast is the distance between them changing **(a)** after 3 min (0.05 h) and **(b)** after 6 min (0.10 h)?

29. *Electronics* The impedance Z in ohms (Ω) of a circuit is given by

$$Z = \frac{RX}{R+X}$$

If R is constant at $3.00\,\Omega$ and X is increasing at a rate of $1.45\,\Omega$/min, what is the rate at which Z is changing when $X = 1.05\,\Omega$?

30. *Electronics* The impedance Z in a series circuit is given by $Z^2 = R^2 + X^2$, where X is the reactance. If $X = 12\,\Omega$ and R increases at the rate of $3.0\,\Omega$/s, what is the rate at which Z is changing when $R = 6.0\,\Omega$?

31. *Electricity* If two variable resistances R_1 and R_2 are linked in parallel, then the effective resistance R is given by $\frac{1}{R} = \frac{1}{R_1} + \frac{1}{R_2}$. At a certain instant, R_1 is $4.00\,\Omega$ and increasing at the rate of $0.50\,\Omega$/s. At the same time, R_2 is $5.00\,\Omega$ and increasing at the rate of $0.40\,\Omega$/s. What is the rate of change of R?

32. *Electricity* The power P, in watts, in a circuit varies according to $P = RI^2$. If $R = 80.00\,\Omega$ and I varies at the rate of 0.24 A/s, what is the rate of change of P when $I = 2.5$ A?

33. *Electronics* The resistance in a resistor is given by $R = 35.0 + 0.0174T^2$, where R is in ohms and T is in degrees Celcius. How is the resistance changing when the temperature of the resistor is 47.0°C and is decreasing at the rate of 1.25°C/min?

34. *Aeronautical engineering* In the air, the speed of sound v, in m/s, is given by the formula $v = 331\sqrt{\frac{T}{273}}$, where T is the temperature in degrees Kelvin. (n degrees Celsius $= n + 273$ degrees Kelvin.) As an airplane increases altitude, the outside temperature decreases at the rate of 5.8°C/km. What is the rate of change in the speed of sound when the temperature outside the airplane is −30°C?

35. *Acoustical engineering* The **Doppler effect** is the apparent change in frequency of a sound source when there is relative motion between the source and the observer. When a person stands still, the apparent change in frequency f in hertz (Hz) of an object moving toward the person is given by the formula

$$f = f_s \frac{v_L}{v_L - v_s}$$

where f_s is the frequency of sound, in Hz, from the source; v_s is the velocity of the source in m/s; and v_L is the speed of sound in m/s. At 21°C, the speed of sound is about 343 m/s. A train is moving at a speed of 29 m/s toward an observer at the instant it sounds it horn at a frequency of 200 Hz. **(a)** What is the frequency of sound to the observer? **(b)** What is the apparent rate of change of sound to the observer?

36. *Medical technology* Blood flows faster the closer it is to the center of a blood vessel. According to **Poiseuille's laws**, the velocity, V, of blood is given by $V = k\left(R^2 - r^2\right)$, where R is the radius of the blood vessel, r is the distance of a layer of blood flow from the center of the vessel, and k is a constant.
   (a) Find $\frac{dV}{dt}$ when $k = 475$. Treat r as a constant.
   (b) Suppose that a cross-country skier's blood vessel has radius $R = 0.015$ mm and that the cold weather is causing the vessel to contract at a rate of $\frac{dR}{dt} = -0.001$ mm/min. How fast is the velocity of the blood changing?

37. *Medical technology* One of the treatments for high blood pressure is to give a patient some medication to dilate the blood vessels. Use **Poiseuille's law** with $k = 637.5$ to find the rate of change of the blood velocity when $R = 0.0250$ mm and R is changing at 0.002 mm/min.

38. *Police science* Crime rates are influenced by temperature. In a certain town with a population of 100,000, the crime rate has been approximated at $C = 0.1\,(T - 47)^2 + 95$, where C is the number of crimes per month and T is the average monthly temperature. The average temperature was 51°F, and by the end of the month the average monthly temperature was rising at the rate of 7°F/mo. How fast is the crime rate rising at the end of April?

39. *Electronics* The impedance of a circuit containing a resistance and an inductance in parallel is given by $Z = \frac{RX}{R+X}$. If $R = 7.5\,\Omega$ and X is decreasing at the rate of $2.5\,\Omega$/s, at what rate is Z changing when $X = 4.0\,\Omega$?

40. *Petroleum engineering* An oil spill on the open sea in calm weather spreads in a circular pattern. If the radius of the oil slick increases at the rate of 0.9 m/s, how fast is the area of the spill increasing when its radius is 225 m?

41. *Electronics* The impedance of a circuit containing a resistance and an inductance in parallel is given by $Z = \frac{RX}{R+X}$. If $R = 7.5\,\Omega$ and Z is increasing at the rate of $2.0\,\Omega$/s, at what rate is X changing when $X = 3.5\,\Omega$?

42. *Petroleum engineering* Oil is leaking from a tanker at the rate of 5 000 L/min. The leak results in an oil spill that is circular on the ocean's surface. The depth of the spill varies linearly from a maximum of 5.0 cm at the origin of the spill to a minimum of 0.5 cm at the outside edge of the oil slick. How fast is the radius of the slick increasing 4 h after the tanker started leaking?

43. *Heating technology* A boiler pipe has a circular cross-section. The inner radius of the pipe was originally 1 in. but is decreasing at the rate of 0.1 in./yr due to mineralization. At what rate is the cross-sectional area of its opening decreasing at the instant when the radius is 0.5 in.?

44. *Construction technology* A concrete pier is being constructed to support a bridge. The pier is in the shape of the frustum of a square pyramid. The bottom base of the pier measures 3 m on each side, the top base measures 2 m on a side, and the height is 2 m. If the concrete is poured at a constant rate of 0.01 m^3/s, at what rate is the height h of the poured concrete rising when $h = 1$ m?

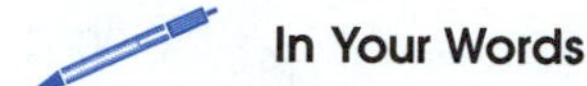

In Your Words

45. Without looking in the book, describe the steps used to solve a related rate problem.

46. Explain how you can tell a problem involves related rates.

5.6 NEWTON'S METHOD

One method for finding the roots of a polynomial isby using linear interpolation. In this section, we will learn a faster method that uses derivatives. This new technique is known as **Newton's method** or the **Newton-Raphson method** for finding roots.

In using the linear interpolation method, we used the locator theorem to determine that a root existed between the real numbers a and b. We will begin Newton's method the same way. For a continuous function f over an interval, we have found two real numbers, a and b, in the interval, such that one of the values of $f(a)$ and $f(b)$ is positive and the other is negative.

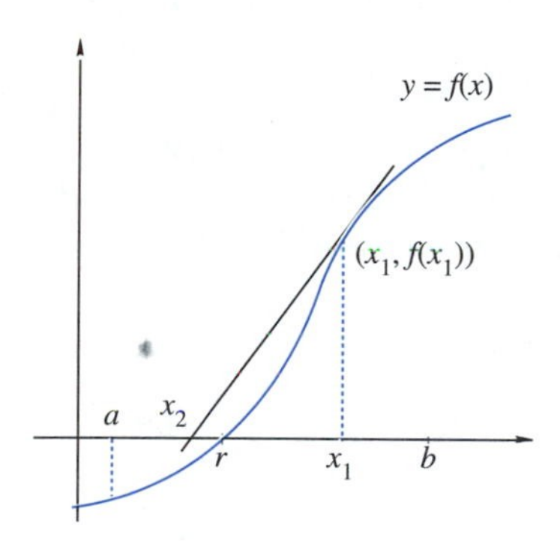

FIGURE 5.28

Assume that $a < b$. We will select some point x_1 in the interval $[a, b]$. We know that the tangent to the graph of f at x_1 is given by $f'(x_1)$. We also know that the equation of this tangent line is $y - f(x_1) = f'(x_1)(x - x_1)$ or $y = f(x_1) + f'(x_1)(x - x_1)$. Suppose that x_2 is the x-intercept of this tangent line as shown in Figure 5.28. Then we know that $(x_2, 0)$ is a point on the tangent line. Thus, the tangent line satisfies the equation

$$0 = f(x_1) + f'(x_1)(x_2 - x_1)$$

If $f'(x_1) \neq 0$, we can rewrite this equation as

$$x_2 = x_1 - \frac{f(x_1)}{f'(x_1)}$$

The process can be continued. We can use x_2 to determine a new point x_3, and so on until we get a point x_n, where $f(x_n)$ is within some desired distance of zero. In general, if x_n is the nth approximation, then

$$x_{n+1} = x_n - \frac{f(x_n)}{f'(x_n)}$$

provided $f'(x_n) \neq 0$.

EXAMPLE 5.22

Use Newton's method to approximate the root between -4 and -3, if $P(x) = x^3 + x^2 - 7x + 3$.

Solution We will let our first guess, x_1, be -3. The results of the successive approximations are given in the following table. $P'(x) = 3x^2 + 2x - 7$

i	x_i	$P(x_i)$	$P'(x_i)$	$x_{i+1} = x_i - \frac{P(x_i)}{P'(x_i)}$	$P(x_{i+1})$
1	−3.0000	6.0000	14.0000	−3.4286	−1.5487
2	−3.4286	−1.5487	21.4087	−3.3563	−0.0490
3	−3.3563	−0.0490	20.0816	−3.3539	−0.0009

■

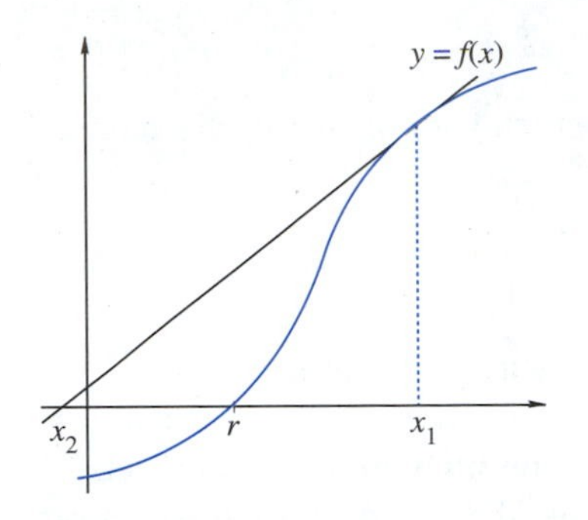

FIGURE 5.29

Linear interpolation takes five steps to get $x_6 = -3.3535$ and $P(x_6) = 0.0062$. Newton's method took three steps to get $x_4 = -3.3539$ and $P(x_4) = -0.0009$. It appears that Newton's method is almost twice as fast as the linear interpolation method.

Newton's method does not always work. In particular, if the first choice of x_1 is not sufficiently close to the root r, then the second approximation, x_2, will be farther away from r instead of closer. This is indicated in Figure 5.29. In general, you do not want to select a point where the derivative is close to zero. As a rule of thumb, whenever $|f'(x_1)| \leq 1$, then select a different choice for x_1.

EXAMPLE 5.23

Use Newton's method to approximate $\sqrt{11}$.

Solution This is equivalent to approximating the positive real root of $f(x) = x^2 - 11$. Since $f(3) = -2$ and $f(4) = 5$, we know there is a root between 3 and 4. Using $f'(x) = 2x$ and selecting $x_1 = 3$, we get the following results.

i	x_i	$P(x_i)$	$P'(x_i)$	$x_{i+1} = x_i - \dfrac{P(x_i)}{P'(x_i)}$	$P(x_{i+1})$
1	3.0000	−2.0000	6.0000	3.3333	0.1109
2	3.3333	0.1109	6.6667	3.3167	0.0005
3	3.3167	0.0005	6.6334	3.3166	−0.0002

Thus, $\sqrt{11} \approx 3.3166$.

Exercise Set 5.6

In Exercises 1–7, use Newton's method to approximate the real roots to four decimal places.

1. $x^3 + 5x - 8 = 0$
2. $x^3 - 4x + 2 = 0$
3. $x^4 - x - 4 = 0$
4. $x^4 - 2x^3 - 3x + 2 = 0$
5. $x^4 = 125$
6. $x^3 = 91$
7. $x^5 = 111$

Solve Exercises 8–9.

8. *Sheet metal technology* An open-topped barrel in the shape of a right circular cylinder is to be fabricated from sheet metal. The manufacturer has specified that it holds a volume of $V = 10\,000\,\text{cm}^3$ and has a total surface area of $A = 2\,400\,\text{cm}^2$.
   (a) Show that the radius r of the barrel must satisfy the equation $2V + \pi r^3 - Ar = 0$, where V and A are the area and volume given above.
   (b) Use Newton's method to find the radius of the barrel accurate to 4 decimal places.

9. *Air traffic control* One method for clearing fog from an airport runway is to ignite long troughs filled with fuel that have been placed along the sides of the runway. A trough is to be formed from a rectangular sheet of metal that measures $20\,\text{m} \times 0.7\,\text{m}$. The cross-section of the trough is an arc of a circle with radius r and subtended angle θ. If each trough is to hold $0.85\,\text{m}^3$ of fuel, use Newton's method to determine the size of θ.

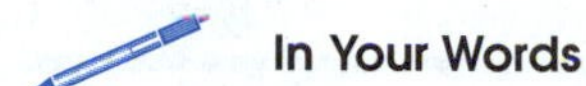

In Your Words

10. What is Newton's method and how is it used?

11. What are the advantages and disadvantages of Newton's method compared to using a graphing calculator?

5.7 DIFFERENTIALS

In studying the derivative of a function $y = f(x)$, we used the notations $f'(x)$ and dy/dx to represent the derivative. The symbols dy and dx are called **differentials**. They may be given their own meaning as we'll see in this section.

Since $f'(x) = \dfrac{dy}{dx}$, we can solve this equation for dy and get $dy = f'(x)dx$, which is called the **differential of** f. In this equation, $dx \neq 0$.

EXAMPLE 5.24

Find the differential of the function $f(x) = 3 + x + 14x^{2/7}$.

Solution Here $f'(x) = 1 + 4x^{-5/7}$. Thus, the differential is

$$dy = f'(x)\,dx = \left(1 + 4x^{-5/7}\right) dx$$

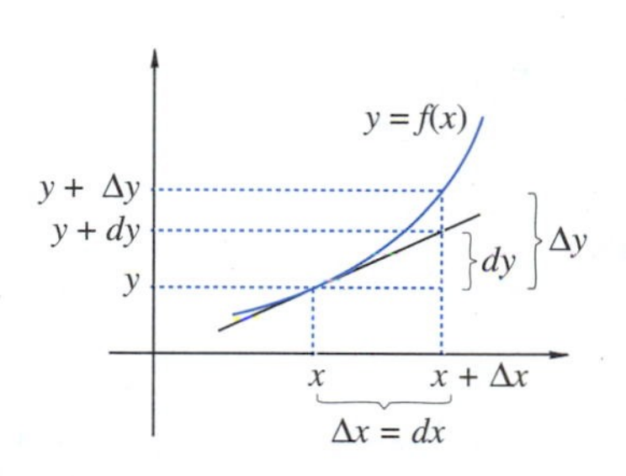

FIGURE 5.30

Figure 5.30 shows the change in height of a point that moves along the tangent line to the function $y = f(x)$ at the point $(x, f(x))$, as x changes from x to $x + \Delta x$. This change in height is called dy. The actual change in the value of y is the increment Δy. If we think of x as a fixed point, then dy is a linear function of the increment Δx. For this reason, dy is called the **linear approximation** to the actual increment Δy. We can approximate $f(x + \Delta x)$ by writing dy in place of Δy. Thus, we see that

$$f(x + \Delta x) = y + \Delta y \approx y + dy$$

Since $y = f(x)$ and $dy = f'(x)\Delta x$, we get the **linear approximation formula**:

$$f(x + \Delta x) \approx f(x) + f'(x)\Delta x$$

EXAMPLE 5.25

If $y = x^3$, determine Δy and dy, if $x = 2$ and $\Delta x = 0.01$.

Solution Since $f(x + \Delta x) = y + \Delta y$ and $y = f(x)$, we can see that $\Delta y = f(x + \Delta x) - f(x)$. Thus, when $x = 2$ and $\Delta x = 0.01$, we have

$$\begin{aligned}\Delta y &= (x + \Delta x)^3 - x^3\\ &= (2.01)^3 - 2^3\\ &= 8.1206 - 8 = 0.1206\end{aligned}$$

We also know that $dy = f'(x)dx$. If $f(x) = x^3$, then $f'(x) = 3x^2$. Since $dx = \Delta x$, we have

$$dy = 3(2)^2(.01) = 0.12$$

As you can see, dy is a good approximation for Δy.

Application

EXAMPLE 5.26

A spherical ball-bearing has a radius of 8 mm when it is new. What is the approximate volume of metal lost after it wears down to a radius of 7.981 mm?

Solution The exact volume of the metal lost is the change in volume ΔV of the sphere, where $V = \frac{4}{3}\pi r^3$. The change in radius is $\Delta r = 7.981 - 8.000 = -0.019$ mm. The change in volume is approximated by dV, where

$$\begin{aligned} dV &= 4\pi r^2 dr \\ &= 4\pi(8^2)(-0.019) \approx -15.281 \end{aligned}$$

The approximate loss in volume is 15.281 mm^3.

Application

EXAMPLE 5.27

A hemispherical dome has a radius of 25 ft. It is going to be given a coat of paint 0.01 in. thick. What is the approximate volume of paint needed for this job?

Solution What we are looking at is the change in volume of this hemisphere when the radius changes from 25.000 ft to 25.00083 ft (0.01 in. $\approx$ 0.00083 ft) or $\Delta r = 0.00083$ ft. The volume of a hemisphere is one-half that of a sphere, or $V = \frac{2}{3}\pi r^3$. Again, using differentials we see that $dV = 2\pi r^2 dr$. When $r = 25$ and $dr = 0.00083$, we have

$$\begin{aligned} dv &= 2\pi(25^2)(0.00083) \\ &\approx 3.2594 \end{aligned}$$

About 3.26 ft^3 of paint are needed.

Differentials can be used to estimate errors in results.

Application

EXAMPLE 5.28

A cylindrical can with an open top has an inside height of exactly 100 mm and a radius of 25 mm with a maximum error of 0.1 mm. Estimate the maximum error in the calculated volume of this can.

Solution If r denotes the radius and h the height, then $V = \pi r^2 h$. If the maximum error in the radius is denoted as dr, then we can see that $dV = 2\pi rh\,dr$. If $r = 25$, $h = 100$, and $dr = 0.1$, then

$$\begin{aligned} dV &= 2\pi(25)(100)(0.1) \\ &= 500\pi \approx 1\,570.8 \text{ mm}^3 \end{aligned}$$

Thus, the maximum error in volume due to the error on the radius is 1 570.8 mm^3.

If there is an error in measurement, we define the **relative** or **average error** as the ratio of the actual error to the actual size of the quantity. This is often written in differential form as

$$\text{relative error} = \frac{dy}{y}$$

The relative error is usually expressed as a percentage.

In Example 5.28, the actual error in the radius was 0.1 mm when the radius was 25 mm. Thus the relative error in measuring the radius is $\frac{dr}{r} = \frac{0.1}{25} = 0.004 = 0.4\%$. The actual volume of the can is $62\,500\pi$ with a possible error of 500π. The relative error is $\frac{500\pi}{62\,500\pi} = 0.008 = 0.8\%$.

Exercise Set 5.7

In Exercises 1–6, find the differential of the given functions.

1. $f(x) = x^4 - x^2$
2. $g(x) = 3x^2 + 2x$
3. $h(x) = 5x^3 - x^2 + x$
4. $j(x) = (x^2 - 4)^3$
5. $k(x) = \sqrt[3]{4 - 2x}$
6. $m(x) = \frac{x}{x - 4}$

In Exercises 7–12, find Δy and dy for the given values of x and Δx.

7. $y = x^2 - x$, $x = 3$, $\Delta x = 0.2$
8. $y = x^3 + x$, $x = 2$, $\Delta x = 0.1$
9. $f(x) = 3x^2 - 4x + 1$, $x = 5$, $\Delta x = 0.15$
10. $g(x) = x^4$, $x = 4.1$, $\Delta x = 0.1$
11. $h(x) = \frac{1}{x^3}$, $x = 3$, $\Delta x = 0.2$
12. $j(x) = \frac{x}{x+1}$, $x = 0.5$, $\Delta x = 0.1$

Solve Exercises 13–36.

13. *Physics* A tank filled with water is in the shape of a right-circular cone. The radius of the cone is measured to be 2 m with a maximum error of 0.01 m. The height is exactly 4 m. **(a)** What is the approximate volume of the water in the tank? **(b)** What is the maximum error in this volume? **(c)** What is the relative error in the volume?
14. *Physics* The radius of a circular disk is estimated to be 200 mm with a maximum error of 1.5 mm. **(a)** What is the maximum error in calculating the area of one side of the disc? **(b)** What is the relative error in the radius? **(c)** What is the relative error in the area?
15. The side of a square is measured to be 24 in. with a possible error of 0.02 in. **(a)** What is the maximum error in the area of the square? **(b)** What is the relative error?
16. A cube has an edge measuring 1.452 m. If the error in measurement is ± 0.5 mm, what is the approximate error in the volume?
17. The radius of a hemispherical dome is measured as 100 m with a maximum error of 10 mm. What is the maximum error in calculating its surface area?
18. With what accuracy must the radius of the dome in Exercise 17 be measured in order to have a percentage error of at most 0.01% in its calculated surface area?

19. The diameter of a sphere is measured to be 12.4 m. If measurement error is plus or minus half the unit of the most precise digit, that is ± 0.05 m, **(a)** what is the approximate error in the volume of the sphere due to measurement error? **(b)** What is the approximate relative error?

20. If the diameter in Exercise 19 is given as 12.40 m, what is the approximate error in the volume and the relative error in volume due to measurement error?

21. *Energy technology* The **Stefan-Bolzmann law** for the emission of radiant energy from the surface of a body is given by $R = \sigma T^4$, where R is the rate of emission per unit area, T is the temperature in degrees Kelvin, and σ is the Stefan-Bolzmann constant. If a relative error of 0.02 is made in T, what is the maximum error in R?

22. *Electricity* A wire has a resistance of R ohms given by $R = \frac{250}{r^2}$, where r is the radius of the wire. There is a possible error in the radius of $\pm 0.4\%$.

(a) What is the approximate error in the resistance? (Express your answer in terms of r.)

(b) What is the relative error?

23. *Electricity* In a dc circuit, $V = IR$. If a circuit has a constant voltage of 30 V, what is the approximate change in the current if the resistance changes from 10.0 to 10.1 Ω?

24. *Electricity* The equivalent resistance R when a resistance of 20 Ω and a variable resistance R_v are connected in parallel is

$$R = \frac{20R_v}{20 + R_v}$$

If the variable resistance is changed from 10.0 to 10.1 Ω, what is the approximate change in R?

25. *Electricity* The resistance in a resistor is given by $R = 35.0 + 0.0174T^2$, where R is in ohms (Ω) and T is the temperature in degrees Celsius (°C). If the error in reading the temperature at $T = 125°\text{C}$ is $\pm -0.5°\text{C}$, what is the approximate error in the resistance?

26. *Construction* A coat of paint 0.5 mm thick is applied to the surface of a spherical storage tank with a radius of 20 m. How many liters of paint are needed?

27. *Mechanical engineering* A company manufactures stainless-steel, spherical ball-bearings with a radius of 8 mm. The percentage error in the radius must be no more than 0.5%. What is the approximate percentage error for the surface area of the ball bearing?

28. *Material science* Stainless steel has a mass of 0.00794 g/mm^3.

(a) What is the mass of one of the ball-bearings in Exercise 27?

(b) What is the maximum error in the mass?

29. *Industrial technology* In a precision manufacturing process, ball bearings must be made with a radius of 0.6 mm, with a maximum error in the radius of ± 0.015 mm. Estimate the maximum error in the volume of the ball bearing.

30. *Pharmacology* The concentration of a certain drug in the bloodstream x hours after being administered is approximately $C(x) = \frac{5x}{9+x^2}$. Use differentials to approximate the changes in concentration as x changes from **(a)** 1 h to 1.5 h, **(b)** 2 h to 2.25 h, and **(c)** 3 h to 3.25 h.

31. *Medical technology* A tumor is approximately spherical in shape. If the radius of the tumor changes from 15 mm to 15.5 mm, use differentials to find the approximate change in volume of the tumor.

32. *Petroleum engineering* The top of an oil slick is in the shape of a circle. Find the approximate increase in the area of the slick's top if its radius changes from 2.1 km to 2.3 km.

33. *Business* A company estimates that the revenue in dollars from the sale of n computers is given by

$$R(n) = 2345 + 0.02n^2 + 0.00015n^3$$

Use differentials to approximate the change in revenue from the sale of one more computer when 500 have been sold.

34. *Meteorology* A spherical weather balloon is being inflated. Use differentials to approximate the change in the volume of the balloon as the radius changes from 20.0 in. to 21.5 in.

35. *Machine technology* A plate in the shape of a right circular cylinder 3.2 cm thick was originally made 30.0 cm in diameter. If 0.2 cm is trimmed all around its edge, use differentials to approximate the volume of the metal that was trimmed.

36. *Acoustical engineering* The frequency, f, in Hertz, of the fundamental note on a guitar string is related to the tension, T, in the string; the length, L, of the string; and the linear density, μ, of the string according to the formula

$$f = \frac{1}{2L}\sqrt{\frac{T}{\mu}}$$

Estimate the amount by which the frequency will change if the tension is adjusted 1% during tuning. Let $L = 1.15$ m, $\mu = 0.0036$ kg/m, and $T = 620$ N.

In Your Words

37. What is meant by the differential terms dy and dx?

38. Explain the difference between dy and Δy.

5.8 ANTIDERIVATIVES

Until now, we have always differentiated a function to arrive at a solution. For example, if we were given a position function, we took its first derivative to get its velocity function and its second derivative to gets its acceleration function. Is it possible to reverse this process? For example, if we know the velocity function for an object, could we determine its position function? As we will see in this section, it is sometimes possible to reverse the differentiation process.

Suppose we have a function F and the derivative of F is f. Thus, $F' = f$. In this case, F is called the **antiderivative** of f.

EXAMPLE 5.29

Find an antiderivative of $f(x) = 5x^3$.

Solution We want to find $F(x)$, where $F'(x) = f(x) = 5x^3$. When we differentiate a polynomial, we reduce the power of x by 1. We also multiply the coefficient by the power of x. Thus, the derivative of x^n is nx^{n-1}. Since the power of f is 3, then this must be one less than the power of F. Thus, F is of the form $F(x) = kx^4$. Now $F'(x) = 4kx^3 = 5x^3$ and so $k = \frac{5}{4}$. Thus $F(x) = \frac{5}{4}x^4$.

But, we have not found a unique solution. For example, the derivative of $\frac{5}{4}x^4 + 9$ is also $5x^3$, and the derivative of $\frac{5}{4}x^4 - 7$ is $5x^3$. In fact, since the derivative of a constant C is 0, the antiderivative of $5x^3$ is $\frac{5}{4}x^4 + C$.

What we have discovered is that, in general, if $f(x) = kx^n$, then

$$F(x) = \frac{k}{n+1}x^{n+1} + C, \text{ if } n \neq -1$$

Notice that

$$F'(x) = \frac{k(n+1)}{n+1}x^{n+1-1} = kx^n = f(x).$$

So F is the antiderivative of f. A special case occurs when $n = 0$. Then $f(x) = k$ and $F(x) = kx + C$.

Antiderivatives Differ by a Constant

If G and F are both antiderivatives of the same function, f, then

$$G(x) = F(x) + C \qquad \text{where } C \text{ is a constant.}$$

EXAMPLE 5.30

Find the antiderivative of $g(x) = 3x^2 - 2x + 5$.

Solution Using this new method to take the antiderivative of each term, we get

$$G(x) = x^3 - x^2 + 5x + C$$

You should check to see that $G'(x) = g(x)$.

EXAMPLE 5.31

Find the antiderivative of $h(x) = 4x^2 + 2x^{-2}$.

Solution The antiderivative of $4x^2$ is $\frac{4}{3}x^3$ and of $2x^{-2}$ is $-2x^{-1}$, so $H(x) = \frac{4}{3}x^3 - 2x^{-1} + C$.

We can apply antiderivatives to many problems. For example, we know that the acceleration due to the Earth's gravity is about 32 ft/s^2 or 9.8 m/s^2 downward. We also know that the acceleration is the derivative of the velocity or that the velocity function is an antiderivative of the acceleration. Similarly, the position function is an antiderivative of the velocity function. With enough information we can determine the value of the constant of the antiderivative. This is demonstrated in the next example.

Application

EXAMPLE 5.32

An object is thrown straight downward from the top of a building with an initial speed of 28 ft/s. The object strikes the ground with a speed of 220 ft/s. How tall is the building?

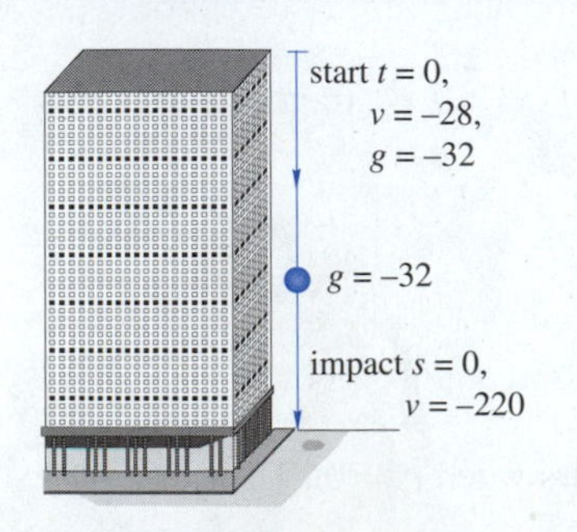

FIGURE 5.31

Solution First, let's set up a coordinate system with the ground level at $y = 0$, as shown in Figure 5.31. We know that acceleration is given by $a(t) = -32$. The velocity is the antiderivative of the acceleration and so $v(t) = -32t + C$. When the object is thrown, $t = 0$ and we are told that $v(0) = -28$. (It is negative because the velocity is downward.) Thus, we can see that $C = -28$ and $v(t) = -32t - 28$.

The object strikes the ground with a velocity of 220 ft/s downward or -220. Thus, when it strikes the ground, $v(t) = -32t - 28 = -220$. Solving for t, we see that it strikes the ground when $t = 6$ s.

The height of the object is given by the position function $s(t)$, and is the antiderivative of the velocity. Thus, $s(t) = -16t^2 - 28t + k$. When the object

EXAMPLE 5.32 (Cont.) hits the ground, $t = 6$ and $s(t) = 0$, so $-16(6^2) - 28(6) + k = 0$. Solving, we get $k = 744$ and $s(t) = -16t^2 - 28t + 744$. At $t = 0$ when the object is released, $s(0) = 744$ and so the height of the building is 744 ft.

Exercise Set 5.8

In Exercises 1–16, find the antiderivative of the given function.

1. $f(x) = 7$
2. $g(x) = 4x$
3. $h(x) = 2 - 3x^2$
4. $j(x) = x^2 - 3x + 5$
5. $k(x) = 4x^3 - 3x^2 + 2x + 9$
6. $m(x) = \frac{4}{5}x^3 - 6x^2$
7. $f(x) = x^{-4} + 2x^{-3} - x^{-2} + 5$
8. $g(x) = \dfrac{1}{x^3} + 2$
9. $h(x) = \sqrt{x} + x - \dfrac{1}{x^2}$
10. $j(x) = 3x^2 - \dfrac{1}{\sqrt[3]{x}} - \dfrac{4}{x^2}$
11. $k(x) = 2x^{-1.4} + 3.5x^{-6} + \dfrac{1.2}{x^{1.3}}$
12. $m(x) = \dfrac{2}{\sqrt{x}} + 3x^2\sqrt{x}$
13. $s(t) = t^2 + 2t$
14. $a(t) = \frac{2}{3}t - \frac{3}{4}t^2 + \sqrt{t^3}$
15. $v(t) = 42t - 5$
16. $p(v) = \frac{1}{3}v^3 - \dfrac{3}{v^4}$

Solve Exercises 17–36.

17. *Physics* A ball is dropped from a building that is 600 ft high. **(a)** How long does it take it to reach the ground? **(b)** What is its velocity when it hits the ground?

18. *Physics* A ball is dropped from a building that is 140 m high. **(a)** How long does it take to reach the ground? **(b)** What is its velocity when it hits the ground?

19. *Physics* A ball is thrown straight down from a building with an initial speed of 25 ft/s. It hits the ground with a speed of 185 ft/s. **(a)** How long was the ball in the air? **(b)** How high is the building?

20. *Physics* A ball is thrown straight down from a building with an initial speed of 12.1 m/s. It hits the ground with a speed of 66 m/s. **(a)** How long was the ball in the air? **(b)** How high is the building?

21. *Physics* A ball is thrown straight upward from the top of a building with an initial speed of 160 ft/s. It strikes the ground with a speed of 384 ft/s. **(a)** How long is it in the air? **(b)** How high is the building?

22. *Physics* A ball is thrown straight upward from the top of a building with an initial speed of 45 m/s. It strikes the ground with a speed of 120 m/s. **(a)** How long is it in the air? **(b)** How high is the building?

23. *Transportation* A car starting from rest has a constant acceleration of 3.2 m/s^2. How far has the car traveled when the speed is 32 m/s?

24. *Automotive engineering* A car's brakes are applied when it is going 45 mph and provides a constant deceleration of 20 ft/s^2. **(a)** How long does it take the car to stop? **(b)** How far does it travel before it stops? (Change 45 mph to ft/s.)

25. *Aeronautical engineering* A runway is 500 m long. If a plane lands at an end of the runway with a speed of 50 m/s, what is the constant acceleration that would bring the plane to a stop at the other end of the runway?

26. *Automotive engineering* The angular acceleration of a wheel is given by $\alpha(t) = 4 - t$ rad/s^2. Angular velocity is given by $\omega(t)$ and angular displacement by $\theta(t)$. If $\theta(0) = 0$ rad, and $\omega(0) = 10$ rad/s, find
(a) an expression for angular displacement,
(b) the angular displacement when the wheel turns in the opposite direction, and
(c) the time when the angular velocity is -14 rad/s.

27. *Machine technology* When the motor turning a flywheel is shut off, the angular velocity of the wheel is given by $\omega(t) = 12.4 - 3.4t + 0.30t^2$ rad/s.
(a) During the first 3 s after the motor has been shut off, how many revolutions does the wheel complete?
(b) When the angular acceleration is $\alpha(t) = -1$, how many revolutions has the wheel completed?

28. *Physics* The acceleration of a ball starting from rest and moving in a straight line is proportional to the time t in seconds (s). If the ball moves 24 m in the first 4 s, find an expression for the displacement $s(t)$.

29. *Electronics* An expression for the current i in a circuit in amperes (A) is given by $i = 4.4t - 2.1t^2$, where t is in seconds. If $i = \dfrac{dq}{dt}$, where q is the charge in coulombs (C), find **(a)** an expression for q at $t = 0.0$ s, $q = 5.0$ C, and **(b)** the charge at $t = 3.2$ s.

30. *Electronics* The current $i(t)$, in amperes (A), flowing to a capacitor is given by $i(t) = 6\sqrt{t}$, where t is time in seconds (s). If $i(t) = q'(t)$, where q is the charge in coulombs (C), find the charge on the capacitor at $t = 0.25$ s, if $q(0.16) = 0.347$ C.

31. *Electronics* The voltage in volts (V) across a coil is given by $v(t) = -N\phi'(t)$, where N is the number of turns in the coil, $\phi(t)$ is the magnetic flux in webers (Wb), and t is the time in seconds (s). If $v(t) = 2t - 4t^{1/3}$ and $\phi(0) = 0.020$ Wb when N is 200 turns, find **(a)** a general expression for $\phi(t)$ and **(b)** $\phi(0.729)$.

32. A general expression for the slope of a tangent at any point on a certain curve is given by $m = 4x + 2$. If the curve goes through the point $(-1, 5)$, state the equation of the curve.

33. The slope of a normal line to a curve $y = f(x)$ is given by $m = \sqrt{x}$. Find $f(x)$, if $f(4) = -3$.

34. *Industrial engineering* A stamping machine produces computer cases at a rate that varies over the working day according to the equation $\dfrac{dN}{dt} = 175 + 2t - 0.2t^2$, where t is the time in hours. **(a)** Find an expression for N and **(b)** calculate N for an 8-hour day if $N = 0$ at $t = 0$.

35. *Thermodynamics* The temperature, T, in °C in an industrial furnace varies from its center to a point outside. The rate of temperature change is given by the equation

$$\frac{dT}{dx} = -\frac{5750}{(x+1)^3}$$

Find an expression for T at a distance of x meters from the center if $T = 2900$°C at $x = 0$.

36. *Ecology* Biologists are treating a stream contaminated with bacteria. The level of contamination is changing at a rate of $\dfrac{dN}{dt} = -\dfrac{870}{t^2} - 210$ bacteria/cm^3/day, where t is the number of days since the treatment began. Find a function $N(t)$ to estimate the level of contamination if the level after 1 day was about 7320 bacteria/cm^3.

In Your Words

37. Explain how an antiderivative differs from a derivative.

38. A function has an infinite number of antiderivatives. How can you determine which is the "correct" antiderivative for a particular function?

CHAPTER 5 REVIEW

Important Terms and Concepts

Acceleration
Antiderivative
Concave downward
Concave upward
Concavity
Critical point
Critical value
Decreasing function
Differentials
Extrema
First derivative test
Increasing function
Inflection point
Linear approximation formula
Maximum
  Absolute
  Local
  Relative
Minimum
  Absolute
  Local
  Relative
Newton's method
Position function
Rectilinear motion
Related rates
Second derivative test
Speed
Velocity
  Average velocity
  Instantaneous velocity

Review Exercises

In Exercises 1–10, find critical values, the extrema, the intervals on which the curve is concave up and concave down, the inflection points, and the asymptotes. Sketch the curve.

1. $f(x) = x^4 - x^2$
2. $g(x) = x^4 - 32x$
3. $h(x) = 18x^2 - x^4$
4. $j(x) = 3x^4 - 4x^3 + 1$
5. $k(x) = \sqrt[5]{x}$
6. $f(x) = x\sqrt[3]{4-x}$
7. $g(x) = \dfrac{x^2+1}{x^2-4}$
8. $h(x) = \dfrac{x-1}{x+2}$
9. $j(x) = \dfrac{x}{x^2+x-2}$
10. $k(x) = \dfrac{x^3}{x^2-9}$

In Exercises 11 and 12, position functions of points moving rectilinearly are given. Find the function for the velocity and acceleration at time *t*. When are the position, velocity, and acceleration maximum and minimum for the given interval?

11. $s(t) = t^3 - 9t^2 + t,\ [-4, 4]$
12. $s(t) = t + \dfrac{4}{t},\ [1, 4]$

In Exercises 13–16, find the antiderivative of each function.

13. $f(x) = 3x^2 - 4x$
14. $g(x) = 2x^{-3} + 5x^6$
15. $h(x) = \sqrt{x} + x^2 - \dfrac{3}{x^2}$
16. $s(t) = 4.9t^2 - 3.6t + 14$

Solve Exercises 17–29.

17. *Physics* An object is thrown directly upward with a velocity of 288 ft/s. Its height $s(t)$ in feet above the ground after t s is given by $s(t) = 288t - 16t^2$. **(a)** What are the velocity and acceleration after t s? **(b)** What are the height, velocity, and acceleration after 4 s? **(c)** What is the maximum height? **(d)** At what time is the height a maximum? **(e)** When does the object strike the ground? **(f)** What is its velocity when it strikes the ground?

18. *Physics* An object rolls down a ramp such that the distance in centimeters after t s is given by $s(t) = 15t^2 + 5$. **(a)** What is its velocity after t s? **(b)** What is its velocity after 2 s? **(c)** When will its velocity be 105 cm/s?

19. *Business* Suppose it costs $1 + 0.00058v^{3/2}$ dollars/mi to operate a truck at v mph. Additional costs, including the driver's salary, amount to \$25 per hour. What speed will minimize the total cost of a 1,000 mi trip?

20. *Construction* A cylindrical barrel is to be constructed to hold 246π ft^3 of liquid. The cost per ft^2 of constructing the side of the barrel is three times the cost of constructing the top and the bottom. What are the dimensions of the barrel that will cost the least to construct?

21. *Agriculture* What is the maximum rectangular area that can be enclosed with 1 200 m of fencing?

22. *Petroleum engineering* A storage tank is in the shape of a sphere with a radius of 8 ft. If oil is pumped into the tank at the rate of 20 gal/min, how fast is the oil level rising when it has reached a height of 11 ft? (1 gal ≈ 0.1335805 ft^3.)

23. The radius of a sphere is 1.5 m with a possible error in measurement of 0.01m=1 cm. What is the maximum error you would expect in **(a)** the calculation of the volume? **(b)** the calculation of the surface area?

24. *Sheet metal technology* A gutter is to be formed by bending up equal widths of the opposite sides of a sheet of metal that is 380 mm wide. If the bent sides are perpendicular to the base, what dimensions will give the largest capacity?

25. *Thermodynamics* The voltage of a certain thermocouple as a function of temperature is given by $v = 4.5T + 0.0003T^3$. What is the approximate change in voltage when the temperature changes from 100°C to 101°C?

26. *Navigation* Two ships leave the same port at the same time. One travels west at the rate of 12 km/h and the other travels south at the rate of 5 km/h. At what rate is the distance between them changing 3 h after they leave the port?

27. A right-circular cylinder is to have a volume of 20 kL. What dimensions will make the total surface area a minimum?

28. *Navigation* A motorist is in a desert 8 mi from point A, the nearest point on a long straight road. The motorist wants to get to an aid station at point B on the road 10 mi from A. The car can travel 15 mph in the desert and 55 mph on the road. At what point should the motorist meet the road to get to B in the shortest possible time?

29. *Physics* A ball is dropped from the top of a building. It takes 7 s for the ball to reach the ground. **(a)** How high is the building? **(b)** What is its speed when the ball strikes the ground?

CHAPTER 5 TEST

1. For the function $f(x) = x^4 - 8x^2$, determine **(a)** all critical values, **(b)** all extrema, **(c)** intervals on which f is concave upward, **(d)** intervals on which f is concave downward, and **(e)** sketch the curve.

2. For the function $g(x) = \dfrac{x^2}{9 - x^2}$, determine **(a)** all critical values, **(b)** all extrema, **(c)** intervals on which f is concave upward, **(d)** intervals on which f is concave downward, **(e)** asymptotes, and **(f)** sketch the curve.

3. Use Newton's method to approximate the positive root of $x^2 - 7 = 0$.
4. Determine the antiderivative of $f(x) = 5x^4 - 7x$.
5. The position function of points moving rectilinearly is given by the function $s(t) = t^3 - 12t^2 + 5$. **(a)** Find the function for the velocity and acceleration at time t. **(b)** When are the position, velocity, and acceleration maximum and minimum for the interval $[-2, 4]$?
6. A rectangular page is to contain 24 in.^{2} of print. The margins at the top and bottom of the page are $1\frac{1}{2}$ in., the margin toward the binding is $1\frac{1}{4}$ in., and the margin on the outside edge is $\frac{3}{4}$ in. What should the dimensions of the page be so that the least amount of paper is used?
7. A farm silo consists of a right circular cylinder of radius r and height h topped by a hemisphere. The volume of the silo is $V = \pi r^2 h + \frac{2}{3}\pi r^3$. If $h = 20$ m and $V = 800\,\text{m}^3$, use Newton's method to find r to two decimal places.

CHAPTER

6

Integration

Integration is used to determine area and volume. In Section 6.5, you will learn a numerical method for determining the cross-sectional area of the amount of dirt that must be excavated at a building site.

Courtesy of Mike Nelson

In Chapters 4 and 5, we looked at one of the two main parts of calculus—differential calculus. We have not finished looking at differential calculus, but it is time to take a look at the other main part of calculus—integral calculus.

In Chapter 3, we spent time looking at some of the foundations of calculus. As we discovered, calculus was developed in order to solve two types of problems. One of the problems, called the tangent question, formed the basis for our study of differential calculus. The other type of problem, which we called the area question, gives the foundation for our study of integral calculus.

6.1 THE AREA QUESTION AND THE INTEGRAL

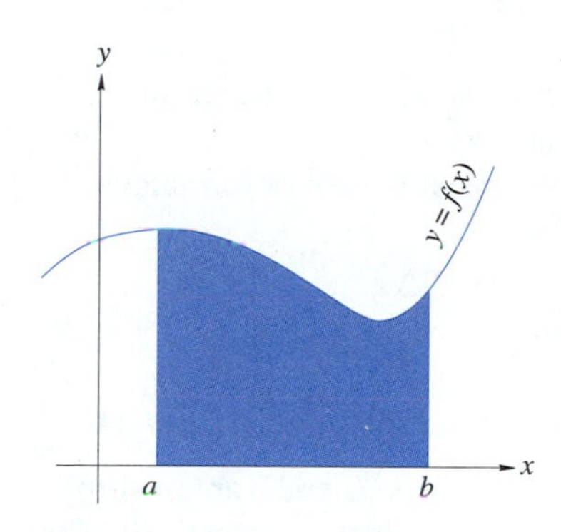

FIGURE 6.1

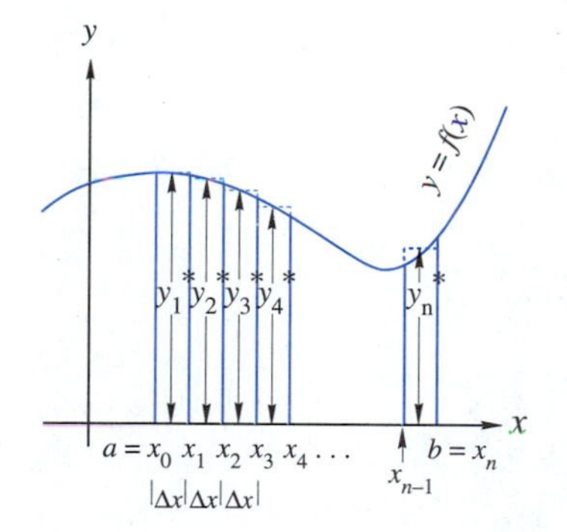

FIGURE 6.2

In Section 3.2, we looked at the area question. We began by looking at the shaded area in Figure 6.1. The area was enclosed by the curve $f(x)$, the x-axis, and the vertical lines $x = a$ and $x = b$.

We began our effort to find the area by dividing the interval $[a, b]$ into n equal segments. We let $a = x_0$ and $b = x_n$ with the first segment $[x_0, x_1]$, the second $[x_1, x_2]$, the third $[x_2, x_3]$, and so on. The midpoint of the ith interval was designated $x_i{}^*$ and we located the height of the curve at that point, labeling it $y_i{}^* = f(x_i{}^*)$. Thus, in the interval $[x_0, x_1]$, the midpoint was $x_1{}^*$ and the height was $y_1{}^* = f(x_1{}^*)$, as shown in Figure 6.2.

These $y_i{}^*$ were used as altitudes of rectangles. The width of each rectangle was $\Delta x = \dfrac{x_n - x_0}{n}$. So, the area of the first rectangle is $(y_1{}^*)(\Delta x)$, the area of the second rectangle is $(y_2{}^*)(\Delta x)$, the area of the third rectangle is $(y_3{}^*)(\Delta x)$, and so on. Thus, the area A under the curve was approximated by

$$\begin{aligned} A &\approx (y_1{}^*)(\Delta x) + (y_2{}^*)(\Delta x) + (y_3{}^*)(\Delta x) + \cdots + (y_n{}^*)(\Delta x) \\ &= (y_1{}^* + y_2{}^* + y_3{}^* + \cdots + y_n{}^*)\Delta x \\ &= [f(x_1{}^*) + f(x_2{}^*) + f(x_3{}^*) + \cdots + f(x_n{}^*)]\Delta x \end{aligned}$$

Using summation or sigma notation, we have

$$A = \sum_{i=1}^{n} f(x_i{}^*)\Delta x$$

In this problem, we made things easy by letting each interval be the same size and by selecting the midpoints of the intervals. This did not have to be done. We can let each interval be a different size. The width of the ith interval would be $\Delta x_i = x_i - x_{i-1}$. We can also pick any point in the interval, say $x_i{}'$, and let $f(x_i{}')$ be the altitude of the rectangle at that point. The area would then be approximated by the formula

$$A \approx \sum_{i=1}^{n} f(x_i{}')\Delta x_i.$$

In the first method, the more intervals that are selected the closer we will get to the actual area. Taking this further, we can see that

$$A = \lim_{n\to\infty} \sum_{i=1}^{n} f(x_i{}^*)\Delta x$$

In the second method, as the widths of the intervals get smaller, we get a closer approximation to the actual area. If we let $\|\Delta x\|$ represent the width of the largest Δx_i, then we can see that

$$A = \lim_{\|\Delta x\|\to 0} \sum_{i=1}^{n} f(x_i{}')\Delta x_i$$

In fact, a combination of these two is the easiest to use. We will let the number of equal-sized intervals increase and we will select any convenient point in each interval to get

$$A = \lim_{n\to\infty} \sum_{i=1}^{n} f(x_i') \Delta x$$

Notice that, since each interval is the same size, increasing the number of intervals has the effect of decreasing the width of the interval.

We will adopt a more convenient notation to represent the above formula:

$$A = \int_a^b f(x)\,dx = \lim_{n\to\infty} \sum_{i=1}^{n} f(x_i') \Delta x$$

The middle expression, $\int_a^b f(x)\,dx$, is called the **definite integral of f from a to b**. The function $f(x)$ is known as the **integrand** and the numbers a and b are the **limits of integration**. More specifically, a is the **lower limit** and b is the **upper limit**. The symbol $\int$ is an elongated "S" for "sum" and is called the **integral sign**.

The following summation formulas will be helpful in working the examples and exercises in this section.

Summation Formulas

$$\sum_{i=1}^{n} c = nc$$

$$\sum_{i=1}^{n} cf(x) = c\sum_{i=1}^{n} f(x)$$

$$\sum_{i=1}^{n} i = 1+2+3+4+\cdots+n = \frac{n(n+1)}{2}$$

$$\sum_{i=1}^{n} i^2 = 1^2+2^2+3^2+4^2+\cdots+n^2$$

$$= 1+4+9+16+\cdots+n^2 = \frac{n(n+1)(2n+1)}{6}$$

$$\sum_{i=1}^{n} i^3 = 1^3+2^3+3^3+4^3+\cdots+n^3$$

$$= 1+8+27+64+\cdots+n^3 = \left[\frac{n(n+1)}{2}\right]^2$$

EXAMPLE 6.1

Evaluate $\int_0^4 x^3\,dx$.

Solution The interval is $[0,4]$ and we will divide it into n subintervals of equal length. Thus, each subinterval has a length $\Delta x = \frac{4}{n}$. While we can select any point in the subinterval, we will select the right-hand endpoint. The values of x_i, as shown in Figure 6.3, are $x_1 = 0 + \frac{4}{n} = \frac{4}{n}$, $\left(x_1 = 0 + \frac{4}{n}\right.$, because 0 is the left endpoint of the interval $\left.[0,4]\right)$, $x_2 = 0 + 2\left(\frac{4}{n}\right) = \frac{8}{n}$, $x_3 = 0 + 3\left(\frac{4}{n}\right) = \frac{12}{n}, \ldots,$ $x_i = 0 + i\left(\frac{4}{n}\right) = \frac{4i}{n}, \ldots, x_n = 0 + n\left(\frac{4}{n}\right) = 4.$

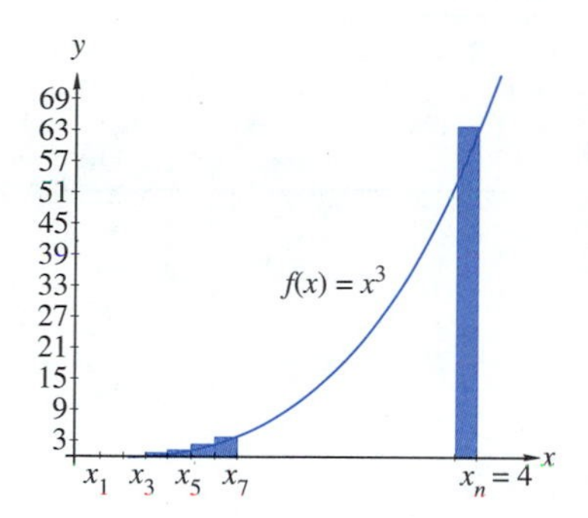

FIGURE 6.3

The altitude of each rectangle is computed using $f(x) = x^3$, and so

$$f(x_1) = \left(\frac{4}{n}\right)^3 = \frac{64}{n^3}$$

$$f(x_2) = \left[2\left(\frac{4}{n}\right)\right]^3 = 2^3 \cdot \frac{64}{n^3}$$

$$f(x_3) = \left[3\left(\frac{4}{n}\right)\right]^3 = 3^3 \cdot \frac{64}{n^3}$$

$$\vdots$$

$$f(x_i) = \left[i\left(\frac{4}{n}\right)\right]^3 = i^3 \cdot \frac{64}{n^3}$$

$$\vdots$$

and $\quad f(x_n) = 4^3 = 64$

The sum of the areas of the rectangles can be written as

$$\sum_{i=1}^{n} f(x_i)\,\Delta x = \sum_{i=1}^{n} i^3 \frac{64}{n^3} \cdot \frac{4}{n} = \frac{256}{n^4} \sum_{i=1}^{n} i^3$$

since $\frac{256}{n^4}$ is a constant. Now $\sum_{i=1}^{n} i^3 = \left[\frac{n(n+1)}{2}\right]^2$, and so

$$\begin{aligned}
\frac{256}{n^4} \sum_{i=1}^{n} i^3 &= \frac{256}{n^4}\left[\frac{n(n+1)}{2}\right]^2 \\
&= \frac{256}{n^4}\frac{n^2(n+1)^2}{4} \\
&= \frac{64(n+1)^2}{n^2} \\
&= \frac{64(n^2+2n+1)}{n^2}
\end{aligned}$$

EXAMPLE 6.1 (Cont.)

We are now ready to find the integral.

$$\begin{aligned}\int_0^4 x^3\,dx &= \lim_{n\to\infty} \frac{64\left(n^2+2n+1\right)}{n^2}\\ &= \lim_{n\to\infty} \frac{64\left(1+\frac{2}{n}+\frac{1}{n^2}\right)}{1}\\ &= 64\end{aligned}$$

Thus, the area of the region is 64 square units.

EXAMPLE 6.2

Evaluate $\int_{-1}^{2} \left(x^2+1\right) dx$.

Solution The interval $[-1,2]$ has length 3. Using equal subintervals, each subinterval will have length $\Delta x = \frac{3}{n}$.

Again, using the right-hand endpoint of each subinterval as shown in Figure 6.4, $x_i = -1 + \frac{3i}{n} = \frac{3i-n}{n}$. Since $f(x) = x^2+1$, we see that $f(x_i) = \left(\frac{3i-n}{n}\right)^2 + 1 = \frac{9i^2-6in+n^2}{n^2} + 1 = \frac{9i^2}{n^2} - \frac{6i}{n} + 2$.

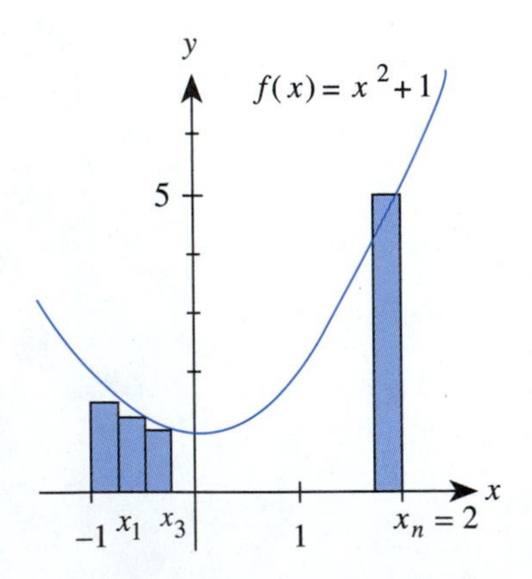

FIGURE 6.4

Thus,

$$\begin{aligned}\sum_{i=1}^{n} f(x_i)\,\Delta x &= \sum_{i=1}^{n}\left(\frac{9i^2}{n^2} - \frac{6i}{n} + 2\right)\left(\frac{3}{n}\right)\\ &= \sum_{i=1}^{n}\left(\frac{27i^2}{n^3} - \frac{18i}{n^2} + \frac{6}{n}\right)\\ &= \frac{27}{n^3}\sum_{i=1}^{n} i^2 - \frac{18}{n^2}\sum_{i=1}^{n} i + \frac{6}{n}\sum_{i=1}^{n} 1\\ &= \frac{27}{n^3}\left[\frac{n(n+1)(2n+1)}{6}\right] - \frac{18}{n^2}\left[\frac{n(n+1)}{2}\right] + \frac{6}{n}\cdot n\\ &= \frac{27}{n^3}\left(\frac{2n^3+3n^2+n}{6}\right) - \frac{18}{n^2}\left(\frac{n^2+n}{2}\right) + 6\\ &= \left(9 + \frac{27}{2n} + \frac{9}{2n^2}\right) - \left(9 + \frac{9}{n}\right) + 6\end{aligned}$$

EXAMPLE 6.2 (Cont.)

We find the integral by taking the limit of this sum as follows:

$$\int_{-1}^{2}(x^2+1)\,dx = \lim_{n\to\infty}\left[\left(9+\frac{27}{2n}+\frac{9}{2n^2}\right)-\left(9+\frac{9}{n}\right)+6\right]$$
$$=(9+0+0)-(9+0)+6=6$$

Thus, the area of the region is 6 square units.

Exercise Set 6.1

In Exercises 1–18, use summation notation and limits to evaluate the definite integrals.

1. $\int_1^3 x\,dx$
2. $\int_0^2 2x\,dx$
3. $\int_0^4 (3x+1)\,dx$
4. $\int_1^3 (2x+2)\,dx$
5. $\int_1^3 (2x-2)\,dx$
6. $\int_0^4 (1-3x)\,dx$
7. $\int_1^4 (3-2x)\,dx$
8. $\int_0^4 (x^2)\,dx$
9. $\int_0^2 (x^2-1)\,dx$
10. $\int_0^4 (x^2+2)\,dx$
11. $\int_0^2 (2x^2+1)\,dx$
12. $\int_0^4 (2-x^2)\,dx$
13. $\int_1^3 x^3\,dx$
14. $\int_1^3 (1-x^3)\,dx$
15. $\int_0^4 (x^3+2)\,dx$
16. $\int_0^3 (2x^3-1)\,dx$
17. $\int_0^2 (x^2+x)\,dx$
18. $\int_0^1 x^4\,dx$

Solve Exercise 19.

19. Use the computer program from Exercise 11 of Exercise Set 3.2 to solve Exercises 1–18. Input different values of n to see how close the computer approximations get to the actual value.

In Your Words

20. In the expression $\int_a^b f(x)\,dx$, what is the integrand, the lower limit of integration, and the upper limit of integration? What does each of them tell you?
21. What are the differences in the meaning between $\sum_1^{10} f(x)\,\Delta x$ and $\int_1^{10} f(x)\,dx$?

6.2 THE FUNDAMENTAL THEOREM OF CALCULUS

In this section, we will examine the relationship between the derivative and the definite integral. We will then look at three properties of integrals. One of the most important tools we will attain in this section will be a general method for evaluating a definite integral.

Figure 6.5 shows the graph of $y=f(x)$. We will assume that $f(x)$ is continuous on the closed interval $[a,b]$ and that its graph does not fall below the x-axis. From Section 6.1, we know that the area of the shaded region is given by $\int_a^b f(x)\,dx$. We

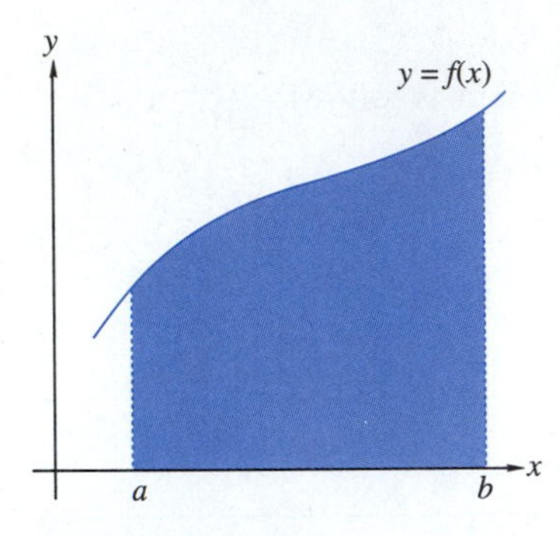

FIGURE 6.5

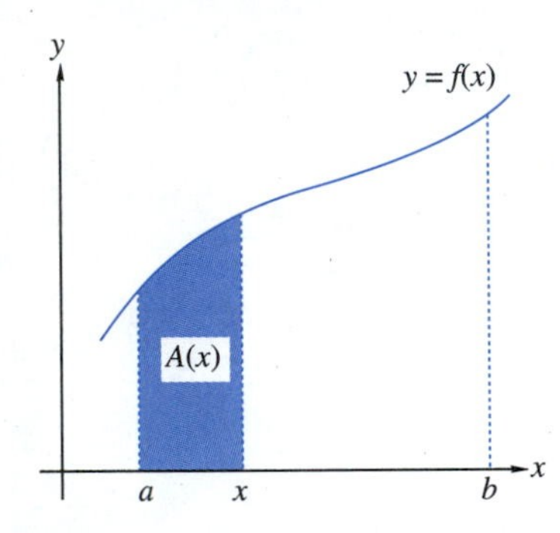

FIGURE 6.6

found that area by dividing the region into smaller and smaller rectangles, finding the sum of the areas of these rectangles, and taking the limit of this sum as the number of rectangles became infinitely large. We will now look at another way to determine the area.

Suppose there is a function $A(x)$, which we will refer to as an "area" function. The function $A(x)$ gives the area of the region below the graph of f and above the x-axis from a to x, where a is a fixed point and $a \leq x \leq b$. The region is shaded in Figure 6.6.

We already know two properties of $A(x)$:

1. $A(a) = 0$, since there is no area from a to a.
2. $A(b) = \int_a^b f(x)\,dx$.

Now, suppose that x is increased by h units as shown in Figure 6.7 and $A(x + h)$ is the area of the new shaded region. The difference in the area of the new region and the area of the original region is shown in Figure 6.8, and is written as $A(x+h) - A(x)$.

The area of the region in Figure 6.9 is the same as the area of a rectangle whose base is h and whose height is some value $\overline{y}$. Since the value of $\overline{y}$ depends on h, $\overline{y}$ is a function of h. The area of the rectangle is $A(x+h) - A(x)$ and is also represented as $\overline{y}h$. Thus, we have the equation

$$A(x+h) - A(x) = \overline{y}h$$

Dividing by h:
$$\frac{A(x+h) - A(x)}{h} = \overline{y}$$

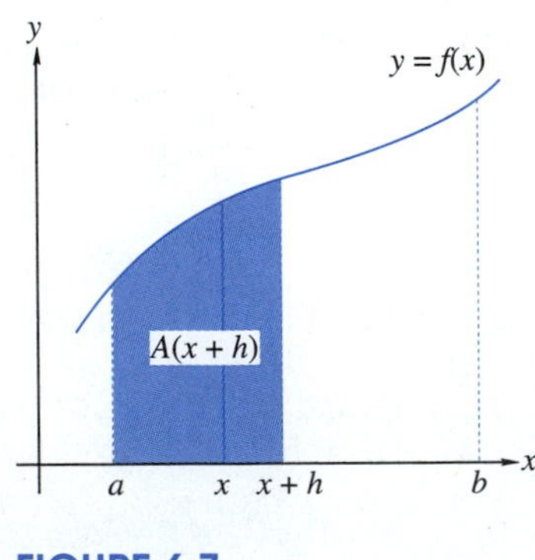

FIGURE 6.7

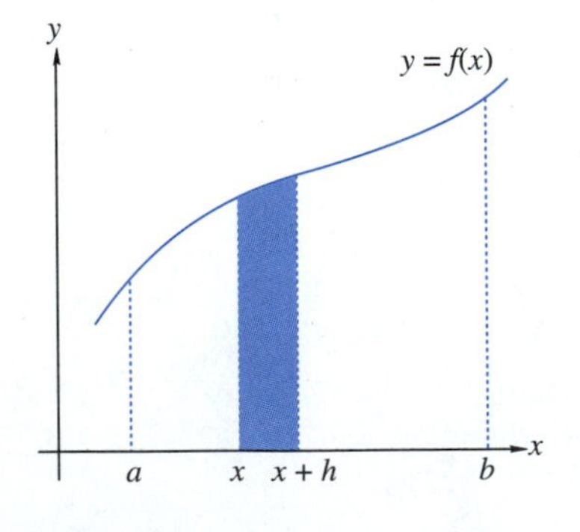

FIGURE 6.8

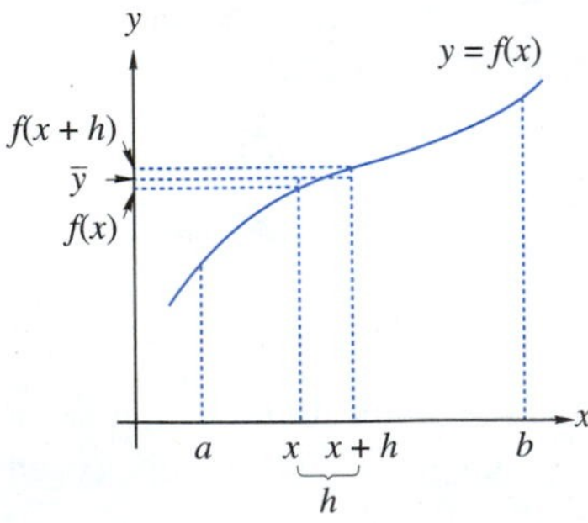

FIGURE 6.9

Now, what happens as the value of h approaches zero? One thing that will happen is that the value of $\overline{y}$ will get closer to the value of $f(x)$, and so

$$\lim_{h\to 0}\frac{A(x+h)-A(x)}{h}=\lim_{h\to 0}\overline{y}=f(x)$$

But, the left-hand side of this equation is the derivative of $A(x)$, thus we have

$$A'(x)=f(x)$$

What this means is that the area function $A(x)$ has the property that its derivative A' is the function f. That means that A is an antiderivative of f.

In Section 5.7, we found that the antiderivative of $f(x)$ was $F(x)+C$, where C is a constant. Thus, we know that $A(x)=F(x)+C$. Now, $A(a)=0$, so when $x=a$, we have $0=F(a)+C$ or $C=-F(a)$. Thus, we have the equation

$$A(x)=F(x)-F(a)$$

If $x=b$, then this equation becomes

$$A(b)=F(b)-F(a)$$

But, $A(b)=\int_a^b f(x)\,dx$. Combining these last two equations we get

$$\int_a^b f(x)\,dx=F(b)-F(a)$$

As we can see, there is a relationship between a definite integral and differentiation. All we need to do to find $\int_a^b f(x)\,dx$ is find an antiderivative of f, which we call F, and subtract the value of F at the lower limit from its value at the upper limit. This result is known as the **Fundamental Theorem of Calculus**. It was discovered independently by Newton and Leibnitz.

Fundamental Theorem of Calculus

If f is continuous on the interval $[a,b]$, then

$$\int_a^b f(x)\,dx=F(x)\Big|_a^b=F(b)-F(a)$$

where F is a function such that $F'=f$.

In stating the Fundamental Theorem of Calculus, we introduced some new notation, $F(x)\,\Big|_a^b$. This notation is often used to help keep track of the limits of integration after we have integrated but before the antiderivative is evaluated.

EXAMPLE 6.3

Evaluate $\int_0^3 x^3\,dx$.

Solution This is the same function we evaluated in Example 25.1. We know that the antiderivative of x^3 is $\frac{x^4}{4}+C$, so

$$\int_0^3 x^3\,dx = \left.\frac{x^4}{4}\right|_0^4$$
$$= \frac{4^4}{4} - \frac{0^4}{4} = 64 - 0 = 64$$

Note When using antiderivatives with the Fundamental Theorem of Calculus, the constant is not used. If the antiderivative of $f(x)$ is $F(x)+C$, then

$$\int_a^b f(x)\,dx = F(x)+C\,\Big|_a^b$$
$$= [F(b)+C] - [F(a)+C]$$
$$= F(b) - F(a)$$

As you can see, the constant C does not have to be included since $C - C = 0$. From now on we will call the process of finding an antiderivative by its more common title of **integration**.

EXAMPLE 6.4

Find the area under the curve of $f(x) = \frac{1}{x^3}$ from $x = 1$ to $x = 3$.

Solution The desired area is shown in Figure 6.10. Note that f is continuous from $x = 1$ to $x = 3$. Since the curve lies above the x-axis in the interval $[-1, 3]$, the area can be found by evaluating $\int_1^3 f(x)\,dx$. Using the Fundamental Theorem of Calculus, we have

$$\int_1^3 \frac{1}{x^3}\,dx = \int_1^3 x^{-3}\,dx$$
$$= \left.\frac{1}{-2}x^{-2}\right|_1^3$$
$$= \left(\frac{1}{-2\,(3^2)}\right) - \left(\frac{1}{-2\cdot 1^2}\right)$$
$$= \frac{-1}{18} + \frac{1}{2}$$
$$= \frac{8}{18} = \frac{4}{9} \text{ square units}$$

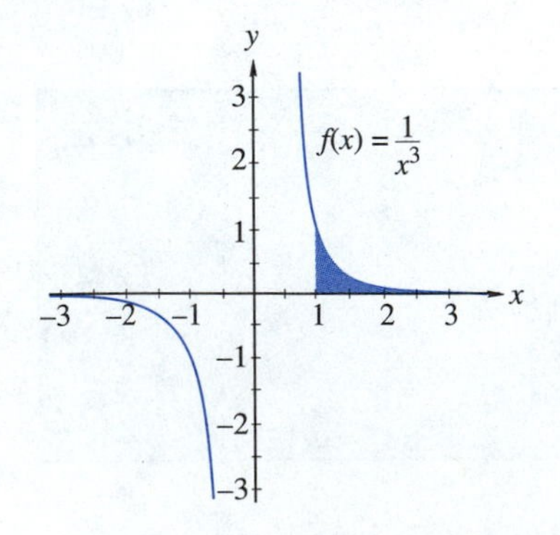

FIGURE 6.10

At the present time we can only find the area when a curve is above the x-axis.

EXAMPLE 6.5

Evaluate $\int_1^3 (3x-2)\,dx$.

Solution

$$\int_1^3 (3x-2)\,dx = \left(\frac{3}{2}x^2 - 2x\right)\Bigg|_1^3$$
$$= \left[\frac{3}{2}(3^2) - 2(3)\right] - \left[\frac{3}{2}(1^2) - 2(1)\right]$$
$$= \left(\frac{27}{2} - 6\right) - \left(\frac{3}{2} - 2\right) = 8$$

The following are three properties, or rules, of definite integrals.

Properties of Definite Integrals

Rule 1: $\int_a^b kf(x)\,dx = k\int_a^b f(x)\,dx$, where k is a constant

Rule 2: $\int_a^b [f(x) \pm g(x)]\,dx = \int_a^b f(x)\,dx \pm \int_a^b g(x)\,dx$

Rule 3: If f is continuous on an interval containing a, b, and c, then

$$\int_a^b f(x)\,dx = \int_a^c f(x)\,dx + \int_c^b f(x)\,dx$$

EXAMPLE 6.6

Evaluate $\int_0^2 (3x^2 - \sqrt{x})\,dx$.

Solution

$$\int_0^2 (3x^2 - \sqrt{x})\,dx = \int_0^2 (3x^2 - x^{1/2})\,dx$$
$$= \left(x^3 - \frac{2}{3}x^{3/2}\right)\Bigg|_0^2$$
$$= \left[2^3 - \frac{2}{3}(2)^{3/2}\right] - \left[0^3 - \frac{2}{3}(0)^{3/2}\right]$$
$$= 8 - \frac{4\sqrt{2}}{3}$$

EXAMPLE 6.7

Evaluate $\int_1^2 4x^2\,dx$.

Solution According to rule 1,

$$\int_1^2 4x^2\,dx = 4\int_1^2 x^2\,dx = 4\left(\frac{1}{3}x^3\Big|_1^2\right)$$
$$= 4\left(\frac{8}{3} - \frac{1}{3}\right)$$
$$= \frac{28}{3}$$

EXAMPLE 6.8

Evaluate $\int_1^2 (x^2 - x)\,dx$.

Solution Using rule 2, we see that

$$\int_1^2 (x^2 - x)\,dx = \int_1^2 (x^2)\,dx - \int_1^2 x\,dx$$
$$= \frac{1}{3}x^3\Big|_1^2 - \frac{1}{2}x^2\Big|_1^2$$
$$= \frac{7}{3} - \frac{3}{2} = \frac{5}{6}$$

EXAMPLE 6.9

Evaluate $\int_{-1}^1 3x^2\,dx$.

Solution Using rule 3, we know that

$$\int_{-1}^1 3x^2\,dx = \int_{-1}^0 3x^2\,dx + \int_0^1 3x^2\,dx$$
$$= x^3\Big|_{-1}^0 + x^3\Big|_0^1$$
$$= (0+1) + (1-0) = 2$$

At present these three rules are not particularly needed. But, we are setting the foundation for later, when they will play an important part in our continued development and use of calculus.

Application

EXAMPLE 6.10

From the top of a building 192 ft high, a ball is thrown upward with an initial velocity of 64 ft/s. **(a)** Find the position function giving the height s as a function of the time t. **(b)** When does the ball strike the ground?

Solutions

(a) We will let $t = 0$ be the initial time. We are given the initial conditions that

$$s(0) = 192 \quad \text{(initial height is 192 ft)}$$
$$s'(0) = 64 \quad \text{(initial velocity is 64 ft/s)}$$

Since acceleration due to gravity is -32 ft/s^2, we have

$$s''(t) = -32$$

and integrating, we obtain

$$\begin{aligned} s'(t) &= \int s''(t)\,dt \\ &= \int -32\,dt = -32t + C_1 \end{aligned}$$

Since $s'(0) = 64 = -32(0) + C_1$, we have $C_1 = 64$. Thus, $s'(t) = -32t + 64$. Now, by integrating $s'(t)$, we have

$$s(t) = \int s'(t)\,dt = \int (-32t + 64)\,dt = -16t^2 + 64t + C_2$$

Using the initial height $s(0) = 192 = -16(0^2) + 64(0) + C_2$, we obtain $C_2 = 192$. We have now found the position function

$$s(t) = -16t^2 + 64t + 192$$

(b) The ball will hit the ground when $s(t) = 0$.

$$\begin{aligned} s(t) = -16t^2 + 64t + 192 &= 0 \\ -16(t^2 - 4t - 12) &= 0 \\ -16(t-6)(t+2) &= 0 \\ t &= -2, 6 \end{aligned}$$

Since t must be positive, we conclude that the ball hits the ground 6 s after it was thrown.

Exercise Set 6.2

In Exercises 1–37, use the Fundamental Theorem of Calculus to evaluate the given definite integral.

1. $\int_0^5 6\,dx$

2. $\int_1^2 (x-6)\,dx$

3. $\int_1^5 2x\,dx$

4. $\int_{-1}^2 (6-2x)\,dx$

5. $\int_{-1}^3 \frac{5}{3}x^3\,dx$

6. $\int_0^5 4x^2\,dx$

7. $\int_1^2 \frac{8}{5}x^3\,dx$

8. $\int_{-2}^2 6x\,dx$

9. $\int_0^2 (4x+5)\,dx$
10. $\int_1^3 (-4)\,dt$
11. $\int_2^0 (4x+5)\,dx$
12. $\int_{-1}^2 (2t-t^2)\,dt$
13. $\int_1^3 (-4x)\,dx$
14. $\int_2^5 (5w^2+2w)\,dw$
15. $\int_0^3 (t^2+2t)\,dt$
16. $\int_{-2}^0 (5-t^3)\,dt$
17. $\int_{-1}^3 (3y^2-2y)\,dy$
18. $\int_0^4 (s^2-4s+16)\,ds$
19. $\int_1^4 (4-2w^2)\,dw$
20. $\int_{-2}^0 -\,dt$
21. $\int_2^3 (x^2+2x+1)\,dx$
22. $\int_{-3}^0 (3y-y^3)\,dy$
23. $\int_3^2 (x^2+2x+1)\,dx$
24. $\int_1^4 \left(\sqrt{x}-\frac{1}{\sqrt{x}}\right)dx$
25. $\int_6^8 dx$
26. $\int_{1/2}^1 \frac{4\,dx}{x^2}$
27. $\int_{-4}^{-2} (3y^2-y-1)\,dy$
28. $\int_1^4 \left(\frac{3}{x^3}-\frac{2}{x^2}\right)dx$
29. $\int_2^4 (x^2-2x+1)\,dx$
30. $\int_0^4 \left(\sqrt{x}\right)^3 dx$
31. $\int_1^2 -3x^{-4}\,dx$
32. $\int_0^1 3x^{5/6}\,dx$
33. $\int_{-2}^{-1} \frac{t^{-2}}{3}\,dt$
34. $\int_0^2 \left(x^4-x^3\right)dx$
35. $\int_0^2 \sqrt{x}\,dx$
36. $\int_1^3 x\sqrt[3]{x}\,dx$
37. $\int_{1/2}^3 \left(\frac{1}{\sqrt{x}}-2\right)dx$

Solve Exercises 38–48.

38. The Fundamental Theorem of Calculus *seems* to say that

$$\int_{-1}^1 x^{-2}\,dx = -x^{-1}\Big|_{-1}^1 = -2$$

But x^{-2} is always positive and we would expect the area under this curve to be positive. Why does the Fundamental Theorem of Calculus appear to give a wrong answer?

39. *Physics* A ball is projected upward along a frictionless inclined plane. The velocity of the ball is given by $v(t) = 4.00t - 1.00t^2$ in meters per second (m/s). What is the change in displacement in the first 4.00 s?

40. *Mechanical technology* The angular velocity in radians per second (rad/s) for a gear is given by $\omega(t) = 1.00t^3 - 27.0t$. What is the change in angular displacement between $t = 1.00$ s and $t = 3.00$ s?

41. *Dynamics* From the top of a building 180 m high, a ball is thrown down with an initial velocity of 10 m/s. **(a)** How fast is the ball moving after 3 s? **(b)** How far does the ball move in the first 3 s? **(c)** How long does it take the ball to hit the ground?

42. *Aeronautics* A hot-air balloon is rising slowly, so the balloonist drops a sandbag from the bottom of the balloon. If the sandbag was dropped at an elevation of 840 ft, what was the speed of the balloon at the time the sandbag was dropped if it took 8 s for the sandbag to hit the ground?

43. *Nuclear technology* A charged particle enters a linear accelerator. Its initial velocity is 500 m/s and its velocity increases with a constant acceleration to a velocity of 10 500 m/s in 0.01 s. **(a)** What is the acceleration? **(b)** How far does the particle travel in that 0.01 s?

44. *Electricity* The current i in amperes (A) to a capacitor is given by $i = 1.00t + 1.00\sqrt{t}$, where t is time in seconds (s). If $i = \frac{dq}{dt}$, where q is the charge in coulombs, find the change in the charge on the capacitor from $t = 1.00$ s to $t = 9.00$ s.

45. *Environmental science* The manager of a wildlife preserve has started a management program to control the preserve's moose population. It is estimated that the population will continue to grow according to the function $N'(t) = 25 - 6t^{1/2}$ per year, where t is the number of years since the plan was implemented. Find the change in the population during the first 9 years of the program.

46. *Industrial engineering* A stamping machine produces computer cases at a rate that varies over the working day according to the equation $\frac{dN}{dt} = 175 + 2t - 0.2t^2$, where t is the time in h.
(a) Find an expression for $N(t)$ if $N(0) = 0$.
(b) Find the number of computer cases the machine produces in the first 4 hours of the day.
(c) Find the number of computer cases the machine produces in the last 4 hours of the 8-hour day.

47. *Forestry* A certain type of hardwood tree will grow at the rate of $G'(t) = 0.5 + 4t^{-3}$ ft/yr, where t is time in years after the tree is 20 years old.
(a) How much will the tree grow in the second year after it is 20 years old?
(b) How much will the tree grow in the fifth year after it is 20 years old?
(c) How much will the tree grow in years 1 through 5 after it is 20 years old?

48. *Ecology* Pollution from a factory is entering a lake. The rate of concentration of the pollutant at time t is given by $P'(t) = 91t^{5/2}$, where t is the number of years since the factory began introducing pollutants into the lake. Ecologists estimate that the lake can accept a total level of pollution of 5,720 units before all the fish are killed. How long can the factory pollute the lake before all the fish are killed?

In Your Words

49. Explain the fundamental theorem of calculus.

50. Three properties of definite integrals were given in the text. Explain what each one means and how you can use it.

6.3 THE INDEFINITE INTEGRAL

The Fundamental Theorem of Calculus showed us a relationship between definite integrals and antiderivatives. In fact, we found that we were able to evaluate $\int_a^b f(x)\,dx$ by finding an antiderivative of f. The general form of the antiderivative is called the **indefinite integral**. The integral symbol, $\int$, without any limits is used to indicate an indefinite integral. Thus, the notation $\int f(x)\,dx$ is used to indicate all the antiderivatives of f.

EXAMPLE 6.11

$$\int x^2\,dx = \frac{1}{3}x^3 + C$$

$$\int 4x^3\,dx = x^4 + C$$

$$\int (2x^{-3} + 5)\,dx = -x^{-2} + 5x + C$$

In each of these examples, C represents an arbitrary constant called the **constant of integration**.

It is important to point out the difference between a definite integral and an indefinite integral. The definite integral is a number; the indefinite integral is a family of functions.

The value of the definite integral depends on the limits of integration. For example,

$$\int_1^2 3x^2\,dx = x^3 \Big|_1^2 = 7 \qquad \int_1^3 3x^2\,dx = x^3 \Big|_1^3 = 26$$

Notice that in each case the definite integral is a number.

However, the indefinite integral has no limits to affect the answer. Thus,

$$\int 3x^2\,dx = x^3 + C$$

and C, or the constant of integration, is used to show that this integral represents a family of functions and not just one particular value.

Properties of Indefinite Integrals

At the end of Section 6.2, we gave three rules for definite integrals. We are now going to give several rules for indefinite integrals. Two of these rules are the indefinite integral versions of previous rules. All of them are based on the six rules for derivatives we gave in Section 4.2.

Because an indefinite integral of f is defined as an antiderivative of f, we know that we can differentiate the indefinite integral and get the original function. Thus,

Rule 1

$$\frac{d}{dx}\int f(x)\,dx = f(x)$$

EXAMPLE 6.12

$$\frac{d}{dx}\int \sqrt[3]{x^2-5}\,dx = \sqrt[3]{x^2-5}$$

The next rule is similar to one we had for definite integrals.

Rule 2

$$\int kf(x)\,dx = k\int f(x)\,dx, \text{ where } k \text{ is a real number}$$

This rule states that we can, in effect, factor a constant out of a function before we integrate it.

The next rule is also similar to one we had for definite integrals.

Rule 3

$$\int [f(x) \pm g(x)]\, dx = \int f(x)\, dx \pm \int g(x)\, dx$$

Thus, the indefinite integral of a sum (or difference) is equal to the sum (or difference) of the indefinite integrals.

The next rule is the antiderivative of rule 3 from Section 4.2.

Rule 4

$$\int x^n\, dx = \frac{1}{n+1} x^{n+1} + C, n \neq -1$$

EXAMPLE 6.13

$$\begin{aligned} \int (3x^2 + 4x - 5)\, dx &= \int 3x^2\, dx + \int 4x\, dx - \int 5\, dx && \text{By rule 3} \\ &= 3\int x^2 dx + 4\int x\, dx - 5\int dx && \text{By rule 2} \\ &= 3\left(\frac{x^3}{3}\right) + 4\left(\frac{x^2}{2}\right) - 5x + C && \text{By rule 4} \\ &= x^3 + 2x^2 - 5x + C \end{aligned}$$

We used three of the four rules in Example 6.13.

Method of Substitution

All of the rules just given are useful, but they do not show us how to integrate $\int (3x-7)^{10}\, dx$, $\int (x^4+6)^3 4x^3\, dx$, or $\int \sqrt[3]{4x^3 - 8}x^2\, dx$. The first integral could be found by expanding $(3x-7)^{10}$ and then integrating, but it would be very long and inefficient. Also, this method will not work on the last two integrals.

Some integrals that cannot be evaluated directly by using the previously stated rules can sometimes be evaluated using the method of substitution. This method involves the introduction of a function that changes the integrand, such that the given rules will work when integrating. Examples 6.14 and 6.15 show how to use the method of substitution on the last two integrals given.

EXAMPLE 6.14

Use the method of substitution to determine

$$\int (x^4+6)^3 4x^3\,dx$$

Solution In this integral, we will first use the substitution $u = x^4+6$. Differentiating with respect to x, we get $\frac{du}{dx} = 4x^3$ or $du = 4x^3\,dx$. We then substitute u for x^4+6 and du for $4x^3\,dx$ in the original integral, with the following result:

$$\int (x^4+6)^3 4x^3\,dx = \int u^3\,du \qquad \text{Let } x^4+6 = u \text{ and } 4x^3\,dx = du.$$

$$= \frac{1}{4}u^4 + C \qquad \text{Integrating with respect to } u.$$

$$= \frac{1}{4}\left(x^4+6\right)^4 + C \qquad \text{Substituting for } u\text{, we obtain this.}$$

EXAMPLE 6.15

Use the method of substitution to find $\int \sqrt[3]{4x^3-8}\,x^2\,dx$.

Solution We will let $u = 4x^3-8$, and so $\frac{du}{dx} = 12x^2$. Thus, $du = 12x^2\,dx$ and $x^2\,dx = \frac{1}{12}\,du$. Substituting these values in the given integral, we get the following solution:

$$\int \sqrt[3]{4x^3-8}\,x^2\,dx = \int \left(4x^3-8\right)^{1/3} x^2\,dx \qquad \text{Let } 4x^3-8 = u \text{ and } x^2\,dx = \frac{1}{12}du$$

$$= \int \left(u^{1/3}\right)\frac{1}{12}du$$

$$= \frac{1}{12}\int u^{1/3}\,du$$

$$= \frac{1}{12}\frac{u^{4/3}}{4/3} + C$$

$$= \frac{1}{16}u^{4/3} + C$$

$$= \frac{1}{16}\left(4x^3-8\right)^{4/3} + C$$

We substituted not only for the integrand, but for the differential. The most difficult part is selecting the correct substitution. Once a substitution is chosen, its differential is determined. In all cases, we are attempting to rewrite the integral so as to get a more general form of rule 4:

$$\int u^n\,du = \frac{u^{n+1}}{n+1}, \qquad n \neq -1$$

The substitution we select must allow us to rewrite $f(x)\,dx$ as $u^n\,du$ with the possible exception of a multiplicative constant.

EXAMPLE 6.16

Evaluate $\displaystyle\int \frac{dx}{\sqrt{x^2+5}}$.

Solution If we let $u = x^2+5$, then $du = 2x\,dx$. But $2x$ is not a constant, so we cannot write this integral as a constant multiple of $u^n\,du$. Thus, we cannot use the substitution method and we cannot integrate this problem using the present methods.

EXAMPLE 6.17

Determine $\displaystyle\int (x^3+5)^2\,dx$.

Solution If we let $u = x^3+5$, then $du = 3x^2\,dx$. Again the substitution method cannot be used because the given integrand is not a constant multiple of $u^n\,du$. However, we can expand the given problem and solve it.

$$\begin{aligned}\int (x^3+5)^2\,dx &= \int (x^6+10x^3+25)\,dx \\ &= \frac{1}{7}x^7 + \frac{5}{2}x^4 + 25x + C\end{aligned}$$

Application

EXAMPLE 6.18

The work, W, done by a variable force, F, over a distance, x, is given by $W = \int F\,dx$. What is the formula for the work done if $F = 15(4x+1)^2$ and $W = 4$ ft·lb when $x = 0$ ft?

EXAMPLE 6.18 (Cont.)

Solution Since $W = \int F\,dx$, we have $W = \int 15(4x+1)^2\,dx$. Let $u = 4x+1$ and then $\frac{du}{dx} = 4$ and $4\,dx = du$. Making these substitutions, we obtain the following solution. Notice in the second line of the solution that we rewrite 15 as $\frac{15}{4} \cdot 4$ in order to get $4\,dx$.

$$\begin{aligned} W &= \int 15(4x+1)^2\,dx \\ &= \int \frac{15}{4}(4x+1)^2 4\,dx \qquad \text{Let } 4x+1 = u \text{ and } 4\,dx = du \\ &= \frac{15}{4}\int u^2\,du \\ &= \frac{15}{4}\frac{u^3}{3} + C \\ &= 1.25\,(4x+1)^3 + C \end{aligned}$$

Thus, the formula for the work is $W = 1.25\,(4x+1)^3 + C$. But, we are given some additional information. We are told that $W = 4$ ft·lb when $x = 0$ ft. Substituting $x = 0$ in our formula for W produces $W = 1.25\,(4(0)+1)^3 + C = 1.25(0+1)^3 + C = 1.25 + C$. Since $W = 4$ when $x = 0$, then $1.25 + C = 4$ and $C = 4 - 1.25 = 2.75$. With this value for C we get the desired formula for the work as:

$$W = 1.25\,(4x+1)^3 + 2.75$$

Exercise Set 6.3

In Exercises 1–56, find the given indefinite integral.

1. $\int 9\,dx$
2. $\int 3x\,dx$
3. $\int 6x^2\,dx$
4. $\int \sqrt{x}\,dx$
5. $\int x^2\sqrt{x}\,dx$
6. $\int x^{-2/3}\,dx$
7. $\int (t^3+1)\,dt$
8. $\int (2-4s^3)\,ds$
9. $\int (y^2+4y-3)\,dy$
10. $\int \left(4\sqrt{y} + \frac{1}{4\sqrt{y}}\right)\,dy$
11. $\int \left(\sqrt[3]{x} - \frac{3}{\sqrt[3]{x^2}}\right)\,dx$
12. $\int \left(\frac{1}{x^3} - \frac{3}{\sqrt{x}} + \frac{\sqrt{x}}{3}\right)\,dx$
13. $\int (x^2+3)2x\,dx$
14. $\int (3x^2-5)6x\,dx$
15. $\int (4-2x^2)4x\,dx$
16. $\int (2x^3+1)^4 6x^2\,dx$
17. $\int (3-x^2)^3 2x\,dx$
18. $\int (x^2-3)^4 x\,dx$
19. $\int \sqrt{x^2+4}x\,dx$
20. $\int (x^3+1)^5 x^2\,dx$
21. $\int \frac{(\sqrt{x}-1)^3}{\sqrt{x}}\,dx$
22. $\int (x^4-5)^7 x^3\,dx$

23. $\int (3x^3+1)^4 x^2\,dx$

24. $\int (3-6x^2)^{3/2} 2x\,dx$

25. $\int \frac{x\,dx}{\sqrt{x^2+3}}$

26. $\int \frac{x\,dx}{\sqrt{(4x^2-1)^3}}$

27. $\int \frac{x\,dx}{(x^2+3)^3}$

28. $\int (x^2+3)^2\,dx$

29. $\int (3x^2-1)^2\,dx$

30. $\int \frac{x^2-1}{x+1}\,dx$

31. $\int \frac{x^2-x-6}{x+2}\,dx$

32. $\int \frac{3x\,dx}{(x^2+4)^5}$

33. $\int \frac{5x\,dx}{\sqrt[3]{x^2-1}}$

34. $\int \frac{dx}{\sqrt{4-x}}$

35. $\int (1+3x^2)^2 x\,dx$

36. $\int 2x\sqrt{5x^2+3}\,dx$

37. $\int 3x^2(4x^3-5)^{2/3}\,dx$

38. $\int \frac{2x^2\,dx}{(5-3x^3)^{2/3}}$

39. $\int (1+3x^2)^2\,dx$

40. $\int (2x^2+4x)^3(4x+4)\,dx$

41. $\int \sqrt{4x^2+2x}(4x+1)\,dx$

42. $\int (x^2+x-5)^4(2x+1)\,dx$

43. $\int \frac{x-1}{(x-1)^3}\,dx$

44. $\int \frac{2x+1}{(x^2+x-5)^3}\,dx$

45. $\int (2x^4-3)^2 8x\,dx$

46. $\int \frac{x+3}{\sqrt[3]{x^2+6x}}\,dx$

47. $\int (x^3-3x)^{2/3}(x^2-1)\,dx$

48. $\int (x^3-2)\sqrt{x^4-8x}\,dx$

49. $\int (3x+1)^3\,dx$

50. $\int 7x(x^3-2)^2\,dx$

51. $\int (2x^3-1)\sqrt[4]{x^4-2x}\,dx$

52. $\int \frac{x^2}{(4-x^3)^{3/2}}\,dx$

53. $\int (t+7)^{1/2}\,dt$

54. $\int \frac{x^4-1}{x-1}\,dx$

55. $\int \frac{y^3-8}{y-2}\,dy$

56. $\int (6x-7)^8\,dx$

In Exercises 57–66, evaluate the given definite integral.

57. $\int_0^2 (x^2-4)2x\,dx$

58. $\int_0^4 x(4x^2+2)^3\,dx$

59. $\int_1^2 6x(3x^2-7)\,dx$

60. $\int_1^3 2x^2(3x^3-1)^2\,dx$

61. $\int_0^2 2x(20-3x^2)\,dx$

62. $\int_0^5 x\sqrt{x^2+144}\,dx$

63. $\int_0^3 (x^3+2)x^2\,dx$

64. $\int_1^2 \frac{dx}{(x+4)^2}$

65. $\int_1^4 (3x^2-2)^4 x\,dx$

66. $\int_{-1}^0 (6x^2-1)^3 x\,dx$

Solve Exercises 67–72.

67. *Environmental science* The air quality control office estimates that for a population of x thousand people, the level of pollutant in the air is increasing at a rate of $L'(x)=0.3+0.002x$ parts per million (ppm) per thousand people. Find a function, $L(x)$, that estimates the level of the pollutants if the level is 7.2 ppm when the population is 20,000 people.

68. *Ecology* Biologists are treating a stream contaminated with bacteria. The level of contamination is changing at the rate of $N'(t)=-\frac{750}{t^2}-120$ bacteria/cm^3/day, where t is the number of days since the treatment began. Find a function, $N(t)$, to estimate the level of contamination if the level after 1 day was about 7,320 bacteria/cm^3.

69. *Business* The marginal profit of a fast-food restaurant is given by $P'(x) = 4x + 35$, where x is the sales volume in thousands of hamburgers. The "profit" is $\$-150$ when no hamburgers are sold. Determine the profit function, $P(x)$.

70. *Medical technology* A medical laboratory estimates that t hours after some certain bacteria are introduced into a culture, the population will be increasing at the rate of $P'(t) = \dfrac{1800}{(18-0.5t)^{1/2}}$ bacteria per hour. Find the increase in the population during the first 8 hours.

71. *Thermodynamics* The temperature, T, in °C, in an industrial furnace varies from its center to a point outside. The rate of temperature change is given by the equation

$$\frac{dT}{dx} = -\frac{5750}{(x+1)^3}$$

Find an expression for T at a distance of x m from the center if $T(0) = 2900°\text{C}$.

72. *Electronics* The current, I, and the charge, q, are related by $I = \dfrac{dq}{dt}$. At $t = 0$ s, the current to a discharged capacitor is $I = 0.005$ A. If the current remains constant, what is the charge, q, in coulombs on the capacitor?

In Your Words

73. What are the differences in the meaning of $\int f(x)\,dx$ and $\int_a^b f(x)\,dx$?

74. Describe how to use the method of substitution.

≡ 6.4 THE AREA BETWEEN TWO CURVES

We introduced the definite integral as an answer to one of the two basic questions that lead to the development of calculus—the area question. In this section, we will continue our study of area.

Until now, we have only been concerned with finding the area of a curve that is above the x-axis. Suppose we want to find the area for a curve that goes below the x-axis, such as the one shown in Figure 6.11. If we were to compute its area, since $f(x) < 0$, we would get $\int_a^b f(x)\,dx < 0$ or a negative area. This does not make sense. The idea of area means that the area of something is a nonnegative number. Thus, we need a better idea for finding the area. This will become clearer as we look at the next example.

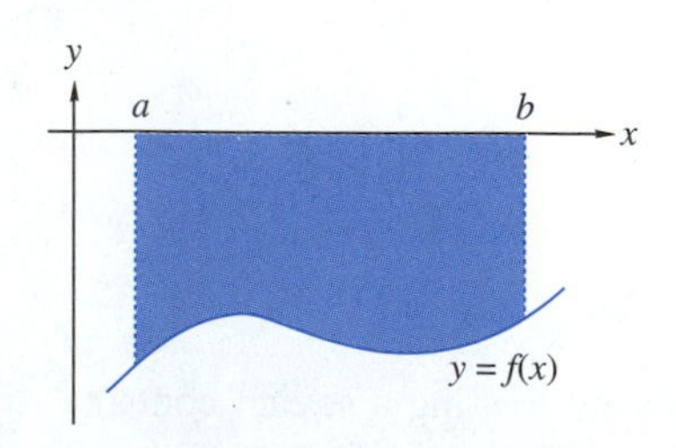

FIGURE 6.11

EXAMPLE 6.19

Find the area between the curve $f(x)=x^3$ and the x-axis from $x=-2$ to $x=2$.

Solution The graph of the desired region is shaded in Figure 6.12. Using the definite integral we see that

$$\begin{aligned}\int_{-2}^{2} x^3\,dx = \frac{1}{4}x^4\Big|_{-2}^{2} &= \frac{1}{4}(2^4)-\frac{1}{4}(-2)^4\\ &= \frac{1}{4}(16)-\frac{1}{4}(16)\\ &= 4-4\\ &= 0\end{aligned}$$

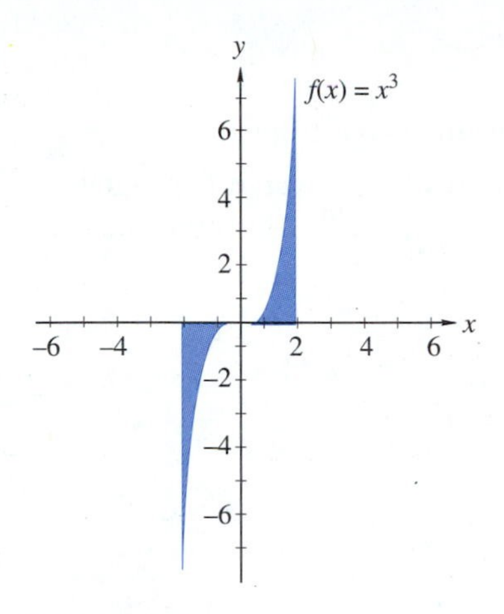

FIGURE 6.12

This does not make sense. As we can see, the shaded area of Figure 6.12 is not zero. Because this shaded area is symmetric about the origin, we can see that the area above the graph is the same as the area below the graph. Thus, we can see that the total area is $2\int_{0}^{2} x^3\,dx = 2(4) = 8$.

The integral $\int_a^b f(x)\,dx$ defines the area between the curve $y=f(x)$ and the x-axis only when $f(x)$ is a nonnegative continuous function for all x in the interval $[a,b]$. If $f(x)$ is continuous but $f(x)<0$ for all x in the interval $[a,b]$, then $\int_a^b f(x)\,dx<0$ and we define the area to be $\left|\int_a^b f(x)\,dx\right|$. If $f(x)$ is continuous and assumes both positive and negative values on the interval $[a,b]$, then the area is found by separating the integral into two or more parts.

One of the rules for definite integrals states that if a, b, and c are three points in an interval and if f is continuous on that interval, then

$$\int_a^b f(x)\,dx = \int_a^c f(x)\,dx + \int_c^b f(x)\,dx$$

Until now we have used this rule only once, in Example 6.9. Here, in Example 6.19, we have the opportunity to use it again. Since the curve crosses the x-axis at $x=0$, we can compute the area as

$$\begin{aligned}\int_{-2}^{2} x^3\,dx &= \left|\int_{-2}^{0} x^3\,dx\right| + \int_{0}^{2} x^3\,dx\\ &= \left|\frac{1}{4}x^4\Big|_{-2}^{0}\right| + \left.\frac{1}{4}x^4\right]_{0}^{2}\\ &= \left|\frac{1}{4}0^4-\frac{1}{4}(-2)^4\right| + \left[\frac{1}{4}2^4-\frac{1}{4}0^4\right]\\ &= |0-4|+(4-0)\\ &= 4+4=8\end{aligned}$$

Caution

As you can see, it is very important that you sketch the graph when you are doing an area problem.

EXAMPLE 6.20

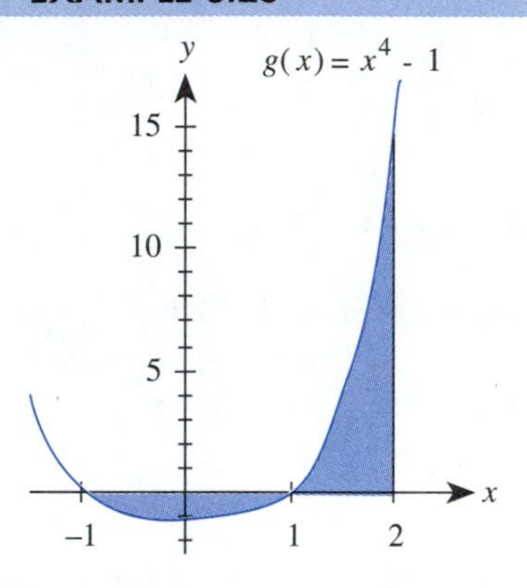

FIGURE 6.13

Find the area between the curve $g(x) = x^4 - 1$ and the x-axis from $x = -1$ to $x = 2$.

Solution The desired area is shown in Figure 6.13. The curve lies below the x-axis for the interval $[-1, 1]$ and above the x-axis for $[1, 2]$. The total area is the sum of these two separate areas. Thus,

$$\text{Area} = \left|\int_{-1}^{1} (x^4 - 1)\, dx\right| + \int_{1}^{2} (x^4 - 1)\, dx$$

Now,

$$\int_{-1}^{1} (x^4 - 1)\, dx = \left[\frac{1}{5}x^5 - x\right]_{-1}^{1}$$
$$= \left(\frac{1}{5} - 1\right) - \left(-\frac{1}{5} + 1\right) = -\frac{8}{5}$$

and
$$\int_{1}^{2} (x^4 - 1)\, dx = \left[\frac{1}{5}x^5 - x\right]_{1}^{2}$$
$$= \left(\frac{32}{5} - 2\right) - \left(\frac{1}{5} - 1\right) = \frac{26}{5}$$

Thus, the total area of the region is $\left|-\frac{8}{5}\right| + \frac{26}{5} = \frac{34}{5} = 6.8$ square units.

EXAMPLE 6.21

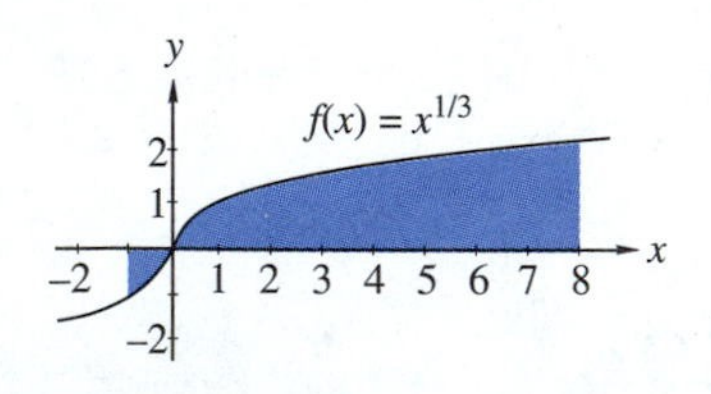

FIGURE 6.14

Find the area between the curve $f(x) = x^{1/3}$ and the x-axis from $x = -1$ to $x = 8$.

Solution The desired area is the shaded region of Figure 6.14. The graph of $f(x) = x^{1/3}$ crosses the x-axis at $x = 0$, so the area A will be

$$A = \left|\int_{-1}^{0} x^{1/3}\, dx\right| + \int_{0}^{8} x^{1/3}\, dx$$
$$= \left|\frac{3}{4}x^{4/3}\Big|_{-1}^{0}\right| + \frac{3}{4}x^{4/3}\Big|_{0}^{8}$$
$$= \left|\frac{3}{4}(0)^{4/3} - \frac{3}{4}(-1)^{4/3}\right| + \left[\frac{3}{4}(8)^{4/3} - \frac{3}{4}(0)^{4/3}\right]$$
$$= \left|0 - \frac{3}{4}\right| + (12 - 0)$$
$$= \frac{3}{4} + 12 = 12\frac{3}{4}$$

Note

The direct evaluation of $\int_{-1}^{8} x^{1/3}\,dx = \frac{3}{4}x^{4/3}\Big|_{-1}^{8} = 12 - \frac{3}{4} = 11\frac{1}{4}$. This is not the area of the shaded region in Figure 6.14. It is, however, the value of the definite integral $\int_{-1}^{8} x^{1/3}\,dx$. You will need to be careful when evaluating a definite integral to proceed using the method that is correct for the desired meaning.

Let's expand our thinking to include a different type of area. Until now, each area has been for some region between a curve and the x-axis. Now we want to find the area between two curves.

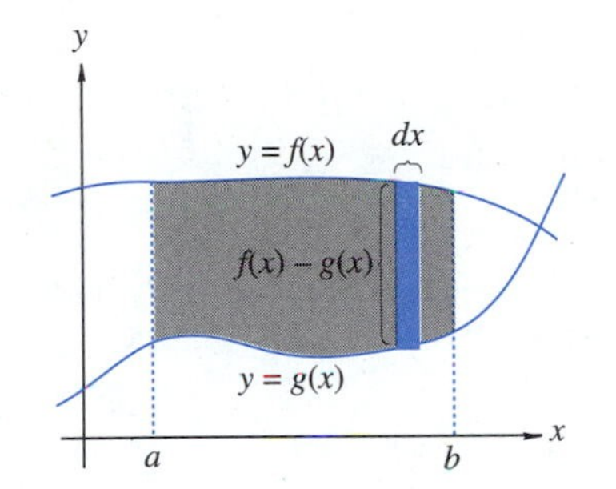

FIGURE 6.15

Consider the functions f and g as shown in Figure 6.15. We want to find the area between the curves $y = f(x)$ and $y = g(x)$ and the lines $x = a$ and $x = b$. We will return to finding the area by dividing the region into rectangular regions of thickness dx. The height of each of these rectangles is $f(x) - g(x)$, regardless of the position of f and g. As a result, the area is given by

$$\int_{a}^{b} [f(x) - g(x)]\,dx$$

The result works whether the graphs lie above the x-axis, partially above and partially below the x-axis, or both lie below the x-axis.

EXAMPLE 6.22

Find the area enclosed by the graphs of $f(x) = 12x - 3x^2$ and $g(x) = 3x - 12$.

Solution The graph of the two functions is shown in Figure 6.16. Before we can compute the area, we need to locate the points of intersection of the two graphs. Thus, we need to determine when $f(x) = g(x)$ or when

$$12x - 3x^2 = 3x - 12$$
$$3x^2 - 9x - 12 = 0$$
$$3(x-4)(x+1) = 0$$

and so the graphs intersect when $x = -1$ and $x = 4$.

From Figure 6.16, we can see that $f(x) \geq g(x)$ on $[-1, 4]$, so the area we want is

$$\begin{aligned}\int_{-1}^{4} [f(x) - g(x)]\,dx &= \int_{-1}^{4} [(12x - 3x^2) - (3x - 12)]\,dx \\ &= \int_{-1}^{4} (9x - 3x^2 + 12)\,dx \\ &= \left(\frac{9}{2}x^2 - x^3 + 12x\right)\Big|_{-1}^{4} \\ &= (72 - 64 + 48) - \left(\frac{9}{2} + 1 - 12\right) \\ &= 56 - (-6.5) = 62.5\end{aligned}$$

The desired area is 62.5 square units.

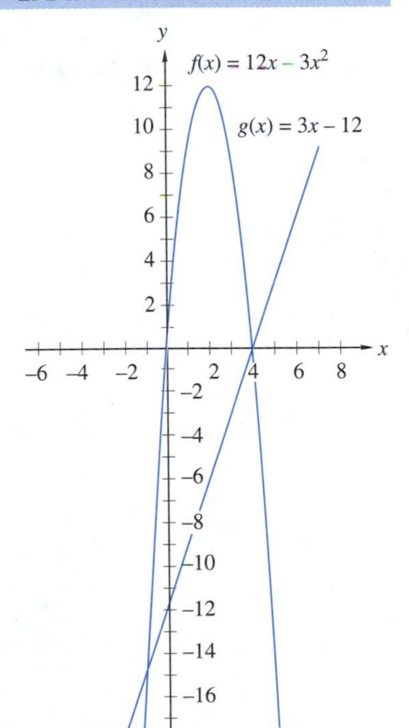

FIGURE 6.16

In the next example, we will combine the techniques that we used in the previous two examples.

EXAMPLE 6.23

Find the area enclosed by the graphs of $f(x) = x^3$ and $g(x) = x$.

Solution The graph of the two functions is shown in Figure 6.17. The points of intersection occur when

$$f(x) = g(x)$$

or when

$$x^3 = x$$

$$x^3 - x = 0$$

$$x(x^2 - 1) = x(x-1)(x+1) = 0$$

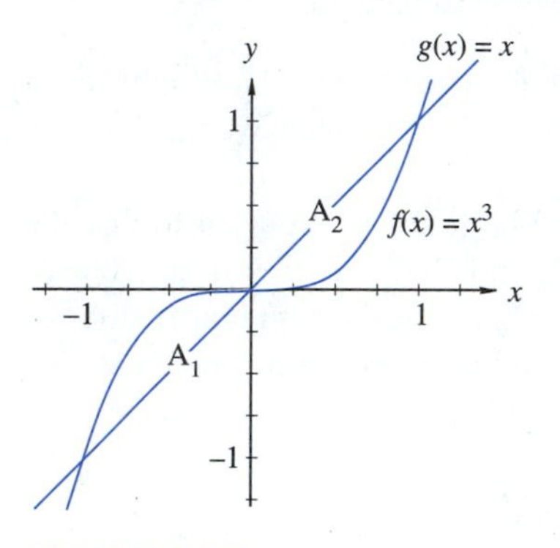

FIGURE 6.17

The two graphs intersect when $x = 0$, $x = 1$, and $x = -1$.

Study the graph of the two curves. On the interval $[-1, 0]$, $f(x) \geq g(x)$, and so the area A_1 enclosed by the curves in this interval is

$$\begin{aligned} A_1 &= \int_{-1}^{0} [f(x) - g(x)]\,dx = \int_{-1}^{0} (x^3 - x)\,dx \\ &= \left(\frac{1}{4}x^4 - \frac{1}{2}x^2\right)\Bigg|_{-1}^{0} \\ &= 0 - \left(\frac{1}{4} - \frac{1}{2}\right) = \frac{1}{4} \end{aligned}$$

On the interval $[0, 1]$, $g(x) \geq f(x)$, and so the area A_2 enclosed by the curves in this interval is

$$\begin{aligned} A_2 &= \int_{0}^{1} [g(x) - f(x)]\,dx = \int_{0}^{1} (x - x^3)\,dx \\ &= \left(\frac{1}{2}x^2 - \frac{1}{4}x^4\right)\Bigg|_{0}^{1} \\ &= \left(\frac{1}{2} - \frac{1}{4}\right) - 0 = \frac{1}{4} \end{aligned}$$

The total area is $A_1 + A_2 = \frac{1}{4} + \frac{1}{4} = \frac{1}{2}$.

Note Again, we need to point out the importance of graphing. If you had applied the rule for the area between two curves without graphing, you could easily have reached the incorrect result shown below:

$$\begin{aligned}\int_{-1}^{1}[f(x)-g(x)]\,dx &= \int_{-1}^{1}(x^3-x)\,dx\\ &= \left(\frac{1}{4}x^4-\frac{1}{2}x^2\right)\Bigg|_{-1}^{1}\\ &= \left(\frac{1}{4}-\frac{1}{2}\right)-\left(\frac{1}{4}-\frac{1}{2}\right)\\ &= 0\end{aligned}$$

EXAMPLE 6.24

Find the area enclosed by the graphs of $f(x)=\sqrt{9-9x}$, $g(x)=\sqrt{9-x}$, and the x-axis.

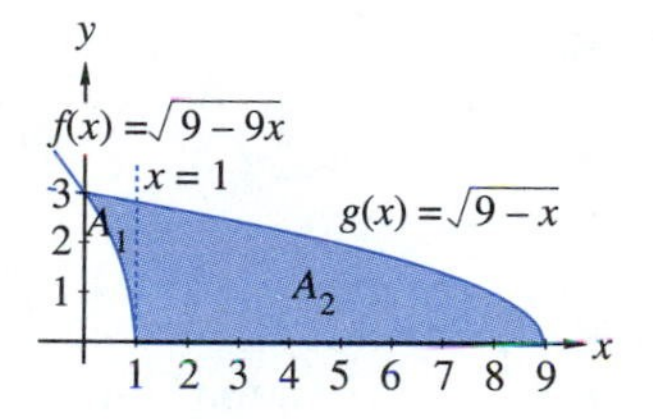

FIGURE 6.18

Solution The graphs of these functions are shown in Figure 6.18. The shading indicates the area we want to measure.

Again, we need to exercise caution. In the interval $[0, 1]$, the shaded region A_1 is between $f(x)$ and $g(x)$ with $g(x) \geq f(x)$. Thus

$$\begin{aligned}A_1 &= \int_0^1 [g(x)-f(x)]\,dx\\ &= \int_0^1\left(\sqrt{9-x}-\sqrt{9-9x}\right)dx\\ &= \int_0^1\sqrt{9-x}\,dx-\int_0^1\sqrt{9-9x}\,dx\\ &= -\frac{2}{3}(9-x)^{3/2}\Bigg|_0^1-\frac{-2}{27}(9-9x)^{3/2}\Bigg|_0^1\\ &= \left[-\frac{2}{3}(8)^{3/2}+\frac{2}{3}(9)^{3/2}\right]-\left[\frac{-2}{27}(0)+\frac{2}{27}(9)^{3/2}\right]\\ &= \left(-\frac{16}{3}\sqrt{8}+18\right)-(0+2)\\ &= 16-\frac{16}{3}\sqrt{8}\end{aligned}$$

In the interval $[1, 9]$, the shaded region A_2 is between $g(x)$ and the x-axis, with $g(x) \geq 0$, and so

$$A_2=\int_1^9\sqrt{9-x}\,dx=-\frac{2}{3}(9-x)^{3/2}\Bigg|_1^9=\frac{16}{3}\sqrt{8}$$

The total area is $A_1+A_2=16-\frac{16}{3}\sqrt{8}+\frac{16}{3}\sqrt{8}=16$.

Partitions Along the y-axis

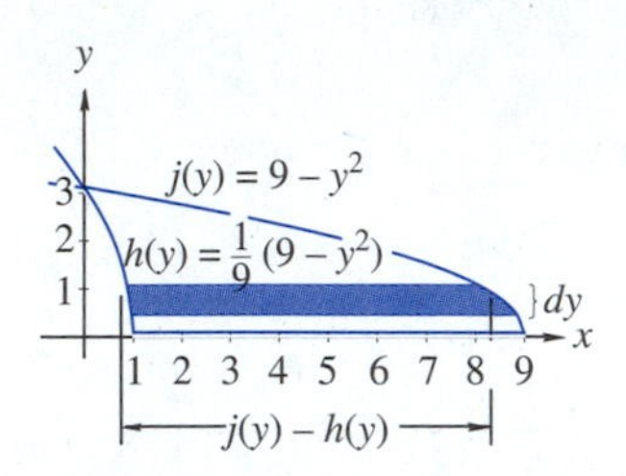

FIGURE 6.19

Until now, we have always integrated with respect to x. We have done this because the functions have always been given in terms of x and the x-axis. Let's look again at Example 6.24 and rewrite the functions in terms of y.

We had $f(x) = \sqrt{9-9x}$ or $y = \sqrt{9-9x}$. If we solve this for x, we get $y^2 = 9 - 9x$ or $x = \frac{1}{9}(9-y^2)$. Let's call this function $h(y) = \frac{1}{9}(9-y^2)$. In the same way, if we solve $g(x)$ for y, we get $y = \sqrt{9-x}$ or $x = 9 - y^2$. We will call this function $j(y)$, and so $j(y) = 9 - y^2$.

Look at Figure 6.19. This time we have partitioned the y-axis and we get rectangles of width dy and length $j(y) - h(y)$. We want the area A between these curves and the x-axis (or when $y = 0$). Thus,

$$\begin{aligned} A &= \int_0^3 [j(y) - h(y)]dy \\ &= \int_0^3 \left[(9-y^2) - \frac{1}{9}(9-y^2)\right] dy \\ &= \int_0^3 \left(8 - \frac{8}{9}y^2\right) dy \\ &= \left(8y - \frac{8}{27}y^3\right)\Big|_0^3 \\ &= 8(3) - \frac{8}{27}(3)^3 = 24 - 8 = 16 \end{aligned}$$

As you can see in this case, a partition of the y-axis made the integration simpler. There is no clear method to tell which way to partition the graph. Usually, the graph will give you some indication.

Improper Integrals

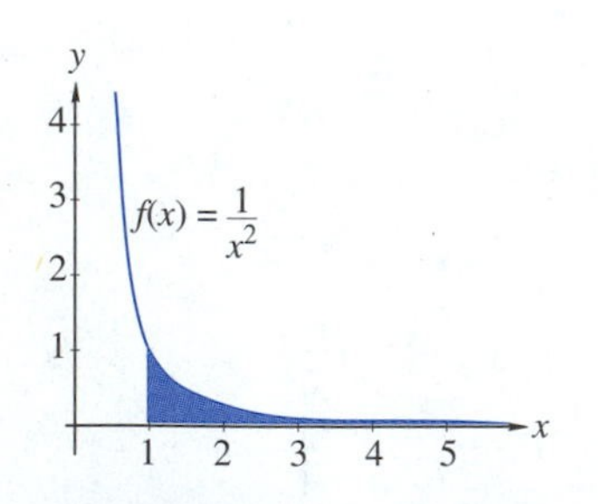

FIGURE 6.20

In the next example, we want to determine the area between a curve and the x-axis. In this case, we want the area between $f(x) = \frac{1}{x^2}$, the x-axis, and to the right of $x = 1$, as shown in Figure 6.20. We know that the x-axis is an asymptote of $\frac{1}{x^2}$, so the curve never intersects the x-axis. Thus, it looks as if the area is determined by

$$\int_1^\infty \frac{1}{x^2}\,dx$$

When we defined the definite integral at the beginning of this chapter, it was over a closed interval $[a, b]$. This integral is on the half-open interval $[a, \infty)$. To get around this problem, we will define

$$\int_1^\infty \frac{1}{x^2}\,dx = \lim_{b\to\infty} \int_1^b \frac{1}{x^2}\,dx, \text{ if the limit exists}$$

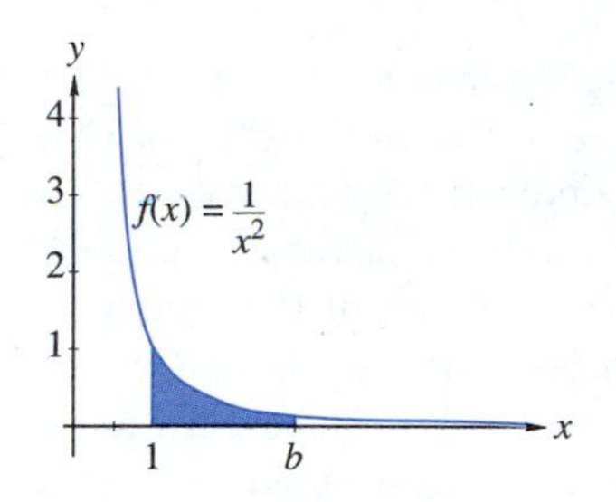

FIGURE 6.21

This is an example of an **improper integral**. Geometrically, we are finding the area under the graph on the interval $[1, b]$, as shown in Figure 6.21, and then we let $b \to \infty$. If the limit exists, the limit is defined to be the area. For this problem,

$$\begin{aligned}\int_1^{\infty} \frac{1}{x^2}\,dx &= \lim_{b\to\infty} \int_1^{b} \frac{1}{x^2}\,dx \\ &= \lim_{b\to\infty} \left(-\frac{1}{x}\right)\Bigg|_1^b \\ &= \lim_{b\to\infty} \left(-\frac{1}{b}+1\right) = 1\end{aligned}$$

Exercise Set 6.4

In Exercises 1–42, find the area of the region bounded by the given curves. Sketch the graphs and shade the desired area first.

1. $f(x) = x$, $x = 4$, x-axis
2. $f(x) = -2x$, $x = -2$, x-axis
3. $f(x) = x^2$, $x = 4$, x-axis
4. $h(x) = x^2 + 4$, $x = -1$, $x = 1$, x-axis
5. $g(x) = 3x + 1$, $x = 10$, x-axis
6. $h(x) = 10 - 2x$, $x = 0$, x-axis
7. $k(x) = 2x + 5$, $x = -2$, $x = 1$, x-axis
8. $j(x) = 2x^2$, x-axis, $x = 1$, $x = 2$
9. $k(x) = 2x^2 - x$, x-axis, $x = -2$
10. $j(x) = x^3 + 1$, $x = 1$, x-axis
11. $m(x) = x^2 - 4x$, x-axis
12. $n(x) = x^2 - 4$, x-axis
13. $f(x) = 9 - x^2$, x-axis
14. $h(x) = 6 - x - x^2$, x-axis
15. $g(x) = x^3 - 4x$, x-axis
16. $f(x) = x^3 - x^2$, x-axis
17. $h(x) = \sqrt{x+9}$, x-axis, $x = 0$
18. $j(x) = \sqrt{x}$, $x = 9$, x-axis
19. $f(x) = x^2$, $g(x) = 2x$
20. $g(x) = 4 - x^2$, $h(x) = x + 2$
21. $f(x) = x^2 + 3$, $g(x) = 9$

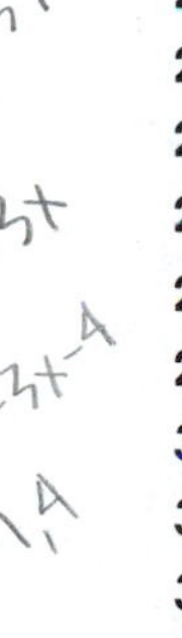

22. $f(x) = x^2 + 4$, $x = -2$, $x = 2$, $y = -1$
23. $f(x) = x - 4$, $g(x) = \sqrt{2x}$
24. $f(x) = \sqrt{x}$, $x = 0$, $y = 4$
25. $g(x) = 4 - x^2$, $h(x) = -3x$
26. $f(x) = x^2 - 4$, $g(x) = 8 - 2x^2$
27. $h(x) = 2x - \frac{1}{2}x^2$, $j(x) = \frac{1}{2}x - 2$
28. $m(x) = x^4$, $n(x) = x^2$
29. $j(x) = 8 - x^2$, $k(x) = x^2$, $x = -2$, $x = 1$
30. $f(x) = 2\sqrt{x}$, $y = 4$, $x = -2$, x-axis
31. $f(x) = x^3$, $g(x) = 4x$
32. $h(x) = x + 3$, $k(x) = 9 - x^2$, $x = 3$
33. $g(x) = 4x - x^2 + 8$, $h(x) = x^2 - 2x$
34. $f(x) = \sqrt{x}$, $g(x) = -x$, $h(x) = x - 2$
35. $f(x) = \sqrt{4x}$, $y = 3$, $x = 0$
36. $h(x) = x^3 - 3x + 2$, $k(x) = x + 2$
37. $y = \sqrt{16 - 2x}$, $y = \sqrt{16 - 4x}$, x-axis
38. $f(x) = x^{1/3}$, $g(x) = -2$, $x = 8$
39. $f(x) = x^{-5/3}$, x-axis, to the right of $x = 8$
40. $f(x) = \frac{1}{x^2}$, $g(x) = 4$, $x = -2$, $x = 2$, x-axis
41. $x^4 y = 1$, x-axis, to the right of $x = 1$
42. $g(x) = x^4 - x^2$, x-axis

Solve Exercises 43–50.

43. *Product design* A tile manufacturing company plans to produce enamel tiles with the design and color scheme shown in Figure 6.22. Will more blue or white enamel be needed?

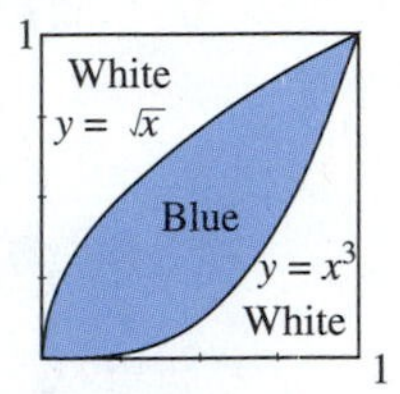

FIGURE 6.22

44. *Environmental science* A new smog-control device will reduce the output of sulfur oxides from automobile exhausts but increase the production of nitrous oxides. It is estimated that using this device will produce a rate of savings to the community of $S(x) = -x^2 + 11x + 15$, in millions of dollars after x years of using the device. The increase in the production of nitrous oxides will result in a rate of additional costs to the community of $C(x) = \frac{5}{4}x^2$, in millions of dollars. If $S(x) - C(x)$ represents the net savings, **(a)** For how many years will it pay to use this new device? **(b)** What will be the net total savings for this period of time?

45. *Industrial design* A double-channel trough used on an assembly line is constructed with end plates defined by the curve $y = x^4 - 2x^2 + 1$ and the line $y = 9$, where x and y are measured in cm. What is the volume of a trough that is 10 m long and has these end plates?

46. *Machine technology* Find the area of the cam outlined by the curves $y = 10x - x^2$ and $4y = x^2$, where x and y are measured in cm.

47. *Automotive engineering* The work cycle, or **Otto cycle**, of an internal combustion engine may be approximated by the pressure-volume (or pV) diagram shown in Figure 6.23. The area of the region between the curves is proportional to the work done by the engine during one cycle. The curves represent adiabatic, or zero, heat input/output processes, where $pV^\gamma = k$, where k is a constant and $\gamma \approx 1.33$ for gasoline vapor. Find the work done by the cycle shown in Figure 6.23. (Note: the answer will be in British thermal units, Btu.)

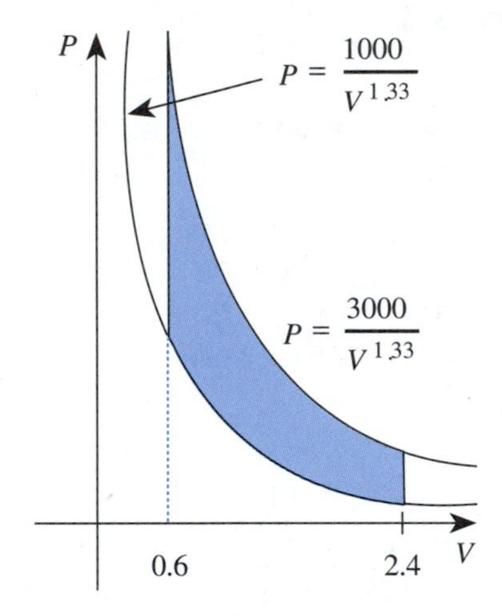

FIGURE 6.23

48. *Automotive engineering* The work cycle, or **Diesel cycle**, of a diesel engine may be approximated by the pV-diagram shown in Figure 6.24. The area of the region between the curves is proportional to the work done by the engine during one cycle. The curves represent adiabatic, or zero, heat input/output processes, where $pV^\gamma = k$, where k is a constant, and $\gamma \approx 1.40$ for diesel vapor. Find the work done by the cycle shown in Figure 6.24. (Note: the answer will be in British thermal units, Btu.)

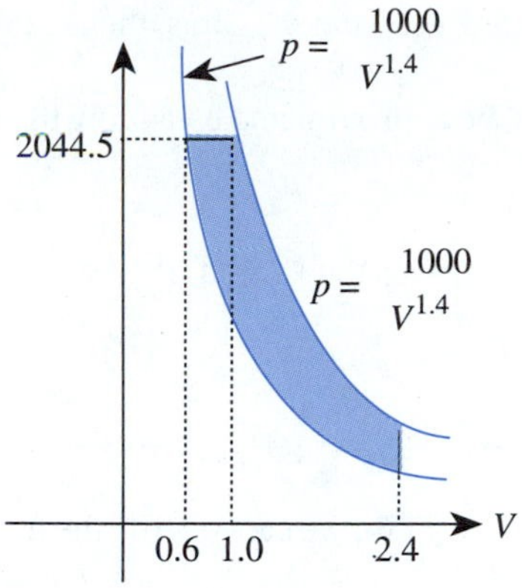

FIGURE 6.24

49. *Mechanical engineering* A heavy flywheel initially rotating on its axis is slowing because of friction in its bearings. If its angular acceleration, α, is given by $\alpha = \dfrac{d\omega}{dt} = \dfrac{100}{\sqrt{(t+1)^3}}$, find its angular velocity, ω, at time $t = 10$ s, if $\omega = 0$ at $t = 0$.

50. *Electronics* The magnetic potential, V, at a point on the axis of a certain coil is given by an equation of the form $V = l\displaystyle\int_a^{\infty} \frac{x}{(R^2+x^2)^{3/2}}\,dx$, where k and R are constants. Integrate in order to find an expression for V.

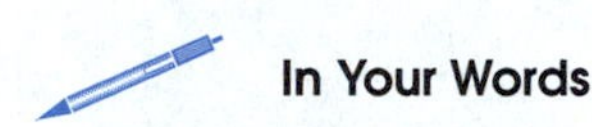

In Your Words

51. Describe the process used to find the area between two curves.

52. What is an indefinite integral?

6.5 NUMERICAL INTEGRATION

The Fundamental Theorem of Calculus provides a useful method for finding the area under a curve. It is not always an easy task to find the antiderivative of a function. It is also possible that we do not know how to find the antiderivative of a certain function. In this section, we will look at two methods for approximating the value of a definite integral to as many decimal places as desired. All techniques of this type come under the heading of **numerical integration**. The two methods that we will examine are called the **trapezoidal rule** and **Simpson's rule**. As with most numerical techniques, your work with these can be eased by the use of a calculator or a computer.

The Trapezoidal Rule

In both the trapezoidal rule and Simpson's rule, we will use some of the work we did in developing the definite integral back in Section 6.1. First we will assume that we want to find the area under some curve, which we will describe by $y = f(x)$. We will also assume that we want to find the area under the curve $f(x)$, above the x-axis, and between the lines $x = a$ and $x = b$, with $a < b$.

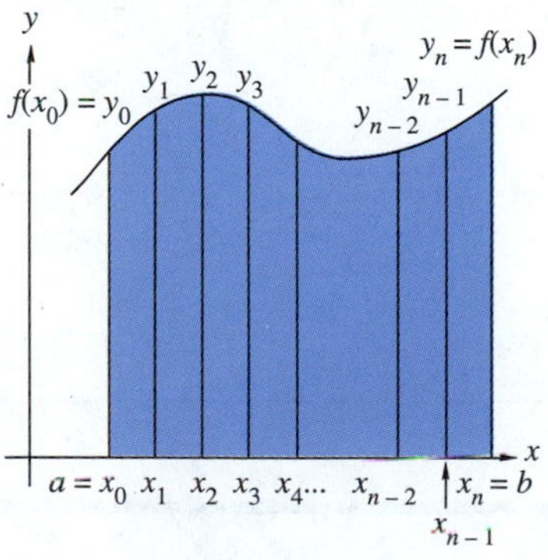

We will divide the interval $[a, b]$ into n equal parts of length $dx = \dfrac{b-a}{n}$, where $a = x_0 < x_1 < x_2 < x_3 < \cdots < x_{n-1} < x_n = b$. The corresponding points on the curve are $y_0 = f(x_0)$, $y_1 = f(x_1)$, $y_2 = f(x_2)$, $y_3 = f(x_3)$, $\ldots$, $y_{n-1} = f(x_{n-1})$, $y_n = f(x_n)$. If we connect these points, we obtain n trapezoids as shown in Figure 6.25.

FIGURE 6.25

The area of a trapezoid is equal to one-half the product of the altitude and the sum of its bases. A typical trapezoid is shown in Figure 6.26a. If you draw the trapezoid as shown in Figure 6.26b, you will find it easier to see that the altitude of

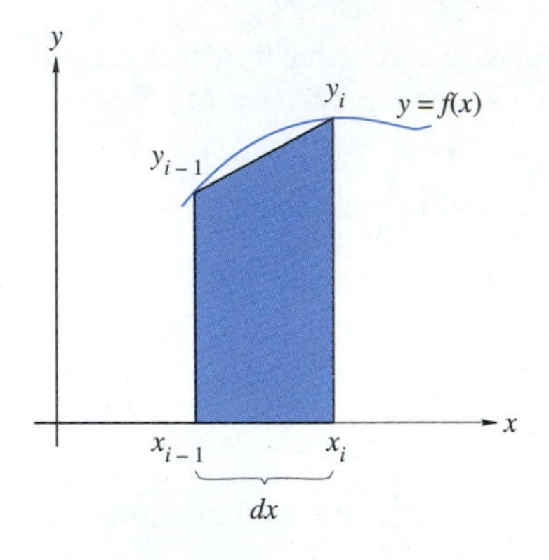

FIGURE 6.26a

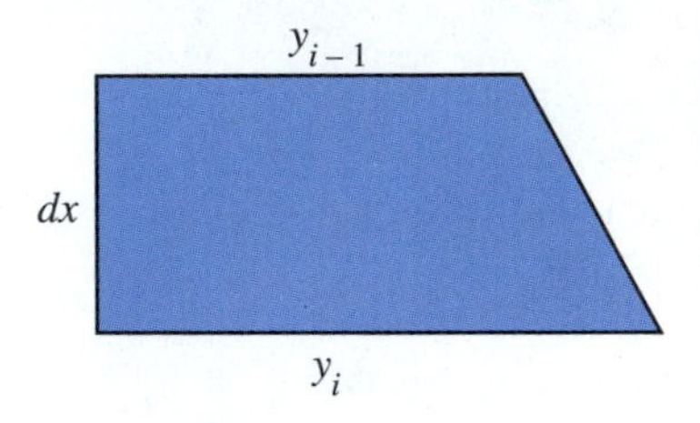

FIGURE 6.26b

this trapezoid is dx and the lengths of its bases are y_{i-1} and y_i. Thus, the area of the trapezoid in Figure 6.26a is $\frac{1}{2}(y_{i-1}+y_i)\,dx$.

The area under the curve in Figure 6.25 can be approximated by adding the areas of all the trapezoids. Thus,

$$\begin{aligned}\int_a^b f(x)\,dx &\approx \frac{1}{2}(y_0+y_1)\,dx+\frac{1}{2}(y_1+y_2)\,dx+\frac{1}{2}(y_2+y_3)\,dx+\cdots\\&\quad+\frac{1}{2}(y_{n-1}+y_n)\,dx\\&=\frac{1}{2}(y_0+2y_1+2y_2+2y_3+\cdots+2y_{x-1}+y_n)\,dx\end{aligned}$$

Using the more conventional function notation, $y=f(x)$, we can rewrite the formula as follows:

> **Trapezoidal Rule**
> $$\begin{aligned}\int_a^b f(x)\,dx &\approx \frac{dx}{2}[f(x_0)+2f(x_1)+2f(x_2)+2f(x_3)+\cdots\\&\quad+2f(x_{n-1})+f(x_n)]\end{aligned}$$
> where $dx=\dfrac{b-a}{n}$.

EXAMPLE 6.25

Use the trapezoidal rule with $n=6$ to approximate the value of $\int_0^2 x^3\,dx$.

Solution For $n=6$, we have $dx=\frac{2-0}{6}=\frac{1}{3}$. Thus, $x_0=0$, $x_1=\frac{1}{3}$, $x_2=\frac{2}{3}$, $x_3=1$, $x_4=\frac{4}{3}$, $x_5=\frac{5}{3}$, and $x_6=2$. Using the trapezoidal rule, we have

$$\begin{aligned}\int_0^2 x^3\,dx &\approx \frac{1/3}{2}\left[0^3+2\left(\frac{1}{3}\right)^3+2\left(\frac{2}{3}\right)^3+2(1)^3+2\left(\frac{4}{3}\right)^3+2\left(\frac{5}{3}\right)^3+2^3\right]\\&=\frac{1}{6}\left(0+\frac{2}{27}+\frac{16}{27}+2+\frac{128}{27}+\frac{250}{27}+8\right)\\&=\frac{37}{9}\\&\approx 4.111\end{aligned}$$

The exact value obtained by integration is 4.

Simpson's Rule

For the second approximation technique, we will use the same curve. Again, the interval $[a,b]$ is divided into n equal parts, but this time n must be an even number. Instead of the straight lines we drew connecting y_0, y_1, y_2, y_3, and so on, we are

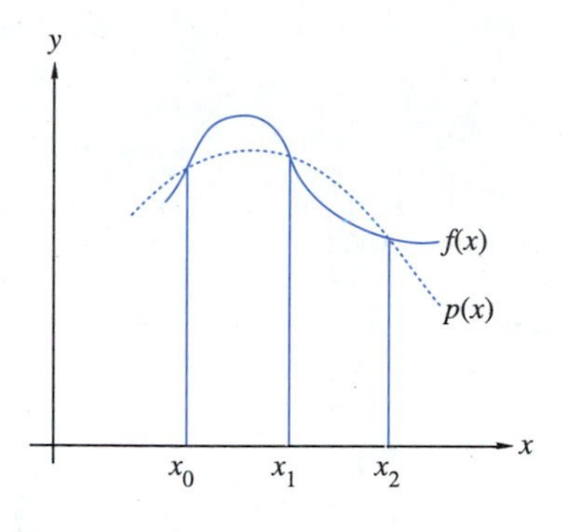

FIGURE 6.27

going to draw parabolas through each group of three consecutive points, as shown in Figure 6.27. A lengthy algebraic manipulation will show that

$$\int_{x_0}^{x_2} f(x)\,dx = \frac{dx}{3}[f(x_0) + 4f(x_1) + f(x_2)]$$

Now, if we did this for each subinterval $[x_0, x_2]$, $[x_2, x_4]$, $[x_4, x_6]$, ..., $[x_{n-2}, x_n]$, we would get the result known as Simpson's rule.

Simpson's Rule

$$\int_a^b f(x)\,dx \approx \frac{dx}{3}[f(x_0) + 4f(x_1) + 2f(x_2) + 4f(x_3) + 2f(x_4) + \cdots + 2f(x_{n-2}) + 4f(x_{n-1}) + f(x_n)]$$

where n is an even number and $dx = \dfrac{b-a}{n}$.

EXAMPLE 6.26

Use Simpson's rule with $n = 6$ to approximate the value of $\int_0^2 x^3\,dx$.

Solution Since $n = 6$, n is an even number. This is the same integral we approximated in Example 6.25, so the values of x and dx are the same.

$$\begin{aligned}\int_0^2 x^3\,dx &\approx \frac{1/3}{3}\left[0^3 + 4\left(\frac{1}{3}\right)^3 + 2\left(\frac{2}{3}\right)^3 + 4(1)^3 + 2\left(\frac{4}{3}\right)^3 + 4\left(\frac{5}{3}\right)^3 + 2^3\right] \\ &= \frac{1}{9}\left(0 + \frac{4}{27} + \frac{16}{27} + 4 + \frac{128}{27} + \frac{500}{27} + 8\right) \\ &= \frac{1}{9}(36) \\ &= 4\end{aligned}$$

This happens to be a better approximation than we got using the trapezoidal rule. In fact, for this function it gives the exact value of the integral.

Both the trapezoidal rule and Simpson's rule are much easier to use if one can take advantage of a calculator or computer.

Some scientific calculators use Simpson's rule to evaluate numeric integrals of functions. The next five examples demonstrate how to do this. Examples 6.27 and 6.28 use a Casio fx-7700G and Examples 6.29, 6.30, and 6.31 use a TI-82.

EXAMPLE 6.27

Use a Casio fx-7700G to evaluate $\int_0^2 x^3\,dx$.

Solution Press the following sequence of keys:

SHIFT ∫dx X,θ,T x^y 3 SHIFT , 0 SHIFT , 2) EXE

After a slight pause, the calculator will return the solution 4, the same as that which we got in Example 6.26.

EXAMPLE 6.28

Use a Casio fx-7700G to evaluate $\int_1^5 (2x^5 - 3x)\,dx$.

Solution Press:

SHIFT ∫dx 2 X,θ,T x^y 5 − 3 X,θ,T SHIFT , 1 SHIFT , 5) EXE

After a fairly long pause, we get the result 5,172. So, according to the calculator, $\int_1^5 (2x^5 - 3x)\,dx = 5{,}172$. How does this compare with the actual value?

$$\begin{aligned}\int_1^5 (2x^5 - 3x)\,dx &= \frac{1}{3}x^6 - \frac{3}{2}x^2 \Big|_1^5 \\ &= \left(\frac{15{,}625}{3} - \frac{75}{2}\right) - \left(\frac{1}{3} - \frac{3}{2}\right) \\ &= 5{,}172\end{aligned}$$

EXAMPLE 6.29

Use a TI-82 to evaluate $\int_0^2 x^3\,dx$.

Solution There are two ways to use a TI-82 to evaluate a definite integral. We will show the first method in this example and the other in the next example.

In the first method you press the following sequence of keys:

MATH 9 X,T,θ ∧ 3 , X,T,θ , 0 , 2) ENTER

After a slight pause, the calculator will return the solution 4, the same as we got in Example 6.27. (Note that pressing MATH 9 accesses the "`fnInt(`" option on the calculator.)

EXAMPLE 6.30

Use a TI-82 to evaluate $\int_1^5 (2x^5 - 3x)\,dx$.

Solution In this method you begin as if your are going to graph the integrand. So, for this integral begin by entering the integrad as $y_1 = 2x^5 - 3x$. Then, after you access the `fnInt(` option on the calculator, you recall the name of the function, in this case, y_1. The remainder of the procedure is exactly the same as shown in

EXAMPLE 6.30 (Cont.)

Example 6.28. Press:

MATH 9 2nd VARS 1 1 , X,T,θ , 1 , 5) ENTER

After a brief pause, we get the result 5172.

According to the calculator, $\int_1^5 (2x^5 - 3x)\,dx = 5172$. How does this compare with the actual value? As we showed in Example 25.27, $\int_1^5 (2x^5 - 3x)\,dx = 5{,}172$, so the calculator gave the exact value of this definite integral.

It is also possible to get the calculator to both shade the region being studied and evaluate the integral. Since an integral is a measure of the area under a curve, this can provide a graphical and numerical interpretation of an integral.

EXAMPLE 6.31

Use a TI-82 to graph the region, and determine the area under the curve of $f(x) = x^2 + \frac{1}{x^3} - 4x + 6$ from $x = 1$ to $x = 4.2$.

Solution As in the previous method, you begin as if you are going to graph the integrand. So, for this integral begin by entering the integrand as $y_1 = x^2 + \frac{1}{x^3} - 4x + 6$. You may want to graph the curve, in order to select a viewing window that shows the graph over the entire interval from $x = 1$ to $x = 4.2$. In this example, a viewing window of $[0, 4.7] \times [-1, 8]$ was selected.

Now access 2nd CALC. (On a TI-82, the CALC key is above the TRACE key.) Select the ∫f(x)dx option on the calculator, by pressing 7. The result in Figure 6.28a shows the graph of the function, puts a blinking cursor on the curve at the x-value in the middle of the window, asks for a Lower Limit?, and gives the coordinates of the cursor. Use the left or right arrow key, ◄ or ►, to move the cursor to the lower limit, in this case $x = 1$, and press the ENTER key. The result is shown in Figure 6.28b. Notice that the calculator is now asking for the Upper Limit?. Use the ► key to move the cursor to the x-value that indicates the upper limit as shown in Figure 6.28c. Press ENTER. You should see the calculator shade the area of the region being intergrated and then, at the bottom of the screen, give an approximation of the area as 10.754322, as shown in Figure 6.28d.

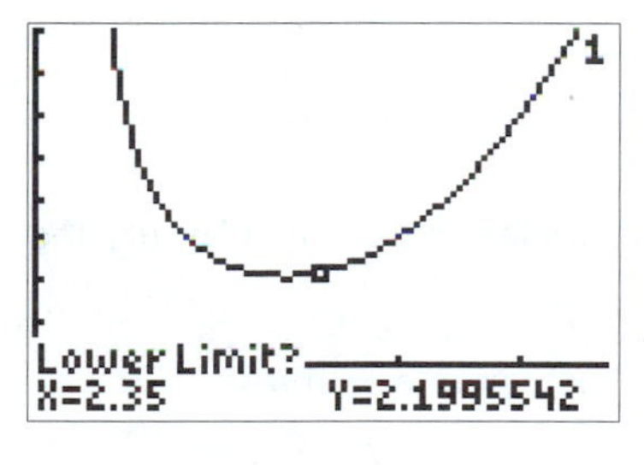

[0, 4.7, 1] x [−1, 8, 1]

FIGURE 6.28a

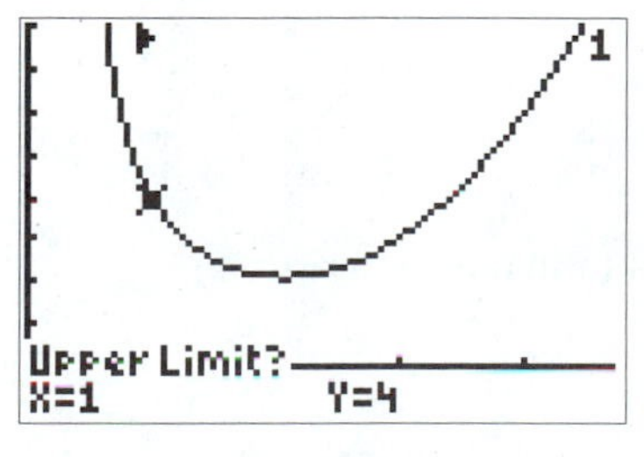

[0, 4.7, 1] x [−1, 8, 1]

FIGURE 6.28b

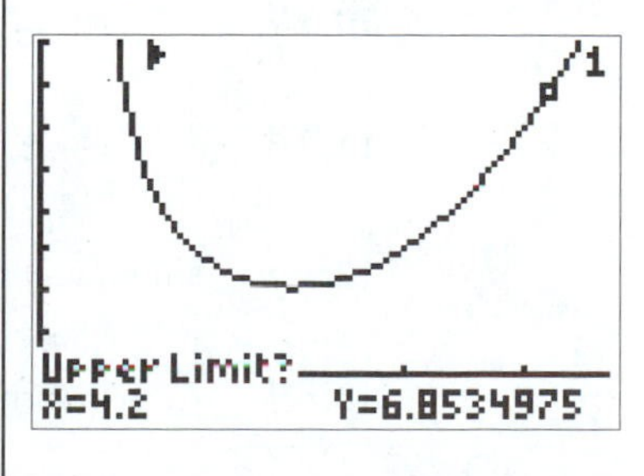

[0, 4.7, 1] x [−1, 8, 1]

FIGURE 6.28c

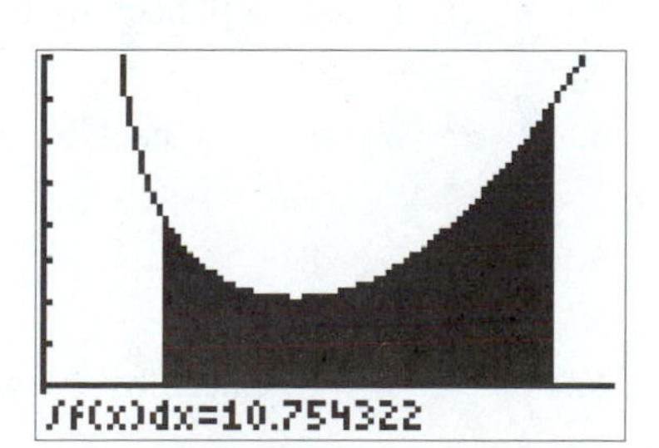

[0, 4.7, 1] x [−1, 8, 1]

FIGURE 6.28d

It is also possible to use both of these formulas to obtain integrals of functions that are available only in graphical or tabular form, as shown by the next example.

Application

EXAMPLE 6.32

The following table gives the results of a series of drillings meant to determine the depth of the bedrock at a building site. These drillings were taken along a straight line down the middle of the lot where the building will be placed. In the table, x is the distance from the front of the parking lot and y is the corresponding depth. Both x and y are given in feet.

x	0	20	40	60	80	100	120	140	160
y	33	35	40	45	42	38	46	40	48

Approximate the area of this cross-section by using Simpson's rule.

Solution We first determine that $n = 8$, an even number, and $dx = 20$. By Simpson's rule,

$$\begin{aligned}\int_0^{160} f(x)\,dx &\approx \frac{20}{3}[33+4(35)+2(40)+4(45)+2(42)+4(38)\\&\quad +2(46)+4(40)+48]\\&= \frac{20}{3}(969)\\&= 6{,}460\,\text{ft}^2\end{aligned}$$

Exercise Set 6.5

In Exercises 1–16, find the approximate value of the given integrals with the specified value of *n* by (a) the trapezoidal rule and (b) Simpson's rule.

1. $\int_1^2 x^4\,dx,\ n=6$ (Check by integration.)
2. $\int_0^2 2x^5\,dx,\ n=8$ (Check by integration.)
3. $\int_0^1 \sqrt{x}\,dx,\ n=4$ (Check by integration.)
4. $\int_1^2 x^{1/3}\,dx,\ n=4$ (Check by integration.)
5. $\int_1^5 \frac{dx}{x},\ n=8$
6. $\int_1^4 \frac{dx}{x^2},\ n=6$ (Check by integration.)
7. $\int_1^5 \frac{dx}{1+x},\ n=8$
8. $\int_0^2 x^{\pi}\,dx,\ n=4$ (Check by integration.)
9. $\int_0^1 \frac{4}{1+x^2},\ n=10$
10. $\int_0^2 e^x\,dx,\ n=4$
11. $\int_0^2 \sqrt{1+x^3}\,dx,\ n=8$
12. $\int_0^{0.2} \sin x\,dx,\ n=4$ (x in radian measure)
13. $\int_{-2}^2 \sqrt{x+4}\,dx,\ n=8$
14. $\int_1^3 \ln x\,dx,\ n=8$

15. $\int_0^4 \frac{\sqrt{x}}{1+\sqrt{x}} dx, n = 8$

16. $\int_0^1 \frac{e^x + e^{-x}}{2} dx, n = 4$

In Exercises 17–19, use the trapezoidal rule and Simpson's rule to find the approximate area under the curve defined by the given sets of experimental data.

17.

x	5.0	6.0	7.0	8.0	9.0	10.0
y	4.2	3.9	3.8	4.0	3.5	3.4

x	11.0	12.0	13.0
y	3.9	4.1	4.3

18.

x	2.00	2.10	2.20	2.30	2.40	2.50	2.60
y	4.32	4.57	5.14	5.78	6.84	6.62	6.51

19.

x	1.00	1.25	1.50	1.75	2.00
y	16.32	16.48	16.73	16.42	16.38

x	2.25	2.50
y	16.29	16.25

Solve Exercises 20–26.

20. *Environmental science* Find the area of the pond in Figure 6.29.

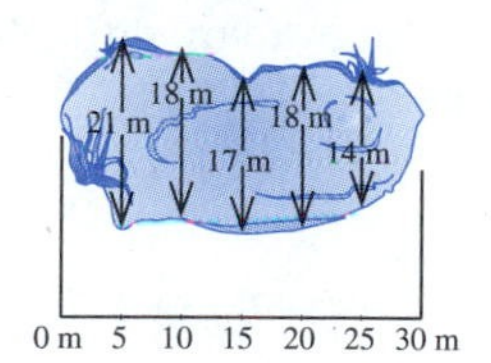

FIGURE 6.29

21. *Electronics* The current i in amperes (A) to a capacitor is given by $i = 4.0e^{-10t} - 4.0e^{-20t}$, where t is time in seconds (s). If $i = \frac{dq}{dt}$ where q is the charge in coulombs (C), use **(a)** the trapezoid rule and **(b)** Simpson's rule to find the approximate change in charge on the capacitor from $t = 1.0$ s to $t = 2.0$ s, using four intervals.

22. *Light* The graph in Figure 6.30 shows the light in lumens given off by the flash of a flashbulb. A typical focal plane shutter is open from 10 milliseconds to 60 milliseconds after the camera button is pressed. What is the total amount of light given by the flashbulb during this interval?

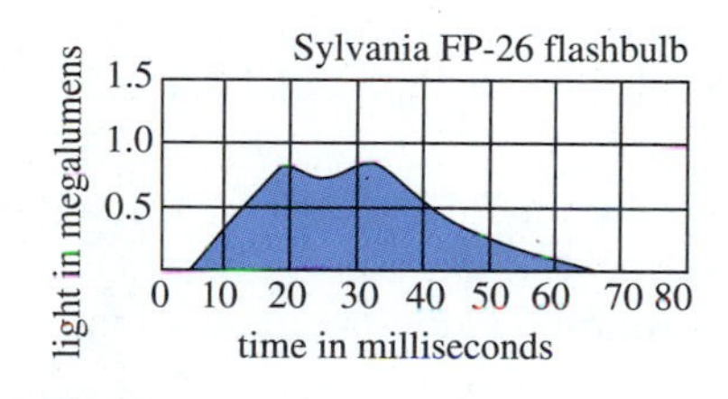

FIGURE 6.30

23. Write a computer program to determine the approximate area under a curve. The program should allow you to select whether to approximate the area by **(a)** the trapezoidal rule, **(b)** Simpson's rule, or **(c)** both rules. Your program should also allow for the input of the endpoints and the number of intervals. If Simpson's rule is used, the program should check to make sure that the number of intervals in even. The program should also allow for either a function $y = f(x)$ to be given or a set of observed data points to be used. Check your program by comparing results with Examples 6.25 and 6.26.

24. *Business* The marginal revenue for the sale of computers by a store is given by the function $R'(x) = 25\sqrt{x}e^{-0.004x}$, where $0 \le x \le 600$. Use the trapezoidal rule with $n = 10$ to determine the approximate revenue from the sale of the first 600 computers sold by the store.

25. *Agriculture* To determine the amount of fertilizer needed for an irregularly shaped field, a farmer needed to estimate the number of acres in the field.. The farmer drew a line across the field and then, at 100 ft intervals, measured across the land perpendicular to the farmer's line. The farmer's measurements are shown in the following table:

x	0	100	200	300	400	500
y	300	300	350	400	500	600

x	600	700	800	900	1000
y	500	600	550	650	600

Use both the trapezoidal rule and Simpson's rule to estimate the number of acres if 1 acre $= 43{,}500\,\text{ft}^2$.

26. *Pharmacology* The reaction rate to a new drug is given by $r = e^{-t^2} + \frac{1}{t}$ where t is the time measured in hours after the drug is administered. Use both the trapezoidal rule and Simpson's rule with $n = 8$ to estimate the total reaction to the drug from $t = 1$ to $t = 9$.

In Your Words

27. Explain the purpose of the trapezoidal rule and Simpson's rule.

28. Write an application in your technology area of interest that requires you the use of either the trapezoidal rule or Simpson's rule for its solution. Give your problem to a classmate and see if he or she understands and can solve your problem. Rewrite the problem as necessary to remove any difficulties encountered by your classmate.

CHAPTER 6 REVIEW

Important Terms and Concepts

Area between curves
Definite integral
  Properties of
Fundamental Theorem of Calculus
Improper integrals
Indefinite integrals
  Properties of
Numerical integration
Simpson's rule
Substitution method
Summation formulas
Trapezoidal rule

Review Exercises

In Exercises 1–30, evaluate the given definite or indefinite integral.

1. $\int x^5\,dx$

2. $\int \left(\sqrt[3]{x}+4\right)dx$

3. $\int_0^3 (x^2+4x-6)\,dx$

4. $\int_1^4 (x+4)^2\,dx$

5. $\int_1^8 \frac{dt}{\sqrt[3]{t}}$

6. $\int_1^4 \left(\sqrt{x}+4\right)dx$

7. $\int \frac{dt}{(t+5)^4}$

8. $\int \frac{x\,dx}{\sqrt[3]{x^2+4}}$

9. $\int_1^8 x\sqrt[3]{x}\,dx$

10. $\int_1^4 \frac{4\,dx}{3\sqrt{x^3}}$

11. $\int (u^4-u^{-4})\,du$

12. $\int \frac{x^4-1}{x^3}\,dx$

13. $\int 2x\sqrt{x^2-5}\,dx$

14. $\int_0^4 x\sqrt{x^2+9}\,dx$

15. $\int x\sqrt[3]{6-x^2}\,dx$

16. $\int 3x\left(4-x^2\right)^3\,dx$

17. $\int_{-1}^{2} (x^2+4)(x^3-3)\,dx$

18. $\int_1^2 x\,(x+2)^2\,dx$

19. $\int \frac{y^2\,dy}{(y^3-5)^{2/3}}$

20. $\int \frac{5-6x+x^2}{1-x}\,dx$

21. $\int_0^1 (1-x^2)^3\,dx$

22. $\int \frac{x+2}{(x^2+4x)^2}\,dx$

23. $\int \left(\sqrt{2x}-\frac{1}{\sqrt{2x}}\right)\,dx$

24. $\int \left(x^3-4x\right)\left(3x^2-4\right)\,dx$

25. $\int_1^4 \frac{9-t^3}{3t^2}\,dt$

26. $\int_0^1 \sqrt{x}\,(3\sqrt{x}-5)\,dx$

27. $\int_{-1}^{8} \frac{4\sqrt[3]{u}\,du}{(1+u^{4/3})^3}$

28. $\int_1^2 \frac{3x^3-2x^{-3}}{x^2}\,dx$

29. $\int_1^{\infty} \frac{dx}{x^4}$

30. $\int_0^{\infty} \frac{4x\,dx}{(3x^2+1)^5}$

In Exercises 31–38, find the areas bounded by the given curves.

31. $f(x)=4x-6$, x-axis, $x=-3$, $x=4$
32. $g(x)=2-x-x^2$, x-axis
33. $f(x)=x^2-x-12$, $g(x)=-6$
34. $h(x)=\sqrt{x+2}$, x-axis, $x=7$
35. $f(x)=4x^3$, $g(x)=x$
36. $f(x)=x^3-3x^2$, $g(x)=x-3$
37. $y^2+x=0$, $y=2x+1$
38. $f(x)=2x^3$, $g(x)=x^4$

In Exercises 39–44, find the approximate value of the given integrals with the specified values of *n* by (a) the trapezoidal rule and (b) Simpson's rule.

39. $\int_0^1 \sqrt{1-x^2}\,dx,\ n=10$

40. $\int_1^4 \frac{1}{3x}\,dx,\ n=6$

41. $\int_2^{2.5} \sqrt{x^3-1}\,dx,\ n=10$

42. $\int_{-1}^{4} \sqrt{1+x^5}\,dx,\ n=10$

43. $\int_0^1 \left(e^x-e^{-x}\right)\,dx,\ n=4$

44.

x	1.00	1.25	1.50	1.75	2.00
y	4.25	5.72	5.13	3.19	2.10

x	2.25	2.50	2.75	3.00
y	0.15	1.65	3.10	3.70

Solve Exercises 45–48.

45. *Dynamics* An object is moving rectilinearly with a velocity $v(t)=t+\frac{1}{\sqrt{1+t}}$ m/s. **(a)** If the object began at the origin, what is its position when $t=15$? **(b)** What is its acceleration when $t=15$?

46. *Electricity* The power P in watts (W) in a circuit is given by $P=\frac{dw}{dt}$, where w is dissipated energy in joules (J) and t is time in seconds (s). If the power in a circuit is given by $P=10t^3$, find the energy dissipated from $t=3.0$ s to $t=5.0$ s.

47. *Electricity* The induced voltage v in volts (V) in an inductor is given by $v=-L\frac{di}{dt}$, where i is the current in amperes (A) and L is the inductance in henries (H). If v is given by $v=2.0-1.0t^2$, where t is time in seconds (s) and $L=2$ H, find the change in the current to the inductor from $t=1.0$ s to $t=6.0$ s.

48. *Highway safety* A car moving initially at 88 ft/s, decelerates at the constant rate of 4 ft/s^2. **(a)** How long does it take to stop the car? **(b)** How far does it travel from the time the brakes are applied until it stops?

CHAPTER 6 TEST

In Exercises 1–7, evaluate the given integral.

1. $\int_0^1 2x\,dx$

2. $\int_1^4 (x+1)(x^2+1)\,dx$

3. $\int (x^2+3)^2\,dx$

4. $\int 4(x^4+7)^5 x^3\,dx$

5. $\int \frac{x^2+2}{x^2}\,dx$

6. $\int (x^2+8x)^3(x+4)\,dx$

7. $\int_1^2 (3x^2\sqrt{x^3+x}+\sqrt{x^3+x})\,dx$

Solve Exercises 8–11.

8. Find the total area of the region between the curve $y=4x^3-4$ and the x-axis over the interval $[-1,2]$.

9. Find the area of the region enclosed by the curve $y=x^2-2$ and the line $x+y=4$.

10. Use the trapezoidal and Simpson's rules to determine $\int_2^4 \frac{1}{x}\,dx$ with $n=8$.

11. For a particular circuit, the current, in amperes, after time t, in seconds, at a certain point, P, is given by $i=0.0004(t^{-1/2}+t^{1/2})$. Find the charge, in C, that passes point P during the first second.

CHAPTER

7

Applications of Integration

A tank truck, like the milk truck shown here, is used to carry liquids. In Section 7.7, you will learn how to determine the force on one end of this truck when it is full of milk.

Courtesy of H.P. Hood Inc.

We introduced the integral as a method for determining the area under a curve. Later, we showed that integrals could be used to determine the velocity and position functions from the acceleration. We also used integrals to find the area between two curves. In this chapter, we will see some of the many other applications for the integral.

We begin with the average value and root mean square of a function. The average value will be used to determine the amount of heating oil that will be used during one winter day. Examples will use the root mean square to determine the effective current and voltage in an electric circuit.

Next, we look at methods for determining the volume of a three-dimensional object. Two different methods will be examined to allow you to select the better method for a particular problem. Then we will look at methods for finding the length of a curve and the area of a surface formed by revolving that curve around a line.

We will end the chapter with three sections that look at several physical applications—an object's mass, its center of gravity, its moment of inertia, the amount of work needed to move something, and the pressure exerted by a liquid.

7.1 AVERAGE VALUES AND OTHER ANTIDERIVATIVE APPLICATIONS

In this section, we will look at two ideas: One is the idea of the average value of a function; the other idea looks at some additional applications of the antiderivative.

Average Value of a Function

The concept of an average is a common one to most of us. Sometimes the term is applied to a finite number of items such as the average grade on a test by a group of students. Sometimes it is based on a continuous function. There are times when that function is known and times when it is not.

One example that most of us hear every day is the reference to the average temperature. The weather bureau assumes that there is some function f that will give the temperature T at some time t. Thus, $T = f(t)$. Normally, what is done is to take a temperature reading every hour. The average or mean temperature $\bar{T}$ is the sum of these temperatures divided by the number of hours. Thus, for one day, we have

$$\bar{T} = \frac{1}{24}\sum_{t=1}^{24} f(t)$$

If we divide the day into smaller intervals and take a temperature reading at each of those intervals, we get a more accurate reading. So, if we take n temperature readings we have

$$\bar{T} = \frac{1}{n}\sum_{i=1}^{n} f(t_i)$$

where t_i represents the time of the ith temperature reading.

Naturally, the larger that n becomes the more accurate is the average. If we wanted the true average, we would take the limit of $\frac{1}{n}\sum_{i=1}^{n} f(t_i)$ as n approaches infinity or

$$\bar{T} = \lim_{n\to\infty} \frac{1}{n}\sum_{i=1}^{n} f(t_i)$$

The right-hand side of this equation looks familiar. If dt represents the width of a time interval, $\frac{b-a}{n}$, where a is the beginning of the first interval and b the end of the last interval, and if we write $\frac{1}{n}$ as $\frac{b-a}{n}\cdot\frac{1}{b-a}$, then

$$\begin{aligned}
\bar{T} &= \lim_{n\to\infty} \frac{1}{b-a}\sum_{n=1}^{n} f(t_i)\frac{b-a}{n} \\
&= \frac{1}{b-a}\lim_{n\to\infty}\sum_{i=1}^{n} f(t_i)dt \\
&= \frac{1}{b-a}\int_a^b f(t)dt
\end{aligned}$$

If we assume that f is continuous, then we can use this idea for the following definition.

Average Value

If f is a continuous function over the interval $[a,b]$, then the **average value** $\bar{y}$ is

$$\bar{y} = \frac{1}{b-a}\int_a^b f(x)dx$$

Application

EXAMPLE 7.1

A ball is dropped from a height of 490 m. Find its average velocity between the time it is dropped and the time it hits the ground.

Solution If acceleration is $-9.8\,\text{m/s}^2$ and, since $v(0)=0$, then we have $v(t) = -9.8t$. We also have $s(0)=490$, and so $s(t) = -4.9t^2+490$. Solving for t, we see that it takes 10 s to hit the ground. The average velocity is

$$\begin{aligned}\bar{v} &= \frac{1}{10-0}\int_0^{10} v(t)dt \\ &= \frac{1}{10}[s(t)]\Big|_0^{10} \\ &= \frac{1}{10}(-4.9t^2+490)\Big|_0^{10} \\ &= \frac{1}{10}(-490) = -49\end{aligned}$$

The average velocity is 49 m/s downward.

Application

EXAMPLE 7.2

The amount of heating oil required to heat a house for one day during the winter can be approximated by multiplying the difference between 20°C and the average daily temperature by 0.8 gal. During one particular cold spell, the temperature T, at any time of the day t h after midnight, was given by $T(t) = 2 - \frac{1}{15}(t-15)^2$. Determine the level of oil consumption during one of these days.

Solution If we let A represent the amount of oil required and $\bar{T}$ the average temperature during a day, then we have

$$A = (20°\text{C} - \bar{T})(0.8\,\text{gal/°C})$$

We will first compute the average temperature during one of these days. We will let t range from midnight at the start of the day, $t=0$, to midnight at the end of the

EXAMPLE 7.2 (Cont.)

day, $t = 24$. Thus, for our formula, $a = 0$ and $b = 24$

$$\begin{aligned}\bar{T} &= \frac{1}{24-0}\int_0^{24} T(t)dt \\ &= \frac{1}{24-0}\int_0^{24}\left[2-\frac{1}{15}(t-15)^2\right]dt \\ &= \frac{1}{24}\left[2t-\frac{1}{45}(t-15)^3\right]_0^{24} \\ &= \frac{1}{24}(31.80-75) = \frac{1}{24}(-43.2) = -1.8\end{aligned}$$

So, the average temperature during this particular 24-hour period was $-1.8°C$.

We can now determine the amount of fuel oil consumed during this period.

$$\begin{aligned}A &= (20°C - \bar{T})(0.8\,\text{gal/°C}) \\ &= [20°C - (-1.8°C)](0.8\,\text{gal/°C}) \\ &= (21.8°C)(0.8\,\text{gal/°C}) \\ &= 17.44\,\text{gal}\end{aligned}$$

A total of 17.44 gal of heating oil were used during this 24-h period.

Root Mean Square

The **root mean square**, rms, of a function $y = f(x)$ over the interval $[a, b]$ is defined to be the square root of the average value of y^2 on the interval.

Root Mean Square

If a function f is continuous on $[a, b]$, then the **root mean square**, f_{rms}, is

$$f_{\text{rms}} = \sqrt{\frac{1}{b-a}\int_a^b [f(x)]^2\,dx}$$

EXAMPLE 7.3

Find the root mean square of $f(x) = x^4$ on $[-1, 3]$.

Solution

$$\begin{aligned}f_{\text{rms}} &= \sqrt{\frac{1}{3-(-1)}\int_{-1}^{3}(x^4)^2\,dx} \\ &= \sqrt{\frac{1}{4}\int_{-1}^{3}x^8\,dx}\end{aligned}$$

EXAMPLE 7.3 (Cont.)

$$= \sqrt{\frac{1}{4}\cdot\frac{1}{9}x^9\Big|_{-1}^{3}}$$

$$= \sqrt{\frac{1}{36}[19{,}683-(-1)]}$$

$$= \sqrt{\frac{19{,}684}{36}} \approx 23.38$$

Because an alternating current changes continuously, its maximum value, $\pm I_{\max}$, does not indicate its ability to do work or to produce heat as does the magnitude of a direct current. For this reason, it is customary to refer to the **effective current**. The same is true for the voltage, where it is more common to refer to the **effective voltage** rather than $V_{\max}$.

Note The effective current and the effective voltage are the same as the root mean square of the current and voltage. In symbols, this means that $i_{\text{eff}} = i_{\text{rms}}$ and $V_{\text{eff}} = V_{\text{rms}}$.

If the resistance is constant, then the average or mean power, in watts, produced by the current is

$$P = i_{\text{rms}}^2 \cdot R$$

Application

EXAMPLE 7.4

If a current $i = t - 2t^2$ A flows through a resistor of $20\,\Omega$ from $t = 0$ to $t = 3$ s, find i_{eff} and the power generated.

Solution

$$i_{\text{eff}} = i_{\text{rms}} = \sqrt{\frac{1}{3}\int_0^3 (t-2t^2)^2\,dt}$$

$$= \sqrt{\frac{1}{3}\int_0^3 (t^2-4t^3+4t^4)dt}$$

$$= \sqrt{\frac{1}{3}\left[\frac{1}{3}t^3 - t^4 + \frac{4}{5}t^5\right]_0^3}$$

$$= \sqrt{\frac{1}{3}(9-81+194.4)}$$

$$= \sqrt{40.8} \approx 6.39\,\text{A}$$

and

$$P = \left(\sqrt{40.8}\right)^2 \cdot 20$$

$$\approx 816\,\text{W}$$

Current and Power

The current i, in amperes, in an electric circuit is defined as the time rate of change of the charge q, in coulombs, which passes a given point in the circuit. This is written as

$$i = \frac{dq}{dt}$$

When this is placed in the differential form, we get $dq = i\,dt$. Integrating both side of this equation, we get

$$q = \int i\,dt$$

Thus we can find the total charge transmitted in a circuit.

The charge in a capacitor is given by

$$q = CV_c$$

where C is the capacitance in farads and V_c represents the voltage between the capacitor terminals. Substituting for q from the previous formula and solving for V_c, we get

$$V_c = \frac{1}{C}q$$

$$\text{or} \quad V_c = \frac{1}{C}\int i\,dt$$

Application

EXAMPLE 7.5

The current in a circuit is given by the equation $i = 4t^3 + 2$ A. How many coulombs are transmitted in 3 s?

Solution We have $q = \int i\,dt$ and from the information given we can see that

$$\begin{aligned} q &= \int_0^3 (4t^3 + 2)dt \\ &= (t^4 + 2t)\big|_0^3 \\ &= 87\,\text{C} \end{aligned}$$

Note The C at the end of Example 7.5 stands for coulombs. Do not confuse it with the C for capacitance, with the constant of integration, or with °C for degrees Celsius.

Application

EXAMPLE 7.6

A 1-μF capacitor receives a charge such that it has a terminal-to-terminal voltage of 33 V. We connect this capacitor at $t = 0$ s to a source that sends a current $i = t^2$ A into it. What is the voltage across the capacitor when $t = 0.2$ s?

EXAMPLE 7.6 (Cont.)

Solution We will need the formula $V_c = \frac{1}{C}\int i\,dt$. We are given $C = 1\,\mu\text{F} = 10^{-6}\,\text{F}$ and $i = t^2$, so

$$V_c = \frac{1}{10^{-6}}\int t^2\,dt = 10^6 \cdot \frac{1}{3}t^3 + k$$

(We will let k be the constant of integration.) Now when $t = 0$, $V_c = 33$ and $k = 33$ and so

$$V_c = 10^6 \cdot \frac{1}{3}t^3 + 33$$

When $t = 0.2$, we get

$$\begin{aligned} V_c &= 10^6 \cdot \frac{1}{3}(0.2)^3 + 33 \\ &\approx 2{,}699.67\,\text{V} \\ &\approx 2{,}700\,\text{V} \end{aligned}$$

Application

EXAMPLE 7.7

A certain capacitor has a voltage of 150 V across it. A current given by $i = 0.08\sqrt{t}$ is sent through the circuit. After 0.36 s, the voltage across the capacitor is measured at 200 V. What is the capacitance of this capacitor?

Solution We will use the equation $V_c = \frac{1}{C}\int i\,dt$. Since $i = 0.08\sqrt{t}$, we have

$$\begin{aligned} V_c &= \frac{1}{C}\int 0.08t^{1/2}\,dt \\ &= \frac{0.08}{C}\int t^{1/2}\,dt \\ &= \frac{0.16}{3C}t^{3/2} + k \end{aligned}$$

When $t = 0$, we know that $V_c = 150$ and so $k = 150$. Thus,

$$V_c = \frac{0.16}{3C}t^{3/2} + 150$$

Now $V_c = 200\,\text{V}$ when $t = 0.36$ s, and so

$$\begin{aligned} 200 &= \frac{0.16}{3C}(0.36)^{3/2} + 150 \\ 200 &= \frac{0.01152}{C} + 150 \end{aligned}$$

Solving for C, we get

$$\begin{aligned} C &= 0.0002304\,\text{F} \\ &= 2.304 \times 10^{-4}\,\text{F} \\ &= 230.4\,\mu\text{F} \end{aligned}$$

As a last application in this section, the amount of energy (**work**), W, expended in an electrical system is given by

$$W = \int P\,dt$$

Here W is in joules, the power P is in watts, and t is in seconds.

Application

EXAMPLE 7.8

The power in a certain electrical system is changed according to $P = 5t^{1/4}$ W. What is the formula for energy in this circuit?

Solution Using the formula $W = \int P\,dt$ with $P = 5t^{1/4}$, we have

$$\begin{aligned} W &= \int 5t^{1/4}\,dt \\ &= 4t^{5/4} + k\,\text{J} \end{aligned}$$

The **instantaneous power** in a circuit can be represented by $P = Vi$, $P = i^2R$, or $P = V^2/R$, and any of these can be substituted for P in the formula $W = \int P\,dt$.

Exercise Set 7.1

In Exercises 1–12, find the average value and the root mean square of the given function over the indicated interval.

1. $f(x) = x^2$, over $[0, 1]$
2. $g(x) = 3x^2$, over $[2, 4]$
3. $h(x) = x^2 - 1$, over $[1, 3]$
4. $j(x) = 9 - x^2$, over $[-3, 3]$
5. $f(x) = \sqrt{x+4}$, over $[0, 5]$
6. $k(x) = x^3 - 1$, over $[1, 3]$
7. $f(t) = t^2 + 4$, over $[0, 4]$
8. $g(t) = 16t^2 - 25$, over $[2, 4]$
9. $f(x) = 6x - x^2$, over $[1, 2]$
10. $h(t) = 16t^2 - 8t + 1$, over $[0, 3]$
11. $j(t) = 4.9t^2 - 2.8t + 4$, over $[0, 2.5]$
12. $k(t) = 4.9t^2 - 4.2t + 9$, over $[0, 1.6]$

Solve Exercises 13–32.

13. *Meteorology* The rainfall per day, measured in centimeters, x days after the beginning of the year is $0.0002(4991 + 366x - x^2)$. What is the average rainfall for the first 90 days? For the year (365 days)?
14. *Material science* The mass density of a metal bar of length 4 m is given by $\rho(x) = 1 + x - \sqrt{x}$ kg/m^3, where x is the distance in meters from one end of the bar. What is the average mass density over the length of the bar?
15. *Physics* If a ball is dropped from a height of 360 ft, find its average height and average velocity between the time it is dropped and the time it hits the ground.
16. A 20000-L water tank takes 10 min to drain. After t min, the amount of water left in the tank is given by $V(t) = 200(100 - t^2)$. What is the average amount of water in the tank while it is draining?

17. *Meteorology* One day the temperature of the air t h after noon was given by $60+4t-t^{2/3}$°F.
    (a) What was the average temperature between noon and 6 p.m.?
    (b) What was the average temperature between 3 p.m. and 9 p.m.?
18. *Business* A laboratory receives a shipment of 900 cases of test tubes every 30 days. The number of cases on hand t days after the shipment arrives is given by $I(t)=900-15\sqrt{120t}$. What is the average daily inventory?
19. *Electricity* Find the average current from $t=0$ to $t=3$ s, if $i=1.0t+\sqrt{t}$ A.
20. *Electricity* Find the effective current for the current in Exercise 19.
21. *Electricity* Find the average current from $t=0$ to $t=4$ s, if $i=4t\sqrt{t^2+1}$ A.
22. *Electricity* What is the effective current for the current in Exercise 21?
23. *Electricity* If the current in Exercise 21 flows through a 30-Ω resistor, what is its average power?
24. *Electricity* A current $i=t^3-2t^2$ A flows through a resistance of 5.0 Ω from $t=0.0$ to $t=0.25$ s. Find the average power developed over that interval.
25. *Electricity* The voltage across a 10-Ω resistor is given by $V=4\sqrt{t}-2t$. **(a)** What is the effective voltage during the time interval from $t=0$ to $t=6$ s?
    (b) What is the average power generated during this time interval?
26. *Electricity* A voltage of $v=8.0t^{1/2}-16t^{3/2}$ V is applied across a resistance of 9.75 Ω. Find the average power developed during the period when v is greater than 0?
27. *Electricity* A capacitor contains a charge of 0.01 C. If it is supplied with a current of $i=2t$ A, what charge resides in the capacitor after 0.1 s?
28. *Electricity* An 80-μF capacitor is charged to 100 V. The capacitor is then supplied a current $i=0.04t^3$ A. After what time interval does the capacitor voltage reach 225 V?
29. *Electricity* The voltage applied to a circuit was $V=2t+1$ V. If the current was $i=0.03t$ A, what was the energy delivered from $t=0$ to $t=50$ s?
30. *Electricity* The voltage across a 90-μF capacitor is zero. What is the voltage after 0.001 s, if a current of 0.20 A charges the capacitor?
31. *Electricity* The voltage across a 7.5-μF capacitor is zero. What is the voltage after 0.005 s, if a current of $i=0.4t$ A charges the capacitor?
32. *Electricity* The current in a certain wire changes with time according to the relation $i=\sqrt[3]{1+5t}$. How many coulombs of charge pass a certain point in the first 3 s?

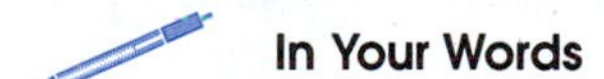

In Your Words

33. Describe how to calculate the average value of a function.
34. What is the root mean square? Give three applications of the root mean square.

7.2 VOLUMES OF REVOLUTION: DISK AND WASHER METHODS

Consider the area under the graph of a continuous nonnegative function, $y=f(x)$, over the interval $[a,b]$. If this area is revolved around the x-axis, then we get a solid of revolution such as the one shown in Figure 7.1. Common examples of solids of revolution include a cone, a cylinder, a soft drink bottle, and a table leg formed by turning a piece of wood on a lathe.

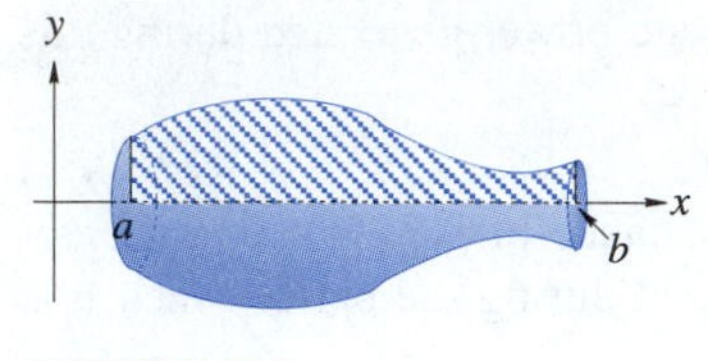

FIGURE 7.1

A cone is formed by revolving the area under a line of the form $f(x) = mx$ from 0 to h around the x-axis, as shown in Figure 7.2a. A cylinder is formed when the area under a horizontal line from a to b is revolved around the x-axis as shown in Figure 7.2b. A soft drink bottle is formed by revolving the area under a curve, such as the one shown in Figure 7.2c.

Disks

In this section and in Section 7.3, we will develop ways for finding the volume of solids of revolution such as these. As when we developed the definite integral, we will let $y = f(x)$ be a continuous function with $f(x) \geq 0$ on the interval $[a, b]$. If we rotate the area under this curve around the x-axis, we generate a solid of revolution. If we cut this solid with parallel planes all perpendicular to the x-axis, we will see that a typical cross-section is a circle and a typical slice is a disk as shown in Figure 7.3a.

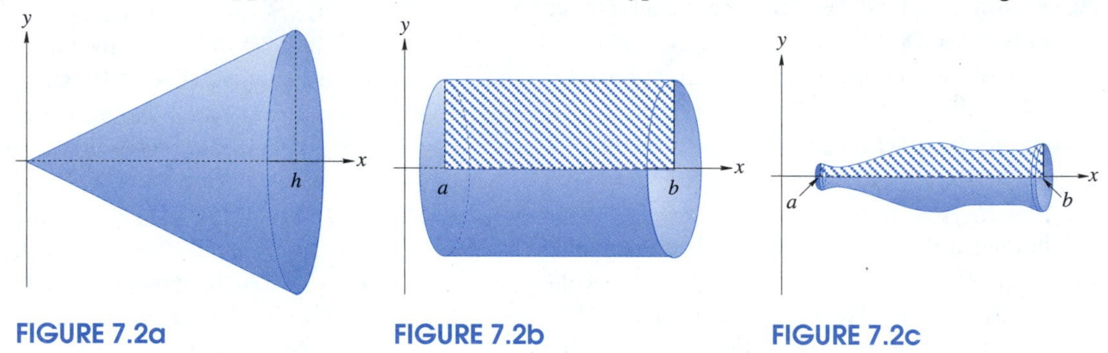

FIGURE 7.2a FIGURE 7.2b FIGURE 7.2c

If we divide $[a, b]$ into subintervals and let x_i be a point in one of the subintervals, then a typical disk is a cylinder with height dx and radius $f(x_i)$ as shown in Figure 7.3b. The volume of a cylinder is $\pi r^2 h$, so the volume of this disk is $\pi[f(x_i)]^2\,dx$. Summing the volume of all of these disks over the interval $[a, b]$, we get the total volume as shown in the following box.

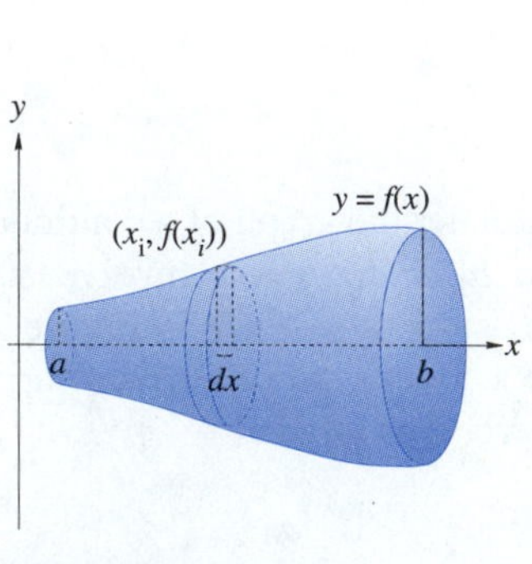

FIGURE 7.3a

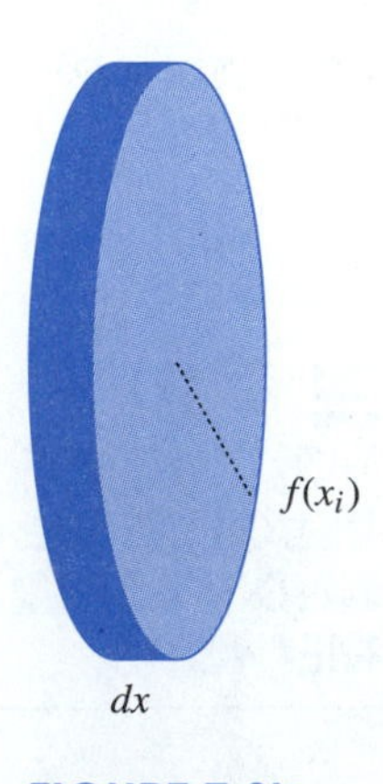

FIGURE 7.3b

Disk Method for Finding a Volume of Revolution (x-Axis)

The volume V of the solid generated by revolving the area between the graph of $y = f(x)$ and the x-axis from $x = a$ to $x = b$ around the x-axis is

$$V = \pi \int_a^b [f(x)]^2 \, dx$$

It is often easier to remember the disk method by using the following guidelines:

1. Sketch the graph over the limits of integration.
2. Draw a typical disk by slicing the solid perpendicular to the axis of revolution.
3. Determine the radius and thickness of the disk.
4. Determine the volume of a typical disk by using $\pi\,(\text{radius})^2 \cdot (\text{thickness})$.
5. Use the following integration formula to find the total volume:

$$V = \pi \int_a^b (\text{radius})^2 \cdot (\text{thickness})$$

EXAMPLE 7.9

Find the volume of the solid of revolution generated by revolving the area under the graph of $f(x) = \sqrt{x}$ from $x = 0$ to $x = 4$ around the x-axis.

Solution The graph of $f(x) = \sqrt{x}$ on $[0, 4]$ is given in Figure 7.4a. A rectangle has been drawn to show a typical element.

The solid that results from revolving the graph of f around the x-axis is shown in Figure 7.4b, with a disk formed by the typical element. The radius of the disk is $y = \sqrt{x}$, and its thickness is dx. A typical disk is shown in Figure 7.4c. Its volume is

$$\pi\,(\text{radius})^2 \cdot (\text{thickness}) = \pi \left(\sqrt{x}\right)^2 dx$$

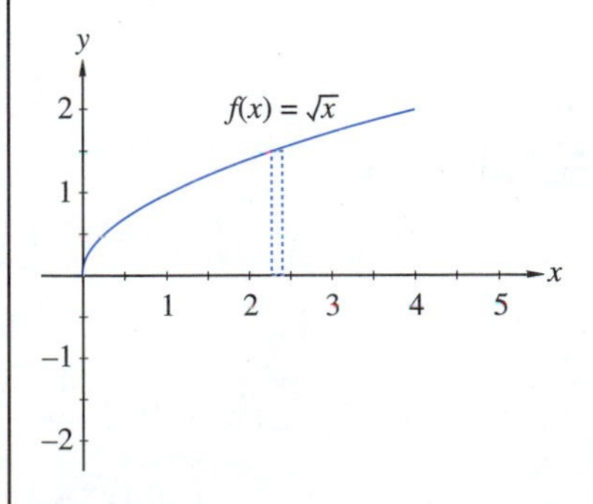

FIGURE 7.4a

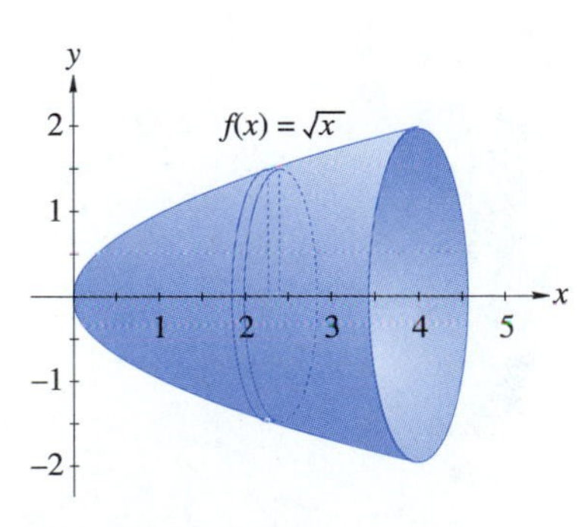

FIGURE 7.4b

EXAMPLE 7.9 (Cont.)

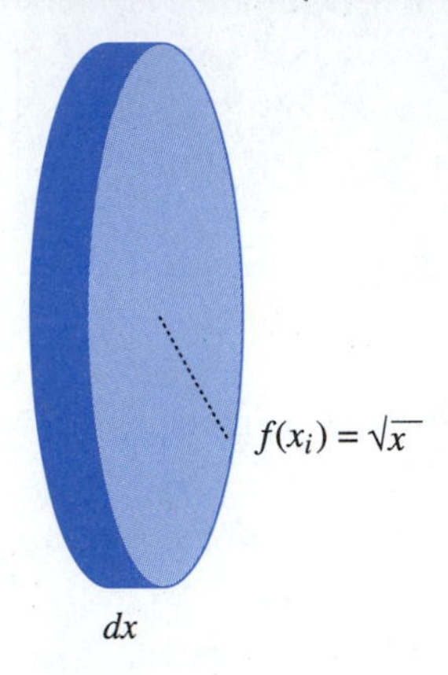

FIGURE 7.4c

Thus, we get the following integral.

$$
\begin{aligned}
V &= \pi \int_0^4 (\sqrt{x})^2\,dx \\
&= \pi \int_0^4 x\,dx \\
&= \pi \left(\frac{1}{2}x^2\right)\Bigg|_0^4 = 8\pi
\end{aligned}
$$

The volume is 8π units^3.

Application

EXAMPLE 7.10

A wing tank for an airplane is a solid of revolution formed by rotating the region bounded by the graph of $y = \frac{1}{16}x\sqrt{4-x}$ and the x-axis from $x = 0$ to $x = 4$ around the x-axis, where x and y are measured in meters, as shown in Figure 7.5a. Find the volume of the tank.

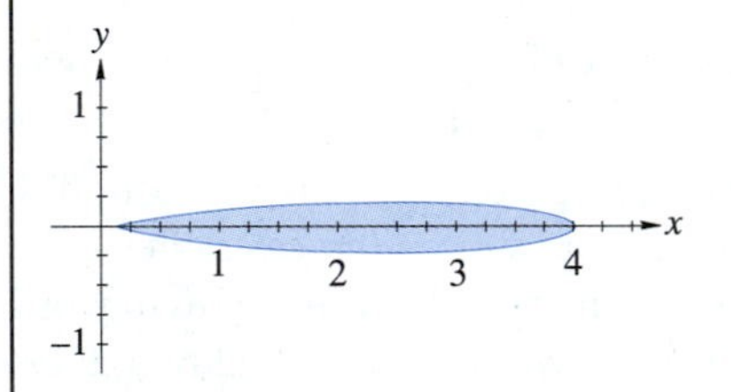

FIGURE 7.5a

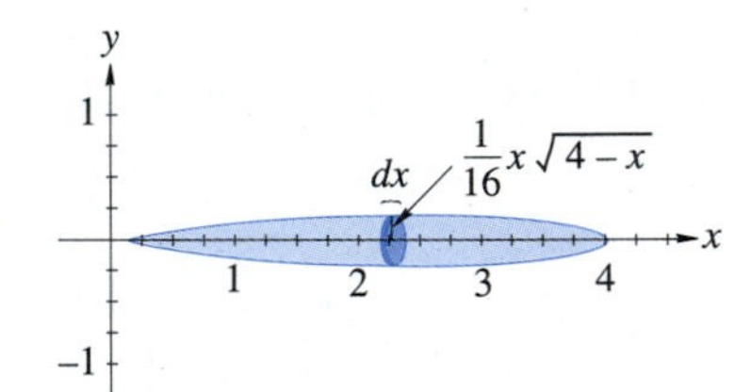

FIGURE 7.5b

Solution A disk formed from a typical element is shown in Figure 7.5b. The radius of the disk is $\frac{1}{16}x\sqrt{4-x}$ and its thickness is dx. Thus the volume of a typical disk is

$$\pi\,(\text{radius})^2 \cdot (\text{thickness}) = \pi\left(\frac{1}{16}x\sqrt{4-x^2}\right)^2 dx$$

and the total volume is

$$
\begin{aligned}
V &= \pi \int_0^4 \left(\frac{1}{16}x\sqrt{4-x}\right)^2 dx \\
&= \pi \int_0^4 \frac{1}{256}x^2(4-x)\,dx \\
&= \frac{\pi}{256}\int_0^4 (4x^2 - x^3)\,dx
\end{aligned}
$$

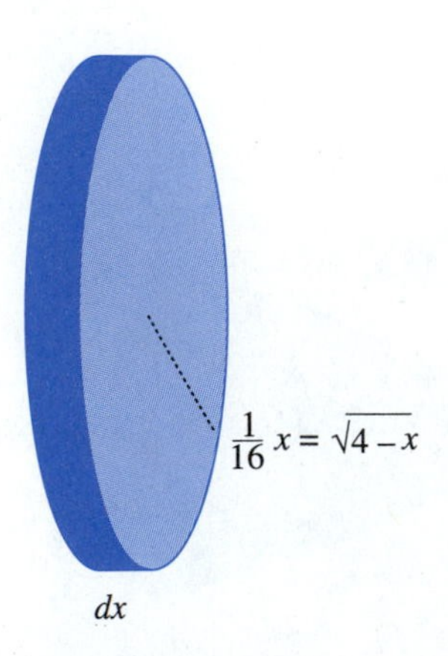

FIGURE 7.5c

EXAMPLE 7.10 (Cont.)

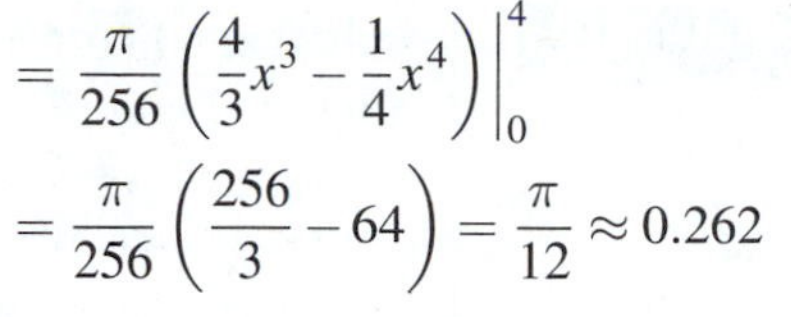

$$= \frac{\pi}{256}\left(\frac{4}{3}x^3 - \frac{1}{4}x^4\right)\Bigg|_0^4$$

$$= \frac{\pi}{256}\left(\frac{256}{3} - 64\right) = \frac{\pi}{12} \approx 0.262$$

The volume of this wing tank is approximately $0.262\,\text{m}^3$.

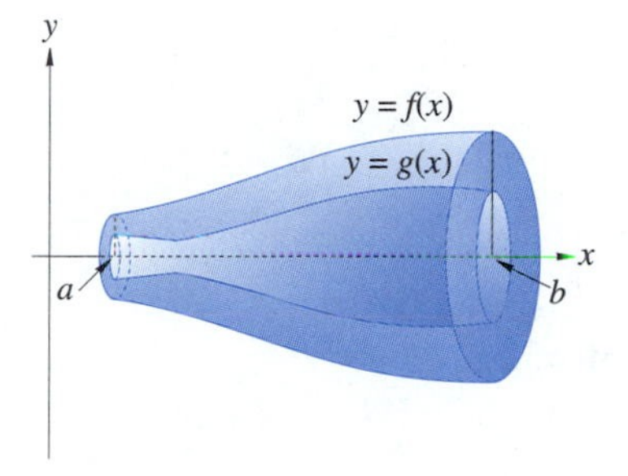

FIGURE 7.6a

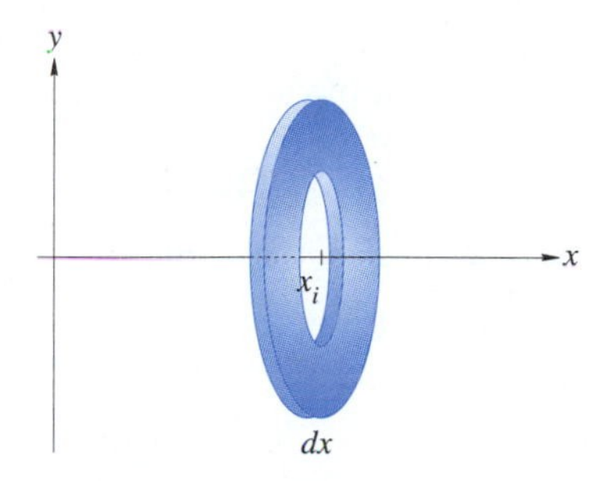

FIGURE 7.6b

Washers

In the disk method, we revolved the area between one curve and the x-axis around the x-axis. But, what happens if we revolve the area between two curves around the x-axis? In Figure 7.6a, we revolved the area enclosed by $y = f(x)$ and $y = g(x)$, where $f(x) \geq g(x)$, and between the lines $x = a$ and $x = b$.

If we sliced this as we did in the disk method, we would get a disk with a hole in the middle. Thus, the slice would look like a washer, as shown in Figure 7.6b. As before, the height of this washer is dx. Its volume could be determined by finding the volume of the solid disk formed by $y = f(x)$ and subtracting the disk that forms the hole in the washer, $y = g(x)$. The area of the entire solid disk is $\pi[f(x_i)]^2\,dx$ and of the hole is $\pi[g(x_i)]^2\,dx$. Thus, the area of the washer is

$$\pi[f(x)]^2\,dx - \pi[g(x)]^2\,dx = \pi[f(x)]^2 - [g(x)]^2\,dx$$

Summing this over the entire interval, we get a second method for finding a volume of revolution.

Washer Method for Finding a Volume of Revolution

The volume V of the solid generated by revolving the area between the graph of $y = f(x)$ and the graph of $y = g(x)$ from $x = a$ to $x = b$ around the x-axis is

$$V = \pi \int_a^b \left\{[f(x)]^2 - [g(x)]^2\right\} dx$$

where $f(x) =$ the outer radius and $g(x) =$ the inner radius of the washer.

EXAMPLE 7.11

Find the volume of the solid of revolution generated by revolving the area enclosed by $y = x^2$ and $y = 4x$ around the x-axis.

EXAMPLE 7.11 (Cont.)

Solution Let $f(x) = x^2$ and $g(x) = 4x$. The graph of these two functions is shown in Figure 7.7. The two graphs intersect when $f(x) = g(x)$ or when $x^2 = 4x$, which is at $x = 0$ and $x = 4$. On the interval $[0, 4]$, $4x \geq x^2$, the outer radius is $f(x) = x^2$, and the inner radius is $g(x) = 4x$. We determine the volume as

$$\begin{aligned} V &= \pi \int_0^4 \left[(4x)^2 - (x^2)^2\right] dx \\ &= \pi \int_0^4 (16x^2 - x^4) dx \\ &= \pi \left(\frac{16}{3}x^3 - \frac{1}{5}x^5\right)\Bigg|_0^4 \\ &= \pi \left(\frac{2{,}048}{15}\right) \approx 136.53\pi \end{aligned}$$

The volume is about 136.53π units^3.

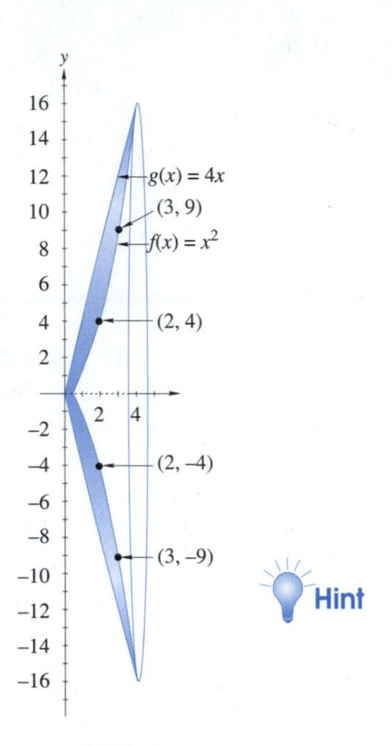

FIGURE 7.7

Hint

The following guidelines are a variation of those for the disk method. Following them should make it easier to remember the washer method.

1. Sketch the graphs of the two functions and find the limits of integration.
2. Draw a typical washer, as in Figure 7.8, and remember to slice the solid perpendicular to the axis of revolution.
3. Find the outer and inner radii for the washer.
4. Determine the thickness of the washer.
5. The volume of a typical washer is

$$\pi\left[(\text{outer radius})^2 - (\text{inner radius})^2\right] \cdot (\text{thickness})$$

6. Use the following integration formula to find the total volume:

$$V = \pi \int_a^b \left[(\text{outer radius})^2 - (\text{inner radius})^2\right] \cdot (\text{thickness})$$

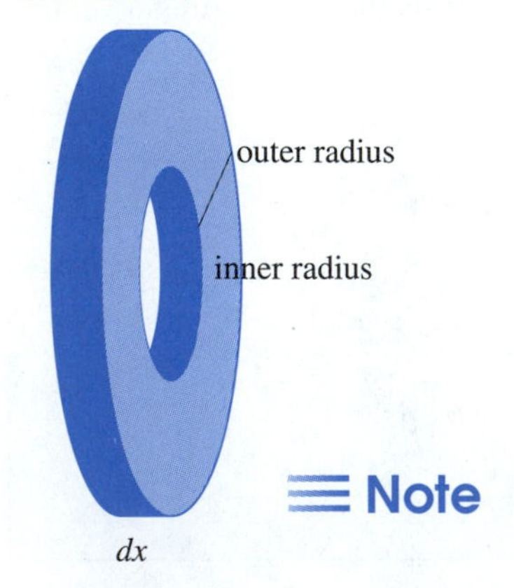

FIGURE 7.8

Note

Be sure to label the volume in appropriate cubic units.

Application

EXAMPLE 7.12

A drafter is designing a plastic bead that can be strung to help form a necklace. The bead is in the shape of a sphere with a hole drilled through the center, as shown in Figure 7.9a. If the sphere has a radius of 0.25 in. and the hole a radius of 0.07 in., how much plastic will be needed to make one bead?

EXAMPLE 7.12 (Cont.)

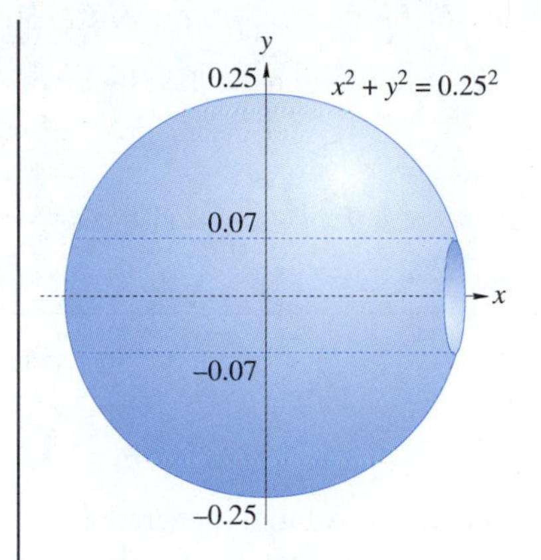

FIGURE 7.9a

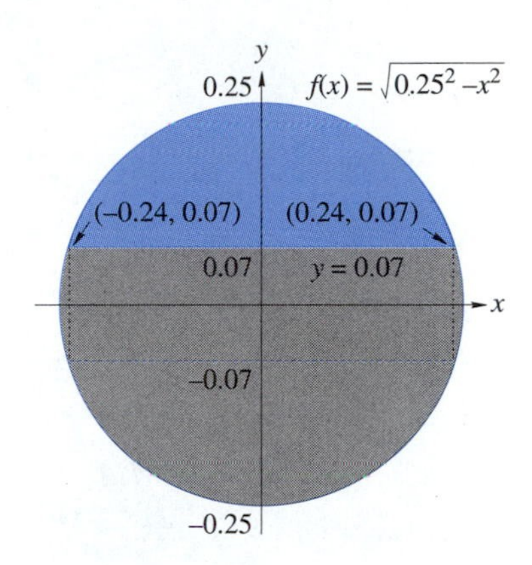

FIGURE 7.9b

Solution We will sketch this so the origin is at the center of the sphere. A cross-section of the bead, as in Figure 7.9b, shows that the sphere is formed by rotating the top half of the circle $x^2 + y^2 = 0.25^2$, so the outer radius is $f(x) = \sqrt{0.25^2 - x^2}$. The hole is formed by rotating the region formed by the line $y = 0.07$, so the inner radius is $g(x) = 0.07$. If we let $y = 0.7$ and we solve the equation $x^2 + y^2 = 0.25^2$, we get the points of intersection at $x = \pm 0.24$ in. Since the thickness is dx, we have all the information we need, and the volume is given by

$$\begin{aligned} V &= \pi \int_{-0.24}^{0.24} \left[\left(\sqrt{0.25^2 - x^2} \right)^2 - (0.07)^2 \right] dx \\ &= \pi \int_{-0.24}^{0.24} (0.0625 - x^2 - 0.0049)\, dx \\ &= \pi \int_{-0.24}^{0.24} (0.0576 - x^2)\, dx \\ &= \pi \left[0.0576x - \frac{x^3}{3} \right]_{-0.24}^{0.24} \\ &= \frac{\pi}{3}(0.055296) = 0.018432\pi \approx 0.0579 \end{aligned}$$

It will take about 0.0579 in.^{3} of plastic to make each bead.

Revolving Around the *y*-Axis

Until now we have found the volumes by revolving regions around the x-axis. It is possible to revolve a region around any line. We will look at revolving the area enclosed by $x = g(y)$, $y = c$, and $y = d$, where $g(y) \geq 0$, around the y-axis, as shown in Figure 7.10. The process is the same as when a region is revolved around the x-axis. Again, slices are taken, only this time they are perpendicular to the y-axis. A typical disk is shown in Figure 7.10. The height of the disk is dy and its radius is $g(y_i)$, so the volume of the disk is $\pi[g(y_i)]^2\, dy$ and the total volume is given in the following box.

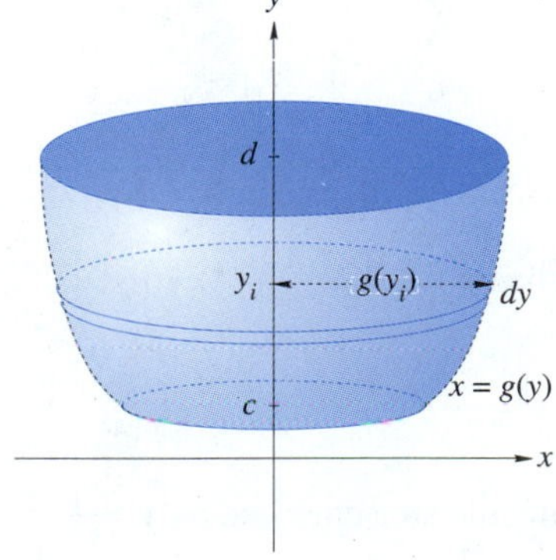

FIGURE 7.10

Disk Method for Finding a Volume of Revolution (y-Axis)

The volume V of the solid generated by revolving the area between the graph of $x = g(y)$ and the y-axis from $y = c$ to $y = d$ around the y-axis is

$$V = \pi \int_c^d (\text{radius})^2 \cdot (\text{thickness}) = \pi \int_c^d [g(y)]^2 \, dy$$

EXAMPLE 7.13

Find the volume of revolution generated by revolving the area enclosed by $y = \frac{1}{2}x^2$, $y = 0$, and $y = 8$ around the y-axis.

Solution Since the region is to be revolved around the y-axis, we will draw, horizontally, the rectangle that forms a typical disk as shown in Figure 7.11. Since this is revolved around the y-axis, we need to describe the curve as a function of y. Solving $y = \frac{1}{2}x^2$ for x, we get $x = g(y) = \sqrt{2y}$. Thus, the volume is

$$V = \pi \int_0^8 (\sqrt{2y})^2 \, dy = \pi \int_0^8 2y \, dy = \pi y^2 |_0^8 = 64\pi$$

The volume is 64π units³.

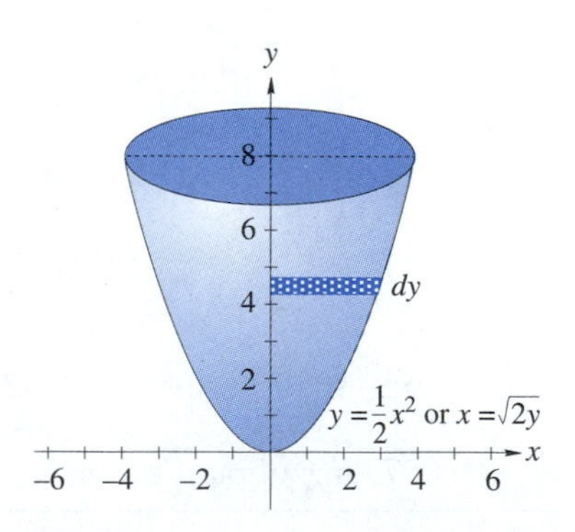

FIGURE 7.11

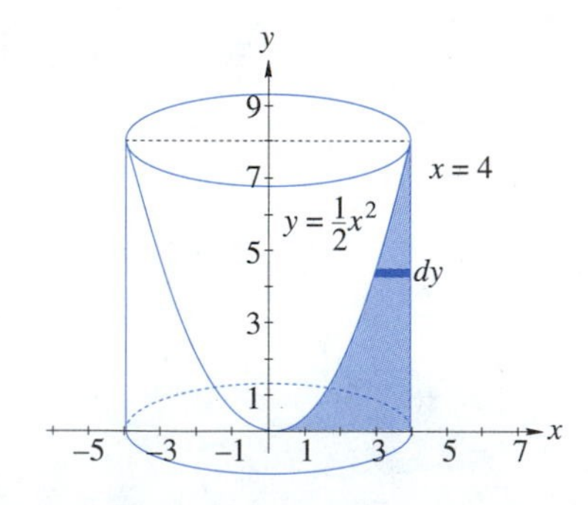

FIGURE 7.12

EXAMPLE 7.14

Find the volume of revolution generated by revolving the area enclosed by $y = \frac{1}{2}x^2$, $x = 0$, $x = 4$, and the x-axis around the y-axis.

Solution The graph of the region that is being revolved is shown in Figure 7.12. The point of intersection of $x = 4$ and $y = \frac{1}{2}x^2$ is $y = 8$. A typical element would

EXAMPLE 7.14 (Cont.)

form a washer, so its volume is $[4^2 - (\sqrt{2y})^2]dy$ and the total volume is

$$V = \pi \int_0^8 \left[(4)^2 - (\sqrt{2y})^2\right] dy = \pi \int_0^8 (16 - 2y)dy$$
$$= \pi \left(16y - y^2\right)\Big|_0^8 = 64\pi$$

The volume is 64 units³.

Application

EXAMPLE 7.15

A hemispherical water tank, as shown in Figure 7.13a, has a radius of 4 m. The water is 1.5 m deep at the center. How much water is in the tank?

Solution We will sketch a cross-section of the tank with the center at the origin of the quarter-circle used to generate the hemisphere, as shown in Figure 7.13b. A typical element has been sketched in the figure. The length of this element is determined by solving the equation for the circle, $x^2 + y^2 = 4^2$ for x, with the result $x = \sqrt{4^2 - y^2}$. The thickness of the element is dy. Since the elements will move from the bottom of the tank, when $y = -4$, to the top of the water, when $y = -2.5$, we have the limits of integration. We see that the volume is

$$V = \pi \int_{-4}^{-2.5} \left(\sqrt{4^2 - y^2}\right)^2 dy$$
$$= \pi \int_{-4}^{-2.5} \left(16 - y^2\right) dy$$
$$= \pi \left[16y - \frac{y^3}{3}\right]_{-4}^{-2.5}$$
$$= \pi \left(-\frac{835}{24} + \frac{128}{3}\right) = \left(\frac{189}{24}\right)\pi = \frac{63}{8}\pi \approx 24.74$$

There is about 24.74 m³ of water in the tank.

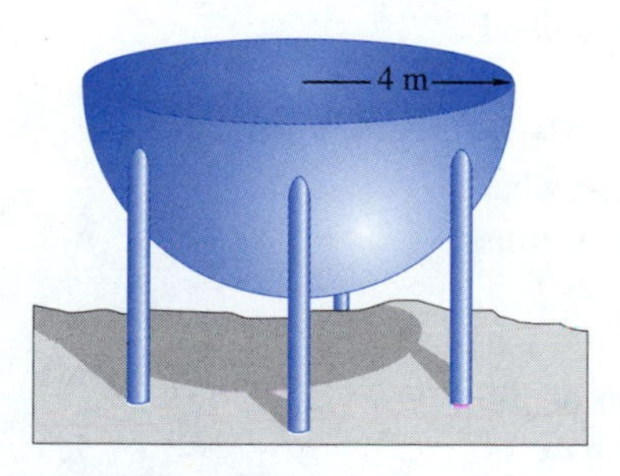

FIGURE 7.13a

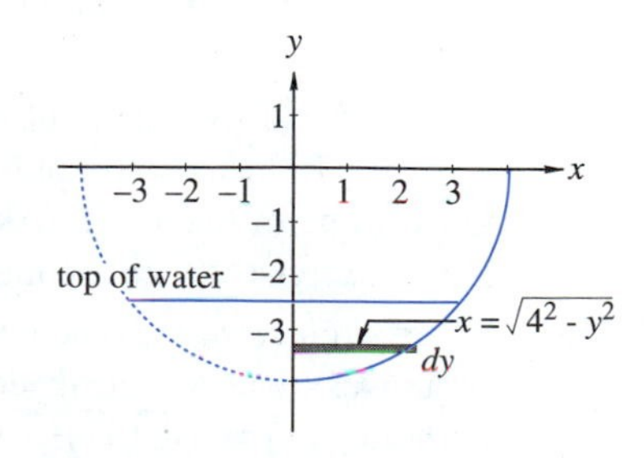

FIGURE 7.13b

Exercise Set 7.2

In Exercises 1–19, find the volume of the solid of revolution generated by revolving the enclosed region about the indicated axis.

1. $y = 4x$, $x = 1$, $x = 5$, $y = 0$, about the x-axis
2. $y = x^2$, $x = 0$, $x = 2$, $y = 0$, about the x-axis
3. $y = x$, $x = 4$, $y = 0$, about the x-axis
4. $y = 6 - x$, $x = 0$, $y = 0$, about the x-axis
5. $y = \sqrt{x+1}$, $x = 1$, $x = 7$, $y = 0$, about the x-axis
6. $y = x^3$, $x = 2$, $y = 1$, about the x-axis
7. $y = x^2$, $y = 4$, $x = 0$, about the y-axis
8. $x + y = 5$, $y = 0$, $x = 0$, about the y-axis
9. $y = x$, $y = 4 - x$, $x = 0$, about the x-axis
10. $y = \sqrt[3]{x}$, $x = 0$, $y = 8$, about the y-axis
11. $y = \sqrt[3]{x}$, $x = 8$, $y = 0$, about the y-axis
12. $y = \sqrt{x}$, $y = x^2$, about the y-axis
13. $y = 4 - x^2$, $y = 8 - 2x^2$, about the x-axis
14. $y = x^2$, $y = x^3$, about the y-axis
15. $y = x^2$, $y = x^3$, about the x-axis
16. $y = x^2$, $y = 4x$, about the x-axis
17. $y = x^2$, $y = 4x$, about the y-axis
18. $y = \sqrt{8 - x^3}$, $x = 0$, $y = 0$, about the x-axis
19. $y = \sqrt{8 - x^2}$, $x = 0$, $y = 0$, about the y-axis

Solve Exercises 20–28.

20. Derive the formula for the volume of a cone of radius r and height h.
21. Derive the formula for the volume of a sphere of radius r.
22. The equation of an ellipse centered at the origin is
$$\frac{x^2}{a^2} + \frac{y^2}{b^2} = 1$$
Calculate the volume of the solid generated by revolving this ellipse around the x-axis.
23. *Electronics* A parabolic reflector is formed by rotating a parabola around its axis. Suppose a satellite television antenna is a parabolic reflector with a diameter of 3 m and a depth of 1 m. **(a)** Find the equation of this parabola. **(b)** What is the volume contained by the reflector?
24. *Electronics* If the "skin" of the parabolic reflector in Exercise 19 has a constant 0.4 cm thickness, what is the volume of the skin? (Assume the dimensions given in Exercise 18 are the interior dimensions.)
25. *Nuclear engineering* The cooling towers at a nuclear power plant are designed in the shape of the hyperbola $g(y) = \sqrt{(147)^2 + 0.16y^2}$. The base of the tower is 442 ft below the vertex and the top is 123 ft above the vertex as shown in Figure 7.14. If we assume the hyperbola is rotated around the y-axis, what is the volume of the cooling tower?
26. *Nuclear engineering* If the walls of the cooling tower in Exercise 25 are a constant 6 in. thick, then the interior walls are in the shape of a hyperbola with the equation $x = \sqrt{146.5^2 + 0.16y^2}$ that is rotated around the y-axis. What is the volume of the walls of the cooling tower?
27. *Wastewater technology* The tank on a water tower is a sphere of radius 18 m.
   (a) Determine the volume of water when the water is 4 m deep at the center.
   (b) Determine the volume of water when the water is 24 m deep at the center.

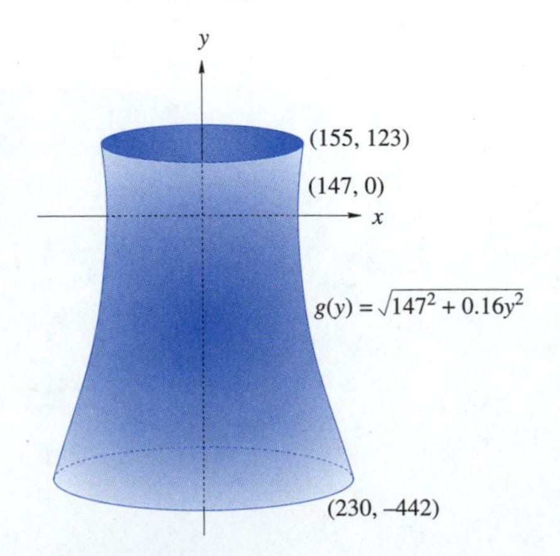

FIGURE 7.14

28. *Mechanical technology* A flange is formed in the shape of a solid of revolution generated by rotating about the x-axis the region bounded by a graph of $y = x^2 + 5$, where x and y are in centimeters; the y-axis; and the lines $y = 2$ and $x = 4$. Determine the mass of this flange if it is made of a material of density $0.016\,\text{kg/cm}^3$. (Remember: mass is the product of the density and volume.)

In Your Words

29. Describe how to find the volume of a solid of revolution by using the disk method.
30. Describe how to find the volume of a solid of revolution by using the washer method.
31. Describe the changes that have to be made when a volume of revolution is formed by revolving a region around the y-axis rather than around the x-axis.

7.3 VOLUMES OF REVOLUTION: SHELL METHOD

In Section 7.2, we found the volume of a solid of revolution by the disk and washer methods. In this section, we will learn how to find these volumes by another method—the shell method. The shell method is based on finding the volume of cylindrical shells and adding those volumes to get the total volume.

An example of a cylindrical shell is a water pipe. A cylindrical shell has some thickness to it. In order to find its volume, we need to find the inner radius r_1 of the shell and the outer radius r_2, as shown in Figure 7.15. If the shell has height h, then its volume is

$$V = \pi(r_2)^2h - \pi(r_1)^2h = \pi h(r_2{}^2 - r_1{}^2)$$

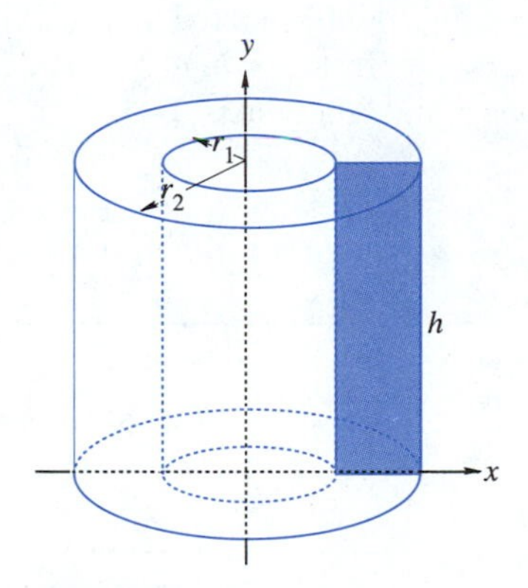

FIGURE 7.15

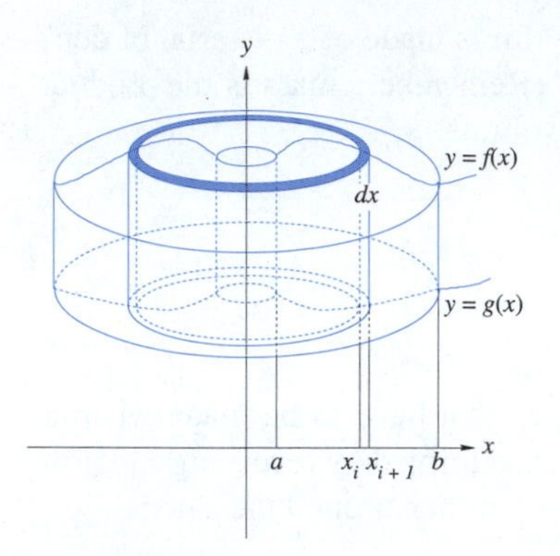

FIGURE 7.16

Some extra algebraic work will change this to a more useful formula.

$$V = \pi h(r_2{}^2 - r_1{}^2) = \pi h(r_2 + r_1)(r_2 - r_1) = 2\pi h\left(\frac{r_2 + r_1}{2}\right)(r_2 - r_1)$$

This formula is very involved, but we will show you a way to remember it after we work Example 7.16. In this last statement, $\frac{r_2 + r_1}{2}$ represents the average radius of the shell and $r_2 - r_1$ represents the thickness.

If we have two curves, $y = f(x)$ and $y = g(x)$, with $f(x) \geq g(x)$, that are continuous over the interval $[a, b]$ with $a \geq 0$, and we revolve the area between these curves around the y-axis, we get a solid of revolution similar to the one in Figure 7.16. Next we divide the interval $[a, b]$ into subintervals. Each subinterval will have the width $r_2 - r_1 = dx$, height $f(x_i) - g(x_i)$, and average radius $\frac{x_{i-1} + x_i}{2}$. Thus, when this subinterval is revolved around the y-axis, it has a volume of

$$2\pi[f(x_i) - g(x_i)]\left(\frac{x_{i-1} + x_i}{2}\right)dx$$

If dx is small enough $\left(\frac{x_{i-1} + x_i}{2} \approx x_i\right)$, and if we sum all the partitions over the interval $[a, b]$, we get the following formula.

Shell Method for Finding a Volume of Revolution (y-Axis)

The volume V of the solid generated by revolving the area between the graph of $y = f(x)$ and $y = g(x)$ from $x = a$ to $x = b$, where a and b are both on the same side of the y-axis around the y-axis, is

$$V = 2\pi \int_a^b x[f(x) - g(x)]\,dx$$

EXAMPLE 7.16

Find the volume of the solid generated by revolving around the y-axis the region between $y = 5x^3 - x^4$ and the x-axis.

Solution The graph of this region is shown in Figure 7.17. (Notice that the scale is not the same on the x- and y-axes.) These two curves, $y = 0$ and $y = 5x^3 - x^4$,

EXAMPLE 7.16 (Cont.)

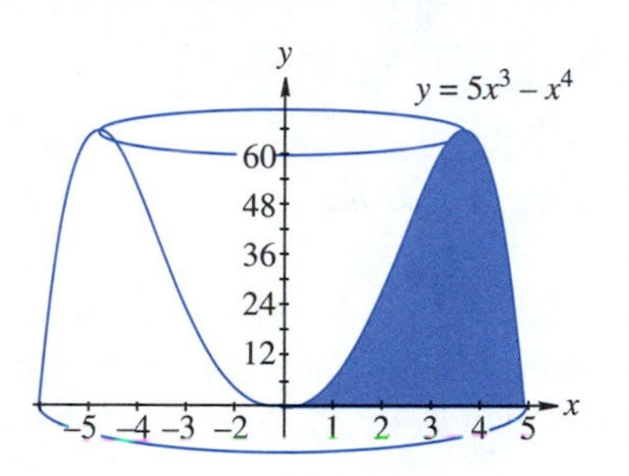

FIGURE 7.17

intersect at $x = 0$ and at $x = 5$; so, $a = 0$ and $b = 5$. Using the new formula, we see that

$$\begin{aligned} V &= 2\pi \int_0^5 x[(5x^3 - x^4) - 0]\,dx \\ &= 2\pi \int_0^5 (5x^4 - x^5)dx = 2\pi \left(x^5 - \frac{1}{6}x^6\right)\Bigg|_0^5 \\ &= 2\pi \left(3{,}125 - \frac{15{,}625}{6}\right) = 1{,}041\frac{2}{3}\pi \end{aligned}$$

The volume is $1{,}041\frac{2}{3}\pi$ units^3.

It would have been possible to solve Example 7.16 by using the washer method. But, that would have meant solving the equation $y = 5x^3 - x^4$ for x in order to get a function of y. This would have been a very difficult and complicated task. The shell method provides a much easier alternative.

It is easier to remember the shell method by using the following guidelines:

1. Sketch the graph over the limits of integration.
2. Draw a typical shell parallel to the axis of revolution.
3. Determine the radius, height, and thickness of the shell.
4. The volume of a typical shell is 2π(radius) · (height) · (thickness).
5. Use the following integration formula to determine the total volume:

$$V = 2\pi \int_a^b (\text{radius}) \cdot (\text{height}) \cdot (\text{thickness})$$

EXAMPLE 7.17

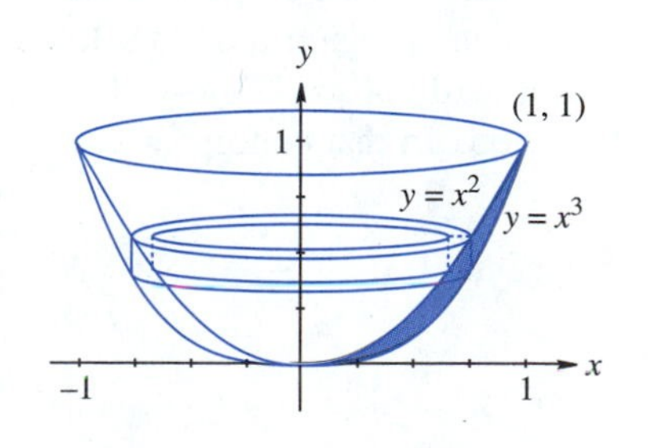

FIGURE 7.18

Find the volume of the solid generated by revolving around the y-axis the region in the first quadrant that is between $y = x^2$ and $y = x^3$.

Solution These two curves intersect at $(0, 0)$ and $(1, 1)$ as shown in Figure 7.18, so $a = 0$ and $b = 1$. On this interval, $x^2 \geq x^3$ and so, using the shell method, the volume is

$$\begin{aligned} V &= 2\pi \int_0^1 x(x^2 - x^3)dx \\ &= 2\pi \int_0^1 (x^3 - x^4)dx \\ &= 2\pi \left(\frac{1}{4}x^4 - \frac{1}{5}x^5\right)\Bigg|_0^1 \\ &= 2\pi \left(\frac{1}{4} - \frac{1}{5}\right) = \frac{\pi}{10} \end{aligned}$$

The volume is $\frac{\pi}{10}$ units^3.

We can just as easily use the shell method to revolve a region around the x-axis. When this is done, the function will have to be written to ensure that it is a function of y.

EXAMPLE 7.18

Find the volume of the solid generated by revolving around the x-axis the region bounded by $y=\sqrt{x}$, $y=0$, and $x=4$.

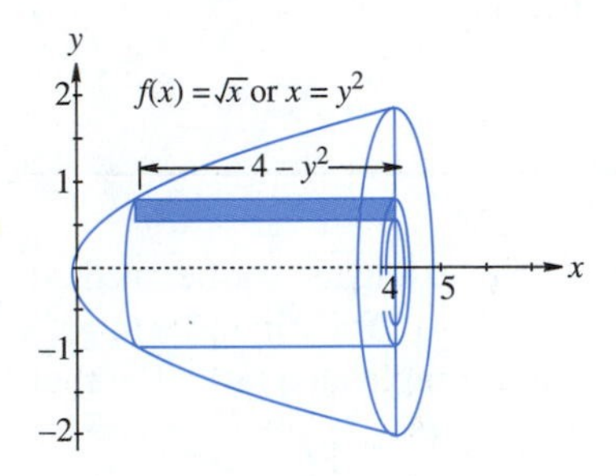

FIGURE 7.19

Solution The graph of this region is shown in Figure 7.19. In this instance, we have two curves that must be written as functions of y. The first is $x=4$ and the other, $y=\sqrt{x}$, is $x=y^2$. Thus, the height of each shell is $4-y^2$ and the radius is y. Then the volume is

$$\begin{aligned} V &= 2\pi \int_0^2 y(4-y^2)dy \\ &= 2\pi \int_0^2 (4y-y^3)dy \\ &= 2\pi \left(2y^2 - \frac{1}{4}y^4\right)\Big|_0^2 \\ &= 2\pi(8-4) = 8\pi \end{aligned}$$

The volume is 8π units^3.

Compare the result of Example 7.18 to that of Example 7.9, where we found the volume of the same solid by using the disk method.

Rotation about a Noncoordinate Axis

As we said, it is possible to rotate a region around any line in order to get a desired volume of revolution. In the next example, we will rotate a region around a line parallel to the y-axis. The formula for the shell method will not work. Instead, we need to rely on the basic concept behind the shell method. In that concept we saw that

$$V = 2\pi(\text{radius})(\text{height})(\text{thickness})$$

EXAMPLE 7.19

Find the volume of the solid formed by revolving the region bounded by $y=x^2+2$ and $y=x+8$ around the line $x=4$.

Solution The region is shown in Figure 7.20. Since this region is revolved around the line $x=4$, the radius of a typical shell is $4-x$. The two curves, $y=x+8$ and $y=x^2+2$, intersect at $(-2,6)$ and $(3,11)$. In the interval $[-2,3]$ we can see that

EXAMPLE 7.19 (Cont.)

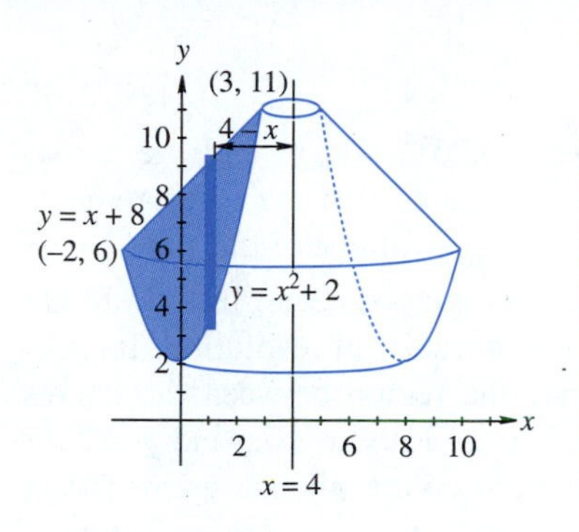

FIGURE 7.20

$x+8 \geq x^2+2$ and so the height of a typical shell is $(x+8)-(x^2+2)=6+x-x^2$. Thus, the volume of a typical shell is

$$\begin{aligned} V &= 2\pi \int_{-2}^{3} (6+x-x^2)(4-x)dx \\ &= 2\pi \int_{-2}^{3} (24-2x-5x^2+x^3)dx \\ &= 2\pi \left(24x - x^2 - \frac{5}{3}x^3 + \frac{1}{4}x^4 \right)\Big|_{-2}^{3} \\ &\approx 2\pi\,[38.25-(-34.67)] = 145.84\pi \end{aligned}$$

The volume is 145.84π units^3. ▪

Summary of Disk and Shell Methods

In Sections 7.2 and 7.3, we have shown two methods for finding the volume of a solid of revolution. These two methods are summarized in Table 26.1.

TABLE 7.1 Summary of Disk and Shell Methods

	Disk or Washer Method	Shell Method
Revolution about x-axis	Partition the x-axis. Use vertical strips of width dx. Integrate with respect to x.	Partition the y-axis. Use horizontal strips of width dy. Integrate with respect to y.
Revolution about y-axis	Partition the y-axis. Use horizontal strips of width dy. Integrate with respect to y.	Partition the x-axis. Use vertical strips of width dx. Integrate with respect to x.

Exercise Set 7.3

In Exercises 1–18, use the shell method to obtain the volume of the solid of revolution generated by revolving the enclosed region about the indicated axis.

1. $y=x^2$, $y=0$, $x=2$, about the y-axis
2. $x=y^3$, $x=8$, about the y-axis
3. $y=x^2-1$, $x=1$, $y=8$, about the y-axis
4. $y=x^3+x$, $x=0$, $x=2$, about the y-axis
5. $y=25-x^2$, $y=0$, about the y-axis
6. $y=x^2$, $y=8-x^2$, about the y-axis
7. $y=x^3$, $x=0$, $y=8$, about the x-axis
8. $y=x^4$, $y=x^3$, about the x-axis
9. $y=\sqrt[3]{x}$, $x=0$, $y=2$, about the x-axis
10. $x=\sqrt[3]{y^2}$, $x=0$, $y=8$, about the x-axis
11. $y=x^{1/2}-x^{3/2}$, $y=0$, about the y-axis
12. $x=y$, $x+2y=3$, $y=0$, about the x-axis
13. $y=2x^3$, $y^2=4x$, about the x-axis
14. $x=4y-y^3$, $x=0$, about the x-axis

15. $y = x^2$, $y = 2x$, about the line $y = 6$

16. $y = 2 - x$, $y = x^2$, $y = 1$, about the x-axis (the area above $y = 1$)

17. $y = 2 - x$, $y = x^2$, about $x = 1$

18. $y = 4 - x^2$, $y = 0$, about the line $x = 4$

Solve Exercises 19–22.

19. A hole of radius 2 is drilled vertically through the center of a sphere of radius 5. Find the volume removed by the drilling.

20. *Civil engineering* A water tank is in the shape of a sphere with a radius of 10 m. **(a)** What is the volume of water in the tank if it is filled to a depth of 3 m? **(b)** If the density of water is 1 000 kg/m^3, what is the mass of the water in the tank?

21. *Sheet metal technology* The inside of a bundt pan, used for baking, is designed by revolving the region between $y = -(x - 3.25)^4 + \left(\frac{5}{4}\right)^4$ and the x-axis around the y-axis. Assume that the measurements are in inches, and find the volume of this pan.

22. *Industrial design* The glass reflector for a flashlight is made in the shape of a solid of revolution. It is obtained by revolving the region between the curves $x = 0.5y^2$, $x = 0.5y^2 + 0.1$, the x-axis, and $y = 2$ for $0 \le x \le 2$ around the x-axis. If all measurements are in centimeters, what is the volume of the reflector?

In Your Words

23. Describe how to find the volume of a solid of revolution by using the shell method.

24. Describe the changes that have to be made when a volume of revolution is formed using the shell method rather than either the disk or washer method.

25. Draw an example of a region for which the washer method is more appropriate than the shell method when the region is revolved around the y-axis. Explain why the washer method is more appropriate.

26. Draw an example of a region for which the shell method is more appropriate than the washer method when the region is revolved around the y-axis. Explain why the shell method is more appropriate.

7.4 ARC LENGTH AND SURFACE AREA

In this section, we will look at two related ideas. One is the length of a curve; the second is the area of the surface that is formed when that curve is rotated around an axis.

One useful application for the length of a curve is in determining the lengths of the cables that are used to support a suspension bridge. The cables between the towers are parabolic in shape and so this results in finding the length of a parabola between two values.

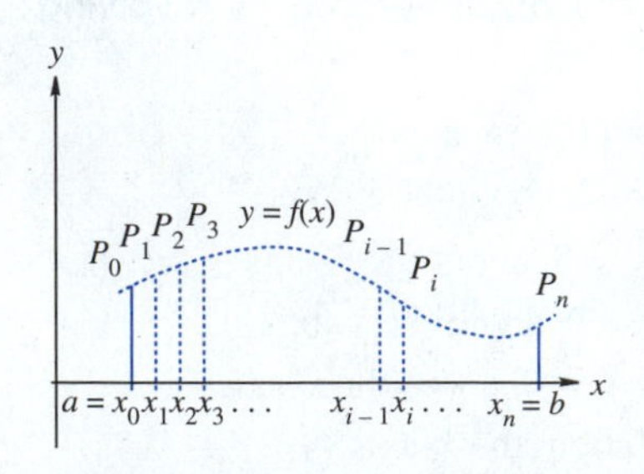

FIGURE 7.21

Consider a curve such as $y = f(x)$, as shown in Figure 7.21. We would like to find the length of this curve from $x = a$ to $x = b$. As before, we will partition the interval $[a, b]$ into n subintervals. With each subinterval $x_0, x_1, x_2, \ldots, x_n$ there is a corresponding point on the graph $P_0, P_1, P_2, \ldots, P_n$. Each point P_i has the coordinates $(x_i, f(x_i))$. If we connect these points in succession, we get a series of line segments that approximate the curve. The sum of the lengths of these segments will approximate the length of the curve. This sum L can be written as

$$L = d(P_0, P_1) + d(P_1, P_2) + d(P_2, P_3) + \cdots + d(P_{n-1}, P_n)$$
$$= \sum_{i=1}^{n} d(P_{i-1}, P_i)$$

where $d(P_{i-1}, P_i)$ is the length of the segment from P_{i-1} to P_i.

Using the distance formula, we get

$$d(P_{i-1}, P_i) = \sqrt{(x_i - x_{i-1})^2 + [f(x_i) - f(x_{i-1})]^2}$$
$$= \sqrt{(\Delta x_i)^2 + (\Delta y_i)^2}$$

Factoring out $(\Delta x_i)^2$ produces

$$= \sqrt{1 + \left(\frac{\Delta y_i}{\Delta x_i}\right)^2}\,\Delta x_i$$

where $\Delta x_i = x_i - x_{i-1}$, and $\Delta y_i = f(x_i) - f(x_{i-1})$. Thus, we can rewrite the formula for L as

$$L = \sum_{i=1}^{n} \sqrt{1 + \left(\frac{\Delta y_i}{\Delta x_i}\right)^2}\,\Delta x_i$$

We are going to assume that f has a derivative; then $\lim\limits_{x_i \to 0} \dfrac{\Delta y_i}{\Delta x_i} = \dfrac{dy_i}{dx_i} = f'(x_i)$. With this change, the length of the curve L becomes

$$L = \sum_{i=1}^{n} \sqrt{1 + [f'(x_i)]^2}\,dx_i$$

Now, as the length of each subinterval gets smaller and smaller, we get the desired formula as a limit.

Length of a Graph

If f is a function with a continuous derivative on $[a, b]$, then the length of the graph of $y = f(x)$ on $[a, b]$ is

$$L = \int_a^b \sqrt{1 + [f'(x)]^2}\,dx$$

EXAMPLE 7.20

Find the length of $y = 3 + x^{2/3}$ from $x = 1$ to $x = 8$.

Solution If $y = f(x) = 3 + x^{2/3}$, then $f'(x) = \dfrac{2}{3x^{1/3}}$. Thus,

$$\begin{aligned} L &= \int_1^8 \sqrt{1 + \left[\frac{2}{3x^{1/3}}\right]^2}\,dx \\ &= \int_1^8 \sqrt{1 + \frac{4}{9x^{2/3}}}\,dx \\ &= \int_1^8 \sqrt{\frac{9x^{2/3} + 4}{9x^{2/3}}}\,dx \\ &= \int_1^8 \frac{\sqrt{9x^{2/3} + 4}}{3x^{1/3}}\,dx \end{aligned}$$

We will use the substitution method to solve this. If $u = 9x^{2/3} + 4$, then $du = 6x^{-1/3}dx$ and so $\frac{1}{3}x^{-1/3}\,dx = \frac{1}{18}\,du$. Thus,

$$\begin{aligned} L &= \int \frac{\sqrt{u}\,du}{18} \\ &= \frac{1}{18}\int \sqrt{u}\,du \\ &= \frac{1}{18} \cdot \frac{2}{3}u^{3/2} = \frac{1}{27}u^{3/2} \end{aligned}$$

Since $u = 9x^{2/3} + 4$, we have

$$\begin{aligned} L &= \frac{1}{27}\left(9x^{2/3} + 4\right)^{3/2}\Big|_1^8 \\ &= \frac{1}{27}\left(40^{3/2} - 13^{3/2}\right) \\ &\approx 7.63 \end{aligned}$$

We should point out that if the curve is a function of y, such as $x = g(y)$, where g' is continuous on $[c, d]$, then the length of L, of the graph of $x = g(y)$ from c to d, is given by

$$L = \int_c^d \sqrt{1 + [g'(y)]^2}\,dy$$

EXAMPLE 7.21

Find the length of $x = y^{3/2}$ from $y = 1$ to $y = 4$.

Solution Since $x' = \frac{3}{2}y^{1/2}dy$,

$$L = \int_1^4 \sqrt{1+\left(\frac{3}{2}y^{1/2}\right)^2}\,dy$$

$$= \int_1^4 \sqrt{1+\frac{9}{4}y}\,dy$$

$$= \frac{4}{9}\int_1^4 \frac{9}{4}\sqrt{1+\frac{9}{4}y}\,dy$$

$$= \frac{2}{3}\cdot\frac{4}{9}\left(1+\frac{9}{4}y\right)^{3/2}\Bigg|_1^4$$

$$= \frac{8}{27}\left[(10)^{3/2} - \left(\frac{13}{4}\right)^{3/2}\right] \approx 7.63$$

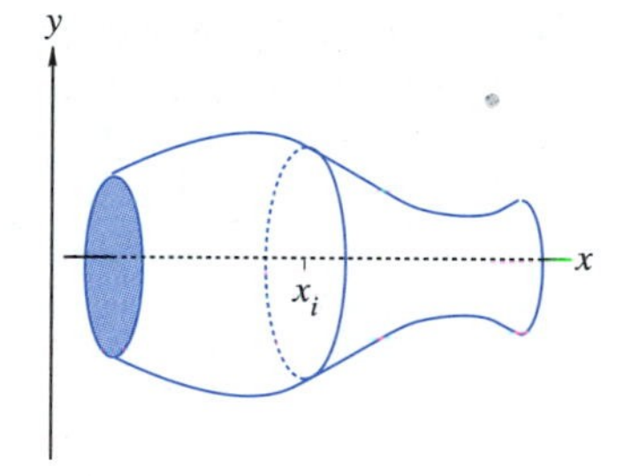

FIGURE 7.22

Surface Area

Now suppose that the curve in Figure 7.21 is rotated around the x-axis. Then a surface is formed. If we want to find the surface area of this figure, we need to think of it as a series of circles. In Figure 7.22, the circle with center at x_i has a circumference of $2\pi f(x_i)$, or $2\pi y_i$, where $y_i = f(x_i)$. If we add the circumferences of all the circles over the entire length of the curve from a to b, we get a total surface area, S.

Surface Area

If f is a function with a continuous derivative on $[a, b]$, then the area of the surface generated by revolving the graph of $y = f(x)$ about the x-axis is

$$S = \int_a^b 2\pi y\sqrt{1+[f'(x)]^2}\,dx$$

EXAMPLE 7.22

Find the surface area of the solid generated by rotating $y = \frac{1}{3}x^3$ from $x = 0$ to $x = 2$ around the x-axis.

Solution The formula for the surface is

$$S = \int_a^b 2\pi y\sqrt{1+[f'(x)]^2}\,dx$$

EXAMPLE 7.22 (Cont.)

Here, $y = \frac{1}{3}x^3$ and $f'(x) = x^2$, so we have

$$S = \int_0^2 \frac{2}{3}\pi x^3 \sqrt{1+(x^2)^2}\,dx = \frac{2}{3}\pi \int_0^2 x^3\sqrt{1+x^4}\,dx$$

If we let $u = 1+x^4$, then $du = 4x^3dx$, and we have

$$\begin{aligned}\frac{2}{3}\pi \cdot \frac{1}{4}\int \sqrt{u}\,du &= \frac{\pi}{6}\cdot\frac{2}{3}u^{3/2}\\ &= \frac{\pi}{9}(1+x^4)^{3/2}\Big|_0^2 = \frac{\pi}{9}(17^{3/2}-1^{3/2})\\ &\approx 7.68\pi\end{aligned}$$

The surface area is about 7.68π units^2.

At present, our ability to solve many of these problems is limited by our lack of ability in integration techniques. We will be using arc length and surface areas as we improve our skills at integration.

Exercise Set 7.4

In Exercises 1–6, find the lengths of the curves over the indicated interval.

1. $y = \frac{1}{3}(x^2+2)^{3/2}$, over $[0,3]$
2. $y = x^{3/2}$, over $[0,8]$
3. $9x^2 = 4y^3$, from the point $(0,0)$ to the point $(2\sqrt{3},3)$
4. $y = \dfrac{x^4}{4} + \dfrac{1}{8x^2}$, over $[1,2]$
5. $y = \dfrac{x^3}{6} + \dfrac{1}{2x}$, over $[1,3]$
6. $x = \dfrac{y^3}{3} + \dfrac{1}{4y}$, over $[1,3]$

In Exercises 7–12, find the surface area of the solid generated by rotating the given curves around the indicated axis.

7. $y = 4x$ from $x = 0$ to $x = 2$ about the x-axis
8. $y = \frac{12}{5}x$ from $x = 0$ to $x = 8$ about the x-axis
9. $y = x$ from $(0,0)$ to $(2,2)$ about the x-axis
10. $y^2 = 4x$ from $x = 0$ to $x = 8$ about the x-axis
11. $x = 4y$ from $y = 1$ to $y = 3$ about the x-axis
12. $x = \sqrt{y}$ from $y = 0$ to $y = 6$ about the y-axis

Solve Exercises 13 and 14.

13. *Construction* A suspension bridge has its roadbed supported by vertical cables that are connected to a cable that is strung between two towers and from each tower to the ground. Each cable from the top of the tower to the ground fits the equation $y = \frac{2}{75}x^{3/2}$. Find the length of one of the cables, if it meets the ground 225 m from the base of the tower.
14. *Interior design* A lamp shade is formed by rotating the graph of $y = (4-x)^3$ over $[2,4]$, around the x-axis, where x and y are in inches. What is the surface area of the lamp shade?

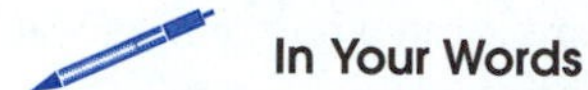

In Your Words

15. Explain how to find the length of a graph.
16. Write an application in your technology area of interest that requires you to find either the arc length of a (nonlinear) curve or the surface area of an irregular shape. Give your problem to a classmate and see if he or she understands and can solve your problem. Rewrite the problem as necessary to remove any difficulties encountered by your classmate.

7.5 CENTROIDS

Thus far we have used integration to study moving objects as if they were particles that had mass but no size. Objects, however, are made up of many particles. If an object is moved, then its particles may have different kinds of behavior. This type of situation can be handled by studying what happens to one particular point of the object. This point is known as the center of mass or the centroid. In this section, we will study how to find the centroid for areas and for solids of revolution.

Suppose we had a metal bar 4 m long that we place along the x-axis with the center of the bar at the origin, one end at 2, and the other end at -2. Next, suppose we put a 10-kg mass at $x=2$ and a 4-kg mass at $x=-2$. We would like to find the center of mass for this system. The **center of mass** of a system is the point where all the mass seems to be concentrated. If the system has a constant density, then the center of mass is called the **centroid**.

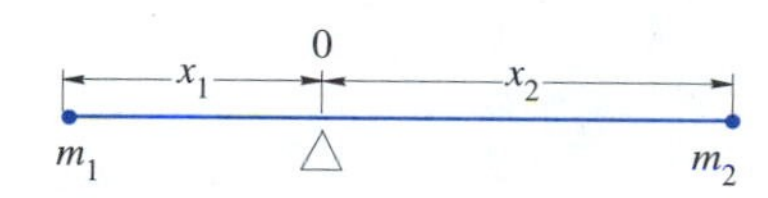

FIGURE 7.23

Before we solve this example, we will look at the idea behind the centroid. Suppose we had a system much like the one in Figure 7.23. The point 0 indicates the origin of the system. A mass m_1 is located x_1 units to the left of the origin. A second mass m_2 is located x_2 units to the right of the origin.

Associated with each mass is its moment with respect to a certain axis. In Figure 7.23, the axis is at the origin. Each **moment** is the product of the mass and its distance from the axis. For the system in Figure 7.23, the moment for mass 1 is m_1d_1 and for mass 2 the moment is m_2d_2. The **first moment** of this system is the sum of the individual moments. In this case, the first moment for the system is $m_1x_1+m_2x_2$. We now select a new point $\bar{x}$ with the property that $(m_1+m_2)\bar{x}$ is equal to the first moment of the system. The location of $\bar{x}$ is the center of mass. Since $(m_1+m_2)\bar{x}=m_1x_1+m_2x_2$, we have

$$\bar{x}=\frac{m_1x_1+m_2x_2}{m_1+m_2}$$

Continuing this process for more than two points gives the general formula for n points:

$$\bar{x}=\frac{\sum_{i=1}^{n} m_ix_i}{\sum_{i=1}^{n} m_i}$$

In this equation, the numerator is the first moment of the system and the denominator is the total mass of the system.

EXAMPLE 7.23

Four particles of masses 2, 4, 3, and 5 kg, respectively, are located on the x-axis at coordinates 3, −2, −3, and 6 m. Where is the center of mass of this system?

Solution

$$\bar{x} = \frac{m_1x_1 + m_2x_2 + m_3x_3 + m_4x_4}{m_1 + m_2 + m_3 + m_4}$$

$$= \frac{2(3) + 4(-2) + 3(-3) + 5(6)}{2 + 4 + 3 + 5} = \frac{19}{14}\text{ m}$$

The center of mass is at the coordinate $\frac{19}{14}$ m on the x-axis.

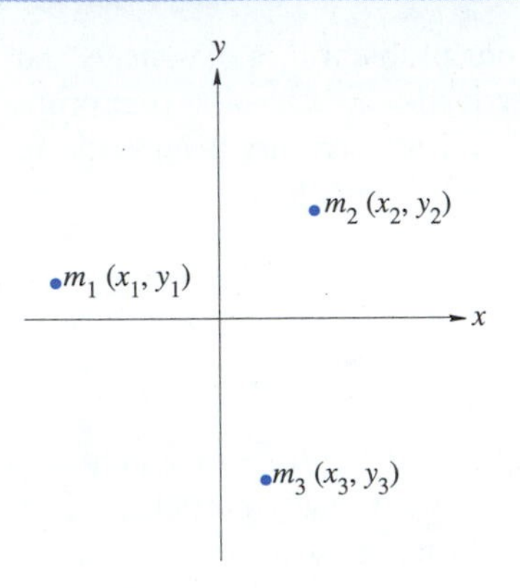

FIGURE 7.24

If we have several point masses located in a plane, then we locate the center of mass by separately finding the center of mass for the x-coordinates and for the y-coordinates. Thus, for the system in Figure 7.24, the center of mass would have coordinates

$$\bar{x} = \frac{m_1x_1 + m_2x_2 + m_3x_3}{m_1 + m_2 + m_3} \text{ and } \bar{y} = \frac{m_1y_1 + m_2y_2 + m_3y_3}{m_1 + m_2 + m_3}$$

In general, we have the formulas in the box for the center of mass for n points.

Center of Mass for n Points

If the point masses $m_1, m_2, \ldots, m_n$ are located at $(x_1, y_1), (x_2, y_2), \ldots, (x_n, y_n)$, respectively, then the center of mass has the coordinates

$$\bar{x} = \frac{\sum_{i=1}^{n} m_ix_i}{m} \text{ and } \bar{y} = \frac{\sum_{i=1}^{n} m_iy_i}{m}$$

or

$$\bar{x} = \frac{M_y}{m} \text{ and } \bar{y} = \frac{M_x}{m}$$

where

$M_y = \sum_{i=1}^{n} m_ix_i$ is the first moment around the y-axis,

$M_x = \sum_{i=1}^{n} m_iy_i$ is the first moment around the x-axis, and

$m = \sum_{i=1}^{n} m_i$ is the total mass of the system.

EXAMPLE 7.24

Calculate the center of mass of the system with four particles of masses 10 g, 5 g, 3 g, and 8 g located respectively at the points $(2,-1)$, $(3,2)$, $(-4,1)$, and $(-3,4)$ (measured in cm).

Solution

$$\begin{aligned}\bar{x} &= \frac{10(2)+5(3)+3(-4)+8(-3)}{10+5+3+8} \\ &= \frac{-1}{26}\text{ cm} \\ \bar{y} &= \frac{10(-1)+5(2)+3(1)+8(4)}{10+5+3+8} \\ &= \frac{35}{26}\text{ cm}\end{aligned}$$

The center of mass is at $\left(-\frac{1}{26}, \frac{35}{26}\right)$.

EXAMPLE 7.25

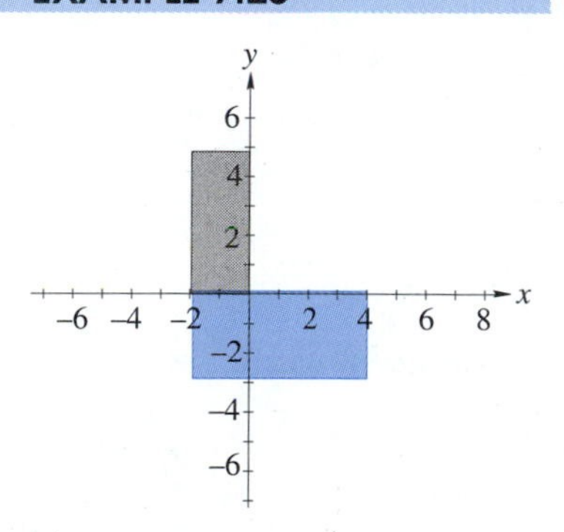

FIGURE 7.25

Find the center of mass of the area in Figure 7.25 if it has uniform density ρ.

Solution We will first divide the area into two rectangles. This division will be along the x-axis with region I above the x-axis and region II below. The geometric center of region I is $\left(-1, \frac{5}{2}\right)$ and of region II, $\left(1, -\frac{3}{2}\right)$. The mass of each region is the product of the density ρ (mass/unit area) and the area. The area of region I is 10 units and so its mass is 10ρ. The area of region II is 18 units and so its mass is 18ρ. The total mass of this figure is

$$m = 10\rho + 18\rho = 28\rho$$

The coordinates of the center of mass will be determined by adding the product of the coordinate of the centroid for each region by its mass. Thus, the x-coordinate is

$$\bar{x} = \frac{(-1)(10\rho)+(1)(18\rho)}{28\rho} = \frac{8\rho}{28\rho} = \frac{2}{7}$$

and the y-coordinate is

$$\bar{y} = \frac{\left(\frac{5}{2}\right)(10\rho)+\left(-\frac{3}{2}\right)(18\rho)}{28\rho} = \frac{-2\rho}{28\rho} = -\frac{1}{14}$$

The center of mass is at $\left(\frac{2}{7}, -\frac{1}{14}\right)$.

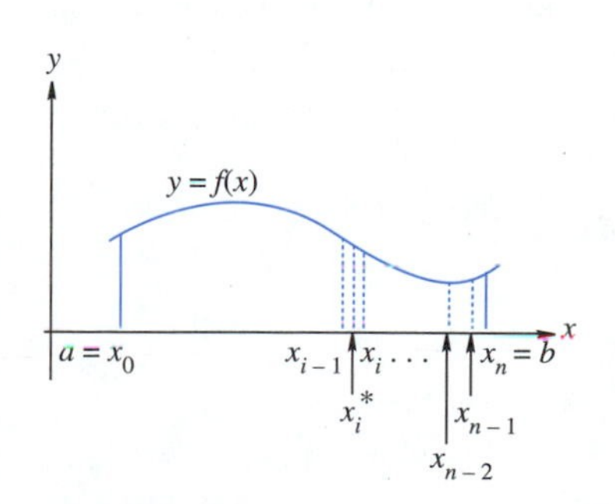

FIGURE 7.26

It is possible for the center of mass to be outside the system. You may have noticed that the density ρ was not used to locate the center of mass. This will happen as long as the object has uniform density, as will be true throughout the remainder of this text.

Now let's apply this above technique to an arbitrary region. To begin, we will let the region be found by $y = f(x)$, $x = a$, $x = b$, and the x-axis, as shown in Figure 7.26. As we have done before, we will partition the interval $[a, b]$ into n subintervals as indicated in Figure 7.26.

Look at the shaded region in Figure 7.26. It is almost rectangular in shape and so its center is the midpoint of the interval $[x_{i-1}, x_i]$, which we will call $x_i{}^*$, and half the height of the rectangle, or $\frac{1}{2}f(x_i{}^*)$. The area of this rectangle is $f(x_i{}^*)(x_i - x_{i-1}) = f(x_i{}^*)\Delta x$, and so the mass of this rectangle is $\rho f(x_i{}^*)\Delta x$, where ρ is the density. The first moment of this shaded region is given by $\rho x_i{}^* f(x_i{}^*)\Delta x$. The first moment for all of these rectangular strips with respect to the y-axis is given by

$$\sum_{i=1}^{n} \rho x_i{}^* f(x_i{}^*)\Delta x$$

So, if the number of intervals increases, we get a first moment of

$$M_y = \lim_{\Delta x \to 0} \sum_{i=1}^{n} \rho x_i{}^* f(x_i{}^*)\Delta x$$

$$\text{or} \qquad M_y = \int_a^b \rho x f(x) dx$$

In a similar way, we can show that the first moment with respect to the x-axis is given by

$$M_x = \frac{1}{2}\int_a^b \rho[f(x)]^2 dx$$

The mass m of the region is

$$m = \int_a^b \rho f(x) dx$$

Combining these, we get the coordinate of the centroid as $(\bar{x}, \bar{y})$ where

$$\bar{x} = \frac{M_y}{m} \text{ and } \bar{y} = \frac{M_x}{m}$$

If $\rho = 1$, then the mass will be the same as the area.

EXAMPLE 7.26

Find the centroid of the plane region bounded by $y = 4 - x^2$ and $y = 0$.

Solution These two curves intersect at $x = -2$ and $x = 2$. Thus, we have

$$\begin{aligned} M_y &= \int_{-2}^{2} \rho x \left(4 - x^2\right) dx \\ &= \int_{-2}^{2} \rho \left(4x - x^3\right) dx \\ &= \rho \left(2x^2 - \frac{1}{4}x^4\right)\Big|_{-2}^{2} \\ &= \rho[(8-4) - (8-4)] = 0 \end{aligned}$$

$$M_x = \frac{1}{2}\int_{-2}^{2} \rho \left(4 - x^2\right)^2 dx$$

EXAMPLE 7.26 (Cont.)

$$= \frac{1}{2}\int_{-2}^{2} \rho\left(16 - 8x^2 + x^4\right) dx$$

$$= \frac{\rho}{2}\left(16x - \frac{8}{3}x^3 + \frac{x^5}{5}\right)\Big|_{-2}^{2}$$

$$= \frac{\rho}{2}\left(\frac{256}{15} + \frac{256}{15}\right)$$

$$= \frac{256}{15}\rho$$

$$m = \int_{-2}^{2} \rho(4 - x^2)dx$$

$$= \rho\left(4x - \frac{x^3}{3}\right)\Big|_{-2}^{2}$$

$$= \rho\left(\frac{16}{3} + \frac{16}{3}\right) = \frac{32}{3}\rho$$

Thus we have the following coordinates:

$$\bar{x} = \frac{M_y}{m} = \frac{0}{32\rho/3} = 0$$

$$\bar{y} = \frac{M_x}{m} = \frac{256\rho/15}{32\rho/3} = \frac{8}{5}$$

The centroid of this region is $\left(0, \frac{8}{5}\right)$.

If the region is between two curves, $f(x)$ and $g(x)$, with $f(x) \geq g(x)$, we have the following, more general, formulas. Notice that if $g(x) = 0$, we have the previous equations.

Moments, Mass, and Centroid of a Plane Region

If f and g are continuous functions with $f(x) \geq g(x)$ on the interval $[a, b]$, then the moments, mass, and centroid of the region of uniform density ρ bounded by the graphs of $y = f(x)$ and $y = g(x)$, on $[a, b]$, are

$$M_x = \frac{1}{2}\int_{a}^{b} \rho([f(x)]^2 - [g(x)]^2)dx$$

$$M_y = \int_{a}^{b} \rho x[f(x) - g(x)]dx$$

$$m = \int_{a}^{b} \rho[f(x) - g(x)]dx$$

$$\bar{x} = \frac{M_y}{m} \text{ and } \bar{y} = \frac{M_x}{m}$$

EXAMPLE 7.27

Find the centroid of the region bounded by $y = x^3$ and $y = \sqrt{x}$.

Solution These two curves intersect at $x = 0$ and $x = 1$. As we can see in Figure 7.27, $\sqrt{x} \geq x^3$ in this interval. Using the new formulas, we get

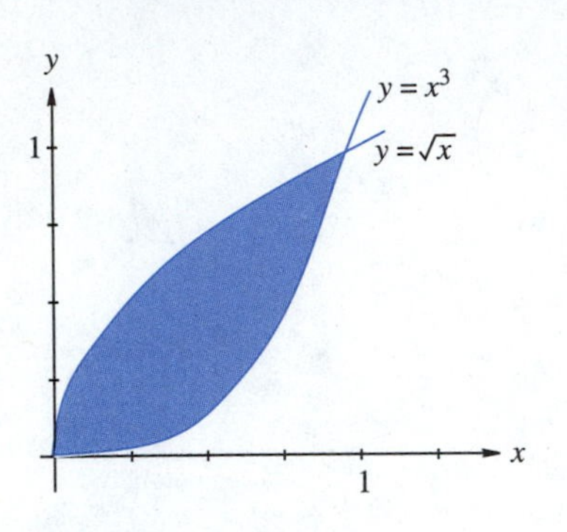

FIGURE 7.27

$$
\begin{aligned}
M_y &= \int_0^1 \rho x \left(\sqrt{x} - x^3\right) dx \\
&= \int_0^1 \rho \left(x^{3/2} - x^4\right) dx \\
&= \rho \left[\frac{2}{5}x^{5/2} - \frac{1}{5}x^5\right]_0^1 = \frac{1}{5}\rho \\
M_x &= \frac{1}{2}\int_0^1 \rho \left[\left(\sqrt{x}\right)^2 - \left(x^3\right)^2\right] dx \\
&= \frac{1}{2}\int_0^1 \rho \left(x - x^6\right) dx \\
&= \rho \frac{1}{2}\left(\frac{1}{2}x^2 - \frac{1}{7}x^7\right)\Big|_0^1 = \frac{5}{28}\rho \\
m &= \int_0^1 \rho \left(\sqrt{x} - x^3\right) dx \\
&= \rho \left(\frac{2}{3}x^{3/2} - \frac{1}{4}x^4\right)\Big|_0^1 = \frac{5}{12}\rho
\end{aligned}
$$

Thus, we have

$$
\begin{aligned}
\bar{x} &= \frac{M_y}{m} = \frac{1/5\rho}{5/12\rho} = \frac{12}{25} \\
\bar{y} &= \frac{M_x}{m} = \frac{5/28\rho}{5/12\rho} = \frac{12}{28} = \frac{3}{7}
\end{aligned}
$$

The centroid is at $\left(\frac{12}{25}, \frac{3}{7}\right)$.

EXAMPLE 7.28

A thin plate covers a region bounded by the x-axis, $y = x^2 + 1$, and $x = 2$. The plate's density at the point (x, y) is $\rho = 5x$. Find the total mass and the center of mass of the plate.

Solution

$$
\begin{aligned}
M_y &= \int_0^2 \rho x \left(x^2 + 1\right) dx \\
&= \int_0^2 (5x)x\left(x^2 + 1\right) dx \\
&= \int_0^2 \left(5x^4 + 5x^2\right) dx = x^5 + \frac{5}{3}x^3\Big]_0^2 = \frac{136}{3}
\end{aligned}
$$

EXAMPLE 7.28 (Cont.)

$$M_x = \frac{1}{2}\int_0^2 \rho\left(x^2+1\right)^2 dx$$
$$= \frac{1}{2}\int_0^2 5x\left(x^4+2x^2+1\right) dx$$
$$= \frac{1}{2}\int_0^2 \left(5x^5+10x^3+5x\right) dx$$
$$= \frac{1}{2}\left(\frac{5}{6}x^6+\frac{5}{2}x^4+\frac{5}{2}x^2\right)\bigg]_0^2 = \frac{153}{3}$$
$$m = \int_0^2 \rho\left(x^2+1\right) dx$$
$$= \int_0^2 \left(5x^3+5x\right) dx = \frac{5}{4}x^4+\frac{5}{2}x^2\bigg]_0^2 = 30$$
$$\bar{x} = \frac{M_y}{m} = \frac{136/3}{30} = \frac{136}{90} = \frac{68}{45}$$
$$\bar{y} = \frac{M_x}{m} = \frac{155/3}{30} = \frac{31}{18}$$

Thus, the total mass is 30 units and the center of mass is at $\left(\frac{68}{45}, \frac{31}{18}\right)$.

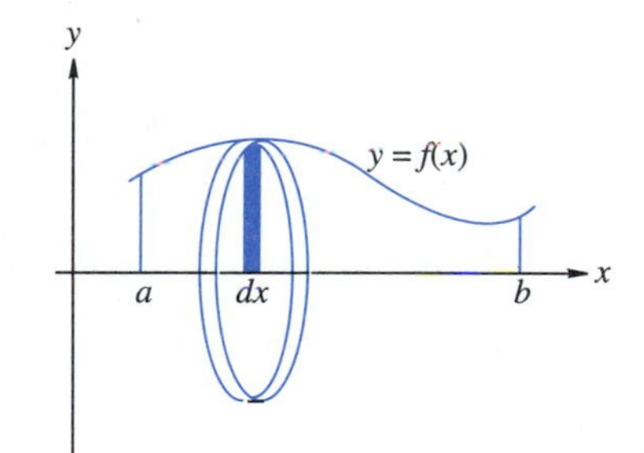

FIGURE 7.28

Centroid of a Solid of Revolution

This method can be extended to solids of revolution. As in the past, we will revolve the region bounded by the curve $y = f(x)$, $x = a$, $x = b$, and the x-axis around the x-axis, as shown in Figure 7.28. Because of the symmetry of the rotation, we know that the centroid is on the axis of rotation. In this case $\bar{y} = 0$. We need only locate $\bar{x}$.

The value of $\bar{x}$ will be $\frac{M_y}{m}$, where M_y is the first moment of the solid with respect to the y-axis and m is the mass. We already know that

$$m = \rho V = \pi \int_a^b \rho[f(x)]^2 dx$$

where V is the volume.

The moment of a typical disk, as shown in Figure 7.28, is the product of x, ρ, and the volume of the disk, $\pi[f(x)]^2 dx$. If we sum these for all the disks, we obtain

$$M_y = \pi \int_a^b \rho x[f(x)]^2 dx$$

Putting these together produces the following result.

Centroid of a Solid of Revolution

If f is a continuous function on the interval $[a, b]$, then the centroid of the solid generated by revolving the region bounded by $y = f(x)$, $x = a$, $x = b$, and the x-axis about the x-axis is at $(\bar{x}, 0)$, where

$$\bar{x} = \frac{M_y}{m}$$

when $M_y = \pi \int_a^b \rho x\,[f(x)]^2\,dx$ and $m = \rho V = \pi \int_a^b \rho[f(x)]^2\,dx$.

EXAMPLE 7.29

Find the centroid of the solid formed by revolving the region bounded by $y = x^2$, $y = 0$, and $x = 2$ about the x-axis, as shown in Figure 7.29.

Solution Because this is revolved around the x-axis, $\bar{y} = 0$. To determine $\bar{x}$, we need m and M_y. We get m from V.

$$\begin{aligned}
V &= \pi \int_0^2 [x^2]^2\,dx \\
&= \pi \int_0^2 x^4\,dx = \pi \frac{x^5}{5}\bigg|_0^2 = \frac{32}{5}\pi \\
m &= \rho V = \frac{32}{5}\rho\pi \\
M_y &= \pi \int_0^2 \rho x\,[x^2]^2\,dx \\
&= \pi \int_0^2 \rho x^5\,dx = \rho\pi\frac{x^6}{6}\bigg|_0^2 = \frac{32}{3}\rho\pi \\
\bar{x} &= \frac{M_y}{m} = \frac{32/3\rho\pi}{32/5\rho\pi} = \frac{5}{3}
\end{aligned}$$

The centroid is at $\left(\frac{5}{3}, 0\right)$.

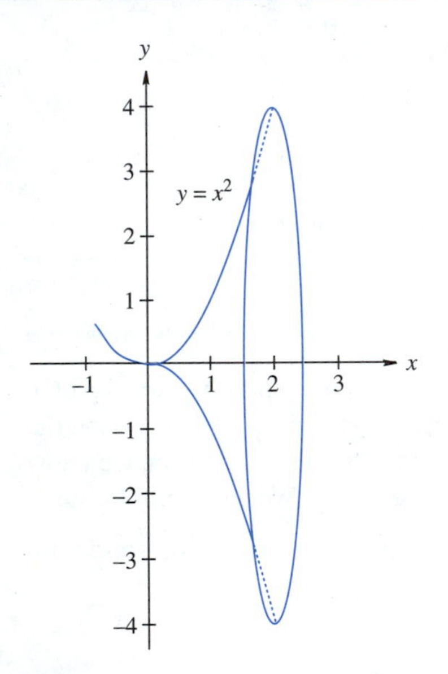

FIGURE 7.29

Note If a solid is revolved around the y-axis, we would follow very similar procedures, only the centroid would be at $(0, \bar{y})$.

EXAMPLE 7.30

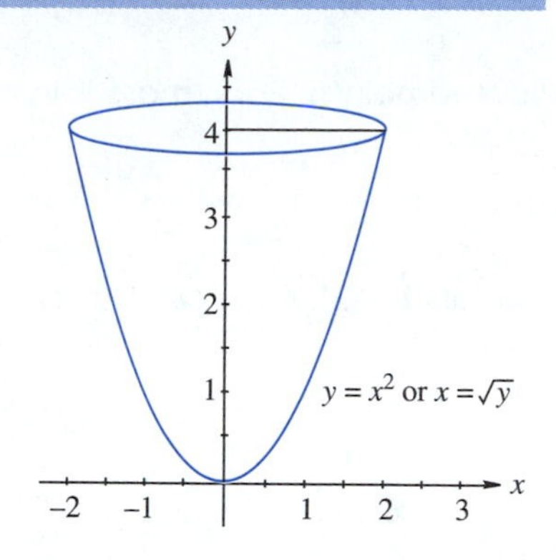

FIGURE 7.30

Find the centroid of the solid formed by revolving the region bounded by $y = x^2$, $y = 4$, and the y-axis around the y-axis as in Figure 7.30.

Solution Because it is revolved around the y-axis, $\bar{x} = 0$. To find $\bar{y}$, we need M_x and m. Again, we get m from V.

$$V = \pi \int_0^4 [\sqrt{y}]^2\, dy$$

$$= \pi \int_0^4 y\, dy = \pi \frac{y^2}{2}\bigg|_0^4 = 8\pi$$

$$m = \rho V = 8\rho\pi$$

$$M_x = \pi \int_0^4 \rho y\, [\sqrt{y}]^2\, dy$$

$$= \pi \int_0^4 \rho y^2\, dy = \pi\rho \frac{y^3}{3}\bigg]_0^4 = \frac{64}{3}\rho\pi$$

$$\bar{y} = \frac{M_x}{m} = \frac{64/3\rho\pi}{8\rho\pi} = \frac{8}{3}$$

The centroid is at $\left(0, \frac{8}{3}\right)$.

Note If a region is bounded by two functions $f(x)$ and $g(x)$ or $h(y)$ and $k(y)$, then the washer method should be used.

EXAMPLE 7.31

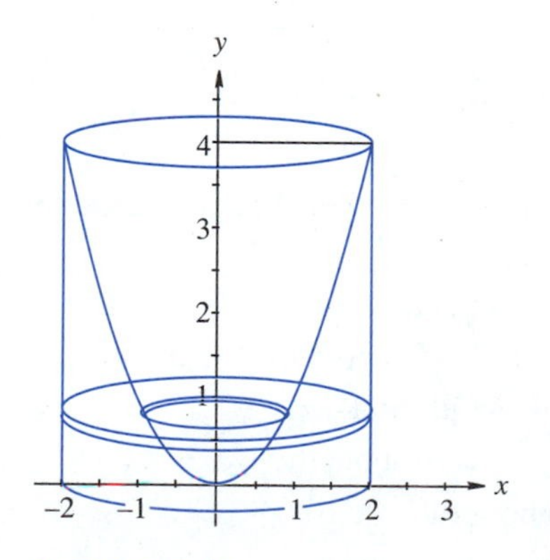

FIGURE 7.31

Find the centroid of the solid formed by revolving the region in Example 7.29 around the y-axis.

Solution Because it is revolved around the y-axis, we know that $\bar{x} = 0$.

We will use the washer method, as shown in Figure 7.31. To determine $\bar{y}$, we need M_x and m.

$$V = \pi \cdot \int_0^4 \left[2^2 - (\sqrt{y})^2\right] dy = \pi \int_0^4 (4 - y)dy$$

$$= \pi \left(4y - \frac{y^2}{2}\right)\bigg|_0^4 = 8\pi$$

$$m = \rho V = 8\pi\rho$$

$$M_x = \pi \int_0^4 \rho y \left[2^2 - (\sqrt{y})^2\right] dy = \pi \int_0^4 \rho(4y - y^2)dy$$

$$= \rho\pi \left(2y^2 - \frac{y^3}{3}\right)\bigg|_0^4 = \frac{32}{3}\pi\rho$$

$$\bar{y} = \frac{M_x}{m} = \frac{32/3\rho\pi}{8\pi\rho} = \frac{4}{3}$$

The centroid is at $\left(0, \frac{4}{3}\right)$.

Exercise Set 7.5

In Exercises 1–4, find the centroid of the system with masses located at the indicated points. Each mass is in grams and the units are in centimeters.

1. $m_1 = 4$ at $(5, 2)$ and $m_2 = 6$ at $(-3, 7)$
2. $m_1 = 3$ at $(-1, 4)$, $m_2 = 4$ at $(-2, -5)$, and $m_3 = 13$ at $(1, 2)$
3. $m_1 = 2$ at $(1, 5)$, $m_2 = 3$ at $(-1, 4)$, and $m_3 = 5$ at $(6, -4)$
4. $m_1 = 1$ at $(1, 1)$, $m_2 = 2$ at $(2, -2)$, $m_3 = 3$ at $(-3, 3)$, and $m_4 = 4$ at $(-4, -4)$

In Exercises 5–8, find the coordinates of the centroid for the indicated figures.

5.
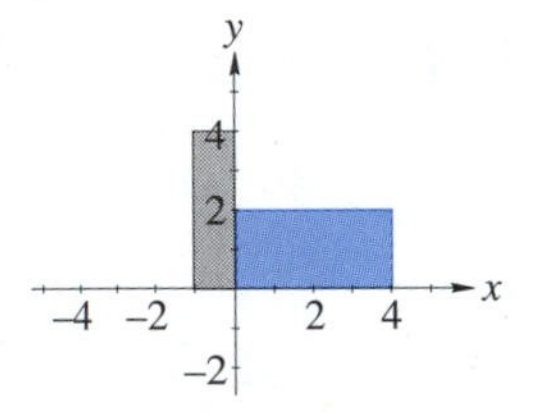

6.
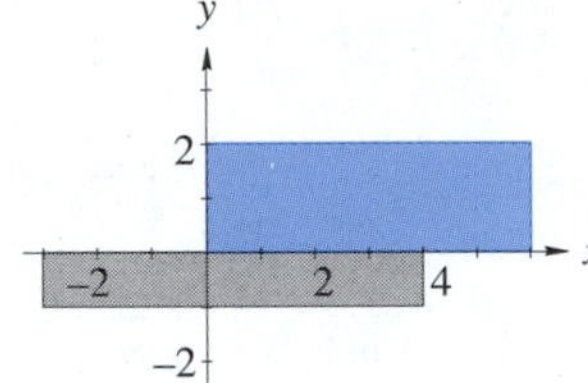

7. 8.
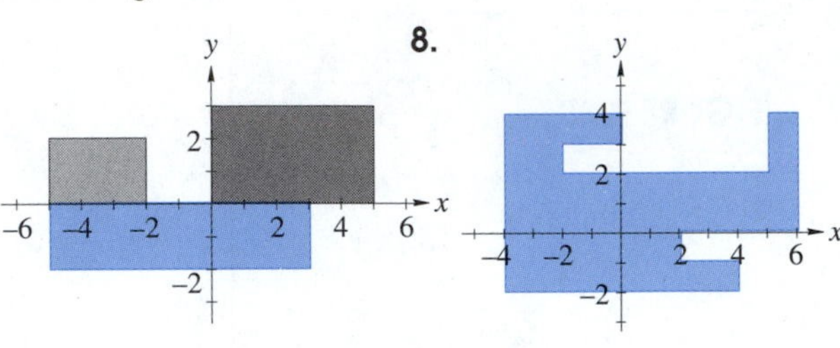

In Exercises 9–14, calculate the centroid of the plane region bounded by the given curve and the *x*-axis over the indicated interval.

9. $y = 2x + 3$, $[0, 3]$
10. $y = x^3$, $[0, 2]$
11. $y = x^{1/3}$, $[0, 8]$
12. $y = x^4$, $[-1, 2]$
13. $y = \sqrt{x + 4}$, $[0, 5]$
14. $y = x^2 + 16$, $[0, 4]$

In Exercises 15–22, find the centroids of the plane regions bounded by the given curves and lines.

15. $y = x^2$, $y = 4x$, $x = 0$, $x = 2$
16. $y = x^3$, $y = 2x$, $x = 0$, $x = 1$
17. $y = 2 - x$, $y = x^2$
18. $y = x^{3/2}$, $y = x$
19. $y = x^2$, $y = 18 - x^2$
20. $y = \sqrt{x}$, $y = x$
21. $y = x$, $y = 12 - x^2$
22. $y = x^2$, $y = x^3$

In Exercises 23–34, find the centroids of the solid of revolution formed by rotating the enclosed region around the indicated axis.

23. $y = x^3$, $y = 0$, $x = 2$, about the x-axis
24. $y = x^3$, $x = 0$, $y = 8$, about the y-axis
25. $y = x^4$, $y = 1$, about the y-axis
26. $y = (x - 1)^2$, $y = 1$, about the y-axis
27. $x + y = 6$, $x = 3$, $y = 0$, $x = 0$, about the x-axis
28. $x + y = 8$, $x = 4$, $y = 0$, $x = 8$, about the x-axis
29. $x + y = 8$, $x = 4$, $y = 0$, about the y-axis
30. $x^2 + y^2 = 16$, $x = 0$, $y = 0$, about the x-axis
31. $x^2 - y^2 = 1$, $x = 3$, about the x-axis
32. $x^2 - y^2 = 1$, $x = 3$, $y = 0$, about the y-axis
33. $y = x^2$, $y = \sqrt{x}$, about the x-axis
34. $y = x^4$, $y = x^2$, about the y-axis

Solve Exercises 35–38.

35. *Construction* The ends of two thin steel rods of equal lengths are welded together to make a right-angle frame, as shown in Figure 7.32.

FIGURE 7.32

(a) Locate the frame's centroid, if the rods are each 1 m long and 1 cm wide.
(b) Locate the frame's centroid, if the rods are each h units long and w units wide.

36. *Product design* A plastic pylon used at work sites is in the shape of a frustum of a cone on a square base, as shown in Figure 7.33. The radii of the bases of the cone are $2\frac{1}{2}$ in. and 7 in. and the height between the bases is 28 in.

(a) Determine the centroid of the portion of the pylon that is in the shape of a frustum of a cone.
(b) If the square base is 1 in. thick and 18 in. on each side, determine the centroid of the entire pylon.

37. Determine the mass and the center of mass of a rod of length 40 cm, where the density ρ of the rod at a point x cm from one end is $5x + 1$ g/cm.

38. Determine the mass and the center of mass of a rod of length 8 m, where the density ρ of the rod at a point x m from one end is $x + 5$ kg/m.

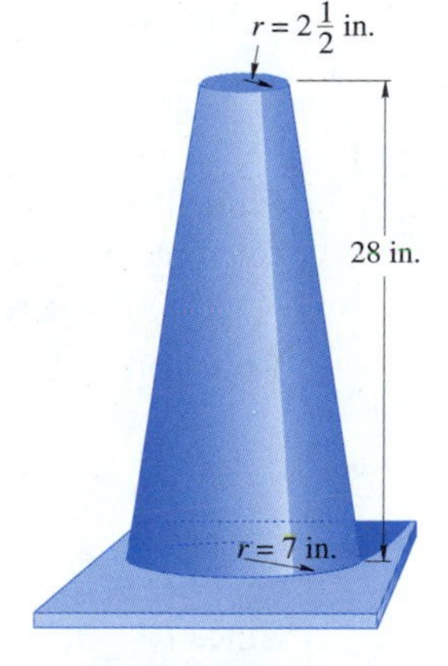

FIGURE 7.33

In Your Words

39. Explain the meaning of the moments, mass, and centroid of a plane region.

40. Describe how to determine the centroid of an irregular-shaped plane region.

7.6 MOMENTS OF INERTIA

In Section 7.5, the first moment around the y-axis (the line $x = 0$) of n masses was defined as

$$M_y = \sum_{i=1}^{n} m_i x_i$$
$$= m_1x_1 + m_2x_2 + m_3x_3 + \cdots + m_nx_n$$

For other problems, other kinds of moments are more useful. In this section, we will study the moments of inertia. The **moment of inertia** is the measure of the tendency of an object to resist a change in motion. The moment of inertia is also called the **second moment** and for a system of n masses around the y-axis it is defined as

$$\begin{aligned} I_y &= \sum_{i=1}^{n} m_i {x_i}^2 \\ &= m_1 {x_1}^2 + m_2 {x_2}^2 + m_3 {x_3}^2 + \cdots + m_n {x_n}^2 \end{aligned}$$

If all of these masses were at the same distance r from the axis of rotation, we would have

$$\begin{aligned} I_y &= m_1 r^2 + m_2 r^2 + m_3 r^2 + \cdots + m_n r^2 \\ &= (m_1 + m_2 + m_3 + \cdots + m_n) r^2 \\ &= m r^2 \end{aligned}$$

where m is the total mass of the system. The number r is called the **radius of gyration**.

EXAMPLE 7.32

Find the moment of inertia and the radius of gyration about the x- and y-axes for three masses of 5 kg, 2 kg, and 4 kg located, respectively, at $(2,1)$, $(4,0)$, and $(-3,-5)$, where the units are in meters.

Solution

$$I_y = \sum_{i=1}^{3} m_i x_i^2 = 5(2)^2 + 2(4)^2 + 4(-3)^2 = 88\,\mathrm{kg}\cdot\mathrm{m}^2$$

Since $I_y = m{r_y}^2$, we know that ${r_y}^2 = \frac{I_y}{m}$ or $r_y = \sqrt{\frac{I_y}{m}}$. The total mass of the system is $m = 5+2+4 = 11$ kg. Thus, $r_y = \sqrt{\frac{88}{11}} = \sqrt{8} = 2\sqrt{2} \approx 2.83$ m.

With respect to the x-axis, we get the results

$$I_x = 5(1)^2 + 2(0)^2 + 4(-5)^2 = 105\,\mathrm{kg}\cdot\mathrm{m}^2$$

and $\qquad r_x = \sqrt{\dfrac{105}{11}} \approx 3.09\,\mathrm{m}$

These answers mean that a mass of 11 kg placed on the line $x = 2.83$ (or $x = -2.83$) will have the same rotational inertia about the y-axis as the three objects. Similarly, placing an 11-kg mass on the line $y = 3.09$ (or $y = -3.09$) will produce the same rotational inertia about the x-axis as the three objects. ▪

It is often of interest to determine the radius of gyration for a system that is rotated about the origin, r_0. In this case, $r_0 = \sqrt{{r_x}^2 + {r_y}^2}$. In Example 7.32, the radius of gyration with respect to the origin is $r_0 = \sqrt{8 + \frac{105}{11}} \approx 4.19$. Thus, placing an 11-kg mass at 4.19 m from the origin will produce the same rotational inertia about the origin as the three objects in Example 7.32.

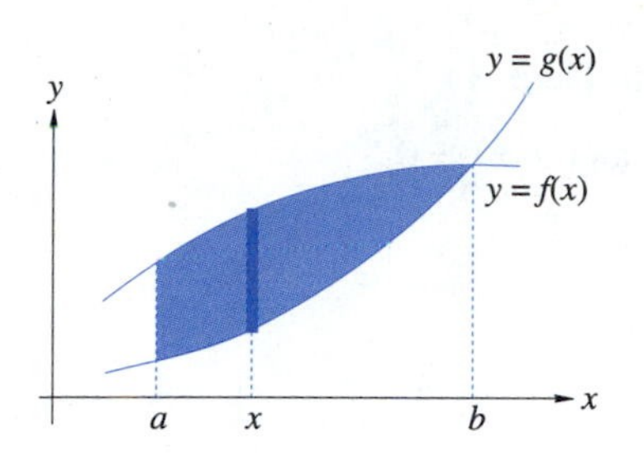

FIGURE 7.34

Moments of Inertia for a Region

Suppose an area is bounded by the curves of two functions such as $y = f(x)$ and $y = g(x)$ and the lines $x = a$ and $x = b$ with $f(x) \geq g(x)$, as shown in Figure 7.34. The moment of inertia for this region with respect to the y-axis is the sum of the moments of inertia for the individual elements of the region. The mass of each element is $\rho[f(x) - g(x)]\,dx$, where ρ is the density (mass per unit area) and $[f(x) - g(x)]\,dx$ is the area of the element. If this element is x-units from the y-axis, then the moment of inertia for the element is $\rho x^2[f(x) - g(x)]\,dx$. If we did this for all the elements, then their sum would give us the moment of inertia.

Moment of Inertia

If a region is bounded by the curves of two functions such as $y = f(x)$ and $y = g(x)$ and by the lines $x = a$ and $x = b$ with $f(x) \geq g(x)$, then the **moment of inertia** of this area with respect to the y-axis, I_y, is

$$I_y = \rho \int_a^b x^2[f(x) - g(x)]\,dx$$

where ρ is the density of the region.

If we have the moment of inertia with respect to the y-axis and the mass of the region, then we can determine the radius of gyration of the region as shown in the following box.

Radius of Gyration

The radius of gyration r_y with respect to the y-axis is

$$r_y = \sqrt{\frac{I_y}{m}}$$

where m is the mass of the region. ($m = \rho A$ where A is the area of the region.)

EXAMPLE 7.33

Find the moment of inertia and the radius of gyration for the region bounded by $y = \sqrt[3]{x}$, $x = 8$, and the x-axis with respect to the y-axis, if the region has a constant density of 3 kg/m^2. Assume all dimensions are meters.

EXAMPLE 7.33 (Cont.)

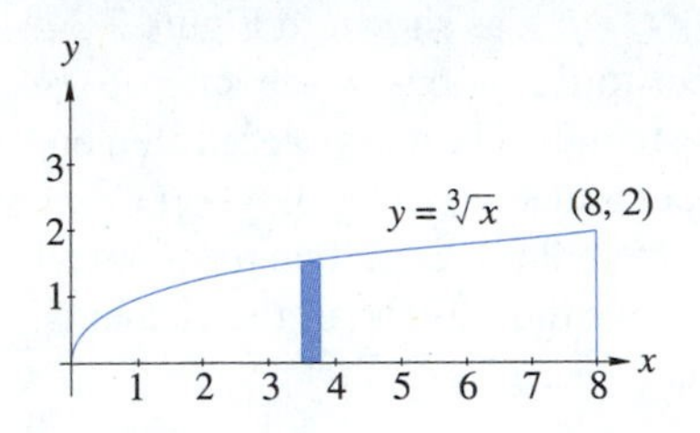

FIGURE 7.35

Solution The region is sketched in Figure 7.35. Here

$$\rho = 3\,\text{kg/m}^2,\ f(x) = \sqrt[3]{x},\ \text{and}\ g(x) = 0$$

$$\begin{aligned} I_y &= 3\int_0^8 x^2(\sqrt[3]{x} - 0)dx \\ &= 3\int_0^8 x^{7/3}\,dx \\ &= 3\left(\frac{3}{10}\right)x^{10/3}\Big|_0^8 = 921.6\,\text{kg}\cdot\text{m}^2 \end{aligned}$$

$$\begin{aligned} A &= \int_0^8 \sqrt[3]{x}\,dx \\ &= \left(\frac{3}{4}\right)x^{4/3}\Big|_0^8 = 12\,\text{m}^2 \end{aligned}$$

Since $m = \rho A$, we have $m = 3A = 36\,\text{kg}$.

$$\begin{aligned} r_y &= \sqrt{\frac{I_y}{m}} \\ &= \sqrt{\frac{921.6}{36}} = \sqrt{25.6} \approx 5.06\,\text{m} \end{aligned}$$

EXAMPLE 7.34

Find the moment of inertia and the radius of gyration of the region in Example 7.33 with respect to the x-axis.

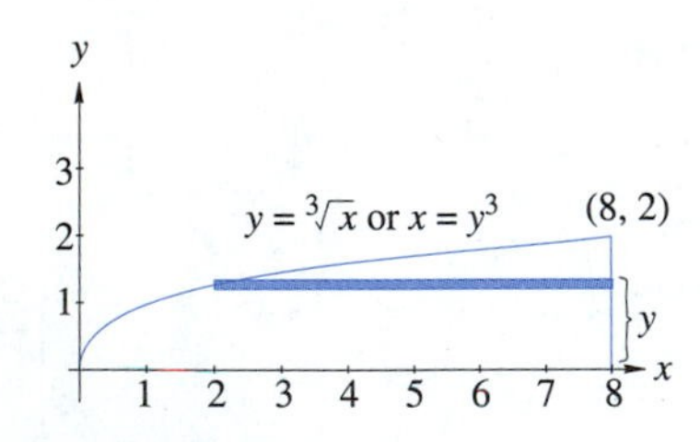

FIGURE 7.36

Solution The region is the same as before, only this time a typical element is parallel to the x-axis, as shown in Figure 7.36. Each element is y units from the x-axis and has the area $(8 - y^3)dy$. Thus, we have

$$\begin{aligned} I_x &= 3\int_0^2 y^2(8 - y^3)dy = 3\int_0^2 (8y^2 - y^5)dy \\ &= 3\left(\frac{8}{3}y^3 - \frac{1}{6}y^6\right)\Big|_0^2 \\ &= 3\left(\frac{64}{3} - \frac{64}{6}\right) = 32\,\text{kg}\cdot\text{m}^2 \end{aligned}$$

$$\begin{aligned} A &= \int_0^2 (8 - y^3)dy \\ &= \left(8y - \frac{1}{4}y^4\right)\Big|_0^2 = (16 - 4) = 12\,\text{kg} \end{aligned}$$

$$m = \rho A = 3A = 36\,\text{kg}$$

$$\begin{aligned} r_x &= \sqrt{\frac{I_x}{m}} \\ &= \sqrt{\frac{32}{36}} = \frac{\sqrt{8}}{3} \approx 0.94\,\text{m} \end{aligned}$$

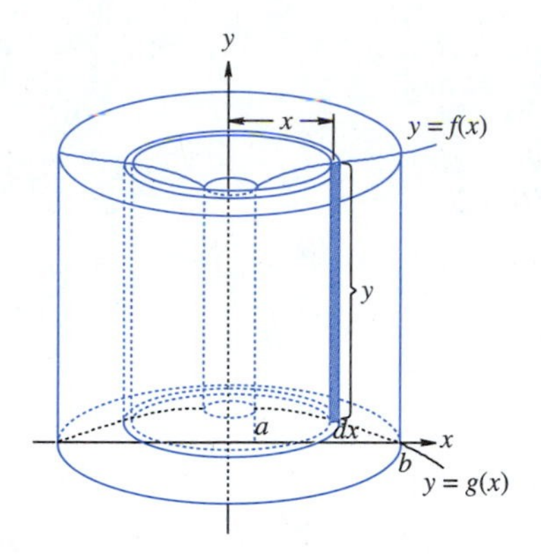

FIGURE 7.37

Moments of Inertia of a Solid of Revolution

We next move to moments of inertia and radii of gyration for solids of revolution. For this we will use the shell method. Consider a region bounded by the curve of the function $y = f(x)$, $x = a$, $x = b$, and $y = g(x)$. The volume of a typical element, such as the one in Figure 7.37, is $2\pi xy\,dx$, where x represents the radius of the cylindrical shell, y, or $f(x) - g(x)$ [assuming $f(x) \geq g(x)$], is its height, and dx is its thickness. The second moment of this shell is $\rho(2\pi xy\,dx)x^2$. Expressing the sums of the second moments as an integral, we get the moment of inertia with respect to the y-axis, I_y, as shown in the following box.

Moment of Inertia; Radius of Gyration

The **moment of inertia** of a solid of revolution formed by generating a region around the y-axis is

$$I_y = 2\pi\rho \int_a^b x^3 y\,dx = 2\pi\rho \int_a^b x^3 [f(x) - g(x)]\,dx$$

and the **radius of gyration** is

$$r_y = \sqrt{\frac{I_y}{m}}$$

where m is the mass of the volume of the solid of revolution.

A similar formula exists for the moment of inertia with respect to the x-axis. This second moment, I_x, is

$$I_x = 2\pi\rho \int_c^d y^3 x\,dy = 2\pi\rho \int_c^d y^3 [h(y) - k(y)]\,dy$$

where the region is bounded by $x = h(y)$, $x = k(y)$, $y = c$, and $y = d$, and where $h(y) \geq k(y)$ on $[c, d]$.

EXAMPLE 7.35

Find the moment of inertia and the radius of gyration of the solid formed by rotating the region bounded by $y = \sqrt[3]{x}$, $x = 8$ cm, and $y = 0$ around the y-axis. This solid will be made from aluminum, which has a density of 2.70 g/cm^3.

Solution For this problem $a = 0$, $b = 8$, $f(x) = \sqrt[3]{x}$, $g(x) = 0$, and $\rho = 2.70$.

$$\begin{aligned} I_y &= 2\pi\rho \int_0^8 x^3 \left(\sqrt[3]{x} - 0\right) dx \\ &= 5.40\pi \int_0^8 x^{10/3}\,dx \end{aligned}$$

EXAMPLE 7.35 (Cont.)

$$= 5.40\pi \left(\frac{3}{13}\right) x^{13/3}\Big|_0^8$$
$$= \frac{16.2\pi}{13}(8192) \approx 32\,070.924\,\text{g}\cdot\text{cm}^2$$

The mass is

$$m = \rho V = 2\pi\rho \int_0^8 x\left(\sqrt[3]{x} - 0\right) dx$$
$$= 5.40\pi \int_0^8 x^{4/3} dx$$
$$= 5.40\pi \left(\frac{3}{7}\right) x^{7/3}\Big|_0^8$$
$$= \frac{16.2\pi}{7}(128) \approx 930.630\,\text{g}$$

The radius of gyration is

$$r_y = \sqrt{\frac{I_y}{m}}$$
$$= \sqrt{\frac{32\,070.924}{930.630}} \approx 5.87\,\text{cm}$$

EXAMPLE 7.36

Find the moment of inertia and the radius of gyration of the solid formed by rotating the region bounded by $y = \sqrt[3]{x}$, $x = 8$ cm, and $y = 0$ around the x-axis. This solid will be made from nylon, which has a density of 1.14 g/cm^3.

Solution This is rotated around the x-axis, so we will use functions of y. Here $h(y) = 8$, $k(y) = y^3$, $c = 0$, $d = 2$, and $\rho = 1.14$. We will not use the value of ρ until we have finished.

$$I_x = 2\pi\rho \int_0^2 y^3\left(8 - y^3\right) dy$$
$$= 2\pi\rho \int_0^2 (8y^3 - y^6) dy$$
$$= 2\pi\rho \left(2y^4 - \frac{1}{7}y^7\right)\Big|_0^2$$
$$\approx 27.43\pi\rho \approx 86.17\rho\,\text{g}\cdot\text{cm}^2$$

Since $\rho = 1.14$, $I_x = 98.23\,\text{g}\cdot\text{cm}^2$. The mass is ρV and so

$$m = 2\pi\rho \int_0^2 y(8 - y^3) dy$$

EXAMPLE 7.36 (Cont.)

$$= 2\pi\rho \int_0^2 (8y - y^4)dy$$

$$= 2\pi\rho \left(4y^2 - \frac{1}{5}y^5\right)\Big|_0^2 = 19.2\pi\rho \approx 60.32\rho\,\text{g}$$

The radius of gyration is

$$r_x = \sqrt{\frac{I_x}{m}}$$

$$= \sqrt{\frac{27.43\pi\rho}{19.2\pi\rho}} = \sqrt{\frac{27.43}{19.2}} \approx 1.20\,\text{cm}$$

Exercise Set 7.6

In Exercises 1–4, find the moment of inertia and the radius of gyration with respect to the origin, r_0, of the given masses at the given points.

1. 3 g at $(4, 0)$ and 5 g at $(-3, 0)$ cm
2. 6 g at $(0, 2)$, 3 g at $(0, -5)$, and 1 g at $(0, 4)$
3. 4 kg at $(2, 1)$ and 3 kg at $(-1, 4)$ m
4. 3 g at $(2, 3)$, 4 g at $(-2, -4)$, and 3 g at $(-4, 5)$

In Exercises 5–14, find the moment of inertia and radius of gyration about the given axis. Unless specified, let ρ represent the density.

5. The region bounded by $y = x^2$, the y-axis, $y = 2$ about y-axis
6. The region of Exercise 5 about x-axis
7. The region bounded by $x = 0$, $y = 0$, $x = 3$, $y = 5$ about the x-axis
8. The region in Exercise 7 about the y-axis
9. The region bounded by $y = x^3$ and $y = x^2$ about the x-axis
10. The region in Exercise 9 about the y-axis
11. The region bounded by $y = 4x - x^2$ and the x-axis, $\rho = 5\ \text{g/cm}^2$ about y-axis
12. The region bounded by $y^3 = x^2$, $y = 4$ and the y-axis, $\rho = 4\ \text{g/cm}^2$ about the x-axis
13. The region bounded by $y = \frac{1}{x^2}$, $y = x^2$, $x = 2$, $x = 1$, $\rho = 8\ \text{g/cm}^2$ about the y-axis
14. The region in Exercise 13 about the x-axis

In Exercises 15–24, find the moment of inertia and the radius of gyration for the solid formed by rotating the region about the given axis. Unless specified, let ρ represent the density.

15. The region bonded by $y = x^2$, the y-axis, and $y = 2$ about the y-axis
16. The region in Exercise 15, but rotated about the x-axis
17. The region bounded by $y = x^2$ and $y = 4x$ about the y-axis
18. The region in Exercise 17 about the x-axis
19. The region bounded by $x = 0$, $y = 0$, $x = 4$ and $y = 6$ about the x-axis
20. The region in Exercise 19, but rotated about the y-axis
21. The region bounded by $y = 4x - x^2$ and $x + y = 4$, $\rho = 3\ \text{g/cm}^3$ about y-axis
22. The region bounded by $y = \sqrt{4 - x}$ and $x + 2y = 4$, $\rho = 5\ \text{g/cm}^3$, about the x-axis.

23. The region bounded by $y = \dfrac{1}{x}$, $y = x^3$, $x = 1$, $x = 2$, $\rho = 5\text{ g/cm}^3$, about the y-axis

24. The region in Exercise 23 about the x-axis

In Your Words

25. Without looking in the text, explain the concept of moment of inertia.

26. What is meant by a radius of gyration?

7.7 WORK AND FLUID PRESSURE

The last applications of the integral that we will consider in this chapter deal with the topic of work. **Work** is defined as the product of the force exerted on an object and the distance the object is moved by that force. For example, if a force of 10 lb is lifted a distance of 5 ft, the work performed is $(10\text{ lb}) \times (5\text{ ft}) = 50\text{ ft}\cdot\text{lb}$. In the same way, if a force of 20 dynes is lifted 9 cm the worked performed is $(20\text{ dynes}) \times (9\text{ cm}) = 180\text{ dyne}\cdot cm = 180$ ergs. Just as an erg is a dyne·cm, the joule is the equivalent of the Newton-meter (N · m).

Most work problems are not as simple as these two. If they were, we certainly would not be discussing them in a chapter about applications of integral calculus.

Hooke's Law

Consider the work done to stretch a spring. When you begin to stretch the spring, it takes very little force. The further the spring is stretched, more force is required. Hooke's law states that the force F required to stretch a spring a distance x is proportional to x. Thus, if k is the constant of proportionality, $F = kx$ and so the force is a function of x or $F(x)$.

We want to determine the amount of work that is done to move the spring from point a to point b. As usual, we partition the interval $[a, b]$ into n subintervals of lengths Δx. If $x_i{}^*$ is the midpoint of one of the intervals, then the work done to move the spring through that interval is approximately $\Delta W = F(x_i{}^*)\Delta x$. Adding the work over all of the intervals and taking the limits as $\Delta x \to 0$, we get the following formula.

> **Work**
>
> The **work** W done by a continuous force directed along the x-axis from $x = a$ to $x = b$ is
>
> $$W = \int_a^b F(x)dx$$

Application

EXAMPLE 7.37

A spring is stretched 10 cm by a force of 3 N, as shown in Figure 7.38. How much work is needed to stretch the spring 40 cm?

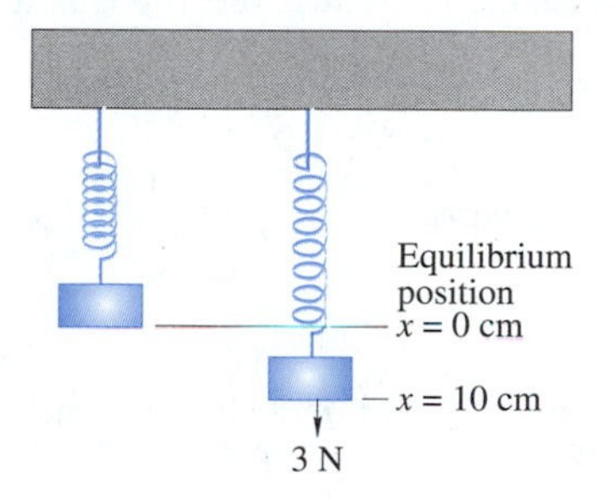

FIGURE 7.38

Solution According to Hooke's law, $F = kx$. We are given $F = 3\,\text{N}$ when $x = 10\,\text{cm} = 0.1\,\text{m}$. Substituting these into the formula for Hooke's law produces

$$\begin{aligned} F &= kx \\ 3 &= k(0.1) \\ 30 &= k \end{aligned}$$

Thus, for this particular spring, $F = 30x$ or the function $F(x) = 30x$. The amount of work needed to stretch this spring $40\,\text{cm} = 0.4\,\text{m}$ is

$$\begin{aligned} W &= \int_0^{0.4} F(x)\,dx \\ &= \int_0^{0.4} 30x\,dx = 15x^2\Big|_0^{0.4} = 2.4\,\text{N}\cdot\text{m} \end{aligned}$$

The work needed to stretch the spring is $2.4\,\text{N}\cdot\text{m} = 2.4\,\text{J}$. (The symbol "J" is used for joule or joules.)

Application

EXAMPLE 7.38

Find the work done in winding 30 m of a 50-m cable, if the cable has a density of 4.5 kg/m.

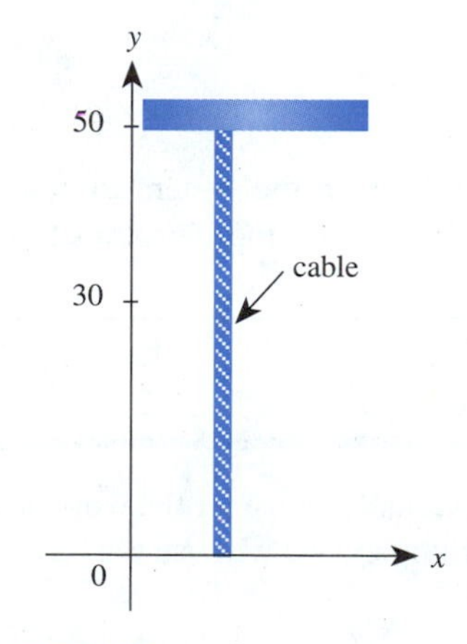

FIGURE 7.39

Solution First we let y denote the length of the cable that has been wound up at any time. The amount of cable that has yet to be wound is $50 - y$. As usual, we will subdivide the interval $[0, 30]$ into n subdivisions. A typical subdivision is shown in Figure 7.39. This subdivision has a length dy and is y units from the top. The mass of a subdivision is $\rho\,dy$ and its weight is $\rho g\,dy$, where ρ is the density and g the gravitational constant. In this example, ρ was given as 4.5 kg/m and, since this problem is in the metric system, $g = 9.8\,\text{m/s}^2$. Thus, the force needed at any point is

$$\begin{aligned} (4.5)(9.8)(\text{amount of cable left to be wound}) &= (4.5)(9.8)(50 - y) \\ &= 44.1(50 - y) \end{aligned}$$

The work needed to move this element from its initial position to the top is $44.1(50 - y)dy$. Since only 30 m are going to be wound, the work is

$$\begin{aligned} W &= \int_0^{30} 44.1(50 - y)dy \\ &= 44.1\left(50y - \frac{1}{2}y^2\right)\Big|_0^{30} \\ &= 44.1(1050)\,\text{J} \\ &= 46\,305\,\text{J} \end{aligned}$$

Electrical Charges

According to Coulomb's law, the electric force between two charges, q_1 and q_2, in a vacuum is directly proportional to the product of these two changes and inversely proportional to the square of the distance r between them. Thus, the force is a function of r and we have

$$F(r) = k\frac{q_1 q_2}{r^2}$$

where $k = 8.988 \times 10^9\,\text{N}\cdot\text{m}^2/\text{C}^2$. Using the formula for work

$$W = \int F(r)\,dr$$

we can determine the work done when two charges move toward or away from each other.

Application

EXAMPLE 7.39

Find the work done when two protons, $q = 1.6 \times 10^{-19}$ C each, move until they are 100 nm apart, if they were originally separated by 1 m.

Solution We have $q_1 = q_2 = 1.6 \times 10^{-19}$. The particles move from 1 m apart to $100\,\text{nm} = 100 \times 10^{-9}\,\text{m} = 10^{-7}$ m apart.

$$\begin{aligned} W &= \int_1^{10^{-7}} \frac{(8.988 \times 10^9)(1.6 \times 10^{-19})(1.6 \times 10^{-19})}{r^2}\,dr \\ &= 2.301 \times 10^{-28} \int_1^{10^{-7}} \frac{dr}{r^2} \\ &= 2.301 \times 10^{-28} \left(\frac{-1}{r}\right)\Bigg|_1^{10^{-7}} \\ &= -2.301 \times 10^{-21}\,\text{J} \end{aligned}$$

The negative sign means that the force must be done *on* the system to move the particles together. Thus, it takes 2.301×10^{-21} J acting on these protons to move them from 1 m apart to 100 nm apart.

Pumping Liquids

As a third example of a problem where the force is variable, consider the work needed to pump a liquid out of, or into, a tank. In these examples, we will use the basic idea that work = (weight of object) × (distance moved).

Application

EXAMPLE 7.40

An oil tank is in the shape of a right-circular cylinder with a height of 40 ft, a radius of 10 ft, and is filled with oil to a depth of 30 ft. How much work is required to pump the oil out of the top of the tank, if oil weighs 54.8 lb/ft^3?

Solution We will position the tank as shown in Figure 7.40. The position of the top of the tank is at $x = 0$ and the bottom at $x = 40$. The work needed to pump a

EXAMPLE 7.40 (Cont.)

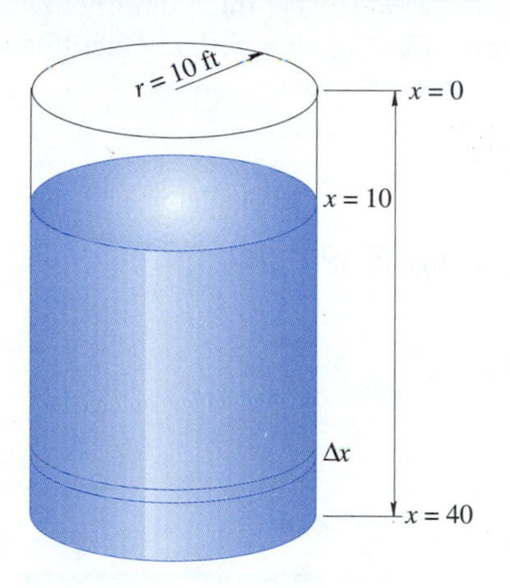

FIGURE 7.40

certain volume of oil over the top depends on the weight of this volume of oil and its distance from the top.

The oil fills the tank from $x = 10$ to $x = 40$. We will partition the interval $[10, 40]$ into subintervals and let Δx be the length of one interval. The volume at this interval is $\pi r^2 h$, and since $r = 10$ and $h = \Delta x$, we have the volume as $100\pi\,\Delta x$. We are told that oil weighs 54.8 lb/ft^3, so the weight of this layer is $(54.8)(100\pi\,\Delta x)$. If the interval is located at $x_i{}^*$, then the work done to lift this layer is $x_i{}^*(54.8)(100\pi\,\Delta x)$.

The layers of oil occur from 10 ft to 40 ft, thus the work required is given by

$$\begin{aligned}\int_{10}^{40} (54.8)(100\pi)x\,dx &= 5480\pi\left(\frac{1}{2}x^2\right)\Bigg|_{10}^{40}\\ &= 4{,}110{,}000\pi\ \text{ft}\cdot\text{lb}\\ &\approx 12{,}911{,}946\ \text{ft}\cdot\text{lb}\end{aligned}$$

It takes a total of 12,911,946 ft·lb to pump the oil out of the top of the tank.

The procedure used in Example 7.40 follows the following guidelines for finding the work needed to pump a liquid out of a tank.

Guidelines for Finding Work Done During Pumping

1. Draw a picture with a coordinate system.
2. Determine the mass of a thin horizontal slab of the liquid.
3. Find an expression for the work needed to lift this slab to its destination.
4. Integrate the work expression from step 3 from the bottom of the liquid to the top.

Application

EXAMPLE 7.41

Suppose a hemispherical tank of radius 10 ft is on a platform so that the bottom is 30 ft off the ground. How much work is done filling the tank from a source of water at ground level, if the water is pumped through a hole in the bottom?

Solution We will put the origin of the vertical axis at ground level. Then the bottom of the tank is at 30 ft and the top is at 40 ft, as shown in Figure 7.41a.

Divide the interval $[30, 40]$ into n subintervals. Suppose a typical interval is x_i units from the top of the tank, as shown in Figure 7.41b. The slice of water taken at this point has a radius y_i. Because the tank is a hemisphere, we know that the radius r of the tank is the hypotenuse of the right triangle formed by x_i and y_i. This means that $x^2 + y^2 = r^2$. The volume of this slice of water is $\Delta V = \pi y^2 \Delta x = \pi(r^2 - x^2)\Delta x$. The force required to lift this slice is

$$\Delta w = 62.4\pi(r^2 - x^2)\Delta x$$

EXAMPLE 7.41 (Cont.)

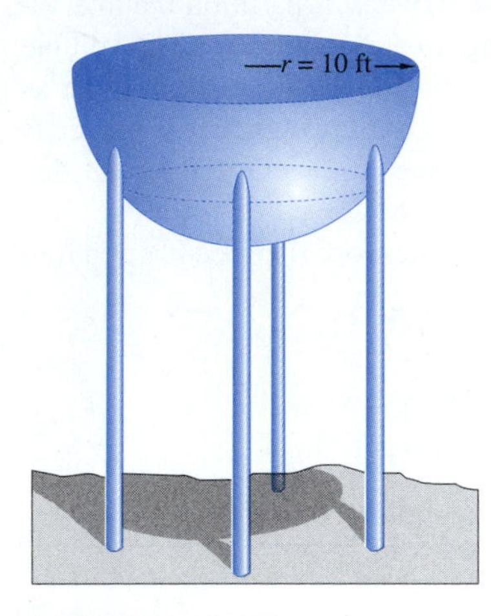

FIGURE 7.41a

where the specific weight of water is $w = 62.4\,\text{lb/ft}^3$. Now all we need is the distance this slice was lifted. Since it was lifted from $x = 0$ to a point x_i units below $x = 40$, it was lifted a total of $40 - x_i$ units. Putting this together with the fact that $r = 10$, we get

$$\begin{aligned} W &= \int_0^{10} 62.4\pi(10^2 - x^2)(40 - x)dx \\ &= 62.4\pi \int_0^{10} (4{,}000 - 100x - 40x^2 + x^3)dx \\ &= 62.4\pi \left(4{,}000x - 50x^2 - \frac{40}{3}x^3 + \frac{1}{4}x^4\right)\Bigg|_0^{10} \\ &\approx 4{,}737{,}521.7\,\text{ft}\cdot\text{lb} \end{aligned}$$

Fluid Pressure

When a container is built to hold a fluid, it is important to know the force F caused by the liquid pressure on the sides of the container. The pressure P exerted by a liquid of mass density ρ at depth h units below the surface of the liquid is defined as

$$P = \rho g h$$

where g is the acceleration of gravity, $32\,\text{ft/s}^2 \approx 9.8\,\text{m/s}^2$.

In the English system we are given the weight density w as lb/ft^3. Here $w = \rho g$.

If a flat plate is submerged in a fluid and placed in a horizontal position, the pressure on the plate is the same at all points. The total force on one side of the plate is the product of the pressure P and the area A of the side $F = PA = \rho g h A$.

Application

EXAMPLE 7.42

A tank 3 m long and 2 m wide is filled with 1.5 m of water. What is the force exerted on the bottom of the tank?

Solution The density of water is about $\rho = 1\,000\,\text{kg/m}^3$. The area of the bottom of the tank is $A = (3\,\text{m})(2\,\text{m}) = 6\,\text{m}^2$. The pressure is $P = (1\,000)(9.8)(1.5) = 14\,700\,\text{N/m}^2$. $F = PA = (14\,700)(6) = 88\,200\,\text{N}$.

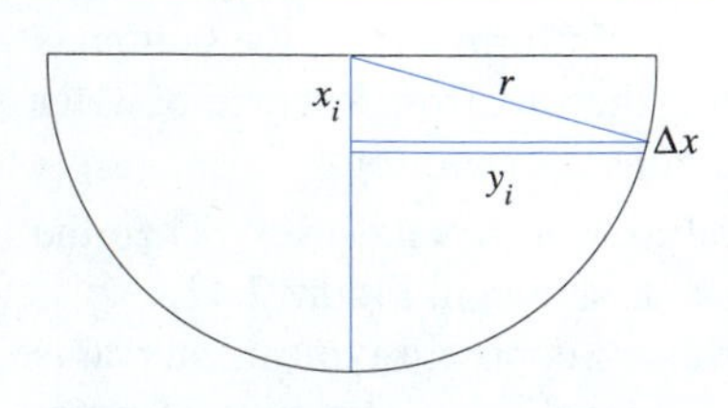

FIGURE 7.41b

If a flat plate submerged in a fluid of density ρ is placed in a vertical position, then the pressure varies as the depth changes. If the plate is subdivided into n horizontal strips, as shown by the typical strip in Figure 7.42, then the pressure is almost constant on each strip.

As shown in Figure 7.42, each horizontal strip has a width dy and length $L(y)$. So, the area is $L(y)\,dy$. We introduce a coordinate system, as shown in the figure. If the plate extends over the interval $[c, d]$ on the y-axis, then for each y in this interval, the depth of each strip is $h(y)$. If dy is small, the pressure at any point in the horizontal

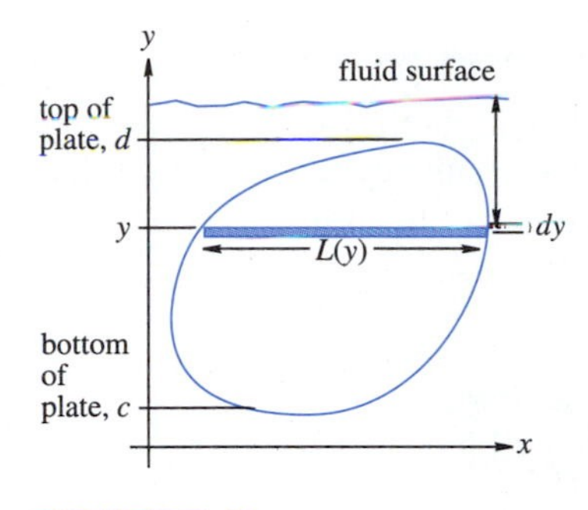

FIGURE 7.42

strip is $\rho h(y)$. Thus, the force on each rectangular strip is given by the product of the pressure on any point in the strip and the area of the strip, or $\rho h(y)L(y)dy$. Integrating this over the depth of the plate, we get the following formula.

Fluid Force

The total force F exerted by a fluid of constant density ρ on a submerged vertical plate from depth $y = c$ to $y = d$ is given by

$$F = \rho \int_c^d h(y)L(y)dy$$

where $h(y)$ is the depth of the fluid at y and $L(y)$ is the horizontal length of the plate at y.

Application

EXAMPLE 7.43

A tank has a cross-section in the shape of a trapezoid with a lower base of 4 ft, and upper base 8 ft. It is filled with 2.5 ft of water. What is the total pressure due to the water on one end of the tank?

Solution A sketch of the tank is shown in Figure 7.43. An xy-coordinate system has been placed on the figure to help in the calculations. The y-axis is partitioned into n subintervals. A typical strip formed by this partition is shaded in the figure. The force against this ith strip is $P_i A_i = 62.4(2.5 - y_i)A_i$, where $h(y_i) = 2.5 - y_i$ is the depth of the strip and A_i is its area.

The length of the strip is approximately $2x_i$. But the edge of the tank, line BC, passes through $(2, 0)$ and $(4, 2.5)$. This line has the equation $y = \frac{2.5}{2}x - 2.5$ and so $x = \frac{2}{2.5}y + 2 = \frac{4}{5}y + 2$. Thus the length of the ith strip is $2\left(\frac{4}{5}y_i + 2\right)$ and its area $A_i = 2\left(\frac{4}{5}y_i + 2\right)\Delta y = \left(\frac{8}{5}y_i + 4\right)\Delta y_i$.

We know that the force F_i on this ith strip is

$$F_i = 62.4\,(2.5 - y_i)\left(\frac{8}{5}y_i + 4\right)\Delta y_i$$

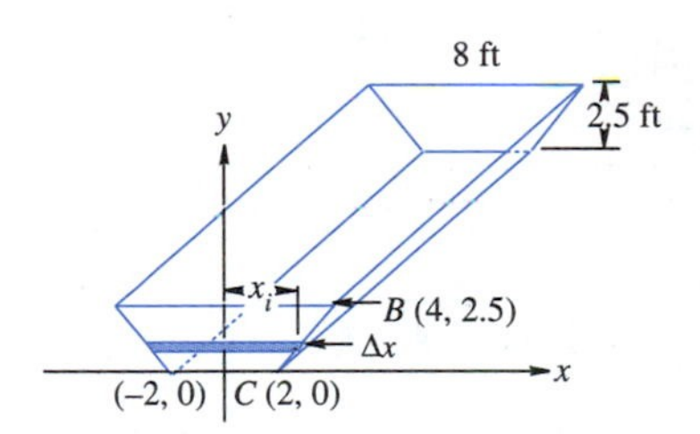

FIGURE 7.43

Integrating, we get the total force

$$\begin{aligned} F &= \int_0^{2.5} 62.4\,(2.5 - y)\left(\frac{8}{5}y + 4\right)dy \\ &= 62.4 \int_0^{2.5}\left(10 - \frac{8}{5}y^2\right)dy \\ &= 62.4\left(10y - \frac{8}{15}y^3\right)\Bigg|_0^{2.5} \\ &= 62.4(16.67) \approx 1{,}040\,\text{lb} \end{aligned}$$

Application

EXAMPLE 7.44

A tank truck, like the milk truck in the photograph in Figure 7.44a, is used to carry liquids. If the tank on this truck is in the shape of a circular cylinder with a radius of 3 ft and length 35 ft, determine the force on one end of this truck when it is full of milk, if milk has a density of 64.5 lb/ft^3.

Courtesy of H.P. Hood Inc.

FIGURE 7.44a

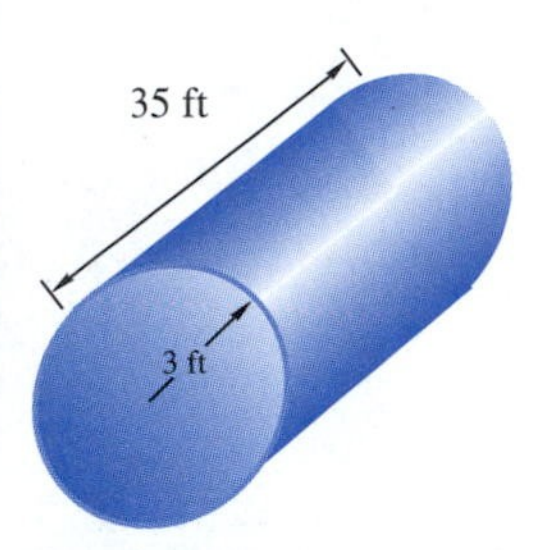

FIGURE 7.44b

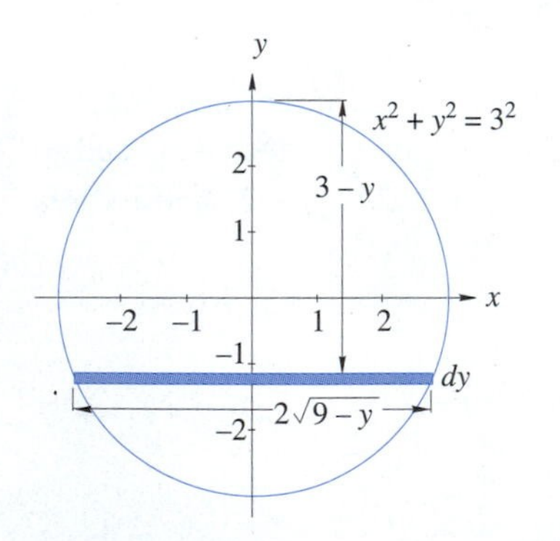

FIGURE 7.44c

Solution A sketch of the tank is shown in Figure 7.44b. A cross-section of an end of the tank is shown in Figure 7.44c. A coordinate system and a typical horizontal strip have been drawn in the figure. Since this is a circular cylinder and the origin of the coordinate system is placed at the center of the base, the circle is described by the equation $x^2 + y^2 = 3^2$, so the length of the strip is $x = 2\sqrt{9-y^2}$. Since the width of the strip is dy, its area is $A_i = 2\sqrt{9-y^2}\,dy$. As can be seen in Figure 7.44c, the depth of this strip is $3-y$. Thus, the force on this strip is $F_i = \rho(3-y)2\sqrt{9-y^2}\,dy$ and the total force on the end of the tank is

$$
\begin{aligned}
F &= 2\rho\int_{-3}^{3}(3-y)\sqrt{9-y^2}\,dy \\
&= 6\rho\int_{-3}^{3}\sqrt{9-y^2}\,dy + 2\rho\int_{-3}^{3}(-y)\sqrt{9-y^2}\,dy
\end{aligned}
$$

While we do not yet know how to integrate the left-hand integral, we do know that its integral will be the area of one-half a circle of radius 3. From geometry, we know that the area of a complete circle of radius 3 is $\pi r^2 = \pi 3^2 = 9\pi$ and so the desired integral will be one-half that amount, or $\frac{9}{2}\pi$. Substituting this in the previous equation, we have

$$
\begin{aligned}
F &= 6\rho\frac{9}{2}\pi + 2\rho\int_{-3}^{3}(-y)(9-y^2)^{1/2}\,dy \\
&= 27\pi\rho + \frac{2}{3}\,\rho(9-y^2)^{3/2}\Big|_{-3}^{3} = 27\pi\rho
\end{aligned}
$$

Since milk has a density of $\rho = 64.5$ lb/ft^3, the total force on the end of the tank is $1{,}741.5\pi \approx 5{,}471.08$ lb/ft^3. Notice that the length of the tank had no effect on the force exerted on the ends. ▪

Exercise Set 7.7

Solve Exercises 1–36.

1. *Physics* A force of 3 lb stretches a spring 4 in. How much work is done in stretching the spring 10 in.?
2. *Physics* If the spring in Exercise 1 is stretched from 1 ft to 2 ft, how much work is done?
3. *Physics* A force of 6 N stretches a spring 20 cm. How much work is done to stretch the spring 1 m?
4. *Physics* How much work is done to stretch the spring in Exercise 3 from 10 cm to 20 cm?
5. *Physics* If the work done in stretching a spring from rest at 0.50 m to a length of 0.60 m is 0.20 J, find the force needed to keep the spring at this length.
6. *Physics* A spring has a natural (or unstretched) length of 10 in. A force of 5 lb compresses the spring to a length of 8 in. Find the work needed to compress it from 9 in. to 6 in.
7. *Physics* If 105 ft·lb of work was needed to stretch the spring in Exercise 1 from its equilibrium position, how far was it stretched?
8. *Mechanical engineering* A cable 50 ft long and weighting 2 lb/ft is hanging from a winch. How much work is needed to wind 20 ft of it up?
9. *Mechanical engineering* A cable 30 m long weighing 2.5 N/m is hanging from a winch. How much work is needed to wind all of the cable onto the winch?
10. *Mechanical engineering* A chain with a mass of 4.7 kg/m has a length of 15 m. If there is a load of 87 kg at the end of the chain, what work is done in winding the chain and load up 10 m?
11. *Mechanical engineering* A chain 12 ft long is lying on the floor. If the chain weighs 0.75 lb/ft, how much work is needed to raise one end to a height of 20 ft?
12. *Mechanical engineering* A chain 5 m long with a density of 0.4 kg/m is lying on the floor. How much work is needed to raise one end to a height of 10 m?
13. *Mechanical engineering* A tank with a mass of 45 kg is filled with 1000 L of water. The water is used to spray chemical cleaner off the side of a building. If the water is used at the rate of 20 L/min and the tank is raised at 0.50 m/min, how much work is done in raising the tank 20 m up the wall?
14. *Nuclear physics* Find the work needed to move a charge of $+5 \times 10^{-7}$ C and a charge of -2×10^{-7} C until they are 1 cm apart, if they were originally 3 mm apart. (Note: Change all distances to meters.)
15. *Nuclear physics* A charge of $+4 \times 10^{-19}$ C is 5 cm from a charge of $+5 \times 10^{-8}$ C. How much work is needed to move them until they are 10 nm apart? ($1\text{ nm} = 10^{-9}$ m.)
16. *Nuclear physics* A proton has a charge of $+1.6 \times 10^{-18}$ C and an electron has a charge of -1.6×10^{-19} C. How much work is needed to move them from 530 nm apart to 1 mm apart?
17. *Physics* According to Newton's law of gravitation, the gravitational force of attraction between two objects that are r units apart is given by $F(r) = \frac{k}{r^2}$. If two objects are 1 cm apart, how much work is needed to separate them by a distance of 1 m? (Leave answer in terms of k.)
18. *Physics* If two objects are 1 ft apart, how much work is needed to separate them until they are 20 ft apart? (Leave answer in terms of k.)
19. *Space technology* How much work is required to lift a 1,000-lb satellite from the surface of the Earth to an orbit of 1,000 mi above the Earth's surface? (See Exercise 17 with $k = 16 \times 10^9$ mi^2·lb and the radius of the Earth as 4,000 mi.)
20. *Petroleum engineering* A cylindrical tank 4 ft in radius and 8 ft high is filled with petroleum. How much work is needed to pump the oil out over the top? (The density of oil is about 54.8 lb/ft^3.)
21. *Petroleum engineering* How much work is required to pump all the oil over the top of a full cylindrical tank that is 4 m in diameter and 8 m high? (The mass density of petroleum varies. For this, and subsequent exercises, let the mass density $\rho = 880$ kg/m^3.)

22. *Recreation* A swimming pool is 2 m deep, 10 m long, and 7 m wide. If it is filled with water, how much work is needed to empty the pool, if the water is pumped over the top?
23. *Recreation* How much work is needed to empty the pool in Exercise 22, if the water was only 1.5 m deep?
24. *Petroleum engineering* The oil in a cylindrical tank that is 10 ft high and has a diameter of 6 ft is to be pumped to a height 5 ft above the top of the tank. How much work is required?
25. *Petroleum engineering* An upright conical tank is filled with oil from a reservoir that is 15.0 m below the bottom of the tank. If the top of the tank has a radius of 2.40 m and the height of the tank is 3.60 m, what work is done in filling the tank when the density of the oil is 847 kg/m^3?
26. *Wastewater technology* A water tank is in the shape of a hemisphere with a radius of 4 m. If the tank is already filled with water, how much work is needed to pump all the water over the top of the tank? (The density of water is 1000 kg/m^3.)
27. *Wastewater technology* How much work is required to pump the top 2 m of water out the top of the tank in Exercise 26?
28. *Wastewater technology* If the tank in Exercise 26 is placed on a platform so that the bottom of the tank is 10 m above the ground, how much work is required to fill the tank with water to the depth of 3 m from ground level?
29. *Recreation* **(a)** What is the pressure of the water on the bottom of the tank in Exercise 22? **(b)** What is the force of the water on the bottom of this tank?
30. *Petroleum engineering* **(a)** What is the pressure of the petroleum on the bottom of the tank in Exercise 24? **(b)** What is the force of the petroleum on the bottom of this tank?
31. *Recreation* If the pool in Exercise 22 is full of water, what is the total force due to water pressure on one short side of the pool?
32. *Recreation* For the pool in Exercise 22, what is the total force due to water pressure on a long side of the pool?
33. *Civil engineering* A vertical flood gate is in the shape of an isosceles triangle which has a base of 3 m and an altitude of 2 m. If the base of the floodgate is on the surface of the water, what is the force on its face?
34. *Recreation* A rectangular swimming pool is 40.0 m long, 1.00 m deep at the shallow end, and 3.00 m deep at the deep end. Assuming that all of the sides are vertical, find the total force on the long side of the swimming pool when it is full.
35. *Transportation* A tanker truck is in the shape of a right-circular cylinder. The radius of the cylinder is 1.6 m and the density of gasoline is about $\rho = 680\,\text{kg/m}^3$. If the tanker is half filled, what is the total force due to the pressure of the gasoline on one end of the cylinder?
36. *Transportation* A tanker truck is in the shape of a right-circular cylinder with a radius of 4 ft. If the tank is half full of gasoline, and the weight of gasoline is about 42.5 lb/ft^3, what is the total force due to the pressure of the gasoline on one end of the tanker?

In Your Words

37. Explain what is meant by the concept of work as used in this text.
38. Without looking in the text, describe the four guidelines for finding the work done during pumping.

CHAPTER 7 REVIEW

Important Terms and Concepts

Arc length
Average value
Centroid
Fluid pressure
Hooke's law
Mass
Moment of inertia
Moments
  First
  Second
Radius of gyration
Root mean square
Surface area
Volumes of revolution
  Disk method
  Shell method
  Washer method
Work

Review Exercises

In Exercises 1–4, find the average value and the root mean square of the function over the given interval.

1. $f(x)=4x$, over $[1,6]$
2. $g(x)=x^3$, over $[0,2]$
3. $h(x)=x^2-4$, over $[2,4]$
4. $k(t)=4.9t^2-2.8t-4$, over $[0,2.5]$

In Exercises 5–12, find (a) the volume and (b) the centroid of the solid of revolution generated by revolving the enclosed region about the indicated axis.

5. $y=6x$, $x=1$, $x=5$, $y=0$, about the x-axis
6. $y=-x$, $x=1$, $x=4$, $y=0$, about the y-axis
7. $y=x^4$, $x=1$, $x=2$, $y=0$, about the x-axis
8. $y=\sqrt[3]{x^2}$, $x=0$, $x=1$, $y=0$, about the x-axis
9. $y=x^3$, $y=\sqrt[3]{x}$, about the y-axis
10. $y=x^3$, $y=\sqrt[3]{x}$, about the x-axis
11. $y=x^2$, $y=9-x^2$, about the y-axis
12. $y=x^3$, $y=4x$, about the line $x=5$

In Exercises 13–14, find the lengths of the curves over the indicated interval.

13. $x=\sqrt{y^3}$ from $y=0$ to $y=4$
14. $y=\sqrt{8x^3}$ from $x=0$ to $x=2$

In Exercises 15–16, find (a) the lengths of the curves over the indicated interval and (b) the surface area of the solid generated by rotating the curve around the indicated axis.

15. $y=\dfrac{x^4}{4}+\dfrac{1}{8x^2}$ from $x=1$ to $x=3$, about the x-axis
16. $x=\dfrac{y^6}{6}+\dfrac{1}{16y^4}$ from $y=1$ to $y=2$, about the y-axis

Solve Exercises 17–8.

17. *Physics* An unstretched spring 0.6 m long is stretched to a length of 0.8 m by a force of 5 N. How much work is needed to stretch the spring to a length of 1.2 m?
18. *Electricity* The voltage across a 60-μF capacitor is zero. What is the voltage after 0.001 s, if a current of 0.40 A charges the capacitor?
19. *Electricity* The electric current, as a function of time, for a certain circuit is $I=4t-t^3$. What is the average value of the current, with respect to time, for the first 2 s?
20. *Physics* If a ball is dropped from the top of the Washington Monument, which is 555 ft high, what is its average height and average velocity between the time it is dropped and the time it strikes the ground?

21. *Construction* A steel beam weighing 1,000 lb hangs from a 50-ft cable. The cable weighs 5 lb/ft. How much work is done in winding 30 ft of cable with a winch?

22. *Civil engineering* A dam is shaped like a trapezoid with a height of 100 ft. It is 400 ft long at the top and 200 ft long at the bottom. When the water level behind the dam is level with the top, what is the total force that the water exerts on the dam?

23. *Petroleum engineering* A tank is in the shape of a right-circular cylinder with a radius of 6 m and a height of 16 m. If it is filled with petroleum, how much work is needed to pump it out the top? (Let $\rho = 880\,\text{kg/m}^3$.)

24. *Waste technology* A tank is in the shape of an inverted cone with a radius of 6 ft and a height of 16 ft. If it is filled with water, how much work is needed to pump it out the top?

CHAPTER 7 TEST

1. Find the average value of $f(x) = 3x^2 + 1$ over the interval $[1, 5]$.

2. Determine the root mean square of $g(x) = x^3 - 1$ over $[1, 3]$.

3. Find the volume of the solid generated by revolving the region bounded by $y = \sqrt{9x+1}$, $x = 0$, $x = 7$, and the x-axis about the x-axis.

4. Find the centroid of the solid of revolution generated by revolving the region bounded by $y = x^2 + 1$, $x = 0$, $x = 4$, and $y = 0$, about the x-axis.

5. Find the length of the curve $y = 1 + x^{3/2}$ from $x = 1$ to $x = 6$.

6. Find the surface area of the solid generated by rotating the curve $y = \frac{1}{6}x^3$ from $x = 1$ to $x = 3$ around the x-axis.

7. How much work does it take to pump the water from a full upright circular-cylindrical tank of radius 5 m and height 10 m to a point 4 m above the top of the tank?

8. A vertical gate is in the shape of a right triangle with its base on top (and the right angle at the bottom), as shown in Figure 7.45. If the base is 5 ft below the surface of the water and the altitude of the gate is 3 ft, what is the force against the gate?

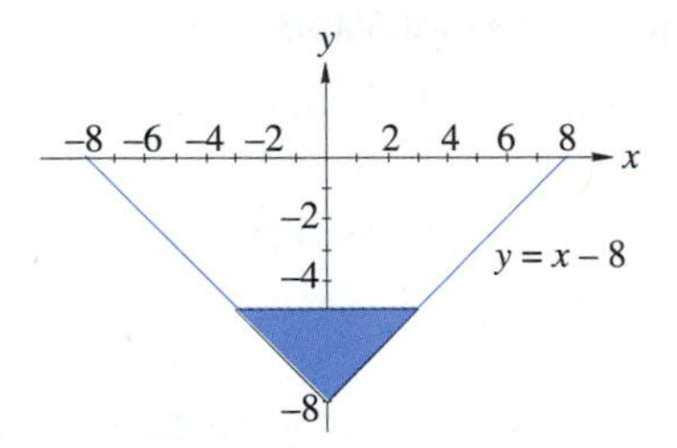

FIGURE 7.45

CHAPTER

8

Derivatives of Transcendental Functions

How far back from a movie screen like the one shown should you sit in order to have the best viewing angle? In Section 8.4, we will show how to use calculus to answer this question.

We have learned how to differentiate and integrate polynomials and certain other algebraic functions. A function that is not algebraic is called transcendental. The transcendental functions include the trigonometric, logarithmic, exponential, and inverse trigonometric functions. In this chapter, we will learn how to differentiate these four types of transcendental functions.

In Sections 8.1 and 8.2 we develop the derivatives of the six trigonometric functions and in Section 8.3 we develop the derivatives of the inverse trigonometric functions. Section 8.4 looks at some applications that use these derivatives.

Sections 8.5–8.6 examine derivatives of exponential and logarithmic functions, and in Section 8.7, we consider several applications that involve the derivatives of these functions.

8.1 DERIVATIVES OF THE SINE AND COSINE FUNCTIONS

In Chapter 4, we developed some rules for differentiating. These rules allowed us to find derivatives much faster than using the four-step method. In this section, we are going to develop rules for the derivatives of the sine and cosine functions. In order to do this, we return to the four-step method.

Before we develop derivatives for the sine and cosine, we must review three ideas. In Section 3.3 we showed that

$$\lim_{x \to 0} \frac{\sin x}{x} = 1$$

where x is measured in radians.

In Section 3.3 we showed that

$$\sin(A + B) = \sin A \cos B + \cos A \sin B$$

and

$$\sin(A - B) = \sin A \cos B - \cos A \sin B$$

There is one additional limit that we will need: $\lim\limits_{x \to 0} \dfrac{\cos x - 1}{x}$. We will determine its limit using algebra.

$$\begin{aligned}
\lim_{x \to 0} \frac{\cos x - 1}{x} &= \lim_{x \to 0} \frac{\cos x - 1}{x} \cdot \frac{\cos x + 1}{\cos x + 1} \\
&= \lim_{x \to 0} \frac{\cos^2 x - 1}{x(\cos x + 1)} \\
&= \lim_{x \to 0} \frac{-\sin^2 x}{x(\cos x + 1)} \\
&= \lim_{x \to 0} \frac{\sin x}{x} \cdot \lim_{x \to 0} \frac{-\sin x}{\cos x + 1} = 1 \cdot \frac{0}{2} = 0
\end{aligned}$$

Thus, we have shown the following:

$$\lim_{x \to 0} \frac{\cos x - 1}{x} = 0$$

Derivative of Sine Function

We are now ready to find the derivative of $f(x) = \sin x$ by the four-step method.

Step 1: $f(x+h) = \sin(x+h)$

Step 2:
$$\begin{aligned} f(x+h) - f(x) &= \sin(x+h) - \sin x \\ &= \sin x \cos h + \cos x \sin h - \sin x \\ &= \sin x(\cos h - 1) + \cos x \sin h \end{aligned}$$

Step 3:
$$\begin{aligned} \frac{f(x+h) - f(x)}{h} &= \frac{\sin x\,(\cos h - 1)}{h} + \frac{\cos x \sin h}{h} \\ &= (\sin x)\left(\frac{\cos h - 1}{h}\right) + (\cos x)\left(\frac{\sin h}{h}\right) \end{aligned}$$

Step 4:
$$\begin{aligned} \lim_{h \to 0} \frac{f(x+h) - f(x)}{h} &= \lim_{h \to 0}(\sin x)\left(\frac{\cos h - 1}{h}\right) + \lim_{h \to 0}(\cos x)\left(\frac{\sin h}{h}\right) \\ &= (\sin x)(0) + (\cos x)(1) = \cos x \end{aligned}$$

Since, $\lim_{h \to 0} \frac{f(x+h) - f(x)}{h} = (\sin x)(0) + (\cos x)(1) = \cos x$, we have the following derivative for the sine function.

Derivative of the Sine Function	$\frac{d}{dx}(\sin x) = \cos x$

A more general form of the derivative of the sine function is developed by using the chain rule. In Section 4.4, we found that if $y = f(u)$ and $u = g(x)$, then

$$y' = f'(u)g'(x)$$

or, in differential notation,

$$\frac{dy}{dx} = \frac{dy}{du} \cdot \frac{du}{dx}$$

Thus, we have the following.

Derivative of the Sine Function	$\frac{d}{dx}(\sin u) = (\cos u)\frac{du}{dx}$

We will use the chain rule in the following examples.

EXAMPLE 8.1

If $y = \sin x^2$, what is y'?

Solution First, we need to remember that $\sin x^2 = \sin(x^2)$. We will use the chain rule with $y = \sin u$ and $u = x^2$. Then

$$y' = f'(u)g'(x) = (\cos u)(2x)$$

Since $u = x^2$, we have

$$\begin{aligned} y' &= (\cos x^2)(2x) \\ \text{or} \quad &= 2x\cos x^2 \end{aligned}$$

EXAMPLE 8.2

If $y = (\sin x)^3$, find y'.

Solution We can solve this by either using the general power rule or the chain rule. We will show how to do it both ways.
Power rule:

$$\begin{aligned} y' &= 3(\sin x)^2 \left(\frac{d}{dx}\sin x\right) \\ &= 3(\sin x)^2(\cos x) \\ &= 3\sin^2 x\cos x \end{aligned}$$

[Remember, $\sin^2 x = (\sin x)^2$.]
Chain rule: Let $u = \sin x$, so $y = u^3$

$$\begin{aligned} y' &= f'(u)g'(x) \\ &= 3u^2\cos x \\ &= 3(\sin x)^2\cos x = 3\sin^2 x\cos x \end{aligned}$$

EXAMPLE 8.3

Differentiate $f(x) = \sin\sqrt{3x^2+1}$.

Solution We let $u = \sqrt{3x^2+1} = (3x^2+1)^{1/2}$. So,

$$\frac{du}{dx} = \frac{1}{2}(3x^2+1)^{-1/2}(6x) = \frac{3x}{\sqrt{3x^2+1}}$$

Now $f(u) = \sin u$ and so

$$\begin{aligned} f'(x) &= \cos u\frac{du}{dx} \\ &= \left(\cos\sqrt{3x^2+1}\right)\left(\frac{3x}{\sqrt{3x^2+1}}\right) \\ &= \frac{3x}{\sqrt{3x^2+1}}\cos\sqrt{3x^2+1} \\ &= \frac{3x\cos\sqrt{3x^2+1}}{\sqrt{3x^2+1}} \end{aligned}$$

Derivative of Cosine Function

Next we find the derivative of $y = \cos x$. We will use the identity $\cos x = \sin(\frac{\pi}{2} - x)$ and the chain rule with $y = \sin u$ and $u = \frac{\pi}{2} - x$.

$$\begin{aligned} y &= \cos x = \sin\left(\frac{\pi}{2} - x\right) = \sin u \\ y' &= \frac{dy}{du} \cdot \frac{du}{dx} = (\cos u)(-1) = -\cos u \\ &= -\cos\left(\frac{\pi}{2} - x\right) = -\sin x \end{aligned}$$

Again, using the chain rule, we get the more general form.

Derivative of the Cosine Function $\quad \dfrac{d}{dx}(\cos u) = -\sin u \dfrac{du}{dx}$

EXAMPLE 8.4

If $y = \cos(3x + 1)$, determine y'.

Solution If we let $u = 3x + 1$, then $y = \cos u$ and

$$\begin{aligned} y' &= f'(u)g'(x) \\ &= (-\sin u)(3) = -3\sin u \\ &= -3\sin(3x + 1) \end{aligned}$$

EXAMPLE 8.5

Find the derivative of $y = x^2 \cos(3x + 1)$.

Solution Using the product rule we have

$$y' = x^2 \frac{d}{dx}[\cos(3x + 1)] + [\cos(3x + 1)] \frac{d}{dx}(x^2)$$

Since $\dfrac{d}{dx}[\cos(3x + 1)] = -3\sin(3x + 1)$, we have

$$y' = -3x^2 \sin(3x + 1) + 2x\cos(3x + 1)$$

EXAMPLE 8.6

Differentiate: $f(x) = \dfrac{\sin 3x}{\cos^2 x}$.

Solution We will use the quotient rule.

$$f'(x) = \frac{\cos^2 x \frac{d}{dx}(\sin 3x) - \sin 3x \frac{d}{dx}\cos^2 x}{(\cos^2 x)^2}$$

$$= \frac{\cos^2 x(3\cos 3x) - \sin 3x(2\cos x)(-\sin x)}{\cos^4 x}$$

$$= \frac{3\cos^2 x \cos 3x + 2\sin x \sin 3x \cos x}{\cos^4 x}$$

$$= \frac{3\cos x \cos 3x + 2\sin x \sin 3x}{\cos^3 x}$$

Application

EXAMPLE 8.7

The voltage, V, in volts, available from the wall outlet of a residential house is given by $V = 120\sin 377t$, where t is the time, in seconds. Find the rate of change of the voltage with respect to time when $t = 4.50$ ms.

Solution The rate of change of the voltage with respect to time means that we want to find $\frac{dV}{dt}$. Differentiating, we obtain $\frac{dV}{dt} = V' = 120(377)\cos 377t = 45240\cos 377t$.

Converting $t = 4.50$ ms to seconds, we obtain $t = 4.50\text{ ms} = 0.00450\text{ s}$. Substituting this value for t in our derived formula for V', we obtain $V' = 45240 \times \cos 377(0.0045) = -5671.869$. The voltage is changing at the rate of -5671.869 V/s.

Exercise Set 8.1

Differentiate the functions in Exercises 1–30.

1. $y = \sin 3x$
2. $y = \cos 4x$
3. $y = 3\cos 2x$
4. $y = 5\sin 6x$
5. $y = \sin(x^2 + 1)$
6. $y = \cos(x^3 - 5)$
7. $y = 4\sin^2 3x$
8. $y = 5\cos^3 2x$
9. $y = \cos(3x^2 - 2)$
10. $y = \sin(2x^3 + 1)$
11. $y = \sin\sqrt{x}$
12. $y = \cos x^{3/2}$
13. $y = \cos\sqrt{2x^3 - 4}$
14. $y = \sin^2\sqrt{x+1}$
15. $y = x^2 + \sin^2 x$
16. $y = \cos x + \sin x$
17. $y = \sin x \cos x$
18. $y = (\sin x - \cos x)^2$
19. $y = \dfrac{2\cos x}{\sin 2x}$
20. $y = \dfrac{\sin 2x}{2x}$
21. $y = x^2 \sin x$
22. $y = x^3 \cos x$
23. $y = \sqrt{x}\sin x$
24. $y = \dfrac{\sin^3 x}{x}$
25. $y = (\sin 2x)(\cos 3x)$
26. $y = x^2\cos(3x^2 - 1)$
27. $y = \sin^3(x^4)$
28. $y = \sqrt{\cos\sqrt{x}}$
29. $y = \sin^2 x + \cos^2 x$
30. $y = \sin^2 x - 2\cos^2 x$

Solve Exercises 31–48.

31. If $f(x) = \sin x$, what is $f''(x)$?

32. If $g(x) = \cos x$, what is $g''(x)$?

33. Find y''' if $y = \sin x$.

34. Find y''' if $y = \cos x$.

35. Find $\dfrac{d^4}{dx^4}(\sin x)$.

36. Find $\dfrac{d^4}{dx^4}(\cos x)$.

37. **(a)** Determine the equation of the line tangent to $f(x) = \sqrt{2}\cos x$ at $x = \frac{\pi}{4}$.

(b) Graph f and the tangent line to f at $x = \dfrac{\pi}{4}$ on the same coordinate system.

38. **(a)** Determine the equation of the line tangent to $g(x) = 2\cos^2 x$ at $x = \dfrac{2\pi}{3}$.

(b) Graph g and the tangent line to g at $x = \dfrac{2\pi}{3}$ on the same coordinate system.

39. **(a)** Determine the equation of the normal line to $h(x) = \sin 4x$ at $\left(\dfrac{\pi}{2}, 0\right)$.

(b) Graph h and the normal line to h at $\left(\dfrac{\pi}{2}, 0\right)$ on the same coordinate system.

40. **(a)** Determine the equation of the normal line to $j(x) = \sin x \cos 2x$ at $\left(\dfrac{\pi}{4}, 0\right)$.

(b) Graph j and the normal line to j at $\left(\dfrac{\pi}{4}, 0\right)$ on the same coordinate system.

41. *Electronics* The voltage, V, in volts, available from the wall outlet of a residential house is given by $V = 120\sin 377t$, where t is the time, in seconds. Find the rate of change of the voltage with respect to time when $t = 1.571$ s.
-3470 V

42. *Automotive engineering* The motion of an automobile engine piston approximates simple harmonic motion given by $s = 4.8\cos 75t$, where y is the displacement in centimeters and t is in seconds. Find the velocity and acceleration when $t = 0.01$ s.

43. *Electronics* The charge across a capacitor is $q(t) = 0.25\cos(t - 1.45)$. Find the current to the capacitor at $t = 7.6$ s.

44. *Navigation* A nautical mile varies with latitude. The equation $l = 6077 - 31\cos 2\theta$ gives the length of a nautical mile in feet at latitude θ, in radians. Find the rate at which l changes with respect to θ when $\theta = \frac{\pi}{3}$.

45. *Optics* According to Malus's law the amount of light transmitted, I, in terms of the angle of incidence, θ, and the maximum intensity of light transmitted, M, is given by $I = M\cos^2\theta$. Find the rate at which I changes with respect to θ when $\theta = \frac{\pi}{3}$.

46. *Lighting technology* The intensity of reflected light, I_{L_r}, increases with the angle of incidence, θ, as given by

$$I_{L_r} = I_{L_i} - I_{L_i}(1 - \mu_r)\cos\theta_i$$

Find the rate at which I_{L_r} changes with respect to θ_i when $\theta_i = \dfrac{\pi}{6}$, $\mu_r = 0.6$, and $I_{L_i} = 100$ cd.

47. *Electricity* The current in an electric circuit is given by $I = 2\sin t + \cos t$. Find the maximum and minimum values of the current for one cycle.

48. *Electronics* The magnetic field, B, in tesla (T), produced by a rotating armature varies with the angle of rotation according to the equation

$$B = \frac{2\sin\theta}{1 + \cos^2(2\theta)}$$

Find the rate at which B changes with respect to θ when $\theta = \dfrac{\pi}{6}$.

In Your Words

49. The derivatives of the sine and cosine functions are very similar. Explain how you can tell which derivative has the minus sign.

50. Many objects move in circles or, like pistons, have linear motion that is the result of moving something in a circular motion. Think of a place in your technology area of interest where this is the case. Write an application that uses derivatives based on your observations. Give your problem to a classmate and see if he or she understands and can solve your problem. Rewrite the problem as necessary to remove any difficulties encountered by your classmate.

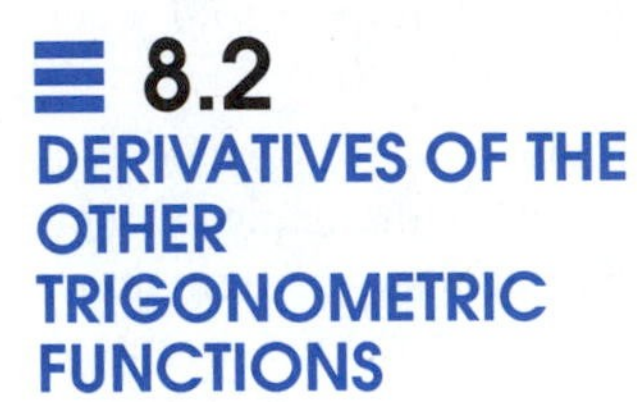

8.2 DERIVATIVES OF THE OTHER TRIGONOMETRIC FUNCTIONS

In Section 8.1, we developed the formulas for the derivatives of the sine and cosine functions. In this section, we will use those formulas to develop the derivatives for the other trigonometric functions.

Derivative of Tangent Function

In order to find the derivatives of the tangent function, we use one of the basic trigonometric identities.

$$\tan u = \frac{\sin u}{\cos u}$$

Using this, we get

$$\begin{aligned}
\frac{d}{dx}\tan u &= \frac{d}{dx}\left(\frac{\sin u}{\cos u}\right) \\
&= \frac{\cos u \dfrac{d}{dx}\sin u - \sin u \dfrac{d}{dx}\cos u}{\cos^2 u} \\
&= \frac{\cos^2 u \dfrac{du}{dx} + \sin^2 u \dfrac{du}{dx}}{\cos^2 u} \\
&= \frac{(\cos^2 u + \sin^2 u)\dfrac{du}{dx}}{\cos^2 u}
\end{aligned}$$

Since $\cos^2 u + \sin^2 u = 1$, we have

$$\begin{aligned}
\frac{d}{dx}\tan u &= \frac{1}{\cos^2 u}\frac{du}{dx} \\
&= \sec^2 u \frac{du}{dx}
\end{aligned}$$

So, we have the derivative of the tangent function.

Derivative of Tangent Function $\qquad \dfrac{d}{dx}\tan u = \sec^2 u \dfrac{du}{dx}$

Derivatives of Secant, Cosecant, and Cotangent Functions

In a similar manner, using $\sec u = \dfrac{1}{\cos u}$, we have

$$\begin{aligned}\frac{d}{dx}\sec u &= \frac{d}{dx}\left(\frac{1}{\cos u}\right)\\ &= \frac{\cos u \frac{d}{dx}(1) - (1)\frac{d}{dx}\cos u}{\cos^2 u}\\ &= \frac{0-(-\sin u)\frac{du}{dx}}{\cos^2 u}\\ &= \frac{\sin u}{\cos^2 u}\frac{du}{dx}\\ &= \frac{1}{\cos u}\frac{\sin u}{\cos u}\frac{du}{dx}\\ &= \sec u \tan u \frac{du}{dx}\end{aligned}$$

Thus, we have the derivative of the secant function.

Derivative of Secant Function $\qquad \dfrac{d}{dx}\sec u = \sec u \tan u \dfrac{du}{dx}$

The development of the derivatives for the remaining two trigonometric functions will be done as exercises.

Derivative of Cotangent and Cosecant Functions

$$\frac{d}{du}\cot u = -\csc^2 u \frac{du}{dx}$$

$$\frac{d}{du}\csc u = -\csc u \cot u \frac{du}{dx}$$

EXAMPLE 8.8

Differentiate $y = \sqrt{\tan 2x}$.

Solution We rewrite this as $y = (\tan 2x)^{1/2}$ and use the general power rule:

$$\begin{aligned}y' &= \frac{1}{2}(\tan 2x)^{-1/2}\frac{d}{dx}(\tan 2x)\\ &= \frac{1}{2}(\tan 2x)^{-1/2} 2\sec^2 2x\\ &= \frac{\sec^2 2x}{\sqrt{\tan 2x}}\end{aligned}$$

EXAMPLE 8.9

Differentiate $y = x\csc^2(4x)$.

Solution We will use the product rule.

$$\begin{aligned} y' &= x\frac{d}{dx}[\csc^2(4x)] + \csc^2(4x)\frac{d}{dx}(x) \\ &= x(2\csc 4x)\frac{d}{dx}(\csc 4x) + \csc^2 4x \\ &= 2x\csc 4x(-\csc 4x\cot 4x)4 + \csc^2 4x \\ &= -8x\csc^2 4x\cot 4x + \csc^2 4x \\ &= (\csc^2 4x)(1 - 8x\cot 4x) \end{aligned}$$

EXAMPLE 8.10

Differentiate $f(x) = \sec^3\sqrt{x}$.

Solution $$\begin{aligned} f'(x) &= 3\sec^2\sqrt{x}\frac{d}{dx}\left(\sec\sqrt{x}\right) \\ &= 3\sec^2\sqrt{x}\left(\sec\sqrt{x}\tan\sqrt{x}\right)\frac{d}{dx}\sqrt{x} \\ &= \frac{3}{2\sqrt{x}}\sec^3\sqrt{x}\tan\sqrt{x} \end{aligned}$$

EXAMPLE 8.11

Differentiate $g(x) = \sin 2x\tan x^2$.

Solution We will use the product rule.

$$\begin{aligned} g'(x) &= \sin 2x\frac{d}{dx}\tan x^2 + \tan x^2\frac{d}{dx}\sin 2x \\ &= \sin 2x\sec^2 x^2\frac{d}{dx}x^2 + \tan x^2\cos 2x\frac{d}{dx}(2x) \\ &= 2x\sin 2x\sec^2 x^2 + 2\tan x^2\cos 2x \end{aligned}$$

EXAMPLE 8.12

Use implicit differentiation to find y', if $y = x + \cot(xy)$.

Solution Using the implicit differentiation and the chain rule produces

$$\begin{aligned} y' &= 1 + [-\csc^2(xy)]\frac{d}{dx}(xy) \\ &= 1 - [\csc^2(xy)](xy' + y) \\ y'[1 + x\csc^2(xy)] &= 1 - y\csc^2(xy) \\ y' &= \frac{1 - y\csc^2(xy)}{1 + x\csc^2(xy)}, \text{ if } x\csc^2 xy \neq -1 \end{aligned}$$

The derivatives of the six trigonometric functions are summarized below.

Derivatives of the Trigonometric Functions

$$\frac{d}{dx}\sin u = \cos u\frac{du}{dx} \qquad \frac{d}{du}\csc u = -\csc u\cot u\frac{du}{dx}$$

$$\frac{d}{dx}\cos u = -\sin u\frac{du}{dx} \qquad \frac{d}{du}\sec u = \sec u\tan u\frac{du}{dx}$$

$$\frac{d}{dx}\tan u = \sec^2 u\frac{du}{dx} \qquad \frac{d}{du}\cot u = -\csc^2 u\frac{du}{dx}$$

Exercise Set 8.2

Differentiate each of the functions in Exercises 1–30.

1. $y = \tan^2\sqrt{x}$
2. $y = \tan 4x$
3. $y = \sec 5x$
4. $y = \cot(1+2x)$
5. $y = \csc(2x-1)$
6. $y = \tan\frac{x}{2}$
7. $y = \sin x\tan x$
8. $y = \sin x\cot x$
9. $y = \tan^3 x$
10. $y = \cot^3 4x$
11. $y = \sec^4(x^2)$
12. $y = \csc^3\sqrt{x^3}$
13. $y = \tan x\cot x$
14. $y = \sec x\csc x$
15. $y = \sin^2 x\cot x$
16. $y = \sin 5x^2$
17. $y = \tan\frac{1}{x}$
18. $y = \cot\sqrt{x^2+1}$
19. $y = \sqrt{1+\tan x^2}$
20. $y = \sqrt{\tan x+\cot x}$
21. $y = \frac{\tan x}{1+\sec x}$
22. $y = (\csc x+\cot x)^3$
23. $y = \frac{\cot x}{1-\csc x}$
24. $y = \sqrt{x}+\tan\sqrt{x}$
25. $y = (\csc x+2\tan x)^3$
26. $y = \sqrt{1+\cot^2 x}$
27. $y = (\tan 2x)^{3/5}$
28. $y = x\csc x$
29. $y = \frac{\sec x}{1+\tan x}$
30. $y = \frac{\cos x}{1+\sec^2 x}$

Solve Exercises 31–32.

31. Develop the formula $\frac{d}{dx}\cot u = -\csc^2 u\frac{du}{dx}$.
32. Develop the formula $\frac{d}{dx}\csc u = -\csc u\cot u\frac{du}{dx}$.

In Exercises 33–36, find the second derivative of the given function.

33. $y = \tan 2x$
34. $y = \sec 3x$
35. $y = x\tan x$
36. $y = \frac{\cos x}{x}$

In Exercises 37–40, use implicit differentiation to find y'.

37. $y = x\sin y$
38. $\sin xy + xy = 0$
39. $x+y = \sin(x+y)$
40. $y^2 = \tan 4xy$

Solve Exercises 41–44.

41. *Optics* The index of refraction of the medium that light enters is given by $n = \csc\theta$, where θ is the critical angle. Find the rate of change of n with respect to θ when $\theta = \frac{\pi}{8}$.
42. *Aeronautics* An airplane that is flying horizontally at a constant speed of 360 mph is approaching a control tower at a height of 6,250 ft above an observer in the tower. At what rate is the angle of elevation changing when $\theta = 37°$?

43. *Electrical engineering* The strength of a radar signal, in millivolts, varies with the orientation of the antenna according the the formula $I = 16\cot^2(2\theta)$. Find the rate of change of signal strength when $\theta = \frac{7\pi}{16}$.

44. *Physics* If the displacement in meters of an object at time t in seconds is given by the equation $s = 8t^2 \tan t^2$, $0 \le t \le 1.5$, find its velocity and acceleration when $t = 0.8$ s.

In Your Words

45. Describe how you can tell the difference between the derivative of the tangent and the derivative of cotangent functions.

46. Describe how you can tell the difference between the derivative of the tangent and the derivative of secant functions.

8.3 DERIVATIVES OF INVERSE TRIGONOMETRIC FUNCTIONS

In an earlier course you studied the inverse trigonometric functions. As a typical example, the inverse sine function was written as either arcsin or $\sin^{-1}$. If $y = \arcsin x$, then $\sin y = x$. Because the trigonometric functions are periodic, the values of y lie in a certain interval. In review, we have the following inverse trigonometric functions.

Inverse Trigonometric Functions

Function	Domain	Range
$y = \sin^{-1} x$	$-\frac{\pi}{2} \le y \le \frac{\pi}{2}$	$-1 \le x \le 1$
$y = \cos^{-1} x$	$0 \le y \le \pi$	$-1 \le x \le 1$
$y = \tan^{-1} x$	$-\frac{\pi}{2} < y < \frac{\pi}{2}$	x is any real number.
$y = \cot^{-1} x$	$0 < y < \pi$	x is any real number.
$y = \sec^{-1} x$	$0 \le y < \frac{\pi}{2}$ or $\pi \le y \le \frac{3\pi}{2}$	$\lvert x\rvert \ge 1$
$y = \csc^{-1} x$	$0 < y \le \frac{\pi}{2}$ or $\pi < y \le \frac{3\pi}{2}$	$\lvert x\rvert \ge 1$

To find the derivative of $y = \sin^{-1} u$, we write it as $u = \sin y$ and differentiate implicitly.

$$\frac{du}{dx} = \left(\frac{d}{dy}\sin y\right)\frac{dy}{dx}$$

$$\frac{du}{dx} = \cos y\,\frac{dy}{dx}$$

Solving for $\frac{dy}{dx}$ we get

$$\frac{dy}{dx} = \frac{1}{\cos y}\frac{du}{dx}$$

One of the Pythagorean identities states that $\sin^2 y + \cos^2 y = 1$. Solving this for $\cos y$, we get $\cos y = \pm\sqrt{1-\sin^2 y}$. Since y is in $\left[-\frac{\pi}{2}, \frac{\pi}{2}\right]$, we know $\cos y \geq 0$, and so we select the positive square root. Thus, $\cos y = \sqrt{1-\sin^2 y}$. But, $u = \sin y$ and so $\cos y = \sqrt{1-u^2}$. Thus, we have the derivative of the inverse sine:

Derivative of Sin^{-1} $\quad \frac{d}{dx}\sin^{-1} u = \frac{1}{\sqrt{1-u^2}}\frac{du}{dx}, \quad (|u| < 1)$

EXAMPLE 8.13

Differentiate $y = \sin^{-1}(3x^2)$.

Solution We will use the chain rule with $y = \sin^{-1} u$ and $u = 3x^2$. Then

$$\begin{aligned} y' &= \frac{1}{\sqrt{1-(3x^2)^2}}\left[\frac{d}{dx}(3x^2)\right] \\ &= \frac{6x}{\sqrt{1-9x^4}} \end{aligned}$$

EXAMPLE 8.14

Find y', if $y = \sqrt{\sin^{-1}(4x-3)}$.

Solution We will use an expanded version of the chain rule where $y' = \frac{dy}{dx} = \frac{dy}{du}\cdot\frac{du}{dv}\cdot\frac{dv}{dx}$ with $y = \sqrt{u} = u^{1/2}$, $u = \sin^{-1} v$, and $v = 4x-3$.

$$\begin{aligned} y' &= \left(\frac{d}{du}u^{1/2}\right)\left(\frac{d}{dv}\sin^{-1} v\right)\left[\frac{d}{dx}(4x-3)\right] \\ &= \left(\frac{1}{2}u^{-1/2}\right)\left(\frac{1}{\sqrt{1-v^2}}\right)(4) \\ &= \frac{2}{\sqrt{u}\sqrt{1-v^2}} \\ &= \frac{2}{\sqrt{\sin^{-1}(4x-3)}\sqrt{1-(4x-3)^2}} \end{aligned}$$

Through a process similar to the one we used to develop the derivative of $y = \sin^{-1} x$, we can determine the following derivative of $\cos^{-1}$.

Derivative of Cos^{-1} $\quad \dfrac{d}{dx}\cos^{-1} u = \dfrac{-1}{\sqrt{1-u^2}}\dfrac{du}{dx}, \quad (|u| < 1)$

Note The derivative of the arccos function is the negative of the derivative of the arcsin function.

EXAMPLE 8.15

Find y', if $y = \cos^{-1}(2x^3 + x)$.

Solution Using the chain rule, we let $y = \cos^{-1} u$ and $u = 2x^3 + x$. Then,

$$\begin{aligned} y' &= \frac{dy}{du} \cdot \frac{du}{dx} \\ &= \frac{-1}{\sqrt{1-u^2}}(6x^2+1) \\ &= \frac{-(6x^2+1)}{\sqrt{1-(2x^3+x)^2}} \end{aligned}$$

To find a formula for the derivative of $y = \tan^{-1} u$, we write $u = \tan y$ and differentiate implicitly with respect to x.

$$\begin{aligned} \frac{du}{dx} &= \frac{d}{dy}\tan y \frac{dy}{dx} \\ &= \sec^2 y \frac{dy}{dx} \end{aligned}$$

Solving for $\dfrac{dy}{dx}$, we get

$$\frac{dy}{dx} = \frac{1}{\sec^2 y}\frac{du}{dx}$$

Since $\sec^2 y = 1 + \tan^2 y$ and $u = \tan y$, we have

$$\frac{dy}{dx} = \frac{1}{1+\tan^2 y}\frac{du}{dx} = \frac{1}{1+u^2}\frac{du}{dx}$$

We conclude the following.

Derivative of Tan^{-1} $$\frac{d}{dx}\tan^{-1}u = \frac{1}{1+u^2}\frac{du}{dx}$$

In a similar manner, we can show the following.

Derivative of Cot^{-1} $$\frac{d}{dx}\cot^{-1}u = \frac{-1}{1+u^2}\frac{du}{dx}$$

Note The derivative of the arccot function is the negative of the derivative of the arctan function.

EXAMPLE 8.16

Differentiate $y = (\tan^{-1} 3x)^2$.

Solution We will use the chain rule with $y = u^2$, $u = \tan^{-1} v$, and $v = 3x$. With these substitutions,

$$\begin{aligned}
\frac{dy}{dx} &= \frac{dy}{dy}\cdot\frac{du}{dv}\cdot\frac{dv}{dx} \\
&= \frac{d}{du}(u^2)\cdot\frac{d}{dv}(\tan^{-1} v)\frac{d}{dx}(3x) \\
&= 2u\left(\frac{1}{1+v^2}\right)3 = \frac{6u}{1+v^2} \\
&= \frac{6(\tan^{-1} 3x)}{1+9x^2}
\end{aligned}$$

EXAMPLE 8.17

Differentiate $y = x^4 \arctan x^2$.

Solution We need to use the product rule to solve this problem. Remember that $\arctan x^2 = \tan^{-1} x^2$.

$$\begin{aligned}
y' &= x^4\frac{d}{dx}\arctan x^2 + \left(\frac{d}{dx}x^4\right)\arctan x^2 \\
&= x^4\left(\frac{1}{1+x^4}\right)2x + 4x^3 \arctan x^2 \\
&= \frac{2x^5}{1+x^4} + 4x^3 \arctan x^2
\end{aligned}$$

We will have less opportunity to use the derivatives of the other two inverse trigonometric functions. However, we present them here in a summary of the derivatives of all six inverse trigonometric functions.

Derivatives of Inverse Trigonometric Functions

$$\frac{d}{dx}\sin^{-1}u = \frac{1}{\sqrt{1-u^2}}\frac{du}{dx}, \qquad |u| < 1$$

$$\frac{d}{dx}\cos^{-1}u = \frac{-1}{\sqrt{1-u^2}}\frac{du}{dx}, \qquad |u| < 1$$

$$\frac{d}{dx}\tan^{-1}u = \frac{1}{1+u^2}\frac{du}{dx}$$

$$\frac{d}{dx}\cot^{-1}u = \frac{-1}{1+u^2}\frac{du}{dx}$$

$$\frac{d}{dx}\sec^{-1}u = \frac{1}{u\sqrt{u^2-1}}\frac{du}{dx}, \qquad |u| > 1$$

$$\frac{d}{dx}\csc^{-1}u = \frac{-1}{u\sqrt{u^2-1}}\frac{du}{dx}, \qquad |u| > 1$$

Exercise Set 8.3

Differentiate the given function in Exercises 1–28.

1. $y = \sin^{-1} 2x$
2. $y = \cos^{-1} 4x$
3. $y = \tan^{-1} \frac{x}{2}$
4. $y = \sin^{-1}\left(\frac{x}{3}\right)$
5. $y = \cos^{-1}(1-x^2)$
6. $y = \sin^{-1}\sqrt{x}$
7. $y = \sin^{-1}(1-2x)$
8. $y = \tan^{-1}(4x-1)$
9. $y = \cos^{-1}(x^3-x)$
10. $y = \sin^{-1}(x-3)^2$
11. $y = \sec^{-1}(4x+2)$
12. $y = \csc^{-1}(3x+1)$
13. $y = x\sin^{-1}x$
14. $y = \sin^{-1}\sqrt{x+1}$
15. $y = x^2\cos^{-1}x$
16. $y = \cos^{-1}\sqrt{1-x^2}$
17. $y = \tan^{-1}\left(\frac{2x-1}{2x}\right)$
18. $y = \csc^{-1}\sqrt{x}$
19. $y = x\tan^{-1}(x+1)$
20. $y = (1+\tan^{-1}x)^2$
21. $y = \sqrt{\sin^{-1}(1-x^2)}$
22. $y = \sqrt{1-x^2}\sin^{-1}x$
23. $y = x\cot^{-1}(1+x^2)$
24. $y = \sin^{-1}\left(\frac{x-1}{x+1}\right)$
25. $y = \frac{\sin^{-1}x}{x}$
26. $y = \frac{\arccos x^2}{x}$
27. $y = \sqrt[3]{\arcsin x}$
28. $y = \sqrt[3]{\arctan x^2}$

Solve Exercises 29–32.

29. Show that $\frac{d}{du}(\cos^{-1}u) = \frac{-1}{\sqrt{1-u^2}}\frac{du}{dx}$.

30. Show that $\frac{d}{du}(\cot^{-1}u) = \frac{-1}{1+u^2}\frac{du}{dx}$.

31. *Photography* A race car is traveling at a speed of 320 ft/s as it passes a television camera at trackside 42 ft away. Find $\frac{d\theta}{dt}$, the rate at which the camera must turn to follow the car when $\theta = 15°$.

32. *Aeronautics* The angle of elevation between an observer and an airplane that is flying horizontally directly toward the observer is changing. If the airplane is flying at a constant speed of 270.0 mph at a height of 7,920 ft above an observer, at what rate is the angle of elevation changing when the plane is at a horizontal distance of 6 mi from the observer?

In Your Words

33. The answer to Exercise 31 is 7.1086 rad/s. Is it practical to place a television camera at this location? If not, what are some alternative locations? Justify your answers by determining the rate at which the camera must turn to follow the car when it covers $\theta = 15°$ at each of these locations.

34. The derivatives for the inverse trigonometric functions seem naturally to group themselves into three groups of two. Which derivatives would you group? Why did you decide to put these in one group? How are you going to remember the derivatives of the functions in the other groups? Are there any other ways you could group these functions that might help you remember their derivatives?

8.4 APPLICATIONS

In this section, we have the opportunity to apply the derivatives of the trigonometric and inverse trigonometric functions. The manner in which these functions are applied is essentially the same as for algebraic functions. We will use derivatives to find slopes, tangents and normal lines, maxima and minima, related rates, and differential problems.

Application

EXAMPLE 8.18

If $f(x) = 2\sqrt{3}\sin x + \cos 2x$, find the extrema and sketch the graph of one complete cycle of f.

Solution Since the period of $\sin x$ is 2π and the period of $\cos 2x$ is π, the period of f is 2π. We will graph one period from $x = 0$ to $x = 2\pi$. Differentiating, we get $f'(x) = 2\sqrt{3}\cos x - 2\sin 2x$.

We know that $\sin 2x = 2\sin x \cos x$. Making this substitution, we get

$$\begin{aligned} f'(x) &= 2\sqrt{3}\cos x - 4\sin x \cos x \\ &= \cos x(2\sqrt{3} - 4\sin x) \end{aligned}$$

The critical values occur when $f'(x) = 0$. This happens when $\cos x = 0$ or when $2\sqrt{3} - 4\sin x = 0$. Now, $\cos x = 0$ at $\frac{\pi}{2}$ and $\frac{3\pi}{2}$ on $[0, 2\pi]$, and $2\sqrt{3} - 4\sin x = 0$ if $\sin x = \dfrac{\sqrt{3}}{2}$ or when $x = \frac{\pi}{3}$ or $\frac{2\pi}{3}$.

The second derivative is $f''(x) = -2\sqrt{3}\sin x - 4\cos 2x$. If we evaluate the second derivative at each of the critical values, we find that $f''\left(\frac{\pi}{2}\right) > 0$ and $f''\left(\frac{3\pi}{2}\right) > 0$ while $f''\left(\frac{\pi}{3}\right) < 0$ and $f''\left(\frac{2\pi}{3}\right) < 0$. Thus, we have minima at $\frac{\pi}{2}$ and $\frac{3\pi}{2}$, and maxima at $\frac{\pi}{3}$ and $\frac{2\pi}{3}$.

EXAMPLE 8.18 (Cont.)

In order to determine inflection points, we determine when the second derivative is 0. If we substitute $\cos 2x = 1 - 2\sin^2 x$, we get

$$-2\sqrt{3}\sin x - 4\cos 2x = -2\sqrt{3}\sin x - 4 + 8\sin^2 x = 0$$

Using the quadratic formula, we obtain

$$\sin x = \frac{2\sqrt{3} \pm \sqrt{12+128}}{16} = \frac{2\sqrt{3} \pm \sqrt{140}}{16} = \frac{\sqrt{3} \pm \sqrt{35}}{8}$$

Solving this for x over $[0, 2\pi]$, we obtain $x \approx 1.28$, 1.87, 3.69, and 5.73 rad as possible critical values. (In degrees, these values for x are approximately 72.9°, 107.1°, 211.5°, and 328.5°.)

Summarizing we get the following results.

Minima: $\left(\frac{\pi}{2}, 2.46\right)$ and $\left(\frac{3\pi}{2}, -4.46\right)$

Maxima: $\left(\frac{\pi}{3}, 2.5\right)$ and $\left(\frac{2\pi}{3}, 2.5\right)$

Inflection points: $(1.27, 2.48)$, $(1.87, 2.48)$, $(3.69, -1.35)$, and $(5.73, -1.37)$ or, in degrees, $(73°, 2.48)$, $(107°, 2.48)$, $(211.5°, -1.35)$, and $(328.5°, -1.37)$

The graph of this function over $[0, 2\pi]$ is shown in Figure 8.1.

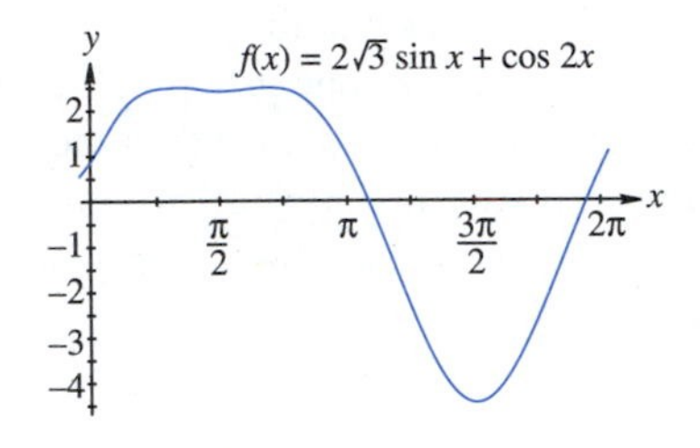

FIGURE 8.1

Application

EXAMPLE 8.19

Find the equation of the tangent and normal lines to $f(x) = x^2 + \sin x$ at $(1, 1.84)$.

Solution The slopes of the tangent and normal lines are found by evaluating the derivative of $f'(x)$ at $x = 1$. (Even though it is not stated, it should be understood that all trigonometric functions will be evaluated in radians.) Taking the derivative of f, we obtain

$$f'(x) = 2x + \cos x$$

and when $x = 1$ rad, $f'(x) \approx 2.54$.

Using the point-slope form for the equation of a line, we have $y - y_1 = f'(x)(x - x_1)$. We know that $(x_1, y_1) = (1, 1.84)$ and so $y - 1.84 = 2.54(x-1)$ or $y = 2.54x - 0.70$.

A normal line is perpendicular to a tangent to $f(x)$; its slope is $\dfrac{-1}{f'(x)}$.

Again using the point-slope form for the equation of a line, we have

$$\begin{aligned} y - 1.84 &= \frac{-1}{2.54}(x-1) \\ &= -0.39(x-1) \\ \text{or} \quad y &= -0.39x + 2.23 \end{aligned}$$

We have shown that the equation of the tangent to $f(x) = x^2 + \sin x$ at $(1, 1.84)$ is $y = 2.54x - 0.70$ and the equation of the normal line at this same point is $y = -0.39x + 2.23$.

Application

EXAMPLE 8.20

A balloon is 500 m above the ground. It is moving horizontally at the rate of 10 m/s away from an observer. If the distance from the observer to a point directly below the balloon is 100 m, how quickly is the angle of elevation changing?

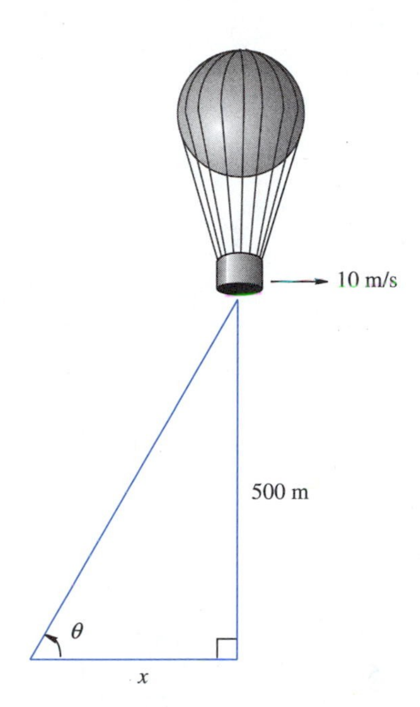

FIGURE 8.2

Solution A sketch of the problem is shown in Figure 8.2. We have $\tan\theta = \frac{500}{x}$, where x is the distance from the observer to the point directly under the balloon and θ is the angle of elevation. Because we are asking "how quickly," we will differentiate θ with respect to time t. That means we want to determine $\frac{d\theta}{dt}$. Taking the derivative, we get $\sec^2\theta\frac{d\theta}{dt} = \frac{-500}{x^2}\cdot\frac{dx}{dt}$ and so, solving for $\frac{d\theta}{dt}$, we obtain

$$\begin{aligned}\frac{d\theta}{dt} &= \frac{-500}{x^2\sec^2\theta}\cdot\frac{dx}{dt}\\ &= \frac{-500}{x^2}\cos^2\theta\cdot\frac{dx}{dt}\end{aligned}$$

We let $\tan\theta = \frac{500}{x}$, and so $\theta = \arctan\frac{500}{x}$. When $x = 100$ m, $\theta = \arctan\frac{500}{x} = \arctan 5 \approx 1.37$ (78.7°), so $\cos^2\theta = 0.038$. Now, since we were given $\frac{dx}{dt} = 10$ m/s, we have

$$\frac{d\theta}{dt} = \frac{-500}{(100)^2}(0.038)(10) = -0.019$$

The angle of elevation is decreasing at a rate of 0.019 rad/s $\approx$ 1.09 degrees/s.

Application

EXAMPLE 8.21

The screen in a movie theater, like the one in Figure 8.3a, is 30 ft high and its bottom edge is 10 ft above the level of a seated person's eyes. How far from a point directly below the screen should a person sit in order for the angle formed by the top of the screen, the person's eyes, and the bottom of the screen to be a maximum?

Solution A sketch of the problem is shown in Figure 8.3b. The desired angle that we want maximized is θ. We have let x represent the distance from the viewer to the point on the floor below the screen. From the figure, we can see that $\tan(\theta+\phi) = \frac{40}{x}$ and $\tan\phi = \frac{10}{x}$. This means that

$$\tan\theta = \tan[(\theta+\phi)-\phi]$$

EXAMPLE 8.21 (Cont.)

$$= \frac{\tan(\theta+\phi)-\tan\phi}{1+\tan(\theta+\phi)\tan\phi}$$

$$= \frac{\frac{40}{x}-\frac{10}{x}}{1+\frac{40}{x}\cdot\frac{10}{x}} = \frac{\frac{30}{x}}{1+\frac{400}{x^2}}$$

$$= \frac{30x}{x^2+400}$$

FIGURE 8.3a

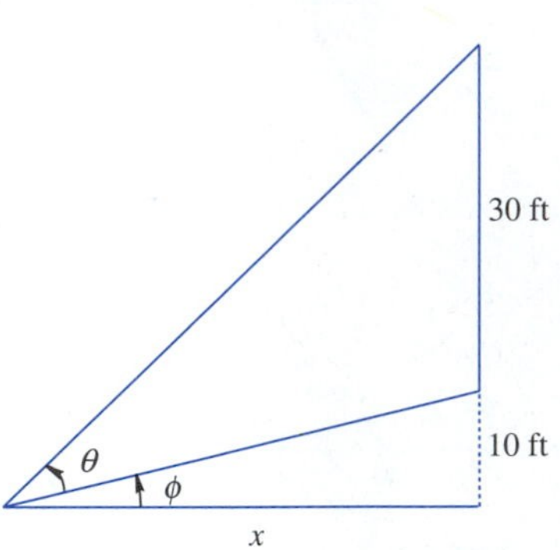

FIGURE 8.3b

Thus, $\theta = \arctan\left(\frac{30x}{x^2+400}\right)$. We want θ to be a maximum, so we take the derivative of arctan:

$$\frac{d\theta}{dx} = \frac{1}{1+\left(\frac{30x}{x^2+400}\right)^2}\left[\frac{d}{dx}\left(\frac{30x}{x^2+400}\right)\right]$$

$$= \left[\frac{(x^2+400)^2}{x^4+800x^2+900x^2+160{,}000}\right]\left[\frac{(x^2+400)30-30x(2x)}{(x^2+400)^2}\right]$$

$$= \frac{12{,}000-30x^2}{x^4+1700x^2+160{,}000}$$

The derivative is 0 when $12{,}000-30x^2 = 30(400-x^2) = 0$. The critical value is $x = 20$ and, by the first derivative test, it is a maximum. Thus, the viewer should sit 20 ft from the front of the screen.

Exercise Set 8.4

1. Show that $\sin x$ is increasing on the interval $\left[-\frac{\pi}{2}, \frac{\pi}{2}\right]$.
2. Show that $\tan x$ is increasing for all values where it is defined.
3. Show that $\arccos x$ is decreasing for all $|x| \leq 1$.
4. Show that $\arctan x$ is increasing for all values of x.

In Exercises 5–14 find all extrema and points of inflection and sketch the graph for $0 \leq x \leq 2\pi$.

5. $f(x) = 2\sin x + \cos 2x$
6. $g(x) = 2\cos x + \sin 2x$
7. $h(x) = \cos x - \sin x$
8. $j(x) = \cos^2 x - \sin x$
9. $k(x) = \cos^2 x + \sin x$
10. $m(x) = \sec x - \tan x$
11. Find an equation of the line tangent to $y = \sin x$ at $x = \frac{\pi}{6}$.
12. Find an equation of the line normal to $y = \sin x$ at $x = \frac{\pi}{6}$.
13. Find an equation of the line tangent to $y = \tan^3 x$ at $x = \frac{\pi}{4}$.
14. Find an equation of the line normal to $y = \tan^3 x$ at $x = \frac{\pi}{4}$.

Solve Exercises 15–26.

15. *Navigation* A searchlight is following an airplane flying at an altitude of 6,000 ft in a straight line over the light. The plane's velocity is 400 mph. At what rate is the searchlight turning when the distance between the light and the plane is 10,000 ft?

16. *Navigation* A ship is moving in a straight line past a lighthouse. At its closest it is 300 m from the lighthouse. The lighthouse keeper is watching the ship through a telescope. How fast does the telescope have to turn if the ship's velocity is 10 km/h and it is 400 m from the ship to the lighthouse?

17. *Navigation* The lighthouse in Exercise 16 is located 500 m from the base of a cliff that runs along a straight shoreline. The searchlight turns at the rate of 1 rpm. How fast is the light beam moving along the cliff when it is at a point on the cliff that is 500 m from the point that is closest to the lighthouse?

18. *Electronics* The power P in a certain ac circuit is to remain constant. As the impedance phase angle θ is varied, the **apparent power**, P_{app}, is defined as $P_{\text{app}} = P\sec\theta$. If $P = 15$ W, find the time-rate change of P_{app}, if θ is changing at the rate of 0.6 rad/min when $\theta = \frac{\pi}{3}$.

19. *Navigation* A radar antenna is rotating at a rate of 30 rpm. If the antenna is aboard a ship that is 8 mi from a straight shore line, how fast does the radar beam travel along the shore when the beam makes an angle of 40° with the shore?

20. *Physics* A weight hangs on a spring that is 2 m long when it is at rest with the weight attached. The weight is pulled down 0.75 m and then released. The weight oscillates up and down. The length x of the spring at any time after t s have elapsed is given by $x = 2 + 0.75\cos 2\pi t$.
   (a) What is the length of the spring when $t = 0, \frac{1}{2}$, and 1?
   (b) What is the velocity function of the weight?
   (c) What is the weight's velocity when $t = 0, \frac{1}{2}$, and 1?
   (d) What is the acceleration function of the weight?
   (e) When is the velocity a maximum?

21. *Environmental science* The amount of fluid flowing across a weir (or dam) with a V-shaped notch is approximately

$$\bar{Q} = 1.33h^{5/2}\tan\frac{\theta}{2}$$

where $\bar{Q}$ is the volume of flow given in m^3/s, h is the height in meters, and θ is the angle of the notch, as shown in Figure 8.4. Assuming that h is kept constant at 2 m while θ is increasing, how fast is $\bar{Q}$ changing when **(a)** $\theta = 30°$? **(b)** $\theta = 40°$? **(c)** At what angle θ is $\bar{Q}$ a maximum, if h is constant?

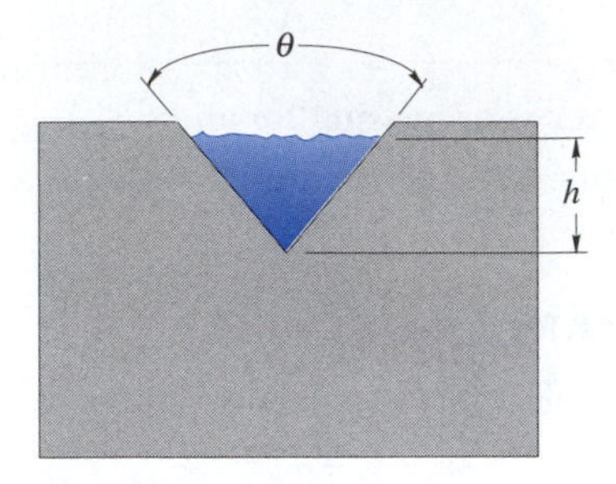

FIGURE 8.4

22. *Civil engineering* A guy wire is attached to a ring 12 m above the ground on an antenna pole. If the length s of the wire is changed, the angle between the pole and the wire is also changed. If the wire is shortened at the rate of 0.4 m/min, what is the rate of change of this angle when s is 20 m?

23. *Electricity* The current in a certain electric circuit is given by $I = \sin^2 3t$ A. At what rate is I changing when $t = 1.5$ s?

24. *Interior design* A 4-ft high picture is hung with the bottom 2.5 ft above eye level. How far away should a person stand so that the angle formed by the top of the picture, the eye, and the bottom of the picture is a maximum?

25. *Physics* A 2-kg mass is attached to a spring and is pulled out and released. At any time t, the length of the spring is $x(t) = \frac{1}{4}\cos 2t$ m. What is the maximum speed of this mass?

26. *Sheet metal technology* A gutter is to be made from a sheet of metal 30 cm wide by turning up strips of width 10 cm along each side that make equal angles θ, as shown in Figure 8.5. What angle θ will make the cross-sectional area a maximum?

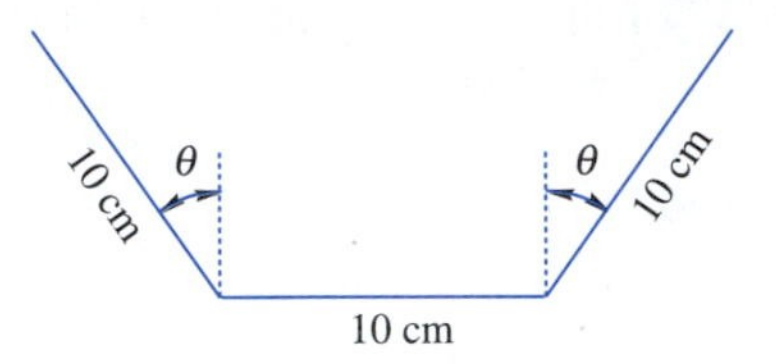

FIGURE 8.5

Use Newton's method to approximate the real roots to four decimal places in Exercises 27–30.

27. $f(x) = \cos x + x - 2$

28. $f(x) = \sin x - x^2$

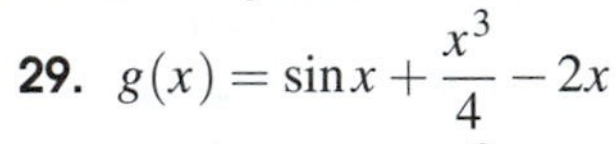

29. $g(x) = \sin x + \dfrac{x^3}{4} - 2x$

30. $h(x) = \cos x + x^2 - 3$

In Your Words

31. Write an application in your technology area of interest that requires you to use derivatives of a trigonometric function. Give your problem to a classmate and see if he or she understands and can solve your problem. Rewrite the problem as necessary to remove any difficulties encountered by your classmate.

32. As in Exercise 31, write a word problem in your technology area that requires you to use an inverse trigonometric function. This problem should use a different trigonometric function than the one in Exercise 31.

8.5 DERIVATIVES OF LOGARITHMIC FUNCTIONS

In Sections 8.5 and 8.6, we will study the derivatives of two other transcendental functions—the logarithmic and exponential functions.

The idea of a limit is often used to develop

$$\lim_{n\to\infty}\left(1+\frac{1}{n}\right)^n = e$$

or, similarly, letting $x = \frac{1}{n}$

$$\lim_{x\to 0}(1+x)^{1/x} = e$$

Any symbol can be used in place of x. For example, if we used $\frac{h}{x}$, then we would have

$$\lim_{\frac{h}{x}\to 0}\left(1+\frac{h}{x}\right)^{x/h} = e$$

Now let's look at the derivative of $f(x) = \log_a x$. Using the four-step method produces the following result.

Step 1: $f(x+h) = \log_a(x+h)$

Step 2: $f(x+h) - f(x) = \log_a(x+h) - \log_a x$

$$= \log_a\left(\frac{x+h}{x}\right)$$

$$= \log_a\left(1+\frac{h}{x}\right)$$

Step 3: $\dfrac{f(x+h)-f(x)}{h} = \dfrac{\log_a(1+h/x)}{h}$

$$= \frac{1}{h}\log_a\left(1+\frac{h}{x}\right)$$

This last expression does not appear to be of much help since it cannot be simplified by our usual methods. But, we can rewrite $\frac{1}{h}$ as $\frac{1}{x}\cdot\frac{x}{h}$, and this gives us the help we need.

$$\frac{1}{h}\log_a\left(1+\frac{h}{x}\right) = \frac{1}{x}\cdot\frac{x}{h}\log_a\left(1+\frac{h}{x}\right) = \frac{1}{x}\log_a\left(1+\frac{h}{x}\right)^{x/h}$$

We are now ready for the last step.

Step 4: $\displaystyle\lim_{h\to 0}\frac{f(x+h)-f(x)}{h} = \lim_{h\to 0}\frac{1}{x}\log_a\left(1+\frac{h}{x}\right)^{x/h}$

$$= \frac{1}{x}\lim_{h\to 0}\log_a\left(1+\frac{h}{x}\right)^{x/h}$$

Since x is a fixed number, if $h \to 0$, then $\frac{h}{x} \to 0$ and we have

$$\begin{aligned}\frac{1}{x}\lim_{h\to 0}\log_a\left(1+\frac{h}{x}\right)^{x/h} &= \frac{1}{x}\lim_{\frac{h}{x}\to 0}\log_a\left(1+\frac{h}{x}\right)^{x/h}\\ &= \frac{1}{x}\log_a e\end{aligned}$$

From the change of base formula for logarithms, we have $\log_a x = \frac{\ln x}{\ln a}$, so $\log_a e = \frac{\ln e}{\ln a} = \frac{1}{\ln a}$. Thus we have the following results for the derivative of the logarithm for the general base a.

Derivative of $y = \log_a x$

$$\begin{aligned}\frac{d}{dx}\log_a x &= \frac{1}{x}\log_a e\\ &= \frac{1}{x}\cdot\frac{1}{\ln a}\end{aligned}$$

Now $\ln x = \log_e x$ and, since $\log_e e = 1$, we have

$$\frac{d}{dx}\ln x = \frac{1}{x}\log_e e = \frac{1}{x}$$

Using the chain rule, we have the following two derivatives.

Derivative of Log Function

$$\begin{aligned}\frac{d}{du}\log_a u &= \frac{1}{u}\log_a e\frac{du}{dx}\\ &= \frac{1}{u}\cdot\frac{1}{\ln a}\frac{du}{dx}\end{aligned}$$

and

$$\frac{d}{du}\ln u = \frac{1}{u}\frac{du}{dx}$$

EXAMPLE 8.22

Differentiate $y = \log_3 x^2$.

Solution We have $a = 3$, and if we use properties of logarithms, then we can rewrite this as $y = \log_3 x^2 = 2\log_3 x$. Differentiating, we obtain $y' = 2\left(\frac{1}{x\ln 3}\right) = \frac{2}{x\ln 3}$.

EXAMPLE 8.23

Differentiate $y = \log_4(x^2+1)$.

Solution Here $a = 4$ and $u = x^2+1$, so

$$\begin{aligned} y &= \frac{1}{u}\left(\frac{1}{\ln 4}\right)\frac{d}{dx}(x^2+1) \\ &= \frac{1}{x^2+1}\left(\frac{1}{\ln 4}\right)(2x) \\ &= \frac{2x}{(x^2+1)\ln 4} \end{aligned}$$

EXAMPLE 8.24

Differentiate $y = \ln\sqrt{x}$.

Solution We will find this derivative by two different methods. The second method uses our definition for the derivative of $\ln u$.

Method 1: In the first method we will use some of the properties of logarithms.

$$y = \ln\sqrt{x} = \ln(x^{1/2}) = \frac{1}{2}\ln x$$

Differentiating, we get

$$y' = \frac{1}{2}\cdot\frac{1}{x} = \frac{1}{2x}$$

Method 2: If we let $u = \sqrt{x}$, then the given equation is $y = \ln u$, and we have

$$\begin{aligned} y' &= \frac{1}{u}\frac{du}{dx} \\ &= \frac{1}{\sqrt{x}}\cdot\frac{1}{2\sqrt{x}} \\ &= \frac{1}{2x} \end{aligned}$$

The second method was not much simpler than the first, but it gave us a method that we can use on more difficult problems.

EXAMPLE 8.25

Differentiate $y = \ln\sqrt{\dfrac{\sin^2 x}{x^3}}$.

Solution We first simplify this by using the properties of logarithms.

$$\begin{aligned} y &= \ln\sqrt{\frac{\sin^2 x}{x^3}} \\ &= \ln\left(\frac{\sin^2 x}{x^3}\right)^{1/2} = \frac{1}{2}\ln\left(\frac{\sin^2 x}{x^3}\right) \\ &= \frac{1}{2}(\ln\sin^2 x - \ln x^3) \\ &= \frac{1}{2}(2\ln\sin x - 3\ln x) \\ y &= \ln\sin x - \frac{3}{2}\ln x \end{aligned}$$

Now, we will differentiate the function $y = \ln\sin x - \frac{3}{2}\ln x$.

$$\begin{aligned} y' &= \frac{1}{\sin x}\frac{d}{dx}(\sin x) - \frac{3}{2}\left(\frac{1}{x}\right) \\ &= \frac{\cos x}{\sin x} - \frac{3}{2x} \\ &= \cot x - \frac{3}{2x} \end{aligned}$$

Exercise Set 8.5

In Exercises 1–20, differentiate the given function and simplify the resulting derivative.

1. $y = \log 5x$
2. $y = \log_5 x^3$
3. $y = \ln(x^2 + 4x)$
4. $y = \ln\sqrt{x^2 + 4x}$
5. $y = \sqrt{\ln(x^2 + 4x)}$
6. $y = \sqrt{\ln x^2 + 4x}$
7. $y = \ln\dfrac{1}{x}$
8. $y = \ln\cos x$
9. $y = \ln\tan x$
10. $y = \log_4\sqrt{x}$
11. $y = \dfrac{\ln x}{x}$
12. $y = \ln\left(\dfrac{x}{1+x}\right)$
13. $y = \sqrt{\ln\left(\dfrac{1+x}{1-x}\right)}$
14. $y = \frac{1}{2}\ln\left(\dfrac{1+x^2}{1-x^2}\right)$
15. $y = \ln\dfrac{4x^3}{\sqrt{x^2+4}}$
16. $y = \dfrac{\ln x}{x^3}$
17. $y = (\ln x)^2$
18. $y = \sin(\ln x)$
19. $y = \ln(\ln x)$
20. $y = \dfrac{x}{\ln x}$

In Exercises 21–30 find the second derivative of the given function.

21. $y = \ln x$
22. $y = x\ln x$
23. $y = \dfrac{1}{x}\ln x$
24. $y = \sqrt{\ln x}$
25. $y = \ln\sqrt{\dfrac{\sin^2 x}{x^3}}$
26. $y = \ln\sqrt[3]{\dfrac{\sin^3 x}{x}}$
27. $y = \sqrt{\ln x^2}$
28. $y = \ln(x + \cos x)$
29. $y = x^2\ln(\sin 2x)$
30. $y = (\cos x)(\ln x)$

Solve Exercises 31–34.

31. *Agriculture* Mediterranean fruit flies, better known as medflies, have been discovered in a citrus orchard. Based on their experience, the Department of Agriculture believes that the population of medflies t hours after the orchard has been sprayed with a pesticide is about $N(t) = 72 - 5t\ln(0.045t) - 0.5t$, where $0 \le t \le 36$.
(a) What will be the maximum number of medflies in the orchard?
(b) Will the medfly infestation be eradicated within the 36 h period? If so, how long did it take to eliminate all the medflies? If not, how many flies remain at the end of the 36 h period?

32. *Computer technology* The speed of the signal, s, in an ethernet line is given by $s = -kt^2\ln t$, where k is a constant and the variable, t, is a function of the thickness of the line.
(a) Determine $\frac{ds}{dt}$.
(b) What is the maximum value of s in terms of k?

33. *Environmental science* The level of ozone in a major city is a function of the time of the day or night. It has been estimated that t h after midnight in a major city the ozone level P in parts per million (ppm) in the air is about $P(t) = 0.014 - 0.008t\ln(0.025t)$.
(a) At what time is the maximum ozone level?
(b) What is the maximum level?

34. *Electronics* A cylindrical capacitor consists of a pair of oppositely charged coaxial cylinders. Its capacitance is given by the equation $C = \frac{kL}{\ln(r_b/r_a)}$ where r_b is the radius of the outer cylinder and r_a is the radius of the inner cylinder. Find $\frac{dC}{dr_a}$ when $r_b = 1.5$ cm, $r_a = 1.0$ cm, $L = 6.0$ m, and $k = 55.6$ pF/m. (Note: 1 pF $= 10^{-12}$ F.)

In Your Words

35. Describe how to differential $\log_b x$.

36. If you use the change of base formula $\log_b x = \frac{\ln x}{\ln b}$, then you only need one rule for differentiating a logarithmic function. Explain how this is done.

8.6 DERIVATIVES OF EXPONENTIAL FUNCTIONS

The logarithmic and exponential functions are inverses of each other. As a result, we can use our knowledge of differentiating logarithmic functions to show how to find the derivatives of exponential functions.

Suppose we have the exponential function $y = f(x) = a^u$. If we take the logarithm of both sides we have

$$\ln y = \ln a^u = u\ln a$$

Using implicit differentiation, we have

$$\frac{d}{dx}\ln y = \frac{du}{dx}\ln a$$

$$\frac{1}{y}\frac{dy}{dx} = \frac{du}{dx}\ln a$$

Solving for $\frac{dy}{dx}$, we get

$$\frac{dy}{dx} = y\frac{du}{dx}\ln a$$

But $y = a^u$, and so

$$\frac{dy}{dx} = a^u \frac{du}{dx} \ln a,$$

or

Derivative of $y = a^u$ $$\frac{d}{dx}a^u = a^u \frac{du}{dx} \ln a$$

EXAMPLE 8.26

Differentiate $y = 3^x$.

Solution Here $a = 3$. Notice that $u = x$ and so $\frac{du}{dx} = \frac{dx}{dx} = 1$.

Then $y' = 3^x \cdot 1 \cdot \ln 3 = 3^x \ln 3$.

EXAMPLE 8.27

Differentiate $y = 5^{x^2+1}$.

Solution Here $a = 5$ and $u = x^2 + 1$, so $\frac{du}{dx} = 2x$.

$$y' = 5^{x^2+1}(2x)\ln 5$$

Now lets look at the special case when $y = e^u$. According to the rule for differentiating $y = a^u$, we obtain

$$y = e^u \frac{du}{dx} \ln e$$

But, $\ln e = 1$ and so

Derivative of $y = e^u$ $$\frac{d}{dx}e^u = e^u \frac{du}{dx}$$

EXAMPLE 8.28

Differentiate $y = e^x$.

Solution Here, $u = x$, so $\frac{du}{dx} = \frac{dx}{dx} = 1$

$$y' = e^x \frac{du}{dx} = e^x$$

EXAMPLE 8.29

Differentiate $y = e^{\sqrt{x}}$.

Solution We have $u = \sqrt{x}$ and so $\dfrac{du}{dx} = \dfrac{1}{2\sqrt{x}}$. Differentiating, we get

$$\begin{aligned} y' &= e^{\sqrt{x}}\left(\frac{d}{dx}\sqrt{x}\right) \\ &= e^{\sqrt{x}}\frac{1}{2\sqrt{x}} = \frac{e^{\sqrt{x}}}{2\sqrt{x}} \end{aligned}$$

EXAMPLE 8.30

Differentiate $y = xe^x$.

Solution We will have to use the product rule.

$$\begin{aligned} y' &= x\frac{d}{dx}e^x + e^x\frac{d}{dx}x \\ &= xe^x + e^x \\ &= e^x(x+1) \end{aligned}$$

EXAMPLE 8.31

Differentiate $y = x^3 3^x$.

Solution Again, we use the product rule.

$$\begin{aligned} y' &= x^3\frac{d}{dx}3^x + 3^x\frac{d}{dx}x^3 \\ &= x^3 \cdot 3^x \ln 3 + 3^x 3x^2 \\ &= x^2 \cdot 3^x(x\ln 3 + 3) \end{aligned}$$

EXAMPLE 8.32

Differentiate $y = e^{\sin x}$.

Solution

$$\begin{aligned} y' &= e^{\sin x}\frac{d}{dx}(\sin x) \\ &= \cos x e^{\sin x} \end{aligned}$$

EXAMPLE 8.33

Differentiate $y = \ln\tan e^{5x}$.

Solution Using the derivative of the logarithm, we get

$$\begin{aligned} y' &= \frac{1}{\tan e^{5x}}\frac{d}{dx}(\tan e^{5x}) \\ &= \frac{1}{\tan e^{5x}}\sec^2 e^{5x}\frac{d}{dx}e^{5x} \\ &= \frac{\sec^2 e^{5x}}{\tan e^{5x}}e^{5x}\frac{d}{dx}(5x) \\ &= \frac{5e^{5x}\sec^2 e^{5x}}{\tan e^{5x}} \end{aligned}$$

Logarithmic Differentiation

In developing the rule for the derivative of the exponential function, we took the logarithm of both sides. This method, known as **logarithmic differentiation**, can often simplify derivatives when both the base and the power are variables, as in the next example.

EXAMPLE 8.34

Differentiate $y = x^{\sin x}$.

Solution Taking the logarithm of both sides and simplifying gives

$$\ln y = \sin x \ln x$$

Using implicit differentiation, we get

$$\begin{aligned} \frac{1}{y}y' &= \frac{1}{x}\sin x + (\ln x)(\cos x) \\ y' &= y\left[\frac{\sin x}{x} + (\ln x)(\cos x)\right] \\ &= x^{\sin x}\left[\frac{\sin x}{x} + (\ln x)(\cos x)\right] \end{aligned}$$

Hyperbolic Functions

Some combinations of functions happen often enough in applications that they are given special names. The **hyperbolic functions** form such a group. Each hyperbolic function is abbreviated by placing an h at the end of the symbol representing its corresponding circular trigonometric function. Thus hyperbolic sine is abbreviated sinh, hyperbolic cosine is abbreviated cosh, and so on. The hyperbolic functions are defined as follows.

Definitions of Hyperbolic Functions

Hyperbolic sine	$\sinh x = \dfrac{e^x - e^{-x}}{2}$
Hyperbolic cosine	$\cosh x = \dfrac{e^x + e^{-x}}{2}$
Hyperbolic tangent	$\tanh x = \dfrac{\sinh x}{\cosh x} = \dfrac{e^x - e^{-x}}{e^x + e^{-x}}$
Hyperbolic cotangent	$\coth x = \dfrac{\cosh x}{\sinh x} = \dfrac{e^x + e^{-x}}{e^x - e^{-x}}$
Hyperbolic secant	$\text{sech } x = \dfrac{1}{\cosh x} = \dfrac{2}{e^x + e^{-x}}$
Hyperbolic cosecant	$\text{csch } x = \dfrac{1}{\sinh x} = \dfrac{2}{e^x - e^{-x}}$

Just as with the trigonometric functions, hyperbolic functions satisfy their own set of identities. As you can see in the following box, except for differences in sign, these are similar to the identities for the trigonometric functions. While we will not develop these, some are stated here because they are especially useful to help solve problems involving hyperbolic functions.

Identities for Hyperbolic Functions

Relations Between Squares of Functions

$$\cosh^2 x - \sinh^2 x = 1$$

$$\tanh^2 x = 1 - \text{sech}^2 x$$

$$\coth^2 x = 1 + \text{csch}^2 x$$

Half-Angle Formulas

$$\sinh \frac{x}{2} = \sqrt{\frac{\cosh x - 1}{2}}$$

$$\cosh \frac{x}{2} = \sqrt{\frac{\cosh x + 1}{2}}$$

Double-Angle Formulas

$$\sinh 2x = 2 \sinh x \cosh x$$

$$\cosh 2x = \cosh^2 x + \sinh^2 x$$

We are now ready to examine derivatives of hyperbolic functions. We will develop the derivative of the hyperbolic sine function, and then state the derivatives and integrals of the remaining functions.

EXAMPLE 8.35

Differentiate $y = \sinh x$.

Solution

$$y = \sinh x = \frac{1}{2}(e^x - e^{-x})$$

$$y' = \frac{1}{2}\left(e^x - e^{-x}\frac{d}{dx}(-x)\right)$$

$$= \frac{1}{2}(e^x + e^{-x})$$

$$= \cosh x$$

Derivatives of Hyperbolic Functions

$$\frac{d}{dx}(\sinh u) = \cosh u\frac{du}{dx} \qquad \frac{d}{dx}(\coth u) = -\text{csch}^2 u\frac{du}{dx}$$

$$\frac{d}{dx}(\cosh u) = \sinh u\frac{du}{dx} \qquad \frac{d}{dx}(\text{sech}\, u) = -\text{sech}\, u \tanh u\frac{du}{dx}$$

$$\frac{d}{dx}(\tanh u) = \text{sech}^2 u\frac{du}{dx} \qquad \frac{d}{dx}(\text{csch}\, u) = -\text{csch}\, u \coth u\frac{du}{dx}$$

It can be shown that any cable that hangs freely between two supports hangs in the shape of a hyperbolic cosine curve. Examples of such curves include telephone lines and electric power cables strung from one pole or tower to another. In Section 9.3, we will examine applications involving hyperbolic functions.

Inverse Hyperbolic Functions

Since $\sinh x$ and $\cosh x$ are defined in terms of e^x, we might expect that their inverse functions are expressed in terms of the natural logarithm function. As the following box shows, this is indeed the case. As with the inverse trigonometric functions, you need to recall that

$$y = \sinh^{-1} x, \text{ if and only if } \sinh y = x$$

Since each of the inverse hyperbolic functions can be defined in terms of its natural logarithm, we should have little difficulty finding their derivatives.

Formulas for Inverse Hyperbolic Functions

$$\sinh^{-1} x = \ln\left(x + \sqrt{x^2+1}\right), \qquad -\infty < x < \infty$$

$$\cosh^{-1} x = \ln\left(x + \sqrt{x^2-1}\right), \qquad x \geq 1$$

$$\tanh^{-1} x = \frac{1}{2}\ln\frac{1+x}{1-x}, \qquad |x| < 1$$

$$\coth^{-1} x = \frac{1}{2}\ln\frac{1+x}{1-x}, \qquad |x| > 1$$

$$\operatorname{sech}^{-1} x = \ln\left(\frac{1+\sqrt{1-x^2}}{x}\right), \qquad 0 < x \leq 1$$

$$\operatorname{csch}^{-1} x = \ln\left(\frac{1}{x} + \frac{\sqrt{1-x^2}}{|x|}\right), \qquad x \neq 0$$

EXAMPLE 8.36

If $y = \sinh^{-1} x$, determine y'.

Solution Rewriting $y = \sinh^{-1} x$ as $y = \ln(x + \sqrt{x^2+1})$ and using the derivative of the natural logarithm produces

$$\begin{aligned} y' &= \left(\frac{1}{x+\sqrt{x^2+1}}\right)\frac{d}{dx}\left(x+\sqrt{x^2+1}\right) \\ &= \frac{1}{x+\sqrt{x^2+1}}\left(1+\frac{x}{\sqrt{x^2+1}}\right) \\ &= \frac{\sqrt{x^2+1}+x}{\left(x+\sqrt{x^2+1}\right)\sqrt{x^2+1}} \\ &= \frac{1}{\sqrt{x^2+1}} \end{aligned}$$

Using similar procedures, we can develop all of the following derivatives.

Derivatives of Inverse Hyperbolic Functions

$$\frac{d}{dx}\left(\sinh^{-1} u\right) = \frac{1}{\sqrt{u^2+1}}\frac{du}{dx}$$
$$\frac{d}{dx}\left(\cosh^{-1} u\right) = \frac{1}{\sqrt{u^2-1}}\frac{du}{dx}, \quad u \geq 1$$
$$\frac{d}{dx}\left(\tanh^{-1} u\right) = \frac{1}{1-u^2}\frac{du}{dx}, \quad |u| < 1$$
$$\frac{d}{dx}\left(\coth^{-1} u\right) = \frac{1}{1-u^2}\frac{du}{dx}, \quad |u| > 1$$
$$\frac{d}{dx}\left(\text{sech}^{-1} u\right) = \frac{-1}{u\sqrt{1-u^2}}\frac{du}{dx}, \quad 0 < u < 1$$
$$\frac{d}{dx}\left(\text{csch}^{-1} u\right) = \frac{-1}{|u|\sqrt{1+u^2}}\frac{du}{dx}, \quad u \neq 1$$

EXAMPLE 8.37

If $f(x) = \sinh^{-1}\sqrt{x^4-9}$, determine $f'(x)$.

Solution Using the formula for the derivative of $\sinh^{-1}$ and the chain rule, we obtain

$$\begin{aligned} f'(x) &= \frac{1}{\sqrt{\left(\sqrt{x^4-9}\right)^2+1}}\left[\frac{1}{2}(x^4-9)^{-1/2}(4x^3)\right] \\ &= \frac{1}{\sqrt{x^4-9+1}}\left[\frac{2x^3}{\sqrt{x^4-9}}\right] \\ &= \frac{2x^3}{\sqrt{x^4-8}\sqrt{x^4-9}} \end{aligned}$$

In Section 9.5, we will examine applications involving the inverse hyperbolic functions.

Exercise Set 8.6

In Exercises 1–26, differentiate and simplify the given function.

1. $y = 4^x$
2. $y = e^{x^2}$
3. $y = 5^{\sqrt{x}}$
4. $y = 6^{x^2+x}$
5. $y = 2^{\sin x}$
6. $y = e^{5x+3}$
7. $y = e^{x^2+x}$
8. $y = e^{2x}$
9. $y = 4^{x^4}$
10. $y = x + e^x$
11. $y = \dfrac{e^x}{x^2}$
12. $y = x^3 e^x$
13. $y = \dfrac{1+e^x}{x^2}$
14. $y = \dfrac{1+e^x}{e^x}$
15. $y = e^{\tan x}$
16. $y = e^{\cos 3x}$
17. $y = \sin e^{x^2}$
18. $y = \cos 2^{x^2}$
19. $y = 3^x(x^3 - 1)$

20. $y = e^{x^2 - \ln x}$

21. $y = x^{\cos x}$

22. $y = x^{x^3}$

23. $y = (\sin x)^x$

24. $y = e^{\arctan x}$

25. $y = \ln \sin e^{3x}$

26. $y = e^{\sin x} \ln \sqrt{x}$

Solve Exercises 27–32.

27. Find y' implicitly, if $y = e^y + y + x$.

28. Find y' implicitly, if $ye^x - xe^y = 1$

29. Show that $\frac{d}{dx}(\cosh x) = \sinh x$.

30. Verify that $\cosh^2 x - \sinh^2 x = 1$.

31. Verify that $\cosh x + \sinh x = e^x$.

32. Verify that $\sinh 2x = 2 \sinh x \cosh x$.

In Exercises 33–42, find the derivative of the given function.

33. $y = e^{\sinh 3x}$

34. $f(x) = \sinh(2x + 3)$

35. $g(x) = \tanh \sqrt{x^3 + 4}$

36. $h(x) = \sinh \ln (x^2 + 3x)$

37. $j(x) = \cosh^3 (x^4 + \sin x)$

38. $k(x) = \tanh^2 x + \operatorname{sech}^2 x$

39. $f(x) = \sinh^{-1} 7x$

40. $g(x) = \cosh^{-1} \sqrt{x}$

41. $j(x) = x \sinh^{-1} \frac{1}{x}$

42. $k(x) = \tanh^{-1}(\sin x)$, $-\frac{\pi}{2} < x < \frac{\pi}{2}$

Solve Exercises 43–48.

43. *Advertising* An electronics dealer estimates that the total number people who have heard one of her radio advertisements is approximated by $N(t) = 615\left(1 - e^{-0.02t}\right)$, where t is the number of days since the advertisement began. How fast is N growing after 7 days?

44. *Medical technology* At the beginning of an experiment, a culture of 100 bacteria is growing in a medium that will allow a maximum of 10,000 bacteria to survive. The population, N, at time t, in hours, is estimated to be $N(t) = \frac{10{,}000}{1 + 99e^{-0.14t}}$ after the experiment has begun.
 (a) What was the initial population?
 (b) Find a formula for the rate of change of the bacteria population.
 (c) How fast is the population changing at the end of 5 h?
 (d) When is the population growing the fastest?
 (e) What is the fastest rate of population growth?

45. *Electronics* The charge across a capacitor is $q(t) = 5e^{-t} \cos 2.5t$.
 (a) What is the current in the capacitor $I = q'(t)$?
 (b) What is the current at $t = 0.45$ s?

46. *Meteorology* If we assume that the temperature of the atmosphere is a constant, then the formula $14.7e^{-0.0003655h}$ gives the atmospheric pressure, P, in pounds per square inch, at altitude h, in feet above sea level, for $0 \le h \le 50$ mi.
 (a) Find the rate of change of P with respect to h.
 (b) What is $\frac{dP}{dh}$ for an airplane that is flying at 35,000 ft?

47. *Biology* Consider a culture of bacteria growing in a glass jar. The number of bacteria present after t h is estimated to be $N(t) = 10^5 + 10^6 \tanh(0.1t)$.
 (a) What is the rate of growth of these bacteria?
 (b) What is the rate of growth of these bacteria at $t = 6.93$ h?
 (c) How many bacteria are there at $t = 6.93$ h?

48. *Environmental science* Radon gas can readily diffuse through solid materials such as brick and concrete. If the direction of diffusion in a basement wall is perpendicular to the surface, then the radon concentration $C(x)$ in J/cm^3 in the air-filled pores within the wall at distance x from the outside surface, can be approximated by

$$C(x) = A\sinh(qx) + B\cosh(qx) + k$$

where constant q depends on the porosity of the wall, the half-life of radon, and a diffusion coefficient; constant k is the maximum radon concentration in the air-filled pores; and A and B are constants that depend on the initial conditions. Find the rate at which the concentration is changing with respect to distance x cm.

In Your Words

49. You can use the change of base formula $\log_b x = \dfrac{\ln x}{\ln b}$ so that you need but one rule for differentiating a logarithmic function. Will this also work to learn the derivatives of e^x and a^x? If you answered yes, explain how it is done. If you answered no, explain why it will not work.

50. Explain how the derivatives of the sine and cosine functions differ from those of the hyperbolic sine and hyperbolic cosine functions, respectively.

8.7 APPLICATIONS

In this section, we will look at applications that require the derivative of logarithmic and exponential functions.

Application

EXAMPLE 8.38

Sketch the graph of $f(x) = \dfrac{x}{e^x}$. Locate all extrema and inflection points.

Solution One thing that we notice before we even begin with calculus is that the function $\dfrac{1}{e^x} = e^{-x}$ is always positive. This means that when $x < 0$, $f(x) = \dfrac{x}{e^x} < 0$ and when $x > 0$, $f(x) > 0$.

Taking the derivative of $f(x) = \dfrac{x}{e^x} = xe^{-x}$, we get $f'(x) = -xe^{-x} + e^{-x} = e^{-x}(1-x)$. The only critical value occurs when $x = 1$. Since $e^{-x} > 0$, the sign of $f'(x)$ depends on the sign of $1-x$. When $x < 1$, we see that $1-x > 0$ and when $x > 1$, then $1-x < 0$. Thus, by the first derivative test, there is a maximum when $x = 1$ or at the point $\left(1, \dfrac{1}{e}\right) \approx (1, 0.368)$.

Taking the second derivative, we get $f''(x) = xe^{-x} - e^{-x} - e^{-x} = (x-2)e^{-x}$. The second derivative is zero when $x = 2$. If $x > 2$, then $f''(x) > 0$, so $f(x)$ is concave up on $(2, \infty)$. If $x < 2$, then $f''(x) < 0$, and $f(x)$ is concave down on $(-\infty, 2)$. Thus, the point $\left(2, \dfrac{2}{e^2}\right) \approx (2, 0.271)$ is a point of inflection.

A sketch of this curve is shown in Figure 8.6.

FIGURE 8.6

Application

EXAMPLE 8.39

Find an equation of the tangent and the normal lines to the graph of $y = \ln x^3$ at $(1,0)$.

Solution In order to determine these lines, we need the slope, which means we need to find the first derivative and evaluate it when $x = 1$. Using properties of logarithms, we have $y = \ln x^3 = 3\ln x$. Thus, $y' = \dfrac{3}{x}$.

When $x = 1$, $y' = 3$ and so the slope of the tangent line is 3 and the slope of the normal line is $\frac{-1}{3}$. Using the slope-intercept form for the equation of a line, we get the tangent line

$$\begin{aligned} y - 0 &= 3(x-1) \\ y &= 3x - 3 \end{aligned}$$

Similarly, since the slope of the normal line is $-\frac{1}{3}$, the equation of the normal line is

$$\begin{aligned} y - 0 &= -\frac{1}{3}(x-1) \\ y &= -\frac{1}{3}x + \frac{1}{3} \\ \text{or} \qquad 3y + x &= 1 \end{aligned}$$

Application

EXAMPLE 8.40

A 1-kg object suspended from a spring stretches it 15 cm. An unknown velocity is given to the object at $t = 0$. Because of a damping factor, the position s of the object at time t is given by $s(t) = -2e^{-2t}\sin 4t$. **(a)** What is the velocity given the object? **(b)** What is the maximum displacement of the object? **(c)** Sketch the graph of this function.

Solution In order to find the velocity, we need to differentiate $s(t)$.

$$\begin{aligned} v(t) = s'(t) &= -2(4e^{-2t}\cos 4t - 2e^{-2t}\sin 4t) \\ &= -4e^{-2t}(2\cos 4t - \sin 4t) \end{aligned}$$

We can now answer the first two parts. **(a)** The initial velocity occurs when $t = 0$ and $v(0) = -8$, so the initial velocity is 8 cm/s downward. **(b)** The maximum displacement will occur when $s'(t) = 0$. Since $e^{-2t} > 0$ for all values of t, $s'(t) = 0$ when $2\cos 4t - \sin 4t = 0$, or when $\sin 4t = 2\cos 4t$. Dividing both sides by $\cos 4t$, we get $\tan 4t = 2$ or $4t = \tan^{-1} 2$. Using a calculator, we see that $4t = 1.107 + \pi n$, so $t = 0.277 + \frac{\pi}{4}n$ rad. When $n = 0$, we have $t = 0.277$ and $s(t) \approx -1.028$ cm. When $n = 1$, we obtain $t = 0.277 + \frac{\pi}{4}$ and we get $s(t) = 0.214$ cm. The maximum displacement is at $(0.277, -1.028)$. **(c)** The graph of this function is shown in Figure 8.7.

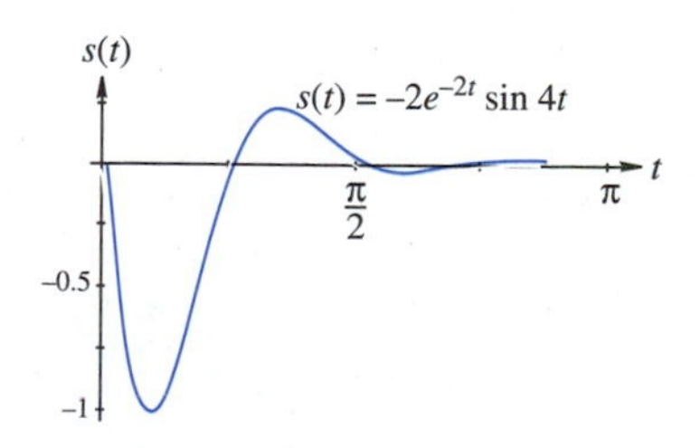

FIGURE 8.7

Application

EXAMPLE 8.41

A submarine telephone cable consists of a conducting circular core surrounded by a layer of insulation. If x is the ratio of the radius of the core to the thickness of the insulation, then the speed v of the signal is proportional to $x^2 \ln\left(\frac{1}{x}\right)$. Thus $v(x) = kx^2 \ln\frac{1}{x}$, where k is the constant of proportionality. For what value of x is the speed of the signal a maximum?

Solution We take the first derivative of $v(x)$.

$$\begin{aligned} v'(x) &= kx^2\left(\frac{-1/x^2}{1/x}\right) + 2kx\ln\frac{1}{x} \\ &= -kx + 2kx\ln\frac{1}{x} \\ &= kx\left(2\ln\frac{1}{x} - 1\right) \end{aligned}$$

We see that $v'(x)$ is undefined when $x = 0$, and that $v'(x) = 0$ when $\ln\frac{1}{x} = \frac{1}{2}$ or when $\ln x = \frac{-1}{2}$. Solving for x, we see that $v'(x) = 0$ when $x = e^{-1/2} \approx 0.607$. For x in the interval $(0, e^{-1/2})$, we have $v'(x) > 0$, and for x in $(e^{-1/2}, \infty)$, we see that $v'(x) < 0$. Thus, $s(x)$ is a maximum when $x = e^{-1/2}$.

Exercise Set 8.7

In Exercises 1–10, find all extrema and points of inflection and sketch the graph.

1. $y = x\ln x$
2. $y = \dfrac{\ln x}{x}$
3. $y = xe^x$
4. $y = x^2e^x$
5. $y = \ln\dfrac{1}{x^2+1}$
6. $y = \dfrac{4\ln^2 x}{x}$
7. $y = \sinh x$
8. $y = \cosh x$
9. $y = e^{-x}\cos x$ for $-2\pi \le x \le 2\pi$
10. $y = \dfrac{\sin x}{\ln(x+2)}$ for $0 \le x \le 4\pi$

Solve Exercises 11–24.

11. Find an equation of the line tangent to $y = x^3\ln x$ at $(1, 0)$.
12. Find an equation of the line normal to $y = x^3\ln x$ at $(1, 0)$.
13. Find an equation of the line tangent to $y = x^2e^x$ at $(1, e)$.
14. Find an equation of the line normal to $y = x^2e^x$ at $(1, e)$.

15. *Physics* A particle moves on a coordinate line with its position at t s given by $s(t) = e^{-3t}$.
(a) What is its velocity function?
(b) When is its velocity a maximum?

16. *Physics* A particle moves along a coordinate line with its position given by $s(t) = t^2 + 4\ln(t+1)$, where t is in seconds.
(a) What is the velocity function?
(b) What is the acceleration function?
(c) When is the velocity a minimum?
(d) What is the minimum velocity?

17. *Physics* A particle moves on a coordinate line with its position given by $s(t) = \sin(e^t)$.
(a) What are the velocity and acceleration functions?
(b) When does the particle first have a velocity of 0 and what is its acceleration at that time?

18. *Biology* The number of bacteria in a particular colony at time t hours is given by

$$x = 5000e^{0.5t}$$

What is the rate of growth of this colony?

19. *Material science* The density of a 5-m bar is given by $\rho(x) = xe^{-x^{2/3}}$ kg/m, where x is in meters from the left of the bar. At what point is the bar the most dense?

20. *Meteorology* The atmospheric pressure of a height of x m above sea level is given by $P(x) = 10^4e^{-0.0012x}$ kg/m^2. What is the rate of change of the pressure with respect to the height?

21. *Electricity* The current in a series RL circuit is given by $i = 1 - e^{-t^2/2L}$. When is the current a maximum?

22. *Electronics* The instantaneous charge on a capacitor for a certain electric circuit is given by $Q(t) = e^{-6t}(4\cos 8t + 3\sin 8t) - 0.4\cos 10t$.
(a) What is the rate of change of this charge?
(b) What is the rate of change when $t = \frac{\pi}{16}$.

23. *Space technology* The power supply P in watts in a satellite is given by $P = 100e^{-0.015t}$, where t is measured in days. What is the time-rate of change of power when $t = 50$ days?

24. *Refrigeration technology* A warm object is placed in a refrigerator to cool. The temperature in degrees Celsius, at any given time t in minutes, is given by $T = 10 + 15e^{-0.875t}$. What is the rate of change of the temperature when $t = 30$ min?

In Exercises 25–28, use Newton's method to approximate the real roots of the given function.

25. $f(x) = x - \ln x - 2$ **26.** $g(x) = \ln x - \sin x$ **27.** $h(x) = e^x \cdot \cos x$ **28.** $j(x) = e^x + 2x - 2$

In Your Words

29. Write an application in your technology area of interest that requires you to use derivatives of a logarithmic function. Give your problem to a classmate and see if he or she understands and can solve your problem. Rewrite the problem as necessary to remove any difficulties encountered by your classmate.

30. Write an application in your technology area of interest that requires you to use derivatives of an exponential function. Give your problem to a classmate and see if he or she understands and can solve your problem. Rewrite the problem as necessary to remove any difficulties encountered by your classmate.

CHAPTER 8 REVIEW

Important Terms and Concepts

Hyperbolic functions
Inverse hyperbolic functions
Logarithmic differentiation

Review Exercises

In Exercises 1–24, differentiate and simplify the given functions.

1. $y = \sin 2x + \cos 3x$
2. $y = \tan 3x^2$
3. $y = \tan^2 3x$
4. $y = \sqrt{\sin 2x}$
5. $y = x^2 \sin x$
6. $y = \dfrac{\cos 3x}{3x}$
7. $y = \sin^{-1}(3x - 2)$
8. $y = \cos^{-1} x^2$
9. $y = \arctan 3x^2$
10. $y = \arctan\left(\dfrac{1+x}{1-x}\right)$
11. $y = \arcsin\sqrt{1-x^2}$
12. $y = \arccos\sqrt{-3+4x-x^2}$
13. $y = \log_4(3x^2 - 5)$
14. $y = \ln(x+4)^3$
15. $y = \ln^2(x^3+4)$
16. $y = \ln\cos x$
17. $y = \ln\dfrac{x^3}{(4x-3)^2}$
18. $y = \ln\arctan x$
19. $y = e^{4x^2}$
20. $y = e^{\ln x^2}$
21. $y = e^{\sin x^2}$
22. $y = xe^{x^2} + \sin x$
23. $y = e^{-x}\sin 5x$
24. $y = e^{-x}\ln x$

In Exercises 25–30, locate all the extrema and the points of inflection for the given function. Sketch the graph.

25. $y = x^2 e^x$
26. $y = 4e^{-9x^2}$
27. $y = e^{-1/2x}\sin 2x$ on $[0, 2\pi]$
28. $y = e\sin 3x - e^{-1}\cos x$ on $[0, 2\pi]$
29. $y = x^3\ln x, x > 0$
30. $y = 2\arccos\dfrac{x}{4}$

In Exercises 31–36, find an equation of (a) the line tangent and (b) the line normal to the given curve at the indicated point.

31. $y = \arctan x$ at $\left(1, \frac{\pi}{4}\right)$
32. $y = \sin^3 x$ at $\left(\dfrac{\pi}{4}, \dfrac{\sqrt{2}}{4}\right)$
33. $y = \ln 3x$ at $(1, 0)$
34. $y = \dfrac{e^x}{x}$ at $(1, e)$
35. $y = \ln(\sin x)$ at $\left(\frac{\pi}{4}, -\ln\sqrt{2}\right)$
36. $y = \sin(\ln x)$ at $(1, 0)$

Solve Exercises 37–40.

37. *Thermodynamics* An object is heated to 95°C and allowed to cool in a room with air temperature of 20°C. The temperature T after t min is given by the equation $T = 75e^{-0.05109t} + 20$.
    (a) What is the temperature of the object after 10 min?
    (b) What is the rate of change of the temperature after 10 min?
38. *Physics* A person parachutes from an airplane flying at 10,000 ft and opens the parachute immediately. The height of the parachutist t s after jumping is given by $s(t) = 10{,}000 - \frac{400}{3}\ln(\cosh 1.6t)$.
    (a) What is the velocity of the parachutist?
    (b) What is the velocity when $t = 60$ s?
39. *Advertising* A billboard parallel to a highway is to be 40 ft high and its bottom will be 8 ft above the eye of a passing motorist. How far from the highway should the billboard be placed in order to maximize the angle it subtends at the motorist's eyes?
40. *Meteorology* A weather balloon is rising from the ground at the rate of 2 m/s from a point 100 m from an observer. Determine the rate of increase of the angle of inclination of the observer's line of sight when the balloon is 30 m high.

CHAPTER 8 TEST

In Exercises 1–10, differentiate the given functions.

1. $f(x) = \sin 7x$
2. $g(x) = \tan(3x^2 + 2x)$
3. $h(x) = e^{2x}$
4. $j(x) = \ln 5x^2$
5. $k(x) = \cos^{-1}\left(e^{x^2}\right)$
6. $f(x) = \dfrac{\sin 4x}{1 + e^{2x}}$
7. $g(x) = (7x^2 + 3x)\tan^2 5x$
8. $h(x) = e^{\sin x} \ln \sqrt{x}$
9. The charge across a capacitor is $q(t) = 3e^{-t}\cos 2.5t$. Find the current to the capacitor when $t = 0.65$ s.
10. A particle is moving along the x-axis so that its distance x (in meters) from the origin at time t s is $x(t) = 3.2\sqrt{\tan 5.0t}$. Find the velocity at time $t = 1.5$ s.

CHAPTER

9 Techniques of Integration

The main cable joining two towers of a suspension bridge, like the one shown, takes the shape of a parabola. In Section 9.7, we will learn how to determine the length of this cable.

Courtesy of New York Convention and Visitors Bureau Inc.

In Chapters 4 and 8, we learned to differentiate almost any function. What is perhaps even more important is that we have developed a method that will help us find the derivative of almost any conceivable function. The same cannot be said about integration.

In this chapter, we will learn more about integration. We will begin by expanding our use of the power formula to include transcendental functions. Next we will look at the integrals for some logarithmic, exponential, and trigonometric functions. Finally, we will learn some techniques that can be used to integrate some additional functions.

When you have finished this chapter, you will have a better idea of which functions can be integrated and how to proceed in order to find the indefinite integral.

9.1 THE GENERAL POWER FORMULA

In Section 6.3, the fourth rule of integration stated that, for any variable x, $\int x^n\,dx = \frac{1}{n+1}x^{n+1} + C$. If we let u represent any function, then this can be restated as the **general power formula**.

General Power Formula for Integration

$$\int u^n\,du = \frac{1}{n+1}u^{n+1} + C, n \neq -1$$

Until now we have only applied the general power formula to algebraic functions, but it can be applied to transcendental functions as well, as is shown in the next five examples.

EXAMPLE 9.1

Find $\int \sin^2 x \cos x\,dx$.

Solution If we let $u = \sin x$, then $du = \cos x\,dx$ and the integral is as follows.

$$\begin{aligned}\int \sin^2 x \cos x\,dx &= \int u^2\,du \\ &= \frac{1}{3}u^3 + C \\ &= \frac{1}{3}\sin^3 x + C\end{aligned}$$

EXAMPLE 9.2

Find $\int \frac{\ln^4 x}{x}\,dx$.

Solution If $u = \ln x$, then $du = \frac{1}{x}\,dx$ and we have

$$\begin{aligned}\int \frac{\ln^4 x}{x}\,dx &= \int (\ln x)^4 \left(\frac{1}{x}\right)\,dx \\ &= \int u^4\,du \\ &= \frac{1}{5}u^5 + C \\ &= \frac{1}{5}\ln^5 x + C\end{aligned}$$

Note Recall that integration can be checked by differentiation. Thus,

$$D_x\left(\frac{1}{5}\ln^5 x + C\right) = \frac{1}{5}5\ln^4 x\left(\frac{1}{x}\right) = \frac{\ln^4 x}{x}$$

EXAMPLE 9.3

Determine $\int \sec^5 x \tan x \, dx$.

Solution Let $u = \sec x$ and then $du = \sec x \tan x \, dx$. If we write $\sec^5 x \tan x \, dx$ as $\sec^4 x(\sec x \tan x \, dx)$, then we have

$$\begin{aligned}\int \sec^5 x \tan x \, dx &= u^4 \, du \\ &= \frac{1}{5}u^5 + C \\ &= \frac{1}{5}\sec^5 x + C\end{aligned}$$

EXAMPLE 9.4

Determine $\int \frac{e^{3x}\,dx}{\sqrt{1+e^{3x}}}$.

Solution We will let $u = 1 + e^{3x}$ and so $du = 3e^{3x}\,dx$. We do not have du in our problem, but we do have $\frac{1}{3}\,du$. We will multiply the problem by $\frac{3}{3}$ in order to get du into the problem.

$$\begin{aligned}\int \frac{e^{3x}\,dx}{\sqrt{1+e^{3x}}} &= \frac{3}{3}\int \frac{e^{3x}\,dx}{\sqrt{1+e^{3x}}} \\ &= \frac{1}{3}\int \frac{3e^{3x}\,dx}{\sqrt{1+e^{3x}}} \\ &= \frac{1}{3}\int \frac{du}{\sqrt{u}} \\ &= \frac{1}{3}\int u^{-1/2}\,du = \left(\frac{1}{3}\right)\frac{2}{1}u^{1/2} + C \\ &= \frac{2}{3}u^{1/2} = \frac{2}{3}\left(1+e^{3x}\right)^{1/2} + C\end{aligned}$$

EXAMPLE 9.5

Evaluate $\int_0^1 \frac{\arcsin x}{\sqrt{1-x^2}}\,dx$.

Solution We first determine the improper integral by letting $u = \arcsin x$ and $du = \frac{dx}{\sqrt{1-x^2}}$. Then

$$\int \frac{\arcsin x}{\sqrt{1-x^2}}\,dx = \int u\,du$$
$$= \frac{1}{2}u^2 + C$$
$$= \frac{1}{2}(\arcsin x)^2 + C$$

Evaluating this, we get the proper integral

$$\int_0^1 \frac{\arcsin x}{\sqrt{1-x^2}\,dx} = \frac{1}{2}(\arcsin x)^2\Big|_0^1$$
$$= \frac{1}{2}\left(\frac{\pi}{2}\right)^2 = \frac{\pi^2}{8}$$

Application

EXAMPLE 9.6

The charge q stored in a capacitor decreases at the rate of

$$\frac{dq}{dt} = e^{-t}\left(e^{-t} - 4\right)^2$$

What is the charge dissipated from $t = 0.5$ s to $t = 2$ s?

Solution As before, we will first determine the improper integral. If $u = e^{-t} - 4$, then $du = -e^{-t}\,dt$. Substituting these into the original integral, produces

$$\int e^{-t}\left(e^{-t} - 4\right)^2 dt = -\int u^2\,du$$
$$= -\frac{1}{3}u^3 + C$$
$$= -\frac{1}{3}\left(e^{-t} - 4\right)^3 + C$$

Now we can evaluate this over the time interval form $t = 0.5$ s to $t = 2$ s.

$$\int_{0.5}^2 e^{-t}\left(e^{-t} - 4\right)^2 dt = -\frac{1}{3}\left(e^{-t} - 4\right)^3\Big|_{0.5}^2$$
$$\approx 2.1611 - .9019 = 1.2592$$

About 1.26 C is dissipated during this time interval.

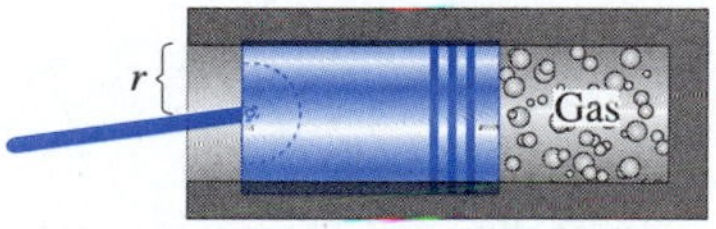

FIGURE 9.1

Consider a piston of radius r in a cylindrical casing, such as the one shown in Figure 9.1. As the gas in the cylinder expands, the piston moves and work is done. If we let P represent the pressure of the gas against the piston head and V the volume of the gas, then as the volume of the gas expands from V_0 to V_1, the temperature remains constant. The work done in moving the piston is given by the formula

$$W = \int_{V_0}^{V_1} P\,dV$$

If we assume that the pressure of the gas is inversely proportional to its volume, then $P = \frac{k}{V}$ and the work done in moving the piston is expressed as

$$W = \int_{V_0}^{V_1} \frac{k}{V}\,dV$$

Application

EXAMPLE 9.7

A quantity of gas with an initial volume of $0.100\,\text{m}^3$ and pressure 300 kPa expands to a volume of $0.200\,\text{m}^3$. Find the work done by the gas, assuming that the pressure of the gas is inversely proportional to its volume.

Solution Since the pressure of the gas is inversely proportional to its volume, we know that $P = \frac{k}{V}$. We are given that when the volume is $0.100\,\text{m}^3$, the pressure is 300 kPa, which means that $300 = \frac{k}{0.100}$. Solving for k, we find that $k = 30$. Thus, the work is

$$\begin{aligned} W &= \int_{V_0}^{V_1} \frac{k}{V}\,dV \\ &= \int_{0.100}^{0.200} \frac{30}{V}\,dV \\ &= 30 \ln V \Big]_{0.100}^{0.200} \\ &= 30[\ln 0.2 - \ln 0.1] = 30 \ln\left(\frac{0.2}{0.1}\right) \\ &= 30 \ln 2 \approx 20.7944 \end{aligned}$$

So, the work is about 20.7944 N·m.

Exercise Set 9.1

In Exercises 1–28, find the given integral.

1. $\int \sin^3 x \cos x\,dx$
2. $\int \cos^4 x \sin x\,dx$
3. $\int \sqrt{\sin^3 x} \cos x\,dx$
4. $\int \sin 2x \cos 2x\,dx$
5. $\int x \sec^2 x^2\,dx$
6. $\int \cos^3 x \sin x\,dx$
7. $\int \sec^2 x \tan x\,dx$
8. $\int \frac{e^{2x}}{(1+e^{2x})^{1/2}}\,dx$
9. $\int \frac{\arccos x}{\sqrt{1-x^2}}\,dx$

10. $\int \frac{\arctan x}{1+x^2}\,dx$

11. $\int \frac{1}{x}\sqrt{\frac{\operatorname{arccsc} x}{x^2-1}}\,dx$

12. $\int \frac{\arctan 6x}{1+36x^2}\,dx$

13. $\int \frac{(\arcsin 2x)^3}{\sqrt{1-4x^2}}\,dx$

14. $\int \frac{(\ln x)^3}{x}\,dx$

15. $\int \frac{[\ln(x+4)]^4}{x+4}\,dx$

16. $\int \frac{[5+2\ln x]^3}{x}\,dx$

17. $\int \frac{e^x\,dx}{(e^x+4)^2}$

18. $\int (e^x+e^{-x})^{1/3}(e^x-e^{-x})\,dx$

19. $\int (e^{2x}-1)^3 e^{2x}\,dx$

20. $\int (3e^{3x}+1)^{2/3} e^{3x}\,dx$

21. $\int_0^{\pi/4} \sin^3 x\cos x\,dx$

22. $\int_{\pi/4}^{\pi/3} \tan^3 x\sec^2 x\,dx$

23. $\int_1^2 \frac{x[\ln(x^2+1)]^3}{x^2+1}\,dx$

24. $\int_1^{\sqrt{3}} \frac{(\arctan x)^3}{1+x^2}\,dx$

25. $\int_1^2 \frac{e^{2x}\,dx}{(e^{2x}-1)^2}$

26. $\int_0^2 8(e^{2x}-1)^3 e^{2x}\,dx$

Solve Exercises 27–32.

27. Find the area under the curve $y=\sin^2 x\cos x$ from $x=0$ to $x=\pi$.

28. Find the area under the curve $y=(e^{3x}+1)e^{3x}$ from $x=0$ to $x=1$.

29. *Electricity* The charge q stored in a capacitor decreases at the rate of

$$\frac{dq}{dt}=e^{-t}\left(e^{-t}-2\right)^2$$

What is the charge dissipated from $t=0.7$ s to $t=3$ s?

30. *Electricity* The charge q stored in a capacitor decreases at the rate of

$$\frac{dq}{dt}=e^{-t}\left(e^{-t}-5\right)^2$$

What is the charge dissipated from $t=0.25$ s to $t=2.50$ s?

31. *Ecology* Ecologists are treating a stream contaminated with bacteria. The level of contamination is changing at the rate of $\frac{dN}{dt}=-\frac{1020}{t^2}-340$ bacteria/cm^3/day, where t is the number of days since the treatment began.
   (a) Find a function $N(t)$ that will estimate the level of contamination if $N(1)=9{,}750$ bacteria/cm^3.
   (b) Determine $N(10)$.

32. *Quality control* A quality control engineer has determined that the rate of expenditure in dollars/year for maintaining a certain machine at a specified level is given by $M'(t)=\sqrt{3t^2+18t}(6t+18)$, where t is time measured in years. Maintenance expenditures for the third year were \$945.
   (a) Find the total maintenance function, $M(t)$.
   (b) According to the engineer, the machine should be replaced when the total expenses reach \$7,200. How long should the company keep this machine?

In Your Words

33. In the general power formula for integration, what is the role of du?

34. In the general power formula for integration, how do you determine which part of the integrand is u and which part is du?

9.2 BASIC LOGARITHMIC AND EXPONENTIAL INTEGRALS

The general power formula from Section 9.1 is valid for all values of n except $n=-1$. This gap can be filled by taking advantage of the derivative of the natural logarithm function and the fact that the integral is the inverse operation of the derivative.

We know from Chapter 8 that

$$\frac{d}{dx}\ln u = \frac{1}{u}\frac{du}{dx}$$

In differential form this is written as

$$d(\ln u) = \frac{1}{u}\,du$$

Thus, we see that by using antiderivatives, we obtain

$$\int \frac{1}{u}\,du = \ln u + C,\ u > 0$$

The logarithmic function is defined for positive values of u. What do we do if u is negative? Well, if u is negative, then $u < 0$ and $-u > 0$, so

$$\frac{d}{dx}\ln(-u) = \frac{-1}{u}\left(-\frac{du}{dx}\right) = \frac{1}{u}\frac{du}{dx}$$

From this we see that

$$\int \frac{1}{u}\,du = \int \frac{-1}{u}(-du) = \ln(-u) + C, \text{ if } u < 0$$

We can combine these two cases if we remember that

$$|u| = \begin{cases} -u & \text{if } u < 0 \\ u & \text{if } u > 0 \end{cases}$$

This means that

$$\int \frac{du}{u} = \ln|u| + C$$

We have now filled the gap in the general power formula.

EXAMPLE 9.8

Determine $\int \frac{x\,dx}{x^2+4}$.

Solution If we let $u = x^2+4$, then $du = 2x\,dx$. If we multiply the above integral by $\frac{2}{2}$, we get

$$\begin{aligned}\int \frac{x\,dx}{x^2+4} &= \frac{2}{2}\int \frac{x\,dx}{x^2+4} \\ &= \frac{1}{2}\int \frac{2x\,dx}{x^2+4}\end{aligned}$$

EXAMPLE 9.8 (Cont.)

$$= \frac{1}{2}\int \frac{du}{u}$$
$$= \frac{1}{2}\ln|u| + C$$
$$= \frac{1}{2}\ln|x^2+4| + C$$

In this case, since $x^2 \geq 0$, $x^2+4 > 0$, so $|x^2+4| = x^2+4$ and we can write the answer as $\frac{1}{2}\ln(x^2+4)+C$.

We also learned in Chapter 8 that $\frac{d}{dx}e^u = e^u\frac{du}{dx}$. Reversing this, we get the integral of $e^u\,du$.

$$\int e^u\,du = e^u + C$$

EXAMPLE 9.9

Find $\int e^{\sin x}\cos x\,du$.

Solution If $u = \sin x$, then $du = \cos x\,du$. Thus

$$\int e^u\,du = e^u + C$$
$$= e^{\sin x} + C$$

If we are working with bases other than e, we get the following integral.

$$\int a^u\,du = a^u \log_a e + C$$
$$= \frac{a^u}{\ln a} + C$$

EXAMPLE 9.10

Find $\int 3^x\,dx$.

Solution We let $u = x$ and so $du = dx$.

$$\begin{aligned}\int 3^x\,dx &= \int 3^u\,du\\ &= \frac{3^u}{\ln 3} + C\\ &= \frac{3^x}{\ln 3} + C\end{aligned}$$

EXAMPLE 9.11

Determine $\int x^2 e^{x^3}\,dx$.

Solution If $u = x^3$, then $du = 3x^2\,dx$. We will need to multiply by $\frac{3}{3}$ to get

$$\begin{aligned}\int x^2 e^{x^3}\,dx &= \frac{1}{3}\int 3x^2 e^{x^3}\,dx\\ &= \frac{1}{3}\int e^u\,du\\ &= \frac{1}{3}e^u + C\\ &= \frac{1}{3}e^{x^3} + C\end{aligned}$$

When integrating e raised to a function, the substitution $u =$ the exponent of e usually leads to a recognizable integration.

EXAMPLE 9.12

Evaluate $\int_1^2 \frac{e^{1/x}}{x^2}\,dx$.

Solution If $u = \frac{1}{x} = x^{-1}$, then $du = -x^{-2}\,dx$ and $\frac{1}{x^2} = -du$. This produces

$$\int \frac{e^{1/x}\,dx}{x^2} = -\int e^u\,du = -e^u + C$$

Since $u = \frac{1}{x}$, we have

$$\begin{aligned}\int_1^2 \frac{e^{1/x}}{x^2}\,dx &= -e^{1/x}\Big|_1^2\\ &= -e^{1/2} - (-e^1) = e - \sqrt{e}\\ &\approx 1.0696\end{aligned}$$

Application

EXAMPLE 9.13

An object weighing 128 lb is thrown from a height of 200 ft with an initial velocity of 16 ft/s. If the air resistance is proportional to the velocity of the body and the limiting velocity is 128 ft/s, the velocity of the object at time t is given by $v(t) = 112e^{-t/4} - 128$. What is the position of the object at any time t?

Solution The velocity function is the derivative of the position function, $s(t)$. This means that the position function is

$$\begin{aligned} s(t) &= \int v(t)\,dt \\ &= \int (112e^{-t/4} - 128)\,dt \\ &= -448e^{-t/4} - 128t + C \end{aligned}$$

Now at $t = 0$, we know that the object is 200 ft above the ground. So, $s(0) = 200$. Since $s(0) = -448e^0 - 128(0) + C = -448 + C$, we find that $C = 648$. Thus, the position function of this object at time t is $s(t) = -448e^{-t/4} - 128t + 648$ ft.

Application

EXAMPLE 9.14

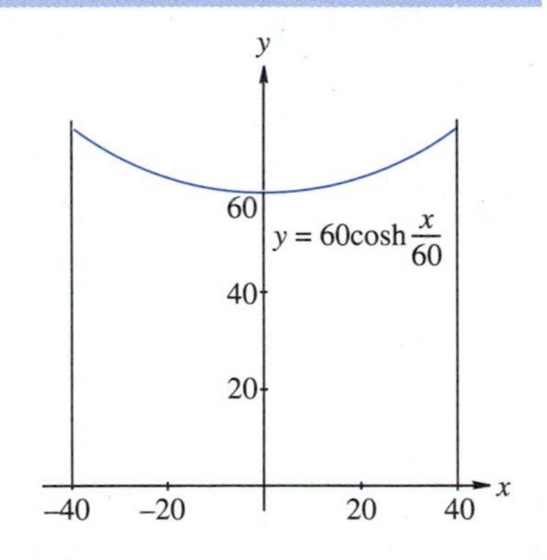

FIGURE 9.2

Any free hanging cable suspended from two towers, like the one shown in Figure 9.2, takes the shape of a **catenary** whose equation is

$$y = a\cosh\frac{x}{a}$$

for some real number a. If an electric cable is hung between two towers that are 80.00 m apart so that the lowest part of the cable is $a = 60.00$ m above the ground, the cable takes the equation

$$y = 60\cosh\frac{x}{60}$$

Find the arc length of the cable between the two towers.

Solution A sketch of the cable with a coordinate system has been drawn in Figure 9.2. Since $y = 60\cosh\frac{x}{60}$, we know from Section 8.6 that $y = 60\cosh\frac{x}{60} = 30\left(e^{x/60} + e^{-x/60}\right)$. To find the arc length, we need $(y')^2$. We can determine that $y' = \frac{1}{2}\left(e^{x/60} - e^{-x/60}\right)$ and so

$$\begin{aligned} (y')^2 &= \left[\frac{1}{2}\left(e^{x/60} - e^{-x/60}\right)\right]^2 \\ &= \frac{1}{4}\left(e^{x/30} - 2 + e^{-x/30}\right) \end{aligned}$$

EXAMPLE 9.14 (Cont.)

We are now ready to determine the arc length L of the cable.

$$\begin{aligned}
L &= \int_a^b \sqrt{1+(y')^2}\,dx \\
&= \int_{-40}^{40} \sqrt{1+\frac{1}{4}\left(e^{x/30}-2+e^{-x/30}\right)}\,dx \\
&= \int_{-40}^{40} \sqrt{\frac{1}{4}\left(e^{x/30}+2+e^{-x/30}\right)}\,dx \\
&= \int_{-40}^{40} \sqrt{\left[\frac{1}{2}\left(e^{x/60}+e^{-x/60}\right)\right]^2}\,dx \\
&= \int_{-40}^{40} \frac{1}{2}\left(e^{x/60}+e^{-x/60}\right)dx \\
&= 30\left(e^{x/60}-e^{-x/60}\right)\Big|_{-40}^{40} \\
&= 60\left(e^{2/3}-e^{-2/3}\right) \approx 86.0590
\end{aligned}$$

The cable is about 86.06 m long.

Exercise Set 9.2

Evaluate the integrals in Exercises 1–28.

1. $\displaystyle\int \frac{dx}{4x+1}$
2. $\displaystyle\int \frac{dx}{9x-5}$
3. $\displaystyle\int e^{-3x}\,dx$
4. $\displaystyle\int e^{6x}\,dx$
5. $\displaystyle\int \frac{\sin x}{\cos x}\,dx$
6. $\displaystyle\int \frac{10x+3}{5x^2+3x-7}\,dx$
7. $\displaystyle\int \frac{2\sec^2 x}{\tan x}\,dx$
8. $\displaystyle\int \frac{2e^x}{e^x+4}\,dx$
9. $\displaystyle\int 4^x\,dx$
10. $\displaystyle\int \left(e^x-e^{-x}\right)dx$
11. $\displaystyle\int \left(e^x+e^{-x}\right)dx$
12. $\displaystyle\int \sec^2 x\, e^{\tan x}\,dx$
13. $\displaystyle\int \frac{e^{2x}}{1+e^{2x}}\,dx$
14. $\displaystyle\int \frac{x}{1+x^2}\,dx$
15. $\displaystyle\int \frac{\ln(1/x)}{x}\,dx$
16. $\displaystyle\int e^{\frac{\sqrt{x}}{8\sqrt{x}}}\,dx$
17. $\displaystyle\int e^{\frac{\sqrt[3]{x}}{x^{2/3}}}\,dx$
18. $\displaystyle\int \frac{(\ln x)^{3/4}}{x}\,dx$
19. $\displaystyle\int_0^4 \frac{x}{x^2+1}\,dx$
20. $\displaystyle\int_0^1 \frac{dx}{e^{2x}}$
21. $\displaystyle\int_1^3 \frac{1}{\ln e^x}\,dx$
22. $\displaystyle\int_0^3 \frac{x^2}{x^3+1}\,dx$
23. $\displaystyle\int_0^1 x^3 e^{x^4}\,dx$
24. $\displaystyle\int_{\pi/4}^{\pi/3} \frac{dx}{(x^2+1)\arctan x}$
25. $\displaystyle\int_2^4 \left(\tfrac{1}{5}\right)^{x/2}\,dx$
26. $\displaystyle\int_{\ln 2}^{\ln 5} \frac{e^x}{e^x+1}\,dx$
27. Find the area between the curve $y=e^{x+1}$, the y-axis, $y=x^2$, and $x=1$.
28. Find the area bounded by $y=0$, $x=0$, $y=e^{-x}$, and $x=4$.

Solve Exercises 29–38.

29. *Physics* An object of mass 24 kg is thrown vertically upward from the ground into the air with an initial velocity of 10 m/s. If the object encounters air resistance proportional to its velocity, the velocity in m/s of the object at any time t is given by

$$v(t) = 394e^{0.025t} - 384$$

(a) What is the position of the object at any time t?
(b) At what time does the object reach its maximum height?
(c) What is the maximum height?

30. *Electricity* Find the average value of the current $i = e^{(-2/3)t}$ in a circuit from $t = 0$ to $t = 2$.

31. Find the volume of the solid generated by rotating the region bounded by $y = e^{-x}$, $y = 0$, $x = 0$, and $x = 2$ around the x-axis.

32. *Electricity* The voltage across an inductor varies according to $v = 200e^{-50t}$.
(a) Find the average voltage over the inductor from $t = 0$ to $t = 0.02$ s.
(b) Determine the effective voltage from $t = 0$ to $t = 0.02$ s.

33. *Electronics* The signal voltage V across a telephone circuit changed with the distance s mi along the circuit at the rate $\dfrac{dV}{ds} = -0.45e^{-0.15s}$. If the applied voltage at the sending end was 3 V, at what distance along the line is the voltage 1.5 V?

34. *Physics* A quantity of gas with an initial volume of $0.200\ m^3$ and pressure 500 kPa expands to a volume of $0.800\ m^3$. Find the work done by the gas, assuming that the pressure of the gas is inversely proportional to its volume.

35. *Physics* A quantity of gas with an initial volume of $0.400\ ft^3$ and pressure 20 psi expands to a volume of $1.500\ ft^3$. Find the work done by the gas, assuming that the pressure of the gas is inversely proportional to its volume.

36. *Pharmacology* If x mg of a certain drug is administered to a person, the rate of change in the person's temperature, in °F, with respect to the dosage is given by $T'(x) = \frac{4}{2x+9}$. A dose of 1 mg raises the person's temperature 2.4°F. Determine the function that gives total change in body temperature.

37. *Nuclear engineering* The integral $F(x) = \displaystyle\int_0^x 12te^{-3t^2}\,dt$ appears in the analysis of failure rates and reliability in nuclear reactor design. Integrate to find $F(x)$.

38. *Medical technology* The rate of change of the concentration $C(t)$ at time t of a radioactive tracer drug is $C'(t) = 2^{-t}$, where t is measured in hours. If the initial concentration is 1 μg/L, determine the concentration at time t.

In Your Words

39. You can use the change of base formula $\log_b x = \dfrac{\ln x}{\ln b}$ so that you need but one rule for differentiating a logarithmic function. Will this also work to integrate $\ln u$ and $\log_b u$? If you answered yes, explain how it is done. If you answered no, explain why it will not work.

40. You can use the change of base formula $\log_b x = \dfrac{\ln x}{\ln b}$ so that you need but one rule for differentiating a logarithmic function. Will this also work to integrate e^x and a^x? If you answered yes, explain how it is done. If you answered no, explain why it will not work.

9.3 BASIC TRIGONOMETRIC AND HYPERBOLIC INTEGRALS

In this section, we will discuss the basic trigonometric integrals. This will include the integrals of the six trigonometric functions. Integration is the inverse of differentiation. Integrating the derivatives of the six trigonometric functions gives us the following formulas.

Basic Trigonometric Integrals

$$\int \cos u \, du = \sin u + C$$

$$\int \sin u \, du = -\cos u + C$$

$$\int \sec^2 u \, du = \tan u + C$$

$$\int \csc^2 u \, du = -\cot u + C$$

$$\int \sec u \tan u \, du = \sec u + C$$

$$\int \csc u \cot u \, du = -\csc u + C$$

The substitution $u =$ the angle will usually allow an easier integration.

EXAMPLE 9.15

Determine $\int 2\cos 2x \, dx$.

Solution Here the angle is $2x$, so we will let $u = 2x$. Then $du = 2\,dx$ and

$$\begin{aligned}\int 2\cos 2x \, dx &= \int \cos u \, du \\ &= \sin u + C \\ &= \sin 2x + C\end{aligned}$$

EXAMPLE 9.16

Find $\int \sec(3x+4)\tan(3x+4)\,dx$.

Solution If we let $u = 3x + 4$, then $du = 3\,dx$ and $dx = \frac{1}{3}du$. We get the following result:

$$\begin{aligned}\int \sec(3x+4)\tan(3x+4)\,dx &= \frac{1}{3}\int \sec u \tan u \, du \\ &= \frac{1}{3}\sec u + C \\ &= \frac{1}{3}\sec(3x+4) + C\end{aligned}$$

EXAMPLE 9.17

Evaluate $\int \tan^2 x\,dx$.

Solution This does not seem to fit in one of the six basic equations. But, if we remember that $\tan^2 x = \sec^2 x - 1$, we can see that

$$\begin{aligned}\int \tan^2 x\,dx &= \int (\sec^2 x - 1)\,dx \\ &= \int \sec^2 x\,dx - \int dx \\ &= \tan x - x + C\end{aligned}$$

We have yet to determine the integrals for four of the trigonometric functions—tangent, cotangent, secant, and cosecant. We will develop them next.

For the tangent and cotangent functions, we use the fact that $\tan x = \dfrac{\sin x}{\cos x}$ and $\cot x = \dfrac{\cos x}{\sin x}$.

$\int \tan u\,du = \int \dfrac{\sin u}{\cos u}\,du$. If we let $v = \cos u$, then $dv = -\sin u\,du$. Multiplying the integral by $\frac{-1}{-1}$, we get $\int \dfrac{\sin u}{\cos u}\,du = -\int \dfrac{-\sin u}{\cos u}\,du = -\int \dfrac{dv}{v} = -\ln|v| + C$. But $v = \cos u$, so

$$\int \tan u\,du = -\ln|\cos u| + C$$

Remembering that $-\ln|\cos u| = \ln(|\cos u|)^{-1} = \ln\left|\dfrac{1}{\cos u}\right| = \ln|\sec u|$, we get the following two alternative results for $\int \tan u\,du$.

$$\begin{aligned}\int \tan u\,du &= -\ln|\cos u| + C \\ &= \ln|\sec u| + C\end{aligned}$$

In the exercises you will be asked to show that

$$\int \cot u\,du = \ln|\sin u| + C$$

Developing the integral of the secant and cosecant functions are a little more complicated. We will develop the integral of the secant function and let you show that our integral of the cosecant function is true. The integral of the secant function requires us to multiply the function by an unusual form of the number 1: $\dfrac{\sec u + \tan u}{\sec u + \tan u}$. We now proceed as follows.

$$\begin{aligned}\int \sec u\,du &= \int \frac{(\sec u)(\sec u + \tan u)}{\sec u + \tan u}\,du \\ &= \int \frac{\sec^2 u + \sec u \tan u}{\sec u + \tan u}\,du\end{aligned}$$

If we let $v = \sec u + \tan u$, then $dv = (\sec u \tan u + \sec^2 u)\,du$. This means that our integral can be rewritten as

$$\int \frac{dv}{v} = \ln|v| + C = \ln|\sec u + \tan u| + C$$

Thus, we have the result we were seeking. A similar effort would produce the integral for the cosecant function.

Integrals of Secant and Cosecant

$$\int \sec u\,du = \ln|\sec u + \tan u| + C$$

$$\int \csc u\,du = \ln|\csc u - \cot u| + C$$

EXAMPLE 9.18

Find $\int \tan(7x-5)\,dx$.

Solution If we let $u = 7x - 5$, then $du = 7\,dx$ and $dx = \frac{1}{7}\,du$. This produces

$$\begin{aligned}\int \tan(7x-5)\,dx &= \frac{1}{7}\int \tan u\,du \\ &= \frac{1}{7}\ln|\sec u| + C \\ &= \frac{1}{7}\ln|\sec(7x-5)| + C\end{aligned}$$

EXAMPLE 9.19

Evaluate $\int x \sec 3x^2\,dx$.

Solution Let $u = 3x^2$, then $du = 6x\,dx$ and $x\,dx = \frac{1}{6}\,du$. These substitutions produce

$$\begin{aligned}\int x\sec 3x^2\,dx &= \frac{1}{6}\int \sec u\,du \\ &= \frac{1}{6}\ln|\sec u + \tan u| + C \\ &= \frac{1}{6}\ln|\sec 3x^2 + \tan 3x^2| + C\end{aligned}$$

Application

EXAMPLE 9.20

The current in a particular circuit is given by the equation $i = 100\cos(120t - \sqrt{2})$. What is the average current as t ranges from 0 to 0.5 s?

Solution The average current $\bar{i}$ is given by

$$\begin{aligned}
\bar{i} &= \frac{1}{b-a}\int_a^b i(t)\,dt \\
&= \frac{1}{0.5-0}\int_0^{0.5} 100\cos(120t - \sqrt{2})\,dt \\
&= \frac{1}{0.5}\frac{100}{120}\sin\left(120t - \sqrt{2}\right)\Big|_0^{0.5} \\
&\approx \frac{5}{3}[0.8932 - (-0.9878)] \\
&\approx 3.135
\end{aligned}$$

The average current is 3.135 A.

Integrals of Hyperbolic Functions

In Section 8.6, we defined and found derivatives of the hyperbolic and inverse hyperbolic functions. We will next examine the integrals of the hyperbolic functions. In Section 9.5, we will integrate the inverse hyperbolic functions.

Based on the derivatives of the hyperbolic functions, we have the following integrals.

Integrals Involving Hyperbolic Functions

$$\int \sinh u\,du = \cosh u + C \qquad \int \operatorname{csch}^2 u\,du = -\coth u + C$$

$$\int \cosh u\,du = \sinh u + C \qquad \int \operatorname{sech} u \tanh u\,du = -\operatorname{sech} u + C$$

$$\int \operatorname{sech}^2 u\,du = \tanh u + C \qquad \int \operatorname{csch} u \coth u\,du = -\operatorname{csch} u + C$$

EXAMPLE 9.21

Evaluate $\int \cosh 9x\,dx$.

Solution If we let $u = 9x$, then $du = 9\,dx$. We multiply the integral by $\frac{9}{9}$, and obtain the result

$$\begin{aligned}\int \cosh 9x\,dx &= \frac{9}{9}\int \cosh 9x\,dx \\ &= \frac{1}{9}\int 9\cosh 9x\,dx \\ &= \frac{1}{9}\sinh 9x + C\end{aligned}$$

EXAMPLE 9.22

Evaluate $\int \coth 17x\,dx$.

Solution Here we rewrite $\coth 17x$ as $\dfrac{\cosh 17x}{\sinh 17x}$. Now, let $u = \sinh 17x$, then $du = 17\cosh 17x\,dx$. We use these results as we evaluate the following integral.

$$\begin{aligned}\int \coth 17x\,dx &= \int \frac{\cosh 17x}{\sinh 17x}\,dx \\ &= \frac{1}{17}\int \frac{du}{u} \\ &= \frac{1}{17}\ln|u| + C \\ &= \frac{1}{17}\ln|\sinh 17x| + C\end{aligned}$$

In Example 9.14, we went through a rather long procedure to solve a problem similar to this. Now that we are able to integrate $\cosh x$, we can use the much simpler process demonstrated by the next example.

Application

EXAMPLE 9.23

A cable suspended between two poles takes the shape $y = 5\cosh\frac{x}{5}$ for $-10.00 \le x \le 10.00$. Find the length of the cable, if x and y are in meters.

Solution Recalling from Section 7.4, the formula for the arc length is $L = \int_a^b \sqrt{1 + [f'(x)]^2}\,dx$. We begin by differentiating the given function, with the result

$$y' = \sinh\left(\frac{x}{5}\right)$$

EXAMPLE 9.23 (Cont.)

Substituting this into the arc length formula, we have

$$\begin{aligned}
L &= \int_{-10}^{10} \sqrt{1 + \left[\sinh\left(\frac{x}{5}\right)\right]^2}\, dx \\
&= \int_{-10}^{10} \sqrt{\cosh^2\left(\frac{x}{5}\right)}\, dx \\
&= \int_{-10}^{10} \cosh\left(\frac{x}{5}\right) dx \\
&= 5 \sinh \frac{x}{5}\Big]_{-10}^{10} \\
&\approx 5[3.63 - (-3.63)] \\
&= 36.30
\end{aligned}$$

This cable is about 36.30 m long.

Exercise Set 9.3

In Exercises 1–40, determine the given integral.

1. $\int \sin \frac{x}{2}\, dx$
2. $\int \cos 3x\, dx$
3. $\int \tan 5x\, dx$
4. $\int x \csc x^2\, dx$
5. $\int \frac{1}{\cos x}\, dx$
6. $\int \frac{1}{\sin x}\, dx$
7. $\int \frac{1}{\cos^2 x}\, dx$
8. $\int \frac{\csc x}{\sin x}\, dx$
9. $\int \frac{\tan \sqrt{x}}{\sqrt{x}}\, dx$
10. $\int \tan^2 6x\, dx$
11. $\int \sin(4x - 1)\, dx$
12. $\int \frac{\sin x + \cos x}{\cos x}\, dx$
13. $\int \frac{\sin x}{\cos^2 x}\, dx$
14. $\int \frac{1 + \sin x}{\cos x}\, dx$
15. $\int \frac{\sec 3x \tan 3x}{5 + 2 \sec 3x}\, dx$
16. $\int \cos^2 x \sin x\, dx$
17. $\int \sec^4 3x \tan 3x\, dx$
18. $\int \tan 3x \sec^2 3x\, dx$
19. $\int \frac{3\, dx}{\tan 3x}$
20. $\int \frac{5\, dx}{\sin^2 5x}$
21. $\int (\sec x + 2)^2\, dx$
22. $\int (\tan x + 3)^2\, dx$
23. $\int_0^{\pi/2} (\sec 0.5x + 5)^2\, dx$
24. $\int_{\pi/12}^{\pi/6} (\cot 3x - 4)^2\, dx$
25. $\int \frac{1 + \cos 4x}{\sin^2 4x}\, dx$
26. $\int x \csc^2 x^2\, dx$
27. $\int \tan \frac{x}{4}\, dx$
28. $\int x \sec^2(x^2 + 1) \tan(x^2 + 1)\, dx$
29. $\int_{\pi/4}^{\pi/2} \frac{1 + \cot^2 x}{\csc^2 x}\, dx$
30. $\int_0^{\pi/2} \frac{\cos x}{1 + \sin x}\, dx$
31. $\int_{\pi/4}^{\pi/2} \frac{\csc \sqrt{x} \cot \sqrt{x}}{\sqrt{x}}\, dx$
32. $\int_0^{\pi/8} \sec^5 2x \tan 2x\, dx$
33. $\int_{\pi/6}^{\pi/2} \frac{\cos^2 x}{\sin x}\, dx$
34. $\int_0^{5\pi/12} \frac{\sec^2 x}{2 \tan x + 4}\, dx$

35. $\int x \sinh x^2\, dx$

36. $\int \frac{\sinh x}{1+\cosh x}\, dx$

37. $\int 3x \cosh x^2 \sqrt{\sinh x^2}\, dx$

38. $\int \operatorname{sech}^2 5x\, dx$

39. $\int \sinh^3 x \cosh^2 x\, dx$

40. $\int \tanh 3x \operatorname{sech} 3x\, dx$

Solve Exercises 41–52.

41. Show that $\int \cot u\, du = \ln \sin u + C$

42. Show that $\int \csc u\, du = \ln \csc u - \cot u + C$

43. Find the area of the region bounded by $y = \sin 2x$, $y = 0$, $x = 0$, and $x = \frac{\pi}{2}$.

44. Find the area of the region bounded by $y = \sec x$, $y = x$, $x = 0$, and $x = \frac{\pi}{4}$.

45. Find the average value of $y = \tan x$ on the interval $\left[0, \frac{\pi}{4}\right]$

46. *Physics* A certain particle has its velocity described by $v(t) = 5 \sin 2t$. If we know that at $t = 0$ it was located at $s = 4$, what is its distance function $s(t)$?

47. *Electricity* The electromotive force E of a particular electrical circuit is given by

$$E = 5 \sin 4t$$

where E is measured in volts and t is in seconds. Find the average value of E during the first second.

48. *Electricity* A 60-Hz alternator generates a current given by $i = 6.5 \sin(377t)$ A and flows through a 40-Ω resistor. The rate P at which heat is produced in the resistor is given by $P = 40i^2$. Determine the average rate of heat produced over one complete cycle. (Hint: Use the half-angle identity for the sine.)

49. *Construction* A power cable suspended between two poles takes the shape $y = 12 \cosh \frac{x}{12}$ for $-18.0 \le x \le 18.0$. Find the length of the cable, if x and y are in meters.

50. *Construction* Electric wires suspended between two towers form a catenary with the equation $y = 60 \cosh \frac{x}{60}$, where x is measured in feet. Find the length of the suspended wire, if the towers are 120 ft apart.

51. *Construction* Electric wires suspended between two towers have the equation $y = 80 \cosh \frac{x}{80}$, where x is measured in feet. Find the length of the suspended wire, if the towers are 300 ft apart.

52. *Construction* The Gateway Arch in St. Louis has the shape of an inverted catenary. The shape of the arch can by approximated by

$$y = -127.7 \cosh \frac{x}{127.7} + 757.7$$

for $-315.0 \le x \le 315.0$, where all units are in feet.

(a) Determine the maximum height of the arch.

(b) Determine the width of the opening of the arch at its base.

(c) Approximate the total open area under the arch.

(d) Approximate the total length of the arch.

In Your Words

53. Explain how the integrals of the sine and cosine functions differ from those of the hyperbolic sine and hyperbolic cosine functions, respectively.

54. Explain how to use the method for developing $\int \tan u\, du$ and your knowledge of $\int \sinh u\, du$ and $\int \cosh u\, du$ to develop $\int \tanh u\, du$.

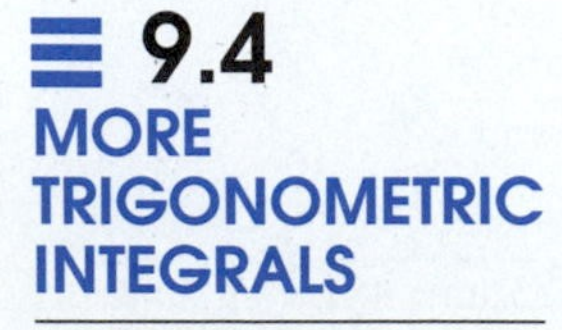

9.4 MORE TRIGONOMETRIC INTEGRALS

In Section 9.3, we learned the basic trigonometric integrals. In Section 9.1, we learned how to integrate some trigonometric functions by using the general power formula. In this section, we will learn some additional methods for integrating trigonometric functions. These methods will use the Pythagorean identities and half-angle formulas. Those equations are given below, so that you will be able to use them more easily.

Pythagorean Identities

$$\sin^2\theta + \cos^2\theta = 1$$
$$\tan^2\theta + 1 = \sec^2\theta$$
$$1 + \cot^2\theta = \csc^2\theta$$

Half-Angle Formulas

$$\sin^2\theta = \frac{1-\cos 2\theta}{2}$$
$$\cos^2\theta = \frac{1+\cos 2\theta}{2}$$

The integrals discussed in this section fall into four types. We will give examples on integrating each type.

Type I: $\int \sin^n x\,dx \qquad \int \cos^n x\,dx, \qquad n$ a positive integer

This first type is really a special case of the second type. However, we will use this as a good learning experience. We will consider two cases: **(a)** n is odd and **(b)** n is even.

When n is an odd positive integer such as 3, 5, or 15, then $n-1$ is an even integer. We will first rewrite the function as $\sin^n x = \sin^{n-1} x \sin x$. Since $n-1$ is even, we can use the identity $\sin^2 x = 1 - \cos^2 x$ and then use the general power formula.

EXAMPLE 9.24

Evaluate $\int \sin^7 x\,dx$.

Solution

$$\begin{aligned}\int \sin^7 x\,dx &= \int \sin^6 x \sin x\,dx \\ &= \int (\sin^2 x)^3 \sin x\,dx \\ &= \int (1-\cos^2 x)^3 \sin x\,dx\end{aligned}$$

If we let $u = \cos x$, then $du = -\sin x\,dx$ and the integral becomes

$$\begin{aligned}\int \sin^7 x\,dx &= \int (1-\cos^2 x)^3 \sin x\,dx \\ &= -\int (1-u^2)^3\,du \\ &= -\int (1-3u^2+3u^4-u^6)\,du \\ &= -\left(u - u^3 + \frac{3}{5}u^5 - \frac{1}{7}u^7\right) + C \\ &= -\cos x + \cos^3 x - \frac{3}{5}\cos^5 x + \frac{1}{7}\cos^7 x + C\end{aligned}$$

We would use the same technique to find $\int \cos^n x\,dx$ if n was odd, except that we would use the identity $\cos^2 x = 1 - \sin^2 x$ and the substitution $u = \sin x$.

If n is an even positive integer, we will use the half-angle formulas to help simplify the integral.

EXAMPLE 9.25

Determine $\int \sin^2 3x\,dx$.

Solution Use the half-angle formula $\sin^2\theta = \frac{1}{2}(1-\cos\theta)$ with $\theta = 2x$. Then

$$\begin{aligned}\int \sin^2 3x\,dx &= \frac{1}{2}\int (1-\cos 6x)\,dx \\ &= \frac{1}{2}x - \frac{1}{2}\int \cos 6x\,dx \\ &= \frac{1}{2}x - \frac{1}{12}\sin 6x + C\end{aligned}$$

EXAMPLE 9.26

Find $\int \cos^4 5x\,dx$.

Solution

$$\begin{aligned}\int \cos^4 5x\,dx &= \int (\cos^2 5x)^2\,dx \\ &= \int \left[\frac{1}{2}(1+\cos 10x)\right]^2 dx \\ &= \frac{1}{4}\int (1+2\cos 10x+\cos^2 10x)\,dx \\ &= \frac{1}{4}\int \left[1+2\cos 10x+\frac{1}{2}(1+\cos 20x)\right] dx \\ &= \frac{1}{4}x+\frac{1}{20}\sin 10x+\frac{1}{8}x+\frac{1}{160}\sin 20x+C \\ &= \frac{3}{8}x+\frac{1}{20}\sin 10x+\frac{1}{160}\sin 20x+C\end{aligned}$$

Application

EXAMPLE 9.27

Find the area under the graph of $y = \cos^4 5x$ and above the x-axis from $x = 0$ to $x = 2\pi$.

EXAMPLE 9.27 (Cont.)

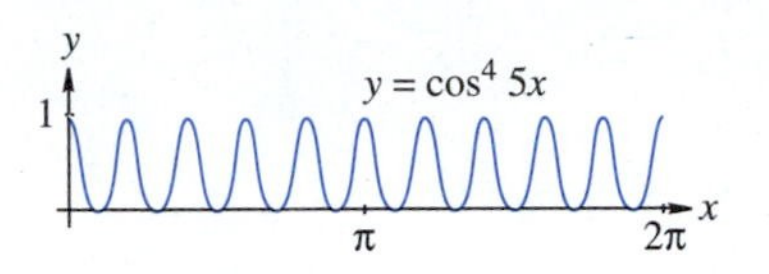

FIGURE 9.3

Solution The graph of this curve is shown in Figure 9.3. In Example 9.26, we found that $\int \cos^4 5x\,dx = \frac{3}{8}x + \frac{1}{20}\sin 10x + \frac{1}{160}\sin 20x + C$. Thus, the area under this curve over the indicated interval is

$$\begin{aligned} A &= \int_0^{2\pi} \cos^4 5x\,dx \\ &= \left[\frac{3}{8}x + \frac{1}{20}\sin 10x + \frac{1}{160}\sin 20x\right]_0^{2\pi} \\ &\approx 2.3562 - 0 = 2.3562 \end{aligned}$$

The desired area is about 2.3562 square units. ▪

Type II: $\int \sin^m x \cos^n x\,dx$

Again, we will consider two cases: **(a)** either m or n is an odd positive integer and **(b)** both m and n are even positive integers. Since Type I is a special case of Type II, where $m = 0$ or $n = 0$, we can use the same techniques we used on Type I integrals.

EXAMPLE 9.28

Find $\int \sin^{1/3} x \cos^5 x\,dx$.

Solution The power of the cosine function is an odd positive integer. We will rewrite $\cos^5 x$ as $(\cos^2 x)^2 \cos x$ and let $u = \sin x$.

$$\begin{aligned} \int \sin^{1/3} x \cos^5 x\,dx &= \int \sin^{1/3} x (\cos^2 x)^2 \cos x\,dx \\ &= \int \sin^{1/3} x (1 - \sin^2 x)^2 \cos x\,dx \\ &= \int u^{1/3}(1 - u^2)^2\,du \\ &= \int u^{1/3}(1 - 2u^2 + u^4)\,du \\ &= \int (u^{1/3} - 2u^{7/3} + u^{13/3})\,du \\ &= \frac{3}{4}u^{4/3} - \frac{3}{5}u^{10/3} + \frac{3}{16}u^{16/3} + C \\ &= \frac{3}{4}\sin^{4/3} x - \frac{3}{5}\sin^{10/3} x + \frac{3}{16}\sin^{16/3} x + C \end{aligned}$$

▪

If both m and n are even numbers, then we can use the Pythagorean identity $\sin^2 x + \cos^2 x = 1$.

EXAMPLE 9.29

Determine $\int \sin^4 x \cos^2 x\,dx$.

Solution Substituting $1 - \sin^2 x$ for $\cos^2 x$, we obtain

$$\begin{aligned}\int \sin^4 x \cos^2 x\,dx &= \int \sin^4 x(1 - \sin^2 x)\,dx \\ &= \int (\sin^4 x - \sin^6 x)\,dx\end{aligned}$$

The integrand is now in the form of a Type I integral and we can use those techniques to solve this integral.

Application

EXAMPLE 9.30

Find the volume of the solid of revolution generated by rotating the region bounded by $y = \sin^{1/6} x \cos^{5/2} x$, $y = 0$, $x = 0$, and $x = \dfrac{\pi}{2}$ around the x-axis.

Solution The graph of this region is shown in Figure 9.4a and the solid of revolution formed by rotating it around the x-axis is shown in Figure 9.4b. A typical disk has been drawn in the solid. The volume of this solid is

$$\begin{aligned}V &= \pi \int_0^{\pi/2} \left(\sin^{1/6} x \cos^{5/2} x\right)^2 dx \\ &= \pi \int_0^{\pi/2} \sin^{1/3} x \cos^5 x\,dx\end{aligned}$$

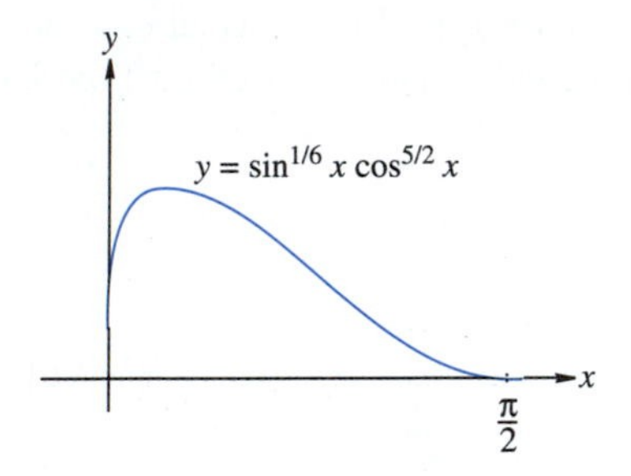

FIGURE 9.4a

From Example 9.28, we know that

$$\int \sin^{1/3} x \cos^5 x\,dx = \frac{3}{4}\sin^{4/3} x - \frac{3}{5}\sin^{10/3} x + \frac{3}{16}\sin^{16/3} x + C$$

Thus,

$$\begin{aligned}V &= \pi \left[\frac{3}{4}\sin^{4/3} x - \frac{3}{5}\sin^{10/3} x + \frac{3}{16}\sin^{16/3} x\right]_0^{\pi/2} \\ &= \pi\left(\frac{3}{4} - \frac{3}{5} + \frac{3}{16}\right) \\ &= \frac{27}{80}\pi\end{aligned}$$

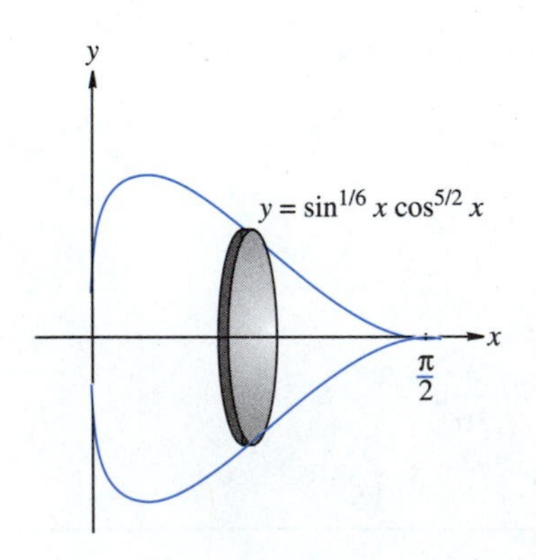

FIGURE 9.4b

Type III: $\int \tan^m x \sec^n x\,dx \qquad \int \cot^m x \csc^n x\,dx$

In this type, we will look first at the case where n is even. We will factor a $\sec^2 x$ out of the integrand and rewrite the remainder of the integrand in terms of the tangent by using the identity $\sec^2 x = 1 + \tan^2 x$. (If the integral involves the cotangent and cosecant functions, you would factor a $\csc^2 x$ out of the integrand and use $\csc^2 x = \cot^2 x + 1$.)

EXAMPLE 9.31

Find $\int \tan^3 x \sec^4 x \, dx$.

Solution

$$\begin{aligned}\int \tan^3 x \sec^4 x \, dx &= \int \tan^3 x (\sec^2 x) \sec^2 x \, dx \\ &= \int \tan^3 x (\tan^2 x + 1) \sec^2 x \, dx \\ &= \int (\tan^5 x + \tan^3 x) \sec^2 x \, dx\end{aligned}$$

If we let $u = \tan x$, then $du = \sec^2 x \, dx$, and we can rewrite this as

$$\begin{aligned}\int (u^5 + u^3) \, du &= \frac{1}{6}u^6 + \frac{1}{4}u^4 + C \\ &= \frac{1}{6}\tan^6 x + \frac{1}{4}\tan^4 x + C\end{aligned}$$

If m is an odd number, then we will factor $\tan x \sec x$ out of the integrand and once again use the identity $\tan^2 x = \sec^2 x - 1$ (or factor $\csc x \cot x$ out of the integrand and use $\cot^2 x = \csc^2 x - 1$).

EXAMPLE 9.32

Evaluate $\int \tan^3 x \sec^3 x \, dx$.

Solution

$$\begin{aligned}\int \tan^3 x \sec^3 x \, dx &= \int \tan^2 x \sec^2 x (\tan x \sec x) \, dx \\ &= \int (\sec^2 x - 1) \sec^2 x (\tan x \sec x) \, dx \\ &= \int (\sec^4 x - \sec^2 x)(\tan x \sec x) \, dx\end{aligned}$$

If $u = \sec x$, then $du = \tan x \sec x \, du$, and we get

$$\begin{aligned}\int (u^4 - u^2) \, du &= \frac{1}{5}u^5 - \frac{1}{3}u^3 + C \\ &= \frac{1}{5}\sec^5 x - \frac{1}{3}\sec^3 x + C\end{aligned}$$

Similar methods can be used for $\int \cot^m x \csc^n x \, dx$.

Type IV: $\int \tan^n x \, dx \qquad \int \sec^n x \, dx, \qquad n$ even

Both of these require the use of the identity $\tan^2 x = \sec^2 x - 1$. In Section 9.7, we will learn how to integrate $\sec^n x$ when n is odd.

EXAMPLE 9.33

Find $\int \tan^4 x\,dx$.

Solution

$$\begin{aligned}\int \tan^4 x\,dx &= \int \tan^2 x \tan^2 x\,dx \\ &= \int \tan^2 x(\sec^2 x - 1)\,dx \\ &= \int (\tan^2 x\sec^2 x - \tan^2 x)\,dx \\ &= \int (\tan^2 x\sec^2 x - \sec^2 x + 1)\,dx \\ &= \frac{1}{3}\tan^3 x - \tan x + x + C\end{aligned}$$

EXAMPLE 9.34

Evaluate $\int \sec^4 3x\,dx$.

Solution

$$\begin{aligned}\int \sec^4 3x\,dx &= \int \sec^2 3x \sec^2 3x\,dx \\ &= \int (1+\tan^2 3x)\sec^2 3x\,dx \\ &= \int (\sec^2 3x + \tan^2 3x\sec^2 3x)\,dx \\ &= \frac{1}{3}\tan 3x + \frac{1}{9}\tan^3 3x + C\end{aligned}$$

Exercise Set 9.4

In Exercises 1–30, evaluate the given integral.

1. $\int \sin^2 3x\cos 3x\,dx$
2. $\int \sin 2x\cos^2 2x\,dx$
3. $\int \cos x\sin^3 x\,dx$
4. $\int \sin^3 4x\cos^2 4x\,dx$
5. $\int \cos 5x\sin^4 5x\,dx$
6. $\int \cos^2 y\sin^5 y\,dy$
7. $\int \sin^2 x\cos^4 x\,dx$
8. $\int \sin^8 x\,dx$
9. $\int \cos^6 3x\,dx$
10. $\int \frac{\cos x}{\sin^3 x}\,dx$
11. $\int \sin^2 2\theta\cos^4 2\theta\,d\theta$
12. $\int \sin^4 3\theta\cos^2 3\theta\,d\theta$
13. $\int \sec^2 x\tan^2 x\,dx$
14. $\int \sec^4 y\tan^3 y\,dy$
15. $\int \csc^4 x\cot x\,dx$
16. $\int \frac{\sin^2\theta}{\cos^4\theta}\,d\theta$
17. $\int \sin^{1/2} 3\theta\cos^3 3\theta\,d\theta$
18. $\int \csc^6 2x\cot^2 2x\,dx$
19. $\int \csc x\cot^3 x\,dx$
20. $\int \sec t\tan^5 t\,dt$
21. $\int \tan^3 x\sec^2 x\,dx$
22. $\int \cot 2x\csc^4 2x\,dx$
23. $\int \tan^6 x\sec^2 x\,dx$
24. $\int \tan^5 x\,dx$
25. $\int \cot^6 x\,dx$
26. $\int \sec^6 2\theta\,d\theta$
27. $\int_0^{\pi/4} \tan^2 x\,dx$

28. $\int_{\pi/6}^{\pi/3} \tan^3 2t\,dt$

29. $\int_0^{\pi/2} \sin^4 x\,dx$

30. $\int_0^{\pi/6} \sec^3 2x \tan 2x\,dx$

Solve Exercises 31–36.

31. Find the volume of the solid of revolution generated by rotating the region bounded by $y = \cos x$, $y = 0$, $x = 0$, and $x = \frac{\pi}{3}$ around the x-axis.

32. *Electricity* Find the root mean square of the voltage for one period, if the voltage is given by $V = 80 \sin 120\pi t$.

33. *Physics* The integral $I = \frac{mr^2}{4}\int_0^{2\pi} \sin^2\theta\,d\theta$ gives the moment of inertia about a diameter of a circular disk of radius r and mass m.
(a) Integrate to find I.
(b) Find the moment of inertia for a disk of radius 0.1 m and mass 12.4 kg.

34. *Physics* Under certain conditions, the rate of radiation by an accelerated charge is given by $R = \int \cos^3\theta\,d\theta$. Find an expression for the rate of radiation.

35. *Electronics* Find the effective current from $t = 0$ to $t = 0.5$ s, if $i = 2 \sin t\sqrt{\cos t}$.

36. *Electronics* Find the rms of the current $i = 4 \sin t \cos t$ from $t = 0$ to $t = \pi$ s.

In Your Words

37. Explain how to evaluate $\int \sin^m x \cos^n x\,dx$ when m is an odd positive integer and n is an even positive integer.

38. Explain how to evaluate $\int \sin^m x \cos^n x\,dx$ when m is even and n is odd.

39. Explain how to evaluate $\int \sin^m x \cos^n x\,dx$ when m and n are even positive integers.

9.5 INTEGRALS RELATED TO INVERSE TRIGONOMETRIC AND INVERSE HYPERBOLIC FUNCTIONS

In this section, we will discuss the integrals of certain algebraic functions that result from derivatives of inverse trigonometric functions.

Integrals Related to Inverse Trigonometric Functions

In Section 8.3, we learned the derivatives of inverse trigonometric functions. In particular, we learned that

$$\frac{d}{dx}\sin^{-1}u = \frac{1}{\sqrt{1-u^2}}\frac{du}{dx}, \quad |u| < 1$$

$$\frac{d}{dx}\tan^{-1}u = \frac{1}{1+u^2}\cdot\frac{du}{dx}$$

$$\frac{d}{dx}\sec^{-1}u = \frac{1}{u\sqrt{u^2-1}}\frac{du}{dx}, \quad |u| > 1$$

The derivatives for the $\operatorname{arccos} x$, $\operatorname{arccot} x$, and $\operatorname{arccsc} x$ were similar except for the signs of the derivatives. We will look at ways of reversing these differentiation formulas.

The first integration formula we will consider is

$$\int \frac{du}{\sqrt{a^2-u^2}} = \arcsin\frac{u}{a} + C$$

In this formula, we assume that a is a constant.

EXAMPLE 9.35

Find $\int \frac{dx}{\sqrt{16-x^2}}$.

Solution

$$\int \frac{dx}{\sqrt{16-x^2}} = \int \frac{dx}{\sqrt{4^2-x^2}} = \arcsin\frac{x}{4} + C$$

The second integration formula we will consider is

$$\int \frac{du}{a^2+u^2} = \frac{1}{a}\arctan\frac{u}{a} + C$$

EXAMPLE 9.36

Find $\int \frac{3\,dx}{7+9x^2}$.

Solution $7+9x^2$ does not seem to be in the form a^2+x^2. If we let $u=3x$, then $u^2=9x^2$ and $du=3\,dx$, and if $a=\sqrt{7}$, then $a^2=7$. Thus $7+9x^2=(\sqrt{7})^2+(3x)^2$ and

$$\int \frac{du}{7+9x^2} = \frac{1}{\sqrt{7}}\arctan\frac{3x}{\sqrt{7}} + C$$

The last integral is

$$\int \frac{du}{u\sqrt{u^2-a^2}} = \frac{1}{a}\operatorname{arcsec}\frac{u}{a} + C$$

EXAMPLE 9.37

Evaluate $\int \frac{dx}{x\sqrt{16x^2-9}}$.

Solution If $u = 4x$ and $a = 3$, then $du = 4\,dx$ and

$$\begin{aligned}\int \frac{dx}{x\sqrt{16x^2-9}} &= \int \frac{4\,dx}{4x\sqrt{16x^2-9}}\\ &= \int \frac{du}{u\sqrt{u^2-a^2}}\\ &= \frac{1}{a}\operatorname{arcsec}\frac{u}{a}+C\\ &= \frac{1}{3}\operatorname{arcsec}\frac{4x}{3}+C\end{aligned}$$

EXAMPLE 9.38

Find $\int \frac{dx}{x^2+2x+10}$.

Solution This does not appear to fit any of the methods we have yet seen. But if we complete the square, we will see that

$$\begin{aligned}x^2+2x+10 &= x^2+2x+1+9\\ &= (x+1)^2+3^2\end{aligned}$$

If we let $u = x+1$ and $du = dx$, then

$$\begin{aligned}\int \frac{dx}{x^2+2x+10} &= \int \frac{du}{u^2+3^2}\\ &= \frac{1}{3}\arctan\frac{u}{3}+C\\ &= \frac{1}{3}\arctan\frac{x+1}{3}+C\end{aligned}$$

EXAMPLE 9.39

Determine $\int \frac{2x-7}{4x^2+9}\,dx$.

Solution We will rewrite this as the sum of two integrals:

$$\int \frac{2x-7}{4x^2+9}\,dx = \int \frac{2x}{4x^2+9}\,dx - \int \frac{7}{4x^2+9}\,dx$$

EXAMPLE 9.39 (Cont.)

The first integral is in logarithmic form with $u = 4x^2 + 9$ and $du = 8x\,dx$. Multiplying by $\frac{4}{4}$, we get $\frac{1}{4}\int \frac{8x\,dx}{4x^2+9} = \frac{1}{4}\int \frac{du}{u}$. The second integral is of the form $\int \frac{dv}{a^2+v^2}$ with $v = 2x$, $dv = 2\,dx$, and $a = 3$. Thus we have

$$\begin{aligned}\int \frac{2x-7}{4x^2+9}\,dx &= \int \frac{2x\,dx}{4x^2+9} - 7\int \frac{dx}{4x^2+9}\\ &= \frac{1}{4}\int \frac{du}{u} - 7\int \frac{dv}{v^2+3^2}\\ &= \frac{1}{4}\ln|u| - \frac{7}{3}\arctan\frac{v}{3} + C\\ &= \frac{1}{4}\ln(4x^2+9) - \frac{7}{3}\arctan\frac{2x}{3} + C\end{aligned}$$

Caution

Many integrals may look like they are of the inverse trigonometric form when they are not. Some integrals that use the general power formula or are of a logarithmic form look very similar to the integrals we studied in this chapter. It is important that you recognize the correct integration form to use. The exercises in this section will include integral forms from previous sections.

As an example of some of the confusion that is possible, look at the following integrals.

(a) $\int \frac{dx}{1+x^2}$ **(b)** $\int \frac{x\,dx}{1+x^2}$ **(c)** $\int \frac{dx}{\sqrt{1-x^2}}$ **(d)** $\int \frac{x\,dx}{\sqrt{1-x^2}}$

Integrals **(a)** and **(b)** are *almost* identical. So are integrals **(c)** and **(d)**. Look carefully! Integral **(a)** is an inverse tangent integral, but integral **(b)** is in the logarithmic form. Integral **(c)** is an inverse sine integral, but integral **(d)** uses the general power formula.

Integrals Related to Inverse Hyperbolic Functions

We present some integrals of the inverse hyperbolic functions in the box on the facing page. While it is not a complete list, it is sufficient for our purposes. You can verify each integral by differentiating the right-hand side of each equation.

EXAMPLE 9.40

Evaluate $\int_{-1}^{4} \frac{1}{\sqrt{9+4x^2}}\,dx$.

Solution This is in the form $\int \frac{1}{\sqrt{u^2+a^2}}\,du$ with $u = 2x$, $a = 3$, and $du = 2\,dx$. Thus, we have

$$\begin{aligned}\int_{-1}^{4} \frac{1}{\sqrt{9+4x^2}}\,dx &= \frac{1}{2}\int_{-1}^{4} \frac{2}{\sqrt{9+4x^2}}\,dx \\ &= \frac{1}{2}\sinh^{-1}\frac{2x}{3}\Big]_{-1}^{4} \\ &\approx \frac{1}{2}[1.7074-(-0.6251)] \approx 1.1663\end{aligned}$$

Integrals Involving Inverse Hyperbolic Functions

$$\int \frac{1}{\sqrt{u^2+a^2}}\,du = \sinh^{-1}\frac{u}{a}+C$$

$$\int \frac{1}{\sqrt{u^2-a^2}}\,du = \cosh^{-1}\frac{u}{a}+C$$

$$\int \frac{1}{a^2-u^2}\,du = \frac{1}{a}\tanh^{-1}\frac{u}{a}+C$$

$$\int \frac{1}{u\sqrt{a^2-u^2}}\,du = \frac{1}{a}\operatorname{sech}^{-1}\frac{u}{a}+C$$

Exercise Set 9.5

In Exercises 1–30, find the given integral.

1. $\int \frac{dx}{\sqrt{4-x^2}}$
2. $\int \frac{dx}{\sqrt{1-9x^2}}$
3. $\int \frac{dx}{\sqrt{4-9x^2}}$
4. $\int \frac{8\,dx}{1+16x^2}$
5. $\int \frac{-x\,dx}{\sqrt{9-x^2}}$
6. $\int \frac{dx}{x\sqrt{x^2-16}}$
7. $\int \frac{dx}{3x\sqrt{9x^2-4}}$
8. $\int \frac{x\,dx}{\sqrt{1-25x^2}}$
9. $\int \frac{dx}{1+(3-x)^2}$
10. $\int \frac{dx}{(2x+1)\sqrt{(2x+1)^2-4}}$
11. $\int \frac{4x-6}{4x^2+25}\,dx$
12. $\int \frac{dx}{(x+3)\sqrt{(x+3)^2-1}}$
13. $\int \frac{dx}{x^2+6x+10}$
14. $\int \frac{dx}{\sqrt{-4x^2+4x+15}}$
15. $\int \frac{x\,dx}{1+x^4}$
16. $\int \frac{\sec^2 x\,dx}{4+\tan^2 x}$
17. $\int \frac{\sin x\,dx}{\sqrt{1-\cos^2 x}}$

18. $\int \frac{x-5}{\sqrt{x^2-10x+16}}\,dx$

19. $\int_0^{\pi/4} \frac{\cos x\,dx}{1+\sin^2 x}$

20. $\int_3^4 \frac{dx}{\sqrt{-x^2+8x-15}}$

21. $\int \frac{dx}{\sqrt{x^2-25}}$

22. $\int \frac{dx}{\sqrt{(x-3)^2+16}}$

23. $\int \frac{dx}{\sqrt{25+9x^2}}$

24. $\int \frac{e^{2x}}{25-e^{4x}}\,dx$

25. $\int_0^{0.5} \frac{dx}{\sqrt{1-x^2}}$

26. $\int_0^{\sqrt{3}} \frac{x\,dx}{\sqrt{1+x^2}}$

27. $\int_1^5 \frac{dx}{x^2-4x+13}$

28. $\int_{-5}^{-1} \frac{dx}{\sqrt{-x^2-7x-6}}$

29. $\int_0^{\pi/6} \frac{\sec^2 x\,dx}{1+16\tan^2 x}$

30. $\int_0^1 \frac{x\,dx}{\sqrt{1+x^4}}$

Solve Exercises 31–38.

31. Find the area enclosed by $y=\frac{1}{1+x^2}$, $y=0$, $x=0$, and $x=1$.

32. Find the area enclosed by $y=\frac{1}{\sqrt{1-4x^2}}$, $x=-\frac{1}{4}$, $x=\frac{1}{4}$, and $y=0$.

33. The region bounded by the graphs of $y=e^x$, $y=\frac{1}{\sqrt{x^2+1}}$, and $x=1$ is rotated around the x-axis. Find the volume of the solid that is generated.

34. Find the moment M_y with respect to the y-axis of the region bounded by $y=\frac{1}{1+x^4}$, $y=0$, $x=1$, and $x=2$. Assume $\rho=1$.

35. Find the length of the curve $y=\sqrt{4-x^2}$ from $x=-1$ to $x=1$.

36. Find the centroid of the region bounded by $y=\frac{4}{\sqrt{4-x^2}}$, $x=0$, $x=1$, and $y=0$.

$\left[\text{Hint: } \int \frac{1}{4-x^2}\,dx = \frac{1}{4}\ln\left|\frac{x+2}{x-2}\right| + C\right]$

37. *Transportation* When a tractor trailer turns a corner at the intersection of two perpendicular streets, its rear wheels follow a curve known as a **tractrix**. If the origin is the middle of the intersection, the tractix can be shown to be the graph of $y=f(x)$, where

$$f'(x) = -\frac{1}{x\sqrt{1-x^2}} + \frac{x}{\sqrt{1-x^2}}$$

Find an equation for y, given the fact that $f(1)=0$.

38. *Robotics* During one cycle, the velocity in m/s of a robotic welding device is given by $v(t)=4t-\frac{12}{4-t^2}$, where t is time in seconds since the cycle began. Find an expression for the position $s(t)$, in meters, at time t if $s(0)=0$.

In Your Words

39. In the caution in this section, you were warned that many integrals look alike. In particular, you were shown

(a) $\int \frac{dx}{1+x^2}$ (b) $\int \frac{x\,dx}{1+x^2}$

(c) $\int \frac{dx}{\sqrt{1-x^2}}$ (d) $\int \frac{x\,dx}{\sqrt{1-x^2}}$

and told that integrals a and b are *almost* identical as are integrals c and d. Explain how you determine which type of integral each one is.

40. Describe how to determine the integrals of $\int \frac{dx}{1+x^2}$ and $\int \frac{dx}{1-x^2}$.

9.6 TRIGONOMETRIC SUBSTITUTION

In Section 9.5, we looked at integrals that had square roots in the denominator. In this section, we will look at integrals that contain one of the forms of $\sqrt{a^2-x^2}$, $\sqrt{a^2+x^2}$, or $\sqrt{x^2-a^2}$ in the integrand. In each case, we will use an appropriate trigonometric substitution to eliminate the radical and, hopefully, change the integrand into a trigonometric integral that we have already studied.

For each of the cases in this section, we will make a different substitution. The substitutions can be memorized, but you will probably prefer to learn the technique and derive the necessary relationships. Before we present the three substitutions, let's work an example.

EXAMPLE 9.41

Evaluate $\int \frac{1}{(x^2+25)^{3/2}}\,dx$.

Solution If we use the substitution $x = 5\tan\theta$, then $dx = 5\sec^2\theta\,d\theta$. Making these substitutions in the original integral we get

$$\begin{aligned}\int \frac{dx}{(x^2+25)^{3/2}} &= \int \frac{5\sec^2\theta\,d\theta}{[(5\tan\theta)^2+25]^{3/2}}\\ &= \int \frac{5\sec^2\theta\,d\theta}{(25\tan^2\theta+25)^{3/2}}\end{aligned}$$

But, $25\tan^2\theta+25 = 25(\tan^2\theta+1) = 25\sec^2\theta$, and so

$$\begin{aligned}\int \frac{5\sec^2\theta\,d\theta}{(25\sec^2\theta)^{3/2}} &= \int \frac{5\sec^2\theta\,d\theta}{(5\sec\theta)^3}\\ &= \int \frac{d\theta}{25\sec\theta}\\ &= \frac{1}{25}\int \cos\theta\,d\theta\\ &= \frac{1}{25}\sin\theta + C\end{aligned}$$

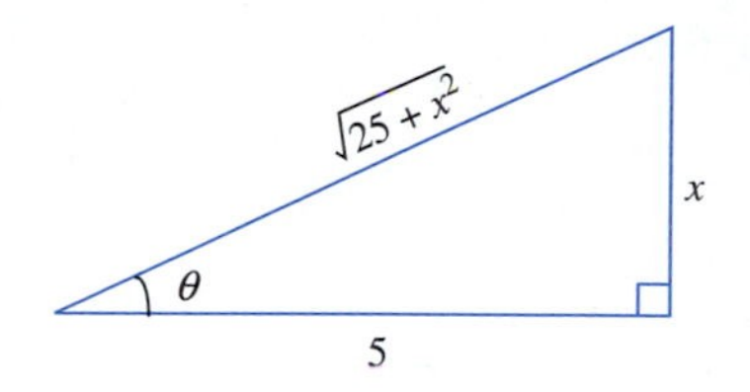

FIGURE 9.5

This result is in terms of θ and not in terms of x. In order to get it in terms of x, we need to return to our substitution $x = 5\tan\theta$. In Figure 9.5, we have drawn a right triangle with an acute angle θ. Two sides are marked so that $\tan\theta = \frac{x}{5}$. The length of the third side was determined by using the Pythagorean theorem. Studying this triangle, we can see that $\sin\theta = \frac{x}{\sqrt{25+x^2}}$. Putting this in the previous answer changes

$$\frac{1}{25}\sin\theta + C \text{to} \frac{x}{25\sqrt{25+x^2}} + C$$

Thus, we get the result

$$\int \frac{1}{(x^2+25)^{3/2}}\,dx = \frac{x}{25\sqrt{25+x^2}} + C$$

▪

Note This process had two parts. One part was the substitution to get the integrand written in terms of θ. The second part involved a right triangle with acute angle θ and the sides in terms of x. The three cases and the proper substitutions are given in Table 9.1.

TABLE 9.1

If the integrand involves	Use the substitution	Use the identity	And use the right triangle
$\sqrt{a^2-u^2}$	$u=a\sin\theta$	$\cos^2\theta=1-\sin^2\theta$	a, u, θ, $\sqrt{a^2-u^2}$
$\sqrt{a^2+u^2}$	$u=a\tan\theta$	$\sec^2\theta=1+\tan^2\theta$	$\sqrt{a^2+u^2}$, u, θ, a
$\sqrt{u^2-a^2}$	$u=a\sec\theta$	$\tan^2\theta=\sec^2\theta-1$	u, $\sqrt{u^2-a^2}$, θ, a

EXAMPLE 9.42

Find $\displaystyle\int\frac{dx}{x^2\sqrt{4-9x^2}}$.

Solution This integral involves $\sqrt{a^2-u^2}$, where $u=3x$ and $a=2$. We will substitute $u=2\sin\theta$ for $3x$. Thus $x=\frac{2}{3}\sin\theta$ and $dx=\frac{2}{3}\cos\theta\,d\theta$, so we have

$$\begin{aligned}\int\frac{dx}{x^2\sqrt{4-9x^2}}&=\int\frac{\frac{2}{3}\cos\theta\,d\theta}{\left(\frac{2}{3}\sin\theta\right)^2\sqrt{4-(2\sin\theta)^2}}\\&=\int\frac{\frac{2}{3}\cos\theta\,d\theta}{\left(\frac{4}{9}\sin^2\theta\right)\sqrt{4(1-\sin^2\theta)}}\\&=\int\frac{\cos\theta\,d\theta}{\frac{2}{3}\sin^2\theta\sqrt{4\cos^2\theta}}\\&=\int\frac{\cos\theta\,d\theta}{\frac{2}{3}(\sin^2\theta)(2\cos\theta)}\end{aligned}$$

EXAMPLE 9.42 (Cont.)

$$= \int \frac{3\,d\theta}{4\sin^2\theta}$$
$$= \int \frac{3}{4}\csc^2\theta = \frac{-3}{4}\cot\theta + C$$

From the triangle in Figure 9.6, we see that

$$\cot\theta = \frac{\sqrt{4-9x^2}}{3x}$$

and
$$-\frac{3}{4}\cot\theta + C = \frac{-3}{4}\frac{\sqrt{4-9x^2}}{3x} + C$$
$$= -\frac{\sqrt{4-9x^2}}{4x} + C$$

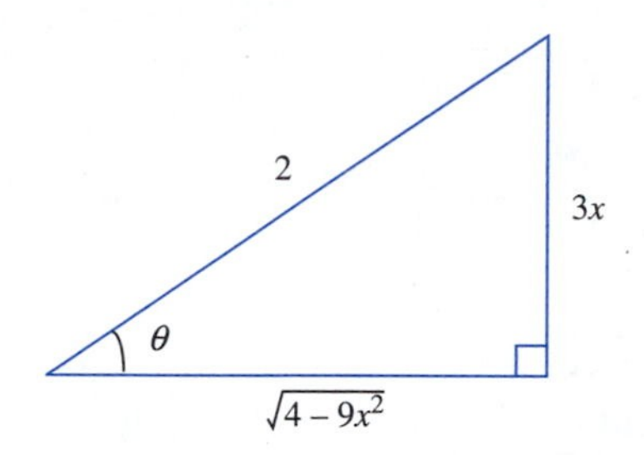

FIGURE 9.6

EXAMPLE 9.43

Determine $\int_0^4 \sqrt{16-x^2}\,dx$.

Solution This is in the form $\sqrt{a^2-u^2}$, so we'll let $u = x = 4\sin\theta$ and $dx = 4\cos\theta\,d\theta$. Substituting, we obtain

$$\int_0^4 \sqrt{16-x^2}\,dx = \int_{x=0}^{x=4} \sqrt{16-16\sin^2\theta}\,(4\cos\theta)\,d\theta$$
$$= \int_{x=0}^{x=4} 4\sqrt{1-\sin^2\theta}\,(4\cos\theta)\,d\theta$$
$$= 16\int_{x=0}^{x=4} \sqrt{\cos^2\theta}\cos\theta\,d\theta$$
$$= 16\int_{x=0}^{x=4} \cos^2\theta\,d\theta$$

From Section 9.5, we know that to integrate this we let $\cos^2\theta = \frac{1}{2}(1+\cos 2\theta)$ and get

$$16\int_{x=0}^{x=4} \frac{1}{2}(1+\cos 2\theta)\,d\theta = 8\left(\theta + \frac{1}{2}\sin 2\theta\right)\Bigg|_{x=0}^{x=4}$$

From the triangle in Figure 9.7, we see that if $\theta = \arcsin\frac{x}{4}$ and since $\sin 2\theta = 2\sin\theta\cos\theta = \frac{x\sqrt{16-x^2}}{8}$, we have

$$8\left(\theta + \frac{1}{2}\sin 2\theta\right)\Bigg|_{x=0}^{x=4} = 8\left(\arcsin\frac{x}{4} + \frac{x\sqrt{16-x^2}}{16}\right)\Bigg|_0^4$$
$$= 8\left(\frac{\pi}{2}+0\right) - (0+0) = 4\pi$$

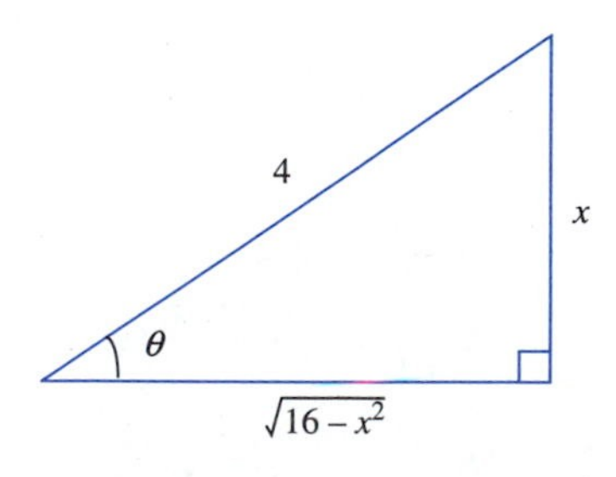

FIGURE 9.7

EXAMPLE 9.44

Find $\displaystyle\int \frac{dx}{x^2\sqrt{x^2+49}}$.

Solution We have the term $\sqrt{x^2+a^2}$ with $a=7$. We will use the substitution $x=7\tan\theta$ with $dx=7\sec^2\theta\,d\theta$. The problem then becomes

$$\begin{aligned}
\int \frac{dx}{x^2\sqrt{x^2+49}} &= \int \frac{7\sec^2\theta\,d\theta}{(7\tan\theta)^2\sqrt{49\tan^2\theta+49}} \\
&= \int \frac{7\sec^2\theta\,d\theta}{49\tan^2\theta\sqrt{49(\tan^2\theta+1)}} \\
&= \int \frac{7\sec^2\theta\,d\theta}{49\tan^2\theta\sqrt{49\sec^2\theta}} \\
&= \int \frac{7\sec^2\theta\,d\theta}{49(\tan^2\theta)(7\sec\theta)} \\
&= \int \frac{\sec\theta\,d\theta}{49\tan^2\theta} \\
&= \frac{1}{49}\int \cot^2\theta\sec\theta\,d\theta \\
&= \frac{1}{49}\int \frac{\cos^2\theta}{\sin^2\theta}\frac{1}{\cos\theta}\,d\theta \\
&= \frac{1}{49}\int \frac{\cos\theta}{\sin^2\theta}\,d\theta \\
&= \frac{1}{49}\left(\frac{-1}{\sin\theta}\right)+C = \frac{-1}{49}\csc\theta + C
\end{aligned}$$

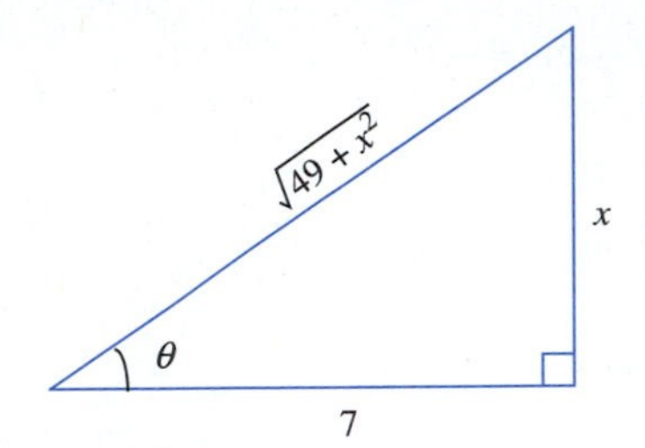

FIGURE 9.8

From the triangle in Figure 9.8, we see that $\csc\theta = \dfrac{\sqrt{49+x^2}}{x}$, and so

$$\int \frac{dx}{x^2\sqrt{x^2+49}} = \frac{-\sqrt{49+x^2}}{49x} + C$$

Exercise Set 9.6

In Exercises 1–36, determine the given integral.

1. $\displaystyle\int \frac{x}{\sqrt{9-x^2}}\,dx$
2. $\displaystyle\int \frac{dx}{\sqrt{9-x^2}}$
3. $\displaystyle\int \frac{x^2}{\sqrt{9-x^2}}\,dx$
4. $\displaystyle\int (9-x^2)^{3/2}\,dx$
5. $\displaystyle\int x^3\sqrt{9-x^2}\,dx$
6. $\displaystyle\int \frac{x}{\sqrt{x^2-9}}\,dx$
7. $\displaystyle\int \frac{dx}{\sqrt{x^2-9}}$
8. $\displaystyle\int \frac{x^3}{\sqrt{x^2-9}}\,dx$
9. $\displaystyle\int (x^2-9)^{1/2}\,dx$
10. $\displaystyle\int x^3\sqrt{x^2-9}\,dx$
11. $\displaystyle\int \frac{x}{\sqrt{x^2+9}}\,dx$
12. $\displaystyle\int \frac{dx}{\sqrt{x^2+9}}$
13. $\displaystyle\int \frac{dx}{x^2+9}$
14. $\displaystyle\int \frac{x^2}{x^2+9}\,dx$
15. $\displaystyle\int x^3\sqrt{9+x^2}\,dx$
16. $\displaystyle\int \frac{\sqrt{1-x^2}}{x^2}\,dx$
17. $\displaystyle\int \frac{\sqrt{4-3x^2}}{x^4}\,dx$

18. $\int \frac{\sqrt{4x^2-9}}{x^3}\,dx$
19. $\int \frac{x^3}{\sqrt{9x^2+4}}\,dx$
20. $\int \frac{dx}{(3x^2+6)^{3/2}}$
21. $\int \frac{\sqrt{x^2+1}}{x^2}\,dx$
22. $\int \frac{dx}{\sqrt{4-(x-1)^2}}$
23. $\int \sqrt{4-(x-1)^2}\,dx$
24. $\int \frac{dx}{\sqrt{4+(x-1)^2}}$
25. $\int \frac{dx}{\sqrt{(x-1)^2-4}}$
26. $\int \sqrt{x^2-2x+5}\,dx$
27. $\int \sqrt{x^2-2x-3}\,dx$
28. $\int \frac{dx}{(x^2-2x+10)^{3/2}}$
29. $\int \frac{dx}{(x-3)\sqrt{x^2-6x+25}}$
30. $\int \frac{x-3}{\sqrt{x^2-6x+25}}\,dx$
31. $\int_1^3 \frac{\sqrt{9-x^2}}{x^2}\,dx$
32. $\int_1^2 \frac{dx}{(8-x^2)^{3/2}}$
33. $\int_4^8 \frac{dx}{x\sqrt{x^2-4}}$
34. $\int_1^4 \frac{dx}{x^4\sqrt{x^2+4}}$
35. $\int_0^6 x^2\sqrt{36-x^2}\,dx$
36. $\int_0^{1.5\sqrt{3}} \frac{x^2\,dx}{\sqrt{9-x^2}}$

Solve Exercises 37–44.

37. Find the area under the graph of $y = \frac{x^3}{\sqrt{16-x^2}}$ from $x = 0$ to $x = 3$.
38. Find the arc length of the portion of the parabola $y = 10x - x^2$ that is above the x-axis.
39. Find the area bounded by the graph of $y = \sqrt{9+x^2}$, $y = 0$, $x = 0$ and $x = 4$.
40. *Civil engineering* The density of a 6-m bar is given by $\rho(x) = \sqrt{108-3x^2}$. What is the total mass of the bar?
41. An ellipsoid of revolution is obtained by revolving the ellipse $\frac{x^2}{25} + \frac{y^2}{4} = 1$ around the x-axis. What is the surface area of this ellipsoid?
42. *Electricity* The current in a circuit varies according to $i = t\sqrt{t+2}$ A. Find the charge transmitted during the interval from $t = 0$ to $t = 1$ s.
43. *Electricity* The current in a transformer is given by $i = \int \frac{\sqrt{t^2+1}}{9t^2}\,dt$. Find the current from $t = 0.5$ to $t = 1$ s.
44. *Petroleum engineering* A 10-ft-long cylindrical oil tank is lying on its side. The base of the tank is an ellipse described by the equation $x^2 + 4y^2 = 36$. If all measurements are in feet, find the volume of oil in the tank when the depth of the oil in the center is 2.5 ft.

In Your Words

45. What trigonometric substitution do you use for integrands of the form $\sqrt{a^2+u^2}$? Explain how this substitution is used.
46. In Exercise 45, you stated the trigonometric substitution you would use for integrands of the form $\sqrt{a^2+u^2}$. Will this same substitution work for integrands of the form $(a^2+u^2)^n$? If so, describe how you would use it. If not, tell how you would determine $\int (a^2+u^2)^n\,du$.

9.7 INTEGRATION BY PARTS

One of the most useful integration techniques is a direct result of the product rule of differentiation. You should remember that the product rule states that if f and g are two functions whose derivatives exist, then

$$D_x[f(x)g(x)] = f(x)g'(x) + g(x)f'(x)$$

Rewriting this, we get

$$f(x)g'(x) = D_x[f(x)g(x)] - g(x)f'(x)$$

Integrating both sides, we have

$$\int f(x)g'(x)\,dx = \int D_x[f(x)g(x)]\,dx - \int g(x)f'(x)\,dx$$

Since $\int D_x[f(x)g(x)]\,dx = f(x)g(x)$, this can be written as

$$\int f(x)g'(x)\,dx = f(x)g(x) - \int g(x)f'(x)\,dx$$

It is customary in this equation to let $u = f(x)$ and $v = g(x)$, so $du = f'(x)\,dx$ and $dv = g'(x)\,dx$. The formula then becomes what is called the integration by parts formula.

Integration by Parts

$$\int u\,dv = uv - \int v\,du$$

In order to apply this formula, we need to split the integrand into two parts. One part is labeled u and the other part is labeled dv. Unfortunately, there are no firm rules for selecting the part that is labeled u and the part that is dv. In some problems, there are several possible choices and only experience or trial and error will help you with the correct choices. There are, however, three general guidelines that should be followed.

Guidelines for Integrating by Parts

1. dx is always part of dv.
2. You must be able to integrate dv.
3. $v\,du$ is easier to integrate than $u\,dv$.

Note

As an aid, we will make a table for each problem that uses integration by parts. The table will be in the following form:

u		v
du		dv

Two of the cells in this table will be filled in when we select u and dv. We will have to determine the values for the other two cells before we can apply the integration by parts formula.

EXAMPLE 9.45

Determine $\int xe^x\,dx$.

Solution We will select $u = x$ and $dv = e^x\,dx$. So, the table looks like

u	x	v	
du		dv	$e^x\,dx$

We will complete the table by taking the derivative of u and by integrating dv. The completed table now looks like this:

u	x	v	e^x
du	dx	dv	$e^x\,dx$

The integration by parts formula states that $\int u\,dv = uv - \int v\,du$. Using the above values produces

$$\begin{aligned}\int xe^x\,dx &= xe^x - \int e^x\,dx \\ &= xe^x - e^x + C \\ &= (x-1)e^x + C\end{aligned}$$

When deciding which function to select for u, remember the *LIATE* rule: u should be the first available type of function from the following list:

Logarithmic
Inverse trigonometric
Algebraic
Trigonometric
Exponential

EXAMPLE 9.46

Evaluate $\int x\cos x\,dx$.

Solution We set up our table with $u = x$ and $dv = \cos x\,dx$.

u	x	v	$\sin x$
du	dx	dv	$\cos x\,dx$

Then

$$\begin{aligned}\int x\cos x\,dx &= uv - \int v\,du \\ &= x\sin x - \int \sin x\,dx \\ &= x\sin x - (-\cos x) + C \\ &= x\sin x + \cos x + C\end{aligned}$$

EXAMPLE 9.47

Find $\int x^3 e^{x^2}\,dx$.

Solution The natural split is to let $u = x^3$ and $dv = e^{x^2}\,dx$. The difficulty with this is that we cannot integrate $e^{x^2}\,dx$. If we let $dv = xe^{x^2}\,dx$, then dv can be integrated. This, of course, means that $u = x^2$.

u	x^2	v	$\frac{1}{2}e^{x^2}$
du	$2x\,dx$	dv	$xe^{x^2}\,dx$

$$\begin{aligned}\int x^3 e^{x^2}\,dx &= \frac{1}{2}x^2 e^{x^2} - \int xe^{x^2}\,dx \\ &= \frac{1}{2}x^2 e^{x^2} - \frac{1}{2}e^{x^2} + C \\ &= \frac{1}{2}e^{x^2}(x^2 - 1) + C\end{aligned}$$

There are times when it is necessary to use integration by parts more than once on the same problem. This is demonstrated in the next example.

EXAMPLE 9.48

Evaluate $\int x^2 e^x\,dx$.

Solution If we let $u = x^2$ and $dv = e^x\,dx$, then the table is as follows.

u	x^2	v	e^x
du	$2x\,dx$	dv	$e^x\,dx$

$$\begin{aligned}\int x^2 e^x\,dx &= x^2 e^x - \int 2xe^x\,dx \\ &= x^2 e^x - 2\int xe^x\,dx\end{aligned}$$

The integral on the right-hand side can be obtained by integration by parts. (In fact, we did this in Example 9.45.) Using integration by parts, we obtain

$$\begin{aligned}\int x^2 e^x\,dx &= x^2 e^x - 2(xe^x - e^x) + C \\ &= e^x(x^2 - 2x + 2) + C\end{aligned}$$

There are times that integration by parts seem to be leading us in the wrong direction when, in fact, it is not. This next example illustrates one of these times.

EXAMPLE 9.49

Find $\int e^x \sin x\, dx$.

Solution We will let $u = e^x$ and $dv = \sin x\, dx$. The table is

u	e^x	v	$-\cos x$
du	$e^x\, dx$	dv	$\sin x\, dx$

$$\int e^x \sin x\, dx = -e^x \cos x - \int -e^x \cos x\, dx$$
$$= -e^x \cos x + \int e^x \cos x\, dx$$

We will use integration by parts a second time.

u	e^x	v	$\sin x$
du	$e^x\, dx$	dv	$\cos x\, dx$

$$\int e^x \sin x\, dx = -e^x \cos x + e^x \sin x - \int e^x \sin x\, dx$$

It looks as if we have gone around in a circle. In some respects we have, because we have the same integral on the right-hand side that we started with. Adding this integral, $\int e^x \sin x\, dx$, to both sides of the previous equation produces

$$2\int e^x \sin x\, dx = -e^x \cos x + e^x \sin x$$
$$= -e^x(\cos x - \sin x)$$

Dividing both sides by 2 gives us the answer we are after.

$$\int e^x \sin x\, dx = -\frac{1}{2}e^x(\cos x - \sin x) + C$$
$$= \frac{1}{2}e^x(\sin x - \cos x) + C$$

EXAMPLE 9.50

Find $\int \sec^3 \theta\, d\theta$.

Solution We will let $u = \sec\theta$ and $dv = \sec^2\theta\, d\theta$. The table is

u	$\sec\theta$	v	$\tan\theta$
du	$\sec\theta \tan\theta\, d\theta$	dv	$\sec^2\theta\, d\theta$

EXAMPLE 9.50 (Cont.)

Applying integration by parts, we obtain

$$\begin{aligned}\int \sec^3\theta\,d\theta &= \sec\theta\tan\theta - \int \sec\theta\tan^2\theta\,d\theta \\ &= \sec\theta\tan\theta - \int \sec\theta\left(\sec^2\theta - 1\right)\,d\theta \\ &= \sec\theta\tan\theta - \int \sec^3\theta\,d\theta + \int \sec\theta\,d\theta \\ &= \sec\theta\tan\theta - \int \sec^3\theta\,d\theta + \ln|\sec\theta + \tan\theta| + C\end{aligned}$$

Adding $\int \sec^3\theta\,d\theta$ to both sides, we get

$$2\int \sec^3\theta\,d\theta = \sec\theta\tan\theta + \ln|\sec\theta + \tan\theta| + C$$

and dividing by 2 produces

$$\int \sec^3\theta\,d\theta = \frac{1}{2}\sec\theta\tan\theta + \frac{1}{2}\ln|\sec\theta + \tan\theta| + C$$

Application

EXAMPLE 9.51

Find the volume of the region generated by rotating the area between the curves $y = \sqrt{\ln x}$, $y = x$, and the lines $x = 1$ and $x = e$ around the x-axis.

Solution A sketch of this region has been shaded in Figure 9.9. As you can see, $x > \sqrt{\ln x}$ for all values of $x > 1$. So, using the washer method, we obtain

$$\begin{aligned}V &= \pi\int_1^e \left[x^2 - \left(\sqrt{\ln x}\right)^2\right] dx \\ &= \pi\int_1^e \left(x^2 - \ln x\right) dx \\ &= \pi\int_1^e x^2\,dx - \int_1^e \ln x\,dx\end{aligned}$$

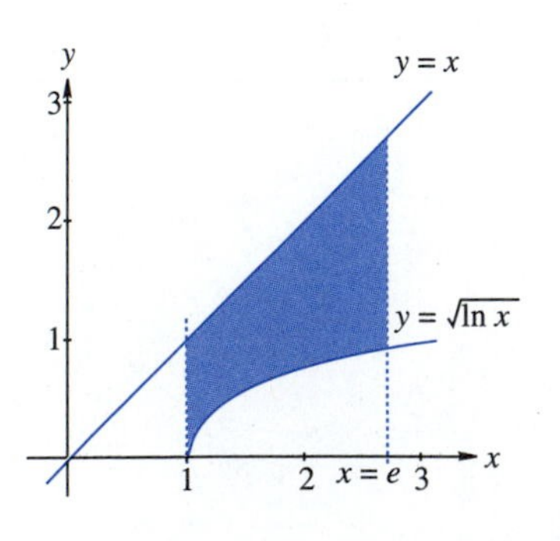

FIGURE 9.9

We will need to use integration by parts on $\int \ln x\,dx$. If we let $u = \ln x$ and $dv = dx$, then we can complete the table as follows.

u	$\ln x$	v	x
du	$\frac{1}{x}\,dx$	dv	dx

EXAMPLE 9.51 (Cont.)

Thus, $\int \ln x\,dx = x\ln x - \int dx = x(\ln x - 1)$, so we are ready to complete the work for determining the volume.

$$\begin{aligned} V &= \pi \int_1^e x^2\,dx - \int_1^e \ln x\,dx \\ &= \pi\left[\frac{1}{3}x^3 - x(\ln x - 1)\right]_1^e \\ &= \pi\frac{e^3-4}{3} \approx 16.8447 \end{aligned}$$

The volume of this region when it is rotated around the x-axis is approximately 16.8447 cubic units.

When a problem involves repeated integration by parts, a tabular method can often be used to organize the work. The technique works well for integrals of the form

$$\int P(x)\sin ax\,dx, \qquad \int P(x)\cos ax\,dx, \qquad \text{or} \qquad \int P(x)e^{ax}\,dx$$

where $P(x)$ is a polynomial. In each case, left $u = P(x)$. The next examples will demonstrate how this method can be used.

EXAMPLE 9.52

Find $\int x^3 \cos 2x\,dx$.

Solution Here $P(x) = x^3$, so we will let $u = x^3$ and $dv = \cos 2x\,dx$. Next, we construct a table with three columns, as follows:

Alternate signs	u and its derivatives	dv and its antiderivatives
$+$	x^3	$\cos 2x$
$-$	$3x^2$	$\frac{1}{2}\sin 2x$
$+$	$6x$	$-\frac{1}{4}\cos 2x$
$-$	6	$-\frac{1}{8}\sin 2x$
$+$	0	$\frac{1}{16}\cos 2x$

↑
Differentiate until you obtain a 0 as a derivative.

EXAMPLE 9.52 (Cont.)

The solution is given by adding the *signed* products of the diagonal entries.

$$\int x^3 \cos 2x\,dx = (x^3)\frac{1}{2}\sin 2x - (3x^2)\left(-\frac{1}{4}\cos 2x\right) + (6x)\left(-\frac{1}{8}\sin 2x\right) - (6)\frac{1}{16}\cos 2x + C$$

$$= \frac{1}{2}x^3 \sin 2x + \frac{3}{4}x^2 \cos 2x - \frac{3}{4}x \sin 2x - \frac{3}{8}\cos 2x + C$$

EXAMPLE 9.53

Find $\int (5x^2 - 3x + 7)e^{4x}\,dx$.

Solution Here $P(x) = 5x^2 - 3x + 7$, so $u = 5x^2 - 3x + 7$ and $dv = e^{4x}\,dx$. Again, we construct a table with three columns.

Alternate signs	u and its derivatives	dv and its antiderivatives
$+$	$5x^2 - 3x + 7$	e^{4x}
$-$	$10x - 3$	$\frac{1}{4}e^{4x}$
$+$	10	$\frac{1}{16}e^{4x}$
$-$	0	$\frac{1}{64}e^{4x}$

The solution is given by adding the *signed* products of the diagonal entries.

$$\int (5x^2 - 3x + 7)e^{4x}\,dx = (5x^2 - 3x + 7)\frac{1}{4}e^{4x} - (10x - 3)\frac{1}{16}e^{4x} + (10)\frac{1}{64}e^{4x} + C$$

$$= \left(\frac{5}{4}x^2 - \frac{11}{8}x + \frac{67}{32}\right)e^{4x} + C$$

Application

EXAMPLE 9.54

A suspension bridge, like the one shown in Figure 9.10a, has its roadbed supported by vertical cables that are connected to a main cable strung between two towers. The main cable does not take the shape of a catenary, but takes instead the shape of a parabola. If the towers are 1,000 ft apart and the main cable is connected to the towers 100 ft above the roadbed, determine the arc length of the main cable.

Solution We will put the origin at the roadbed midway between the towers; that is, at the vertex of the parabola, as shown in Figure 9.10b. This means that the cables are attached to the towers at the points $(-500, 100)$ and $(500, 100)$. By placing the vertex at the origin, we know that this parabola is of the form $y = 4px^2$. Since $(500, 100)$ is a point on this parabola, we see that $100 = 4p(500)^2$, so $p = \dfrac{1}{10{,}000}$

EXAMPLE 9.54 (Cont.)

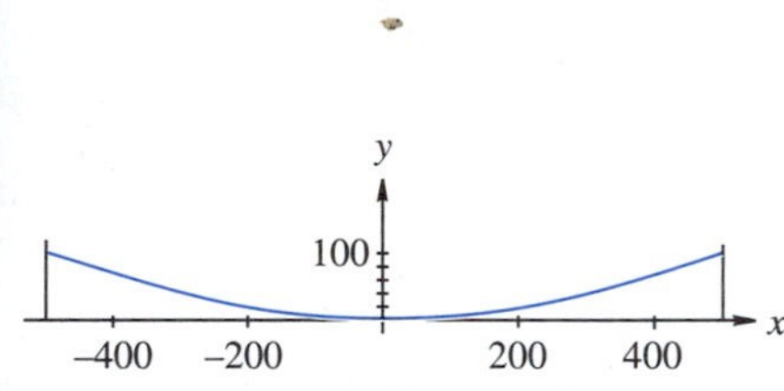

Courtesy of New York Convention and Visitors Bureau Inc.

FIGURE 9.10a **FIGURE 9.10b**

and the parabola has the equation $y = \frac{4}{10{,}000}x^2 = \frac{1}{2{,}500}x^2$. For the arc length we need the derivative of y, which is $y' = \frac{2}{2{,}500}x = \frac{1}{1{,}250}x$. So, the length of the cable is

$$L = \int_{-500}^{500} \sqrt{1 + \left(\frac{1}{1{,}250}x\right)^2}\,dx = \frac{1}{1{,}250}\int_{-500}^{500} \sqrt{1{,}250^2 + x^2}\,dx$$

Using the trigonometric substitutions of $x = 1{,}250\tan\theta$ and $dx = 1{,}250\sec^2\theta\,d\theta$, the indefinite integral becomes

$$\begin{aligned}
&\frac{1}{1{,}250}\int \sqrt{1{,}250^2 + x^2}\,dx \\
&\quad = \frac{1}{1{,}250}\int \sqrt{1{,}250^2 + 1{,}250^2\tan^2\theta}\,\left(1{,}250\sec^2\theta\right)d\theta \\
&\quad = 1{,}250\int \sqrt{1 + \tan^2\theta}\sec^2\theta\,d\theta \\
&\quad = 1{,}250\int \sec^3\theta\,d\theta
\end{aligned}$$

From Example 9.50, we know that

$$\int \sec^3\theta\,d\theta = \frac{1}{2}\sec\theta\tan\theta + \frac{1}{2}\ln|\sec\theta + \tan\theta| + C$$

and so

$$\begin{aligned}
L &= \frac{1{,}250}{2}\left[\sec\theta\tan\theta + \ln|\sec\theta + \tan\theta|\right] + C \\
&= 625\left[\sec\theta\tan\theta + \ln|\sec\theta + \tan\theta|\right] + C
\end{aligned}$$

EXAMPLE 9.54 (Cont.)

Now, since $\tan\theta = \dfrac{x}{1{,}250}$, we determine that $\sec\theta = \dfrac{\sqrt{1{,}250^2 + x^2}}{1{,}250}$ and we have

$$\begin{aligned} L &= 625\left[\frac{\sqrt{1{,}250^2+x^2}}{1{,}250}\frac{x}{1{,}250} + \ln\left|\frac{\sqrt{1{,}250^2+x^2}}{1{,}250} + \frac{x}{1{,}250}\right|\right]_{-500}^{500} \\ &= 625\left[\frac{2\sqrt{29}}{25} + \ln\left|\frac{\sqrt{29}+2}{5}\right| - \left(-\frac{2\sqrt{29}}{25} + \ln\left|\frac{\sqrt{29}-2}{5}\right|\right)\right] \\ &= 625\left(\frac{4\sqrt{29}}{25} + \ln\left|\frac{\sqrt{29}+2}{\sqrt{29}-2}\right|\right) \approx 1{,}026.0606 \end{aligned}$$

Thus, the length of the main cable is approximately 1,026.0606 ft.

Exercise Set 9.7

In Exercises 1–24, find the given integral.

1. $\int x\ln x\,dx$
2. $\int \tan^{-1}x\,dx$
3. $\int x\sin 2x\,dx$
4. $\int x^5\ln x\,dx$
5. $\int x^2\ln x\,dx$
6. $\int x\sqrt{1+x}\,dx$
7. $\int x\sin^{-1}x\,dx$
8. $\int x^2\sin x\,dx$
9. $\int x^3e^{2x}\,dx$
10. $\int x\tan^{-1}x\,dx$
11. $\int x\sqrt{4x+1}\,dx$
12. $\int x^2\cos 3x\,dx$
13. $\int x^2e^{x/4}\,dx$
14. $\int e^{4x}\sin 2x\,dx$
15. $\int e^x\cos x\,dx$
16. $\int x^4e^{2x}\,dx$
17. $\int x^3\cos x^2\,dx$
18. $\int x^2e^{-x}\,dx$
19. $\int x^3e^{-x}\,dx$
20. $\int \sin^2x\,dx$
21. $\int e^{-x}\cos x\,dx$
22. $\int x4^x\,dx$
23. $\int x\cos 4x\,dx$
24. $\int e^{4x}\cos 2x\,dx$

Solve Exercises 25–38.

25. The region bounded by $y = \cos x$, $y = 0$, $x = 0$, and $x = \dfrac{\pi}{2}$ is rotated around the x-axis. What is the volume of the solid that is generated?
26. Find the volume of the solid generated by rotating the region in Exercise 25 around the y-axis.
27. *Electricity* During a certain interval the current in a circuit varied according to $i = 4t\sin 2t$ A. What was the charge transferred during the period from $t = 0$ to $t = 4$ s?
28. The density of a 6-m bar is given by $\rho = xe^{-x}$ kg/m, where x is measured in meters. What is the total mass of the bar?
29. Find the centroid of the area bounded by $y = x^2e^x$, $y = 0$, $x = -1$ and $x = 2$.
30. Find the root mean square of $y = \sqrt{\cos^{-1}x}$ from $x = 0$ to $x = 1$.
31. *Physics* The force acting on a point on a coordinate axis is given by $F(x) = x^3\cos x\,dx$. Find the work required to move the point from $x = 0$ to $x = \dfrac{\pi}{2}$.
32. *Physics* A particle moves along a coordinate axis with a velocity of $v(t) = t^5(1-t^3)^{1/2}$. How far does it travel from the time $t = 0$ to $t = 1$, where t is in seconds?

33. *Construction* Determine the length of the main cable of a suspension bridge, if the towers are 1,000 ft apart and the main cable is connected to the towers 200 ft above the roadbed.

34. *Pharmacology* The rate of reaction to a drug is given by $r'(t) = 3t^2e^{-t}$, where t is the time in hours since the drug was administered.

(a) Find an expression for $r(t)$, the total reaction to the drug, if $r(0) = 0$.

(b) Find the total reaction to the drug for the first 8 hours.

35. *Automotive technology* Friction between spinning circular surfaces is important in the design of clutch mechanisms. In one type of clutch mechanism, two flat circular disks of radius R are mounted so they can be brought into contact causing a net frictional force. In one case the contact pressure can be given by $P(r) = P_0e^{-kr}$ where k is a positive constant, P_0 is the initial pressure, and r is the distance from the center of the discs.

(a) Evaluate the integral $F = 2\pi \int_0^R r \cdot P(r)\,dr$ in order to determine the total contact force between the discs.

(b) Evaluate the integral $T = 2\pi\mu \int_0^R r^2 \cdot P(r)\,dr$ in order to determine the total torque force between the discs, where μ is the coefficient of friction.

36. *Petroleum engineering* The production of a new oil field in thousands of barrels per month is estimated to be $R(t) = 10te^{-0.1t}$, where t is in months.

(a) Integrate R in order to determine the total production, P, of the well for the first t months.

(b) Estimate the total production in the first year of operation.

(c) Estimate the total production in the second year of operation.

37. *Environmental science* The concentration of particulate matter in ppm t hours after a factory ceases operation for the day is given by $C(t) = \dfrac{40\ln(t+2)}{(t+2)^2}$. Find the average concentration for the time period from $t = 0$ to $t = 6$.

38. *Electronics* When a periodic electromotive force is applied to an RL circuit, the current can be calculated using $i(t) = \dfrac{V}{L}\int e^{kt}\sin\omega t\,dt$. Determine the current.

In Your Words

39. Without looking in the text, write the integration by parts formula and describe how to use it.

40. Without looking in the text, summarize the guidelines for integrating by parts and explain how to use them.

41. A table was given in the note following the the guidelines for integrating by parts. Explain how to use this table.

9.8 USING INTEGRATION TABLES

There are many types of integrals that we encounter in applications. It would be almost impossible for anyone to remember how to find each of these integrals. For this reason, extensive tables of integrals have been prepared. One book contains a list of over 700 integrals. A much shorter list is given in Appendix C at the back of this book.

In the first seven sections of this chapter, we learned how to integrate many different types of functions. All of the integrals in Appendix C can be obtained from one of these methods, but you can often save a lot of time by using the tables for one of these integrals.

EXAMPLE 9.55

$\int 5\cos^2 5x\,dx.$

Solution This integral involves a trigonometric function, so we look in the section of the table headed Trigonometric Forms. Note that this fits Formula 57 if we let $u = 5x$ and $du = 5\,dx$. Then by direct substitution, we have

$$\int 5\cos^2 5x\,dx = \frac{5x}{2} + \frac{1}{2}\sin 5x\cos 5x + C$$

Notice that Formula 57 has a second version, so we could have also written

$$\int 5\cos^2 5x\,dx = \frac{5x}{2} + \frac{1}{4}\sin 10x + C$$

Not all integrals can be solved by using direct substitution. Sometimes we have to multiply by a form of 1 in order to get the problem in the correct form.

EXAMPLE 9.56

Solve $\int \frac{2\,dx}{x(25x^2-9)^{3/2}}$.

Solution The closest version to this form is Formula 45. Here $u = 5x$, $a = 3$, and $du = 5\,dx$. With these substitutions, Formula 45 becomes

$$\int \frac{du}{u(u^2-a^2)^{3/2}} = \int \frac{5\,dx}{5x(25x^2-9)^{3/2}}$$
$$= \int \frac{dx}{x(25x^2-9)^{3/2}}$$

This is almost the integral we have been asked to solve. If we multiply by $\frac{2}{2}$, we get

$$\frac{1}{2}\int \frac{2\,dx}{x(25x^2-9)^{3/2}} = \frac{1}{2}\left(\frac{-1}{9\sqrt{25x^2-9}} - \frac{1}{27}\operatorname{arcsec}\frac{5x}{3}\right) + C$$

by Formula 45.

The integral you are trying to solve does not always bear a great resemblance to the integrals in the table. You will have to recognize the necessary substitutions.

EXAMPLE 9.57

Find $\int \frac{\tan 2x\,dx}{3+5\cos 2x}$.

Solution A look through all of the integrals in the Trigonometric Forms section does not reveal any formula that even remotely resembles this one. In fact, there are no formulas that contain both the tangent and the cosine functions. That is a clue. Let's rewrite this entirely in terms of sine and cosine.

$$\int \frac{\tan 2x\,dx}{3+5\cos 2x} = \int \frac{\sin 2x\,dx}{\cos 2x(3+5\cos 2x)}.$$

EXAMPLE 9.57 (Cont.)

If we let $u = \cos 2x$, then $du = -2\sin 2x$ and we can rewrite this as

$$\frac{-1}{2}\int \frac{-2\sin 2x\,dx}{\cos 2x(3+5\cos 2x)} = \frac{-1}{2}\int \frac{du}{u(3+5u)}$$

This looks like Formula 48 with $a = 3$ and $b = 5$, and we see that

$$\int \frac{\tan 2x\,dx}{3+5\cos 2x} = \frac{-1}{2}\left(\frac{1}{3}\ln\left|\frac{\cos 2x}{3+5\cos 2x}\right|\right)+C$$
$$= \frac{-1}{6}\ln\left|\frac{\cos 2x}{3+5\cos 2x}\right|+C$$

As the last example, we will consider one of the formulas that has an integral in the answer.

EXAMPLE 9.58

Determine $\int \sin^4 x\,dx$.

Solution This is definitely of the form in Formula 64 with $n = 4$. According to that formula,

$$\int \sin^4 x\,dx = -\frac{1}{4}\sin^3 x\cos x + \frac{3}{4}\int \sin^2 x\,dx$$

The integral in the answer is also of the form of formula 64 with $n = 2$. It also fits formula 56. In this case, we have a choice. We will use formula 56 to get the final answer.

$$\int \sin^4 x\,dx = -\frac{1}{4}\sin^3 x\cos x + \frac{3}{4}\left(\frac{1}{2}x - \frac{1}{4}\sin 2x\right)+C$$
$$= -\frac{1}{4}\sin^3 x\cos x + \frac{3}{8}x - \frac{3}{16}\sin 2x + C$$

Any formula for an integral that contains an integral on the right-hand side of the formula is known as a **reduction formula**. You will notice that the function being integrated on the right-hand side in a reduction formula is the same as the original function, but is of a lower degree.

Exercise Set 9.8

In Exercises 1–24, use the table of integrals in Appendix C to integrate the given function.

1. $\int (1+\tan 3x)^2\,dx$
2. $\int \frac{\sqrt{16+25x^2}}{x}\,dx$
3. $\int \frac{dx}{x(4x-3)}$
4. $\int \frac{5x\,dx}{3+7x}$
5. $\int \frac{x^2\,dx}{x^6\sqrt{16+x^6}}$
6. $\int (25-4x^2)^{3/2}\,dx$
7. $\int \frac{x^2\,dx}{\sqrt{9-x^2}}$
8. $\int x^2\sqrt{9x^2-49}\,dx$

9. $\int \cos^3 x\, dx$

10. $\int \sin 5x \sin 2x\, dx$

11. $\int \sin^6 3x\, dx$

12. $\int e^{-6x} \sin 10x\, dx$

13. $\int e^{10x} \cos 6x\, dx$

14. $\int x^2 \tan^{-1} x\, dx$

15. $\int x^7 \ln x\, dx$

16. $\int \frac{\sqrt{7-9x^2}}{x}\, dx$

17. $\int \arcsin 4x\, dx$

18. $\int \frac{dx}{(4-3x^2)^{3/2}}$

19. $\int \frac{\sqrt{9+x^2}}{x}\, dx$

20. $\int e^{\sin x} \sin x \cos x\, dx$

21. $\int x^3 e^{2x}\, dx$

22. $\int x^4 \ln 2x\, dx$

23. $\int x^3 \sin 2x\, dx$

24. $\int \frac{\sqrt{\tan^2 2x - 9}}{\cos^2 2x}\, dx$

Solve Exercises 25–28.

25. *Electronics* Find the average value of the voltage $V(t) = t^2 e^{5t}$ from $t = 0$ s to $t = 2.5$ s.

26. *Robotics* The angular velocity, $\omega = \frac{d\theta}{dt}$, of a certain rotating system depends on the time according to the equation $\omega(t) = te^{0.25t}$, where t is in seconds. The number of revolutions, R, during the first n s is given by $R = \frac{1}{2\pi}\int_0^n \omega(t)\, dt$.
    (a) Determine the number of revolutions during the first n s.
    (b) Determine the number of revolutions during the first 10 s.

27. *Mechanical engineering* The force in N acting on an object is given by $F = \frac{1}{49-9x^2}$, where x is the distance in meters from the initial position. Find the work done in moving the object from $x = 0$ to $x = 2.00$.

28. *Electronics* The current across capacitance $C = 0.02$ F in an electric circuit is given by $I = \frac{1}{t^2+t}$, where $t \geq 0$ and t in s. The voltage across the capacitance at any time t is given by $V = \frac{1}{C}\int I\, dt$. If the voltage across this capacitance is 0 when $t = 1$, find the voltage across the capacitance at any time t.

In Your Words

29. Find a function in an earlier section that you were not able to integrate. Describe how you would use integration tables to now complete this integral and then use the tables to integrate the function.

30. Write an application in your technology area of interest that requires you to integrate a function. Give your problem to a classmate and see if he or she understands and can solve your problem using any of the techniques in this chapter. Rewrite the problem as necessary to remove any difficulties encountered by your classmate.

CHAPTER 9 REVIEW

Important Terms and Concepts

Catenary
Integration
- By parts
- By general power rule
- Of exponential functions
- Of inverse trigonometric functions
- Of logarithmic functions
- Of trigonometric functions
- By trigonometric substitution
- Using tables

Trigonometric substitution

Review Exercises

In Exercises 1–44, integrate the given function without the use of the table of integrals in Appendix C.

1. $\int xe^{3x}\,dx$
2. $\int \sin^3 x\cos^2 x\,dx$
3. $\int \frac{x}{\sqrt{25-x^2}}dx$
4. $\int x^3\sqrt{25-x^2}\,dx$
5. $\int \sin^4 2x\cos 2x\,dx$
6. $\int 2(e^x-e^{-x})\,dx$
7. $\int \frac{dx}{9x+5}$
8. $\int \tan^2 8x\,dx$
9. $\int \sin(7x+2)\,dx$
10. $\int \sin^3 2x\cos 2x\,dx$
11. $\int \sin^5 3x\cos^2 3x\,dx$
12. $\int \frac{dx}{x^2+4x+20}$
13. $\int \frac{dx}{(4x^2+49)^{3/2}}$
14. $\int x^3\ln x\,dx$
15. $\int x^2e^{x^3}\,dx$
16. $\int \frac{dx}{\sqrt{4x^2+49}}$
17. $\int \frac{x\,dx}{\sqrt{4x^2+49}}$
18. $\int \cot 5x\csc^4 5x\,dx$
19. $\int \frac{\sec 4x\tan 4x}{9+2\sec 4x}dx$
20. $\int \sin^6\frac{3x}{2}dx$
21. $\int \frac{[\ln(2x+1)]^5}{2x+1}dx$
22. $\int \frac{\arctan 7x}{1+49x^2}dx$
23. $\int \frac{x^2}{x^3+4}dx$
24. $\int 4x^3e^{x^4}\,dx$
25. $\int \tan\frac{x}{5}dx$
26. $\int x\sin^3 x\,dx$
27. $\int e^{\cos x}\sin x\,dx$
28. $\int x^4e^{-x}\,dx$
29. $\int \frac{\cos x\,dx}{\sin^2 x+9}$
30. $\int \frac{e^x\,dx}{e^x+16}$
31. $\int_1^e x^3\ln x^2\,dx$
32. $\int e^{8x}\cos 2x\,dx$
33. $\int \sin^{1/3}4x\cos^5 4x\,dx$
34. $\int \frac{\sin^3 x\,dx}{\cos^4 x}$
35. $\int \frac{\cos x\,dx}{\sqrt{16-4\sin^2 x}}$
36. $\int \frac{e^{5x}}{4-e^{5x}}dx$
37. $\int x^5e^{x^2}\,dx$
38. $\int \frac{e^{3x}}{(e^{3x}-1)^2}dx$
39. $\int \frac{(\arctan 2x)^4}{1+4x^2}dx$
40. $\int \frac{\sec^2 5x\,dx}{2\tan 5x+9}$
41. $\int_0^{\ln 49}\sqrt{9+e^x}\,dx$
42. $\int_0^1 \arcsin\left(\frac{x}{2}\right)dx$
43. $\int_0^3 \frac{x^3\,dx}{\sqrt{9+x^2}}$
44. $\int_0^{\pi/2}\sin^3 x\cos^3 x\,dx$

CHAPTER 9 TEST

In Exercises 1–6, integrate the given function.

1. $\int \frac{e^x\,dx}{\sqrt{9-e^x}}$
2. $\int \tan^2 4x\cos^4 4x\,dx$
3. $\int \frac{5\,dx}{x^2+1}$
4. $\int \frac{4x\,dx}{(x^2+1)^3}$
5. $\int xe^{4x}\,dx$
6. $\int \frac{e^{\tan x}}{\cos^2 x}dx$

Solve Exercises 7 and 8.

7. Find the volume of the solid generated by revolving about the x-axis the region bounded by the x-axis and the curve $y=\sqrt{x}\,e^x$ from $x=0$ to $x=1$.
8. Use the table of integrals in Appendix C to find the arc length of $y=x^2$ from $x=0$ to $x=2$.

CHAPTER

10

Parametric Equations and Polar Coordinates

The position, velocity, and acceleration of a projectile, like this football, can be described by pairs of parametric equations. In Section 10.2, we will learn how to use these equations to describe the flight of such a projectile.

Courtesy of Eric Lars Bakke/Denver Broncos

In Sections 10.1 and 10.5, we introduce parametric equations and polar coordinates. Later, we will use polar coordinates in our work with complex numbers. In this chapter, we will look at the calculus of these two types of equations.

10.1 PARAMETRIC EQUATIONS

Graphing an equation in terms of two variables x and y is often done by setting up a table of values. We usually select a value for x or y and solve for the other variables. In the case of a function such as $y = x^2$ or $y = x^3 + 2x$, this has not been a problem. We have had difficulty with equations that were not functions, such as $x^2 + y^2 = 9$.

Parametric Equations

One solution is to rewrite this equation by expressing x and y as functions of a third variable, called a **parameter**. These equations are called **parametric equations**.

EXAMPLE 10.1

Describe and sketch the curve represented by the parametric equations

$$x = 2t,\ y = t^2 - 4$$

Solution The parameter for these equations is t. We will set up a table of values for t, x, and y. The graph is shown in Figure 10.1. We connect the points in order of the increasing values of t, as indicated by the arrows in Figure 10.1.

t	−4	−3	−2	−1	0	1	2	3	4
x	−8	−6	−4	−2	0	2	4	6	8
y	12	5	0	−3	−4	−3	0	5	12

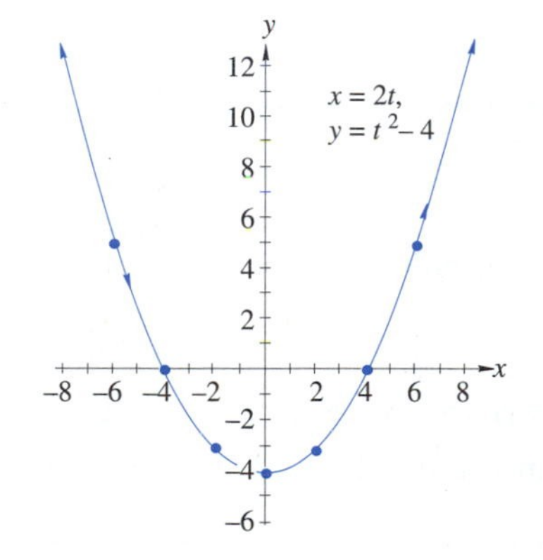

FIGURE 10.1

Does the curve look familiar? It looks very much like some of the equations that we graphed earlier. It is possible to eliminate the parameter and write the equation in rectangular form. In these two equations, since $x = 2t$, then $t = \frac{x}{2}$. Substituting this value for t in the equation $y = t^2 - 4$, we get $y = \frac{x^2}{4} - 4$ or $y = \frac{1}{4}x^2 - 4$. ▪

When eliminating the parameter, you must be very careful about the domain. Consider the following example.

EXAMPLE 10.2

Describe and sketch the curve represented by the parametric equations $x = 3\cos t$ and $y = \cos 2t$.

Solution We will set up a table of values for t, x, and y.

t	0	$\frac{\pi}{6}$	$\frac{\pi}{4}$	$\frac{\pi}{3}$	$\frac{\pi}{2}$	$\frac{2\pi}{3}$	$\frac{3\pi}{4}$	$\frac{5\pi}{6}$	π	$\frac{7\pi}{6}$	$\frac{5\pi}{4}$	$\frac{4\pi}{3}$	$\frac{3\pi}{2}$
x	3	2.6	2.1	1.5	0	−1.5	−2.1	−2.6	−3	−2.6	−2.1	−1.5	0
y	1	0.5	0	−0.5	−1	−0.5	0	0.5	1	0.5	0	−0.5	−1

FIGURE 10.2

Notice that these values begin to repeat once we get to $t = \pi$. The sketch of this curve is shown in Figure 10.2. Again, we get a shape similar to the one in Example 10.1,

EXAMPLE 10.2 (Cont.)

except that the curve does not continue in each direction. It oscillates back and forth. Using techniques we will develop later, we can eliminate the parameter to form the rectangular equation $y = \frac{2x^2}{9} - 1$, where $-3 \le x \le 3$. Notice the restriction on the domain of x.

A simpler example demonstrating how the domain is often restricted when parametric equations are written in the rectangular form is shown by the next example.

EXAMPLE 10.3

Describe and sketch the curve represented by the parametric equations $x = t^2$ and $y = 2t^2$.

Solution Since $x = t^2$ and $y = 2t^2$, we see that the rectangular form is $y = 2x$. This is the equation of a straight line. Notice, however, that for all values of t, $x \ge 0$ and $y \ge 0$. The graph formed by these parametric equations is shown in Figure 10.3 as a solid line. The remainder of the curve $y = 2x$ is indicated by the dashed line.

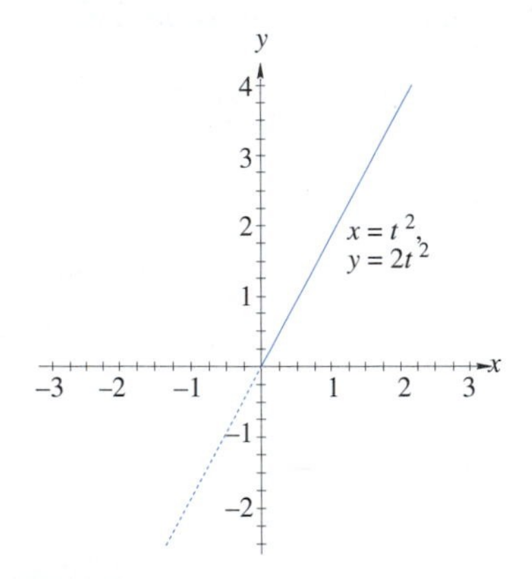

FIGURE 10.3

Lissajous Figures

When the parametric equations of a point describe simple harmonic motion, the resulting curve is called a **Lissajous figure**. When voltages of different frequencies are applied to the vertical and horizontal plates of an oscilloscope, a Lissajous figure results.

EXAMPLE 10.4

Sketch the graph of the parametric equations $x = \cos t$ and $y = \sin 3t$.

Solution The Lissajous curve for these parametric equations is shown in Figure 10.4.

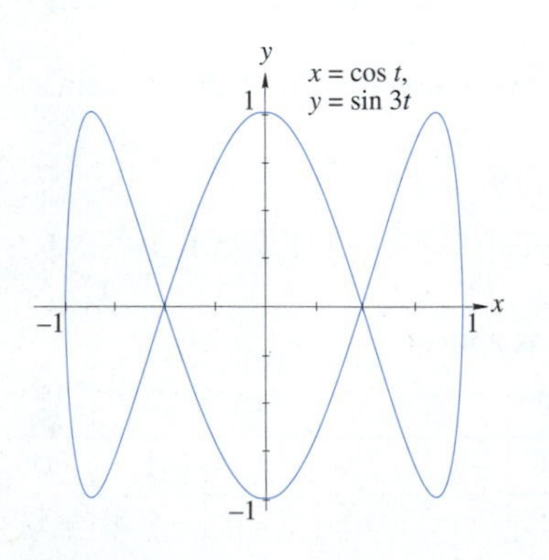

FIGURE 10.4

Examine the Lissajous figure in Figure 10.4. There are three loops along the top edge of the figure and one loop along the side. This is a ratio of 3:1. Now look at the frequencies of the parametric equations that generated this curve. The frequency of $x = \cos t$ is $\frac{1}{2\pi}$ and the frequency of $y = \sin 3t$ is $\frac{3}{2\pi}$. The ratio of the frequencies is 3:1. In Exercise Set 10.1 we will predict the number of loops on the top and side from the parametric equations and then graph the curve. While this is only an exercise in this text, it is a technique that can be used to calibrate signal generators.

Using a Calculator to Graph Parametric Equations

Before you can use a graphing calculator to draw the graphs of parametric equations, you need to put the calculator in parametric mode. On a TI-82 this is done by pressing MODE ▼ ▼ ▼ ► ENTER. The screen on a TI-82 should look like the one shown in Figure 10.5. On a Casio fx-7700G, press the key sequence MODE SHIFT ×, to set the calculator in parametric mode.

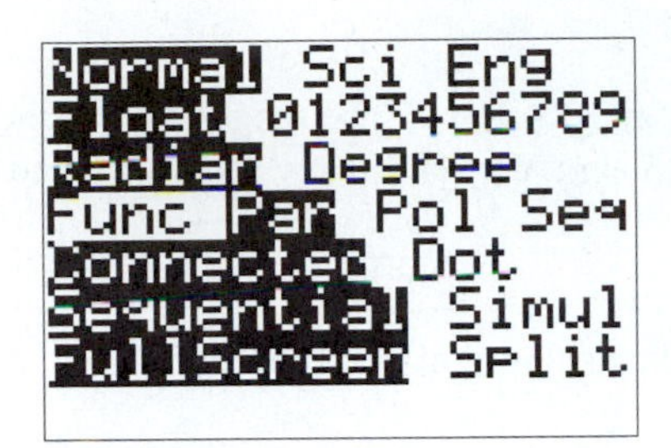

FIGURE 10.5

EXAMPLE 10.5

Use a graphing calculator to sketch the curve represented by the parametric equations $x = 2t$ and $y = t^2 - 4$.

Solution These are the same parametric equations we graphed in Example 10.1. We will use the following table from that example to help graph these equations.

t	−4	−3	−2	−1	0	1	2	3	4
x	−8	−6	−4	−2	0	2	4	6	8
y	12	5	0	−3	−4	−3	0	5	12

Press WINDOW. On a TI-82, you are first asked for Tmin, Tmax, and Tstep. Based on the table, we will let Tmin = −4 and Tmax = 4. The value of Tstep determines how much the value of t should be increased before calculating new values for x and y. We will pick Tstep = 0.5. You may want to try different values. Based on this table, let Xmin = −8, Xmax = 8, Xscl = 1, Ymin = −4, Ymax = 4, and Yscl = 1. A Casio fx-7700G calculator asks for these same values, but Tmin, Tmax, and Tptch are requested at the end, rather than at the beginning.

Now press Y=. On the first line, enter the right-hand side of the parametric equation for x. Press

2 X,T,θ

Notice that when you pressed the X,T,θ key a T is displayed on the screen. This will always happen when the calculator is in parametric mode.

Next, enter the parametric equation for y. Press ENTER to move the cursor to the line labeled Y1T and press the key sequence

X,T,θ x^2 − 4

The result in displayed in Figure 10.6a. Now press GRAPH and you should see the result in Figure 10.6b. ■

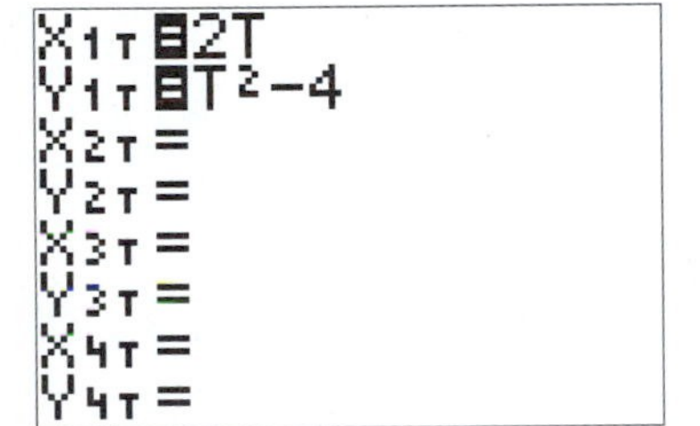

FIGURE 10.6a

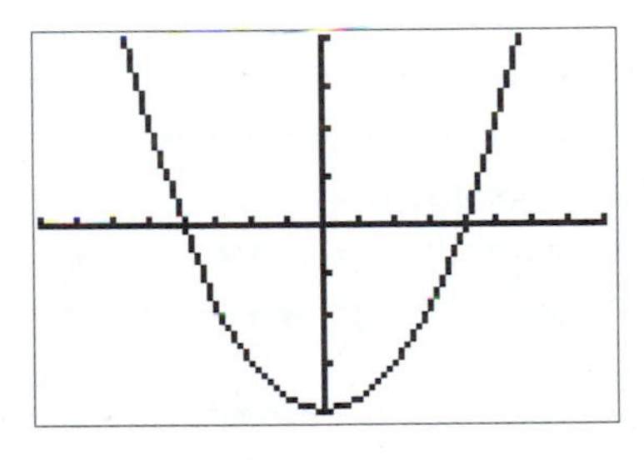

[−8, 8, 1] × [−4, 4, 1]

FIGURE 10.6b

EXAMPLE 10.6

Use a graphing calculator to sketch the curve represented by the parametric equations $x = 2\cos t$ and $y = \sin t$.

Solution Because the periods of sin and cos are 2π, we will let Tmin = 0, Tmax = 6.3, and Tstep = 0.1. Since the range of $x = 2\cos t$ is $[-2, 2]$, we choose Xmin = −2,

EXAMPLE 10.6 (Cont.)

Xmax = 2, and Xscl = 1. The range of $y = \sin t$ is $[-1, 1]$. We let Ymin = −1.5, Ymax = 1.5, and Yscl = 1. Now press [Y=] and enter the parametric equations by pressing

2 [COS] [X,T,θ] [ENTER] [SIN] [X,T,θ] [GRAPH]

The result is displayed in Figure 10.7.

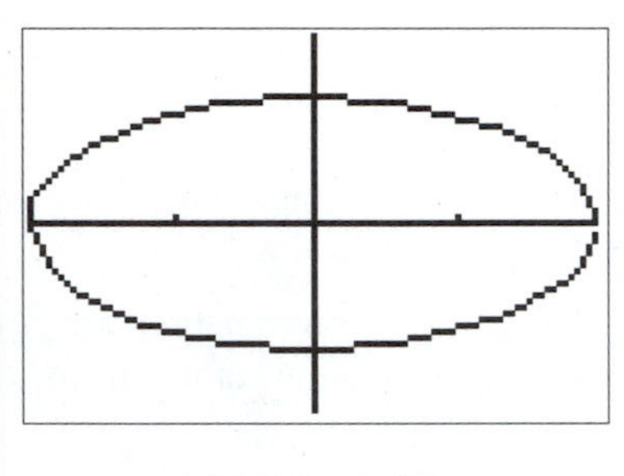

$[-2, 2, 1] \times [-1.5, 1.5, 1]$

FIGURE 10.7

Exercise Set 10.1

Graph each curve given by the parametric equations in Exercises 1–14. Make a table of values for at least six values of *t*. Show how the figure was drawn by drawing arrows on it. Eliminate the parameters in Exercises 1–6 and write the equation as a function of *y*.

1. $x = t,\ y = 3t$
2. $x = 2t,\ y = 4t + 1$
3. $x = t,\ y = \dfrac{1}{t}$
4. $x = t + 5,\ y = 3t - 2$
5. $x = 3 - t,\ y = t^2 - 9$
6. $x = t + 2,\ y = t^2 - t$
7. $x = 2\sin t,\ y = 2\cos t$
8. $x = 5\sin t,\ y = 2\cos t$
9. $x = 5\sin t,\ y = 3\sin t$
10. $x = 2\cos t,\ y = 6\cos t$
11. $x = t - \sin t,\ y = 1 - \cos t$
12. $x = \tan t,\ y = 6\cot t$
13. $x = \sec t,\ y = 2\csc t$
14. $x = 3\sec t,\ y = \tan t$

Examine the pairs of parametric equations in Exercises 15–24. For each pair, (a) predict the ratio of the loops along the top to the number of loops along the side, then (b) graph each of these curves. If possible, use a graphing calculator or computer graphing program to graph the curves.

15. $x = \sin t,\ y = \cos t$
16. $x = \sin 2t,\ y = \cos t$
17. $x = \sin t,\ y = \cos 2t$
18. $x = \sin 3t,\ y = \cos t$
19. $x = \sin 4t,\ y = \cos t$
20. $x = \sin 4t,\ y = \cos 2t$
21. $x = \sin 3t,\ y = \cos 2t$
22. $x = \sin 4t,\ y = \cos 3t$
23. $x = \sin 5t,\ y = \cos 2t$
24. $x = \sin 5t,\ y = \cos 3t$

Exercises 25–30 are Lissajous curves. For these curves, we have changed the amplitude. Graph each curve.

25. $x = 2\sin t,\ y = \cos t$
26. $x = 4\sin t,\ y = \cos t$
27. $x = 3\sin 2t,\ y = 2\cos t$
28. $x = 3\sin 3t,\ y = \cos t$
29. $x = \sin 3t,\ y = 4\cos t$
30. $x = 2\sin 2t,\ y = 5\cos 3t$

Solve Exercises 31 and 32.

31. *Automotive technology* The motion of a piece of gravel leaving a (rear) wheel at an angle α with speed V, in ft/s, can be described by

$$x = (V\cos\alpha)t$$
$$y = (V\sin\alpha)t - \frac{1}{2}gt^2$$

where $g \approx 32$ ft/s^2.

(a) Assume that a car is traveling 30 mph (44 ft/s) and that three pieces of gravel leave its rear tire, one at $\alpha = 30°$, one at $\alpha = 45°$, and one at $\alpha = 50°$. Graph the path of each of these three pieces of gravel on the same set of axes.

(b) Rewrite these parametric equations as one equation in rectangular form.

(c) Use your answer to (b) to determine how far the gravel travels before hitting the road.

32. *Business* After a new consumer electronics product is introduced, sales rise quickly and the price gradually decreases. Let t be the number of years since a product was introduced (so, $t = 0$ is when the product was introduced). Suppose the unit price at time t, in hundreds of dollars, is $p(t) = \frac{t^2+20}{t^2+5}$, and the monthly sales, in 100,000 units, are $s(t) = \frac{t^2+3t}{t^2+1}$.

(a) Graph this pair of parametric equations for 5 years with $s(t)$ on the horizontal axis and $p(t)$ on the vertical axis.

(b) What was the price when the product was introduced? After 1 year? After 5 years?

(c) What were the monthly sales during the 12th month? After 5 years?

In Your Words

33. Distinguish between rectangular equations and parametric equations.

34. Describe situations in which it is more helpful to use parametric equations than rectangular equations for graphing equations.

10.2 DERIVATIVES OF PARAMETRIC EQUATIONS

The graph of a function such as $y = f(x)$ is intersected no more than once by a vertical line. To study more complicated graphs that are not graphs of functions, we need to use a different method. One of these methods is to introduce a third variable and to use a pair of equations. In this way, we let the equations

$$x = f(t) \quad \text{and} \quad y = g(t)$$

define a curve in the plane in terms of the variable t. These equations are called **parametric equations** and for each value of t they give us a value of x and a value of y.

EXAMPLE 10.7

Graph the parametric equations $x = \sin t$ and $y = 2\cos t$ for $0 \le t \le 2\pi$.

Solution A table of values follows with the graph shown in Figure 10.8. ▪

t	0	$\frac{\pi}{6}$	$\frac{\pi}{3}$	$\frac{\pi}{2}$	$\frac{2\pi}{3}$	$\frac{5\pi}{6}$	π	$\frac{7\pi}{6}$	$\frac{4\pi}{3}$	$\frac{3\pi}{2}$	$\frac{5\pi}{3}$	$\frac{11\pi}{6}$	2π
$x = f(t)$	0	0.500	0.866	1	0.866	0.500	0	−0.500	−0.866	−1	−0.866	−0.500	0
$y = g(t)$	2	1.732	1.000	0	−1.000	−1.732	−2	−1.732	−1.000	0	1.000	1.732	2

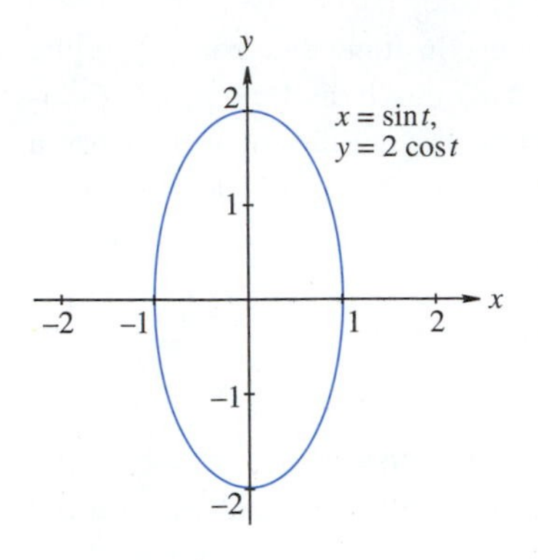

FIGURE 10.8

As we see, the curve in Example 10.7 is an ellipse. In order to eliminate the parameter t and write this as an equation in terms of x and y, we use the fact that

$$\sin^2 t + \cos^2 t = 1$$

From the equation $x = \sin t$ and $y = 2\cos t$, we obtain $x = \sin t$ and $\frac{y}{2} = \cos t$. Substituting these into $\sin^2 t + \cos^2 t = 1$ produces

$$x^2 + \frac{y^2}{4} = 1$$

which is indeed the equation of an ellipse centered at the origin with a vertical major axis of 4 and minor axis of 2.

Now, suppose that we are interested in finding the slope of this graph, or the slope of the graph of any curve with equations in parametric form. This means we want the derivative $\frac{dy}{dx}$. The following statement shows how to find the derivative of the parametric equations for a curve.

Derivative of Parametric Equations

If a curve is described by the parametric equations $x = f(t)$ and $y = g(t)$, and if $\frac{dx}{dt} \neq 0$, then the derivative of these parametric equations is

$$\frac{dy}{dx} = \frac{dy/dt}{dx/dt}$$

EXAMPLE 10.8

Find the derivative of the parametric equations in Example 10.7: $x = \sin t$, $y = 2\cos t$.

Solution To find this derivative, we need to recall that $\frac{d}{dt}(\sin t) = \cos t$ and $\frac{d}{dt}(\cos t) = -\sin t$.

$$\frac{dy}{dx} = \frac{dy/dt}{dx/dt} = \frac{-2\sin t}{\cos t} = -2\tan t$$

If you want to find a second derivative, then you take the derivative of the first derivative. Because the first derivative is a function of t and we want to take the derivative with respect to x, we must use the chain rule.

$$\frac{d}{dx}\left(\frac{dy}{dx}\right) = \frac{d}{dt}\left(\frac{dy/dt}{dx/dt}\right)$$

$$= \frac{\frac{d}{dx}\left(\frac{dy}{dt}\right)}{dx/dt}$$

$$= \frac{\frac{d}{dt}\left(\frac{dy}{dx}\right)}{dx/dt}$$

Thus, we have the following rule for $\frac{d^2y}{dx^2}$, the second derivative of parametric equations.

Second Derivative of Parametric Equations

If a curve is described by the parametric equations $x = f(t)$ and $y = g(t)$, and if $\frac{dx}{dt} \neq 0$, then the second derivative of these parametric equations is

$$\frac{d^2y}{dx^2} = \frac{\frac{d}{dt}\left(\frac{dy}{dx}\right)}{\frac{dx}{dt}}$$

Caution

The notation $\frac{d}{dt}\left(\frac{dy}{dx}\right)$ is used to indicate that we are finding the derivative of $\frac{dy}{dx}$ with respect to t, *not* $\frac{d}{dt}$ multiplied by $\frac{dy}{dx}$.

Similarly, the third derivative is $\frac{d^3y}{dx^3} = \frac{\frac{d}{dt}\left(\frac{d^2y}{dx^2}\right)}{\frac{dx}{dt}}$. Notice that the denominator for each derivative is $\frac{dx}{dt}$.

EXAMPLE 10.9

Find the second derivative of the parametric equations $x = \sin t$ and $y = 2\cos t$.

Solution In Example 10.8, we found that $\frac{dx}{dt} = \cos t$ and $\frac{dy}{dx} = -2\tan t$. We need to first determine $\frac{d}{dt}\left(\frac{dy}{dx}\right) = \frac{d}{dt}(-2\tan t) = -2\sec^2 t$. Thus, the second derivative is

$$\frac{d^2y}{dx^2} = \frac{\frac{d}{dt}\left(\frac{dy}{dx}\right)}{\frac{dx}{dt}}$$

$$= \frac{-2\sec^2 t}{\cos t} = -2\sec^3 t$$

EXAMPLE 10.10

Determine the first and second derivatives of $x = 4(t - \sin t)$ and $y = 4(1 - \cos t)$, then sketch the graph of the curve described by these parametric equations.

Solution The first derivative is

$$\frac{dy}{dx} = \frac{dy/dt}{dx/dt} = \frac{4\sin t}{4(1-\cos t)} = \frac{\sin t}{1-\cos t}$$

For the second derivative, we need $\frac{d}{dt}\left(\frac{dy}{dx}\right) = \frac{d}{dt}\left(\frac{\sin t}{1-\cos t}\right)$. We use the quotient rule to obtain this derivative. Thus, the second derivative is

$$\frac{d^2y}{dx^2} = \frac{\frac{d}{dt}\left(\frac{dy}{dx}\right)}{dx/dt} = \frac{\frac{(1-\cos t)\cos t - \sin t \sin t}{(1-\cos t)^2}}{4(1-\cos t)}$$
$$= \frac{\cos t - \cos^2 t - \sin^2 t}{4(1-\cos t)^3}$$
$$= \frac{\cos t - 1}{4(1-\cos t)^3}$$
$$= \frac{-1}{4(1-\cos t)^2}$$

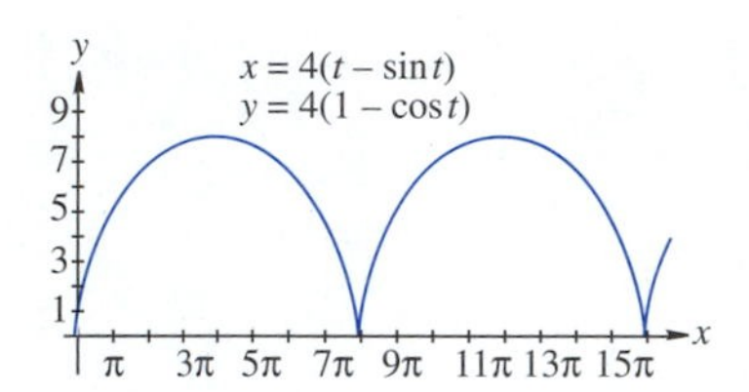

FIGURE 10.9

You might have noticed that, except when t is an odd multiple of π, the second derivative is always negative. This indicates that, except for odd multiples of π, this curve is always concave downward. Maximum points occur when t is an odd multiple of π (π, 3π, etc.). Minima occur when t is an even multiple of π (0, 2π, 4π, etc.). When $t = \pi$, we get the maximum at $x = 4\pi$, $y = 8$. The graph is shown in Figure 10.9.

EXAMPLE 10.11

Consider the curve defined by the parametric equations $x = 4\cos t + 1$ and $y = 2\sin t$. **(a)** Find the equation for the line tangent to this curve at $t = \frac{\pi}{4}$, **(b)** find the equation for the line normal to this curve at $t = \frac{\pi}{4}$, **(c)** find all the points in the interval $[0, 2\pi]$ where the curve has a horizontal or vertical tangent line, and **(d)** sketch the graph of this curve and the tangent to the curve at $t = \frac{\pi}{4}$.

Solution

(a) To find the equation for the tangent to this curve, we evaluate the derivative at $t = \frac{\pi}{4}$.

$$\frac{dy}{dx} = \frac{dy/dt}{dx/dt}$$
$$= \frac{2\cos t}{-4\sin t} = -\frac{1}{2}\cot t$$

EXAMPLE 10.11 (Cont.)

When $t=\frac{\pi}{4}$, the slope of the tangent line is $-\frac{1}{2}\cot\left(\frac{\pi}{4}\right)=-\frac{1}{2}$. The x- and y-coordinates of this point are

$$x=4\cos\left(\frac{\pi}{4}\right)+1=4\left(\frac{\sqrt{2}}{2}\right)+1=2\sqrt{2}+1\approx 3.828$$

$$y=2\sin\left(\frac{\pi}{4}\right)=2\left(\frac{\sqrt{2}}{2}\right)=\sqrt{2}\approx 1.414$$

Thus, the equation of the tangent line at this point is

$$\begin{aligned} y-y_0&=m(x-x_0)\\ y-\sqrt{2}&=-\frac{1}{2}\left[x-\left(2\sqrt{2}+1\right)\right]\\ &=-\frac{1}{2}x+\sqrt{2}+\frac{1}{2}\\ y&=-\frac{1}{2}x+2\sqrt{2}+\frac{1}{2}\end{aligned}$$

(b) Since the tangent line to the curve at $t=\frac{\pi}{4}$ has a slope of $-\frac{1}{2}$, the slope of the normal line to the curve at this point is $\dfrac{-1}{-1/2}=2$. The equation of the normal line is

$$\begin{aligned} y-y_0&=m(x-x_0)\\ y-\sqrt{2}&=2\left[x-\left(2\sqrt{2}+1\right)\right]\\ &=2x-4\sqrt{2}-2\\ y&=2x-3\sqrt{2}-2\end{aligned}$$

(c) This curve has horizontal tangent lines when $\dfrac{dy}{dx}=0$, and it has vertical tangent lines when $\dfrac{dy}{dx}$ is undefined. In (a), we found that $\dfrac{dy}{dx}=-\frac{1}{2}\cot t$.

Now, $\cot t=0$ when $t=\frac{\pi}{2}$ and when $t=\frac{3\pi}{2}$. Thus, this curve has a horizontal tangent line when $t=\frac{\pi}{2}$, which is the point $x=4\cos\left(\frac{\pi}{2}\right)+1=1$, $y=2\sin\left(\frac{\pi}{2}\right)=2$ or $(1,2)$. The other horizontal tangent line is when $t=\frac{3\pi}{2}$, or at the point $(1,-2)$.

The curve has vertical tangent lines when $t=0$, or at the point $(5,0)$, and when $t=\pi$ at the point $(-3,0)$.

(d) The graph of the curve and the tangent when $t=\frac{\pi}{4}$, as drawn on a TI-82, are shown in Figure 10.10. The graph of the tangent line was drawn by using the parametric equations $x=t$, $y=-\frac{1}{2}t+2\sqrt{2}+\frac{1}{2}$. ▪

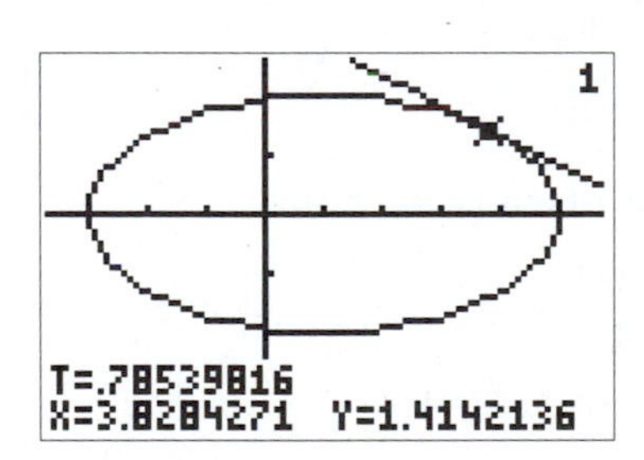

[−3.7, 5.7,1] x [−3.6, 2.6, 1]

FIGURE 10.10

Application

EXAMPLE 10.12

A certain projectile has its position at time t described by the parametric equations $s_x(t) = 6{,}250t\cos\frac{7\pi}{15}$ and $s_y(t) = 6{,}250t\sin\frac{7\pi}{15} - 16t^2$, where $s_x(t)$ represents the horizontal component of its position in feet at time t and $s_y(t)$ is the corresponding vertical component. When $t = 6$ s, determine **(a)** the horizontal and vertical components of the projectile's position, **(b)** its horizontal and vertical velocity, and **(c)** its horizontal and vertical acceleration.

Solutions **(a)** When $t = 6$, the horizontal and vertical components are $s_x(6) = 6{,}250(6)\cos\frac{7\pi}{15} \approx 3{,}919.82$ ft and $s_y(6) = 6{,}250(6)\sin\frac{7\pi}{15} - 16(6)^2 \approx 36{,}718.57$ ft. **(b)** The horizontal and vertical velocities are found by taking the derivatives of the horizontal and vertical position components with respect to time. Thus, the horizontal velocity is $v_x(t) = s_x'(t) = 6{,}250\cos\frac{7\pi}{15}$. When $t = 6$, we get $v_x(6) = 6{,}250\cos\frac{7\pi}{15} \approx 653.30$ ft/s. The vertical velocity is $v_y(t) = s_y'(t) = 6{,}250\sin\frac{7\pi}{15} - 32t$, and when $t = 6$, we obtain $v_y(6) = 6{,}250\sin\frac{7\pi}{15} - 32(6) \approx 6{,}023.76$ ft/s. **(c)** The horizontal and vertical accelerations are found by taking the derivatives of the horizontal and vertical velocity components with respect to time. Thus, the horizontal acceleration is $a_x(t) = v_x'(t) = 0$. The vertical acceleration is $a_y(t) = v_y'(t) = -32$ ft/s^2.

Exercise Set 10.2

In Exercises 1–10, find the first and second derivatives of the given parametric equations. Locate all extrema and any inflection points.

1. $x = t^2 + t,\ y = t + 1$
2. $x = t + 3,\ y = t^2 + t$
3. $x = t^2 - 6t + 12,\ y = t + 4$
4. $x = t^2 + 6t + 12,\ y = t + 4$
5. $x = t^2 + t,\ y = t^2 - t$
6. $x = 4t + 1,\ y = 9t^2$
7. $x = 3 + 4\cos t,\ y = -1 + \cos t$
8. $x = 3 + 4\cos t,\ y = 1 - \sin t$
9. $x = 2 + \sin t,\ y = -1 + \cos t$
10. $x = -2 + 4e^t,\ y = 3 + 2e^{-t}$

In Exercises 11–16, find an equation of the lines tangent and normal to the curve at the given point.

11. $x = 2t - 1,\ y = 4t^2 - 2t,\ t = 1$
12. $x = t - 4,\ y = t^3 + 2t^2 - 5t - 2,\ t = 1$
13. $x = t^3,\ y = t^2,\ t = -3$
14. $x = 2\cos t,\ y = 3\sin t,\ t = \frac{\pi}{4}$
15. $x = 2 + \cos t,\ y = 2\sin t,\ t = \frac{\pi}{2}$
16. $x = e^t + 1,\ y = e^t + e^{-t},\ t = 1$

In Exercises 17–20, find all the points where the curve has a horizontal or vertical tangent.

17. $x = t + 3,\ y = t^2 - 4t$
18. $x = t - 4,\ y = (t^2 + t)^2$
19. $x = 3\cos t,\ y = 5\sin t$
20. $x = t^2 + 1,\ y = \cos t$

Solve Exercises 21–26.

21. *Physics* A certain projectile has its position, in feet, at time t described by the parametric equations $s_x(t) = 3{,}780t\cos\frac{4\pi}{15}$ and $s_y(t) = 3{,}780t\sin\frac{4\pi}{15} - 16t^2$. When $t = 5$ s, determine

(a) the horizontal and vertical components of the projectile's position,

(b) its horizontal and vertical velocity, and

(c) its horizontal and vertical acceleration.

22. *Physics* A certain projectile has its position, in meters, at time t described by the parametric equations $s_x(t) = 1{,}250t\cos\frac{3\pi}{11}$ and $s_y(t) = 1{,}250t\sin\frac{3\pi}{11} - 4.9t^2$. When $t = 8$ s, determine

(a) the horizontal and vertical components of the projectile's position,

(b) its horizontal and vertical velocity, and

(c) its horizontal and vertical acceleration.

23. *Electronics* An electron in an electric field moves in a path described by the parametric equations

$$x = \frac{100}{\sqrt{t^2+1}} \qquad \text{and} \qquad y = \frac{100t}{\sqrt{t^2+1}}$$

where x and y are in kilometers and t is in seconds.

(a) Find the magnitude and direction of this electron when $t = 3.0$ s.

(b) Find the horizontal and vertical components of this electron's velocity.

(c) Find the magnitude and direction of the velocity of this electron when $t = 3.0$ s.

(d) Find the horizontal and vertical components of this electron's acceleration.

(e) Find the magnitude and direction of this electron's acceleration when $t = 3.0$ s.

24. *Automotive technology* The motion of a piece of gravel leaving a (rear) wheel at angle α with speed V, in feet per second, can be described by

$$x = (V\cos\alpha)t$$

$$y = (V\sin\alpha)t - \frac{1}{2}gt^2$$

where $g \approx 32$ ft/s^2. Assume that a car is traveling 30 mph (44 ft/s) and that three pieces of gravel leave its rear tire, one at $\alpha = 30°$, one at $\alpha = 45°$, and one at $\alpha = 50°$.

(a) Determine the horizontal and vertical components of each piece's velocity.

(b) Find the magnitude and direction of the velocity of each piece of gravel at $t = 1$ s.

25. *Space technology* During the first 100 s after launch, a spacecraft moves in a path described by the parametric equations

$$x = 10\sqrt{t^4+1} - 1 \qquad \text{and} \qquad y = 40t^{3/2}$$

where x and y are in meters and t is in seconds.

(a) Find the horizontal and vertical components of this spacecraft's velocity.

(b) Determine the equations for the magnitude and direction of this spacecraft's velocity at time t.

(c) Find the magnitude and direction of the velocity of this spacecraft when $t = 5.0$ s.

26. *Space technology* Consider the spacecraft in Exercise 25.

(a) Find the horizontal and vertical components of this spacecraft's acceleration.

(b) Determine the equations for the magnitude and direction of this spacecraft's acceleration at time t.

(c) Find the magnitude and direction of the acceleration of this spacecraft when $t = 5.0$ s.

In Your Words

27. Describe a situation when you might want to know the derivatives of the horizontal and vertical components of a function.

28. Without looking in the text, explain how to take the first and second derivatives of a function described by the parametric equations $x = f(t)$ and $y = g(t)$.

10.3 INTRODUCTION TO VECTORS

Our study will begin with the introduction of two quantities: scalars and vectors. After we finish this introductory material, you will be ready to learn some of the basic operations with scalars and vectors. We will use these operations to help solve some applied problems.

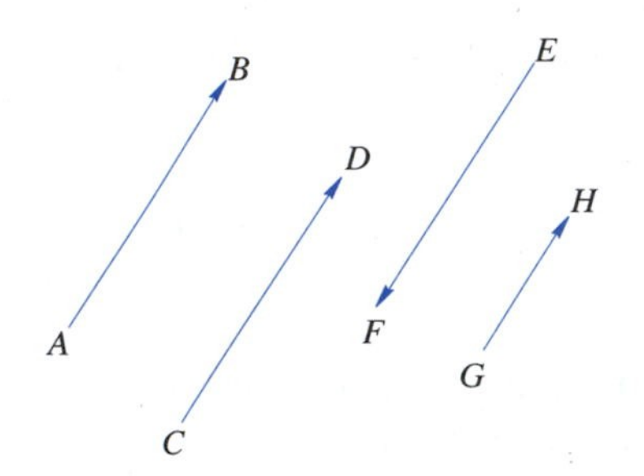

FIGURE 10.11

Scalars

Quantities that have size, or magnitude, but no direction are called **scalars**. Time, volume, mass, speed, distance, and temperature are some examples of scalars.

Vectors

A **vector** is a quantity that has both magnitude and direction. The magnitude of the vector is indicated by its length. The direction is often given by an angle. A vector is usually pictured as an arrow with the arrowhead pointing in the direction of the vector. Force, velocity, acceleration, torque, and electric and magnetic fields are all examples of vectors.

Vectors are usually represented with boldface letters such as **a** or **A** or by letters with arrows over them, $\overrightarrow{a}$ or $\overrightarrow{A}$. If a vector extends from a point A, called the **initial point**, to a point B, called the **terminal point**, then the vector can be represented by $\overrightarrow{AB}$ with the arrowhead over the terminal point B. In written work, it is hard to write in boldface, so you should use $\overrightarrow{a}$ or $\overrightarrow{b}$ or $\overrightarrow{AB}$.

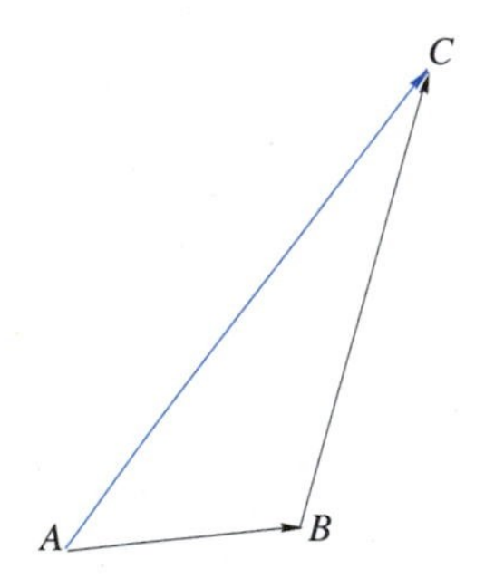

FIGURE 10.12a

The magnitude of a vector **A** is usually denoted by $|\mathbf{A}|$ or A. The magnitude of a vector is never negative. Two vectors, $\overrightarrow{AB}$ and $\overrightarrow{CD}$, are **equal**, or **equivalent**, if they have the same magnitude and direction, and we write $\overrightarrow{AB} = \overrightarrow{CD}$. In Figure 10.11, $\overrightarrow{AB} = \overrightarrow{CD}$ but $\overrightarrow{AB} \neq \overrightarrow{EF}$ because, while they have the same magnitude, they are not in the same direction. Also, $\overrightarrow{AB} \neq \overrightarrow{GH}$, because they do not have the same magnitude, even though they are in the same direction.

Resultant Vectors

If $\overrightarrow{AB}$ is the vector from A to B and $\overrightarrow{BC}$ is the vector from B to C, then the vector $\overrightarrow{AC}$ is the **resultant vector** and represents the sum of $\overrightarrow{AB}$ and $\overrightarrow{BC}$.

$$\overrightarrow{AC} = \overrightarrow{AB} + \overrightarrow{BC}$$

The sum of two vectors is shown geometrically in Figure 10.12a. It is also possible to add vectors such as $\overrightarrow{AB} + \overrightarrow{AD}$, as shown in Figure 10.12b. Here $\overrightarrow{AC} = \overrightarrow{AB} + \overrightarrow{AD}$. As you can probably guess from the shape of Figure 10.12b, this method of adding vectors is called the **parallelogram method**.

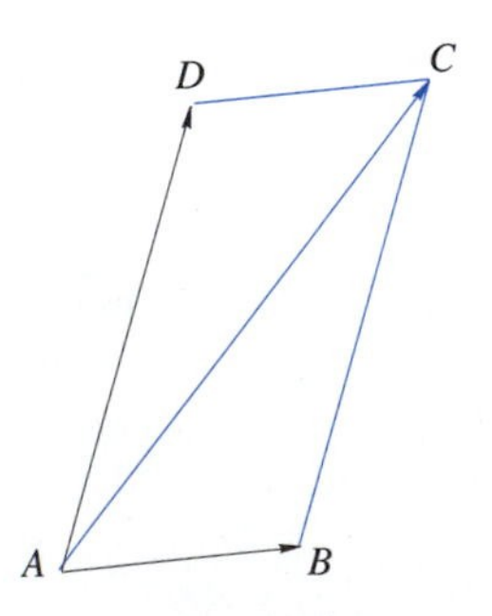

FIGURE 10.12b

The vectors that are being added do not need an endpoint in common. example, in Figure 10.13 to find $\overrightarrow{AB} + \overrightarrow{EF}$ we would find the vector $\overrightarrow{BC}$, whic equivalent to $\overrightarrow{EF}$. That is, $\overrightarrow{BC} = \overrightarrow{EF}$. We are then back to the first method an $\overrightarrow{AB} + \overrightarrow{EF} = \overrightarrow{AB} + \overrightarrow{BC} = \overrightarrow{AC}$.

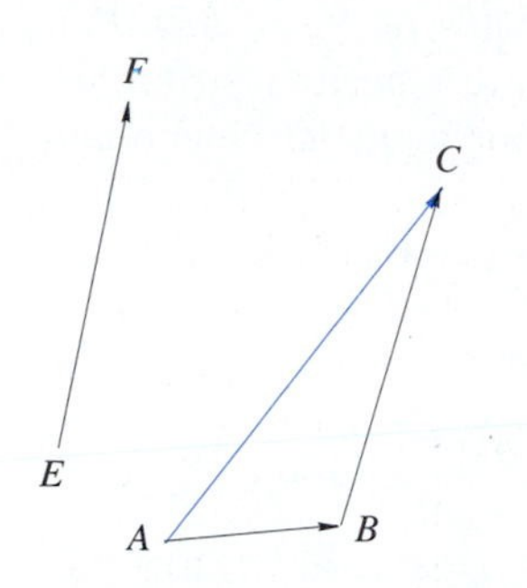

FIGURE 10.13

Vectors can also be subtracted. Here you need to realize that the vector $-\mathbf{V}$ has the opposite direction of the vector $\mathbf{V}$. An example of vector subtraction, $\overrightarrow{AB} - \overrightarrow{CD} = \overrightarrow{R}$ is shown in Figure 10.14.

Another way to represent subtraction of vectors is shown in Figure 10.15. If we want to subtract $\overrightarrow{AB} - \overrightarrow{CD}$ as in Figure 10.14, we could draw $\overrightarrow{AE}$, where $\overrightarrow{AE} = \overrightarrow{CD}$. Then $\overrightarrow{EB}$ would be the vector that represents $\overrightarrow{AB} - \overrightarrow{CD} = \overrightarrow{AB} - \overrightarrow{AE}$. Notice that in Figure 10.15, $\overrightarrow{EB}$ is drawn from the terminal point of the second vector in the difference $(\overrightarrow{AE})$ to the terminal point of the first vector $(\overrightarrow{AB})$.

A vector can be multiplied by a scalar. Thus, $2\mathbf{a}$ has twice the magnitude but the same direction as $\mathbf{a}$ and $5\mathbf{a}$ has five times the magnitude and the same direction as $\mathbf{a}$. (See Figure 10.16.)

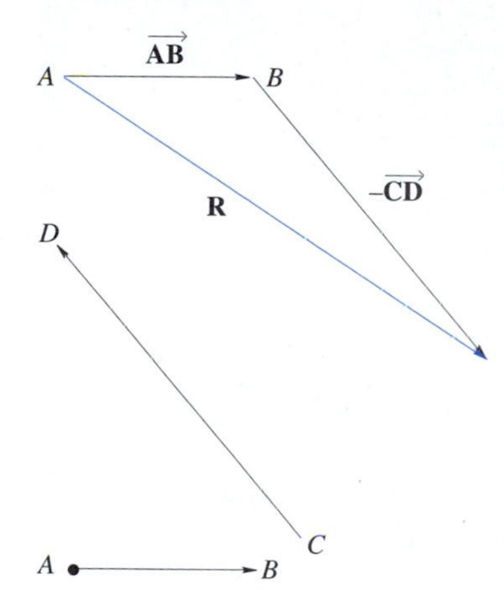

FIGURE 10.14

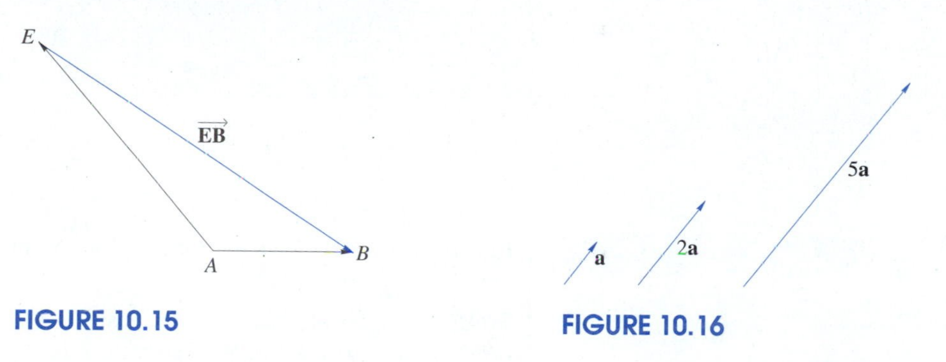

FIGURE 10.15

FIGURE 10.16

Component Vectors

In addition to adding two vectors together to find a resultant vector, we often need to reverse the process and think of a vector as the sum of two other vectors. Any two vectors that can be added together to give the original vector are called **component vectors**. To resolve a vector means to replace it by its component vectors. Usually a vector is resolved into component vectors that are perpendicular to each other.

Position Vector

If P is any point in a coordinate plane and O is the origin, then $\overrightarrow{OP}$ is called the **position vector** of P. It is relatively easy to resolve a position vector into two component vectors by using the x- and y-axes. These are called the **horizontal** (or x-)**component** and the **vertical** (or y-)**component**. In Figure 10.17, $\mathbf{P}_x$ is the horizontal component of $\overrightarrow{OP}$ and $\mathbf{P}_y$ is the vertical component.

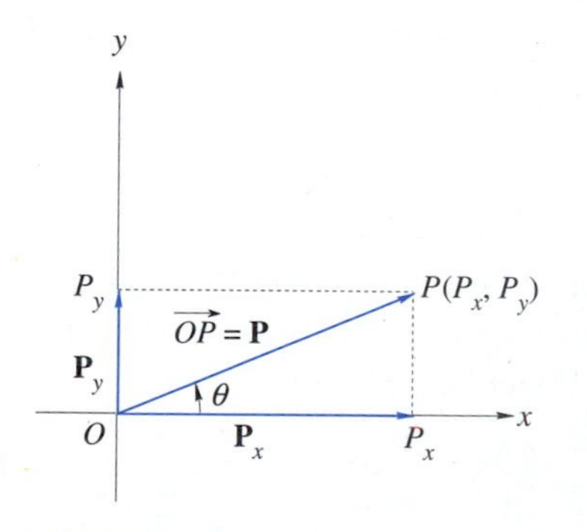

FIGURE 10.17

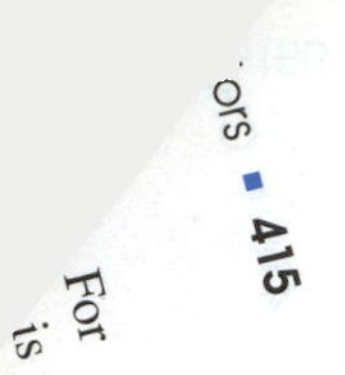

Finding Component Vectors

If you study Figure 10.17, you can see that the coordinates of P are (P_x, P_y). (Remember that P_x is the magnitude of vector $\mathbf{P}_x$.) Since every position vector has O as its initial point, it is easier to refer to a position vector by its terminal point. From now on we will refer to $\overrightarrow{OP}$ as **P**.

If the angle that a position vector **P** makes with the positive x-axis is θ, then the components of **P** are found as follows.

Components of a Vector

A position vector **P** that makes an angle θ with the positive x-axis can be resolved into component vectors $\mathbf{P}_x$ and $\mathbf{P}_y$ along the x- and y-axis respectively, with magnitudes P_x and P_y, where

$$\mathbf{P}_x = P\cos\theta$$

and

$$\mathbf{P}_y = P\sin\theta$$

EXAMPLE 10.13

Resolve a vector 12.0 units long and at an angle of 150° into its horizontal and vertical components.

Solution Consider this to be a position vector and put the initial point at the origin. The vector will look like vector **P** in Figure 10.18. We are told that $P = 12$ and $\theta = 150°$, so

$$\begin{aligned}\mathbf{P}_x &= 12\cos 150° \\ &\approx 12(-0.866) \\ &= -10.392\end{aligned}$$

and

$$\begin{aligned}\mathbf{P}_y &= 12\sin 150° \\ &= 12(0.5) \\ &= 6\end{aligned}$$

FIGURE 10.18

We have resolved **P** into two component vectors. One component is along the negative x-axis and has approximate magnitude 10.4 units. The other component is along the positive y-axis and has magnitude 6.00 units.

EXAMPLE 10.14

Vector **P** is shown in Figure 10.19a. Resolve **P** into its horizontal and vertical components.

EXAMPLE 10.14 (Cont.)

Solution This vector is in Quadrant III, so both components will be negative. The reference angle is 67°. We want to know the angle that this vector makes with the positive x-axis. Since the vector is in Quadrant III, $\theta = 180° + 67° = 247°$. We see that $P = 130$, so we determine

$$\begin{aligned}\mathbf{P}_x &= 130\cos 247° \\ &\approx 130(-0.3907) \\ &\approx -50.7950 \\ &\approx -50.8 \\ \mathbf{P}_y &= 130\sin 247° \\ &\approx 130(-0.9205) \\ &\approx -119.6656 \\ &\approx -120\end{aligned}$$

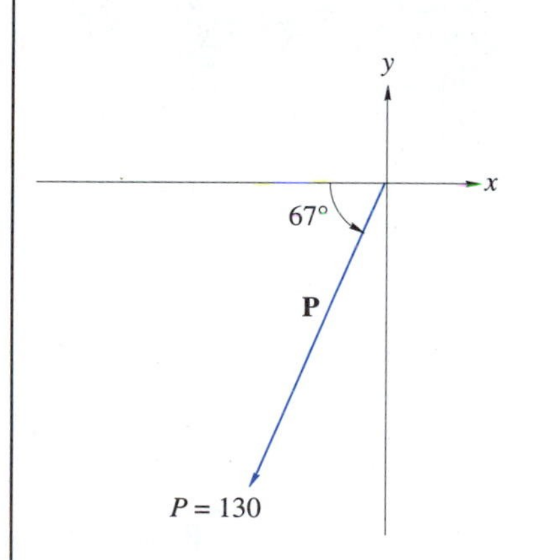

FIGURE 10.19a

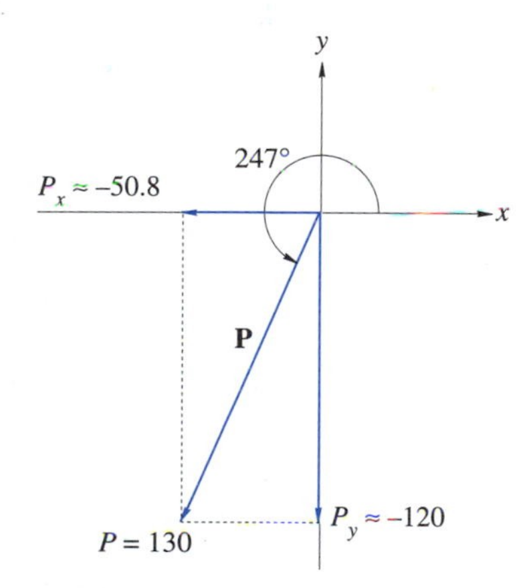

FIGURE 10.19b

Thus, **P** has been resolved into component vectors $\mathbf{P}_x$ of magnitude 50.8 units along the negative x-axis and $\mathbf{P}_y$ of magnitude 120 units along the negative y-axis, as shown in Figure 10.19b.

Application

EXAMPLE 10.15

The police officer in Figure 10.20a is measuring skid marks at an accident scene by pushing a wheel tape with a force of 10 lb and holding the handle at an angle of 46° with the ground. Resolve this into its horizontal and vertical component vectors.

EXAMPLE 10.15 (Cont.)

Solution A vector diagram has been drawn over the photograph (Figure 10.20a) of the police officer operating the wheel tape. The vector diagram is shown alone in Figure 10.20b. The initial point of the vector is at the officer's hand and the terminal point is at the hub, or axle, of the wheel tape. We have placed the origin of our

Courtesy of Michael A. Gallitelli, Metroland Photo Inc.

FIGURE 10.20a

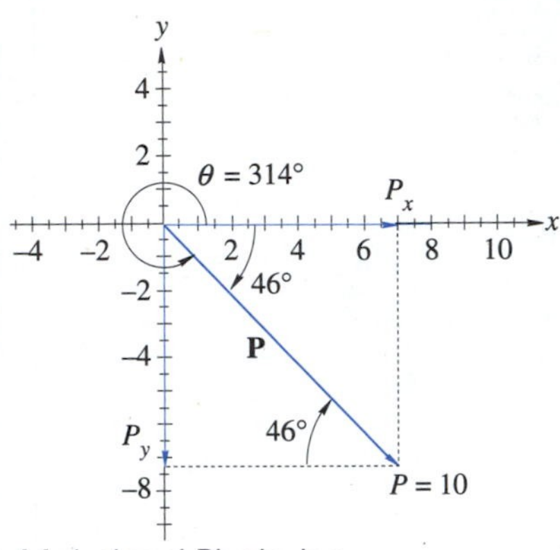

FIGURE 10.20b

coordinate system at the initial point of the vector, and the horizontal, or x-axis, is parallel to the ground.

The vector makes an angle of $360° - 46° = 314°$ with the positive x-axis. Thus, we have $P = 10$ and $\theta = 314°$, so

$$\begin{aligned} \mathbf{P}_x &= 10\cos 314° \\ &\approx -6.9466 \\ \text{and} \quad \mathbf{P}_y &= 10\sin 314° \\ &\approx -7.1934 \end{aligned}$$

The police officer exerts a horizontal force of about 6.9 lb and a vertical force of approximately 7.2 lb.

Application

EXAMPLE 10.16

A cable supporting a television tower exerts a force of 723 N at an angle of 52.7° with the horizontal, as shown in Figure 10.21. Resolve this force into its vertical and horizontal components.

EXAMPLE 10.16 (Cont.)

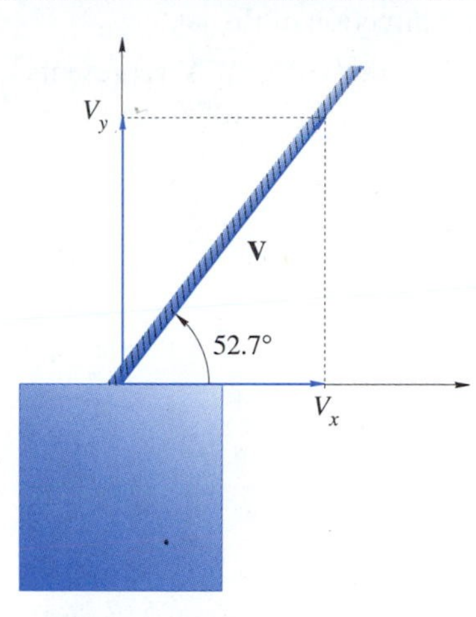

FIGURE 10.21

Solution A vector diagram has been drawn in Figure 10.21. If the cable is represented by vector **V**, then the horizontal component is $\mathbf{V}_x$ and the vertical component is $\mathbf{V}_y$. In this example, $\theta = 52.7°$, and so,

$$\begin{aligned} V_x &= 723\cos 52.7° \\ &\approx 438.1296 \\ \text{and} \quad V_y &= 723\sin 52.7° \\ &\approx 575.1273 \end{aligned}$$

We see that this cable exerts a horizontal force of approximately 438 N and a vertical force of about 575 N.

Finding the Magnitude and Direction of Vectors

If we have the horizontal and vertical components of a vector **P**, then we can use the components to determine the magnitude and direction of the resultant vector.

Magnitude and Direction of a Vector

If $\mathbf{P}_x$ is the horizontal component of vector **P** and $\mathbf{P}_y$ is its vertical component, then

$$|\mathbf{P}| = P = \sqrt{P_x^{\,2} + P_y^{\,2}}$$

$$\text{and} \qquad \theta_{\text{Ref}} = \tan^{-1}\left|\frac{P_y}{P_x}\right|$$

EXAMPLE 10.17

A position vector **V** has its horizontal component $\mathbf{V}_x = -24$ and its vertical component $\mathbf{V}_y = -32$. What are the direction and magnitude of **V**?

Solution A sketch of this problem in Figure 10.22 shows that **V** is in the third quadrant. Since $\theta_{\text{Ref}} = \tan^{-1}\left|\frac{-32}{-24}\right| \approx 53.13°$, we see that $\theta \approx 180° + 53.13° = 233.13°$.

We know that

$$\begin{aligned} V &= \sqrt{V_x^{\,2} + V_y^{\,2}} \\ &= \sqrt{(-24)^2 + (-32)^2} = \sqrt{1{,}600} \\ &= 40 \end{aligned}$$

So, the magnitude of **V** is 40 and **V** is at an angle of 233.13°.

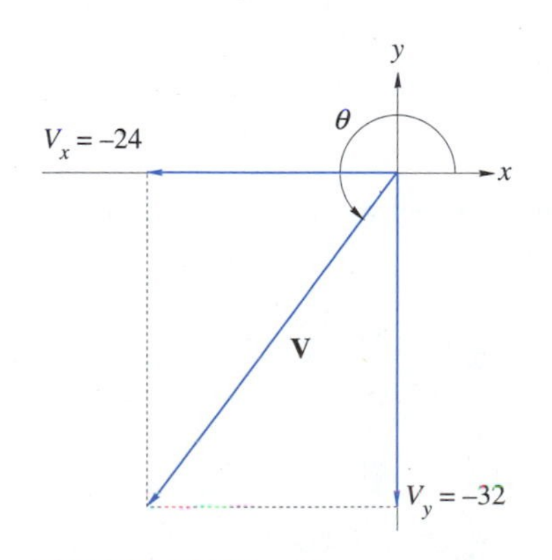

FIGURE 10.22

Application

EXAMPLE 10.18

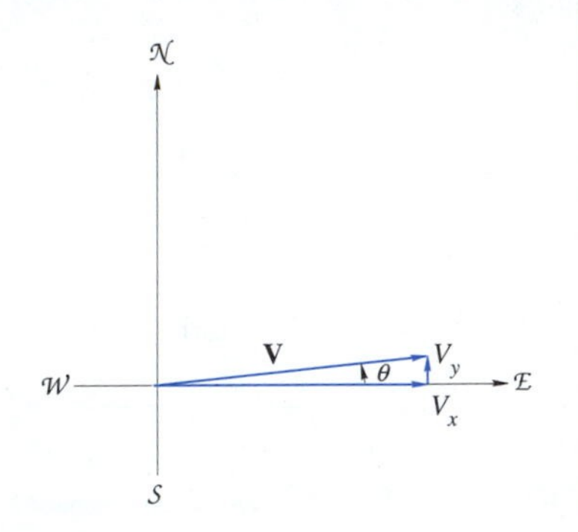

FIGURE 10.23

A pilot heads a jet plane due east at a ground speed of 425.0 mph. If the wind is blowing due north at 47 mph, find the true speed and direction of the jet.

Solution A sketch of this situation is shown in Figure 10.23. If **V** represents the vector with components $\mathbf{V}_x$ and $\mathbf{V}_y$, then we are given $V_x = 425.0$ and $V_y = 47$. Thus,

$$\begin{aligned} V &= \sqrt{V_x^2 + V_y^2} \\ &= \sqrt{425.0^2 + 47^2} \\ &\approx 427.59 \end{aligned}$$

$$\text{and} \qquad \begin{aligned} \theta &= \tan^{-1}\left(\frac{47}{425}\right) \\ &= \tan^{-1} 0.110588 \end{aligned}$$

$$\text{and so,} \qquad \theta = 6.31^\circ$$

The jet is flying at a speed of approximately 427.6 mph in a direction 6.31° north of due east.

Exercise Set 10.3

In Exercises 1–4, add the given vectors by drawing the resultant vector.

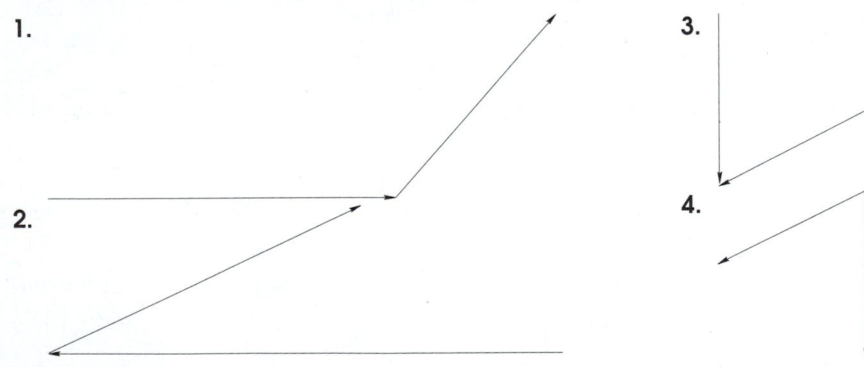

For Exercises 5–20, trace each of the vectors A–D in Figure 10.24. Use these vectors to find each of the indicated sums or differences.

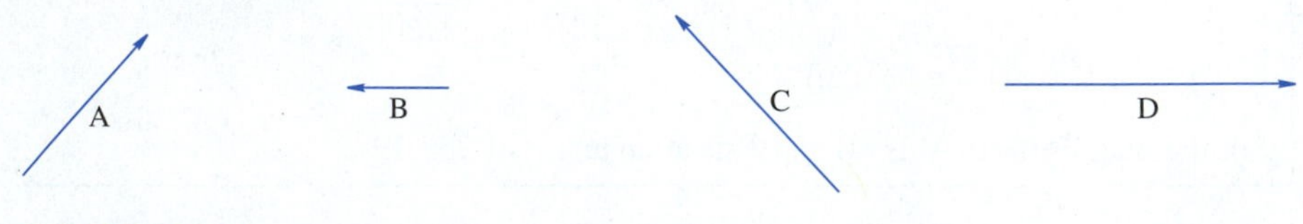

FIGURE 10.24

5. $\mathbf{A}+\mathbf{B}$
6. $\mathbf{B}+\mathbf{D}$
7. $\mathbf{A}+\mathbf{B}+\mathbf{C}$
8. $\mathbf{B}+\mathbf{C}+\mathbf{D}$
9. $\mathbf{A}-\mathbf{B}$
10. $\mathbf{C}-\mathbf{D}$
11. $\mathbf{B}-\mathbf{A}$
12. $\mathbf{D}-\mathbf{C}$
13. $\mathbf{A}+3\mathbf{B}$
14. $\mathbf{C}+2\mathbf{D}$
15. $2\mathbf{C}-\mathbf{D}$
16. $3\mathbf{B}-\mathbf{A}$
17. $\mathbf{A}+2\mathbf{B}-\mathbf{C}$
18. $\mathbf{A}+3\mathbf{C}-\mathbf{D}$
19. $2\mathbf{B}-\mathbf{C}+2.5\mathbf{D}$
20. $4\mathbf{C}-3\mathbf{B}-2\mathbf{D}$

In Exercises 21–26, use trigonometric functions to find the horizontal and vertical components of the given vectors.

21. Magnitude 20, $\theta = 75°$
22. Magnitude 16, $\theta = 212°$
23. Magnitude 18.4, $\theta = 4.97$ rad
24. Magnitude 23.7, $\theta = 2.22$ rad
25. $V = 9.75, \theta = 16°$
26. $P = 24.6, \theta = 317°$

In Exercises 27–30, the horizontal and vertical components are given for a vector. Find the magnitude and direction of each resultant vector.

27. $\mathbf{A}_x = -9; \mathbf{A}_y = 12$
28. $\mathbf{B}_x = 10; \mathbf{B}_y = -24$
29. $\mathbf{C}_x = 8; \mathbf{C}_y = 15$
30. $\mathbf{D}_x = -14; \mathbf{D}_y = -20$

Solve Exercises 31–38.

31. *Navigation* A ship heads into port at 12.0 km/h. The current is perpendicular to the ship at 5 km/h. What is the resultant velocity of the ship?
32. *Navigation* A pilot heads a jet plane due east at a ground speed of 756.0 km/h. If the wind is blowing due north at 73 km/h, find the true speed and direction of the jet.
33. *Construction* A cable supporting a tower exerts a force of 976 N at an angle of 72.4° with the horizontal. Resolve this force into its vertical and horizontal components.
34. *Construction* A sign of mass 125.0 lb hangs from a cable. A worker is pulling the sign horizontally by a force of 26.50 lb. Find the force and the angle of the resultant force on the sign.
35. *Medical technology* A ramp for the physically challenged makes an angle of 12° with the horizontal. A woman and her wheelchair weigh 153 lb. What are the components of this weight parallel and perpendicular to the ramp?
36. *Electricity* If a resistor, a capacitor, and an inductor are connected in series to an ac power source, then the effective voltage of the source is given by the vector $\mathbf{V}$, where $\mathbf{V}$ is the sum of the vector quantities $\mathbf{V}_R$ and $\mathbf{V}_L - \mathbf{V}_C$, as shown in Figure 10.25. The phase angle, ϕ, is the angle between $\mathbf{V}$ and $\mathbf{V}_R$, where $\tan\phi = \dfrac{\mathbf{V}_L - \mathbf{V}_C}{\mathbf{V}_R}$. If the effective voltages across the circuit components are $V_R = 12$ V, $V_C = 10$ V, and $V_L = 5$ V, determine the effective voltage and the phase angle.

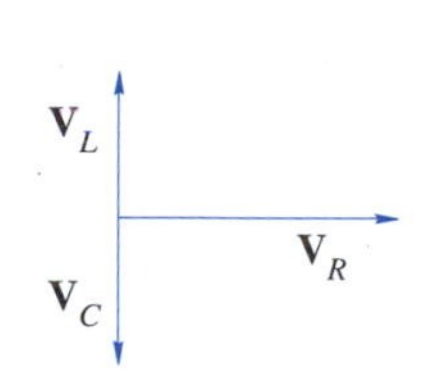

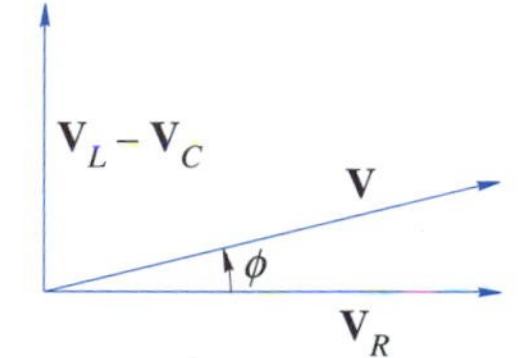

FIGURE 10.25

37. *Electricity* A resistor, capacitor, and inductor are connected in series to an ac power source. If the effective voltages across the circuit components are $V_R = 15.0$ V, $V_C = 17.0$ V, and $V_L = 8.0$ V, determine the effective voltage and the phase angle.
38. *Electricity* A resistor, capacitor, and inductor are connected in series to an ac power source. If the effective voltages across the circuit components are $V_R = 22.6$ V, $V_C = 15.2$ V, and $V_L = 28.3$ V, determine the effective voltage and phase angle.

In Your Words

39. **(a)** Distinguish between scalars and vectors. How are they alike and how are they different?

    (b) What does it mean for two vectors to be equal?

40. **(a)** Describe how to use the components of a vector to determine the vector's magnitude and direction.

    (b) Explain how to use the magnitude and direction of a vector to determine its component vectors.

10.4 DERIVATIVES OF VECTORS

Vectors are particularly useful in the study of motion and forces. In this section, we will combine the idea of parametric equations with the concept of vectors. This will allow us to better describe motion along a curve.

We know that any vector **P** can be described in terms of its components $\mathbf{P}_x$ and $\mathbf{P}_y$. In particular, $\mathbf{P} = \mathbf{P}_x + \mathbf{P}_y$. Often we express a vector in terms of the **unit vectors** **i** and **j**. The unit vector **i** has magnitude equal to 1, and its direction is along the positive x-axis. In much the same manner, the unit vector **j** has magnitude equal to 1 and direction is along the positive y-axis. Using **i** and **j**, any vector can be uniquely written as

$$\mathbf{P} = P_x\mathbf{i} + P_y\mathbf{j}$$

for suitable values of P_x and P_y. Notice from this definition that the magnitude of **P** is still

$$|P| = \sqrt{P_x{}^2 + P_y{}^2}$$

and the direction θ of **P** is determined from

$$\begin{aligned} \tan\theta &= \frac{P_y}{P_x} \\ \text{or} \quad \theta_{\text{Ref}} &= \tan^{-1}\left|\frac{P_y}{P_x}\right| \end{aligned}$$

This definition of **P** in terms of the unit vectors **i** and **j** allows us to use parametric equations with vectors for a new type of function, a **vector-valued function** or simply, a **vector function**.

> **Vector Functions and Parametric Equations**
>
> If a curve is described by the parametric equations $x = f(t)$ and $y = g(t)$, then the curve can be expressed as the vector-valued function $\mathbf{P}(t)$, where
>
> $$\mathbf{P}(t) = f(t)\mathbf{i} + g(t)\mathbf{j}$$

EXAMPLE 10.19

Determine the vector-valued function of the curve described by the parametric equations $x = \ln t^2$ and $y = \cos\sqrt{1-t^2}$.

Solution In this situation we are given parametric equations for x and y. If we let $x = f(t) = \ln t^2$ and $y = g(t) = \cos\sqrt{1-t^2}$, then the desired vector-valued

EXAMPLE 10.19 (Cont.)

function is

$$\mathbf{P}(t) = \ln t^2\mathbf{i} + \cos\sqrt{1-t^2}\,\mathbf{j}$$

If parameter t is time, then a specific value of t indicates a specific point on the curve $(x, y) = (f(t), g(t))$. The parametric equation can also be considered as describing the curvilinear motion of a particle. Here t not only indicates the point on the curve, but the position of the particle at time t and the direction in which it is moving.

If a vector is used to describe the motion of a particle, then the particle's velocity and acceleration can be determined by taking appropriate derivatives.

Derivatives of Vector-Valued Function

If $\mathbf{p}(t) = f(t)\mathbf{i} + g(t)\mathbf{j}$ is a vector-valued function and f and g are differentiable, then the first derivative of $\mathbf{p}$ is

$$\mathbf{p}'(t) = f'(t)\mathbf{i} + g'(t)\mathbf{j}$$

and the second derivative of $\mathbf{p}$ is

$$\mathbf{p}''(t) = f''(t)\mathbf{i} + g''(t)\mathbf{j}$$

As usual, the first derivative of the position function is the velocity function and the second derivative of the position function is the acceleration function.

EXAMPLE 10.20

The motion of a particle is given by the parametric equations $x = 3 - t^2$ and $y = \frac{1}{t}$, where x and y are in meters and t is in seconds.

(a) Sketch a graph of the particle's position for $-3 \le t \le 3$.

(b) Write the vector-valued function that describes the motion of this particle.

(c) Find the velocity vector for the particle.

(d) Determine the velocity and direction when $t = 1$.

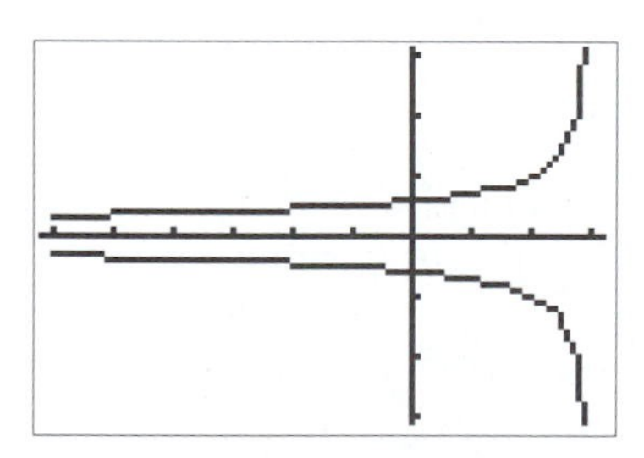

[−6.2, 3.2, 1] x [−3.1, 3.1, 1]

$-3 \le t \le 3$

FIGURE 10.26a

Solutions

(a) The graph of these parametric equations, as drawn on a TI-82, is shown in Figure 10.26a. Notice that when $t = 0$, there is no point on the graph.

(b) Since $x = f(t) = 3 - t^2$ and $y = g(t) = \frac{1}{t}$, the vector-valued function for this particle is

$$\mathbf{p}(t) = (3 - t^2)\,\mathbf{i} + \left(\frac{1}{t}\right)\mathbf{j}$$

EXAMPLE 10.20 (Cont.)

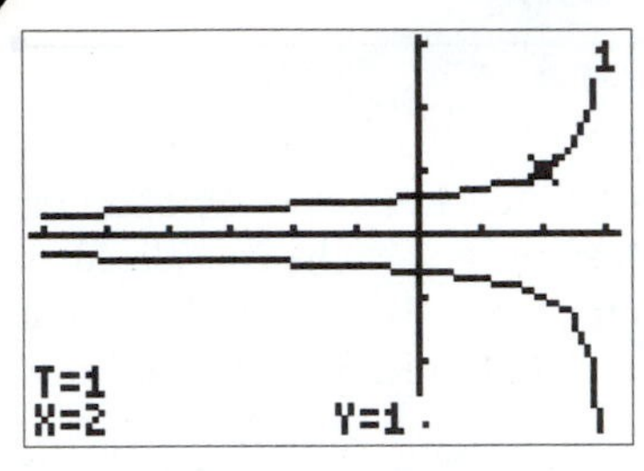

[–6.2, 3.2, 1] x [–3.1, 3.1, 1]

$-3 \leq t \leq 3$

FIGURE 10.26b

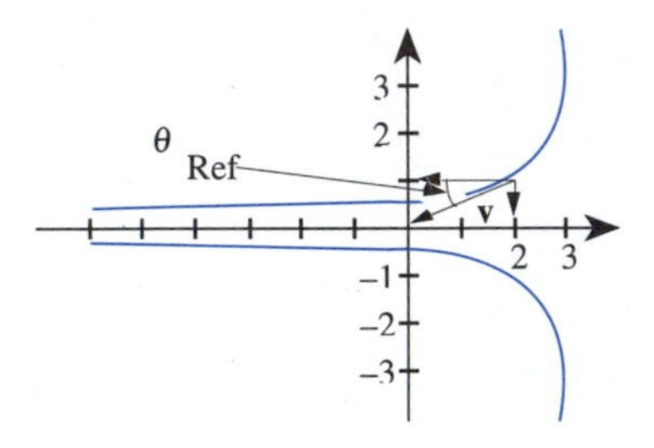

FIGURE 10.26c

(c) Here $f'(t) = -2t$ and $g'(t) = -\dfrac{1}{t^2} = -t^{-2}$. As a result, the velocity vector is

$$\mathbf{v}(t) = (-2t)\mathbf{i} - \left(\frac{1}{t^2}\right)\mathbf{j}$$

(d) When $t = 1$, the particle is located at the point with coordinates $(2, 1)$. This is the point highlighted in Figure 10.26b. At this time, when $t = 1$, the velocity vector is

$$\mathbf{v}(1) = -2\mathbf{i} - \mathbf{j}$$

The original curve and its velocity vector when $t = 1$ are shown in Figure 10.26c. The magnitude, or speed, of the velocity vector is

$$|\mathbf{v}| = \sqrt{(-2)^2 + (-1)^2} = \sqrt{5} \approx 2.236$$

For the angle θ in Figure 10.26c, we have

$$\begin{aligned} \tan\theta &= \frac{-1}{-2} = \frac{1}{2} \\ \text{or} \quad \theta_{\text{Ref}} &\approx 26.6^\circ \end{aligned}$$

Hence, the direction is

$$\theta = 180^\circ + \theta_{\text{Ref}} \approx 180^\circ + 26.6^\circ = 206.6^\circ$$

EXAMPLE 10.21

The motion of a particle is given by the parametric equations $x = \cos t$ and $y = \sin 2t$, where x and y are in meters and t is in seconds.

(a) Determine the velocity vector, $\mathbf{v}$, and the acceleration vector, $\mathbf{a}$, for the particle.

(b) Determine the velocity and acceleration when $t = \dfrac{4\pi}{3}$.

Solutions

1. Since $x = f(t) = \cos t$, we have $v_x = f'(t) = -\sin t$ and $a_x = f''(t) = -\cos t$. Also, since $y = g(t) = \sin 2t$, we have $v_y = g'(t) = 2\cos 2t$ and $a_y = g''(t) = -4\sin 2t$. Thus, the velocity vector is given by

$$\begin{aligned} \mathbf{v}(t) &= v_x\mathbf{i} + v_y\mathbf{j} \\ &= (-\sin t)\mathbf{i} + (2\cos 2t)\mathbf{j} \end{aligned}$$

and the acceleration vector is

$$\begin{aligned} \mathbf{a}(t) &= a_x\mathbf{i} + a_y\mathbf{j} \\ &= (-\cos t)\mathbf{i} - (4\sin 2t)\mathbf{j} \end{aligned}$$

EXAMPLE 10.21 (Cont.)

2. At $t = \frac{4\pi}{3}$, then $v_x = -\sin\left(\frac{4\pi}{3}\right) = \frac{\sqrt{3}}{2} \approx 0.866$ and $v_y = 2\cos\left(2 \cdot \frac{4\pi}{3}\right) = 2\cos\left(\frac{8\pi}{3}\right) = -1$. Thus, when $t = \frac{4\pi}{3}$, the velocity vector is

$$\mathbf{v}\left(\frac{4\pi}{3}\right) = \frac{\sqrt{3}}{2}\mathbf{i} - \mathbf{j}$$

So, the magnitude of the velocity at $t = \frac{4\pi}{3}$ is

$$|\mathbf{v}| = \sqrt{\left(\frac{\sqrt{3}}{2}\right)^2 + (-1)^2} = \sqrt{\frac{7}{4}} \approx 1.32 \text{ m/s}$$

and the direction is

$$\begin{aligned}\theta &= \tan^{-1}\left(\frac{v_y}{v_x}\right)\\ &= \tan^{-1}\left(\frac{-1}{\sqrt{3}/2}\right)\\ &= \tan^{-1}\left(\frac{-2}{\sqrt{3}}\right) \approx -49.1^\circ\end{aligned}$$

When $t = \frac{4\pi}{3}$, we have $a_x = 0.5$ and $a_y = -2\sqrt{3} \approx -3.464$. Thus, the acceleration vector is

$$\mathbf{a}\left(\frac{4\pi}{3}\right) = 0.5\mathbf{i} - 2\sqrt{3}\mathbf{j} \text{ m/s}^2$$

The magnitude of the acceleration is

$$|\mathbf{a}| = \sqrt{0.5^2 + \left(-2\sqrt{3}\right)^2} = \sqrt{12.25} = 3.5 \text{ m/s}^2$$

while the direction is

$$\theta = \tan^{-1}\left(\frac{-2\sqrt{3}}{0.5}\right) = \tan^{-1}\left(-4\sqrt{3}\right) \approx -81.8^\circ$$

Application

EXAMPLE 10.22

Vertical wind shear in the lowest 300 ft above the ground is very important to aircraft during takeoffs and landings. Vertical wind shear is defined as $\mathbf{v}'(h)$, where $\mathbf{v}$ is the wind velocity and h is the height above the ground. During strong wind gusts at one airport, the wind velocity (in mi/hr) for altitudes h between 0 and 200 ft is estimated to be

$$\mathbf{v}(h) = \left(14 + 0.008h^{3/2}\right)\mathbf{i} + \left(12 + 0.006h^{3/2}\right)\mathbf{j}$$

EXAMPLE 10.22 (Cont.)

Calculate the magnitude of the vertical wind shear 125 ft above the ground.

Solution Since the vertical wind shear is the derivative of the wind's velocity, we want to find the derivative of $\mathbf{v}(h)$ and evaluate it when $h = 125$.

$$\begin{aligned}\mathbf{v}'(h) &= \frac{3}{2}\left(0.008h^{1/2}\right)\mathbf{i} + \frac{3}{2}\left(0.006h^{1/2}\right)\mathbf{j}\\ &= \left(0.012h^{1/2}\right)\mathbf{i} + \left(0.009h^{1/2}\right)\mathbf{j}\end{aligned}$$

At $h = 125$, we have

$$\begin{aligned}\mathbf{v}'(h) &= 0.012\left(125^{1/2}\right)\mathbf{i} + 0.009\left(125^{1/2}\right)\mathbf{j}\\ &= 0.060\sqrt{5}\mathbf{i} + 0.045\sqrt{5}\mathbf{j}\\ &\approx 0.134\mathbf{i} + 0.101\mathbf{j}\end{aligned}$$

The magnitude of this wind shear is given by

$$\sqrt{\left(0.060\sqrt{5}\right)^2 + \left(0.045\sqrt{5}\right)^2} \approx \sqrt{0.018 + 0.010125} \approx 0.168$$

Thus, the wind shear has a magnitude of 0.168 mi/hr/ft.

Exercise Set 10.4

In Exercises 1–6, (a) graph each curve given by the parametric equation in terms of *t*, (b) find the velocity vector v, and (c) find v, the magnitude of v, and its direction at the indicated value of *t*.

1. $x = t^2 + t$ and $y = t + 1$ at $t = 1$
2. $x = t^3 - t$ and $y = 4t - 3t^2$ at $t = 1$
3. $x = \cos t$ and $y = 2 + \sin 3t$ at $t = \frac{2\pi}{3}$
4. $x = \cos^2 t$ and $y = \sin^2(2t) - \cos t$ at $t = \frac{3\pi}{4}$
5. $x = e^{\sin t}$ and $y = -e^{\cos t}$ at $t = \pi$
6. $x = 4\ln t$ and $y = 6\sqrt{10 - t}$ at $t = 1$

In Exercises 7–10, find the magnitude and direction of the velocity and acceleration at the given value(s) of *t*.

7. $x = 0.4t^{5/2}$ and $y = \frac{0.5}{(t-3)^2}$ at $t = 4$ where distance is in feet and t in seconds
8. $x = \frac{5}{3t^2 + 1}$ and $y = t^2 - t$ at $t = 0$ where distance is in feet and t in seconds
9. $x = \sin t$ and $y = \frac{1}{2}\ln t$ at $t = -\frac{3\pi}{2}$ and $t = \pi$ where distance is in meters and t in seconds
10. $x = \cos^2 t$ and $y = \sin t + \cos t$ at $t = \frac{\pi}{2}$ and $t = \frac{3\pi}{4}$ where distance is in meters and t in seconds
11. *Space technology* A spacecraft moves according to the equations $x = 10(\sqrt{1 + t^4} + 1)$, $y = 50t^{3/2}$ for the first two minutes after launch, where x and y are in meters and t is in seconds. Find the speed and direction of the velocity of the spacecraft 20.0 s and 100 s after launch.
12. *Meteorology* A weather balloon is released from the ground and rises in the atmosphere. Its distance, in meters, above the ground from the moment it is released is given by $y = 360t - 9t^2$, where t varies from 0 to 15 minutes. Because the wind is blowing in gusts, the position of the balloon with respect to the ground is given by $x = 0.8t^2 + 0.9\sin^2 t$, with a positive direction towards the east.
    (a) Determine the velocity of the balloon at any time t.

(b) After 90 s, determine the position, speed, and direction of the balloon.

13. *Fire science* In fighting forest fires, airplanes are often used to drop water or a fire suppressant on the fire. The drop has to be timed so that the liquid will hit the fire and not a part of the forest that has already burned. Suppose that a plane at an altitude of 320 ft is traveling at 150 mi/h = 220 ft/s in level flight when it drops its load of water. The trajectory of the water is given by the equations $x = 220(t - 0.025t^2 + e^{-t} - 1)$ and $y = 32\left(1610 - 40t - 1600e^{-t/40}\right)$.
(a) Determine the velocity vector.
(b) Determine the magnitude and direction of the velocity 3 s after the water is dropped.

14. *Electronics* An electron in an electric field moves in a path described by the parametric equations $x = \dfrac{100}{\sqrt{t^2+1}} = 100\left(t^2+1\right)^{-1/2}$ and $y = \dfrac{100t}{\sqrt{t^2+1}} = 100t\left(t^2+1\right)^{-1/2}$ where x and y are in km and t is in seconds.
(a) Find the velocity vector of this electron.
(b) Find the magnitude and direction of the velocity of this electron when $t = 3.0$ s.
(c) Find the acceleration vector of this electron.
(d) Find the magnitude and direction of this electron's acceleration when $t = 3.0$ s.

In Your Words

15. Describe a situation in which you might want to use a vector-valued function rather than parametric equations.

16. Without looking in the text, explain how to take the first and second derivatives of the vector-valued function $\mathbf{p}(t) = f(t)\mathbf{i} + g(t)\mathbf{j}$.

10.5 POLAR COORDINATES

Every time we have represented a point in a plane, we used the rectangular coordinate system. Each point has an x- and a y-coordinate. There is another type of coordinate system that is used to represent points. This system is called the **polar coordinate system**.

To introduce a system of polar coordinates in a plane, we begin with a fixed point O called the **pole** or **origin**. From the pole, we shall draw a half-line that has O as its endpoint. This half-line will be called the **polar axis**.

Polar Coordinates

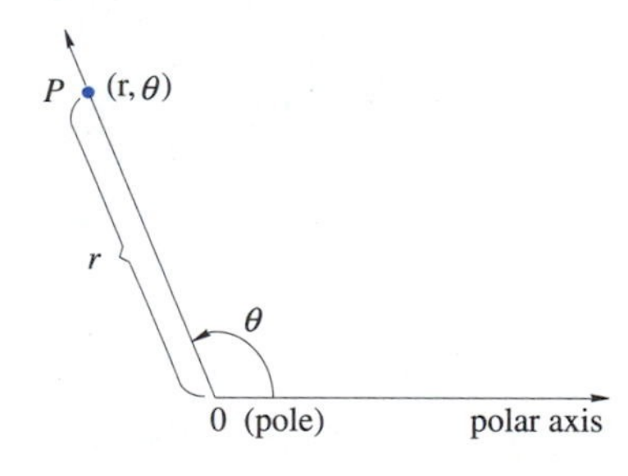

FIGURE 10.27

Consider any point in the plane different from point O. Call this point P. Let the polar axis form the initial side of an angle and $\overrightarrow{OP}$ the terminal side. This angle has measure θ. If the distance from O to P is r, we can say that the **polar coordinates** of P are (r, θ), as shown in Figure 10.27.

We have some of the same understanding about polar coordinates as we do about the angles from trigonometry. If the angle is generated by a counterclockwise rotation of the polar axis, the angle is positive. If it is generated by a clockwise rotation, θ is negative. If r is negative, the terminal side of the angle is extended in the opposite direction through the pole and is measured off r units on this extended side.

A special type of graph paper, called **polar coordinate paper**, is used to graph polar coordinates. This paper has concentric circles with their centers at the pole. The distance between any two consecutive circles is the same. Lines are drawn through the pole and correspond to some of the common angles. An example of polar coordinate paper is shown in Figure 10.28.

EXAMPLE 10.23

Plot the points with the following polar coordinates: $(5, \frac{\pi}{4})$, $(-5, \frac{\pi}{4})$, $(3, 150°)$, $(3, -70°)$, $(6, \pi)$, $(4, 0°)$, and $(6, 450°)$.

Solution The points are plotted in Figure 10.29. ▪

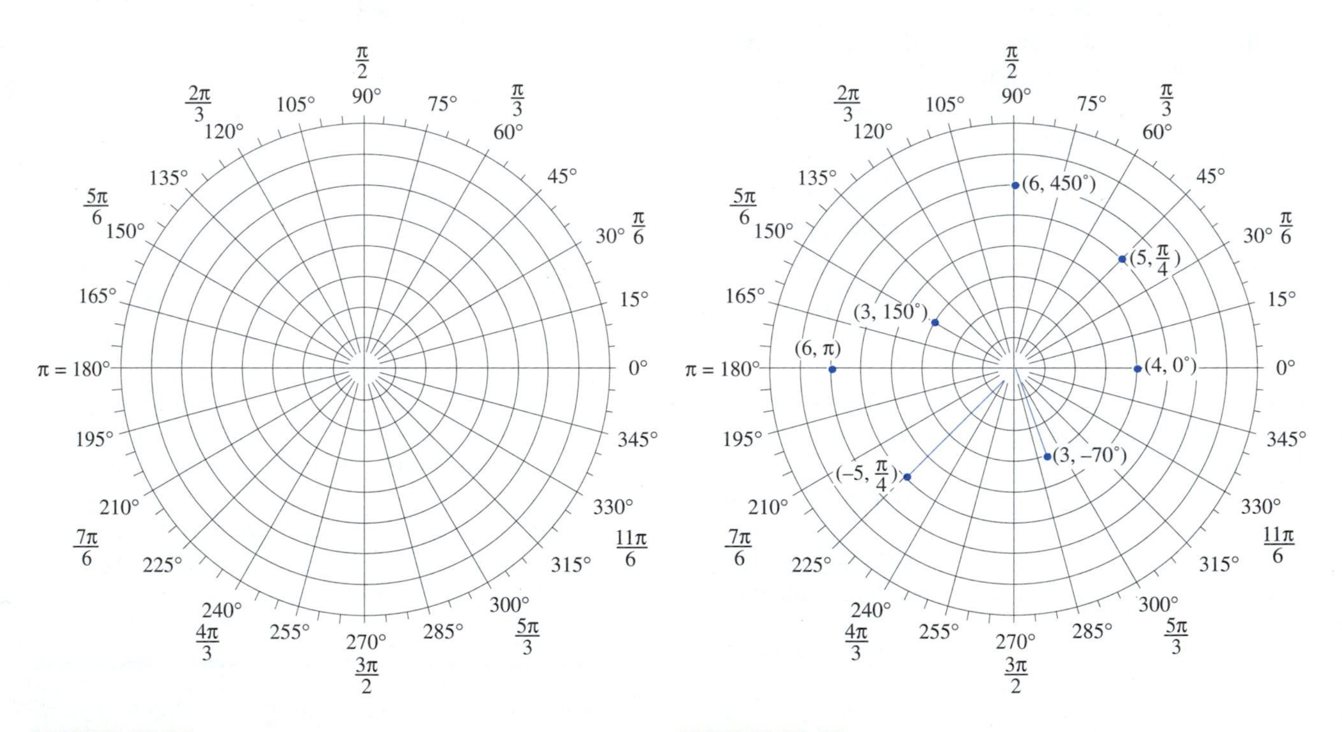

FIGURE 10.28

FIGURE 10.29

Notice that there is nothing unique about these points. For example, the point $(6, 450°)$ would be the same as the points $(6, 90°)$, $(6, -270°)$, or $(-6, 270°)$. The pole has the polar coordinate $(0, \theta)$, where θ can be any angle.

Converting Between Polar and Rectangular Coordinates

Converting between the polar coordinate system and the rectangular coordinate system requires the use of trigonometry.

Converting Polar Coordinates to Rectangular Coordinate

If the point P has polar coordinates (r, θ) and rectangular coordinates

$$x = r\cos\theta$$
$$y = r\sin\theta$$

This converts the polar coordinates to rectangular coordinates.

EXAMPLE 10.24

Find the rectangular coordinates of the point with the polar coordinates $\left(8, \frac{5\pi}{6}\right)$.

Solution From the equations previously given, we have $x = r\cos\theta$ and $y = r\sin\theta$. In this example, $r = 8$ and $\theta = \frac{5\pi}{6}$. So, $x = 8\left(\frac{-\sqrt{3}}{2}\right) = -4\sqrt{3} \approx -6.93$ and $y = 8\left(\frac{1}{2}\right) = 4$. The rectangular coordinates are $(-4\sqrt{3}, 4)$.

To convert from rectangular to polar coordinates, we need to use the Pythagorean theorem.

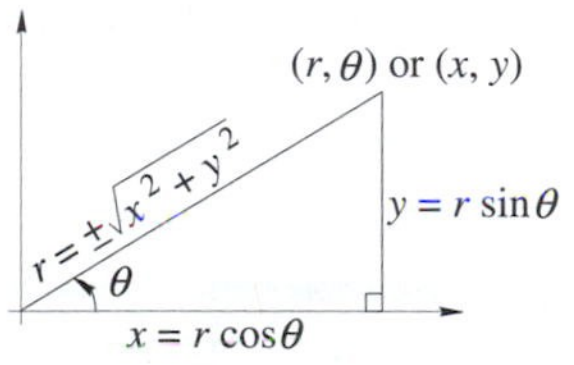

FIGURE 10.30

Converting Rectangular Coordinates to Polar Coordinates

The equations

$$r^2 = x^2 + y^2 \quad \text{or} \quad r = \pm\sqrt{x^2 + y^2}$$
$$\tan\theta = \frac{y}{x}, x \neq 0$$

will convert rectangular coordinates to polar coordinates. (See Figure 10.30.)

EXAMPLE 10.25

Find polar coordinates of $(-4\sqrt{3}, 4)$.

Solution This is the reverse of Example 10.24, so we know that the answer should be $\left(8, \frac{5\pi}{6}\right)$. Let's practice using the two conversion formulas $r^2 = x^2 + y^2$ and $\tan\theta = \frac{y}{x}$.

EXAMPLE 10.25 (Cont.)

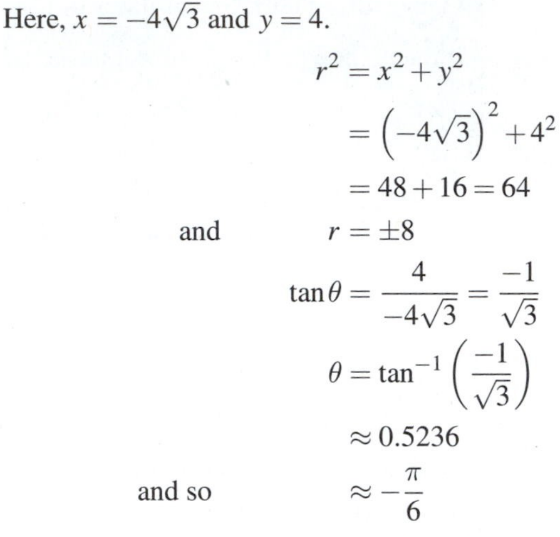

Here, $x = -4\sqrt{3}$ and $y = 4$.

$$\begin{aligned} r^2 &= x^2 + y^2 \\ &= \left(-4\sqrt{3}\right)^2 + 4^2 \\ &= 48 + 16 = 64 \\ \text{and} \qquad r &= \pm 8 \\ \tan\theta &= \frac{4}{-4\sqrt{3}} = \frac{-1}{\sqrt{3}} \\ \theta &= \tan^{-1}\left(\frac{-1}{\sqrt{3}}\right) \\ &\approx 0.5236 \\ \text{and so} \qquad &\approx -\frac{\pi}{6} \end{aligned}$$

It looks as if the answer is $\left(8, -\frac{\pi}{6}\right)$ or $\left(-8, -\frac{\pi}{6}\right)$. Now $\left(-8, -\frac{\pi}{6}\right)$ is correct, but $\left(8, -\frac{\pi}{6}\right)$ is not correct. What happened? Plot $\left(-4\sqrt{3}, 4\right)$. It is in Quadrant II. Now plot $= \left(8, \frac{-\pi}{6}\right)$. It is in Quadrant IV. Remember that $\tan^{-1}$ will only give angles in Quadrants I and IV. You should first determine which quadrant a point is in. If it is in Quadrants II or III, you will need to add π (or 180°) to the answer you get using the . conversion formula. $\Big($Because θ may be in Quadrant II or III, we write $\tan\theta = \frac{y}{x}$, rather than $\tan^{-1}\frac{y}{x} = \theta.\Big)$ If we add π to $\frac{-\pi}{6}$, we get $\frac{5\pi}{6}$. Thus, two possible answers are $\left(-8, -\frac{\pi}{6}\right)$ and $\left(8, \frac{5\pi}{6}\right)$. Both of these are polar coordinates of the point with Cartesian coordinates $(-4\sqrt{3}, 4)$.

Polar Equations

Equations can also be written using the variables r and θ. These are called **polar equations**. A polar equation states a relationship between all the points (r, θ) that satisfy the equation. In the remaining part of this section, we will graph some polar equations.

As we usually do when we graph an equation, we will use a table of values, plot the points in the table, then connect these points in order as the values of θ increase.

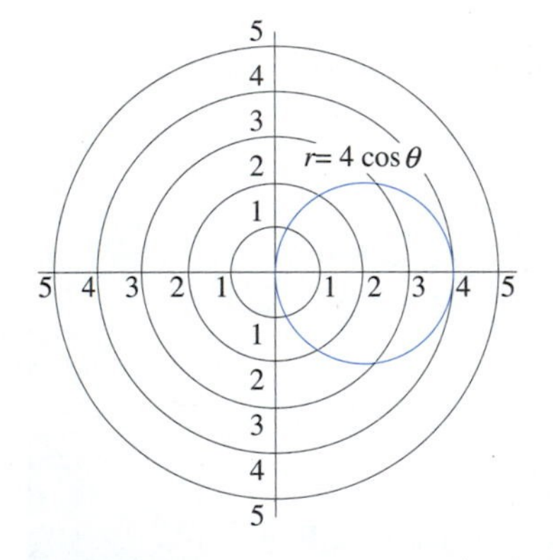

FIGURE 10.31

EXAMPLE 10.26

Graph the function $r = 4\cos\theta$.

Solution A table of values follows. The graph of the points is shown in Figure 10.31.

θ	0°	30°	60°	90°	120°	150°	180°	210°	240°	270°	300°	330°	360°
r	4	3.46	2	0	−2	−3.46	−4	−3.46	−2	0	2	3.46	4

Notice that this is a circle with radius 2 centered at $(2, 0°)$.

Circles

The graph of any equation of the type $r = a\cos\theta$ is a circle of radius $\left|\frac{a}{2}\right|$ and center $\left(\frac{a}{2}, 0°\right)$. An equation of the type $r = a\sin\theta$ is a circle with radius $\left|\frac{a}{2}\right|$ and center $\left(\frac{a}{2}, 90°\right)$.

EXAMPLE 10.27

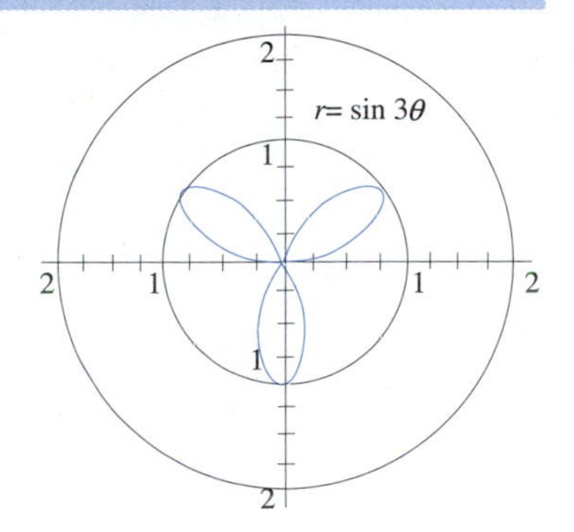

FIGURE 10.32

Graph the function $r = \sin 3\theta$.

Solution The table of values follows. The graph of these points is in Figure 10.32.

θ	0°	10°	20°	30°	40°	50°	60°	70°	80°	90°
$r = \sin 3\theta$	0	0.5	0.87	1	0.87	0.5	0	−0.5	−0.87	−1

θ	90°	100°	110°	120°	130°	140°	150°	160°	170°	180°
$r = \sin 3\theta$	−1	−0.87	−0.5	0	0.5	0.87	1	0.87	0.5	0

A curve of this type is called a **rose**. This rose has three petals.

Roses

Any polar equation of the form $r = \sin n\theta$ or $r = \cos n\theta$, where n is a positive integer, is a rose. If n is an odd number, the rose will have n petals. If n is an even number, the rose will have $2n$ petals.

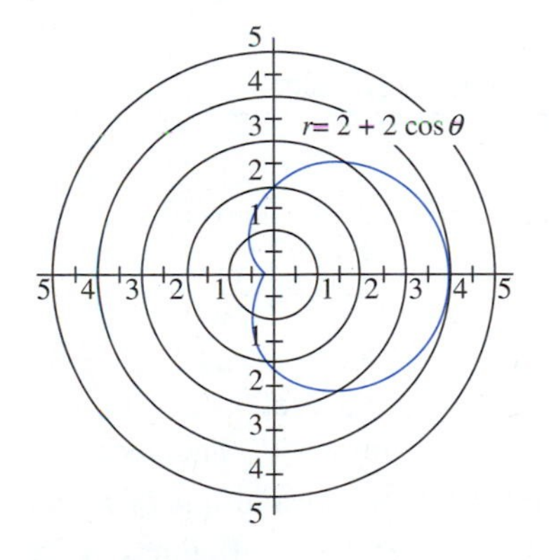

FIGURE 10.33

For many functions, polar equations are much simpler to work with than are rectangular equations (and vice versa). A good example of this will be seen in a later chapter.

Another type of polar graph is demonstrated by the graph of the polar equation $r = 2 + 2\cos\theta$. The graph of this equation is shown in Figure 10.33.

The curve in Figure 10.33 is called a **cardioid** because of its heart shape. In Exercise Set 10.5, we will graph other polar equations. If the curve has a special name, we will give that name.

Using a Calculator to Graph Polar Equations

You can use a graphing calculator or a computer to help you graph many of these curves. We will describe how it is done using a graphics calculator. The procedures for a TI-82 differ quite a bit from those for a Casio fx-7700G. We will use a TI-82 for the first example and a Casio fx-7700G for the second example.

EXAMPLE 10.28

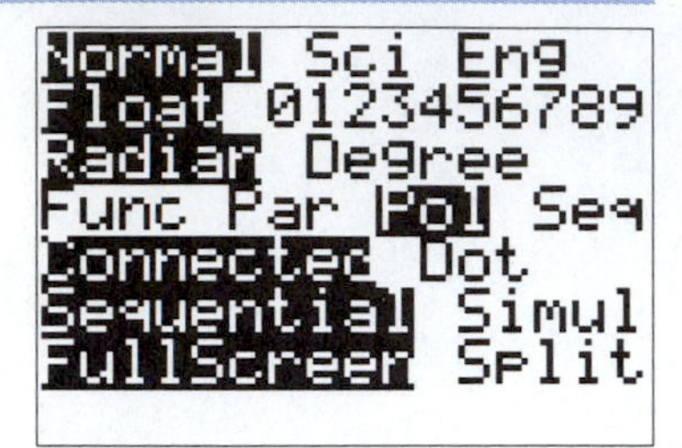

FIGURE 10.34a

Use a TI-82 graphing calculator to graph $r = \sin^2 3\theta$.

Solution First, put your calculator in polar mode. On a TI-82 this is done by pressing [MODE] [▼] [▼] [▼] [▶] [▶] [ENTER]. The screen on a TI-82 should look like the one shown in Figure 10.34a.

Press [WINDOW] to set the size of the viewing window. The function $r = \sin^2 3\theta$ has a period of 2π, so it will take no more than from $\theta = 0$ to $\theta = 2\pi$ to completely sketch this graph, so set $\theta\text{min} = 0$, $\theta\text{max} = 2\pi$, and $\theta\text{step} = \pi/24$. Since the function has an amplitude of 1, a window of $[-1.5, 1.5] \times [1, 1]$ should be right. You are now ready to enter the function into the calculator.

Now press [Y=]. On the first line, enter the right-hand side of the parametric equation for r. Press

[(] [sin] 3 [X,T,θ] [)] [x^2]

Notice that when you pressed the [X,T,θ] key a θ was displayed on the screen. This will always happen when the calculator is in polar mode. Remember, that $\sin^2\theta$ is entered into a calculator or computer as $(\sin\theta)^2$. The result in displayed in Figure 10.34b. Now press [GRAPH] and you should see the result in Figure 10.34c.

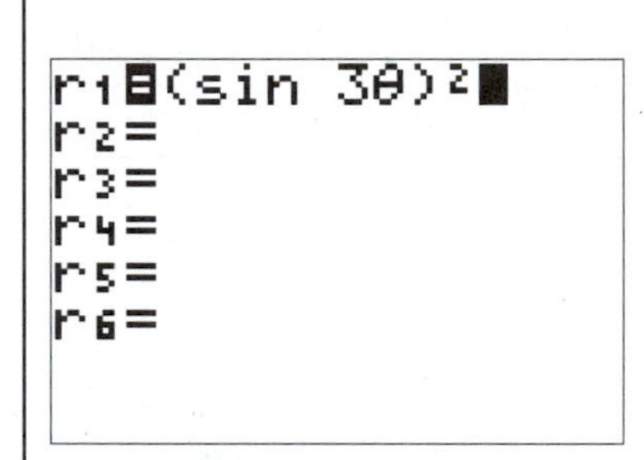

FIGURE 10.34b

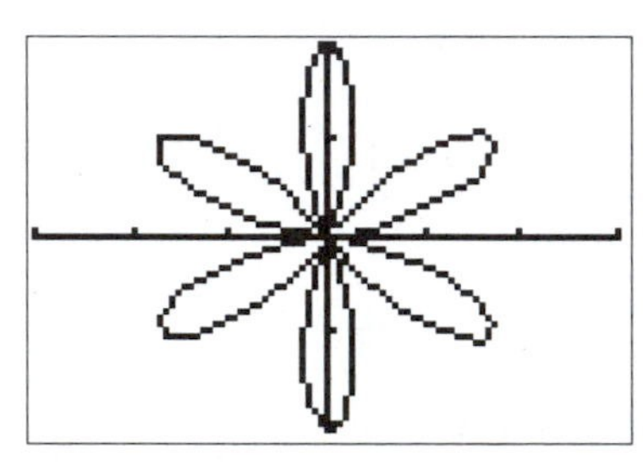

$[-1.5,1.5,0.5]\times[-1,1,0.5]$

FIGURE 10.34c

EXAMPLE 10.29

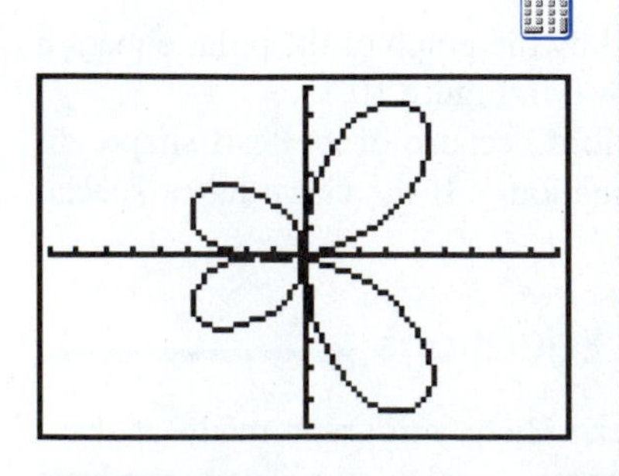

FIGURE 10.35

Use a Casio fx-7700G graphing calculator to graph $r = 5\sin 2\theta - 2\cos 3\theta$.

Solution First, put your calculator in polar mode. Press [MODE] [SHIFT] [−]. Next, set the range. Since the ranges of sine and cosine are both $[-1, 1]$, we know that $-7 \le r \le 7$. Because the screen is longer horizontally than it is vertically, we set Xmin -9, Xmax $= 9$, Xscl $= 1$, Ymin $= -6$, Ymax $= 6$, and Yscl $= 1$. After you press [EXE], you get a new screen, where you enter the values of θ. The period of $\sin 2\theta$ is π; the period of $\cos 3\theta$ is $\frac{2}{3}\pi$. Since the least common multiple of π and $\frac{2}{3}\pi$ is 2π, the period of r is a multiple of 2π. We will set $\theta\text{min} = 0$, $\theta\text{max} = 2\pi$ (press 2 [SHIFT] [π]), and $\theta\text{ptch} = 0.1$. Press [EXE] [Graph] and then the following:

5 [SIN] 2 [X,θ,T] [−] 3 [X,θ,T] [EXE]

The result is a butterfly-shaped curve as shown in Figure 10.35.

Exercise Set 10.5

In Exercises 1–12, plot the points with the given polar coordinates.

1. $\left(2, \frac{\pi}{4}\right)$
2. $(4, 60°)$
3. $(3, 90°)$
4. $(5, \pi)$
5. $\left(-4, \frac{2\pi}{3}\right)$
6. $(-6, 30°)$
7. $(-4, 0°)$
8. $\left(-2, \frac{5\pi}{6}\right)$
9. $(5, -30°)$
10. $\left(3, -\frac{3\pi}{4}\right)$
11. $\left(-7, \frac{11\pi}{6}\right)$
12. $(-2, -270°)$

In Exercises 13–24, convert each polar coordinate to its equivalent rectangular coordinates.

13. $\left(4, \frac{\pi}{3}\right)$
14. $(5, 75°)$
15. $(2, 135°)$
16. $\left(3, \frac{3\pi}{2}\right)$
17. $(-6, 20°)$
18. $\left(-3, \frac{5\pi}{3}\right)$
19. $(-2, 4.3)$
20. $(-5, 255°)$
21. $(3, -170°)$
22. $\left(4, -\frac{\pi}{8}\right)$
23. $(-6, -2.5)$
24. $(-3, -195°)$

In Exercises 25–36, convert each rectangular coordinate to an equivalent polar coordinate with $r > 0$.

25. $(4, 4)$
26. $(3, 6)$
27. $(4, 3)$
28. $(5, 12)$
29. $(-20, 21)$
30. $(-12, 5)$
31. $(-3, 4)$
32. $(9, -5)$
33. $(-7, -10)$
34. $(-8, -3)$
35. $(2, 9)$
36. $(-6, 1)$

In Exercises 37–58, graph the polar equations.

37. $r = 4$
38. $r = -6$
39. $r = 3\sin\theta$
40. $r = -5\cos\theta$
41. $r = 4 - 4\sin\theta$ (cardioid)
42. $r = 1 + 3\cos\theta$ (limacon)
43. $r = 3\cos 5\theta$ (five-petaled rose)
44. $r = 3\sin 2\theta$ (four-petaled rose)
45. $r = 5\sec\theta$
46. $r = -7\csc\theta$
47. $r = \theta$ (Let θ get larger than 4π.)
48. $r = 3^\theta$ (spiral)
49. $r = 4 + 4\sec\theta$
50. $r = \dfrac{1}{\theta}, \theta > 0$
51. $r = 3 + \cos\theta$
52. $r^2 = 16\sin 2\theta$
53. $r = 2 + 5\sin\theta$
54. $r = 1 + 4\sec\theta$
55. $r = \dfrac{6}{3 + 2\sin\theta}$
56. $r = \dfrac{6}{1 + 3\cos\theta}$
57. $r = \dfrac{3}{2 + 2\cos\theta}$
58. $r = \dfrac{4\sec\theta}{2\sec\theta - 1}$

Solve Exercises 59–61.

59. *Space technology* The polar equation for a certain satellite's orbit is given by

$$r = \frac{6000}{1.4} - 0.25\cos\theta$$

Sketch the graph of this satellite.

60. *Architecture* In the design of a geodesic dome, an architect uses the equation

$$r^2 = \frac{E^2}{E^2\cos^2\theta + \sin^2\theta}$$

where E is a constant.

(a) What is the rectangular form of this equation?

(b) Let $E = 0.5$ and graph this function on your calculator.

61. *Electronics* The field strength, r, in μV/m, of a broadcast station 1 mi from the antenna is given by

$$r = 2 + 5\cos 2\theta + \sin\theta$$

Use your calculator to sketch the graph of this antenna pattern.

In Section 10.5, we studied graphs in the polar coordinate system. Some of the special curves we were able to draw were the rose, the cardioid, and the limaçon. In this section, we will return to the polar coordinate system and see how we can graph the conic sections using polar coordinates.

As we progressed through this chapter, you saw that the equations of the conic sections have very simple forms if the center or vertex is at the origin. But, we had to learn a different form of each conic section. Polar coordinates will allow us to use one equation to represent each conic except the circle. Each of these conics will have a focus at the origin and one axis will be a coordinate axis.

When we defined the parabola, we said that it was the set of points that was an equal distance from the focus and the directrix. The ratio of these two distances, since the distances are the same, is 1. This ratio is called the **eccentricity**. In the problems for the ellipse and the hyperbola, we also used the eccentricity. Each conic can be defined in terms of a point on the curve and the ratio of its distance from a focus and a line called the directrix. This ratio is the eccentricity, e.

Do not confuse the e used to denote the eccentricity with the number $e \approx 2.71828182846$.

Each type of conic is determined by the eccentricity and leads to the general definition of a conic section that follows.

General Definition of a Conic Section

Let ℓ be a fixed line (the directrix) and F a fixed point (**focus**) not on ℓ. A **conic section** is the set of all points P in the plane such that

$$\frac{d(P, F)}{d(P, \ell)} = e$$

where $d(P, F)$ is the distance from P to F, and $d(P, \ell)$ is the perpendicular distance from P to ℓ. The constant e is called the **eccentricity** and if

$0 < e < 1$, the conic is an ellipse

$e = 1$, the conic is a parabola

$e > 1$, the conic is a hyperbola

Since $d(P, F)$ and $d(P, \ell)$ are distances, they are not negativ
negative.

Now, if $P(r, \theta)$ is a point on a conic with focus O, directr...
eccentricity e (see Figure 10.36), then

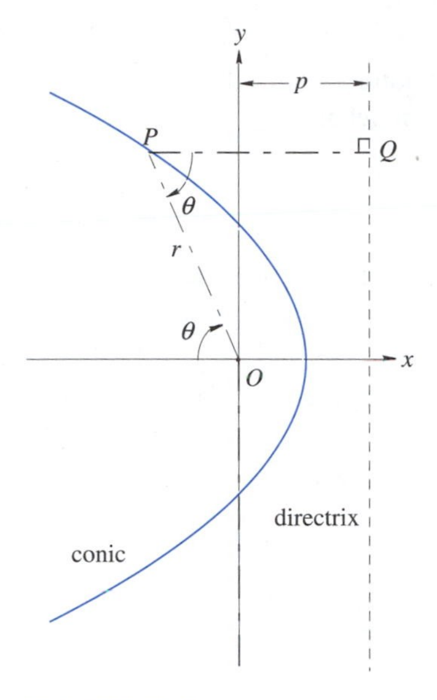

FIGURE 10.36

$$\frac{d(P, O)}{d(P, Q)} = e$$

where Q is a point on the directrix.

But, $d(P, O) = r$ and $d(P, Q) = p + r\cos\theta$, and so

$$\frac{r}{p + r\cos\theta} = e$$

Solving this for r, we obtain

$$r = \frac{pe}{1 - e\cos\theta}$$

If we had chosen the directrix as $x = -p$ $(p > 0)$, we would have obtained the equation

$$r = \frac{pe}{1 + e\cos\theta}$$

If the directrix is $y = \pm p(p > 0)$, the equations would be

$$r = \frac{pe}{1 \pm e\sin\theta}$$

Thus, we have two sets of equations, and by determining e, we can determine the type of curve. The previous results are summarized in the following box.

Polar Equations of Conic Sections

A polar equation that has one of the four forms

$$r = \frac{pe}{1 \mp e\cos\theta} \qquad r = \frac{pe}{1 \pm e\sin\theta}$$

is a conic section. The conic is a parabola if $e = 1$, an ellipse if $0 < e < 1$, or a hyperbola if $e > 1$.

EXAMPLE 10.30

Describe and sketch the graph of the equation

$$r = \frac{10}{5 + 3\cos\theta}$$

Solution Since the constant term in the denominator must be 1, we divide both numerator and denominator by 5. The equation then becomes

$$r = \frac{2}{1 + \frac{3}{5}\cos\theta}$$

From this, we see that $e = \frac{3}{5}$. Since $\frac{3}{5} < 1$, the conic is an ellipse. The denominator contains the cosine function and so the major axis of this conic section is horizontal. The vertices can be determined by setting θ equal to 0 and π. When $\theta = 0$, $r = \dfrac{2}{8/5} = \dfrac{10}{8} = 1.25$ and when $\theta = \pi$, $r = \dfrac{2}{2/5} = 5$. So $2a = 5 + 1.25 = 6.25$, or $a = 3.125$. The eccentricity $e = \dfrac{c}{a}$, so $\dfrac{3}{5} = \dfrac{c}{3.125}$ and we get $c = 1.875$. Finally, $b^2 = a^2 - c^2 = 6.25$, thus $b = 2.5$. The sketch of this ellipse is given in Figure 10.37.

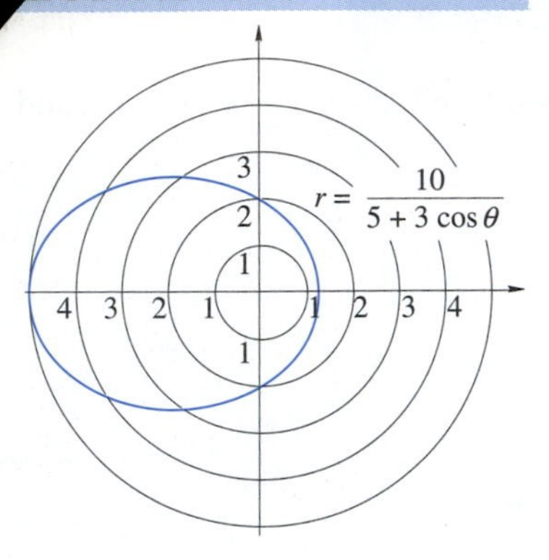

FIGURE 10.37

EXAMPLE 10.31

Describe and sketch the graph of the equation

$$r = \frac{8}{2 + 2\sin\theta}$$

Solution We divide the numerator and denominator by 2, obtaining

$$r = \frac{4}{1 + 1\sin\theta}$$

From this we see that $e = 1$, and the curve is a parabola. Also, since $pe = 4$, $p = 4$. This curve has a vertical axis, so the directrix is $y = 4$. If we plot the points that correspond to the x- and y-intercepts, we get the following table and the curve in Figure 10.38.

θ	0	$\frac{\pi}{2}$	π	$\frac{3\pi}{2}$
r	4	2	4	Not defined

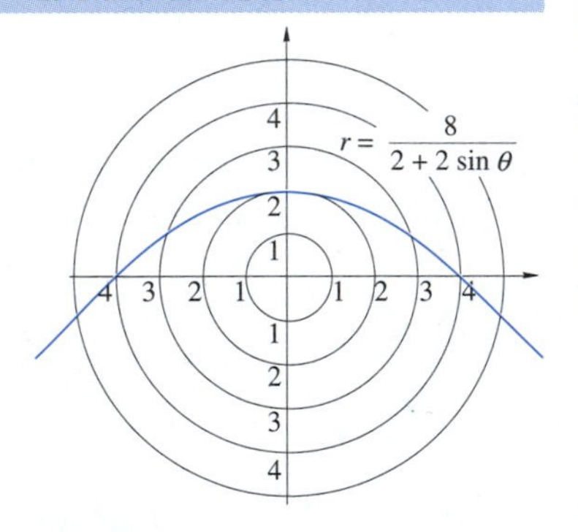

FIGURE 10.38

In Section 2.7, we learned a very complicated process for rotating conic sections on a rectangular coordinate system. The polar coordinate equation of a rotated conic

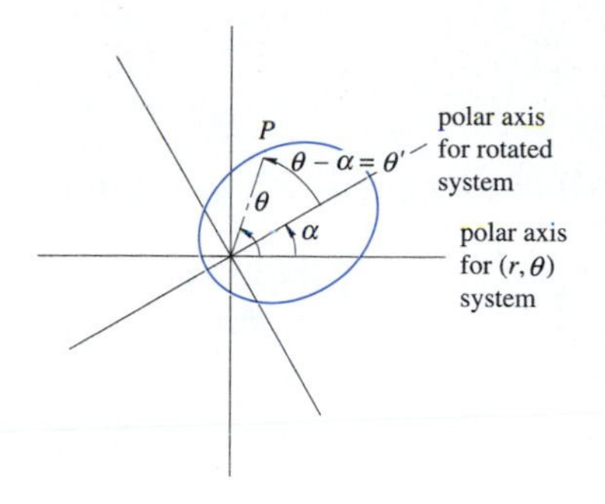

FIGURE 10.39

is quite simple. Consider the ellipse in Figure 10.39. It has been rotated through a positive angle α about its focus at the origin O. In the rotated system with polar coordinates (r', θ'), the ellipse has the equation

$$r' = \frac{pe}{1 - e\cos\theta'}$$

But, $r' = r$ and $\theta' = \theta - \alpha$, so the equation of this ellipse in the original (unrotated) polar coordinate system is

$$r = \frac{pe}{1 - e\cos(\theta - \alpha)}$$

Since, in this example, the conic is an ellipse, $0 < e < 1$. But if $e = 1$, the conic would be a parabola, and if $e > 1$, the conic would be a hyperbola.

EXAMPLE 10.32

Discuss and sketch the graph of the equation

$$r = \frac{1.30}{1 + 0.65\cos\left(\theta + \frac{\pi}{6}\right)}$$

Solution This equation has a denominator that begins with the number 1, so we can immediately see that $e = 0.65$. Since $e < 1$, the conic is an ellipse. The angle is $\theta + \frac{\pi}{6}$, so the major axis is rotated $\alpha = -\dfrac{\pi}{6}$. The denominator contains the cosine function, thus the major axis is the horizontal axis, which has been rotated $-\frac{\pi}{6}$ rad. Again, the vertices can be determined by setting $\theta = 0$ and $\theta = \pi$. This gives $V = 0.83$ and $V' = 2.97$, so $2a = 3.80$ and $a = 1.90$. The eccentricity $e = \dfrac{c}{a}$, so $c = ea = (0.65)(1.90) = 1.235$. Thus, $b^2 = a^2 - c^2 = 2.085$ or $b \approx 1.444$. The sketch of this ellipse is shown in Figure 10.40.

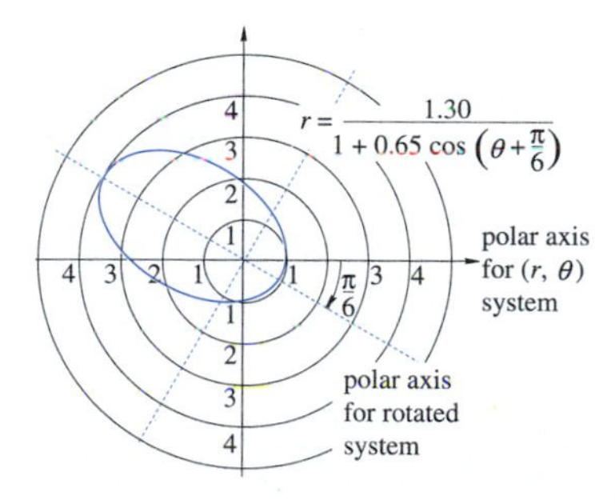

FIGURE 10.40

We will see some applications of this polar equation for a rotated conic section in the following exercise set.

Using a Graphing Calculator

In order to graph a conic section in polar coordinates, you must make sure that your calculator is in polar mode. (Not all graphing calculators have a polar mode, so check your user's manual.) Once your calculator is in polar mode, you graph a conic section by entering the function $r(\theta)$ into the calculator, setting an appropriate viewing window, and graphing the function.

EXAMPLE 10.33

Use a graphing calculator to graph $r = \dfrac{1.30}{1+0.65\cos\left(\theta+\dfrac{\pi}{6}\right)}$.

Solution This is the same function we graphed in Example 10.32, so the result should look much like the graph in Figure 10.40.

Make sure that the calculator is in both radian and polar modes. On a TI-82 press [Y=] and enter the right-hand side of the equation, as shown in Figure 10.41a. Using the window settings θmin $= 0$, θmax $= 2\pi$, θstep $= 0.1$, Xmin $= -4.7$, Xmax $= 4.7$, Xscl $= 1$, Ymin $= -3.1$, Ymax $= 3.1$, Yscl $= 1$, you should get the result in Figure 10.41b.

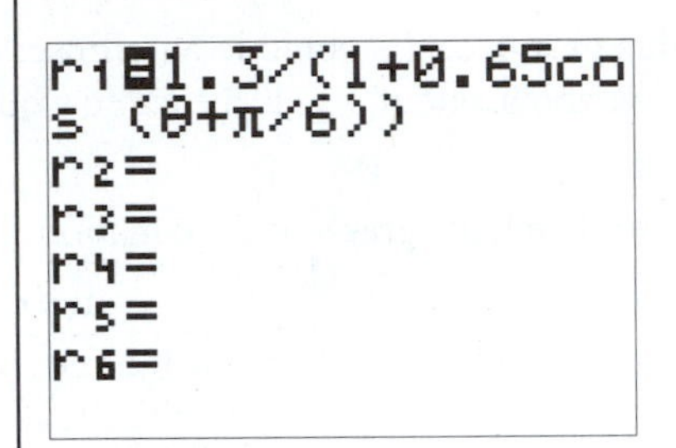

FIGURE 10.41a

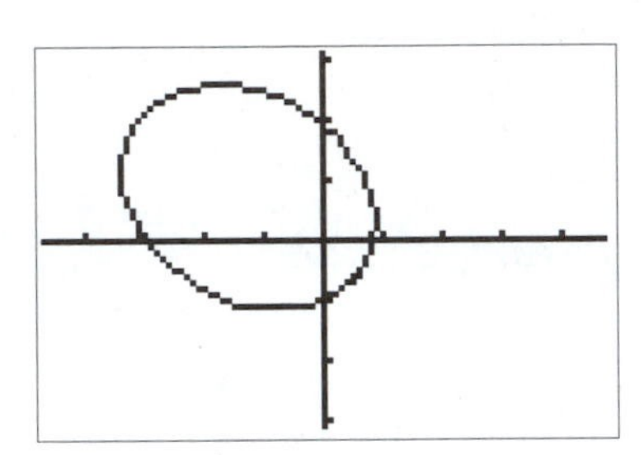

FIGURE 10.41b

Exercise Set 10.6

In Exercises 1–10, identify and sketch the conic section with the given equation.

1. $r = \dfrac{6}{1+3\cos\theta}$
2. $r = \dfrac{8}{1-2\sin\theta}$
3. $r = \dfrac{12}{3-\cos\theta}$
4. $r = \dfrac{15}{3+9\sin\theta}$
5. $r = \dfrac{12}{4-4\cos\theta}$
6. $r = \dfrac{8}{6+6\sin\theta}$
7. $r = \dfrac{12}{3+2\cos\theta}$
8. $r = \dfrac{12}{2-3\cos\theta}$
9. $r = \dfrac{12}{2-4\cos\left(\theta+\frac{\pi}{3}\right)}$
10. $r = \dfrac{6}{2+\sin\left(\theta+\frac{\pi}{6}\right)}$

In Exercises 11–16, find a polar equation of the conic with focus at the origin and the given eccentricity and directrix.

11. Directrix $x = 4$; $e = \frac{3}{2}$
12. Directrix $x = -2$; $e = \frac{3}{4}$
13. Directrix $y = -5$; $e = 1$
14. Directrix $y = 3$; $e = 2$
15. Directrix $x = 1$; $e = \frac{2}{3}$
16. Directrix $x = 5$; $e = 1$

Solve Exercises 17–20.

17. *Astronomy* The planet Mercury travels around the sun in an elliptical orbit given approximately by

$$r = \frac{3.442 \times 10^7}{1-0.206\cos\theta}$$

where r is measured in miles and the sun is at the pole. Determine Mercury's greatest and shortest distance from the sun.

18. *Mechanical engineering* An engine gear is tested for dynamic balance by rotating it and tracing a polar graph of a point on its rim. Draw the graph given by the equation $r = \dfrac{4}{1 - 0.05\sin\theta}$ and determine the state of balance of the gear.

19. *Mechanical engineering* A cam is shaped such that the equation of the upper half is given by $r = 2 + \cos\theta$ and the equation of the lower half is given by $r = \dfrac{3}{2 - \cos\theta}$. Sketch the shape of this cam.

20. *Astronomy* A certain comet is following a parabolic path, with the sun as the focus. When the comet is 50 000 000 km from the center of the sun, the line from the comet to the sun makes an angle of 45° with the axis of the parabola.
    (a) Write an equation that describes the path of this comet.
    (b) Sketch the path of this comet.
    (c) How close will the comet come to the center of the sun, that is, how far is the vertex of this parabola from the sun?

In Your Words

21. Describe how to tell if a polar equation of a conic section is a parabola, ellipse, or hyperbola.

22. Explain the differences between the e used to denote the eccentricity and the number $e \approx 2.71828182846$.

10.7 DIFFERENTIATION IN POLAR COORDINATES

Parametric equations use the rectangular or xy-coordinate system. Polar coordinates do not. They use the polar coordinate system, where every point in the plane can be represented as a pair of numbers (r, θ) with $r \geq 0$ and $0 \leq \theta \leq 2\pi$. The number represented by r is the distance of the point from the origin (or pole) and the number represented by θ is the direction.

The polar and rectangular coordinate systems are related by the equations

$$x = r\cos\theta \text{ and } y = r\sin\theta$$

Similarly, we have

$$r = \sqrt{x^2 + y^2} \text{ and } \tan\theta = \frac{y}{x}, \text{ if } x \neq 0$$

If we are given a polar equation, $r = f(\theta)$, it is a simple matter to find the derivative $\dfrac{dr}{d\theta}$. The problem is that this does not represent the same thing as the derivative $\dfrac{dy}{dx}$, where $y = g(x)$. As you remember, $\dfrac{dy}{dx}$ represents the slope of the tangent to the curve $y = g(x)$. The expression $\dfrac{dr}{d\theta}$ represents the change in the radius of the curve with respect to angle θ.

Let us return to our original equation, $r = f(\theta)$. We want to be able to use the derivative as a measure of the slope of the graph of this equation. This means we need to express $r = f(\theta)$ in terms of x and y or $x = r\cos\theta$ and $y = r\sin\theta$. Since $r = f(\theta)$, we get the equations

$$x = f(\theta)\cos\theta \text{ and } y = f(\theta)\sin\theta$$

These are parametric equations in terms of θ rather than t. Thus, using the parametric differentiation from Section 10.2,

$$\begin{aligned}\frac{dy}{dx} &= \frac{dy/d\theta}{dx/d\theta} \\ &= \frac{f'(\theta)\sin\theta + f(\theta)\cos\theta}{f'(\theta)\cos\theta - f(\theta)\sin\theta} \\ &= \frac{r'\sin\theta + r\cos\theta}{r'\cos\theta - r\sin\theta}\end{aligned}$$

This provides the following definition.

Derivative of an Equation in Polar Form

If $r = f(\theta)$ is a differentiable function, then the slope of the tangent line to the graph of r at the point (r, θ) is

$$\frac{dy}{dx} = \frac{r'\sin\theta + r\cos\theta}{r'\cos\theta - r\sin\theta}$$

provided $\dfrac{dx}{d\theta} \neq 0$ at (r, θ).

EXAMPLE 10.34

Find the derivative $\dfrac{dy}{dx}$ of $r = 1 + 2\cos\theta$.

Solution First we determine that $r' = \dfrac{dr}{d\theta} = -2\sin\theta$. Then

$$\begin{aligned}\frac{dy}{dx} &= \frac{r'\sin\theta + r\cos\theta}{r'\cos\theta - r\sin\theta} \\ &= \frac{(-2\sin\theta)(\sin\theta) + (1+2\cos\theta)\cos\theta}{(-2\sin\theta)(\cos\theta) - (1+2\cos\theta)\sin\theta} \\ &= \frac{-2\sin^2\theta + \cos\theta + 2\cos^2\theta}{-4\sin\theta\cos\theta - \sin\theta} \\ &= \frac{2(\cos^2\theta - \sin^2\theta) + \cos\theta}{-4\sin\theta\cos\theta - \sin\theta} \\ &= \frac{2\cos 2\theta + \cos\theta}{-2\sin 2\theta - \sin\theta}\end{aligned}$$

That is a rather messy formula. Fortunately, we will usually want to know the value of the derivative at a specific point, as shown in the next example.

EXAMPLE 10.35

Find the slope of the curve $r = 1 + \sin\theta$ at $\theta = \frac{2\pi}{3}$.

Solution $r' = \cos\theta$ and so

$$\begin{aligned}\frac{dy}{dx} &= \frac{r'\sin\theta + r\cos\theta}{r'\cos\theta - r\sin\theta}\\ &= \frac{\cos\theta\sin\theta + (1+\sin\theta)\cos\theta}{\cos\theta\cos\theta - (1+\sin\theta)\sin\theta}\\ &= \frac{\cos\theta(1+2\sin\theta)}{\cos^2\theta - \sin\theta - \sin^2\theta}\end{aligned}$$

When $\theta = \frac{2\pi}{3}$, $\sin\theta = \dfrac{\sqrt{3}}{2}$ and $\cos\theta = \frac{-1}{2}$, thus

$$\begin{aligned}\frac{dy}{dx} &= \frac{\dfrac{-1}{2}\left[1+2\left(\dfrac{\sqrt{3}}{2}\right)\right]}{\left(\frac{-1}{2}\right)^2 - \dfrac{\sqrt{3}}{2} - \left(\dfrac{\sqrt{3}}{2}\right)^2}\\ &= \frac{\dfrac{-1}{2}(1+\sqrt{3})}{\frac{1}{4} - \dfrac{\sqrt{3}}{2} - \frac{3}{4}} = \frac{\dfrac{-1}{2}(1+\sqrt{3})}{\dfrac{-\sqrt{3}}{2} - \frac{1}{2}} = \frac{1+\sqrt{3}}{1+\sqrt{3}} = 1\end{aligned}$$

The slope of $r = 1 + \sin\theta$ at $\theta = \frac{2\pi}{3}$ is 1.

EXAMPLE 10.36

Find an equation of the normal and tangent lines to $r = 2\sin 2\theta$ at $\theta = \frac{\pi}{4}$.

Solution $r' = 4\cos 2\theta$ and so

$$\begin{aligned}\frac{dy}{dx} &= \frac{r'\sin\theta + r\cos\theta}{r'\cos\theta - r\sin\theta}\\ &= \frac{4\cos 2\theta\sin\theta + 2\sin 2\theta\cos\theta}{4\cos 2\theta\cos\theta - 2\sin 2\theta\sin\theta}\\ &= \frac{2\cos 2\theta\sin\theta + \sin 2\theta\cos\theta}{2\cos 2\theta\cos\theta - \sin 2\theta\sin\theta}\end{aligned}$$

When $\theta = \frac{\pi}{4}$, we have

$$\begin{aligned}\frac{dy}{dx} &= \frac{2\left(\cos\dfrac{\pi}{2}\right)\left(\sin\dfrac{\pi}{4}\right) + \left(\sin\dfrac{\pi}{2}\right)\left(\cos\dfrac{\pi}{4}\right)}{2\left(\cos\dfrac{\pi}{2}\right)\left(\cos\dfrac{\pi}{4}\right) - \left(\sin\dfrac{\pi}{2}\right)\left(\sin\dfrac{\pi}{4}\right)}\\ &= \frac{0 + \dfrac{\sqrt{2}}{2}}{0 - \dfrac{\sqrt{2}}{2}} = -1\end{aligned}$$

So, the slope of the tangent line is -1.

EXAMPLE 10.36 (Cont.)

Next, we determine the x- and y-coordinates when $\theta = \frac{\pi}{4}$. When $\theta = \frac{\pi}{4}$, we have $r = 2\sin 2\theta = 2\sin\frac{\pi}{2} = 2$. So, the point $\left(2, \frac{\pi}{4}\right)$ is equivalent to the point (x, y), where $x = r\cos\theta = 2\cos\frac{\pi}{4} = \sqrt{2}$ and $y = r\sin\theta = \sqrt{2}$. Thus, the tangent line passing through the point with x- and y-coordinates $(\sqrt{2}, \sqrt{2})$ and with a slope of -1, has the equation $y - \sqrt{2} = -(x - \sqrt{2})$ or $y = -x + 2\sqrt{2}$.

The normal line has a slope of 1 and has the equation

$$y - \sqrt{2} = x - \sqrt{2}$$

or

$$y = x$$

EXAMPLE 10.37

Find the extrema of $r = 1 + \sin\theta$.

Solution In Example 10.35 we found that

$$\frac{dy}{dx} = \frac{\cos\theta(1 + 2\sin\theta)}{\cos^2\theta - \sin\theta - \sin^2\theta}$$

The numerator is 0 if $\cos\theta = 0$ or if $\sin\theta = \frac{-1}{2}$. Thus, $\theta = \frac{\pi}{2}, \frac{3\pi}{2}, \frac{7\pi}{6}$, and $\frac{11\pi}{6}$ are critical values.

The denominator is 0 if

$$\cos^2\theta - \sin\theta - \sin^2\theta = 0$$

Since $\cos^2\theta = 1 - \sin^2\theta$, we can write this as

$$1 - \sin^2\theta - \sin\theta - \sin^2\theta = 0$$

or

$$2\sin^2\theta + \sin\theta - 1 = 0$$

$$(2\sin\theta - 1)(\sin\theta + 1) = 0$$

This is a true equation when $\sin\theta = \frac{1}{2}$ or $\sin\theta = -1$, so $\frac{\pi}{6}, \frac{5\pi}{6}$, and $\frac{3\pi}{2}$ are critical values.

Checking these by the first derivative test, we see that $(r, \theta) = \left(2, \frac{\pi}{2}\right)$ and $\left(0, \frac{3\pi}{2}\right)$ are maxima and that $(r, \theta) = \left(\frac{1}{2}, \frac{7\pi}{6}\right)$ and $\left(\frac{1}{2}, \frac{11\pi}{6}\right)$ are minima.

As you can see from the graph of this curve in Figure 10.42, $\frac{5\pi}{6}$ and $\frac{\pi}{6}$ are not where r (or y) is a maximum or minimum. The first derivative test does indicate that they are extreme points and indeed the maximum value of x occurs when $\theta = \frac{\pi}{6}$ and the minimum x-value occurs when $\theta = \frac{5\pi}{6}$. The tangent lines are vertical at the points $\left(\frac{3}{2}, \frac{\pi}{6}\right)$ and $\left(\frac{3}{2}, \frac{5\pi}{6}\right)$.

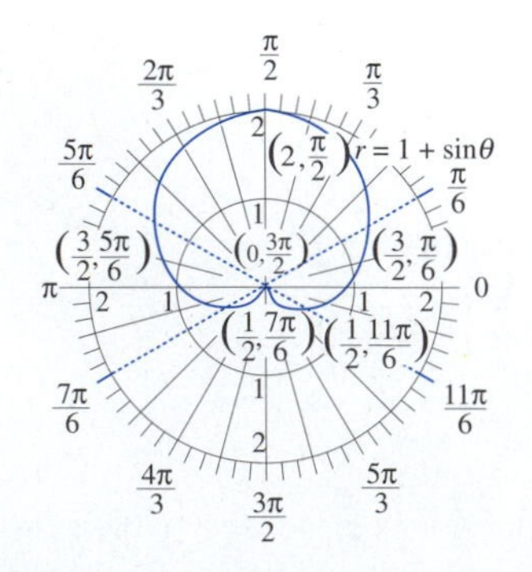

FIGURE 10.42

Exercise Set 10.7

In Exercises 1–8, find $\frac{dy}{dx}$.

1. $r = 3\sin\theta$
2. $r = -2\cos\theta$
3. $r = 1+\cos\theta$
4. $r = \cos 3\theta$
5. $r = 1+\cos 3\theta$
6. $r = \csc\theta$
7. $r = \tan\theta$
8. $r = e^{\theta}$

In Exercises 9–16, find the equations (in rectangular coordinates) of the tangent and normal lines to the given curve at the given point. The coordinates of the given points are in polar coordinates, (r, θ).

9. $r = \sin\theta$, $\left(1, \frac{\pi}{2}\right)$
10. $r = 2\cos\theta$, $\left(\sqrt{2}, \frac{\pi}{4}\right)$
11. $r = 5\sin 3\theta$, $\left(\frac{5}{\sqrt{2}}, \frac{\pi}{12}\right)$
12. $r = 2-3\sin\theta$, $\left(\frac{1}{2}, \frac{5\pi}{6}\right)$
13. $r = 6\sin^2\theta$, $\left(4.5, \frac{2\pi}{3}\right)$
14. $r = 2+3\sec\theta$, $\left(-4, \frac{2\pi}{3}\right)$
15. $r = e^{\theta}$, $\left(2.8497, \frac{\pi}{3}\right)$
16. $r = \tan\theta$, $\left(-1, \frac{3\pi}{4}\right)$

In Exercises 17–20, find all extrema.

17. $r = 3\cos 2\theta$
18. $r = 1-\cos\theta$
19. $r = 1+2\cos\theta$
20. $r = \sin^2\theta$

Solve Exercises 21 and 22.

21. *Robotics* A robot arm joint moves in the elliptical path described by the equation $r = \frac{10}{5+3\cos\theta}$, where r is in meters and $0 \le \theta < 2\pi$.
   (a) Find $\frac{dr}{d\theta}$.
   (b) Determine the value of θ where r is a maximum.
22. *Robotics* Consider the robot arm joint described in Exercise 21.
   (a) Find expressions for the x- and y-coordinates of r in terms of θ.
   (b) Determine $\frac{dy}{dx}$.
   (c) Determine where the graph has horizontal tangents and where it has vertical tangents.

In Your Words

23. What is a physical interpretation of the expression $\frac{dr}{d\theta}$?
24. What is a physical interpretation of the expression $\frac{dy}{dx}$ when $x = r\cos\theta$ and $y = r\sin\theta$ and $r = f(\theta)$ is a polar equation?

10.8 ARC LENGTH AND SURFACE AREA REVISITED

In Section 7.4, we studied the arc length of a curve and how the arc length could be used to find the surface area of a solid generated around a line. The curves we studied before were all expressed in rectangular coordinates. In this section, we will learn how to determine and use the arc length of a curve that is in polar form or given as parametric equations.

Arc Length of Parametric Equations

If we have a curve described by the function $y = f(x)$, then the length L of the curve from $x = a$ to $x = b$ is given by the equation

$$L = \int_a^b \sqrt{1 + [f'(x)]^2}\, dx = \int_a^b \sqrt{1 + \left(\frac{dy}{dx}\right)^2}\, dx$$

If we differentiate this equation with respect to x, we get

$$\frac{dL}{dx} = \sqrt{1 + \left(\frac{dy}{dx}\right)^2}$$

Squaring both sides results in

$$\left(\frac{dL}{dx}\right)^2 = 1 + \left(\frac{dy}{dx}\right)^2$$

or

$$(dL)^2 = (dx)^2 + (dy)^2$$

Now, suppose that x and y are both functions of a third variable t. Then they are parametric equations of the form $x = f(t)$ and $y = g(t)$. Differentiating the previous equation with respect to t we obtain

$$\left(\frac{dL}{dt}\right)^2 = \left(\frac{dx}{dt}\right)^2 + \left(\frac{dy}{dt}\right)^2$$

or

$$\frac{dL}{dt} = \sqrt{\left(\frac{dx}{dt}\right)^2 + \left(\frac{dy}{dt}\right)^2}$$

Integrating from $t = t_1$ to $t = t_2$, we get the formula for the arc length of a curve represented by parametric equations.

Arc Length of a Curve Represented by Parametric Equations

$$L = \int_{t_1}^{t_2} \sqrt{\left(\frac{dx}{dt}\right)^2 + \left(\frac{dy}{dt}\right)^2}\, dt$$

EXAMPLE 10.38

Find the length of the curve $x = 2t^2 + 5$, $y = 3t^3 - 1$ from $t = 0$ to $t = 2$.

Solution To use the new formula, we need to first differentiate x and y. $\frac{dx}{dt} = 4t$ and $\frac{dy}{dt} = 9t^2$. Thus, we have

$$\begin{aligned} L &= \int_0^2 \sqrt{(4t)^2 + (9t^2)^2}\,dt \\ &= \int_0^2 \sqrt{16t^2 + 81t^4}\,dt \\ &= \int_0^2 t\sqrt{16 + 81t^2}\,dt \end{aligned}$$

If we let $u = 16 + 81t^2$, then $du = 162t\,dt$ and we have

$$\begin{aligned} \frac{1}{162}\int \sqrt{u}\,du &= \frac{1}{243}u^{3/2} = \frac{1}{243}\left(16 + 81t^2\right)^{3/2}\Big|_0^2 \\ &= \frac{1}{243}\left(340\sqrt{340} - 64\right) \approx 25.536 \end{aligned}$$

The arc length of this curve is about 25.536 units.

EXAMPLE 10.39

Find the length of $x = e^t \sin t$, $y = e^t \cos t$ from $t = 0$ to $t = \pi$.

Solution Again, we begin by finding

$$\frac{dx}{dt} = e^t \cos t + e^t \sin t = e^t(\cos t + \sin t)$$

and

$$\frac{dy}{dt} = -e^t \sin t + e^t \cos t = e^t(\cos t - \sin t)$$

Thus

$$\begin{aligned} L &= \int_0^\pi \sqrt{[e^t(\cos t + \sin t)]^2 + [e^t(\cos t - \sin t)]^2}\,dt \\ &= \int_0^\pi \sqrt{e^{2t}(\cos^2 t + 2\cos t \sin t + \sin^2 t) + e^{2t}(\cos^2 t - 2\cos t \sin t + \sin^2)}\,dt \\ &= \int_0^\pi \sqrt{e^{2t}(1 + 2\cos t \sin t) + e^{2t}(1 - 2\cos t \sin t)}\,dt \\ &= \int_0^\pi \sqrt{2e^{2t}}\,dt = \sqrt{2}\int_0^\pi e^t\,dt = \sqrt{2}e^t\Big|_0^\pi = \sqrt{2}(e^\pi - 1) \approx 31.312 \end{aligned}$$

So, the arc length of this curve is approximately 31.312 units.

Arc Length of a Polar Equation

Suppose we have a curve given by the polar equation $r = f(\theta)$, where f is continuous on the interval $a \le \theta \le b$. We know that we can express this in rectangular coordinates using

$$\begin{aligned} x &= r\cos\theta \\ &= f(\theta)\cos\theta \end{aligned} \qquad \text{and} \qquad \begin{aligned} y &= r\sin\theta \\ &= f(\theta)\sin\theta \end{aligned}$$

We can find the arc length of this curve using the form for parametric equations. We have $\dfrac{dx}{d\theta} = f'(\theta)\cos\theta - f(\theta)\sin\theta$ and $\dfrac{dy}{d\theta} = f'(\theta)\sin\theta + f(\theta)\cos\theta$. Squaring these and adding produces the following. (Notice that we reversed the order in which the terms were written.)

$$\left(\frac{dx}{d\theta}\right)^2 = [f(\theta)]^2\sin^2\theta - 2f'(\theta)f(\theta)\sin\theta\cos\theta + [f'(\theta)]^2\cos^2\theta$$
$$\left(\frac{dy}{d\theta}\right)^2 = [f(\theta)]^2\cos^2\theta + 2f'(\theta)f(\theta)\sin\theta\cos\theta + [f'(\theta)]^2\sin^2\theta$$

Adding, we get

$$\begin{aligned} \left(\frac{dx}{d\theta}\right)^2 + \left(\frac{dy}{d\theta}\right)^2 &= [f(\theta)]^2(\sin^2\theta + \cos^2\theta) + [f'(\theta)]^2(\cos^2\theta + \sin^2\theta) \\ &= [f(\theta)]^2 + [f'(\theta)]^2 \end{aligned}$$

Thus, the length L of a curve in polar form $r = f(\theta)$ from $\theta = a$ to $\theta = b$ is given by the following formula.

Arc Length of a Curve in Polar Form: $r = f(\theta)$

If $r = f(\theta)$ has a continuous first derivative for $a \le \theta \le b$ and if $r = f(\theta)$ is traced exactly once for $a \le \theta \le b$, then the length of the curve is

$$\begin{aligned} L &= \int_a^b \sqrt{[f(\theta)]^2 + [f'(\theta)]^2}\, d\theta \\ &= \int_a^b \sqrt{r^2 + \left(\frac{dr}{d\theta}\right)^2}\, d\theta \end{aligned}$$

EXAMPLE 10.40

Find the arc length of $r = 4\cos\theta$ from $\theta = 0$ to $\theta = \pi$.

Solution A sketch of this curve is shown in Figure 10.43. We need $r' = -4\sin\theta$, then

$$\begin{aligned} L &= \int_0^\pi \sqrt{(4\cos\theta)^2 + (-4\sin\theta)^2}\, d\theta \\ &= \int_0^\pi \sqrt{16\cos^2\theta + 16\sin^2\theta}\, d\theta \end{aligned}$$

EXAMPLE 10.40 (Cont.)

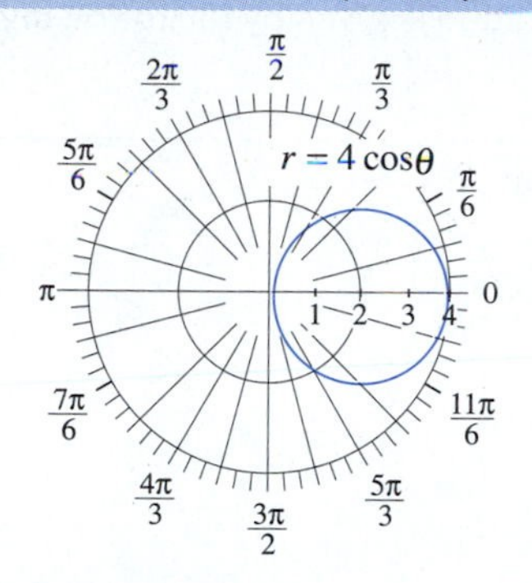

FIGURE 10.43

$$= \int_0^{\pi} \sqrt{16(\cos^2\theta + \sin^2\theta)}\, d\theta = \int_0^{\pi} \sqrt{16} d\theta$$

$$= \int_0^{\pi} 4\, d\theta = 4\theta \Big|_0^{\pi} = 4\pi$$

The arc length of $r = 4\cos\theta$ from $\theta = 0$ to $\theta = \pi$ is 4π units. As you can see from the graph in Figure 10.43, this curve is a circle of radius 2. Thus, the arc length is its circumference, $C = 2\pi r = 2\pi(2) = 4\pi$ units.

EXAMPLE 10.41

Find the arc length of $r = 1 + \sin\theta$ from $\theta = 0$ to $\theta = 2\pi$.

Solution The sketch of this curve is in Figure 10.44.

Here $r' = \cos\theta$, and so the arc length is

$$L = \int_0^{2\pi} \sqrt{(1+\sin\theta)^2 + (\cos\theta)^2}\, d\theta$$

$$= \int_0^{2\pi} \sqrt{1 + 2\sin\theta + \sin^2\theta + \cos^2\theta}\, d\theta$$

$$= \int_0^{2\pi} \sqrt{2 + 2\sin\theta}\, d\theta$$

Using the trigonometric identity from Section 20.3, $1 + \cos\theta = 2\cos^2\left(\frac{\theta}{2}\right)$, this becomes

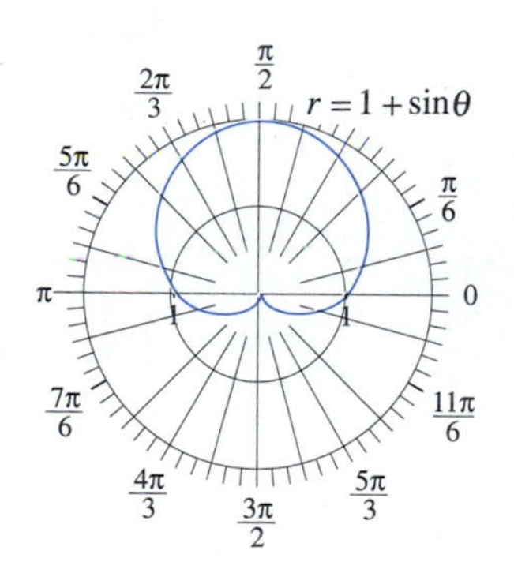

FIGURE 10.44

$$L = \int_0^{2\pi} \sqrt{4\cos^2\left(\frac{\theta}{2}\right)}\, d\theta$$

$$= \int_0^{2\pi} 2\left|\cos\left(\frac{\theta}{2}\right)\right| d\theta$$

Now, for $0 \le \theta \le \pi$, we know that $\cos\left(\frac{\theta}{2}\right) \ge 0$ and for $\pi \le \theta \le 2\pi$, we know $\cos\left(\frac{\theta}{2}\right) \le 0$ which means that for $\pi \le \theta \le 2\pi$, we have $\left|\cos\left(\frac{\theta}{2}\right)\right| = -\cos\left(\frac{\theta}{2}\right)$. We are now ready to evaluate the integral.

$$L = \int_0^{\pi} 2\cos\left(\frac{\theta}{2}\right) d\theta - \int_{\pi}^{2\pi} 2\cos\left(\frac{\theta}{2}\right) d\theta$$

$$= 4\sin\left(\frac{\theta}{2}\right)\Big]_0^{\pi} - 4\sin\left(\frac{\theta}{2}\right)\Big]_{\pi}^{2\pi}$$

$$= 4[1-0] - 4[0-1] = 8$$

The arc length of this cardioid is 8 units.

Surface Area

As we saw in Section 7.4, the area of a surface of revolution is given by the following formulas.

Surface Area of a Curve Revolved around an Axis

If a curve, written in rectangular form $y = f(x)$ from $x = a$ to $x = b$, is revolved around the x-axis, the area of its surface of revolution is

$$S = 2\pi \int_a^b y\sqrt{1+[f'(x)]^2}\,dx$$

If it is revolved around the y-axis, then

$$S = 2\pi \int_a^b x\sqrt{1+[f'(x)]^2}\,dx$$

The corresponding formulas for parametric and polar equations are given as follows.

Surface Area of a Curve Revolved around an Axis (Parametric Form)

If the curve, written in parametric form, $x = f(t)$, $y = g(t)$, $a \le t \le b$, is revolved around the x-axis, the surface area is

$$S = 2\pi \int_a^b y\sqrt{\left(\frac{dx}{dt}\right)^2 + \left(\frac{dy}{dt}\right)^2}\,dt$$

where $y \ge 0$. If it is revolved around the y-axis, then

$$S = 2\pi \int_a^b x\sqrt{\left(\frac{dx}{dt}\right)^2 + \left(\frac{dy}{dt}\right)^2}\,dt$$

where $x \ge 0$.

EXAMPLE 10.42

Find the surface area of the curve with the parametric equations $x = 2\cos^3 t$, $y = 2\sin^3 t$ from $0 \le t \le 2\pi$ when it is revolved around the x-axis.

Solution The graph of $x = 2\cos^3 t$, $y = 2\sin^3 t$ for $0 \le t \le 2\pi$ is shown in Figure 10.45a. Since we are required to have $y \ge 0$, we will restrict this to $0 \le t \le \pi$. We will also take advantage of the symmetry of the figure and find the surface area when the portion in the first quadrant is revolved around the x-axis and then double that answer. Restricting t to the interval $\left[0, \frac{\pi}{2}\right]$ produces the desired graph, as shown in Figure 10.45b.

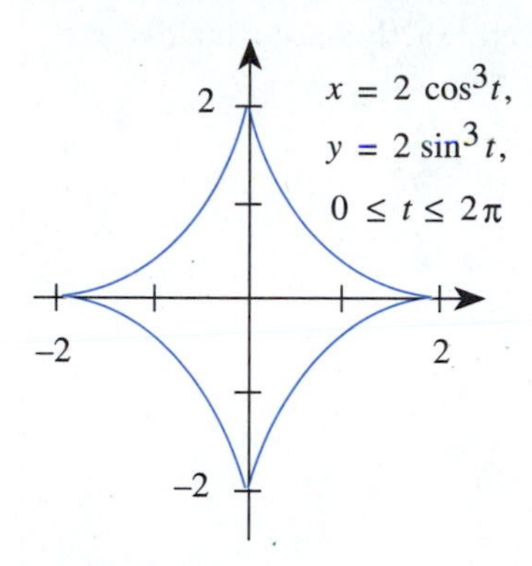

FIGURE 10.45a

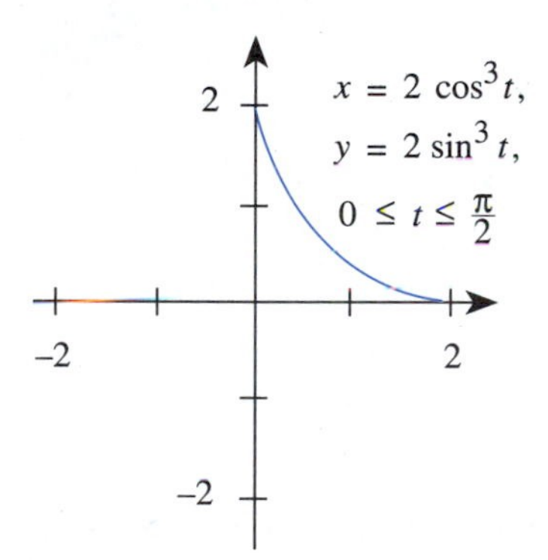

FIGURE 10.45b

When the curve in Figure 10.45b is revolved around the x-axis, the surface area is

$$
\begin{aligned}
S &= 2\pi \int_0^{\pi/2} y\sqrt{\left(\frac{dx}{dt}\right)^2 + \left(\frac{dy}{dt}\right)^2}\, dt \\
&= 2\pi \int_0^{\pi/2} \left(2\sin^3 t\right) \sqrt{\left(-6\cos^2 t \sin t\right)^2 + \left(6\sin^2 t \cos t\right)^2}\, dt \\
&= 2\pi \int_0^{\pi/2} \left(2\sin^3 t\right) \sqrt{36\cos^4 t \sin^2 t + 36\sin^4 t \cos^2 t}\, dt \\
&= 2\pi \int_0^{\pi/2} \left(2\sin^3 t\right) \sqrt{36\cos^2 t \sin^2 t \left(\cos^2 t + \sin^2 t\right)}\, dt \\
&= 2\pi \int_0^{\pi/2} \left(2\sin^3 t\right) \sqrt{36\cos^2 t \sin^2 t}\, dt \\
&= 2\pi \int_0^{\pi/2} \left(2\sin^3 t\right) \left(6\cos t \sin t\right) dt \\
&= 2\pi \int_0^{\pi/2} 12\sin^4 t \cos t\, dt \\
&= 2\pi \frac{12}{5} \sin^5 t \Big]_0^{\pi/2} = \frac{24\pi}{5}
\end{aligned}
$$

Doubling this result produces a surface area of $2\left(\frac{24\pi}{5}\right) = \frac{48\pi}{5} \approx 30.1593$ units^2.

Surface Area of a Curve Revolved around an Axis (Polar Form)

If the curve, written in polar form $r = f(\theta)$ is traced exactly once from $\theta = a$ to $\theta = b$, and is revolved around the polar (or x) axis, then

$$S = 2\pi \int_a^b y\sqrt{[f(\theta)]^2 + [f'(\theta)]^2}\,d\theta$$

$$= 2\pi \int_a^b y\sqrt{r^2 + \left(\frac{dr}{d\theta}\right)^2}\,d\theta$$

where $y = r\sin\theta$.

If the curve is revolved around the $\frac{\pi}{2}$ (or y) axis, then

$$S = 2\pi \int_a^b x\sqrt{[f(\theta)]^2 + [f'(\theta)]^2}\,d\theta$$

$$= 2\pi \int_a^b x\sqrt{r^2 + \left(\frac{dr}{d\theta}\right)^2}\,d\theta$$

where $x = r\cos\theta$.

EXAMPLE 10.43

The curve $r = e^\theta$ for $0 \le \theta \le \pi$ is revolved around the polar axis. Find the surface area of the generated figure.

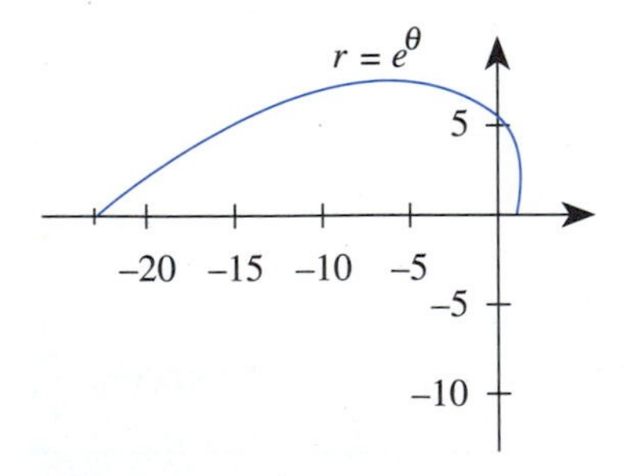

FIGURE 10.46a

Solution The sketch of this curve is shown in Figure 10.46a. When it is revolved around the polar axis, the surface shown in Figure 10.46b results. This curve is in polar form. According to the formula, the surface area is given by

$$S = 2\pi \int_0^\pi y\sqrt{[f(\theta)]^2 + [f'(\theta)]^2}\,d\theta$$

where $y = r\sin\theta$.

We have $y = e^\theta \sin\theta$ and $f'(\theta) = e^\theta$, so

$$S = 2\pi \int_0^\pi e^\theta \sin\theta\sqrt{(e^\theta)^2 + (e^\theta)^2}\,d\theta$$

$$= 2\pi \int_0^\pi e^\theta \sin\theta\sqrt{2e^{2\theta}}\,d\theta$$

$$= 2\pi \int_0^\pi e^\theta \sin\theta(e^\theta\sqrt{2})\,d\theta$$

$$= 2\sqrt{2}\pi \int_0^\pi e^{2\theta} \sin\theta\,d\theta$$

EXAMPLE 10.43 (Cont.)

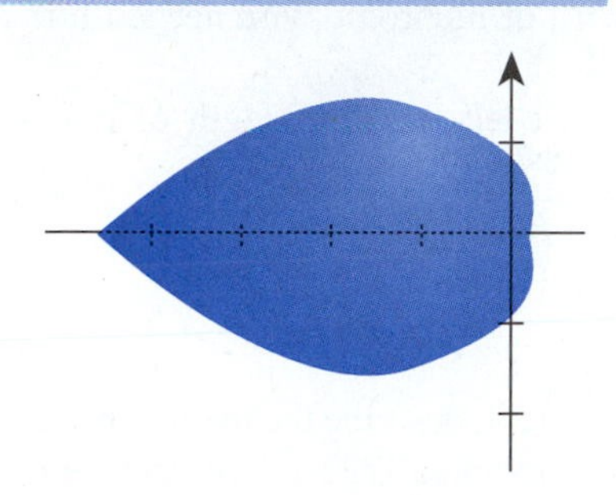

FIGURE 10.46b

Integration by parts (or Formula 87 in Appendix C) produces

$$= 2\sqrt{2}\pi \left[\frac{e^{2\theta}}{5}(2\sin\theta - \cos\theta)\right]\Bigg|_0^{\pi}$$

$$= \frac{2}{5}\sqrt{2}\pi[e^{2\pi} + e^0]$$

$$= \frac{2}{5}\sqrt{2}(e^{2\pi} + 1)\pi \approx 953.428$$

This surface has an area of approximately 953.428 square units.

Exercise Set 10.8

In Exercises 1–10, find the lengths of the given curves for the given intervals.

1. $x = 4t^3$, $y = 3t^2$ from $t = 0$ to $t = 1$
2. $x = \sin t$, $y = \cos t$ from $t = 0$ to $t = \pi$
3. $x = 3\cos t$, $y = 3\sin t$ from $t = 0$ to $t = 2\pi$
4. $x = \cos^3 t$, $y = \sin^3 t$ from $t = 0$ to $t = \frac{\pi}{2}$
5. $x = \cos t + t\sin t$, $y = \sin t - t\cos t$ from $t = 0$ to $t = \pi$
6. $r = \theta$ from $\theta = 0$ to $\theta = \frac{\pi}{2}$
7. $r = 1 + \cos\theta$ from $\theta = 0$ to $\theta = \pi$
8. $r = \cos^2\theta$ from $\theta = 0$ to $\theta = \frac{\pi}{2}$
9. $r = \sin^2\theta$ from $\theta = 0$ to $\theta = \frac{\pi}{2}$
10. $r = e^{\theta/2}$ from $\theta = 0$ to $\theta = 4$

In Exercises 11–18, find the area of the surface of revolution.

11. $x = t + 4$, $y = t^3$ from $t = 0$ to $t = 2$ around the x-axis
12. $x = t$, $y = 4 - t^2$ from $t = 0$ to $t = 2$ around the y-axis
13. $x = \cos t$, $y = \sin t$ from $t = 0$ to $t = \frac{\pi}{2}$ around the x-axis
14. $x = 1 + \sin t$, $y = \cos t$ from $t = 0$ to $t = \frac{\pi}{2}$ around the y-axis
15. $x = 1 + \sin t$, $y = \cos t$ from $t = 0$ to $t = \frac{\pi}{2}$ around the x-axis
16. $r = \sin\theta$, from $\theta = 0$ to $\theta = \frac{\pi}{2}$ around the polar axis
17. $r = 1 + \cos\theta$, from $\theta = 0$ to $\theta = \pi$ around the polar axis
18. $r = e^{\theta/2}$ from $\theta = 0$ to $\theta = \pi$ around the polar axis

Solve Exercises 19 and 20.

19. *Environmental science* An **Archimedean spiral** has a polar equation of the form $r = a + b\theta$. Each successive loop of this spiral is the same distance $d = 2\pi b$ from the adjacent loops. One brand of mosquito coil is designed as an Archimedean spiral described by the equation $r = b\theta$, where r is in cm and $0 \le \theta < 2n\pi$. Here n determines the number of complete coils.
   (a) What is the length of this coil if there are four complete coils that are 1 cm apart?
   (b) If the coil burns at the rate of 5 cm/h, how long before the coil is all burned?

20. *Electronics* An electric heater coil is constructed from nichrome wire of resistance $4\,\Omega/\text{m}$. If L is the length of nichrome used, the total resistance, R, of the heating element is $R = 4L\,\Omega$. The shape of the coil is described by the equation $r = \frac{\theta}{2\pi}$, where $2\pi \le \theta \le 8\pi$. If the heater is designed to be connected across a voltage, V, of 110 V

(a) Find the length, L, of nichrome wire needed for this coil.

(b) Find the power needed for this length of coil. (Remember, $P = \frac{V^2}{R}$.)

In Your Words

21. Without looking in the text, describe the method used to determine the arc length of a curve in either polar or parametric form.

22. Without looking in the text, describe the method used to determine the surface area of a curve revolved around an axis if the equation of the curve is in either polar or parametric form.

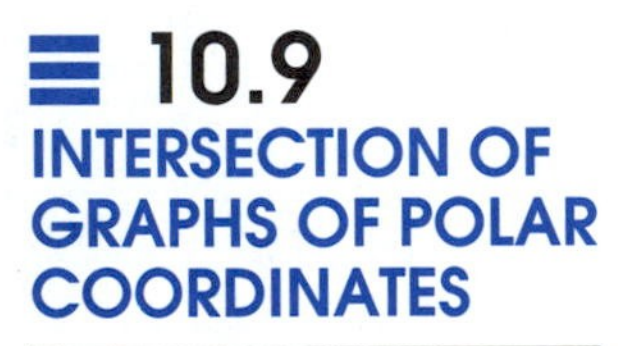

10.9 INTERSECTION OF GRAPHS OF POLAR COORDINATES

In Section 10.10, we will be finding the area of a region of the plane enclosed by one or more graphs of polar equations. In order to do, this we will need to know where these graphs intersect. Previously, when we needed to find where the graphs of two equations in rectangular coordinates intersected, we solved the equations simultaneously. This does not always work when we are working in polar coordinates.

EXAMPLE 10.44

Find the points of intersection of $r = 1$ and $r = 2\cos\theta$.

Solution Substituting $r = 1$ from the first equation into the second equation, we get

$$1 = 2\cos\theta \text{ or } \cos\theta = \frac{1}{2}$$

Solving this for θ, we get $\theta = \frac{\pi}{3}$ or $\frac{5\pi}{3}$. Thus the points of intersection are $\left(1, \frac{\pi}{3}\right)$ and $\left(1, \frac{5\pi}{3}\right)$. The graphs of these two curves, and their points of intersection, are shown in Figure 10.47.

We solved these two equations simultaneously and got the points of intersection. But, as we said, this does not always work. Consider the next example.

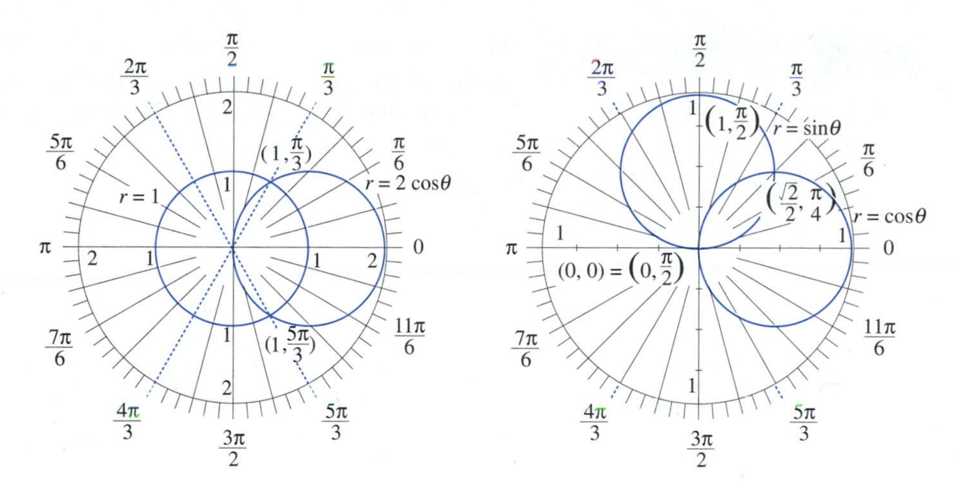

FIGURE 10.47

FIGURE 10.48

EXAMPLE 10.45

Find the points of intersection of $r = \sin\theta$ and $r = \cos\theta$.

Solution If we solve these simultaneously, we get $\sin\theta = \cos\theta$ or $\frac{\sin\theta}{\cos\theta} = \tan\theta = 1$. This is true when $\theta = \frac{\pi}{4} + n\pi$ or, in the range $0 \leq \theta < 2\pi$, when $\theta = \frac{\pi}{4}$ or $\frac{5\pi}{4}$. Thus, we get the points $\left(\frac{\sqrt{2}}{2}, \frac{\pi}{4}\right)$ and $\left(-\frac{\sqrt{2}}{2}, \frac{5\pi}{4}\right)$. We have a problem. These are not different points. They are just different ways of writing the same point.

Now, look at the graphs of these two functions as shown in Figure 10.48. As you can see, they do intersect in two points. One point we found, $\left(\frac{\sqrt{2}}{2}, \frac{\pi}{4}\right)$, the other point is the pole. The reason we did not find the pole when we solved the equations is that for the curve $r = \sin\theta$, the pole is at the points $(0, \pi + n\pi)$. For the curve $r = \cos\theta$, the pole is at $(0, \frac{\pi}{2} + n\pi)$. Both curves pass through the pole but not for the same values of θ. Because they intersect for different values of θ, we cannot find these points by using simultaneous equations. This point is at $(0, 0) = (0, \frac{\pi}{2})$.

In order to find the points of intersection of two graphs in polar form, we need to do two things. First sketch a graph of the curves. While your curve does not have to be precise it must be good enough to see how many points of intersection exist. A graphing calculator will be very helpful here.

Second, solve the polar equations simultaneously. Check your solutions to make sure you have all the points of intersection with the other curve and that you do not have any duplicates.

EXAMPLE 10.46

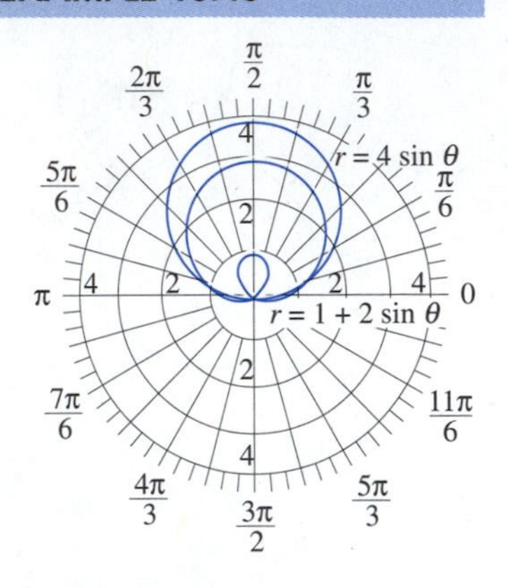

FIGURE 10.49

Find the points of intersection of $r = 4\sin\theta$ and $r = 1 + 2\sin\theta$.

Solution The sketch of these two curves is shown in Figure 10.49. As you can see, they both pass through the pole, so we know that $(0,0)$ is a point of intersection.

There are two other points of intersection. Solving the equations simultaneously, we get

$$\begin{aligned} 4\sin\theta &= 1 + 2\sin\theta \\ 2\sin\theta &= 1 \\ \sin\theta &= \frac{1}{2}, \text{ so } \theta = \frac{\pi}{6} \text{ or } \frac{5\pi}{6} \end{aligned}$$

The points of intersection are $(0,0)$, $\left(2, \frac{\pi}{6}\right)$, and $\left(2, \frac{5\pi}{6}\right)$.

EXAMPLE 10.47

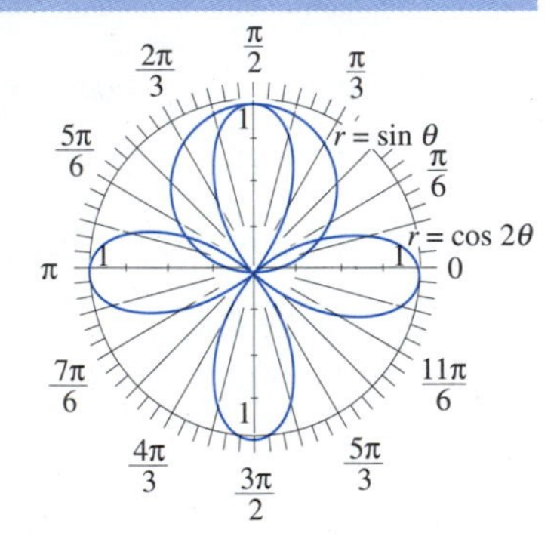

FIGURE 10.50

Find the points of intersection of $r = \sin\theta$ and $r = \cos 2\theta$.

Solution The graphs of these two curves are shown in Figure 10.50. As you can see, there are four points of intersection. Solving the equations simultaneously, we get

$$\sin\theta = \cos 2\theta$$

Using the identity $\cos 2\theta = 1 - 2\sin^2\theta$, we see that

$$\begin{aligned} \sin\theta &= 1 - 2\sin^2\theta \\ \text{or} \qquad 2\sin^2\theta + \sin\theta - 1 &= 0 \\ (2\sin\theta - 1)(\sin\theta + 1) &= 0 \\ \sin\theta &= \frac{1}{2} \text{ or } \sin\theta = -1 \\ \text{and} \qquad \theta &= \frac{\pi}{6}, \frac{5\pi}{6}, \text{ or } \frac{3\pi}{2} \end{aligned}$$

Thus, three points of intersection for these two polar curves are $\left(\frac{1}{2}, \frac{\pi}{6}\right)$, $\left(\frac{1}{2}, \frac{5\pi}{6}\right)$, and $\left(-1, \frac{3\pi}{2}\right) = \left(1, \frac{\pi}{2}\right)$. Checking the pole, we see that $(0,0)$ is also a solution.

The methods that we have shown will allow you to find all of the points of intersection for some equations. However, there are other pairs of simultaneous equations that can be used to locate other points of intersection. We will not use these other pairs of equations in this text.

Exercise Set 10.9

In Exercises 1–12, find all points of intersection of the graphs of the two polar equations. Grap

1. $r = 3\theta,\ r = \frac{\pi}{2}$
2. $r = \dfrac{\theta}{4},\ r = \frac{\pi}{6}$
3. $r = \frac{1}{2},\ r = \cos\theta$
4. $r = 2 - 2\sin\theta,\ r = 2 - 2\cos\theta$
5. $r = 1 - \sin\theta,\ r = 1 + \cos\theta$
6. $r = -4 + 4\cos\theta,\ r = -4 + 4\sin\theta$

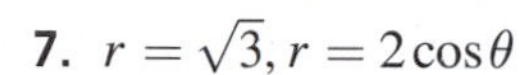

7. $r = \sqrt{3},\ r = 2\cos\theta$
8. $r = \sin 2\theta,\ r = \sin\theta$
9. $r = \sin 2\theta,\ r = \sqrt{2}\sin\theta$
10. $r = 2 + 2\cos\theta,\ r = \dfrac{1}{1 - \cos\theta}$
11. $r = 1 - \sin\theta,\ r = \dfrac{1}{1 - \sin\theta}$
12. $r = \sin 2\theta,\ r = \cos 2\theta$

In Your Words

13. Describe how you would find the intersection of two graphs in polar coordinates.
14. Why do you think it is important to find the points where the graphs of two equations in polar coordinates intersect?

10.10 AREA IN POLAR COORDINATES

In Section 10.9, we learned how to find the points of intersection for the graphs of two polar equations. In this section, we will consider the problem of finding the area of a region in a plane that has been enclosed by the graph of a polar equation and two rays that have the pole as a common vertex. Next, we will find the area enclosed by the graph of two polar equations.

Also in this section, we will use the following formula for the area of a sector of a circle.

> **Area of a Sector of a Circle**
>
> If a sector has angle θ (in radians) and radius r, then its area is
>
> $$A = \frac{1}{2}\theta r^2$$

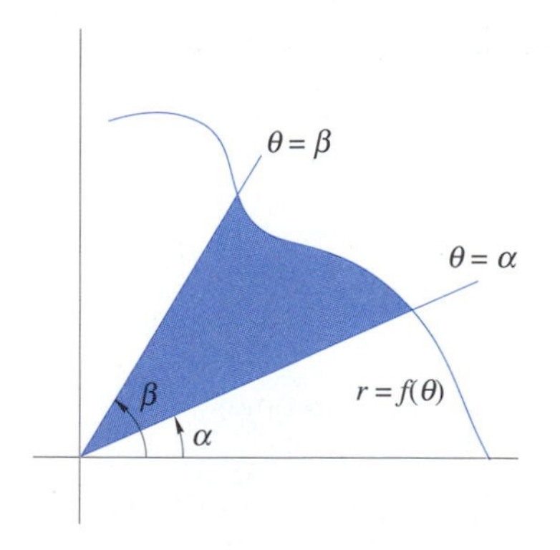

FIGURE 10.51

Now, suppose we have a region bounded by $r = f(\theta)$ and the two rays $\theta = \alpha$ and $\theta = \beta$, where $\alpha < \beta$, as shown in Figure 10.51. We want to find the area and so we will subdivide something into n intervals. With rectangular coordinates we divided the interval $[a, b]$. We will do a similar thing here. We will divide the angular interval from α to β into n subintervals where

$$\alpha = \theta_0 < \theta_1 < \theta_2 < \cdots < \theta_n = \beta$$

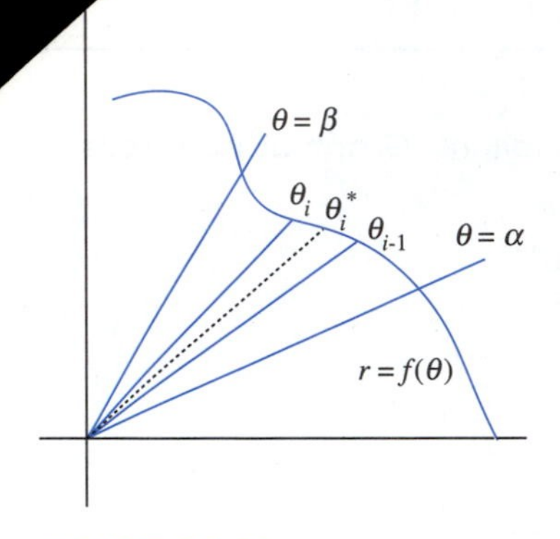

FIGURE 10.52

We will select the ray $\theta_i{}^*$ at the midpoint of the ith interval, as shown in Figure 10.52. The area of this interval is approximately

$$A \approx \frac{1}{2}(\theta_i - \theta_{i-1})[f(\theta_i{}^*)]^2$$

But, what is $\theta_i - \theta_{i-1}$? Nothing more than $\Delta\theta_i$. So, the area becomes

$$A \approx \frac{1}{2}[f(\theta_i{}^*)]^2 \Delta\theta_i$$

Adding all of these intervals and taking the limit, we get the following formula for the area of a region.

Area of a Sector for a Polar Curve

If f is continuous and $f(\theta) \geq 0$ on $[\alpha, \beta]$ with $0 \leq \alpha < \beta \leq 2\pi$, then area A of the region bounded by the graphs of $r = f(\theta)$, $\theta = \alpha$, and $\theta = \beta$ is

$$A = \frac{1}{2}\int_{\alpha}^{\beta} [f(\theta)]^2 \, d\theta$$

or

$$A = \frac{1}{2}\int_{\alpha}^{\beta} r^2 \, d\theta$$

EXAMPLE 10.48

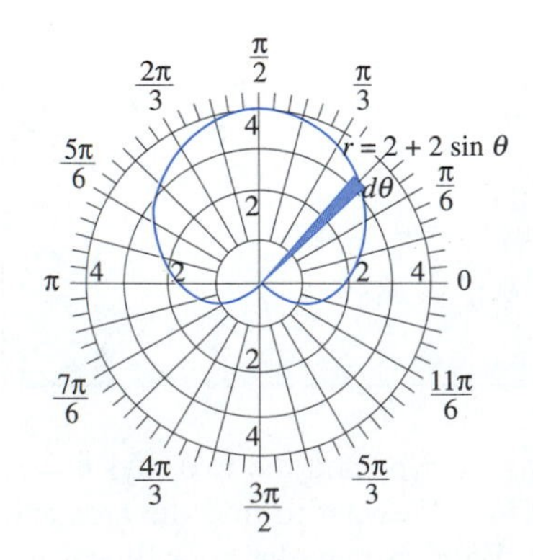

FIGURE 10.53

Find the area enclosed by the cardioid $r = 2 + 2\sin\theta$ from $0 \leq \theta \leq 2\pi$.

Solution The graph of this curve is shown in Figure 10.53. Here, we have $\alpha = 0$ and $\beta = 2\pi$. A typical circular sector has been shaded in the figure and its central angle has been labeled $d\theta$.

$$\begin{aligned} A &= \frac{1}{2}\int_0^{2\pi} (2 + 2\sin\theta)^2 \, d\theta \\ &= \frac{1}{2}\int_0^{2\pi} (4 + 8\sin\theta + 4\sin^2\theta) \, d\theta \end{aligned}$$

Using the half-angle formula $\sin^2\theta = \frac{1}{2}(1 - \cos 2\theta)$, the area becomes

$$\begin{aligned} A &= \frac{1}{2}\int_0^{2\pi} [4 + 8\sin\theta + 2(1 - \cos 2\theta)] \, d\theta \\ &= \frac{1}{2}(4\theta - 8\cos\theta + 2\theta - \sin 2\theta)\Big|_0^{2\pi} \\ &= \frac{1}{2}[(8\pi - 8 + 4\pi) - (-8)] = 6\pi \end{aligned}$$

The area enclosed by this cardioid is 6π square units.

The following guidelines will help you to find the area of a polar region. Look back over Example 10.48 to see if we followed these guidelines.

Guidelines for Finding the Area of a Polar Region

1. Sketch the region, labeling the graph of $r = f(\theta)$. Find the smallest value $\theta = \alpha$ and the largest value of $\theta = \beta$ for points (r, θ) in the region.
2. Sketch a typical circular sector and label its central angle $d\theta$.
3. Express the area of the sector in Step 2 as $\frac{1}{2}r^2\,d\theta$.
4. Integrate the expression in Step 3 over the limits α to β.

EXAMPLE 10.49

Find the area enclosed by the outer loop of the limaçon $r = 1 + 2\sin\theta$.

Solution The graph of this curve is shown in Figure 10.54. The curve intersects the pole when $1 + 2\sin\theta = 0$ or $\sin\theta = -\frac{1}{2}$. This happens twice—at $\theta = \frac{7\pi}{6}$ and $\frac{11\pi}{6}$. Between these two values of θ, we get the inner loop. So, what we want is the area of the curve from 0 to $\frac{7\pi}{6}$ and then again from $\frac{11\pi}{6}$ to 2π. We can get that same area by integrating from $\dfrac{-\pi}{6}$ to $\frac{7\pi}{6}$, since, on a circle, the angle $\frac{11\pi}{6} = \dfrac{-\pi}{6}$.

$$\begin{aligned}
A &= \frac{1}{2}\int_{-\pi/6}^{7\pi/6} (1 + 2\sin\theta)^2\,d\theta \\
&= \frac{1}{2}\int_{-\pi/6}^{7\pi/6} (1 + 4\sin\theta + 4\sin^2\theta)\,d\theta \\
&= \frac{1}{2}\int_{-\pi/6}^{7\pi/6} (1 + 4\sin\theta + 2 - 2\cos 2\theta)\,d\theta \\
&= \frac{1}{2}(3\theta - 4\cos\theta - \sin 2\theta)\Big|_{-\pi/6}^{7\pi/6} \\
&= \frac{1}{2}\left[\left(\frac{7\pi}{2} + 2\sqrt{3} - \frac{\sqrt{3}}{2}\right) - \left(\frac{-\pi}{2} - 2\sqrt{3} + \frac{\sqrt{3}}{2}\right)\right] \\
&= 2\pi + 2\sqrt{3} - \frac{\sqrt{3}}{2} \\
&= 2\pi + \frac{3}{2}\sqrt{3} \approx 8.88
\end{aligned}$$

The area enclosed by this region is about 8.88 square units.

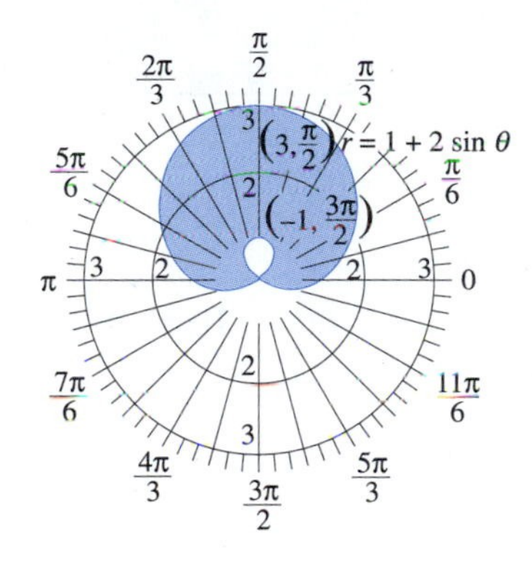

FIGURE 10.54

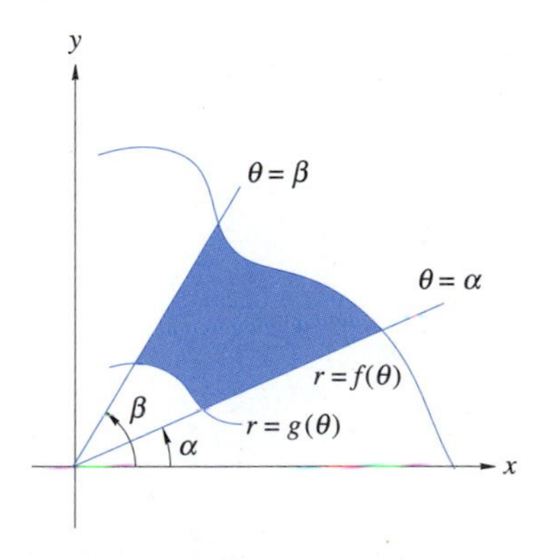

FIGURE 10.55

Now suppose we have two curves, $r = f(\theta)$ and $r = g(\theta)$, with $f(\theta) \geq g(\theta)$. We want to find the area of the region between these curves and bounded by $\theta = \alpha$ and $\theta = \beta$, as shown in Figure 10.55. We do this, as we did with rectangular coordinates, by subtracting the area bounded by the inner curve from the area bounded by the outer curve. This gives us the formula for the area enclosed between two polar curves in the following box.

Area Enclosed Between Two Polar Curves

If f and g are continuous and $f(\theta) \geq g(\theta) \geq 0$ on $[\alpha, \beta]$ with $0 \leq \alpha < \beta \leq 2\pi$, then area A of the region bounded by the graphs of $r = f(\theta)$, $r = g(\theta)$, $\theta = \alpha$, and $\theta = \beta$ is

$$A = \frac{1}{2}\int_{\alpha}^{\beta} [f(\theta)]^2\, d\theta - \frac{1}{2}\int_{\alpha}^{\beta} [g(\theta)]^2\, d\theta$$

$$= \frac{1}{2}\int_{\alpha}^{\beta} \left([f(\theta)]^2 - [g(\theta)]^2\right) d\theta$$

If you let $r_2 = f(\theta)$ and $r_1 = g(\theta)$, then this can be written as

$$A = \frac{1}{2}\int ({r_2}^2 - {r_1}^2)\, d\theta$$

EXAMPLE 10.50

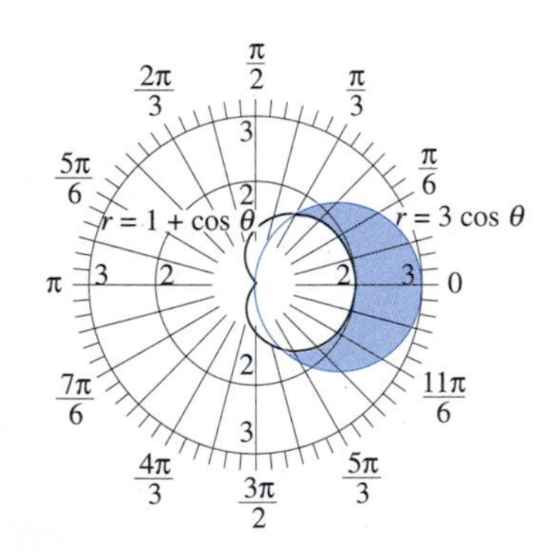

FIGURE 10.56

Find the area inside $r = 3\cos\theta$ and outside $r = 1 + \cos\theta$.

Solution The area we want to integrate has been shaded in Figure 10.56. We will need to find the points of intersection for these two curves.

Solving the equations simultaneously, we get

$$3\cos\theta = 1 + \cos\theta$$
$$2\cos\theta = 1$$
$$\cos\theta = \frac{1}{2}$$

So, the curves intersect when

$$\theta = \frac{\pi}{3} \text{ and } \frac{-\pi}{3}$$

The area is

$$A = \frac{1}{2}\int_{-\pi/3}^{\pi/3} \left[(3\cos\theta)^2 - (1+\cos\theta)^2\right] d\theta$$

$$= \frac{1}{2}\int_{-\pi/3}^{\pi/3} \left[(9\cos^2\theta) - (1 + 2\cos\theta + \cos^2\theta)\right] d\theta$$

$$= \frac{1}{2}\int_{-\pi/3}^{\pi/3} (8\cos^2\theta - 1 - 2\cos\theta)\, d\theta$$

$$= \frac{1}{2}\int_{-\pi/3}^{\pi/3} \left[8\left(\frac{1+\cos 2\theta}{2}\right) - 1 - 2\cos\theta\right] d\theta$$

EXAMPLE 10.50 (Cont.)

$$= \frac{1}{2}\int_{-\pi/3}^{\pi/3} (3+4\cos 2\theta - 2\cos\theta)\,d\theta$$

$$= \frac{1}{2}(3\theta + 2\sin 2\theta - 2\sin\theta)\Big|_{-\pi/3}^{\pi/3}$$

$$= \frac{1}{2}\left[(\pi + \sqrt{3} - \sqrt{3}) - (-\pi - \sqrt{3} + \sqrt{3})\right]$$

$$= \pi \text{ square units}$$

We could have made our calculations easier in the last problem if we had taken advantage of the fact that the region was symmetric with respect to the polar axis. Notice that half of the area is above the x-axis and half is below. Thus, we could have found the area from 0 to $\frac{\pi}{3}$ and then doubled it as follows.

$$A = 2\cdot\frac{1}{2}\int_0^{\pi/3} [(3\cos\theta)^2 - (1+\cos\theta)^2]\,d\theta$$

$$= 3\theta + 2\sin 2\theta - 2\sin\theta\Big|_0^{\pi/3}$$

$$= (\pi + \sqrt{3} - \sqrt{3}) - (0) = \pi$$

Exercise Set 10.10

In Exercises 1–10, sketch the graph of the equation and find the area of the region enclosed by the graph.

1. $r = 4\sin\theta$
2. $r = 5\cos\theta$
3. $r = 1 - \cos\theta$
4. $r = 4 + 4\sin\theta$
5. $r = \sin 2\theta$
6. $r = \cos 3\theta$
7. $r = 4 + \cos\theta$
8. $r = 4 + 3\cos\theta$
9. $r^2 = \sin\theta$
10. $r^2 = 4\cos 2\theta$

In Exercises 11–16, find the area enclosed by the given polar equation and the given rays.

11. $r = 3\cos\theta$, $\theta = 0$, $\theta = \frac{\pi}{4}$
12. $r = 4\sin\theta$, $\theta = 0$, $\theta = \frac{\pi}{3}$
13. $r = \cos 2\theta$, $\theta = 0$, $\theta = \frac{\pi}{4}$
14. $r = \sin 4\theta$, $\theta = 0$, $\theta = \frac{\pi}{8}$
15. $r = e^{2\theta}$, $\theta = 0$, $\theta = \frac{\pi}{2}$
16. $r = 5\theta$, $\theta = 0$, $\theta = \frac{\pi}{6}$

In Exercises 17–20, find the area of the region bounded by one loop of the graph of the given equation.

17. $r = 4\cos 2\theta$
18. $r^2 = 4\cos 2\theta$
19. $r = 2\sin 3\theta$
20. $r = \sin 6\theta$

In Exercises 21–26, sketch the graph and find the area of the region described.

21. Inside $r = 1 + \cos\theta$ and outside $r = 1$
22. Inside $r = 2\sin\theta$ and outside $r = 1$
23. Inside $r = 1 - \sin\theta$ and outside $r = 2\cos\theta$
24. Inside $r^2 = 8\sin\theta$ and outside $r = 2$
25. Inside $r = \cos\theta$ and outside $r = 1 - \sin\theta$
26. Inside $r = \sin\theta$ and inside $r = \sqrt{3}\cos\theta$

In Your Words

27. Without looking in the text, describe the guidelines for finding the area of a polar region.

28. Describe how to find the area between two polar curves.

CHAPTER 10 REVIEW

Important Terms and Concepts

Arc length
- Of parametric equations
- Of polar equations

Area
- Between two polar curves
- Of a polar region
- Of a sector of a circle

Eccentricity

Parametric equations
- First derivative
- Higher order derivatives

Polar equations
- Derivative
- For conic sections
- Intersection of graphs

Sector of a circle
- Area

Surface area
- Generated by parametric equations
- Generated by polar equations

Vector-valued functions
- Derivative

Review Exercises

In Exercises 1–4, (a) find the first and second derivative of the given function and (b) find the slope of the tangent line at the point corresponding to $t = 1$.

1. $x = t^2,\ y = t^3 + t$

2. $x = \frac{1}{t},\ y = \frac{2}{t}$

3. $x = \cos^2 t,\ y = \sin t$

4. $x = \sqrt{t},\ y = e^{-t}$

In Exercises 5–8, (a) find the first derivative of the given function and (b) find the slope of the tangent line at the point where $\theta = \frac{\pi}{4}$.

5. $r = 3 + 2\cos\theta$

6. $r = 6\cos 2\theta$

7. $r^2 = \sin\theta$

8. $r = \dfrac{8}{3 + \cos\theta}$

In Exercises 9–18, sketch the curves of the given equation.

9. $x = 3t,\ y = 5t - 2$

10. $x = t + 2,\ y = t^2 + t$

11. $x = 4\sin t,\ y = 3\cos t$

12. $x = \sin t,\ y = \cos 4t$

13. $r = \cos 7\theta$

14. $r = 3 + 2\sin\theta$

15. $r = \dfrac{16}{5 - 3\cos\theta}$

16. $r = \dfrac{9}{3 - 5\cos\theta}$

17. $r = \dfrac{9}{3 + 3\sin\theta}$

18. $r = \dfrac{2}{1 + \cos\theta}$

In Exercises 19–24, find the arc length of the given curve over the given interval.

19. $x = t^2,\ y = t^3,\ 0 \le t \le 2$

20. $x = t^2,\ y = 2t,\ 0 \le t \le 3$

21. $x = 4\sin t,\ y = 4\cos t,\ 0 \le t \le \frac{\pi}{2}$

22. $r = e^{2\theta},\ 0 \le \theta \le \pi$

23. $r = \cos^2\left(\dfrac{\theta}{2}\right),\ 0 \le \theta \le \pi$

24. $r = 1 - \cos\theta,\ -\pi \le \theta \le 0$

In Exercises 25–30, find the area of the surface generated by revolving the given curve around the indicated axis.

25. $x = t^2,\ y = t,\ 0 \le t \le 2$ around the x-axis

26. $x = t^2,\ y = t - \dfrac{t^3}{3},\ 0 \le t \le 1$ around the x-axis

27. $x = t^2,\ y = t^3,\ 0 \le t \le 2$ around the y-axis

28. $x = e^t \sin t,\ y = e^t \cos t,\ 0 \le t \le \frac{\pi}{2}$ around the y-axis

29. $r = 6\sin\theta,\ 0 \le \theta \le \pi$ around the polar axis

30. $r = 4 + 4\cos\theta,\ 0 \le \theta \le \pi$ around the polar axis

In Exercises 31–34, find the points of intersection of the graph of the given pair of polar equations.

31. $r = 4,\ r = 4 + 4\sin\theta$

32. $r = 3,\ r = 6\sin\theta$

33. $r = 2,\ r = 4 + 4\cos\theta$

34. $r = 1,\ r = 2\sin 2\theta$

In Exercises 35–40, find the indicated area and sketch the graph.

35. Inside $r = 1 + \cos\theta$

36. Inside $r = 4 + 4\sin\theta$

37. Inside $r = 4 + 4\sin\theta$ and outside $r = 4$

38. Inside $r = 6\sin\theta$ and outside $r = 3$

39. Inside $r = \sin\theta$ and outside $r = 1 + \cos\theta$

40. Inside $r = 2\sin 2\theta$ and outside $r = 1$

CHAPTER 10 TEST

1. Sketch $x = 2t,\ y = t^2 + 1$.
2. Sketch $r = 1 + 2\cos\theta$.
3. Find the first and second derivative of $x = t^3,\ y = t^2 + 5t$.
4. Find the equation of the tangent line to the graph of $x = \sin t,\ y = \cos^2 t$ at $\left(\dfrac{\sqrt{2}}{2}, \frac{1}{2}\right)$, when $t = \frac{\pi}{4}$.
5. Find the derivative of $r = 4 - 2\sin\theta$.
6. Find the equation of the normal line to the graph of $r = 8\sin 3\theta$ at the point $\left(4\sqrt{2}, \frac{\pi}{4}\right)$.
7. Find the arc length of the curve described by the parametric equations $x = e^t,\ y = 2e^t$ from $0 \le t \le 2$.
8. Find the surface area of the surface generated by revolving the graph of $r = 4\sin\theta$ from $0 \le \theta \le \frac{\pi}{2}$ around the polar axis.
9. Find the points of intersection of the curves $r = 4\cos\theta$ and $r = 1 - \cos\theta$.
10. Find the area of the region that lies inside the circle $r = 1$ and outside the cardioid $r = 1 - \cos\theta$.
11. Identify and sketch the graph of the polar equation $r = \dfrac{2}{1 + 4\cos\theta}$.

CHAPTER

11

Partial Derivatives and Multiple Integrals

Robots, like the one shown, use a three-dimensional coordinate system. In Sections 11.1 and 11.6, we will study the coordinate systems used by robots like this.

Courtesy of Motoman Inc.

Until now, all of our work has been in the plane. The derivatives and integrals that we have learned have all applied to curves in the plane. We have used two types of coordinate systems—the rectangular and the polar—but we live in a three-dimensional world, so it is important that we study three-dimensional mathematics.

In this chapter, we will learn about measuring and graphing objects in three dimensions. We will learn about three new coordinate systems and see how they are used in robotics, and we will learn how to take derivatives and integrals in three dimensions.

11.1 FUNCTIONS IN TWO VARIABLES

The functions that we have used until now have always been functions of one variable. The typical function, $y = f(x)$, had x as the variable. Parametric equations of the form $x = f(t)$ and $y = g(t)$, had t as the variable. Polar equations, such as $r = f(\theta)$, had θ as the variable.

We have, however, worked with functions of more than one variable. For example, the formula for the volume of a cylinder, $V = \pi r^2 h$, is a function of two variables, r and h. As you can see, the volume of a cylinder depends on both the radius of the base and the height of the cylinder. To indicate that it is a function of two variables, we could write

$$V = f(r, h) = \pi r^2 h$$

EXAMPLE 11.1

For the function $V = f(r, h) = \pi r^2 h$, find **(a)** $f(5, 4)$, **(b)** $f(10, 4)$, and **(c)** $f(5, 8)$.

Solutions

(a) Here, $r = 5$ and $h = 4$, so we have $f(5, 4) = \pi \cdot 5^2 \cdot 4 = 100\pi$.

(b) In this example, we have $r = 10$ and $h = 4$, and so $f(10, 4) = \pi \cdot 10^2 \cdot 4 = 400\pi$.

(c) Here, $r = 5$ and $h = 8$, with the result $f(5, 8) = \pi \cdot 5^2 \cdot 8 = 200\pi$.

We define a **real function of two variables** as any rule that assigns a unique real number to each ordered pair of real numbers (x, y) in a certain set D of the xy-plane.

We often write a real function of two variables in the form $z = f(x, y)$. Here, z represents a **dependent variable** and x and y are the **independent variables**. The numbers in D of the definition form the **domain** of the function f. The **range** of f is all the real numbers $f(x, y)$, where (x, y) is in the domain D. As you can see, this is very similar to what we did with a function of one variable. Of course, $f(a, b)$ means the value of the function f when $x = a$ and $y = b$.

EXAMPLE 11.2

If $f(x, y) = \dfrac{x^2 - y^2}{x^2 + y^2}$, find $f(2, 1)$

Solution $f(2, 1) = \dfrac{2^2 - 1^2}{2^2 + 1^2} = \dfrac{4 - 1}{4 + 1} = \dfrac{3}{5}$.

EXAMPLE 11.3

If $f(x, y) = e^{x^2 - xy}$, find $f(3, 2)$.

Solution Here, $x = 3$ and $y = 2$, so $f(3, 2) = e^{3^2 - 3 \cdot 2} = e^{9-6} = e^3$.

EXAMPLE 11.4

If $f(x,y)=\sqrt{x}+x\sqrt{y}$, find $f(x,2x)-f(x,x^2)$.

Solution This is a little different because we want to subtract the second value of the function from the first. Also, notice that in both cases the value of x is the same. The value of the y-variable changes.

$$f(x,2x)=\sqrt{x}+x\sqrt{2x}$$

$$f(x,x^2)=\sqrt{x}+x\sqrt{x^2}=\sqrt{x}+x|x| \qquad \text{Remember, } \sqrt{x^2}=|x|.$$

The difference is

$$\begin{aligned} f(x,2x)-f(x,x^2) &= (\sqrt{x}+x\sqrt{2x})-(\sqrt{x}+x|x|) \\ &= x\sqrt{2x}-x|x| \\ &= x\left(\sqrt{2x}-|x|\right) \end{aligned}$$

Domain

As with functions of one variable, we must be careful that we do not divide by zero. We must also be sure that the values of x and y lead to real number answers. For example, in Example 11.4, if either $x<0$ or $y<0$, then $f(x,y)$ would produce an imaginary value. But, this is a *real* value function and imaginary numbers are not allowed. The domain consists of all values of $x\geq 0$ and of $y\geq 0$.

In a similar manner, the domain of Example 11.2 includes all real values of (x,y) except the origin $(0,0)$, because this gives a zero in the denominator.

Functions of More Than Two Variables

We have been concentrating on functions of two variables. However, we could just as easily have a function of three variables, such as $w=f(x,y,z)$, or of four variables, such as $u=f(x,y,z,w)$, or of as many variables as we like.

EXAMPLE 11.5

If $f(x,y,z,w)=xy^2+z\tan w-e^{yz}$, find $f\left(3,1,2,\frac{\pi}{4}\right)$.

Solution Here $x=3$, $y=1$, $z=2$, and $w=\frac{\pi}{4}$, so

$$\begin{aligned} f(x,y,z,w) &= 3\cdot 1^2+2\tan\frac{\pi}{4}-e^{1\cdot 2} \\ &= 3+2-e^2 \\ &= 5-e^2 \end{aligned}$$

The domain of this function consists of all real numbers for x, y, and z. However, the values of $w\neq\frac{\pi}{2}(2n+1)$ where n is an integer. Thus, w cannot be an odd multiple of $\frac{\pi}{2}$ such as $\frac{\pi}{2}$, $\frac{3\pi}{2}$, $\frac{5\pi}{2}$, or $-\frac{\pi}{2}$, because the tangent function is not defined at these points.

Example 11.5 shows how to work with functions of four variables. As we saw, these functions are treated in the same manner as those of two variables. For now, we will concentrate on functions of two variables. Using a function of two variables to express a problem is no easier or more different than it is for one variable. No general rules exist that you have not already encountered while working with just one variable. As before, you may originally write the problem with more than two variables. You should then use the relationships between variables to eliminate some of them.

Application

EXAMPLE 11.6

An open rectangular box is constructed out of different materials. The material for the bottom costs \$1.25/ft^2. The material for the sides costs \$1.50/ft^2. The volume of the box is to be 4 ft^3. Express the cost of the material as a function of the length and width of the box.

Solution Let's use l for the length, w for the width, and h for the height of this box. Then

$$V = lwh$$

The cost of the material depends on the surface area. The area of the sides is $(2l+2w)h$. The area of the bottom is lw, so the cost of the material is

$$C = 1.50(2l+2w)h + 1.25lw$$

This is not a function of the length and width because it includes the variable h.

But $V = lwh = 4$, and so $h = \frac{4}{lw}$. With this substitution, we get the desired result

$$\begin{aligned} V &= 1.50(2l+2w)\frac{4}{lw} + 1.25lw \\ &= \frac{12(l+w)}{lw} + 1.25lw \end{aligned}$$

▪

In this section, we will only be writing equations as functions of two variables. Later we will differentiate and integrate these functions in order to solve some problems.

Exercise Set 11.1

Solve Exercises 1–10.

1. Let $f(x, y) = 3x + 4y - xy$. Find: **(a)** $f(1,0)$, **(b)** $f(0,1)$, **(c)** $f(2,1)$, **(d)** $f(x+h, y)$, **(e)** $f(x, y+h)$.
2. Let $g(x, y) = x^2y + \sin x - \cos y$. Find: **(a)** $g\left(\frac{\pi}{2}, \pi\right)$, **(b)** $g\left(\pi, \frac{\pi}{2}\right)$, **(c)** $g\left(\frac{\pi}{2}, \frac{\pi}{4}\right)$, **(d)** $g(x+h, y)$, **(e)** $g(x, y+h)$.
3. Let $j(x, y) = \sqrt{xy} - x + \frac{4}{y}$. Find: **(a)** $j(-1,-2)$, **(b)** $j(-1,-4)$, **(c)** $j(4,1)$, **(d)** $j(x+h, y)$, **(e)** $j(x, y+h)$.

4. Let $k(x, y) = e^x + e^{xy} - y^2$. Find: **(a)** $k(1, 2)$, **(b)** $k(2, 1)$, **(c)** $k(2, 3)$, **(d)** $k(x+h, y)$, **(e)** $k(x, y+h)$.

5. Let $f(x, y) = \dfrac{2xy - x^2}{y - x}$. Find: **(a)** $f(1, 0)$, **(b)** $f(0, 1)$, **(c)** $f(2, 1)$, **(d)** $f(1, 2)$, **(e)** the domain of f.

6. Let $g(x, y) = \dfrac{\ln(x+y)}{y}$. Find: **(a)** $g(1, 1)$, **(b)** $g(2, 1)$, **(c)** $g(3, 6)$, **(d)** $g(4, -3)$, **(e)** the domain of g.

7. Express the volume of a cone as a function of the radius of the base and the height.

8. Express the volume of a prism as a function of the area of its base and its height.

9. Express the lateral surface area of a right circular cylinder as a function of the radius of its base and its height.

10. Express the total surface area of a cone as a function of its slant height and the radius of its base.

Solve Exercises 11–24.

11. *Petroleum engineering* The cost of the bottom and top of a cylindrical petroleum storage tank is \$200/m^2 and the cost of the side is \$1,000/m^2. Write the total cost of constructing such a tank as a function of the radius and height.

12. *Wastewater technology* A conical sewage tank is being constructed. The base costs \$1,500/ft^2 and the side costs \$200/ft^2. Write the total cost as a function of the radius r and slant height s.

13. *Medical technology* According to **Poiseuille's law**, the velocity v in cm/min of blood flow r units from the center of an artery of radius R is given by

$$v(R, r) = c\left(R^2 - r^2\right)$$

where c is a constant based on the length of the vessel, the viscosity of the blood, and the blood pressure. If $c = 1$, find $v(0.0075, 0.0045)$.

14. *Meteorology* Under certain conditions, the wind speed in mile per hour of a tornado at a distance d feet from its center can be approximated by the function

$$S(a, V, d) = \frac{aV}{0.51d^2}$$

where a is a constant that depends on certain atmospheric conditions and V is the volume of the tornado in cubic feet. A certain tornado has a volume of 2,400,000 ft^3 and $a = 0.62$. Approximate the wind speed **(a)** 100 ft from the center of this tornado and **(b)** 200 ft from the center of this tornado.

15. *Meteorology* Wind speed affects the temperature a person feels and often makes it feel colder than it is. The **Wind Chill Index** (*WCI*) represents the equivalent temperature in degrees Fahrenheit, that exposed skin would feel if there was little or no wind. If F represents the actual temperature in degrees Fahrenheit as measured by a thermometer and v is the velocity of the wind in miles per hour, then

$$WCI(v, F) = \begin{cases} F, & 0 \le v \le 4 \\ 91.4 - \dfrac{(10.45 + 6.69\sqrt{v} - 0.447v)(91.4 - F)}{22}, & 4 < v \le 45 \\ 1.60F - 55, & v > 45 \end{cases}$$

Find the wind chill index to the nearest 1 degree when **(a)** $T = 20°$F and $v = 20$ mph, **(b)** $T = 10°$F and $v = 20$ mph, **(c)** $T = 10°$F and $v = 4$ mph, and **(d)** $T = 20°$F and $v = 25$ mph.

16. *Meteorology* Humidity affects the temperature a person feels and, in the summer, often makes it feel hotter than it actually is. The **Apparent Temperature Index** (*ATI*) attempts to measure what hot weather "feels like" to the average person under various conditions of temperature and relative humidity. If F represents the actual temperature in degrees Fahrenheit as measured by a thermometer and RH is the relative humidity expressed as a decimal, then

$$ATI(F, RH) = 0.885F - 78.7RH + 1.20F(RH) + 2.70$$

Find the apparent temperature index to the nearest 1 degree, when **(a)** $T = 95°\text{F}$ and $RH = 30\%$, **(b)** $T = 95°\text{F}$ and $RH = 60\%$, **(c)** $T = 95°\text{F}$ and $RH = 90\%$, and **(d)** $T = 25°\text{F}$ and $RH = 75\%$.

17. *Package design* The packaging department in a company has been asked to design a rectangular box with no top and a partition down the middle, as in Figure 11.1. The dimensions of the box, in inches, are x, y, and z.

(a) Write a function of three variables for the number of cubic inches in the volume of the box.

(b) Write a function of three variables for the number of square inches in the material needed to construct the box.

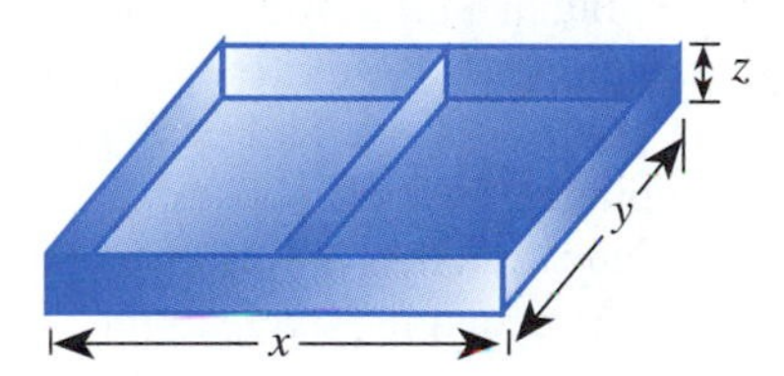

FIGURE 11.1

18. *Construction* A farmer wants to build a rectangular pen and then divide it with two interior fences, as shown in Figure 11.2. The exterior fence costs \$24/ft, and the interior fence costs \$18.50/ft. Write a function of two variables for the total cost of the material needed to construct the pen.

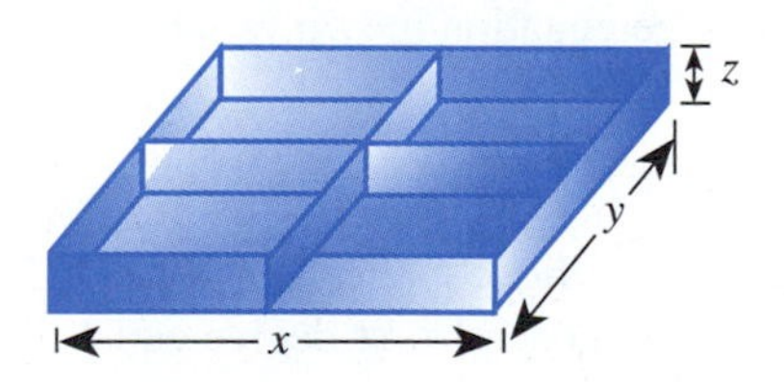

FIGURE 11.2

19. *Police science* The stopping distance, in feet, for a car after the brakes are applied is given by $S(w, r) = kwr^2$, where k is a constant, w the car's weight in lb, and r its speed in mph. If $k = 0.000\,02$, determine the stopping distance for a car weighing 2,400 lb and traveling 65 mph.

20. *Recreation* In using scuba diving gear, a diver estimates the time of a dive according to the equation $t(V, x) = \dfrac{33V}{x + 33}$, where t is the length of the dive in minutes; V the volume of air at sea level pressure, compressed into the tanks, and x the depth of the dive in feet. Find **(a)** $t(75, 67)$ and **(b)** $t(60, 32)$.

21. *Medical technology* The surface area in square meters of a human is approximated by $A(w, h) = 0.18215w^{0.425}h^{0.725}$, where w is the weight of the person in kilograms and h is the height of the person in meters.

(a) Determine the surface area of a person with a mass of 84 kg and a height of 1.80 m.

(b) Write a function of two variables to determine the surface area in square inches of a person based on the person's weight in pounds and height in inches.

(c) Determine the surface area of a person with a weight of 185 lb and a height of 5 ft 11 in.

22. *Advertising* A company spends x on newspaper advertising and y on television advertising, where x and y are in thousands of dollars per week. It has found that its weekly sales, in thousands of dollars, can be given by $S(x, y) = 5x^2y^3$. **(a)** Find $S(4, 3)$ and **(b)** $S(3, 4)$.

23. *Economics* The **Cobb-Douglas production formula** is used by economists to model the production level of a company. The formula is $P(x, y) = kx^a y^{1-a}$, where P is the total units produced, x is the measure of labor units, y is a measure of capital invested, and k is a constant that varies from product to product. Suppose $P(x, y) = 400x^{0.35}y^{0.65}$ represents the number of units produced by a company, with x units of labor and y units of capital.

(a) How many units of a product will be manufactured if 300 units of labor and 50 units of capital are used?

(b) How many units of a product will be manufactured if the units of labor and capital are both doubled?

24. *Agriculture* The number of cattle that can graze off a certain ranch without causing overgrazing is approximated by $C(x, y) = 9x + 6y - 7$, where x is the number of acres of grass and y is the number of acres of alfalfa.

(a) How many cattle can graze if there are 80 acres of grass and 20 acres of alfalfa?

(b) How many cattle can graze if there are 60 acres of grass and 40 acres of alfalfa?

In Your Words

25. What are the advantages of expressing something as a function of two variables rather than as a function of one variable?

26. Write an application in your technology area of interest that requires that you describe something as a function of two or three variables.

11.2 SURFACES IN THREE DIMENSIONS

Graphing on the rectangular coordinate system in two dimensions meant using two perpendicular axes–the x-axis and the y-axis. Graphing in a **rectangular coordinate system in three dimensions** involves using three perpendicular axes–an x-axis, a y-axis, and a z-axis. Each pair of axes forms a plane in space and each plane is perpendicular to the other two planes. The three planes are called by the axes that define them. Thus, we have an xy-plane, an xz-plane, and a yz-plane. The planes divide space into eight regions called the **octants**. The octant in which all the values are positive is known as the **first octant**. The other octants are not numbered. The coordinate system looks something like the one shown in Figure 11.3. Normally, only the axes are drawn and not the planes.

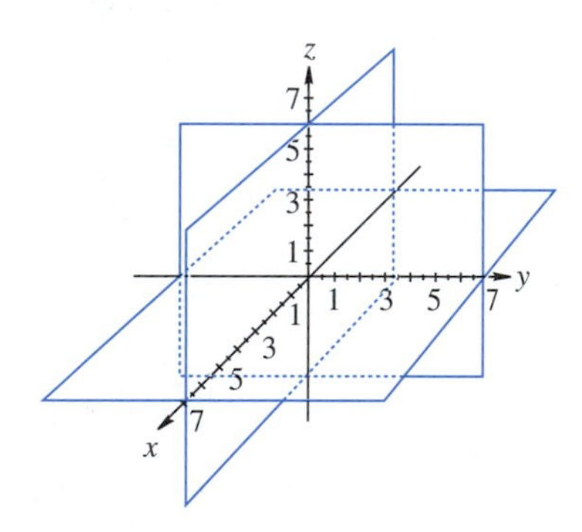

FIGURE 11.3

Locating a point is done by the following procedures and is shown in Figure 11.4. Suppose we want to locate the point $(2, 3, 5)$. First we locate the x- and y-coordinates on their respective axes. The x-coordinate is 2, so we locate the point $(2, 0, 0)$. Notice that the x-axis "comes out" towards you. Next we locate the y-coordinate at $(0, 3, 0)$ on the y-axis. From each of these points we draw a line in the xy-plane perpendicular to the axis. This locates the point $(2, 3, 0)$. From this point we draw a line straight up and 5 units long. We have now located the point $(2, 3, 5)$.

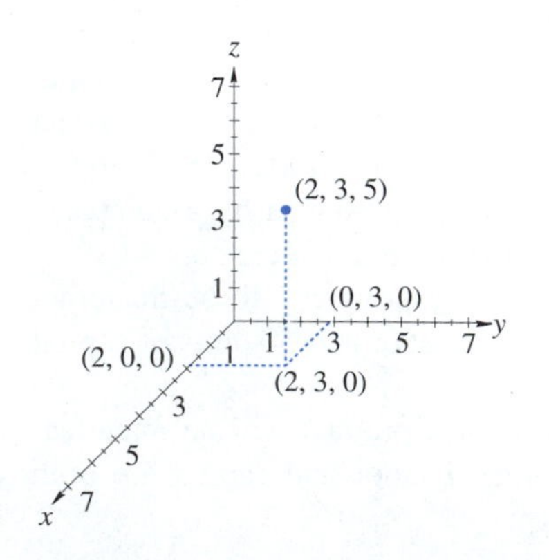

FIGURE 11.4

In Chapter 2, we showed that a second-degree equation in two variables was of the form

$$Ax^2 + Bxy + Cy^2 + Dx + Ey + F = 0$$

where A, B, and C were not all 0. The graph of this equation was one of four types—circle, ellipse, hyperbola, or parabola. The general first-degree equation in two variables $Ax + By + C = 0$ graphed a straight line.

A second-degree equation in three variables is of the form

$$Ax^2 + By^2 + Cz^2 + Dxy + Exz + Fyz + Gx + Hy + Iz + J = 0$$

This equation graphs one of nine types, called the **quadric surfaces**.

Planes

A general first-degree equation, $Ax + By + Cz + D = 0$, will graph a plane in three dimensions.

EXAMPLE 11.7

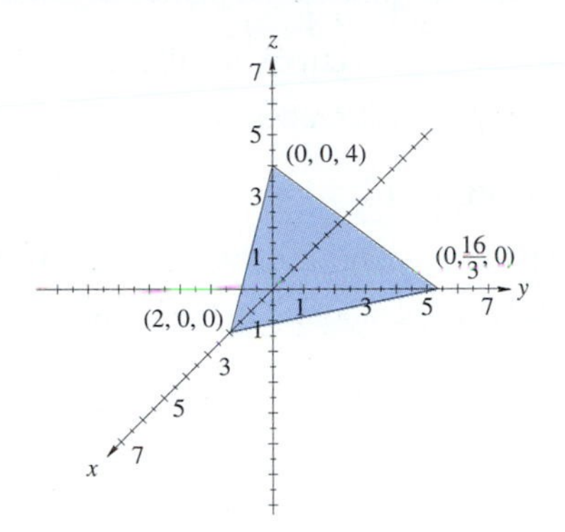

FIGURE 11.5

Sketch the graph of $4x + \frac{3}{2}y + 2z - 8 = 0$.

Solution The intercepts occur where the graph crosses the respective axes. By finding the intercepts, we get three points. It only takes three points to define a plane, so the intercepts are enough to do so.

If we let two of the variables have a value of 0 and solve the remaining equation, we will get the intercepts. If $x = y = 0$, then $z = 4$ and $(0, 0, 4)$ is one point. If $x = z = 0$, then $y = \frac{16}{3}$ and $(0, \frac{16}{3}, 0)$ is the y-intercept. If $y = z = 0$, then $x = 2$ and $(2, 0, 0)$ is the x-intercept. Plotting these three points and connecting them as in Figure 11.5 gives us a representation of the plane.

The shaded region in Figure 11.5 represents only part of the plane. In this particular case, we get the part of the plane that is in the first octant. Figure 11.6 shows the graph of $4x - \frac{3}{2}y - 2z + 8 = 0$. Here again, we only show the part of the plane that is in one octant.

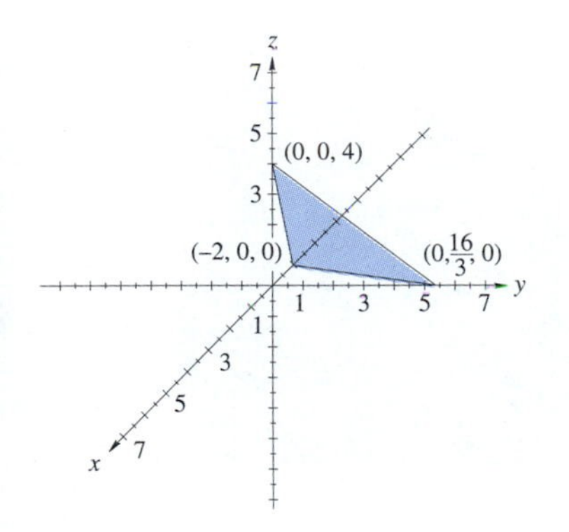

FIGURE 11.6

Quadric Surfaces

The previous two graphs are of planes. In general, the graph of an equation in three variables, which is equivalent to a function in two independent variables, is a **surface** in space. So far, we have looked at plane surfaces. In the remainder of this section, we will present the nine quadric surfaces, but before we present them we need to learn some terminology.

The intersection of two surfaces in space is a **curve**. The **traces** of a surface are the curves formed when the surface intersects a coordinate plane. A **section** is formed by the intersection of the surface with a plane other than a coordinate plane. Usually, a section is formed when the surface intersects a plane parallel to a coordinate plane.

In the list of quadric surfaces that follows, we will present the "standard equations." If the surface has a center, it will be at the origin. If the surface has symmetry, it will be symmetric with a coordinate axis. For each surface, we will give the intercepts, a description of the traces, and a sketch of the surface. In all cases, the letters a, b, and c represent positive real numbers.

1. **Ellipsoid:** $\frac{x^2}{a^2} + \frac{y^2}{b^2} + \frac{z^2}{c^2} = 1$ (Figure 11.7)

   The intercepts are $(\pm a, 0, 0)$, $(0, \pm b, 0)$, and $(0, 0, \pm c)$. The traces are ellipses, as are all sections parallel to the coordinate plane.

2. **Elliptic Cone:** $z^2 = \frac{x^2}{a^2} + \frac{y^2}{b^2}$ (Figure 11.8)

   The only intercept is at the origin $(0, 0, 0)$ and is called the **vertex**. The trace in the xy-plane is the origin. The trace in the yz-plane is the intersecting lines

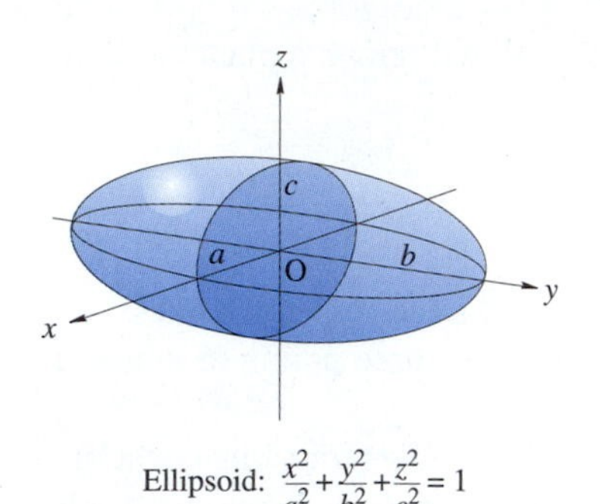

FIGURE 11.7

$z = \pm\frac{y}{b}$. The trace in the xz-plane is the intersecting lines $z = \pm\frac{x}{a}$. Sections parallel to the xy-plane are ellipses. If $a = b$, the elliptic cone become a **circular cone**.

3. Elliptic Paraboloid: $z = \frac{x^2}{a^2} + \frac{y^2}{b^2}$ (Figure 11.9)

The only intercept is at the origin and is called the **vertex**. The trace in the xy-plane is the origin. The trace in each of the other two planes is a parabola—in the xz-plane it is $z = \frac{x^2}{a^2}$ and in the yz-plane $z = \frac{y^2}{b^2}$. Sections parallel to the xy-plane are ellipses and sections parallel to the other two planes are parabolas.

4. Hyperbolic Paraboloid: $z = \frac{y^2}{b^2} - \frac{x^2}{a^2}$ (Figure 11.10)

The only intercept is at the origin. The traces in the xy-plane are the intersecting lines $\frac{y}{b} = \pm\frac{x}{a}$. The traces in the other two planes are parabolas. In the xz-plane it is $z = \frac{-x^2}{a^2}$ and in the yz-plane, $z = \frac{y^2}{b^2}$. Sections parallel to the xy-plane are hyperbolas. Sections parallel to the other two planes are parabolas.

The origin plays a unique role in this figure. It is a minimum point for the trace in the yz-plane and is a maximum point for the trace in the xy-plane. It is known as a **saddle point**.

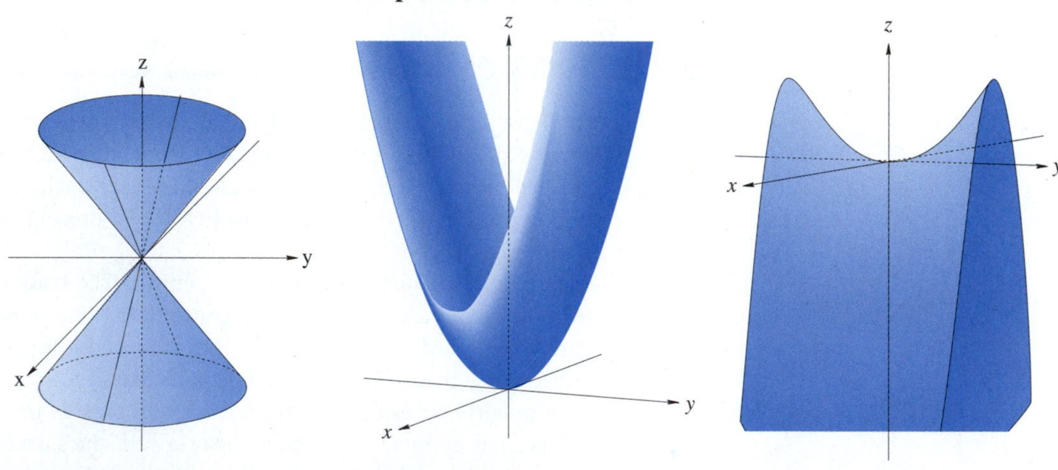

FIGURE 11.8 **FIGURE 11.9** **FIGURE 11.10**

5. Hyperboloid of One Sheet: $\frac{x^2}{a^2} + \frac{y^2}{b^2} - \frac{z^2}{c^2} = 1$ (Figure 11.11)

The intercepts are $(\pm a, 0, 0)$ and $(0, \pm b, 0)$. The trace in the xy-plane is the ellipse $\frac{x^2}{a^2} + \frac{y^2}{b^2} = 1$. The traces in the other two planes are hyperbolas. In the xz-plane, the trace is $\frac{x^2}{a^2} - \frac{z^2}{c^2} = 1$ and in the yz-plane it is $\frac{y^2}{b^2} - \frac{z^2}{c^2} = 1$.

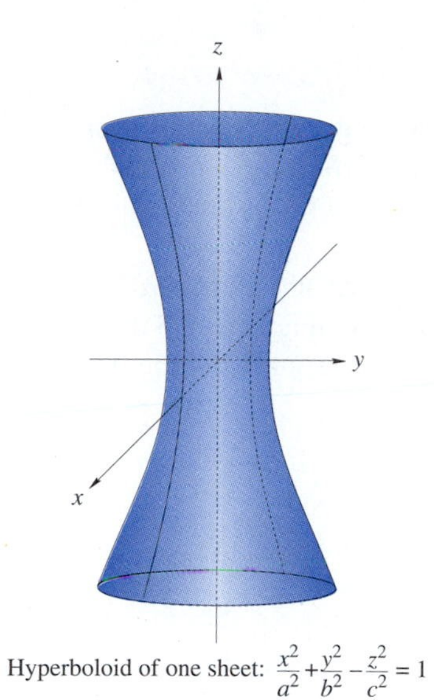

Hyperboloid of one sheet: $\frac{x^2}{a^2}+\frac{y^2}{b^2}-\frac{z^2}{c^2}=1$

FIGURE 11.11

Sections parallel to the xy-plane are ellipses and sections parallel to the other coordinate planes are hyperbolas. If $a = b$, the surface is called a **hyperboloid of revolution**.

6. Hyperboloid of Two Sheets: $\frac{x^2}{a^2} + \frac{y^2}{b^2} - \frac{z^2}{c^2} = -1$ (Figure 11.12)

The intercepts are the two points $(0, 0, \pm c)$. The trace in the xz-plane is the hyperbola $\frac{z^2}{c^2} - \frac{x^2}{a^2} = 1$ and in the yz-plane, the hyperbola $\frac{z^2}{c^2} - \frac{y^2}{b^2} = 1$. There is no trace in the xy-plane. Sections parallel to the xy-plane are ellipses and sections parallel to the other two planes are hyperbolas.

The last three quadric surfaces are called **cylinders**. A cylinder is generated by a line moving along a curve while remaining parallel to a fixed line. Notice that for each cylinder, one variable is missing.

7. Parabolic Cylinder: $x^2 = 4ay$ (Figure 11.13)

This surface is formed by a line parallel to the z-axis moving along the parabola $x^2 = 4ay$. One way to visualize this is to take a piece of paper and fold it so that two opposite edges form a parabola.

8. Elliptic Cylinder: $\frac{x^2}{a^2} + \frac{y^2}{b^2} = 1$ (Figure 11.14)

This surface is generated by a line parallel to the z-axis moving along the ellipse satisfying the given standard equation. You can visualize this by taking a sheet of paper and folding it until opposite edges meet. The other two edges should be in the shape of an ellipse.

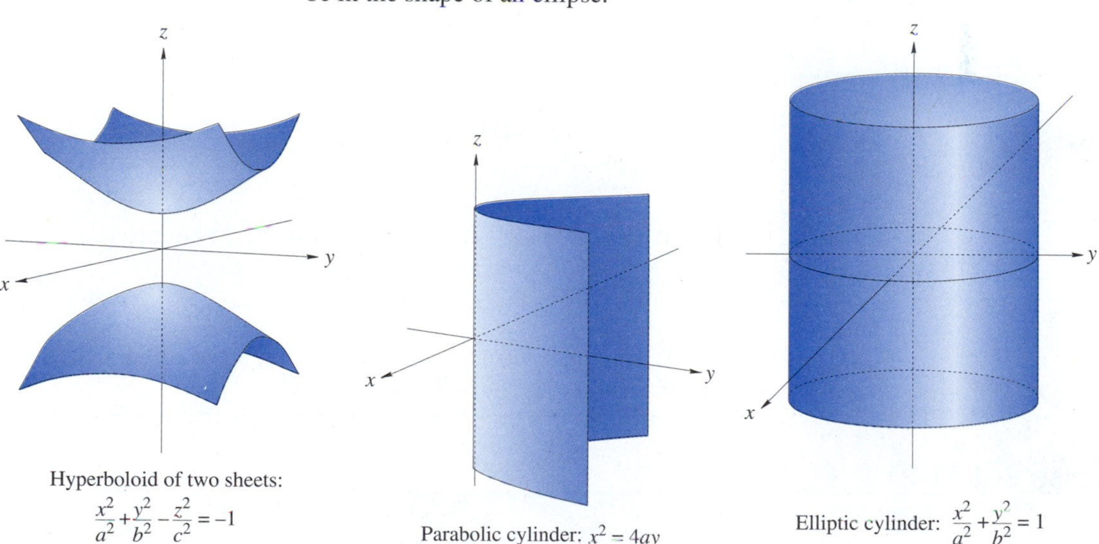

Hyperboloid of two sheets: $\frac{x^2}{a^2}+\frac{y^2}{b^2}-\frac{z^2}{c^2}=-1$

FIGURE 11.12

Parabolic cylinder: $x^2 = 4ay$

FIGURE 11.13

Elliptic cylinder: $\frac{x^2}{a^2}+\frac{y^2}{b^2}=1$

FIGURE 11.14

9. Hyperbolic Cylinder: $\dfrac{x^2}{a^2} - \dfrac{y^2}{b^2} = 1$ (Figure 11.15)

This is generated by a line parallel to the z-axis moving along the hyperbola with the given standard equation. This surface has two parts. You can visualize this surface by folding two sheets of paper where each sheet traces one branch of the hyperbola.

Remember that the nine types of quadric surfaces are all in standard position. It is certainly possible to interchange the variables. This will not change the type of surface, only its orientation. You may have to complete the square in order to get an equation into a recognizable form.

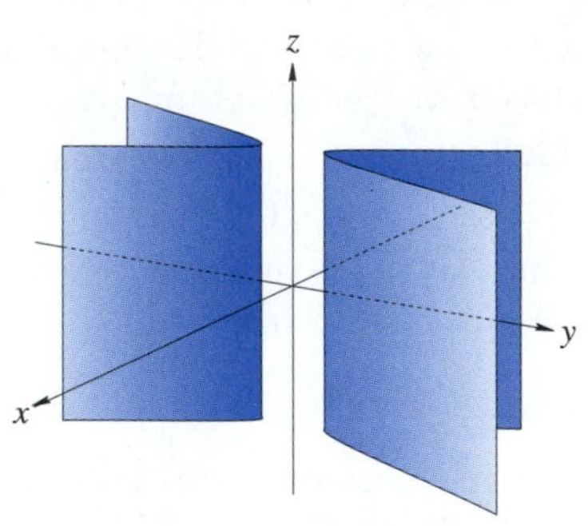

FIGURE 11.15

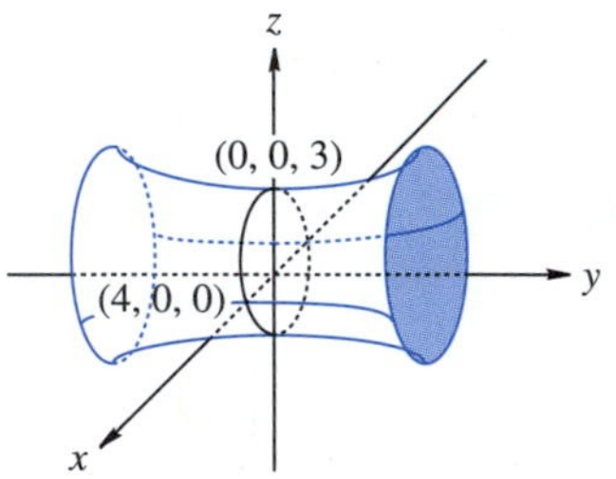

FIGURE 11.16

EXAMPLE 11.8

Identify the surface and sketch the graph of

$$9x^2 - 36y^2 + 16z^2 = 144.$$

Solution Divide both sides of the equation by 144 to get

$$\frac{x^2}{16} - \frac{y^2}{4} + \frac{z^2}{9} = 1$$

This is in the form of a hyperboloid of one sheet whose axis is the y-axis. The sketch of this curve is shown in Figure 11.16.

Exercise Set 11.2

In Exercises 1–20, identify each surface and sketch the graph in the rectangular coordinate system in three dimensions.

1. $x + 2y + 3z - 6 = 0$
2. $3x + 2y - z - 6 = 0$
3. $x = 8y^2$
4. $x^2 + y^2 + z^2 = 9$
5. $x^2 + 2y^2 + z^2 = 4$
6. $4x^2 + 9y^2 = 36$
7. $2x^2 + y^2 + 2z^2 = 8$
8. $4x^2 + y^2 - z^2 = 1$
9. $4x^2 + y^2 - z^2 = -1$
10. $4x^2 - 9y^2 = 1$
11. $3x - 4y - 6z = 12$
12. $xy = 4$

13. $z = x^2 + 4y^2 - 16$
14. $2x^2 + y^2 - 2z^2 = -8$
15. $36z = 4x^2 + 9y^2$
16. $36z^2 = 4x^2 + 9y^2$
17. $36z = 4x^2 - 9y^2$
18. $36z^2 = 1 - 4x^2 - 9y^2$
19. $2x = y^2$
20. $3x + 2y - 3z = 6$

In Your Words

21. Describe how to sketch a three-dimensional rectangular coordinate system.
22. What are quadric surfaces? Name each one and sketch its basic appearance.

11.3 PARTIAL DERIVATIVES

Derivatives have been a useful tool for us; but the derivatives that we have taken have always been with functions of one variable. Now we are working with functions of two or more independent variables. In this section, we are going to learn a method of taking derivatives that we can use with functions of several variables. This method is called **partial derivatives**.

Suppose we have a function of two variables, such as $z = f(x, y)$. If we treat y as a constant, this acts as a function in just one variable, x. We know how to take the derivative of a function in one variable. If we take the derivative of this "one-variable" function, we will call it a **partial derivative of f with respect to x**. We use the notation $\frac{\partial z}{\partial x}$, $\frac{\partial f}{\partial x}$, or f_x to indicate a partial derivative with respect to x. In the same way, if we take the derivative of $f(x, y)$ with respect to y (by treating x as a constant), then we have taken the **partial derivative of f with respect to y**. The notation $\frac{\partial z}{\partial y}$, $\frac{\partial f}{\partial y}$, or f_y is used for a partial derivative of f with respect to y. The formal definitions for partial derivatives are given in the following box.

Partial Derivatives of a Function of Two Variables

If $z = f(x, y)$, then the **first partial derivatives of** f with respect to x and y are

$$\frac{\partial f}{\partial x} = f_x = \lim_{h \to 0} \frac{f(x+h, y) - f(x, y)}{h}$$

$$\frac{\partial f}{\partial y} = f_y = \lim_{h \to 0} \frac{f(x, y+h) - f(x, y)}{h}$$

provided the limits exist.

In these definitions, if $z = f(x, y)$, then to find f_x you need to consider y a constant and differentiate with respect to x. Similarly, to find f_y, consider x a constant and differentiate with respect to y.

EXAMPLE 11.9

If $f(x, y) = x^2 + 2xy$, find $\dfrac{\partial f}{\partial x}$, $\dfrac{\partial f}{\partial y}$, $f_x(1, 2)$, and $f_y(1, 2)$.

Solution In finding the partial derivative of f with respect to x, we treat y as a constant. Thus

$$\frac{\partial f}{\partial x} = 2x + 2y$$

For the partial derivative of f with respect to y, we treat x as a constant and get

$$\frac{\partial f}{\partial y} = 0 + 2x = 2x$$

$$f_x(1, 2) = 2 + 4 = 6$$

$$f_y(1, 2) = 2$$

EXAMPLE 11.10

If $z = \sqrt{x^2 + y^3}$, calculate $\dfrac{\partial z}{\partial x}$ and $\dfrac{\partial z}{\partial y}$.

Solution First we write $z = (x^2 + y^3)^{1/2}$.

$$\frac{\partial z}{\partial x} = \frac{1}{2}(x^2 + y^3)^{-1/2}(2x + 0) = \frac{x}{\sqrt{x^2 + y^3}}$$

$$\frac{\partial z}{\partial y} = \frac{1}{2}(x^2 + y^3)^{-1/2}(0 + 3y^2) = \frac{3y^2}{2\sqrt{x^2 + y^3}}$$

EXAMPLE 11.11

If $z = \dfrac{x}{y}\sin(x^2y^3)$, determine $\dfrac{\partial z}{\partial x}$ and $\dfrac{\partial z}{\partial y}$.

Solution We need to use the product and chain rules of differentiation.

$$\begin{aligned}
\frac{\partial z}{\partial x} &= \frac{x}{y}\left[\frac{\partial}{\partial x}\sin(x^2y^3)\right] + \sin(x^2y^3)\left[\frac{\partial}{\partial x}\left(\frac{x}{y}\right)\right] \\
&= \frac{x}{y}\left[\cos(x^2y^3)\right](2xy^3) + \left[\sin(x^2y^3)\right]\left(\frac{1}{y}\right) \\
&= 2x^2y^2\cos(x^2y^3) + \frac{1}{y}\sin(x^2y^3)
\end{aligned}$$

$$\begin{aligned}
\frac{\partial z}{\partial y} &= \frac{x}{y}\left[\frac{\partial}{\partial y}\sin(x^2y^3)\right] + \sin(x^2y^3)\left[\frac{\partial}{\partial y}\left(\frac{x}{y}\right)\right] \\
&= \frac{x}{y}\left[\cos(x^2y^3)\right](3x^2y^2) + \left[\sin(x^2y^3)\right]\left(\frac{-x}{y^2}\right) \\
&= 3x^3y\cos(x^2y^3) - \frac{x}{y^2}\sin(x^2y^3)
\end{aligned}$$

Geometric Interpretation of Partial Derivatives

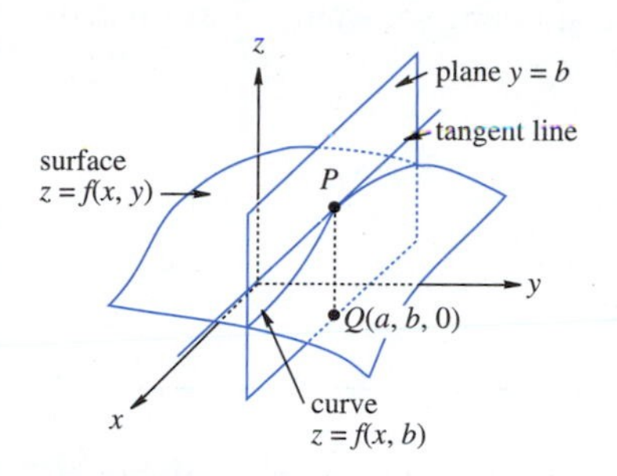

FIGURE 11.17

The partial derivatives f_x and f_y are the slopes of tangent lines to certain curves on the surface of $z = f(x, y)$. Consider the point $P(a, b, f(a, b))$ on this surface. This point lies directly above the point $Q(a, b, 0)$ in the xy-plane, as shown in Figure 11.17. The vertical plane $y = b$ is parallel to the xz-plane. (If $b = 0$, the plane $y = b$ is the xz-plane.) The plane $y = b$ intersects the surface in the curve $z = f(x, b)$ through the point P. Because this curve is in the plane $y = b$, the x values can change, but the y values are always b. The partial derivative of z with respect to x represents the slope of a line tangent to this curve $z = f(x, b)$.

In the same way, the partial derivative of z with respect to y represents the slope of a line tangent to the curve $z = f(a, y)$. This curve is formed by the vertical plane $x = a$, which is parallel to the yz-plane, intersecting the surface as shown in Figure 11.18.

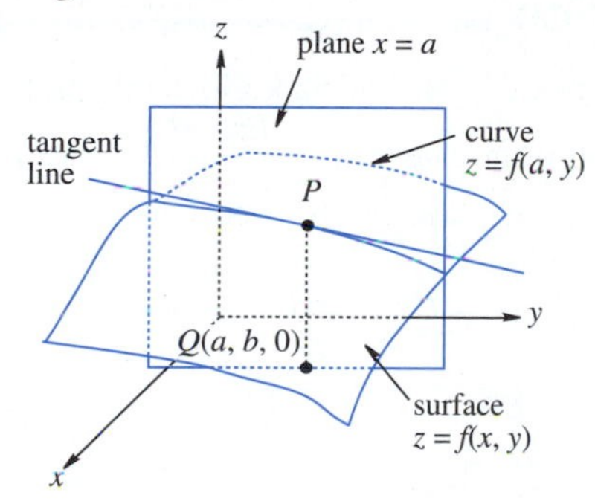

FIGURE 11.18

To find the slope of a line tangent to surface $z = f(x, y)$ and parallel to the xz-plane at the point $P(a, b, f(a, b))$, calculate $\dfrac{\partial z}{\partial x}$.

Similarly, to find the slope of a line tangent to surface $z = f(x, y)$ and parallel to the yz-plane at the point $P(a, b, f(a, b))$, calculate $\dfrac{\partial z}{\partial y}$.

EXAMPLE 11.12

(a) Find the slope of a line tangent to the surface $z = x^2 + 3y^2$ and parallel to the xz-plane at the point $(2, -1, 7)$. **(b)** Find the slope of the line that is tangent to this surface and parallel to the yz-plane at the same point.

Solutions **(a)** We want to find $\dfrac{\partial z}{\partial x}$ and then evaluate it at $(2, -1, 7)$:

$$\frac{\partial z}{\partial x} = 2x$$

EXAMPLE 11.12 (Cont.)

So, at $(2,-1,7)$, $\left.\dfrac{\partial z}{\partial x}\right|_{(2,-1,7)} = 4$. **(b)** For the second part, we want $\dfrac{\partial z}{\partial y}$ evaluated at the same point:

$$\frac{\partial z}{\partial y} = 6y$$

$$\text{So,} \qquad \left.\frac{\partial z}{\partial y}\right|_{(2,-1,7)} = -6.$$

Applications of partial derivatives are found in many ways. The following is but one example. Others are shown in the exercises and in the Section 11.4.

Application

EXAMPLE 11.13

A light source I of a candelabra located r m from a surface inclined at an angle θ provides an illuminance $E = \dfrac{I\cos\theta}{r^2}$ lumens per square meter (lm/m^2). Find the rate at which E varies with respect to I and evaluate it for $\theta = \frac{\pi}{6}$, $r = 2$ m, and $I = 100$ cd.

Solution

$$\frac{\partial E}{\partial I} = \frac{\cos\theta}{r^2}$$

$$\left.\frac{\partial E}{\partial I}\right|_{\substack{r=2\\ \theta=\frac{\pi}{6}}} = \frac{\cos(\frac{\pi}{6})}{2^2}$$

$$= \frac{\sqrt{3}/2}{4} = \frac{\sqrt{3}}{8} \approx 0.217\,\text{lm/m}^2$$

Higher Partial Derivatives

We know that if we have a function in two independent variables, such as $z = f(x, y)$, we can take the partial derivative with respect to either variable: $\dfrac{\partial z}{\partial x}$ and $\dfrac{\partial z}{\partial y}$. We can just as easily differentiate each of these derivatives with respect to either variable. Thus, for $\dfrac{\partial z}{\partial x}$ we find the derivative with respect to x, $\dfrac{\partial}{\partial x}\left(\dfrac{\partial z}{\partial x}\right) = \dfrac{\partial^2 z}{\partial x^2}$, or we can take it with respect to y, $\dfrac{\partial}{\partial y}\left(\dfrac{\partial z}{\partial x}\right) = \dfrac{\partial^2 z}{\partial y\,\partial x}$. In the same way, we can differentiate $\dfrac{\partial z}{\partial y}$ with respect to x, $\dfrac{\partial}{\partial x}\left(\dfrac{\partial z}{\partial y}\right) = \dfrac{\partial^2 z}{\partial x\,\partial y}$, and with respect to y, $\dfrac{\partial}{\partial y}\left(\dfrac{\partial z}{\partial y}\right) = \dfrac{\partial^2 z}{\partial y^2}$. Thus, we have the following second partial derivatives for a function of two variables.

Second Partial Derivatives of a Function of Two Variables

If $z = f(x, y)$, then there are four **second partial derivatives** of f:

1. Differentiate twice with respect to x:

$$\frac{\partial}{\partial x}\left(\frac{\partial z}{\partial x}\right) = \frac{\partial^2 f}{\partial x^2} = f_{xx}$$

2. Differentiate twice with respect to y:

$$\frac{\partial}{\partial y}\left(\frac{\partial z}{\partial y}\right) = \frac{\partial^2 f}{\partial y^2} = f_{yy}$$

3. Differentiate first with respect to x and then with respect to y:

$$\frac{\partial}{\partial y}\left(\frac{\partial z}{\partial x}\right) = \frac{\partial^2 f}{\partial y\,\partial x} = f_{xy}$$

4. Differentiate first with respect to y and then with respect to x:

$$\frac{\partial}{\partial x}\left(\frac{\partial z}{\partial y}\right) = \frac{\partial^2 f}{\partial x\,\partial y} = f_{yx}$$

Note The last two second-order partial derivatives

$$\frac{\partial^2 z}{\partial x\,\partial y} = f_{yx}(x, y) \text{ and } \frac{\partial^2 z}{\partial y\,\partial x} = f_{xy}(x, y)$$

are called **mixed partial derivatives**. Notice the differences in these two equations. The notation f_{xy} means that we should first differentiate with respect to x and then differentiate that result with respect to y.

In Section 11.4, we shall see how to use second partial derivatives to determine relative maximum, relative minimum, concavity, and saddle points, a new property.

EXAMPLE 11.14

Find all second-order partial derivatives of $f(x, y) = \sin(xy)$.

Solution We first find the first-order partial derivatives.

$$f_x = y\cos(xy)$$
$$f_y = x\cos(xy)$$

EXAMPLE 11.14 (Cont.)

Then, we get the following second-order partial derivatives.

$$f_{xx} = \frac{\partial}{\partial x}(f_x) = -y^2 \sin(xy)$$
$$f_{xy} = \frac{\partial}{\partial y}(f_x) = -xy\sin(xy) + \cos(xy)$$
$$f_{yx} = \frac{\partial}{\partial x}(f_y) = -xy\sin(xy) + \cos(xy)$$
$$f_{yy} = \frac{\partial}{\partial y}(f_y) = -x^2 \sin(xy)$$

Note In this example, $f_{xy} = f_{yx}$ for all (x, y). As it turns out, this will be the case for most functions.

Exercise Set 11.3

In Exercises 1–16, find f_x and f_y.

1. $f(x, y) = x^2y + y^2x$
2. $f(x, y) = x^2y + 7y^2$
3. $f(x, y) = 3x^2 + 6xy^3$
4. $f(x, y) = \dfrac{x}{y^2}$
5. $f(x, y) = \dfrac{x^2 + y^2}{y}$
6. $f(x, y) = \dfrac{x + y}{\sqrt{xy}}$
7. $f(x, y) = e^y \cos x + e^x \sin y$
8. $f(x, y) = x^2 \cos y + y^2 \cos x$
9. $f(x, y) = e^{2x+3y}$
10. $f(x, y) = e^{x^2+y^2}$
11. $f(x, y) = \sin(x^2y^3)$
12. $f(x, y) = \ln(x^2y^3)$
13. $f(x, y) = \ln\sqrt{x^2 + y^2}$
14. $f(x, y) = \tan^{-1}\dfrac{y}{x}$
15. $f(x, y) = \sin^2(3xy)$
16. $f(x, y) = e^{\sqrt{x^2+y^2}}$

In Exercises 17–24, compute all second-order partial derivatives.

17. $f(x, y) = x^3y^2 - xy^5$
18. $f(x, y) = \sin xy^3$
19. $f(x, y) = 3e^{xy^3}$
20. $f(x, y) = \sin(x + y^2)$
21. $f(x, y) = \dfrac{2x}{y^5}$
22. $f(x, y) = e^x \tan y$
23. $f(x, y) = \ln(x^3y^5)$
24. $f(x, y) = \sqrt{x^2y + 3y^2}$

Solve Exercises 25–36.

25. *Physics* The **ideal gas law** states that $P = \frac{nRT}{V}$, where n is the number of moles of an ideal gas, T is the absolute temperature, and R is the universal gas constant. If $n = 20$ and $T = 22°C = 295°K$, what is the rate of change of the pressure with respect to the volume, when the volume is 3 L?

26. *Thermodynamics* The temperature distribution T of a heated plate located in the xy-plane is given by

$$T = \frac{100}{\ln 2}\ln(x^2 + y^2), \quad \text{for } 1 \le x^2 + y^2 \le 4$$

Find the rate of change of T in a direction parallel to the x-axis at the point $(1, 0)$ and at the point $(0, 1)$.

27. *Physics* The volume V (in cubic centimeters) of one mole of an ideal gas is given by

$$V = \frac{82.06T}{P}$$

where P is the pressure in atmospheres (atm) and T is the absolute temperature. Find the rate of change of the volume of 1 mole with respect to pressure, when $T = 300°K$ and $P = 5$ atm.

28. *Physics* Find the rate of change of the volume of 1 mole of the ideal gas in Exercise 27 with respect to temperature, when $T = 300°K$ and $P = 5$ atm.

29. Let C be the trace of the paraboloid $z = 16 - x^2 - y^2$ on the plane $x = 1$. Find the slope of the tangent line to C at the point $(1, 3, 6)$.

30. *Physics* For the ideal gas equation $PV = nRT$, show that

$$\left(\frac{\partial V}{\partial T}\right)\left(\frac{\partial T}{\partial P}\right)\left(\frac{\partial P}{\partial V}\right) = -P$$

31. *Medical technology* One of **Poiseuille's laws** states that the resistance, R, for blood flowing in a blood vessel depends on the length, L, and radius, r, of the vessel according to the formula $R = k\frac{L}{r^4}$, where k is a constant.

(a) What is the rate of change in the resistance with respect to the length?

(b) What is the rate of change in the resistance with respect to the radius?

32. *Economics* The production of a computer manufacturing company is given approximately by the Cobb-Douglas production formula $P(x, y) = 65x^{0.3}y^{0.7}$, where x is the number of units of labor and y is the number of units of capital.

(a) Find $P_x(x, y)$, the **marginal productivity of labor**.

(b) If the company is currently using 1,500 units of labor and 5,400 units of capital, find the marginal productivity of labor.

(c) Find $P_y(x, y)$, the **marginal productivity of capital**.

(d) If the company is currently using 1,500 units of labor and 5,400 units of capital, find the marginal productivity of capital.

(e) For the greatest increase in productivity, should the management of the company encourage increased use of labor or capital?

33. *Medical technology* The surface area in square inches of a human is approximated by $A(w, h) = 14.085w^{0.425}h^{0.725}$, where w is the weight of the person in pounds and h is the height of the person in inches.

(a) Find $A_w(w, h)$ and $A_h(w, h)$.

(b) Compute $A_w(w, h)$ and $A_h(w, h)$ for a person with a weight of 185 lb and a height of 5 ft 11 in.

(c) Interpret your answers from (b).

34. *Medical technology* The surface area in square meters of a human is approximated by $A(w, h) = 0.18215w^{0.425}h^{0.725}$, where w is the weight of the person in kilograms and h is the height of the person in meters.

(a) Find $A_w(w, h)$ and $A_h(w, h)$.

(b) Compute $A_w(w, h)$ and $A_h(w, h)$ for a person with a weight of 84 kg and a height of 1.80 m.

35. *Police science* The stopping distance for a car after the brakes are applied is given by $S(w, r) = 0.000\,02wr^2$, where w is the car's weight in pounds and r its speed in miles per hour.
(a) Determine $S_w(w, r)$.
(b) Calculate $S_w(w, r)$ for a car weighing 2,400 lb and traveling 65 mph.
(c) Determine $S_r(w, r)$.
(d) Calculate $S_r(w, r)$ for a car weighing 2,400 lb and traveling 65 mph.
(e) How much will the stopping distance increase for this car if its velocity is increased from 65 mph to 66 mph?

36. *Meteorology* Wind speed affects the temperature a person feels and often makes it feel colder than it is. The wind chill index (WCI) represents the equivalent temperature, in degrees Fahrenheit, that exposed skin

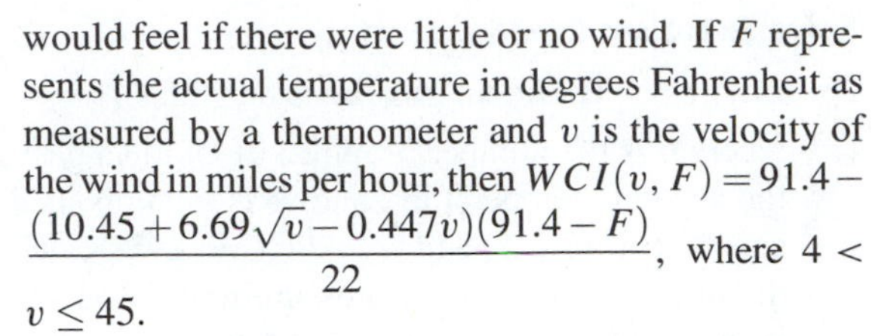

would feel if there were little or no wind. If F represents the actual temperature in degrees Fahrenheit as measured by a thermometer and v is the velocity of the wind in miles per hour, then $WCI(v, F) = 91.4 - \dfrac{(10.45 + 6.69\sqrt{v} - 0.447v)(91.4 - F)}{22}$, where $4 < v \leq 45$.
(a) Evaluate $WCI(v, F)$ for wind velocity of 30 mph and temperature of −4°F.
(b) Determine $WCI_v(v, F)$.
(c) Calculate $WCI_v(v, F)$ for wind velocity of 30 mph and temperature of −4°F.
(d) Determine $WCI_F(v, F)$.
(e) Calculate $WCI_F(v, F)$ for wind velocity of 30 mph and temperature of −4°F.
(f) Would you feel more comfortable if the wind velocity increased 1 mph or if temperature dropped 1°F?

In Your Words

37. Without looking in the text, describe how to find the partial derivatives of a function of two variables.

38. Without looking in the text, **(a)** list all the second partial derivatives of a function of two variables, and **(b)** state how each second partial derivative is found.

11.4 SOME APPLICATIONS OF PARTIAL DERIVATIVES

Some applications of partial derivatives were discussed in Section 11.3. You may remember that we applied derivatives to such problems as related rates, maxima, and minima. Partial derivatives can be used to solve similar problems. In this section, we will learn how to calculate the differential of a function of two variables and how to apply it. We will also learn how to apply maxima and minima for functions of two variables.

Differentials

If $y = f(x)$ is a function in one variable and if $f'(x)$ exists, then the change in y, Δy, from x to $x + h$, is approximated by the differential of f

$$dy = f'(x)\,dx$$

We discussed this idea in Section 5.7. We can define a differential for a function of two variables in much the same manner.

Differential for a Function of Two Variables

If $z = f(x, y)$ is a function of two variables, then the **differential** dz, usually referred to as the **total differential** of z, is defined as

$$dz = \frac{\partial z}{\partial x} dx + \frac{\partial z}{\partial y} dy$$

EXAMPLE 11.15

Find the differential dz, if $z = x^2 + 3xy - 2y^2$.

Solution We will first compute $\frac{\partial z}{\partial x}$ and $\frac{\partial z}{\partial y}$.

$$\frac{\partial z}{\partial x} = 2x + 3y$$

$$\frac{\partial z}{\partial y} = 3x - 4y$$

Using the formula $dz = \frac{\partial z}{\partial x} dx = \frac{\partial z}{\partial y} dy$, we get the total differential of z.

$$dz = (2x + 3y)\,dx + (3x - 4y)\,dy$$

EXAMPLE 11.16

For the function in Example 11.15, estimate the change in z when x changes from 3.0 to 3.1 and y changes from 6.0 to 5.8.

Solution From Example 11.15, we have $dz = (2x + 3y)\,dx + (3x - 4y)\,dy$. We have the values $x = 3.0$, $y = 6.0$, $dx = 0.1$, and $dy = -0.2$. Notice that dy is negative because the value of y is decreasing. Substituting these values into the equation for dz, we get

$$\begin{aligned} dz &= [2(3.0) + 3(6.0)]0.1 + [3(3.0) - 4(6.0)](-0.2) \\ &= 24(0.1) + (-15)(-0.2) \\ &= 2.4 + 3.0 \\ &= 5.4 \end{aligned}$$

The answer in Example 11.16 is an approximation to the actual change. We could compute the actual change by finding

$$\begin{aligned} f(3.1, 5.8) - f(3.0, 6.0) &= \left[(3.1)^2 + 3(3.1)(5.8) - 2(5.8)^2\right] \\ &\quad - \left[(3.0)^2 + 3(3.0)(6.0) - 2(6.0)^2\right] \\ &= (9.61 + 53.94 - 67.28) - (9 + 54 - 72) \\ &= -3.73 - (-9) = 5.27 \end{aligned}$$

Application

EXAMPLE 11.17

A bottling company requires cans in the shape of a right circular cylinder of height 11.7 cm and radius 3.1 cm. The can manufacturer claims a percentage error of no more than 0.2% in the height and 0.1% in the radius. What is the maximum error in the volume?

Solution The volume of a right circular cylinder is given by $V = \pi r^2 h$. The total differential dV is

$$\begin{aligned} dV &= \frac{\partial V}{\partial r}dr + \frac{\partial V}{\partial h}dh \\ &= 2\pi rh\,dr + \pi r^2\,dh \end{aligned}$$

The relative error of the radius is $\frac{|\Delta r|}{r} = \frac{|dr|}{r} = 0.001$ since $0.1\% = 0.001$ and the relative error of the height is $\frac{|\Delta h|}{h} = \frac{|dh|}{h} = 0.002$. Thus, we have $|dr| = 0.001r$ and $|dh| = 0.002h$. The maximum error in the volume is then given by

$$\begin{aligned} dV &= 2\pi rh|dr| + \pi r^2|dh| \\ &= 0.002\pi r^2 h + .002\pi r^2 h \\ &= 0.004\pi r^2 h \\ &= 0.004V \end{aligned}$$

Thus, we see that the change in volume is at most $0.004 = 0.4\%$.

Since

$$\begin{aligned} V &= \pi r^2 h = (3.1)^2(11.7)\pi \\ &\approx 353.23 \end{aligned}$$

$dV = 0.004V \approx 1.41$ and so the actual volume lies between 351.82 and 354.64 cm^3.

Note The definition of the total differential can be extended to any number of variables. For example, if you have a function of three variables

$$w = f(x, y, z)$$

then $$dw = \frac{\partial w}{\partial x}dx + \frac{\partial w}{\partial y}dy + \frac{\partial w}{\partial z}dz$$

Chain Rule

If x and y are parametric equations, where $x = g(t)$ and $y = h(t)$, and you have a function of two variables, $z = f(x, y)$, then you can use the following version of the **chain rule**.

Chain Rule for a Function of Two Variables $$\frac{dz}{dt} = \frac{\partial z}{\partial x}\cdot\frac{dx}{dt} + \frac{\partial z}{\partial y}\cdot\frac{dy}{dt}$$

EXAMPLE 11.18

Find $\frac{dw}{dt}$, where $w = e^{-x^2-y^2}$, $x = t$, and $y = \sqrt{t}$.

Solution
$$\frac{dw}{dt} = \frac{\partial w}{\partial x} \cdot \frac{dx}{dt} + \frac{\partial w}{\partial y} \cdot \frac{dy}{dt}$$
$$= \left(-2xe^{-x^2-y^2}\right)(1) + \left(-2ye^{-x^2-y^2}\right)\left(\frac{1}{2}t^{-1/2}\right)$$

Substituting for x and y, we get

$$= \left(-2te^{-t^2-t}\right) + \left(-2\sqrt{t}e^{-t^2-t}\right)\left(\frac{1}{2\sqrt{t}}\right)$$
$$= -(2t+1)e^{-(t^2+t)}$$

Chain Rule and Related Rates

The chain rule can be used with problems of related rates, such as the one in the next example.

Application

EXAMPLE 11.19

The radius of a right circular cone is increasing at the rate of 3 cm/s, while its height is increasing at the rate of 5 cm/s. How fast is the volume changing, when $r = 18$ cm and $h = 30$ cm?

Solution We are asked to find $\frac{dV}{dt}$, where $V = \frac{1}{3}\pi r^2 h$. Using the chain rule, we obtain

$$\frac{dV}{dt} = \frac{\partial V}{\partial r} \cdot \frac{dr}{dt} + \frac{\partial V}{\partial h} \cdot \frac{dh}{dt}$$

Since $\frac{\partial V}{\partial r} = \frac{2}{3}\pi rh$ and $\frac{\partial V}{\partial h} = \frac{1}{3}\pi r^2$, we have

$$\frac{dV}{dt} = \frac{2}{3}\pi rh\frac{dr}{dt} + \frac{1}{3}\pi r^2\frac{dh}{dt}$$

When $r = 18$, $h = 30$, $\frac{dr}{dt} = 3$, and $\frac{dh}{dt} = 5$, we get

$$\frac{dV}{dt} = \frac{2}{3}\pi(18)(30)(3) + \frac{1}{3}\pi(18)^2(5)$$
$$= 1\,620\pi \text{ cm}^3/\text{s}$$

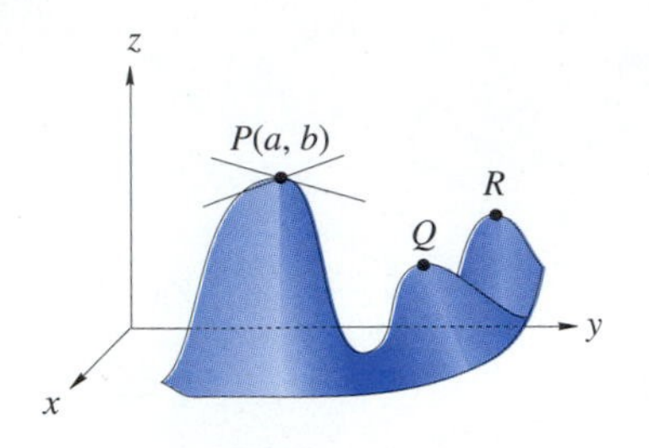

FIGURE 11.19

Maxima and Minima

Our last topic in this section deals with the extrema; that is, the maximum and minimum points on a surface. Suppose we have a surface, $z = f(x, y)$, such as the one in Figure 11.19. We have several points, P, Q, and R, that all seem to be local or relative maximum points. If one of these is a maximum point, then every tangent line is horizontal, as is shown at point P. In particular, we have

$$\frac{\partial z}{\partial x} = 0 = \frac{\partial z}{\partial y}$$

or

$$f_x(a, b) = 0 = f_y(a, b)$$

if P is the point (a, b). The same would be true if P was a minimum point. Thus, in order for $z = f(x, y)$ to have a local maximum or minimum, both partial derivatives of f must exist and have a value of 0. We state this formally as the following first derivative test.

First Derivative Test for Local Extrema

If $z = f(x, y)$ has a local maximum or minimum value at an interior point (a, b) of its domain, where f_x and f_y are both defined, then

$$f_x(a, b) = 0 = f_y(a, b)$$

or

$$\frac{\partial z}{\partial x} = \frac{\partial z}{\partial y} = 0 \text{ at } (a, b)$$

This provides us with a list of critical values but no guarantee that a maximum or minimum exists. The **second derivative test for functions of two variables** provides us with this information. Unfortunately, the second derivative test is fairly complicated.

An important consideration in the determination of relative maxima and minima of $z = f(x, y)$ is the determinant given below, which we have called Δ (delta).

Note Remember that the determinant $\begin{vmatrix} a & b \\ c & d \end{vmatrix} = ad - bc$.

$$\Delta = \begin{vmatrix} \dfrac{\partial^2 z}{\partial x^2} & \dfrac{\partial^2 z}{\partial y\,\partial x} \\ \dfrac{\partial^2 z}{\partial x\,\partial y} & \dfrac{\partial^2 z}{\partial y^2} \end{vmatrix} = \frac{\partial^2 z}{\partial x^2} \cdot \frac{\partial^2 z}{\partial y^2} - \left(\frac{\partial^2 z}{\partial y\,\partial x}\right)^2$$

(Notice that we are assuming that $\dfrac{\partial^2 z}{\partial x\,\partial y} = \dfrac{\partial^2 z}{\partial y\,\partial x}$.)

Second Derivative Test

If $z = f(x, y)$, $\dfrac{\partial z}{\partial x} = \dfrac{\partial z}{\partial y} = 0$ at (a, b), and

(a) if $\Delta > 0$ and $\dfrac{\partial^2 z}{\partial x^2} > 0$, then $f(a, b)$ is a **relative** or **local minimum**;

(b) if $\Delta > 0$ and $\dfrac{\partial^2 z}{\partial x^2} < 0$, then $f(a, b)$ is a **relative** or **local maximum**;

(c) If $\Delta < 0$ at (a, b), then $(a, b, f(a, b))$ is a **saddle point**;

(d) If $\Delta = 0$ at (a, b), the test fails,

$$\text{where} \quad \Delta = \begin{vmatrix} \dfrac{\partial^2 z}{\partial x^2} & \dfrac{\partial^2 z}{\partial y\, \partial x} \\ \dfrac{\partial^2 z}{\partial x\, \partial y} & \dfrac{\partial^2 z}{\partial y^2} \end{vmatrix} = \frac{\partial^2 z}{\partial x^2} \cdot \frac{\partial^2 z}{\partial y^2} - \left(\frac{\partial^2 z}{\partial y\, \partial x} \right)^2$$

It turns out that if $\Delta > 0$, then $\dfrac{\partial^2 z}{\partial x^2} \cdot \dfrac{\partial^2 z}{\partial y^2} > 0$, so these two second-order partial derivatives have the same sign. Thus, you can replace $\dfrac{\partial^2 z}{\partial x^2}$ with $\dfrac{\partial^2 z}{\partial y^2}$ in (a) and (b).

In (c), $\Delta < 0$. This point (a, b) was neither a maximum nor a minimum, but it is a special point called a **saddle point**. An example of a saddle point is at the origin in Figure 11.10. This surface is a minimum for the trace in the yz-plane and is a maximum for the trace in the xy-plane.

EXAMPLE 11.20

If $f(x, y) = 3x - x^3 - 3xy^2$, find the extrema of f.

Solution We have $f_x(x, y) = 3 - 3x^2 - 3y^2$ and $f_y(x, y) = -6xy$. Critical values will occur when $f_x = f_y = 0$. The fact that $f_y = -6xy = 0$ implies that either $x = 0$ or $y = 0$. The fact that $f_x = 0$ when $x = 0$ implies that $y = \pm 1$, and when $y = 0$, $x = \pm 1$. Thus, we have four critical points: $(1, 0)$, $(0, 1)$, $(-1, 0)$, and $(0, -1)$.

The second-order partial derivatives are $\dfrac{\partial^2 f}{\partial x^2} = -6x$, $\dfrac{\partial^2 f}{\partial y^2} = -6x$, $\dfrac{\partial^2 f}{\partial y\, \partial x} = -6y$, $\dfrac{\partial^2 f}{\partial x\, \partial y} = -6y$, so

$$\begin{aligned} \Delta &= (-6x)(-6x) - (-6y)^2 = 36x^2 - 36y^2 \\ &= 36(x^2 - y^2) \end{aligned}$$

at each of the critical points.

Now if $x = 0$, then $x^2 - y^2 = -1$ and so $\Delta < 0$; thus we do not have an extremum. [But we do have a saddle point at $(0, 1)$ and $(0, -1)$.]

If $y = 0$, then $x^2 - y^2 = 1$ and $\Delta > 0$. If $x = 1$, then $\dfrac{\partial^2 f}{\partial x^2} = -6$ and so $(1, 0)$ is a relative maximum. If $x = -1$, $\dfrac{\partial^2 f}{\partial x^2} = 6$ and so $(-1, 0)$ is a relative minimum. ▪

Exercise Set 11.4

In Exercises 1–8, find the total differential of each function.

1. $z = x^2 + y^2$
2. $z = 2x^2 + xy - y^2$
3. $z = 3x^2 + 4xy - 2y^3$
4. $z = xye^{x+y}$
5. $z = \arctan\left(\frac{y}{x}\right)$
6. $z = \ln(x^2 + y^2)$
7. $z = x \tan yx$
8. $z = \ln\left(\frac{x}{y}\right)$

In Exercises 9–12, find $\frac{dz}{dt}$ by using the chain rule.

9. $z = x^2 + y^2,\ x = te^t,\ y = t^2e^t$
10. $z = \frac{1}{x^2 + y^2},\ x = \cos 2t,\ y = \sin 2t$
11. $z = e^u \sin v,\ u = \sqrt{t},\ v = \pi t$
12. $z = x^3 - y,\ x = te^{-t},\ y = \sin t$

In Exercises 13–20, find all critical points and test for relative extrema.

13. $z = x^2 + 4y^2 + x + 8y + 1$
14. $z = x^2 + y^2 - 2x + 4y + 2$
15. $z = 20 + 12x - 12y - 3x^2 - 2y^2$
16. $z = xy + 3x - 2y + 4$
17. $z = x^2 + 2xy - y^2$
18. $z = x^2 - 3xy - y^2$
19. $z = x^3 + x^2y + y^2$
20. $z = x^3 + y^3 + 3xy$

Solve Exercises 21–36.

21. The base and height of a rectangle are measured as 10 cm, and 25 cm, respectively, with a possible error of 0.1 cm in each. Use differentials to determine the maximum possible error in the area of the rectangle.
22. The radius of the base of a right circular cone is measured as 4 in.; its height is measured as 12 in. A possible error of $\frac{1}{16}$ in. was made in measuring each dimension. Use differentials to estimate the maximum error that might occur in computing **(a)** the volume of the cone and **(b)** the total surface area of the cone.
23. *Electricity* The total resistance R of two resistors R_1 and R_2 connected in parallel is

$$R = \frac{R_1 R_2}{R_1 + R_2}$$

If $R_1 = 100\,\Omega$ and $R_2 = 400\,\Omega$ with a maximum error of 1% in each measurement, what is the maximum error in the value of R?

24. *Physics* The equation relating pressure P, volume V, and temperature T of a confined gas is $PV = kT$, where k is a constant. If $P = 0.5$ psi when $V = 66$ in.^{3} and $T = 110$°F, approximate the change in P, if V and T change to 70 in.^{3} and 112°F.
25. *Physics* If P and V in Exercise 24 are changing at the rate of $\frac{dP}{dt}$ and $\frac{dV}{dt}$, respectively, use the chain rule to find a formula for $\frac{dT}{dt}$.
26. *Physics* A confined gas satisfies the equation $PV = kT$, as described in Exercises 24 and 25, with $k = 0.4$. If the pressure is increasing at the rate of 0.05 psi/min and the volume is decreasing at the rate of 1 in.^{3}/min, find the rate of change of the temperature at the moment $P = 0.5$ psi and $V = 66$ in.^{3}
27. *Electronics* The energy stored in an inductor is $w = \frac{Li^2}{2}$. If L changes from 3.0 to 3.02 H while i changes from 1.0 to 0.9 A, what is the approximate change in w?
28. A vertical line is moving to the right at 2 cm/min, and a horizontal line is moving upward at 3 cm/min. What is the rate of change in the area of the rectangle formed by the coordinate axes and these two lines when $x = 6$ and $y = 7$?

29. *Sheet metal technology* A manufacturer wishes to make an open rectangular box of volume $V = 500\,\text{cm}^3$. Find the dimensions of the box that will use the least possible amount of material.

30. *Sheet metal technology* Rework Exercise 29 if the box has a lid.

31. *Sheet metal technology* An open rectangular box is constructed out of different materials. The material for the bottom costs $\$1.25/\text{ft}^2$. The material for the sides costs $\$1.50/\text{ft}^2$. The volume of the box is to be $4\,\text{ft}^3$. **(a)** What are the dimensions of the box that will cost the least to build? (See Example 11.6.) **(b)** What is this cost?

32. *Thermodynamics* A flat plate is heated such that the temperature T at any point (x, y) is given by $T = x^2 + 2y^2 - x$. What is the temperature at the coldest point?

33. *Pharmacology* When x units of drug A and y units of drug B are used in the treatment of one type of cancer, the reaction of the patient 12 h after their administration is given by $R(x, y) = 15xy(2 - x - 4y)$. Find x and y in order to maximize this reaction.

34. *Business* The total cost in dollars to produce x units of electrical tape and y units of packing tape is given by

$$C(x, y) = 2x^2 + 3y^2 - 2xy + 2x - 125y + 4500$$

(a) Find the number of units of each kind of tape that should be produced so that the total cost is a minimum.

(b) Find the minimum total cost.

35. *Electronics* Electric charge is distributed uniformly around a thin ring of radius a, with total charge Q. The potential at a point P, a distance x from the center of a ring on a line through the center of the ring and perpendicular to the plane of the ring, is

$$V(x) = \frac{1}{4\pi\epsilon_0}\frac{Q}{\sqrt{x^2 + a^2}}$$

where ϵ_0 is a constant. Find $E_x = -\dfrac{\partial V}{\partial x}$, the rate of change of V for a displacement parallel to the x-axis.

36. *Electronics* Charge Q is uniformly distributed along a rod with length $2a$. The potential at a point on the perpendicular bisector of such a rod at distance x from its center is

$$V(x) = \frac{1}{4\pi\epsilon_0}\frac{Q}{2a}\ln\frac{\sqrt{a^2 + x^2} + a}{\sqrt{a^2 + x^2} - a}$$

Find $E_x = -\dfrac{\partial V}{\partial x}$.

In Your Words

37. Describe the chain rule for a function of two variables.

38. Explain the first derivative test for local extrema.

≡ 11.5 MULTIPLE INTEGRALS

We have been looking at ways to differentiate functions of two variables and ways to use those derivatives. We will now look at the inverse process—integration of functions of two variables. The process of integrating a function of two variables is very similar to that of differentiating a function of two variables. In each case, an operation (differentiation or integration) is performed while holding one of the independent variables constant.

Let $z = f(x, y)$ be a continuous function over a bounded region R in the xy-plane, as shown in Figure 11.20. We want to find the volume between the surface and R.

We will assume that R is rectangular and that it is bounded by $x = a$, $x = b$, $y = c$, and $y = d$. We will divide the intervals $[a, b]$ and $[c, d]$ into partitions, much

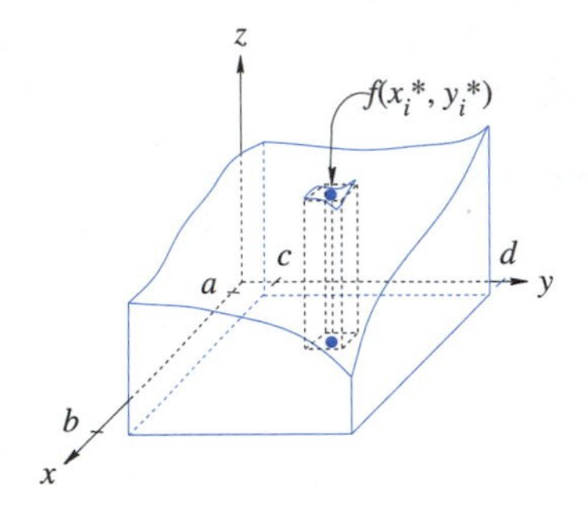

FIGURE 11.20

as we did when we were developing the integral in Chapter 8. We will divide $[a, b]$ into n intervals $a = x_0 < x_1 < x_2 < \cdots < x_i < \cdots < x_n = b$ and we will divide $[c, d]$ into m intervals $c = y_0 < y_1 < y_2 < \cdots < y_j < \cdots y_m = d$. It is possible that $n = m$, but it is not necessary. Thus, we have $\Delta x = \dfrac{b-a}{n}$ and $\Delta y = \dfrac{d-c}{m}$.

Now, select the ith interval on the x-axis, $[x_{i-1}, x_i]$ and the jth interval on the y-axis, $[y_{j-1}, y_j]$. This gives a rectangular region in the xy-plane. From each vertex of this rectangle, erect a vertical line that intersects the surface. What we have approximates a rectangular parallelepiped. We want to get a measure of the volume of that parallelepiped. As before, we will select the midpoint of each interval and call them $x_i{}^*$ and $y_j{}^*$, and this point $(x_i{}^*, y_j{}^*)$ is the center of the rectangle. We will erect a vertical line from that point until it meets the surface at $f(x_i{}^*, y_j{}^*)$. If we assume that the height of the rectangular parallelepiped has a height of $f(x_i{}^*, y_j{}^*)$, then it has a volume of

$$V_{ij} = f(x_i{}^*, y_j{}^*), \Delta x\, \Delta y$$

We now add all of the approximate volumes to get an approximation of the total volume. The total volume is

$$\begin{aligned} V \approx & V_{11} + V_{12} + V_{13} + \cdots + V_{1m} \\ & + V_{21} + V_{22} + \cdots + V_{2m} \\ & \vdots \\ & + V_{n1} + V_{n2} + \cdots + V_{nm} \end{aligned}$$

If we use summation notation, we need to use two summation signs, because we are summing over two variables i and j.

$$V = \sum_{i=1}^{n}\sum_{j=1}^{m} V_{ij} = \sum_{i=1}^{n}\sum_{j=1}^{m} f(x_i{}^*, y_j{}^*)\, \Delta x\, \Delta y$$

When we take the limits of each of these as both Δx and Δy approach zero, we get

$$V = \iint_R f(x, y)\, dx\, dy$$

We now give a more formal definition of a double integral.

Double Integral

If the rectangular region, R, is defined by $a \le x \le b$ and $c \le y \le d$, then we can calculate this double integral as

$$\begin{aligned} \iint_R f(x, y)\, dy\, dx &= \int_a^b \int_c^d f(x, y)\, dy\, dx \\ &= \int_c^d \int_a^b f(x, y)\, dx\, dy \end{aligned}$$

EXAMPLE 11.21

Calculate the volume of the region beneath the surface $z = xy^2 + 2y$ and the rectangle where $0 \leq x \leq 2$ and $1 \leq y \leq 4$.

Solution Using the new equation, we have

$$\int_0^2 \int_1^4 (xy^2 + 2y)\,dy\,dx = \int_0^2 \left\{ \int_1^4 (xy^2 + 2y)\,dy \right\} dx$$

We first perform the integration inside the braces. Since this integral is to be taken with respect to y, we treat x as a constant. We will evaluate this definite inner integral:

$$\begin{aligned} \int_0^2 \left\{ \int_1^4 (xy^2 + 2y)\,dy \right\} dx &= \int_0^2 \left(\frac{1}{3}xy^3 + y^2 \right) \Bigg|_{y=1}^{y=4} dx \\ &= \int_0^2 \left[\left(\frac{64}{3}x + 16 \right) - \left(\frac{1}{3}x + 1 \right) \right] dx \\ &= \int_0^2 (21x + 15)\,dx \end{aligned}$$

We now have an ordinary single variable integral that we can evaluate easily.

$$\begin{aligned} \int_0^2 (21x + 15)\,dx &= \frac{21}{2}x^2 + 15x \Bigg|_0^2 \\ &= \frac{21}{2} \cdot 4 + 15 \cdot 2 \\ &= 42 + 30 = 72 \end{aligned}$$

It is also possible to first integrate this over x and then over y.

EXAMPLE 11.22

Evaluate $\int_1^4 \int_0^2 (xy^2 + 2y)\,dx\,dy$.

Solution

$$\begin{aligned} \int_1^4 \int_0^2 (xy^2 + 2y)\,dx\,dy &= \int_1^4 \left\{ \int_0^2 (xy^2 + 2y)\,dx \right\} dy \\ &= \int_1^4 \left[\frac{1}{2}x^2y^2 + 2xy \right] \Bigg|_{x=0}^{x=2} dy \\ &= \int_1^4 (2y^2 + 4y)\,dy \\ &= \left(\frac{2}{3}y^3 + 2y^2 \right) \Bigg|_1^4 \\ &= \left(\frac{128}{3} + 32 \right) - \left(\frac{2}{3} + 2 \right) = 72 \end{aligned}$$

Average Value Over Rectangular Regions

In Section 7.1, we defined the average value of continuous function f over interval $[a,b]$. The **average value**, $\bar{y}$, was

$$\bar{y} = \frac{1}{b-a}\int_a^b f(x)\,dx$$

This definition can be extended to the function of two variables over rectangular regions as shown in the following box. Notice that the denominator in the expression is the area of the rectangle, R.

> **Average Value Over a Rectangular Region**
>
> The **average value** of the function $f(x,y)$ over the rectangular region of width x, with $a \le x \le b$, length y, and with $c \le y \le d$, is
>
> $$\begin{aligned}\bar{f} &= \frac{1}{(b-a)(d-c)}\int\int_R f(x,y)\,dy\,dx \\ &= \frac{1}{(b-a)(d-c)}\int_a^b\int_c^d f(x,y)\,dy\,dx \\ &= \frac{1}{(b-a)(d-c)}\int_c^d\int_a^b f(x,y)\,dx\,dy\end{aligned}$$

EXAMPLE 11.23

Find the average value of $f(x,y) = 4 - \frac{3}{2}x - \frac{1}{2}y$ over the rectangular region $R = x \times y$ where $0 \le x \le 2$ and $0 \le y \le 4$.

Solution Here $a = 0$, $b = 2$, $c = 0$, and $d = 4$. Using these substitutions, we obtain

$$\begin{aligned}\bar{f} &= \frac{1}{(b-a)(d-c)}\int_a^b\int_c^d f(x,y)\,dy\,dx \\ &= \frac{1}{(2-0)(4-0)}\int_0^4\int_0^2 \left(4 - \frac{3}{2}x - \frac{1}{2}y\right)dx\,dy \\ &= \frac{1}{8}\int_0^4 \left[\left(4x - \frac{3}{4}x^2 - \frac{1}{2}yx\right)\right]_0^2 dy \\ &= \frac{1}{8}\int_0^4 (5-y)\,dy \\ &= \frac{1}{8}\left[5y - \frac{1}{2}y^2\right]_0^4 \\ &= \frac{1}{8}[20-8] = \frac{3}{2}\end{aligned}$$

Double Integration Over a Simple Region

Now, suppose that the region is not a rectangle. Then it may be a vertically simple region or a horizontally simple region, as shown in Figure 11.21. A region R is **vertically simple**, if it is described by the inequalities

$$a \leq x \leq b, \quad g_1(x) \leq y \leq g_2(x)$$

where g_1 and g_2 are continuous functions on $[a,b]$. A region R is **horizontally simple**, if it is described by the inequalities

$$c \leq y \leq d, \; h_1(y) \leq x \leq h_2(y)$$

where h_1 and h_2 are continuous on $[c,d]$.

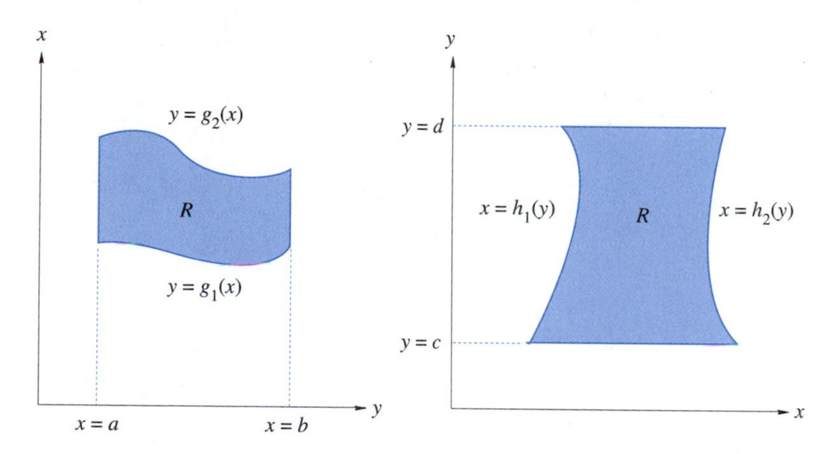

FIGURE 11.21

We can compute a double integral over a region R that is either horizontally or vertically simple, using the following rule.

Double Integration Over a Simple Region

If $z = f(x,y)$ is continuous on the region R and if R is a vertically simple region, then

$$\iint_R f(x,y)\,dA = \int_a^b \int_{g_1(x)}^{g_2(x)} f(x,y)\,dy\,dx$$

If R is a horizontally simple region, then

$$\iint_R f(x,y)\,dA = \int_c^d \int_{h_1(y)}^{h_2(y)} f(x,y)\,dx\,dy$$

The symbol dA represents the change in area of R.

EXAMPLE 11.24

Evaluate $\iint_R (x^3+4y)\,dA$, where R is the region in the xy-plane bounded by $y=x^2$ and $y=2x$.

Solution The region R is shown in Figure 11.22. Notice that this is both horizontally and vertically simple. We can use either rule. Suppose we assume it is vertically simple with $y=x^2$ the lower boundary (g_1) and $y=2x$ the upper boundary (g_2). We have $0 \le x \le 2$. By the rule for double integration over a simple region,

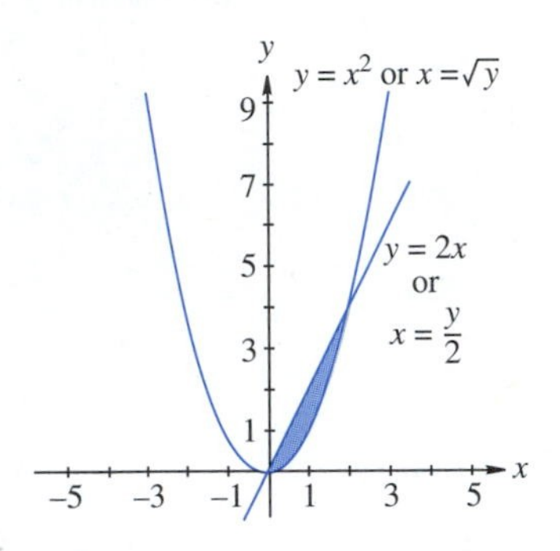

FIGURE 11.22

$$
\begin{aligned}
\iint_R (x^3+4y)\,dA &= \int_0^2 \int_{x^2}^{2x} (x^3+4y)\,dy\,dx \\
&= \int_0^2 (x^3y+2y^2)\Big|_{y=x^2}^{y=2x} dx \\
&= \int_0^2 \left\{\left[x^3(2x)+2(2x)^2\right]-\left[x^3(x^2)+2(x^2)^2\right]\right\} dx \\
&= \int_0^2 (2x^4+8x^2)-(x^5+2x^4)\,dx \\
&= \int_0^2 (8x^2-x^5)\,dx \\
&= \left(\frac{8}{3}x^3-\frac{1}{6}x^6\right)\Big|_0^2 \\
&= \frac{8}{3}(8)-\frac{1}{6}(64) \\
&= \frac{32}{3}
\end{aligned}
$$

If we assumed that the region was horizontally simple, then we would use $x=\frac{1}{2}y$ as the lower boundary (h_1) and $x=\sqrt{y}$ as the upper boundary (h_2). The integral would then be

$$
\begin{aligned}
\iint_R (x^3+4y)\,dA &= \int_0^4 \int_{1/2y}^{\sqrt{y}} (x^3+4y)\,dx\,dy \\
&= \int_0^4 \left(\frac{1}{4}x^4+4xy\right)\Big|_{x=1/2y}^{x=\sqrt{y}} \\
&= \int_0^4 \left[\left(\frac{1}{4}y^2+4y^{3/2}\right)-\left(\frac{1}{64}y^4-2y^2\right)\right] dy \\
&= \frac{32}{3}
\end{aligned}
$$

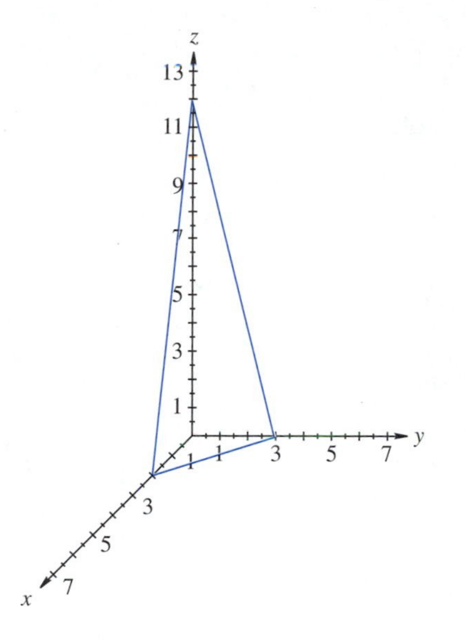

FIGURE 11.23

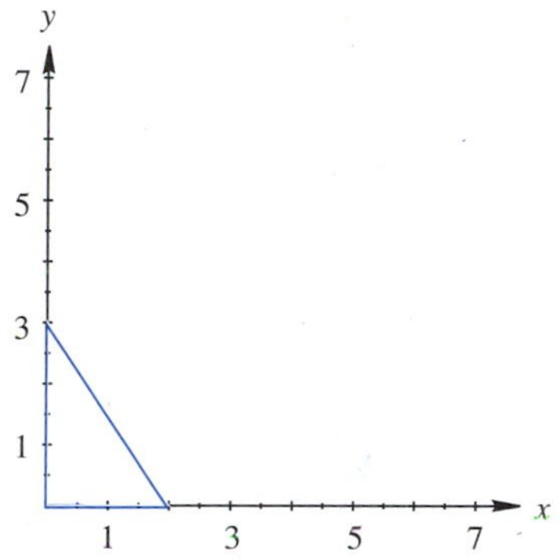

FIGURE 11.24

EXAMPLE 11.25

Find the volume of the solid bounded by the plane $6x+4y+z=12$ and the coordinate planes, as shown in Figure 11.23.

Solution When finding a volume, the integrand is always the function that represents the surface. In this case, $z=f(x,y)=12-6x-4y$. The limits of integration are found by examining the region R in the xy-plane, as shown in Figure 11.24. Notice that the xy-trace in Figure 11.24 is the line $y=-\frac{3}{2}(x-2)$. Thus, we have $g_1(x)=0$ and $g_2(x)=-\frac{3}{2}(x-2)$, and the outer limits are $x=0$ and $x=2$. The double integral becomes

$$\begin{aligned}
V &= \int_0^2 \int_0^{-3(x-2)/2} (12-6x-4y)\,dy\,dx \\
&= \int_0^2 (12y-6xy-2y^2)\Big|_0^{-\frac{3}{2}(x-2)}\,dx \\
&= \int_0^2 -18(x-2)+9x(x-2)+\frac{-9}{2}(x-2)^2\,dx \\
&= \int_0^2 \left(\frac{9}{2}x^2-18x+18\right)dx \\
&= \left(\frac{3}{2}x^3-9x^2+18x\right)\Big|_0^2 \\
&= 12
\end{aligned}$$

Double integrals can also be performed in polar coordinates. As we will see after Section 11.6 they can be extended to triple integrals for functions of three variables.

Exercise Set 11.5

In Exercises 1–16, evaluate the double integrals.

1. $\int_0^1 \int_0^{x^2} xy\,dy\,dx$
2. $\int_{-1}^1 \int_{-2}^2 (2xy - 3y^2)\,dy\,dx$
3. $\int_0^{\pi} \int_{-\pi/2}^{\pi/2} \sin x \cos y\,dy\,dx$
4. $\int_0^1 \int_0^2 \sqrt{x+y}\,dy\,dx$
5. $\int_0^{\ln 2} \int_0^{\ln 5} e^{x+y}\,dy\,dx$
6. $\int_0^1 \int_y^{y^2} (x+y)\,dx\,dy$
7. $\int_0^1 \int_x^{\sqrt{x}} (x+y)\,dy\,dx$
8. $\int_0^4 \int_{-\sqrt{2x}}^{\sqrt{2x}} (2x+y)\,dy\,dx$
9. $\int_0^1 \int_{x^3}^{x} (y-x)\,dy\,dx$
10. $\int_0^4 \int_y^{\sqrt{y}} (3x+2y)\,dx\,dy$
11. $\int_0^2 \int_0^3 (xy + x - y)\,dy\,dx$
12. $\int_0^1 \int_0^x xy\,dy\,dx$
13. $\int_0^1 \int_0^{1-x} (x^2y + xy^2)\,dy\,dx$
14. $\int_0^2 \int_0^{x^2} (x^2 - y^2)\,dy\,dx$
15. $\int_0^1 \int_{y^2}^{y} (xy+1)\,dx\,dy$
16. $\int_1^2 \int_y^{y^2} \frac{x}{y}\,dx\,dy$

In Exercises 17–26, evaluate the given integral.

17. $\iint_R (x^2 - y^2)\,dA$, where R is the region bounded by $x = 0$, $y = 1$, and $y = x$.
18. $\iint_R x^2 y\,dA$, where R is the region bounded by $y = x^2$, $y = 0$, and $x = 2$.
19. $\iint_R (x^3 - y^3)\,dA$, where R is the region bounded by $y = x^3$, $x = 0$, and $y = 8$.
20. $\iint_R xy\,dA$, where R is the region bounded by $y = x$, and $y = x^4$.
21. $\iint_R (x+y)\,dA$, where R is the region bounded by $xy = 8$ and $x + y = 9$.
22. $\iint_R (x^2 + y^2)\,dA$, where R is the region bounded by $x = 1$, $x = 4$, $y = 0$, and $y = 5$.
23. $\iint_R (x+y)^2\,dA$, where R is the region in the first quadrant bounded by $y = x^3$ and $y = x$.
24. $\iint_R xy\,dA$, where R is the region bounded by $x = y^2$ and $x = y + 4$.
25. $\iint_R xy\,dA$, where R is the region bounded by $x = 0$ and $x = y^2 - 9$.
26. $\iint_R \sqrt{xy}\,dA$, where R is the region bounded by $y = 1$ and $y = x^2$.

In Exercises 27–30, find the average value for each of the functions over regions *R* having the given boundaries.

27. $f(x, y) = x^2 + y^2;\ 0 \le x \le 2,\ 0 \le y \le 3$
28. $f(x, y) = (x+y)^2;\ 1 \le x \le 5,\ -1 \le y \le 1$
29. $f(x, y) = \frac{x}{y};\ 1 \le x \le 4,\ 2 \le y \le 7$
30. $f(x, y) = e^{2x+y};\ 1 \le x \le 2,\ 2 \le y \le 3$

In Exercises 31–34, find the volume of the given solid.

31. The solid bounded by the plane $x+2y+z=4$ and the three coordinate planes

32. The solid bounded by the cylinder $x^2+z^2=4$ and the planes $y=0$ and $y=2$

33. The solid bounded by the plane $y=0$, $y=x$, and the cylinder $x^2+z^2=2$

34. The solid bounded by the surface $z=e^{-(x+y)}$ and the three coordinate planes

Solve Exercises 35–38.

35. *Business* A manufacturing company has two plants. The weekly cost function for the first plant at production level x is $C_1(x)=0.2x^2+40x+3600$, and at the second plant at production level y the cost is $C_2(y)=0.4y^2+24y+6{,}000$. Find the average weekly cost of the company for $300 \leq x \leq 400$ and $350 \leq y \leq 550$, that is, find the average cost of $C_1(x)+C_2(y)$.

36. *Business* If a company invests x thousand labor-hours and y million dollars in the production of N thousand units of a certain item, then according to the Cobb-Douglas production formula, $N(x,y)=x^{0.75}y^{0.25}$, where $10 \leq x \leq 20$ and $1 \leq y \leq 2$. What is the average number of units produced for these ranges of labor-hour and capital expenditures?

37. *Environmental science* An industrial plant is located in the center of a rectangular-shaped small town that is 4 mi long and 2 mi wide. The plant emits particulate matter into the atmosphere. The concentration of particulate matter in parts per million at a point d mi from the plant is given by $C=100-15d^2$.

(a) Express C as a function of x and y, where x and y are the horizontal and vertical distance of d from the plant.

(b) Determine the average concentration of particulate matter throughout the town.

38. *Police science* Under ideal conditions, if a person driving a car brakes hard, the car will travel distance L in feet before it stops, where $L=0.000\,02xy^2$, x is the weight of the car in pounds, and y is the speed of the car in miles per hour. What is the average distance that a car weighing between 2,000 and 3,000 lb will travel at speeds between 50 and 60 mph?

In Your Words

39. Without looking in the text, describe how to determine the double integral of a function defined over a region.

40. How does a simple region differ from a region?

41. Without looking in the text, explain how to determine the average value of a function over a rectangular region.

42. Describe the geometric meaning of the double integral of a function of two variables.

11.6 CYLINDRICAL AND SPHERICAL COORDINATES

Until now, all of our three-dimensional work has been done with rectangular coordinates. However, there are several ways to represent points in space. In this section, we will look at two common ways to do this. Both of these methods are used to describe the motion of robots.

Cylindrical Coordinate System

In the cylindrical coordinate system, a point P is identified by $P=(r,\theta,z)$. In this identification, r and θ are the polar coordinates of P as it is projected on the xy-plane. Thus, $r \geq 0$ and $0 \leq \theta \leq 2\pi$. The value z represents the distance of the xy-plane from P. In Figure 11.25, P is projected onto the xy-plane at Q. Thus, the distance from P to Q is z; that is, $d(P, Q) = z$.

EXAMPLE 11.26

Discuss the graphs of each of the following equations in cylindrical coordinates: **(a)** $r = 5$, **(b)** $\theta = \frac{\pi}{3}$, **(c)** $z = 4$, **(d)** $3 \le r \le 5$, $\frac{\pi}{4} \le \theta \le \frac{\pi}{3}$, $4 \le z \le 7$.

Solutions

(a) If $r = 5$ (or any positive constant), then θ and z can assume any value. As a result, we obtain a right circular cylinder with radius r, as shown in Figure 11.26a.

(b) If $\theta = \frac{\pi}{3}$, we obtain a half-plane through the z-axis, as shown in Figure 11.26b.

(c) If $z = 4$, we get a plane parallel to the xy-plane (the polar plane), as shown in Figure 11.26c.

(d) First, the values $3 \le r \le 5$ give the region between the cylinders $r = 3$ and $r = 5$. The values $\frac{\pi}{4} \le \theta \le \frac{\pi}{3}$ define the "wedge-shaped" region between half-planes $\theta = \frac{\pi}{4}$ and $\theta = \frac{\pi}{3}$. Finally, $4 \le z \le 7$ is the "slice" (or "slab") of space between the planes $z = 4$ and $z = 7$. Putting this together, we get the shaded solid shown in Figure 11.26d. ▪

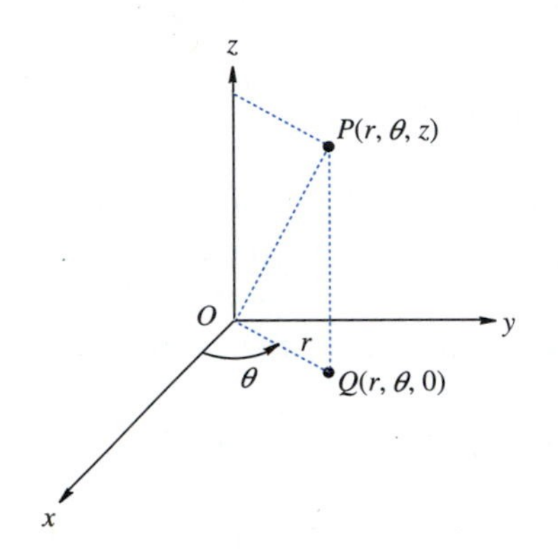

FIGURE 11.25

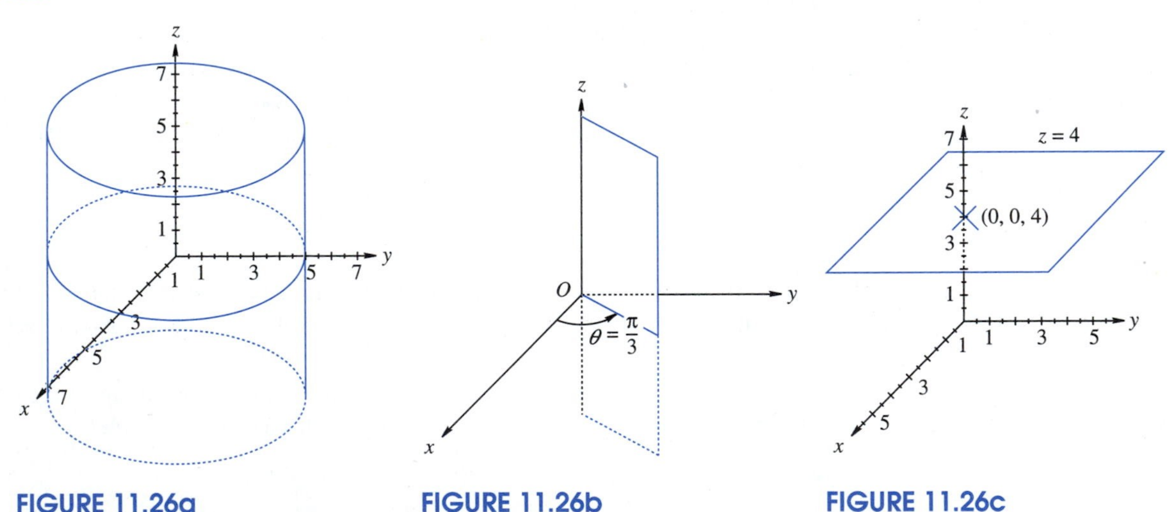

FIGURE 11.26a FIGURE 11.26b FIGURE 11.26c

Changing Between Cylindrical and Rectangular Coordinates

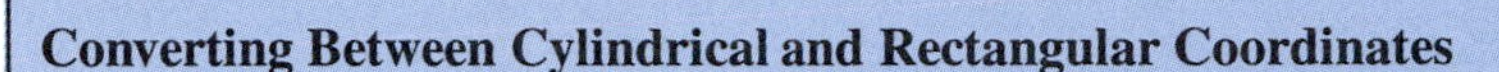

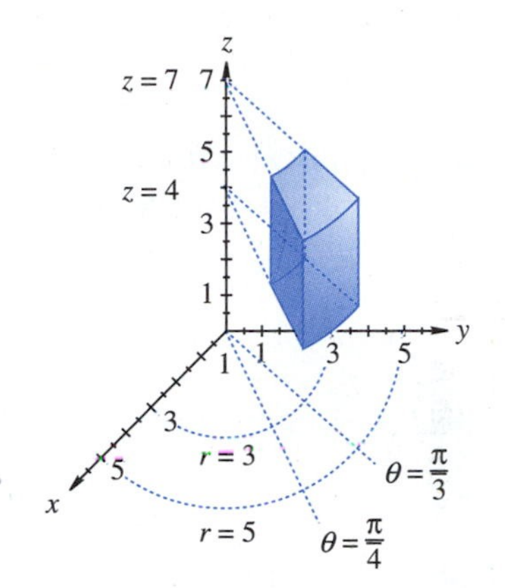

FIGURE 11.26d

We often find it necessary to change from cylindrical coordinates to rectangular coordinates or vice versa. The following conversion formulas are very similar to the ones we used with polar coordinates.

Converting Between Cylindrical and Rectangular Coordinates

A point with the cylindrical coordinates (r, θ, z) has the rectangular coordinates (x, y, z), where

$$x = r\cos\theta, \qquad y = r\sin\theta, \qquad z = z \tag{1}$$

Similarly, a point with the rectangular coordinates (x, y, z) has the cylindrical coordinates (r, θ, z), where

$$r = \sqrt{x^2 + y^2}, \qquad \tan\theta = \frac{y}{x}, \qquad z = z \tag{2}$$

EXAMPLE 11.27

Convert $P\left(4, \frac{5\pi}{6}, 7\right)$ from cylindrical to rectangular coordinates.

Solution We have $r = 4$, $\theta = \frac{5\pi}{6}$, and $z = 7$. Substituting these into the equations in (1), we obtain

$$x = r\cos\theta = 4\left(\frac{-\sqrt{3}}{2}\right) = -2\sqrt{3} \approx -3.464$$
$$y = r\sin\theta = 4\left(\tfrac{1}{2}\right) = 2$$
$$z = z = 7$$

So in rectangular coordinates, $P = \left(-2\sqrt{3}, 2, 7\right)$.

EXAMPLE 11.28

Convert $(5, -12, 2)$ from rectangular to cylindrical coordinates.

Solution Here, we substitute into the equations in (2), with the result

$$r = \sqrt{x^2 + y^2} = \sqrt{5^2 + (-12)^2} = \sqrt{25 + 144} = 13$$

$\tan\theta = \frac{y}{x} = \frac{-12}{5} = -2.4$. So, $\theta = \tan^{-1}(-2.4) = 5.107$ rad [since the point $(5, -12)$ is in the fourth quadrant], and $z = 2$.

Thus, the rectangular coordinates $(5, -12, 2)$ are $(13, 5.107, 2)$ in cylindrical coordinates.

Spherical Coordinates

The other new coordinate system that we are going to learn is the spherical coordinate system. A typical point P in space is represented as

$$P = (\rho, \theta, \phi)$$

where $\rho \geq 0$, $0 \leq \theta \leq 2\pi$, $0 \leq \phi \leq \pi$, and where ρ is the positive distance from P to the origin, θ is the same as in cylindrical coordinates, and ϕ is the angle between $\overrightarrow{OP}$ and the positive z-axis, as shown in Figure 11.27.

EXAMPLE 11.29

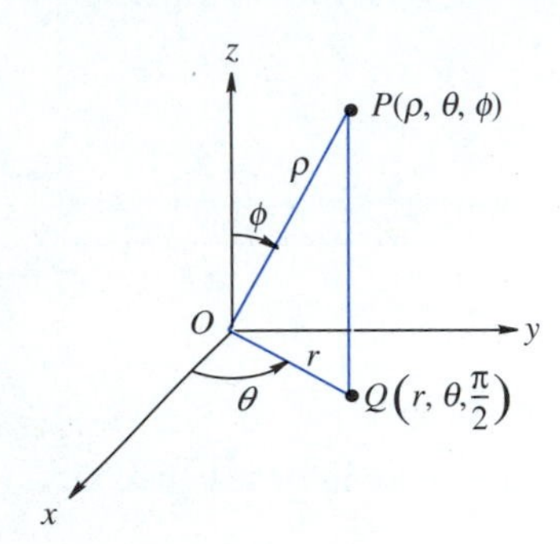

FIGURE 11.27

Discuss the graphs of the following equations in spherical coordinates: **(a)** $\rho = 5$, **(b)** $\theta = \frac{\pi}{3}$, **(c)** $\phi = \frac{\pi}{6}$, **(d)** $3 \leq \rho \leq 5$, $\frac{\pi}{4} \leq \theta \leq \frac{\pi}{3}$, $\frac{\pi}{6} \leq \phi \leq \frac{\pi}{4}$.

Solution

(a) If $\rho = 5$ (or any positive constant), we get all the points in space that are 5 units from the origin. This is a sphere centered at the origin with radius 5, as shown in Figure 11.28a.

(b) If $\theta = \frac{\pi}{3}$, we get a half-plane containing the z-axis, as shown in Figure 11.28b. This is the same graph we would get in cylindrical coordinates, if $\theta = \frac{\pi}{3}$.

(c) If $\phi = \frac{\pi}{6}$, the graph would be obtained by rotating the vector $\overrightarrow{OP}$ around the z-axis. This yields a circular cone, as shown in Figure 11.28c. If $\phi > \frac{\pi}{2}$, the cone would be below the xy-plane, as shown in Figure 11.28d. If $\phi = \frac{\pi}{2}$, we would get the xy-plane.

(d) Here, $3 \leq \rho \leq 5$ gives the region between the spheres $\rho = 3$ and the sphere $\rho = 5$. Next, $\frac{\pi}{4} \leq \theta \leq \frac{\pi}{3}$ is, as in the cylindrical coordinates, a wedge-shaped region between $\theta = \frac{\pi}{4}$ and $\theta = \frac{\pi}{3}$. Finally, $\frac{\pi}{6} \leq \phi \leq \frac{\pi}{4}$ describes the region between the cone $\phi = \frac{\pi}{6}$ and the cone $\phi = \frac{\pi}{4}$. Combining these we would get the **spherical wedge** shown in Figure 11.28e.

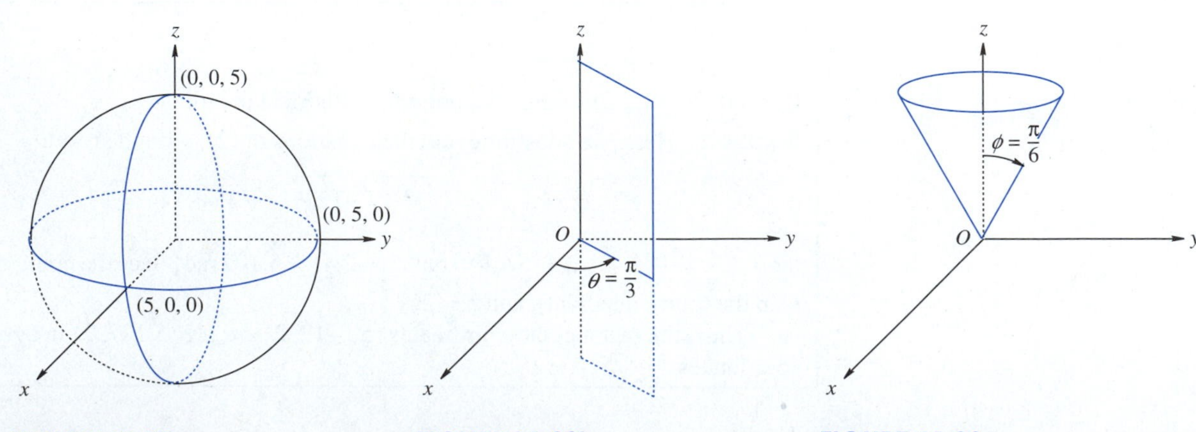

FIGURE 11.28a FIGURE 11.28b FIGURE 11.28c

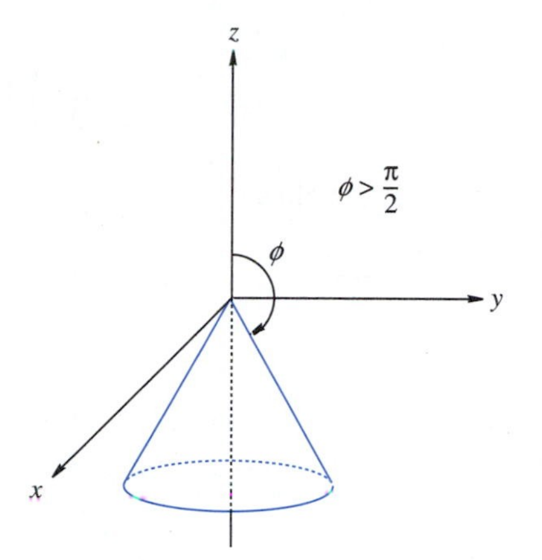

FIGURE 11.28d

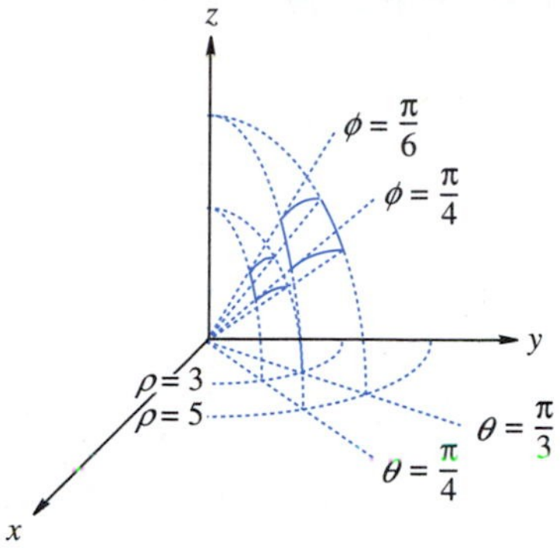

FIGURE 11.28e

Changing Between Spherical and Rectangular Coordinates

To convert from spherical to rectangular coordinates, we use the following formulas:

Converting Between Spherical and Rectangular Coordinates

A point with the spherical coordinates (ρ, θ, ϕ) has the rectangular coordinates (x, y, z), where

$$x = \rho \sin\phi \cos\theta, \qquad y = \rho \sin\phi \sin\theta, \qquad z = \rho \cos\phi \tag{3}$$

Similarly, a point with the rectangular coordinates (x, y, z) has the spherical coordinates (ρ, θ, ϕ), where

$$\rho = \sqrt{x^2 + y^2 + z^2}, \qquad \tan\theta = \frac{y}{x}, \qquad \cos\phi = \frac{z}{\rho} \tag{4}$$

EXAMPLE 11.30

Convert $\left(8, \frac{\pi}{6}, \frac{3\pi}{4}\right)$ from spherical to rectangular coordinates.

Solution We have $\rho = 8$, $\theta = \frac{\pi}{6}$, and $\phi = \frac{3\pi}{4}$. Using the formulas in (3), we have

$$\begin{aligned} x = \rho \sin\phi \cos\theta &= 8 \sin\frac{3\pi}{4} \cos\frac{\pi}{6} \\ &= 8\left(\frac{\sqrt{2}}{2}\right)\left(\frac{\sqrt{3}}{2}\right) = 2\sqrt{6} \end{aligned}$$

$$y = \rho \sin\phi \sin\theta = 8\left(\frac{\sqrt{2}}{2}\right)\left(\frac{1}{2}\right) = 2\sqrt{2}$$

EXAMPLE 11.30 (Cont.)

$$z = \rho \cos \phi = 8\left(-\frac{\sqrt{2}}{2}\right) = -4\sqrt{2}$$

So, the spherical coordinates of $\left(8, \frac{\pi}{6}, \frac{3\pi}{4}\right)$ are, in rectangular coordinates, $(2\sqrt{6}, 2\sqrt{2}, -4\sqrt{2})$.

EXAMPLE 11.31

Convert $(1, -\sqrt{3}, \sqrt{5})$ from rectangular to spherical coordinates.

Solution Here, $x = 1$, $y = -\sqrt{3}$, and $z = \sqrt{5}$. Using the formulas in (4) produces

$$\begin{aligned}
\rho &= \sqrt{x^2 + y^2 + z^2} \\
&= \sqrt{(1)^2 + (-\sqrt{3})^2 + (\sqrt{5})^2} \\
&= \sqrt{1 + 3 + 5} = 3 \\
\tan\theta &= \frac{y}{x} \\
\theta_{\text{Ref}} &= \tan^{-1}\frac{-\sqrt{3}}{1} \\
&= -\frac{\pi}{3}
\end{aligned}$$

Since $x > 0$ and $y < 0$, θ is between $\frac{3\pi}{2}$ and 2π. So,

$$\begin{aligned}
\theta &= \frac{5\pi}{3} \\
\cos\phi &= \frac{z}{\rho} = \frac{\sqrt{5}}{3} \\
\phi &= \cos^{-1}\frac{\sqrt{5}}{3} \approx 0.730
\end{aligned}$$

The rectangular coordinates $(1, -\sqrt{3}, \sqrt{5})$ have the spherical coordinates $\left(3, \frac{5\pi}{3}, 0.730\right)$.

Industrial Robots and Coordinate Systems

Industrial robots are currently used to weld, cast, form, assemble, paint, transfer, inspect, and load or unload parts into and out of many different machines. The industrial robot can be thought of as a mechanical arm that is in a fixed location. In fact, it is often bolted to the floor or wall.

The work envelope of a robot consists of all the points in space that can be touched by the end of the robot's arm. This work envelope is a three-dimensional volume of points. Three major axes of motion provide the largest portion of the robot's work envelope. The coordinate system used for these three major axes of motion provides one means of classifying robots. The three major axes of motion are: **(a)** a vertical lift stroke; **(b)** an in-and-out or horizontal reach stroke, and **(c)** a rotational, traverse, or swing motion about the vertical axis.

A cylindrical coordinate robot has a horizontal arm assembled to a vertical axis. The vertical axis is mounted on a rotating base. The horizontal arm can move in and out and move vertically up and down on the vertical axis. This arm assembly can also rotate left and right about the vertical axis. The motions of the three major axes form a portion of a cylinder as the work envelope of the robot, as shown in Figure 11.29.

A spherical coordinate robot consists of an arm that moves in and out in a reach stroke. However, the arm uses a pivoting vertical motion instead of a true vertical stroke and the arm can move left or right about the vertical pivot axis. These motions form a portion of a sphere as a work envelope. (See Figure 11.30.)

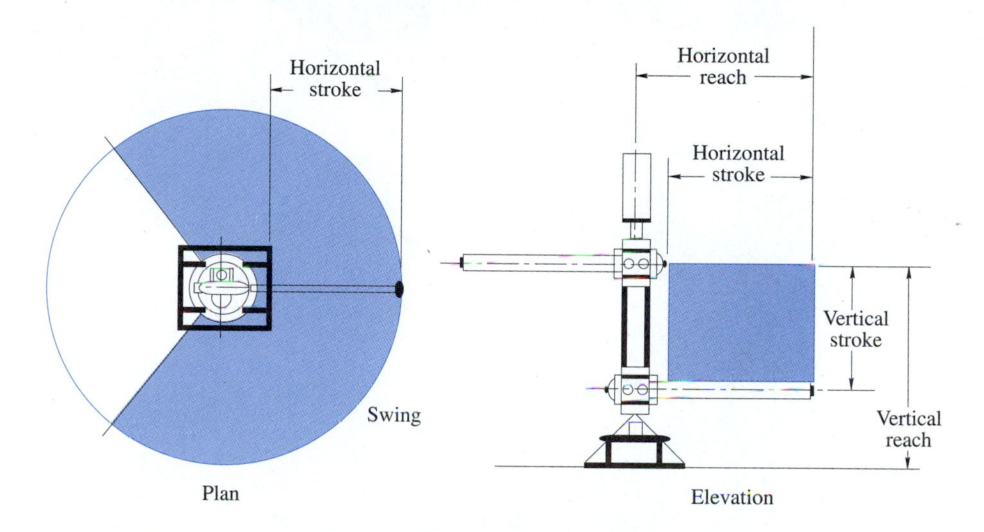

FIGURE 11.29

A third work-envelope structure is a rectangular coordinate robot. This type of robot has a horizontal arm assembled to a vertical lift axis. The vertical axis is mounted on a linear traverse base. The work envelope covered by these motions is illustrated in Figure 11.31.

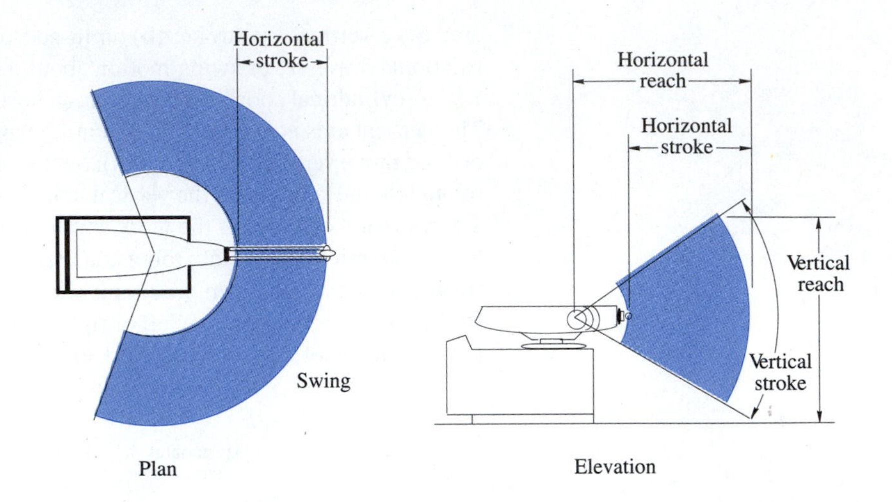

Work Envelope:
Spherical Coordinate Robot

FIGURE 11.30

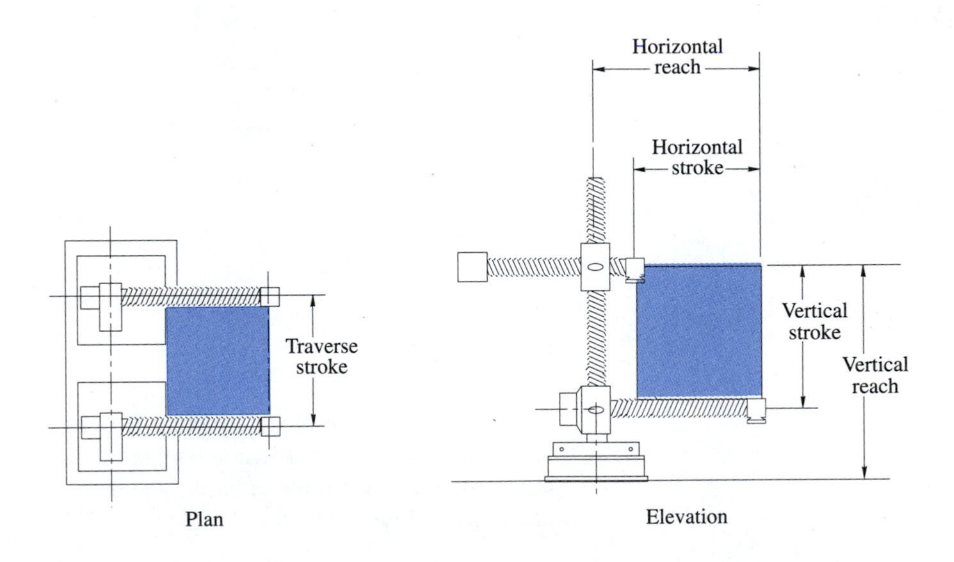

Work Envelope:
Rectangular Coordinate Robot

FIGURE 11.31

Exercise Set 11.6

In Exercises 1–6, convert the cylindrical coordinate to its equivalent rectangular coordinate.

1. $(2, \frac{\pi}{4}, 2)$
2. $(3, \frac{2\pi}{3}, -2)$
3. $(2, 0, 4)$
4. $(0, \frac{5\pi}{4}, -5)$
5. $(5, \frac{5\pi}{4}, 0)$
6. $(4, \frac{5\pi}{3}, 7)$

In Exercises 7–12, convert the rectangular coordinate to its equivalent cylindrical coordinate.

7. $(2, 2, 5)$
8. $(3, -4, -5)$
9. $(-4, 3, 2)$
10. $(-1, -\sqrt{3}, 4)$
11. $(12, -5, -3)$
12. $(6, 8, -4)$

In Exercises 13–18, convert the spherical coordinate to its equivalent rectangular coordinate.

13. $\left(4, \frac{\pi}{4}, \frac{\pi}{6}\right)$
14. $\left(1, \frac{\pi}{6}, \frac{\pi}{2}\right)$
15. $\left(3, \frac{\pi}{2}, \frac{5\pi}{3}\right)$
16. $\left(2, \frac{5\pi}{6}, \frac{3\pi}{4}\right)$
17. $\left(5, \frac{7\pi}{6}, \frac{2\pi}{3}\right)$
18. $\left(5, \frac{5\pi}{3}, \frac{\pi}{4}\right)$

In Exercises 19–24, convert the rectangular coordinate to its equivalent spherical coordinate.

19. $(4, 3, 0)$
20. $(4, \sqrt{5}, -2)$
21. $(2, 1, -2)$
22. $(4, -\sqrt{3}, 2)$
23. $(1, 1, \sqrt{2})$
24. $(-\sqrt{3}, \sqrt{3}, -\sqrt{3})$

In Exercises 25–30, describe and sketch the curves and regions determined by the given equations and inequalities.

25. $\theta = \frac{\pi}{4}, \phi = \frac{\pi}{4}$
26. $\theta = \frac{3\pi}{4}$
27. $3 \leq \rho \leq 5$
28. $\rho = 5, \theta = \frac{\pi}{3}$
29. $\rho \leq 4, \phi \leq \frac{\pi}{6}$
30. $0 \leq \theta \leq \frac{\pi}{4}, 0 \leq \phi \leq \frac{\pi}{4}, 0 \leq \rho \leq 4$

31. Write a computer program to convert a coordinate in the rectangular, cylindrical, or spherical systems to its equivalent in the other two systems. Check your program on the answers to Exercises 1–24.

In Your Words

32. Explain the cylindrical, spherical, and rectangular three-dimensional coordinate system.
33. Why do we need three different three-dimensional coordinate systems?

11.7 MOMENTS AND CENTROIDS

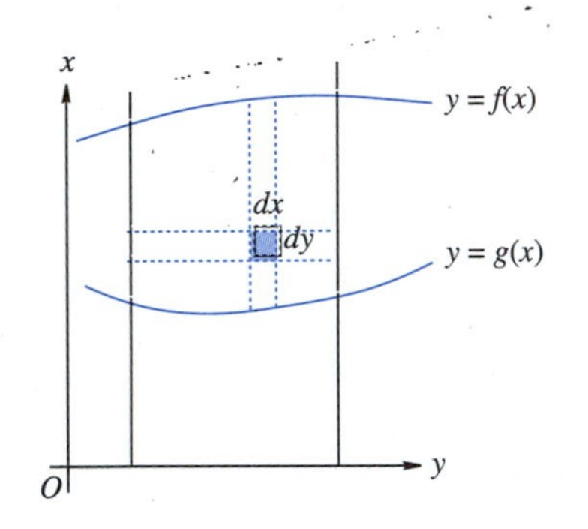

FIGURE 11.32

In Chapter 7, we looked at ways in which integration could be used to determine the centroids and the moments of inertia of a figure. In this section, we will see how to use double integration to find the centroids and moments of inertia of these same figures.

Consider the region shown in Figure 11.32. We can determine the area of this figure by using the integral $A = \int_a^b [f(x) - g(x)]\,dx$. But, $f(x) - g(x) = \int_{g(x)}^{f(x)} dy$, so we can write the area as a double integral:

$$A = \int_a^b \int_{g(x)}^{f(x)} dy\,dx$$

In Section 7.5, we showed that the moment of an area with respect to the y-axis, M_y, is $\rho \int_a^b x\,dA$, where ρ was the mass per unit area. In symbols we wrote this as

$$M_y = \rho \int_a^b x[f(x) - g(x)]\,dx$$

But, since $f(x) - g(x) = \int_{g(x)}^{f(x)} dy$, we can write this as

$$M_y = \rho \int_a^b \int_{g(x)}^{f(x)} x\,dy\,dx$$

Because ρ was constant, its effect was cancelled in locating the coordinates of the centroid. The total mass m was given by ρA.

Thus, the x-coordinate $\bar{x}$ of the centroid is given by

$$\bar{x} = \frac{M_y}{m} = \frac{\rho \int_a^b \int_{g(x)}^{f(x)} x\,dy\,dx}{m}$$

In a similar manner, we can show that the moment of the area with respect to the x-axis is given by

$$M_x = \rho \int_a^b \int_{g(x)}^{f(x)} y\,dy\,dx$$

As a result, the y-coordinate $\bar{y}$ of the centroid is given by

$$\bar{y} = \frac{M_x}{m} = \frac{\rho \int_a^b \int_{g(x)}^{f(x)} y\,dy\,dx}{m}$$

The formulas for determining the mass, moments, and centroid of a region in the xy-plane are summarized in the following box. The examples following the box show how to use these formulas.

Area, Mass, Moments, and Centroid of a Region in the xy-Plane

If ρ is a continuous density function on a plane region R, then the area, mass, moments, and centroid of R are as follows:

Area: $A = \iint_R dy\,dx$

Mass: $m = \iint_R \rho(x, y)\,dy\,dx$

First Moments:
$$M_x = \iint_R y\rho(x, y)\,dy\,dx$$
$$M_y = \iint_R x\rho(x, y)\,dy\,dx$$

Centroid: $(\bar{x}, \bar{y})$, where $\bar{x} = \dfrac{M_y}{m}$ and $\bar{y} = \dfrac{M_x}{m}$

EXAMPLE 11.32

Find the centroid of the region bounded by $y = x$, $y = 6 - 2x$, and the y-axis, as shown in Figure 11.33a.

Solution The two functions intersect at the point $(2, 2)$. So, we have $f(x) = 6 - 2x$, $g(x) = x$, $a = 0$, and $b = 2$.

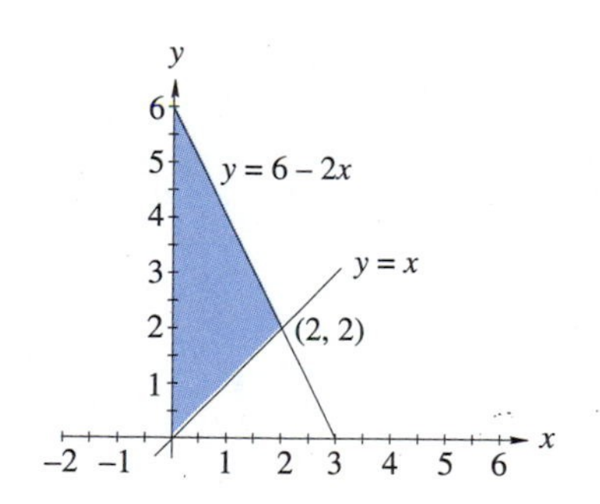

FIGURE 11.33a

$$\begin{aligned}
A &= \int_0^2 \int_x^{6-2x} dy\,dx = \int_0^2 y\Big|_x^{6-2x} dx \\
&= \int_0^2 [(6-2x) - x]\,dx = \int_0^2 (6-3x)\,dx \\
&= 6x - \frac{3}{2}x^2\Big|_0^2 = 6
\end{aligned}$$

$$m = \rho A = 6\rho$$

$$\begin{aligned}
M_y &= \rho \int_0^2 \int_x^{6-2x} x\,dy\,dx = \rho \int_0^2 y\Big|_x^{6-2x} x\,dx \\
&= \rho \int_0^2 (6-3x)x\,dx = \rho \int_0^2 (6x - 3x^2)\,dx \\
&= \rho(3x^2 - x^3)\Big|_0^2 = 4\rho
\end{aligned}$$

$$M_x = \rho \int_0^2 \int_x^{6-2x} y\,dy\,dx = \rho \int_0^2 \frac{1}{2}y^2\Big|_x^{6-2x} dx$$

EXAMPLE 11.32 (Cont.)

$$= \frac{1}{2}\rho \int_0^2 [(6-2x)^2 - x^2]\,dx = \frac{1}{2}\rho \int_0^2 (36 - 24x + 3x^2)\,dx$$

$$= \frac{1}{2}(36x - 12x^2 + x^3)\rho\Big|_0^2 = 16\rho$$

Thus, $\bar{x} = \dfrac{M_y}{m} = \dfrac{4\rho}{6\rho} = \dfrac{2}{3}$ and $\bar{y} = \dfrac{M_x}{m} = \dfrac{16\rho}{6\rho} = \dfrac{8}{3}$.

The centroid is at $\left(\frac{2}{3}, \frac{8}{3}\right)$, as indicated in Figure 11.33b.

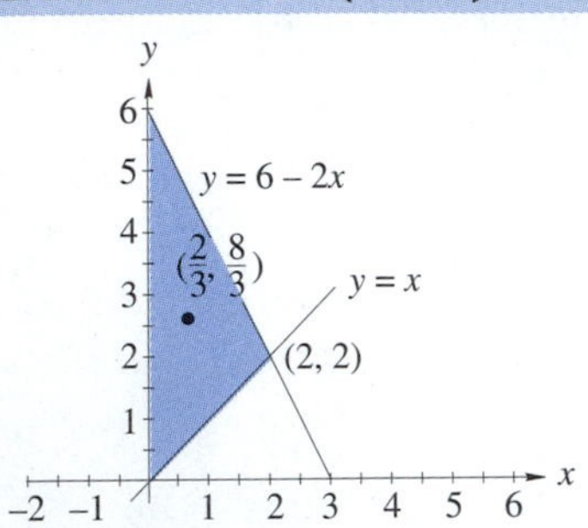

FIGURE 11.33b

EXAMPLE 11.33

Find the centroid of the region bonded by $y = x^2$ and $y = x + 2$.

Solution The region is shaded in Figure 11.34a. The curves intersect at $(2, 4)$ and $(-1, 1)$. We have $f(x) = x + 2$, $g(x) = x^2$, $a = -1$, and $b = 2$.

$$A = \int_{-1}^2 \int_{x^2}^{x+2} dy\,dx = \int_{-1}^2 y\Big|_{x^2}^{x+2} dx = \int_{-1}^2 (x + 2 - x^2)\,dx$$

$$= \left(\frac{1}{2}x^2 + 2x - \frac{1}{3}x^3\right)\Bigg|_{-1}^2 = 4.5$$

$$m = \rho A = 4.5\rho$$

$$M_y = \rho \int_{-1}^2 \int_{x^2}^{x+2} x\,dy\,dx = \rho \int_{-1}^x y\Big|_{x^2}^{x+2} x\,dx$$

$$= \rho \int_{-1}^2 [(x+2) - x^2]\,x\,dx = \rho \int_{-1}^2 (x^2 + 2x - x^3)\,dx$$

$$= \rho\left(\frac{1}{3}x^3 + x^2 - \frac{1}{4}x^4\right)\Bigg|_{-1}^2 = 2.25\rho$$

$$M_x = \rho \int_{-1}^2 \int_{x^2}^{x+2} y\,dy\,dx = \frac{1}{2}\rho \int_{-1}^2 y^2\Big|_{x^2}^{x+2} dx$$

$$= \frac{1}{2}\rho \int_{-1}^2 \left[(x+2)^2 - \left(x^2\right)^2\right] dx$$

$$= \frac{1}{2}\rho \int_{-1}^2 (x^2 + 4x + 4 - x^4)\,dx$$

$$= \frac{1}{2}\rho\left(\frac{1}{3}x^3 + 2x^2 + 4x - \frac{x^5}{5}\right)\Bigg|_{-1}^2 = 7.2\rho$$

$$\bar{x} = \frac{M_y}{m} = \frac{2.25\rho}{4.5\rho} = 0.5,\ \bar{y} = \frac{M_x}{m} = \frac{7.2\rho}{4.5\rho} = 1.6$$

The centroid is at (0.5, 1.6), as shown in Figure 11.34b.

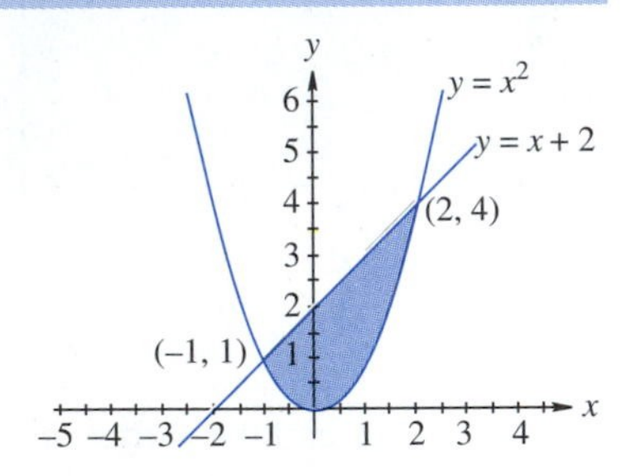

FIGURE 11.34a

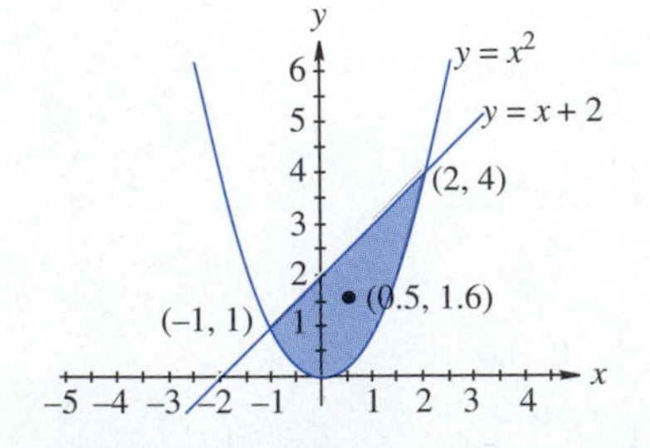

FIGURE 11.34b

In Section 7.6, we showed that the **moment of inertia** of a region with respect to the y-axis is

$$I_y = \rho \int_a^b x^2[f(x) - g(x)]\,dx$$

Again, since $\int_{g(x)}^{f(x)} dy = f(x) - g(x)$, we can write

$$I_y = \rho \int_a^b \int_{g(x)}^{f(x)} x^2\,dy\,dx$$

Similarly, we can show that the moment of inertia with respect to the x-axis is

$$I_x = \rho \int_a^b \int_{g(x)}^{f(x)} y^2\,dy\,dx$$

As before, the radius of gyration with respect to the y-axis is $r_y = \sqrt{\frac{I_y}{m}}$ and with respect to the x-axis is

$$r_x = \sqrt{\frac{I_x}{m}}$$

These results are summarized in the box on page 508.

EXAMPLE 11.34

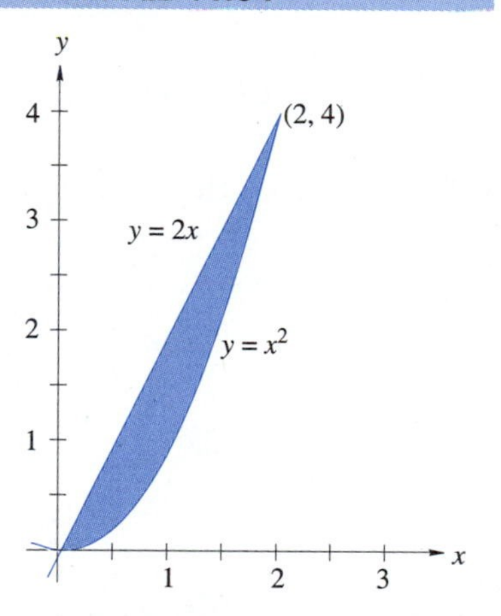

FIGURE 11.35

Find I_x and I_y for the region bounded by $y = 2x$ and $y = x^2$.

Solution The region is shaded in Figure 11.35. Here, we have $f(x) = 2x$, $g(x) = x^2$, $a = 0$, and $b = 2$.

$$\begin{aligned}
I_y &= \rho \int_0^2 \int_{x^2}^{2x} x^2\,dy\,dx = \rho \int_0^2 y\Big|_{x^2}^{2x} x^2\,dx \\
&= \rho \int_0^2 (2x - x^2)x^2\,dx \\
&= \rho \int_0^2 (2x^3 - x^4)\,dx = \rho \left(\frac{1}{2}x^4 - \frac{1}{5}x^5\right)\Big|_0^2 = 1.6\rho \\
I_x &= \rho \int_0^2 \int_{x^2}^{2x} y^2\,dy\,dx = \rho \int_0^2 \frac{1}{3}y^3\Big|_{x^2}^{2x} dx \\
&= \frac{1}{3}\rho \int_0^2 (8x^3 - x^6)\,dx \\
&= \frac{1}{3}\left(2x^4 - \frac{1}{7}x^7\right)\rho\Big|_0^2 = \frac{32}{7}\rho
\end{aligned}$$

The moments of inertia for the region bounded by $y = 2x$ and $y = x^2$ are $I_x = \frac{32}{7}\rho$ and $I_y = 1.6\rho$.

Moments of Inertia and Radii of Gyration of a Region in the xy-Plane

If ρ is a continuous density function on a plane region R, then the moments of inertia and radii of gyration of R are as follows:

Moments of Inertia (First Moments):

About the x-axis: $$I_x = \iint_R y^2 \rho(x, y)\, dy\, dx$$

About the y-axis: $$I_y = \iint_R x^2 \rho(x, y)\, dy\, dx$$

About the origin: $$I_0 = \iint_R (x^2 + y^2)\, \rho(x, y)\, dy\, dx$$

Radii of Gyration:

About the x-axis: $$r_x = \sqrt{\frac{I_x}{m}}$$

About the y-axis: $$r_y = \sqrt{\frac{I_y}{m}}$$

About the origin: $$r_0 = \sqrt{\frac{I_0}{m}}$$

where m is the mass of the region.

Note The number I_0 is a measure of the **polar moment of inertia**.

EXAMPLE 11.35 Find the moments of inertia and the radius of gyration for the region bounded by $y = \sqrt[3]{x}$, $x = 8$, and the x-axis with respect to the y-axis.

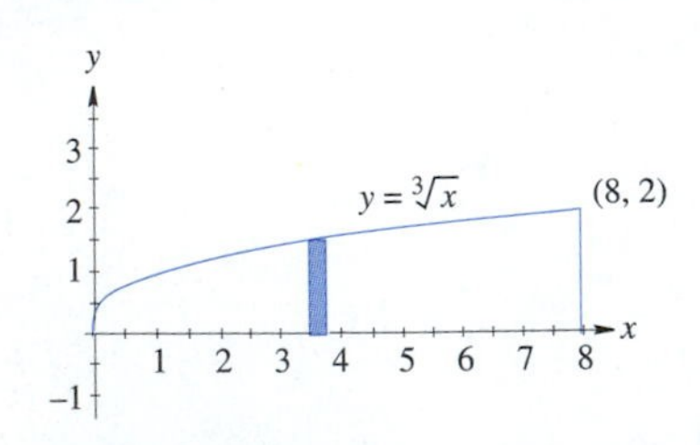

FIGURE 11.36

Solution A sketch of this region is shown in Figure 11.36. Here, we have $f(x) = \sqrt[3]{x}$, $g(x) = 0$, $a = 0$, and $b = 8$.

$$A = \int_0^8 \int_0^{\sqrt[3]{x}} dy\, dx = \int_0^8 y\Big|_0^{\sqrt[3]{x}} dx$$

$$= \int_0^8 x^{1/3}\, dx = \frac{3}{4}x^{4/3}\Big|_0^8 = 12$$

$$m = \rho A = 12\rho$$

$$I_y = \rho \int_0^8 \int_0^{\sqrt[3]{x}} x^2\, dy\, dx = \rho \int_0^8 y\Big|_0^{\sqrt[3]{x}} x^2\, dx = \rho \int_0^8 x^{7/3}\, dx$$

$$= \frac{3}{10}\rho x^{10/3}\Big|_0^8 = 307.2\rho$$

$$I_x = \rho \int_0^8 \int_0^{\sqrt[3]{x}} y^2\, dy\, dx = \rho \int_0^8 \frac{1}{3}y^3\Big|_0^{\sqrt[3]{x}} dx$$

EXAMPLE 11.35 (Cont.)

$$= \frac{\rho}{3}\int_0^8 x\,dx = \frac{1}{6}x^2\rho\Big|_0^8 = \frac{32}{3}\rho$$

$$r_y = \sqrt{\frac{I_y}{m}} = \sqrt{\frac{307.2\rho}{12\rho}} = \sqrt{25.6} \approx 5.06$$

For this region, the moments of inertia are $I_x = \frac{32}{3}\rho$ and $I_y = 307.2\rho$ and the radius of gyration around the y-axis is 5.06. ■

EXAMPLE 11.36

Find the moment of inertia and the radius of gyration for the region bounded by $y = \sqrt[3]{x}$, $x = 8$, and the x-axis with respect to the origin.

Solution This is the same region as that used in Example 11.35, so we know that its mass is $m = 12\rho$.

We first determine its moment of inertia about the origin.

$$\begin{aligned}
I_0 &= \rho\int_0^8\int_0^{\sqrt[3]{x}} (x^2+y^2)\,dy\,dx \\
&= \rho\int_0^8 \left(x^2y+\frac{1}{3}y^3\right)\Big|_0^{\sqrt[3]{x}}\,dx \\
&= \rho\int_0^8 \left(x^{7/3}+\frac{1}{3}x\right)dx \\
&= \rho\left[\frac{3}{10}x^{10/3}+\frac{1}{6}x^2\right]_0^8 = \frac{4{,}768}{15}\rho
\end{aligned}$$

Using this result, we determine the radius of gyration about the origin

$$\begin{aligned}
r_0 &= \sqrt{\frac{I_0}{m}} \\
&= \sqrt{\frac{4{,}768\rho/15}{12\rho}} = \sqrt{\frac{1{,}192}{45}} \approx 5.1467
\end{aligned}$$

■

Exercise Set 11.7

In Exercises 1–6, calculate the centroids of the plane region bounded by the given curve and the x-axis over the indicated interval.

1. $y = 2x+3$, $[0,3]$
2. $y = x^3$, $[0,2]$
3. $y = x^{1/3}$, $[0,8]$
4. $y = x^4$, $[-1,2]$
5. $y = \sqrt{x+4}$, $[0,5]$
6. $y = \sqrt{x^2+16}$, $[0,4]$

In Exercises 7–12, find the centroids of the plane regions bounded by the given curves and lines.

7. $y = x^2$, $y = 4x$, $x = 0$, $x = 2$

8. $y = x^3$, $y = 8x$, $x = 0$, $x = 2$

9. $y = x^{3/2}$, $y = x$

10. $y = x^2$, $y = 18 - x^2$

11. $y = x$, $y = 12 - x^2$

12. $y = x^2$, $y = x^3$

In Exercises 13–20, find the moment of inertia and radius of gyration about the given axis.

13. The region bounded by $y = x^2$, $x = 0$, and $y = 2$, about the y-axis

14. The region in Exercise 13 about the x-axis

15. The region bounded by $x = 0$, $y = 0$, $x = 3$, and $y = 5$, about the x-axis

16. The region in Exercise 15 about the y-axis

17. The region bounded by $y = 4x - x^2$ and $y = 0$, about the x-axis

18. The region in Exercise 17 about the y-axis

19. The region bounded by $y = \frac{1}{x}$, $y = x^2$, $x = 2$, and $x = 1$, about the y-axis

20. The region in Exercises 19 about the x-axis

In Exercises 21–24, for each of the plane regions, find (a) the moment of inertia about the origin and (b) the radius of gyration about the origin.

21. The region bounded by $y = x^2$, $x = 0$, $y = 2$. (This is the same region as in Exercise 13.)

22. The region bounded by $x = 0$, $y = 0$, $x = 3$, $y = 5$. (This is the same region as in Exercise 15.)

23. The region bounded by $y = 4x - x^2$, $y = 0$. (This is the same region as in Exercise 17.)

24. The region bounded by $y = \frac{1}{x}$, $y = x^2$, $x = 2$, $x = 1$. (This is the same region as in Exercise 19.)

In Your Words

25. Give physical explanations of what is meant by the moments and centroid of a region in the plane.

26. Give physical explanations of what is meant by the radius of gyration of a region in the plane.

CHAPTER 11 REVIEW

Important Terms and Concepts

Centroids
Chain rule for a function of two variables
Curve
Cylindrical coordinate system
Differentials
Functions of three variables
Functions of two variables
    Domain
    Range
Integration
    Double
    Multiple
Moments
Octant
Partial derivatives
    First derivative test
    Second derivative test
Planes
Quadric surfaces
    Ellipsoid
    Elliptic cone
    Elliptic cylinder
    Elliptic paraboloid
    Hyperbolic cylinder

Hyperbolic paraboloid
Hyperboloid of one sheet
Hyperboloid of two sheets
Rectangular coordinate system
Related rates
Spherical coordinate system
Surfaces in three dimensions

Intersection of
Curve
Section
Trace
Maxima
Minima
Saddle point

Review Exercises

In Exercises 1–8, identify and sketch the given surface.

1. $x+3y+2z=6$
2. $4x-3y-2z=12$
3. $4y^2+z^2=4$
4. $x^2+y^2+z^2=16$
5. $y=4x^2$
6. $9x^2-4y^2=1$
7. $x^2+4y^2-z^2=16$
8. $36z^2=9x^2+4y^2$

In Exercises 9–12, find $\frac{\partial z}{\partial x}, \frac{\partial z}{\partial y}, \frac{\partial^2 z}{\partial x^2}, \frac{\partial^2 z}{\partial y^2}$, and $\frac{\partial^2 z}{\partial x \partial y}$.

9. $z=3x^2+6xy^3$
10. $z=\dfrac{x^2+y^2}{y}$
11. $z=e^x\cos y-e^y\sin x$
12. $z=\ln\sqrt{x^2+y^3}+\sin^2(3xy)$

In Exercises 13–18, evaluate the given integral.

13. $\displaystyle\int_0^2\int_0^{x^3} xy\,dy\,dx$
14. $\displaystyle\int_0^{\pi/2}\int_0^{\pi} \sin x\cos y\,dy\,dx$
15. $\displaystyle\int_0^1\int_0^4 \sqrt{x+y}\,dy\,dx$
16. $\displaystyle\int_0^4\int_0^9 xy\sqrt{x^2+y^2}\,dy\,dx$
17. $\displaystyle\int_0^{\ln 4}\int_0^{\ln 10} e^{x+2y}\,dy\,dx$
18. $\displaystyle\int_0^1\int_0^{x^2} x\,dy\,dx$

In Exercises 19 and 20, convert the cylindrical coordinate to its equivalent rectangular coordinate and equivalent spherical coordinate.

19. $(4, \frac{\pi}{6}, 2)$
20. $(9, \frac{5\pi}{6}, -3)$

In Exercises 21–22, convert the spherical coordinate to its equivalent rectangular coordinate and equivalent cylindrical coordinate.

21. $(4, \frac{3\pi}{4}, \frac{\pi}{6})$
22. $(5, \frac{7\pi}{6}, \frac{2\pi}{3})$

In Exercises 23 and 24, find the average value for each of the functions over regions *R* having the given boundaries.

23. $f(x,y)=x^2+y^4, 0\le x\le 4, -2\le y\le 2$
24. $f(x,y)=x^3+2y, 1\le x\le 5, 0\le y\le 4$

Solve Exercises 25–30.

25. *Drafting* An open-topped rectangular box is to have a volume of 300 in.^{3} Find the dimensions that minimize its surface area.

26. *Electricity* The total resistance R of two resistances R_1 and R_2 connected in parallel is given by the formula

$$\frac{1}{R} = \frac{1}{R_1} + \frac{1}{R_2} = \frac{R_1 + R_2}{R_1 R_2}$$

Suppose that R_1 and R_2 are measured to be $300\,\Omega$ and $600\,\Omega$, respectively, with a maximum error of 1% in each measurement. Use differentials to estimate the maximum error (in ohms) in the calculated value of R.

27. *Electricity* Ohm's law states that $R = \frac{V}{I}$. Measurements are $V = 116\,\text{V}$ and $I = 2\,\text{A}$ with a possible error of 0.2 V and 0.01 A, respectively. Use differentials to approximate the maximum error in the calculated value of R (in ohms).

28. What is the smallest amount of material needed to build an open-top rectangular box enclosing a volume of $25\,\text{m}^3$?

29. *Physics* According to the ideal gas law, the pressure P, volume V, and absolute temperature T of n moles of an ideal gas are related by

$$PV = nRT$$

where R is a constant. The pressure of 8 moles of an ideal gas is increasing at a rate of $0.4\,\text{N/cm}^2\text{/min}$, while the temperature is increasing at a rate of $0.5°\text{K/min}$. How fast is the volume of the gas changing when $V = 1\,000\,\text{m}^3$ and $P = 4\,\text{N/cm}^2$?

30. The radius of a right circular cone is increasing at a rate of 10 cm/s, while its height is increasing at a rate of 15 cm/s. How is the volume changing when $r = 180\,\text{cm}$ and $h = 270\,\text{cm}$?

In Exercises 31–34, find the centroid of the indicated region.

31. $y = 6x,\ x = 1,\ x = 5,\ y = 0$

32. $y = x^4,\ x = 1,\ x = 2,\ y = 0$

33. $y = x^3,\ y = \sqrt[3]{x}$

34. $y = x^2,\ y = 9 - x^2$

In Exercises 35–38, find the moment of inertia and radius of gyration about the given axis.

35. The region bounded by $y = 6x$, $x = 0$, $x = 5$, and $y = 0$, about the x-axis

36. The region bounded by $y = 4 - x$, $x = 0$, $x = 4$, and $y = 0$, about the y-axis

37. The region bounded by $y = \sqrt[3]{x^2}$, $x = 0$, $x = 1$, and $y = 0$, about the x-axis

38. The region in the first quadrant bounded by $y = x^3$ and $y = \sqrt[3]{x}$, about the y-axis

Solve Exercises 39–42.

39. Find **(a)** the moment of inertia and **(b)** the radius of gyration about the origin for the region bounded by $y = 10x$, $x = 0$, $x = 8$, and $y = 0$.

40. Find **(a)** the moment of inertia and **(b)** the radius of gyration about the origin for the region in the first quadrant bounded by $y = 6x$, $y = x^2$.

41. Evaluate $\iint_R (x^2 - y)\,dA$, where R is the region bounded by $y = x$, $x + y = 4$, and $y = 0$.

42. Evaluate $\iint_R (x - 4y)\,dA$, where R is the region bounded by $y = x$ and $y = x^2 + x - 5$.

CHAPTER 11 TEST

In Exercises 1–4, identify and sketch the given surfaces.

1. $2x - y + 4z = 8$
2. $x^2 + y^2 + z^2 = 9$
3. $\frac{x^2}{4} + \frac{y^2}{9} - \frac{z^2}{4} = 1$
4. $z = 9x^2 + 4y^2 - 36$

Solve Exercises 5–14.

5. If $z = 4x^3 - 5x^2y^4$, determine $\frac{\partial z}{\partial x}$ and $\frac{\partial z}{\partial y}$.
6. If $z = \sqrt{x^2 + 4y^3} + \ln(x^2 y)$, determine $\frac{\partial^2 z}{\partial x^2}$.
7. Evaluate $\int_0^4 \int_0^{\pi/4} x \sin y \, dy \, dx$.
8. Evaluate $\int_0^2 \int_0^{x^2} 3xy^2 \, dy \, dx$.
9. Convert the cylindrical coordinates $\left(3, \frac{5\pi}{6}, 8\right)$ to its equivalent rectangular coordinates and equivalent spherical coordinates.
10. Convert the spherical coordinate $\left(4, \frac{7\pi}{6}, \frac{3\pi}{4}\right)$ to its equivalent rectangular coordinates and equivalent cylindrical coordinates.
11. Find the volume of the solid region bounded by the paraboloid $z = 4 - x^2 - 4y^2$ and the xy-plane.
12. Determine the centroid of the region in the xy-plane bounded by the graphs of $y = (x-1)^2 = x^2 - 2x + 1$ and $y = 4$.
13. Find the average value of $f(x, y) = x^2y^3$ for $-1 \le x \le 2, 0 \le y \le 4$.
14. Find **(a)** the moment of inertia and **(b)** the radius of gyration about the origin for the region in the first quadrant bounded by $x = 0$, $y = x$, $y = 6 - x^2$.

CHAPTER

12

Infinite Series

Calculators and computers use infinite series to calculate the values of many functions. In Section 12.3, you will learn how to use the Maclaurin series to perform such calculations.

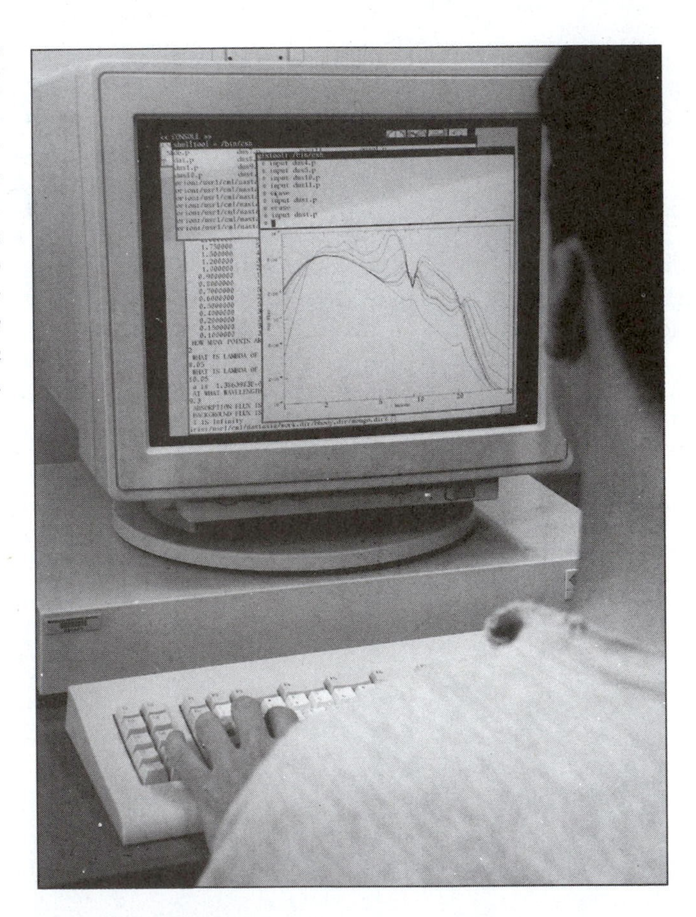

In this chapter, we begin by studying sequences and series. An infinite series is an indicated sum of infinitely many terms. We will look at some of the cases when an infinite series converges and when it does not. We are able to tell when a geometric series converges and what its sum will be. In this chapter, we will discuss methods for computing the sum of an infinite series. Infinite series arise very naturally in science and mathematics because many functions occur naturally in the form of infinite series. And, we will see that some functions can be written as infinite series in order to help with numerical computations.

12.1 SEQUENCES

A **sequence** is a set of numbers arranged in some order. Each number in the sequence is labeled with a variable, such as a. The variable is indexed with a natural number that tells its position in the sequence. The numbers a_1, a_2, a_3, and so on are the **terms** of the sequence. So, the first term in the sequence is a_1, the second term a_2, the third term a_3, and so on.

EXAMPLE 12.1

The sequence $3, 7, 11, 15, 19, 23$ has $a_1 = 3$, $a_2 = 7$, $a_3 = 11$, $a_4 = 15$, $a_5 = 19$, and $a_6 = 23$. This sequence has six terms.

Many sequences follow some sort of pattern. The pattern is usually described by the nth term of the sequence. This term, a_n, is called the **general term** of the sequence. A **finite sequence** has a specific number of terms and so it has a last term. An **infinite sequence** does not have a last term. The notation $\{a_n\}$ is often used to present the nth term of a sequence. The $\{\ \}$ indicate that it is a sequence.

EXAMPLE 12.2

Find the first six terms of the sequence $a_n = 4n + 1$.

Solution

$$a_1 = 4(1) + 1 = 5$$
$$a_2 = 4(2) + 1 = 9$$
$$a_3 = 4(3) + 1 = 13$$
$$a_4 = 4(4) + 1 = 17$$
$$a_5 = 4(5) + 1 = 21$$
$$a_6 = 4(6) + 1 = 25$$

The first six terms of this sequence are 5, 9, 13, 17, 21, and 25.

EXAMPLE 12.3

Find the first five terms of the sequence $\{n^2 - 3\}$.

Solution

$$a_1 = 1^2 - 3 = 1 - 3 = -2$$
$$a_2 = 2^2 - 3 = 4 - 3 = 1$$
$$a_3 = 3^2 - 3 = 9 - 3 = 6$$
$$a_4 = 4^2 - 3 = 16 - 3 = 13$$
$$a_5 = 5^2 - 3 = 25 - 3 = 22$$

The first five terms of this sequence are -2, 1, 6, 13, and 22.

EXAMPLE 12.4

Find the first seven terms of the sequence $\left\{\frac{n}{n+2}\right\}$.

Solution

$$a_1 = \frac{1}{1+2} = \frac{1}{3}$$
$$a_2 = \frac{2}{2+2} = \frac{2}{4} = \frac{1}{2}$$
$$a_3 = \frac{3}{3+2} = \frac{3}{5}$$
$$a_4 = \frac{4}{4+2} = \frac{4}{6} = \frac{2}{3}$$
$$a_5 = \frac{5}{5+2} = \frac{5}{7}$$
$$a_6 = \frac{6}{6+2} = \frac{6}{8} = \frac{3}{4}$$
$$a_7 = \frac{7}{7+2} = \frac{7}{9}$$

The first seven terms of this sequence are $\frac{1}{3}$, $\frac{1}{2}$, $\frac{3}{5}$, $\frac{2}{3}$, $\frac{5}{7}$, $\frac{3}{4}$, and $\frac{7}{9}$.

Recursion Formula

A **recursion formula** defines a sequence in terms of one or more previous terms. A sequence that is specified by giving the first term, or the first few terms, and a recursion formula is said to be **defined recursively**.

EXAMPLE 12.5

Find the first five terms of the sequence defined recursively by $a_1 = 2$ and $a_n = na_{n-1}$.

Solution We are told that $a_1 = 2$. By the recursion formula

$$a_2 = 2(a_1) = 2(2) = 4$$
$$a_3 = 3(a_2) = 3(4) = 12$$
$$a_4 = 4(a_3) = 4(12) = 48$$
$$a_5 = 5(a_4) = 5(48) = 240$$

The first five terms are 2, 4, 12, 48, and 240.

EXAMPLE 12.6

Give the first six terms of the sequence defined by $a_1 = 1$, $a_n = -2a_{n-1}$.

Solution

$$a_1 = 1$$
$$a_2 = -2(a_1) = -2(1) = -2$$
$$a_3 = -2(a_2) = -2(-2) = 4$$
$$a_4 = -2(a_3) = -2(4) = -8$$
$$a_5 = -2(a_4) = -2(-8) = 16$$
$$a_6 = -2(a_5) = -2(16) = -32$$

The first six terms are 1, −2, 4, −8, 16, and −32.

Application

EXAMPLE 12.7

A contractor is preparing a bid for constructing an office building. The first floor will cost $275,000. Each floor above the first will cost $15,000 more than the floor below it. **(a)** How much will the fifth floor cost? **(b)** What is the total cost for the first five floors?

Solutions

(a) Notice that this is a recursive sequence with $a_1 = 275{,}000$, $a_n = a_{n-1} + 15{,}000$. The first five terms of the sequence are

$$\begin{aligned} a_1 &= 275{,}000 \\ a_2 &= a_1 + 15{,}000 = 275{,}000 + 15{,}000 = 290{,}000 \\ a_3 &= a_2 + 15{,}000 = 290{,}000 + 15{,}000 = 305{,}000 \\ a_4 &= a_3 + 15{,}000 = 305{,}000 + 15{,}000 = 320{,}000 \\ a_5 &= a_4 + 15{,}000 = 320{,}000 + 15{,}000 = 335{,}000 \end{aligned}$$

The cost of the fifth floor is $335,000.

(b) The total cost for the first 5 floors is

$$\begin{aligned} a_1 + a_2 + a_3 + a_4 + a_5 &= 275{,}000 + 290{,}000 + 305{,}000 \\ &\quad + 320{,}000 + 335{,}000 \\ &= 1{,}525{,}000 \end{aligned}$$

The cost for the first five floors in $1,525,000.

Exercise Set 12.1

In Exercises 1–12, find the first six terms with the given specified general term.

1. $a_n = \dfrac{1}{n+1}$
2. $b_n = \dfrac{(-1)^n}{n}$
3. $a_n = (n+1)^2$
4. $a_n = \dfrac{1}{n(n+1)}$
5. $a_n = (-1)^n n$
6. $a_n = \dfrac{1+(-1)^n}{1+4n}$
7. $\left\{\dfrac{2n-1}{2n+1}\right\}$
8. $\left\{\left(\frac{-1}{2}\right)^n\right\}$
9. $\left\{\left(\frac{2}{3}\right)^{n-1}\right\}$
10. $\{n \cos n\pi\}$
11. $\left\{\left(\dfrac{n-1}{n+1}\right)^2\right\}$
12. $\left\{\dfrac{n^2-1}{n^2+1}\right\}$

In Exercises 13–24, find the first six terms of the recursively defined sequence.

13. $a_1 = 1,\ a_n = na_{n-1}$
14. $a_1 = 3,\ a_n = a_{n-1} + n$
15. $a_1 = 5,\ a_n = a_{n-1} + 3$
16. $a_1 = 2,\ a_n = (a_{n-1})^n$
17. $a_1 = 2,\ a_n = (a_{n-1})^{n-1}$
18. $a_1 = 1,\ a_n = \left(\dfrac{-1}{n}\right) a_{n-1}$
19. $a_1 = 1,\ a_n = n^{a_{n-1}}$
20. $a_1 = \frac{1}{2},\ a_n = (a_{n-1})^{-n}$
21. $a_1 = 1,\ a_2 = \frac{1}{2},\ a_n = (a_{n-1})(a_{n-2})$
22. $a_1 = 1,\ a_2 = 1,\ a_n = a_{n-2} + a_{n-1}$
23. $a_1 = 0,\ a_2 = 2,\ a_n = a_{n-1} - a_{n-2}$
24. $a_1 = 1,\ a_2 = 2,\ a_n = (a_{n-1})(a_{n-2})$

Solve Exercises 25–28.

25. *Environmental Science* An ocean beach is eroding at the rate of 3 in. per year. If the beach is currently 25 ft wide, how wide will the beach be in 25 y? (Hint: $a_1 = 25\text{ ft} - 3\text{ in.}$; find a_{25}.)

26. *Construction* A contractor is preparing a bid for constructing an office building. The foundation and basement will cost \$750,000. The first floor will cost \$320,000. The second floor will cost \$240,000. Each floor above the second will cost \$12,000 more than the floor below it. **(a)** How much will the 10th floor cost? **(b)** How much will the 20th floor cost?

27. *Business* You are offered a job with a starting salary of \$26,000 per year with a guaranteed raise of \$1,100 per year. What can you expect as an annual salary in your **(a)** 5th year? **(b)** 10th year?

28. *Architecture* The first row of an auditorium has 60 seats. Each row after the first has 4 more seats than the row in front of it. How many seats are in the 10th row?

In Your Words

29. What is a recursion formula? Support your explanation by giving an example different from those in the text.

30. What is a sequence? How does a finite sequence differ from an infinite sequence?

12.2 SERIES

There are many times when we need to add the terms of a sequence. The sum of the terms of a sequence is called a **series**. The series

$$a_1 + a_2 + a_3 + a_4 + a_5$$

is a **finite series** with five terms. A finite series can have $1, 2, 5, 19, 437$, or any number of terms as long as the number is a natural number. A series of the form

$$a_1 + a_2 + a_3 + a_4 + \cdots$$

is an **infinite series**. An infinite series has an infinite or endless number of terms.

Summation Notation

A more compact notation is often used to indicate a series. This notation is referred to as **summation** or **sigma notation** because it uses the capital Greek letter sigma ($\sum$). In general, the sigma notation means

$$\sum_{k=1}^{n} a_k = a_1 + a_2 + a_3 + \cdots + a_n$$

Here $\sum$ indicates a sum. The letter k is called an **index of summation**. The summation begins with $k = 1$ as is indicated below the $\sum$ and ends with $k = n$ as is indicated above the $\sum$.

Sometimes it is useful to indicate that a series begins with the zeroth, second, or other term. If it starts with the 0th term, then

$$\sum_{k=0}^{n} a_k = a_0 + a_1 + a_2 + \cdots + a_n$$

and if it starts with the second term,

$$\sum_{k=2}^{n} a_k = a_2 + a_3 + a_4 + \cdots + a_n$$

The numbers below and above the $\sum$ are the **limits of summation**. Here they are 2 and n.

EXAMPLE 12.8

Evaluate $\sum_{k=1}^{5} (-3)k$.

Solution

$$\begin{aligned}\sum_{k=1}^{5}(-3)k &= (-3)1+(-3)2+(-3)3+(-3)4+(-3)5\\ &= -3+(-6)+(-9)+(-12)+(-15)\\ &= -45\end{aligned}$$

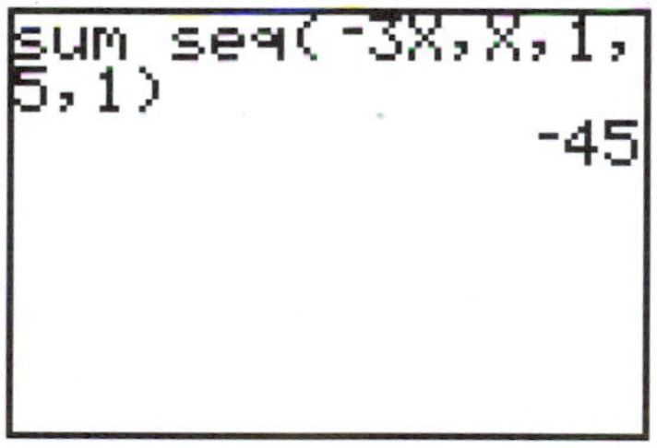

sum seq(-3X,X,1,5,1)
-45

FIGURE 12.1

You can use some calculators to add sequences. For instance, you can use a TI-82 to evaluate $\sum_{k=1}^{5}(-3)k$ by using the keystrokes [2nd] [LIST] [MATH] 5(sum) [2nd] [LIST] 5(seq() [(−)] 3 [X,T,θ] [,] [X,T,θ] [,] 1 [,] 5 [,] 1 [)] [ENTER]. The result, −45, as shown in Figure 12.1, is the same result we obtained in Example 12.8.

EXAMPLE 12.9

Evaluate $\sum_{k=1}^{4} (2k-3)$.

Solution

$$\begin{aligned}\sum_{k=1}^{4}(2k-3) &= (2\cdot 1-3)+(2\cdot 2-3)+(2\cdot 3-3)+(2\cdot 4-3)\\ &= (2-3)+(4-3)+(6-3)+(8-3)\\ &= -1+1+3+5\\ &= 8\end{aligned}$$

EXAMPLE 12.10

Evaluate $\sum_{k=0}^{3} \frac{2^k}{3k-5}$.

Solution

$$\begin{aligned}\sum_{k=0}^{3} \frac{2^k}{3k-5} &= \frac{2^0}{3\cdot 0-5} + \frac{2^1}{3\cdot 1-5} + \frac{2^2}{3\cdot 2-5} + \frac{2^3}{3\cdot 3-5} \\ &= \frac{1}{0-5} + \frac{2}{3-5} + \frac{4}{6-5} + \frac{8}{9-5} \\ &= \frac{1}{-5} + \frac{2}{-2} + \frac{4}{1} + \frac{8}{4} \\ &= 4\tfrac{4}{5}\end{aligned}$$

Partial Sums

Informally speaking, the expression $\sum_{k=0}^{n} a_k$ directs us to find the "sum" of the terms $a_1, a_2, a_3, \ldots, a_n$. To carry out this process, we proceed as follows. We let S_n denote the sum of the first n terms of the series. Thus,

$$\begin{aligned}S_1 &= a_1 \\ S_2 &= a_1 + a_2 \\ S_3 &= a_1 + a_2 + a_3 \\ &\vdots \\ S_n &= a_1 + a_2 + a_3 + \cdots + a_n = \sum_{k=1}^{n} a_k\end{aligned}$$

The number S_n is called the n**th partial sum** of the series. The sequence S_1, S_2, $S_3, \ldots, S_n$ is called the **sequence of partial sums**.

EXAMPLE 12.11

Find the first five partial sums of the series $\sum_{k=1}^{n} k^2$.

Solution

$$\begin{aligned}S_1 &= 1^2 = 1 \\ S_2 &= 1^2 + 2^2 = 5 \\ S_3 &= 1^2 + 2^2 + 3^2 = 14 \\ S_4 &= 1^2 + 2^2 + 3^2 + 4^2 = 30 \\ S_5 &= 1^2 + 2^2 + 3^2 + 4^2 + 5^2 = 55\end{aligned}$$

Arithmetic Series

Now let's look at series formed from two special sequences. An arithmetic sequence has a first term of a_1 and a common difference d. The sum of the first n terms of an arithmetic sequence is the nth partial sum of the sequence:

$$S_n = a_1 + (a_1 + d) + (a_1 + 2d) + (a_1 + 3d) + \cdots + [a_1 + (n-1)d] \quad (1)$$

If we write the nth term a_n first, then S_n could be written as

$$S_n = a_n + (a_n - d) + (a_n - 2d) + (a_n - 3d) + \cdots + [a_n - (n-1)d] \quad (2)$$

If we add the corresponding terms of equations (1) and (2), we get

$$2S_n = (a_1 + a_n) + (a_1 + a_n) + (a_1 + a_n) + (a_1 + a_n) + \cdots + (a_1 + a_n)$$

The right-hand side has n terms that are all the same, $a_1 + a_n$. Thus, we have the following sum S_n of the first n terms of an arithmetic sequence.

Sum of First n Terms of an Arithmetic Sequence

The sum S_n of the first n terms of an arithmetic sequence is

$$S_n = \frac{n(a_1 + a_n)}{2}$$

where a_1 is the first term and a_n is the nth term.

EXAMPLE 12.12

Find the sum of the first eight terms of the arithmetic sequence 3, 8, 13, 18, 23, 28, 33, 38, ...

Solution We will use the formula $S_n = \frac{n(a_1 + a_n)}{2}$ with $n = 8$, $a_1 = 3$, and $a_8 = 38$.

$$S_8 = \frac{8(3 + 38)}{2} = 164$$

EXAMPLE 12.13

Find the sum of the first 14 terms of the arithmetic sequence $9, 3, -3, \ldots$.

Solution In this sequence, $d = -6$, $a_1 = 9$, and $n = 14$. We know that

$$\begin{aligned} a_{14} &= a_1 + (14 - 1)d \\ &= 9 + 13(-6) \\ &= -69 \end{aligned}$$

EXAMPLE 12.13 (Cont.)

We can now find the sum of the first 14 terms.

$$\begin{aligned} S_{14} &= \frac{14(a_1 + a_n)}{2} \\ &= \frac{14(9 - 69)}{2} \\ &= -420 \end{aligned}$$

The sum of the first 14 terms of this sequence is -420.

Geometric Series

Now let's see if we can develop a similar formula for the first n terms of a geometric series. The sum of the first n terms of a geometric sequence is

$$S_n = a_1 + a_1 r + a_1 r^2 + a_1 r^3 + \cdots + a_1 r^{n-1} \tag{3}$$

Multiplying both sides of equation (3) by r, we get

$$rS_n = a_1 r + a_1 r^2 + a_1 r^3 + a_1 r^4 + \cdots + a_1 r^{n-1} + a_1 r^n \tag{4}$$

If we subtract equation (4) from equation (3), we get

$$S_n - rS_n = a_1 - a_1 r^n$$

Factoring each side $S_n(1-r) = a_1(1-r^n)$ and solving for S_n, we have the following formula for the first n terms of a geometric sequence.

Sum of First n Terms of a Geometric Sequence

The sum S_n of the first n terms of a geometric sequence is

$$S_n = a_1 \frac{1 - r^n}{1 - r}$$

where a_1 is the first term and r is the common ratio.

EXAMPLE 12.14

Find the sum of the first 10 terms of the geometric sequence $4, -8, 16, -32, \ldots$.

Solution In this sequence, $a_1 = 4$, $r = -2$, and $n = 10$.

$$\begin{aligned} S_{10} &= a_1 \frac{1 - r^{10}}{1 - r} \\ &= 4\left(\frac{1 - (-2)^{10}}{1 - (-2)}\right) \end{aligned}$$

EXAMPLE 12.14 (Cont.)

$$= 4\left(\frac{1-1{,}024}{3}\right)$$
$$= 4\left(\frac{-1{,}023}{3}\right)$$
$$= -1{,}364$$

The sum of the first 10 terms of $4-8+16-32+\cdots$ is $-1{,}364$.

EXAMPLE 12.15

Find the sum of the geometric series $\frac{1}{2}+\frac{1}{3}+\frac{2}{9}+\cdots+\frac{1}{2}(\frac{2}{3})^8$.

Solution In this series, $a_1 = \frac{1}{2}$, $r = \frac{2}{3}$, and $n = 9$.

$$S_9 = \frac{1}{2}\left(\frac{1-(\frac{2}{3})^9}{1-\frac{2}{3}}\right) = \frac{1}{2}\left(\frac{1-\frac{512}{19{,}683}}{1-\frac{2}{3}}\right) = \frac{1}{2}\left(\frac{\frac{19{,}171}{19{,}683}}{\frac{1}{3}}\right) = \frac{57{,}513}{39{,}366}$$
$$\approx 1.46$$

An Application of the Geometric Series

One application of the geometric series deals with compound interest. Suppose you deposit \$100 in a savings account that pays 8% compounded annually. The amount of money after 1 year would be

$$100 + 100(0.08) = 100 + 8$$
$$= \$108$$

After 2 years the amount would be

$$108 + 108(0.08) = 108 + 8.64$$
$$= \$116.64$$

As you can see, the amount after each year forms a geometric sequence

$$100 + 108 + 116.64 + \cdots$$

where $a_1 = 100$ and $r = 1.08$. Thus $r = 1 + i$, where i is the interest rate. From this, we get the formula for compound interest.

Compound Interest

$$A = P(1+i)^n$$

where A is the amount after n interest periods, P is the principal or initial amount invested, and i is the interest rate per interest period expressed as a decimal.

Thus A is similar to S_n. Since the initial investment is P, this formula begins with $n = 0$.

The interest rate is given for the interest period. This means that in order to determine i, you must divide the annual interest rate by the number of interest periods. An 8% interest rate compounded quarterly (four times a year) would have $i = \frac{0.08}{4} = 0.02$. (Remember $8\% = 0.08$.) If the 8% interest rate was compounded monthly, or 12 times a year, then $i = \frac{0.08}{12} = 0.0067$.

Application

EXAMPLE 12.16

If \$100 is deposited in a savings account paying 6% annually, what is the amount after 1 year if interest is compounded **(a)** quarterly, or **(b)** monthly?

Solutions We will use $A = P(1+i)^n$ in both parts with $P = 100$.

(a) Since interest is compounded quarterly

$$\begin{aligned} i &= \frac{0.06}{4} = 0.015 \text{ and } n = 4 \\ A &= 100(1+0.015)^4 \\ &= 106.14 \end{aligned}$$

The total is \$106.14 at the end of 1 year.

(b) In this example, the interest is compounded monthly, so

$$\begin{aligned} i &= \frac{0.06}{12} = 0.005 \text{ and } n = 12 \\ A &= 100(1.005)^{12} \\ &= 106.17 \end{aligned}$$

The total is \$106.17 at the end of 1 year.

Application

EXAMPLE 12.17

Suppose the money in Example 12.16 were left in the savings account for 4 more years (a total of 5). What would it then be worth?

Solutions

(a) At quarterly compounding, i is still 0.015, but n is now 4×5 (4 times a year for 5 years).

$$\begin{aligned} A &= 100(1.015)^{20} \\ &= 134.69 \end{aligned}$$

(b) At monthly compounding, $i = 0.005$ and $n = 12 \times 5 = 60$.

$$\begin{aligned} A &= 100(1.005)^{60} \\ &= 134.89 \end{aligned}$$

The totals are \$134.69 if the money is compounded quarterly for 5 years and \$134.89 if it is compounded monthly.

Some older calculators have a $\boxed{\Sigma}$ or $\boxed{\Sigma+}$ key. This key is used to sum a group of numbers. The total is placed in memory and the number of terms is displayed. On some calculators, it is possible to recall this total from memory and on others it is not. The $\boxed{\Sigma+}$ key on a calculator is primarily used for statistical problems.

Caution Check the owner's manual for your calculator to see if you can use your calculator to sum a sequence.

Exercise Set 12.2

In Exercises 1–4, write the first four terms of the indicated series.

1. $\sum_{k=1}^{20} 3\left(\frac{1}{2}\right)^k$
2. $\sum_{n=1}^{50} [4+(n-1)3]$
3. $\sum_{n=0}^{20} (-1)^n \frac{3^n}{n+1}$
4. $\sum_{k=1}^{60} (-1)^k$

In Exercises 5–12, evaluate the given sum.

5. $\sum_{k=1}^{5} (k-3)$
6. $\sum_{k=0}^{4} (2k+1)$
7. $\sum_{k=1}^{6} k^2$
8. $\sum_{k=2}^{8} \frac{k+2}{(k-1)k}$
9. $\sum_{n=1}^{5} n^3$
10. $\sum_{n=1}^{8} \frac{(-1)^n(n^2+1)}{n}$
11. $\sum_{i=3}^{8} \frac{2^i}{3i+1}$
12. $\sum_{i=0}^{7} \frac{(-1)^{i+1}(i+1)^2}{2i+1}$

In Exercises 13–28, determine whether the terms of the given series form an arithmetic or geometric series and find the indicated sum.

13. $3+6+9+12+\cdots;\ S_{10}$
14. $5+1-3+\cdots;\ S_{12}$
15. $1+2+4+\cdots;\ S_8$
16. $\frac{1}{2}+2+8+\cdots;\ S_9$
17. $-6-2+2+\cdots;\ S_{12}$
18. $1-\frac{1}{2}+\frac{1}{4}+\cdots;\ S_{10}$
19. $0.5+0.75+1.0+\cdots;\ S_{20}$
20. $0.5+0.2+0.08+\cdots;\ S_{15}$
21. $\frac{3}{4}-\frac{1}{4}+\frac{1}{12}+\cdots;\ S_{16}$
22. $0.4+0.04+0.004+\cdots;\ S_6$
23. $0.4+1.6+2.8+\cdots;\ S_{14}$
24. $16+8+4+\cdots;\ S_{10}$
25. $2\sqrt{3}+6+6\sqrt{3}+\cdots;\ S_{15}$
26. $\sqrt{5}+10+20\sqrt{5}+\cdots;\ S_{12}$
27. $3+9+27+\cdots;\ S_8$
28. $5-25+125+\cdots;\ S_{10}$

Solve Exercises 29–36.

29. *Physics* A ball is dropped to the ground from a height of 80 m, and each time it bounces 80% as high as it did on the previous bounce. How far has it traveled when it hits the ground the fourth time? The tenth time?

30. *Physics* A pendulum swings a distance of 50 m initially from one side to the other. After the first swing, each swing is only 0.8 of the distance of the previous swing. What is the total distance covered by the pendulum in 10 swings?

31. *Construction* The end of a gable roof is in the shape of a triangle. If the span is 32 ft (as shown in Figure 12.2), the height of the gable is 8 ft, and the studs are placed every 16 in., what is the total length of the studs?

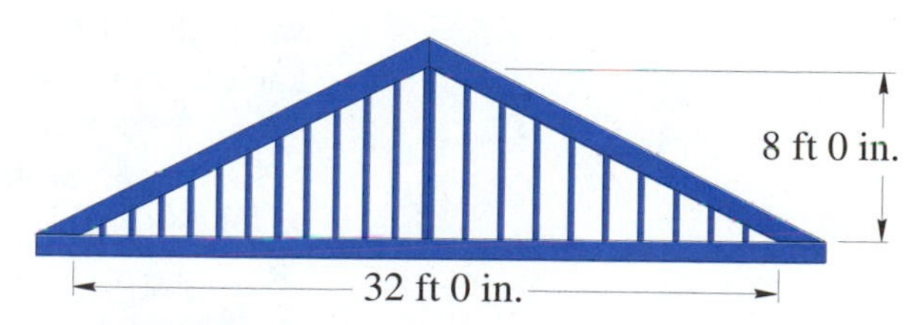

FIGURE 12.2

32. *Finance* How much will \$2,000 earn in 3 years if it is invested in a savings account that pays 8% compounded quarterly?

33. *Finance* You have the opportunity to invest the \$2,000 in Exercise 32 in a savings account that pays 8% compounded monthly. How much would this total after 3 years? How much more was this than the amount in Exercise 32?

34. *Finance* Your final chance is to invest the money in Exercise 32 at 8% compounded continuously. What would this total after 3 years?

35. *Construction* In Exercise 26, Exercise Set 12.1, a contractor was preparing a bid for the construction of an office building. The foundation and basement will cost \$750,000 and the first floor will cost \$320,000. The second floor will cost \$240,000 and each floor above the second will cost \$12,000 more than the floor below it. **(a)** If the 10th floor will cost \$336,000, how much will a 10-story building cost? **(b)** If the 20th floor will cost \$456,000, how much will it cost to build a 20-story building?

36. *Finance* At the beginning of each year, the owner of a automotive repair shop deposits \$3,000 into a retirement account paying 5.25% compounded annually. How much money will be in the account when the owner retires 25 years later? (Note: The owner retires at the end of the 25th year.)

In Your Words

37. Explain how a series and a sequence are different.

38. What is a partial sum? How you you determine the sequence of partial sums?

39. Describe how to determine the sum of the first n terms of an arithmetic sequence.

40. Describe how to determine the sum of the first n terms of a geometric sequence.

12.3 POWER AND MACLAURIN SERIES

Until now, the series that we have considered have all had terms that are constants. In this section, the terms of the series will be a function. This means that the series is of the form $\sum f_n(x)$.

Power Series

We will begin by looking at any series of the form

$$\sum_{k=0}^{\infty} a_k x^k = a_0 + a_1 x + a_2 x^2 + a_3 x^3 + \cdots$$

where the coefficients a_0, a_1, a_2, a_3, ... are all constants. A series of this type is known as a **power series in** x. In a power series, if $x = 0$, then the series is equal to a_0.

A variation on the power series in x is the power series in $x - c$, which is defined as follows.

Power Series in $x - c$

A series of the form

$$\sum_{k=0}^{\infty} a_k(x-c)^k = a_0 + a_1(x-c) + a_2(x-c)^2 + a_3(x-c)^3 + \cdots$$

is called a **power series in $x - c$.**

Notice that the power series in x,

$$\sum_{k=0}^{\infty} a_k x^k = a_0 + a_1 x + a_2 x^2 + a_3 x^3 + \cdots$$

is a special case of the power series in $x - c$, where $c = 0$.

EXAMPLE 12.18

Write the first five terms of each of the following power series: **(a)** $\sum_{k=0}^{\infty} \frac{x^k}{k!}$, **(b)** $\sum_{k=0}^{\infty} \frac{kx^k}{3^k}$, and **(c)** $\sum_{k=0}^{\infty} \frac{x^k}{k+2}$.

Solutions

(a) $$\sum_{k=0}^{\infty} \frac{x^k}{k!} = \frac{x^0}{0!} + \frac{x^1}{1!} + \frac{x^2}{2!} + \frac{x^3}{3!} + \frac{x^4}{4!} + \cdots \qquad \text{Recall: } 0! = 1$$

$$= 1 + x + \frac{x^2}{2} + \frac{x^3}{6} + \frac{x^4}{24} + \cdots$$

(b) $$\sum_{k=0}^{\infty} \frac{kx^k}{3^k} = \frac{0 \cdot x^0}{3^0} + \frac{1 \cdot x^1}{3^1} + \frac{2 \cdot x^2}{3^2} + \frac{3 \cdot x^3}{3^3} + \frac{4 \cdot x^4}{3^4} + \cdots$$

$$= 0 + \frac{x}{3} + \frac{2x^2}{9} + \frac{x^3}{9} + \frac{4x^4}{81} + \cdots$$

(c) $$\sum_{k=0}^{\infty} \frac{x^k}{k+2} = \frac{x^0}{0+2} + \frac{x^1}{1+2} + \frac{x^2}{2+2} + \frac{x^3}{3+2} + \frac{x^4}{4+2} + \cdots$$

$$= \frac{1}{2} + \frac{x}{3} + \frac{x^2}{4} + \frac{x^3}{5} + \frac{x^4}{6} + \cdots$$

A power series converges for some values of x and diverges for others. The values of x for which the series converges differ from series to series. For example, the series in Example 12.18(a) converges for all values of x, the series in Example 12.18(b) converges if $|x| < 3$, and the series in Example 12.18(c) converges if $-1 \leq x < 1$. The values of x for which a series converges is called the **interval of convergence**.

Note It is not our purpose to determine the interval of convergence for a series. We will assume that any series that we work with will converge for all values of x or for some given interval.

Properties of Convergent Series

If a power series has an interval of convergence that is greater than zero, then three properties are true for all values of x within that interval. If $f(x) = \sum_{k=0}^{\infty} a_k x^k = a_0 + a_1x + a_2x^2 + a_3x^3 + a_4x^4 + \cdots$, then for all values of x within the interval of convergence, **(a)** f is continuous, **(b)** f is differentiable, and **(c)** f can be integrated.

Maclaurin Series

We will immediately use the first two of these properties. Let's assume that for some interval $-R < x < R$, the function f is represented by the power series just given. We may obtain the coefficients $a_0, a_1, a_2, \ldots$ in terms of f and its derivatives by using part (b) from the previous paragraph.

$$\begin{aligned}
f(x) &= a_0 + a_1x + a_2x^2 + a_3x^3 + a_4x^4 + a_5x^5 + \cdots \\
f'(x) &= a_1 + 2a_2x + 3a_3x^2 + 4a_4x^3 + 5a_5x^4 + \cdots \\
f''(x) &= 2a_2 + 3\cdot 2a_3x + 4\cdot 3a_4x^2 + 5\cdot 4a_5x^3 + \cdots \\
f'''(x) &= 3\cdot 2a_3 + 4\cdot 3\cdot 2a_4x + 5\cdot 4\cdot 3a_5x^2 + \cdots \\
f^{(4)}(x) &= 4\cdot 3\cdot 2a_4 + 5\cdot 4\cdot 3\cdot 2a_5x + \cdots \\
f^{(5)}(x) &= 5\cdot 4\cdot 3\cdot 2a_5 + \cdots
\end{aligned}$$

If we let $x = 0$, then we get a value of 0 for each term in these equations, except for the first term. So, we have $f(0) = a_0$, $f'(0) = a_1$, $f''(0) = 2a_2$, $f'''(0) = 3\cdot 2a_3 = 6a_3$, $f^{(4)}(0) = 4\cdot 3\cdot 2a_4 = 24a_4$, $f^{(5)}(0) = 5\cdot 4\cdot 3\cdot 2a_5 = 120a_5$, and so on. Look carefully at each of these terms. In general, we have $f^{(n)}(0) = n(n-1)(n-2)(n-3)\cdots(2)\cdot 1a_n$. Since $n(n-1)(n-2)(n-3)\cdots 2\cdot 1 = n!$, we can write $f^{(n)}(0) = n!a_n$. Solving for a_n, we see that

$$a_n = \frac{f^{(n)}(0)}{n!}$$

Remember that the notation $f^{(n)}(x)$ indicates the nth derivative of $f(x)$ and *not* the nth power of $f(x)$.

Substituting these values into the equation for the function, we get the **Maclaurin series**.

Maclaurin Series

$$\begin{aligned} f(x) &= f(0) + f'(0)x + \frac{f''(0)}{2!}x^2 + \frac{f'''(0)}{3!}x^3 + \frac{f^{(4)}(0)}{4!}x^4 + \cdots \\ &\quad + \frac{f^{(n)}(0)}{n!}x^n + \cdots \\ &= \sum_{n=0}^{\infty} \frac{f^{(n)}(0)}{n!}x^n \end{aligned}$$

where $f^{(0)}(x) = f(0)$.

EXAMPLE 12.19

Find the Maclaurin series of $f(x) = e^x$.

Solution To use our definition, we need to evaluate f and its derivatives at 0.

$$\begin{aligned} f(x) &= e^x & f(0) &= 1 \\ f'(x) &= e^x & f'(0) &= 1 \\ f''(x) &= e^x & f''(0) &= 1 \end{aligned}$$

In general, $f^{(n)}(x) = e^x$ and $f^{(n)}(0) = 1$. Thus, the Maclaurin series of $f(x) = e^x$ is

$$\begin{aligned} e^x &= 1 + x + \frac{x^2}{2!} + \frac{x^3}{3!} + \frac{x^4}{4!} + \cdots + \frac{x^n}{n!} + \cdots \\ &= \sum_{n=0}^{\infty} \frac{x^n}{n!} \end{aligned}$$

Notice that this was the same series we used in Example 12.18(a).

EXAMPLE 12.20

Find the Maclaurin series of $f(x) = \cos x$.

Solution Again, we evaluate f and its derivatives at 0.

$$\begin{aligned} f(x) &= \cos x & f(0) &= 1 \\ f'(x) &= -\sin x & f'(0) &= 0 \\ f''(x) &= -\cos x & f''(0) &= -1 \\ f'''(x) &= \sin x & f'''(0) &= 0 \\ f^{(4)}(x) &= \cos x & f^{(4)}(0) &= 1 \end{aligned}$$

As you can see, we are back at the beginning: $f^{(5)}(x) = f'(x)$, $f^{(6)}(x) = f''(x)$, and so on. Substituting these values yields the Maclaurin series.

$$\cos x = 1 - \frac{x^2}{2!} + \frac{x^4}{4!} - \frac{x^6}{6!} + \frac{x^8}{8!} + \cdots + (-1)^n \frac{x^{2n}}{(2n)!} + \cdots$$

EXAMPLE 12.21

Find the Maclaurin series of $f(x) = \sin 2x$.

Solution

$$\begin{aligned} f(x) &= \sin 2x & f(0) &= 0 \\ f'(x) &= 2\cos 2x & f'(0) &= 2 = 2^1 \\ f''(x) &= -4\sin 2x & f''(0) &= 0 \\ f'''(x) &= -8\cos 2x & f'''(0) &= -8 = -2^3 \\ f^{(4)}(x) &= 16\sin 2x & f^{(4)}(0) &= 0 \\ f^{(5)}(x) &= 32\cos 2x & f^{(5)}(0) &= 32 = 2^5 \end{aligned}$$

We seem to have a pattern. If n is an even number, $f^{(n)}(0) = 0$. If n is an odd number, then $n = 2k+1$, and $f^{(n)}(0) = f^{(2k+1)}(0) = (-1)^k 2^{2k+1}$.

The series is

$$\sin 2x = 2x - \frac{8x^3}{3!} + \frac{2^5x^5}{5!} - \cdots + \frac{(-1)^k(2x)^{2k+1}}{(2k+1)!}.$$

The expansions in the box below are important and have been listed to help you. In Examples 12.19 and 12.20, we developed the first and third expressions. The rest are Exercises 1–4 in Exercise Set 12.3.

Some Important Maclaurin Series

$$e^x = 1 + x + \frac{x^2}{2!} + \frac{x^3}{3!} + \frac{x^4}{4!} + \frac{x^5}{5!} + \cdots \quad \text{, for all } x$$

$$\sin x = x - \frac{x^3}{3!} + \frac{x^5}{5!} - \frac{x^7}{7!} + \frac{x^9}{9!} - \frac{x^{11}}{11!} + \cdots \quad \text{, for all } x$$

$$\cos x = 1 - \frac{x^2}{2!} + \frac{x^4}{4!} - \frac{x^6}{6!} + \frac{x^8}{8!} - \frac{x^{10}}{10!} + \cdots \quad \text{, for all } x$$

$$\sinh x = x + \frac{x^3}{3!} + \frac{x^5}{5!} + \frac{x^7}{7!} + \cdots \quad \text{, for all } x$$

$$\cosh x = 1 + \frac{x^2}{2!} + \frac{x^4}{4!} + \frac{x^6}{6!} + \cdots \quad \text{, for all } x$$

$$\ln(1+x) = x - \frac{x^2}{2} + \frac{x^3}{3} - \frac{x^4}{4} + \frac{x^5}{5} - \frac{x^6}{6} + \cdots \quad \text{, for } -1 < x \le 1$$

Application

EXAMPLE 12.22

A certain type of skin wound heals at the rate of

$$A'(t) = \frac{-50}{t^2+20} \text{ cm}^2/\text{day}$$

EXAMPLE 12.22 (Cont.)

Suppose that initially a wound had an area of $15\,\text{cm}^2$. Use a Maclaurin series to approximate the area of the wound after 2 days.

Solution Here, if we let $f(t) = A'(t) = \dfrac{-50}{t^2+20}$, then we have

$$f(t) = \frac{-50}{t^2+20} \qquad f(0) = -\frac{5}{2}$$

$$f'(t) = \frac{100t}{(t^2+20)^2} \qquad f'(0) = 0$$

$$f''(t) = \frac{100(20-3t^2)}{(t^2+20)^3} \qquad f''(0) = \frac{1}{4}$$

Thus, we can approximate $A'(t)$ with the Maclaurin series

$$\begin{aligned} A'(t) &\approx -\frac{5}{2} + \frac{1/4}{2!}t^2 \\ &= -\frac{5}{2} + \frac{1}{8}t^2 \end{aligned}$$

Since this is the rate at which the wound is healing, the size of the wound in square centimeters at the end of day t is

$$\begin{aligned} A(t) &= \int A'(t)\,dt \\ &= -\frac{5}{2}t + \frac{1}{24}t^3 + C \end{aligned}$$

We are given that the wound was initially $15\,\text{cm}^2$, so $A(0) = 15$. Thus,

$$A(t) = -\frac{5}{2}t + \frac{1}{24}t^3 + 15$$

At the end of day 2, the wound has an area of

$$\begin{aligned} A(2) &= -5 + \frac{1}{3} + 15 \\ &= 10\frac{1}{3}\,\text{cm}^2 \end{aligned}$$

Exercise Set 12.3

In Exercises 1–16, find the first four nonzero terms of the Maclaurin series expansion for the given function.

1. $\sin x$
2. $\sinh x$ $\left(\text{Remember: } \sinh x = \dfrac{e^x - e^{-x}}{2}.\right)$
3. $\cosh x$ $\left(\text{Remember: } \cosh x = \dfrac{e^x + e^{-x}}{2}.\right)$
4. $\ln(1+x)$
5. e^{3x}
6. $\sin^3 x$
7. $\ln(1+x^2)$
8. e^{-x}
9. $\cos x^2$

10. e^{-x^2}

11. $e^x \sin x$

12. xe^x

13. $x^2 e^{-x^2}$

14. $e^{\sin x}$

15. $x \sin 3x$

16. $e^{-x} \cos x$

Solve Exercises 17–20.

17. *Medical technology* A certain type of skin wound heals at the rate of

$$A'(t) = \frac{-60}{t^2+30} \text{ cm}^2\text{/day}$$

Suppose that initially a wound had an area of 12 cm^2. Use a second-degree Maclaurin series to approximate the area of the wound after 2 days.

18. *Electronics* The current in amperes in a certain circuit is given by

$$i(\alpha) = 2 + \cos\alpha + 2\sin\alpha$$

Find the first four terms of the Maclaurin series expansion of this function.

19. *Hydraulics* The function $E(x) = \sec x$ is used in the study of fluid flow to define the Euler numbers. Find the first three nonzero terms of the Maclaurin series expansion of this function.

20. *Medical technology* The rate of healing for one type of skin wound in square centimeters per day is given by

$$A'(t) = \frac{-75}{t^2+25}$$

Suppose that initially a wound had an area of 16 cm^2.

(a) Use a Maclaurin series to approximate the area of the wound after t days.

(b) What is the approximate area of the wound after 3 days?

In Your Words

21. Describe how to determine a Maclaurin series.

22. Why are series important? What are some "real world" applications of infinite series?

12.4 OPERATIONS WITH SERIES

In Section 12.3, we learned how to use Maclaurin series to expand several functions. In this section, we will learn some operations with series that will provide quicker ways to expand these, and other, series. In Section 12.5, we will see how to make use of these series to develop tables.

In Section 12.3, we developed the Maclaurin series for $f(x)$. This same method could be used to expand $f(u)$. Thus, we would have

$$f(u) = \sum_{n=0}^{\infty} \frac{f^{(n)}(0)}{n!}u^n = f(0) + f'(0)u + \frac{f''(0)}{2!}u^2 + \frac{f'''(0)}{3!}u^3 + \cdots$$

By letting u assume values of $2x$, x^2, $-x$, x^{-2}, and so on, we can evaluate $f(2x)$, $f(x^2)$, $f(-x)$, $f(x^{-2})$, and so on.

EXAMPLE 12.23

Find the Maclaurin expansion of e^{3x}.

Solution From the Section 12.3, we know that

$$e^u = 1 + u + \frac{u^2}{2!} + \frac{u^3}{3!} + \cdots$$

We want the expansion of e^{3x}, so if we let $u = 3x$, we have

$$\begin{aligned} e^{3x} &= 1 + 3x + \frac{(3x)^2}{2!} + \frac{(3x)^3}{3!} + \frac{(3x)^4}{4!} + \cdots \\ &= 1 + 3x + \frac{9}{2}x^2 + \frac{9}{2}x^3 + \frac{27}{8}x^4 + \cdots \end{aligned}$$

EXAMPLE 12.24

Find the Maclaurin expansion of $\cos x^2$.

Solution We know that

$$\cos u = 1 - \frac{u^2}{2!} + \frac{u^4}{4!} - \frac{u^6}{6!} + \cdots$$

Thus, if $u = x^2$, we have

$$\begin{aligned} \cos x^2 &= 1 - \frac{(x^2)^2}{2!} + \frac{(x^2)^4}{4!} - \frac{(x^2)^6}{6!} + \cdots \\ &= 1 - \frac{x^4}{2!} + \frac{x^8}{4!} - \frac{x^{12}}{6!} + \cdots \end{aligned}$$

Compare this to the amount of work that we had to do when we worked Exercise 9 in Exercise Set 12.3.

Basic Operations with Series

The basic operations can be used with series in order to obtain other series. Thus, you can add, subtract, multiply, or divide series in the same way as you do polynomials. The next example will show you how this can be done.

EXAMPLE 12.25

Determine the Maclaurin series expansion of $e^x \sin x$.

Solution We know that if $f(x) = e^x$, then $f(x) = 1 + x + \frac{x^2}{2!} + \frac{x^3}{3!} + \frac{x^4}{4!} + \cdots$, and if $g(x) = \sin x$, then $g(x) = x - \frac{x^3}{3!} + \frac{x^5}{5!} - \frac{x^7}{7!} + \cdots$.

We want $e^x \sin x = f(x) \cdot g(x)$. Thus, we can multiply these two series term by term:

$$\begin{aligned} e^x \sin x &= \left(1 + x + \frac{x^2}{2!} + \frac{x^3}{3!} + \frac{x^4}{4!} + \cdots\right)\left(x - \frac{x^3}{3!} + \frac{x^5}{5!} - \frac{x^7}{7!} + \cdots\right) \\ &= x + x^2 + \frac{x^3}{3} - \frac{4x^5}{5!} + \cdots \end{aligned}$$

Differentiation and Integration of Series

It is also possible to differentiate and integrate functions by differentiating or integrating the terms of their series expansion. We used this procedure to help solve Example 12.22.

EXAMPLE 12.26

Show that $\frac{d}{dx}e^x = e^x$.

Solution We know that $e^x = 1+x+\frac{x^2}{2!}+\frac{x^3}{3!}+\frac{x^4}{4!}+\cdots$, so

$$\begin{aligned}\frac{d}{dx}e^x &= \frac{d}{dx}\left(1+x+\frac{x^2}{2!}+\frac{x^3}{3!}+\frac{x^4}{4!}+\cdots\right)\\ &= 0+1+\frac{2x}{2!}+\frac{3x^2}{3!}+\frac{4x^3}{4!}+\cdots\\ &= 1+x+\frac{x^2}{2!}+\frac{x^3}{3!}+\cdots\\ &= e^x\end{aligned}$$

Application

EXAMPLE 12.27

Approximate the area under the graph of $y = \cos x^2$ to four decimal places.

Solution In order to find this area, we need to evaluate $\int_0^1 \cos x^2\,dx$. In Example 12.24, we found that

$$\cos x^2 = 1-\frac{x^4}{2!}+\frac{x^8}{4!}-\frac{x^{12}}{6!}+\cdots$$

Integrating the individual terms of this series allows us to obtain the desired integral.

$$\begin{aligned}\int_0^1 \cos x^2\,dx &= \int_0^1\left(1-\frac{x^4}{2!}+\frac{x^8}{4!}-\frac{x^{12}}{6!}+\cdots\right)dx\\ &= x-\frac{x^5}{5\cdot 2!}+\frac{x^9}{9\cdot 4!}-\frac{x^{13}}{13\cdot 6!}+\cdots\Big|_0^1\\ &= 1-\frac{1}{5\cdot 2}+\frac{1}{9\cdot 4!}-\frac{1}{13\cdot 6!}+\cdots\\ &\approx 1-0.1+0.00463-0.00010\\ &= 0.9045\end{aligned}$$

The last example of this section will show a relationship between three transcendental functions. This result makes use of the imaginary number $j=\sqrt{-1}$. In this relationship, we will need to assume that the Maclaurin expansions of e^x, $\sin x$, and $\cos x$ are valid for complex numbers.

Look back at the Maclaurin expansions of $\sin x$ and $\cos x$. The expansion of the sine function has only odd powers. The expansion of the cosine function has only even powers. On the other hand, the expansion of e^x contains all powers of x. Now, if we expand e^{jx}, we get

$$\begin{aligned} e^{jx} &= 1 + jx + \frac{(jx)^2}{2!} + \frac{(jx)^3}{3!} + \frac{(jx)^4}{4!} + \frac{(jx)^5}{5!} + \frac{(jx)^6}{6!} + \cdots \\ &= 1 + jx + \frac{j^2x^2}{2!} + \frac{j^3x^3}{3!} + \frac{j^4x^4}{4!} + \frac{j^5x^5}{5!} + \frac{j^6x^6}{6!} + \cdots \end{aligned}$$

We need to remember that $j = \sqrt{-1}$, $j^2 = -1$, $j^3 = -j$, $j^4 = 1$, and then the pattern repeats. We get

$$\begin{aligned} e^{jx} &= 1 + jx - \frac{x^2}{2!} - \frac{jx^3}{3!} + \frac{x^4}{4!} + \frac{jx^5}{5!} - \frac{x^6}{6!} - \cdots \\ &= \left(1 - \frac{x^2}{2!} + \frac{x^4}{4!} - \frac{x^6}{6!}\right) + j\left(x - \frac{x^3}{3!} + \frac{x^5}{5!} - \cdots\right) \\ &= \cos x + j \sin x \end{aligned}$$

This relationship, $e^{jx} = \cos x + j\sin x$, is known as **Euler's identity**.

A complex number in rectangular form is expressed as $a + bj$, where a and b are real numbers. This same number, $a + bj$, is written in polar form as $r(\cos\theta + j\sin\theta) = r\operatorname{cis}\theta$. In exponential notation, $a + bj$ is written as $re^{j\theta}$, where $r = \sqrt{a^2 + b^2}$.

EXAMPLE 12.28

Express $4e^{5.4j}$ in rectangular form.

Solution We know that $a + bj = re^{j\theta}$. Thus, $r = 4$ and $\theta = 5.4$ rad and we have

$$\begin{aligned} a + bj &= 4(\cos 5.4 + j\sin 5.4) \\ &\approx 4(0.6347 - 0.7728j) \\ &= 3.9822 + 0.3764j \end{aligned}$$

One final point about Euler's identity: Some of the most familiar and, some people say, interesting numbers in mathematics are e, π, j, 0, and -1. Using Euler's identity, we see that

$$\begin{aligned} e^{j\pi} &= \cos\pi + j\sin\pi \\ &= -1 \end{aligned}$$

Thus, we have

$$e^{j\pi} = -1$$

$$\text{or} \qquad e^{j\pi} + 1 = 0$$

As we will see, Euler's identity is used in differential equations and in the theory of electrical circuits.

Exercise Set 12.4

In Exercises 1–12, find the first four nonzero terms of the Maclaurin expansion of the given function.

1. $f(x) = e^{3x}$
2. $g(x) = e^{-4x}$
3. $h(x) = \cos\frac{x}{2}$
4. $j(x) = \sin x^3$
5. $k(x) = \cos x^3$
6. $f(x) = \sin 3x$
7. $g(x) = \sin 2x^2$
8. $h(x) = \ln(1-x)$
9. Show that $\frac{d}{dx}\sin x = \cos x$ by using Maclaurin series.
10. Use Maclaurin series to show $\frac{d}{dx}\cos x^2 = -2x\sin x^2$.
11. Use Maclaurin series to show that $\frac{d}{dx}e^{2x} = 2e^{2x}$.
12. Use Maclaurin series to show that $\frac{d}{dx}\sin 3x = 3\cos 3x$.

In Exercises 13–16, evaluate the given integrals by use of three terms of the appropriate series.

13. $\int_0^1 \sin x^2\,dx$
14. $\int_0^{0.1} \frac{e^x - 1}{x}dx$
15. $\int_0^{0.2} \sin\sqrt{x}\,dx$
16. $\int_0^1 \frac{\sin x}{x}dx$

Solve Exercises 17–20.

17. Find the Maclaurin expansion of $e^{-x}\cos x$ using the method of Example 12.25.
18. Find the Maclaurin expansion of $\frac{1}{2}(e^x - e^{-x})$ by adding the terms of the appropriate series. Compare this to your results to Exercise 2, from Exercise Set 11.1.
19. Find the Maclaurin expansion of $x^2e^{-x^2}$ using the method of Example 12.25.
20. Find the Maclaurin expansion of $\tan x$ by dividing $\sin x$ by $\cos x$.

In Exercises 21–24, change the given complex number to the indicated form.

21. $5e^{0.8j}$ (rectangular)
22. $5-12j$ (polar)
23. $6\text{ cis }\frac{4\pi}{3}$ (exponential)
24. $-3+4j$ (exponential)

In Exercises 25–28, find the approximate area bounded by the given curves or lines by using the first three nonzero terms of the appropriate Maclaurin series.

25. $y = e^{x^2}, x = 0, x = 1, y = 0$
26. $y = \cos x^2, x = 0, x = \frac{\pi}{2}, y = 0$
27. $y = x^2e^x, x = 0.2, x = 0, y = 0$
28. $y = e^{-x^2}, x = 0, x = 1, y = 0$

Solve Exercises 29–34.

29. *Electricity* Use the first five nonzero terms of the Maclaurin expansion of $\cos x$ to obtain an approximation of the voltage $V = 15\cos\Omega t$, when $\Omega t = 0.1$ rad.
30. Use the Maclaurin series expansion of $\sinh x$ and $\cosh x$ to show that $e^x = \sinh x + \cosh x$.
31. *Electricity* **(a)** Express as a Maclaurin series the current $i = \sin t^2$ A. **(b)** What is the amount of charge transmitted by this current from $t = 0$ to $t = 0.02$ s?
32. Find the volume generated by rotating the area bounded by $y = e^{-x}$, $y = 0$, $x = 0$, and $x = 0.1$ about the y-axis, by using the first three terms of the appropriate series.

33. *Environmental technology* A train derailment has resulted in the spill of a certain pollutant. The amount of pollutant from the spill in the local groundwater decreases according to the equation $P = P_0\left(e^{-2t} + e^{-t}\right)$, where t is in weeks and P is measured in ppm.
(a) Find the first four terms of the Maclaurin series expansion of P
(b) Use your answer to (a) to find $\frac{dP}{dt}$.
(c) Use your answer to (b) to calculate the value of $\frac{dP}{dt}$ at $t = 1.4$.

34. *Quality control* The manager of a computer manufacturing company guarantees a certain model of computer for two years. The probability that a computer will last t years without service is given by $p(t) = 0.05e^{-0.05t}$.
(a) Find the first four terms of the Maclaurin series expansion of p.
(b) Use your answer to (a) to find the probability that a computer will need service before it is 2 years old by evaluating $\int_0^2 p(t)\,dt$.

In Your Words

35. What is the advantage of using a series expansion of a function to differentiate or integrate the function?

36. Which of the operations of addition, subtraction, multiplication, and division can be used to combine two series? What precautions do you need to take when you do this? What are the advantages or disadvantages of using these operations to combine two series?

12.5 NUMERICAL TECHNIQUES USING SERIES

Power series expansions can be used to compute numerical values of transcendental functions. If you use enough terms, you can get the value of the function to any degree of accuracy. Calculations such as these are used to make tables of values.

One question often arises when using a method such as this to approximate a value. That question concerns the accuracy of the approximation.

Alternating Series

Most of the time, we will use a series in which the signs alternate from positive to negative and back to positive. This is called an **alternating series** as defined in the following box.

Alternating Series

An **alternating series** is either of the form

$$\sum_{k=0}^{\infty}(-1)^k a_k = a_0 - a_1 + a_2 - a_3 + a_4 - a_5 + \cdots$$

or of the form

$$\sum_{k=0}^{\infty}(-1)^{k+1} a_k = -a_0 + a_1 - a_2 + a_3 - a_4 + \cdots$$

where each number a_k is positive for all values of k.

If an alternating series converges, then the sum can be obtained to any degree of accuracy by adding the first n terms of the series where the $(n+1)$st term is less than the desired accuracy. This means that if you want an approximation within some value, say E, of the true value, then you need to find the first term of the series, a_{n+1}, when $a_{n+1} < E$. Then $\sum_{n=0}^{n} a_n$ will be within the desired amount of accuracy.

EXAMPLE 12.29

Compute the value of $e^{-0.2}$ with an accuracy within 0.0001.

Solution Since $e^x = 1 + x + \frac{x^2}{2!} + \frac{x^3}{3!} + \frac{x^4}{4!} + \cdots$, then

$$e^{-x} = 1 - x + \frac{x^2}{2!} - \frac{x^3}{3!} + \frac{x^4}{4!} - \frac{x^5}{5!} + \cdots$$

We want to approximate $e^{-0.2}$ to within 0.0001.

$$\begin{aligned} e^{-0.2} &= 1 - 0.2 + \frac{(0.2)^2}{2!} - \frac{(0.2)^3}{3!} + \frac{(0.2)^4}{4!} - \frac{(0.2)^5}{5!} + \cdots \\ &= 1 - 0.2 + 0.02 - 0.00133 + 0.000067 - \cdots \end{aligned}$$

Since $0.000067 < 0.0001$, we need only add the first four terms to get within 0.0001 of the actual value.

$$e^{-0.2} \approx 1 - 0.2 + 0.02 - 0.0013 = 0.8187$$

The value by using a calculator is approximately 0.81873.

Using Series to Approximate Values

EXAMPLE 12.30

Calculate the value of $\ln 1.1$ with an accuracy within 0.0001.

Solution We have $\ln(1+x) = x - \frac{x^2}{2} + \frac{x^3}{3} - \frac{x^4}{4} + \cdots$, so

$$\begin{aligned} \ln 1.1 = \ln(1+0.1) &= 0.1 - \frac{(0.1)^2}{2} + \frac{(0.1)^3}{3} - \frac{(0.1)^4}{4} + \cdots \\ &= 0.1 - 0.005 + 0.00033 - 0.000025 + \cdots \end{aligned}$$

When we add the first three terms we get $\ln 1.1 \approx 0.09533$. Since $0.000025 < 0.0001$, we do not need to add the fourth term to get the desired accuracy. The actual value (using a calculator) is approximately 0.09531.

EXAMPLE 12.31

Calculate the value of $\cos 3°$ accurate to within 0.0001.

Solution We know that

$$\cos x = 1 - \frac{x^2}{2!} + \frac{x^4}{4!} - \frac{x^6}{6!} + \cdots$$

EXAMPLE 12.31 (Cont.)

This is for a value of x in radians. Since $3° = \frac{\pi}{60}$ rad we have

$$\begin{aligned}\cos 3° = \cos\frac{\pi}{60} &= 1 - \frac{(\pi/60)^2}{2!}\\ &= 1 - \frac{\pi^2}{7{,}200}\\ &= 1 - 0.0014\\ &= 0.9986\end{aligned}$$

Since the next nonzero term, $\frac{(\pi/60)^4}{4!} \approx 0.0000003 < 0.0001$, we needed only the first two terms.

EXAMPLE 12.32

Approximate $\int_0^{0.5} e^{-x^2}\,dx$ accurately to within 0.001.

Solution The Maclaurin series for e^{-x^2} is

$$1 - x^2 + \frac{x^4}{2!} - \frac{x^6}{3!} + \cdots$$

Integrating, we get

$$\begin{aligned}\int_0^{0.5} e^{-x^2}\,dx &= \int_0^{0.5}\left(1 - x^2 + \frac{x^4}{2!} - \frac{x^6}{3!} + \cdots\right)dx\\ &= x - \frac{x^3}{3} + \frac{x^5}{5\cdot 2!} - \frac{x^7}{7\cdot 3!} + \cdots\Big|_0^{0.5}\\ &= \frac{1}{2} - \frac{(1/2)^3}{3} + \frac{(1/2)^5}{5\cdot 2!} - \frac{(1/2)^7}{7\cdot 3!} + \cdots\\ &= \frac{1}{2} - \frac{1}{3\cdot 2^3} + \frac{1}{5\cdot 2!(2)^5} - \frac{1}{7\cdot 3!(2)^7} + \cdots\\ &= 0.5 - 0.04167 + 0.00313 - 0.00019\end{aligned}$$

This is an alternating series. The last term, $0.00019 < 0.001$, so the sum of the first three terms gives the desired accuracy:

$$\int_0^{0.5} e^{-x^2}\,dx \approx 0.5 - 0.04167 + 0.00313 = 0.46146$$

Application

EXAMPLE 12.33

The current supplied to an initially discharged 1-μF capacitor varies according to $i = \frac{0.1(1-\cos t)}{t}$ A, where t is in seconds. What is the voltage across the capacitor when $t = 0.1$ s?

EXAMPLE 12.33 (Cont.)

Solution The voltage across a capacitor at any instant is given by

$$V_C = \frac{1}{C}\int i\,dt$$

where C is the capacitance in farads and V_C is in volts. Here, we have $C = 1\,\mu\text{F} = 10^{-6}\,\text{F}$, so

$$V_C = \frac{1}{10^{-6}}\int_0^{0.1} \frac{0.1(1-\cos t)}{t}\,dt = 10^5 \int_0^{0.1} \frac{1-\cos t}{t}\,dt$$

None of our integration techniques from Chapter 9 and none of the integration formulas in Appendix C show us how to complete this integral. However, using the series for $\cos x$, we can integrate this expression.

We know that $\cos t = 1 - \frac{t^2}{2!} + \frac{t^4}{4!} - \frac{t^6}{6!} + \cdots$, and so $1 - \cos t = \frac{t^2}{2!} - \frac{t^4}{4!} + \frac{t^6}{6!} - \frac{t^8}{8!} + \cdots$. Dividing by t produces $\frac{1-\cos t}{t} = \frac{t}{2!} - \frac{t^3}{4!} + \frac{t^5}{6!} - \frac{t^7}{8!} + \cdots$.

Substituting this series into the integral, we get

$$\begin{aligned}
V_C &= 10^5 \int_0^{0.1} \frac{1-\cos t}{t}\,dt \\
&= 10^5 \int_0^{0.1} \left(\frac{t}{2!} - \frac{t^3}{4!} + \frac{t^5}{6!} - \frac{t^7}{8!} + \cdots\right) dt \\
&= 10^5 \left(\frac{t^2}{2!\cdot 2} - \frac{t^4}{4!\cdot 4} + \frac{t^6}{6!\cdot 6} - \frac{t^8}{8!\cdot 8} + \cdots\right)\Bigg|_0^{0.1} \\
&= 10^5 \left(\frac{(0.1)^2}{2!\cdot 2} - \frac{(0.1)^4}{4!\cdot 4} + \frac{(0.1)^6}{6!\cdot 6} - \frac{(0.10)^8}{8!\cdot 8} + \cdots\right) \\
&= 10^5(0.0025 - 0.000001 + \cdots) \\
&\approx 250\,\text{V}
\end{aligned}$$

There are 250 V across this capacitor when $t = 0.1$ s.

Exercise Set 12.5

In Exercises 1–10, calculate the value of the given function accurately to within 0.0001.

1. $e^{-0.3}$
2. $\sin(0.1)$
3. $\cos(0.1)$
4. $e^{0.2}$
5. $\sin 5°$
6. $\ln 1.5$
7. $\ln 0.97$
8. $\sqrt{e}$
9. $\ln 0.5$
10. $\cos 5°$

Evaluate the integrals in Exercises 11–16 to an accuracy of 0.0001.

11. $\int_0^{0.5} \frac{1-\cos x}{x}\,dx$

12. $\int_0^{0.5} \frac{x-\sin x}{x}\,dx$

13. $\int_0^{1} \frac{\sin x - x}{x^2}\,dx$

14. $\int_0^{0.5} \sqrt{x}\cos\sqrt{x}\,dx$

15. $\int_0^{0.2} \frac{e^x-1}{x}\,dx$

16. $\int_{0.5}^{1} \frac{\ln(1+x)}{x}\,dx$

Solve Exercises 17–24.

17. **(a)** Use the Maclaurin expansion to show that, for small values of θ, $\sin\theta = \theta$. **(b)** Determine the values of θ for which this is true.

18. *Electricity* The current delivered to an initially discharged 1-μF capacitor varies according to $i = \frac{0.1(t-\sin t)}{t}$ A, where t is in seconds. What is the voltage across the capacitor when $t = 0.2$ s?

19. *Electricity* The charge q, in microcolumbs (μC), on a certain capacitor is

$$q = \int_0^{0.5} \left(1 - e^{-0.05t^2}\right) dt$$

Determine this charge accurately to within 0.001 μC.

20. *Electricity* The charge q, in microcoulombs, on a capacitor is given by

$$q = \int_0^{1.5} \left(1.0 - e^{0.25t^2}\right) dt$$

Use a Maclaurin series to approximate this integral.

21. *Electricity* The voltage in a circuit is given by $V = 0.75e^{0.75t}$. **(a)** Express the voltage as a Maclaurin series. **(b)** Determine an approximation of the voltage when $t = 0.45$ s.

22. *Construction technology* The **Fresnel integral**

$$c(x) = \int_0^x \cos t^2\,dt$$

is used to describe the longitudinal displacement of a beam subject to a periodic force. This integral cannot be found directly.

(a) Find the first four terms of the Maclaurin series expansion of $\cos t^2$ by substituting t^2 for x in the series for $\cos x$.

(b) Use your answer to (a) to approximate this integral.

(c) Use your answer to (b) to evaluate $c(0.3)$.

23. *Electrical engineering* The integral

$$V(x) = V_0 \int_0^x e^{-t^2}\,dt$$

is used to calculate the voltage $V(x)$ along a very long transmission line of negligible inductance and capacitance per unit length, where V_0 is the voltage applied to the transmitting end of the line.

(a) Find the first four terms of the Maclaurin series expansion of e^{-t^2} by substituting $-t^2$ for x in the series for e^x.

(b) Evaluate this integral using the first four nonzero terms of the Maclaurin series for this function.

(c) Find $V(1.4)$.

(d) Estimate the maximum error of your approximation in (c).

24. *Electronics* The charge, q, in millicoulombs (mC), on a certain capacitor is

$$q = \int_0^{1.0} \left(1 - e^{0.1t^2}\right) dt$$

Evaluate this integral using the first four terms of the Maclaurin series for this function.

In Your Words

25. Without looking in the text, describe an alternating series.

26. How do you know when an alternating series will be within a desired amount of accuracy?

12.6 TAYLOR SERIES

Until now, we have used Maclaurin series to evaluate functions around zero. Let's assume that we have a power series whose interval of convergence is $(c-R, c+R)$, where R is a positive real number. This power series will be of the form

$$f(x) = \sum_{k=0}^{\infty} a_k(x-c)^k = a_0 + a_1(x-c) + a_2(x-c)^2 + \cdots$$

Again, in taking successive derivatives of f, we get

$$\begin{aligned}
f(x) &= a_0 + a_1(x-c) + a_2(x-c)^2 + a_3(x-c)^3 + a_4(x-c)^4 + \cdots \\
f'(x) &= a_1 + 2a_2(x-c) + 3a_3(x-c)^2 + 4a_4(x-c)^3 + \cdots \\
f''(x) &= 2a_2 + 3\cdot 2a_3(x-c) + 4\cdot 3a_4(x-c)^2 + \cdots \\
f'''(x) &= 3\cdot 2a_3 + 4\cdot 3\cdot 2a_4(x-c) + \cdots \\
f^{(4)}(x) &= 4\cdot 3\cdot 2a_4 + \cdots
\end{aligned}$$

and, for any positive integer n,

$$f^{(n)}(x) = n!a_n + (n+1)!\,a_{n+1}(x-c) + \cdots$$

If we evaluate each of these equations when $x = c$, then every term, except the first, will have a value of 0. Thus, we get

$$\begin{aligned}
f(c) &= a_0 & \text{or} \quad a_0 &= f(c) \\
f'(c) &= a_1 & a_1 &= f'(c) \\
f''(c) &= 2!a_2 & a_2 &= \frac{1}{2!}f''(c) \\
f'''(c) &= 3!a_3 & a_3 &= \frac{1}{3!}f'''(c) \\
&\vdots & &\vdots \\
f^{(n)}(c) &= n!a_n & a_n &= \frac{1}{n!}f^{(n)}(c)
\end{aligned}$$

Substituting these values into the series expansion for $f(x)$, we get the **Taylor series** for f about c.

> **Taylor series**
>
> The **Taylor Series** for f about c is
>
> $$f(x) = f(c) + f'(c)(x-c) + \frac{f''(c)(x-c)^2}{2!} + \frac{f'''(c)(x-c)^3}{3!} + \cdots + \frac{f^{(n)}(c)(x-c)^n}{n!} + \cdots$$

Note When $c = 0$, the Taylor series is equivalent to the Maclaurin series.

EXAMPLE 12.34

Expand e^x in a Taylor series around 1.

Solution

$$\begin{aligned} f(x) &= e^x & f(1) &= e \\ f'(x) &= e^x & f'(1) &= e \\ f''(x) &= e^x & f''(1) &= e \\ &\vdots & &\vdots \end{aligned}$$

$$\begin{aligned} f(x) &= e + e(x-1) + \frac{e}{2!}(x-1)^2 + \frac{e}{3!}(x-1)^3 + \frac{e}{4!}(x-1)^4 + \cdots \\ &= e\left[1 + (x-1) + \frac{(x-1)^2}{2!} + \frac{(x-1)^3}{3!} + \frac{(x-1)^4}{4!} + \cdots\right] \\ &= e\left[x + \frac{(x-1)^2}{2!} + \frac{(x-1)^3}{3!} + \frac{(x-1)^4}{4!} + \cdots\right] \end{aligned}$$

This example gives us a method to evaluate e^x for values of x that are close to 1. ▪

EXAMPLE 12.35

Use a Taylor series to evaluate $e^{0.9}$.

Solution From Example 12.34, we know that for values of x close to 1,

$$e^x = e\left[x + \frac{(x-1)^2}{2!} + \frac{(x-1)^3}{3!} + \frac{(x-1)^4}{4!} + \cdots\right]$$

In this example, $x = 0.9$, so

$$\begin{aligned} e^{0.9} &= e\left[0.9 + \frac{(0.9-1)^2}{2!} + \frac{(0.9-1)^3}{3!} + \frac{(0.9-1)^4}{4!} + \cdots\right] \\ &= e\left[0.9 + \frac{(-0.1)^2}{2!} + \frac{(-0.1)^3}{3!} + \frac{(-0.1)^4}{4!} + \cdots\right] \\ &= e(0.9 + 0.005 - 0.000167 + 0.000004 - \cdots) \\ &= 0.904837e \\ &\approx 2.4596 \end{aligned}$$

▪

EXAMPLE 12.36

Expand $\sin x$ in a Taylor series about $\frac{\pi}{3}$.

Solution

$$\begin{aligned} f(x) &= \sin x & f\left(\frac{\pi}{3}\right) &= \frac{\sqrt{3}}{2} \\ f'(x) &= \cos x & f'\left(\frac{\pi}{3}\right) &= \frac{1}{2} \\ f''(x) &= -\sin x & f''\left(\frac{\pi}{3}\right) &= \frac{-\sqrt{3}}{2} \\ f'''(x) &= -\cos x & f'''\left(\frac{\pi}{3}\right) &= -\frac{1}{2} \end{aligned}$$

$$\sin x = \frac{\sqrt{3}}{2} + \frac{1}{2}\left(x - \frac{\pi}{3}\right) - \frac{\sqrt{3}}{2}\cdot\frac{1}{2!}\left(x - \frac{\pi}{3}\right)^2 - \frac{1}{2}\cdot\frac{1}{3!}\left(x - \frac{\pi}{3}\right)^3 + \cdots$$

▪

EXAMPLE 12.37

Calculate the approximate value of $\sin 61^\circ$ by using a Taylor series.

Solution We have the Taylor expansion of $\sin x$ around $\frac{\pi}{3} = 60^\circ$ from Example 12.36. We will need to express 61° in radians. Since $x = 61^\circ = \frac{61\pi}{180}$ and $c = 60^\circ = \frac{\pi}{3} = \frac{60\pi}{180}$, we have $x - c = \frac{61\pi}{180} - \frac{60\pi}{180} = \frac{\pi}{180}$. Thus, from Example 12.36, we obtain

$$\begin{aligned}\sin 61^\circ &= \frac{\sqrt{3}}{2} + \frac{1}{2}\left(\frac{\pi}{180}\right) - \frac{\sqrt{3}}{2}\cdot\frac{1}{2!}\left(\frac{\pi}{180}\right)^2 - \frac{1}{2}\cdot\frac{1}{3!}\left(\frac{\pi}{180}\right)^3 + \cdots \\ &\approx 0.8660 + 0.0087 - 0.0001 - \cdots \\ &= 0.8746\end{aligned}$$

Application

EXAMPLE 12.38

The damped current in an electronic circuit has the form

$$i = \frac{\sin \pi t}{t}$$

Calculate the approximate value of this function at $t = 1.9$ by using a Taylor series about $t = 2$.

Solution Using a procedure similar to the one used in Example 12.36, we determine the following fundamental information needed for the Taylor expansion of $\sin \pi t$ around $t = 2$.

$$\begin{aligned} f(x) &= \sin \pi x & f(2) &= \sin 2\pi = 0 \\ f'(x) &= \pi \cos \pi x & f'(2) &= \pi \cos 2\pi = \pi \\ f''(x) &= -\pi^2 \sin \pi x & f''(2) &= -\pi^2 \sin 2\pi = 0 \\ f'''(x) &= -\pi^3 \cos \pi x & f'''(2) &= -\pi^3 \cos 2\pi = -\pi^3 \end{aligned}$$

Using these values, we obtain

$$\begin{aligned}\sin \pi t &= 0 + \pi(t-2) - 0\cdot\frac{1}{2!}(t-2)^2 - \pi^3\frac{1}{3!}(t-2)^3 + \cdots \\ &= \pi(t-2) - \frac{\pi^3}{3!}(t-2)^3 + \cdots\end{aligned}$$

Now dividing by t, we obtain

$$\frac{\sin \pi t}{t} = \pi\left(\frac{t-2}{t}\right) - \frac{\pi^3}{3!}\frac{(t-2)^3}{t} + \cdots$$

and when we let $t = 1.9$, we get

$$\begin{aligned}\frac{\sin \pi t}{t} &= \pi\left(\frac{-0.1}{1.9}\right) - \frac{\pi^3}{3!}\frac{(-0.1)^3}{1.9} + \cdots \\ &\approx -0.165347 - (-0.002720) + \cdots \\ &\approx -0.162627\end{aligned}$$

A calculator gives $\dfrac{\sin(1.9\pi)}{1.9} \approx -0.162641$.

Exercise Set 12.6

In Exercises 1–12, expand the function in a Taylor series about the given value of c.

1. $f(x) = e^x$, $c = 2$
2. $g(x) = e^{2x}$, $c = -1$
3. $h(x) = \sin x$, $c = \frac{\pi}{4}$
4. $j(x) = \cos x$, $c = \frac{\pi}{4}$
5. $k(x) = \dfrac{1}{x}$, $c = 2$
6. $m(x) = \sqrt{x}$, $c = 4$
7. $f(x) = \dfrac{1}{(x+1)^2}$, $c = 0$
8. $g(x) = e^{1+x}$, $c = 1$
9. $h(x) = e^{-x}$, $c = 1$
10. $j(x) = \ln x$, $c = 4$
11. $k(x) = \sqrt[3]{x}$, $c = 8$
12. $m(x) = \tan x$, $c = \frac{\pi}{4}$

In Exercises 13–24, evaluate the given function by use of the series expansion you used in Exercises 1 through 12.

13. $e^{2.1}$
14. $e^{-0.8}$
15. $\sin 43°$
16. $\cos 43°$
17. $\frac{1}{1.8}$
18. $\sqrt{3.9}$
19. $\dfrac{1}{1.12^2}$
20. $e^{1.9}$
21. $e^{-0.8}$
22. $\ln 4.1$
23. $\sqrt[3]{7.8}$
24. $\tan 43°$

Solve Exercises 25–26.

25. *Electricity* The damped current in an electronic circuit has the form
$$i = \frac{\sin \pi t}{t}$$
Calculate the approximate value of this function at $t = 2.1$ by using the Taylor series about $t = 2$.

26. *Electricity* The damped current in an electronic circuit has the form
$$i = \frac{\sin 2\pi t}{t}$$
Calculate the approximate value of this function at $t = 1.05$ by using the Taylor series about $t = 1$.

In Your Words

27. Without looking in the text, describe a Taylor series.
28. How does a Taylor series differ from a Maclaurin series?

12.7 FOURIER SERIES

So far in this chapter, we have been concerned with functions that can be expanded in power series. In this section, we will consider a different kind of series—a trigonometric series.

Many functions that we use in science and technology are periodic functions. The first periodic functions that we encountered were the trigonometric functions. The **Fourier series** expresses a given function in terms of the sine and cosine functions.

Periodic Function

A **periodic function** is a function f where $f(x+p)=f(x)$. The smallest positive value of p is the period of the function.

The sine and cosine functions both have periods of 2π because $\sin(x+2\pi)=\sin x$ and $\cos(x+2\pi)=\cos x$, and because there is no positive value smaller than 2π for which this is true. In general, $\sin ax$ and $\cos ax$ have periods of $\frac{2\pi}{a}$.

Now, suppose that we have a periodic function f that has a period of $2L$. We will represent f by a series of sines and cosines, where

$$f(x)=a_0+a_1\cos\frac{\pi x}{L}+a_2\cos\frac{2\pi x}{L}+\cdots+a_n\cos\frac{n\pi c}{L}+\cdots$$
$$+b_1\sin\frac{\pi x}{L}+b_2\sin\frac{2\pi x}{L}+\cdots+b_n\sin\frac{n\pi x}{L}+\cdots \qquad (*)$$

Now, $\sin\left(\frac{\pi x}{L}\right)$ and $\cos\left(\frac{\pi x}{L}\right)$ have periods of $\frac{2\pi}{\pi/L}=2L$, $\sin\left(\frac{2\pi x}{L}\right)$ and $\cos\left(\frac{2\pi x}{L}\right)$ have periods of L, and in general $\sin\left(\frac{n\pi x}{L}\right)$ and $\cos\left(\frac{n\pi x}{L}\right)$ have periods of $\frac{2L}{n}$. All the terms have a common period of $2L$. This series is the **Fourier series**.

We need to find the values of the coefficients a_n and b_n. When we developed the Maclaurin and Taylor series, we used derivatives. To find the coefficients of the Fourier series, we will use integrals. In each case, we will integrate over the interval $(-L,L)$.

To find a_0, we will integrate both sides of equation (*) from $-L$ to L.

$$\int_{-L}^{L} f(x)\,dx=\int_{-L}^{L} a_0\,dx+\int_{-L}^{L} a_1\cos\frac{\pi x}{L}\,dx+\cdots+\int_{-L}^{L} b_1\sin\frac{\pi x}{L}\,dx+\cdots$$

Now, $\int_{-L}^{L} f(x)\,dx=a_0x\Big|_{-L}^{L}=2La_0$, since the integrals of all the trigonometric functions are 0. {For example, the integral $\int a_1\cos\frac{\pi x}{L}\,dx=\frac{a_1L}{\pi}\sin\frac{\pi x}{L}\Big|_{-L}^{L}=\frac{a_1L}{\pi}[\sin\pi-\sin(-\pi)]=0.$} Solving for a_0, we get $a_0=\frac{1}{2L}\int_{-L}^{L} f(x)\,dx.$

Next, we will find the value of a_k, $k=1,2,3,\ldots$. To do this, we multiply both sides of equation $(*)$ by $\cos\frac{k\pi x}{L}$ and then integrate from $-L$ to L. For example,

$$\int_{-L}^{L} f(x)\cos\frac{k\pi x}{L}\,dx=\int_{-L}^{L} a_0\cos\frac{k\pi x}{L}\,dx+\int_{-L}^{L} a_1\cos\frac{\pi x}{L}\cos\frac{k\pi x}{L}\,dx+\cdots$$
$$+\int_{-L}^{L} a_k\cos^2\frac{k\pi x}{L}\,dx+\cdots+\int_{-L}^{L} b_1\sin\frac{\pi x}{L}\cos\frac{k\pi x}{L}\,dx+\cdots$$

$$+\int b_k \sin\frac{k\pi x}{L}\cos\frac{k\pi x}{L}\,dx+\cdots \qquad (**)$$

From this, you can see that there are two types of integrals: These are

$$\int_{-L}^{L}\cos\frac{m\pi x}{L}\cos\frac{k\pi x}{L}\,dx \text{ and } \int_{-L}^{L}\sin\frac{m\pi x}{L}\cos\frac{k\pi x}{L}\,dx$$

where m and k are integers. It can be shown that

$$\int_{-L}^{L}\cos\frac{m\pi x}{L}\cos\frac{k\pi x}{L}\,dx=\begin{cases}0 & \text{if } m\neq k\\ L & \text{if } m=k\end{cases}$$

and

$$\int_{-L}^{L}\sin\frac{m\pi x}{L}\cos\frac{k\pi x}{L}\,dx=0$$

Thus, it follows that all the terms on the right-hand side of equation $(**)$ are, with one exception, zero. From this we see that

$$a_k=\frac{1}{L}\int_{-L}^{L}f(x)\cos\frac{k\pi x}{L}\,dx,\ k=1,2,3,\ldots$$

Finally, we will find the values of b_k by using a similar technique. This time we will multiply both sides of (*) by $\sin\frac{k\pi x}{L}$ for $k=1,2,3,\ldots$ and integrate from $-L$ to L. When we do this, we find that

$$b_k=\frac{1}{L}\int_{-L}^{L}f(x)\sin\frac{k\pi x}{L}\,dx,\ k=1,2,3,\ldots$$

In summary, we have the following facts about the Fourier series.

Fourier Series

The Fourier series for a function $f(x)$ of period $2L$ is given by

$$\begin{aligned}f(x)=a_0+a_1\cos\frac{\pi x}{L}+a_2\cos\frac{2\pi x}{L}+\cdots\\ +b_1\sin\frac{\pi x}{L}+b_2\sin\frac{2\pi x}{L}+\cdots\end{aligned}$$

where the coefficients are given by

$$\begin{aligned}a_0&=\frac{1}{2L}\int_{-L}^{L}f(x)\,dx\\ a_k&=\frac{1}{L}\int_{-L}^{L}f(x)\cos\frac{k\pi x}{L}\,dx\\ b_k&=\frac{1}{L}\int_{-L}^{L}f(x)\sin\frac{k\pi x}{L}\,dx\end{aligned}$$

EXAMPLE 12.39

Find a Fourier series for the function

$$f(x) = \begin{cases} -1, & -2 < x < 0 \\ 1, & 0 < x < 2 \end{cases}$$

with a period of 4.

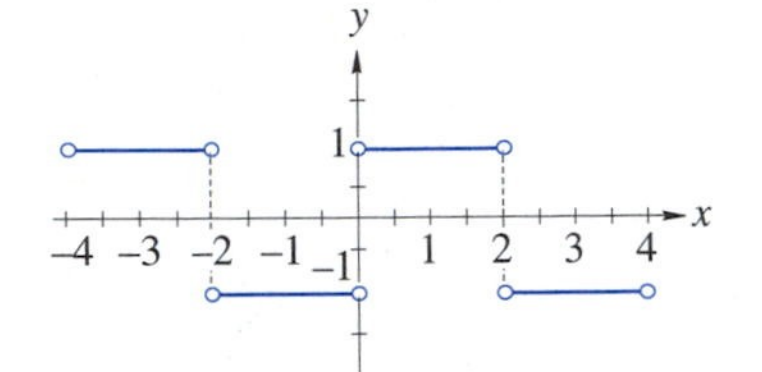

FIGURE 12.3

Solution The graph of this function is given in Figure 12.3. From this graph you can see why this is often called a "square wave." Since the period is 4, we have $2L = 4$ or $L = 2$.

$$\begin{aligned} a_0 &= \frac{1}{4}\int_{-2}^{2} f(x)\,dx = \frac{1}{4}\int_{-2}^{0}(-1)\,dx + \frac{1}{4}\int_{0}^{2}(1)\,dx \\ &= \frac{1}{4}(-x)\Big|_{-2}^{0} + \frac{1}{4}x\Big|_{0}^{2} \\ &= \frac{1}{4}(0-2) + \frac{1}{4}(2-0) = 0 \end{aligned}$$

For $n \geq 1$, we have

$$\begin{aligned} a_n &= \frac{1}{2}\int_{-2}^{2} f(x)\cos\frac{n\pi x}{2}\,dx \\ &= \frac{1}{2}\int_{-2}^{0}\left(-\cos\frac{n\pi x}{2}\right)dx + \frac{1}{2}\int_{0}^{2}\cos\frac{n\pi x}{2}\,dx \\ &= \frac{1}{2}\left(\frac{-2}{\pi n}\right)\sin\frac{n\pi x}{2}\Big|_{-2}^{0} + \frac{1}{2}\left(\frac{2}{\pi n}\right)\sin\frac{n\pi x}{2}\Big|_{0}^{2} \\ &= 0 + \frac{1}{\pi n}\sin(-n\pi) + \frac{1}{\pi n}\sin(n\pi) - 0 \\ &= 0 \end{aligned}$$

since the sine of a multiple of π is 0.

Finally, for $n \geq 1$, we have

$$\begin{aligned} b_n &= \frac{1}{2}\int_{-2}^{2} f(x)\sin\frac{n\pi x}{2}\,dx \\ &= \frac{1}{2}\int_{-2}^{0}\left(-\sin\frac{n\pi x}{2}\right)dx + \frac{1}{2}\int_{0}^{2}\sin\frac{n\pi x}{2}\,dx \\ &= \frac{1}{2}\left(\frac{2}{n\pi}\right)\cos\frac{n\pi x}{2}\Big|_{-2}^{0} + \frac{1}{2}\left(\frac{2}{n\pi}\right)\left(-\cos\frac{n\pi x}{2}\right)\Big|_{0}^{2} \\ &= \frac{1}{n\pi}[\cos 0 - \cos(-n\pi)] - \frac{1}{n\pi}[\cos(n\pi) - \cos 0] \\ &= \frac{1}{n\pi}(1 - \cos n\pi) - \frac{1}{n\pi}(\cos n\pi - 1) \\ &= \frac{2}{n\pi}(1 - \cos n\pi) \\ &= \begin{cases} 0 & \text{, if } n \text{ is even} \\ \dfrac{4}{n\pi} & \text{, if } n \text{ is odd} \end{cases} \end{aligned}$$

EXAMPLE 12.39 (Cont.)

Substituting these values for the coefficients into the series, we get

$$\begin{aligned} f(x) &= \frac{4}{\pi}\sin\frac{\pi x}{2} + \frac{4}{3\pi}\sin\frac{3\pi x}{2} + \frac{4}{5\pi}\sin\frac{5\pi x}{2} + \cdots \\ &= \frac{4}{\pi}\left(\sin\frac{\pi x}{2} + \frac{1}{3}\sin\frac{3\pi x}{2} + \frac{1}{5}\sin\frac{5\pi x}{2} + \cdots\right) \\ &= \frac{4}{\pi}\sum_{k=1}^{\infty}\frac{1}{2k-1}\sin\frac{(2k-1)\pi x}{2} \end{aligned}$$

It is curious that a series of sine and cosine functions are used to approximate a graph of straight lines. But, if we graph the first term

$$\frac{4}{\pi}\sin\frac{\pi x}{2}$$

we get the graph in Figure 12.4a.

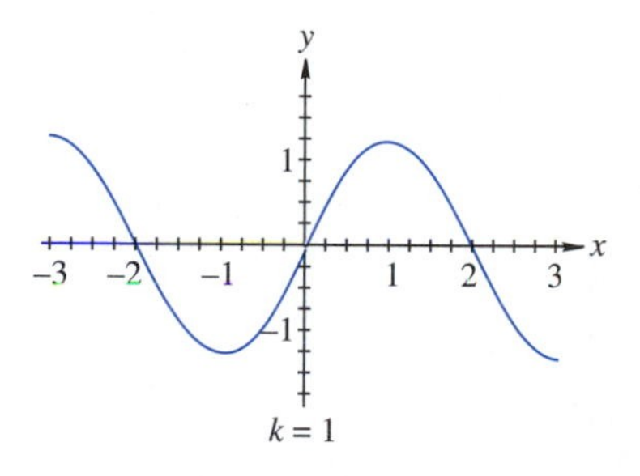

y
1
–1
x
–3
–2
–1
1
2
3
k = 1

FIGURE 12.4a

y
1
x
–3
–2
–1
1
2
3
k = 2

FIGURE 12.4b

If we graph the first two terms of this series, $\frac{4}{\pi}\left(\sin\frac{\pi x}{2} + \frac{1}{3}\sin\frac{3\pi x}{2}\right)$, we obtain the graph in Figure 12.4b. The first three terms give the graph in Figure 12.4c. Graphs for the first 4, 5, 10, 20, and 40 terms are shown in Figure 12.4d–h, respectively.

y
1
x
–3
–2
–1
1
2
3
k = 3

FIGURE 12.4c

y
1
x
–3
–2
–1
1
2
3
k = 4

FIGURE 12.4d

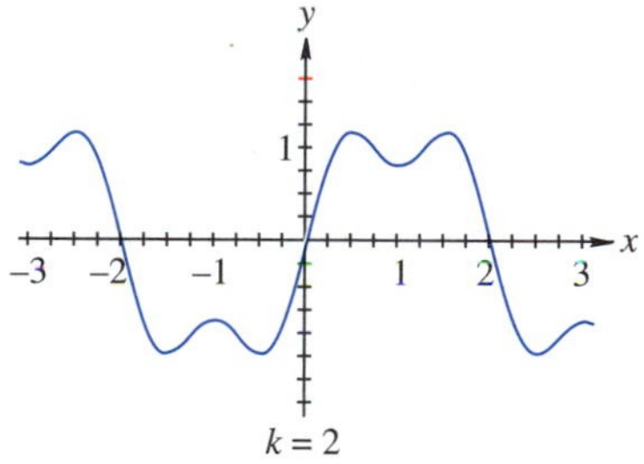

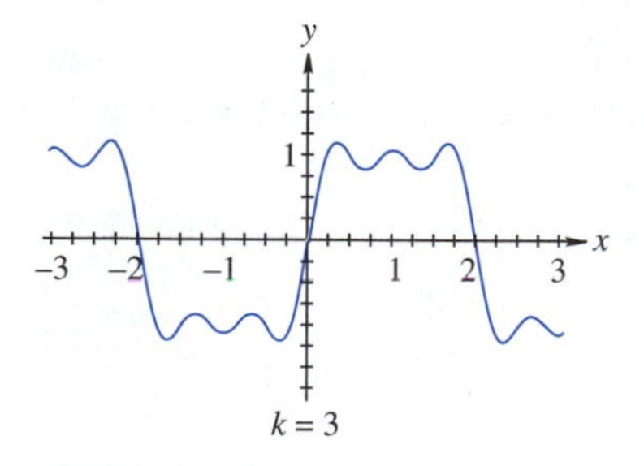

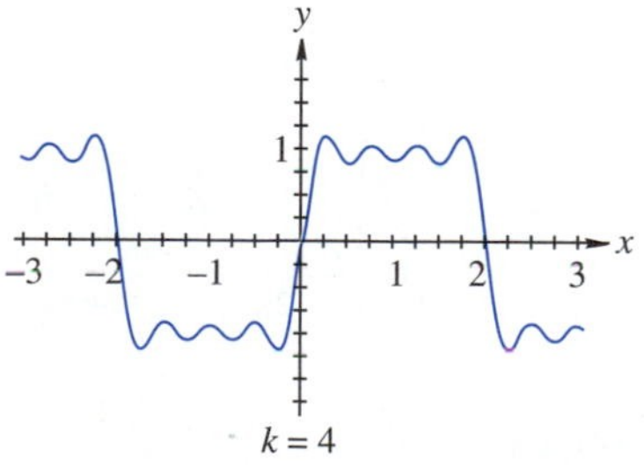

As you can see, with each additional term the graph gets closer to the graph of the actual function. Notice that at the points of discontinuity, the graph passes through the point midway between the ordinates -1 and 1.

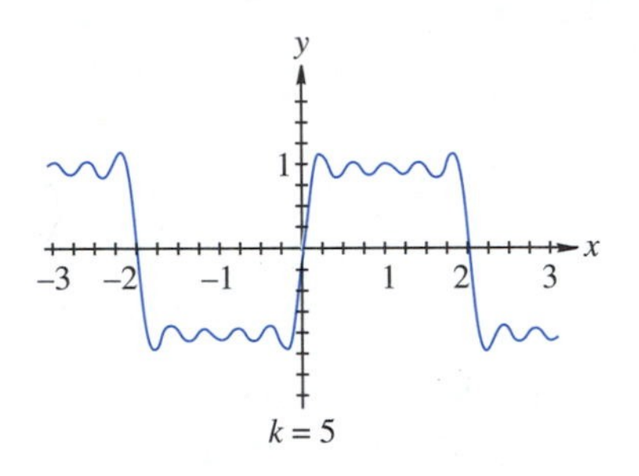

FIGURE 12.4e

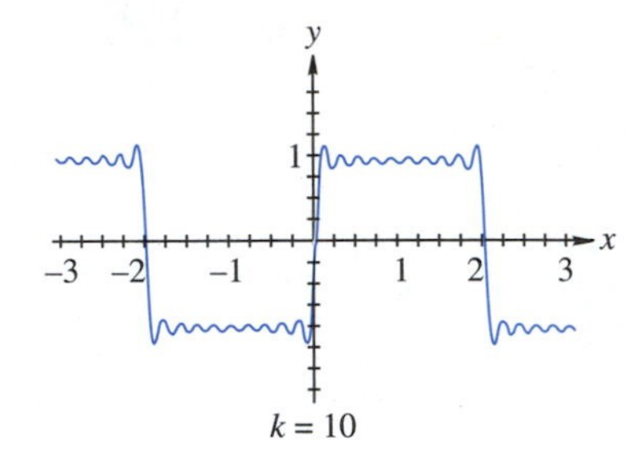

FIGURE 12.4f

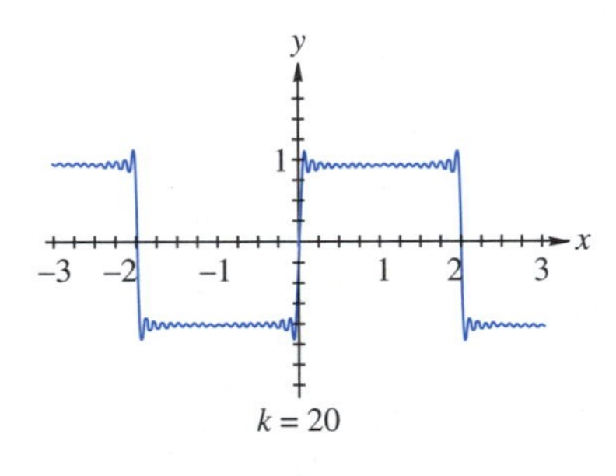

FIGURE 12.4g

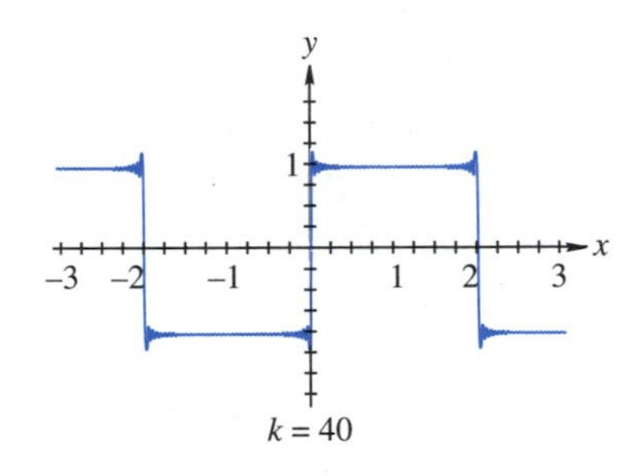

FIGURE 12.4h

Application

EXAMPLE 12.40

The electron beam that forms the picture on a television or computer screen is controlled basically by sawtooth voltages and currents. Find a Fourier series for the sawtooth function $f(x) = x$, $0 < x < 2\pi$, with a period of 2π.

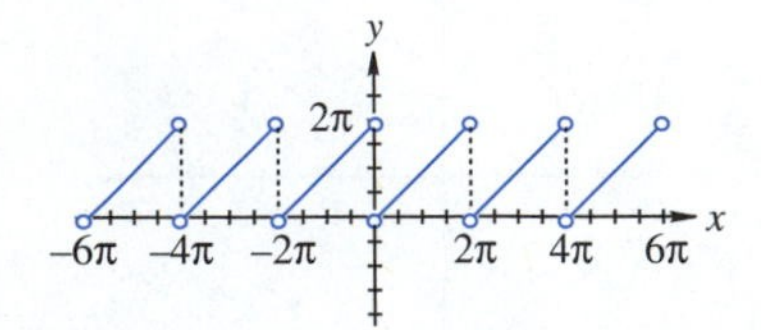

FIGURE 12.5

Solution The graph of this function is the saw tooth function shown in Figure 12.5. Since the period is 2π, we have $2L = 2\pi$ or $L = \pi$.

$$a_0 = \frac{1}{2\pi}\int_0^{2\pi} x\,dx = \pi$$

If $n \geq 1$,

$$\begin{aligned} a_n &= \frac{1}{\pi}\int_0^{2\pi} x\cos\frac{n\pi x}{\pi}\,dx \\ &= \frac{1}{\pi}\int_0^{2\pi} x\cos nx\,dx \\ &= \frac{1}{\pi}\left(\frac{\cos nx}{n^2} + \frac{x\sin x}{n}\right)\Big|_0^{2\pi} \end{aligned}$$

EXAMPLE 12.40 (Cont.)

$$= \frac{1}{\pi}\left[\left(\frac{1}{n^2}+0\right)-\left(\frac{1}{n^2}+0\right)\right]=0$$

$$b_n = \frac{1}{\pi}\int_0^{2\pi} x \sin\frac{n\pi x}{\pi}\,dx$$

$$= \frac{1}{\pi}\int_0^{2\pi} x \sin nx\,dx$$

$$= \frac{1}{\pi}\left(\frac{\sin nx}{n^2}-\frac{x\cos nx}{n}\right)\Big|_0^{2\pi}$$

$$= \frac{1}{\pi}\left[\left(0-\frac{2\pi}{n}\right)-(0-0)\right]=\frac{-2}{n}$$

Thus, the Fourier series is

$$\pi - 2\left(\sin x + \frac{\sin 2x}{2} + \frac{\sin 3x}{3} + \cdots\right)$$

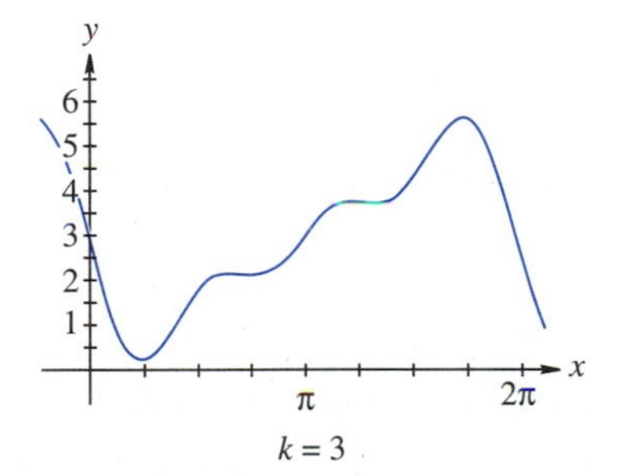

FIGURE 12.6a

The graph of the first three terms of this Fourier expansion is given in Figure 12.6a and of the first five terms is shown in Figure 12.6b. The graphs in Figure 12.6c–e, are for the first 10, 20, and 40 terms, as indicated. Again, you can see that the larger number of terms gives a better approximation.

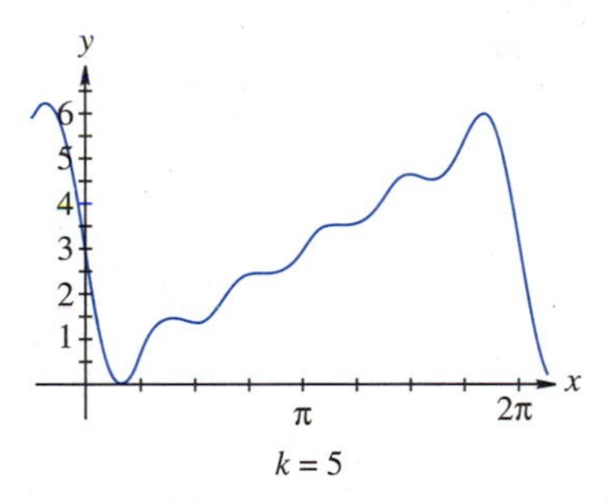

FIGURE 12.6b

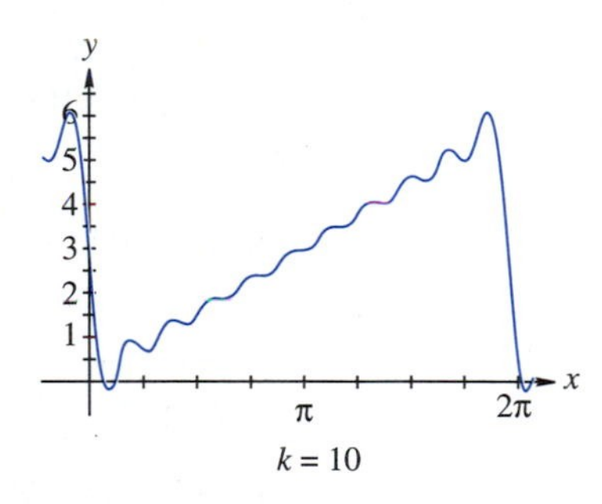

FIGURE 12.6c

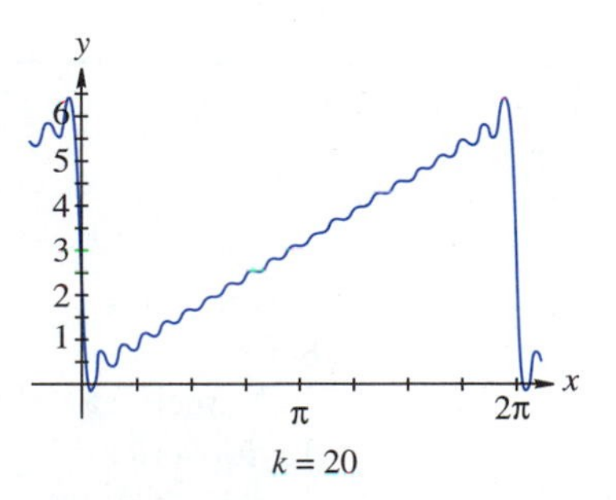

FIGURE 12.6d

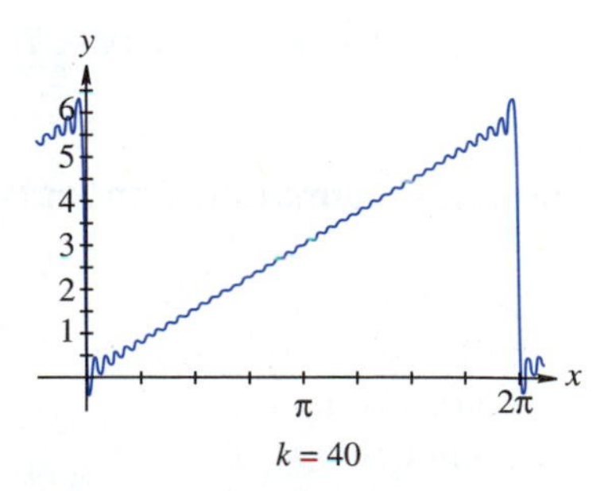

FIGURE 12.6e

Exercise Set 12.7

In Exercises 1–10, obtain Fourier series expansions for the function defined in the indicated interval and with the indicated period outside of the interval. Graph the function.

1. $f(x) = \begin{cases} 2, & -\pi < x < 0 \\ -2, & 0 < x < \pi \end{cases}$, period 2π
2. $g(x) = 1, -2 < x < 2,$ period 4
3. $h(x) = \begin{cases} 0, & -3 < x < 0 \\ 2, & 0 < x < 3 \end{cases}$, period 6
4. $j(x) = \begin{cases} 0, & -1 < x < 0 \\ x, & 0 < x < 1 \end{cases}$, period 2
5. $f(x) = \begin{cases} -x, & -\pi < x < 0 \\ x, & 0 < x < \pi \end{cases}$, period 2π
6. $g(x) = \begin{cases} 3, & 0 < x < 4 \\ -3, & 4 < x < 8 \end{cases}$, period 8
7. $h(x) = x, -\pi < x < \pi,$ period 2π
8. $j(x) = \begin{cases} 0, & -\frac{\pi}{2} < x < 0 \\ 2, & 0 < x < \frac{\pi}{2} \end{cases}$, period π
9. $f(x) = \begin{cases} 1, & 0 < x < \frac{2\pi}{3} \\ 0, & \frac{2\pi}{3} < x < \frac{4\pi}{3}, \\ -1, & \frac{4\pi}{3} < x < 2\pi \end{cases}$ period 2π
10. $g(x) = \begin{cases} x, & 0 < x \le \pi \\ \pi, & \pi < x < 2\pi \end{cases}$, period 2π

Solve Exercises 11–12.

11. *Electronics* Some electronic devices allow current to pass through in only one direction. When an alternating current is used with one of these devices, the current exists for only half the cycle. The equation for such a **half-wave rectifier** is

$$h(x) = \begin{cases} 0, & -\pi < x < 0 \\ \sin x, & 0 < x < \pi \end{cases}, \text{ period } 2\pi$$

Obtain the Fourier series expansion and graph this function.

12. *Electronics* A **full-wave rectifier** is described by

$$f(x) = \begin{cases} \sin x, & 0 < x < \pi \\ -\sin x, & \pi < x < 2\pi \end{cases}, \text{ period } 2\pi$$

Obtain the Fourier series expansion and graph this function.

In Your Words

13. Without looking in the text, describe a Fourier series.
14. How does a Fourier series differ from either a Maclaurin or a Taylor series?

CHAPTER 12 REVIEW

Important Terms and Concepts

Alternating series
Euler's identity
Fourier series
Infinite series
Maclaurin series
Power series
Series
  Numerical techniques
Taylor series

Review Exercises

In Exercises 1–8, find the Maclaurin series for *f*(*x*).

1. $f(x) = \begin{cases} \dfrac{1-\cos x}{x}, & x \neq 0 \\ 0, & x = 0 \end{cases}$
2. $f(x) = xe^{-2x}$
3. $f(x) = \sin x \cos x$
4. $f(x) = \ln(2+x)$
5. $f(x) = (1+x)^{3/4}$
6. $f(x) = e^{-x^4}$
7. $f(x) = \sin x^2$
8. $f(x) = \dfrac{1-e^x}{x}$

Solve Exercises 9–12.

9. Evaluate $\int_0^1 \frac{\sin x}{\sqrt{x}}\,dx$ accurately to 0.0001.
10. Evaluate $\int_0^1 x^3 e^{-x^3}\,dx$ accurately to 0.0001.
11. Evaluate $e^{-0.25}$ accurately to 0.0001.
12. Evaluate $\int_0^1 \cos x^4\,dx$ accurately to 0.0001.

In Exercises 13–16, find the Taylor series expansion about the given value of *c*.

13. $\cos x$, $c = \frac{\pi}{6}$
14. $\sinh x$, $c = \ln 2$
15. $\sqrt{x}$, $c = 9$
16. $\ln x$, $c = 1$

Use the appropriate Taylor series expansions from the previous four problems to evaluate each of the following.

17. $\cos 29°$
18. $\sinh(\ln 1.9)$
19. $\sqrt{9.1}$
20. $\ln(1.1)$

In Exercises 21–24, determine the Fourier series for the given function. Graph the function.

21. $f(x) = \begin{cases} 1, & -\pi < x < 0 \\ 0, & 0 < x < \pi \end{cases}$, period 2π
22. $g(x) = \begin{cases} 1, & -\pi < x < 0 \\ 2, & 0 < x < \pi \end{cases}$, period 2π
23. $h(x) = x^2$, $-\pi < x < \pi$, period 2π
24. $j(x) = 4 - x^2$, $-\pi < x < \pi$, period 2π

CHAPTER 12 TEST

1. Find the Maclaurin series for $f(x) = \sin x^2$.
2. Find the Maclaurin series expansion for $g(x) = \cos\sqrt{x}$.
3. Find the Taylor series expansion of $h(x) = \dfrac{1}{x}$ about $c = 2$.
4. Use your Taylor series expansion from Exercises 3 to evaluate $\frac{1}{2.1}$.
5. Find the Taylor series expansion of $j(x) = \sqrt[3]{x}$ about $c = -1$.

6. Use your Taylor series expansion from Exercises 5 to evaluate $\sqrt[3]{-0.9}$.

7. Determine the Fourier series for $f(x) = x, 0 < x < 2\pi$, period 2π.

8. Determine the Fourier series for

$$g(x) = \begin{cases} \cos x, & 0 < x < \pi \\ -\cos x, & -\pi < x < 0 \end{cases}, \quad \text{period } 2\pi.$$

9. Evaluate $\int_0^1 \cos x^3 \, dx$ accurately to 0.0001.

10. The charge q, in microcoulombs, on a capacitor is given by

$$q = \int_0^{0.15} e^{-0.30t^2} \, dt$$

Use a Maclaurin series to approximate this integral.

CHAPTER

13

First-Order Differential Equations

Building ventilation systems can cause problems for the environment. The systems must be designed to ensure efficiency, safety, and comfort. In Section 13.6, we will solve problems related to efficient ventilation.

For the last ten chapters, we have worked with equations that involved differentials or derivatives. In some cases, we were required to find the corresponding functions that would produce these equations. An equation that involves an unknown function and its derivative is known as a **differential equation**.

If the unknown function depends on only one independent variable, then it is an **ordinary differential equation**. When we studied derivatives, we learned that a derivative of a function with two or more independent variables is a partial derivative. Similarly, a differential equation in which the unknown function involves two or more independent variables is called a **partial differential equation**.

Differential equations are often used in science and technology. They occur in physics, chemistry, and biology. In this chapter, we will examine certain methods for solving differential equations. We will also show how solving differential equations can help solve applied problems.

13.1 SOLUTIONS OF DIFFERENTIAL EQUATIONS

The basic types of differential equation that we will consider are those that involve first and second derivatives. As we mentioned in the chapter introduction, differential equations are often divided into ordinary and partial differential equations. Differential equations are also classified by their order and degree.

Order and Degree

The **order** of a differential equation is the order of the highest derivative appearing in the equation. If an equation contains only first derivatives, it is called a **first-order** differential equation. If the equation contains second derivatives and no other higher-order derivatives, it is a **second-order** differential equation. A second-order differential equation can have both first and second derivatives or just second derivatives. The **degree** of a differential equation that can be written as a polynomial is the highest power of the highest-order derivative.

EXAMPLE 13.1

What are the order and degree of each of the following differential equations?

(a) $\frac{dy}{dx} = 9x - 7 + y^2$

(b) $\frac{d^2y}{dx^2} + 5\left(\frac{dy}{dx}\right)^3 = 2$

(c) $6\frac{d^3y}{dx^3} + (\cos x)\frac{d^2y}{dx^2} + 5x^2y - 3 = y$

(d) $\left(\frac{d^2y}{dx^2}\right)^3 + 2y\left(\frac{dy}{dx}\right)^4 + y^4\left(\frac{dy}{dx}\right)^5 = x$

Solutions Equation (a) is a first-order differential equation.

Equations (b) and (d) are both second-order differential equations. Equation (c) is a third-order differential equation. Note that in equation (d), the highest derivative is the second, so this is a second-order differential equation even though its first derivatives are raised to the fourth and fifth power and the second derivative is raised to the third power.

Equations (a), (b), and (c) are all of degree one, since in each case the highest-order derivative is only raised to the first power. In equation (d), the degree is 3, because the highest order derivative, $\frac{d^2y}{dx^2}$, is raised to the third power.

Don't confuse order and degree. The symbols $\frac{d^2y}{dx^2}$ and $\left(\frac{dy}{dx}\right)^2$ do not mean the same thing. The equation $\frac{d^2y}{dx^2} = 0$ has order 2 and degree 1; the equation $\left(\frac{dy}{dx}\right)^2 = 0$ has order 1 and degree 2.

Solutions of Differential Equations

A **solution** of a differential equation is a relation between the variables that satisfy the equation. This means that when you substitute the solution into the differential equation, you get an algebraic identity. A **particular solution** of a differential equation is any single, or one, solution. The **general solution** of a differential equation is the set of all solutions that satisfy the equation.

EXAMPLE 13.2

Is the equation $y = A\sin 3x + B\cos 3x$, where A and B are constants, a general solution of $y'' + 9y = 0$? What are some particular solutions?

Solution Since $y = A\sin 3x + B\cos 3x$, we need to find y'' and substitute it into the given differential equation.

$$\begin{aligned} y' &= 3A\cos 3x - 3B\sin 3x \\ y'' &= -9A\sin 3x - 9B\cos 3x \end{aligned}$$

Substituting for y'' and y in the given equation, we obtain

$$y'' + 9y = (-9A\sin 3x - 9B\cos 3x) + 9(A\sin 3x + B\cos 3x) = 0$$

and so this is a general solution of the given differential equation.

We can get some particular solutions by substituting values for A and B. If $A = 0$ and $B = 2$, then $y = 2\cos 3x$ is a particular solution of the given equation. If we let $A = 1$ and $B = -4$, then $y = \sin 3x - 4\cos 3x$ is another particular solution.

EXAMPLE 13.3

Show that $y = 2e^{-2x} + xe^{-2x}$ is a solution of the equation $y + 2y' + y'' = xe^{-2x}$.

Solution Differentiating, we get

$$\begin{aligned} y' &= -4e^{-2x} + e^{-2x} - 2xe^{-2x} \\ &= -3e^{-2x} - 2xe^{-2x} \\ y'' &= 6e^{-2x} - 2e^{-2x} + 4xe^{-2x} \\ &= 4e^{-2x} + 4xe^{-2x} \end{aligned}$$

EXAMPLE 13.3 (Cont.)

Substituting for y, y', and y'' in the given equation produces

$$\begin{aligned} y + 2y' + y'' &= (2e^{-2x} + xe^{-2x}) + 2(-3e^{-2x} - 2xe^{-2x}) \\ &\quad + (4e^{-2x} + 4xe^{-2x}) \\ &= 2e^{-2x} + xe^{-2x} - 6e^{-2x} - 4xe^{-2x} \\ &\quad + 4e^{-2x} + 4xe^{-2x} \\ &= (2 - 6 + 4)e^{-2x} + (1 - 4 + 4)xe^{-2x} \\ &= 0e^{-2x} + xe^{-2x} \\ &= xe^{-2x} \end{aligned}$$

So, $y = 2e^{-2x} + xe^{-2x}$ is a particular solution of $y + 2y' + y'' = xe^{-2x}$.

Earlier we solved some elementary differential equations. For example, when we were looking at some applications of derivatives, we found that if we integrated the acceleration function we got the general solution of the velocity. If we were given the velocity at some particular time, then we were able to find a particular solution. If we integrated the velocity function, we found the general solution for the distance.

EXAMPLE 13.4

Find the function y, if $y' = 6x^2$ and $y(0) = 5$.

Solution Since $y' = 6x^2$, then to find y we need to integrate y'.

$$y = \int 6x^2\,dx = 2x^3 + C$$

We are told that when $x = 0$, $y = 5$. Substituting, we see that $y = 2(0)^3 + C = 5$, and so $C = 5$. The particular solution is $y = 2x^3 + 5$.

Notice that we were given the value of y for a specific value of x. This allowed us to find the particular solution to this differential equation. A differential equation, such as the one in Example 13.4, along with some conditions on the unknown function and its derivatives at the same value of the independent variable is called an **initial value problem**. The conditions are called **initial conditions**. If the conditions are given at more than one value of the independent variable, the problem is a **boundary value problem** and the conditions are called **boundary conditions**.

EXAMPLE 13.5

Solve the boundary value problem $y'' = 4x$, $y(0) = -2$, $y'(1) = 3$.

Solution Since $y'' = 4x$, we can find y' by integrating.

$$y' = \int 4x\,dx = 2x^2 + C_1$$

EXAMPLE 13.5 (Cont.)

We know $y'(1) = 2(1)^2 + C_1 = 2 + C_1 = 3$, so $C_1 = 1$. Thus, we have determined that $y' = 2x^2 + 1$.

Integrating y' in order to find y, we have

$$y = \int y'\,dx = \int (2x^2 + 1)\,dx$$
$$= \frac{2}{3}x^3 + x + C_2$$

Since $y(0) = \frac{2}{3}(0)^3 + 0 + C_2 = -2$, we know $C_2 = -2$, thus the desired solution is $y = \frac{2}{3}x^3 + x - 2$.

Application

EXAMPLE 13.6

A driver in a car traveling 60 mph applies the brakes so that the car's braking system produces a maximum deceleration of 32 ft/s^2. If the car maintains that deceleration, **(a)** how long will it take the car to stop and **(b)** how far will it travel before coming to a complete stop?

Solutions We will let s represent the number of feet the car has traveled since the brakes were applied. The velocity of the car at t seconds after the brakes were applied is v and the acceleration a is -32 ft/s^2. (Notice that the deceleration is really a negative acceleration.) So,

$$a(t) = \frac{dv}{dt} = \frac{d^2s}{dt^2} = -32 \tag{1}$$

Integrating equation (1) produces

$$v(t) = \frac{ds}{dt} = -32t + C_1 \tag{2}$$

We were given $v(0) = 60$ mph, so

$$v(0) = -32(0) + C_1 = 60$$

Thus, $C_1 = 60$ mph. Because the acceleration is in ft/s^2, we need to convert C_1 to ft/s.

$$C_1 = \frac{60\,\text{mi}}{1\,\text{h}} = \frac{60 \times 5{,}280\,\text{ft}}{3{,}600\,\text{s}} = 88\,\text{ft/s}$$

So, equation (2) becomes

$$v(t) = -32t + 88 \tag{3}$$

Integrating equation (3), we obtain

$$s = -16t^2 + 88t + C_2$$

EXAMPLE 13.6 (Cont.)

Now at $t = 0$, the car has not traveled, so $s = 0$, so $C_2 = 0$. Thus, the car traveled

$$s(t) = -16t^2 + 88t \text{ ft} \tag{4}$$

t seconds after the brakes were applied.

(a) In order to determine how long it takes to stop the car, we use equation (3). When the car stops, the velocity is 0. So solving $-32t + 88 = 0$ for t, we see that it takes $t = 2.75$ s to stop the car.

(b) Substituting $t = 2.75$ in equation (4), we determine that the car travels $-16(2.75)^2 + 88(2.75) = -121 + 242 = 121$ ft before it stops.

As you can see, in order to solve a differential equation, we need to find some method of changing the equation so that the terms can be integrated. Earlier, with the acceleration and velocity functions, this was an easy task. Not all differential equations are that easy. In the following section we will spend time learning how to transform differential equations so they can be integrated.

Exercise Set 13.1

In Exercises 1–16, show that the each equation is a solution of the indicated differential equation.

1. Equation: $y = e^x$, differential equation: $y'' - y = 0$
2. Equation: $y = e^{2x} + e^x$, differential equation: $y'' - 4y' + 4y = e^x$
3. Equation: $y = 2x^3$, differential equation: $xy' = 3y$
4. Equation: $y = x^3 + 3$, differential equation: $3y - xy' = 9$
5. Equation: $y = 5\cos x$, differential equation: $y' + y\tan x = 0$
6. Equation: $y = e^x - 1$, differential equation: $y' - y = 1$
7. Equation: $y = x^2 + 3x$, differential equation: $xy' - 2y + 3xy'' - 3x = 0$
8. Equation: $y = 4x^2$, Differential Equation: $xy' = 2y$
9. Equation: $y = 2e^x + 3e^{-x} - 4x$, differential equation: $y'' - y = 4x$
10. Equation: $x^2 - y^2 + 2xy = 9$, differential equation: $y' = \dfrac{y+x}{y-x}$
11. Equation: $y = 9x^2 + 6x - 54$, differential equation: $y''' = 0$
12. Equation: $xy^2 = 7x - 3$, differential equation: $x(y')^2 + 2yy' + xyy'' = 0$
13. Equation: $y = x^2 + 4$, differential equation: $y' = 2x$
14. Equation: $2y = \sin x$, differential equation: $y'' + 5y = 2\sin x$
15. Equation: $y = 5e^{-3x} + 2e^{2x}$, differential equation: $y'' + y' - 6y = 0$
16. Equation: $y = 5\cos x$, differential equation: $y'' + y = 0$

In Exercises 17–20, solve the given initial or boundary value problem for *y*.

17. $y'' = 6x^2$, $y'(1) = -2$, $y(1) = 3$
18. $y'' = \sin x$, $y'(0) = 2$, $y(\pi) = \pi$
19. $\dfrac{dy}{dx} = \sec^2 x$, $y = -6$ when $x = \frac{\pi}{4}$
20. $\dfrac{d^2y}{dx^2} = e^{-x}$, $y'(0) = 0$, $y(1) = -1$

Solve Exercises 21–32.

21. *Automotive technology* A driver in a car traveling 65 mph applies the brakes so that the car's braking system produces a maximum deceleration of 30 ft/s^2. If the car maintains this deceleration, **(a)** how long will it take the car to stop and **(b)** how far will it travel before coming to a complete stop?

22. *Automotive technology* A driver in a car traveling 70 mph applies the brakes so that the car's braking system produces a maximum deceleration of 28 ft/s^2. If the car maintains this deceleration, **(a)** how long will it take the car to stop and **(b)** how far will it travel before coming to a complete stop?

23. *Highway safety* Suppose that a car is sitting at a traffic light 151 ft directly ahead of the point where the brakes are applied on the car in Exercise 21. **(a)** How long after the brakes are applied will the cars hit? **(b)** At what speed with the first car strike this stopped car?

24. *Highway safety* Suppose a car is sitting at a traffic light 184.5 ft directly ahead of the point where the brakes are applied on the car in Exercise 22. **(a)** How long after the brakes are applied will the cars hit? **(b)** At what speed will the first car strike this stopped car?

25. *Physics* A ball is thrown upward from a point 192 ft above the ground with an initial velocity of 64 ft/s. **(a)** How long will it take for the ball to hit the ground? **(b)** How fast will it be traveling when it hits the ground?

26. *Physics* A ball is thrown upward from a point 34.3 m above the ground with an initial velocity of 29.4 m/s. **(a)** How long will it take for the ball to hit the ground? **(b)** How fast will it be traveling when it hits the ground? (Use $g = a(t) = -9.8$ m/s^2.)

27. *Medical technology* A woman with a headache takes a 500-mg pain-relief tablet. The tablets are metabolized (used up) by her body at a constant rate of 75 mg/h. **(a)** Find the formula for the amount left in her body t h after the pills are taken. **(b)** How long will it take for the entire amount of pain reliever to be metabolized if no more tablets are taken?

28. *Medical technology* After 3 h, the woman in Exercise 27 still has a headache, so she takes another 500-mg pain-relief tablet. There are still 275 mg of the first tablet left in her system. As before, the tablets are metabolized by her body at a constant rate of 75 mg/h. **(a)** Find the formula for the amount left in her body t h after the first pill is taken. **(b)** How long will it take for the entire amount of pain reliever to be metabolized if no more tablets are taken?

29. *Medical technology* A patient takes a 5-mg dose of cortisone. The patient's body metabolizes the cortisone at the rate of $\frac{1}{2\sqrt{t}}$ mg/h. How long will it take for the entire amount of cortisone to be metabolized if no more is taken?

30. *Environmental science* The population $N(t)$ at time t of a certain species of animal in a controlled environment is assumed to satisfy the logistics growth law

$$\frac{dN}{dt} = \frac{1}{250}N(1{,}000 - N)$$

(a) Show that $N(t) = \frac{1{,}000}{1 + Ce^{-4t}}$ is a general solution of this general differential equation.
(b) If $N(0) = 200$, determine the value of C.

31. *Electricity* In an electric circuit containing constant resistance R and constant inductance L, current i at time t satisfies the differential equation

$$L\frac{di}{dt} + Ri = V$$

where V is the constant voltage. Show that a general solution of this differential equation is $i = \frac{V}{R} + ke^{-Rt/L}$, where k is a constant.

32. *Pharmacology* The number $N(t)$ of bacteria in a culture at time t is assumed to satisfy the differential equation

$$\frac{dN}{dt} = 100 - 0.5N$$

Show that $N(t) = 200 - Ce^{-0.5t}$ is a general solution of this differential equation.

In Your Words

33. What is a differential equation? How can you tell if a differential equation is ordinary or partial?

34. What is the difference between the order and degree of a differential equation? Illustrate your answer with a second-order differential equation that is not second-degree and with a second-degree differential equation that is not second order.

35. How are particular and general solutions to differential equations alike and how are they different?

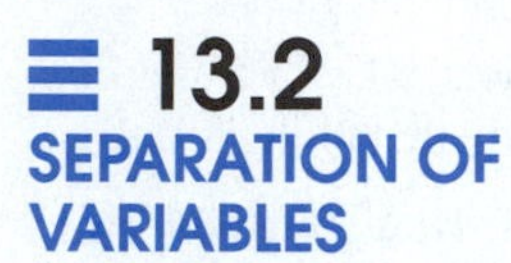

13.2 SEPARATION OF VARIABLES

There is no one method that will allow us to obtain integrable factors of a differential equation. There are some methods that arise often and we will look at these. The first method is the simplest and, as a result, you should always look to see if you can use the method of **separation of variables**

In Section 13.1, we discussed the order and degree of a differential equation. If a differential equation is of the first order and first degree, then it can be written in the form $\frac{dy}{dx} = f(x, y)$. This can also be written in the differential form

$$M(x, y)\,dx + N(x, y)\,dy = 0 \qquad (*)$$

where $M(x, y)$ and $N(x, y)$ may represent constants, functions of either x or y, or functions of x and y.

EXAMPLE 13.7

Write the differential equation $\frac{dy}{dx} = 4x^2$ in the differential form

$$M(x, y)\,dx + N(x, y)\,dy = 0$$

Solution $\frac{dy}{dx} = 4x^2$ can be written as $4x^2\,dx - dy = 0$ by multiplying both sides by dx and then subtracting dy from both sides. Here $M = 4x^2$ and $N = -1$.

EXAMPLE 13.8

Write the differential equation $\frac{dy}{dx} = \frac{x+3y}{x-y}$ in the differential form $M(x, y)\,dx + N(x, y)\,dy = 0$.

Solution Begin by multiplying both sides by the common denominator $dx(x-y)$. Then simplify this result to get it in the desired form.

$$\frac{dy}{dx} = \frac{x+3y}{x-y}$$

$$dy(x-y) = (x+3y)\,dx$$

$$(x+3y)\,dx - (x-y)\,dy = 0$$

or as

$$(x+3y)\,dx + (y-x)\,dy = 0$$

Here $M = x+3y$ and $N = y-x$.

In order to find the solution to an equation of the form $M(x, y)\,dx + N(x, y)\,dy = 0$, it is necessary to integrate. But in order to integrate M, it must be a function of x alone. In order to integrate N, it must be only a function of the variable y.

As you can see, we have a problem if M is a function of y or if it is a function of both x and y. We also have a problem if N is a function of x or a function of both x and y. We need to use some algebraic means to rewrite $M(x, y)\,dx + N(x, y)\,dy = 0$ in the form

$$A(x)dx + B(y)dy = 0$$

where A is a function of x alone and B is a function of y alone. If this is possible, then we can integrate A with respect to x and B with respect to y.

Separation of Variables

A first order differential equation is **separable** if it can be written in the form

$$M(x, y)\,dx + N(x, y)\,dy = 0$$

or in the equivalent form

$$A(x)dx + B(y)dy = 0$$

When we write the equation in this way, we have then **separated the variables**.

In using separation of variables, try to get all the terms involving x and dx on one side of the equation and the terms that involve y and dy on the other side.

EXAMPLE 13.9

Solve $\dfrac{dy}{dx} = \dfrac{2y^2}{x^3}$.

Solution First we rearrange the equation so that the terms involving y and dy are on the left-hand side and the terms involving x and dx are on the right-hand side.

$$\frac{dy}{dx} = \frac{2y^2}{x^3}$$

$$\frac{dy}{y^2} = \frac{2\,dx}{x^3}$$

This can be rewritten as

$$\frac{2\,dx}{x^3} - \frac{dy}{y^2} = 0$$

or $\quad 2x^{-3}\,dx - y^{-2}\,dy = 0$

EXAMPLE 13.9 (Cont.)

This is now in the form $A(x)dx + B(y)dy = 0$ with $A(x) = 2x^{-3}$ and $B(y) = -y^{-2}$.
Integrating, we get

$$\int 2x^{-3}\,dx - \int y^{-2}\,dy = 0$$

$$-x^{-2} + y^{-1} = C$$

or

$$-\frac{1}{x^2} + \frac{1}{y} = C$$

If we can select a function, $I(x, y)$, that we can multiply by the differential equation so that we can integrate the equation, then this function, $I(x, y)$, is known as the **integrating factor**. We need to select the integrating factor so that we remove the x-terms from the coefficient of dy without introducing any factors of y into the coefficient of dx. Similarly, we need to remove the y-terms from the coefficient of dx without introducing any factors of x into the coefficient of dy.

In Section 13.3, we will learn how to find integrating factors, but for now Examples 13.10–13.14 will show the general idea.

EXAMPLE 13.10

Solve $x(y^2 - 3)\,dx + y(x^2 + 4)\,dy = 0$.

Solution The integrating factor is $I(x, y) = \dfrac{1}{(y^2 - 3)(x^2 + 4)}$. Multiplying the differential equation by the integrating factor produces

$$\frac{x\,(y^2 - 3)}{(y^2 - 3)\,(x^2 + 4)}\,dx + \frac{y\,(x^2 + 4)}{(y^2 - 3)\,(x^2 + 4)}\,dy = 0$$

$$\frac{x}{x^2 + 4}\,dx + \frac{y}{y^2 - 3}\,dy = 0$$

Now we know $\displaystyle\int \frac{x}{x^2 + 4}\,dx = \frac{1}{2}\ln\left(x^2 + 4\right) + C$ and $\displaystyle\int \frac{y}{y^2 - 3}\,dy = \frac{1}{2}\ln\left|y^2 - 3\right| + C$, so we have

$$\int \frac{x}{x^2 + 4}\,dx + \int \frac{y}{y^2 - 3}\,dy = 0$$

$$\frac{1}{2}\ln\left(x^2 + 4\right) + \frac{1}{2}\ln\left|y^2 - 3\right| = C$$

or, using the properties of logarithms,

$$\ln\left|\left(x^2 + 4\right)\left(y^2 - 3\right)\right| = 2C$$

If we write the constant of integration $2C$ as $\ln C_1$, then

$$\ln\left|\left(x^2 + 4\right)\left(y^2 - 3\right)\right| = \ln C_1$$

or

$$\left(x^2 + 4\right)\left(y^2 - 3\right) = C_1$$

This is a general solution to this differential equation. Selecting different values for C_1 will give different particular solutions. Thus, $(x^2+4)(y^2-3)=-10$ and $(x^2+4)(y^2-3)=e^3$ are two particular solutions to the differential equation in Example 13.10.

EXAMPLE 13.11

Solve the differential equation

$$3y\sin x\,dx+4y^2\cos x\,dy=0$$

with the initial condition that $y(0)=5$.

Solution From the note just before Example 13.10, the integrating factor seems to be $\frac{1}{y\cos x}$. Multiplying by the integrating factor, we get

$$\frac{3y\sin x}{y\cos x}dx+\frac{4y^2\cos x}{y\cos x}dy=0$$
$$\frac{3\sin x}{\cos x}dx+4y\,dy=0$$

Integrating, we obtain

$$3\int\frac{\sin x}{\cos x}dy+4\int y\,dy=0$$
$$-3\ln|\cos x|+2y^2=C$$

or

$$3\ln|\sec x|+2y^2=C$$

or

$$\ln\left|\sec^3 x\right|+2y^2=C$$

Now, the initial condition states that when $x=0$, $y=5$, so

$$\ln\left|\sec^3(0)\right|+2(5)^2=C$$

or

$$\ln 1+50=C$$

And, since $\ln 1=0$, we have $C=50$. Thus, this particular solution is

$$\ln\left|\sec^3 x\right|+2y^2=50$$

Don't fall into a trap. Not every differential equation needs to be multiplied by an integrating factor. This is shown in Examples 13.12 and 13.13.

EXAMPLE 13.12

Solve: $\frac{dy}{dx}=\frac{4x^3}{y^2}$.

Solution Rewriting, we get

$$4x^3\,dx-y^2\,dy=0$$

EXAMPLE 13.12 (Cont.)

This can be integrated as it is:

$$\int 4x^3\,dx - \int y^2\,dy = 0$$

and so

$$x^4 - \frac{1}{3}y^3 = C$$

EXAMPLE 13.13

Solve: $\dfrac{dy}{dx} = \dfrac{y^2}{x^2+9}$.

Solution We can rewrite this differential equation as

$$\frac{dy}{y^2} = \frac{dx}{x^2+9}$$

Integrating, we get the general solution

$$-\frac{1}{y} = \frac{1}{3}\arctan\frac{x}{3} + C$$

EXAMPLE 13.14

Solve the differential equation

$$y' = \frac{4y}{x(y-5)}$$

Solution We need to recognize that $y' = \dfrac{dy}{dx}$, so the differential equation can be written as

$$x(y-5)\,dy = 4y\,dx$$

The integrating factor is $\dfrac{1}{xy}$. Multiplying by the integrating factor, the equation becomes

$$\frac{y-5}{y}\,dy = \frac{4}{x}\,dx$$

Now, $\dfrac{y-5}{y} = 1 - \dfrac{5}{y}$, so we can integrate as follows.

$$\int\left(1-\frac{5}{y}\right)dy = \int\frac{4}{x}\,dx$$

$$y - 5\ln|y| = 4\ln|x| + \ln C$$

or

$$y = \ln\left|y^5\right| + \ln x^4 + \ln C$$

$$y = \ln\left|Cy^5x^4\right|$$

EXAMPLE 13.14 (Cont.)

This can also be written as

$$e^y = Cy^5x^4$$

$$\text{or} \quad C_1 e^y = y^5 x^4$$

Application

EXAMPLE 13.15

Newton's law of cooling states that the rate of change of the temperature of an object is proportional to the difference between the temperature of the object, T, and the temperature of the surrounding medium, T_m.

(a) Write a differential equation to express Newton's law of cooling.

(b) Use separation of variables to find the general solution to this differential equation.

(c) Find the particular solution if $T = T_0$ at $t = 0$ s.

(d) The temperature of a turkey was 185°F when it was removed from the oven and placed on a carving tray in a room with a constant temperature of 72°F. After 10 min, the temperature of the turkey was 178°F. Find the temperature 15 min after the turkey was removed from the oven.

Solutions

(a) Here T_m is a constant. If $T > T_m$, then $T - T_m$ is positive. However, T is decreasing, so the rate of change of the temperature is negative. Thus, we get $\frac{dT}{dt} = -k(T - T_m)$, where k is a positive constant of proportionality.

(b) Rewriting this differential equation, we get

$$\frac{dT}{T - T_m} + k\,dt = 0$$

Integrating produces

$$\int \frac{dT}{T - T_m} + \int k\,dt = \int 0\,dt$$

$$\ln(T - T_m) + kt = C_1$$

$$\ln(T - T_m) = -kt + C_1$$

$$T - T_m = e^{-kt + C_1} = e^{-kt}e^{C_1} = Ce^{-kt}$$

$$T = T_m + Ce^{-kt}$$

The general solution is $T = T_m + Ce^{-kt}$.

(c) Substituting 0 for t and T_0 for T in the general solution, we obtain $T_0 = T_m + Ce^{-k(0)} = T_m + C$. Solving for C produces $C = T_0 - T_m$. Thus, the particular solution is $T = T_m + (T_0 - T_m)e^{-kt}$.

EXAMPLE 13.15 (Cont.)

(d) Here $T_0 = 185$ and $T_m = 72$. When $t = 10$, we have $T = 178$. These values are substituted into the particular solution, and it is solved for k as follows:

$$\begin{aligned} 178 &= 72 + (185 - 72)e^{-k(10)} \\ 106 &= 113e^{-k(10)} \\ \frac{106}{113} &= e^{-k(10)} \\ \ln\left(\frac{106}{113}\right) &= -k(10) \\ 0.00639487 &\approx k \end{aligned}$$

Using this value for k, the given values of T_0 and T_m, and letting $t = 15$, we obtain

$$\begin{aligned} T &= 72 + 113e^{-0.00639487(15)} \\ &\approx 72 + 102.66 = 174.66 \end{aligned}$$

So, 15 min after it has been out of the oven, the turkey is about 174.66°F.

Exercise Set 13.2

In Exercises 1–20, solve the given differential equation.

1. $y' = \dfrac{2x}{y}$
2. $\dfrac{dy}{dx} = \dfrac{2y}{x}$
3. $5x\,dx + 3y^2\,dy = 0$
4. $5y\,dx + 3y^2\,dy = 0$
5. $5y\,dx + 3\,dy = 0$
6. $4y^5\,dx - 3x^2\,dy = 0$
7. $4y^5\,dx - 3x^2y\,dy = 0$
8. $xyy' + \sqrt{1+y^2} = 0$
9. $x^3\left(y^2+4\right) + yy' = 0$
10. $x^4\left(y^3-3\right)dx + y^2\left(x^5-2\right)dy = 0$
11. $ye^{x^2}\,dy = 2x\left(y^2+4\right)dx$
12. $e^{x^2}\,dy + x\sqrt{4-y^2}\,dx = 0$
13. $2y + e^{-3x}y' = 0$
14. $\sin^2 y\,dx + \cos^2 x\,dy = 0$
15. $\left(y^2-4\right)\cos x\,dx + 2y\sin x\,dy = 0$
16. $(1+x)^2\,y' = 1$
17. $y' = \dfrac{x^2y - y}{y^2+3}$
18. $\dfrac{dy}{dx} = \dfrac{y}{x^2+6x+9}$
19. $\sec 3x\,dy + y^2e^{\sin 3x}\,dx = 0$
20. $2xy + \left(1+x^2\right)y' = 0$

In Exercises 21–26, find the particular solution of the given differential equation for the given conditions.

21. $y' + y^2x^3 = 0$, $y = -1$ when $x = 2$
22. $x\,dy + y\,dx = 0$, $y = 3$ when $x = 2$
23. $\dfrac{dy}{dx} = \dfrac{3x + xy^2}{y + x^2y}$, $y = 3$ when $x = 1$
24. $\sin^2 y\,dx + \cos^2 x\,dy = 0$, $y = \frac{3\pi}{4}$ when $x = \frac{\pi}{4}$
25. $y' = 3x^2y - 2y$, $y = 1$ when $x = 1$
26. $\sqrt{1+9x^2}\,dy = y^4x\,dx$, $y = -1$ when $x = 0$

Solve Exercises 27–32.

27. *Food science* A roast is removed from a freezer where the temperature is 10°F and placed in an oven with a constant temperature of 325°F. After one hour in the oven, the temperature of the roast is 75°F. The roast will be ready to take out of the oven when its temperature reaches 145°F.

(a) Write the particular differential equation to fit this situation.

(b) How long will it take this roast to cook?

28. *Medical technology* Blood plasma is stored at 40°F. Before the plasma can be used, it must be 90°F. When the plasma is placed in a warming oven at 120°F, it takes 45 min for the plasma to warm to 90°F. Assume that Newton's law of cooling applies and that k is not influenced by T_m.

(a) Write the particular differential equation to fit this situation.

(b) How long will it take the plasma to warm to 90°F if the oven temperature is set at 150°F?

(c) How long will it take the plasma to warm to 90°F if the oven temperature is set at 80°F?

29. *Hydrology* One particularly cold morning, the rate of growth of the thickness of a layer of ice on a certain lake is determined by the differential equation $\frac{dx}{dt} = \frac{5}{x}$. If $x = 2$ in. at 6:00 A.M., at what time will the ice be 6 in. thick? (Hint: Let $t = 0$ at 6:00 A.M.)

30. *Pharmacology* The amount, A, of a certain drug remaining in a body t hours after an injection decreases at a rate proportional to the amount present.

(a) Write a differential equation to fit this situation.

(b) Use separation of variables to find the general solution of this differential equation.

(c) Find the particular solution if the amount of a certain drug decreases at a rate of 2% per hour and an injection of 20 cc is administered.

(d) Under the conditions in (c), how much of the drug remains in the body 6 hr after the injection?

31. *Medical technology* An influenza epidemic has spread throughout a city of 500,000 people at a rate proportional to the product of the number of people who have been infected and the number who have not been infected. Suppose that 100 people were initially infected and 750 were infected 10 days later.

(a) Write a differential equation to fit this situation.

(b) Use separation of variables to find the general solution of this differential equation.

(c) Use the given infection data to find the particular solution.

(d) How many will be infected after 20 days?

32. *Advertising* A company decides to use only television commercials to introduce a new product to a city of 250,000 people. Suppose the rate at which people learn about the new product is proportional to the number who have not yet heard of it. Suppose further, that no one knew about the product at the start of the advertising campaign and after 7 days, 40,000 are aware of the product.

(a) How many people will know about it after 14 days?

(b) How long will it take for 125,000 people to become aware of the product?

In Your Words

33. Without looking in the text, write the criteria for telling if a first-order differential equation is separable.

34. What is an integrating factor?

13.3 INTEGRATING FACTORS

In Section 13.2, we often used an integrating factor to allow us to use the separation-of-variables method. An integrating factor is used when separation of variables cannot be done easily. In this section, we will learn how to tell when we need to use an integrating factor and how to determine that factor.

A first-order differential equation of the form $M(x, y)\,dx + N(x, y)\,dy = 0$ is **exact** if there exist a function $f(x, y)$ whose derivative is equal to the left-hand side of the differential equation. Thus,

$$df(x, y) = M(x, y)\,dx + N(x, y)\,dy$$

The general solution of an exact differential equation is $f(x, y) = C$.

Test for Exactness

A first-order differential equation is exact if it passes the following test.

1. Both $M(x, y)$ and $N(x, y)$ are continuous and have continuous first partial derivatives over the defined intervals.
2. The first-order differential equation $M(x, y)\,dx + N(x, y)\,dy = 0$ is exact if and only if

$$\frac{\partial M(x, y)}{\partial y} = \frac{\partial N(x, y)}{\partial x}$$

EXAMPLE 13.16

Test to see if each of the following differential equations are exact.

(a) $(2xy + 4x^2)\,dx + x^2\,dy = 0$

(b) $2y^2 dx - x^3 dy = 0$

(c) $y' = \dfrac{xy^2 - 4}{9 - x^2 y}$

(d) $3x^2 y\,dx + (x^3 + \sin y)\,dy = 0$

Solutions

(a) We can see that $M(x, y) = 2xy + 4x^2$ and $N(x, y) = x^2$. Since M and N are polynomials, they are continuous and have continuous first derivatives. Taking the partial derivatives, we get

$$\frac{\partial M}{\partial y} = \frac{\partial(2xy + 4x^2)}{\partial y} = 2x$$
$$\frac{\partial N}{\partial x} = \frac{\partial x^2}{\partial x} = 2x$$

Since $\dfrac{\partial M}{\partial y} = \dfrac{\partial N}{\partial x}$, this equation is exact.

(b) Here, $M = 2y^2$ and $N = -x^3$. Again, both M and N are continuous and have first partial derivatives that are continuous. Differentiating, we get $\dfrac{\partial M}{\partial y} = 4y$ and $\dfrac{\partial N}{\partial x} = -3x^2$. Since $\dfrac{\partial M}{\partial y} \neq \dfrac{\partial N}{\partial x}$, this equation is not exact.

EXAMPLE 13.16 (Cont.)

(c) $y' = \dfrac{xy^2 - 4}{9 - x^2y}$ is not in the correct form. Solving, we obtain $(xy^2 - 4)\,dx - (9 - x^2y)\,dy = 0$, so $M = xy^2 - 4$ and $N = -(9 - x^2y) = x^2y - 9$. Both M and N are continuous and have continuous first partial derivatives. Differentiating produces $\dfrac{\partial M}{\partial y} = 2xy$ and $\dfrac{\partial N}{\partial x} = 2xy$. These partial derivatives are the same, so this differential equation is exact.

(d) We have $M = 3x^2y$ and $N = x^3 + \sin y$. As before, both M and N are continuous and have continuous first partial derivatives. The partial derivatives are

$$\frac{\partial M}{\partial y} = 3x^2 = \frac{\partial N}{\partial x}$$

so the equation is exact. ▪

Integrating Factors

If an equation is not exact, it can often be made exact by the introduction of an integrating factor. In Section 13.2, we introduced the integrating factor. So far, we have been able to find the integrating factor by inspection and through our experience. In the remainder of this section, we develop some procedures for finding integrating factors. First, we will give a more precise definition.

Integrating Factor

A function $I(x, y)$ is an integrating factor for the differential equation $M(x, y)\,dx + N(x, y)\,dy = 0$ if the equation

$$I(x, y)\,[M(x, y)\,dx + N(x, y)\,dy] = 0$$

is exact.

The following table contains combinations of basic differentials, the integrating factor for each, and the resulting exact differential.

Some terms have more than one integrating factor. The integrating factor that should be used will depend on the nature of the problem.

TABLE 13.1

	Group of terms	Integrating factor $I(x, y)$	Exact differential $dg(x, y)$
(1)	$y\,dx + x\,dy$	$\dfrac{1}{xy}$	$d(\ln xy)$
(2)	$x\,dx + y\,dy$	1	$d\left[\dfrac{1}{2}(x^2+y^2)\right]$
(3)		$\dfrac{1}{x^2+y^2}$	$d\left[\dfrac{1}{2}\ln(x^2+y^2)\right]$
(4)	$y\,dx - x\,dy$	$-\dfrac{1}{x^2}$	$d\left(\dfrac{y}{x}\right)$
(5)		$\dfrac{1}{y^2}$	$d\left(\dfrac{x}{y}\right)$
(6)		$-\dfrac{1}{xy}$	$d\left(\ln\dfrac{y}{x}\right)$
(7)		$\dfrac{-1}{x^2+y^2}$	$d\left(\arctan\dfrac{y}{x}\right)$
(8)	$ay\,dx + bx\,dy$, a, b constants	$x^{a-1}y^{b-1}$	$d\left(x^a y^b\right)$

EXAMPLE 13.17

Solve the differential equation

$$y\,dx - x\,dy + 4x\,dx = 0$$

Solution This equation contains the combination $y\,dx - x\,dy$. We have four integrating factors in Table 32.1 from which to select. If we use integrating factors (5), (6), or (7), the last term in the differential equation, $4x\,dx$, cannot be integrated. We will use integrating factor (4).

Multiplying the given differential equation by the integrating factor produces

$$-\frac{1}{x^2}(y\,dx - x\,dy + 4x\,dx) = 0$$

$$\frac{x\,dy - y\,dx}{x^2} - \frac{4}{x}\,dx = 0$$

According to Table 13.1 , the left-hand combination is the derivative of $\dfrac{y}{x}$. The right-hand combination is the derivative of $4\ln|x|$, and so we have

$$\frac{y}{x} - 4\ln|x| = C$$

or $$\frac{y}{x} - \ln x^4 = C$$

EXAMPLE 13.18

Solve $y^2\,dx - y\,dx + x\,dy = 0$.

Solution Here again, we see the combination of $y\,dx - x\,dy$. Rewriting, we get

$$y^2\,dx - (y\,dx - x\,dy) = 0$$

If we multiply by integrating factor (5) of Table 32.1, we obtain

$$dx - \frac{y\,dx - x\,dy}{y^2} = 0$$

Integrating this equation results in

$$x - \frac{x}{y} = C$$

or $\frac{x}{y} = x + C_1$, where $C_1 = -C$. Solving for y, we get the general solution

$$y = \frac{x}{x + C_1}$$

EXAMPLE 13.19

Find the particular solution of the differential equation $x\,dx + x^2\,dx + y\,dy + y^2\,dx = 0$ that satisfies the condition that $y = 3$ when $x = 0$.

Solution This does not appear to contain any of the terms in Table 32.1. If we rearrange terms, we get

$$(x\,dx + y\,dy) + (x^2 + y^2)\,dx = 0$$

The term in the first group of parentheses is like the second group of terms inTable 13.1 . Since the second group in the differential equation has a factor of $x^2 + y^2$, we will use the integrating factor in (3). Multiplying by $I(x, y) = \frac{1}{x^2 + y^2}$, gives

$$\frac{x\,dx + y\,dy}{x^2 + y^2} + \frac{x^2 + y^2}{x^2 + y^2}\,dx = 0$$

$$\frac{x\,dx + y\,dy}{x^2 + y^2} + dx = 0$$

Integrating this last equation, we get

$$\frac{1}{2}\ln(x^2 + y^2) + x = C_1$$

or $$\ln(x^2 + y^2) = -2x + C_2, \text{ where } C_2 = 2C_1$$

Taking the exponential of both sides, the equation becomes

$$e^{\ln(x^2+y^2)} = e^{-2x+C_2}$$

or $$x^2 + y^2 = e^{-2x}e^{C_2}$$

Since e^{C_2} is a constant, which we will call C, we can rewrite this solution as

$$x^2 + y^2 = Ce^{-2x}$$

EXAMPLE 13.19 (Cont.)

This is the general solution. When $x = 0$ and $y = 3$, we obtain

$$0^2 + 3^2 = Ce^{-2\cdot 0}$$
$$9 = Ce^0 = C$$

Thus $C = 9$, and the particular solution is

$$x^2 + y^2 = 9e^{-2x}$$

Exercise Set 13.3

In Exercises 1–10, test the differential equation to see if it is exact.

1. $(x^2 - y)\,dx + x\,dy = 0$
2. $(x^2 + y^2)\,dx + xy\,dy = 0$
3. $(x^2 + y^2)\,dx + 2xy\,dy = 0$
4. $x\,dx + \sin y\,dx - \sin x\,dy = 0$
5. $x\,dx + y\cos x\,dx + \sin x\,dy = 0$
6. $e^{2x}\,dy + 2ye^{2x}\,dx - dy = 0$
7. $2xy\,dx + y\,dy + x\,dx + x^2\,dy = 0$
8. $4x^3y^2\,dx + (3x^4y + 5x^4)dy = 0$
9. $x\,dx + y\,dy = y\,dx - x\,dy$
10. $(y\sin x + x\cos y)dx + \left(\frac{1}{2}y^2\cos x + \sin y\right)dy = 0$

In Exercises 11–26, solve the given differential equation.

11. $y\,dx - x\,dy = 0$
12. $y\,dx - x\,dy - x^3\,dx = 0$
13. $y\,dx - x\,dy - 5y^4\,dy = 0$
14. $y\,dx - x\,dy = x^2y\,dx$
15. $y\,dx + y^2\,dx = x\,dy - x^2\,dx$
16. $y\,dx = x\,dy + 3\,dx$
17. $y\,dx + x\,dy = 3x^3y\,dx$
18. $x\,dx + y\,dy = 3x^2\,dy + 3y^2\,dy$
19. $xy^2\,dx + x^2y\,dy = 0$
20. $4y\,dx - 3x\,dy - x^{-3}y^2\,dy = 0$
21. $2y\,dx + x\,dy = 3x\,dx$
22. $x^2\,dx + 2xy\,dy = y^2\,dx$
23. $3x^2(x^2 + y^2)dx + y\,dx = x\,dy$
24. $(x^2 + y^2)dx = 4x\,dx + 4y\,dy$
25. $\tan(x^2 + y^2)dy = x\,dx + y\,dy$
26. $8x\,dy + 4y\,dx + 9x^{-3}y\,dy = 0$

In Exercises 27–30, find the particular solution to the given differential equation that satisfies the given boundary values.

27. $2x\,dy + 2y\,dx = 3x^3y\,dx$, $y = 3$ when $x = 1$
28. $y\,dx - x\,dy = x^2\,dx$, $y = 4$ when $x = 2$
29. $(x^2 + y + y^2)\,dx = x\,dy$, $y = \frac{\pi}{3}$ when $x = \frac{\pi}{3}$

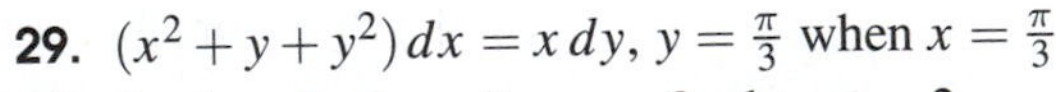

30. $2y\,dx + 3x\,dy = 0$, $y = -2$ when $x = 2$

In Your Words

31. Without looking in the text, describe how to determine if a first-order differential equation is exact.
32. What is an integrating factor? How do you determine if a function is an integrating factor for a differential equation of the form $M(x, y)\,dx + N(x, y)\,dy = 0$?

13.4 LINEAR FIRST-ORDER DIFFERENTIAL EQUATIONS

At this point, we have solved differential equations by using either separation of variables or inspection andTable 13.1 to find an integrating factor. In this section, we will look at the linear first-order differential equation, because it is always possible to find an integrating factor for an equation of this type.

Linear First-Order Differential Equation

A **linear first-order differential equation** is of the form

$$dy + P(x)y\,dx = Q(x)\,dx$$

or

$$\frac{dy}{dx} + P(x)y = Q(x)$$

and has an integrating factor of

$$I(x, y) = e^{\int P(x)\,dx}$$

Notice that P and Q are functions of x only, and that this integrating factor does not depend on y.

EXAMPLE 13.20

Find an integrating factor for $dy = 2xy\,dx + x\,dx$.

Solution We need to write this in the form $dy + P(x)y\,dx = Q(x)\,dx$. Since the term with $P(x)$ also has a factor of y, we will subtract $2xy\,dx$ from (or add $-2xy\,dx$ to) both sides. When we do, we get

$$dy + (-2xy\,dx) = x\,dx$$

which is in the desired form. We can see that $P(x) = -2x$, so the integrating factor is

$$I(x, y) = e^{\int -2x\,dx} = e^{-x^2}$$

The purpose of getting the integrating factor $I(x, y)$ is to multiply the original equation by $I(x, y)$ in order to get an exact differential equation. The left-hand side of the exact differential equation is of the form $\frac{du}{dx}$, where $u = yI(x, y) = ye\int P(x)\,dx$. The right-hand side of the exact differential equation is a function of x only, and can be integrated. Remember that when both sides are integrated, $\int \frac{du}{dx}\,dx = u$. Thus, we can make the following statement about the solution of a linear first-order differential equation.

Solution of a Linear First-Order Differential Equation

The solution of a linear first-order differential equation of the form

$$dy + P(x)y\,dx = Q(x)\,dx$$

or $$\frac{dy}{dx} + P(x)y = Q(x)$$

is

$$ye^{\int P(x)\,dx} = \int Qe^{\int P(x)\,dx}\,dx + C$$

EXAMPLE 13.21

Solve the differential equation in Example 13.20.

Solution The given differential equation was rewritten in the form

$$dy + (-2xy\,dx) = x\,dx$$

with an integrating factor of e^{-x^2}. Multiplying by e^{-x^2}, the equation becomes

$$e^{-x^2}\,dy - 2xye^{-x^2}\,dx = xe^{-x^2}\,dx$$

The left-hand side is the derivative of ye^{-x^2}, so we have

$$\frac{d}{dx}\left(ye^{-x^2}\right) = xe^{-x^2}\,dx$$

Integrating the right-hand side produces

$$\int xe^{-x^2}\,dx = -\frac{1}{2}e^{-x^2} + C$$

so when we integrate the entire equation, we obtain

$$ye^{-x^2} = -\frac{1}{2}e^{-x^2} + C$$

or $$y = Ce^{x^2} - \frac{1}{2}$$

which is the general solution of this differential equation. ▪

EXAMPLE 13.22

Solve the differential equation $y' - 4y = xe^{4x}$.

Solution Since $y' = \frac{dy}{dx}$, this is in the form $\frac{dy}{dx} + P(x)y = Q(x)$ with $P(x) = -4$. Thus, $I(x, y) = e^{\int -4\,dx} = e^{-4x}$. Multiplying by I, we get

$$y'e^{-4x} - 4ye^{-4x} = xe^{4x}e^{-4x}$$

or $$\frac{d}{dx}\left(ye^{-4x}\right) = xe^{4x-4x} = x$$

EXAMPLE 13.22 (Cont.)

Integrating, we obtain

$$ye^{-4x} = \int x\,dx = \frac{1}{2}x^2 + C$$

So,

$$y = \left(\frac{1}{2}x^2 + C\right)e^{4x}$$

is the general solution of this differential equation.

EXAMPLE 13.23

Solve $\cot x\,dy - y\,dx = (\cot x)3e^{\sin x}\,dx$.

Solution This is not in the correct form. If the equation is divided by $\cot x$, we get

$$dy - \frac{y}{\cot x}\,dx = 3e^{\sin x}\,dx$$

or $$dy - y\tan x\,dx = 3e^{\sin x}\,dx$$

Here, $P(x) = -\tan x$ and so the integrating factor is $e^{-\int \tan x\,dx} = e^{\ln|\cos x|} = \cos x$. Multiplying the given equation by the integrating factor, we obtain

$$\cos x\,dy - y\cos x\tan x\,dx = 3(\cos x)e^{\sin x}\,dx$$

Now $\cos x\tan x = \sin x$, so we can rewrite this equation and then integrate the result in order to find the solution:

$$\cos x\,dy - y\sin x\,dx = 3(\cos x)e^{\sin x}\,dx$$

$$\int \frac{d}{dx}(y\cos x)\,dx = 3\int (\cos x)e^{\sin x}\,dx$$

$$y\cos x = 3e^{\sin x} + C$$

EXAMPLE 13.24

Find the particular solution of $dy + y\,dx = \sin x\,dx$, if $y = 1$ when $x = 0$.

Solution Here, $P(x) = 1$ and so $I(x, y) = e^{\int dx} = e^x$. Thus,

$$ye^x = \int e^x \sin x\,dx$$

We can integrate the right-hand side by using integration by parts twice, or by using formula 87 in the Table of Integrals in Appendix C. Either way, we should see that

$$\int e^x \sin x\,dx = \frac{1}{2}e^x(\sin x - \cos x) + C$$

EXAMPLE 13.24 (Cont.)

So, a general solution is

$$ye^x = \frac{1}{2}e^x(\sin x - \cos x) + C$$

or

$$y = \frac{1}{2}\sin x - \frac{1}{2}\cos x + Ce^{-x}$$

To find the particular solution when $x = 0$ and $y = 1$, we see that

$$\begin{aligned} 1 &= \frac{1}{2}\sin(0) - \frac{1}{2}\cos(0) + Ce^{-0} \\ &= \frac{1}{2}(0) - \frac{1}{2}(1) + C \end{aligned}$$

Thus, $C = 1 + \frac{1}{2} = \frac{3}{2}$ and the particular solution is

$$y = \frac{1}{2}\sin x - \frac{1}{2}\cos x + \frac{3}{2}e^{-x}$$

Application

EXAMPLE 13.25

The following model has been proposed for weight loss or gain:

$$\frac{dw}{dt} + 0.005w = \frac{1}{3{,}500}C$$

where $w(t)$ is a person's weight in pounds after t days of consuming exactly C calories/day. Solve this differential equation for $w(t)$.

Solution We first write the differential equation in the form

$$dw + 0.005w\,dt = \frac{1}{3{,}500}C\,dt$$

In this example w replaces y and t replaces x in the linear first-order differential equation. Here $P(t) = 0.005$ and $Q(t) = \frac{1}{3{,}500}C$. This differential equation has an integrating factor of

$$\begin{aligned} I(t,w) &= e^{\int 0.005\,dt} \\ &= e^{0.005t} \end{aligned}$$

Multiplying the (rewritten) given equation by the integrating factor, we obtain

$$e^{0.005t}\,dw + 0.005we^{0.005t}\,dt = \frac{1}{3{,}500}Ce^{0.005t}\,dt$$

Integrating both side of this equation produces

$$we^{0.005t} = \frac{2}{35}Ce^{0.005t} + k$$

EXAMPLE 13.25 (Cont.)

Multiplying this equation by $e^{-0.005t}$, we get the desired result for w.

$$w = \frac{2}{35}C + ke^{-0.005t}$$

Thus, a person's weight in pounds after t days of consuming exactly C calories/day is given by $w = \frac{2}{35}C + ke^{-0.005t}$, where k is a constant.

Exercise Set 13.4

In Exercises 1–20, solve the given differential equation.

1. $dy + 2xy\,dx = 6x\,dx$
2. $y' + y = 2$
3. $y' - 2y = 4$
4. $dy - 3y\,dx = e^{4x}\,dx$
5. $dy + \frac{y}{x}dx = x^2\,dx$
6. $y' - 6y = e^x$
7. $y' - 6y = 12x$
8. $dy + x^2y\,dx = x^2\,dx$
9. $\frac{dy}{dx} = e^x - \frac{y}{x}$
10. $\frac{dy}{dx} = 6xy$
11. $y' - \frac{1}{x^2}y = \frac{1}{x^2}$
12. $y' - 4xy = xe^{x^2}$
13. $y\,dy - 2y^2\,dx = y\sin 2x\,dx$
14. $x\,dy - y\,dx = 4x\,dx$
15. $4x\,dy - y\,dx = 8x\,dx$
16. $\sec x\,dy = (y-1)dx$
17. $dy = \sec^2 x(1+y)\,dx$
18. $xy' = y + (x^2-1)^2$
19. $x\,dy + (1-4x)y\,dx = 4x^2e^{4x}\,dx$
20. $x\,dy - y\,dx = x^3\sin x^2\,dx$

In Exercises 21–24, find the particular solution for the given equation and conditions.

21. $dy + \frac{2y}{x}dx = x\,dx$, $y = 3$ when $x = 2$
22. $dy + 3y\,dx = e^{-2x}dx$, $y = 2$ when $x = 0$
23. $y' - \frac{2y}{x} = x^2\sin 3x$, $y = \pi^2$ when $x = \pi$
24. $\sin x\,dy + (y\cos x - 1)dx = 0$, $y = \pi$ when $x = \frac{\pi}{6}$

Solve Exercises 25 and 26.

25. *Health technology* Use the result in Example 13.25, to determine the following for a person weighing 160 lb who goes on a 2,100 cal/day diet.
    (a) What is the particular solution to the differential equation that meets these conditions?
    (b) How much will this person weigh after 30 days on this diet?
    (c) How long will it take for this person to lose 10 lb?
    (d) Find $\lim_{t\to\infty} w(t)$.
    (e) How do you interpret the results in (d)?
26. *Health technology* Use the result in Example 13.25 to determine the following for a person weighing 200 lb who goes on a 2,100 cal/day diet.
    (a) What is the particular solution to the differential equation that meets these conditions?
    (b) How much will this person weigh after 30 days on this diet?
    (c) How long will it take for this person to lose 10 lb?

In Your Words

27. What is a first-order differential equation? What is its integrating factor?

28. How do you solve a linear first-order differential equation?

13.5 APPLICATIONS

As we stated earlier, differential equations have many applications in science and technology. In this section and in Section 13.6, we will look at some of these applications in such areas as growth and decay, cooling, electric circuits, orthogonal trajectories, falling bodies with air resistance, and dilution. In this section and in Section 13.6, we will do these one area at a time in order to help you to understand them better.

Growth and Decay

Suppose that we have a function $N(t)$ representing the amount (or number) of a substance that is either growing or decaying. $N(t)$ represents the amount that is present at any time t. We will assume that $\frac{dN}{dt}$, the rate of change of this substance at any time t, is proportional to the amount of the substance present at that time. Then, we have $\frac{dN}{dt} = kN$ or

$$\frac{dN}{dt} - kN = 0$$

where k is the constant of proportionality.

Application

EXAMPLE 13.26

A certain radioactive substance is known to decay at a rate proportional to the amount present. An experiment begins with 50 mg of the material. After 8 h it is observed that only 20 mg remain. **(a)** What is the half-life of the substance? **(b)** How much is left after 24 h?

Solutions We begin by determining a formula that describes the amount of radioactive substance present at a given time. Let N denote the amount of the substance present at time t. In particular, when $t = 0$, we know that $N = 50$ and at $t = 8$, we know that $N = 20$.

We have the differential equation $\frac{dN}{dt} - kN = 0$. This differential equation is separable as $\frac{1}{N}\,dN - k\,dt = 0$. Integrating, we get

$$\begin{aligned} \ln N - kt &= C_1 \\ \text{or} \qquad \ln N &= kt + C_1 \end{aligned}$$

which can be simplified as

$$N = Ce^{kt}$$

EXAMPLE 13.26 (Cont.)

Since $N = 50$ when $t = 0$, we have

$$50 = Ce^{k \cdot 0}$$

$$\text{and so} \qquad C = 50$$

and the solution is now of the form $N = 50e^{kt}$.

Since $N = 20$ when $t = 8$, we have

$$\begin{aligned} 20 &= 50e^{8k} \\ \text{or} \qquad \ln 20 &= \ln 50 + \ln e^{8k} \\ &= \ln 50 + 8k \end{aligned}$$

Thus, $8k = \ln 20 - \ln 50 = \ln \frac{20}{50}$ and $k = \frac{1}{8} \ln \frac{2}{5} \approx -0.1145$. Substituting, we find that the amount of the substance at any time t is given by

$$N = 50e^{-0.1145t}$$

where t is measured in hours. **(a)** The half-life is the amount of time it takes for $\frac{1}{2}$ of the original mass to decay. The original mass was 50 mg and $\frac{1}{2}$ of that is 25 mg. So, when $N = 25$, we have

$$\begin{aligned} 25 &= 50e^{-0.1145t} \\ 0.5 &= e^{-0.1145t} \\ -0.1145t &= \ln 0.5 \\ \text{or} \qquad t &= \frac{\ln 0.5}{-0.1145} \approx 6.05\,\text{h} \\ &= 6\,\text{h}\;3\,\text{min} \end{aligned}$$

The half-life of this radioactive substance is 6 h 3 min. **(b)** The amount after 24 h is found from the formula

$$N = 50e^{-0.1145t}$$

when $t = 24$.

$$\begin{aligned} N &= 50e^{(-0.1145)(24)} \\ &\approx 3.20\,\text{mg} \end{aligned}$$

After 24 h, there will be about 3.20 mg of the substance left.

This technique also works on growth problems, particularly those which involve the growth of bacteria as in Exercises 4–6.

Cooling

Newton's law of cooling states that the time rate of change of the temperature of a body is proportional to the temperature difference between the body and the surrounding medium. Let T represent the temperature of the body and T_m represent the temperature of its surrounding medium. The time rate of change of the temperature of the body is $\frac{dT}{dt}$. **Newton's law of cooling** can be written as

$$\frac{dT}{dt} = -k(T - T_m)$$

where k is a positive constant.

Application

EXAMPLE 13.27

A metal object at a temperature of 180°C is placed in a room at a constant temperature of 20°C. After 40 min the temperature of the object is 140°C. Find **(a)** the temperature of the metal after 2 h and **(b)** the time it will take to reach 30°C.

Solutions We begin by developing a formula that describes the temperature of the metal object at any time t. With $T_m = 20°\text{C}$, we have

$$\frac{dT}{dt} + kT = 20k$$

This differential equation is linear and its solution is

$$Te^{kt} = 20e^{kt} + C$$

$$\text{or} \qquad T = 20 + Ce^{-kt}$$

Since $T = 180$ when $t = 0$, we have $C = 160$. Substituting this value for C, we get

$$T = 20 + 160e^{-kt} \qquad (*)$$

At $t = 40$ min, we are given $T = 140$, so

$$140 = 20 + 160e^{-40k}$$

$$\frac{120}{160} = e^{-40k}$$

$$\text{thus,} \qquad k = -\frac{1}{40}\ln\left(\frac{120}{160}\right) \approx 0.0072$$

Substituting this value for k in equation $(*)$, we obtain

$$T = 20 + 160e^{-0.0072t}$$

where t is in minutes. We now have the necessary formula to answer the questions. **(a)** Since 2 h is 120 min, the temperature after 2 h is given by

$$T = 20 + 160e^{(-0.0072)(120)} \approx 87.44°\text{C}$$

After 2 h, the temperature has dropped to about 87.44°C. **(b)** We want to find t when $T = 30°\text{C}$. So,

$$30 = 20 + 160e^{-0.0072t}$$

$$\text{or} \qquad 10 = 160e^{-0.0072t}$$

$$\frac{1}{16} = e^{-0.0072t}$$

$$\text{and so} \qquad -0.0072t = \ln\frac{1}{16}$$

EXAMPLE 13.27 (Cont.)

Solving this for t, we find that

$$t = \frac{\ln 1/16}{-0.0072} \approx 385.08\,\text{min}$$

which is about 6 h 25 min 5 s. So, it will take about 6 h 25 min 5 s for the temperature to drop to 30°C.

Electric Circuits

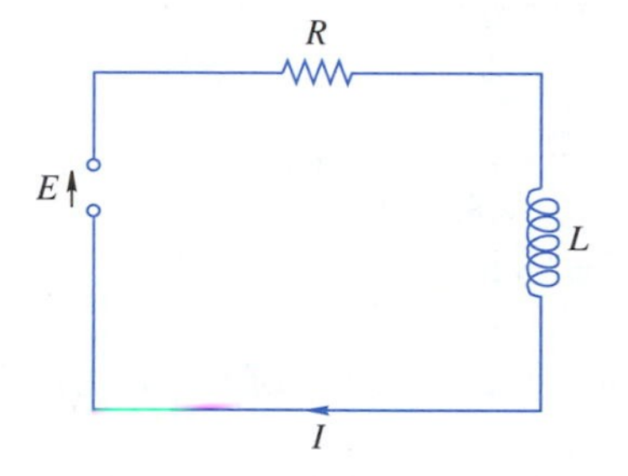

FIGURE 13.1

The amount of current I in an RL circuit as shown in Figure 13.1 is governed by the equation

$$\frac{dI}{dt} + \frac{R}{L}I = \frac{E}{L}$$

where R is the resistance in ohms (Ω), L an inductor in henries (H), and an electromotive force (emf) E is in volts (V).

An RC circuit consists of a resistance, a capacitance C in farads (F), an emf, and no inductance, as shown in Figure 13.2. The equation governing the amount of electrical charge q, in coulombs (C), on the capacitor is

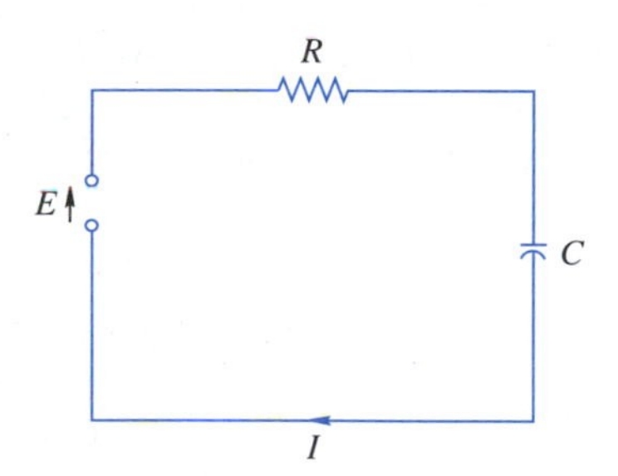

FIGURE 13.2

$$\frac{dq}{dt} + \frac{1}{RC}q = \frac{E}{R}$$

and the relationship between q and I is

$$I = \frac{dq}{dt}$$

Application

EXAMPLE 13.28

An RL circuit has an emf of 10 V, a resistance of 75 Ω, an inductance of 1 H, and no initial current. Find the current in the circuit at any time t.

Solution We have $E = 10$, $R = 75$, and $L = 1$, so

$$\frac{dI}{dt} + 75I = 10$$

This equation is linear. Its solution is

$$I = Ce^{-75t} + \frac{10}{75}$$

At $t = 0$, $I = 0$, and so

$$0 = Ce^{0} + \frac{10}{75}$$

which means that $C = -\frac{10}{75}$. Thus, the current at any time t is given by

$$I = -\frac{10}{75}e^{-75t} + \frac{10}{75}$$

In the previous equation, the quantity $-\frac{10}{75}e^{-75t}$ is called the **transient current**, since this quantity goes to zero as $t \to \infty$. The quantity $C = \frac{10}{75}$ in the same equation is called the **steady-state current**.

Exercise Set 13.5

Solve Exercises 1–16.

1. *Nuclear physics* A radioactive substance decays at a rate proportional to the amount present. If there are 100 g initially and 75 g after 15 days, what is the mass at any time t?
2. *Nuclear physics* What is the half-life of the substance in Exercise 1?
3. *Nuclear physics* A radioactive substance has a half-life of 1,000 years. How long will it take for 10% of the substance to decay?
4. *Medical technology* A bacteria culture is known to grow at a rate proportional to the amount present. After 1 h, 2,000 strands of the bacteria are observed in the culture. After 4 h, there are 6,000 strands. Find an expression for the number of strands of the bacteria present in the culture at any time t.
5. *Medical technology* How many strands were originally in the bacteria culture in Exercise 4?
6. *Medical technology* How frequently does the bacteria culture in Exercise 4 double?
7. *Archeology* The half-life of carbon-14 (C-14) is approximately 5,600 years. When an organism dies, its absorption of C-14, by either breathing or eating, stops. Thus, if we compare the proportional amount of C-14 present in a fossil or other object with the constant ratio in the atmosphere, we can get a close estimate of the object's age. If a fossilized bone is found to contain $\frac{1}{500}$ of the original amount of C-14, what is the age of the fossil?
8. *Optics* When a vertical beam of light passes through a transparent substance, the rate at which its intensity I decreases is proportional to the thickness of the medium in feet. In clear sea water, the intensity 3 ft below the surface of the water is 25% of the initial intensity of the beam. What is the intensity of the beam 18 ft below the surface?
9. *Thermal science* An object at a temperature of 10°C is placed outdoors where the temperature is 30°C. After 10 min the temperature of the object is 15°C.
   (a) Find a formula for the temperature of the object at time t?
   (b) How long does it take the object to reach a temperature of 22°C?
   (c) What is its temperature after 1 h?
10. *Thermal science* An object of unknown temperature is placed in a room of constant temperature 40°F. After 10 min, the temperature of the object is 0°F and after 20 min, the temperature is 10°F. What was the initial temperature of the object?
11. *Food science* A baker removes a cake from a 375°F oven and places it in a 40°F refrigerator to cool. After 15 min, the cake has cooled to 280°F. When the cake cools to 75°F, the icing can be applied. How long will the baker have to wait to apply the icing?
12. *Food science* An object is removed from a freezer with a temperature of −20°C and placed in a room with temperature 20°C. One minute later the object has warmed up to −16°C. How long will it take the object to reach a temperature of 15°C?
13. *Electricity* A generator with an emf of 100 V is connected in series with a $10-\Omega$ resistor and an inductor of 2 H. If the switch is closed at time $t = 0$, determine the current at any time t.
14. *Electricity* Suppose the 100-V generator in Exercise 13 is replaced with one that has an emf of $20\cos 5t$ V. What is the current at any time t?
15. *Electricity* An RL circuit has an emf of 5 V, $R = 50\,\Omega$, $L = 1$ H, and no initial current. What is the current in this circuit at any time t?
16. *Electricity* An RL circuit has an emf of $3\sin 2t$ V, $R = 10\,\Omega$, $L = 0.5$ H, and an initial current of 6 A. What is the current at any time t?

In Your Words

17. Write an application in your technology area of interest that requires you to solve a differential equation that involves growth and decay, cooling, or electric circuits. Give your problem to a classmate and see if he or she understands and can solve your problem using the techniques of this and earlier chapters. Rewrite the problem as necessary to remove any difficulties encountered by your classmate.

18. Write an application in your technology area of interest that requires you to solve a differential equation that involves growth and decay, cooling, or electric circuits. This application should involve a different area than your application in Exercise 17. As before, share your problem with a classmate and rewrite the problem as necessary to remove any difficulties encountered by the classmate.

13.6 MORE APPLICATIONS

In Section 13.5, we looked at applications of differential equations in the areas of growth and decay, cooling, and electric circuits. In this section, we will look at applications in three-additional areas.

Orthogonal Trajectories

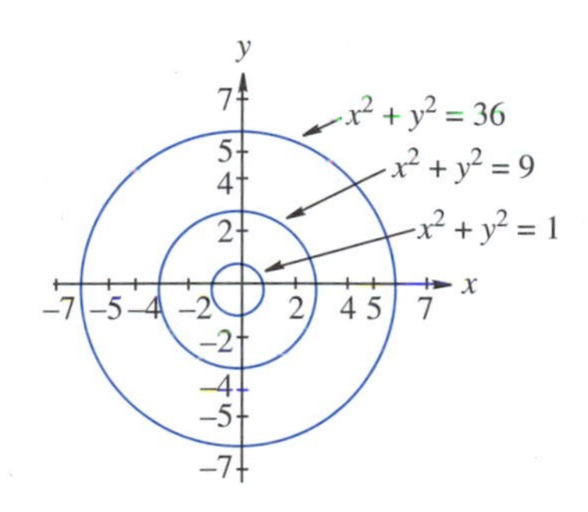

FIGURE 13.3a

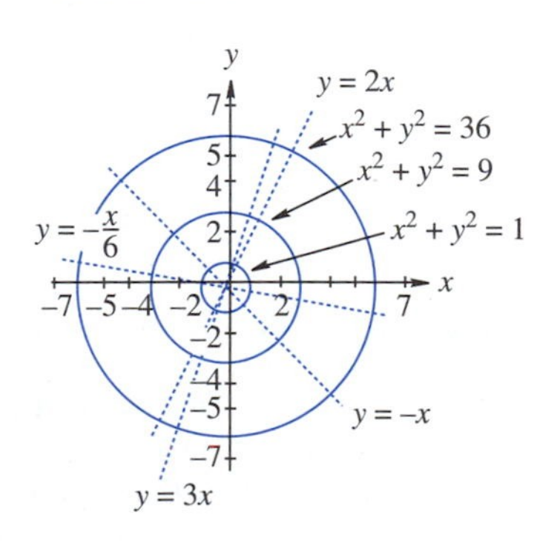

FIGURE 13.3b

A curve that intersects all members of a family of curves at right angles is called an **orthogonal trajectory**. A family of curves is a specified set of curves that satisfy some given conditions.

An example of a family of curves is shown in Figure 13.3a. This is a family of circles that have the origin as their center. The orthogonal trajectories for this family of circles are the family of straight lines through the origin. (Some of the lines are shown as dashed lines in Figure 13.3b). In the same way, the orthogonal trajectories of the family of straight lines passing through the origin are the circles with centers at the origin.

If we have a family of curves of the form $F(x, y, c) = 0$, then to find the orthogonal trajectories we will first implicitly differentiate this equation with respect to x and eliminate the constant c between this derived equation and the original one. We now have an equation in x, y, and y'. Solving this for y', we get an equation of the form

$$\frac{dy}{dx} = f(x, y)$$

The orthogonal trajectories are the solutions of

$$\frac{dy}{dx} = \frac{-1}{f(x, y)}$$

Application

EXAMPLE 13.29

Find the orthogonal trajectories for the family of curves formed by the equation $y = cx^2$.

Solution This family is the set of parabolas shown in Figure 13.4a and can be written as $F(x, y, c) = y - cx^2$. It consists of the parabolas symmetric to the y-axis with vertices at the origin. Using implicit differentiation gives the following results.

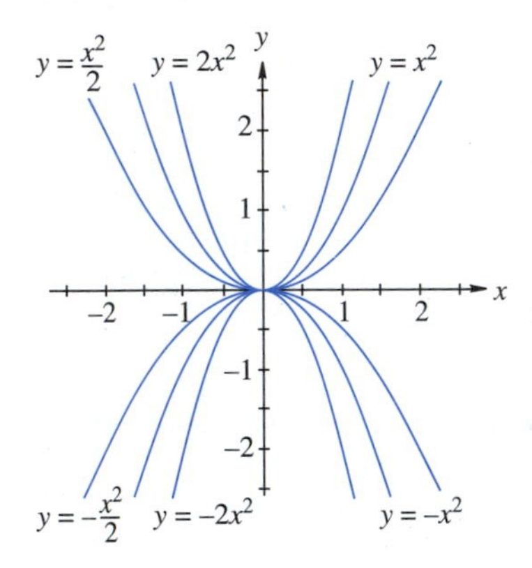

FIGURE 13.4a

$$y - cx^2 = 0$$
$$y' - 2cx = 0$$
$$y' = 2cx \qquad \text{or} \qquad \frac{dy}{dx} = 2cx \qquad (*)$$

Solving $y - cx^2 = 0$ for c, we obtain

$$c = \frac{y}{x^2}$$

Substituting this value for c into $(*)$, we have

$$\frac{dy}{dx} = 2\left(\frac{y}{x^2}\right)x = \frac{2y}{x}$$

Thus, we have $f(x, y) = \dfrac{2y}{x}$. The orthogonal trajectories satisfy the equation $\dfrac{dy}{dx} = \dfrac{-1}{f(x, y)}$, and so

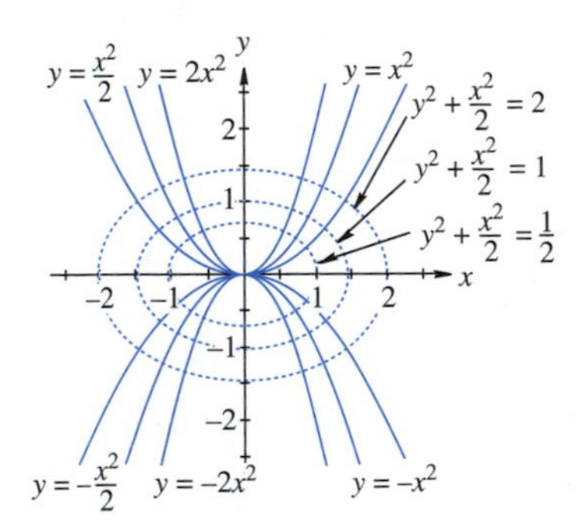

FIGURE 13.4b

$$\frac{dy}{dx} = \frac{-1}{2y/x} = -\frac{x}{2y}$$
$$\text{or} \qquad 2y\,dy + x\,dx = 0$$

The solution of this separable differential equation is

$$y^2 + \frac{1}{2}x^2 = k$$

This is the equation for a family of ellipses. Some members of this family are shown as the dashed lines in Figure 13.4b.

Falling Bodies with Resistance

The force F acting on a freely falling body in a vacuum is given by $F = mg$ were m is the mass and g is the force of gravity. Most of the time when an object falls it is affected by a resisting medium such as air. The resistance depends on the velocity and the size and shape of the object. At relatively low velocities, the resistance appears to be proportional to the velocity, or kv, where $k > 0$.

Newton's second law of motion states that the net force acting on a body is equal to the time rate of change of the momentum of the body. For an object of constant mass, this means that $F = m\frac{dv}{dt}$, where F is the net force of the body and v its velocity, both at time t. Because the resistance opposes the velocity, the net force is $F = mg - kv$. Substituting this into the equation $F = m\frac{dv}{dt}$, we obtain $mg - kv = m\frac{dv}{dt}$, or

$$\frac{dv}{dt} + \frac{kv}{m} = g$$

Application

EXAMPLE 13.30

A body of mass 10 slugs is dropped from a height of 200 ft with no initial velocity. If the retarding force is equal to 0.2 times the velocity, what is the body's velocity after 5 s?

Solution We have $m = 10$ slugs, $g = 32\,\text{ft/s}^2$, and $k = 0.2$, so

$$\frac{dv}{dt} + \frac{0.2v}{10} = 32$$

or

$$\frac{dv}{dt} = 32 - 0.02v$$

Separating variables, we obtain

$$\frac{dv}{32 - 0.02v} = dt$$

Integrating results in

$$-50\ln(32 - 0.02v) = t + C_1$$

For convenience, we will write $C_1 = -50\ln C$ in order to have

$$-50\ln(32 - 0.02v) = t - 50\ln C$$

or

$$\ln(32 - 0.02v) = -\frac{t}{50} + \ln C$$

Taking the exponential of both side results in

$$32 - 0.02v = Ce^{-t/50}$$

EXAMPLE 13.30 (Cont.)

Now, solving for v, we get

$$v = 50\left(32 - Ce^{-t/50}\right)$$

The object started from rest, so when $t = 0$, $v = 0$. Thus

$$0 = 50(32 - C)$$

and so $\quad C = 32$

As a result, the equation for the velocity is given by

$$v = 50(32 - 32e^{-t/50})$$

or $\quad v = 1{,}600\left(1 - e^{-t/50}\right)$

When $t = 5$, we obtain

$$\begin{aligned} v &= 1{,}600\left(1 - e^{-0.1}\right) \\ &= 1{,}600(1 - 0.9048) = 152.32\,\text{ft/s} \end{aligned}$$

So after 5 s, the body's velocity is 152.32 ft/s.

Mixture Problems

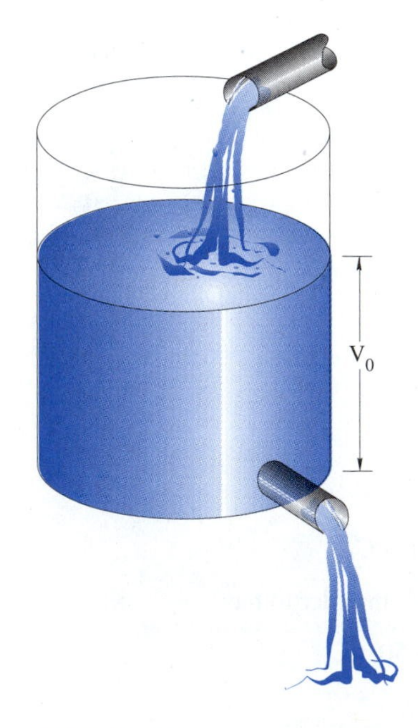

FIGURE 13.5

Suppose you have a tank that originally holds a known quantity V_0 of a solution that contains some quantity a of a dissolved chemical. Another solution containing the quantity b of the same dissolved chemical is added to the tank at the rate of r_1, while at the same time, the well-mixed solution is emptied at the rate r_2, as shown in Figure 13.5. We want to find the quantity of the chemical in the tank at any time t.

If we let $Q(t)$ represent the amount of the chemical at any time, then $\frac{dQ}{dt}$ represents the difference between the rate at which the chemical enters the tank and the rate at which it leaves. The chemical enters the tank at the rate of br_1 per unit of time. It leaves at the rate of ar_2 per unit of time. The volume in the tank at any time is $V_0 + r_1t - r_2t$, where V_0 represents the volume at $t = 0$. Thus, the concentration of the chemical at any time is $\frac{Q}{V_0 + r_1t - r_2t}$ and so it leaves at the rate of $r_2\left(\frac{Q}{V_0 + r_1t - r_2t}\right)$ per unit of time. This means that

$$\frac{dQ}{dt} = br_1 - r_2\left(\frac{Q}{V_0 + r_1t - r_2t}\right)$$

or

$$\frac{dQ}{dt} + \frac{r_2}{V_0 + (r_1 - r_2)t}Q = br_1$$

Application

EXAMPLE 13.31

Initially, 50 lb of salt are dissolved into a large tank that holds 300 gal of water. A brine solution of 2 lb/gal is pumped into the tank at the rate of 3 gal/min. A well-stirred solution is pumped out at the same rate. **(a)** How much salt is in the tank at any one time? **(b)** How much salt is there after 1 h?

Solutions

(a) Here we have $r_1 = r_2 = 3$ gal/min, $V_0 = 300$ gal, $a = 50$ lb, and $b = 2$ lb/gal. Thus,

$$\frac{dQ}{dt} + \frac{3}{300 + (3-3)t} Q = 2 \cdot 3$$

$$\frac{dQ}{dt} + \frac{Q}{100} = 6$$

The integrating factor is $e^{t/100}$, and so

$$\frac{d}{dt}\left(e^{t/100} Q\right) = 6e^{t/100}$$

and $$e^{t/100} Q = 600e^{t/100} + C$$

or $$Q = 600 + Ce^{-t/100}$$

We are given that $Q = 50$ when $t = 0$, so

$$50 = 600 + C$$

or $$C = -550$$

Thus, the quantity of salt that is in the tank at any time t, where t is in minutes, is given by

$$Q(t) = 600 - 550e^{-t/100}$$

(b) After 1 h, or when $t = 60$ min, we have

$$\begin{aligned} Q(60) &= 600 - 550e^{-60/100} \\ &= 600 - 550e^{-0.6} \\ &\approx 298.154 \text{ lb} \end{aligned}$$

Application

EXAMPLE 13.32

An office that measures $15' \times 20' \times 8'$ initially contains 0.1% carbon dioxide. Fresh air containing 0.025% carbon dioxide is pumped into the room at the rate of 100 ft^3/min. The old and new air are well mixed by the ventilation system and the well-mixed air is pumped out at the same rate. **(a)** How much carbon dioxide is in the room at any one time? **(b)** How much carbon dioxide is in the room after 1 h?

EXAMPLE 13.32 (Cont.)

FIGURE 13.6

Solutions

(a) We will let Q represent the amount of carbon dioxide (CO_2) in cubic feet in the room at time t min after beginning to pump fresh air into the room. We are given $r_1 = r_2 = 100\,\text{ft}^3/\text{min}$. The volume of the room is $V_0 = 15 \times 20 \times 8 = 2{,}400\,\text{ft}^3$, and $a = 0.1\%$ of $2{,}400 = 0.001 \times 2{,}400 = 2.4\,\text{ft}^3$ of CO_2. Finally, $br_1 = 0.025\% \times 100 = 0.00025 \times 100 = 0.025\,\text{ft}^3/\text{min}$. Thus, we have the following differential equation.

$$\frac{dQ}{dt} + \frac{100}{2{,}400 + (100 - 100)t} Q = 0.025$$

$$\frac{dQ}{dt} + \frac{Q}{24} = 0.025$$

$$\frac{dQ}{dt} = 0.025 - \frac{Q}{24}$$

$$= \frac{0.6 - Q}{24}$$

Separating variables, we obtain

$$\frac{dQ}{0.6 - Q} = \frac{dt}{24}$$

EXAMPLE 13.32 (Cont.)

Integration yields

$$-\ln(0.6-Q)=\frac{1}{24}t+C_1$$

or

$$\ln(0.6-Q)=-\frac{1}{24}t-C_1$$

So,

$$0.6-Q=e^{-t/24-C_1}$$

$$=Ce^{-t/24}$$

We are given that $Q=2.4\,\text{ft}^3$ when $t=0$, and so

$$0.6-2.4=C$$

or

$$C=-1.8$$

Thus, the quantity of carbon dioxide in the room at any time t, where t is in minutes, is given by

$$Q(t)=0.6+1.8e^{-t/24}$$

(b) After 1 h, or when $t=60$ min, we have

$$Q(60)=0.6+1.8e^{-60/24}$$

$$=0.6+1.8e^{-2.5}$$

$$\approx 0.7478\,\text{ft}^3$$

Dividing this by the volume of the room, we find that this amount of CO_2 represents $0.00031=0.031\%$ of the air in the room.

Exercise Set 13.6

In Exercises 1–4, find the orthogonal trajectories of the given family of curves. In each case, sketch some of the curves and their orthogonal trajectories. Remember to eliminate the constant from $\frac{dy}{dx}$ for the original curves.

1. $x^2+y^2=C$
2. $xy=C$
3. $y=Ce^x$
4. $x^2-y^2=C$

Solve Exercises 5–16.

5. *Energy science* In heat flow, the family of isothermal curves are the curves that join points of equal temperature. If the family of isothermal curves is given by the ellipses $x^2 + \frac{1}{2}y^2 = C^2$, find the curves along which the heat flows. (In other words, find the orthogonal trajectories of the ellipses.)
6. *Energy science* The lines of equal potential in a field of force are all at right angles to the line of force. In an electric field of force caused by charged particles, the lines of force are given by $x^2 + 2y^2 = C$. Find the equation of the line of equal potential.
7. *Physics* A body of mass of 10 kg (weight 98 N) is dropped from a height of 1000 m with no initial velocity. If the retarding force due to air resistance is 0.2 times the velocity, find a formula for the velocity at time t.
8. *Physics* What is the velocity of the object in Exercise 7 after 10 s?
9. *Physics* If the object in Exercise 7 is dropped with an initial velocity of 1 m/s, find **(a)** its velocity at any time t, **(b)** its velocity after 10 s, and **(c)** if the limiting velocity v_L is defined by $v_L = \dfrac{mg}{k}$, what is the limiting velocity?
10. *Physics* An object of mass 10 slugs (weight 320 lb) is dropped from a height of 1,000 ft with no initial velocity. The object encounters air resistance proportional to its velocity. The limiting velocity is known to be 320 ft/s. Find **(a)** the velocity of the object at any time t, **(b)** the position of the object at any time t, and **(c)** the time required for the object to reach a velocity of 160 ft/s.
11. *Environmental science* Fifty gallons of brine originally containing 25 lb of salt are in a tank into which 2 gal of water run each minute with the same amount of the well-stirred mixture running out each minute. How much salt remains in the tank after 15 min?
12. *Environmental science* One hundred gallons of brine originally containing 40 lb of salt are in a tank. Water enters the tank at the rate of 5 gal/min. The same amount of mixture from the tank leaves each minute. How much salt is in the tank after **(a)** 10 min? **(b)** 30 min? **(c)** 1 h?
13. *Environmental science* A tank is filled with 8 gal of brine in which 2 lb of salt is dissolved. Brine with a concentration of 3 lb of salt/gal enters the tank at 4 gal/min, and the well-stirred mixture leaves at the same rate. **(a)** Find the amount of salt at any time t. **(b)** How much salt is there after 8 min?
14. *Environmental science* A tank has 60 gal of brine with 2 lb of salt per gallon. A solution of 3 lb of salt per gallon enters at 2 gal/min and the well-stirred mixture leaves at the same rate. When will there be 150 lb of salt in the tank?
15. *Wastewater technology* A settling tank contains 10 kg of solid industrial waste dissolved in 200 m^3 of water. A solution of 1 kg of solid waste/5 m^3 of water is added at a rate of 10 m^3/hr and mixed to keep the tank's contents at a uniform concentration. **(a)** How much waste is in the tank at any time t, where t is in hr? **(b)** If the contents of the tank is being drained at a rate of 10 m^3/hr, how much dissolved solid will the tank contain after 3 hr?
16. *Medical technology* According to the **Gompertzian relation**, a solid tumor grows more slowly with the passage of time. One differential equation that meets these conditions is

$$\frac{dV}{dt} = \lambda e^{-\alpha t} V$$

where $V(t)$ is the volume of dividing cells at time t and λ and α are positive constants. Solve this differential equation.

In Your Words

17. Write an application in your technology area of interest that requires you to solve a differential equation that involves orthogonal trajectories, falling bodies with resistance, or mixture. Give your problem to a classmate and see if he or she understands and can solve your problem using the techniques of this and earlier chapters. Rewrite the problem as necessary to remove any difficulties encountered by your classmate.

18. Write an application in your technology area of interest that requires you to solve a differential equation that involves orthogonal trajectories, falling bodies with resistance, or mixture. This application should involve an area different from your application in Exercise 17. As before, share your problem with a classmate and rewrite the problem as necessary to remove any difficulties encountered by the classmate.

CHAPTER 13 REVIEW

Important Terms and Concepts

Boundary conditions
Boundary value problem
Cooling
Differential equation
- Degree
- Exact
- General solution
- Order
- Ordinary
- Partial
- Particular solution

Electrical circuits
Falling bodies with resistance
Growth and decay
Initial conditions
Initial value problem
Integrating factor
Linear first-order differential equation
Mixture
Newton's law of cooling
Orthogonal trajectories
Separation of variables

Review Exercises

In Exercises 1–16, find the general solution of the given differential equation.

1. $y' = x + y$
2. $x\,dx - y^2\,dy = 0$
3. $y' = y^2x^3$
4. $y' = 8y$
5. $e^x\,dx - 2y\,dy = 0$
6. $(y^2 - y)\,dx + x\,dy = 0$
7. $(y - xy^3)\,dx + (x - 2x^2y^2)\,dy = 0$
8. $(y+1)\,dx - x\,dy = 0$
9. $y\,dx + (2-x)dy = 0$
10. $(y + x^3y)\,dx + x\,dy = 0$
11. $y^2\,dx + xy\,dy = 0$
12. $dy = 3y\,dx + 6\,dx$
13. $y' - 4xy = x$
14. $y' + \dfrac{12y}{x} = x^{12}$
15. $y' + y = \sin x$
16. $y' - 7y = e^x$

In Exercises 17–20, find the particular solution of the given differential equation for the given boundary values.

17. $\cos x\,dx + y\,dy = 0$, $y = 2$ when $x = 0$
18. $(x^2 + y + y^2)dx - x\,dy = 0$, $y = \frac{\pi}{3}$ when $x = \frac{\pi}{3}$
19. $y' + \dfrac{2}{x}y = x$, $y = 0$ when $x = 1$
20. $dy + 6xy\,dx = 0$, $y = 5$ when $x = \pi$

Solve Exercises 21–30.

21. *Nuclear physics* A certain radioactive substance decays at a rate proportional to the amount present. An experiment finds that after 2 y, 10% of the original mass has decayed. Find **(a)** the expression for the mass at any time t, and **(b)** the half-life of the substance.
22. Find the orthogonal trajectories for the family of curves $y = cx^3$.
23. *Thermal science* An object at a temperature of 0°F is placed in a room with temperature 72°F. After 15 min, the object has reached a temperature of 20°F. Find **(a)** the amount of time it will take the object to reach a temperature of 50°F and **(b)** the temperature of the object after 1 h.
24. *Electricity* A 2-H inductor, a 60-Ω resistor, and a 25-V battery are connected in series. Find the current in the circuit at any time t, if the initial current is 0.
25. *Electricity* An RC circuit has an emf of $400\cos 2t$ V, a resistance of 80 Ω, and a capacitance of 10^{-2} F. When $t = 0$, $q = 0$. What is the change in the circuit at any time t?
26. *Physics* A 192-lb object has a limiting velocity of 16 ft/s when falling in air. The air provides a resisting force proportional to the object's velocity. If $v = 0$ when $t = 0$, what is **(a)** the velocity of the object when $t = 1$ s and **(b)** how long does it take the object to reach a velocity of 15 ft/s?
27. *Environmental science* A room of volume 2,000 ft^3 contains 0.1% carbon dioxide. Fresh air containing 0.035% carbon dioxide is pumped into the room at the rate of 100 ft^3/min and well-mixed air is pumped out at the same rate. **(a)** How much carbon dioxide is in the room at any one time? **(b)** How much carbon dioxide is in the room after 1 h? **(c)** How much carbon dioxide is in the room after 3 h?
28. *Environmental science* A building of volume 25,000 ft^3 contains 0.01% radon. Fresh air containing 0.0005% radon enters the building at the rate of 100 ft^3/min and well-mixed air is pumped out at the same rate. **(a)** How much radon is in the building at any one time? **(b)** How much radon is in the room after 1 h?
29. *Environmental science* A lake of volume 100 000 m^3 contains a 0.01% concentration of a certain pollutant. Because of a chemical spill, a river flowing into the lake becomes polluted. The polluted river contains 0.05% of the pollutant when it flows into the lake at the rate of 25 m^3/day. **(a)** If the water in the lake is well-mixed before it leaves, and it leaves at the same rate as it enters (25 m^3/day), find the concentration of the pollutant at time t. **(b)** When will the concentration of the pollutant be twice the initial concentration?
30. *Environmental science* A lake of volume 10 000 000 m^3 contains a 0.01% concentration of a certain pollutant. Because of a mining operation, a river flowing into the lake becomes polluted. The polluted river contains 0.074% of the pollutant when it flows into the lake at the rate of 500 m^3/day. **(a)** If the water in the lake is well-mixed before it leaves, and if it leaves at the rate of 500 m^3/day, find the concentration of the pollutant at time t. **(b)** When will the concentration of the pollutant be twice the initial concentration? **(c)** What is the maximum amount of pollutant that will be in the lake?

CHAPTER 13 TEST

Solve Exercises 1 and 2.

1. Show that $y = 4e^{2x} + 2e^{-3x}$ is a solution of the differential equation

$$\frac{d^2y}{dx^2} + \frac{dy}{dx} - 6y = 0$$

2. Determine if the equation $y^2\,dx + 2xy\,dy = 0$ is exact.

In Exercises 3–8, solve the given differential equation.

3. $y' = \dfrac{5x^4}{y^3}$
4. $(x^2+1)\,dx + x^2y^2\,dy = 0$
5. $(1+y^2)\,dx + (x^2y + y)\,dy = 0$
6. $\sin x \dfrac{dy}{dx} = 1 - 2y\cos x$
7. $x\,dy - y\,dx - x^6\,dx = 0$
8. $4(x\,dy + y\,dx) + 6x^2y\,dx = 0$, given that $x = 1$ when $y = e$.

Solve Exercises 9–12.

9. Find the orthogonal trajectories of the family $y = ce^{-2x}$.
10. What is the half-life of a radioactive substance, if 30% disappears in 40 y?
11. A mass of 5 slugs is falling under the influence of gravity. Find its velocity after 6 s, if it starts from rest and experiences a retarding force of 0.2 times its velocity.
12. An object cools from 50°C to 45°C in 20 min in air that is maintained at 30°C. What is the object's temperature after 1 h?

CHAPTER

14

Higher-Order Differential Equations

Objects attached to springs are subjected to several forces including retarding forces, such as friction. In Section 14.4, we will learn how to solve this type of harmonic motion problem by using differential equations.

Courtesy of Monroe Muffler/Brake Inc. Photo by John Myers.

In Chapter 13, we limited our discussion to differential equations of the first order. In this chapter, we will look at higher-order differential equations, particularly at second-order, linear, differential equations with constant coefficients.

We will begin by looking at higher-order homogeneous equations that have constant coefficients. Initially these equations will have distinct real roots, but later we will consider those with repeating real roots or with complex roots. The latter are particularly useful in electrical applications.

Next, we will learn how to solve non-homogeneous differential equations with undetermined coefficients. Finally, we will consider instances where the techniques of this chapter can be applied.

14.1 HIGHER-ORDER HOMOGENEOUS EQUATIONS WITH CONSTANT COEFFICIENTS

A linear differential equation of order n has the form

$$a_0(x)\frac{d^n y}{dx^n}+a_1(x)\frac{d^{n-1}y}{dx^{n-1}}+a_2(x)\frac{d^{n-2}y}{dx^{n-2}}+\cdots+a_{n-1}(x)\frac{dy}{dx}+a_n(x)y=F(x)$$

where $a_0(x)$, $a_1(x)$, $a_2(x)$, … , $a_n(x)$, and $F(x)$ all depend only on x and not on y. It is possible that some of the $a_i(x)$ or $F(x)$ elements may be constants. In fact, if $F(x)=0$, the equation is called **homogeneous**. If $F(x)\neq 0$, the equation is **nonhomogeneous**. In this chapter, we will look at methods for solving both homogeneous and nonhomogeneous equations. We will, however, restrict our attention to equations with constant coefficients.

Using the above notation, we have the following definition for a second-order, linear, homogeneous equation with constant coefficients.

Second-Order, Linear, Homogeneous Equation with Constant Coefficients

A second-order, linear, homogeneous equation with constant coefficients is of the form

$$a_0\frac{d^2y}{dx^2}+a_1\frac{dy}{dx}+a_2y=0$$

where a_0, a_1, and a_2 are constants.

It is often more convenient to use the notation D and D^2, to indicate the **operations** of taking the first and second derivative. Thus, $Dy=\frac{dy}{dx}$ and $D^2y=\frac{d^2y}{dx^2}$. The symbols D and D^2 are called the **operators** because they define an operation to be performed.

Operator Notation

A second-order, linear, homogeneous equation with constant coefficients is written in **operator notation** as

$$a_0D^2y+a_1Dy+a_2y=0$$

where a_0, a_1, and a_2 are constants.

This is often written as $\left(a_0D^2+a_1D+a_0\right)y=0$. It could also be written as $a_0y''+a_1y'+a_2y=0$.

We now see how operator notation can be used to solve a differential equation.

EXAMPLE 14.1

Solve the differential equation $(D^2+5D+6)y=0$.

Solution We will begin by factoring the operators as $(D+3)(D+2)y=0$. If we let $z=(D+2)y$, then $(D+3)z=0$. We can solve this equation by using separation of variables. Thus, we have

$$\frac{dz}{dx}+3z=0$$

or $$\frac{dz}{z}+3\,dx=0$$

and so $$\ln z+3x=\ln c_1, \text{ when } c=\ln c_1$$

Solving for z we get

$$\ln z=\ln c_1-3x$$

or
$$\begin{aligned} z&=e^{\ln c_1-3x}\\ &=e^{\ln c_1}e^{-3x}\\ &=c_1e^{-3x} \end{aligned}$$

Replacing z by $(D+2)y$, we have

$$(D+2)y=c_1e^{-3x}$$

This is a linear equation of the first order. Thus, we have

$$dy+2y\,dx=c_1e^{-3x}\,dx$$

The integrating factor is $e^{\int 2\,dx}=e^{2x}$ and so

$$\begin{aligned} ye^{2x}&=\int c_1e^{-3x}e^{2x}\,dx\\ &=\int c_1e^{-x}\,dx\\ &=-c_1e^{-x}+c_2 \end{aligned}$$

Multiplying both sides by e^{-2x}, we get

$$\begin{aligned} y&=-c_1e^{-3x}+c_2e^{-2x}\\ &=ke^{-3x}+c_2e^{-2x}, \text{ where } k=-c_1 \end{aligned}$$

While we have found a general solution for this given differential equation, we would like to find an easier method for determining this solution. It appears that $y=e^{mx}$ is a solution where m is an unknown constant. We substitute this solution into the original equation of Example 14.1 and get

$$\begin{aligned} D^2e^{mx} + 5De^{mx} + 6e^{mx} &= 0 \\ m^2e^{mx} + 5me^{mx} + 6e^{mx} &= 0 \\ (m^2 + 5m + 6)e^{mx} &= 0 \end{aligned}$$

Now, $e^{mx} > 0$ for all values of m. Thus, the equation is true only if $m^2 + 5m + 6 = 0$ or $(m+3)(m+2) = 0$, which occurs when $m = -3$ or $m = -2$.

The equation $a_0m^2 + a_1m + a_2 = 0$ is called the **auxiliary equation** of $a_0D^2y + a_1Dy + a_2y = 0$. Notice the similarity of the two equations.

There are two roots, m_1 and m_2, of the auxiliary equation and there are two arbitrary constants in the solution of $a_0D^2y + a_1Dy + a_2y = 0$. These factors direct us to the general solution of the second-order, linear, homogeneous equation with constant coefficients. This general solution is

$$y = c_1e^{m_1x} + c_2e^{m_2x}$$

where m_1 and m_2 are the solutions to the auxiliary equation. Notice how this corresponds to our solution to Example 14.1.

General Solution of a Second-Order, Linear, Homogeneous Differential Equation

If m_1 and m_2 are distinct real roots of the auxiliary equation $a_0m^2 + a_1m + a_2 = 0$, then

$$y = c_1e^{m_1x} + c_2e^{m_2x}$$

is the **general solution** of the differential equation.

EXAMPLE 14.2

Solve the differential equation $D^2y - 3Dy + 2y = 0$.

Solution The auxiliary equation is

$$m^2 - 3m + 2 = 0$$

Factoring this equation, we get

$$(m-2)(m-1) = 0$$

and so $m_1 = 2$ and $m_2 = 1$

The general solution is

$$y = c_1e^{2x} + c_2e^{x}$$ ▪

EXAMPLE 14.3

Solve the differential equation $D^2y - 5Dy = 0$.

Solution The auxiliary equation is

$$m^2 - 5m = 0$$

Solving this, we get $m(m-5) = 0$, so $m_1 = 0$ and $m_2 = 5$. Thus, the general solution is

$$y = c_1e^{0x} + c_2e^{5x} = c_1 + c_2e^{5x}$$

EXAMPLE 14.4

Solve the differential equation $y'' - y' = 12y$.

Solution We will first rewrite this differential equation using operator notation as follows:

$$D^2y - Dy - 12y = 0$$

and thus the auxiliary equation is

$$m^2 - m - 12 = 0$$

This factors into $(m-4)(m+3) = 0$, so the roots are $m_1 = 4$ and $m_2 = -3$. The general solution becomes

$$y = c_1e^{4x} + c_2e^{-3x}$$

EXAMPLE 14.5

Solve the differential equation $2\frac{d^2y}{dx^2} + 4\frac{dy}{dx} - y = 0$ and find the particular solution that satisfies $\frac{dy}{dx} = 6$ and $y = 0$, when $x = 0$.

Solution We will first write this differential equation in operator notation:

$$2D^2y + 4Dy - y = 0$$

The auxiliary equation is

$$2m^2 + 4m - 1 = 0$$

This equation does not factor. Using the quadratic formula, we get

$$\begin{aligned} m &= \frac{-4 \pm \sqrt{4^2 - 4(2)(-1)}}{2(2)} \\ &= \frac{-4 \pm \sqrt{24}}{4} = \frac{-2 \pm \sqrt{6}}{2} \end{aligned}$$

Thus, the general solution is

$$y = c_1e^{\frac{-2+\sqrt{6}}{2}x} + c_2e^{\frac{-2-\sqrt{6}}{3}x}$$

EXAMPLE 14.5 (Cont.)

Now, let's find the particular solution. When $x = 0$, $y = c_1 + c_2 = 0$, so $c_2 = -c_1$.

Differentiating the general solution, we get

$$y' = \frac{-2+\sqrt{6}}{2} c_1 e^{\frac{-2+\sqrt{6}}{2}x} + \frac{-2-\sqrt{6}}{2} c_2 e^{\frac{-2-\sqrt{6}}{2}x}$$

and when $x = 0$,

$$y' = \frac{-2+\sqrt{6}}{2} c_1 + \frac{-2-\sqrt{6}}{2} c_2$$

Since $c_2 = -c_1$, we have

$$y' = \frac{-2+\sqrt{6}}{2} c_1 + \frac{2+\sqrt{6}}{2} c_1 = \sqrt{6} c_1$$

We are given $\frac{dy}{dx} = 6$ and, since $y' = \sqrt{6}c_1$, we have $\sqrt{6}c_1 = 6$ and so $c_1 = \frac{6}{\sqrt{6}} = \sqrt{6}$ and $c_2 = -\sqrt{6}$. Thus, the particular solution is

$$y = \sqrt{6} e^{\frac{-2+\sqrt{6}}{2}x} - \sqrt{6} e^{\frac{-2-\sqrt{6}}{2}x}$$

This method works for homogeneous linear differential equations of order higher than two. We summarize these results in the following box.

nth-Order Linear Homogeneous Differential Equations

A homogeneous equation of order n with constant coefficients is of the form

$$a_0 \frac{d^n y}{dx^n} + a_1 \frac{d^{n-1} y}{dx^{n-1}} + \cdots + a_{n-1} \frac{dy}{dx} + a_n y = 0$$

and in operator form is

$$a_0 D^n y + a_1 D^{n-1} y + \cdots + a_{n-1} Dy + a_n y = 0$$

The **auxiliary equation** is of the form

$$a_0 m^n + a_1 m^{n-1} + a_2 m^{n-2} + \cdots + a_{n-1} m + a_n = 0$$

If $m_1, m_2, m_3, \ldots, m_n$ are the roots of this auxiliary equation, then

$$y = c_1 e^{m_1 x} + c_2 e^{m_2 x} + \cdots + c_n e^{m_n x}$$

is the **general solution** of the differential equation.

EXAMPLE 14.6

Solve the differential equation

$$D^3y - 3D^2y + 2Dy = 0$$

Solution The auxiliary equation is

$$m^3 - 3m^2 + 2m = 0$$

This factors into $m(m-2)(m-1) = 0$, so $m_1 = 0$, $m_2 = 2$, and $m_3 = 1$. The general solution is

$$\begin{aligned} y &= c_1e^{0x} + c_2e^{2x} + c_3e^x \\ &= c_1 + c_2e^{2x} + c_3e^x \end{aligned}$$

EXAMPLE 14.7

Solve the differential equation

$$(D^3 - 2D^2 - D + 2)y = 0$$

Solution This has the auxiliary equation

$$m^3 - 2m^2 - m + 2 = 0$$

The only possible rational roots are ± 1 and ± 2. Using synthetic division, or one of the other methods we used earlier, we get

$$\begin{array}{r|rrrr} 1 & 1 & -2 & -1 & 2 \\ & & 1 & -1 & -2 \\ \hline & 1 & -1 & -2 & 0 \end{array}$$

and so 1 is a root and the equation factors into

$$\begin{aligned} m^3 - 2m^2 - m + 2 &= (m-1)(m^2 - m - 2) \\ &= (m-1)(m-2)(m+1) \end{aligned}$$

So, $m_1 = 1$, $m_2 = 2$, and $m_3 = -1$. Thus the general solution of the differential equation is

$$y = c_1e^x + c_2e^{2x} + c_3e^{-x}$$

Exercise Set 14.1

In Exercises 1–20, solve the given differential equation.

1. $(D^2 - 7D + 10)y = 0$
2. $(D^2 + 7D + 12)y = 0$
3. $D^2y - 8Dy + 12y = 0$
4. $6D^2y + Dy = y$
5. $2\dfrac{d^2y}{dx^2} + 9\dfrac{dy}{dx} - 5y = 0$
6. $\dfrac{d^2y}{dx^2} = 9y$
7. $4D^2y - 7Dy - 2y = 0$
8. $4y'' + 11y' - 3y = 0$
9. $y'' - 5y' + 4y = 0$

10. $\frac{d^2y}{dx^2}+4\frac{dy}{dx}+3y=0$
11. $y''-y=0$
12. $D^2y=Dy+30y$
13. $y''=7y$
14. $D^2y+Dy-y=0$
15. $2\frac{d^2y}{dx^2}+\frac{dy}{dx}-y=0$
16. $D^2y+2Dy-y=0$
17. $D^3y-6D^2y+11Dy-6y=0$
18. $D^3y-D^2y-4Dy+4y=0$
19. $y'''+6y''+11y'+6y=0$
20. $D^3y-D^2y-17Dy=15y$

In Exercises 21–28, find the particular solutions for the given differential equation.

21. $D^2y+2Dy-15y=0$, $y=2$ and $y'=6$ when $x=0$
22. $3D^2y-14Dy+8y=0$, $y=3$ and $y'=12$ when $x=0$
23. $D^2y+2Dy-8y=0$, $y=0$ and $y'=6$ when $x=0$
24. $D^2y+2Dy-8y=0$, $y=3$ and $y'=-12$ when $x=0$
25. $y''+2y'-15y=0$, $y=5$ and $y'=-1$ when $x=0$
26. $y''-4y'-32y=0$, $y=15$ and $y'=0$ when $x=0$
27. $y'''+y''-4y'-4y=0$, $y=3$, $y'=7$, and $y''=-3$ when $x=0$
28. $y'''+y''-17y'+15y=0$, $y=0$, $y'=-12$, and $y''=0$ when $x=0$

In Your Words

29. Without looking in the text, explain how you can tell if an equation is a second-order, linear, homogeneous differential equation.
30. Without looking in the text, describe how to find the general solution of a second-order, linear, homogeneous differential equation.

14.2 AUXILIARY EQUATIONS WITH REPEATED OR COMPLEX ROOTS

In Section 14.1, all roots of the auxiliary equation were distinct real numbers. In this section, we will look at repeated and complex roots of auxiliary equations.

Repeated Roots

The following example will show what to do when there are multiple (or repeated) roots.

EXAMPLE 14.8

Solve the differential equation $(D^2-6D+9)y=0$.

Solution The auxiliary equation is

$$m^2-6m+9=(m-3)^2=0$$

This equation has the multiple roots of 3 and 3. Using the method of the last section, we would write $y=c_1e^{3x}+c_2e^{3x}=(c_1+c_2)e^{3x}$ or $y=ce^{3x}$. But, this has only one arbitrary constant.

While we will not show it here, you can check that $y=xe^{3x}$ is also a solution. So, the general solution is given by the linear combination

$$y=c_1e^{3x}+c_2xe^{3x}$$

What happens if one root is repeated more than once? A similar thing occurs. For example, if the auxiliary equation is $(m-4)^3 = 0$, then there are three roots that are the same such as 4, 4, and 4. The general solution is given by $y = c_1e^{4x} + c_2xe^{4x} + c_3x^2e^{4x}$.

General Solution with Repeated Roots

If the auxiliary equation has the real root m, which repeats n times, then

$$\begin{aligned} y &= c_1e^{mx} + c_2xe^{mx} + c_3x^2e^{mx} + \cdots + c_nx^{n-1}e^{mx} \\ &= (c_1 + c_2x + c_3x^2 + \cdots + c_nx^{n-1})e^{mx} \end{aligned}$$

is the general solution of the equation.

EXAMPLE 14.9

Solve the differential equation $D^6y + D^5y - 4D^4y - 2D^3y + 5D^2y + Dy = 2y$.

Solution This differential equation has the auxiliary equation $m^6 + m^5 - 4m^4 - 2m^3 + 5m^2 + m - 2 = 0$. This auxiliary equation factors into $(m-1)^3(m+1)^2(m+2) = 0$ and so it has roots of 1, 1, 1, -1, -1, and -2. The general solution of this differential equation is

$$y = (c_1 + c_2x + c_3x^2)e^x + (c_4 + c_5x)e^{-x} + c_6e^{-2x}$$

Complex Roots

Now, suppose that the auxiliary equation $a_0m^2 + a_1m + a_2 = 0$ has real coefficients and complex roots. From the quadratic equation we know that the roots are complex conjugates and are of the form $m = a \pm bj$. From the technique learned in Section 14.1, we can show that the general solution is of the form

$$\begin{aligned} y &= c_1e^{(a+bj)x} + c_2e^{(a-bj)x} \\ &= e^{ax}(c_1e^{bjx} + c_2e^{-bjx}) \end{aligned}$$

The relationship between the exponential and polar form of a complex number states that $re^{j\theta} = r(\cos\theta + j\sin\theta)$. When we apply this to the equation above, we get

$$y = e^{ax}\left(c_1[\cos bx + j\sin bx] + c_2[\cos(-bx) + j\sin(-bx)]\right)$$

But, since $\cos(-\alpha) = \cos\alpha$ and $\sin(-\alpha) = -\sin\alpha$, we can rewrite this as

$$\begin{aligned} y &= e^{ax}[c_1(\cos bx + j\sin bx) + c_2(\cos bx - j\sin bx)] \\ &= e^{ax}(c_1\cos bx + c_2\cos bx + jc_1\sin bx - jc_2\sin bx) \\ &= e^{ax}[(c_1 + c_2)\cos bx + j(c_1 - c_2)\sin bx] \\ &= e^{ax}(c_3\cos bx + c_4\sin bx) \end{aligned}$$

where $\quad c_3 = c_1 + c_2$ and $c_4 = j(c_1 - c_2)$

Complex Roots

If the auxiliary equation $a_0m^2 + a_1m + a_2 = 0$, with real coefficients, has complex roots of the form $a \pm bj$, the general solution of the differential equation is

$$y = e^{ax}(c_1 \cos bx + c_2 \sin bx)$$

(These are different c_1 and c_2 than we used previously. We use them here simply to indicate two arbitrary constants.)

EXAMPLE 14.10

Solve the differential equation

$$D^2y + 4Dy + 13y = 0$$

Solution This has the auxiliary equation $m^2 + 4m + 13 = 0$. Using the quadratic formula, we see that

$$\begin{aligned} m &= \frac{-4 \pm \sqrt{4^2 - 4(13)}}{2} \\ &= \frac{-4 \pm \sqrt{-36}}{2} = \frac{-4 \pm 6j}{2} \\ &= -2 \pm 3j \end{aligned}$$

are the roots of the auxiliary equation. Here, $a = -2$ and $b = 3$, and the general solution is

$$y = e^{-2x}(c_1 \cos 3x + c_2 \sin 3x)$$

■

EXAMPLE 14.11

Solve the differential equation $\frac{d^2y}{dx} - \frac{dy}{dx} + 1 = 0$.

Solution Rewriting this in operator form as

$$D^2y - Dy + y = 0$$

we see that it has the auxiliary equation

$$m^2 - m + 1 = 0$$

Using the quadratic equation, we get

$$m = \frac{1 \pm \sqrt{1-4}}{2} = \frac{1 \pm \sqrt{-3}}{2} = \frac{1}{2} \pm j\frac{\sqrt{3}}{2}$$

and so $a = \frac{1}{2}$ and $b = \frac{\sqrt{3}}{2}$.

EXAMPLE 14.11 (Cont.)

The general solution to this differential equation is

$$e^{x/2}\left(c_1 \cos\frac{\sqrt{3}}{2}x + c_2 \sin\frac{\sqrt{3}}{2}x\right)$$

EXAMPLE 14.12

Solve $y''' - 6y'' + 2y' + 36y = 0$.

Solution This has the auxiliary equation $m^3 - 6m^2 + 2m + 36 = 0$, which has the roots -2 and $4 \pm j\sqrt{2}$. The general solution is

$$y = c_1 e^{-2x} + e^{4x}(c_2 \cos\sqrt{2}x + c_3 \sin\sqrt{2}x)$$

Notice how we were able to combine the real and the complex roots into the general solution.

EXAMPLE 14.13

Solve $y'' + 3y' + 4y = 0$, if $y = 2$ and $y' = 4$ when $x = 0$.

Solution The auxiliary equation is $m^2 + 3m + 4 = 0$ and has roots

$$m = \frac{-3 \pm \sqrt{9-16}}{2} = \frac{-3}{2} \pm j\frac{\sqrt{7}}{2}$$

The general solution is

$$y = e^{-3x/2}\left(c_1 \cos\frac{\sqrt{7}}{2}x + c_2 \sin\frac{\sqrt{7}}{2}x\right)$$

When $x = 0$, this becomes $y = c_1$. Since we were given $y = 2$, then $c_1 = 2$.

The other condition involves y'. If we differentiate the general solution, we get

$$\begin{aligned} y' &= e^{-3x/2}\left(-c_1\frac{\sqrt{7}}{2}\sin\frac{\sqrt{7}}{2}x + c_2\frac{\sqrt{7}}{2}\cos\frac{\sqrt{7}}{2}x\right) \\ &\quad -\frac{3}{2}e^{-3x/2}\left(c_1 \cos\frac{\sqrt{7}}{2}x + c_2 \sin\frac{\sqrt{7}}{2}x\right) \end{aligned}$$

When $x = 0$, this becomes

$$y' = c_2\frac{\sqrt{7}}{2} - \frac{3}{2}c_1$$

We know that $y' = 4$ when $x = 0$ and that $c_1 = 2$. Making these substitutions, we get

$$4 = c_2\frac{\sqrt{7}}{2} - \frac{3}{2}(2)$$

EXAMPLE 14.13 (Cont.)

$$4 = c_2\frac{\sqrt{7}}{2} - 3$$

or $$7 = c_2\frac{\sqrt{7}}{2}$$

and so $$c_2 = \frac{2}{\sqrt{7}}\cdot 7 = 2\sqrt{7}$$

The particular solution to this differential equation is

$$y = e^{-3x/2}\left(2\cos\frac{\sqrt{7}}{2}x + 2\sqrt{7}\sin\frac{\sqrt{7}}{2}x\right)$$

Exercise Set 14.2

In Exercises 1–26, solve the given differential equation.

1. $D^2y + 2Dy + y = 0$
2. $(D^2 - 2D + 1)y = 0$
3. $D^2y - 4Dy + 4y = 0$
4. $y'' - 10y' + 25y = 0$
5. $9y'' - 6y' + y = 0$
6. $16y'' + 8y' + y = 0$
7. $4y'' + 12y' + 9y = 0$
8. $9y'' - 12y' + 4y = 0$
9. $y'' + 4y' + 5y = 0$
10. $y'' + 4y = 0$
11. $D^2y + 9Dy = 0$
12. $D^3y = 0$
13. $D^3y = y$
14. $\frac{d^3y}{dx^2} + y = 0$
15. $\frac{d^3y}{dx^3} + 8y = 0$
16. $\frac{d^3y}{dx^3} - 27y = 0$
17. $y''' - 6y'' + 11y' - 6y = 0$
18. $D^4y - 9D^2y + 20y = 0$
19. $D^4y + 8D^3y + 24D^2y + 32Dy + 16y = 0$
20. $y''' - y'' + y' - y = 0$
21. $(D-1)^2(D+2)^3y = 0$
22. $(D-2)^4(D+5)y = 0$
23. $(D-3)^2(D^2 - 6D - 9)y = 0$
24. $(D+1)^2(D^2 + 4D + 9)y = 0$
25. $(3D^2 - 2D + 1)y = 0$
26. $\frac{d^2y}{dx^2} - 2\frac{dy}{dx} + 3y = 0$

In Exercises 27–34, find the particular solution of the given differential equation with the given conditions.

27. $y'' - 2y' + 5y = 0$, $y = 4$ and $y' = 7$ when $x = 0$
28. $y'' + 8y' + 16y = 0$, $y = 4$ and $y' = -6$ when $x = 0$
29. $(4D^2 - 12D + 9)y = 0$, $y = 1$ when $x = 0$ and $y = 0$ when $x = 1$
30. $(D^2 - 4D + 5)y = 0$, $y = 1$ when $x = 0$ and $y = 3e^{\pi}$ when $x = \frac{\pi}{2}$
31. $y'' + 2y' + y = 0$, $y = 1$ and $y' = -3$ when $x = 0$
32. $y'' - 10y' + 25y = 0$, $y = -4$ and $y' = 4$ when $x = 0$
33. $(D^2 - 6D + 34)y = 0$, $y = 4$ and $y' = -3$ when $x = 0$
34. $(D^2 + 10D + 29)y = 0$, $y = 0.5$ and $y' = 1.5$ when $x = 0$

In Your Words

35. Suppose that the auxiliary equation to a higher-order linear homogeneous differential equation has repeated roots. Explain how you find the general solution to this differential equation.

36. Describe how you would find the general solution to a differential equation if its auxiliary equation has real coefficients and nonreal complex roots.

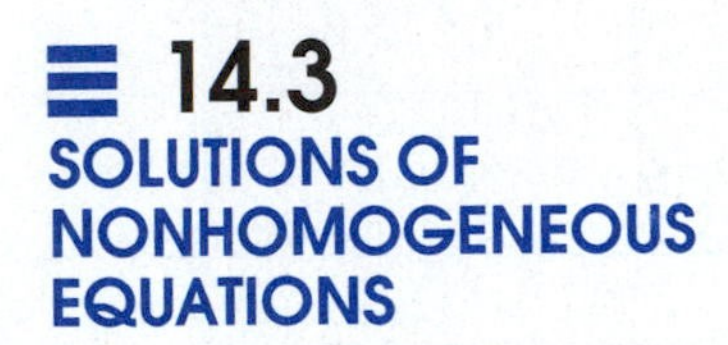

14.3 SOLUTIONS OF NONHOMOGENEOUS EQUATIONS

In Sections 14.1 and 14.2, we focused on homogeneous linear equations of the form

$$a_0D^2y + a_1Dy + a_2y = 0$$

where a_0, a_1, and a_2 were constants. Next, we look at nonhomogeneous linear equations of the form

$$a_0D^2y + a_1Dy + a_2y = f(x)$$

where a_0, a_1, and a_2 are all constants and $f(x)$ is either a constant or a function of x. We will find a way in which to find a general solution, y, which satisfies this equation.

In Sections 14.1 and 14.2, we were able to find a general solution for $a_0D^2y + a_1Dy + a_2y = 0$. This solution is called the **complementary solution** and is denoted y_c. Thus

$$(a_0D^2 + a_1D + a_2)y_c = 0$$

Now, suppose we have found a **particular solution**, y_p, to $a_0D^2y + a_1Dy + a_2y = f(x)$. This means that $(a_0D^2 + a_1D + a_2)y_p = f(x)$. It turns out that $y = y_c + y_p$ is a general solution of the nonhomogeneous linear equation.

Several methods exist for finding y_p. We will use the **method of undetermined coefficients**. This method begins by guessing the form of y_p up to arbitrary multiplicative constants. These constants are then evaluated by substituting the proposed form of y_p into the given differential equation and equating the coefficients of like terms. The guess that we make for y_p is an educated guess. In general, there are the four cases which have been outlined in the following box.

Method of Undetermined Coefficients

Case 1: $f(x) = p_n(x)$, an nth degree polynomial in x. Assume

$$y_p = A_n x^n + A_{n-1}x^{n-1} + \cdots + A_1 x + A_0$$

We will need to determine $A_n, A_{n-1}, \ldots, A_1$, and A_0.

Case 2: $f(x) = e^{ax} p_n(x)$, where a is a known constant and $p_n(x)$ as in case 1. Then, we will assume that

$$y_p = e^{ax}(A_n x^n + A_{n-1}x^{n-1} + \cdots + A_1 x + A_0)$$

As in case 1, we will need to determine the values of $A_n, A_{n-1}, \ldots, A_1$, and A_0.

Case 3: $f(x) = e^{ax} p_n(x) \sin bx$, where a and b are known constants. In this case, we assume that

$$\begin{aligned} y_p &= e^{ax} \sin bx(A_n x^n + A_{n-1}x^{n-1} + \cdots + A_1 x + A_0) \\ &\quad + e^{ax} \cos bx(B_n x^n + B_{n-1}x^{n-1} + \cdots + B_1 x + B_0) \end{aligned}$$

where $A_n, A_{n-1}, \ldots A_1, A_0$, and $B_n, B_{n-1}, \ldots, B_1, B_0$ are values we will need to determine.

Case 4: $f(x) = e^{ax} p_n(x) \cos bx$, where a and b are known constants. In this case, we make the same assumption for y_p as we made in case 3.

Note It is possible for a problem to fit more than one of these cases as we will see in Example 14.17.

Now, this is a very lengthy explanation and is somewhat confusing. Let's work several examples to help make it clear.

EXAMPLE 14.14

Solve the differential equation

$$D^2 y - 3Dy + 2y = 4x^2 - 5$$

Solution First we find the complementary solution. In Example 14.2, we found that $y_c = c_1 e^{2x} + c_2 e^x$.

Here we have $f(x) = 4x^2 - 5$, a second-degree polynomial, and so this is a case 1 problem. We assume that $y_p = A_2 x^2 + A_1 x + A_0$. Thus, $y_p' = 2A_2 x + A_1$ and $y_p'' = 2A_2$. Substituting these results into the differential equation, we have

$$\begin{aligned} (2A_2) - 3(2A_2 x + A_1) + 2(A_2 x^2 + A_1 x + A_0) &= 4x^2 - 5 \\ 2A_2 - 6A_2 x + 3A_1 + 2A_2 x^2 + 2A_1 x + 2A_0 &= 4x^2 - 5 \end{aligned}$$

or

$$2A_2 x^2 + (2A_1 - 6A_2)x + (2A_2 - 3A_1 + 2A_0) = 4x^2 + 0x - 5$$

EXAMPLE 14.14 (Cont.)

Thus, $2A_2x^2 = 4x^2$, so $A_2 = 2$. We also have $(2A_1 - 6A_2)x = 0x$. Since $A_2 = 2$, this means that $2A_1 - 6(2) = 2A_1 - 12 = 0$, and so $A_1 = 6$.

Finally, $2A_2 - 3A_1 + 2A_0 = -5$. Since $A_2 = 2$ and $A_1 = 6$, we have $2(2) - 3(6) + 2A_0 = 4 - 18 + 2A_0 = -5$ or $2A_0 = 9$, so $A_0 = \frac{9}{2}$.

Thus $y_p = 2x^2 + 6x + \frac{9}{2}$, and the general solution is

$$y = c_1e^{2x} + c_2e^x + 2x^2 + 6x + \frac{9}{2}$$

EXAMPLE 14.15

Solve the differential equation $y'' - 5y' = 3e^{7x}$.

Solution In Example 14.3, we found that the complementary solution of this differential equation is $y_c = c_1 + c_2e^{5x}$.

Here we have $f(x) = 3e^{7x}$, which is a case 2 problem with $a = 7$ and $p_n(x) = 3$, a constant polynomial. According to case 2, we assume that

$$y_p = A_0e^{7x}$$

So, we have $y'_p = 7A_0e^{7x}$ and $y''_p = 49A_0e^{7x}$. Substituting these values into the original differential equation gives

$$\begin{aligned} y'' - 5y' &= 3e^{7x} \\ 49A_0e^{7x} - 5\left(7A_0e^{7x}\right) &= \\ 49A_0e^{7x} - 35A_0e^{7x} &= \\ \text{or} \qquad 14A_0e^{7x} &= 3e^{7x} \end{aligned}$$

Thus $A_0 = \frac{3}{14}$, and the general solution is

$$y = c_1 + c_2e^{5x} + \frac{3}{14}e^{7x}$$

EXAMPLE 14.16

Solve the differential equation

$$(D^2 - 6D + 9)y = \cos 2x$$

Solution In Example 14.8, we found that $y_c = c_1e^{3x} + c_2xe^{3x}$. In this example, $f(x) = \cos 2x$, which is the form in case 4 with $a = 0$, $p_n(x) = 1$, and $b = 2$. Thus, we assume the solution to be

$$y_p = A_0 \sin 2x + B_0 \cos 2x$$

Differentiating, we get $y'_p = 2A_0 \cos 2x - 2B_0 \sin 2x$ and $y''_p = -4A_0 \sin 2x - 4B_0 \cos 2x$. Substituting into the differential equation, we obtain

$$\begin{aligned} (-4A_0 \sin 2x - 4B_0 \cos 2x) - 6(2A_0 \cos 2x - 2B_0 \sin 2x) \\ + 9(A_0 \sin 2x + B_0 \cos 2x) = \cos 2x \end{aligned}$$

EXAMPLE 14.16 (Cont.)

Collecting terms, we get

$$(-4A_0+12B_0+9A_0)\sin 2x+(-4B_0-12A_0+9B_0)\cos 2x=\cos 2x$$
$$\text{or} \qquad (5A_0+12B_0)\sin 2x+(5B_0-12A_0)\cos 2x=\cos 2x$$

Thus, $5A_0+12B_0=0$ and $5B_0-12A_0=1$. Solving these simultaneously, we have $A_0=\frac{-12}{169}$ and $B_0=\frac{5}{169}$. This means that $y_p=\frac{-12}{169}\sin 2x+\frac{5}{169}\cos 2x$, and the general solution is

$$y=c_1e^{3x}+c_2xe^{3x}-\frac{12}{169}\sin 2x+\frac{5}{169}\cos 2x$$

EXAMPLE 14.17

Solve $y''+4y=3xe^x+5x-1$.

Solution From Exercise 10 in Exercise Set 14.3, we see that

$$y_c=c_1\cos 2x+c_2\sin 2x$$

In this example, $f(x)=3xe^x+5x-1$. If we write this as $f(x)=g(x)+h(x)$ where $g(x)=3xe^x$ and $h(x)=5x-1$, then h is a case 1 form and g is a case 2 form. Accordingly, we assume that for $g(x)$ we have a solution of the form $e^x(A_1x+A_0)$ and for $h(x)$ the solution is of the form B_1x+B_0. Thus, we try

$$y_p=A_1xe^x+A_0e^x+B_1x+B_0$$

Differentiating, we get $y_p'=A_1e^x+A_1xe^x+A_0e^x+B_1$ and $y_p''=2A_1e^x+A_1xe^x+A_0e^x$. Substituting into the differential equation, we have

$$2A_1e^x+A_1xe^x+A_0e^x+4(A_1xe^x+A_0e^x+B_1x+B_0)=3xe^x+5x-1$$

or

$$(A_1+4A_1)xe^x+(2A_1+A_0+4A_0)e^x+4B_1x+4B_0=3xe^x+5x-1$$

Thus, $5A_1=3$, so $A_1=\frac{3}{5}$ and $2A_1+5A_0=0$, and so $A_0=-\frac{6}{25}$, $B_1=\frac{5}{4}$, and $B_0=-\frac{1}{4}$. The general solution of this differential equation is

$$y=c_1\cos 2x+c_2\sin 2x+\frac{3}{5}xe^x-\frac{6}{25}e^x+\frac{5}{4}x-\frac{1}{4}$$

There are some exceptions to the four cases just given. However, we will not explore them in this book.

Exercise Set 14.3

Solve each of the differential equations in Exercises 1–20.

1. $(D^2 - 10D + 25)y = 4$
2. $(D^2 - 10D + 25)y = 4x^2$
3. $(D^2 - 10D + 25)y = e^{3x}$
4. $(D^2 - 10D + 25)y = 2xe^{3x}$
5. $(D^2 - 10D + 25)y = 10 + e^x$
6. $(D^2 - 10D + 25)y = 3\sin 2x$
7. $(D^2 - 10D + 25)y = 29\sin 2x$
8. $(D^2 - 9)y = 3\cos x$
9. $(D^2 - 4)y = x^2e^x - 3x$
10. $(D^2 + 4)y = 6x + 3$
11. $(D^2 + 9)y = 4\cos x + 2\sin x$
12. $y'' - 5y' + 6y = 9x + 2e^x$
13. $y'' + y = 2e^{3x}$
14. $(D^2 + 2D + 1)y = 4\sin 2x$
15. $y'' - 4y = 8x^2$
16. $y'' + 4y' + 5y = e^{-x} + 10x$
17. $D^2y + y = 4 + \cos 2x$
18. $3D^2y + 2Dy - y = 4 + 2x + 6x^2$
19. $3D^2y + 2Dy + y = 4 + 2x + 6x^2$
20. $y'' + 9y' - y = x^2 + 6e^{2x}$

In Exercises 21–28, find the particular solution of the given differential equation for the given conditions.

21. $y'' - 2y' + y = x^2 - 1$, $y = 7$ and $y' = 15$ when $x = 0$
22. $y'' - 2y' + y = 10$, $y = 20$ and $y' = 5$ when $x = 0$
23. $y'' - y' - 2y = \sin 2x$, $y = 1$ and $y' = \frac{7}{4}$ when $x = 0$
24. $y'' - y' - 2y = e^{3x}$, $y = 2$ and $y' = 11$ when $x = 0$
25. $y'' + 2y' + y = 3x^2 + 2x - 1$, $y = 3$ and $y' = 5$ when $x = 0$
26. $y'' - 5y' - 6y = e^{3x}$, $y = 2$ and $y' = 1$ when $x = 0$
27. $y'' + 9y = 8\cos x$, $y = -1$ and $y' = 1$ when $x = \frac{\pi}{2}$
28. $y'' - 5y' + 6y = e^x(2x - 3)$, $y = 1$ and $y' = 3$ when $x = 0$

In Your Words

29. What are the differences among a particular solution, complementary solution, and general solution to a nonhomogeneous linear equation?
30. What is the method of undetermined coefficients? When do you use it? How do you use it?

14.4 APPLICATIONS

In Sections 14.1–14.3, we learned how to solve some higher-order linear differential equations with constant coefficients. In this section, we will look at two applications of second-order equations.

Simple Harmonic Motion

Perhaps the easiest example of simple harmonic motion is that of a vibrating spring. Consider a spring that is hung from a fixed support as shown in Figure 14.1. Suppose that this spring has an object of mass m suspended from the spring. When the object is at rest we say that it is in the **equilibrium position**. We normally use a vertical coordinate axis with the origin, $y = 0$, at the location of the object in the equilibrium position. This means that y will represent the vertical displacement of the object.

You should remember that, according to Hooke's law, a force F on m due to the stretched or compressed spring is $F = kx$. Newton's second law states that the force acting on a body is equal to its mass times its acceleration. This is equivalent to $F = m\dfrac{d^2x}{dt^2}$. Combining these two values for the force, we have

$$m\frac{d^2x}{dt^2} = -kx$$

where the negative sign is used to indicate that the forces act in opposite directions. An object whose displacement from equilibrium satisfies this equation is said to be executing **simple harmonic motion**.

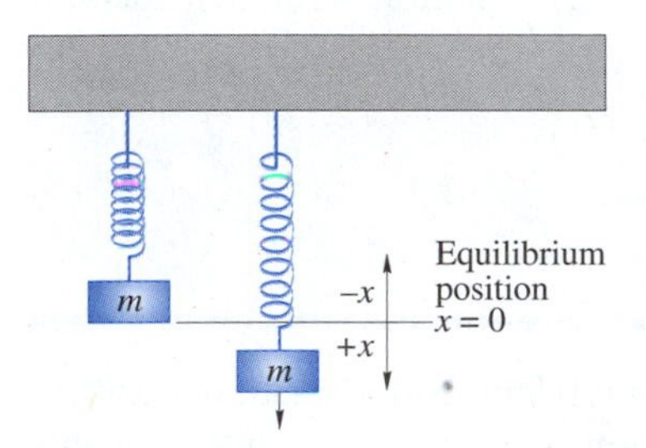

FIGURE 14.1

This, however, is not very practical. The equations that result from solving this differential equation imply that once the object begins to oscillate it will keep moving up and down forever. Our experience tells us that this is not true. What does happen is that the oscillations gradually die down until the object comes to rest at the equilibrium position.

The problem with this new formula is that it ignores the effects of friction. Thus, a reasonable assumption is that there is a retarding force proportional to the velocity. We will represent this force by $-\ell\dfrac{dx}{dt}$, where $\ell > 0$. When this is included, the new formula becomes

$$m\frac{d^2x}{dt} = -kx - \ell\frac{dx}{dt}$$

$$\text{or} \qquad m\frac{d^2x}{dt} + \ell\frac{dx}{dt} + kx = 0$$

If we divide this last equation by m and let $\dfrac{\ell}{m} = 2b$ and $\dfrac{k}{m} = \omega^2$, then we can write the equation in operator notation as

$$(D^2 + 2bD + \omega^2)x = 0$$

The auxiliary equation $m^2 + 2bm + \omega^2 = 0$ has the roots

$$\begin{aligned} m &= \frac{-2b \pm \sqrt{4b^2 - 4\omega^2}}{2} \\ &= -b \pm \sqrt{b^2 - \omega^2} \end{aligned}$$

There are three possible situations depending on the relative sizes of b^2 and ω^2.

If $b^2 > \omega^2$, we say the motion is **overdamped** and the general solution is of the form

$$x = Ae^{-at} + Be^{-ct}$$

where $a = b + \sqrt{b^2 - \omega^2}$ and $c = b - \sqrt{b^2 - \omega^2}$.

If $b^2 = \omega^2$, we say that the motion is **critically damped** and the general solution is

$$x = (A + Bt)e^{-bt}$$

In both overdamped and critically damped motion, the mass of the object m is subject to such a large retarding force that it slows down and returns to equilibrium.

If $b^2 < \omega^2$, then $\sqrt{b^2 - \omega^2}$ is imaginary. If we let $\beta_j = \sqrt{b^2 - \omega^2}$, then the solution is in the form

$$x = e^{-bt}(c_1 \cos \beta t + c_2 \sin \beta t)$$

This motion is known as **underdamped** or **oscillatory motion**. The value $\omega = \sqrt{\frac{k}{m}}$ is defined as the **natural frequency** of the system.

If an external force $f(t)$ acts on the spring, then the equation, written in operator notation, becomes

$$(D^2 + 2bD + \omega^2) = f(t)$$

This is an example of **forced oscillation**.

Application

EXAMPLE 14.18

A 10-kg object is attached to a spring stretching it 0.7 m from its natural position. The object is started in motion from its equilibrium position with an initial velocity of 1 m/s downward. Find the position of the object as a function of time, if the force due to air resistance is 90 times the velocity newtons.

Solution We begin by determining the spring constant. The force stretching the spring is mg. We are given $m = 10$ kg and g, in the metric system, is 9.8 m/s^2. Thus, $F = 10(9.8) = k(0.7)$ and $k = 140$ N/m. We are also given ℓ as 90, so $2b = \frac{\ell}{m}$ or $b = 4.5$, and $\omega^2 = \frac{k}{m} = 14$. Thus, the auxiliary equation is

$$m^2 + 9m + 14 = 0$$

This has the roots 2 and 7, and $b^2 = 20.25 > 14 = \omega^2$, so the general solution is

$$x = Ae^{-2t} + Be^{-7t}$$

Since the object is started in motion from its equilibrium position, we know that when $t = 0$, $x = 0$, and so $A + B = 0$.

The initial velocity is $+1$, so when $t = 0$, $x' = 1 = -2Ae^{-2t} - 7Be^{-7t}$. Thus, $-2A - 7B = 1$. Solving the two equations for A and B, we get $B = -\frac{1}{5}$ and $A = \frac{1}{5}$. Thus,

$$x = \frac{1}{5}(e^{-2t} - e^{-7t})$$

■

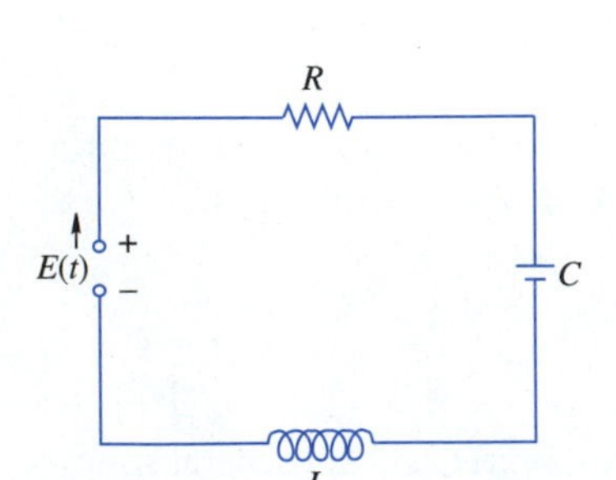

FIGURE 14.2

Electric Circuits

There is an electrical situation that parallels the one just described. Consider an RLC circuit similar to the one shown in Figure 14.2. The voltage across an inductor, a resistor, and a capacitor is given by

$$L\frac{di}{dt}, \quad Ri, \quad \text{and} \quad \frac{q}{C}$$

respectively. We know that $i = \dfrac{dq}{dt}$ and are then able to rewrite these three expressions as

$$L\frac{d^2q}{dt^2}, \quad R\frac{dq}{dt}, \quad \text{and} \quad \frac{q}{C}$$

If these are connected in series with a generator, then the electromotive force (emf) is expressed as $E(t)$. Kirchhoff's second law states that the sum of the voltage around a closed loop equals the sum of the emf of this loop. Applying Kirchhoff's law to the loop in Figure 14.2, we get the result

$$E(t) = L\frac{d^2q}{dt^2} + R\frac{dq}{dt} + \frac{q}{C}$$

Application

EXAMPLE 14.19

An RLC circuit has $R = 16\,\Omega$, $C = 10^{-2}\,\text{F}$, $L = 1\,\text{H}$, and an emf of $36\sin 10t$. If there is no initial charge on the capacitor but an initial current of 5 A when the voltage is first applied $(t = 0)$, what is the expression for the charge on the capacitor?

Solution Using the new formula, we have

$$36\sin 10t = \frac{d^2q}{dt^2} + 16\frac{dq}{dt} + 100q \qquad (*)$$

To find the complementary solution q_c, we must solve the auxiliary equation

$$m^2 + 16m + 100 = 0$$

and so,

$$\begin{aligned} m &= \frac{-16 \pm \sqrt{16^2 - 4(100)}}{2} \\ &= \frac{-16 \pm 12j}{2} = -8 \pm 6j \end{aligned}$$

These are complex roots, so

$$q_c = e^{-8t}(c_1\cos 6t + c_2\sin 6t)$$

For this particular solution we have a case 3 situation from the method of undetermined coefficients (see Section 14.3), and so

$$q_p = A\sin 10t + B\cos 10t$$

with $q_p' = 10A\cos 10t - 10B\sin 10t$ and $q_p'' = -100A\sin 10t - 100B\cos 10t$. Substituting these values into $(*)$, we get

$$\begin{aligned} &(-100A\sin 10t - 100B\cos 10t) + 16(10A\cos 10t - 10B\sin 10t) \\ &\quad + 100(A\sin 10t + B\cos 10t) = 36\sin 10t \end{aligned}$$

EXAMPLE 14.19 (Cont.)

Collecting terms, we obtain

$$(-100A - 160B + 100A)\sin 10t + (-100B + 160A + 100B)\cos 10t = 36\sin 10t$$

Thus, $-160B = 36$ or $B = -0.225$, and $A = 0$.

Then the general solution for $q(t)$ is

$$\begin{aligned} q(t) &= q_c + q_p \\ &= e^{-8t}(c_1\cos 6t + c_2\sin 6t) - 0.225\cos 10t \end{aligned}$$

We are told that there is no initial charge on the capacitor (so $q(0) = 0$) and that the initial current is 5 A (so $q'(0) = 5$). Applying these initial conditions, we obtain

$$q(0) = 0 = c_1 - 0.225 \text{ or } c_1 = 0.225$$

Differentiating produces

$$\begin{aligned} q'(t) = e^{-8t}&(-6c_1\sin 6t + 6c_2\cos 6t) \\ &- 8e^{8t}(c_1\cos 6t + c_2\sin 6t) + 2.25\sin 10t \end{aligned}$$

and when $t = 0$ and $c_1 = 0.225$, we see that

$$\begin{aligned} 5 &= 6c_2 - 8c_1 \\ &= 6c_2 - 8(0.225) \end{aligned}$$

Solving this for c_2, we get $c_2 = \frac{6.8}{6} \approx 1.1333$.

The particular solution for this circuit is

$$q(t) = e^{-8t}(0.225\cos 6t + 1.1333\sin 6t) - 0.225\cos 10t$$

Look more closely at the solution to the last example. The complementary solution, q_c, contains the exponential factor e^{-8t}. As t gets larger this factor approaches 0. After just 2 s, $e^{-8t} = 0.000\,000\,11$. This means that q_c dies out very quickly. In fact, when $t = 2$ s, $q_c = -0.000\,000\,051$. On the other hand, q_p does not contain this factor of e^{-8t}. Thus, since q_c will quickly approach 0, the only significant term left will be q_p. As a result, q_c is called the **transient part** of the solution and q_p is called the **steady-state solution**. The **steady state current** is $i = \dfrac{dq_p}{dt}$.

EXAMPLE 14.20

Find the steady-state current for the RLC circuit in Example 14.19.

Solution The steady-state solution is $q_p = -0.225\cos 10t$. The steady-state current is

$$i = \frac{dq_p}{dt} = 2.25\sin 10t$$

Exercise Set 14.4

Solve Exercises 1–16.

1. *Physics* A 20-lb weight is suspended from a spring. As a result, the spring stretches 5 in. What is the spring constant?
2. *Physics* A 15-lb weight stretches a spring 3 in. What is the spring constant?
3. *Physics* A 5-kg mass is hung from a spring. It stretches the spring 0.2 m. What is the spring constant?
4. *Physics* An 8-kg mass is suspended from a spring and as a result the spring stretches 0.4 m. What is the spring constant?
5. *Physics* The 20-lb weight in Exercise 1 is started in motion from the equilibrium position with an initial velocity of 4 ft/s upward. Find the motion of the weight, if the force due to air resistance is twice the instantaneous velocity.
6. *Physics* The 15-lb weight in Exercise 2 is started in motion from the equilibrium position with an initial velocity of 2 ft/s downward. Find the motion of the weight, if the force due to air resistance is equal to the instantaneous velocity.
7. *Physics* The object in Exercise 3 is set in motion from the equilibrium position with an initial velocity of 2 m/s downward. Find the equation for the motion of the object, if the force due to air resistance is 40 times the instantaneous velocity newtons.
8. *Physics* The object in Exercise 4 is started in motion from the equilibrium position with an initial velocity of 3 m/s upward. Find the equation for the motion of the object, if the force in newtons due to air resistance is 80 times the instantaneous velocity.
9. *Electronics* An RLC circuit with $R = 6\,\Omega$, $C = 10^{-2}$ F, and $L = 0.1$ H has an emf of $50\sin 100t$. Find the charge as a function of time, if $q(0) = 0$ and $i(0) = 0$.
10. *Electronic* An RLC circuit has $R = 5\,\Omega$, $C = 10^{-2}$ F, $L = 0.125$ H, and an emf of $\sin t$. What is the charge as a function of time, if $q(0) = 0$ and $i(0) = 0$.
11. *Electronics* A 0.1-H inductor, a 4×10^{-3}-F capacitor, and a generator with emf of $180\cos 60t$ are connected in series. **(a)** What is the charge as a function of time, if $q(0) = i(0) = 0$? **(b)** What is the steady-state current?
12. *Electronics* An RLC circuit has $R = 10\,\Omega$, $L = 1$ H, $C = 10^{-2}$ F, and an emf of $50\cos 10t$. **(a)** What is the charge as a function of time? **(b)** What is the steady-state current?
13. *Medical technology* Diabetes is usually diagnosed by means of a glucose tolerance test (GTT). In this test the patient comes to the hospital after an overnight fast and is given a large dose of glucose. During the next three to five hours, several measurements are made of the concentration of glucose in the patient's blood, and these measurements are used in the diagnosis of diabetes. If $g(t)$ measures the patient's deviation from an optimum glucose level and if $t = 0$ is the time in minutes at which the glucose load has been completely ingested, then g satisfies

$$\frac{d^2g}{dt^2} + 2\alpha\frac{dg}{dt} + {\omega_0}^2 g = 0$$

where α and ω_0 are positive constants. Solve this differential equation if $\alpha^2 - {\omega_0}^2 < 0$.

14. *Space technology* The differential equation

$$\frac{d^2z}{d\theta^2} + z = k$$

is used in the study of planetary motion to determine r, the distance from the planet to the sun, where $z = \frac{1}{r}$. If k represents a constant and θ is the angle through which the planet has traveled, find the general solution for this differential equation.

15. *Electronics* Find the equation for $q(t)$ in an RLC series circuit with $L = 1.2$ H, $R = 5.0\,\Omega$, $C = 0.025$ F, and $V(t) = 15$ V.
16. *Electronics* A capacitor with capacitance is $\frac{1}{505}$ F, and inductor with coefficient of inductance is $\frac{1}{20}$ H, and a resistor with resistance $1\,\Omega$ are connected in series. If at $t = 0$, $i = 0$ and the charge on the capacitor is 1 C, find **(a)** the charge and **(b)** the current in the circuit due to the discharge of the capacitor when $t = 0.01$ s.

In Your Words

17. Write an application in your technology area of interest that requires you to solve a differential equation that involves simple harmonic motion, electric circuits, or some area that involves setting up and solving higher-order differential equations. Give your problem to a classmate and see if he or she understands and can solve your problem using the techniques of this and earlier chapters. Rewrite the problem as necessary to remove any difficulties encountered by your classmate.
18. Write an application in your technology area of interest that requires you to solve a higher-order differential equation in an area different from your application in Exercise 17. As before, give your problem to a classmate and see if he or she understands and can solve your problem using the techniques of this and earlier chapters. Rewrite the problem as necessary to remove any of your classmate's difficulties.

CHAPTER 14 REVIEW

Important Terms and Concepts

Auxiliary equation
- Complex roots
- Distinct real roots
- Repeated roots

Electric circuits
Homogeneous equation
Method of undetermined coefficients
Nonhomogeneous equation
Operator
Simple harmonic motion
- Critically damped motion
- Damped motion
- Oscillatory motion
- Overdamped motion
- Undamped motion

Review Exercises

Solve each of the following differential equations.

1. $D^2y = 0$
2. $(D^2 - 9)y = 0$
3. $(D^2 - 5D + 6)y = 0$
4. $(D^2 - 5D - 14)y = 0$
5. $(D^2 - 6D + 25)y = 0$
6. $(D^2 + 2D - 2)y = 0$
7. $(D^3 - 8)y = 0$
8. $y''' + y'' - y' - y = 0$
9. $(D^2 + 3D - 4)y = 9x^2$
10. $(D^2 - 4D - 5)y = 3e^{2x}$
11. $(D^2 + 7D + 12)y = \cos 5x$
12. $(D^2 - 6D + 8)y = 6x^2 - 2$
13. $(D^2 + 4D + 4) = 4x + e^{2x}$
14. $(D^2 + 4D - 9)y = x \sin 2x$

Solve Exercises 15 and 16.

15. *Physics* An object of mass 4 kg stretches a spring 0.2 m. The object is set in motion with an initial velocity of 1 m/s downward and the force of the air resistance is $10\frac{dx}{dt}$. What is the equation that describes the motion of the object?

16. *Electronics* An RLC circuit has $R = 6\,\Omega$, $C = 4 \times 10^{-4}$ F, $L = 1$ H, and an emf of $16\cos 10t$. **(a)** If $q(0) = i(0) = 0$, what is the charge as a function of time? **(b)** What is the steady-state current?

CHAPTER 14 TEST

What is the auxiliary equation of the differential equation $(5D^2 + 2D - 3)y = 0$?

In Exercises 1–9, solve the differential equation.

1. $(D^2 - 25)y = 0$
2. $(D^2 + 8D - 20)y = 0$
3. $(D - 5)^3(D^2 + 4)y = 0$
4. $(D^2 + 6D + 13)y = 0$
5. $(D^2 + 6D - 7)y = 0$
6. $(D^2 + 6D - 7)y = 6e^{2x}$
7. $(D^2 + 6D - 7)y = 3\cos 4x$
8. $(D^2 + 6D - 7)y = 8x$
9. $(D^2 + 6D - 7)y = e^{4x}$; if $x = 0$, then $y = \frac{4}{11}$ and $Dy = \frac{5}{11}$.
10. Find an expression for the charge and the steady-state current of the RLC circuit with $L = 0.5$ H, $C = 2 \times 10^{-3}$ F, $R = 30\,\Omega$, and an emf of $5\sin 10t$.

CHAPTER

15

Numerical Methods and Laplace Transforms

Laplace transforms are particularly helpful because they allow the use of algebra to solve differential equations. In Section 15.6, we will use Laplace transforms to solve electric circuit problems.

Courtesy of International Business Machines Corporation

In Chapters 13 and 14, we studied some ways to solve differential equations. We looked at a few of the many kinds of differential equations. There are many others, some of which have important applications. For these equations several numerical methods have been developed.

A numerical method for solving a differential equation is a procedure which gives approximate solutions at certain points. Numerical methods use only the four basic operations of addition, subtraction, multiplication and division or the procedure of evaluating functions. A great deal of time and computation is often needed to get the desired solution using numerical methods. For this reason, numerical methods are ideal for the use of calculators and computers.

After we complete our work with numerical methods we will look at Laplace transforms. These transforms provide an algebraic method of obtaining a particular solution to a differential equation with known initial conditions.

15.1 EULER'S OR THE INCREMENT METHOD

The first numerical method for solving differential equations goes by two names. Some books call it **Euler's method** and some call it the **increment method**.

Suppose we have a first-order differential equations of the form $y' = f(x, y)$. Further, we have an initial value for this equation. This means that the graph of this equation would pass through some known point (x_0, y_0). We will find approximate solutions at $x_1, x_2, x_3, x_4, \ldots$, where the difference between any two successive values of x, such as $x_1 - x_0$, $x_2 - x_1$, or $x_n - x_{n-1}$ is the same value $\Delta x = dx$. In keeping with our work with the definition of the derivative, we will let $dx = h$.

Since $y' = \frac{dy}{dx} = f(x, y)$, we have $dy = f(x, y)\,dx$ or $dy = f(x, y)h$. Now, in order to find the first approximation, y_1, we find $y_0 + dy_0 = y_0 + f(x_0, y_0)h$. Thus,

$$y_1 = y_0 + f(x_0, y_0)h$$

We know that $x_1 = x_0 + h$, and since we just found the approximation y_1, we can approximate $y_2 = y_1 + f(x_1, y_1)h$. In general, we have the following formula for Euler's method.

Euler's Method for Solving Differential Equations

$$y_{n+1} = y_n + f(x_n, y_n)h$$

This is usually set up in table form, as we will show in Example 15.1.

EXAMPLE 15.1

Find an approximate solution to the differential equation $y' = x - y$, if the curve passes through the point $(0, 2)$.

Solution For this problem $x_0 = 0$, $y_0 = 2$, and $f(x, y) = x - y$. We select $h = 0.1$ and obtain

$$\begin{aligned} y_{n+1} &= y_n + f(x_n, y_n)h \\ &= y_n + (x_n - y_n)h \end{aligned}$$

Thus,

$$\begin{aligned} y_1 &= y_0 + (x_0 - y_0)h \\ &= 2 + (0 - 2)(0.1) \\ &= 2 + (-0.2) \\ &= 1.8 \end{aligned}$$

EXAMPLE 15.1 (Cont.)

To find y_2, we let $x_1 = x_0 + h = 0 + 0.1 = 0.1$, and so

$$\begin{aligned} y_2 &= y_1 + (x_1 - y_1)h \\ &= 1.8 + (0.1 - 1.8)(0.1) \\ &= 1.8 + (-0.17) \\ &= 1.63 \end{aligned}$$

In this example, we are able to check our work. We have the differential equation $y' = x - y$. Using the technique from Chapter 32, you can determine that this has the particular solution

$$y = 3e^{-x} + x - 1$$

If we check the values when $x = 0.1$, we see that $y = 1.8145$, and when $x = 0.2$, we get $y = 1.6562$. The following table summarizes our results.

n	x_n	y_n	Correct solution $y = 3e^{-x} + x - 1$
0	0	2	2
1	0.1	1.8	1.8145
2	0.2	1.63	1.6562
3	0.3	1.487	1.5225
4	0.4	1.3683	1.4110
5	0.5	1.27147	1.3196

The approximation we get is directly related to the size of h. In order to show that better values are obtained by using smaller values of h, the following table is given. Three values of h were used for this table: 0.1, 0.05, and 0.01. (The first value for h, 0.1, is the same one we just used.) We have left out the intermediate values of x.

		y		
x	$h = 0.1$	$h = 0.05$	$h = 0.01$	Correct solution $y = 3e^{-x} + x - 1$
0	2	2	2	2
0.1	1.8	1.8075	1.81315	1.81451
0.2	1.63	1.64352	1.65372	1.65619
0.3	1.487	1.50528	1.51910	1.52245
0.4	1.3683	1.39026	1.40692	1.41096
0.5	1.27147	1.29621	1.31502	1.31959
0.6	1.19432	1.22108	1.24147	1.24643

EXAMPLE 15.2

Use Euler's method to solve the differential equation $y' = 3x^2 - y^3$, if the curve passes through the point $(1, 2)$.

Solution This example doesn't specify the size of h. We will first let $h = 0.1$ and then let it be 0.05. If $x_0 = 1$, $y_0 = 2$, and $h = 0.1$, then

$$\begin{aligned} y_1 = y_0 + f(x_0, y_0)h &= 2 + (3 \cdot 1^2 - 2^3)(0.1) \\ &= 2 + (3 - 8)(0.1) \\ &= 1.5 \end{aligned}$$

If $h = 0.05$, then

$$\begin{aligned} y_1 &= 2 + (3 - 8)(0.05) \\ &= 1.75 \end{aligned}$$

Notice that when $h = 0.05$, $x_1 = 1.05$, but when $h = 0.1$, $x_1 = 1.10$. Continuing these two situations, we get the following results.

	y	
x	$h = 0.1$	$h = 0.05$
1	2	2
1.05	—	1.75
1.10	1.5	1.64741
1.15	—	1.60536
1.20	1.5255	1.59687
1.25	—	1.60927
1.30	1.60249	1.63526
1.35	—	1.67012
1.40	1.69798	1.71057
1.45	—	1.75431
1.50	1.79643	1.79973

Euler's method is the easiest of many methods that can be used to obtain numerical approximations of the solution to a differential equation. We will look at one other method in Exercise Set 15.1. This second method, called Heun's method, is much more accurate than Euler's method.

Exercise Set 15.1

1. Write a computer program to use Euler's method for solving differential equations. Check your program on Examples 15.1 and 15.2. Your program should allow you to input the size of h and the values of x_0 and y_0.

In Exercises 2–5, solve the given differential equations by Euler's method for the specified value of *h*. Solve the differential equation exactly and check you approximations against the exact values.

2. $y' = x + 2$, $x_0 = 1$, $y_0 = 4$, $1 \leq x \leq 2$, $h = 0.1$ and $h = 0.05$
3. $y' = 3y + e^x$, $x_0 = 0$, $y_0 = 0$, $0 \leq x \leq 1$, $h = 0.1$ and $h = 0.05$
4. $y' = 2xy + 2x$, $x_0 = 0$, $y_0 = 1$, $0 \leq x \leq 1$, $h = 0.1$
5. $y' = 7y + \sin 2x$, $x_0 = 0$, $y_0 = 1$, $0 \leq x \leq 0.5$, $h = 0.05$

In Exercises 6–12, solve the given differential equation by using Euler's method for the given initial value of *x* and *y* and the given value of *h*.

6. $y' = xy + 5$, $x_0 = 0$, $y_0 = 0$, $0 \leq x \leq 1$, $h = 0.1$
7. $y' = x^2 + y^2$, $x_0 = 0$, $y_0 = 1$, $0 \leq x \leq 1$, $h - 0.1$
8. $y' = x^2 + 4y^2$, $x_0 = 0$, $y_0 = 1$, $0 \leq x \leq 1$, $h = 0.1$
9. $y' = e^{xy}$, $x_0 = 0$, $y_0 = 0$, $0 \leq x \leq 1$, $h = 0.1$
10. $y' = \sqrt{4 + xy}$, $x_0 = 0$, $y_0 = 1$, $0 \leq x \leq 0.5$, $h = 0.05$
11. $y' = \sin(x + y)$, $x_0 = \frac{\pi}{2}$, $y_0 = 1$, $\frac{\pi}{2} \leq x \leq \frac{3\pi}{4}$, $h = 0.05$
12. $y' = y + \cos x$, $x_0 = \pi$, $y_0 = 2$, $\pi \leq x \leq \frac{5\pi}{4}$, $h = 0.05$

Solve Exercises 13–16.

13. Heun's method is based on the formula

$$y_{n+1} = y_n + \frac{h}{2}\left[f(x_n, y_n) + f(x_n + h, y_n + n + hy_n')\right]$$

Write a computer program to use Heun's method for solving differential equations. Check your program on Example 15.1. Your program should allow you to input the size of h and the initial values x_0 and y_0.

14. Use Heun's method to solve Exercises 2–5. Compare your results to the correct solution and to the results you got using Euler's method.

15. *Thermal science* **Stefan's law of radiation** states that the rate of change in temperature of a body at $T(t)$ degrees in a medium at $T_m(t)$ degrees is proportional to ${T_m}^4 - T^4$. Thus, we have

$$\frac{dT}{dt} = k\left({T_m}^4 - T^4\right)$$

where k is a constant. Let $k = 40^{-4}$ and assume the medium temperature is constant, $T_m(t) = 70°$. If $T(0) = 100°$, use Euler's method with $h = 0.1$ to approximate $T(1)$ and $T(2)$.

16. *Thermal science* According to Newton's law of cooling

$$\frac{dT}{dt} = -k(T - T_m)$$

where k is a positive constant. Let $k = 1$ and assume the medium temperature is constant, $T_m(t) = 70°$. If $T(0) = 100°$, use Euler's method with $h = 0.1$ to approximate $T(1)$ and $T(2)$.

In Your Words

17. What is Euler's method for solving differential equations?

18. Why would you use a numerical technique such as Euler's method for solving differential equations?

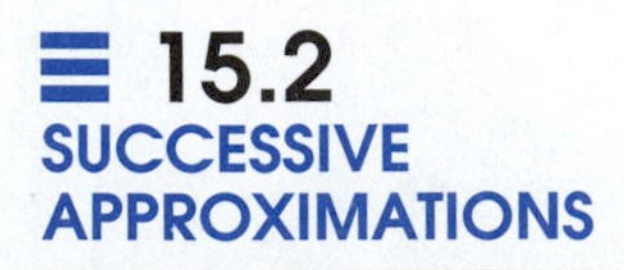

15.2 SUCCESSIVE APPROXIMATIONS

Another method that is used to solve differential equations is based on successive approximations. Using this method, we will develop a function that approximates the actual solution.

As with Euler's method, we begin with a differential equation of the form $y' = f(x, y)$ that has an initial value of (x_0, y_0). Begin by substituting y_0 into

$f(x, y)$. This makes it a function of x only. Then integrate this function, which gives y as a function of x. This we call the first approximation.

If the first approximation is then substituted into the right-hand side of y' and integrated again, a second approximation is obtained. This method can be repeated as long as desired and each time a closer approximation to y is obtained. This method is summarized in the following box.

We will demonstrate this on the differential equations in Examples 15.1 and 15.2 from Section 15.1.

Solving Differential Equations by Successive Approximations

The method of successive approximations to solve the differential equation $y' = f(x, y)$ that has an initial value $f(x_0, y_0)$ uses the following steps:

1. Substitute y_0 into $f(x, y)$ to obtain $f(x, y_0)$.
2. Integrate $f(x, y_0)$ with respect to x.
3. Substitute this result for y in the given equation $y' = f(x, y)$.
4. Integrate this result with respect to x.
5. Repeat steps 3 and 4 to obtain a closer approximation to y.

EXAMPLE 15.3

Obtain the first three approximations to the differential equation $y' = x - y$, if the solution passes through $(0, 2)$.

Solution We are given $y_0 = 2$. We substitute this for y in the differential equation and get $y' = x - 2$. Integrating, we have

$$\int dy = \int (x-2)\,dx$$

$$y = \frac{1}{2}x^2 - 2x + C_1$$

We know that this curve passes through $(0, 2)$, and so $C_1 = 2$. Thus we have

$$y = \frac{1}{2}x^2 - 2x + 2$$

as the first approximation.

Next, we substitute this value for y in the differential equation $y' = x - y$ and get

$$\begin{aligned} y' &= x - \left(\frac{1}{2}x^2 - 2x + 2\right) \\ &= 3x - \frac{1}{2}x^2 - 2 \end{aligned}$$

Integrating, we obtain

$$y = \frac{3}{2}x^2 - \frac{1}{6}x^3 - 2x + C_2$$

EXAMPLE 15.3 (Cont.)

When $x = 0$, $y = 2$, so $C_2 = 2$ and the second approximation is

$$y = \frac{3}{2}x^2 - \frac{1}{6}x^3 - 2x + 2$$

For the third approximation, we use

$$\begin{aligned} y' &= x - y \\ &= x - \left(\frac{3}{2}x^2 - \frac{1}{6}x^3 - 2x + 2\right) \\ &= \frac{1}{6}x^3 - \frac{3}{2}x^2 + 3x - 2 \end{aligned}$$

Integrating both sides produces

$$y = \frac{1}{24}x^4 - \frac{1}{2}x^3 + \frac{3}{2}x^2 - 2x + C_3$$

When $x = 0$, $y = 2$, so $C_3 = 2$. The third approximation is

$$y = \frac{1}{24}x^4 - \frac{1}{2}x^3 + \frac{3}{2}x^2 - 2x + 2$$

How accurate is this? When $x = 0.5$ we get $y = 1.31510$. Euler's method with $h = 0.01$ produced $y = 1.31502$ and the actual value is 1.31959. As you can see, this is slightly more accurate than Euler's method with $h = 0.01$.

EXAMPLE 15.4

Obtain the first two approximations to the differential equation $y' = 3x^2 - y^3$, if the curve passes through the point $(1, 2)$.

Solution As before, we begin by substituting the given value for y into the differential equation, getting

$$y' = 3x^2 - (2)^3 = 3x^2 - 8$$

Integrating both sides produces

$$y = x^3 - 8x + C_1$$

Since $y = 2$ when $x = 1$, we can see that $C_1 = 9$, and so the first approximation is

$$y = x^3 - 8x + 9$$

To get the second approximation, we substitute this into the differential equation with the result

$$\begin{aligned} y' &= 3x^2 - (x^3 - 8x + 9)^3 \\ &= 3x^2 - (x^9 - 24x^7 + 27x^6 + 192x^5 - 432x^4 - 269x^3 \\ &\quad + 1{,}728x^2 - 1{,}944x + 729) \\ &= -x^9 + 24x^7 - 27x^6 - 192x^5 + 432x^4 + 269x^3 - 1{,}725x^2 \\ &\quad + 1{,}944x - 729 \end{aligned}$$

EXAMPLE 15.4 (Cont.)

Integrating both sides and solving for the constant, we have the second approximation

$$y = -\frac{1}{10}x^{10} + 3x^8 - \frac{27}{7}x^7 - 32x^6 + \frac{432}{5}x^5 + \frac{269}{4}x^4 - 575x^3 + 972x^2 - 729x + 213.307$$

As you can see, determining the third approximation would be extremely difficult. This method also has the disadvantage that it can be applied only if the resulting differentials can be integrated. But, as we saw in Example 15.3, this often provides a more accurate solution than does Euler's method.

Exercise Set 15.2

In Exercises 1–4, find the third approximation of the solution of the given differential equations whose solution passes through the given point. Evaluate the solution for the indicated value of *x*. Compare these answers to the exact values and the answers you got using Euler's method in Exercises 2–5 of Exercise Set 15.1.

1. $y' = x + 2$, $(1, 4)$, $x = 2$
2. $y' = 3y + e^x$, $(0, 0)$, $x = 1$
3. $y' = 2xy + 2x$, $(0, 1)$, $x = 1$
4. $y' = 7y + \sin 2x$, $(0, 1)$, $x = 0.5$

In Exercises 5–10, find the second approximation of the solution of the given differential equation whose solution passes through the given point.

5. $y' = x^2 + y^2$, $(0, 1)$
6. $y' = ye^x$, $(0, 1)$
7. $y' = 4 + xy$, $(0, 1)$
8. $y' = x^2 + 4y^2$, $(0, 1)$
9. $y' = \sin 2x + y$, $(\frac{\pi}{2}, 1)$
10. $y' = y + \cos x$, $(\pi, 2)$

Solve Exercises 11–14.

11. Substitute the first few terms of the Maclaurin series for e^{-x} into the expression $3e^{-x} + x - 1$. Compare the results with the solution for Example 15.3.
12. Substitute the first few terms of the Maclaurin series for e^{x^2} into the expression $2e^{x^2} - 1$. Compare the result with the solution for Exercise 3.
13. *Electricity* The current I in an electric circuit at time t is given by

$$I'(t) + 2I(t) = \sin t$$

If $I = 0$ when $t = 0$, calculate I when $t = 0.5$. (Use the third approximation of Euler's method.)

14. *Physics* A 10-kg object is attached to a parachute and dropped from rest. It encounters a force due to air resistance of $4v^2$. The resulting differential equation is

$$\frac{dv}{dt} = 9.8 - 0.4v^2$$

Find the velocity when $t = 0.5$ s by using the third approximation.

In Your Words

15. Describe how the method of successive approximations differs from Euler's method.

16. Is the method of successive approximations or Euler's method easier to use? Why?

15.3 LAPLACE TRANSFORMS

In Section 15.1, we looked at two numerical methods, Euler's method and Heun's method. In the second section, we looked at the method of successive approximations. In the remainder of this chapter, we will look at a third method of solving differential equations. This last method involves **Laplace transforms**. These transforms give us an algebraic method for finding a particular solution of a differential equation from its initial conditions.

> **Laplace Transform**
>
> The **Laplace transform** of a function $f(t)$ is denoted as either $L(f)$ or $F(s)$ and is defined as
>
> $$L(f) = F(s) = \int_0^{\infty} e^{-st} f(t)\,dt$$

As we can see from its notation, the Laplace transform is a function of $f(t)$ and a function of s.

You may remember that in Section 8.4 we learned to evaluate an improper integral, such as one whose upper limit is ∞. In general, we use the method

$$\int_0^{\infty} f(x)\,dx = \lim_{b\to\infty} \int_0^{b} f(x)\,dx$$

if the limit exists.

EXAMPLE 15.5

Find the Laplace transform of the function $f(t) = 1$.

Solution

$$\begin{aligned} L(f) = L(1) &= \int_0^{\infty} 1 \cdot e^{-st} dt \\ &= \lim_{b\to\infty} \int_0^{b} e^{-st} dt \\ &= \lim_{b\to\infty} -\frac{1}{s} e^{-st}\Big|_0^b \\ &= \lim_{b\to\infty} \left(-\frac{1}{s}e^{-sb} + \frac{1}{s}\right) = \frac{1}{s}, \quad s > 0 \end{aligned}$$

So, $L(1) = \frac{1}{s}$, $s > 0$.

Here we assumed $s > 0$ in order to make $\lim_{b\to\infty} e^{-sb} = 0$.

EXAMPLE 15.6

Find the Laplace transform of the function $f(t) = t, t > 0$.

Solution

$$L(f) = \int_0^\infty te^{-st} dt$$

Using integration by parts or Formula 84 from the Table of Integrals in Appendix C, we can evaluate the integral.

$$\begin{aligned} L(f) = L(t) &= \lim_{b\to\infty} \int_0^b te^{-st}\, dt \\ &= \lim_{b\to\infty} \left. \frac{e^{-st}(-st-1)}{s^2} \right|_0^b \\ &= \lim_{b\to\infty} \left[\frac{e^{-sb}(-sb-1)}{s^2} + \frac{1}{s^2} \right] = \frac{1}{s^2} \end{aligned}$$

It can be shown that

$$\lim_{b\to\infty} \frac{e^{-sb}(-sb-1)}{s^2} = 0$$

and so we conclude that

$$L(t) = \frac{1}{s^2}, \quad t > 0$$

EXAMPLE 15.7

Find the Laplace transform of the function $f(t) = \sin at$.

Solution $$\begin{aligned} L(f) = L(\sin at) &= \int_0^\infty e^{-st} \sin at\, dt \\ &= \lim_{b\to\infty} \int_0^b e^{-st} \sin at\, dt \\ &= \lim_{b\to\infty} \left. \left[\frac{e^{-st}(-s\sin at - a\cos at)}{a^2+s^2} \right] \right|_0^b \\ &= \lim_{b\to\infty} \left[\frac{e^{-sb}(-s\sin ab - a\cos ab)}{a^2+s^2} + \frac{a}{a^2+s^2} \right] \\ &= 0 + \frac{a}{a^2+s^2} = \frac{a}{s^2+a^2} \end{aligned}$$

Thus, $L(\sin at) = \dfrac{a}{s^2+a^2}$.

We could continue in this manner to get more Laplace transforms; in fact, you will be asked to do this as part of the exercises. There are some properties of Laplace transforms that you will find helpful.

In particular, the Laplace transform of a sum of two functions is the sum of their Laplace transforms.

Properties of Laplace Transforms

If $f(t)$ and $g(t)$ are functions and k is a constant, then

1. $L[f(t)+g(t)] = L(f)+L(g)$
2. $L[kf(t)] = kL(f)$

These two properties allow us to say that the Laplace transform is **linear** or has the **linearity property**. A short table of Laplace transforms is given in Table 15.1.

TABLE 15.1 Table of Laplace Transforms

	$f(t)=L^{-1}(F)$	$L(f)=F(s)$		$f(t)=L^{-1}(F)$	$L(f)=F(s)$
1.	1	$\frac{1}{s}$	12.	$\sin at - at\cos at$	$\frac{2a^3}{(s^2+a^2)^2}$
2.	t	$\frac{1}{s^2}$	13.	$t\sin at$	$\frac{2as}{(s^2+a^2)^2}$
3.	t^n	$\frac{n!}{s^{n+1}}\ (n=1,2,3,\cdots)$	14.	$\sin at + at\cos at$	$\frac{2as^2}{(s^2+a^2)^2}$
4.	$\frac{1}{\sqrt{\pi t}}$	$\frac{1}{\sqrt{s}}$	15.	$t\cos at$	$\frac{s^2-a^2}{(s^2+a^2)^2}$
5.	e^{at}	$\frac{1}{s-a}$	16.	$\cos at - \cos bt$	$\frac{s(b^2-a^2)}{(s^2+a^2)(s^2+b^2)},\ (a^2 \neq b^2)$
6.	te^{at}	$\frac{1}{(s-a)^2}$	17.	$e^{at}-e^{bt}$	$\frac{a-b}{(s-a)(s-b)}$
7.	$t^{n-1}e^{at}$	$\frac{(n-1)!}{(s-a)^n}$	18.	$ae^{at}-be^{bt}$	$\frac{s(a-b)}{(s-a)(s-b)}$
8.	$\sin at$	$\frac{a}{s^2+a^2}$	19.	$e^{-at}\sin bt$	$\frac{b}{(s+a)^2+b^2}$
9.	$\cos at$	$\frac{s}{s^2+a^2}$	20.	$e^{-at}\cos bt$	$\frac{s+a}{(s+a)^2+b^2}$
10.	$1-\cos at$	$\frac{a^2}{s(s^2+a^2)}$	21.	$\frac{1}{2b^3}e^{-at}[\sin(bt)-bt\cos(bt)]$	$\frac{1}{\left[(s+a)^2+b^2\right]^2}$
11.	$at-\sin at$	$\frac{a^3}{s^2(s^2+a^2)}$	22.	$\frac{1}{2b^3}e^{-at}\left[abt\cos(bt)+b^2t\sin(bt)-a\sin(bt)\right]$	$\frac{s}{\left[(s+a)^2+b^2\right]^2}$

Note: a and b are constants; s and t are variables.

EXAMPLE 15.8

Use Table 15.1 and the linearity property to obtain $L(4\cos 5t + e^{-3t})$.

Solution By the linearity property,

$$L(4\cos 5t + e^{-3t}) = 4L(\cos 5t) + L(e^{-3t})$$

Using transforms 9 and 5 from Table 15.1 , we see that

$$\begin{aligned} 4L(\cos 5t) + L\left(e^{-3t}\right) &= 4\left(\frac{s}{s^2+5^2}\right) + \frac{1}{s+3} \\ &= \frac{4s}{s^2+25} + \frac{1}{s+3} \end{aligned}$$

EXAMPLE 15.9

Use Table 15.1 and the linearity property to obtain $L\left(\sin^2 5t\right)$.

Solution We need to use the trigonometric identity

$$\sin^2\theta = \frac{1}{2}(1-\cos 2\theta)$$

with $\theta = 5t$. This leads to $\sin^2 5t = \frac{1}{2} - \frac{1}{2}\cos 10t$, and so we have

$$L\left(\sin^2 5t\right) = L\left(\frac{1}{2} - \frac{1}{2}\cos 10t\right)$$

By Table 15.1 , $L\left(\frac{1}{2}\right) = \frac{1}{2}L(1) = \frac{1}{2}\left(\frac{1}{s}\right)$ and

$$\begin{aligned} L\left(\frac{1}{2}\cos 10t\right) &= \frac{1}{2}L\left(\cos 10t\right) \\ &= \frac{1}{2}\left(\frac{s}{s^2+10^2}\right) = \frac{1}{2}\left(\frac{s}{s^2+100}\right) \end{aligned}$$

Thus,

$$\begin{aligned} L\left(\sin^2 5t\right) &= L\left(\frac{1}{2} - \frac{1}{2}\cos 10t\right) \\ &= \frac{1}{2}\left(\frac{1}{s}\right) - \frac{1}{2}\left(\frac{s}{s^2+100}\right) \\ &= \frac{50}{s(s^2+100)} \end{aligned}$$

Exercise Set 15.3

In Exercises 1–4, use the definition to find the Laplace transform of *f(t)*. Check your answer with the transform in Table 34.1.

1. $f(t) = n$
2. $f(t) = e^{-at}$
3. $f(t) = \cos at$
4. $f(t) = te^{at}$

In Exercises 5–12, find the transforms of the given functions by using Table 34.1.

5. $f(t) = t^3$
6. $f(t) = e^{2t}$
7. $f(t) = \sin 6t$
8. $f(t) = e^{2t} \sin 6t$
9. $f(t) = t^3 e^{5t}$
10. $f(t) = e^{-5t} \sin 10t$
11. $f(t) = \sin 4t + 4t \cos 4t$
12. $f(t) = t \cos 7t$

In Exercises 13–22, find the transforms of the given functions by using Table 34.1 and the linearity property.

13. $f(t) = \sin 5t + \cos 3t$
14. $f(t) = e^{3t} \sin 5t + 4t^8$
15. $f(t) = 3t \sin 5t - \cos 4t + 1$
16. $f(t) = t^3 e^{4t} + 4e^{3t}$
17. $f(t) = e^{-2t} \cos 3t - 2e^{5t} \sin t$
18. $f(t) = t \cos 3t - 2t \sin 2t$
19. $f(t) = 4 + 3t + 2e^t$
20. $f(t) = 6 - 4t + 2 \sin 10t + e^{5t}$
21. $f(t) = \cos^2 3t$
22. $f(t) = 5e^{2t} - t^3$

In Your Words

23. Without looking in the text, give the definition of the Laplace transform.
24. What are the two properties of Laplace transforms? Give an example showing how each is used.

15.4 INVERSE LAPLACE TRANSFORMS AND TRANSFORMS OF DERIVATIVES

In Section 15.3, we were given a function $f(t)$ and we were able to find its Laplace transform $L(f) = F(s)$. We can reverse this process.

Inverse Laplace Transforms

Suppose we have a function $F(s)$. Can we find the **inverse Laplace transform** $L^{-1}(F) = f(t)$? The table of Laplace transforms (Table 15.1) will often help us find the answer.

EXAMPLE 15.10

If $F(s) = \dfrac{1}{s+9}$, find $L^{-1}(F)$.

Solution If we look in Table 15.1under the column headed $L(f) = F(s)$, we find the fifth entry is $\dfrac{1}{s-a}$. If $a = -9$, this is the desired form of the formula. The column to the left is headed $f(t) = L^{-1}(F)$ and the fifth entry in that column is e^{at}. We know that $a = -9$, so the answer is

$$L^{-1}\left(\frac{1}{s+9}\right) = e^{-9t}$$

▪

EXAMPLE 15.11

If $F(s) = \dfrac{12}{s^2+16}$, what is $f(t)$?

Solution Since $f(t) = L^{-1}(F)$, we want to find the inverse Laplace transform. If we look in Table 15.1 , we do not find any form that fits. There are two forms that have $s^2 + a^2$ for a denominator. If $a = 4$, the denominator is $s^2 + 16$, which is the one in our problem. One of these, in row 9, has the variable s in the numerator. Our problem does not have an s in the numerator, so this must not be the correct form.

The other form, in row 8 of Table 15.1 , has a numerator of a constant, a. We let $a = 4$ and our numerator is $12 = 3 \cdot 4$. We can rewrite this problem as

$$L^{-1}\left(\frac{12}{s^2+16}\right) = 3L^{-1}\left(\frac{4}{s^2+16}\right) = 3\sin 4t$$

▪

EXAMPLE 15.12

Find $L^{-1}\left(\dfrac{4s^2+9s}{s^4+6s^2+9}\right)$.

Solution We notice that $s^4 + 6s^2 + 9 = (s^2+3)^2$, so

$$\frac{4s^2+9s}{s^4+6s^2+9} = \frac{4s^2+9s}{(s^2+3)^2} = \frac{4s^2}{(s^2+3)^2} + \frac{9s}{(s^2+3)^2}$$

Compare $\dfrac{4s^2}{(s^2+3)^2}$ to transpose 14, $\dfrac{2as^2}{(s^2+a^2)^2}$. If $a^2 = 3$, then $a = \sqrt{3}$ and $4s^2 = \left(\dfrac{4}{2\sqrt{3}}\right)2\sqrt{3}s^2 = \left(\dfrac{2}{\sqrt{3}}\right)2\sqrt{3}s^2$. Thus, we can rewrite $\dfrac{4s^2}{(s^2+3)^2} = \dfrac{2}{\sqrt{3}} \cdot \dfrac{2\sqrt{3}s^2}{(s^2+3)^2}$.

Similarly, compare $\dfrac{9s}{(s^2+3)^2}$ to transform 13, $\dfrac{2as}{(s^2+a^2)^2}$. Again, $a = \sqrt{3}$, so

EXAMPLE 15.12 (Cont.)

$\frac{9s}{(s^2+3)^2} = \frac{9}{2\sqrt{3}} \cdot \frac{2\sqrt{3}s}{(s^2+3)^2}$. Thus, we can rewrite the displayed equation as

$$\frac{4s^2+9s}{s^4+6s^2+9} = \frac{2}{\sqrt{3}} \cdot \frac{2\sqrt{3}s^2}{(s^2+3)^2} + \frac{9}{2\sqrt{3}} \cdot \frac{2\sqrt{3}s}{(s^2+3)^2}$$

By using transforms 14 and 13, respectively, we see that

$$L^{-1}\left(\frac{4s^2+9s}{s^4+6s^2+9}\right) = \frac{2}{\sqrt{3}}L^{-1}\left(\frac{2\sqrt{3}s^2}{(s^2+3)^2}\right) + \frac{9}{2\sqrt{3}}L^{-1}\left(\frac{2\sqrt{3}s}{(s^2+3)^2}\right)$$
$$= \frac{2}{\sqrt{3}}(\sin\sqrt{3}t + \sqrt{3}t\cos\sqrt{3}t) + \frac{9}{2\sqrt{3}}t\sin\sqrt{3}t$$

Transforms of Derivatives

When we are solving a differential equation, there is another Laplace transform that is important to finding the solution. This involves finding the transform of the derivative of a function.

We will begin by finding the Laplace transform of the first derivative of a function. If $f(t)$ is a function with first derivative $f'(t)$, then by the definition of a Laplace transform

$$L(f') = \int_0^\infty e^{-st} f'(t)\,dt$$

We can integrate this by parts, if we let $u = e^{-st}$ and $dv = f'(t)\,dt$. Then we have the following table:

$u = e^{-st}$	$v = f(t)$
$du = -se^{-st}dt$	$dv = f'(t)\,dt$

And so,

$$\int_0^\infty e^{-st} f'(t)\,dt = e^{-st} f(t)\big|_0^\infty + s\int_0^\infty e^{-st} f(t)\,dt = 0 - f(0) + sL(f)$$

Notice that the integral of the last term on the right-hand side is the Laplace transform of $f(t)$. So, the Laplace transform of $f'(t)$ is

$$L(f') = sL(f) - f(0)$$

If we followed the same procedure on the second derivative of f, $f''(t)$, we would get the Laplace transform

$$L(f'') = s^2L(f) - sf(0) - f'(0)$$

These results are summarized in the following box.

Laplace Transforms of Derivatives

The Laplace transforms of the first derivative, $f'(t)$, and the second derivative, $f''(t)$, are

$$L(f') = sL(f) - f(0) \qquad (*)$$
$$L(f'') = s^2L(f) - sf(0) - f'(0) \qquad (**)$$

EXAMPLE 15.13

If $f(0) = 2$, express the transform of $f'(t) - 5f(t)$ in terms of s and the transform of $f(t)$.

Solution We will take the Laplace transform of the expression, and by the linearity property we have

$$L(f') - 5L(f)$$

From $(*)$ we have $L(f') = sL(f) - f(0)$, and so, with this substitution, the expression becomes

$$sL(f) - f(0) - 5L(f)$$

We are given the fact that $f(0) = 2$, and, with $L(0) = 0$, the expression becomes

$$sL(f) - 2 - 5L(f)$$

or $$(s - 5)L(f) - 2$$

EXAMPLE 15.14

If $f(0) = 1$ and $f'(0) = 4$, express the transform of $f''(t) - f'(t) - 2f(t)$ in terms of s and the transform of $f(t)$.

Solution Taking the Laplace transform of the expression and using the linearity property, we have

$$L(f'') - L(f') - 2L(f)$$

Using the Laplace transforms of the first and second derivatives for f as given in $(*)$ and $(**)$, we get

$$[s^2L(f) - sf(0) - f'(0)] - [sL(f) - f(0)] - 2L(f)$$

We are given $f(0) = 1$ and $f'(0) = 4$, so the expression becomes

$$[s^2L(f) - s(1) - 4] - [sL(f) - 1] - 2L(f)$$

or

$$(s^2 - s - 2)L(f) - s - 4 + 1$$
$$(s^2 - s - 2)L(f) - (s + 3)$$

After Section 15.5, we will combine our knowledge of Laplace transforms from this and Section 15.3 to solve differential equations.

Exercise Set 15.4

In Exercises 1–10, find the inverse Laplace transforms of the given function.

1. $\frac{4}{s-5}$

2. $\frac{6}{s^2+4}$

3. $\frac{16}{s^3+16s}$

4. $\frac{4}{(s+9)^2}$

5. $\frac{3s}{s^2+9}$

6. $\frac{s+6}{(s+16)^2}$

7. $\frac{5s}{s^2+9s+14}$

8. $\frac{s^2-16}{(s^2+4)^2}$

9. $\frac{2s+6s^2}{(s^2+25)^2}$

10. $\frac{s+5}{s^2+10s+26}$

In Exercises 11–16, express the transforms of the given expression in terms of *s* and *L(f)*.

11. $f''(t)+4f'(t)$, $f(0)=1$, $f'(0)=0$

12. $f''(t)-6f(t)$, $f(0)=3$, $f'(0)=2$

13. $2f''(t)-3f'(t)+f(t)$, $f(0)=1$, $f'(0)=0$

14. $y''-4y'$, $f(0)=2$, $f'(0)=1$

15. $y''+6y'-y$, $f(0)=1$, $f'(0)=-1$

16. $y''-3y'+4y$, $f(0)=-2$, $f'(0)=1$

In Your Words

17. Without looking in the text, write the Laplace transforms of derivatives.

18. What is an inverse Laplace transform? How do you find an inverse Laplace transform?

15.5 PARTIAL FRACTIONS

We are now going to introduce a topic that is often included as an integration technique. Rather than introduce it as an integration method, we chose to let you rely on the table of integrals. But, we have now reached a point where the topic will be valuable and will help us use Laplace transforms to solve differential equations.

It is often necessary to find an inverse Laplace transform, such as

$$L^{-1}\left\{\frac{7}{(s-5)(s+2)}\right\}$$

If you look in Table 15.1 you will not find one for this type of fraction. But, as we will soon show, it is possible to rewrite this fraction as

$$\frac{1}{s-5}-\frac{1}{s+2}$$

We can take the inverse transform of each of these. In fact,

$$L^{-1}\left\{\frac{7}{(s-5)(s+2)}\right\}=L^{-1}\left(\frac{1}{s-5}\right)-L^{-1}\left(\frac{1}{s+2}\right)=e^{5t}-e^{-2t}$$

What we have done is divided the original fraction into **partial fractions**.

If you have an expression of the form $\frac{p(x)}{q(x)}$, where $p(x)$ and $q(x)$ are polynomials, and if the degree of p is less than the degree of q, then it is possible to write

$$\frac{p(x)}{q(x)} = F_1 + F_2 + F_3 + \cdots + F_k$$

where each F_i is of the form

$$\frac{A}{(ax+b)^m} \quad \text{or} \quad \frac{Bx+C}{(ax^2+bx+c)^n}$$

where A, B, and C represent constants and m and n are positive integers. One final condition is that ax^2+bx+c is irreducible in that it has a negative discriminate; that is, $b^2-4ac<0$. The sum on the right-hand side, $F_1+F_2+\cdots+F_k$, is called **partial fraction decomposition** of $\frac{p(x)}{q(x)}$ and each F_i is a **partial fraction.**

The process of finding the partial fraction decomposition is rather straightforward; but, it is also quite lengthy. The process involves four steps, as shown in the following box.

Guidelines for Partial Fraction Decomposition

1. If the degree of the numerator $p(x) \geq$ degree of the denominator $q(x)$, divide $p(x)$ by $q(x)$ and work with the remainder.
2. Factor the denominator $q(x)$ into its linear factors $(ax+b)$ and its irreducible quadratic factors (ax^2+bx+c). If any of these are also factors of the numerator, then they should be cancelled. Collect all repeated factors so that the factors are of the form $(ax+b)^m$ and $(ax^2+bx+c)^n$.
3. For each linear factor of the form $(ax+b)^m$, the decomposition contains a sum of m partial fractions of the form
$$\frac{A_1}{ax+b} + \frac{A_2}{(ax+b)^2} + \frac{A_3}{(ax+b)^3} + \cdots + \frac{A_m}{(ax+b)^m}$$
where $A_1, A_2, A_3, \ldots, A_m$ are constants.
4. For each quadratic factor $(ax^2+bx+c)^n$, the decomposition contains a sum of n partial fractions of the form
$$\frac{A_1x+B_1}{ax^2+bx+c} + \frac{A_2x+B_2}{(ax^2+bx+c)^2} + \cdots + \frac{A_nx+B_n}{(ax^2+bx+c)^n}$$
where $A_1, B_1, A_2, B_2, \ldots, A_n, B_n$ are all constants.

EXAMPLE 15.15

Find the partial fraction decomposition of

$$\frac{x+1}{x^3+x^2-6x}$$

EXAMPLE 15.15 (Cont.)

Solution The denominator factors into $x(x+3)(x-2)$, so the partial fraction decomposition is of the form

$$\frac{x+1}{x^3+x^2-6x} = \frac{A}{x} + \frac{B}{x+3} + \frac{C}{x-2}$$

Multiplying both sides by the common denominator x^3+x^2-6x, we get

$$\begin{aligned} x+1 &= A(x+3)(x-2) + Bx(x-2) + Cx(x+3) \\ &= (A+B+C)x^2 + (A-2B+3C)x - 6A \end{aligned}$$

The coefficients of corresponding terms must be equal. This gives the simultaneous equations

$$\begin{aligned} A+B+C &= 0 && \text{from the } x^2 \text{ terms} \\ A-2B+3C &= 1 && \text{from the } x \text{ terms} \\ -6A &= 1 && \text{from the constant terms} \end{aligned}$$

Solving these simultaneously, we get $A = -\frac{1}{6}$, $B = -\frac{2}{15}$, and $C = \frac{3}{10}$.

The partial fraction decomposition then becomes

$$\frac{x+1}{x^3+x^2-6x} = \frac{-1/6}{x} + \frac{-2/15}{x+3} + \frac{3/10}{x-2}$$

We used simultaneous linear equations to determine the values of the constants A, B, and C. In the next example, we will show a method that can be used to reduce the number of equations that need to be solved.

EXAMPLE 15.16

Find the partial fraction decomposition of

$$\frac{2x+3}{x^3+2x^2+x}$$

Solution The degree of the numerator is less than the denominator. We factor the denominator $x^3+2x^2+x = x(x+1)^2$. These are all linear factors. Thus, we know that the partial fraction decomposition is of the form

$$\frac{2x+3}{x(x+1)^2} = \frac{A}{x} + \frac{B}{x+1} + \frac{C}{(x+1)^2}$$

Notice that the denominator had one factor of x and two factors of $x+1$. The partial fraction decomposition has one fraction with a denominator of x and two with $x+1$ in the denominator.

We need to determine the constants A, B, and C. If we clear the fractions by multiplying both sides of this equation $x(x+1)^2$, we will get

$$2x+3 = A(x+1)^2 + Bx(x+1) + Cx$$

There are a couple of approaches that can be used to solve this for A, B, and C. This equation must be true for all values of x. If $x=0$, two of these terms will

EXAMPLE 15.16 (Cont.)

be 0, and the equation becomes

$$2(0)+3 = A(0+1)^2 + B(0)(0+1) + C(0)$$
$$\text{or} \qquad 3 = A$$

Since $x+1=0$ when $x=-1$, if we let $x=-1$, then two other terms will become 0, and the equation becomes

$$2(-1)+3 = A(-1+1)^2 + B(-1)(-1+1) + C(-1)$$
$$\text{or} \qquad 1 = -C$$

and so $C=-1$.

The equation can now be written as

$$\begin{aligned} 2x+3 &= 3(x+1)^2 + Bx(x+1) - x \\ &= 3(x^2+2x+1) + B(x^2+x) - x \\ &= (3+B)x^2 + (6+B-1)x + 3 \end{aligned}$$

Since there is no x^2 term on the left-hand side, then $3+B=0$ or $B=-3$. We now have the values of A, B, and C and can write

$$\frac{2x+3}{x(x+1)^2} = \frac{3}{x} + \frac{-3}{x+1} + \frac{-1}{(x+1)^2}$$

EXAMPLE 15.17

Find the partial fraction decomposition of

$$\frac{x^2+1}{2x^4-x^3-x}$$

Solution

$$\begin{aligned} \frac{x^2+1}{2x^4-x^3-x} &= \frac{x^2+1}{x(x-1)(2x^2+x+1)} \\ &= \frac{A}{x} + \frac{B}{x-1} + \frac{Cx+D}{2x^2+x+1} \end{aligned}$$

To clear the fractions, we multiply both sides by $x(x-1)(2x^2+x+1)$ and obtain

$$x^2+1 = A(x-1)(2x^2+x+1) + Bx(2x^2+x+1) + (Cx+D)x(x-1)$$

We know that $x(x-1)(2x^2+x+1)=0$ when $x=0$ and when $x=1$. Using each of these in turn, we get the following results:

$$\begin{aligned} &\text{When } x=0: && 1=-A, \text{ so } A=-1 \\ &\text{When } x=1: && 1+1 = B\cdot 1(2+1+1) \\ &\text{And so,} && 2 = 4B \text{ or } B = \frac{1}{2} \end{aligned}$$

The equation can now be written as

$$\begin{aligned} x^2+1 = {} & -1(x-1)(2x^2+x+1) + \frac{x}{2}(2x^2+x+1) \\ & + (Cx+D)x(x-1) \end{aligned}$$

EXAMPLE 15.17 (Cont.)

If we combine the like powers of x on the right-hand side, we have

$$x^2+1=(-2+1+C)x^3+\left(2-1+\frac{1}{2}-C+D\right)x^2$$
$$+\left(1-1+\frac{1}{2}-D\right)x+1$$
$$=(C-1)x^3+\left(\frac{3}{2}-C+D\right)x^2+\left(\frac{1}{2}-D\right)x+1$$

Since there is no x^3-term on the left-hand side, we can see that $C-1=0$ and $C=1$. Likewise, there is no x-term on the left-hand side, so $\frac{1}{2}-D=0$ and $D=\frac{1}{2}$. Thus, the partial fraction decomposition is

$$\frac{x^2+1}{2x^4-x^3+x}=\frac{-1}{x}+\frac{\frac{1}{2}}{x-1}+\frac{x+\frac{1}{2}}{2x^2+x+1}$$

If we wanted to take the inverse Laplace transforms of these partial fractions we would easily find the first two in Table 15.1 . But there is none in the table like the third. The closest are the two that have $(s+a)^2+b^2$ as a denominator. We need to write this partial fraction so its denominator is in this form. To do this, we will complete the square.

$$\frac{x+\frac{1}{2}}{2x^2+x+1}=\frac{x+\frac{1}{2}}{2\left(x^2+\frac{1}{2}x+\frac{1}{2}\right)}$$
$$=\frac{1}{2}\cdot\frac{x+\frac{1}{2}}{\left(x^2+\frac{1}{2}x+\frac{1}{16}\right)+\left(\frac{1}{2}-\frac{1}{16}\right)}$$
$$=\frac{1}{2}\cdot\frac{x+\frac{1}{2}}{\left(x+\frac{1}{4}\right)^2+\frac{7}{16}}$$
$$=\frac{1}{2}\left[\frac{x+\frac{1}{4}}{\left(x+\frac{1}{4}\right)^2+\left(\frac{\sqrt{7}}{4}\right)^2}+\frac{\frac{1}{\sqrt{7}}\cdot\frac{\sqrt{7}}{4}}{\left(x+\frac{1}{4}\right)^2+\left(\frac{\sqrt{7}}{4}\right)^2}\right]$$

Now, if we take the inverse Laplace transform, we obtain

$$L^{-1}\left(\frac{x+\frac{1}{2}}{2x^2+x+1}\right)=\frac{1}{2}L^{-1}\left[\frac{x+\frac{1}{4}}{\left(x+\frac{1}{4}\right)^2+\left(\frac{\sqrt{7}}{4}\right)^2}\right]$$

$$+\frac{1}{2\sqrt{7}}L^{-1}\left[\frac{\frac{\sqrt{7}}{4}}{\left(x+\frac{1}{4}\right)^2+\left(\frac{\sqrt{7}}{4}\right)^2}\right]$$

$$=\frac{1}{2}e^{-t/4}\cos\frac{\sqrt{7}}{4}t+\frac{1}{2\sqrt{7}}e^{-t/4}\sin\frac{\sqrt{7}}{4}t$$

$$=\frac{1}{2}e^{-t/4}\left(\cos\frac{\sqrt{7}}{4}t+\frac{1}{\sqrt{7}}\sin\frac{\sqrt{7}}{4}t\right)$$

You should notice that even after we had completed the square and written the denominator in the correct form, we did not have the numerator in the correct form. We had to rewrite the fraction into two fractions so that the numerators were correct.

EXAMPLE 15.18

Find the partial fraction decomposition of

$$\frac{2x^3+3}{x(x^2+1)^2}$$

Solution We write

$$\frac{2x^3+3}{x(x^2+1)^2}=\frac{A}{x}+\frac{Bx+C}{x^2+1}+\frac{Dx+E}{(x^2+1)^2}$$

Multiplying both sides by $x(x^2+1)^2$ produces the equation

$$2x^3+3=A(x^2+1)^2+(Bx+C)x(x^2+1)+(Dx+E)x$$

If $x=0$, we see that $A=3$, and can write the equation as

$$2x^3+3=3(x^2+1)^2+(Bx+C)x(x^2+1)+(Dx+E)x$$

Rewriting and combining like terms changes the equation to

$$2x^3+3=(3+B)x^4+Cx^3+(6+B+D)x^2+(C+E)x+3$$

From this we see that $B=-3$, $C=2$, $D=-3$, and $E=-2$. Thus, the partial decomposition is

$$\frac{2x^3+3}{x(x^2+1)^2}=\frac{3}{x}+\frac{-3x+2}{x^2+1}+\frac{-3x-2}{(x^2+1)^2}$$

▪

EXAMPLE 15.19

Find

$$L^{-1}\left[\frac{2s^3+3}{s(s^2+1)^2}\right]$$

Solution We found the partial fraction decomposition of this fraction in Example 15.18. Thus

$$\begin{aligned}
L^{-1}\left[\frac{2s^3+3}{s(s^2+1)^2}\right] &= L^{-1}\left(\frac{3}{s}\right) + L^{-1}\left(\frac{-3s+2}{s^2+1}\right) + L^{-1}\left[\frac{-3s-2}{(s^2+1)^2}\right] \\
&= 3 - 3\cos t + 2\sin t - \frac{3}{2}t\sin t - \sin t + t\cos t \\
&= 3 + (t-3)\cos t + \left(1 - \frac{3}{2}t\right)\sin t
\end{aligned}$$

Exercise Set 15.5

In Exercises 1–10, find the partial fraction decomposition of the given fraction.

1. $\dfrac{1}{x(x+1)}$
2. $\dfrac{4}{x(x-3)}$
3. $\dfrac{4}{x(x-1)(x+1)}$
4. $\dfrac{5x}{(x-2)(x+3)}$
5. $\dfrac{3x}{x^2-8x+15}$
6. $\dfrac{4x}{(x+1)(x+2)(x+3)}$
7. $\dfrac{x-1}{x^2+2x+1}$
8. $\dfrac{x}{(x+1)^2(x-2)}$
9. $\dfrac{x^3+x^2+x+2}{(x^2+1)(x^2+2)}$
10. $\dfrac{3x^2+5}{(x^2+1)^2}$

Use the method of partial fractions to find the inverse Laplace transforms of the functions in Exercises 11–26. (For Exercises 11–20, use the decomposition from Exercises 1–10.)

11. $L^{-1}\left[\dfrac{1}{s(s+1)}\right]$
12. $L^{-1}\left[\dfrac{4}{s(s-3)}\right]$
13. $L^{-1}\left[\dfrac{4}{s(s-1)(s+1)}\right]$
14. $L^{-1}\left[\dfrac{5s}{(s-2)(s+3)}\right]$
15. $L^{-1}\left(\dfrac{3s}{s^2-8s+15}\right)$
16. $L^{-1}\left[\dfrac{4s}{(s+1)(s+2)(s+3)}\right]$
17. $L^{-1}\left(\dfrac{s-1}{s^2+s+1}\right)$
18. $L^{-1}\left[\dfrac{s}{(s+1)^2(s-2)}\right]$
19. $L^{-1}\left[\dfrac{s^3+s^2+s+2}{(s^2+1)(s^2+2)}\right]$
20. $L^{-1}\left[\dfrac{3s+5}{(s^2+1)^2}\right]$
21. $L^{-1}\left[\dfrac{2}{s^2(s-1)}\right]$
22. $L^{-1}\left[\dfrac{s}{(s+1)(s^2+1)}\right]$
23. $L^{-1}\left[\dfrac{1}{(s+4)(s^2+9)}\right]$
24. $L^{-1}\left[\dfrac{1}{(s^2+1)(s^2-1)}\right]$
25. $L^{-1}\left[\dfrac{1}{(s^2+1)(s-1)^2}\right]$
26. $L^{-1}\left[\dfrac{1}{(s^2+4s+7)(s+1)}\right]$

In Your Words

27. What is a partial fraction?

28. Without looking in the text, outline the guidelines for partial fraction decomposition.

29. The guidelines for partial fraction decomposition only discuss what to do when $q(x)$ is composed of linear factors of the form $ax+b$ and irreducible quadratic factors of the form ax^2+bx+c. What are you supposed to do if q has a factor of degree higher than 2?

15.6 USING LAPLACE TRANSFORMS TO SOLVE DIFFERENTIAL EQUATIONS

We are now ready to use Laplace transforms to solve differential equations. What we will find are the **particular** solutions of the equation subject to the given initial conditions.

This procedure is relatively direct. It has three parts.

Steps to Solving Differential Equations by Laplace Transforms

1. Find the Laplace transforms of both sides of the differential equation.
2. Solve the resulting algebraic expression for $Y(s)$.
3. Find the inverse transform $y = L^{-1}[Y(s)]$.

Note

We have changed our notation. This is more in keeping with the other differential equations that we have solved. We will let $y = f(t)$. Then $L(f) = L(y) = Y(s)$ and $f(0) = y(0)$, $f'(0) = y'(0)$, and so on. Remember from Section 15.4 that $L(y') = sL(y) - y(0)$ and $L(y'') = s^2L(y) - sy(0) - y'(0)$.

EXAMPLE 15.20

Solve the differential equation $y' - y = 2\sin t$, when $y(0) = 0$.

Solution We take the Laplace transform of both sides:

$$L(y') - L(y) = L(2\sin t)$$

We have $L(y') = sL(y) - y(0) = sY - y(0)$, because $L(y) = Y$. Since $y(0) = 0$, the left-hand side becomes

$$sY - 0 - Y = Y(s-1)$$

On the right-hand side, $L(2\sin t) = \dfrac{2}{s^2+1}$. Thus, we have

$$Y(s-1) = \frac{2}{s^2+1}$$

or

$$Y = \frac{2}{(s-1)(s^2+1)}$$

EXAMPLE 15.20 (Cont.)

and so

$$y = L^{-1}\left[\frac{2}{(s-1)(s^2+1)}\right]$$

To find this inverse transform, we will need to use partial fractions.

$$\frac{2}{(s-1)(s^2+1)} = \frac{A}{s-1} + \frac{Bs+C}{s^2+1}$$

or

$$\begin{aligned} 2 &= A(s^2+1)+(s-1)(Bs+C) \\ &= (As^2+A)+(Bs^2-Bs+Cs-C) \\ &= (A+B)s^2+(C-B)s+(A-C) \end{aligned}$$

Thus,

$$\begin{aligned} A+B &= 0 \\ -B+C &= 0 \\ A-C &= 2 \end{aligned}$$

Solving these simultaneous equations, we get $A = 1$, $B = -1$, and $C = -1$. We conclude that

$$\frac{2}{(s-1)(s^2+1)} = \frac{1}{s-1} + \frac{-s-1}{s^2+1}$$

and

$$\begin{aligned} y &= L^{-1}\left(\frac{1}{s-1} + \frac{-s-1}{s^2+1}\right) \\ &= L^{-1}\left(\frac{1}{s-1}\right) - L^{-1}\left(\frac{s}{s^2+1}\right) - L^{-1}\left(\frac{1}{s^2+1}\right) \\ &= e^t - \cos t - \sin t \end{aligned}$$

The solution to this differential equation is

$$y = e^t - \cos t - \sin t$$

EXAMPLE 15.21

Solve the differential equation $y'' + 2y' + y = e^{-t}$, where $y(0) = 1$ and $y'(0) = 2$.

Solution We begin by taking Laplace transforms of both sides, obtaining

$$L(y'') + 2L(y') + L(y) = L(e^{-t}) \qquad (*)$$

Now, since

$$L(y'') = s^2Y(s) - sy(0) - y(0)$$

and

$$L(y') = sY(s) - y(0)$$

we have

$$L(y'') = s^2Y(s) - s(1) - 2$$

EXAMPLE 15.21 (Cont.)

$$= s^2Y(s) - s - 2$$

and $\quad L(y') = sY(s) - 1$

Substituting these into equation (*) produces

$$[s^2Y(s) - s - 2] + 2[sY(s) - 1] + Y(s) = \frac{1}{s+1}$$

$$(s^2 + 2s + 1)Y(s) - s - 4 = \frac{1}{s+1}$$

or

$$(s^2 + 2s + 1)Y(s) = s + 4 + \frac{1}{s+1}$$

and

$$Y(s) = \frac{s}{s^2 + 2s + 1} + \frac{4}{s^2 + 2s + 1} + \frac{1}{(s+1)(s^2 + 2s + 1)}$$

$$= \frac{s}{(s+1)^2} + \frac{4}{(s+1)^2} + \frac{1}{(s+1)^3}$$

Taking the inverse Laplace transform, we get

$$y = L^{-1}[Y(s)] = L^{-1}\left[\frac{s}{(s+1)^2}\right] + L^{-1}\left[\frac{4}{(s+1)^2}\right] + L^{-1}\left[\frac{1}{(s+1)^3}\right]$$

$$= e^{-t}(1-t) + 4te^{-t} + \frac{1}{2}t^2e^{-t}$$

$$= \left(\frac{1}{2}t^2 + 3t + 1\right)e^{-t}$$

From these two examples, you should see that this method of solving differential equations is basically an algebraic method. This is one of the big advantages of Laplace transforms. We can translate a differential equation into an algebraic equation, which is then changed into the solution to the differential equation.

EXAMPLE 15.22

Solve $y'' - 2y' = 4$, where $y(0) = -1$ and $y'(0) = 2$.

Solution

$$
\begin{aligned}
L(y'') - 2L(y') &= L(4) \\
[s^2Y(s) - sy(0) - y'(0)] - 2[sY(s) - y(0)] &= \frac{4}{s} \\
[s^2Y(s) - s(-1) - 2] - 2[sY(s) - (-1)] &= \frac{4}{s} \\
s^2Y(s) + s - 2 - 2sY(s) - 2 &= \frac{4}{s} \\
(s^2 - 2s)Y(s) + (s - 4) &= \frac{4}{s} \\
(s^2 - 2s)Y(s) &= \frac{4}{s} - s + 4 \\
&= \frac{-s^2 + 4s + 4}{s} \\
Y(s) &= \frac{-s^2 + 4s + 4}{s(s^2 - 2s)} \\
&= \frac{-s^2 + 4s + 4}{s^2(s - 2)}
\end{aligned}
$$

The partial fraction decomposition of the right-hand side is

$$
\frac{-s^2 + 4s + 4}{s^2(s - 2)} = \frac{-3}{s} + \frac{-2}{s^2} + \frac{2}{s - 2}
$$

and so

$$
\begin{aligned}
y = L^{-1}[Y(s)] &= L^{-1}\left[\frac{-s^2 + 4s + 4}{s^2(s - 2)}\right] \\
&= L^{-1}\left(\frac{-3}{s}\right) + L^{-1}\left(\frac{-2}{s^2}\right) + L^{-1}\left(\frac{2}{s - 2}\right) \\
&= -3 - 2t + 2e^{2t}
\end{aligned}
$$

As the last example in this chapter, and in this book, we will use an application from electricity.

Application

EXAMPLE 15.23

Recall that for an RLC series circuit, the differential equation for the instantaneous charge $q(t)$ on the capacitor is

$$
L\frac{d^2q}{dt^2} + R\frac{dq}{dt} + \frac{1}{C}q = E(t)
$$

Determine the charge $q(t)$ and current $i(t)$ for a series with $R = 20\,\Omega$, $L = 1$ H, $C = 0.01$ F, $E(t) = 120\sin 10t$ V, $q(0) = 0$, and $i(0) = 0$.

EXAMPLE 15.23 (Cont.)

Solution Using the given values for R, L, C, and $E(t)$, the differential equation is

$$\frac{d^2q}{dt^2}+20\frac{dq}{dt}+100q=120\sin 10t$$

Taking the Laplace transform of both sides gives

$$L\left(\frac{d^2q}{dt^2}\right)+20L\left(\frac{dq}{dt}\right)+100Lq=120L(\sin 10t)$$

$$[s^2L(q)-sq(0)-q'(0)]+20[sL(q)-q(0)]+100L(q)=120\,L(\sin 10t)$$

Since $q(0)=0$ and $i(0)=q'(0)=0$, we have

$$\begin{aligned}
s^2L(q)+20sL(q)+100L(q)&=120\frac{10}{s^2+100}\\
(s^2+20s+100)L(q)&=\frac{1\,200}{s^2+100}\\
(s+10)^2L(q)&=\frac{1\,200}{s^2+10^2}\\
L(q)&=\frac{1\,200}{(s^2+10^2)(s+10)^2}
\end{aligned}$$

Using partial fractions, we see that

$$\frac{1\,200}{(s^2+10^2)(s+10)^2}=\frac{0.6}{s+10}+\frac{6}{(s+10)^2}+\frac{-0.6s}{s^2+10^2}$$

and so

$$\begin{aligned}
q(t)&=L^{-1}[L(q)]\\
&=L^{-1}\left(\frac{0.6}{s+10}\right)+L^{-1}\left[\frac{6}{(s+10)^2}\right]+L^{-1}\left(\frac{-0.6s}{s^2+10^2}\right)
\end{aligned}$$

and the charge $q(t)$ is

$$q(t)=0.6e^{-10t}+6te^{-10t}-0.6\cos 10t$$

The current $i(t)$ is $q'(t)$, which is

$$\begin{aligned}
i(t)=q'(t)&=-6e^{-10t}+6e^{-10t}-60te^{-10t}+6\sin 10t\\
&=-60te^{-10t}+6\sin 10t
\end{aligned}$$

You can see that the steady-state current is $6\sin 10t$. ▪

Exercise Set 15.6

In Exercises 1–22, use Laplace transforms to solve the given initial value problem.

1. $y' - y = 1$, $y(0) = 0$
2. $y' + 2y = t$, $y(0) = -1$
3. $y' + 6y = e^{-6t}$, $y(0) = 2$
4. $y' - y = \sin 10t$, $y(0) = 0$
5. $y' + 5y = te^{-5t}$, $y(0) = 3$
6. $y' + y = \sin t$, $y(0) = -1$
7. $y'' + 4y = 0$, $y(0) = 2$, $y'(0) = 3$
8. $y'' - 5y' + 4y = 0$, $y(0) = -2$, $y'(0) = 4$
9. $y'' + 4y = e^t$, $y(0) = y'(0) = 0$
10. $y'' + 2y' - 3y = 5e^{2t}$, $y(0) = 2$, $y'(0) = 3$
11. $y'' - 6y' + 9y = t$, $y(0) = 0$, $y'(0) = 1$
12. $y'' + y = 1$, $y(0) = y'(0) = 1$
13. $y'' + 6y' + 13y = 0$, $y(0) = 1$, $y'(0) = -4$
14. $y'' + 2y' + 5y = 8e^t$, $y(0) = y'(0) = 0$
15. $y'' - 4y = 3\cos t$, $y(0) = y'(0) = 0$
16. $y'' - y = e^t$, $y(0) = 1$, $y'(0) = 0$
17. $y'' + 2y' + 5y = 3e^{-2t}$, $y(0) = y'(0) = 1$
18. $y'' - 6y' + 9y = 12t^2e^{3t}$, $y(0) = y'(0) = 0$
19. $y'' + 2y' - 3y = te^{2t}$, $y(0) = 2$, $y'(0) = 3$
20. $y'' + 2y' - 3y = \sin 2t$, $y(0) = y'(0) = 0$
21. $y'' + 2y' + 5y = 10\cos t$, $y(0) = 2$, $y'(0) = 1$
22. $y'' + 2y' + 5y = 10\cos t$, $y(0) = 0$, $y'(0) = 3$

Solve Exercises 23–29.

23. *Electronics* Use Laplace transforms to determine the charge $q(t)$, the current $i(t)$, and the steady-state current of an RLC series circuit, if $R = 20\,\Omega$, $L = 1\,\text{H}$, $C = 0.005\,\text{F}$, $E(t) = 150\,\text{V}$, and $q(0) = i(0) = 0$.
24. *Electronics* The differential equation for the current in an RL series circuit is $L\frac{di}{dt} + Ri = E(t)$, where $E(t)$ is the impressed voltage. If $L = 1\,\text{H}$, $R = 10\,\Omega$, $E(t) = \sin t$, and $i(0) = 0$, find the formula for the current $i(t)$.
25. *Electronics* Determine the charge $q(t)$ and current $i(t)$, of an RLC series circuit, if $R = 6\,\Omega$, $L = 0.1\,\text{H}$, $C = 0.02\,\text{F}$, $E(t) = 6\,\text{V}$, and if $q(0) = i(0) = 0$.
26. *Electronics* An RLC series circuit has values of $R = 20\,\Omega$, $L = 0.05\,\text{H}$, $C = 100\,\mu\text{F}$, and $E(t) = 100\cos 200t$. Determine the charge, current, and steady-state current, if $q(0) = i(0) = 0$.
27. *Physics* A spring hangs in a vertical position with its upper end fixed. We know that the spring constant is $k = 48$ lb/ft. An object weighing 16 lb is attached to the lower end of the spring. After coming to rest, the object is pulled down 2 in. and released. If the air resistance in pounds is $64v$, describe the motion of the object.
28. *Physics* If the support of the spring in Exercise 27 is given a motion of $\cos 4t$ ft, describe the motion of the object.
29. *Automotive technology* A 3,200 lb vehicle rests on a spring-shock absorber system on each of four wheels. The system yields the model

$$y'' + 2y' + 17y = e^{-t}\cos 4t$$

where $e^{-t}\cos 4t$ represents the force in pounds resulting from a bumpy road. If $y = y' = 0$ when $t = 0$, use Laplace transforms to solve this differential equation.

In Your Words

30. Without looking in the text, outline the steps to solving a differential equation by Laplace transforms.
31. Of all the methods used in the text to solve differential equations, which one do you think was the easiest to use? Explain why you think this was the easiest method.

CHAPTER 15 REVIEW

Important Terms and Concepts

Euler's method
Huen's method
Increment method
Inverse Laplace transforms
LaPlace transforms
Numerical methods for solving differential equations
  Euler's method
  Huen's method
  Successive approximations
Partial fractions
Successive approximations

Review Exercises

In Exercises 1–4, solve the given differential equation by using Euler's method for the given initial value of *x* and *y* and the given value of *h*.

1. $y' = -y^2(1+x)$, $x_0 = 0$, $y_0 = 2$, $0 \le x \le 1$, $h = 0.1$, $h = 0.05$
2. $y' = \dfrac{2(x+1)^3}{y}$, $x_0 = 0$, $y_0 = 1$, $0 \le x \le 1$, $h = 0.1$, $h = 0.05$
3. $y' = -0.1y + \dfrac{4}{1+x^2}$, $x_0 = 0$, $y_0 = 60$, $0 \le x \le 1$, $h = 0.05$
4. $y' = \dfrac{6}{\sqrt{1+x^2}}$, $x_0 = 0$, $y_0 = 0$, $0 \le x \le 0.10$, $h = 0.01$

In Exercises 5–8, find the second approximation of the solution of the given differential equations if the solution passes through the given point.

5. $y' = xy + 4$, $(0, 1)$
6. $y' = 2x + y$, $(0, 0)$
7. $y' = y\sqrt{x-9}$, $(10, 1)$
8. $y' = y + \cos x$, $\left(\frac{\pi}{2}, 0\right)$

In Exercises 9–20, solve the given differential equation by use of Laplace transforms, where the solution has the given initial values.

9. $2y' - y = 4$, $y(0) = 1$
10. $3y' + y = t$, $y(0) = 2$
11. $y' + 2y = e^t$, $y(0) = 1$
12. $y' + 5y = 0$, $y(0) = 1$
13. $y'' - y = 0$, $y(0) = y'(0) = 1$
14. $y'' - y = e^t$, $y(0) = y'(0) = 0$
15. $y'' + 2y' + 5y = 0$, $y(0) = 1$, $y'(0) = 0$
16. $y'' + y' + y = 0$, $y(0) = 4$, $y'(0) = -2$
17. $y'' + 2y' + 5y = 3e^{-2t}$, $y(0) = y'(0) = 1$
18. $y'' + 2y' - 3y = 5e^{2t}$, $y(0) = 2$, $y'(0) = 3$
19. $y'' + 4 = \sin t$, $y(0) = y'(0) = 0$
20. $y'' + 4 = t$, $y(0) = -1$, $y'(0) = 0$

Solve Exercises 21 and 22.

21. *Electronics* An RLC series circuit has values of $R = 10\,\Omega$, $L = 0.5\,\text{H}$, $C = 0.01\,\text{F}$, and $E(t) = 10\sin t$ V. If $q(0) = 0$ and $i(0) = 10$ A, find the equation for the charge, the current, and the steady-state current.

22. *Physics* A 20-kg mass is attached to a spring with a spring constant of 700 N/m. The mass is started in motion from an equilibrium position with an initial velocity of 1 m/s upward and an applied external force of $5\sin t$. Find the equation for the motion of this mass, if the force in newtons due to air resistance is 90 times the velocity.

CHAPTER 15 TEST

In Exercises 1–4, use Table 15.1 to find the transform of the given function.

1. $f(t) = 5e^{-3t}$
2. $f(t) = 2t^3 + \sin 8t$
3. $f(t) = e^{-5t}\cos 4t$
4. $f(t) = 8t^9 e^{5t}$

In Exercises 5–7, find the inverse Laplace transform of the given function.

5. $F(s) = \dfrac{s-6}{(s-6)^2+7}$
6. $F(s) = \dfrac{s}{s^2-6s+13}$
7. $F(s) = \dfrac{5}{(s+3)(s^2+1)}$

In Exercises 8–10, use the method of Laplace transforms to solve the given differential equation.

8. $y' + 5y = 0$, $y(0) = 2$
9. $y'' + 2y = e^{-2t}$, $y(0) = 0$, $y'(0) = 0$
10. $y'' + 2y' + y = 0$, $y(0) = 1$, $y'(0) = -1$

Solve Exercises 11 and 12.

11. In an RLC series circuit, $R = 8.0\,\Omega$, $L = 0.20\,\text{H}$, $C = 0.002\,\text{F}$, and $E(t) = 12\sin 20t$ V. Find q as a function of time, if $q(0) = 0$ and $i(0) = q'(0) = 0$.

12. Use Euler's method to find an approximate solution to the differential equation $y' = 2x + y$, if the curve passes through the point $(0, 2)$ and $h = 0.2$.

Appendix

The Electronic Hand-Held Calculator

Until recently you would have had to work all the problems in this book in your head, with pencil and paper, or with a slide rule. There were some calculators available but they were large and bulky—almost the size of a typewriter—and they were very limited in the operations that you could perform with them.

The invention of the integrated circuit made it possible to process and store large amounts of information in a very small space and with very little energy. Later advances in integrated circuits allowed for even more information to be stored on an integrated circuit "chip." The advances in the chip plus the light emitting diode (LED) provided the basis for small, light, inexpensive electronic calculators. More recent calculators use a liquid crystal display (LCD) instead of the LEDs.

A.1 INTRODUCTION

Electronic calculators have quietly and quickly become common tools. People use them at work, at home, and in school. Because we expect that you will use a calculator in your work as a technician, we include this section showing how calculators can and should be used. You are encouraged and expected to use a calculator and/or computer when working problems in this book.

Types of Calculators

What kind of calculator should you get? For the work in this book, you need a "scientific" calculator. This type of calculator naturally has keys for the ten digits, [0] through [9], and for the basic operations ([+], [−], [×], and [÷]). But, the calculator should also have the keys [sin], [cos], [tan], [y^x] or [x^y], [log], and [$\ln x$]. Make sure you have the instruction manual for your calculator. Because not all calculators work the same, the methods described in this book may not work on your calculator. If you have this problem, the instruction manual for your calculator and your teacher should help you learn how to work these problems.

All calculators work in one of two basic ways. Some use **algebraic notation** (**AN**) and the others use **reverse Polish notation** (**RPN**). In the text, most calculator operations are shown using algebraic notation. If it is likely that confusion might result for users of RPN calculators, then an example showing how to use an RPN calculator is provided. It does not matter whether your calculator is an AN or RPN type. Just make sure you read your instruction manual carefully and practice using your calculator.

There are other variations among calculators. Some have one or more addressable memories. Others have the ability to draw graphs.

Another major difference is the order of operations performed by the calculator. A calculator that performs operations according to the usual mathematical order of operations is said to have an **algebraic operating system (AOS)**. Some companies refer to the order in which their calculators perform operations as the **equation operating system (EOS)**. Any calculator that uses the EOS is also using the AOS. If you need to specify that certain operations are performed before others, then you will need to use parentheses.

Consider the problem $9 + 5 \times 6$. Using the order of operations in Section 1.2, you should get an answer of 39. Work the problem on your calculator. Do you get 39 or 84? Some calculators will give one answer; some will give the other. If your calculator gives the answer 84, then you must be very careful to perform the operations in the correct order and either write the intermediate results or, if your calculator has a memory, store the intermediate results in memory. Some calculators have parenthesis keys that can be used to work problems like the one just described.

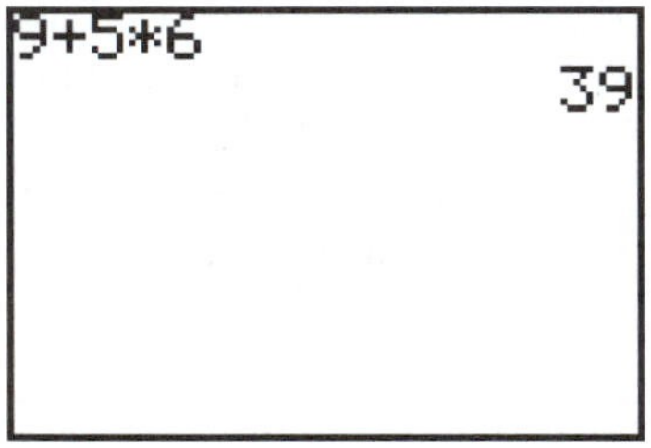

9+5*6
39

FIGURE A.1

All graphing calculators use an AOS. Figure A.1 shows the result of working $9 + 5 \times 6$ on a TI-82.

A Hewlett-Packard HP-48 gives the same result, but you have two ways, the "stack method" and the "equation editor method" to get an HP-48 to calculate $9 + 5 \times 6$. The keystrokes with the two methods are outlined below.

METHOD	PRESS	DISPLAY
stack	9 [ENTER] 5 [ENTER] 6 [×] [+]	39
equation editor	[↰] [ENTER] 9 [+] 5 [×] 6 [ENTER]	'9+5*6'
	[EVAL]	39

If you have any doubts as to which sequence of operations your calculator will follow when you use combined operations, press the [=] key after each individual operation and record, or store in the calculator's memory, the intermediate results. Practice and experience will allow you to become familiar enough with your calculator to use it correctly and to take advantage of its built-in capabilities.

Most calculators display 8 to 10 significant digits and store 2 or 3 more digits for rounding off purposes. However, this is not always the case. Some calculators employ a method called **truncation** in which any digits not displayed are discarded. Thus, 489.781 truncated to tenths is 489.7. This same number rounded off to tenths is 489.8. While you will seldom use numbers with enough significant digits for calculator truncation to be a problem, you should be aware of this possibility.

A **graphics calculator** has a larger screen and can draw and display graphs. One advantage of a graphics calculator is that it displays the numbers and operations and allows you to edit or change them.

One other difference between calculators is how they handle the decimal point. Some calculators are designed to show exactly two decimal places with every computation. These calculators have a **fixed decimal point**. Other calculators only show as many decimal places as result from the computation. These calculators have a **floating decimal point**. For example, a fixed decimal point calculator would show 34.50 when a floating decimal point calculator would show 34.5. Similarly, a floating decimal point calculator would show 47.368 when a fixed decimal point calculator would show 47.37. Most scientific calculators have a [Fix] key that allows you to set the number of digits to the right of the decimal point that will be displayed.

A.2 BASIC OPERATIONS WITH A CALCULATOR

In Section A.1, we examined the differences between calculators. In this section, we will learn how to use the calculator to solve problems such as those in the first few chapters of the text.

Entering Positive and Negative Numbers

Data are entered by pressing the digits in order from left to right. Thus, the number 48.732 would be entered into the calculator by pressing the keys in this order: [4] [8] [·] [7] [3] [2]. If your calculator has a floating decimal point the display should read 48.732; if it has a fixed decimal point, it would read 48.73.

Your calculator probably has a [+/–], [CHS], or [(–)] key. This key allows you to change the sign of the number in the display. Enter 59.638 in your calculator and press the [+/–] or [CHS] key. The display should now read –59.638. Press the [+/–] or [CHS] key again and you should see the calculator return to the original display of 59.638. Notice that the [+/–] or [CHS] key is pressed *after* the number is entered in the calculator.

Some calculators, such as graphics calculators, use a [(–)] key to enter a negative number. The use of the parentheses is to help you tell it apart from the [–] key which is used for subtraction. Here the [(–)] key is pressed *before* the number is entered. Thus, you would press [(–)] 59.638 in order to enter the number –59.638.

The [+/–] or [CHS] key should be pressed after the number has been entered. The [(–)] key should be pressed before the number is entered.

Dual Function Key

Some calculators have a **dual function** key or [2nd] key. In order to save space, calculator makers decided to let some keys perform more than one function. The first function is printed on the key; the second is printed just above the key; and a third function may be printed above or below the key.

To use the first function of a key, all you need to do is press the key. To use the second function, you first press the [2nd] key and then the key below the desired function. For example, on some calculators there is a key labeled [CLR] with CA above it much like this: CA [CLR]. In order to get the CA function, press the [2nd] key and then press the [CLR] key. We will write this as [2nd] [CA] to prevent any possible confusion about which function to use. Some calculators indicate the [2nd] key by [F] or [2nd F] or a blank key [ ] printed in a solid color such as gold or blue.

Graphics calculators have both a second function key and an alpha function key. For example, on the Texas Instruments TI-81, the [x^2] and the [LN] keys each have two symbols above them as shown here:

$\sqrt{\ }$ I e^x S

[x^2]   [LN]

√115
10.72380529

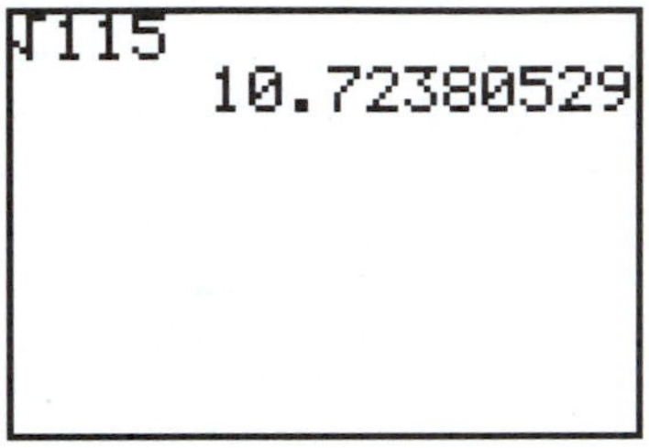

FIGURE A.2

To the left printed in light blue above a key is an operation. For example, to the left above the [x^2] key is the operation $\sqrt{\ }$ and to the left above the [LN] key is the operation e^x. To use an operation, press the [2nd] key and then the key you require. For example, to calculate $\sqrt{115}$, press [2nd] [$\sqrt{\ }$] 115 [ENTER]. The screen displays $\sqrt{\ }115$ on one line and the result, 10.72380529 on the next line, as shown in Figure A.2.

Note To help you use the correct key, the name of the operation and the [2nd] key are in the same color. On a TI-81 and TI-82 calculators, this color is light blue; on a TI-85, it is yellow.

To the right above a key is a letter or other symbol printed in grey. For example, to the right above the [x^2] key is the letter I and to the right above the [LN] key is the letter S. To use these keys, you first press the [ALPHA] key. So, to enter the letter I you press the [ALPHA] [x^2] key. In this book we will indicate that you need to use the [ALPHA] key by placing the letter inside a square. So, if we want you to use the variable I, we will write [ALPHA] [I].

Note Letters and symbols accessed by pressing the [ALPHA] key are color-coordinated. On a TI-8x calculator, this color is grey.

Basic Arithmetic Operations

The basic arithmetic operations are performed using the operation keys [+], [−], [×], and [÷], and the [=], [ENTER], or [EXE] key. The problem $4 \times 8 + 5 \times 6$ would be worked by keying into the calculator as follows:

Algebraic Calculator	RPN Calculator
4 [×] 8 [+] 5 [×] 6 [=]	4 [ENTER] 8 [×] 5 [ENTER] 6 [×] [+]

In each case, the final display is 62.

To find the sum of 14.32 and 9.37 you press

Algebraic Calculator	RPN Calculator
14.32 [+] 9.37 [=]	14.32 [ENTER] 9.37 [+]

and the result 23.69 is displayed.

Subtraction, multiplication, and division are handled in much the same manner.

Note From now on, we will indicate what you should enter and what is displayed in the following manner.

PRESS	DISPLAY
14.32 [+] 9.37 [=]	23.69

There will be times when a third column will be added, so we can include some comments or explanations.

EXAMPLE A.1

Use an algebraic or graphing calculator to evaluate **(a)** $72.1 - 83.12$, **(b)** 4.3×6.1, and **(c)** $7.8 \div 2.4$

Solution

	PRESS	DISPLAY
(a)	72.1 [−] 83.12 [=]	−11.09
(b)	4.3 [×] 6.1 [=]	26.23
(c)	7.8 [÷] 2.4 [=]	3.25

EXAMPLE A.2

Use an RPN calculator to work the problem $43.7 + 56.2$.

Solution

PRESS	DISPLAY
43.7 [ENTER]	43.7
56.2 [+]	99.9

EXAMPLE A.3

Evaluate **(a)** $9.3 + 84.12 - 20.5 - 123.1$ and **(b)** $7.4 \times 2.5 \div 37 \times 8$.

Solution

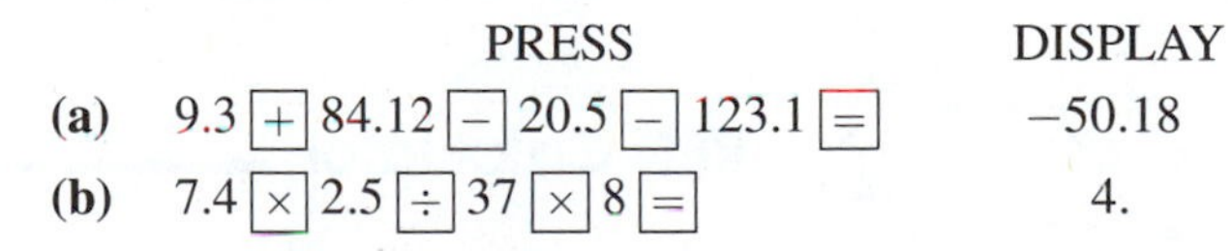

	PRESS	DISPLAY
(a)	9.3 [+] 84.12 [−] 20.5 [−] 123.1 [=]	−50.18
(b)	7.4 [×] 2.5 [÷] 37 [×] 8 [=]	4.

These same two problems would be worked on an RPN calculator as follows:

	PRESS	DISPLAY
(a)	9.3 [ENTER] 84.12 [+] 20.5 [−] 123.1 [−]	−50.18
(b)	7.4 [ENTER] 2.5 [×] 37 [÷] 8 [×]	4.0

Parentheses

Even if your calculator uses the algebraic operating system (AOS) described earlier, there are times when you will need to specify the exact order in which expressions are to be evaluated. There are several different symbols used for grouping—parentheses ( ), brackets [ ], and braces { }—but calculators and computers primarily use parentheses. The parentheses tell the calculator to perform the operations inside the parentheses first. If you are not sure how your calculator will handle an expression, you should use parentheses.

A calculator does not know about implied multiplication. People often write $(7+3)(8-5)$ when they mean $(7+3)\times(8-5)$. You must tell the calculator when quantities are being multiplied, as shown in Example A.4.

EXAMPLE A.4

Multiply $(7+3)\times(8-5)$.

Solution

PRESS	DISPLAY
[(] 7 [+] 3 [)] [×] [(] 8 [−] 5 [)] [=]	30.

A slightly more involved use of parentheses is given in Example A.5. Here you want the calculator to evaluate the entire numerator and then divide by the entire denominator. To make sure that this is what the calculator does, extra sets of parentheses must be added.

EXAMPLE A.5

FIGURE A.3

Evaluate $\dfrac{(8+3)\times(7-21)}{(6+15\div 3)\times 2}$ with your calculator.

Solution Add another set of parentheses around the numerator and the denominator. The problem now looks like $\dfrac{((8+3)\times(7-21))}{((6+15\div 3)\times 2)}$. Now use your calculator to evaluate this expression. The result below, and in Figure A.3, is for a TI-82 graphing calculator.

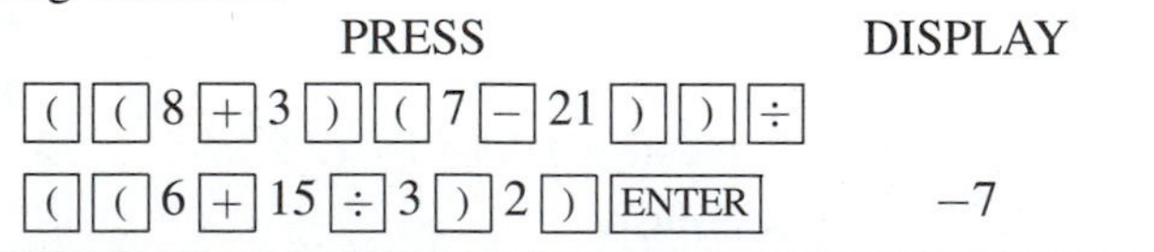

RPN Calculator

Add parentheses to aid in the correct order of operations. It is easiest to work from the inside of parentheses outward, much as you would if you were doing the calculations on paper. Since division is performed before addition, a set of parentheses should be added in the denominator.

$$\frac{(8+3)\times(7-21)}{(6+(15\div 3)\times 2)}$$

You would then use an RPN calculator in the following manner:

PRESS	DISPLAY	COMMENT
8 [ENTER] 3 [+] 7 [ENTER] 21 [−] [×]	−154	(The numerator)
[ENTER]	−154	(Stores the numerator)
15 [ENTER] 3 [÷] 6 [+] 2 [×]	22	(The denominator)
[÷]	−7	(The result)

Note Remember that parentheses come in pairs. Whenever you use a left parenthesis, (, you must also use a right parenthesis,), in that same problem.

Example A.6 provides another chance for you to use parentheses with your calculator.

EXAMPLE A.6

Evaluate $\dfrac{9\times 3+7\times(-8)+(-16)}{((3+6\times 5)+4.5-7.5)\div 1.5}$.

Solution Notice that there are no parentheses in the numerator. If your calculator does not use the AOS, then you need to place parentheses so that you get the right answer for the numerator. Parentheses should be placed as such:

$$(9\times 3)+(7\times -8)+(-16)$$

If your calculator has AOS, then this step is not necessary. You need to rewrite the problem by placing a set of parentheses around the numerator and a set around the denominator. The problem now looks like

$$\frac{(9\times 3+7\times -8+(-16))}{(((3+6\times 5)+4.5-7.5)\div 1.5)}$$

Use a TI-8x calculator to evaluate this expression.

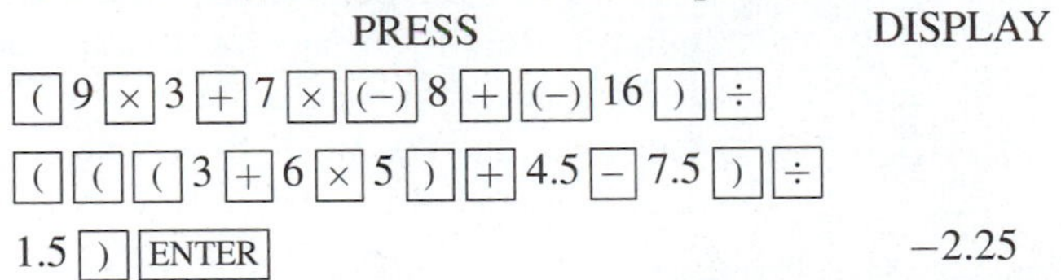

PRESS	DISPLAY
[(] 9 [×] 3 [+] 7 [×] [(−)] 8 [+] [(−)] 16 [)] [÷]	
[(] [(] [(] 3 [+] 6 [×] 5 [)] [+] 4.5 [−] 7.5 [)] [÷]	
1.5 [)] [ENTER]	−2.25

Reciprocals

Many calculators have a [1/x] or [x^{-1}] key that is used to find the reciprocal of a number on display. Remember, your calculator will give the decimal value of the reciprocal. In Example A.7, we use the calculator to evaluate the reciprocals of three numbers.

EXAMPLE A.7

What are the reciprocals of **(a)** 3, **(b)** -19, and **(c)** $\frac{4}{5}$?

Solution

		PRESS	DISPLAY
(a)		3 [1/x]	0.3333333
(b)		19 [+/−] [1/x]	−0.0526316
(c)	Algebraic calculator:	4 [÷] 5 [=] [1/x]	1.25
	RPN calculator:	4 [ENTER] 5 [÷] [1/x]	1.25
	Graphics calculator:	[(] 4 [÷] 5 [)] [x^{-1}] [ENTER]	1.25

Notice that in (c), if you knew that the decimal equivalent of $\frac{4}{5}$ was 0.8, then you could have entered 0.8 [1/x] or 0.8 [x^{-1}] and the display would have read 1.25. But be careful when you try to use a short cut to solve a problem. Remember that the reciprocal of a reciprocal is the original number. We know that the reciprocal of 3 is $\frac{1}{3}$, so the reciprocal of $\frac{1}{3}$ is 3. Now work Example A.8.

EXAMPLE A.8

Use a calculator to **(a)** find the reciprocal of the reciprocal of 3, and **(b)** find the reciprocal of 0.3333333.

Solution

	PRESS	DISPLAY
(a)	3 [1/x]	0.3333333
	[1/x]	3.
(b)	0.3333333 [1/x]	3.0000003

As you can see, the reciprocal of a reciprocal is supposed to return the original number. But, if you enter the decimal equivalent of a number that does not terminate before the eighth place to the right of the decimal point you may not get the exact reciprocal that you want. Sometimes it is worth the few extra steps it takes to enter 1 [÷] 3 instead of 0.3333333.

Exercise Set A.2

Use a calculator to evaluate Exercises 1–24.

1. $28+45$

2. $48+27$

3. $93-47$

4. $86-39$

5. $43-192$

6. $57-216$

7. 21×15

8. 47×12

9. $18 \div 5$

10. $38 \div 4$

11. $15.32+-6.71$

12. $18.71+-7.38$

13. $-14.67 - 8.932$
14. $-21.93 - 6.091$
15. 12.41×-3.4
16. 23.36×-517
17. $-8.35 \div 2.5$
18. $53.72 \div -6.32$
19. $16.37 + 81.4 - 6.437 + -8.5 - -16.372$
20. $-12.4 - 81.37 + -6.9 - 108.4 + -6.43$
21. $8.4 \times 3.7 \div 6.2 \div -4.2 \times 124 \div -0.74$
22. $-165 \div 1.1 \times 2.6 \times -8.5 \div 13 \div -0.015$
23. $\dfrac{(8+7) \times 3 \div ((6.2-5) \times 5)}{18 \div 6 - 3 \times 4.5 - 1.25 \times -8}$
24. $\dfrac{(17.3 + 6.82) \times 4.1 \div (-27.3 \div 3 - 8.9)}{(81.73 - 16.38 \times 220 \div 11 + 40.37) \div (97.6 - 15.4)}$

Use a calculator to determine the reciprocal of the numbers in Exercises 25–30.

25. 4
26. 0.5
27. 4.5
28. 0.1111111
29. 215.3
30. 0.00025

A.3 SOME SPECIAL CALCULATOR KEYS

In Section A.2, we looked at one special key—the [1/x] or reciprocal key and at the grouping keys [(] [)]. In this section you will learn to use eight more special keys.

Keys for Powers or Roots

The first four special keys are used to find decimal values of powers and roots. Two of these should look familiar. These are the [x^2] and [$\sqrt{\ }$] or [$\sqrt{x}$] keys. The [x^2] will calculate the square (or second power) of any number you enter as shown in Example A.9.

EXAMPLE A.9

Evaluate **(a)** 3^2, **(b)** 5^2, **(c)** 4.2^2, **(d)** 135^2, and **(e)** $(-7)^2$.

Solution

	PRESS	DISPLAY
(a)	3 [x^2]	9.
(b)	5 [x^2]	25.
(c)	4.2 [x^2]	17.64
(d)	135 [x^2]	18225.
(e)	Algebraic: 7 [+/–] [x^2]	49.
	RPN: 7 [CHS] [x^2]	49.
	Graphic: [(] [(–)] 7 [)] [x^2] [ENTER]	49.

The [√] key is used to determine the square root of a number. On many calculators you first enter the number and then press the [√] key. (This may require pressing the [2nd] key and then the [√] key.) On a graphics calculator the [√] key is pressed before the numbers.

EXAMPLE A.10

Evaluate **(a)** $\sqrt{25}$, **(b)** $\sqrt{2}$, and **(c)** $\sqrt{797}$.

Solution

	PRESS	DISPLAY
(a)	Algebraic or RPN: 25 [√]	5.
	Graphics: [√] 25	5.
(b)	Algebraic or RPN: 2 [√]	1.4142136
	Graphics: [√] 2	1.4142136
(c)	Algebraic or RPN: 797 [√]	28.231188
	Graphics: [√] 797	28.231188

In Example A.10(a), the displayed number is the exact principal square root of the number. The numbers displayed for $\sqrt{2}$ and $\sqrt{797}$ are approximate values. Both of these are irrational numbers. If you evaluate $(1.4142136)^2$ and $(28.231188)^2$, you may not get 2 and 797. For example, on one calculator $(1.4142136)^2 = 2.0000001$ and $(28.231188)^2 = 796.99998$.

Remember, you can only take square roots of positive real numbers. What happens when you enter 2 [+/−] [√] into your calculator? Some calculators display 1.4142136 in the flashing mode. The flashing indicates that you have made an error. Other calculators display an "E" or the word "Error." A number, such as 0, may be displayed with the "E".

If you want powers larger than two, use the [y^x], [x^y], or [∧] key. This discussion will be written for a calculator with a [y^x] key. If your calculator has an [x^y] or [∧] key, then you should make any necessary changes.

Suppose you want to find 2^3. We know that $2^3 = 2 \times 2 \times 2 = 8$. To get this same answer with your algebraic calculator, enter 2 [y^x] 3 [=]. On an RPN calculator, press 2 [ENTER] 3 [y^x] and on some graphing calculators press 2 [∧] 3 [ENTER].

To evaluate $4^{5/6}$, you need to use parentheses. You should enter 4 [y^x] [(] 5 [÷] 6 [)] [=]. The result is approximately 3.1748. Now work the problems in Example A.11 to see how to use your calculator.

EXAMPLE A.11

Evaluate: **(a)** 4^3, **(b)** $4.3^{2.7}$, **(c)** $(\sqrt{3})^{\sqrt{2}}$, **(d)** 2^{-3}, **(e)** $\sqrt[4]{2}$, and **(f)** $9^{5/3}$.

Solution

	PRESS	DISPLAY
(a)	4 [y^x] 3 [=]	64.
(b)	4.3 [y^x] 2.7 [=]	51.32924
(c)	3 [$\sqrt{\ }$] [y^x] 2.7 [$\sqrt{\ }$] [=]	2.466015575
	TI-81: [$\sqrt{\ }$] 3 [∧] [$\sqrt{\ }$] 2.7 [ENTER]	2.466015575
(d)	2 [y^x] 3 [+/–] [=]	0.125
	Graphics: 2 [y^x] [(–)] 3 [ENTER]	0.125
(e)	2 [y^x] 0.25 [=]	1.1892071
(f)	2 [x^y] [(] 5 [÷] [)] [=]	38.9407384

Memory Keys

Memory keys allow you to store numbers that you plan to use later. These numbers are stored in special places in a calculator. The number of memories available in your calculator may range from as few as one for older calculators to 27 for the TI-82 and TI-83. Calculators such as the TI-85 and TI-92 are limited only by the amount of memory, in bytes, that the calculator can use.

To STOre the number on display, you press the [STO] key. Some calculators use [M] for the Memory key. In calculators with more than one memory, you must tell the calculator in which memory you want to store the number. Calculator memories are numbered. A calculator with ten memories numbers the locations 0–9. If you want to place a displayed number in memory location six, press [STO] [6]. To erase the numbers from all memories press [2nd] [CA].

Calculators that have alphabetical keys have a memory for each letter of the alphabet. For example, to place a number displayed on the TI-81 in memory location B, you press [STO(] and then the [B] key. Do not press the [ALPHA] key. On a Casio fx7700-G, the [→] key acts like the [STO(] key, except that you need to press the [ALPHA] key. For example, to store a number in memory location B, you press [→] [ALPHA] [B].

Once you have stored a number in memory, you need to be able to get it back so you can use it. To do this, use the ReCalL key, [RCL]. On some calculators, this is the Memory Recall, [MR], key. If your calculator has more than one memory, then you must indicate which memory location contains the number. To recall the number in memory location five, press [RCL] [5]. To recall the number in memory location B, press [ALPHA] [B].

Note

Recalling a number from memory does not clear the memory. The number is still stored in that memory, so you can use it again.

Instructions for more special calculator keys are included at various places in the text.

Exercise Set A.3

Use a calculator to evaluate Exercises 1–44.

1. 7^2
2. 11^2
3. -25^2
4. 142^2
5. 3.1^2
6. 7.95^2
7. -18.43^2
8. -26.4^2
9. $/\sqrt{36}$
10. $\sqrt{169}$
11. $\sqrt{1296}$
12. $\sqrt{45796}$
13. $\sqrt{10.24}$
14. $\sqrt{34.81}$
15. $\sqrt{151.5361}$
16. $\sqrt{2631.69}$
17. 9^3
18. 7^4
19. 21^5
20. 53^4
21. 4.7^3
22. 0.24^8
23. 14.1^{22}
24. 23.4^{36}
25. $\sqrt[4]{6561}$
26. $\sqrt[5]{32768}$
27. $\sqrt[7]{62748517}$
28. $\sqrt[6]{11390625}$
29. $\sqrt[7]{340.48254}$
30. $\sqrt[5]{656.75808}$
31. $\sqrt[23]{241}$
32. $\sqrt[62]{11562}$
33. 8^{-3}
34. 9^{-7}
35. $1.1^{4.2}$
36. $2.3^{5.7}$
37. 7^3
38. 12^5
39. 25^{-10}
40. 47^{-8}
41. $\sqrt[3.1]{23}$
42. $\sqrt[2.45]{62}$
43. $\sqrt[3]{\sqrt[4]{9}}$
44. $\sqrt[3]{\sqrt[9]{56}}$

Enter each of the numbers in Exercises 45–52 into your calculator in scientific notation.

45. 6.21×10^{14}
46. 3.781×10^{23}
47. 8.92×10^{-21}
48. 7.631×10^{-46}
49. $8,437,100,000,000$
50. $62,790,000,000$
51. 0.00000000000271
52. 0.0000000000472

Evaluate Exercises 53–60 with the help of a calculator.

53. $4.8 \times 10^{12} \times 5.23 \times 10^{25}$
54. $7.912 \times 10^{35} \times 6.39 \times 10^{24}$
55. $1.73 \times 10^{18} \times 8.42 \times 10^{-21}$
56. $2.71 \times 10^{25} \times 4.32 \times 10^{-16}$
57. $4.31 \times 10^{14} \div 2.31 \times 10^{12}$
58. $8.92 \times 10^{19} \div 4.62 \times 10^{23}$
59. $5.42 \times 10^{21} \div 2.47 \times 10^{-5}$
60. $4.72 \times 10^{31} \div 4.93 \times 10^{-34}$

Use the STO and RCL keys on a calculator to solve Exercises 61–68.

61. $\dfrac{45.37 - 82.91}{6.4 \times 3.7 - 8.9 \times 4}$
62. $\dfrac{(9.31 + 8.72)5.4}{7.3 \times 4.6 - 4 \times 9.1}$
63. $\dfrac{3.7 - 91.4}{(18.32 - 6.7)9.2}$
64. $A = (4.3 + 8.1) \div 7.2$, $B = 5.7A - 34.9$, $C = \dfrac{B}{A} - A$, Find C
65. $A = (4.3 + 8.1) \div 7.2$, $B = 5.7A - 34.9$, $C = \dfrac{B}{A} - A$, Find C
66. $A = 12.4 \div (3.7 - 8.9)$, $B = \sqrt{-A}$, $C = 4B + A$, Find C
67. $A = 4.3 \times 8.7 - 9.1$, $B = A^2 + 7$, $C = (\sqrt{B})A$, Find C
68. $A = (2.7 + 8.9)^2$, $B = \sqrt[3]{4A} - \sqrt{A}$, $C = \dfrac{B^5 + A^3}{AB}$, Find C

≡ A.4 GRAPHING CALCULATORS AND COMPUTER-AIDED GRAPHING

The introduction of microcomputers has allowed people to use inexpensive computers for writing and performing calculations quickly and accurately; but people discovered that words and numbers were not enough, and so computer graphics were introduced.

Some of these graphing capabilities are now available on calculators. Use of a graphing calculator or a computer removes some of the drudgery of plotting equations by hand. In this text, we show how you can use computer and calculator graphing capabilities. However, we will also try to show you how to interpret the information they give you.

The graphics capabilities of computers and calculators vary widely. Graphics examples in this text were run on either a Casio fx-7700G or a Texas Instruments TI-82 graphing calculator or on a Macintosh computer.

Section 1.3 is an optional section that presents the basics of using a graphing calculator. Additional examples that explain more advanced features of graphing calculators are spread throughout the text.

Exercise Set A.4

Review Exercises

Use a calculator to evaluate Exercises 1–20.

1. $43 + -85$
2. $16 - 34$
3. 25×-81
4. $-62 \div 16$
5. 8.32×4.15
6. $-9.41 + 18.301$
7. $4.32 - 1.011$
8. $15.42 \div 0.002$
9. $8.3 + 9.1 - 4.3 \times 4 + 3 \div 5$
10. $-162 \div 1.2 + 4.3 \div 0.4 - 7$
11. $\dfrac{(3+5) \times -2 \div ((6-4.3) \times 7)}{9 \div 3 - 4 \times 2.3 - 5 \times -7}$
12. $\dfrac{(16.3-4) \times 6.2 \div (3.4 \div 18.1 + 3)}{8.1 \div 9 + 10}$
13. 35^2
14. $\sqrt{5}$
15. $(-4)^5$
16. 14^6
17. $\sqrt[7]{813}$
18. $\sqrt[4]{216}$
19. $\sqrt[2.3]{42.1}$
20. $\sqrt[1.3]{25.4}$

In Exercises 21–24, enter the number into your calculator in scientific notation.

21. 4.31×10^{23}
22. 6.42×10^{-16}
23. $34,700,000,000,000,000,000,000,000,000,000$
24. 0.000000000000000000427

Evaluate Exercises 25–26 with the help of a calculator.

25. $2.7 \times 10^{18} \times 3.4 \times 10^{23}$
26. $3.8 \times 10^{-15} \times 4.5 \times 10^{13}$

Appendix

B

The Metric System

The metric system[1] is becoming more important to workers. Almost every major country uses the metric system. Many industries are converting their manufacturing processes and products to metric specifications. In Canada, and every other major industrial country except the United States, the SI metric system is the official measurement system. Competition in the form of international trade has forced many U.S. companies to convert their products to the metric system. The U.S. Congress passed legislation requiring federal agencies to use the metric system in their business activities.

Older technical service publications were printed using the U.S. Customary system, while the latest publications are either in metric or in a combination of both measurement systems. This section gives a brief introduction to the metric system and the different units and their symbols, and tells how to convert within the metric system or between the metric and U.S. Customary system.

Units of Measure

There are seven base units in the metric system and two supplemental units in the SI metric system. All seven base units and the two supplemental units are shown in Table B.1.

All other units are formed from these seven. The base units that you will use most often are those for length (meter), mass or weight (kilogram), time (second), and electric current (ampere). The base unit for temperature is Kelvin, but we will mostly use a variation called the degree Celsius. In addition to the seven base units, there are two supplemental units. We use the supplemental unit for a plane angle (radian).

Each unit can be divided into smaller units or made into larger units. To show that a unit has been made smaller or larger, a prefix is placed in front of the base unit. The prefixes are based on powers of 10. The most common prefixes are shown in Table B.2. For example, 1 kilometer (km) is 1 000 m, 1 milligram (mg) is 0.001 g, 1 megavolt (MV) is 1 000 000 V$=10^6$ V, and 1 nanosecond (ns) is 0.000 000 001 s $=10^{-9}$ s.

[1] The official name for the metric system is The International System of Units (or, in French, Le Système International d'Unités). The official abbreviation for the metric system throughout the world is "SI."

TABLE B.1 SI Metric System Base and Supplemental Units

Quantity	Unit	Symbol
BASE UNITS		
Length	meter	m
Mass	kilogram	kg
Time	second	s
Electric current	ampere	A
Temperature	Kelvin	K
Luminous intensity	candela	cd
Amount of substance	mole	mol
SUPPLEMENTAL UNITS		
Quantity	**Unit**	**Symbol**
Plane angle	radian	rad
Solid angle	steradian	sr

TABLE B.2 Metric Prefixes

Multiple	Prefix	Symbol	
1 000 000 000	$= 10^9$	giga	G
1 000 000	$= 10^6$	mega	M
1 000	$= 10^3$	kilo	k
1			
0.01	$= 10^{-2}$	centi	c
0.001	$= 10^{-3}$	milli	m
0.000 001	$= 10^{-6}$	micro	μ
0.000 000 001	$= 10^{-9}$	nano	n

Many quantities, such as area, volume, speed, velocity, acceleration, and force require a combination of two or more fundamental units. Combinations of these units are referred to as **derived units** and are listed in Table B.3.

Table B.4 shows some of the metric units most commonly used and their uses in various trade and technical areas. This list is not intended to be exhaustive, but to provide you with some idea as to where the different units are used.

Writing Metrics

You should remember some important rules when you use the metric system.

1. The unit symbols that are used are the same in all languages. This means that the symbols used on Japanese cars are the same as those used on German or American cars.
2. The unit symbols are not abbreviations and a period is not put at the end of the symbol.

TABLE B.3 Derived Units for Common Physical Quantities

Quantity	Derived Unit	Symbol	Alternate Symbol
Acceleration	meter per second squared	m/s^2	
Angular acceleration	radian per second squared	rad/s^2	
Angular velocity	radian per second	rad/s	
Area	square meter	m^2	
Concentration, mass (density)	kilogram per cubic meter	kg/m^3	
	gram per liter	g/L	
Capacitance	farad	F	C/V
Electric current	ampere	A	V/Ω
Electric field strength	volt per meter	V/m	
Electric resistance	ohm	Ω	V/A
Electromotive force (emf)	volt	V	J/C
Force	newton	N	$kg \cdot m/s^2$
Frequency	hertz	Hz	s^{-1}
Illuminance	lux	lx	lm/m^2
Inductance	henry	H	$V \cdot s/A$
Light exposure	lumen second	$lm \cdot s$	
Luminance	candela per square meter	cd/m^2	
Magnetic flux	weber	Wb	$V \cdot s$
Moment of force	newton meter	$n \cdot m$	
Moment of inertia, dynamic	kilogram meter squared	$kg \cdot m^2$	
Momentum	kilogram meter per second	$kg \cdot m/s$	
Power	watt	W	J/s
Pressure	pascal	Pa	N/m^2
Quantity of electricity	coulomb	C	
Speed, velocity	meter per second	m/s	
Volume	cubic meter	m^3	
Volume flow rate	cubic meter per second	m^3/s	
	liter per minute	L/min	
Work, energy, quantity of heat	joule	J	$N \cdot m$

3. Unit symbols are shown in lowercase letters except when the unit is named for a person. Examples of unit symbols that are written in capital letters are joule (J), newton (N), pascal (Pa), watt (W), ampere (A), and coulomb (C). The only exception is the use of L for liter. The symbol L is often used for liter to eliminate any confusion between the lowercase unit symbol (l) and the numeral (1). Some people prefer to use a script ℓ for liter.
4. The symbol is the same for both singular and plural (e.g., 1 m and 12 m).
5. Numbers with four or more digits are written in groups of three separated by a space instead of a comma. (The space is optional on four-digit numbers.) This is done because some countries use the comma as a decimal point.

EXAMPLE B.1

Use 2 473 or 2473 instead of 2,473;
45 689 instead of 45,689;
47 398 254.263 72 instead of 47,398,254.26372.

TABLE B.4 Uses of Metric Units in Technical Areas

Quantity	Unit	Symbol	Use
Length	micrometer	μm	paint thickness; surface texture or finish
	millimeter	mm	motor vehicle dimensions; wood; hardware, bolt, and screw dimensions; tool sizes; floor plans
	centimeter	cm	bearing size; length and width of fabric or window; length of weld, channel, pipe, I-beam, or rod
	meter	m	braking distance; turning circle; room size; wall covering; landscaping; architectural drawings; length of pipe or conduit; highway width
	kilometer	km	land distances; maps; odometers
Area	square centimeter	cm^2	piston head surfaces; brake and clutch contact area; glass, tile, or wall covering; area of steel plate
	square meter	m^2	fabric, land, roof, and floor area; room sizes; carpeting; window and/or wall covering
Volume or capacity	cubic centimeter	cm^3	cylinder bore; small engine displacement; tank or container capacity
	cubic meter	m^3	work or storage space; truck body; room or building volume; trucking or shipping space; tank or container capacity; ordering concrete; earth removal
	milliliter	mL	chemicals; lubricant; oils; small liquids; paint;
	liter	L	fuel; large engine displacement; gasoline
Temperature	degree Celsius	°C	thermostats; engine operating temperature; oil or liquid temperature; melting points; welds
Mass or weight	gram	g	tire weights; mailing and shipping packages
	kilogram	kg	batteries; weights; mailing and shipping packages
	metric ton	t	vehicle and load weight; construction material such as sand or cement; crop sales
Bending force, torque, moment of force	newton meter	N · m	torque specifications; fasteners
Pressure/vacuum	kilopascal	kPa	gas, hydraulic, oxygen, tire, air, or air hose pressure; manifold pressure compression; tensile strength
Velocity	kilometers per hour	km/h	vehicle speed; wind speed
	meters per second	m/s	speed of air or liquid through a system
Force, thrust, drag	newton	N	pedal; spring; belt; drawbar
Power	watt	W	air conditioner; heater; engine; alternator
Illumination	lumens per square meter	lm/m^2	intensity of light on a given area

TABLE B.4 (continued)

Quantity	Unit	Symbol	Use
Density	milligrams per cubic meter	mg/m^3	industrial hygiene standards for fumes, mists, and dusts
Flow	cubic meters per second	m^3/s	measure of air exchange in a region; exhaust and air exchange system ratings

6. A zero is placed to the left of the decimal point if the number is less than one (0.52 L, not .52 L).
7. Liter and meter are often spelled litre and metre.
8. The units of area and volume are written by using exponents.

EXAMPLE B.2

5 square centimeters are written as 5 cm^2, not as 5 sq cm;
37 cubic meters are written as 37 m^3, not as 37 cu m

Changing Within the Metric System

Changing units of measurement within a system is called **reduction**. A change from one system to another is called **conversion**. We will now discuss reduction in the metric system, and then conversion between the metric and U.S. customary system afterward.

The metric prefixes in Table B.2 provide a method to help with metric reduction. Using the information in Table B.2, we construct a metric reduction diagram like the one shown in Figure B.1. Starting on the left with the largest prefix shown in Table B.2, we mark each multiple of 10 until we get to the smallest prefix. Notice that not all multiples are labeled. While there is a prefix for each multiple of 10, we have given you only those that you will need.

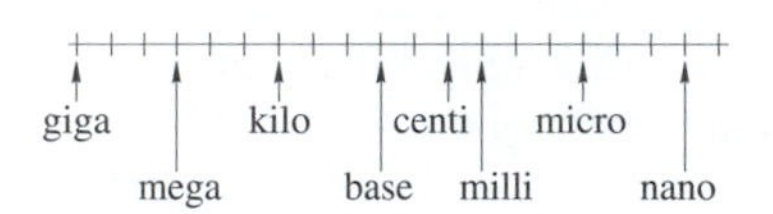

FIGURE B.1

Reduction in the SI Metric System

To change from one metric system unit to another:

1. Mark each unit on the reduction scale in Figure B.1.
2. Move the decimal point as many places as you move along the reduction scale according to the following rules:
   (a) To change to a unit farther to the right, move the decimal point to the right.
   (b) To change to a unit farther to the left, move the decimal point to the left.
3. If you are changing between square units, then move the decimal point *twice* as far as indicated in Step 2.
4. If you are changing between cubic units, then move the decimal point *three times* as far as indicated in Step 2.

EXAMPLE B.3

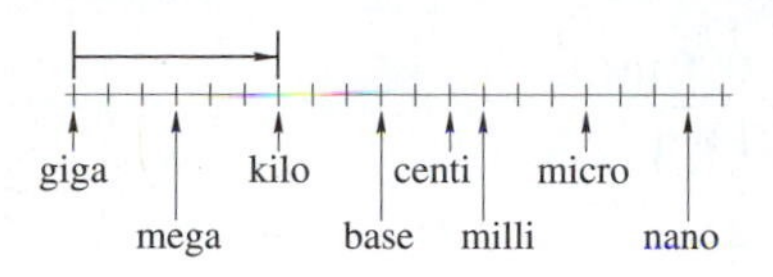

FIGURE B.2

Change 7.35 GV to kilovolts.

Solution Since 7.35 GV represents 7.35 gigavolts, we put a mark at the giga point on the reduction scale in Figure B.2. We want to convert GV to kV, so an arrow is drawn from the first mark to the mark labeled "kilo." The arrow points to the right and is 6 units long, so we move the decimal point 6 units to the right. To do this, we need to insert some zeros, with the result

$$7.35\,\text{GV} = 7\,350\,000\,\text{kV}$$

EXAMPLE B.4

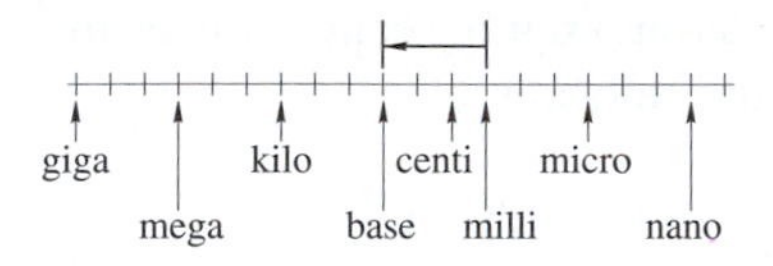

FIGURE B.3

Change 9.6 mm to meters.

Solution Since 9.6 mm represents 9.6 millimeters, we put a mark at the milli point on the reduction scale in Figure B.3. We want to convert mm to m, the base unit, so an arrow is drawn from the first mark to the mark labeled "base." The arrow points to the left and is 3 units long, so we move the decimal point 3 units to the left. Again, we need to insert some zeros, with the result

$$9.6\,\text{mm} = 0.009\,6\,\text{m}$$

EXAMPLE B.5

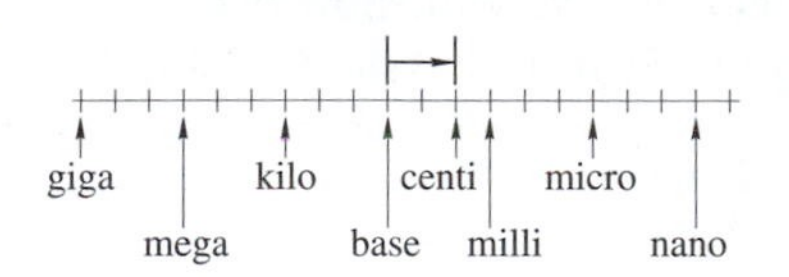

FIGURE B.4

Change 3.7 m^2 to square centimeters.

Solution Since 3.7 m^2 represents 3.7 square meters, we put a mark at the base point on the reduction scale in Figure B.4. We want to convert m^2 to cm^2, so the arrow is drawn from the first mark to the mark labeled "centi." The arrow points to the right and is 2 units long. Since we are converting between square units, we double this number, so we move the decimal point 4 units to the right. Again, we need to insert some zeros, with the result

$$3.7\,\text{m}^2 = 3\,700\,\text{cm}^2$$

To change one set of units into another set of units, you should perform algebraic operations with units to form new units for the derived quantity.

EXAMPLE B.6

Convert a speed of 88.00 km/h to meters per second.

Solution We will use two relationships that will give four possible conversion

EXAMPLE B.6 (Cont.)

factors.

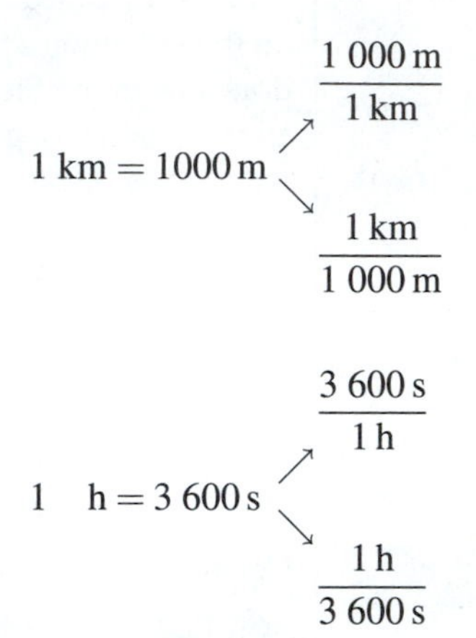

We write the quantity to be changed, then choose the appropriate conversion factors so that all but two of the units "cancel," leaving the units m/s.

$$88\frac{\cancel{\text{km}}}{\cancel{\text{h}}} \times \frac{1\,000\,\text{m}}{1\,\cancel{\text{km}}} \times \frac{1\,\cancel{\text{h}}}{3\,600\,\text{s}} \approx 24.44\,\text{m/s}$$

Changing Between the Metric and Customary Systems

Many technicians are often called upon to use both the SI metric and the U.S. Customary measurement systems. While it is best not to change from either the metric or the customary system to the other, sometimes a worker has to do so. For example, it is possible that the measuring tools available may be different than the measuring system in which the specifications are given. In such a case, the technician will be required to convert from one system to another. We will show one way to change between the U.S. Customary system and the SI metric system. Table B.5 should help you do this.

EXAMPLE B.7

Express 4.3 kg in pounds.

Solution You are changing from the metric (kg) to the customary system.

Look at Table B.5 under the heading, "From Metric to Customary." You are changing a measurement from kg to lb. Look in the column labeled "To change from" until you find the symbol for kilogram (kg). The symbol in the next right-hand column is the symbol for pound (lb).

Now look in the "Multiply by" column just to the right of these two symbols. The number there is 2.205. Multiply the given number of kilograms (4.3) by this number (2.205).

$$4.3 \times 2.205 = 9.4815$$

This means that 4.3 kg is equivalent to 9.4815 lb (or 4.3 kg $\approx$ 9.4815 lb).

TABLE B.5 **Changing Units Between the Metric and Customary Systems**

	FROM METRIC TO CUSTOMARY			FROM CUSTOMARY TO METRIC		
Quantity	**To change from**	**To**	**Multiply by**	**To change from**	**To**	**Multiply by**
Length	μm	mil	0.039 37	mil	μm	25.4
	mm	in.	0.039 37	in.	mm	25.4
	cm	in.	0.393 7	in.	cm	2.54
	m	ft	3.280 8	ft	m	0.304 8
	km	mile	0.621 37	mile	km	1.609 3
Area	cm^2	$in.^2$	0.155	$in.^2$	cm^2	6.451 6
	m^2	ft^2	10.763 9	ft^2	m^2	0.092 9
Volume	cm^3	$in.^3$	0.061	$in.^3$	cm^3	16.387
	m^3	yd^3	1.308	yd^3	m^3	0.764 6
	m^3	gal	264.172	gal	m^3	0.003 785
	mL	fl oz	0.033 8	fl oz	mL	29.574
	L	fl oz	33.814	fl oz	L	0.029 6
	L	pt	2.113	pt	L	0.473 2
	L	qt	1.056 7	qt	L	0.946 4
	L	gal	0.264 2	gal	L	3.785 4
Mass or weight	g	oz	0.035 3	oz	g	28.349 5
	kg	lb	2.205	lb	kg	0.453 6
	t	lb	2205	ton	kg	907.2
Bending moment, torque, moment of force	N · m	lbf · in.	8.850 7	lbf · in.	N · m	0.113
	N · m	lbf · ft	0.737 6	lbf · ft	N · m	1.355 8
Pressure, vacuum	kPa	psi	0.145	psi	kPa	6.894 8
Velocity	km/h	mph	0.621 4	mph	km/h	1.609 3
Force, thrust, drag	N	lbf	0.224 8	lbf	N	4.448 2
Power	W	W	1	W	W	1
Temperature	°C	°F	$\frac{9}{5}(°C)+32$	°F	°C	$\frac{5}{9}(°F-32)$

EXAMPLE B.8

Express 6.5 fluid ounces in liters.

Solution You are changing from the customary (fluid ounces) to the metric (liters) systems.

Look in Table B.5 under the heading "From Customary to Metric." You are changing a measurement from fluid ounces to liters. Look under the column labeled "To change from" until you find the symbol for fluid ounces—fl oz. The fl oz symbol appears twice. Now look in the next right-hand column. Do you see the symbol for liters (L)? It is opposite the second fl oz symbol.

EXAMPLE B.8 (Cont.)

If you look in the "Multiply by" column to the right of these two symbols (fl oz and L), you find 0.029 6. Multiply the number of fluid ounces (6.5) by this number (0.029 6).

$$6.5 \times 0.029\,6 = 0.192\,4$$

This means that 6.5 fluid ounces is equivalent to 0.192 4 L (or 6.5 fl oz $\approx$ 0.192 4 L).

Exercise Set

In Exercises 1–16, reduce the given unit to the indicated unit.

27. 347 g to kilograms

28. 0.26 km to meters

29. 7.92 kW to watts

30. 2.3 μs to seconds

31. 0.023 85 Ω to milliohms

32. 0.000 235 47 MW to milliwatts

33. 9.72 mm to centimeters

34. 0.35 mm to nanometers

35. 835 000 cm^2 to square meters

36. 2.34 km^2 to m^2

37. 4.35 mm^3 to cubic centimeters

38. 91.52 m^3 to cubic millimeters

Solve Exercises 13–16.

39. *Machine technology* A die is 14 mm long. Express this in centimeters.

40. *Law enforcement* A male suspect is 1.97 m tall. Express this in centimeters.

41. *Environmental science* A wastewater treatment plant has 92 600 kg of sludge. Express this in metric tons.

42. *Medical technology* A technician needs 1 125 mL of a sterile solution. Express this in liters.

In Exercises 17–24, change the given units to the indicated unit. (You may want to consult Table B.3.)

43. 45 m/s to km/h

44. 27 mm/s to km/h

45. *Physics* If a force of 126 N gives a body an acceleration of 9 m/s^2, then the body has a mass of $\frac{126\text{ N}}{9\text{ m/s}^2}$. Convert this to kilograms.

46. *Electricity* An ac electric current that is generated by a source of 120 V and with an impedance of 125 Ω has an effective current of $\frac{120\text{ V}}{125\,\Omega}$. Convert this to amperes.

47. *Electricity* The resistance in a heater is $\frac{84\text{ V}}{8\text{ A}}$. Convert this to ohms.

48. *Physics* A technician determines that the absolute pressure in a tank is 112 cm of mercury. The following computation will convert this pressure to kilopascals: (13 600 kg/m^3)(9.8 m/s^2)(112 cm). Convert this to kilopascals.

49. *Automotive technology*
If a 3 250-g piston is held 300 mm above a work surface, then its potential energy relative to the work surface is

$$(3\,250\text{ g})(9.8\text{ m/s}^2)(300\text{ mm})$$

Express this potential energy in joules.

50. *Construction* At the instant a 4-kg sledgehammer reaches a velocity of 26 m/s, it has a kinetic energy of $\frac{1}{2}(4\text{ kg})(26\text{ m/s})^2$. Express this kinetic energy in joules.

In Exercises 25–40, convert the given measurement to the indicated unit.

51. $4\,m =$ ________ ft
52. $16\,L =$ ________ pt
53. $24\,kg =$ ________ lb
54. $27.5\,N \cdot m =$ ________ $lbf \cdot ft$
55. $105\,km/h =$ ________ mph
56. $17\,in. =$ ________ cm
57. $4.5\,qt =$ ________ L
58. $21\,ft^2 =$ ________ m^2
59. $5.8\,lb =$ ________ kg
60. $32\,psi =$ ________ kPa
61. $97\,W =$ ________ W
62. $98.6°F =$ ________ °C
63. *Fire safety* A fire extinguisher contains $2\frac{1}{2}$ gal. Convert this to liters.
64. *Law enforcement* An adult female weighs 120 lb. Convert this to kilograms.
65. *Environmental science* A flow-measuring meter at a wastewater treatment plant recorded $9{,}660{,}\tilde{0}00$ gal in one day. Convert this to liters.
66. *Environmental science* A wastewater treatment plant has a daily flow of $59\,500\,m^3$. Convert this to gallons.

Appendix

Table of Integrals

Basic Forms

1. $\int u\, v = uv - \int v\, u$
2. $\int u^n\, u = \frac{1}{n+1} u^{n+1} + C,\ n \neq -1$
3. $\int \frac{du}{u} = \ln|u| + C$
4. $\int e^u\, u = e^u + C$
5. $\int a^u\, u = \frac{1}{\ln a} a^u + C$
6. $\int \sin u\, u = -\cos u + C$
7. $\int \cos u\, u = \sin u + C$
8. $\int \sec^2 u\, u = \tan u + C$
9. $\int \csc^2 u\, u = -\cot u + C$
10. $\int \sec u \tan u\, u = \sec u + C$
11. $\int \csc u \cot u\, u = -\csc u + C$
12. $\int \tan u\, u = \ln|\sec u| + C$
    $= -\ln|\cos u| + C$
13. $\int \cot u\, u = \ln|\sin u| + C$
14. $\int \sec u\, u = \ln|\sec u + \tan u| + C$
15. $\int \csc u\, u = \ln|\csc u - \cot u| + C$
16. $\int \frac{du}{\sqrt{a^2 - u^2}} = \sin^{-1}\frac{u}{a} + C$
17. $\int \frac{du}{a^2 + u^2} = \frac{1}{a}\tan^{-1}\frac{u}{a} + C$
18. $\int \frac{du}{u\sqrt{u^2 - a^2}} = \frac{1}{a}\sec^{-1}\frac{u}{a} + C$
19. $\int \frac{du}{a^2 - u^2} = \frac{1}{2a}\ln\left|\frac{u+a}{u-a}\right| + C$
20. $\int \frac{du}{u^2 - a^2} = \frac{1}{2a}\ln\left|\frac{u-a}{u+a}\right| + C$

Forms Involving $\sqrt{a^2 - u^2}$

21. $\int \sqrt{a^2 - u^2}\, u = \frac{u}{2}\sqrt{a^2 - u^2} + \frac{a^2}{2}\sin^{-1}\frac{u}{a} + C$
22. $\int u^2\sqrt{a^2 - u^2}\, u =$
    $\frac{u}{8}(2u^2 - a^2)\sqrt{a^2 - u^2} + \frac{a^4}{8}\sin^{-1}\frac{u}{a} + C$
23. $\int \frac{\sqrt{a^2 - u^2}}{u}\, u =$
    $\sqrt{a^2 - u^2} - a\ln\left|\frac{a + \sqrt{a^2 - u^2}}{u}\right| + C$
24. $\int \frac{\sqrt{a^2 - u^2}}{u^2}\, u = -\frac{1}{u}\sqrt{a^2 - u^2} - \sin^{-1}\frac{u}{a} + C$

25. $\displaystyle\int \frac{u^2 u}{\sqrt{a^2-u^2}} = -\frac{u}{2}\sqrt{a^2-u^2}+\frac{a^2}{2}\sin^{-1}\frac{u}{a}+C$

26. $\displaystyle\int \frac{du}{u\sqrt{a^2-u^2}} = -\frac{1}{a}\ln\left|\frac{a+\sqrt{a^2-u^2}}{u}\right|+C$

27. $\displaystyle\int \frac{du}{u^2\sqrt{a^2-u^2}} = -\frac{1}{a^2u}\sqrt{a^2-u^2}+C$

28. $\displaystyle\int (a^2-u^2)^{3/2}\, u = -\frac{u}{8}(2u^2-5a^2)\sqrt{a^2-u^2}+\frac{3a^4}{8}\sin^{-1}\frac{u}{a}+C$

29. $\displaystyle\int \frac{du}{(a^2-u^2)^{3/2}} = \frac{u}{a^2\sqrt{a^2-u^2}}+C$

Forms Involving $\sqrt{u^2 \pm a^2}$

30. $\displaystyle\int \sqrt{u^2\pm a^2}\, u = \frac{u}{2}\sqrt{u^2\pm a^2}\pm \frac{a^2}{2}\ln\left|u+\sqrt{u^2\pm a^2}\right|+C$

31. $\displaystyle\int u^2\sqrt{u^2\pm a^2}\, u = \frac{u}{8}(2u^2\pm a^2)\sqrt{u^2\pm a^2}-\frac{a^4}{8}\ln\left|u+\sqrt{u^2\pm a^2}\right|+C$

32. $\displaystyle\int \frac{\sqrt{u^2+a^2}}{u}\, u = \sqrt{u^2+a^2}-a\ln\left|\frac{a+\sqrt{u^2+a^2}}{u}\right|+C$

33. $\displaystyle\int \frac{\sqrt{u^2-a^2}}{u}\, u = \sqrt{u^2-a^2}-a\cos^{-1}\frac{a}{u}+C$

34. $\displaystyle\int \frac{\sqrt{u^2\pm a^2}}{u^2}\, u = -\frac{\sqrt{u^2\pm a^2}}{u}+\ln\left|u+\sqrt{u^2\pm a^2}\right|+C$

35. $\displaystyle\int \frac{du}{\sqrt{u^2\pm a^2}} = \ln\left|u+\sqrt{u^2\pm a^2}\right|+C$

36. $\displaystyle\int \frac{du}{u\sqrt{u^2+a^2}} = -\frac{1}{a}\ln\left|\frac{a+\sqrt{u^2+a^2}}{u}\right|+C$

37. $\displaystyle\int \frac{du}{u\sqrt{u^2-a^2}} = \frac{1}{a}\cos^{-1}\left|\frac{a}{u}\right|+C = \frac{1}{a}\sec^{-1}\left|\frac{u}{a}\right|$

38. $\displaystyle\int \frac{u^2 u}{\sqrt{u^2\pm a^2}} = \frac{u}{2}\sqrt{u^2\pm a^2}\mp \frac{a^2}{2}\ln\left|u+\sqrt{u^2\pm a^2}\right|+C$

39. $\displaystyle\int \frac{du}{u^2\sqrt{u^2\pm a^2}} = \mp\frac{\sqrt{u^2\pm a^2}}{a^2u}+C$

40. $\displaystyle\int \left(u^2\pm a^2\right)^{3/2} u = \frac{u}{4}\left(u^2\pm a^2\right)^{3/2}\pm \frac{3a^2u}{8}\sqrt{u^2\pm a^2}+\frac{3a^4}{8}\ln\left|u+\sqrt{u^2\pm a^2}\right|+C$

41. $\displaystyle\int \frac{\left(u^2+a^2\right)^{3/2}}{u}\, u = \frac{1}{3}\left(u^2+a^2\right)^{3/2}+a^2\sqrt{u^2+a^2}-a^3\ln\left|\frac{a+\sqrt{u^2+a^2}}{u}\right|+C$

42. $\displaystyle\int \frac{\left(u^2-a^2\right)^{3/2}}{u}\, u = \frac{1}{3}\left(u^2-a^2\right)^{3/2}-a^2\sqrt{u^2-a^2}-a^3\sec^{-1}\frac{u}{a}+C$

43. $\displaystyle\int \frac{du}{\left(u^2\pm a^2\right)^{3/2}} = \frac{\pm u}{a^2\sqrt{u^2\pm a^2}}+C$

44. $\displaystyle\int \frac{du}{u\left(u^2+a^2\right)^{3/2}} = \frac{1}{a^2\sqrt{u^2+a^2}}-\frac{1}{a^3}\ln\left|\frac{a+\sqrt{u^2+a^2}}{u}\right|+C$

45. $\displaystyle\int \frac{du}{u\left(u^2-a^2\right)^{3/2}} = \frac{-1}{a^2\sqrt{u^2-a^2}}-\frac{1}{|a^3|}\sec^{-1}\frac{u}{a}+C$

Forms Involving $a+bu$

46. $\displaystyle\int \frac{u\, u}{a+bu} = \frac{1}{b^2}(a+bu-a\ln|a+bu|)+C$

47. $\displaystyle\int \frac{u^2 u}{a+bu} = \frac{1}{2b^3}\left[(a+bu)^2-4a(a+bu)+2a^2\ln|a+bu|\right]+C$

48. $\displaystyle\int \frac{du}{u(a+bu)} = \frac{1}{a}\ln\left|\frac{u}{a+bu}\right|+C$

49. $\displaystyle\int \frac{u\, u}{(a+bu)^2} = \frac{1}{b^2}\left(\frac{a}{a+bu}+\ln|a+bu|\right)+C$

50. $$\int \frac{du}{u(a+bu)^2} = \frac{1}{a(a+bu)} - \frac{1}{a^2}\ln\left|\frac{a+bu}{u}\right| + C$$

51. $$\int u\sqrt{a+bu}\, u = \frac{2}{15b^2}(3bu-2a)(a+bu)^{3/2}+C$$

52. $$\int \frac{u\, u}{\sqrt{a+bu}} = \frac{1}{3b^2}(bu-2a)\sqrt{a+bu}+C$$

53. $$\int \frac{u^2 u}{\sqrt{a+bu}} = \frac{2}{15b^3}(8a^2+3b^2u^2-4abu)\sqrt{a+bu}+C$$

54. $$\int \frac{du}{u\sqrt{a+bu}} = \begin{cases} \dfrac{1}{\sqrt{a}}\ln\left|\dfrac{\sqrt{a+bu}-\sqrt{a}}{\sqrt{a+bu}+\sqrt{a}}\right| + C & \text{if } a>0 \\ \dfrac{2}{\sqrt{-a}}\tan^{-1}\sqrt{\dfrac{a+bu}{-a}}+C, & \text{if } a<0 \end{cases}$$

55. $$\int \frac{\sqrt{a+bu}}{u}\, u = 2\sqrt{a+bu} + a\int \frac{du}{u\sqrt{a+bu}}$$

Trigonometric Forms

56. $$\int \sin^2 u\, u = \frac{1}{2}u - \frac{1}{4}\sin 2u + C = \frac{1}{2}u - \frac{1}{2}\sin u\cos u + C$$

57. $$\int \cos^2 u\, u = \frac{1}{2}u + \frac{1}{4}\sin 2u + C = \frac{1}{2}u + \frac{1}{2}\sin u\cos u + C$$

58. $$\int \tan^2 u\, u = \tan u - u + C$$

59. $$\int \cot^2 u\, u = -\cot u - u + C$$

60. $$\int \sin^3 u\, u = -\cos u + \frac{1}{3}\cos^3 u + C$$

61. $$\int \cos^3 u\, u = \sin u - \frac{1}{3}\sin^3 u + C$$

62. $$\int \tan^3 u\, u = \frac{1}{2}\tan^2 u + \ln|\cos u| + C$$

63. $$\int \cot^3 u\, u = -\frac{1}{2}\cot^2 u - \ln|\sin u| + C$$

64. $$\int \sin^n u\, u = -\frac{1}{n}\sin^{n-1} u\cos u + \frac{n-1}{n}\int \sin^{n-2} u\, u$$

65. $$\int \cos^n u\, u = \frac{1}{n}\cos^{n-1} u\sin u + \frac{n-1}{n}\int \cos^{n-2} u\, u$$

66. $$\int \tan^n u\, u = \frac{1}{n-1}\tan^{n-1} u - \int \tan^{n-2} u\, u$$

67. $$\int \cot^n u\, u = \frac{-1}{n-1}\cot^{n-1} u - \int \cot^{n-2} u\, u$$

68. $$\int \sec^n u\, u = \frac{1}{n-1}\tan u\sec^{n-2} u + \frac{n-2}{n-1}\int \sec^{n-2} u\, u$$

69. $$\int \csc^n u\, u = \frac{-1}{n-1}\cot u\csc^{n-2} u + \frac{n-2}{n-1}\int \csc^{n-2} u\, u$$

70. $$\int \sin au\sin bu\, u = \frac{\sin(a-b)u}{2(a-b)} - \frac{\sin(a+b)u}{2(a+b)} + C$$

71. $$\int \cos au\cos bu\, u = \frac{\sin(a-b)u}{2(a-b)} + \frac{\sin(a+b)u}{2(a+b)} + C$$

72. $$\int \sin au\cos bu\, u = -\frac{\cos(a-b)u}{2(a-b)} - \frac{\cos(a+b)u}{2(a+b)} + C$$

73. $$\int u\sin u\, u = \sin u - u\cos u + C$$

74. $$\int u\cos u\, u = \cos u + u\sin u + C$$

75. $$\int u^n \sin u\, u = -u^n\cos u + n\int u^{n-1}\cos u\, u$$

76. $$\int u^n \cos u\, u = u^n\sin u - n\int u^{n-1}\sin u\, u$$

77. $$\int \sin^n u\cos^m u\, u = -\frac{\sin^{n-1} u\cos^{m+1} u}{n+m} + \frac{n-1}{n+m}\int \sin^{n-2} u\cos^m u\, u = \frac{\sin^{n+1} u\cos^{m-1} u}{n+m} + \frac{m-1}{n+m}\int \sin^n u\cos^{m-2} u\, u$$

Inverse Trigonometric Forms

78. $\int \sin^{-1} u\, u = u \sin^{-1} u + \sqrt{1-u^2} + C$

79. $\int \cos^{-1} u\, u = u \cos^{-1} u - \sqrt{1-u^2} + C$

80. $\int \tan^{-1} u\, u = u \tan^{-1} u - \frac{1}{2} \ln(1+u^2) + C$

81. $\int u^n \sin^{-1} u\, u =$
$\frac{1}{n+1}\left[u^{n+1} \sin^{-1} u - \int \frac{u^{n+1} u}{\sqrt{1-u^2}}\right], n \neq -1$

82. $\int u^n \cos^{-1} u\, u =$
$\frac{1}{n+1}\left[u^{n+1} \cos^{-1} u + \int \frac{u^{n+1} u}{\sqrt{1-u^2}}\right], n \neq -1$

83. $\int u^n \tan^{-1} u\, u = \frac{1}{n+1}\left[u^{n+1} \tan^{-1} u - \int \frac{u^{n+1} u}{1+u^2}\right],$
$n \neq -1$

Exponential and Logarithmic Forms

84. $\int u e^{au}\, u = \frac{1}{a^2}(au-1)e^{au} + C$

85. $\int u^2 e^{au}\, u = \frac{1}{a^3}(a^2u^2 - 2au + 2)e^{au} + C$

86. $\int u^n e^{au}\, u = \frac{1}{a} u^n e^{au} - \frac{n}{a} \int u^{n-1} e^{au}\, u$

87. $\int e^{au} \sin bu\, u = \frac{e^{au}}{a^2+b^2}(a \sin bu - b \cos bu) + C$

88. $\int e^{au} \cos bu\, u = \frac{e^{au}}{a^2+b^2}(a \cos bu + b \sin bu) + C$

89. $\int \ln u\, u = u \ln u - u + C$

90. $\int u^n \ln u\, u = u^{n+1}\left(\frac{\ln u}{n+1} - \frac{1}{(n+1)^2}\right) + C,$
$n \neq -1$

91. $\int \frac{1}{u \ln u}\, u = \ln|\ln u| + C$

Hyperbolic Forms

92. $\int \sinh u\, u = \cosh u + C$

93. $\int \cosh u\, u = \sinh u + C$

94. $\int \tanh u\, u = \ln \cosh u + C$

95. $\int \coth u\, u = \ln|\sinh u| + C$

96. $\int \operatorname{sech} u\, u = \tan^{-1}|\sinh u| + C$

97. $\int \operatorname{csch} u\, u = \ln\left|\tanh \frac{1}{2} u\right| + C$

98. $\int \operatorname{sech}^2 u\, u = \tanh u + C$

99. $\int \operatorname{csch}^2 u\, u = -\coth u + C$

100. $\int \operatorname{sech} u \tanh u\, u = -\operatorname{sech} u + C$

101. $\int \operatorname{csch} u \coth u\, u = -\operatorname{csch} u + C$

Answers to Odd-Numbered Exercises

ANSWERS FOR CHAPTER 1

Exercise Set 1.1

1.

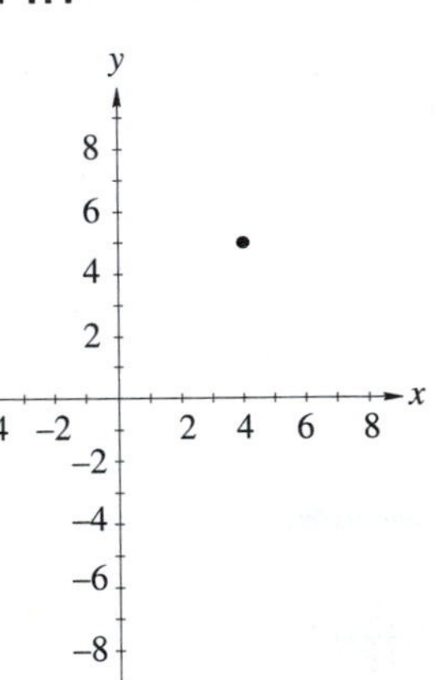

3.

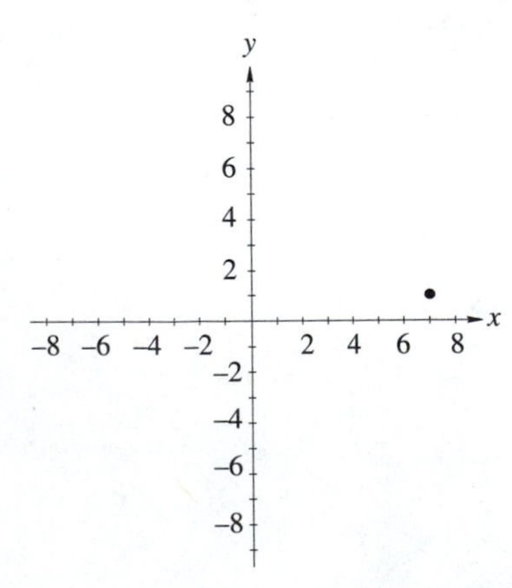

5.

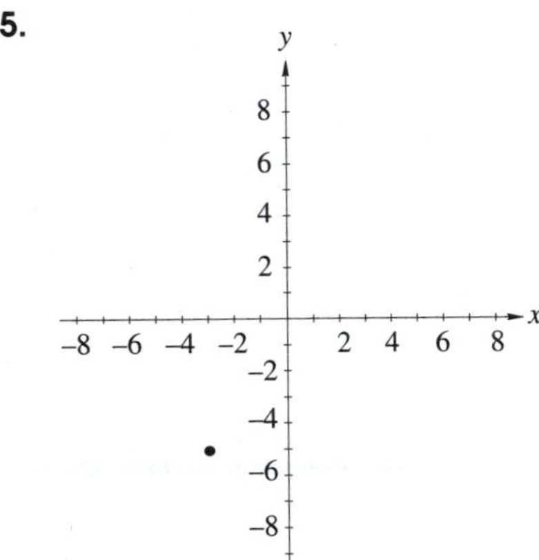

7.

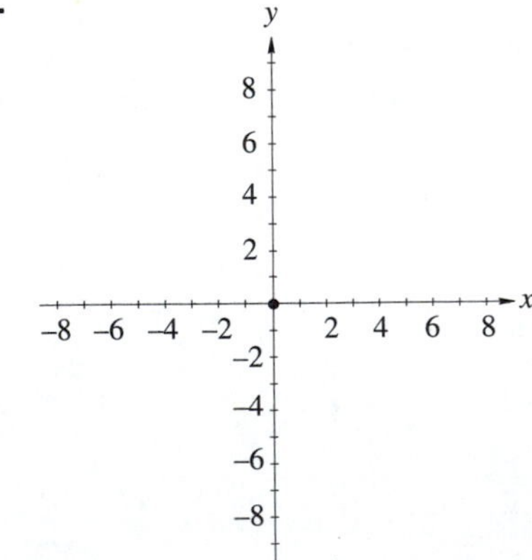

9.

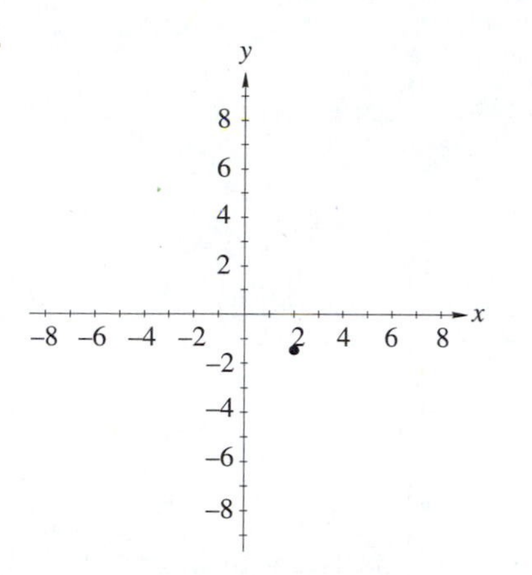

11. $(-1, -4)$

13. They are on a horizontal line and have the same y-coordinate.

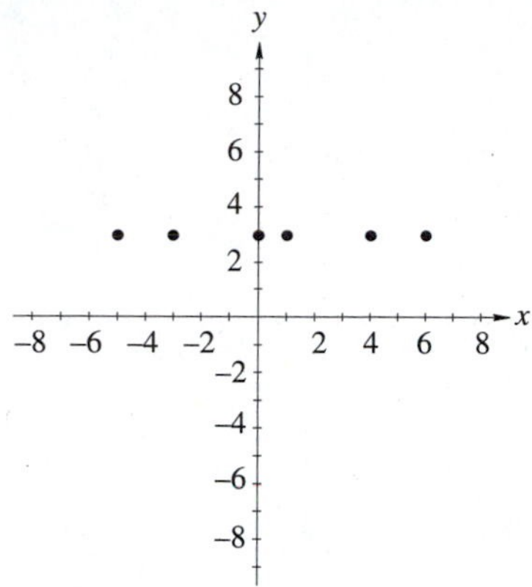

15. The points all lie on the same straight line.

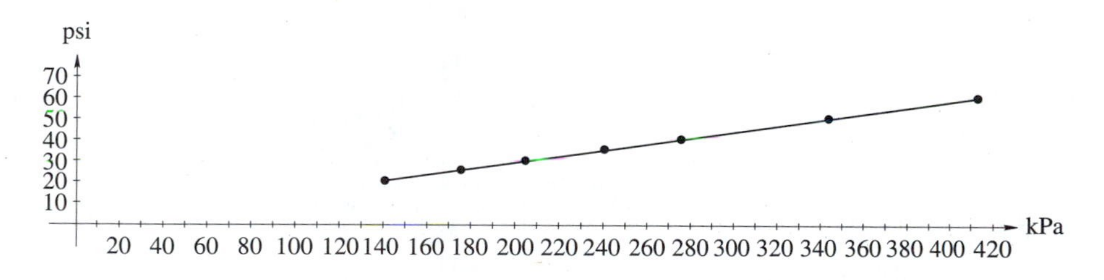

17. On the horizontal line through $y = -2$. **19.** On the vertical line $x = -5$ **21.** To the right of the vertical line through $x = -3$. **23.** To the right of the line through the point $(1, 0)$ and below the horizontal line through the point $(0, -2)$.

25. **(a)**

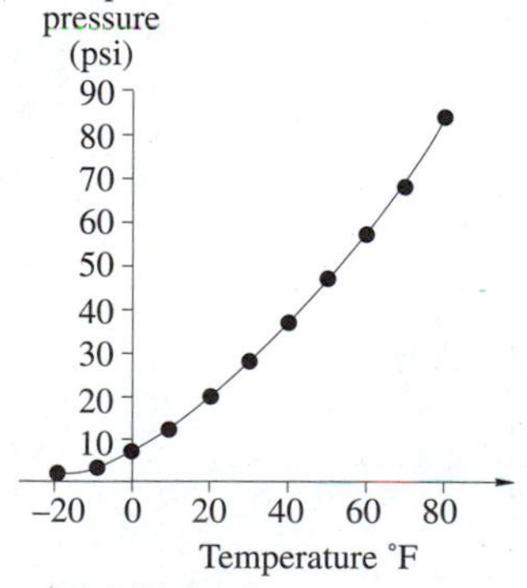

(b) 41.9 psi; **(c)** 99.6 psi

27. **(a)**

(b) Around 317 **(c)** Around 23.5 min

Exercise Set 1.2

1. **(a)**

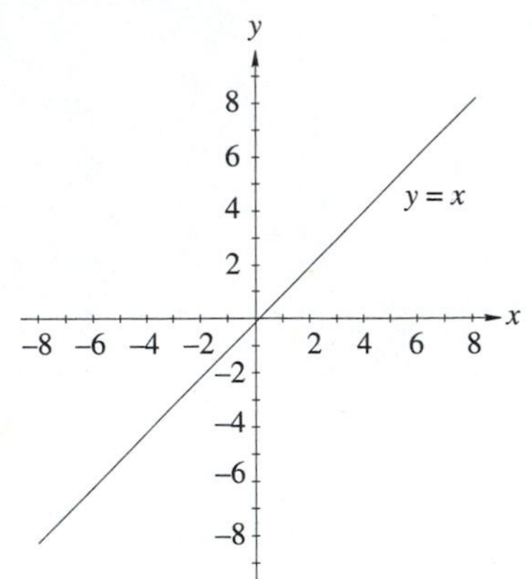

(b) x-intercept is 0; y-intercept is 0; **(c)** $m = 1$

3. **(a)**

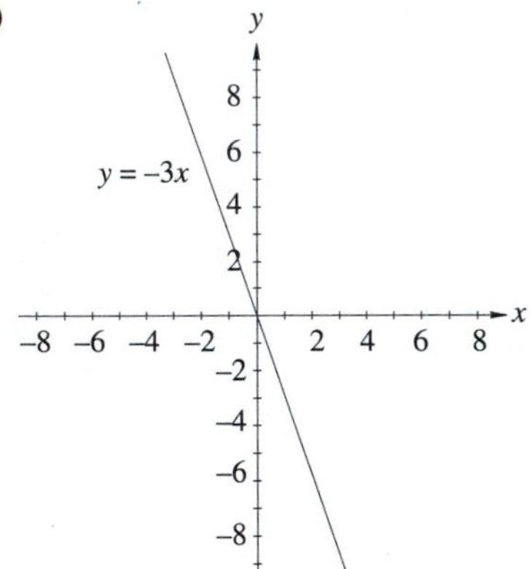

(b) x-intercept is 0; y-intercept is 0; **(c)** $m = -3$

5. **(a)**

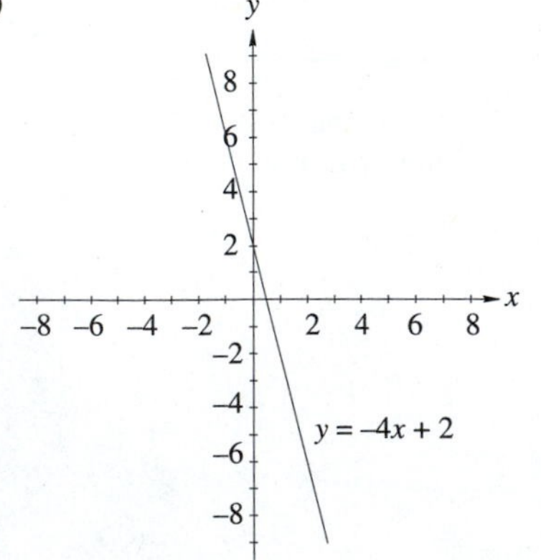

(b) x-intercept is $\frac{1}{2}$; y-intercept is 2; **(c)** $m = -4$

7. **(a)**

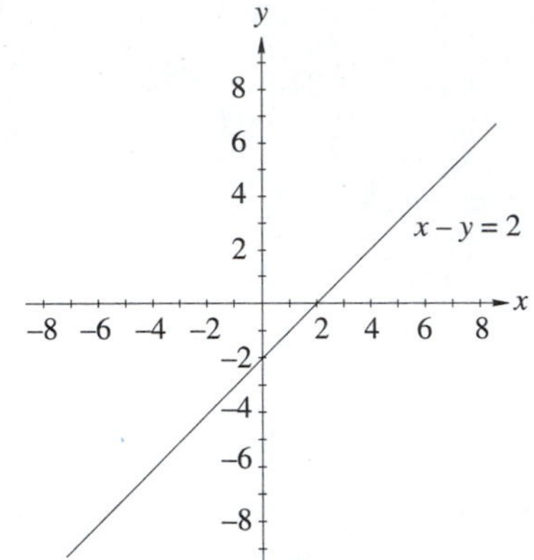

(b) x-intercept is 2; y-intercept is -2; **(c)** $m = 1$

9. **(a)**

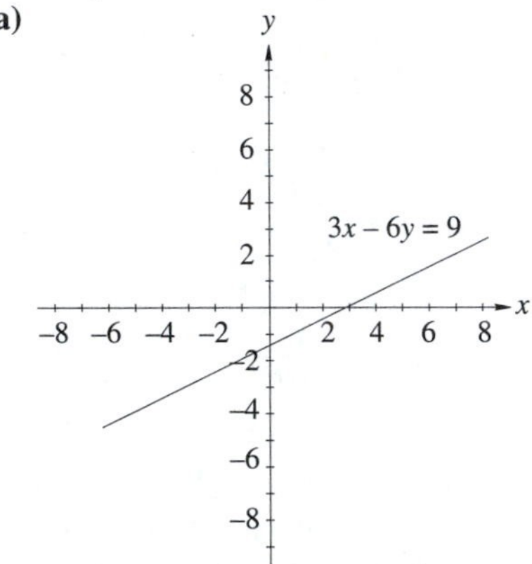

(b) x-intercept is 3; y-intercept is $-\frac{3}{2}$; **(c)** $m = \frac{1}{2}$

11. **(a)**

x	-3	-2	-1	0	1	2	3
y	9	4	1	0	1	4	9

(b)

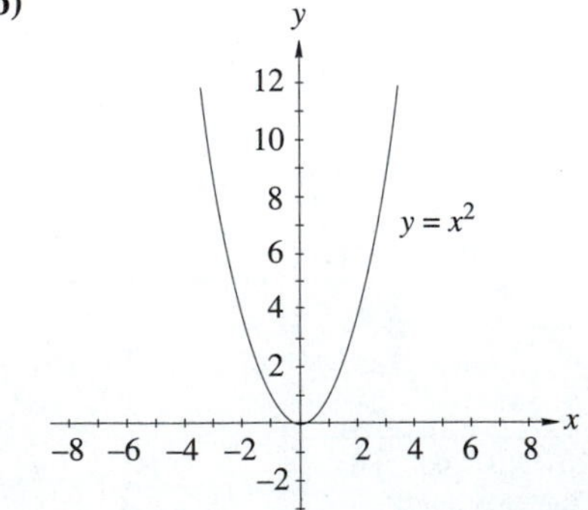

(c) Domain: all real numbers; range: $\{y\colon y \geq 0\}$

13. **(a)**

x	-3	-2	-1	0	1	2	3
y	7	2	-1	-2	-1	2	7

(b)

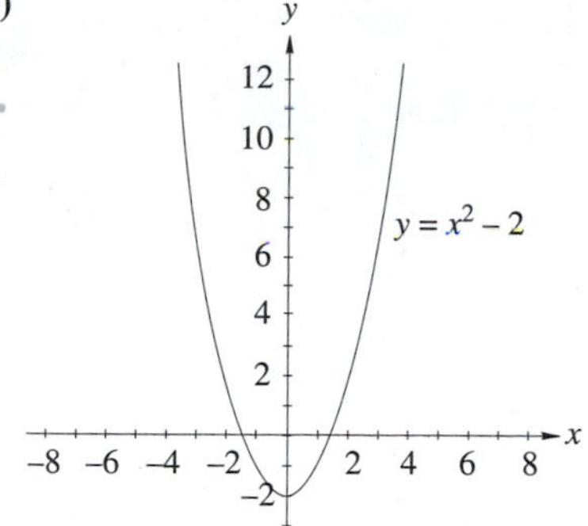

(c) Domain: all real numbers; range: $\{y: y \geq -2\}$

15. (a)

x	-3	-2	-1	0	1	2	3	4	5	6
y	36	25	16	9	4	1	0	1	4	9

(b)

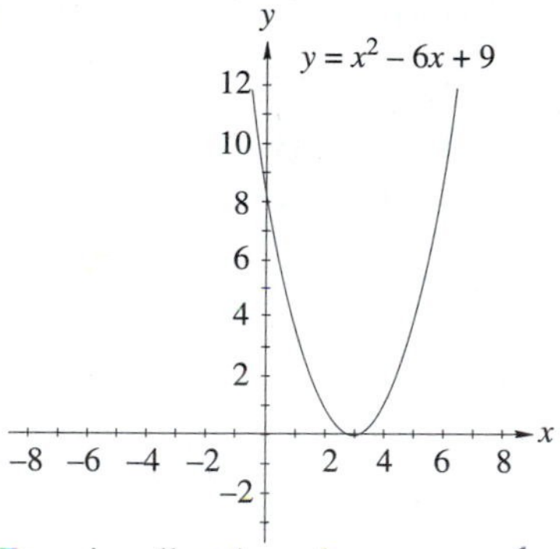

(c) Domain: all real numbers; range: $\{y: y \geq 0\}$

17. (a)

x	-3	-2	-1	$-\frac{1}{2}$	$-\frac{1}{4}$	0
y	$-\frac{1}{3}$	$-\frac{1}{2}$	-1	-2	-4	undefined

x	$\frac{1}{4}$	$\frac{1}{2}$	1	2	3
y	4	2	1	$\frac{1}{2}$	$\frac{1}{3}$

(b)

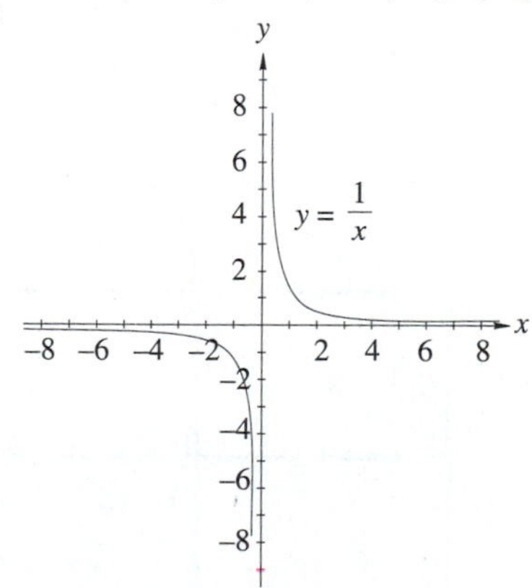

(c) Domain: $\{x: x \neq 0\}$; range: $\{y: y \neq 0\}$

19. (a)

x	-3	-2	-1	0	1	2	3
y	-27	-8	-1	0	1	8	27

(b)

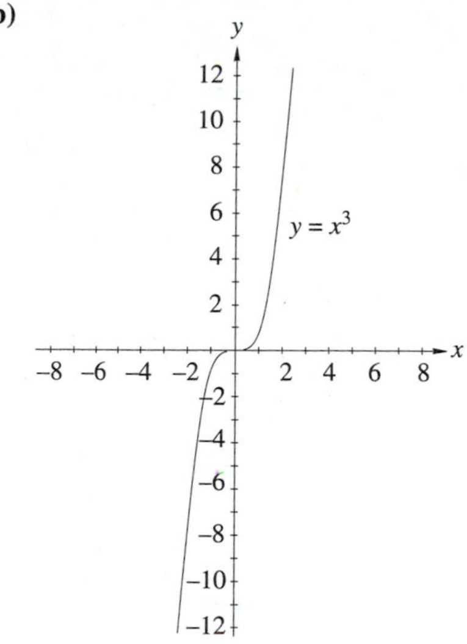

(c) Domain: all real numbers; range: all real numbers

21.

x	0	0	5	-5	3	-3	3	-3	4	4	-4	-4
y	5	-5	0	0	4	4	-4	-4	3	-3	3	-3

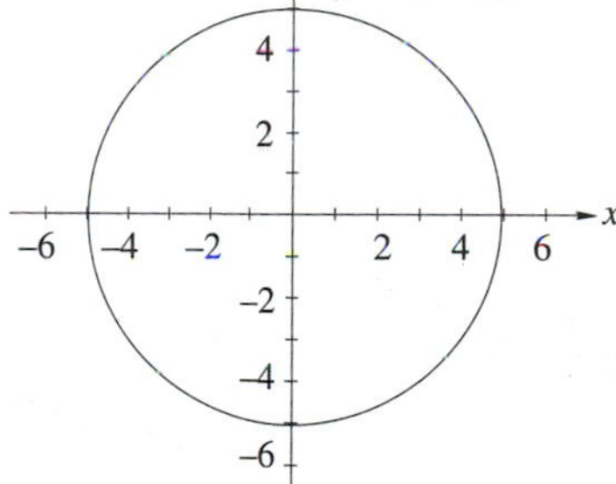

23.

x	-5	-4	-3	-2	-1	0	1	2
y	3	2	1	0	1	2	3	4

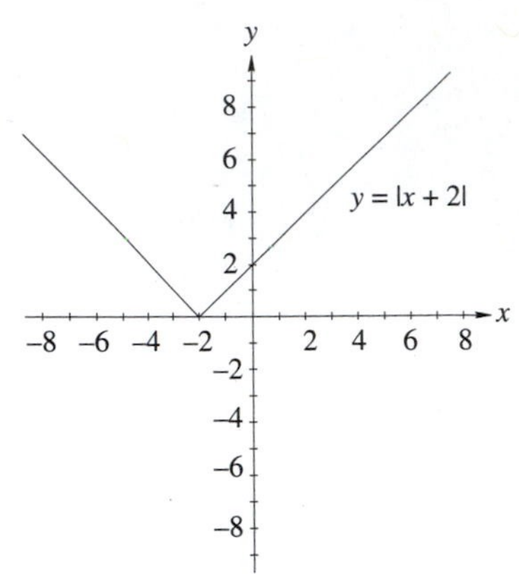

25.

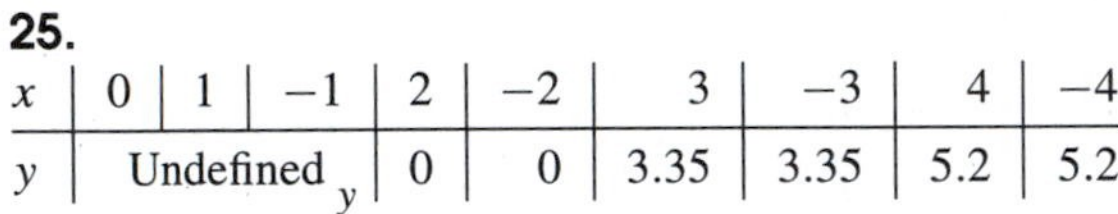

x	0	1	−1	2	−2	3	−3	4	−4
y	Undefined			0	0	3.35	3.35	5.2	5.2

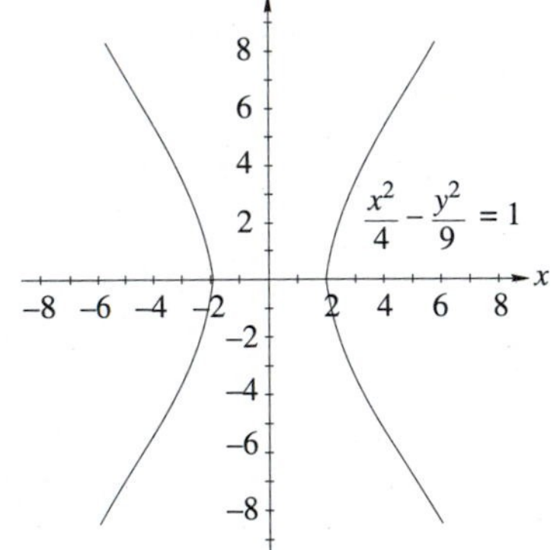

27. Function **29.** Function

31. **(a)**

k	0	1	2	3	4	5
$p(k)$	25.0	8.33	5.00	3.57	2.78	2.27

k	6	7	8	9
$p(k)$	1.92	1.67	1.47	1.31

(b) yes

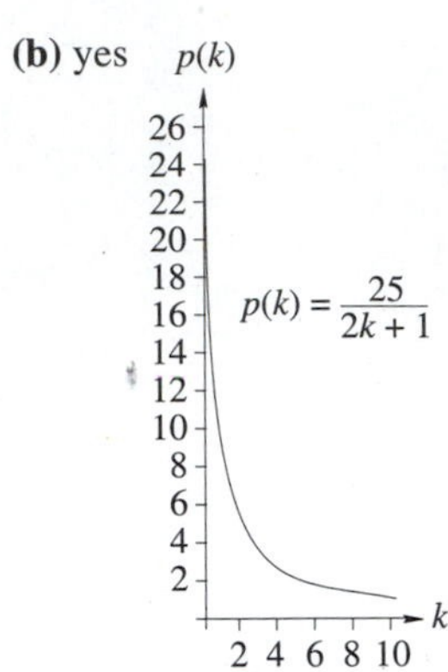

33. **(a)** $0 < p < \$4.65$

(b)

p	0.25	0.50	0.75	1.00
$d(p)$	10.56	11	11.31	11.50

p	1.25	1.50	1.75	2.00
$d(p)$	11.56	11.50	11.31	11

(c)

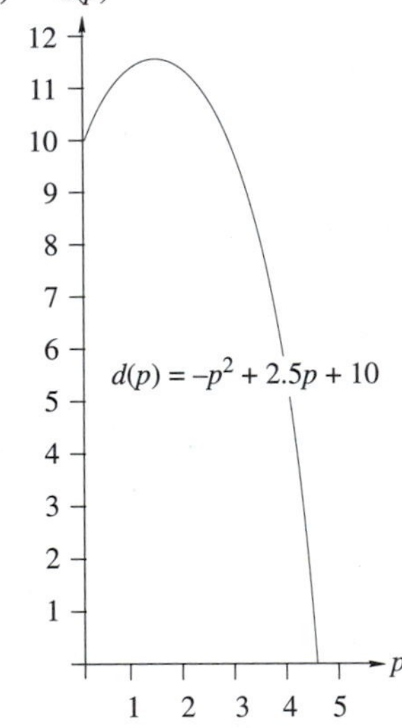

(d) $1.25

Exercise Set 1.3

1.

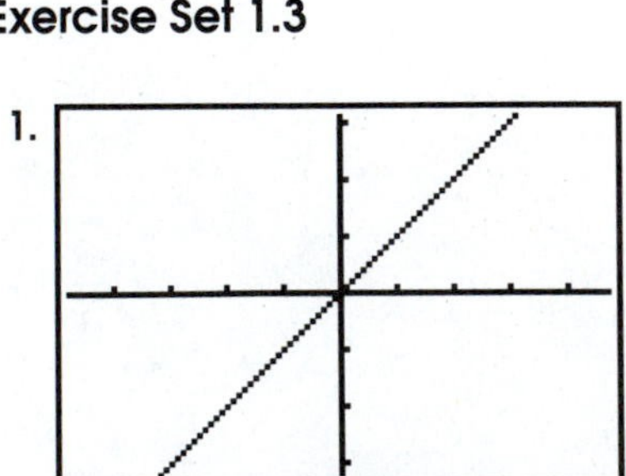

3.

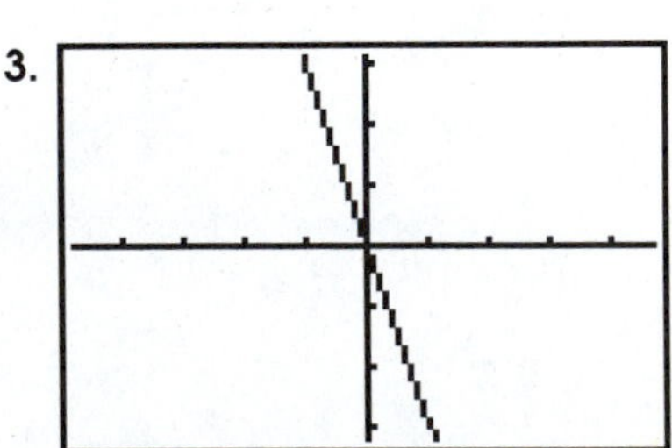

5.

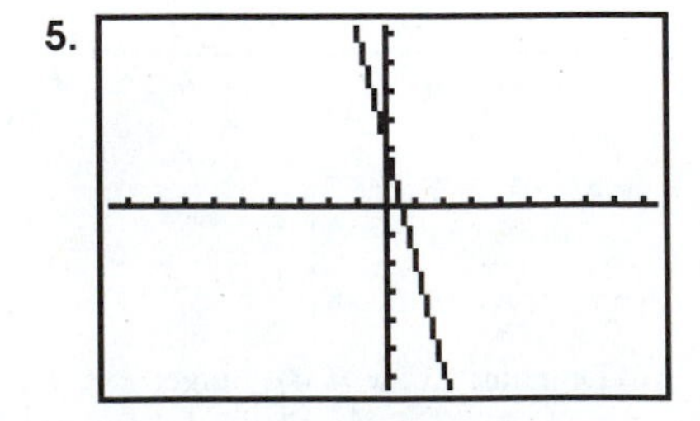

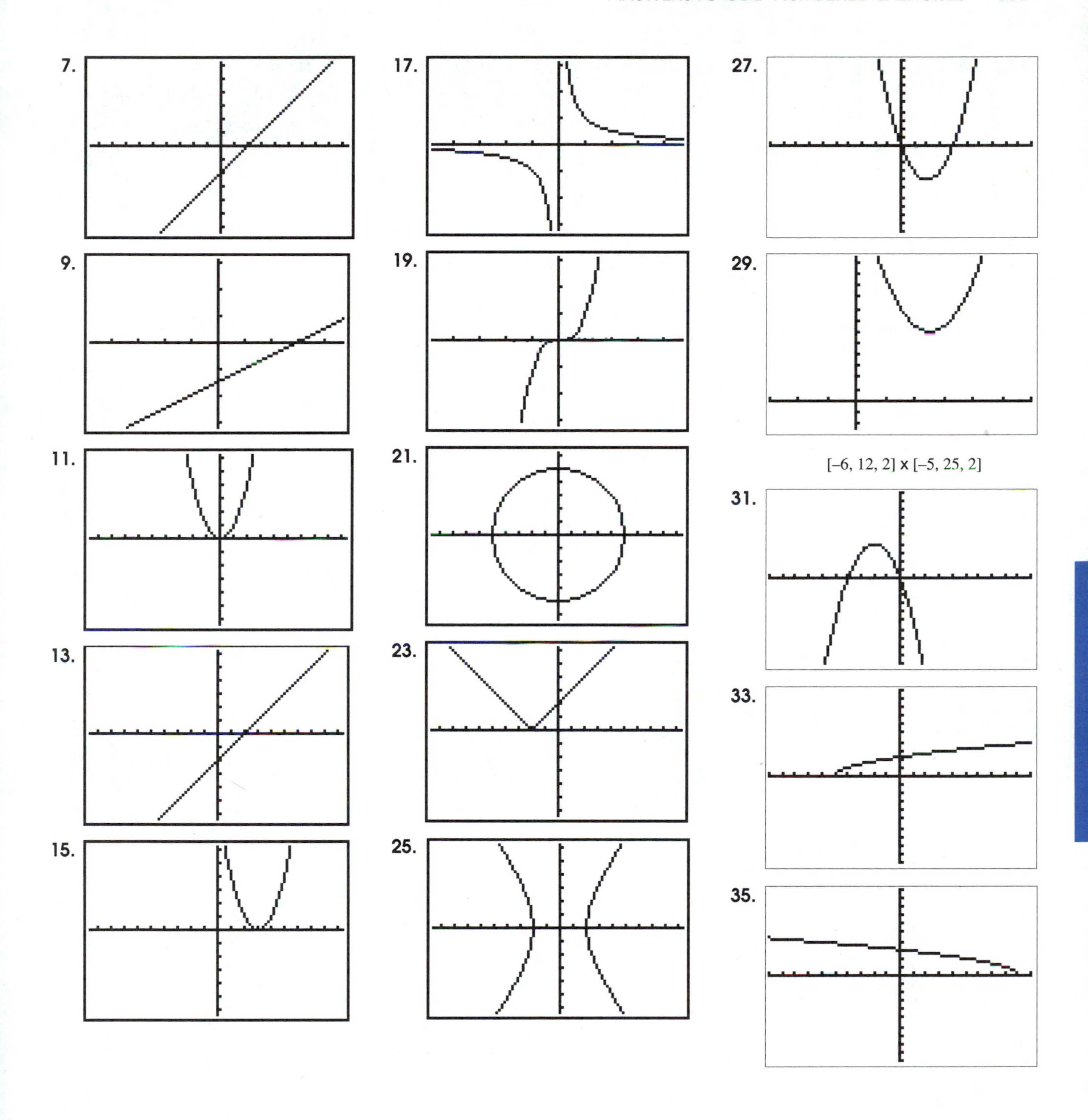
7.
17.
27.
9.
19.
29.
11.
21.
[–6, 12, 2] x [–5, 25, 2]
31.
13.
23.
33.
15.
25.
35.

37.

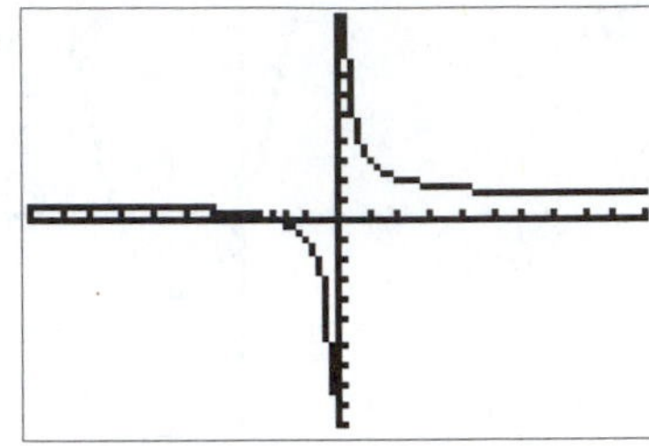

39.

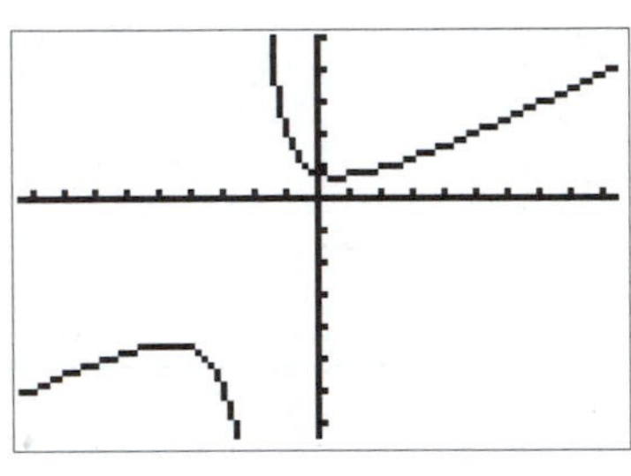

[−9.4, 9.4, 1] x [−10, 32, 2]

41.

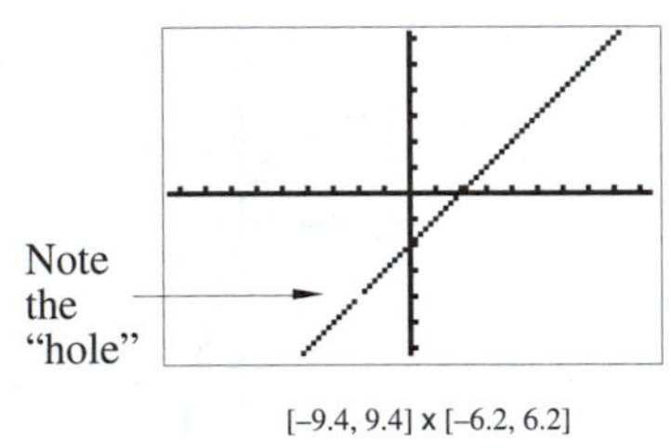

[−9.4, 9.4] x [−6.2, 6.2]

Exercise Set 1.4

1. The root is $x = -\frac{5}{2}$

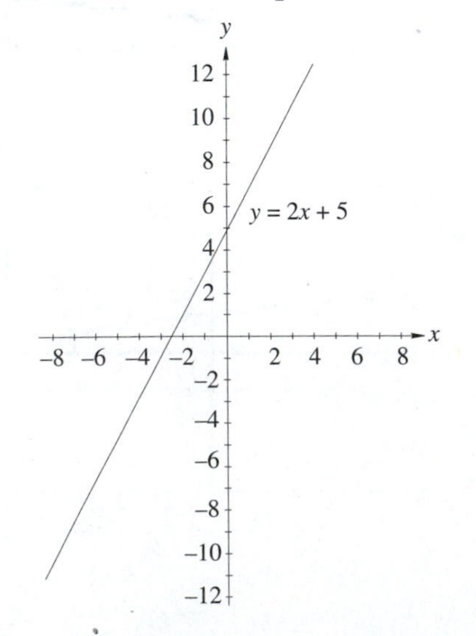

3. The roots are $x = -3$ and $x = 3$

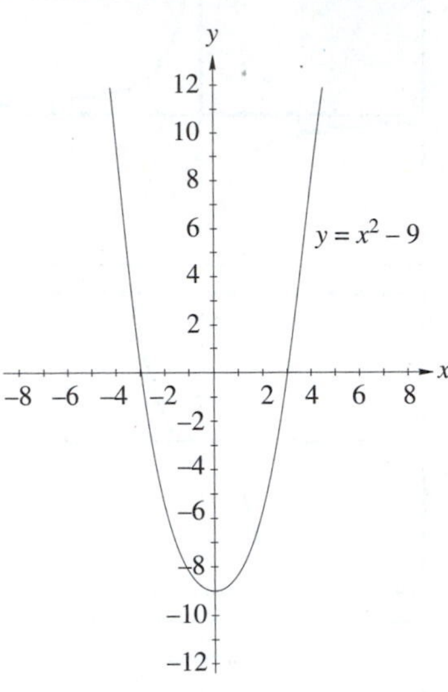

5. The roots are $x = 0$ and $x = 5$

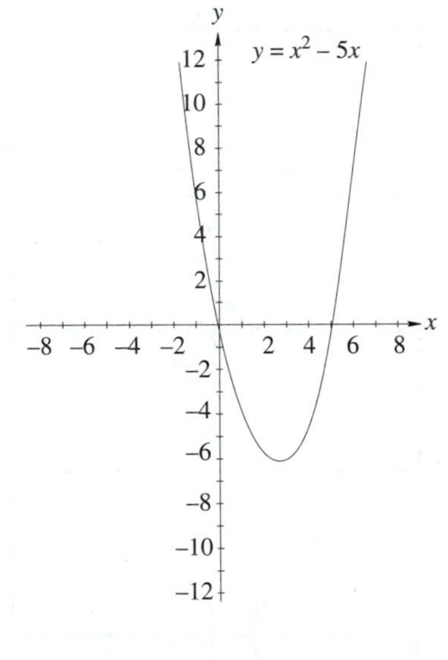

7. The roots are

$$x = -\frac{5}{2} - \frac{\sqrt{37}}{2} \approx -5.5414 \text{ and}$$

$$x = -\frac{5}{2} + \frac{\sqrt{37}}{2} \approx 0.5414$$

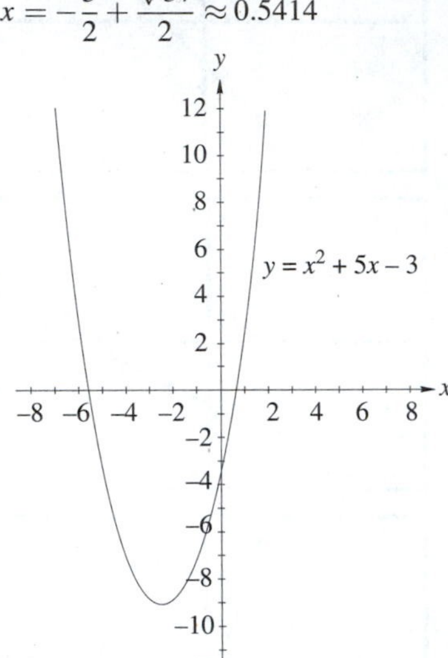

9. The roots are $x = -\frac{9}{4}$ and $x = \frac{6}{5}$

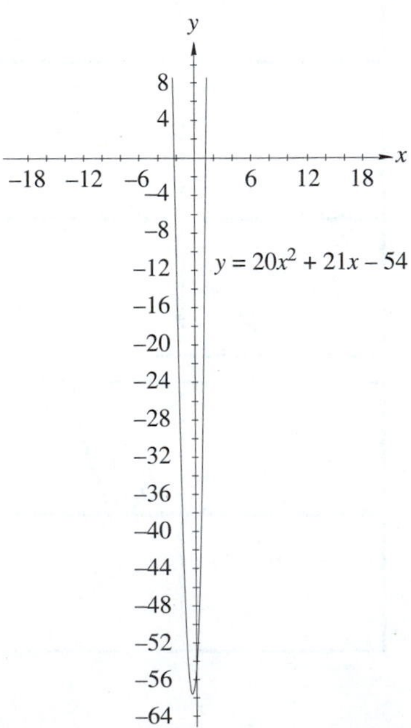

11. The root is $x = -1$.

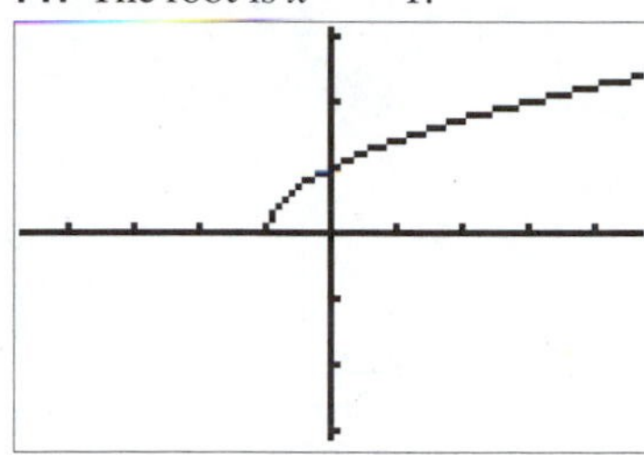

$[-4.7, 4.7] \times [-3.1, 3.1]$

13. The root is $x = 8$.

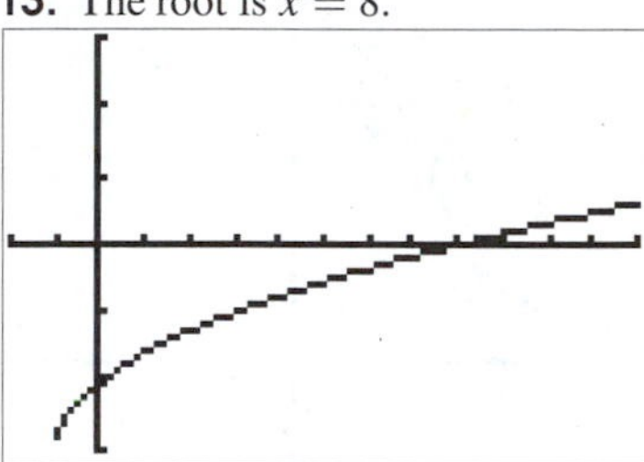

$[-2, 12] \times [-3, 3]$

15. The root is $x = -0.75$.

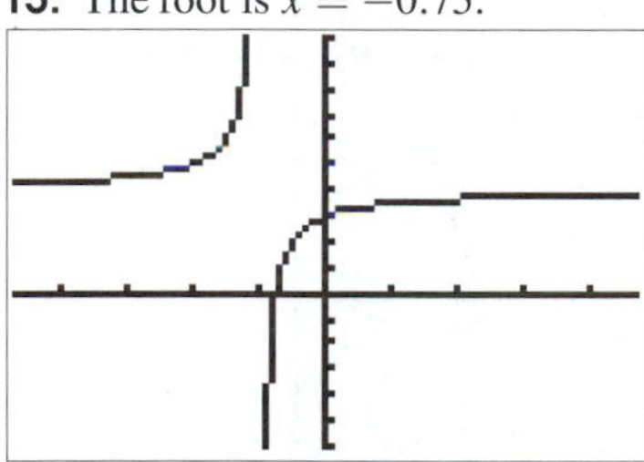

$[-4.7, 4.7] \times [-6, 10]$

17. The root is $x \approx -1.2695$.

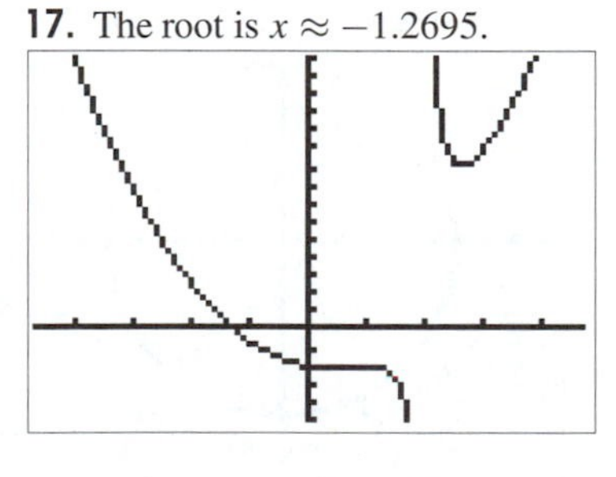

$[-4.7, 4.7] \times [-1, 15]$

19. (a)

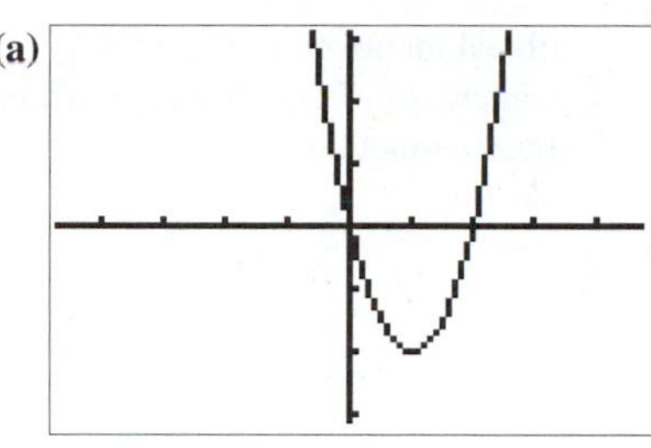

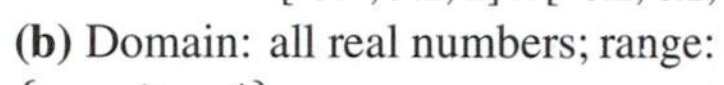

$[-9.4, 9.2, 2] \times [-6.2, 6.2, 2]$

(b) Domain: all real numbers; range: $\{y\colon y \geq -4\}$

21. (a)

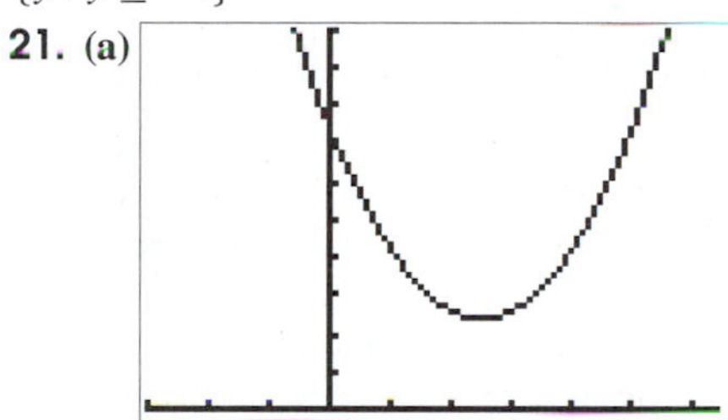

$[-6, 12.8, 2] \times [0, 50, 5]$

(b) Domain: all real numbers; range: $\{y\colon y \geq 12\}$

23. (a)

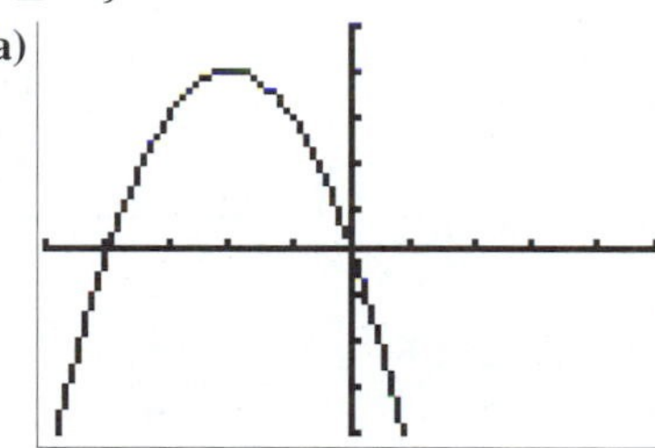

$[-5, 5, 1] \times [-4, 5, 1]$

(b) Domain: all real numbers; range: $\{y\colon y \leq 4\}$

25. (a)

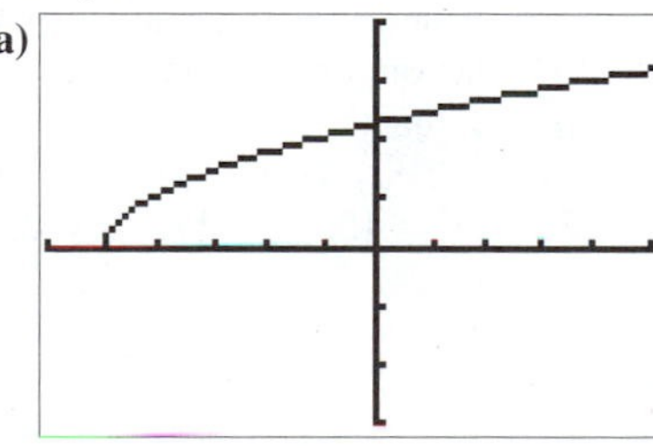

$[-6, 5, 1] \times [-3, 4, 1]$

(b) Domain: $\{x : x \geq -5\}$; range: $\{y\colon y \geq 0\}$

27. (a)

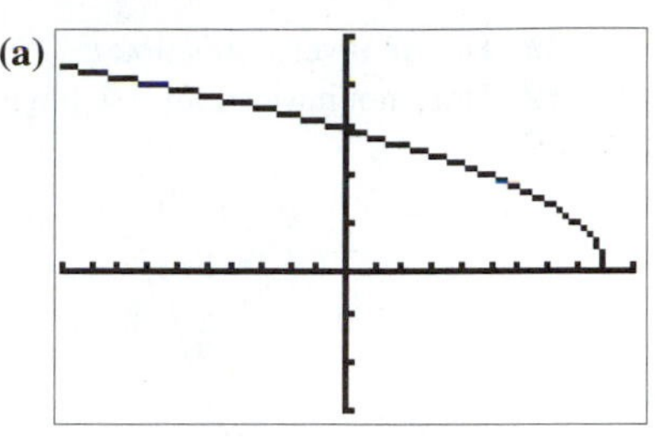

$[-10, 10, 1] \times [-3, 5, 1]$

(b) Domain: $\{x : x \leq 9\}$; range: $\{y\colon y \geq 0\}$

29. (a)

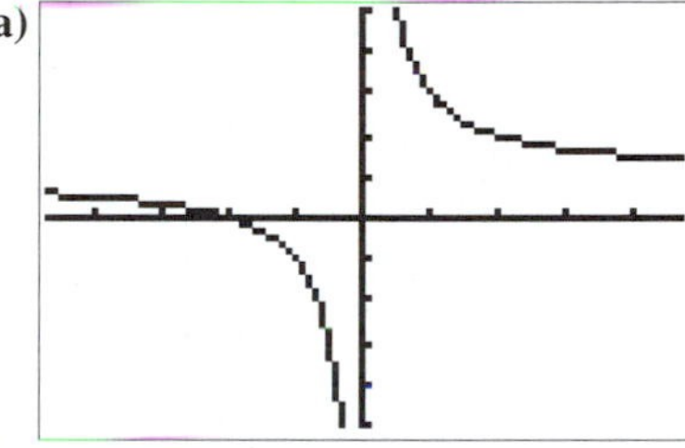

$[-4.7, 4.7, 1] \times [-5, 5, 1]$

(b) Domain: $\{x : x \neq 0\}$; range: $\{y\colon y \neq 1\}$

31. (a)

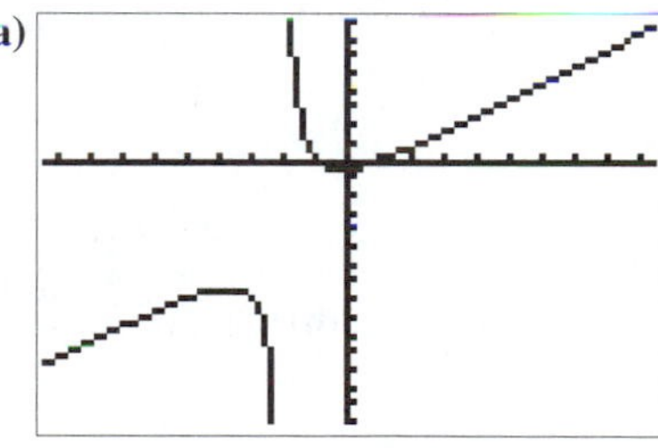

$[-9.4, 9.4, 1] \times [-15, 8, 1]$

(b) Domain: $\{x : x \neq -2\}$; range: $\{y\colon y < -7.46, y > -0.536\}$ (decimal values are approximate)

33. (a)

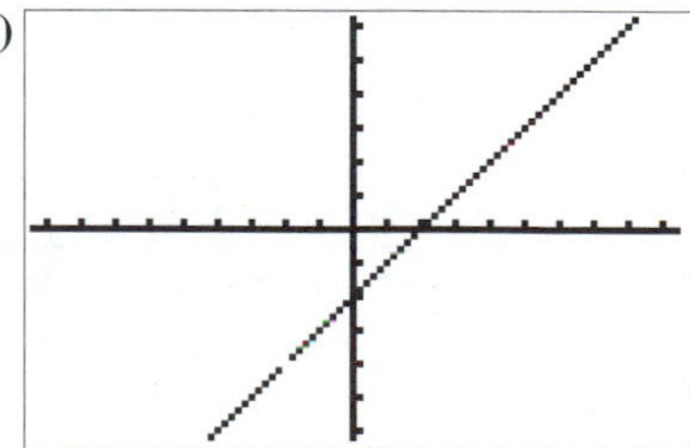

$[-9.4, 9.4, 1] \times [-6.2, 6.2, 1]$

(b) Domain: $\{x : x \neq -2\}$; range: $\{y\colon y \neq -4\}$

35. Has an inverse function
37. Does not have an inverse function
39.

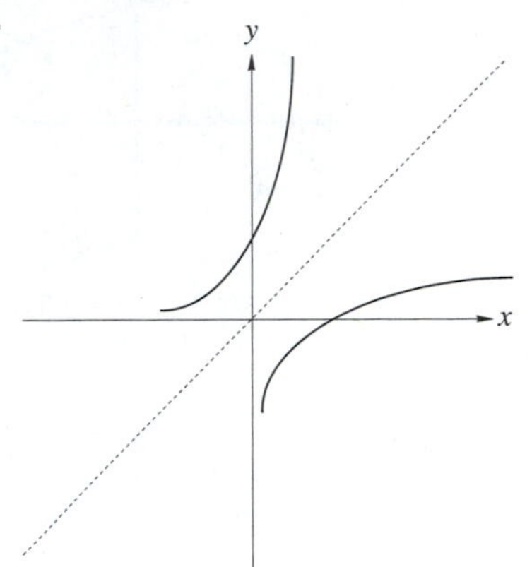

41.

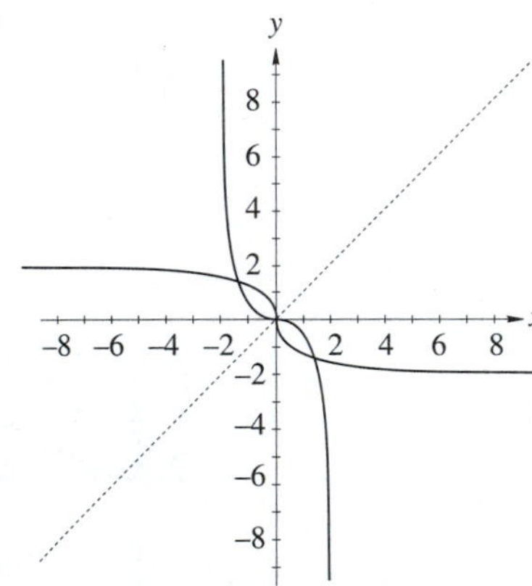

43. **(a)** 30%; **(b)** 60%; **(c)** 73%
45. No

Review Exercises

1. **(a)**

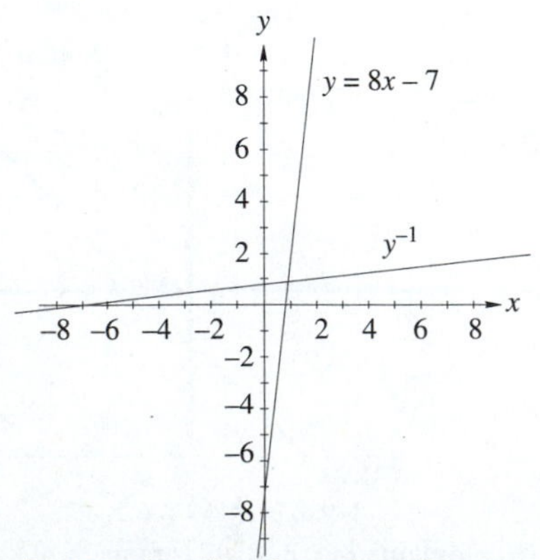

(b) Domain: all real numbers; Range: all real numbers; x-intercept $\frac{7}{8}$; y-intercept -7; **(c)** function; **(d)** has an inverse function

3. **(a)**

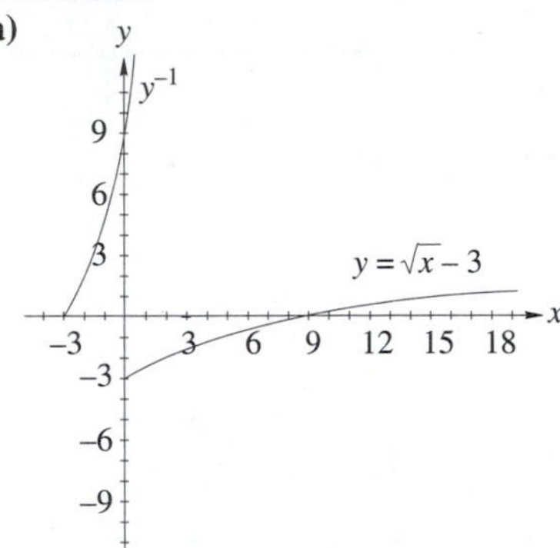

(b) Domain: all non-negative real numbers, $x \geq 0$; Range: all real numbers greater than or equal to -3; x-intercept 9; y-intercept -9; **(c)** function; **(d)** has an inverse function

5. **(a)**

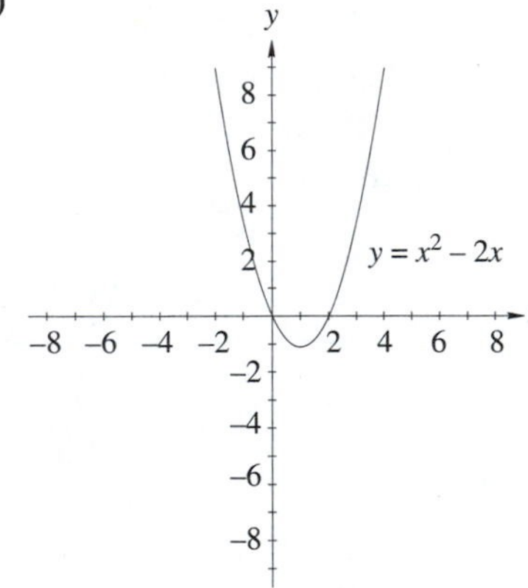

(b) Domain: all real numbers; Range: all real numbers greater than or equal to -1; x-intercepts $0, 2$; y-intercept 0; **(c)** function; **(d)** does not have an inverse function

7. -12 **9.** 0 **11.** $4a - 20$
13. -1 **15.** 0 **17.** $\frac{7}{25} = 0.28$

19.

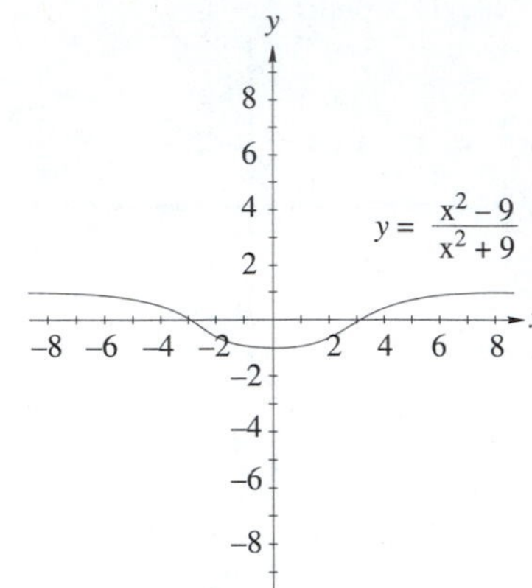

21.

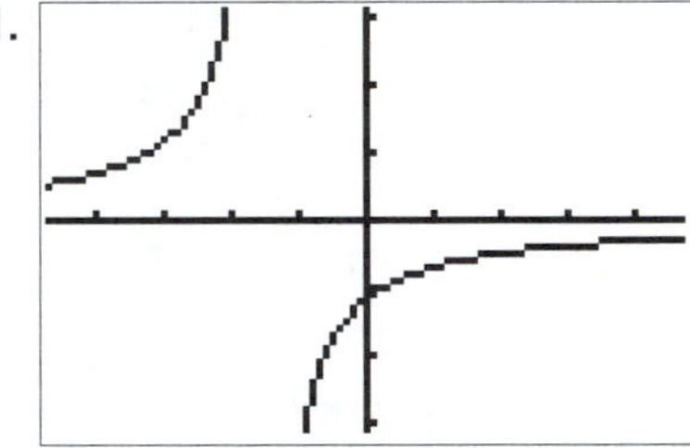

[–9.4, 9.4] x [–6.2, 6.2]

23.

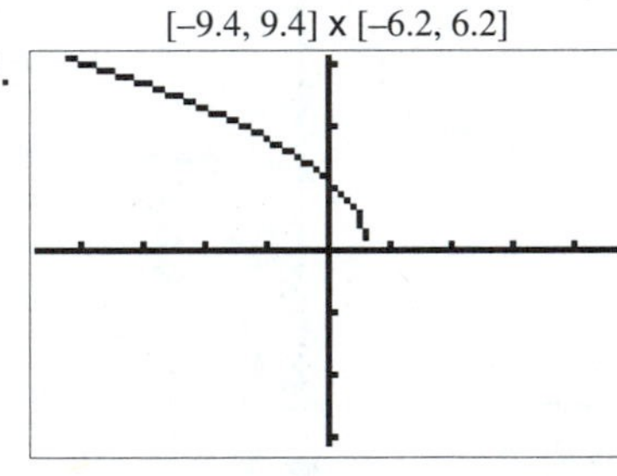

[–9.4, 9.4] x [–6.2, 6.2]

25.

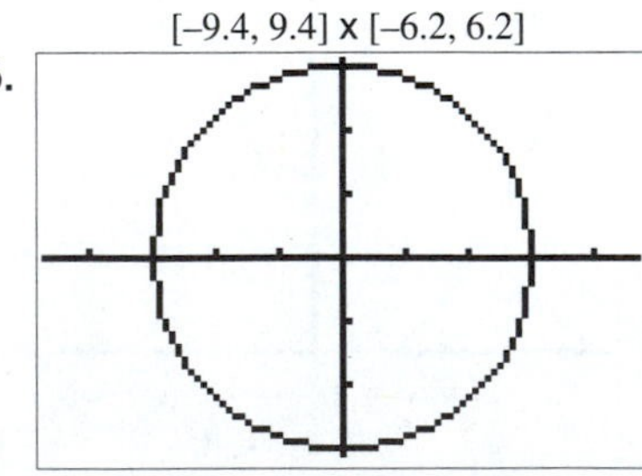

[–4.7, 4.7] x [–3.1, 3.1]

27. $y = -\frac{4}{7}x$; root: $x = 0$

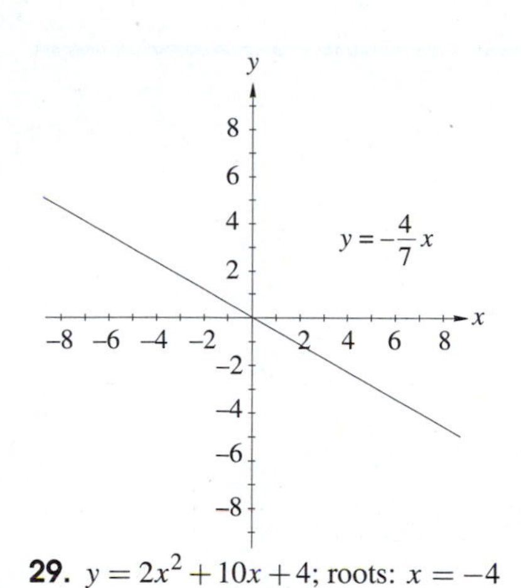

29. $y = 2x^2 + 10x + 4$; roots: $x = -4$ and $x = -1$

31. **(a)** \$29.95, \$17.50, \$15; **(b)** $R(n) = \left(35 - \frac{n}{20}\right)n$; **(c)** \$3024.95, \$6125, \$6000

33.

t	0	1	2	3	4	5
$h(t)$	0	3.08	6.48	10.68	15.68	21.00

t	6	7	8	9	10
$h(t)$	25.68	28.28	26.88	19.08	2.00

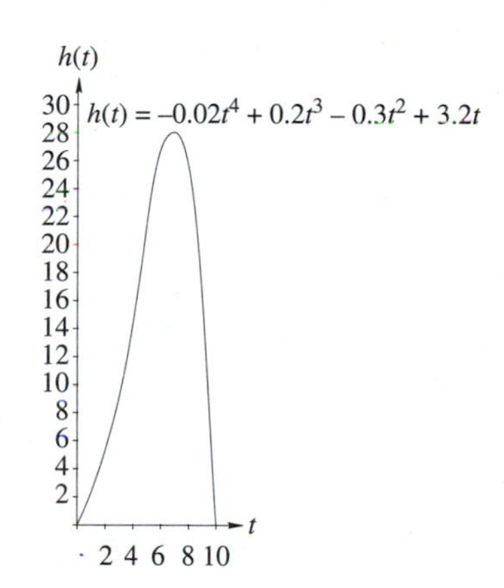

Chapter 1 Test

1. -19 **3.** -5

5. **(a)**

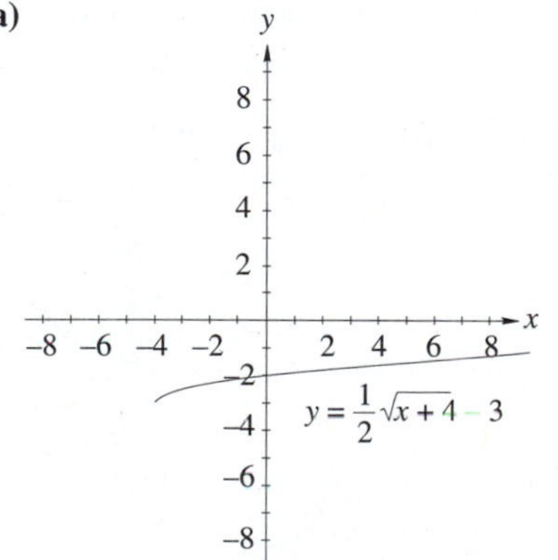

(b) $x \geq -4$; **(c)** $y \geq -2$; **(d)** $x = 32$; **(e)** $y = -2$ **7.** $\frac{3x^2 - x - 70}{x+5}$

9. $3(x+5) = 3x + 15$

11. $\frac{3x - 20}{3x - 10}$

13. / **15.**

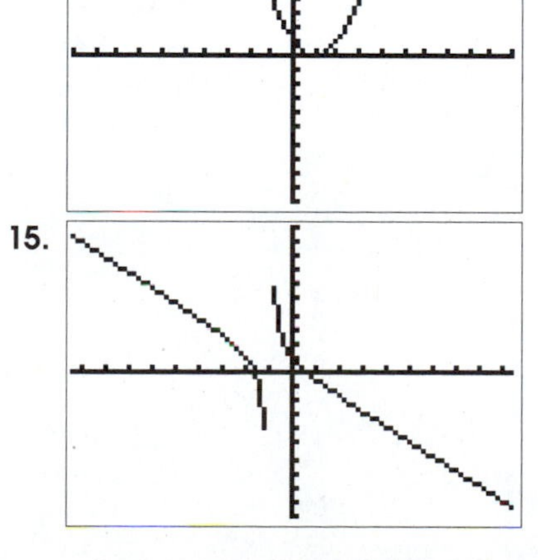

[−9.4, 9.4] x [−10, 10]

17. **(a)**

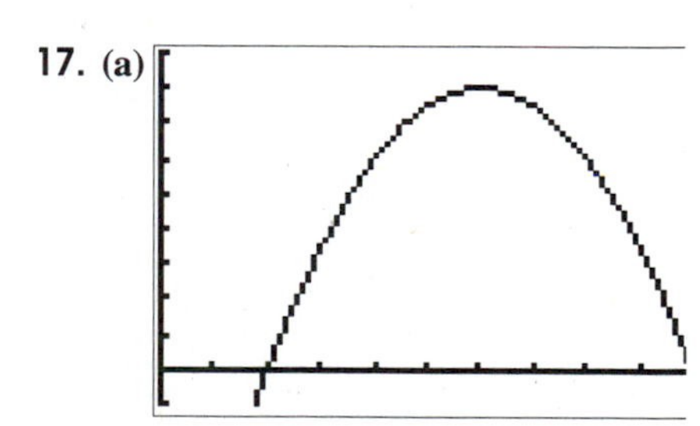

[0, 10, 1] x [−2, 18, 2]

(b) 7, **(c)** 6 ms, **(d)** either 4 ms or 8 ms

ANSWERS FOR CHAPTER 2

Exercise Set 2.1

1. $\sqrt{194}$ **3.** $\sqrt{65}$ **5.** $\sqrt{277}$ **7.** 7
9. $\left(\frac{9}{2}, -\frac{5}{2}\right)$ **11.** $\left(1, -\frac{11}{2}\right)$ **13.** $\left(\frac{15}{2}, -6\right)$
15. $\left(\frac{3}{2}, 5\right)$ **17.** $-\frac{13}{5} = -2.6$ **19.** $-\frac{1}{8}$ **21.** $\frac{14}{9}$
23. 0 **25.** $y - 4 = -2.6(x - 2)$ or $y = -2.6x + 9.2$
27. $y + 6 = -\frac{1}{8}(x - 5)$ or $y = -\frac{1}{8}x - 5\frac{3}{8}$
29. $y - 1 = -\frac{14}{9}(x - 12)$ or $y = \frac{14}{9}x - 17\frac{2}{3}$ **31.** $y = 5$
33. 1 **35.** approximately 0.1511352 **37.** 68.19859°
39. 153.43495° **41.** $-\frac{1}{3}$ **43.** 2 **45.** $y + 5 = 6(x - 2)$ or $y = 6x - 17$ **47.** $y + 2 = -\frac{5}{2}(x - 4)$ or $y = -\frac{5}{2}x + 8$
49. $y + 4 = 1.732(x + 2)$ or $y = 1.732x - 0.536$
51. $y = 0.364x + 3$ **53.** $2x - 3y + 11 = 0$
55. $5x - 2y = 22$ **57.** $m = -\frac{3}{2}$, y-intercept $= 6$, x-intercept $= 4$ **59.** $m = \frac{1}{3}$, y-intercept $= -3$, x-intercept $= 9$ **61.** $v = 2.6 + 0.4t$ **63.** $0.8\,\Omega$
65. **(a)** $C = 1{,}225 + 1.25n$, **(b)** \$26,225
67. **(a)** 362.08 ppm, **(b)** 441.28 ppm, **(c)** Answer varies with year.

Exercise Set 2.2

1. $(x - 2)^2 + (y - 5)^2 = 9$; $x^2 + y^2 - 4x - 10y + 20 = 0$
3. $(x + 2)^2 + y^2 = 16$; $x^2 + y^2 + 4x - 12 = 0$
5. $(x + 5)^2 + (y + 1)^2 = \frac{25}{4}$; $x^2 + y^2 + 10x + 2y + 19.75 = 0$
7. $(x - 2)^2 + (y + 4)^2 = 1$; $x^2 + y^2 - 4x + 8y + 19 = 0$
9. $C = (3, 4)$, $r = 3$

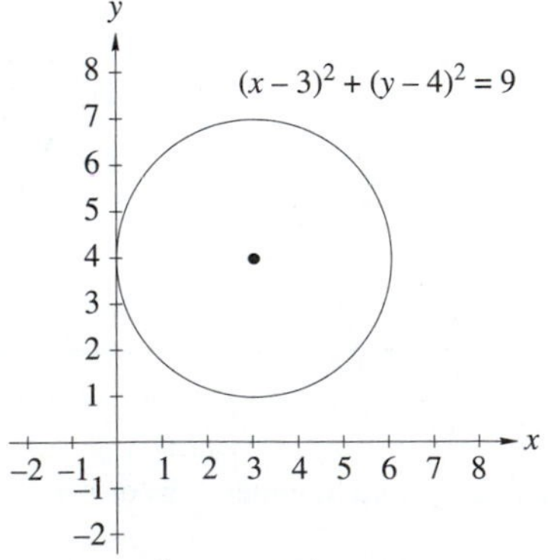

11. $C = \left(-\frac{1}{2}, -\frac{13}{4}\right)$, $r = \sqrt{7}$

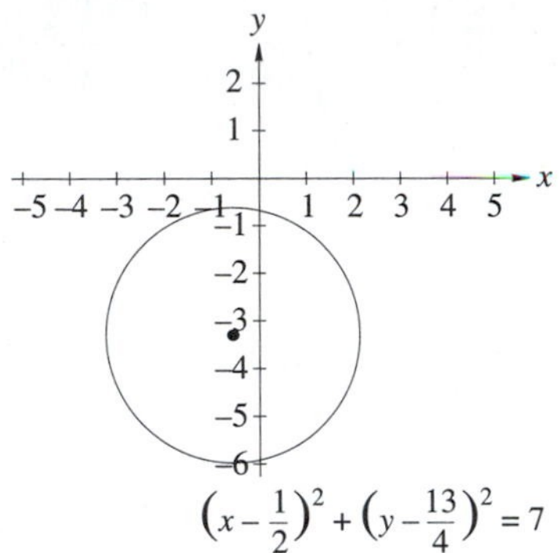

13. $C = \left(0, \frac{7}{3}\right)$, $r = \sqrt{6}$

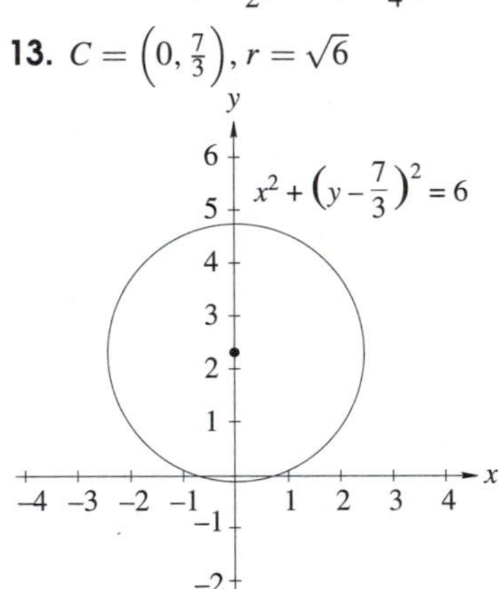

15. Circle, $C = (-2, 3)$, $r = 3$ **17.** Circle, $C = (-5, 3)$, $r = 9$ **19.** Not a circle since $r^2 = -1$ **21.** Circle, $C = (-3, 0)$, $r = 5$ **23.** Circle, $C = \left(-\frac{5}{2}, \frac{9}{2}\right)$, $r = 6$
25. Not a circle since $r^2 = -1.3$ **27.** $x^2 + y^2 = 310.03^2$ (assume circle is at the center of a coordinate system.)
29. $\left(x - 1.495 \times 10^8\right)^2 + y^2 = 1.478 \times 10^{11}$; $x^2 + y^2 + 4x - 12 = 0$ **31.** $x^2 + (y - 0.9)^2 = 0.55^2$
33. **(a)** $x^2 + (y - 5)^2 = 13^2$; **(b)** 28 cm

Exercise Set 2.3

1. $F = (0, 1)$, directrix: $y = -1$, opens upward

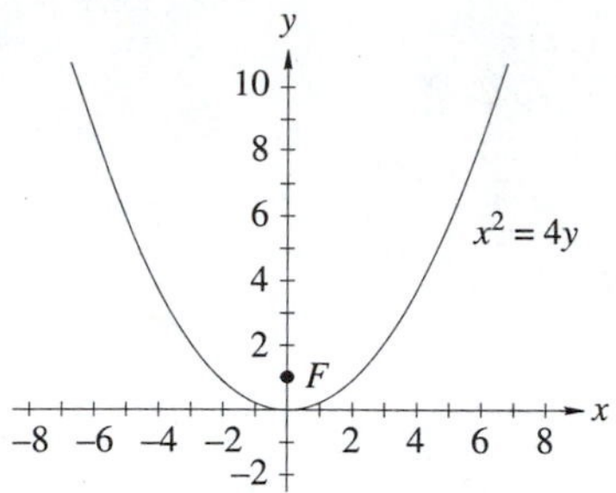

3. $F = (-1, 0)$, directrix: $x = 1$, opens left

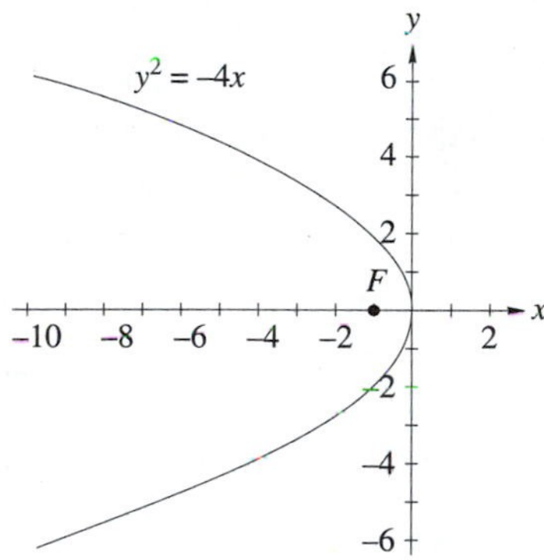

5. $F = (0, -2)$, directrix: $y = 2$, opens downward

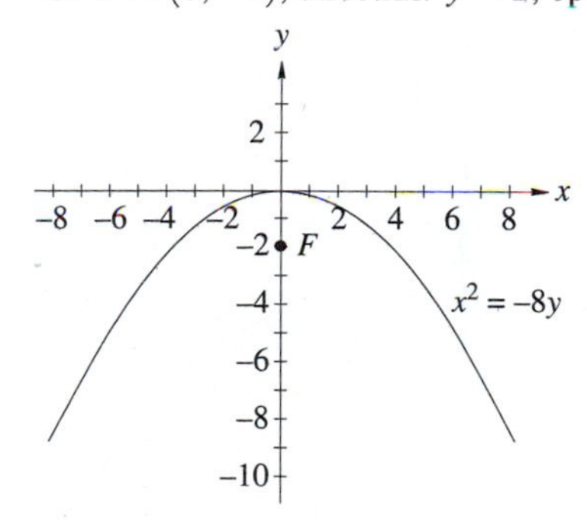

7. $F = \left(\frac{5}{2}, 0\right)$, directrix: $x = -\frac{5}{2}$, opens right

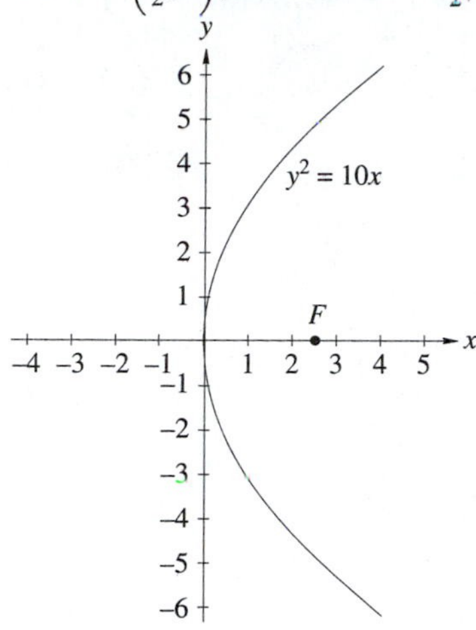

9. $F = \left(0, \frac{1}{2}\right)$, directrix: $x = -\frac{1}{2}$, opens upward

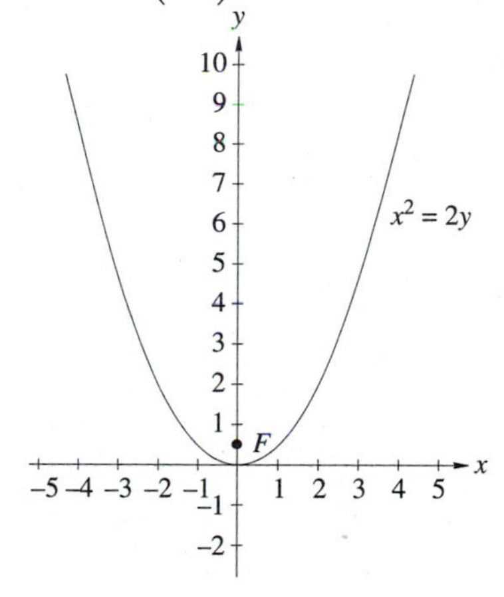

11. $F = \left(-\frac{21}{4}, 0\right)$, directrix: $x = -\frac{21}{4}$, opens left

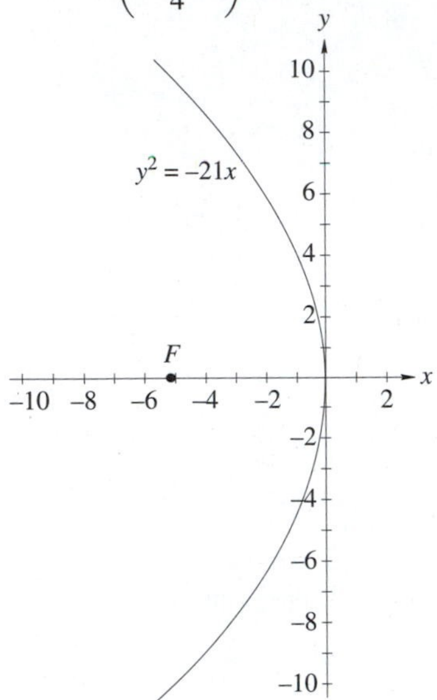

13. $x^2 = 16y$ **15.** $y^2 = -24x$ **17.** $y^2 = 8x$
19. $y^2 = 6x$ **21.** 27.78 m **23.** $\frac{6.25}{6} \approx 1.042$ m from vertex **25.**

27. 3.125 ft

Exercise Set 2.4

1. $\frac{x^2}{36} + \frac{y^2}{20} = 1$ **3.** $\frac{x^2}{12} + \frac{y^2}{16} = 1$ **5.** $\frac{x^2}{25} + \frac{y^2}{16} = 1$
7. $\frac{x^2}{16} + \frac{y^2}{9} = 1$ **9.** $V = (0,3)$, $V' = (0,-3)$, $F = (0,\sqrt{5})$, $F' = (0,-\sqrt{5})$, $M = (2,0)$, $M' = (-2,0)$

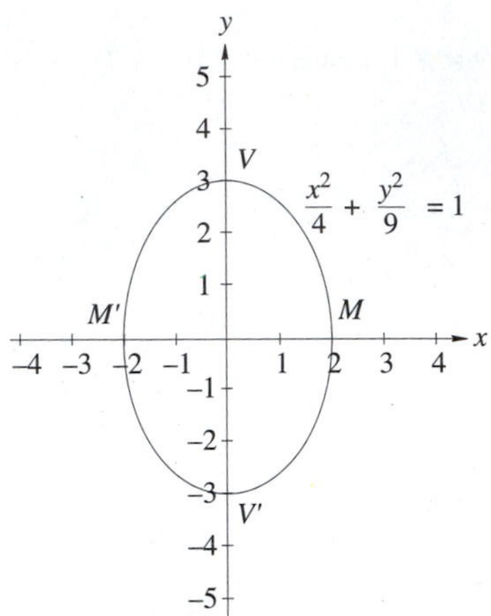

11. $V = (2,0)$, $V' = (-2,0)$, $F = (\sqrt{3},0)$, $F' = (-\sqrt{3},0)$, $M = (0,1)$, $M' = (0,-1)$

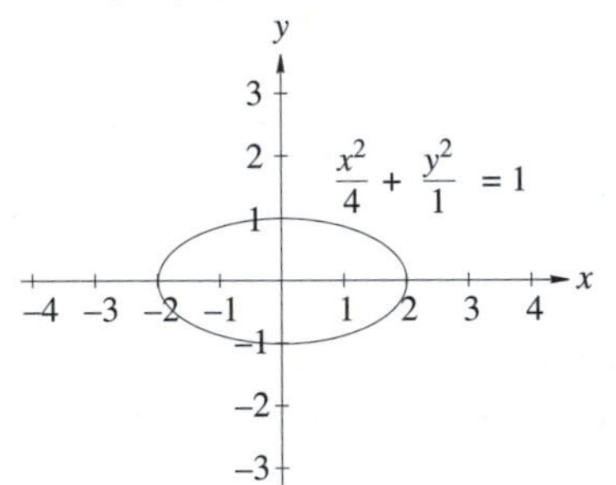

13. $V = (0,2)$, $V' = (0,-2)$, $F = (0,\sqrt{3})$, $F' = (0,-\sqrt{3})$, $M = (1,0)$, $M' = (-1,0)$

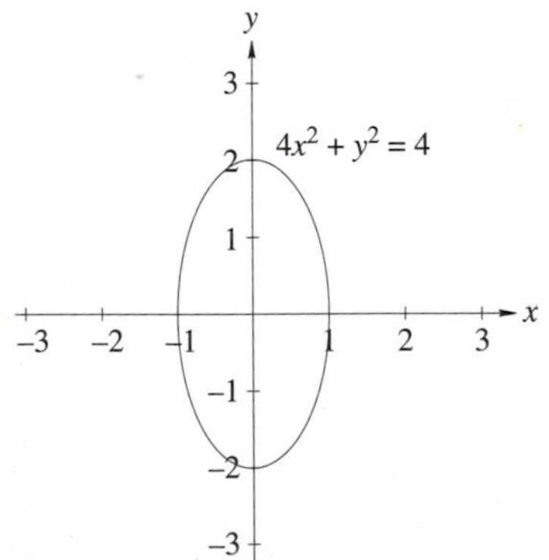

15. $V = (6,0)$, $V' = (-6,0)$, $F = (\sqrt{11},0)$, $F' = (-\sqrt{11},0)$, $M = (0,5)$, $M' = (0,-5)$

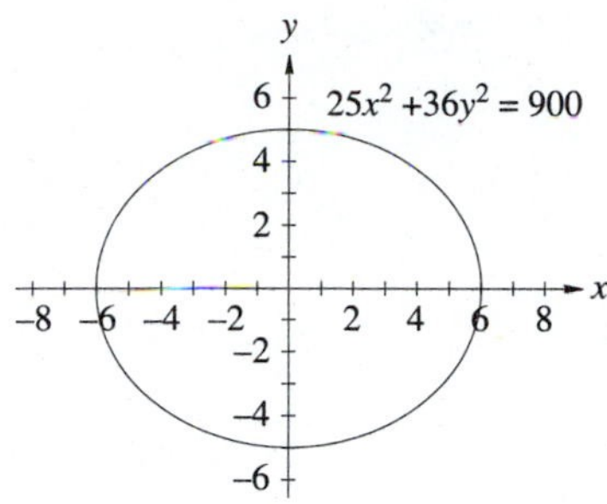

17. 2.992×10^8 km **19.** major axis $= \sqrt{200} \approx 14.14$; minor axis $= 10$ **21.** 0.9999253 **23.** $8\sqrt{2} \approx 11.3$ in.

Exercise Set 2.5

1. $\frac{x^2}{16} - \frac{y^2}{20} = 1$ **3.** $\frac{y^2}{9} - \frac{x^2}{16} = 1$ **5.** $\frac{x^2}{9} - \frac{y^2}{16} = 1$

7. $\frac{x^2}{7} - \frac{y^2}{9} = 1$ **9.** $V = (2, 0)$, $V' = (-2, 0)$, $F = (\sqrt{13}, 0)$, $F' = (-\sqrt{13}, 0)$, $M = (0, 3)$, $M' = (0, -3)$

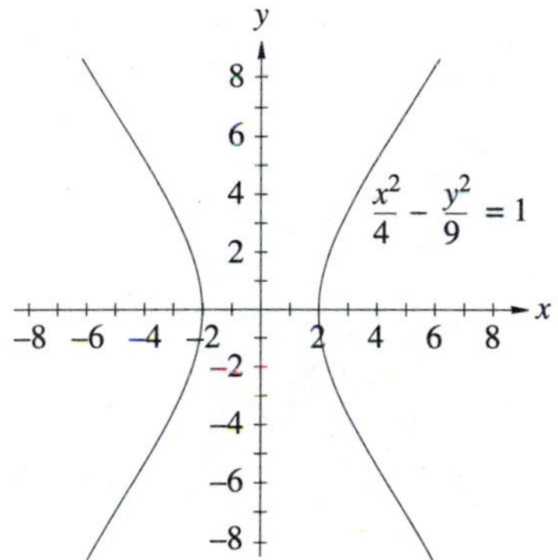

11. $V = (0, 2)$, $V' = (0, -2)$, $F = (0, \sqrt{5})$, $F' = (0, -\sqrt{5})$, $M = (1, 0)$, $M' = (-1, 0)$

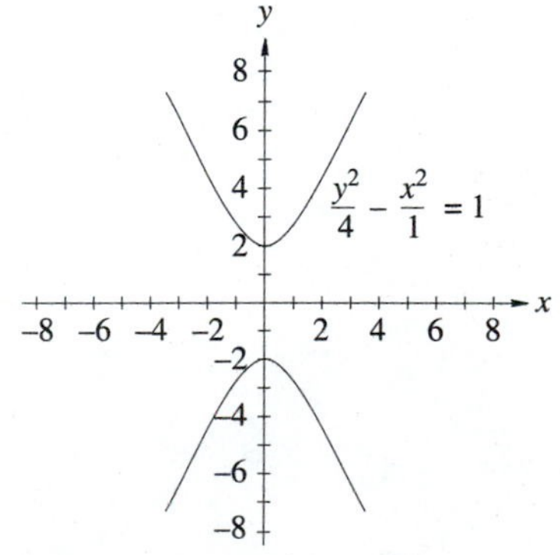

13. $V = (1, 0)$, $V' = (-1, 0)$, $F = (\sqrt{5}, 0)$, $F' = (-\sqrt{5}, 0)$, $M = (0, 2)$, $M' = (0, -2)$

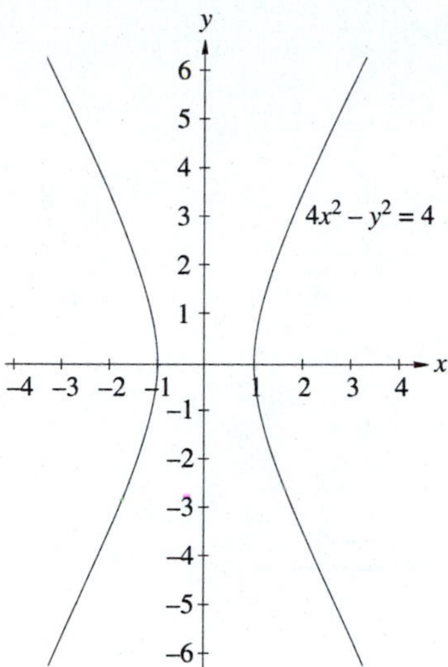

15. $V = (0, 5)$, $V' = (0, -5)$, $F = (0, \sqrt{61})$, $F' = (0, -\sqrt{61})$, $M = (6, 0)$, $M' = (-6, 0)$

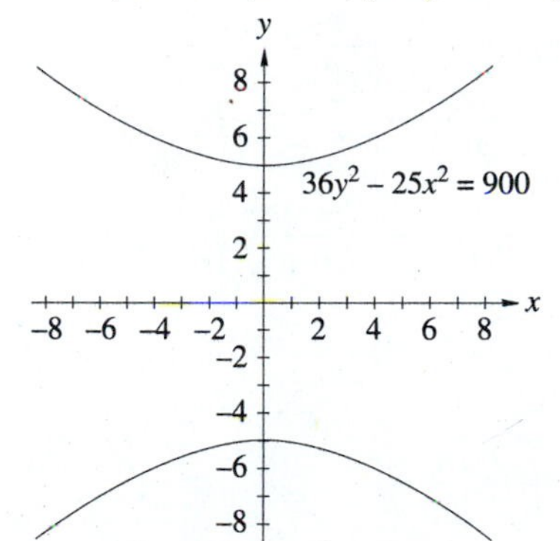

17. $\frac{x^2}{49} - \frac{y^2}{61.25} = 1$ **19.** $\sqrt{2}$

21.

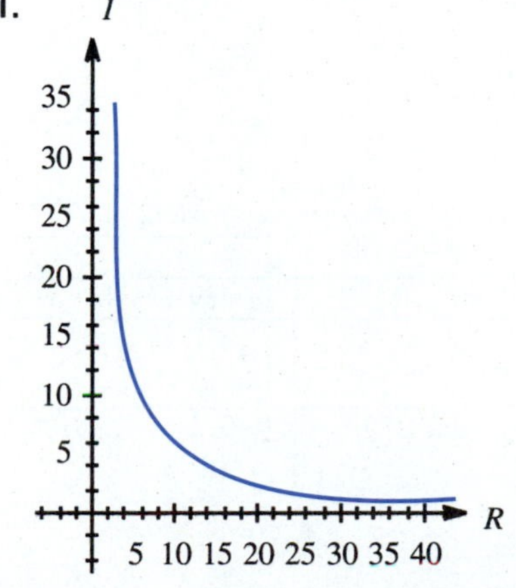

23. 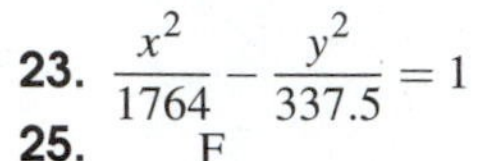
$$\frac{x^2}{1764} - \frac{y^2}{337.5} = 1$$

25.

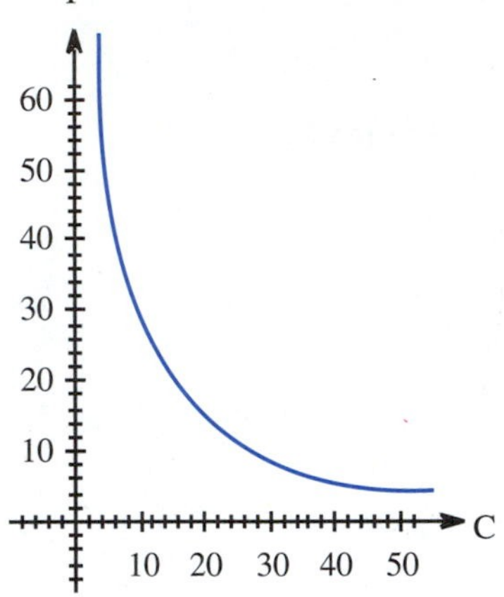

Exercise Set 2.6

1. ellipse;

	$x'y'$-system	xy-system
center	$(0', 0')$	$(4, -3)$
vertices	$(\pm 3', 0')$	$(7, -3)$, $(1, -3)$
foci	$(\pm\sqrt{5}', 0')$	$(\sqrt{5}+4, -3)$, $(-\sqrt{5}+4, -3)$
endpoints of minor axis	$(0', \pm 2')$	$(4, -1)$, $(4, -5)$

$$\frac{(x-4)^2}{9} + \frac{(y+3)^2}{4} = 1$$

3. hyperbola;

	$x'y'$-system	xy-system
center	$(0', 0')$	$(-4, 3)$
vertices	$(0', \pm 6)$	$(-4, 9)$, $(-4, -3)$
foci	$(0', \pm\sqrt{37}')$	$(-4, \sqrt{37}+3)$, $(-4, -\sqrt{37}+3)$
endpoints of conjugate axis	$(\pm 2', 0')$	$(-3, 3)$, $(-5, 3)$

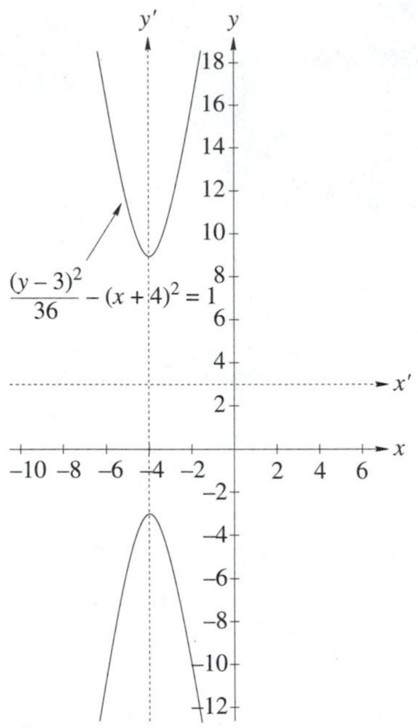

5. hyperbola;

	$x'y'$-system	xy-system
center	$(0', 0')$	$(-5, 3)$
vertices	$(\pm 2', 0')$	$(3, 0)$, $(-7, 0)$
foci	$(2 \pm \sqrt{26}', 0')$	$(2\sqrt{26}-5, 0)$, $(-2\sqrt{26}-5, 0)$
endpoints of conjugate axis	$(0', \pm 10')$	$(-5, 10)$, $(-5, -10)$

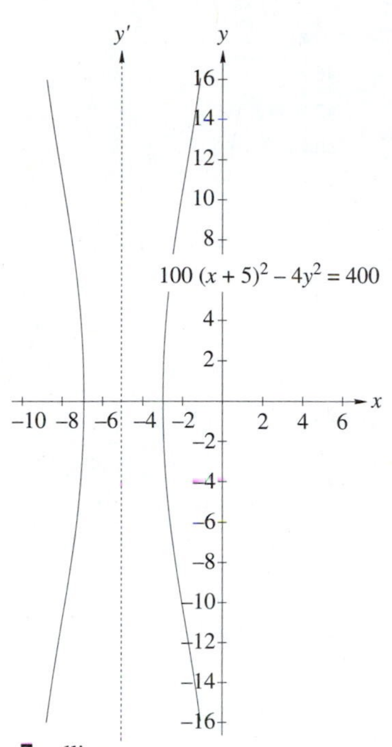

7. ellipse;

	$x'y'$-system	xy-system
center	$(0', 0')$	$(-2, \frac{3}{2})$
vertices	$(0', \pm 3')$	$(-2, \frac{7}{2})$, $(-2, -\frac{1}{2})$
foci	$(0', \pm\sqrt{3}')$	$(-2, \sqrt{3}+\frac{3}{2})$, $(-2, -\sqrt{3}+\frac{3}{2})$
endpoints of minor axis	$(\pm 1', 0')$	$(-1, \frac{3}{2})$, $(-3, \frac{3}{2})$

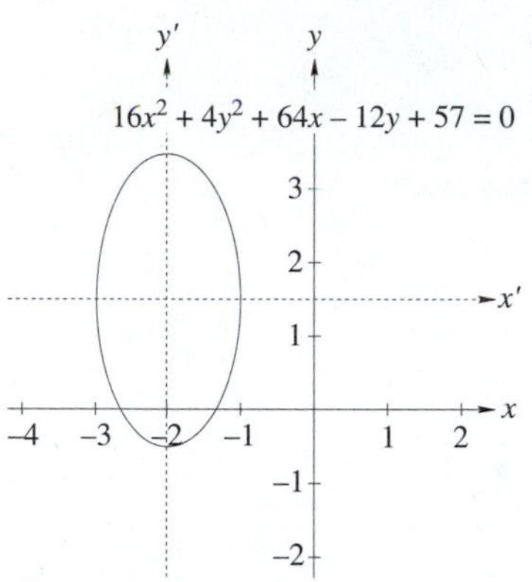

9. ellipse;

	$x'y'$-system	xy-system
center	$(0', 0')$	$(5, 2)$
vertices	$(0', \pm 5')$	$(5, 7)$, $(5, -3)$
foci	$(0', \pm\sqrt{21}')$	$(5, 2+\sqrt{21})$, $(5, 2-\sqrt{21})$
endpoints of minor axis	$(\pm 2', 0')$	$(7, 2)$, $(3, 2)$

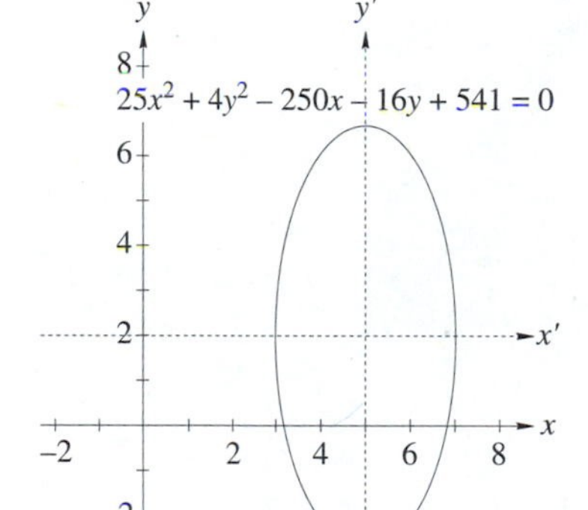

11. hyperbola;

	$x'y'$-system	xy-system
center	$(0', 0')$	$(4, 2)$
vertices	$(\pm\sqrt{2}', 0')$	$(4+\sqrt{2}, 2)$, $(4-\sqrt{2}, 2)$
foci	$(\pm\sqrt{6}', 0')$	$(4+\sqrt{2}, 2)$, $(4-\sqrt{2}, 2)$
endpoints of conjugate axis	$(0', \pm 2')$	$(4, 0)$, $(4, 4)$

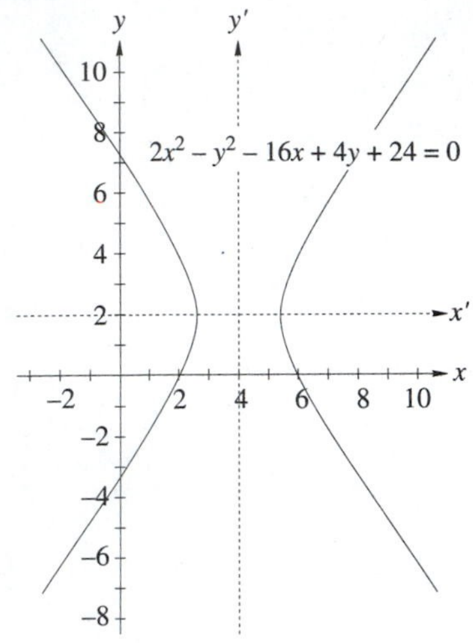

13. $(x-2)^2 = 32(y+3)$ **15.** $\dfrac{(x-4)^2}{36} + \dfrac{(y+3)^2}{20} = 1$

17. $\dfrac{(y-2)^2}{9} - \dfrac{(x+3)^2}{16} = 1$ **19.** $(y-1)^2 = -16(x+5)$

21. $\dfrac{(x+4)^2}{36} + \dfrac{(y-1)^2}{100} = 1$ **23.** This is a parabola with the equation $-4.9(t-3)^2 = s - 44.1$. Maximum height is when $t = 3$ and $s = 44.1$ m.

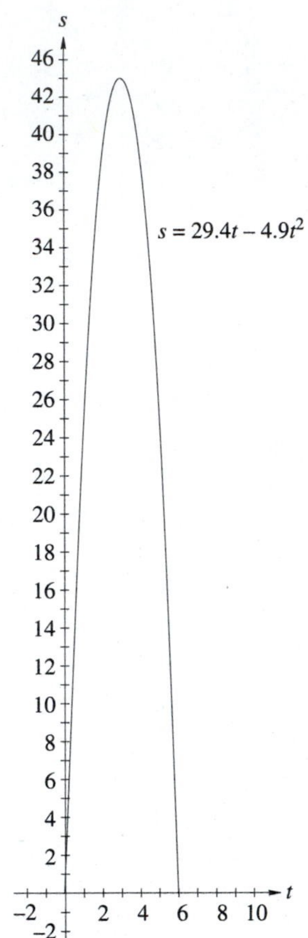

25. If the transverse axis passes through A and B and the conjugate axis passes through the midpoint of $\overline{AB}$, then A is at $(-500, 0)$ and B at $(500, 0)$. The plane is at $(216, 1073.054)$ and lies on the hyperbola $\dfrac{x^2}{8100} - \dfrac{y^2}{241\,900} = 1$

27. **(a)** $x^2 = -200(y-0.02)$; **(b)** 2.00 cm

29. **(a)** $24(y-125)^2 - x^2 = 15{,}000$, upper branch; **(b)** $\left(\dfrac{1{,}200+120\sqrt{5}}{19}, \dfrac{2{,}400+240\sqrt{5}}{19}\right) \approx (77.3, 154, 6)$

31. **(a)** $y = 2\sqrt{0.0625 - 0.25x^2}$ and $y = -2.5\sqrt{0.1 - 0.4x^2}$,

(b)

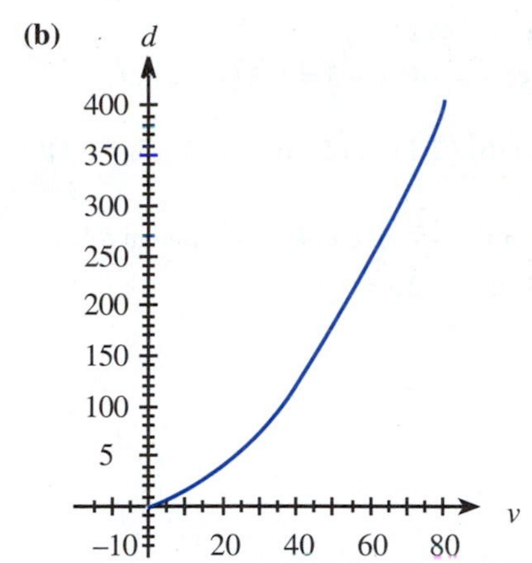

33. **(a)** $c = 1.14x - 0.01x^2$,

(b)

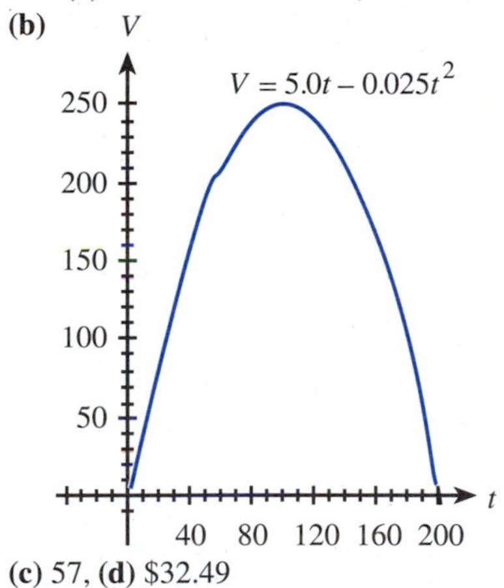

(c) 57, **(d)** \$32.49

Exercise Set 2.7

1. **(a)** discriminant is 1; curve is a hyperbola; **(b)** 45°;

(d) $\dfrac{y''^2}{18} - \dfrac{x''^2}{18} = 1$

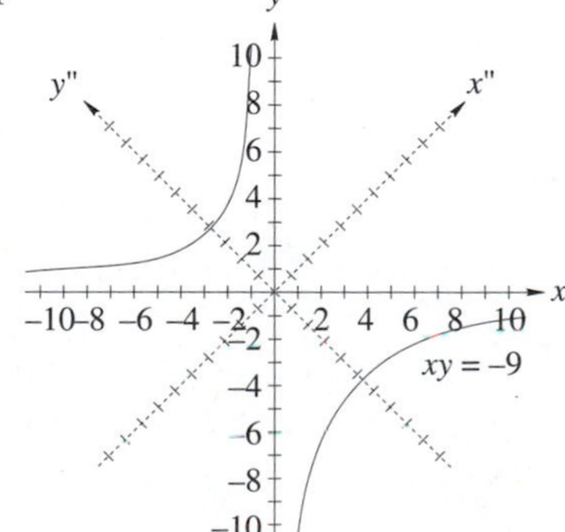

3. **(a)** discriminant is 32; curve is a hyperbola; **(b)** 45°;

(d) $\dfrac{y''^2}{2} - \dfrac{x''^2}{4} = 1$

(e)

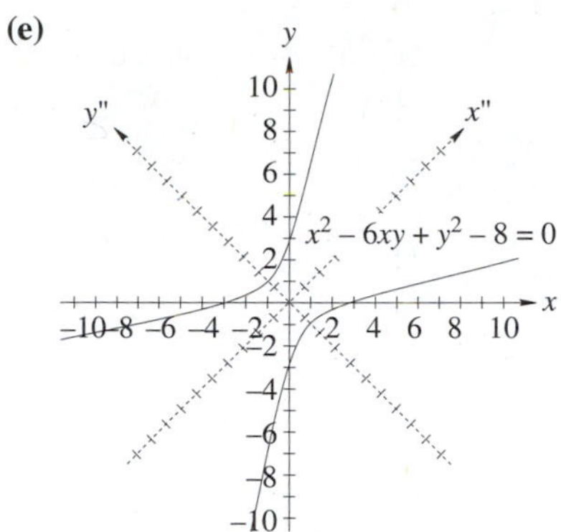

5. **(a)** discriminant is −10,000; curve is an ellipse;

(b) 36.870°; **(d)** $\dfrac{x''^2}{4} + y''^2 = 1$

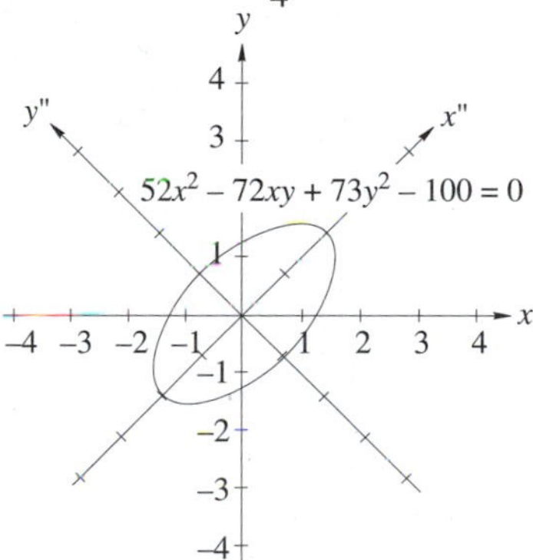

7. **(a)** discriminant is 0; curve is a parabola; **(b)** 45°;

(d) $y''^2 = -\dfrac{\sqrt{2}}{2}x''$

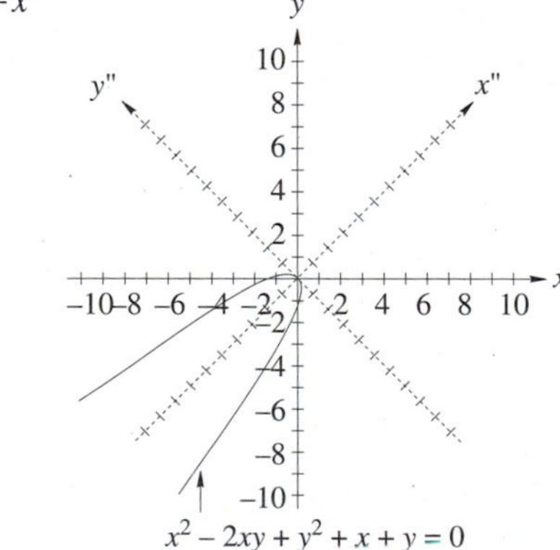

9. **(a)** discriminant is 168; curve is a hyperbola; **(b)** 33.690°;

(d) $\dfrac{x''^2}{7} - \dfrac{y''^2}{6} = 1$

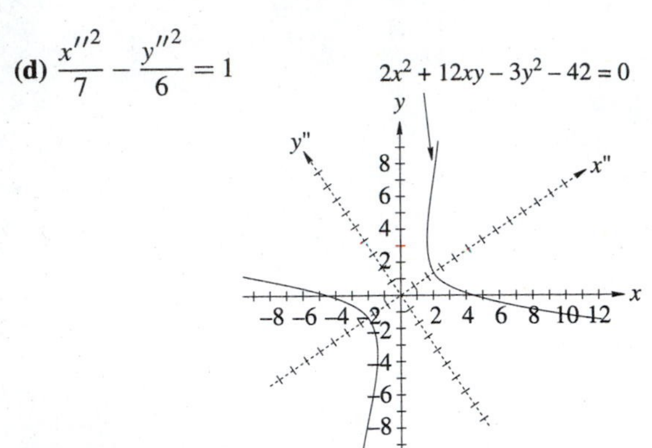

11. **(a)** discriminant is -119; curve is an ellipse; **(b)** $45°$; **(d)** $7x''^2 + 17y''^2 = 52$

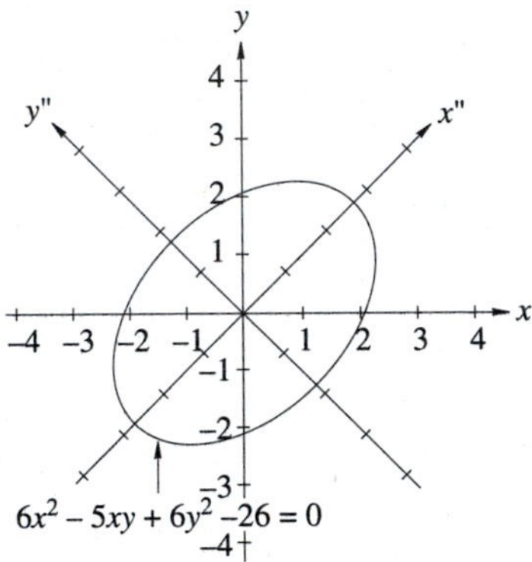

13.

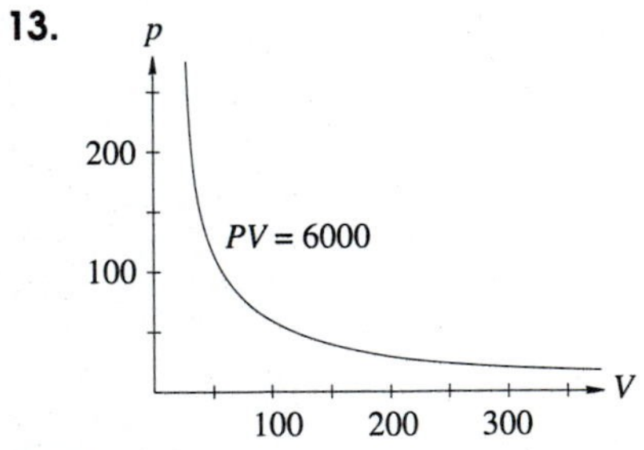

15. Parabola

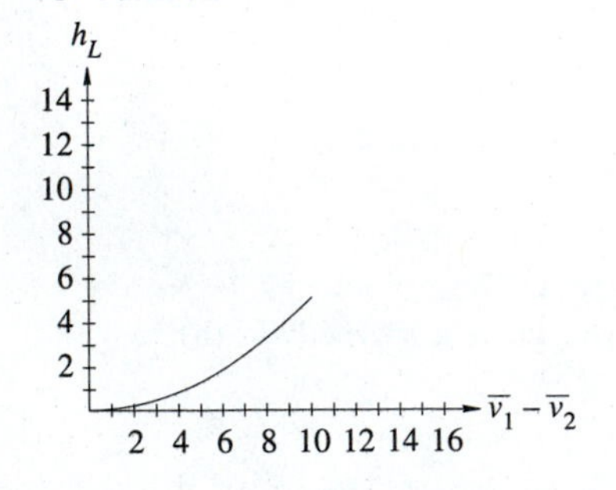

Review Exercises

1. **(a)** 5; **(b)** $(\frac{1}{2}, 7)$; **(c)** $-\frac{4}{3}$; **(d)** $y - 5 = -\frac{4}{3}(x - 2)$ or $4x + 3y = 23$

3. **(a)** $\sqrt{104} = 2\sqrt{26}$; **(b)** $(2, 1)$; **(c)** 5; **(d)** $y + 4 = 5(x - 1)$ or $y - 5x = -10$

5. line in #1: $\frac{3}{4}$; line in #2: $\dfrac{12}{5}$; line in #3: $-\frac{1}{5}$; line in #4: 1

7. $y - 5 = -2(x + 3)$ or $y + 2x = -1$

9.

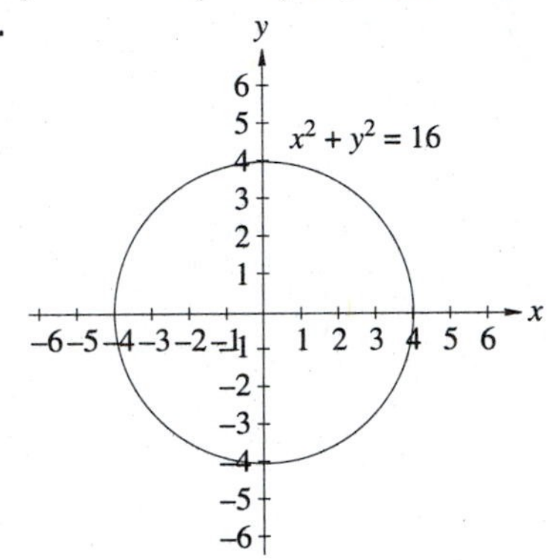

11.

$x^2 + 4y^2 = 16$

13.

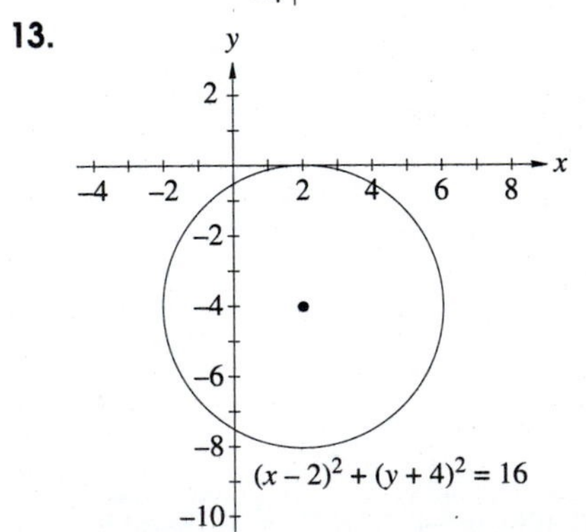

15.

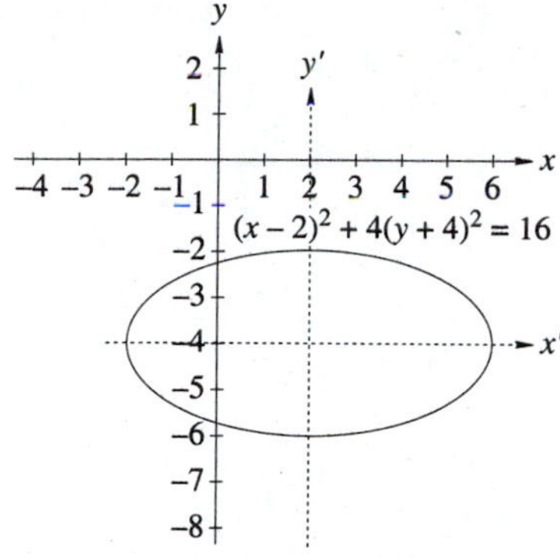

17.

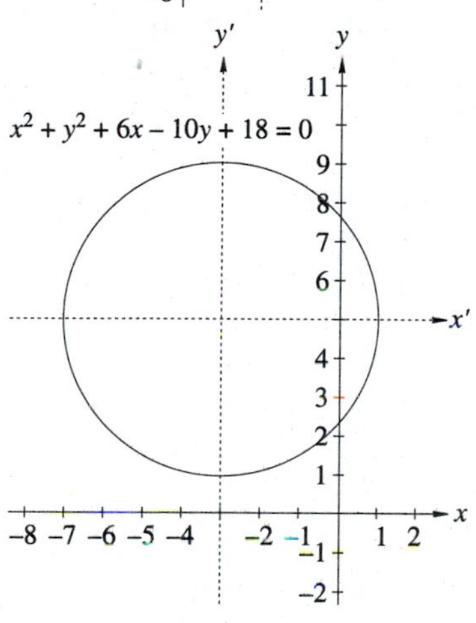

19.

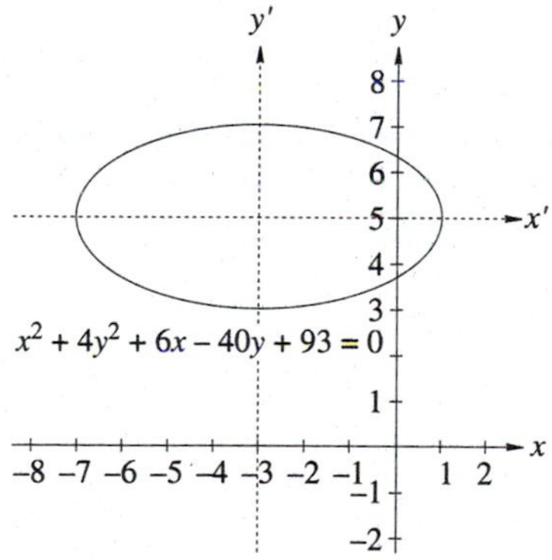

21.

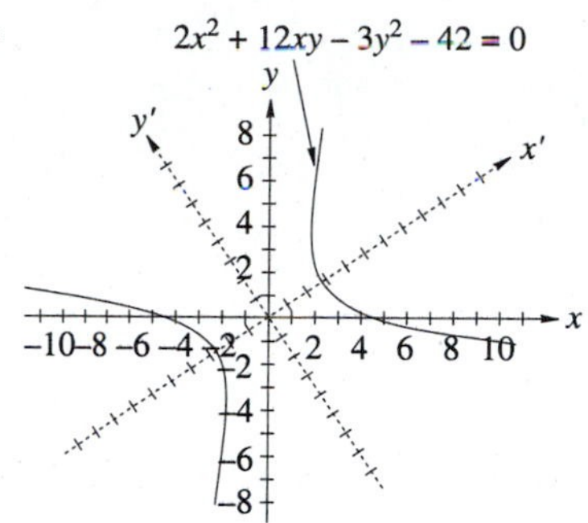

23. $3x^2 + 2\sqrt{3}\,xy + y^2 + 8x - 8\sqrt{3}\,y = 32$

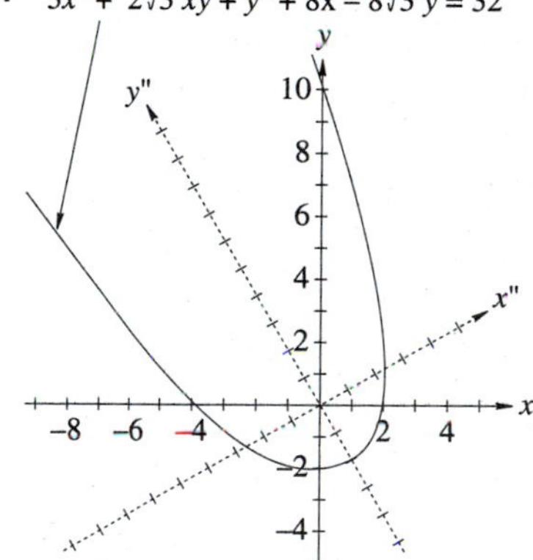

25. **(a)**

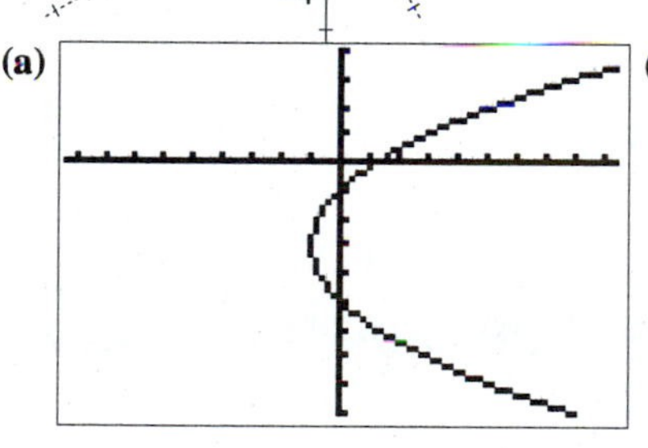

(b) Parabola

[−9.4, 9.4] x [−9, 4]

(c) Horizontal axis, opens to the right, vertex: $(-1, -3)$, directrix: $x = -2$, and focus: $(0, -3)$.

27. $\dfrac{x^2}{50^2} + \dfrac{y^2}{195.84} = 1$

29. If the x-axis passes through A and B and the y-axis through the midpoint of $\overline{AB}$, then B has the coordinates $(200, 0)$ and $A = (-200, 0)$. The plane is at $(91.13, -50)$. The plane lies on the hyperbola $\dfrac{x^2}{6400} - \dfrac{y^2}{33\,600} = 1$.

Chapter 2 Test

1. Focus $(-4.5, 0)$; directrix $x = 4.5$

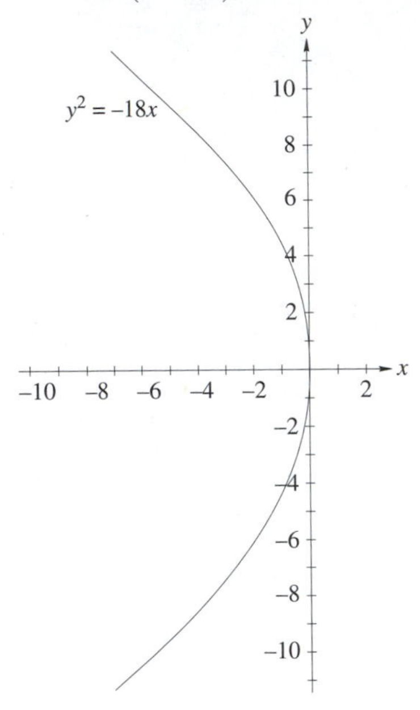

3. (a)

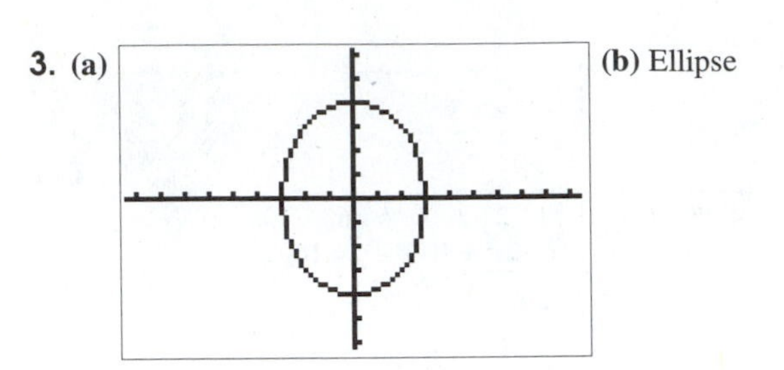

[−9.4, 9.4] x [−6.2, 6.2]

(b) Ellipse **(c)** Center: $(0,0)$, Vertices: $(-3,0)$ and $(3,0)$, Foci: $(-\sqrt{7},0)$ and $(\sqrt{7},0)$; endpoints of minor axis: $(0,4)$ and $(0,-4)$

5. $y + \frac{13}{2} = -\frac{2}{3}(x-3)$ or $6y + 4x + 27 = 0$

7. (a)

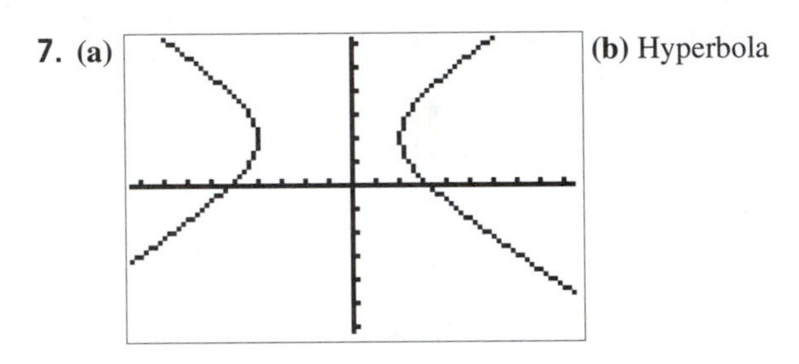

[−9.4, 9.4] x [−6.2, 6.2]

(b) Hyperbola **(c)** Center: $(-1,2)$, Vertices: $(-4,2)$ and $(2,2)$, Foci: $(-1-\sqrt{13},2)$ and $(-1+\sqrt{13},2)$; endpoints of minor axis: $(-1,4)$ and $(-1,0)$ **9. (a)** $\frac{\pi}{4}$; **(b)** an ellipse **11.** 75 ft above its lowest point.

ANSWERS FOR CHAPTER 3

Exercise Set 3.1

1. 9 **3.** 7 **5.** 28 **7.** $-1/2$ **9.** 1

11. $\frac{x_1^2 + x_1 - 2}{x_1 - 1} = x_1 + 2$ **13.** $2x_1 + h$

15. (a) 29 liters/min; **(b)** 39 L/m; **(c)** 34 L/m **17. (a)** 30, **(b)** 27, **(c)** 25.5, **(d)** 24.75, **(e)** 24.3, **(f)** 24.15

21. (a) 36.5 mi, **(b)** 43.8 mph

Exercise Set 3.2

1. 84 **3.** 11.9375 **5.** 181 **7.** 72.28125 **9.** 175.75 **11.** 30.85 **15.** \$2,480 **17.** 1.206 Ω

Exercise Set 3.3

1.

x	0.9	0.99	0.999	0.9999	1.0001	1.001	1.01	1.1
$f(x) = 3x$	2.7	2.97	2.997	2.9997	3.0003	3.003	3.03	3.3

$\lim_{x \to 1} 3x = 3$

3.

x	−1.1	−1.01	−1.001	−1.0001	−0.9999	−0.999	−0.99	−0.9
$h(x) = x^2 + 2$	3.21	3.0201	3.002	3.0002	2.9999	2.998	2.9801	2.81

$\lim_{x \to -1} (x^2 + 2) = 3$

5.

x	-0.1	-0.01	-0.001	-0.0001	0.0001	0.001	0.01	0.1
$f(x) = \frac{\tan x}{x}$	1.0033467	1.0000333	1.0000003	1	1	1.0000003	1.0000333	1.0033467

$\lim_{x\to 0} \frac{\tan x}{x} = 1$

7.

x	0.9	0.99	0.999	0.9999	1.0001	1.001	1.01	1.1
$h(x) = \frac{x}{x-1}$	-9	-99	-999	-9999	10001	1001	101	11

$\lim_{x\to 1} \frac{x}{x-1}$ Does not exist

9. **(a)** 0, both sides of graph near $x = -1$ converge on 0; **(b)** Does not exist, graph goes toward -2 when $x < 2$ and goes toward 1 for $x > 2$; **(c)** 4, both sides of graph near $x = 3$ converge on 4.

11.

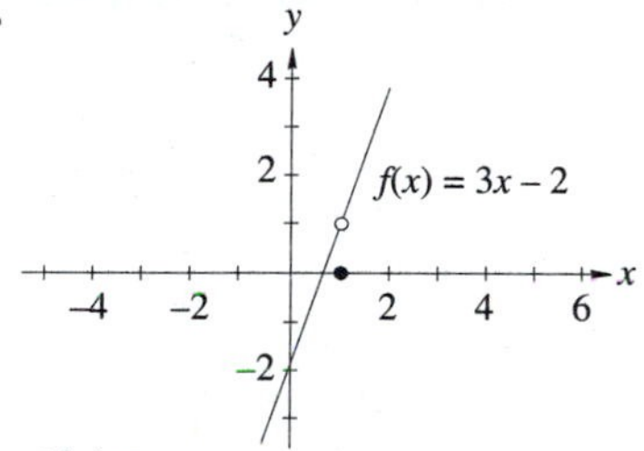

$\lim_{x\to 1} f(x)$ Does not exist

13.

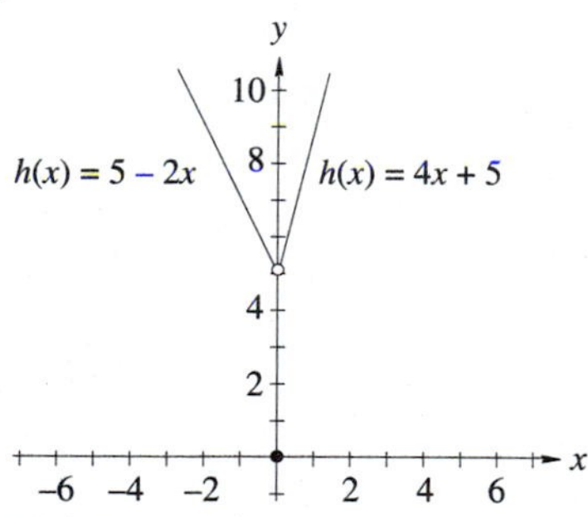

$\lim_{x\to 1} h(x)$ Does not exist

15.

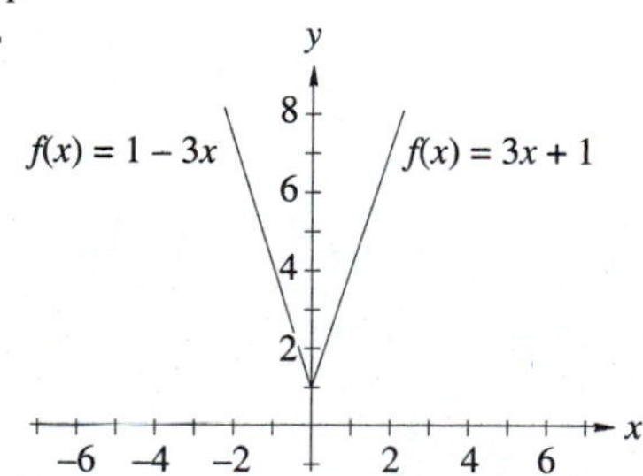

$\lim_{x\to 1} g(x) = 1$

17.

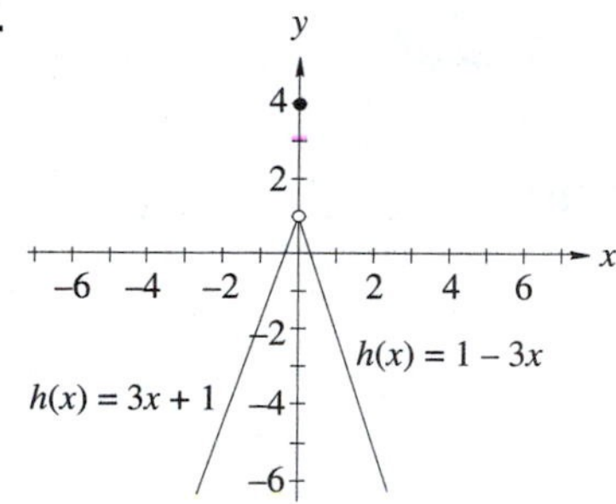

$\lim_{x\to 1} h(x)$ Does not exist

19. -15 **21.** -4 **23.** 8 **25.** 0 **27.** 4

Exercise Set 3.4

1. **(a)** -2; **(b)** 1; **(c)** 4; **(d)** 4 **3.** 3 **5.** ∞ **7.** -4 **9.** -1 **11.** -1 **13.** 0 **15.** 0 **17.** 0 **19.** 2 **21.** 0 **23.** $-\infty$ **25.** **(a)** 0.02381, **(b)** 0.0556, **(c)** 0 **27.** **(a)** \$4.75, **(b)** \$4.75, **(c)** \$4.75

Exercise Set 3.5

1. 8 **3.** 5 **5.** 1 **7.** 6 **9.** 9 **11.** 5 **13.** 8 **15.** 27 **17.** 1 **19.** $\sqrt{6}$ **21.** 2 **23.** -2 **25.** 2 **27.** $\frac{13}{9}$ **29.** 8 **31.** $\frac{2}{5}$ **33.** 0

Exercise Set 3.6

1. Continuous **3.** Not continuous, $\lim_{x\to 1^-} h(x) = 4 \neq 3 = h(1)$ **5.** Not continuous, $k(2)$ is not defined **7.** Not continuous, $\lim_{x\to -2^-} g(x) = 4 \neq g(-2)$ **9.** Not continuous, $\lim_{x\to 2^-} j(x) = 0 = j(2)$ **11.** None **13.** -1 **15.** $-3, 2$ **17.** None **19.** 2 **21.** $(-\infty, -3), (-3, \infty)$ **23.** $(-\infty, -2), (-2, 2), (2, \infty)$ **25.** $[-3, \infty)$ **27.** $(-\infty, -3], [3, \infty)$ **29.** $(-\infty, 0)$, $(0, \infty)$ **31.** **(a)** yes

(b)

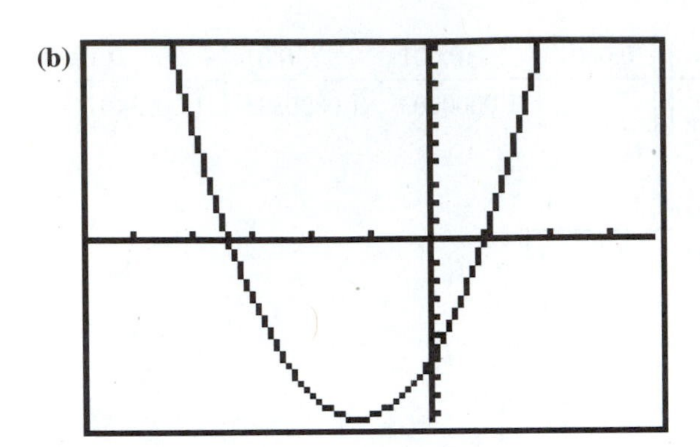

33. **(a)** \$3,585, **(b)** \$3,585, **(c)** \$3,585, **(d)** Yes, **(e)** \$14,155, **(f)** \$14,155, **(g)** \$14,155, **(h)** Yes

Review Exercises

1. 2 **3.** −10 **5.** $b+1$ **7.** 21 **9.** 65.96875 **11.** 0 **13.** 6 **15.** 3 **17.** 0 **19.** 0 **21.** 5 **23.** −3 **25.** $(-\infty,-7)(-7,\infty)$ **27.** $(-\infty,-5),(5,\infty)$

Chapter 3 Test

1. 2 **3.** 13 **5.** 2 **7.** $x=2$ and $x=3$

ANSWERS FOR CHAPTER 4

Exercise Set 4.1

1. 3 **3.** −3 **5.** $6x$ **7.** $4x$ **9.** $-8x$ **11.** $-12x^2$ **13.** $1-9t^2$ **15.** $\dfrac{-1}{(x+1)^2}$ **17.** $\dfrac{8x}{(1-x^2)^2}$ **19.** $32t-6$ **21.** $\dfrac{x}{\sqrt{x^2+4}}$ **23.** $\dfrac{1-x}{\sqrt{2x-x^2}}$ **25.** 3;3 **27.** 98, 66, 2 **29.** 0, 3 **31.** $\dfrac{-10}{21^2}=\frac{-10}{441},\frac{10}{441}$ **33.** $\frac{3}{4}$, $-\frac{3}{4}$ **35.** 0, $\frac{15}{64}$, $-\frac{20}{27}$

Exercise Set 4.2

1. 0 **3.** 7 **5.** $18x$ **7.** $5x^{14}$ **9.** $-10x^{-3}$ **11.** $-x^{1/2}$ **13.** $\dfrac{2\sqrt{3}}{\sqrt{x}}$ **15.** $-x^{-4/3}=\dfrac{-1}{x\sqrt[3]{x}}$ **17.** $18x+3$ **19.** $x^2+x-5-\frac{4}{3}x^{-3}$ **21.** $4x^4-\frac{9}{2}x^2-x^{-3}+2x^{-4}$ **23.** $12\sqrt{3}x^3+\dfrac{3\sqrt{5}}{2\sqrt{x}}+4x$ **25.** $32t-32$ **27.** $-8.0t+t^{-1/2}$ **29.** $\dfrac{\sqrt{6}}{2}t^{-1/2}-6t^{1/2}+\frac{2}{3}t^{-1/3}$ **31.** $\frac{35}{3}x^{4/3}$ **33.** $\dfrac{-8}{x^3}$ **35.** $-\frac{2}{3}x^{-7/6}$ **37.** $\frac{9}{2}x^{1/2}-7-3x^{-5/2}$ **39.** $f'(x)=3x^2+6x$, $f'(x)=0$ when $x=0$ or $x=-2$ **41.** $g'(x)=6x^2+2x-4$, $g'(x)=0$ when $x=-1$ or $x=2/3$ **43.** $j(x)=12t^2-\dfrac{1}{t^2}$, $j(t)=0$ when $t=\dfrac{1}{\sqrt[4]{12}}\approx 0.537285$ **45.** **(a)** 146,382 people, **(b)** 63.52 increase in population per month **47.** **(a)** 172 ft, **(b)** 71 ft/s **49.** **(a)** 133,912 people, **(b)** 1,493 people per year, **(c)** 139,886 people, **(d)** 1,494 people per year **51.** 120 W/A **53.** **(a)** $P(n)=R(n)-C(n)=(78n-0.025n^2)-(9{,}800+22.5n)=55.5n-0.025n^2-9{,}800$, **(b)** $P'(n)=55.5-0.05n$, **(c)** \$53/telephone **55.** **(a)** 605 ft above sea level, **(b)** $t=11$ s, **(c)** $h(2)=621$ ft, $v(2)=48$ ft/s, **(d)** $h(4)=861$ ft, $v(4)=192$ ft/s, **(e)** $h(6)=1{,}325$ ft, $v(6)=240$ ft/s, **(f)** $h(10)=1{,}005$ ft, $v(10)=-720$ ft/s, **(g)** $v(11)=-1{,}320$ ft/s

Exercise Set 4.3

1. $12x-19$ **3.** $18x^2-14x-8$ **5.** $-54x^2+12x+36$ **7.** $16t^3-72t^2+72t$ **9.** $(3w^3-4w^2+2w-5)(2w+w^{-2})+(9w^2-8w+2)(w^2-w^{-1})=15w^4-16w^3+6w^2-16w+4-5w^{-2}$ **11.** $\dfrac{(2x+3)(4)-(4x-1)(2)}{(2x+3)^2}=\dfrac{14}{(2x+3)^2}$ **13.** $\dfrac{(4x-1)(18x)-(9x^2+2)(4)}{(4x-1)^2}=\dfrac{36x^2-18x-4}{(4x-1)^2}$ **15.** $12s+\dfrac{1}{3s^3}$ **17.** $12t^2+\dfrac{4}{(t-2)^2}$ **19.** $2x+1$ **21.** $\dfrac{(3\phi^2-1)(2\phi)-\phi^2(6\phi)}{(3\phi^2-1)^2}$ **23.** $\dfrac{(x^3-3x-1)(9x^2-1)-(3x^3-x+1)(3x^2-3)}{(x^3-3x-1)^2}=\dfrac{-16x^3-12x^2+4}{(x^3-3x-1)^2}$ **25.** $\dfrac{t^{1/3}(6t-1)-(3t^2-t-1)\frac{1}{3}t^{-2/3}}{t^{2/3}}=\dfrac{5t^2-\frac{2}{3}t+1/3}{t^{4/3}}=\dfrac{15t^2-2t+1}{3t^{3/4}}$ **27.** $24w+\dfrac{2}{(w+1)^2}$ **29.** $\dfrac{2x}{(x^2+1)^2}$ **31.** $\dfrac{(2-x^2)(3x-x^3)(-3x^2)-(4-x^3)(5x^4-15x^2+6)}{(2-x^2)^2(3x-x^3)^2}=\dfrac{2x^7-20x^4-12x^3+60x^2-24}{(2-x^2)^2(3x-x^3)^2}$ **33.** $\dfrac{(s^2-1)(s-1)(6s^2)-2s^3[(s^2-1)+(s-1)(2s)]}{(s^2-1)^2(s-1)^2}=\dfrac{12s^5-10s^4-8s^3+6s^2}{(s^2-1)^2(s-1)^2}$

35. $\frac{(3x^3+1)(6x^2-22x+14)-(2x^3-11x^2+14x-3)(9x^2)}{(3x^3+1)^2}$ $=\frac{33x^4-84x^3+33x^2-22x+14}{(3x^3+1)^2}$

37. $\frac{(t^2-1)}{(2t+1)}\left[\frac{(2t+1)-(t-1)(2)}{(2t+1)^2}\right]+$ $\frac{(t-1)}{(2t+1)}\left[\frac{(2t+1)(2t)-(t^2-1)(2)}{(2t+1)^2}\right]=\frac{2t^3+3t^2-5}{(2t+1)^3}$

39. 16 **41.** Tangent: $y-2=-\frac{1}{2}(x-3)$ or $2y+x=7$ Normal: $y-2=2(x-3)$ or $y-2x-4$ **43.** Tangent: $y-3=-\frac{1}{3}(x-1)$ or $3y+x=10$ Normal: $y-3=3(x-1)$ or $y-3x=0$ **45.** Tangent: $y+2=\frac{1}{8}(x+2)$ or $8y-x=-14$ Normal: $y+2=-8(x+2)$ or $y+8x=-18$

47. 8.2 thousand people per year or 8,200 people per year

49. $V'=\frac{6600}{(R+60)^2}$

51. **(a)** $N'(t)=\frac{9t^2+8t^{3/2}-6t-8\sqrt{t}-3}{2\left(3\sqrt{t}+2\right)^2\sqrt{t}}$, **(b)** 777 bacteria/hr

Exercise Set 4.4

1. $12(3x-6)^3$ **3.** $-25(5x-7)^{-6}$

5. $4(2x+3)(x^2+3x)^3$ **7.** $\frac{-4x}{(4x^2+7)^{3/2}}$

9. $3(t^4-t^3+2)^2(4t^3-3t^2)$

11. $3(u^3+2u^{-4})^2(3u^2-8u^{-5})$

13. $2[(x-2)(3x^2-x)]^2[(x-2)(6x-1)+(3x^2-x)]=2[(x-2)(3x^2-x)]^2(9x^2-14x+2)$

15. $6(v^2+1)^2(2v-5)^2+4v(v^2+1)(2v-5)^3$

17. $4\left[\left(x^3-6x^2\right)\left(5x-6x^2+x^3\right)\right]^3\left[\left(x^3-6x^2\right)\left(5-12x+3x^2\right)+\left(5x-6x^2+x^3\right)\left(3x^2-12x\right)\right]=$ $4\left[\left(x^3-6x^2\right)\left(5x-6x^2+x^3\right)\right]^3\left(6x^5-60x^4+164x^3-90x^2\right)$

19. $\frac{(3x^2+2)^3(3x^2-7)}{\sqrt{x^3-7x}}+18x(3x^2+2)^2\sqrt{x^3-7x}=\frac{243x^8-945x^6-1{,}710x^4-732x^2-56}{\sqrt{x^3-7x}}$

21. $\frac{3(v^3-9)^2(v^2-4v)^2(2v-4)-(v^2-4v)(2)(3v^2)(v^3-9)}{(v^3-9)^4}=\frac{12v^7-96v^6+138v^5+540v^4-1728v^3+1728v^2}{(v^3-9)^3}$

23. $6(x^5+4)^5(5x^4)=30x^4(x^5+4)^5$

25. $4x(4x^2-5)^{-1/2}$ **27.** $(14x^2-6)(7x^3-9x)^{-1/3}$

29. $\frac{-6}{(x+3)^3}[\frac{1}{(x+3)^2}+1]^2$

31. $\frac{1}{3}[(x^3+4)^2-2(x^3+4)]^{-2/3}[2(x^3+4)-2](3x^2)=$ $2x^2\left[\left(x^3+4\right)^2-2\left(x^3+4\right)\right]^{-2/3}(x^3+3)$

33. $\left[16\left(3t^2-4t\right)^3-3\right](6t-4)=2{,}592t^7-12{,}096t^6+$ $20{,}736t^5-15{,}360t^4+4{,}096t^3-18t+12$

35. $6(9x^2+4x)^5(18x+4)$

37. $10(11x^5-2x+1)^9(55x^4-2)$ **39.** $\frac{-56(18x)}{(9x^2-4)^9}$

41. $\frac{3x}{\sqrt[4]{2x^2-5}}$ **43.** $\frac{-3072}{x^7}(\frac{128}{x^6}-1)$ **45.** $\frac{-480}{x^{11}}$

47. $y+8=192(x-2)$ or $192x-y=-394$

49. **(a)** $P=E'=\frac{dE}{dt}=144\left(1+4t^2\right)^2t$ W, **(b)** $P=E'=$ $\frac{dE}{dt}=\frac{144\left(1+4t^2\right)^2t}{1000}=0.144\left(1+4t^2\right)^2t$ kW, **(c)** 243.36 kW

51. **(a)** $P'(t)=\frac{250t\left(t^2+30\right)}{(t^2+15)^{3/2}}$, **(b)** 271.8 ppm/day

53. $R'=\frac{4n-4}{(n+1)^3}$

55. **(a)** $I'=-\frac{120X_C}{\left(900+X_C{}^2\right)^{3/2}}$, **(b)** -0.0504 A/Ω

Exercise Set 4.5

1. $-4/5$ **3.** $\frac{1}{2y}$ **5.** x/y **7.** $-9x/16y$

9. $\frac{4x}{2+y}$ **11.** $\frac{2x-y-5y^2}{x+10xy}$

13. $(2x+4y)/(1-4x-2y)$ **15.** $\frac{1+10x}{3+20y}$ **17.** $\frac{-8x}{3y^2}$

19. $\frac{y-3x^2}{18y^2-x}$ **21.** $\frac{1-2xy}{x^2}$ **23.** $-\frac{y^2}{x^2}$

25. $\frac{3(x+1)^{-2}-2xy+3(x+1)^{-2}}{x^2}=\frac{3-2xy(x+1)^2}{x^2(x+1)^2}$

27. $\frac{x^{-2}y-y+1}{x+x^{-1}}=\frac{y-x^2y+x^2}{x^3+x}$ **29.** $\frac{5+6xy^2-2x^3}{y-6x^2y}$

31. $\frac{2y\sqrt{x^2+y^2}+x^3}{2x\sqrt{x^2+y^2}-x^2y}$ **33.** $\frac{6x(x^2+y^2)^2}{1-6y(x^2+y^2)^2}$

35. $y+4=\frac{3}{4}(x-3)$ or $4y-3x+25=0$

37. $y-2=\frac{9}{16}(x+2)$ or $9x-16y+50=0$ **39.** tangent: 1; normal: -1 **41.** **(a)** $\frac{ds}{dt}=\frac{6\sqrt{st}-s}{8s\sqrt{st}+t}$, **(b)** 1.464 mi,

(c) 0.399 mi/min = 23.94 mi/hr **43. (a)** 1.48. **(b)** The store should sell 1.48 television sets for each digital satellite that is sold.

Exercise Set 4.6

1. $24x-12$ **3.** $840x$ **5.** $2x^{-3}$ **7.** $2+\frac{2}{(t+1)^3}$

9. $-6u^{-4}+24u^{-5}=\frac{-6u+24}{u^5}$ **11.** $-\frac{1}{4}(x+1)^{-3/2}$

13. 48 **15.** $\frac{10}{9}x^{-8/3}$

17. $24x(x+1)^{-5}-24(x+1)^{-4}=-24(x+1)^{-5}$

19. $216w^2(3w^2+1)^{-4}-12(3w^2+1)^{-3}=(180w^2-12)(3w^2+1)^{-4}$. **21.** $840x^3-480x$

23. $24x^{-5}$ **25.** $\frac{18x^3+18x^2+6x}{(3x-1)^3}$

27. $\frac{3}{4}x(x+1)^{-5/2}-(x+1)^{-3/2}=\frac{-x-4}{4(x+1)^{5/2}}$

29. $y'=\frac{2x-1}{8y}$, $y''=\frac{16y^2-(2x-1)^2}{64y^3}$

31. $y'=\frac{4}{2xy-1}$; $y''=\frac{-8(4x+2xy^2-y)}{(2xy-1)^2}$

33. $y'=\frac{-2y}{x}$; $y''=\frac{6y}{x^2}$ **35. (a)** $N'(t)=10t-0.2t^{3/2}$, **(b)** $N''(t)=10-0.3t^{1/2}$ **37. (a)** $h'(t)=120-24t$, **(b)** $h''(t)=-24$

Review Exercises

1. $4x-1$ **3.** $1-6x^2$ **5.** $\frac{1}{2\sqrt{x}}$ **7.** $6x+2$

9. $15x^4-\frac{12}{x^4}+\frac{3}{2\sqrt{x}}$ **11.** $\frac{3}{(x+3)^2}$

13. $(x-4)(3x^2-3)+(x^3-3x)=9x^2-30x+12$

15. $\frac{1}{2\sqrt{x+1}}$ **17.** $6x^5-10x^4+12x^3-18x^2-8x+8$

19. $\frac{4x^3+15x^2+7}{2\sqrt{(x^3+7)(x+5)}}$

21. $(x^2+5x)^2(-6x^2)(x^3+1)^{-3}+2(x^2+5x)(2x+5)(x^3+1)^{-2}=\frac{-2x^6-30x^5-100x^4+4x^3+30x^2+50x}{(x^3+1)^3}$

23. $3(x+1)^2(x-1)^2+2(x+1)(x-1)^3=(x-1)^2(x+1)(5x+1)$ **25.** $-30x(x^2+1)^{-4}$

27. $2x(x^2-2)^{-2/3}-3x^3(x^2-2)^{-5/2}=(x^2-2)^{-5/2}(-x^3-4x)$

29. $-4(u^2+\frac{1}{u}-\frac{4}{u^2})^{-5}(2u-u^{-2}+8u^{-3})$

31. $6x^5+12x^3+4x$ **33.** $-\frac{1}{2}$ **35.** $\frac{2y-2x-y^2}{2xy-2x}$

37. $\frac{y^2-2xy^5}{3x^2y^4-1}$ **39.** $42x^5+24x^2$ **41.** $24-\frac{12}{x^4}$

43. $42-420x^{-6}$ **45.** $30+\frac{3}{8}x^{-5/2}$

47. $\frac{6x^2y-8x+2y}{(x^2-1)^2}$ **49.** $y'=\frac{2x-y}{x-2y}$; $y''=\frac{2x^3-8x^2y+4x^2+8xy^2+2xy-2y^2}{(x-2y)^3}$ **51.** 18 2/9

53. Tangent $\frac{17}{6}$; Normal $-\frac{6}{17}$ **55.** 0, −2

57. $y-2=\frac{5}{4}(x+2)$ or $4y-5x-18=0$ **59.** $\pm\frac{\sqrt{3}}{3}$

Chapter 4 Test

1. $15x^2$ **3.** $2x^5(x^6-2)$ **5.** $\frac{2-4x}{(4x^2+1)^{1/2}}$

7. $y'=\frac{3x^2-2y-y^3}{2x+3y^2x}$ **9.** $v(t)=9t+27$

ANSWERS FOR CHAPTER 5

Exercise Set 5.1

1. $v(t)=6t-12$; $a(t)=6$

t	0	1	2	3	4	5
$s(t)$	5	−4	−7	−4	5	20

t = 3 t = 3 t = 5
t = 2 t = 1 t = 0
−10 −7 −4 0 5 10 20

3. $v(t)=3t^3-12$; $a(t)=6t$

t	−4	−3	−2	−1	0	1	2	3	4
$s(t)$	−14	11	18	13	2	−9	−14	−7	18
$v(t)$	36	15	0	−9	−12	−9	0	15	36
$a(t)$	−24	−18	−12	−6	0	6	12	18	24

t = 3
t = 2 t = 1 t = 0 t = −1 t = 4
t = −3 t = −2
t = −4
−14 −9 −5 0 2 5 11 18 25

5. $v(t) = 1 - 4/t^2$; $a(t) = 8/t^3$

t	1	2	3	4
$s(t)$	5	4	17/3	5
$v(t)$	−3	0	0.5555	0.75
$a(t)$	8	1	0.2963	0.125

$t=3$, $t=4$, $t=2$, $t=1$

0 4 5 6

7. $v(t) = t^{-1/2} - \frac{1}{2}t^{-3/2}$; $a(t) = -\frac{1}{2}(t^{-3/2} - \frac{3}{4}t^{-5/2})$

t	1	2	3	4
$s(t)$	3	3.5355	4.0415	4.5
$v(t)$	0.5	0.5303	0.4811	0.4375
$a(t)$	0.25	−0.0442	−0.0481	−0.0391

$t=2$, $t=1$, $t=4$, $t=3$

0 3 4 5 6

9. (a) 9 s; **(b)** $v = 144$ ft/s downward; $a = 32\text{ ft/s}^2$ downward; **(c)** 0 ft (ground level) **11. (a)** 6 s; **(b)** $v - 29.4$ m/s; $a = -9.8\text{ m/s}^2$; **(c)** 0 m **13. (a)** $\frac{dF}{ds} = F(s) = \frac{-100}{s^3}$ dynes/cm; **(b)** $\frac{-100}{27}$ dynes/cm **15. (a)** $\frac{9}{5}$; **(b)** $\frac{5}{9}$; **(c)** $\frac{5}{9}$ **17. (a)** $\frac{dQ}{dt} = -50 + t$ gallons/min; **(b)** −46 gal/min; **(c)** −42 gal/min **19. (a)** 4.0 m/s; **(b)** 8.0 m; **(c)** 4.0 s **21. (a)** $v = -N\frac{d\phi}{dt} = -\frac{60}{t^{1/2}} + 40t$; **(b)** −20 V at $t = 1.0$ s and 340 V at $t = 9.0$ s **23.** $46.24\pi\text{ m}^2\text{/s}$ **25. (a)** $P = \frac{6,41}{V}$; **(b)** $\frac{dP}{dV} = -\frac{6.41}{V^2}$ **27. (a)** $v(t) = s'(t) = 6t - 0.05t^2$, **(b)** $a(t) = v'(t) = s''(t) = 6 - 0.1t$, **(c)** $|v(t)| = |6t - 0.05t^2|$

Exercise Set 5.2

1. Critical values: $x = 4$; increasing on $(4, \infty)$; decreasing on $(-\infty, 4)$; Maximum: none; Minimum: $f(4) = -16$
3. Critical values: $x = 0$; increasing on $(-\infty, 0)$; decreasing on $(0, \infty)$; Maximum: $m(0) = 16$; Minimum: none
5. Critical values: $x = 2$; increasing on $(-\infty, 2)$; decreasing on $(2, \infty)$; Maximum: $j(2) = 31$; Minimum: none
7. Critical values: $x = 2$; increasing on $(2, \infty)$; decreasing on $(-\infty, 2)$; Maximum: none; Minimum: $f(2) = -15$
9. Critical values: $t = \pm 1$; increasing on $(-\infty, -1)$ and $(1, \infty)$; decreasing on $(-1, 1)$; Maximum: $q(-1) = 2$; Minimum: $q(1) = -2$ **11.** Critical values: $x = 0, \pm 6$; increasing on $(-6, 0)$ and $(6, \infty)$; decreasing on $(-\infty, -6)$ and $(0, 6)$; Maximum: $y(0) = 16$; Minimum: $y(-6) = y(6) = -308$ **13.** Critical values: $z = -4$; increasing on $(-\infty, \infty)$ decreasing nowhere; Max: none; Minimum: none. **15.** Critical values: $v = -1, 0, 2$; increasing on $(-1, 0)$ and $(2, \infty)$; decreasing on $(-\infty, -1)$ and $(0, 2)$; Maximum: $g(0) = 12$; Minimum: $g(-1) = 7$ and $g(2) = -20$. **17.** Critical values: none; increasing nowhere; decreasing on $(-\infty, -1)$ and $(-1, \infty)$; Maximum: none; Minimum: none **19.** Critical values: $x = -1$; increasing on $(-1, 0)$ and $(0, \infty)$; decreasing on $(-\infty, -1)$; Maximum: none; Minimum: $g(-1) = 3$ **21.** Critical values: $x = 0$; increasing on $(0, 1)$ and $(1, 2)$; decreasing on $(-\infty, 0)$ and $(2, \infty)$; Maximum: $f(2) = -4$; Minimum: $f(0) = 0$ **23.** Critical values: $t = \pm 1, \pm\frac{\sqrt{2}}{2}$; increasing on $\left(-\frac{\sqrt{2}}{2}, \frac{\sqrt{2}}{2}\right)$; decreasing on $\left(-1, \frac{-\sqrt{2}}{2}\right)$ and $\left(\frac{\sqrt{2}}{2}, 1\right)$; Maximum: $f\left(\frac{\sqrt{2}}{2}\right) = 0.5$; Minimum: $f\left(-\frac{\sqrt{2}}{2}\right) = -0.5$
25. Critical values: $x = 0.6, 1$; increasing on $(-\infty, 0.6)$ and $(1, \infty)$; decreasing on $(0.6, 1)$; Maximum: $(0.6, 0.326)$; Minimum: $(1, 0)$ **27. (a)** 8 s; **(b)** 1274 ft
29. (a) $P = 0.0$ W; **(b)** $t = 2.0$ s **31. (a)** minimum when $R_2 = 0$; no maximum; **(b)** minimum $R = 0$
33. (a) $P'(n) = -n + 8.6$, **(b)** 8.6, **(c)** \$6,498
35. (a) $N'(t) = 2t^{-1/2} + 3t^{1/2} - 2t^{3/2}$, **(b)** $t = 2$ mo

Exercise Set 5.3

1. Maximum: $(2, 12)$, Concave downward: $(-\infty, \infty)$

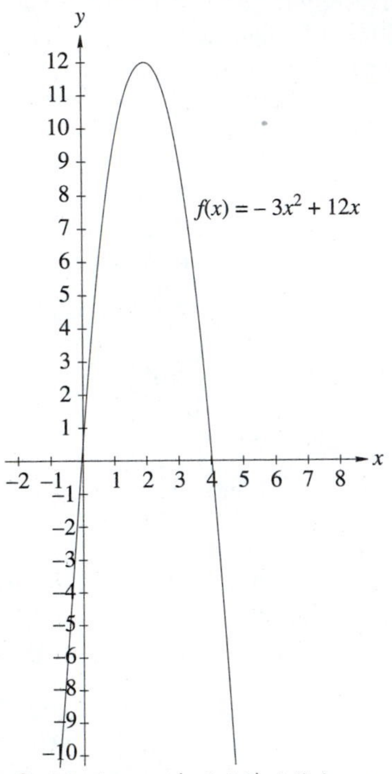

3. Maximum: $(-2, 28)$; Minimum: $(2, -4)$; Concave upward: $(0, \infty)$; Concave downward: $(-\infty, 0)$: Inflection point: $(0, 12)$

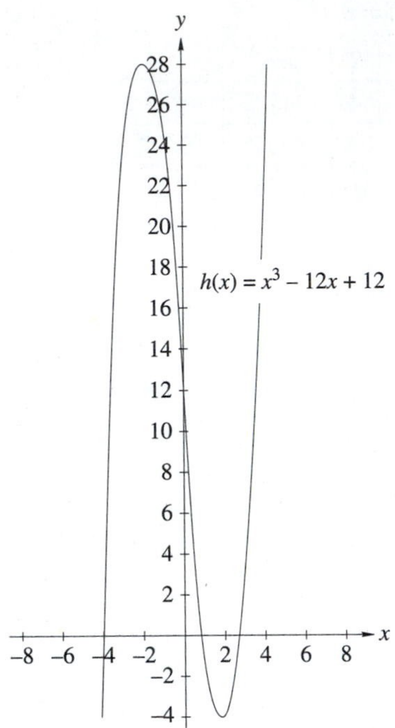

5. Maximum: none; Minimum: none; Concave upward: $(2, \infty)$, Concave downward: $(-\infty, 2)$: Inflection point: $(2, 4)$

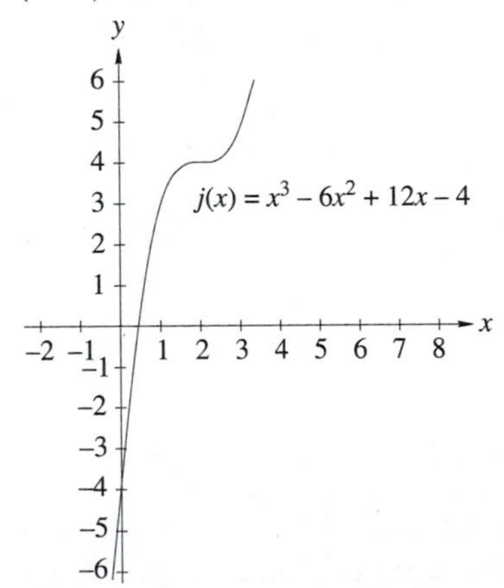

7. Maximum: none, Minimum: none; Concave upward: $\left(-\infty, \frac{1}{2}\right)$; Concave downward: $\left(\frac{1}{2}, \infty\right)$; Inflection point: $\left(\frac{1}{2}, -\frac{1}{2}\right)$

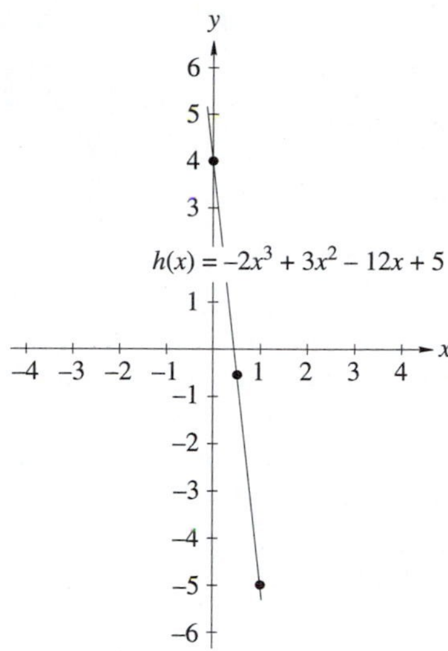

9. Maximum: $(-2, 10.67)$ and $(1, 1.67)$; Minimum: $(0, 0)$; Concave upward: $(-1.22, 0.55)$; Concave downward: $(-\infty, -1.22)$ and $(0.55, \infty)$; Inflection point: $(-1.22, 6.16)$ and $(0.55, 0.90)$

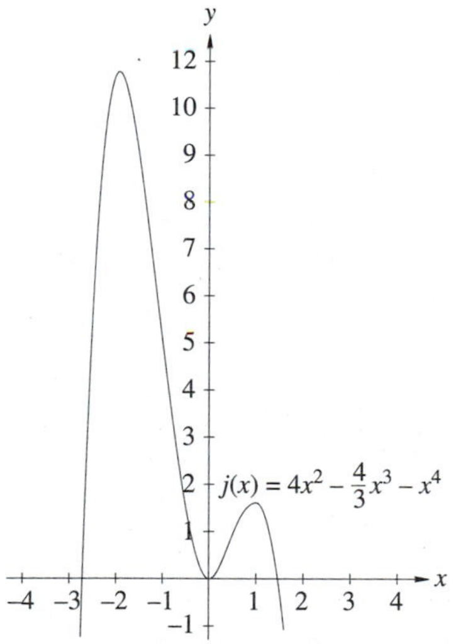

11. Maximum: none; Minimum: none; Concave upward: $(-\infty, 1)$; Concave downward: $(1, \infty)$; Inflection point: $(1, 0)$

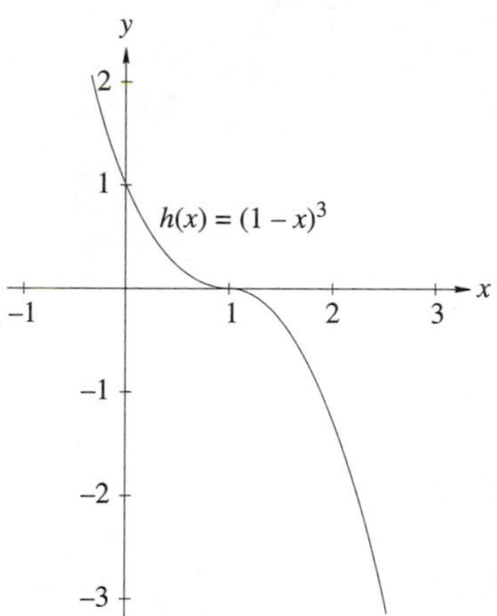

13. Maximum: none; Minimum: none; Concave upward: $(-\infty, 0)$; Concave downward: $(0, \infty)$; Inflection point: $(0, 2)$

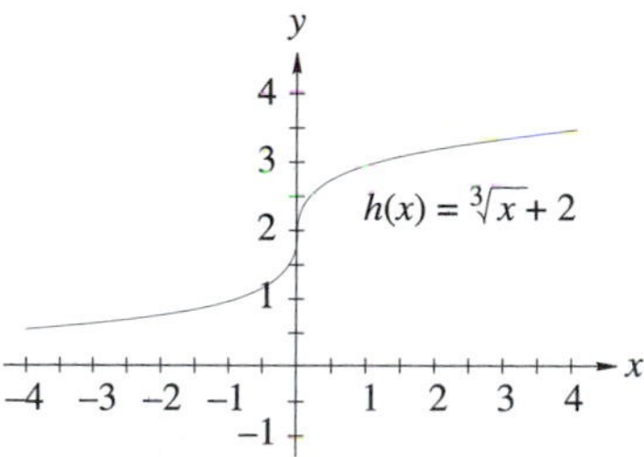

15. Maximum: $(-2, -4)$; Minimum: $(2, 4)$; Concave upward: $(0, \infty)$; Concave downward: $(-\infty, 0)$; Inflection point: none

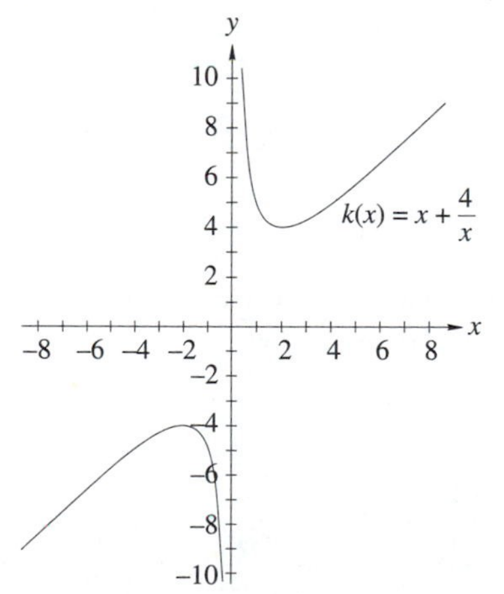

17. Maximum: none; Minimum: $(-1, 3)$; Concave upward: $(-\infty, 0)$ and $\left(\sqrt[3]{2}, \infty\right) \approx (1.26, \infty)$; Concave downward: $\left(0, \sqrt[3]{2}\right) \approx (0, 1.26)$; Inflection point: $\left(\sqrt[3]{2}, 0\right) \approx (1.26, 0)$

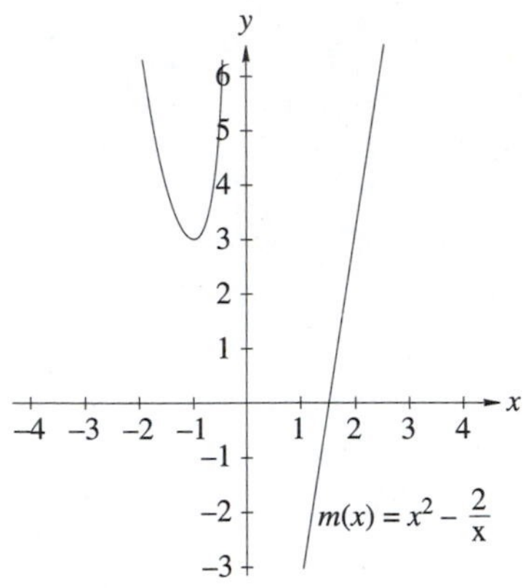

19. Maximum: $(-2, 7.333)$; Minimum: $(3, -13.5)$; Concave upward: $\left(\frac{1}{2}, \infty\right)$; Concave downward: $\left(-\infty, \frac{1}{2}\right)$; Inflection point: $(0.5, -3.083)$

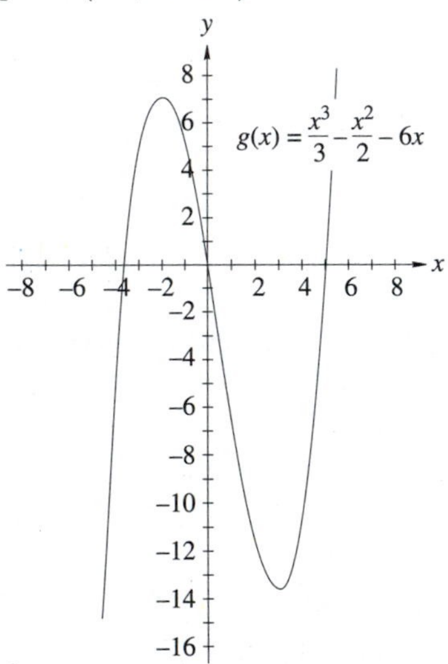

21. Maximum: $(1, 1)$; Minimum: $(0, 0)$; Concave upward: nowhere; Concave downward: $(0, \infty)$; Inflection point: none

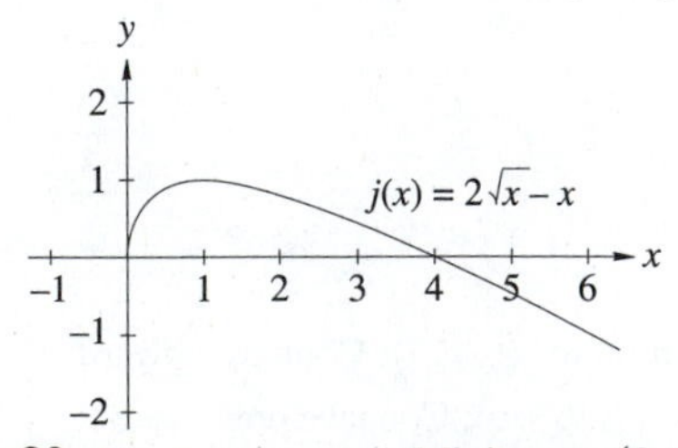

23. Maximum: none; Minimum: $(0, 2)$; Concave upward: nowhere, Concave downward: $(-\infty, \infty)$; Inflection point: none

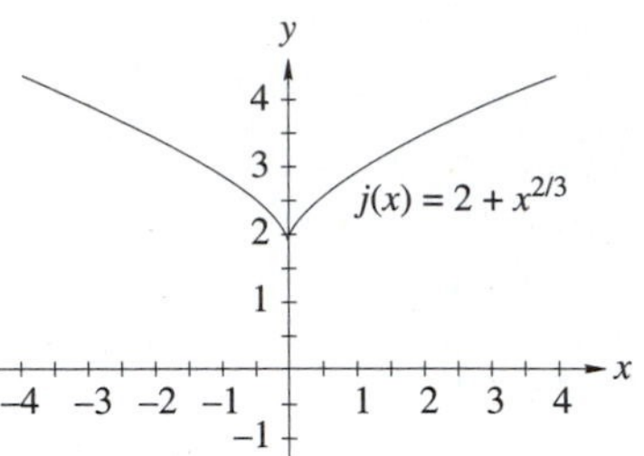

25. Maximum: $(0, 0)$; Minimum: $(4. - 15.119)$; Concave upward: $(-\infty, \infty)$; Concave downward: nowhere; Inflection point: none

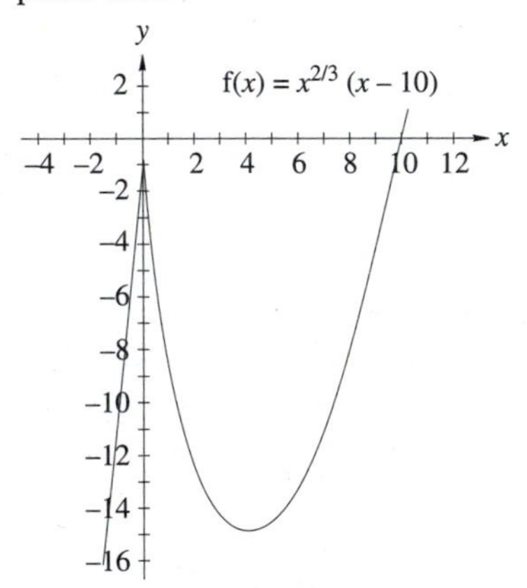

27. Maximum: $\left(1, \frac{1}{2}\right)$; Minimum: $(-1, -\frac{1}{2})$; Concave upward: $(-\sqrt{3}, 0)$, $(-\sqrt{3}, 0)$, $(\sqrt{3}, \infty)$; Concave downward: $(-\infty, -\sqrt{3})$, $(0, \sqrt{3})$; Inflection points: $(-\sqrt{3}, -\sqrt{3/4})$, $(\sqrt{3}, \sqrt{3/4})$

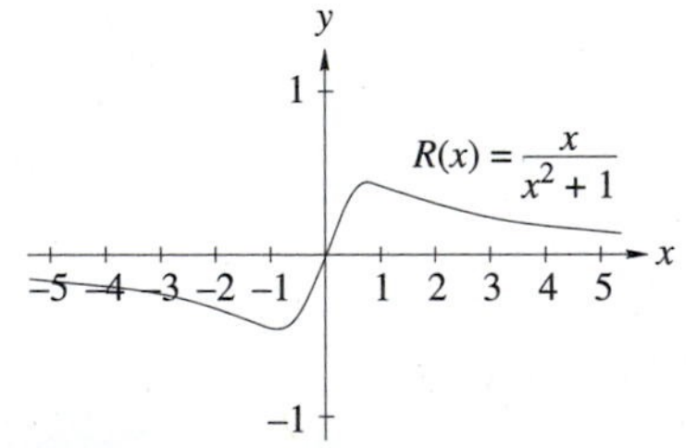

29. $t = 23$ **31.** **(a)** $N(9) = 116$, **(b)** $N'(4) = 8$ crimes/mo

Exercise Set 5.4

1. 2500 ft × 5000 ft; area = 12,500,000 ft^2 **3.** **(a)** 2000 ft along river × 3000 ft; **(b)** cost is \$12,000 **5.** 4 cm **7.** $2\sqrt{2} \approx 2.83$ units horizontally and $\frac{3}{2}\sqrt{2} \approx 2.12$ units vertically **9.** $4'' \times 4'' \times 2''$ **11.** radius = 5.05 cm, height = 10.1 cm **13.** 21.984 cm × 26.381 cm **15.** **(a)** $w = \sqrt{300} \approx 17.32$ in.; $d = \sqrt{600} \approx 24.49$ in. **(b)** $w = 2r\frac{\sqrt{3}}{3}$ in.; $d = 2r\frac{\sqrt{6}}{3}$ in. **17.** 80 mm × 160 mm **19.** 1.25 A **21.** $A = \sqrt{c/b}$ **23.** 2.24 m

25. **(a)** $t = 4.5\,s$; **(b)** $P = 0.0\,W$ **27.** about 3.1 m from the less intense light **29.** $0.578\,L$ **31.** $\frac{h}{d} = \frac{3\pi+2}{4}$ **33.** $12 - \frac{5}{3}\sqrt{3} \approx 9.11$ mi **35.** **(a)** 56.2×74.7 ft, **(b)** \$52,290

Exercise Set 5.5

1. $\frac{400}{3} = 133\frac{1}{3}$ cm^2/min **3.** $\frac{\sqrt[4]{5}}{5} \approx 1.789$ ft/min **5.** 2.9104 km/min = 174.6 km/h **7.** **(a)** $\frac{5}{8\pi} \approx 0.1989$ ft/min; **(b)** $\frac{10}{21\pi} \approx 0.1516$ ft/min; **(c)** $\frac{5}{12\pi} \approx 0.1326$ ft/min **9.** 12.4 mm/min **11.** 15 cm^2/s **13.** 11.25 cm^2/s **15.** 24 mm^2/s **17.** -8.3% of the volume per hour **19.** 387.2 m/s or 1394 km/h **21.** **(a)** -18.0 m/s^2; **(b)** -18.0 m/s^2 **23.** **(a)** Either 5.75 mi west and 5.56 mi north of A or 5.75 mi west and 5.56 mi south of A. **(b)** 14.86 mph **(c)** $\theta = 51.2°$ north of east or 38.8° east of north **25.** **(a)** $\frac{50}{7} \approx 7.143$ ft/s; **(b)** 7.143 ft/s; **(c)** 7.143 ft/s **27.** 0.0064 Ω/s **29.** 0.86 Ω/min **31.** decreasing 0.04725 Ω/s **33.** -2.04 Ω/min **35.** **(a)** 218.47 Hz; **(b)** 20.17 Hz/s **37.** 0.06375 mm/min **39.** -1.06 Ω/s **41.** 4.30 Ω/s **43.** 0.1π in.^{2}/yr ≈ 0.314 in.^{2}/yr

Exercise Set 5.6

1. 1.2289 **3.** -1.2838; 1.5338 **5.** -3.3437; 3.3437 **7.** 2.5649 **9.** 1.1066 rad $\approx 63.40°$

Exercise Set 5.7

1. $dy = (4x^3 - 2x)dx$ **3.** $dy = (15x^2 - 2x + 1)dx$ **5.** $dy = \frac{-2}{3}(4 - 2x)^{-2/3}dx$ **7.** $dy = 1.0, \Delta y = 1.04$ **9.** $dy = 3.9, \Delta y = 3.9675$ **11.** $dy = -0.0074074, \Delta y = -0.0065195$ **13.** **(a)** $\frac{16}{3}\pi$ m^3; **(b)** $\frac{16}{300}\pi m^3 = \frac{4}{75}\pi$ m^3; **(c)** 1% **15.** **(a)** 0.96 in^2; **(b)** $0.00167 = 0.167\%$ **17.** 4π m^2 **19.** **(a)** $\pm$ 24.2 m^3, **(b)** $\pm 2.42\%$ **21.** $0.08\sigma T^4$ **23.** -0.03 A **25.** ± 2.18 Ω **27.** 1.0% **29.** ± 0.07 mm^3 **31.** $450\pi \approx 1414$ mm^3 **33.** \$132.50 **35.** $19.2\pi \approx 60.3$ cm^3.

Exercise Set 5.8

1. $7x + C$ **3.** $2x - x^3 + C$ **5.** $x^4 - x^3 + x^2 + 9x + C$ **7.** $-\frac{1}{3}x^{-3} - x^{-2} + x^{-1} + 5x + C$ **9.** $\frac{2}{3}x^{3/2} + \frac{1}{2}x^2 + \frac{1}{x} + C$ **11.** $-5x^{-0.4} - 0.7x^{-5} - 4x^{-0.3} + C$ **13.** $\frac{1}{3}t^3 + t^2 + C$ **15.** $21t^2 - 5t + C$ **17.** **(a)** 6.12 s; **(b)** 195.96 ft/s **19.** **(a)** 5 s; **(b)** 525 ft **21.** **(a)** 17 s; **(b)** 1904 ft **23.** 160 m **25.** -2.5 m/s^2 **27.** **(a)** 3.9 rev; **(b)** 4.6 rev **29.** **(a)** $q = 2.2t^2 - 0.7t^3 + 5.0$; **(b)** 4.6 C **31.** **(a)** $\phi(t) = \frac{-1}{200}(t^2 - 3t^{4/3} - 4.0)$; **(b)** $\phi(0.729) = 0.027$ Wb **33.** $f(x) = -2\sqrt{x} + 1$ **35.** $T = 2875(x+1)^{-2} + 25$

Review Exercises

1. Critical values: $0, \pm\frac{\sqrt{2}}{2}$; Maximum: $(0,0)$; Minimum: $\left(-\frac{\sqrt{2}}{2}, -0.25\right), \left(\frac{\sqrt{2}}{2}, -0.457\right)$; Inflection point: $\left(\pm\frac{\sqrt{6}}{6}, -0.139\right)$; Concave upward: $(-\infty, -\frac{\sqrt{6}}{6})$, $(\frac{\sqrt{6}}{6}, \infty)$; Concave downward: $(-\frac{\sqrt{6}}{6}, \frac{\sqrt{6}}{6})$

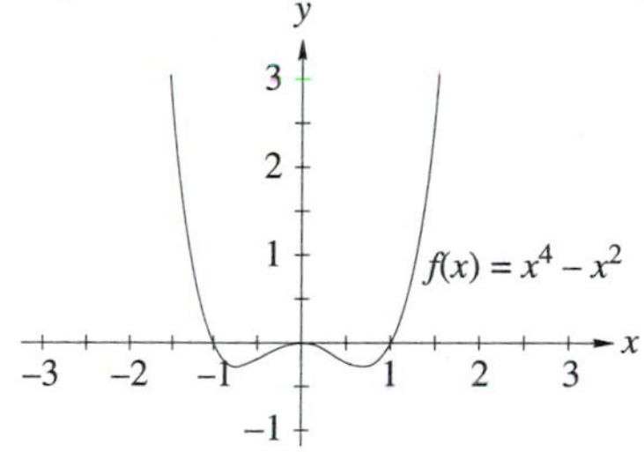

3. Critical values: $0, \pm 3$; Maximum: $(\pm 3, 81)$; Minimum: $(0,0)$; I.P. $(\pm\frac{\sqrt{3}}{3}, 5.889)$; Concave upward: $(-\frac{\sqrt{3}}{3}, \frac{\sqrt{3}}{3})$; Concave downward: $(-\infty, -\frac{\sqrt{3}}{3})$; $(\frac{\sqrt{3}}{3}, \infty)$

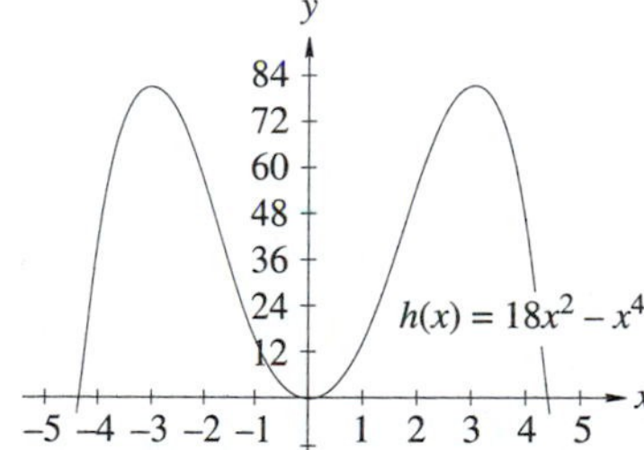

5. Critical value: 0; Maximum: none; Minimum: none; Inflection point: $(0,0)$; Concave upward: $(-\infty, 0)$; Concave downward: $(0, \infty)$

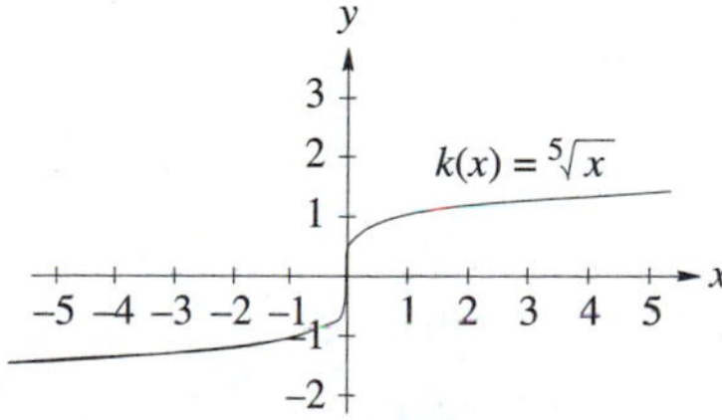

7. Critical value: 0; Maximum: $(0, -\frac{1}{4})$; Minimum: none; Inflection point: none; Asymptotes: $x = \pm 2$; $y = 1$; Concave upward: $(-\infty, -2)$, $(2, \infty)$; Concave downward: $(-2, 2)$

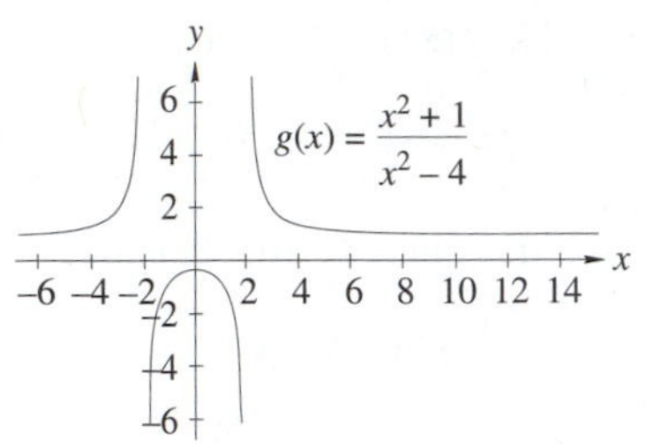

9. Critical values: none; Maximum: none; Minimum: none; Inflection point: $(-0.3275, 0.1475)$; Asymptotes: $x = -2$, $x = 1$, $y = 0$; Concave upward: $(-2, -0.3275)$, $(1, \infty)$; Concave downward: $(-\infty, -2)$, $(-0.3275, 1)$

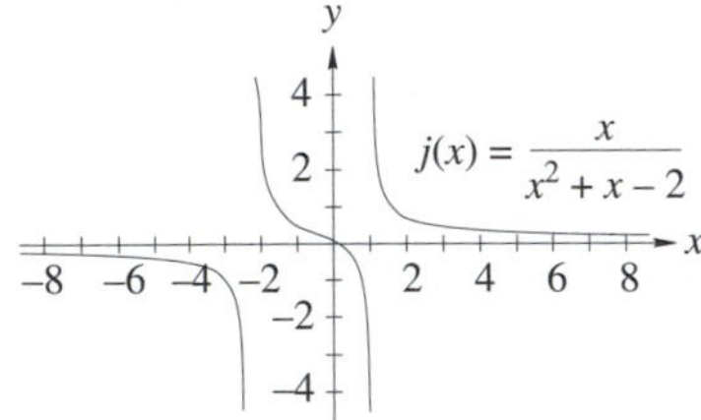

11. distance: Maximum at $(0.056, 0.028)$; Minimum at $(4, -76)$ and $(-4, -212)$; velocity: Maximum at $(4, -23)$ and $(-4, 121)$; Minimum at $(3, -26)$; acceleration: Minimum at $(-4, -42)$ and maximum at $(4, 6)$ **13.** $x^3 - 2x^2 + C$ **15.** $\frac{2}{3}x^{3/2} + \frac{1}{3}x^3 + \frac{3}{x} + C$ **17.** **(a)** $v(t) = 288 - 32t$; $a(t) = -32$; **(b)** height: 896 ft; velocity: 160 ft/s upward; acceleration: -32 ft/s^2; **(c)** 1296 ft; **(d)** 9 s; **(e)** 18 s; **(f)** 288 ft/s downward **19.** 56.98 mph **21.** $300 \times 300 = 90000\text{ m}^2$ **23.** **(a)** $0.09\pi \approx 0.2827\text{ m}^3$; **(b)** $0.12\pi \approx 0.3770\text{ m}^2$ **25.** 13.5 V **27.** $r = \sqrt[3]{\frac{10}{\pi}}\text{ m} \approx 1.4710\text{ m}$, $h = 2\sqrt[3]{\frac{10}{\pi}}\text{ m} \approx 2.942\text{ m}$ **29.** **(a)** 784 ft; **(b)** 244 ft/s

Chapter 5 Test

1. **(a)** $x = -2, 0,$ and 2; **(b)** Maximum $(0, 0)$, Minimum $(-2, -16)$ and $(2, -16)$, **(c)** Concave upward: $\left(-\infty, -\sqrt{\frac{4}{3}}\right)$ and $\left(\sqrt{\frac{4}{3}}, \infty\right)$, **(d)** Concave downward: $\left(-\sqrt{\frac{4}{3}}, \sqrt{\frac{4}{3}}\right)$ **3.** 2.646

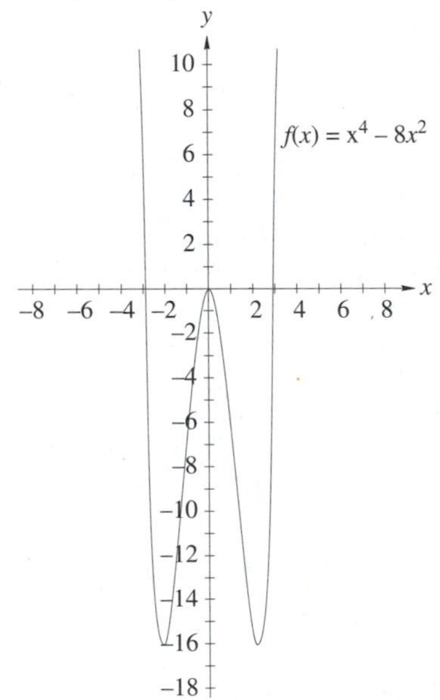

5. **(a)** $v(t) = 3t^2 - 24t$, $a(t) = 6t - 24$; **(b)** Position: Maximum at $(0, 5)$, Minimum at $(4, -123)$, Velocity: Maximum at $(-2, 60)$, Minimum at $(4, -48)$; Acceleration: Maximum at $(4, 0)$, Minimum at $(-2, -36)$ **7.** 3.38 m

ANSWERS FOR CHAPTER 6

Exercise Set 6.1

1. 4 **3.** 28 **5.** 4 **7.** -6 **9.** $\frac{2}{3}$ **11.** $\frac{22}{3}$ **13.** 20 **15.** 72 **17.** $\frac{14}{3}$

Exercise Set 6.2

1. 30 **3.** 24 **5.** $33\frac{1}{3}$ **7.** 6 **9.** 18 **11.** -18 **13.** -16 **15.** 18 **17.** 20 **19.** -30 **21.** $\frac{37}{3}$ **23.** $-\frac{37}{3}$ **25.** 2 **27.** 60 **29.** $\frac{26}{3}$ **31.** $-\frac{7}{8}$ **33.** $\frac{1}{6}$ **35.** $\frac{4}{3}\sqrt{2} \approx 1.8856$ **37.** -2.9501 **39.** 10.7 m **41.** **(a)** 39.4 m/s; **(b)** 74.1 m; **(c)** 5.1258 s **43.** **(a)** 1000000 m/s^2; **(b)** 5005 m **45.** 117 moose **47.** **(a)** 2 ft, **(b)** 0.545 ft, **(c)** 3.92 ft

Exercise Set 6.3

1. $9x + C$ **3.** $2x^3 + C$ **5.** $\frac{2}{7}x^{7/2} + C$ **7.** $\frac{1}{4}t^4 + t + C$ **9.** $\frac{1}{3}y^3 + 2y^2 - 3y + C$ **11.** $\frac{3x\sqrt[3]{x}}{4} - 9\sqrt[3]{x} + C$ **13.** $\frac{1}{2}(x^2+3)^2 + C$ **15.** $-\frac{(4-2x^2)^2}{2} + C$ **17.** $-\frac{1}{4}(3-x^2)^4 + C$ **19.** $\frac{1}{3}(x^2+4)^{3/2} + C$ **21.** $\frac{1}{2}(\sqrt{x}-1)^4 + C$ **23.** $\frac{1}{45}(3x^3+1)^5 + C$ **25.** $\sqrt{x^2+3} + C$ **27.** $-\frac{1}{4}(x^2+3)^{-2} + C$ **29.** $\frac{9}{5}x^5 - 2x^3 + x + C$ **31.** $\frac{1}{2}x^2 - 3x + C$ **33.** $\frac{15}{4}(x^2-1)^{2/3} + C$ **35.** $\frac{1}{18}(1+3x^2)^3 + C$ **37.** $\frac{3}{20}(4x^3-5)^{5/3} + C$

39. $x+2x^3+\frac{9}{5}x^5+C$ **41.** $\frac{1}{3}(4x^2+2x)^{3/2}+C$

43. $-\dfrac{1}{x-1}+C$ **45.** $\frac{16}{5}x^{10}-16x^6+36x^2+C$

47. $\frac{1}{5}(x^3-3x)^{5/3}+C$ **49.** $\frac{1}{12}(3x+1)^4+C$

51. $\frac{2}{5}(x^4-2x)^{5/4}+C$ **53.** $\frac{2}{3}(t+7)^{3/2}+C$

55. $\frac{1}{3}y^3+y^2+4y+C$ **57.** -8 **59.** $\frac{9}{2}$ **61.** $156\frac{1}{3}$

63. 139.5 **65.** 6,865,432.5

67. $L(x)=0.3x+0.001x^2+0.8$

69. $P(x)=2x^2+35x-150$

71. $T(x)=2875(x+1)^{-2}+25$

Exercise Set 6.4

1. 8

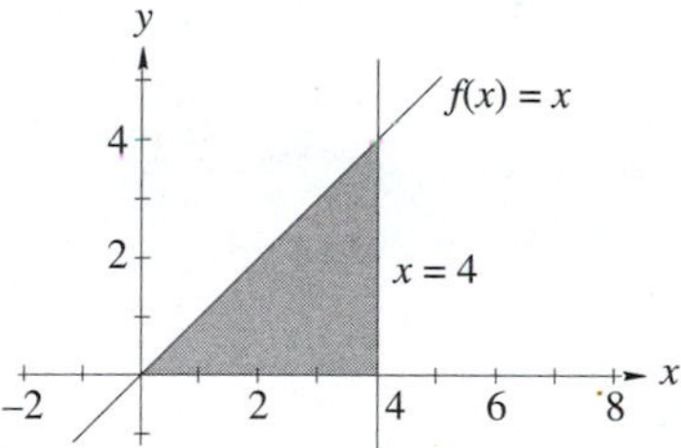

3. $\frac{64}{3}$

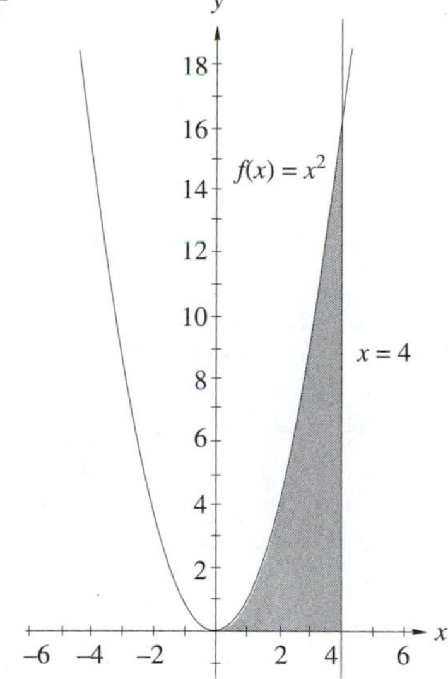

5. $160\frac{1}{6}$

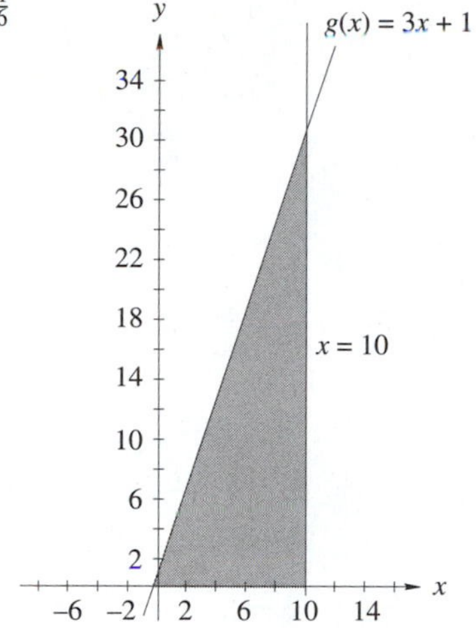

7. 12

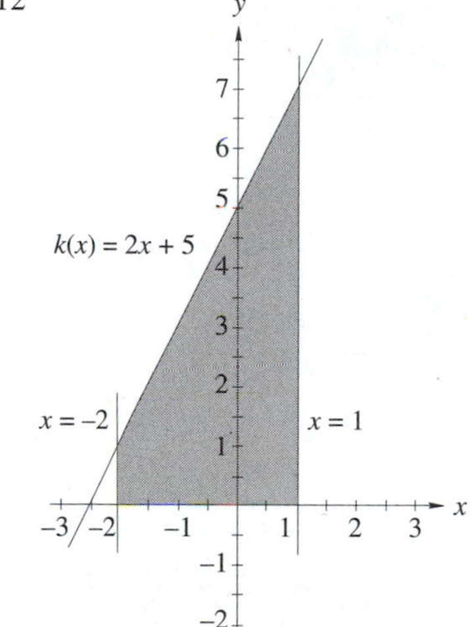

9. $\frac{22}{3}$

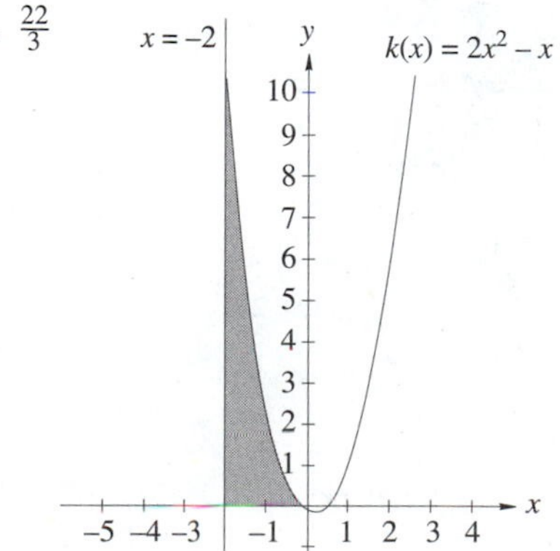

11. $\frac{32}{3}$

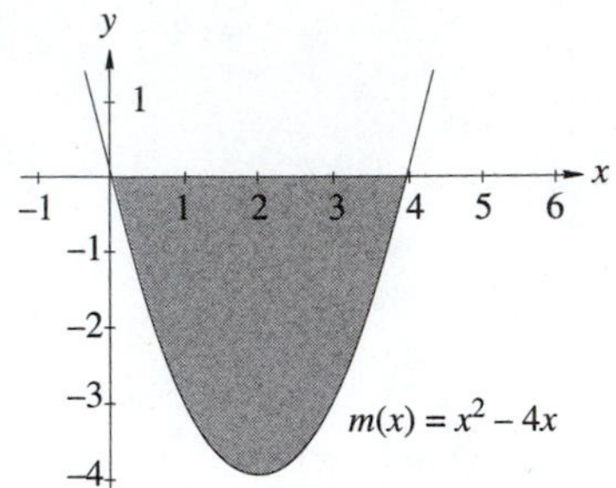

13. 36

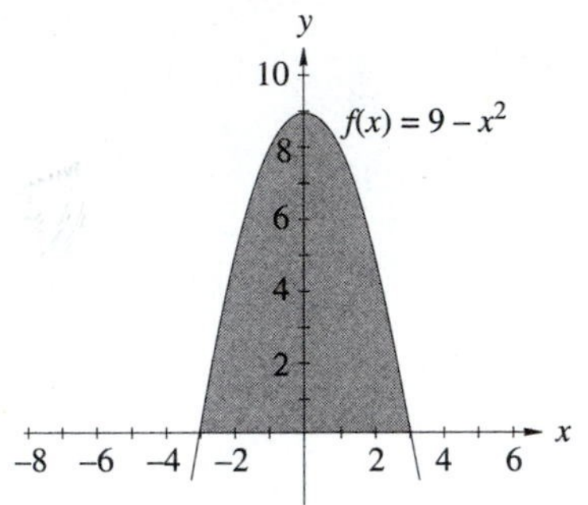

15. 16

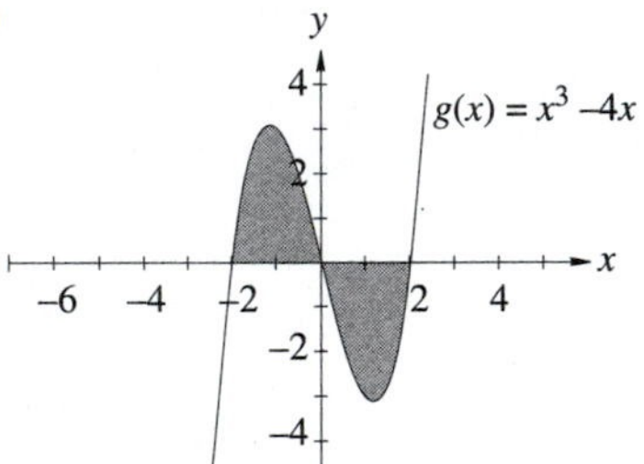

17. 18

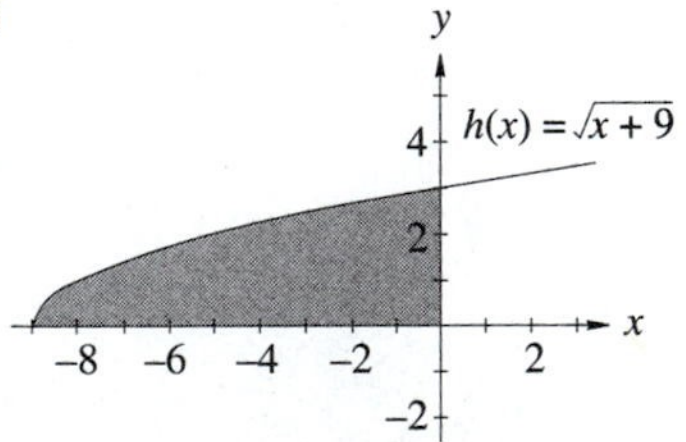

19. $\frac{4}{3}$

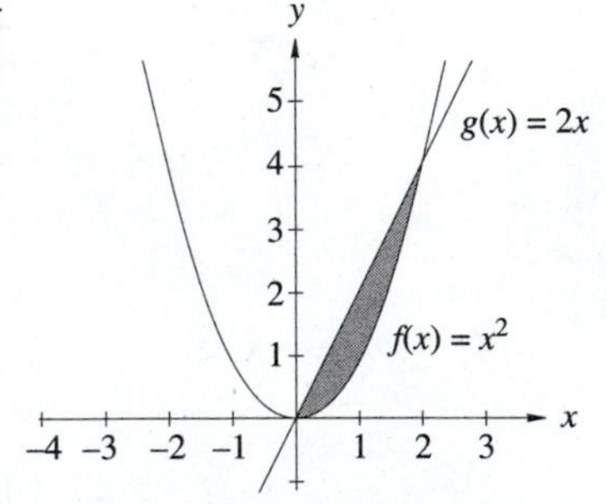

21. $8\sqrt{6} \approx 19.596$

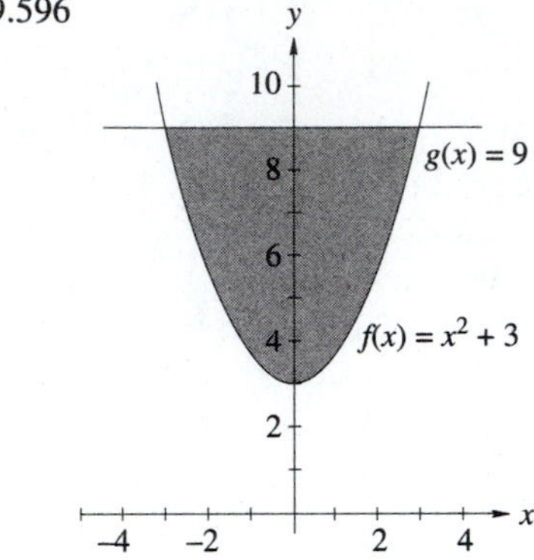

23. 21.333

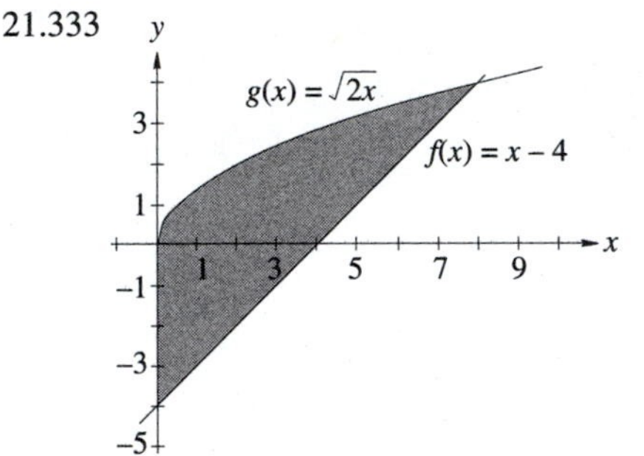

25. 20.8333

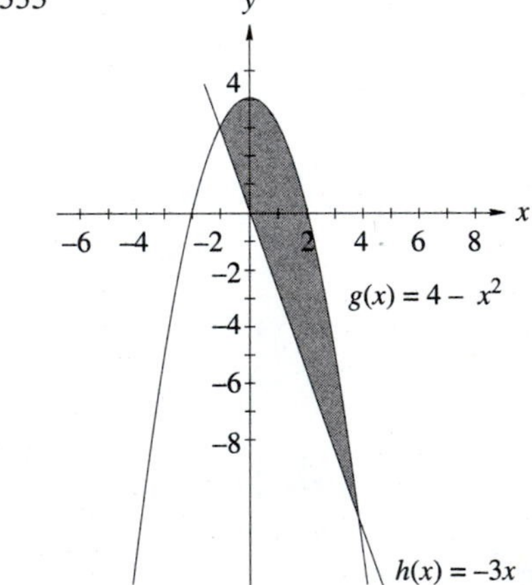

27. $\frac{125}{6} \approx 20.833$

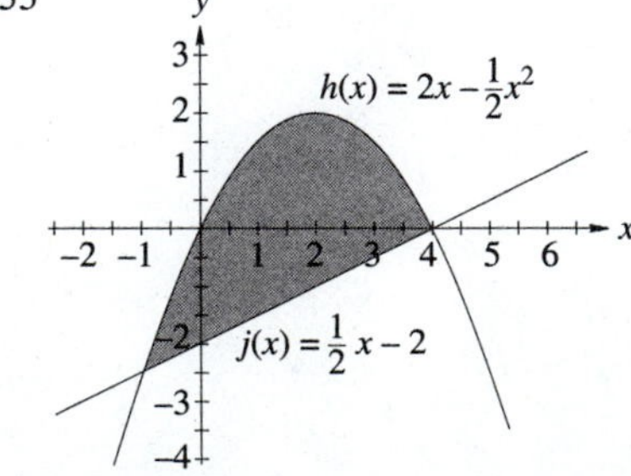

29. 18

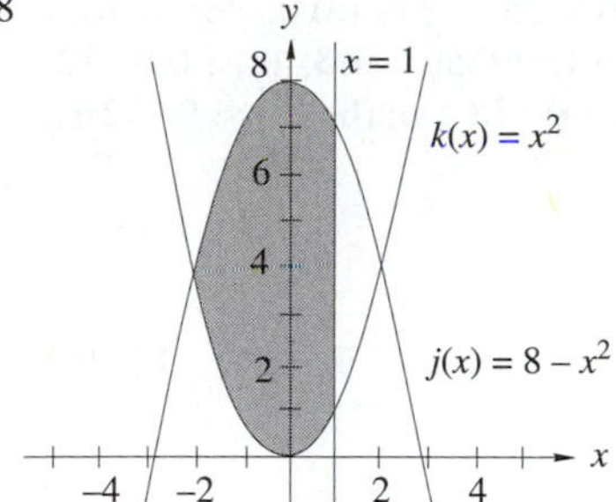

31. 8

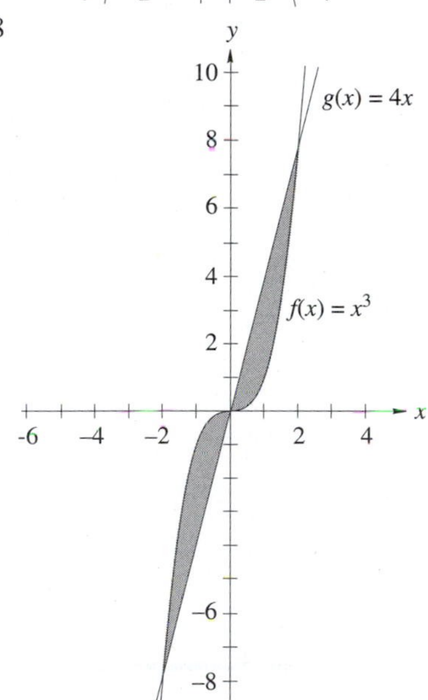

33. $41\frac{2}{3}$

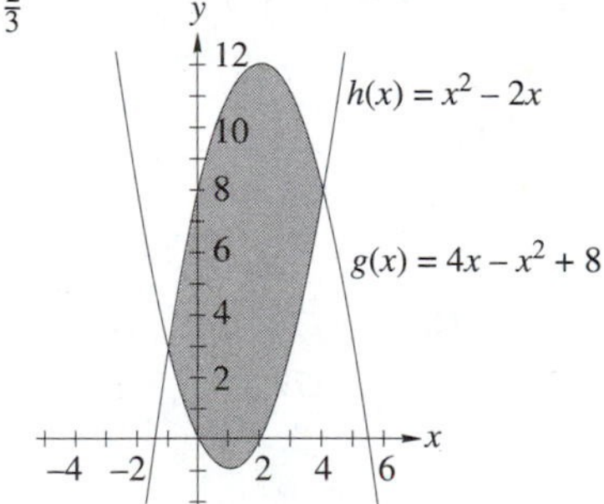

35. 2.25

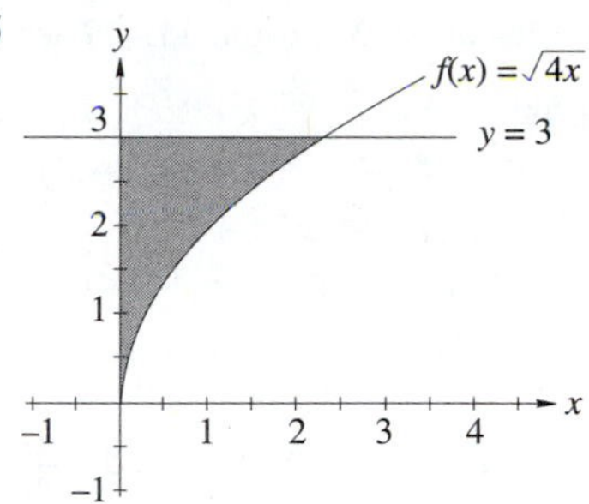

37. $\frac{64}{6} \approx 10.667$

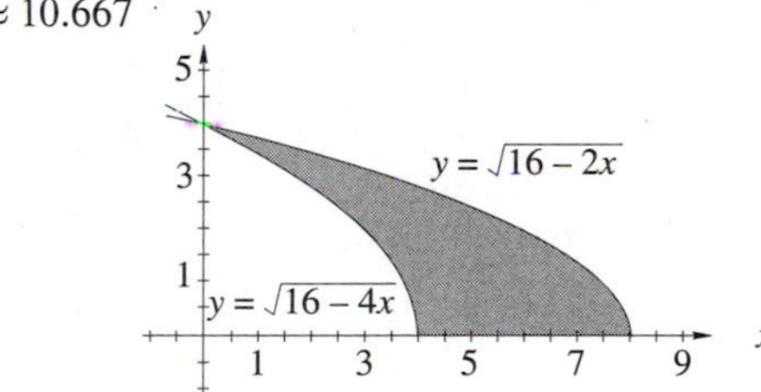

39. 0.375

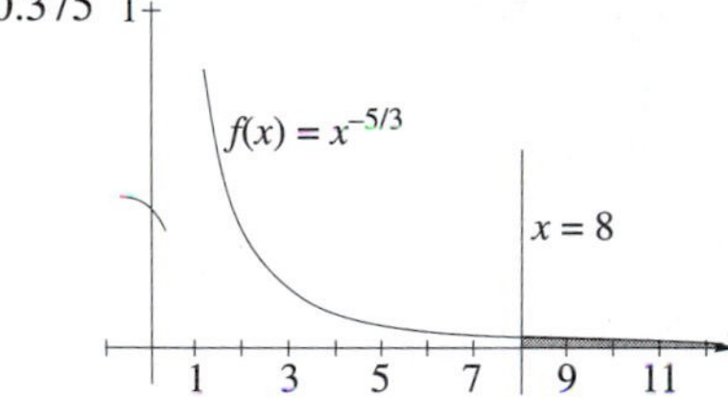

41. $\frac{1}{3}$

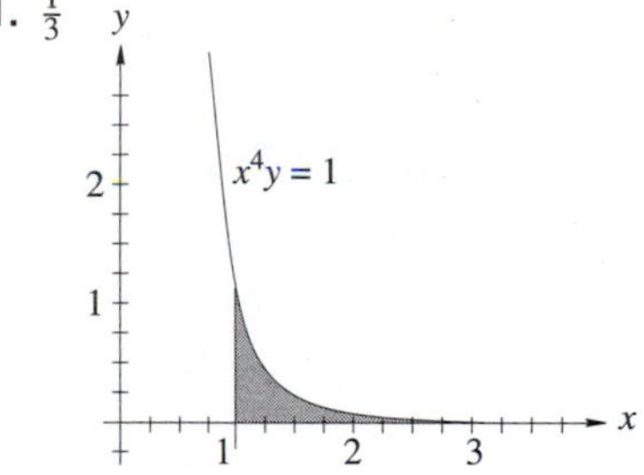

43. White. The blue area is $\frac{5}{12}$ and the white area is $\frac{7}{12}$.
45. $33\,866\frac{2}{3} \approx 33\,866.67 \text{ cm}^3$ **47.** 2630 Btu
49. 139.7 rad/s

Exercise Set 6.5

1. **(a)** 6.264 789 1; **(b)** 6.200 102 86; actual value, 6.2
3. **(a)** 0.643 283; **(b)** 0.656 526 3; actual value $\frac{2}{3}$
5. **(a)** 1.628 968; **(b)** 1.610 847 **7.** **(a)** 1.103 210 7; **(b)** 1.098 726 **9.** **(a)** 3.139 926; **(b)** 3.141 592 6
11. **(a)** 3.251 744; **(b)** 3.241 238 **13.** **(a)** 7.909 233; **(b)** 7.912 321 **15.** **(a)** 2.141 030; **(b)** 2.170 342
17. **(a)** 30.85; **(b)** 30.833 33 **19.** **(a)** 24.646 25; **(b)** 24.629 167
21. **(a)** 26.76 μC; **(b)** 20.34 μC

25. Trapezoidal: 11.264 acres, Simpson: 11.417 acres

Review Exercises

1. $\frac{1}{6}x^6+C$ **3.** 9 **5.** $\frac{9}{2}$ **7.** $-\frac{1}{3}(t+5)^{-3}+C$ **9.** $\frac{381}{7}\approx 54.429$ **11.** $\frac{1}{5}u^5+\frac{1}{3}u^{-3}+C$ **13.** $\frac{2}{3}(x^2-5)^{3/2}+C$ **15.** $-\frac{3}{8}(6-x^2)^{4/3}+C$ **17.** $-\frac{39}{2}$ **19.** $(y^3-5)^{1/3}+C$ **21.** $\frac{16}{35}$ **23.** $\frac{1}{3}(2x)^{3/2}-(2x)^{1/2}+C$ **25.** $-\frac{1}{4}$ **27.** $\frac{855}{2312}\approx 0.36981$ **29.** $\frac{1}{3}$ **31.** 53 **33.** $\frac{125}{6}$

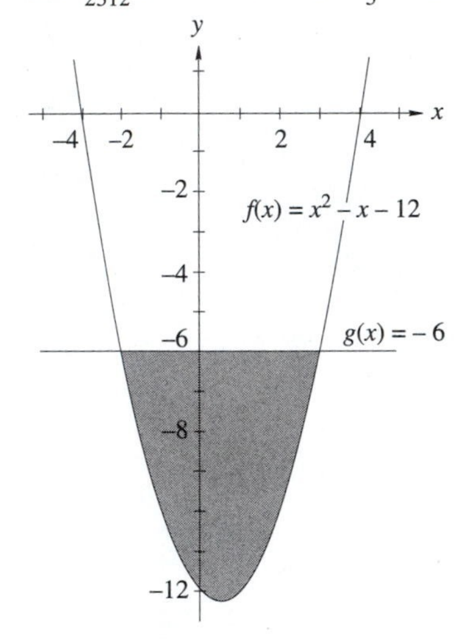

35. $\frac{1}{8}$ **37.** $\frac{9}{16}\approx 0.5625$ **39.** **(a)** 0.776130; **(b)** 0.781752 **41.** **(a)** 1.613695; **(b)** 1.613656 **43.** **(a)** 1.091812; **(b)** 1.086185 **45.** **(a)** 118.5 m; **(b)** $\frac{127}{128}\approx 0.992\,\text{m/s}^2$ **47.** 30.8 A

Chapter 6 Test

1. 1 **3.** $\frac{1}{5}x^5+3x^3+9x+C$ **5.** $x-2x^{-1}+C$ **7.** $\frac{2}{3}(10^{3/2}-2^{3/2})\approx 19.196$ **9.** $\frac{125}{6}$ **11.** 0.00107 C

ANSWERS FOR CHAPTER 7

Exercise Set 7.1

1. $\bar{y}=1/3$, $f_{\text{rms}}=\sqrt{1/5}=\frac{\sqrt{5}}{5}\approx 0.4472$ **3.** $\bar{y}=10/3$, $h_{\text{rms}}=\sqrt{248/14}\approx 4.0661$ **5.** $\bar{y}=2\frac{38}{15}$, $f_{\text{rms}}=\sqrt{\frac{13}{2}}\approx 2.5495$ **7.** $\bar{y}=\frac{28}{3}$, $f_{\text{rms}}=\sqrt{1648/15}\approx 10.4817$ **9.** $\bar{y}=6\frac{2}{3}$, $f_{\text{rms}}=\sqrt{\frac{226}{5}}\approx 6.7231$ **11.** $\bar{y}=10.7083$, $j_{\text{rms}}=12.8992$ **13.** 3.7522 cm for 90 days; 5.4755 cm for the year **15.** $\bar{s}=240$ ft; $\bar{v}=-75.8947$ ft/s **17.** **(a)** 68° F; **(b)** 71° F **19.** 2.6547 A **21.** 23.0309 A **23.** 27136 W **25.** **(a)** 1.3927 V; **(b)** 19.40 W **27.** 0.02 C **29.** 2537.5 J **31.** 0.6667 V

Exercise Set 7.2

1. 661.33π **3.** $64\pi/3$ **5.** 30π **7.** 8π **9.** $16\pi/3$ **11.** $768\pi/7$ **13.** $\frac{512\pi}{5}$ **15.** $2\pi/35$ **17.** $\frac{10\pi}{3}$ **19.** $32\sqrt{2\pi}/3$ **21.** $V=\frac{4}{3}\pi r^3$ **23.** **(a)** $x^2=\frac{2}{3}y$; **(b)** 1.8π **25.** $16{,}913{,}712\pi\,\text{ft}^3\approx 53{,}135{,}993\,\text{ft}^3$ **27.** **(a)** $\frac{800}{3}\pi\approx 837.758\,\text{m}^3$; **(b)** $5760\pi\approx 18095.574\,\text{m}^3$

Exercise Set 7.3

1. 8π **3.** 32π **5.** 625π **7.** $768\pi/7$ **9.** $64\pi/5$ **11.** $\frac{8\pi}{35}$ **13.** $10\pi/7\approx 1.42857\pi$ **15.** $176\pi/15\approx 11.7333\pi$ **17.** $\frac{27\pi}{2}$ **19.** $28\pi\sqrt{21}$ **21.** 99.7 in.^{3}

Exercise Set 7.4

1. 12 **3.** 14/3 **5.** 14/3 **7.** 64.6308π **9.** $4\sqrt{2\pi}$ **11.** $32\sqrt{17}\pi\approx 414.5$ **13.** approximately 244.17 m

Exercise Set 7.5

1. $(0.2, 5)$ **3.** $(2.9, 0.2)$ **5.** $\left(\frac{7}{6}, \frac{4}{3}\right)$ **7.** $\left(\frac{1}{74}, \frac{25}{74}\right)$ **9.** $(1.75, 3.25)$ **11.** $(32/7, 4/5)$ **13.** $(253/95, 195/152)$ **15.** $(5/4, 17/5)$ **17.** $\left(-\frac{1}{2}, \frac{8}{5}\right)$ **19.** $(0, 9)$ **21.** $(0.5, 4.4)$ **23.** $(7/4, 0)$ **25.** $\left(0, \frac{3}{5}\right)$ **27.** $(33/28, 0)$ **29.** $(0, 2.75)$ **31.** $(2.4, 0)$ **33.** $(5/9, 0)$ **35.** **(a)** $\left(\frac{5049.5}{199}, \frac{5049.5}{199}\right) \approx (25.374, 25.374)$; **(b)** $\left(\frac{h^2+hw-w^2}{2(2h-w)}, \frac{h^2+hw-w^2}{2(2h-w)}\right)$ **37.** $m = 4040$ g, $\bar{x} = 26.6$ cm from the less dense end.

Exercise Set 7.6

1. $I_x = 0$; $I_y = 93$; $r_0 = \sqrt{93/8} \approx 3.4095$ **3.** $I_x = 52$; $I_y = 19$; $r_0 = \sqrt{71/7} \approx 3.1847$ **5.** $I_y = 8\rho\sqrt{2/15} \approx 0.7542\rho$; $r_y = \sqrt{0.4} \approx 0.6325$ **7.** $I_x = 125\rho$; $r_x = \sqrt{25/3} \approx 2.8868$ **9.** $I_x \approx 0.014286\rho$; $r_x \approx 0.414$ **11.** $I_y = 256\text{ g}\cdot\text{cm}^2$; $r_y = \sqrt{4.8} \approx 2.1909$ cm **13.** $I_y = 41.6\text{ g}\cdot\text{cm}^2$; $r_y = \sqrt{156/55}\text{ cm} \approx 1.6842$ cm **15.** $I_y = 4\rho\pi/3$; $r_y = \sqrt{2/3} = 0.8165$ **17.** $I_y = 29.867\pi\rho$; $r_y \approx 1.497$ **19.** $I_x = 2592\pi\rho$; $r_x = 3\sqrt{2} \approx 4.2426$ **21.** $I_y = 512\pi\text{ g}\cdot\text{cm}^2$; $r_y = \sqrt{7.76} \approx 2.7852$ cm **23.** $I_x = 158.0952\pi\text{ g}\cdot\text{cm}^2$; $r_y \approx 1.7436$ cm

Exercise Set 7.7

1. $25/8 = 3.125$ ft·lb **3.** 15 J **5.** 4.0 N **7.** 4.8305 ft. **9.** 1125 J **11.** 126 ft·lb. **13.** 127 kJ **15.** 179.76 MJ **17.** 99k J **19.** 80,000 mi-lb. $= 4.224 \times 10^9$ ft·lb. **21.** $1\,238\,720\pi$ J $\approx$ J **23.** 1 286 250 J **25.** 2.866 MJ **27.** 274400π J **29.** **(a)** 19 600 N/m^2; **(b)** 1 372 000 N **31.** 137 200 N **33.** 39 200 N **35.** 24 578 N

Review Exercises

1. $\bar{y} = 14$, $f_{\text{rms}} = 15.1438$ **3.** $\bar{y} = 16/3$, $h_{\text{rms}} = 6.3666$ **5.** $V = 1488\pi$, $\bar{x} = 3.7742$ **7.** $V = 511\pi/9$, $\bar{x} = 1.8018$ **9.** $V = 16\pi/35 \approx 0.4571\pi$, $\bar{y} = 0.5469$ **11.** $V = 20.25\pi$, $\bar{y} = 4.5$ **13.** $L = 9.0734$, $S = 93.9342\pi$ **15.** $L = 20.1111$, $S = 411.0124\pi$ **17.** 4.5 N·m **19.** 2 A **21.** 35,250 ft· lb **23.** $39\,739\,392\pi$ J

Chapter 7 Test

1. 32 **3.** 227.5π **5.** 14.6238 **7.** 2250000π J

ANSWERS FOR CHAPTER 8

Exercise Set 8.1

1. $3\cos 3x$ **3.** $-6\sin 2x$ **5.** $2x\cos(x^2+1)$ **7.** $24\sin 3x\cos 3x$ **9.** $-6x\sin(3x^2-2)$ **11.** $\frac{1}{2\sqrt{x}}\cos\sqrt{x}$ **13.** $-3x^2(2x^3-4)^{-1/2}\sin\sqrt{2x^3-4}$ **15.** $2x + 2\sin x\cos x$ **17.** $\cos^2 x - \sin^2 x = \cos 2x$ **19.** $\frac{-2\sin 2x\sin x - 4\cos x\cos 2x}{\sin^2 2x} = \frac{-2\cos^2 x}{\sin x}$ **21.** $2x\sin x + x^2\cos x$ **23.** $\frac{1}{2}x^{-1/2}\sin x + \sqrt{x}\cos x$ **25.** $2(\cos 2x)(\cos 3x) - 3(\sin 2x)(\sin 3x)$ **27.** $12x^3\sin^2(x^4)\cos(x^4)$ **29.** $2\sin x\cos x - 2\cos x\sin x = 0$ **31.** $-\sin x$ **33.** $-\cos x$ **35.** $\sin x$ **37.** **(a)** $y = -x + 1 + \frac{\pi}{4}$, **(b)**

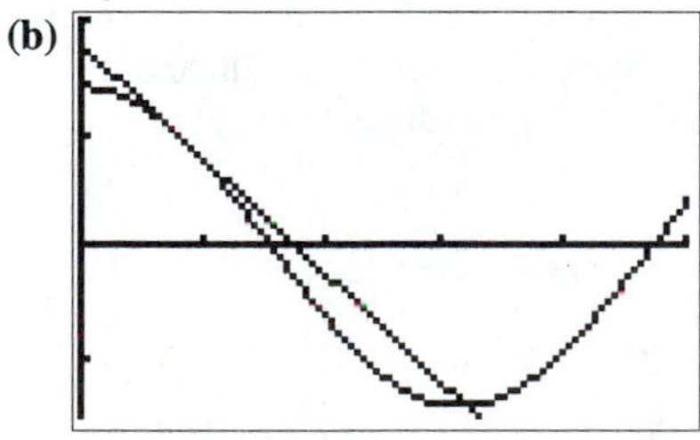

[0, 5, 1] x [−1.5, 2, 1]

39. **(a)** $y = -\frac{1}{4}\left(x - \frac{\pi}{2}\right)$ or $y = -\frac{1}{4}x + \frac{\pi}{8}$, **(b)**

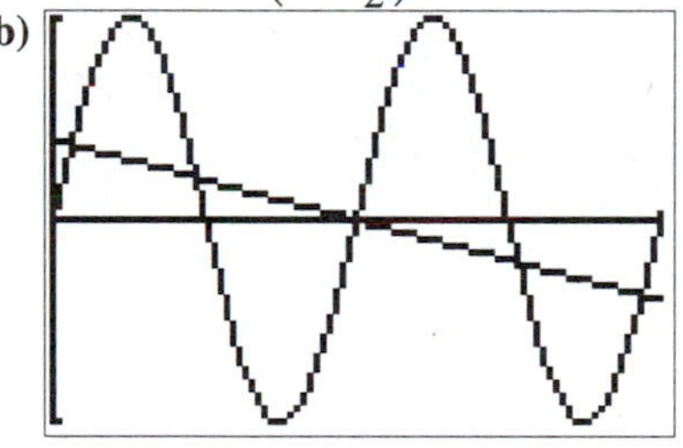

$\left[0, \pi, \frac{\pi}{4}\right] \cdot [-1, 1, 1]$

41. -3470 V **43.** 0.033 A **45.** $-\frac{\sqrt{3}}{2}M \approx -0.866M$ cd/rad **47.** maximum 2.24 A; minimum -2.24 A

Exercise Set 8.2

1. $\frac{\tan\sqrt{x}\sec^2\sqrt{x}}{\sqrt{x}}$ **3.** $5\sec 5x\tan 5x$ **5.** $-2\csc(2x-1)\cot(2x-1)$ **7.** $\sin x\sec^2 x + \cos x\tan x = \sin x(\sec^2 x + 1)$ **9.** $3\tan^2 x\sec^2 x$ **11.** $8x\sec^4(x^2)\tan(x^2)$ **13.** $\sec^2 x\cot x - \tan x\csc^2 x = 0$

15. $2\sin x\cos x\cot x - \sin^2 x\csc^2 x = 2\cos^2 x - 1 = \cos 2x$

17. $\frac{-1}{x^2}\sec^2\frac{1}{x}$ **19.** $\frac{x\sec^2 x^2}{\sqrt{1+\tan x^2}}$

21. $\frac{\sec x(\sec x + \sec^2 x - \tan^2 x)}{(1+\sec x)^2} = \frac{\sec x}{1+\sec x}$

23. $\frac{\csc x(\csc^2 x - \csc x - \cot^2 x)}{(1-\csc x)^2}$

25. $3(\csc x + 2\tan x)^2(2\sec^2 x - \csc x\cot x)$

27. $\frac{6}{5}(\sec^2 2x)(\tan 2x)^{-2/5}$

29. $\frac{\sec x(\tan x + \tan^2 x - sec^2 x)}{(1+\tan x)^2} = \frac{\sec x(\tan x - 1)}{(1+\tan x)^2}$ **31.**

33. $8\sec^2 2x\tan 2x$ **35.** $2\sec^2 x(1 + x\tan x)$

37. $\frac{\sin y}{1 - x\cos y}$ **39.** $\frac{1-\cos(x+y)}{\cos(x+y)-1}$ **41.** -6.31

43. 1055.05 mV/rad

Exercise Set 8.3

1. $\frac{2}{\sqrt{1-4x^2}}$ **3.** $\frac{1/2}{1+x^2/4} = \frac{2}{4+x^2}$

5. $\frac{2x}{\sqrt{1-(1-x^2)^2}} = \frac{2x}{\sqrt{2x^2-x^4}}$ **7.** $\frac{-1}{\sqrt{x-x^2}}$

9. $\frac{1-3x^2}{\sqrt{1-(x^3-x^2)^2}}$ **11.** $\frac{2}{(2x+1)\sqrt{(4x+2)^2-1}}$

13. $\frac{x}{\sqrt{1-x^2}} + \sin^{-1}x$ **15.** $2x\cos^{-1}x - \frac{x^2}{\sqrt{1-x^2}}$

17. $\frac{2}{8x^2-4x+1}$ **19.** $\tan^{-1}(x+1) + \frac{x}{1+(x+1)^2}$

21. $\frac{-1}{\sqrt{2-x^2}\sqrt{\sin^{-1}(1-x^2)}}$

23. $\cot^{-1}(1+x^2) - \frac{2x^2}{2+2x^2+x^4}$

25. $\frac{1}{x}\left(\frac{1}{\sqrt{1-x^2}} - \frac{\sin^{-1}x}{x}\right)$ **27.** $\frac{1}{\sqrt[3]{1-x^2}(\arcsin x)^{2/3}}2$

31. 7.1087 rad/s

Exercise Set 8.4

5. maxima: $(\frac{\pi}{6}, 1.5)$ and $(\frac{5\pi}{6}, 1.5)$; minima: $(\frac{\pi}{2}, 1)$ and $(\frac{3\pi}{2}, -3)$; inflection points: $(1.0030, 1.2646)$, $(2.1386, 1.2646)$, $(3.7765, -0.8896)$, and $(5.6483, -0.8896)$

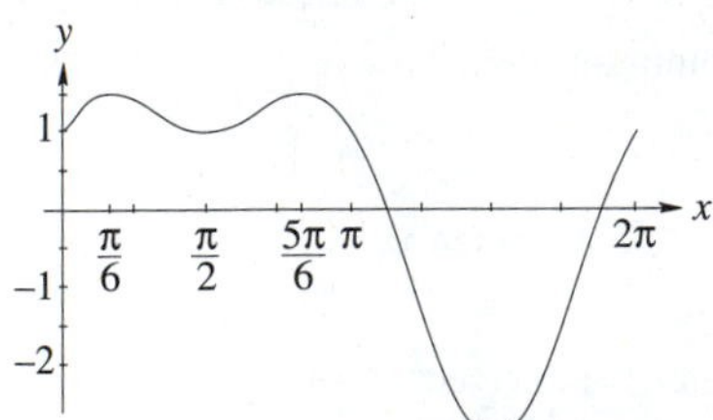

7. maxima: $(\frac{7\pi}{4}, \sqrt{3})$; minima: $(\frac{3\pi}{4}, -\sqrt{3})$; inflection points: $(\frac{\pi}{4}, 0)$, $(5\frac{\pi}{4}, 0)$

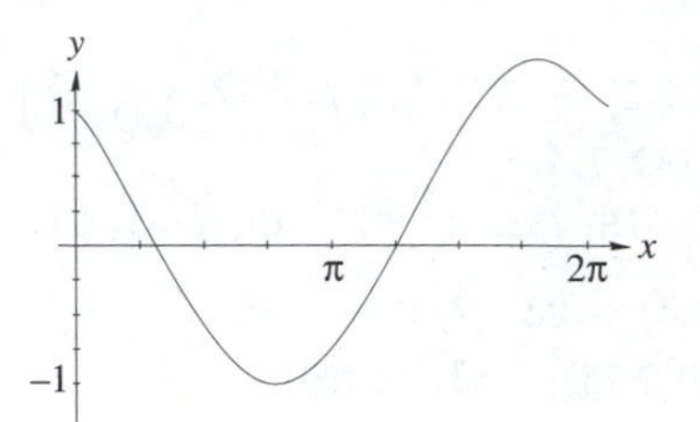

9. maxima: $(\frac{\pi}{6}, 1.25)$, $(\frac{5\pi}{6}, 1.25)$; minima: $(\frac{\pi}{2}, 1)$, $(\frac{3\pi}{2}, -1)$; inflection points: $(0.7297, 1.2222)$, $(2.4119, 1.2222)$, $(\frac{7\pi}{6}, 0)$, $(\frac{11\pi}{6}, 0)$

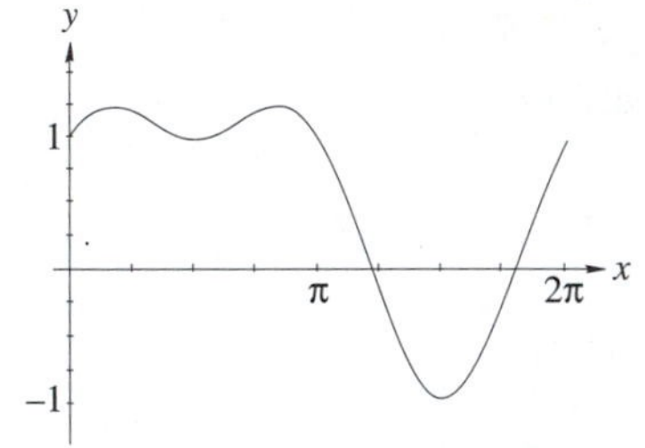

11. $y - 0.5 = \frac{\sqrt{3}}{2}(x - \frac{\pi}{6})$

13. $y - 1 = 6(x - \frac{\pi}{4})$

15. 0.0352 rad/s **17.** 104.71 m/s **19.** 42.8284 mi/sec

21. **(a)** 4.0319 m^3/s; **(b)** 4.2602 m^3/s; **(c)** 180°

23. 1.2364 A/s **25.** 0.5 m/s **27.** 2.9883

29. $0, \pm 2.7200$ **31.**

Exercise Set 8.5

1. $\frac{1}{x}\log e = \frac{1}{x}\frac{1}{\ln 10}$ **3.** $\frac{2x+4}{x^2+4x}$

5. $\frac{x+2}{x^2+4x}[\ln(x^2+4x)]^{-1/2}$ **7.** $-\frac{1}{x}$ **9.** $\sec x\csc x$

11. $\frac{1-\ln x}{x^2}$ **13.** $\frac{1}{1-x^2}[\ln(\frac{1+x}{1-x})]^{-1/2}$ **15.** $\frac{2x^2+12}{x(x^2+4)}$

17. $\frac{2\ln x}{x}$ **19.** $\frac{1}{x\ln x}$ **21.** $\frac{-1}{x^2}$

23. $\frac{2\ln x - 3}{x^3} = \frac{\ln x^2 - 3}{x^3}$ **25.** $\frac{3}{2x^2} - \csc^2 x$

27. $-\frac{\ln x^2 + 1}{x^2(\ln x^2)^{3/2}}$

29. $2\ln(\sin 2x) + 8x\cot 2x - 4x^2\csc^2 2x$ **31.** **(a)** about 109 medflies, **(b)** yes, the medflies will be eradicated in about 31.67 h **33.** **(a)** about 14.72 h, **(b)** about 0.132 ppm

Exercise Set 8.6

1. $4^x\ln 4$ **3.** $\frac{1}{2\sqrt{x}}5^{\sqrt{x}}\ln 5$ **5.** $(\cos x)2^{\sin x}\ln 2$

7. $(2x+1)e^{x^2+x}$ **9.** $4x^3 4^{x^4}\ln 4$ **11.** $e^x(x-2)/x^3$

13. $\frac{(x-2)e^x - 2}{x^3}$ **15.** $\sec^2 x e^{\tan x}$ **17.** $2xe^{x^2}\cos e^{x^2}$

19. $3x^2 3^x + (x^3 - 1)3^x \ln 3$
21. $x^{\cos x}[\frac{\cos x}{x} - (\sin x)(\ln x)]$
23. $(\sin x)^x(\ln \sin x + x \cot x)$ **25.** $3e^{3x} \cot e^{3x}$ **27.** e^{-y}
29. **31.** **33.** $3(\cosh 3x)e^{\sinh 3x}$
35. $g'(x) = 3x^2 \operatorname{sech}^2 \sqrt{x^3+4}$
37. $j'(x) = 3(4x^3 + \cos x)\cosh^2(x^4 + \sin x)\sinh(x^4 + \sin x)$
39. $f'(x) = \dfrac{7}{\sqrt{49x^2+1}}$
41. $j'(x) = \dfrac{-|x|}{x\sqrt{1+x^2}} + \sinh^{-1}\dfrac{1}{x}$ **43.** 10.69 people/day
45. **(a)** $I = q'(t) = -5e^{-t}\cos 2.5t - 12.5e^{-t}\sin 2.5t$, **(b)** $q'(0.45) \approx -3.34$ A
47. **(a)** $N'(t) = 10{,}000 \operatorname{sech}^2(0.1t)$, **(b)** 6401 bacteria/h, **(c)** $699{,}906 \approx 7 \times 10^5$ bacteria

Exercise Set 8.7

1. maxima: none; minima: $(e^{-1}, -e^{-1})$; inflection point: none

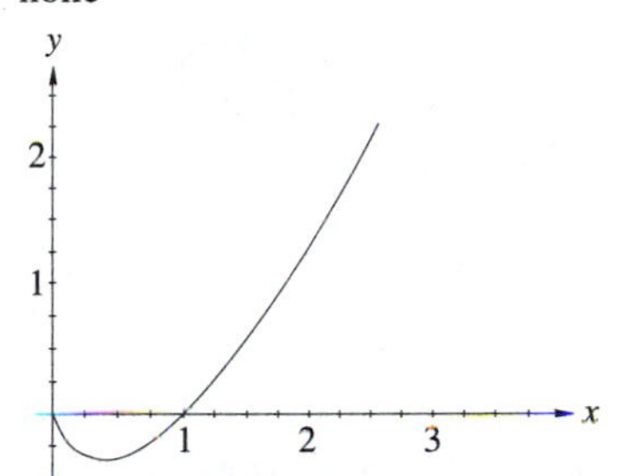

3. maxima: none; minima: $(-1, -e^{-1})$; inflection point: $(-2, -2e^{-2})$

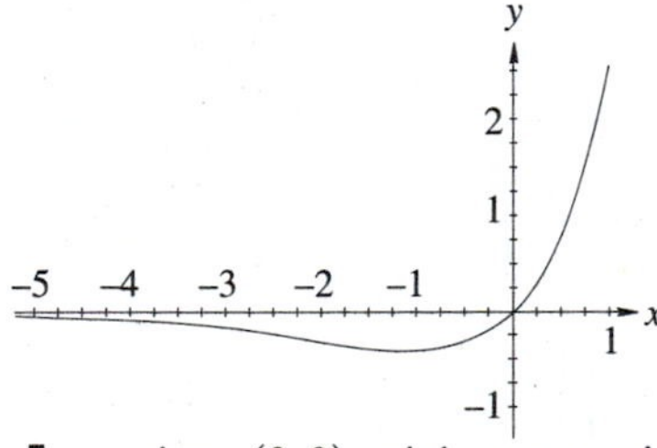

5. maxima: $(0,0)$; minima: none; inflection points: $(\pm 1, \ln \frac{1}{2}) = (\pm 1, -0.6931)$

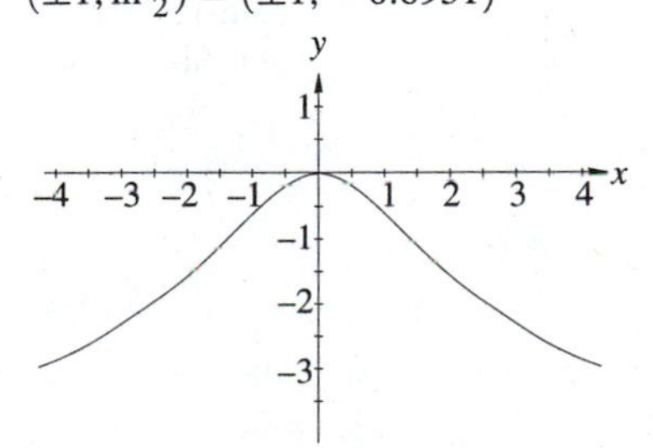

7. maxima: none; minima: none; inflection point: $(0,0)$

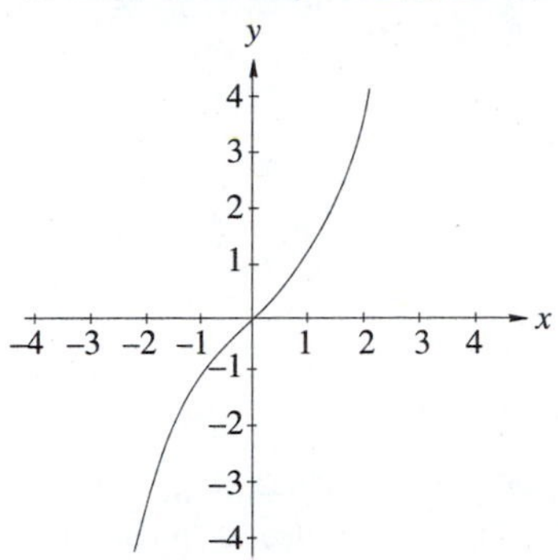

9. maxima: $(-\frac{\pi}{4}, 1.5509)$, $\left(\frac{7\pi}{4}, 0.0029\right)$; minima: $\left(\frac{3\pi}{4}, -0.0670\right)$ and $\left(-\frac{5\pi}{4}, -35.8885\right)$; inflection points: $(-2\pi, 0.0019)$, $(-\pi, -23.1407)$, $(0,1)$, $(\pi, -0.0432)$, $(2\pi, 0.0019)$

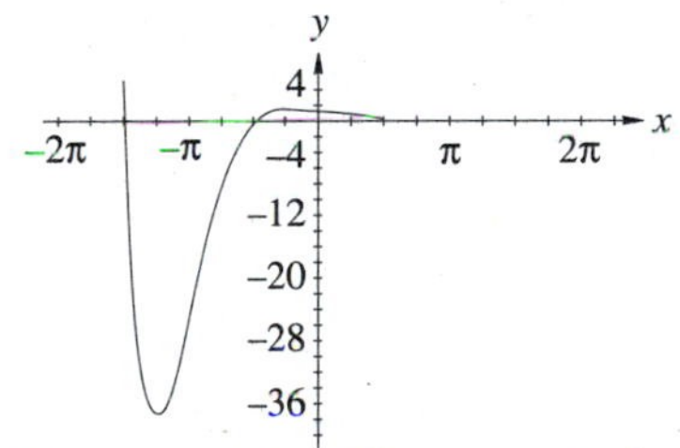

11. $y = x - 1$ **13.** $y - e = 3e(x-1)$
15. **(a)** $v(t) = -3e^{-3t}$; **(b)** never **17.** **(a)** $v(t) = e^t\cos(e^t)$; $a(t) = e^t[\cos(e^t) - e^t\sin(e^t)]$; **(b)** $t = \ln\frac{\pi}{2} \approx 0.4516$ s; $a(t) = -(\frac{\pi}{2})^2 \approx 2.4674$ units/s^2 **19.** $x \approx 1.8371$ m
21. $t = 0$ **23.** -0.7085 W/day **25.** 0.1586; 3.1462
27. $\pm\left(k\pi + \frac{\pi}{2}\right)$, $k = 0, 1, 2, 3, \ldots$

Review Exercises

1. $2\cos 2x - 3\sin 3x$ **3.** $6\tan 3x \sec^2 3x$
5. $2x\sin x + x^2\cos x = x(2\sin x + x\cos x)$
7. $\dfrac{3}{\sqrt{12x - 9x^2 - 3}}$ **9.** $\dfrac{6x}{1+9x^4}$ **11.** $\dfrac{-1}{\sqrt{1-x^2}}$
13. $\dfrac{6x}{3x^2-5}\dfrac{1}{\ln 4}$ **15.** $\dfrac{6x^2\ln(x^3+4)}{x^3+4}$ **17.** $\dfrac{4x-9}{x(4x-3)}$
19. $8xe^{4x^2}$ **21.** $2x(\cos x^2)e^{\sin x^2}$
23. $e^{-x}(5\cos 5x - \sin 5x)$

25. maxima: $(-2, 0.5413)$; minima: $(0,0)$; inflection points: $(-2-\sqrt{2}, 0.3835)$, $(-2+\sqrt{2}, 0.1910)$

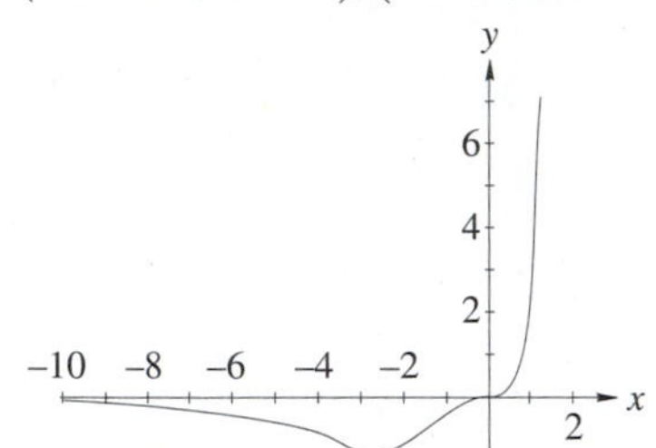

27. maxima: $(0.6629, 0.6964)$, $(3.8045, 0.1448)$; minima: $(2.2337, -0.3175)$, $(5.3753, -0.0660)$; inflection points: $(1.3110, 0.2578)$, $(2.9378, -0.0913)$, $(4.4526, 0.0536)$, $(6.0793, -0.0190)$

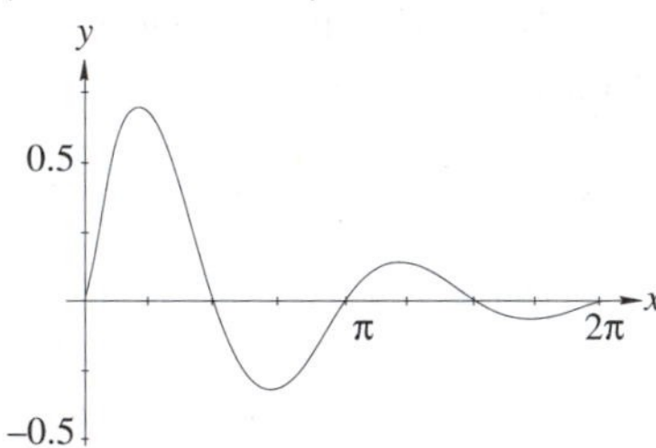

29. maxima: none; minima: $(0.7165, -0.1226)$; inflection point: $(0.4346, -0.0684)$

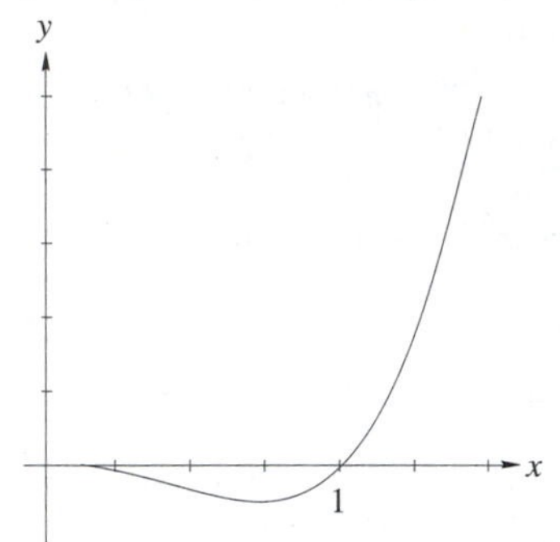

31. **(a)** $y - \frac{\pi}{4} = \frac{1}{2}(x-1)$; **(b)** $y - \frac{\pi}{4} = -2(x-1)$
33. **(a)** $y = x-1$; **(b)** $y = 1-x$
35. **(a)** $y + \ln\sqrt{2} = x - \frac{\pi}{4}$; **(b)** $y + \ln\sqrt{2} = \frac{\pi}{4} - x$
37. **(a)** 64.9967°C; **(b)** −2.2089°C **39.** 19.5959 ft

Chapter 8 Test

1. $7\cos 7x$ **3.** $h'(x) = 2e^{2x}$ **5.** $\dfrac{2xe^{x^2}}{\sqrt{1-e^{2x^2}}}$
7. $(\tan 5x)\left[10(7x^2+3x)\left(\sec^2 5x\right) + (14x+3)\left(\tan^2 5x\right)\right]$
9. -3.82 A

ANSWERS FOR CHAPTER 9

Exercise Set 9.1

1. $\frac{1}{4}\sin^4 x + C$ **3.** $\frac{2}{5}\sin^{5/2} x + C$ **5.** $\frac{1}{2}\tan x^2 + C$
7. $\frac{1}{2}\sec^2 x + C$ or $\frac{1}{2}\tan^2 x + C$ **9.** $-\frac{1}{2}(\arccos x)^2 + C$
11. $-\frac{2}{3}(\text{arccsc}\, x)^{3/2} + C$ **13.** $\frac{1}{8}(\arcsin 2x)^4 + C$
15. $\frac{1}{5}\ln^5(x+4) + C$ **17.** $-(e^x+4)^{-1} + C$
19. $\frac{1}{8}(e^{2x}-1)^4 + C$ **21.** 1/16 **23.** 0.8098
25. 0.0689 **27.** 2/3 **29.** 1.3397 C
31. **(a)** $N(t) = \dfrac{1020}{t} - 340t + 9070$, **(b)** $N(10) = 5772$

Exercise Set 9.2

1. $\frac{1}{4}\ln|4x+1| + C$ **3.** $-\frac{1}{3}e^{-3x} + C$
5. $-\ln|\cos x| + C$ **7.** $2\ln|\tan x| + C$ **9.** $\dfrac{4^x}{\ln 4} + C$
11. $e^x - e^{-x} + C$ **13.** $\frac{1}{2}\ln(1+e^{2x}) + c = \ln\sqrt{1+e^{2x}} + C$
15. $-\frac{1}{2}\ln^2(1/x) + C$ **17.** $3e^{\sqrt[3]{x}} + C$
19. $\frac{1}{2}\ln 17 \approx 1.4166$ **21.** $\ln 3 \approx 1.0986$
23. $\frac{1}{4}(e-1) \approx 0.4296$ **25.** 0.1988
27. $e^2 - (e + 1/3) \approx 4.3374$
29. **(a)** $s(t) = -15{,}760e^{-0.025t} - 38et + 15{,}760$; **(b)** about 1.028 s; **(c)** about 5.1196 m **31.** $\frac{\pi}{2}(1-e^{-4}) \approx 1.5420$
33. $(\ln 0.5) \div (-0.15) \approx 4.6210$ mi **35.** 10.574 ft·lb
37. $F(x) = 2 - 2e^{-3x^2}$

Exercise Set 9.3

1. $-2\cos\dfrac{x}{2} + C$ **3.** $-\frac{1}{5}\ln|\cos x| + C = \frac{1}{5}\ln|\sec x| + C$
5. $\ln|\sec x + \tan x| + C$ **7.** $\tan x + C$
9. $-2\ln|\cos\sqrt{x}| + C = 2\ln|\sec\sqrt{x}| + C$
11. $-\frac{1}{4}\cos(4x-1) + C$ **13.** $\sec x + C$
15. $\frac{1}{6}\ln|5 + 2\sec 3x| + C$ **17.** $\frac{1}{12}\sec^4 3x + C$
19. $\ln|\sin 3x| + C$ **21.** $\tan x + 4\ln|\sec x + \tan x| + 4x + C$
23. $\dfrac{25\pi}{2}$
25. $-\frac{1}{4}(\cot^2 4x + \csc 4x) + C = -\frac{1}{4}\left(\dfrac{\cos 4x + 1}{\sin 4x}\right) + C$
27. $4\ln|\sec x/4| + C$ **29.** $\pi/4$ **31.** 0.4765
33. 0.4509 **35.** $\frac{1}{2}\cosh x^2 + C$ **37.** $\left(\sinh x^2\right)^{3/2} + C$
39. $\frac{1}{5}\cosh^5 x - \frac{1}{3}\cosh^3 x + C$ **41.** Delmar did not provide an answer for this exercise **43.** 1 **45.** .4413.
47. 2.067 V **49.** 51.1027 m **51.** 509.397 ft.

Exercise Set 9.4

1. $\frac{1}{9}\sin^3 3x + C$ **3.** $\frac{1}{4}\sin^4 x + C$ **5.** $\frac{1}{25}\sin^5 5x + C$
7. $\frac{x}{16} + \frac{1}{64}\sin 4x + \frac{1}{48}\sin^3 2x + C$
9. $\frac{5}{16}x + \frac{1}{12}\sin 6x + \frac{1}{64}\sin 12x - \frac{1}{144}\sin^3 6x + C$
11. $\frac{1}{16}\theta - \frac{1}{128}\sin 8\theta + \frac{1}{96}\sin^3 r\theta + C$ **13.** $\frac{1}{3}\tan^3 x + C$
15. $-\frac{1}{2}\cot^2 x - \frac{1}{4}cot^4 x + C$
17. $\frac{2}{9}\sin^{3/2} 3\theta - \frac{2}{21}\sin^{7/2} 3\theta + C$ **19.** $\csc x - \frac{1}{3}\csc^3 x + C$
21. $\frac{1}{4}\tan^4 x + C$ **23.** $\frac{1}{7}\tan^7 x + C$
25. $-\frac{1}{5}\cot^5 x + \frac{1}{3}\cot^3 x - \cot x - x + C$ **27.** $1 - \pi/4$
29. $\frac{3\pi}{16}$ **31.** $\pi\left(\frac{\pi}{6} + \frac{\sqrt{3}}{8}\right) \approx 2.3251$
33. (a) $I = \frac{mr^2\pi}{4}$, (b) $0.0974\,\text{kg}\cdot\text{m}^2$ **35.** 0.5421 A

Exercise Set 9.5

1. $\arcsin\frac{x}{2} + C$ **3.** $\frac{1}{3}\arcsin\frac{3}{2}x + C$ **5.** $\sqrt{9 - x^2} + C$
7. $\frac{1}{2}\operatorname{arcsec}\frac{3x}{2} + C$ **9.** $-\arctan(3 - x) + C$
11. $2\ln(4x^2 + 25) - \frac{3}{5}\arctan\frac{2x}{5} + C$
13. $\arctan(x + 3) + C$ **15.** $\frac{1}{2}\arctan x^2 + C$
17. $-\arcsin(\cos x) + C = x - \frac{\pi}{2} + C$ **19.** 0.6155
21. $\cosh^{-1}\frac{x}{5} + C$ **23.** $\frac{1}{3}\sinh^{-1}\frac{3x}{5} + C$
25. $\frac{\pi}{6} \approx 0.5236$ **27.** 0.36905 **29.** 0.2905 **31.** $\pi/4$
33. $\frac{\pi}{2}(e^2 - 1 - \frac{\pi}{2}) \approx 7.5685$ **35.** $2\pi/3$
37. $y = \operatorname{sech}^{-1} x - \sqrt{1 - x^2} + \operatorname{sech}^{-1} 1$

Exercise Set 9.6

1. $-\sqrt{9 - x^2} + C$ **3.** $-\frac{x}{2}\sqrt{9 - x^2} + \frac{9}{2}\sin^{-1}\frac{x}{3} + C$
5. $\frac{1}{5}(9 - x^2)^{3/2}(6 + x^2) + C$
7. $\ln\left|\frac{x}{3} + \frac{\sqrt{x^2 - 9}}{3}\right| + C = \ln|x + \sqrt{x^2 - 9}| + k$ where $k = C - \ln 3$ **9.** $\frac{1}{2}[x\sqrt{x^2 - 9} - 9\ln(x + \sqrt{x^2 - 9})]$
11. $\sqrt{x^2 + 9} + C$ **13.** $\frac{1}{3}\arctan\frac{x}{3} + C$
15. $(\frac{x^2 - 6}{5})(x^2 + 9)^{3/2} + C$ **17.** $-\frac{1}{12}\frac{(4 - 3x^2)^{3/2}}{x^3} + C$
19. $\frac{1}{243}\sqrt{(9x^2 + 4)^3} - \frac{16}{81}\sqrt{9x^2 + 4} + C$
21. $-\frac{\sqrt{x^2 + 1}}{x} + \ln|x + \sqrt{x^2 + 1}| + C$
23. $\frac{x - 1}{2}\sqrt{4 - (x - 1)^2} + 2\sin^{-1}(\frac{x - 1}{2}) + C$
25. $\ln|x + \sqrt{(x - 1)^2 - 4}| + C$
27. $\frac{1}{2}\left[(x - 1)\sqrt{x^2 - 2x - 3} - 2\ln\left|x - 1 + \sqrt{x^2 - 2x - 3}\right|\right] + C$
29. $\frac{1}{4}\ln\frac{-4 + \sqrt{x^2 - 6x + 25}}{x - 3} + C$ **31.** 1.59747
33. 0.1355 **35.** $81\pi \approx 254.4690$
37. $\frac{-1}{3}(41\sqrt{7} - 128) \approx 6.5080654$
39. $5 + \ln 3^{9/2} \approx 9.9438$
41. $8\pi + \frac{100\pi}{\sqrt{21}}\sin^{-1}\frac{\sqrt{21}}{5} \approx 104.6074$ **43.** ≈ 0.1829 A

Exercise Set 9.7

1. $\frac{1}{2}x^2(\ln x - \frac{1}{2}) + C$ **3.** $\frac{1}{4}\sin 2x - \frac{x}{2}\cos 2x + C$
5. $\frac{x^3}{3}\ln x - \frac{x^3}{9} + C$ **7.** $\frac{1}{2}[x^2\sin^{-1} x + \sqrt{1 - x}] + C$
9. $e^{2x}(\frac{1}{2}x^3 - \frac{3}{4}x^2 + \frac{3}{4}x - \frac{3}{8}) + C$
11. $\frac{1}{60}(6x - 1)(4x + 1)^{3/2} + C$
13. $e^{x/4}(3x^2 - 32x + 128) + C$ **15.** $\frac{1}{2}e^x(\cos x + \sin x) + C$
17. $\frac{1}{2}x^2\sin x^2 + \frac{1}{2}\cos x^2 + C$
19. $-e^{-x}(x^3 + 3x^2 + 6x + 6) + C$
21. $\frac{1}{2}e^{-x}(\sin x - \cos x) + C$ **23.** $\frac{1}{16}\cos 4x + \frac{x}{4}\sin 4x + C$
25. $\pi^2/4$ **27.** $\sin 8 - 8\cos 8 \approx 2.1534$ C
29. $\bar{x} = 1.5972, \bar{y} = 7.8846$ **31.** $\frac{\pi^3}{8} + 3\pi - 6 \approx 7.3006$
33. 1096 ft **35.** (a) $F = -\frac{e^{-kR}2\pi(kR + 1)P_0}{k^2} + \frac{2\pi P_0}{k^2}$, (b) $T = -2\pi\mu\left[\frac{1}{k^3}e^{-kR}\left(k^2R^2 + 2kR + 2\right)\right] + 4\pi\mu\left(\frac{1}{k^3}\right)$
37. 3.0776 ppm

Exercise Set 9.8

1. $\frac{1}{3}(2\ln|\sec 3x| + \tan 3x) + C$ **3.** $-\frac{1}{3}\ln\left|\frac{x}{4x - 3}\right| + C$
5. $\frac{\pm\sqrt{16 + x^6}}{48x^3} + C$ **7.** $-\frac{x}{2}\sqrt{9 - x^2} + \frac{9}{2}\sin^{-1}\frac{x}{3} + C$
9. $\sin x - \frac{1}{3}\sin^3 x + C$ **11.** $\frac{-1}{18}\sin^5 3x\cos 3x - \frac{5}{72}\sin^3 3x\cos 3x - \frac{8}{45}\sin 3x\cos 3x + \frac{5}{48}x + C$
13. $\frac{1}{136}e^{10x}(10\cos 6x + 6\sin 6x) + C$
15. $x^8(\frac{\ln x}{8} - \frac{1}{64}) + C$
17. $\frac{1}{4}(4x\sin^{-1} 4x + \sqrt{1 - 16x^2}) + C$
19. $\sqrt{9 + x^2} - 3\ln\left|\frac{3 + \sqrt{9 + x^2}}{x}\right| + C$
21. $\frac{1}{2}x^3e^{2x} - \frac{3}{4}x^2e^{2x} + \frac{3}{8}(2x - 1)e^{2x} + C = \frac{1}{8}(4x^3 - 6x^2 + 6x - 3)e^{2x} + C$
23. $-\frac{1}{2}(x^3\cos 2x - \frac{3}{2}x^2\sin 2x + \frac{3}{4}\sin 2x - 3x\cos 2x) + C$
25. $\frac{1}{2.5}(286047.5315) \approx 114\,419.0126$ V
27. $\frac{1}{14}\ln 13 \approx 0.1832$ N·m

Review Exercises

1. $\frac{1}{9}(3x - 1)e^{3x} + C$ **3.** $-\sqrt{25 - x^2} + C$
5. $\frac{1}{10}\sin^5 2x + C$ **7.** $\frac{1}{9}\ln|9x + 5| + C$

9. $-\frac{1}{7}\cos(7x+2)+C$
11. $-\frac{1}{21}\cos^7 3x+\frac{2}{15}\cos^5 3x-\frac{1}{9}\cos^3 3x+C$
13. $\dfrac{x}{49\sqrt{4x^2+49}}+C$ **15.** $\frac{1}{3}e^{x^3}+C$
17. $\frac{1}{4}\sqrt{4x^2+49}+C$ **19.** $\frac{1}{8}\ln|9+2\sec 4x|+C$
21. $\frac{1}{12}[\ln(2x+1)]^6+C$ **23.** $\frac{1}{3}\ln|x^3+4|+C$
25. $-5\ln|\cos\dfrac{x}{5}|+C=5\ln|\sec\dfrac{x}{5}|+C$ **27.** $-e^{\cos x}+C$
29. $\frac{1}{3}\tan^{-1}(\dfrac{\sin x}{3})+C$ **31.** $\dfrac{3e^4}{8}+\frac{1}{8}\approx 20.5993$
33. $\frac{3}{320}(\sin^{4/3}4x)(5\sin^4 4x-16\sin^2 4x+20)+C$
35. $\frac{1}{2}\sin^{-1}(\dfrac{\sin x}{2})+C$ **37.** $\frac{1}{2}e^{x^2}(x^4-2x^2+2)+C$
39. $\frac{1}{10}(\arctan 2x)^5+C$ **41.** 12.5861
43. $18-9\sqrt{2}\approx 5.2721\approx 5.2721$

Chapter 9 Test

1. $-2\sqrt{9-e^x}+C$ **3.** $4\arctan x+C$
5. $\frac{1}{4}xe^{4x}-\frac{1}{16}e^{4x}+C=\frac{1}{16}e^{4x}(4x-1)+C$
7. $\dfrac{\pi}{4}(e^2+1)$

ANSWERS FOR CHAPTER 10

Exercise Set 10.1

1.

t	−3	−2	−1	0	1	2	3; $y=3x$
x	−3	−2	−1	0	1	2	3
y	−9	−6	−3	0	3	6	9

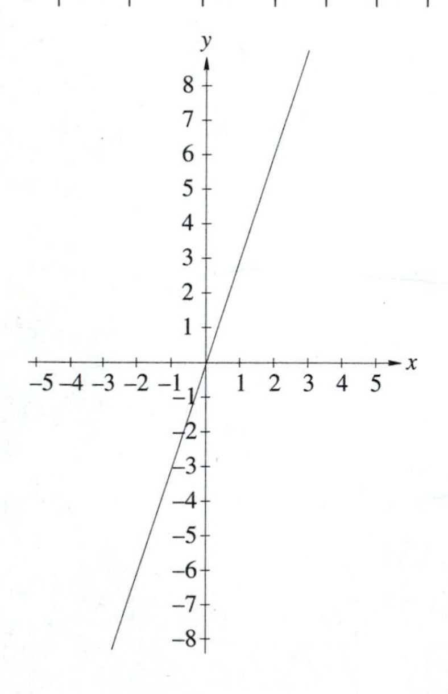

3.

t	−3	−2	−1	0	1	2	3; $y=\dfrac{1}{x}$
x	−3	−2	−1	0	1	2	3
y	$-\frac{1}{3}$	$-\frac{1}{2}$	−1	*	1	$\frac{1}{2}$	$\frac{1}{3}$

*Not defined

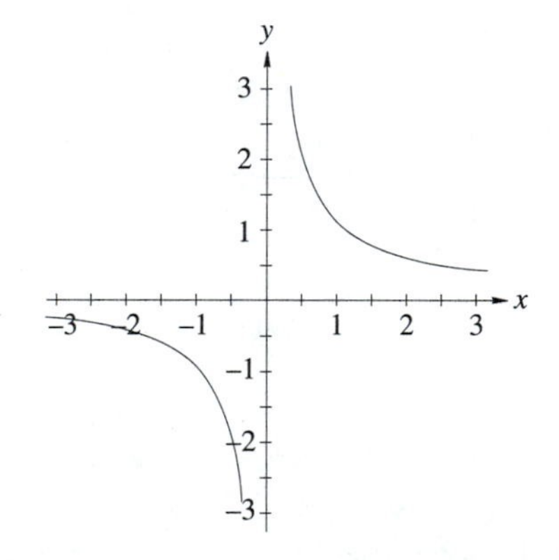

5.

t	-5	-4	-3	-2	-1	0	1	2	3	4	5;
x	8	7	6	5	4	3	2	1	0	-1	-2
y	16	7	0	-5	-8	-9	-8	-5	0	7	16

$y = (x-3)^2 - 9 = x^2 - 6x$

7.

t	0	$\frac{\pi}{4}$	$\frac{\pi}{2}$	$\frac{3\pi}{4}$	π	$\frac{5\pi}{4}$	$\frac{3\pi}{2}$	$\frac{7\pi}{4}$
x	0	$\sqrt{2}$	2	$\sqrt{2}$	0	$-\sqrt{2}$	-2	$-\sqrt{2}$
y	2	$\sqrt{2}$	0	$-\sqrt{2}$	-2	$-\sqrt{2}$	0	$\sqrt{2}$

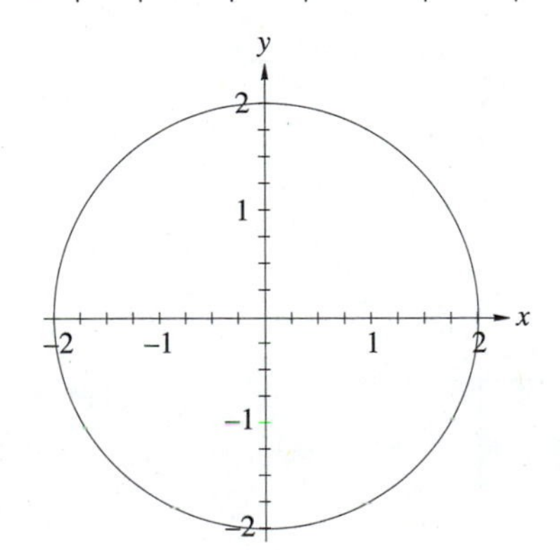

9.

t	0	$\frac{\pi}{4}$	$\frac{\pi}{2}$	$\frac{3\pi}{4}$	π	$\frac{5\pi}{4}$	$\frac{3\pi}{2}$	$\frac{7\pi}{4}$	2π
x	2	$\frac{5\sqrt{2}}{2}$	5	$\frac{5\sqrt{2}}{2}$	0	$-\frac{5\sqrt{2}}{2}$	-5	$-\frac{5\sqrt{2}}{2}$	0
y	0	$\frac{3\sqrt{2}}{2}$	3	$\frac{3\sqrt{2}}{2}$	0	$-\frac{3\sqrt{2}}{2}$	-3	$-\frac{3\sqrt{2}}{2}$	0

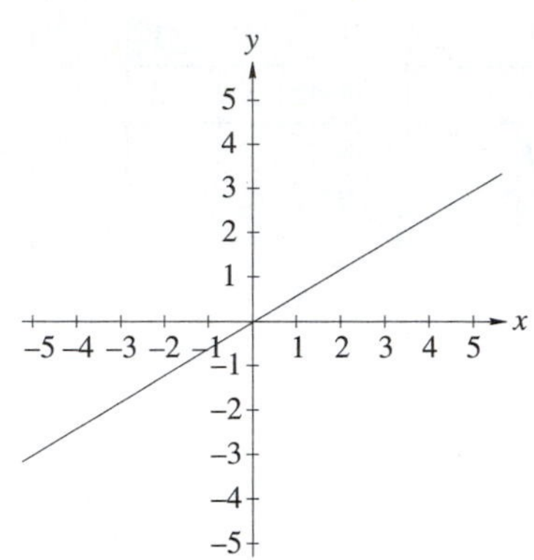

11.

t	-5	-4	-3	-2	-1	0	1	2	3	4	5
x	-5.96	-4.76	-2.86	-1.09	-0.16	0	0.16	1.09	2.86	4.76	5.96
y	0.72	1.65	1.99	1.41	0.46	0	0.46	1.41	1.99	1.65	0.72

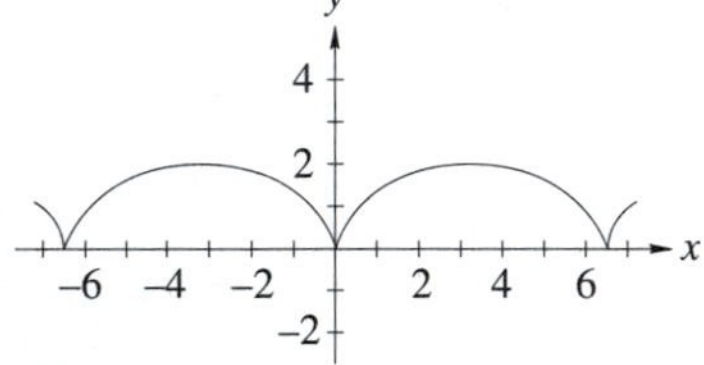

13.

t	0	$\frac{\pi}{6}$	$\frac{\pi}{4}$	$\frac{\pi}{3}$	$\frac{\pi}{2}$	$\frac{2\pi}{3}$	$\frac{3\pi}{4}$	$\frac{5\pi}{6}$	π	$\frac{7\pi}{6}$	$\frac{5\pi}{4}$	$\frac{4\pi}{3}$	$\frac{3\pi}{2}$	$\frac{5\pi}{3}$	$\frac{7\pi}{4}$	$\frac{11\pi}{6}$	2π
x	1	$\frac{2}{\sqrt{3}}$	$\sqrt{2}$	2	*	-2	$-\sqrt{2}$	$-\frac{2}{\sqrt{3}}$	-1	$-\frac{2}{\sqrt{3}}$	$-\sqrt{2}$	-2	*	2	$\sqrt{2}$	$\frac{2}{\sqrt{3}}$	1
y	*	4	$2\sqrt{2}$	$\frac{4}{\sqrt{3}}$	2	$\frac{4}{\sqrt{3}}$	$2\sqrt{2}$	4	*	-4	$-2\sqrt{2}$	$-\frac{4}{\sqrt{3}}$	-2	$-\frac{4}{\sqrt{3}}$	$-2\sqrt{2}$	-4	*

*Not defined

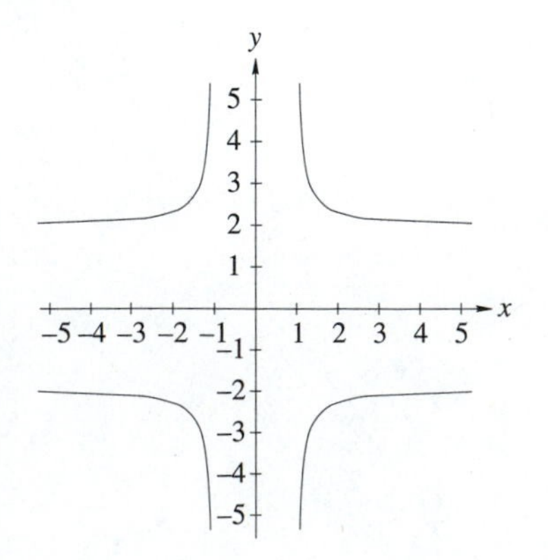

15. **(a)** 1 : 1;

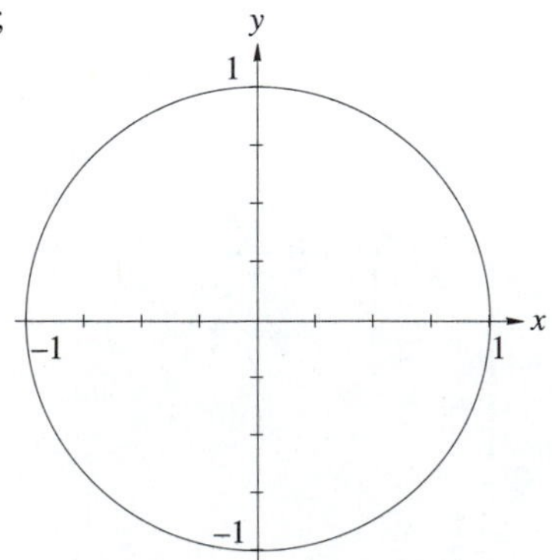

17. **(a)** 2 : 1;

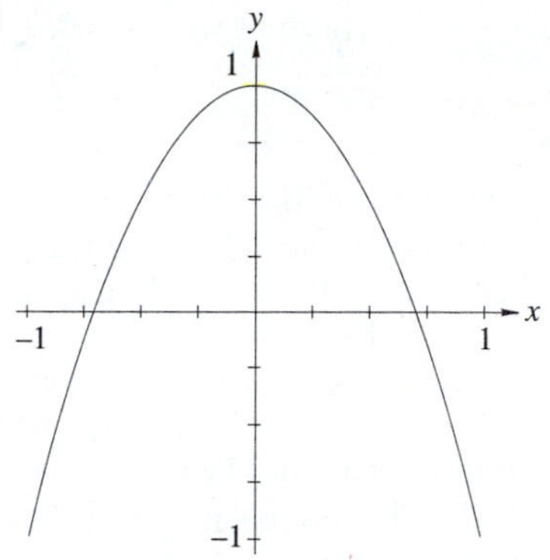

19. **(a)** 1 : 4;

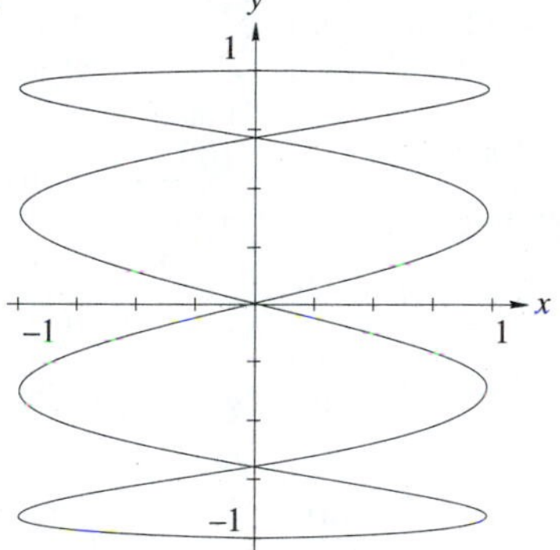

21. **(a)** 2 : 3;

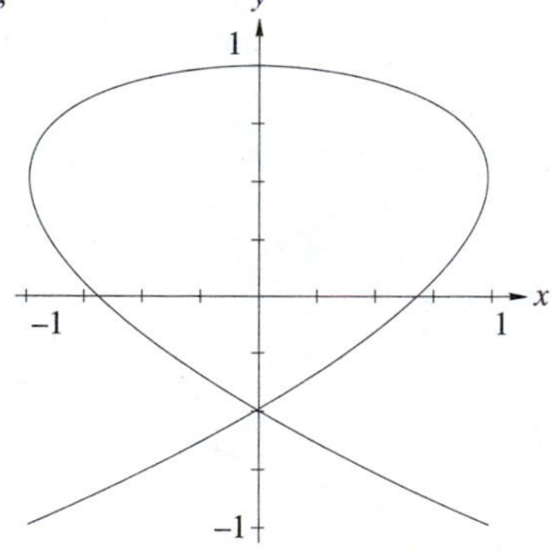

23. **(a)** 2 : 5;

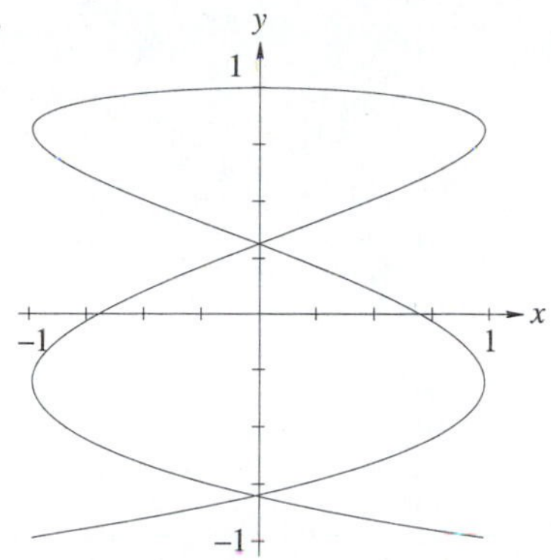

25.

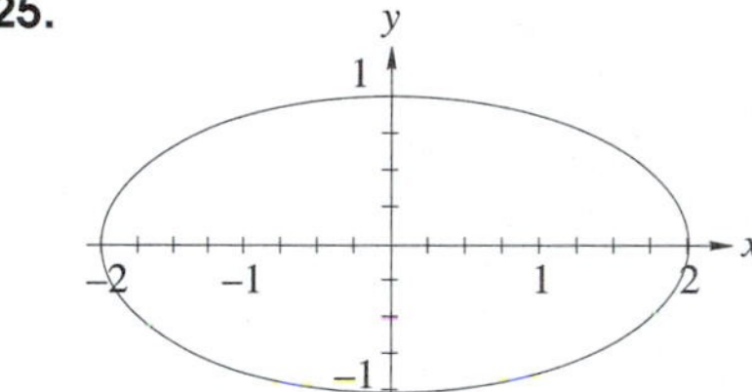

27.

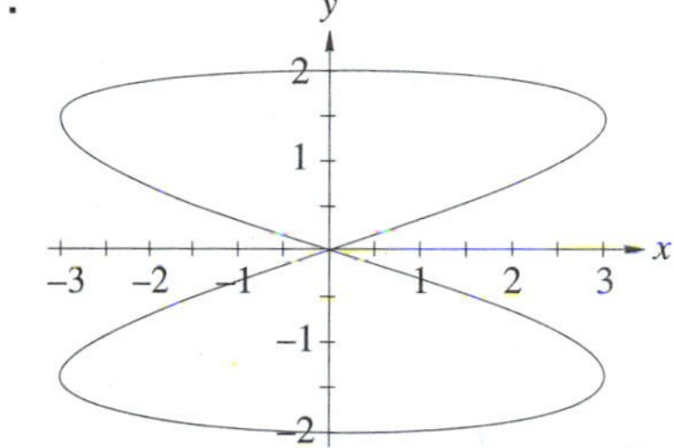

29.

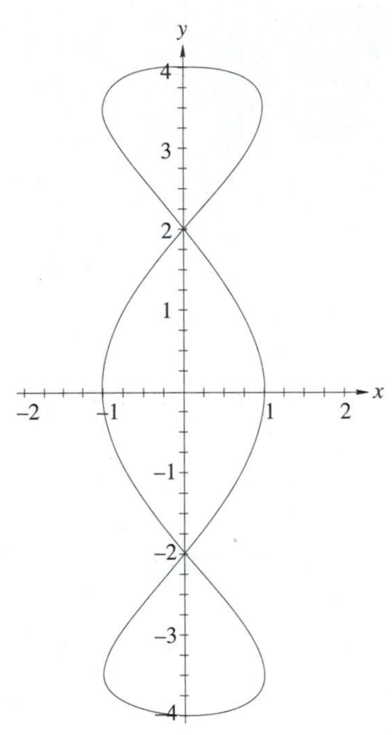

31. **(a)**

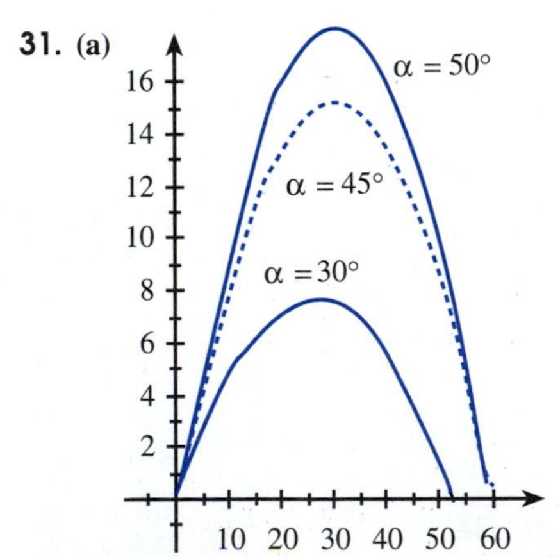

(b) $y = x\tan\alpha - \dfrac{gx^2}{2V^2}\sec^2\alpha$, **(c)** $x = \dfrac{V^2\sin 2\alpha}{g}$

Exercise Set 10.2

1. $\dfrac{dy}{dx} = \dfrac{1}{2t+1}, \dfrac{d^2y}{dx^2} = \dfrac{-2}{(2t+1)^3}$, min at $t = -\frac{1}{2}$: $(-\frac{1}{4}, \frac{1}{2})$

3. $\dfrac{dy}{dx} = \dfrac{1}{2t-6}, \dfrac{d^2y}{dx^2} = \dfrac{-2}{(2t-6)^3}$, min at $t = 3$: $(3, 7)$

5. $\dfrac{dy}{dx} = \dfrac{2t-1}{2t+1}, \dfrac{d^2y}{dx^2} = \dfrac{4}{(2t+1)^3}$, min at $t = 1/2$: $(\frac{3}{4}, -\frac{1}{4})$

7. $\dfrac{dy}{dx} = 1/4; \dfrac{d^2y}{dx^2} = 0$, maxima when $t = 0, 2\pi, \ldots : (7,0)$; min when $t = \pi, 3\pi, \cdots : (-1,-2)$ **9.** $\dfrac{dy}{dx} = -\tan t$: $\dfrac{d^2y}{dx^2} = \dfrac{-1}{\cos^3 t}$; maxima when $t = \pi, 3\pi, \ldots : (2,-2)$ and when $t = 0, 2\pi, \ldots : (2,0)$; min when $t = \pi/2, 5\pi/2, \ldots : (3,-1)$ and $t = 3\pi/2, 7\pi/2, \ldots : (1,-1)$ **11.** tangent: $y = 3x - 1$; normal: $3y + x = 7$ **13.** tangent: $9y + 2x = -36$; normal: $2y - 9x = 261$ **15.** tangent: $y = 2$; normal: $x = 2$ **17.** horizontal tangent when $t = 2 : (5,-4)$; vertical tangents: none **19.** horizontal tangent when $t = \frac{\pi}{2}, \frac{5\pi}{2}, \ldots : (0,5)$ and $t = \frac{3\pi}{2}, \frac{7\pi}{2}, \ldots : (0,5)$; vertical tangents when $t = 0, 2\pi, \ldots : (3,0)$ and $t = \pi, 3\pi, \ldots : (-3,0)$
21. **(a)** $s_x(5) \approx 12{,}646.57$ ft, $s_y(5) \approx 13{,}645.44$ ft; **(b)** $v_x(5) \approx -2809.09$ ft/s, $v_y(5) \approx 2369.32$ ft/s; **(c)** $a_x(5)0$, $a_y(5) = -32$ ft/s^2 **23.** **(a)** 100 km at 71.6°, **(b)** $v_x = -\dfrac{100t}{(t^2+1)^{3/2}}, v_y = \dfrac{100}{(t^2+1)^{3/2}}$, **(c)** magnitude: 10 km/s, direction: −18.4°, **(d)** $a_x = \dfrac{200t^2 - 100}{(t^2+1)^{5/2}}$, $a_y = -\dfrac{300t}{(t^2+1)^{5/2}}$, **(e)** magnitude: $\sqrt{37} \approx 6.08$ km/s^2, direction: −27.9° **25.** **(a)** $v_x = \dfrac{20t^3}{\sqrt{t^4+1}}, v_y = 60\sqrt{t}$, **(b)** magnitude: $20\sqrt{\dfrac{t^6}{t^4+1} + 9t}$ m/s, direction: $\tan^{-1}\left(\dfrac{3\sqrt{t^4+1}}{t^{5/2}}\right)$, **(c)** magnitude: 167.28 m/s, direction: 53.3°

Exercise Set 10.3

1.

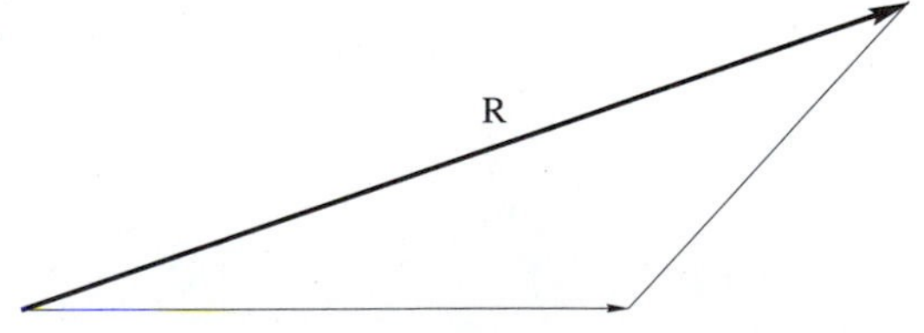

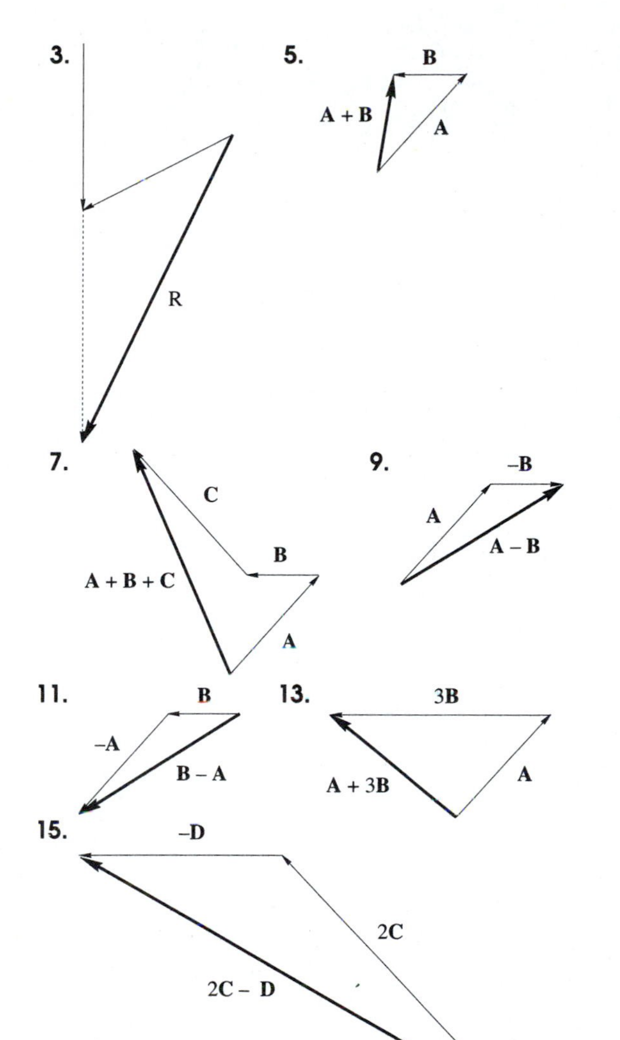

17.

19.

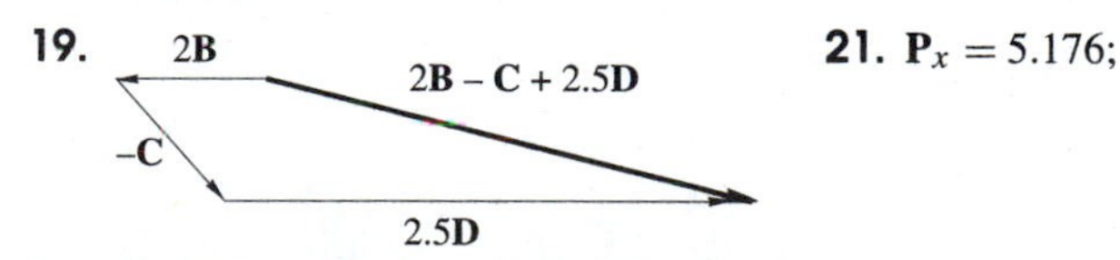

21. $\mathbf{P}_x = 5.176$; $\mathbf{P}_y = 19.319$ **23.** $\mathbf{P}_x = 4.688$; $\mathbf{P}_y = -17.793$ **25.** $\mathbf{V}_x = 9.372$; $\mathbf{V}_y = 2.687$ **27.** $A = 15$ **29.** $C = 17$

31. 13 km/h **33.** $\mathbf{V}_x \approx 295\,\text{N}$, $\mathbf{V}_x \approx 930\,\text{N}$ **35.** parallel to the ramp ≈ 31.8 lb, perpendicular to the ramp ≈ 149.7 lb **37.** $V = 17.5$ V, $\phi \approx 30.96°$

Exercise Set 10.4

1. (a)

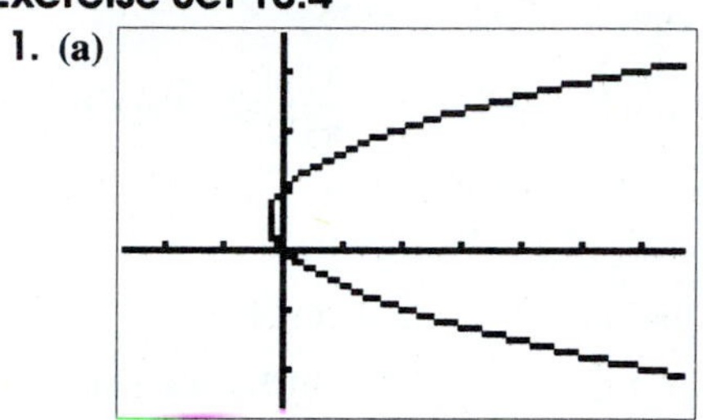

$[-2.7, 6.7, 1] \times [-2.6, 3.6, 1]$

$-4 \le t \le 4$

(b) $\mathbf{v}(t) = (t+1)\mathbf{i} + \mathbf{j}$, **(c)** $\mathbf{v}(1) = 2\mathbf{i} + \mathbf{j}$, with magnitude $\sqrt{5} \approx 2.236$ and direction $\theta \approx 26.5°$

3. (a)

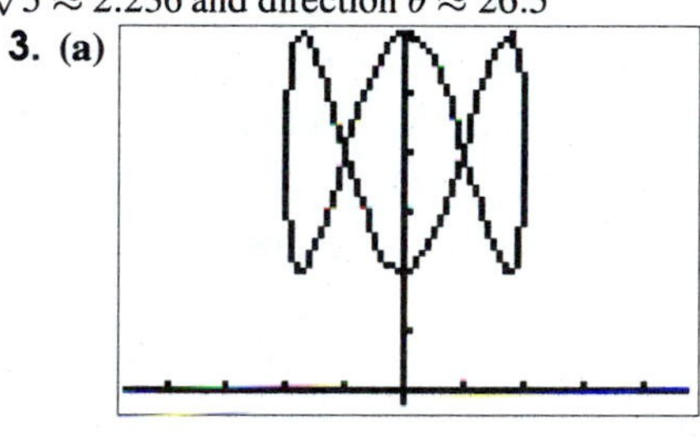

$[-2.35, 2.35, 0.5] \times [-0.1, 3, 0.5]$

$0 \le t \le 6.3$

(b) $\mathbf{v}(t) = (-\sin t)\mathbf{i} + (3\cos 3t)\mathbf{j}$, **(c)** $\mathbf{v}\left(\frac{2\pi}{3}\right) = -\frac{\sqrt{3}}{2}\mathbf{i} + 3\mathbf{j}$, with magnitude $\sqrt{9.75} \approx 3.123$ and direction $\theta \approx 106.1°$

5. (a)

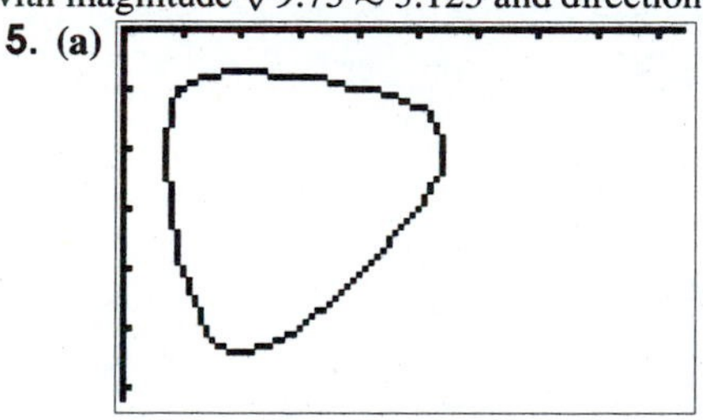

$[0, 4.7, 0.5] \times [-3.1, 0, 0.5]$

$0 \le t \le 2\pi$

(b) $\mathbf{v}(t) = (\cos t)\left(e^{\sin t}\right)\mathbf{i} + (\sin t)\left(e^{\cos t}\right)\mathbf{j}$, (c) $\mathbf{v}(\pi) = -\mathbf{i}$, with magnitude 1 and direction $\theta = 180°$

7. Velocity vector: $\mathbf{v}(4) = 8\mathbf{i} - \mathbf{j}$; magnitude $= \sqrt{65} \approx 8.062$ ft/s; direction $\approx -0.895°$. Acceleration vector: $\mathbf{a}(4) = 3\mathbf{i} + 3\mathbf{j}$; magnitude $= \sqrt{18} \approx 4.243$ ft/s^2; direction $= 45°$

9. When $t=-\dfrac{3\pi}{2}$
Velocity vector: $\mathbf{v}\left(-\dfrac{3\pi}{2}\right)=-\dfrac{2}{3\pi}\mathbf{j}$; magnitude $\dfrac{2}{3\pi}\approx 0.212$ m/s; direction $-90^\circ=270^\circ$
Acceleration vector: $\mathbf{a}\left(-\dfrac{3\pi}{2}\right)=\mathbf{i}-\dfrac{4}{9\pi^2}\mathbf{j}\approx\mathbf{i}-0.045\mathbf{j}$; magnitude ≈ 1.010 m/s^2; direction about -2.58°. When $t=\pi$
Velocity vector: $\mathbf{v}(\pi)=-\mathbf{i}+\dfrac{1}{\pi}\mathbf{j}$; magnitude $\sqrt{1+(1/\pi)^2}\approx 1.049$ m/s; direction about 197.7°
Acceleration vector: $\mathbf{a}(\pi)=-\dfrac{1}{\pi^2}\mathbf{j}\approx -0.101\mathbf{j}$; magnitude $\dfrac{1}{\pi^2}\approx 0.101$ m/s^2; direction $-90^\circ=270^\circ$.

11. At 20.0 s, speed: 522.0 m/s, direction: 40°. At $t=100.0$ s, speed: 2136.0 m/s, direction: 20.56°.

13. **(a)** $\mathbf{v}(t)=$
$220\left(1-0.05t-e^{-1}\right)\mathbf{i}+32\left(-40+40e^{-t/40}\right)\mathbf{j}$
(b) Magnitude: $\sqrt{176^2+(-92.5)^2}\approx 198.8$ ft/s; direction: $\tan^{-1}\left(\dfrac{-92.5}{176}\right)\approx -27.7^\circ$.

Exercise Set 10.5

1.

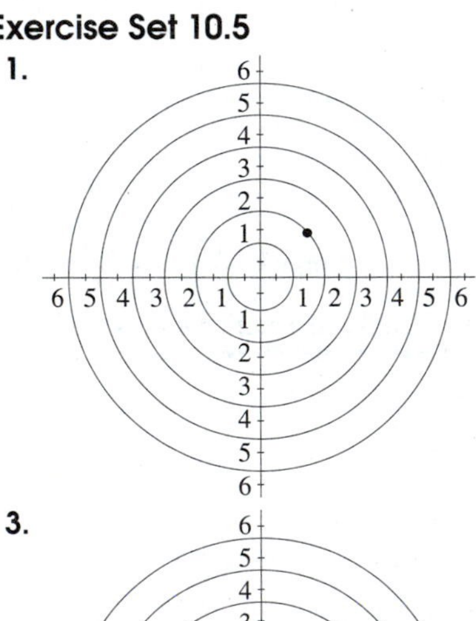

3.

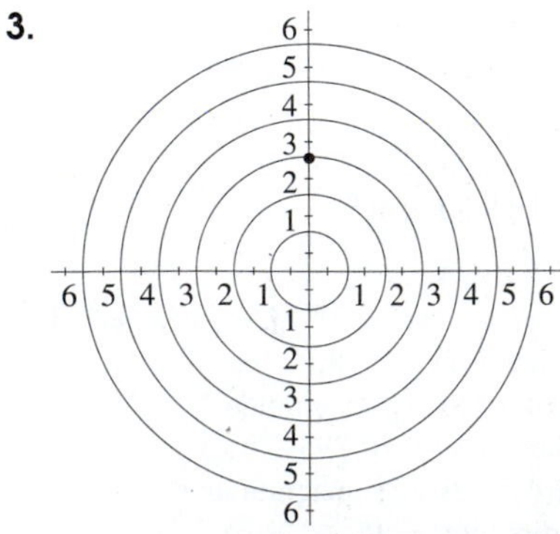

5.

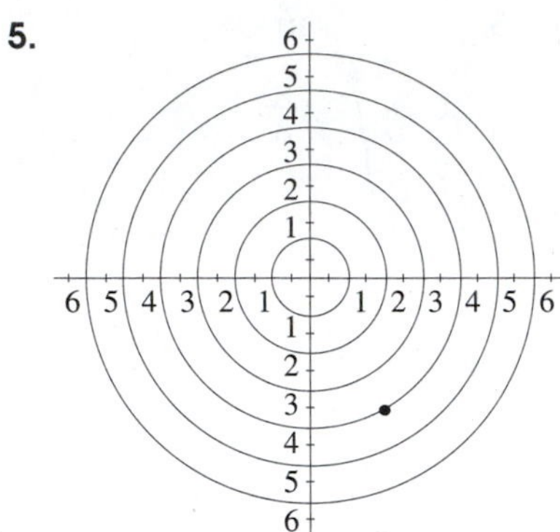

7.

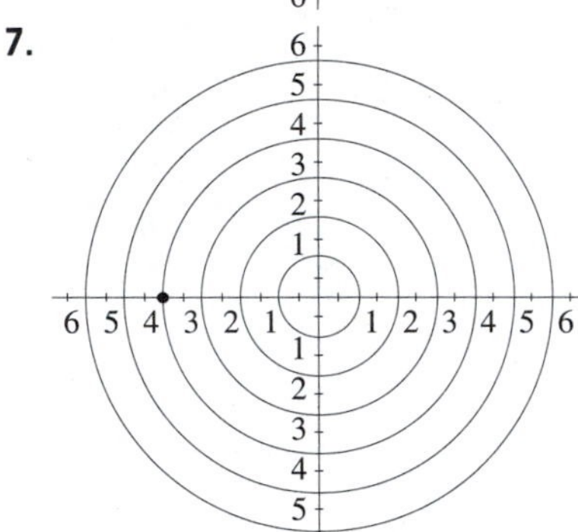

9.

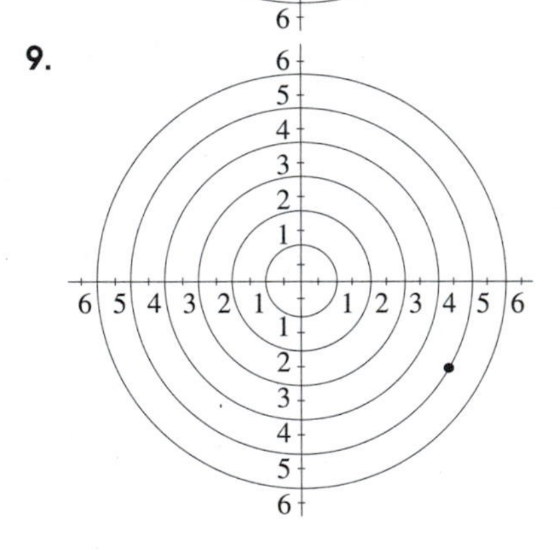

11.

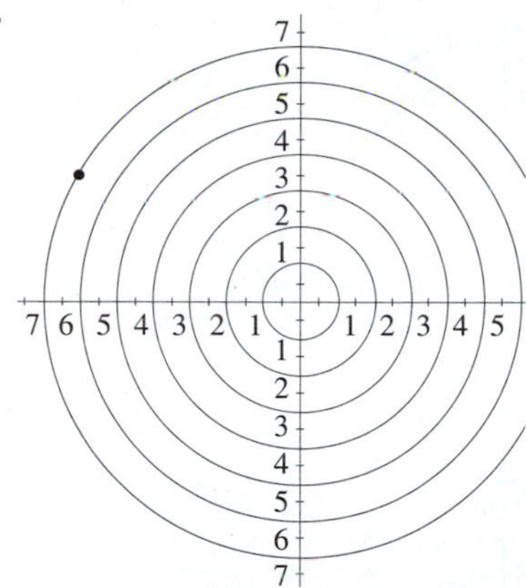

13. $(2, 3.46)$ **15.** $(-1.414, 1.414)$ or $(-\sqrt{2}, \sqrt{2})$
17. $(-5.64, -2.05)$ **19.** $(0.80, 1.83)$ **21.** $(-2.95, 0.52)$
23. $(4.81, 3.59)$
25. $(5.66, 45°) \approx (4\sqrt{2}, 45°) = (4\sqrt{2}, \frac{\pi}{4})$
27. $(5, 35.87°) = (5, 0.64)$
29. $(29, 133.60°) = (29, 2.332)$
31. $(5, 126.87°) = (5, 2.214)$
33. $(12.21, 235.01°) = (12.21, 4.102)$
35. $(9.22, 77.47°) = (9.22, 1.352)$
37.

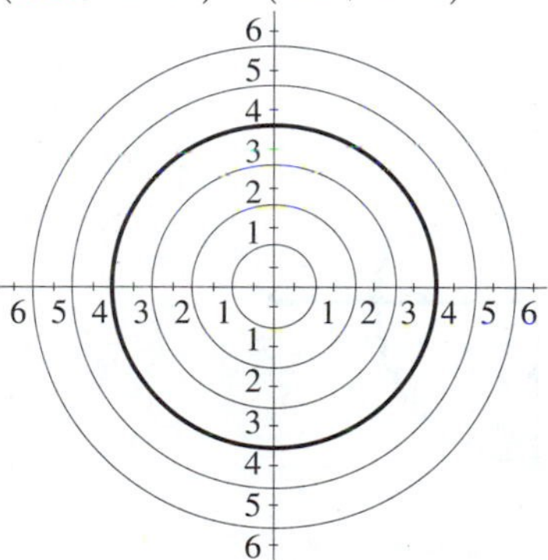

39.

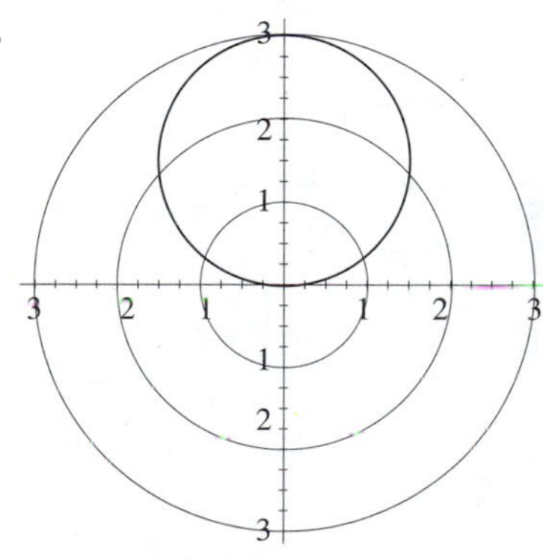

41.

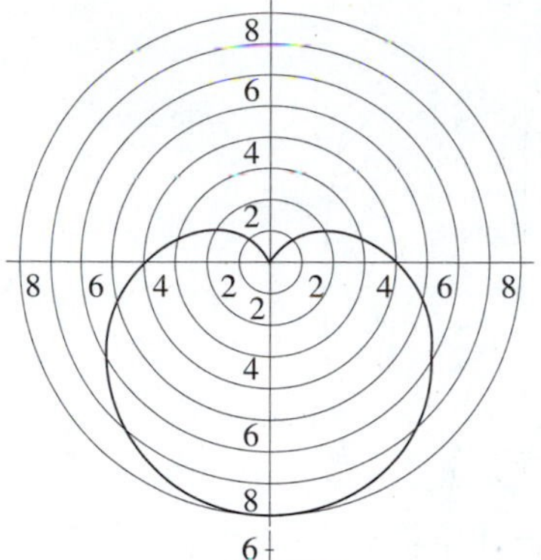

43.

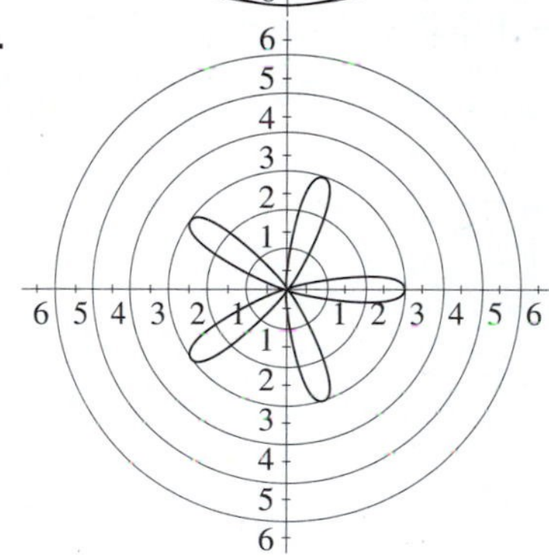

45.

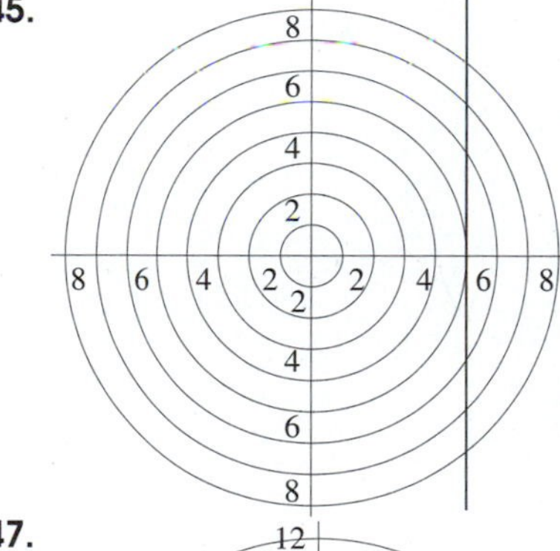

47.

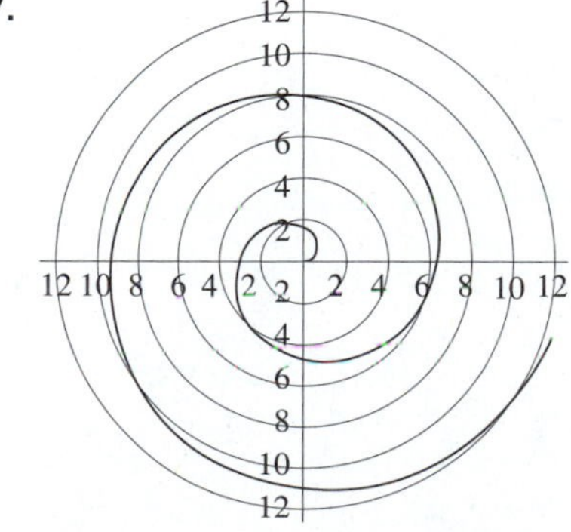

49.

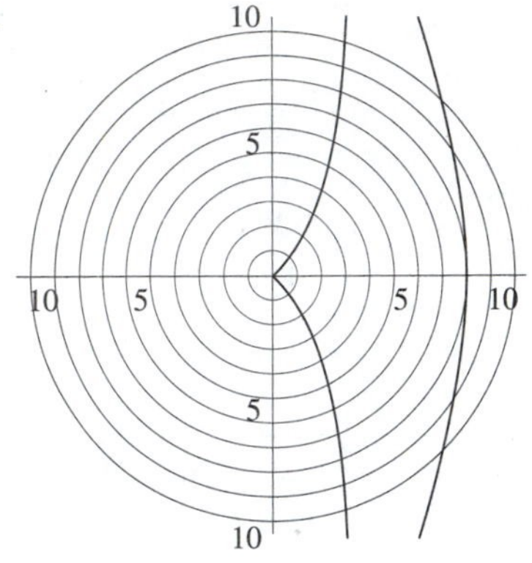

51.

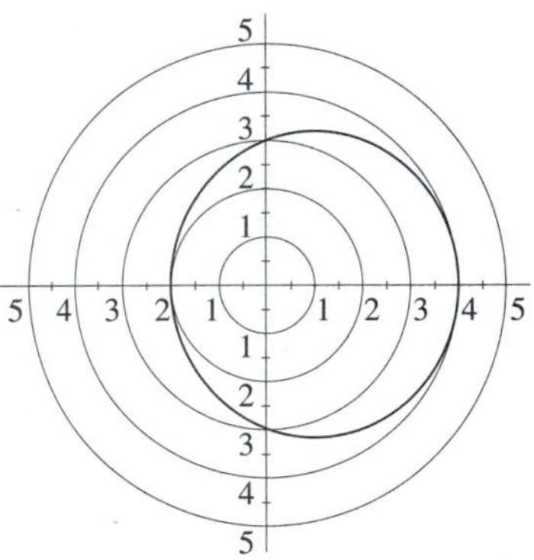

53.

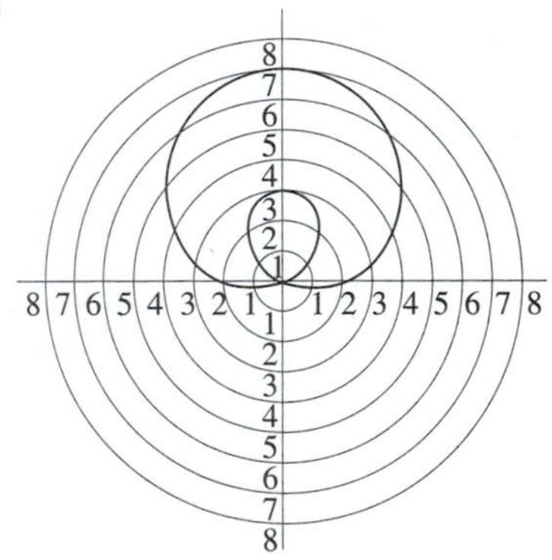

55.

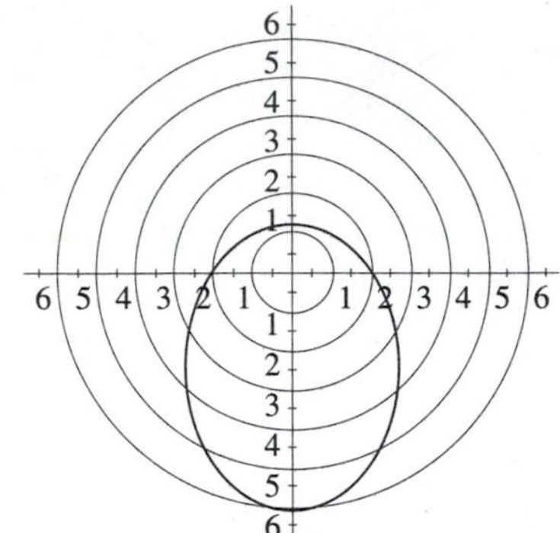

57.

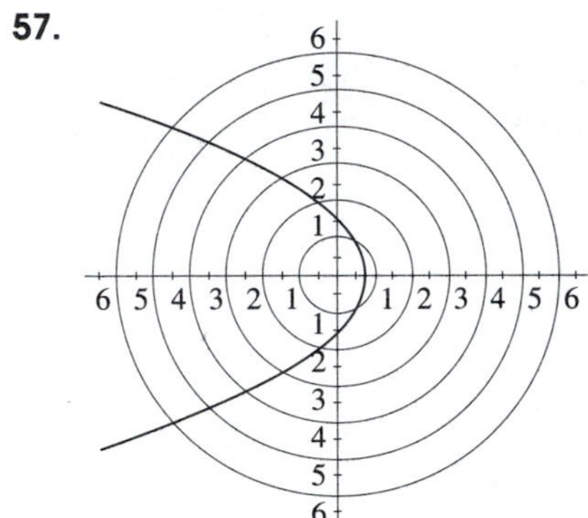

59.

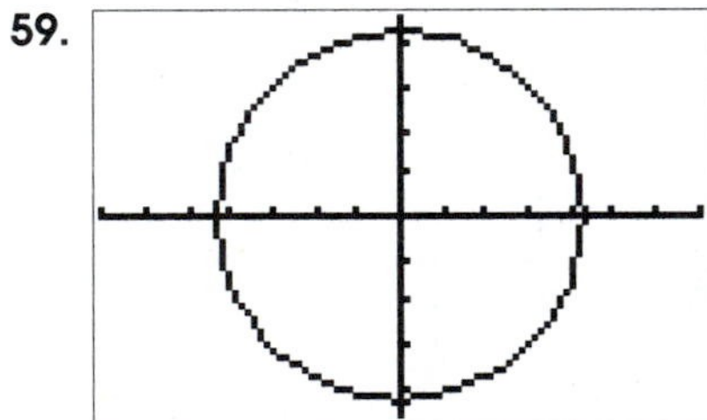

$[-7050, 7050, 1000] \times [-4650, 4650, 1000]$

61.

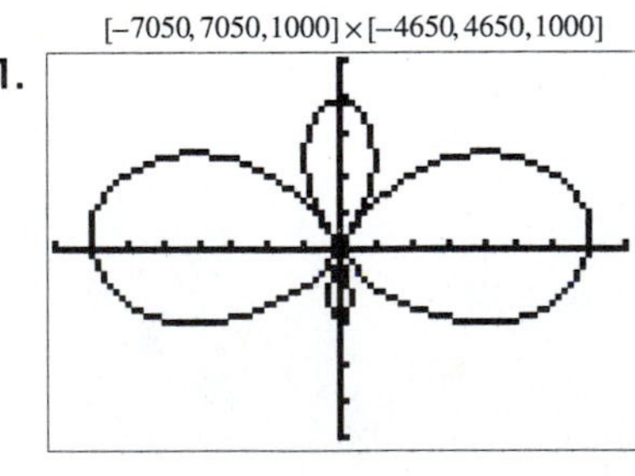

$[-8, 8, 1] \times [-5, 5, 1]$

Exercise Set 10.6

1. hyperbola

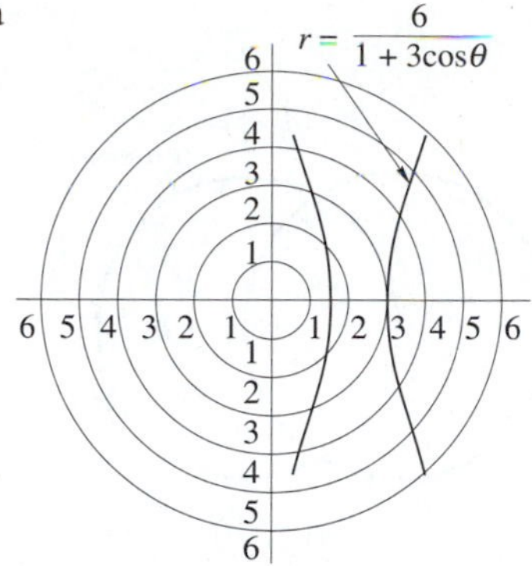

3. ellipse

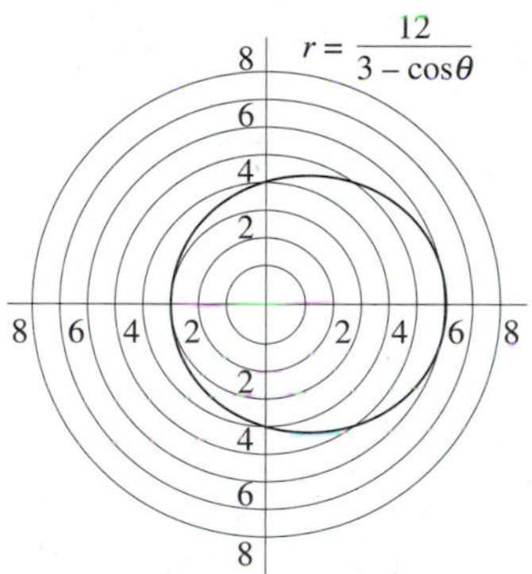

5. parabola

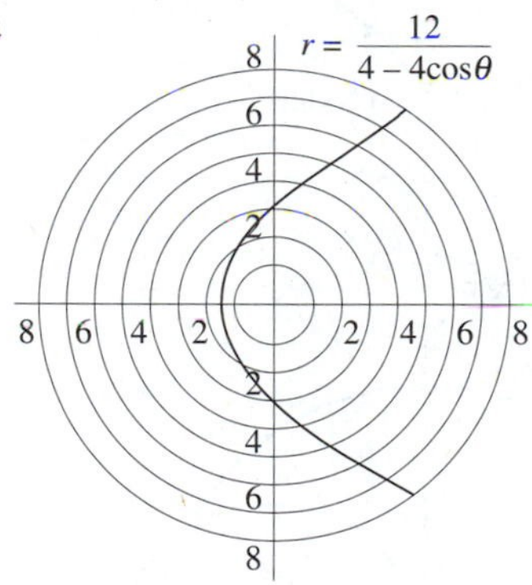

7. ellipse

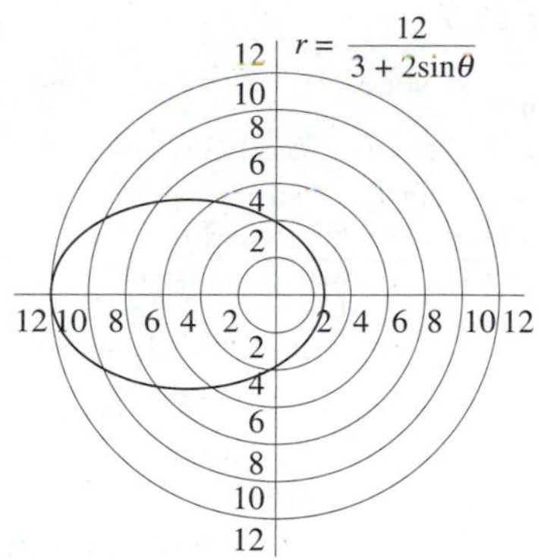

9. hyperbola rotated $\frac{\pi}{3}$ radians

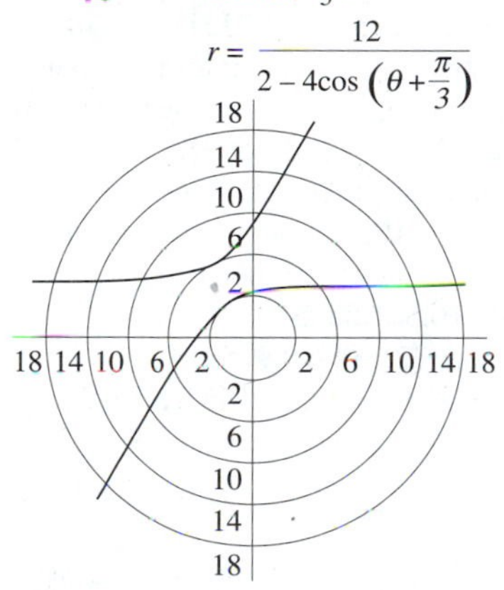

11. $r = \dfrac{6}{1 - \frac{3}{2}\cos\theta} = \dfrac{12}{2 - 3\cos\theta}$

13. $r = \dfrac{5}{1 - \sin\theta}$

15. $r = \dfrac{\frac{2}{3}}{1 - \frac{2}{3}\cos\theta} = \dfrac{2}{3 - 2\cos\theta}$

17. Greatest distance: 4.335×10^7 miles; shortest distance: 2.854×10^7 miles

19.

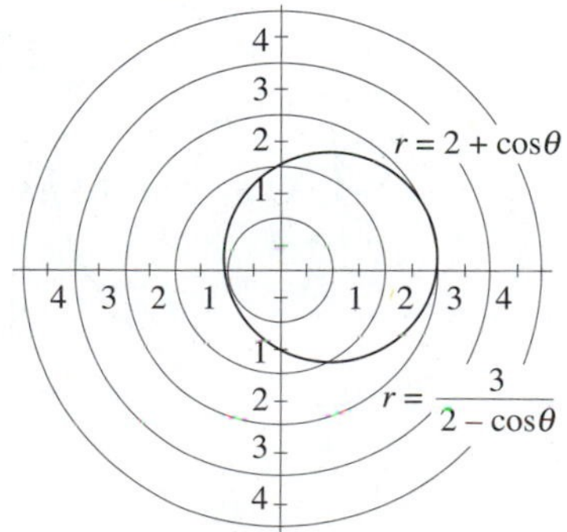

Exercise Set 10.7

1. $\dfrac{2\cos\theta\sin\theta}{\cos^2\theta-\sin^2\theta}=\tan 2\theta$

3. $\dfrac{\cos^2\theta-\sin^2\theta+\cos\theta}{-(2\sin\theta\cos\theta+\sin\theta)}=\dfrac{\cos 2\theta+\cos\theta}{\sin\theta-\sin 2\theta}$

5. $\dfrac{3\sin 3\theta\sin\theta-\cos\theta-\cos\theta\cos 3\theta}{3\sin 3\theta\cos\theta+\sin\theta+\sin\theta+\cos 3\theta}$

7. $\dfrac{\sec^2\theta\sin\theta+\tan\theta\cos\theta}{\sec^2\theta\cos\theta-\tan\theta\sin\theta}=(\sec^2\theta+1)\tan\theta$

9. tangent: $y=1$; normal: $x=0$ **11.** tangent: $y-0.9151=0.6603(x-3.4151)$; normal: $y-0.9151=-1.5146(x-3.4151)$ **13.** tangent: $y-3.8971=5.1962(x+2.25)$ or $y-12\sqrt{3x}=36\sqrt{3}$; normal: $y-3.8971=-0.1925(x+2.25)$ or $36y+4\sqrt{3x}+45\sqrt{3}$ **15.** tangent: $y+3.7321x=7.7875$; normal: $y-0.2679x=2.0861$ **17.** Numerator is 0 when $\cos\theta=0$ or $\sin\theta=\pm\sqrt{6}/6$. Denominator is 0 when $\sin\theta=0$ or $\sin\theta=\pm\sqrt{6}/6$. maxima: $(3,0)$, $(-2,1.1503)$, $(-2.0543,1.9790)$, $(3,\pi)$; minima: $(2,0.4205)$, $(-3,\pi/2)$, $(2,2.7211)$, $(2.0543,3.5498)$ **19.** Numerator $=0$ when $\cos\theta=(-1\pm\sqrt{33}/8)$; denominator $=0$ when $\theta=0$, π, $\cos^{-1}(-\frac{1}{4})$; maxima: $(3,0)$, $(0.5,1.8235)$, $(-1,\pi)$, $(0.5,4.4597)$, $(3,2\pi)$; minima: $(2.1862,0.9359)$, $(-0.6862,2.45738)$, $(-0.6862,3.7094)$, $(2.1862,5.3473)$

21. **(a)** $\dfrac{dr}{d\theta}=\dfrac{30\sin\theta}{(5+3\cos\theta)^2}$, **(b)** $\theta=\pi$

Exercise Set 10.8

1. $\frac{1}{3}(5\sqrt{5}-1)\approx 5.0902$ **3.** 6π **5.** $\frac{1}{2}\pi^2$ **7.** 4 **9.** 1.3802 **11.** $\frac{\pi}{27}(145^{3/2}-1)\approx 203.0436$ **13.** 2π **15.** 2π **17.** $2^5\pi/5=\frac{32}{5}\pi$ **19.** **(a)** 50.56 cm, **(b)** 10^+ h

Exercise Set 10.9

1. $(\frac{\pi}{2},\frac{\pi}{6})$

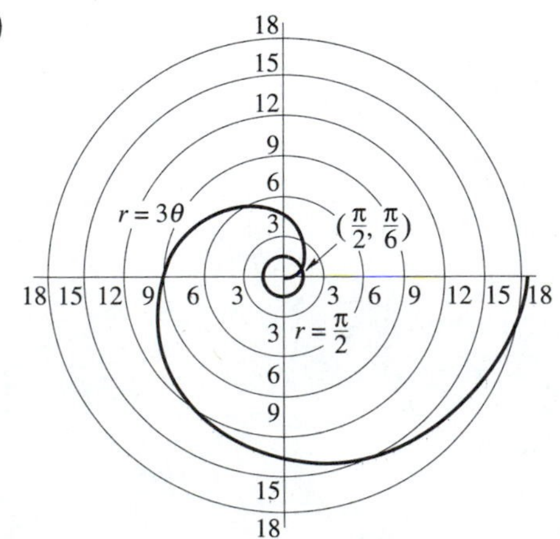

3. $\left(\frac{1}{2},\frac{\pi}{3}\right),\left(\frac{1}{2},\frac{5\pi}{3}\right)$

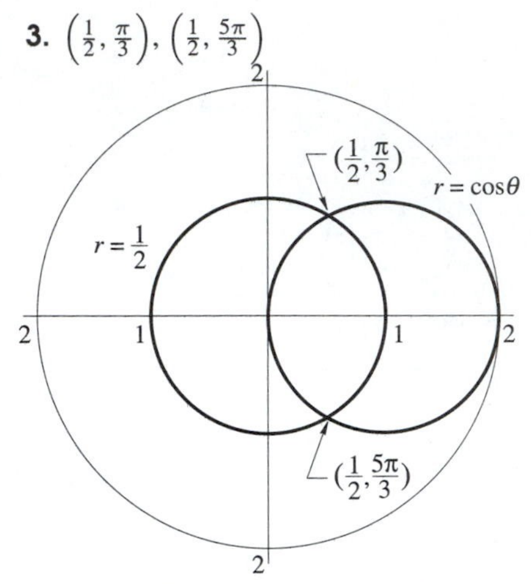

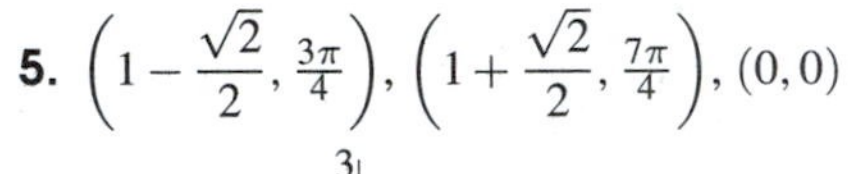

5. $\left(1-\frac{\sqrt{2}}{2},\frac{3\pi}{4}\right),\left(1+\frac{\sqrt{2}}{2},\frac{7\pi}{4}\right),(0,0)$

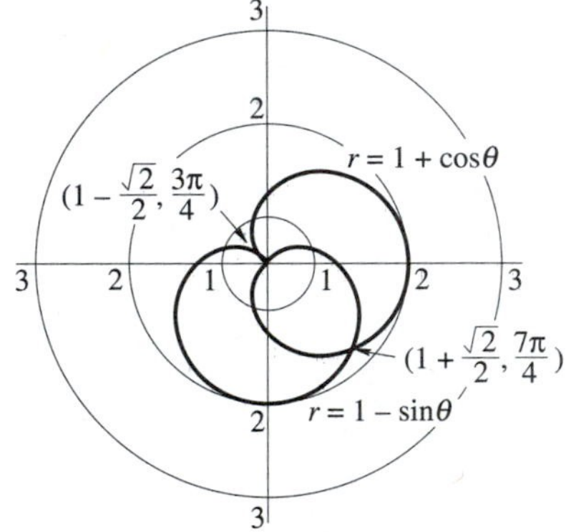

7. $\left(\sqrt{3},\frac{\pi}{6}\right),\left(\sqrt{3},\frac{11\pi}{6}\right)$

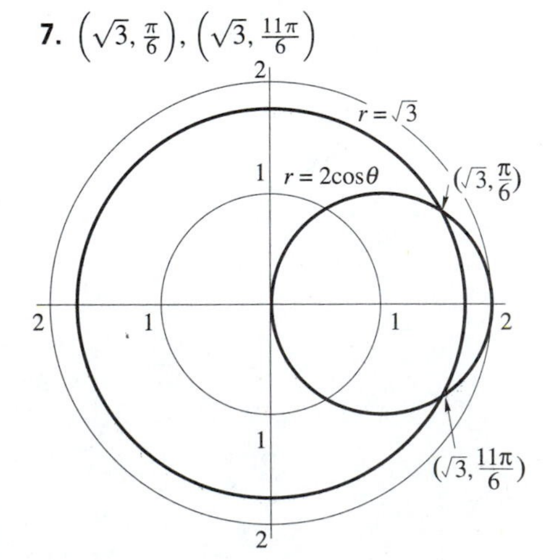

9. $\left(1, \frac{\pi}{4}\right), \left(-1, \frac{7\pi}{4}\right), (0,0)$

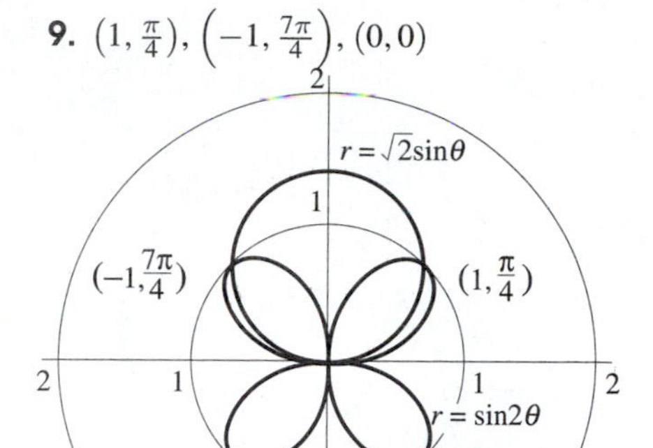

11. $(1,0), (1,\pi)$

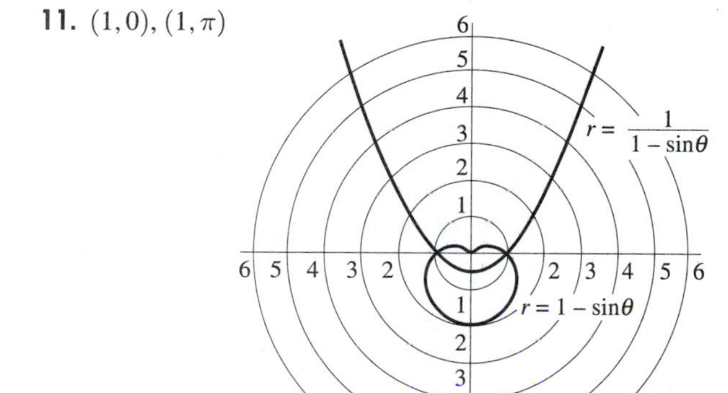

Exercise Set 10.10

1. 4π

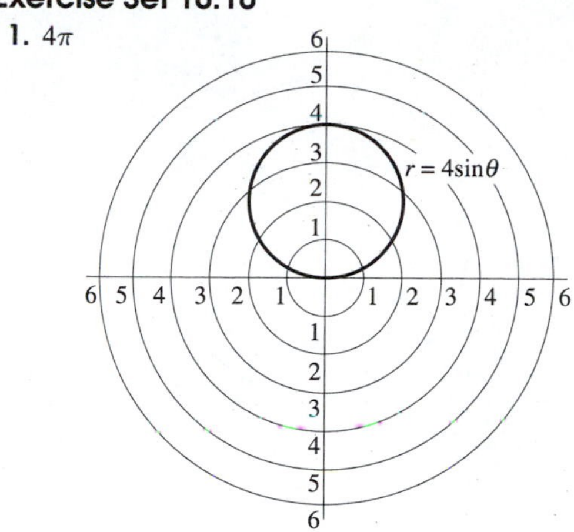

3. $\frac{3}{2}\pi$

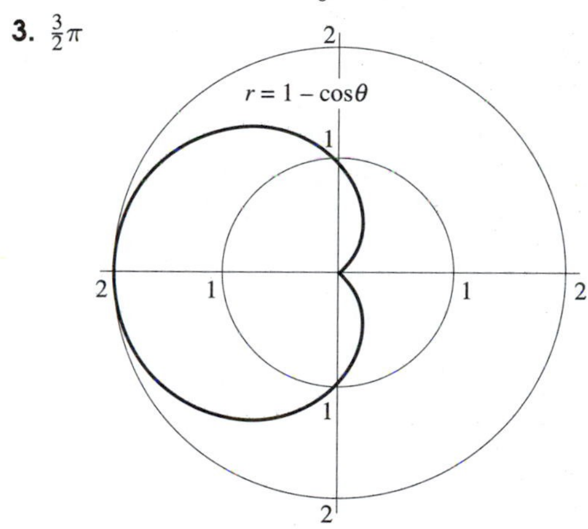

5. $\pi/2$

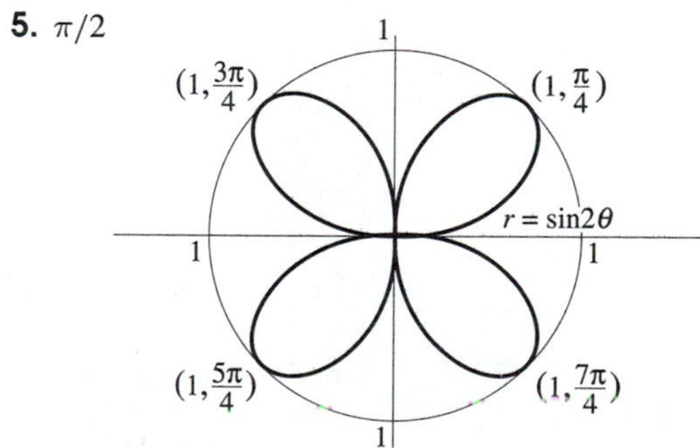

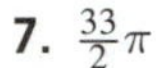

7. $\frac{33}{2}\pi$

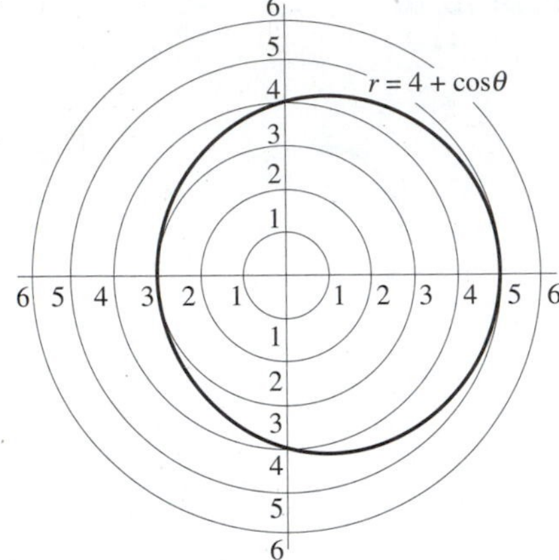

9. 2

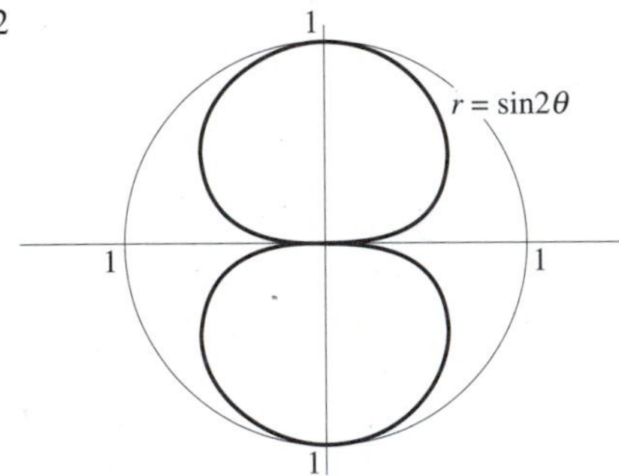

11. $\frac{9}{16}(\pi+2) \approx 2.8921$ **13.** $\pi/16 \approx 0.1963$
15. $\frac{1}{8}(e^{2\pi}-1) \approx 66.8115$ **17.** 2π **19.** $\pi/3$
21. $2+\pi/4 \approx 2.7854$

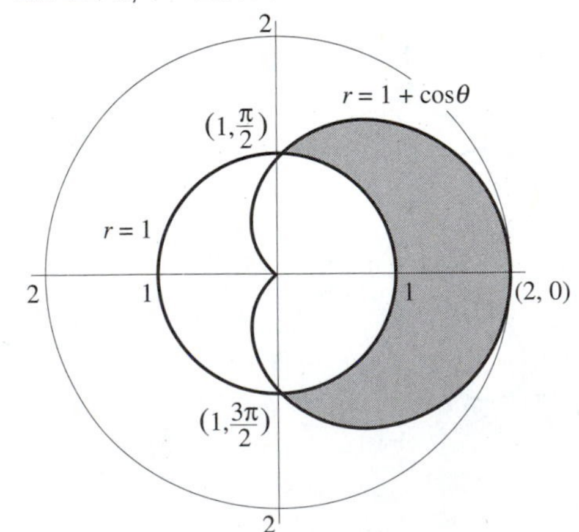

23. 0.382778

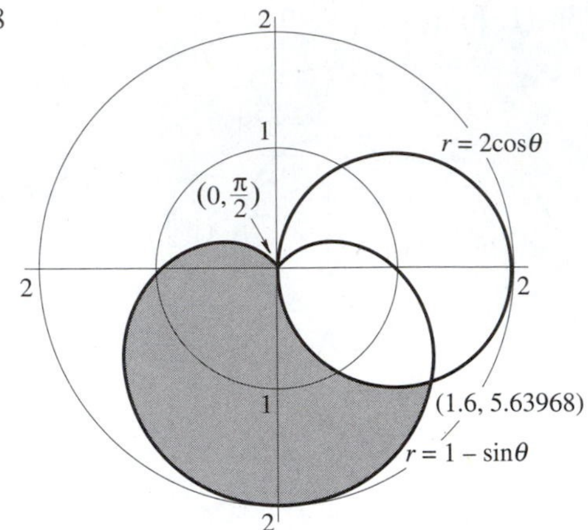

25. $1-\frac{\pi}{4} \approx 0.2146$

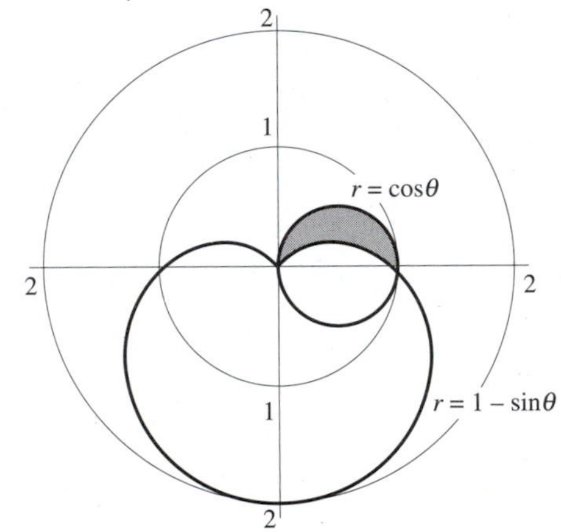

Review Exercises

1. **(a)** $\dfrac{dy}{dx} = \dfrac{3t^2+1}{2t}$; $\dfrac{d^2y}{dx^2} = \dfrac{3t^2-1}{4t^3}$; **(b)** 2

3. **(a)** $\dfrac{dy}{dx} = \dfrac{-1}{2\sin t}$; $\dfrac{d^2y}{dx^2} = \dfrac{-1}{4\sin^3 t}$; **(b)** -0.5942

5. **(a)** $-\dfrac{2\cos 2\theta + 3\cos\theta}{2\sin 2\theta + 3\sin\theta}$; **(b)** -0.5147

7. **(a)** $\dfrac{3\cos\theta\sin\theta}{\cos^2\theta - 2\sin^2\theta}$; **(b)** -3

9.

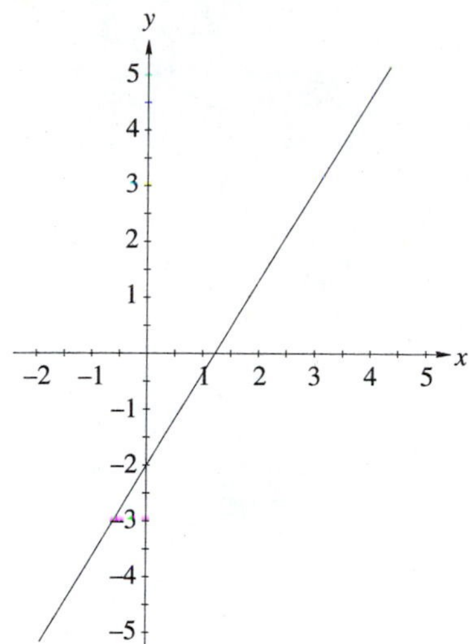

11.

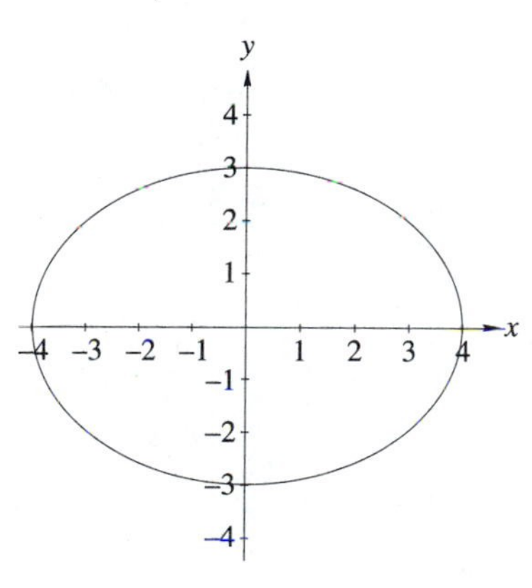

13.

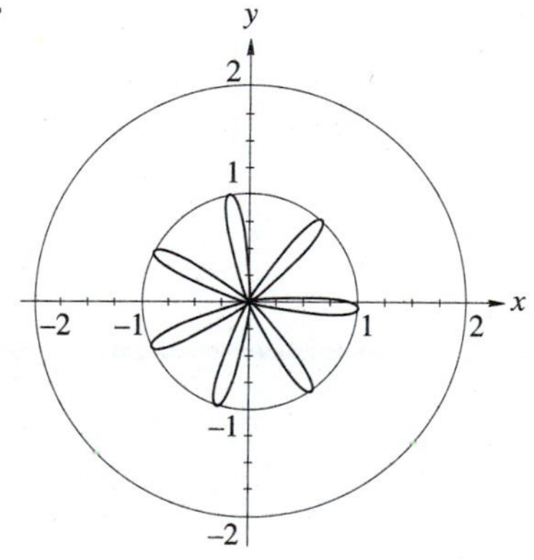

15.

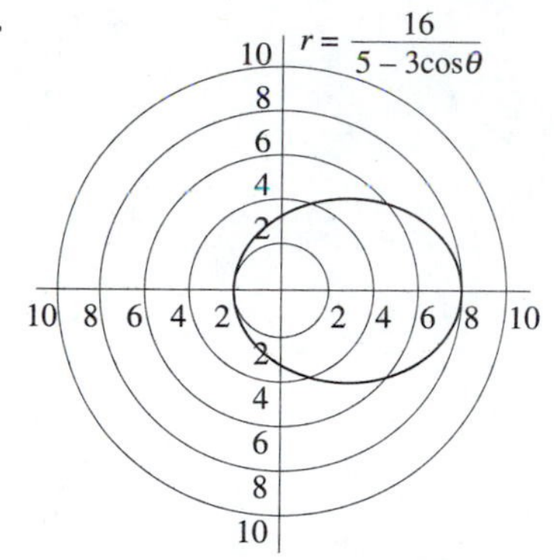

17.

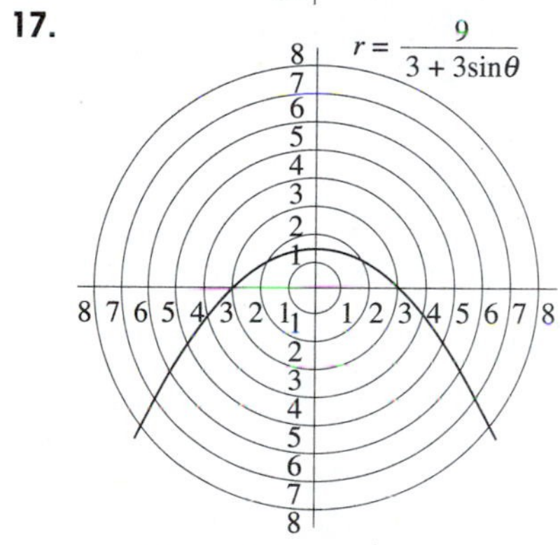

19. $\frac{1}{27}(40^{3/2} - 4^{3/2}) \approx 9.0734$ **21.** 2π **23.** 2

25. $\frac{\pi}{6}(17^{3/2} - 1) \approx 11.5155\pi \approx 36.1769$

27. $41.7485\pi \approx 131.1568$ **29.** $36\pi^2 \approx 355.3058$

31. $0, \pi$ **33.** $2\pi/3, 4\pi/3$

35. $3\pi/2$

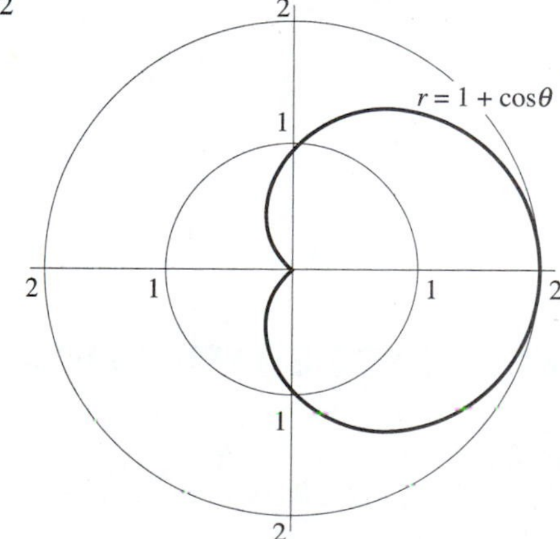

37. $4\pi + 32$

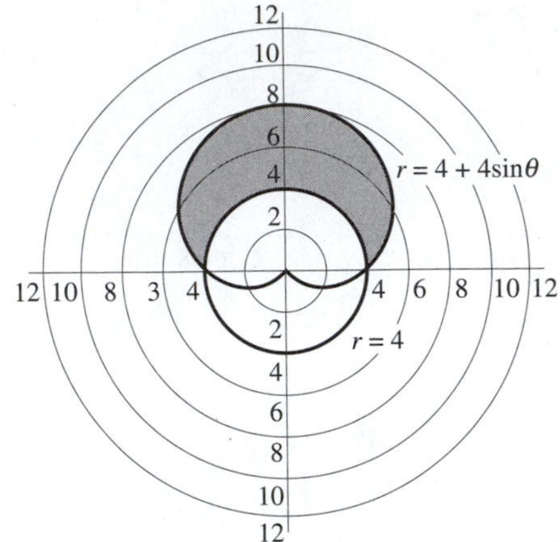

39. $1 - \pi/4 \approx 0.2146$

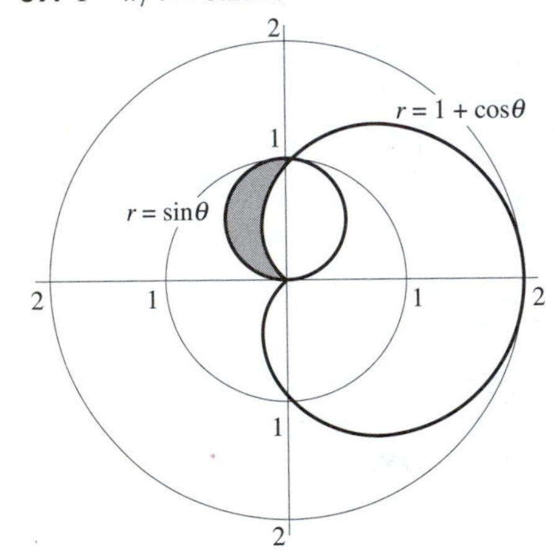

Chapter 10 Test

1.

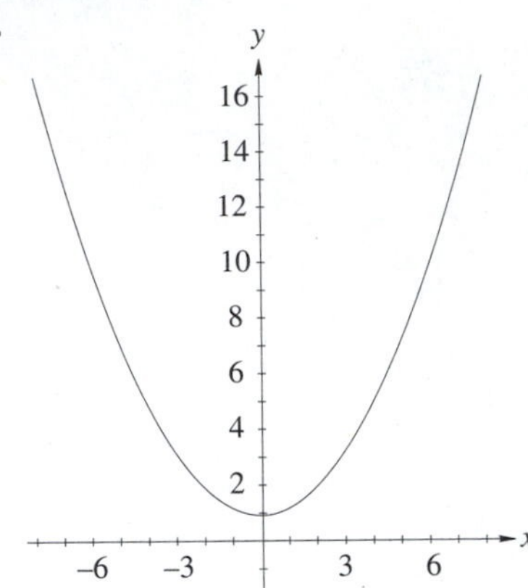

3. $\dfrac{dy}{dx} = \dfrac{3t^2}{2t+5}$; $\dfrac{d^2y}{dx^2} = \dfrac{6t(t+5)}{(2t+5)^3}$

5. $\sin\theta\cos\theta$

7. $\sqrt{5}(e^2 - 1)$

9. $(0.2, 1.3694)$ and $(0.2, -1.3694) = (0.2, -4.9137)$

11. hyperbola

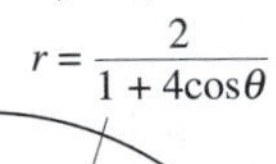

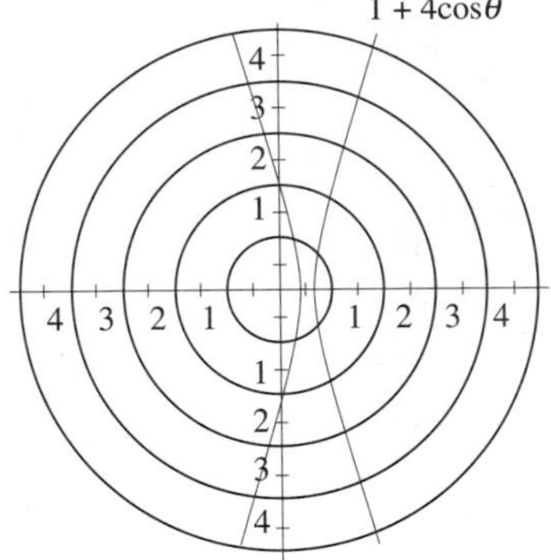

13. **(a)** $\frac{\pi}{4}$; **(b)** an ellipse

ANSWERS FOR CHAPTER 11

Exercise Set 11.1

1. **(a)** 3, **(b)** 4, **(c)** 8, **(d)** $3x + 3h + 4y - xy - hy$, **(e)** $3x + 4y + 4h - xy - xh$

3. **(a)** $\sqrt{2} - 1$, **(b)** 2, **(c)** 2, **(d)** $\sqrt{(x+h)y} - (x+h) + \dfrac{4}{y}$, **(e)** $\sqrt{x(y+h)} - x + \dfrac{4}{y+h}$

5. **(a)** 1, **(b)** 0, **(c)** 0, **(d)** 3, **(e)** all real values of (x, y) except when $x = y$

7. $V = \frac{1}{3}\pi r^2 h$

9. $A = 2\pi rh + 2\pi r^2$

11. $C = 200\pi r(2r + h)$

13. 0.000036 cm/min

15. **(a)** -11°F, **(b)** -25°F, **(c)** 10°F, **(d)** -15°F

17. **(a)** $V(x, y, z) = xyz$, **(b)** $S(x, y, z) = xy + 2xz + 3yz$

19. 202.8 ft

21. **(a)** $1.834\,\text{m}^2$, **(b)** $A(w,h) = 14.085w^{0.425}h^{0.725}$, **(c)** $2{,}847.5\,\text{in.}^2$
23. **(a)** 37,444, **(b)** 74,888

Exercise Set 11.2

1. A plane whose intercepts are the points $(6,0,0)$, $(0,3,0)$, and $(0,0,2)$

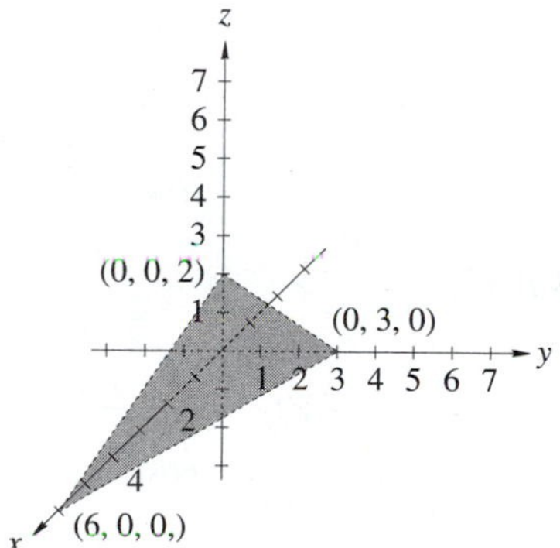

3. Parabolic cylinder

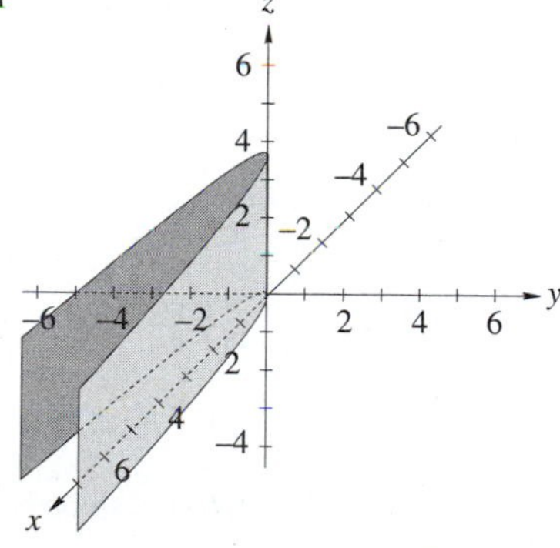

5. Ellipsoid with intercepts at $(\pm 2,0,0)$, $(0,\pm\sqrt{2},0)$, and $(0,0,\pm 2)$

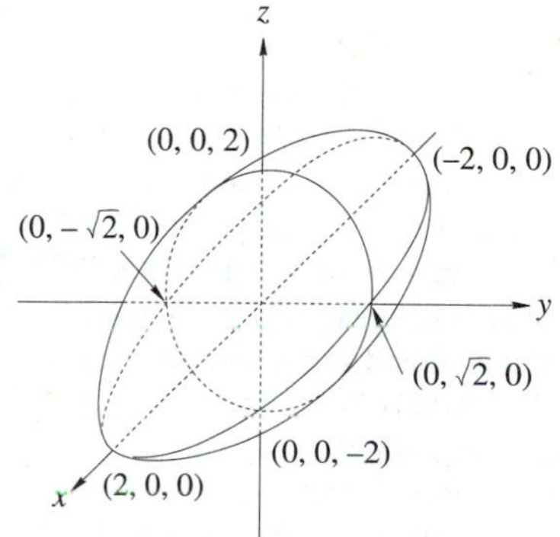

7. Ellipsoid with intercepts at $(\pm 2,0,0)$, $(0,\pm 2\sqrt{2},0)$, and $(0,0,\pm 2)$

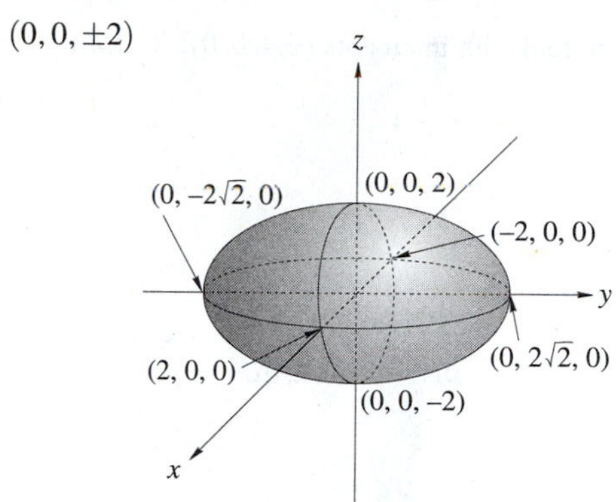

9. Hyperboloid of two sheets with intercepts at $(0,0,\pm 1)$

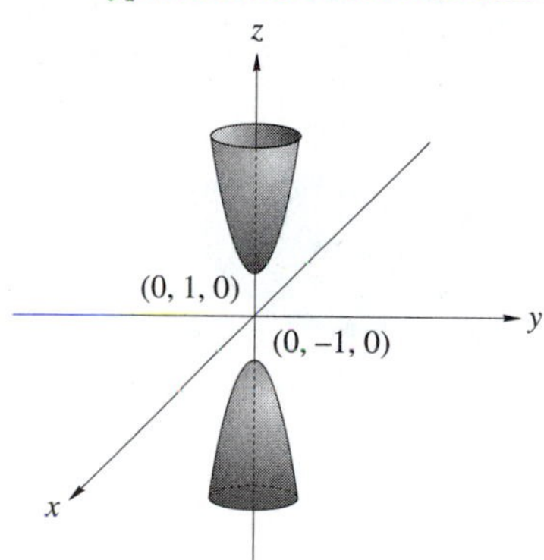

11. Plane with intercepts at $(4,0,0)$, $(0,-3,0)$ and $(0,0,-2)$

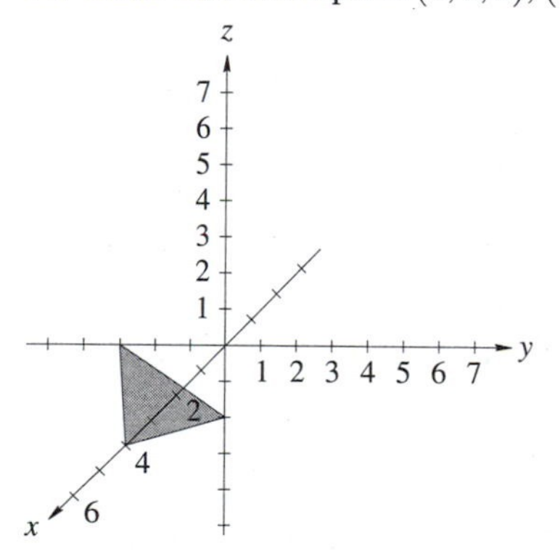

13. Elliptic paraboloid with intercepts $(\pm 4, 0, 0)$, $(0, \pm 2, 0)$, and $(0, 0, -16)$

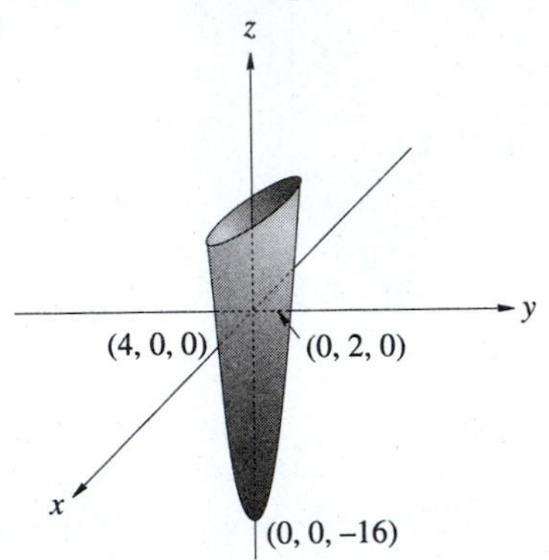

15. Elliptic paraboloid

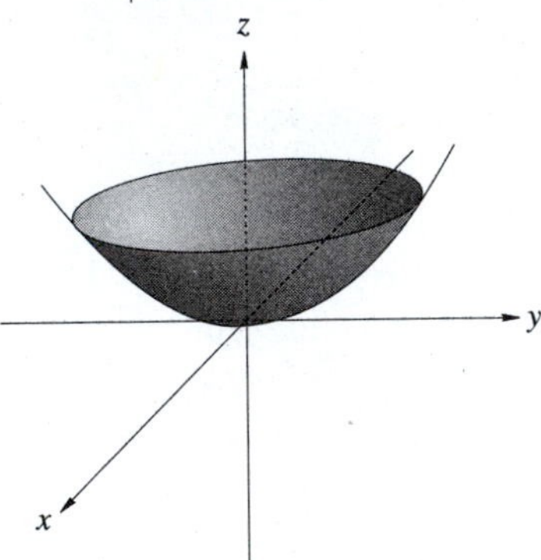

17. Hyperbolic paraboloid

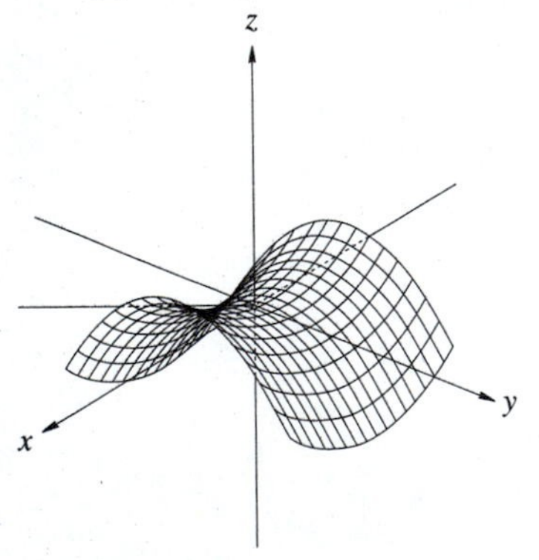

19. Parabolic cylinder

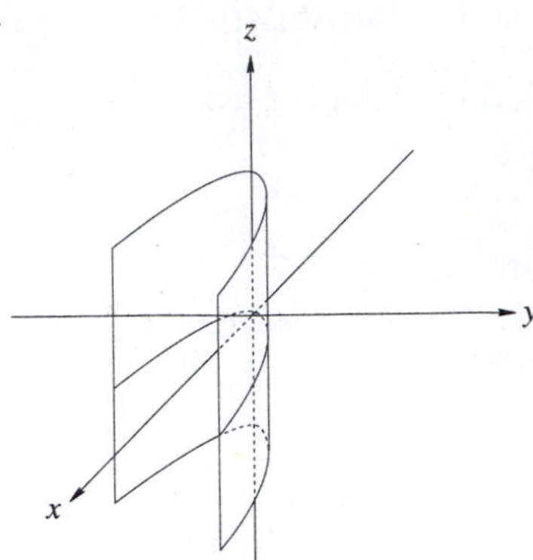

Exercise Set 11.3

1. $f_x = 2xy + y^2$; $f_y = x^2 + 2xy$ **3.** $f_x = 6x + 6y^3$; $f_y = 18xy^2$ **5.** $f_x = \dfrac{2x}{y}$; $f_y = \dfrac{y^2 - x^2}{y^2}$
7. $f_x = e^x \sin y - e^y \sin x$; $f_y = e^y \cos x + e^x \cos y$
9. $f_x = 2e^{2x+3y}$; $f_y = 3e^{2x+3y}$
11. $f_x = 2xy^3 \cos(x^2y^3)$; $f_y = 3x^2y^2 \cos(x^2y^3)$
13. $f_x = \dfrac{x}{x^2+y^2}$; $f_y = \dfrac{y}{x^2+y^2}$
15. $f_x = 6y \sin(3xy)\cos(3xy)$; $f_y = 6x \sin(3xy)\cos(3xy)$
17. $f_{xx} = 6xy^2$; $f_{xy} = f_{yx} = 6x^2y - 5y^4$; $f_{yy} = 2x^3 - 20xy^3$
19. $f_{xx} = 3y^6 e^{xy^3}$; $f_{xy} = f_{yx} = 9y^2 e^{xy^3}(1 + xy^3)$; $f_{yy} = 9xye^{xy^3}(2 + 3xy^3)$ **21.** $f_{xx} = 0$; $f_{xy} = f_{yx} = \dfrac{-10}{y^6}$; $f_{yy} = \dfrac{60x}{y^7}$ **23.** $f_{xx} = \dfrac{-3}{x^2}$; $f_{xy} = f_{yx} = 0$; $f_{yy} = \dfrac{-5}{y^2}$
25. $\dfrac{-5900R}{9}$ kg/m^2 **27.** $-984.72\,\text{cm}^3$/atm
29. $\dfrac{dz}{dy} = -2y$. At $(1, 3, 6)$ the slope is -6. **31.** **(a)** $\dfrac{k}{r^4}$, **(b)** $-\dfrac{4kL}{r^5}$ **33.** **(a)** $A_w(w, h) = 5.9861w^{-0.575}h^{0.725}$ in.^{2}/lb and $A_h(w, h) = 10.2116w^{0.425}h^{-0.275}$ in.^{2}/in.,
(b) $A_w(185, 71) = 6.54$ in.^{2}/lb; $A_h(185, 71) = 29.08$ in.^{2}/in.,
(c) $A_w(185, 71) = 6.54$ in.^{2}/lb means that for a 185 lb person 5′11″ tall, the rate of change in surface area is 6.54 in.^{2} for each pound gained in weight if the height is fixed; $A_h(185, 71) = 29.07$ in.^{2}/in. means that for a 185 lb person 5′11″ tall, the rate of change in surface area is 29.07 in.^{2} for each inch gained in height if the weight remains the same.
35. **(a)** $S_w(w, r) = 0.000\,02r^2$ ft/lb,
(b) $S_w(2400, 65) = 0.0845$ ft/lb,
(c) $S_r(w, r) = 0.000\,04wr$ ft/mph,
(d) $S_r(2400, 65) = 6.24$ ft/mph, **(e)** about 6.24 ft

Exercise Set 11.4

1. $2x\,dx + 2y\,dy$ **3.** $(6x+4y)\,dx + (4x-6y^2)\,dy$
5. $\dfrac{-y}{x^2+y^2}\,dx + \dfrac{x}{x^2+y^2}\,dy$
7. $(\tan yx + yx\sec^2 yx)\,dx + x^2\sec^2 yx\,dy$
9. $2te^{2t}(1+t) + 2t^3e^{2t}(2+t) = 2te^{2t}(1+t+2t^2+t^3)$
11. $\dfrac{e^{\sqrt{t}}}{2\sqrt{t}}\sin\pi t + \pi e^{\sqrt{t}}\cos\pi t$ **13.** minimum at $(-1/2, -1)$
15. minimum at $(2, -3)$ **17.** saddle point at $(0, 0)$
19. saddle point at $\left(3, -\frac{9}{2}\right)$; test fails at $(0, 0)$ **21.** $3.5\,\text{cm}^2$
23. $0.8\,\Omega$ **25.** $\frac{1}{k}(V\cdot\frac{dP}{dt} + P\cdot\frac{dV}{dt})$ **27.** -0.29
29. $h = 5\,\text{cm}$, $l = w = 10\,\text{cm}$ **31.** **(a)** $h = 0.89\,\text{ft}$, $l = w = 2.13\,\text{ft}$; **(b)** \$16.94 **33.** $x = \frac{2}{3}$, $y = \frac{1}{6}$ produces a reaction of $\frac{10}{9}$ **35.** $E_x = \dfrac{Qx}{4\pi\epsilon_0(x^2+a^2)^{3/2}}$

Exercise Set 11.5

1. $\frac{1}{12}$ **3.** 4 **5.** 4 **7.** $\frac{3}{20}$ **9.** $\frac{-4}{105}$ **11.** 6
13. $1/30$ **15.** $\frac{5}{24}$ **17.** $\frac{1}{6}$ **19.** -1876.7473
21. $114\frac{1}{3}$ **23.** $\frac{31}{120} \approx 0.258$ **25.** 121.5 **27.** $\frac{13}{3}$
29. $\frac{1}{2}\ln\frac{7}{2} \approx 0.6264$ **31.** $5\frac{1}{3}$ **33.** $\frac{4}{3}\sqrt{2} \approx 1.8856$
35. \$141,400 **37.** **(a)** $C(x, y) = 100 - 15(x^2+y^2)$, **(b)** 75 ppm

Exercise Set 11.6

1. $(\sqrt{2}, \sqrt{2}, 2)$ **3.** $(2, 0, 4)$
5. $(-03.5355, -3.5355, 0)$
7. $(2.8284, 0.7854, 5) = (2.8284, \pi/4, 5)$
9. $(5, 2.4981, 2)$ **11.** $(13, 5.8884, -3)$
13. $(\sqrt{2}, \sqrt{2}, 3.4641)$ **15.** $(0, -2.5981, 1.5)$
17. $(-3.75, -2.1651, -2.5)$
19. $(5, 0.6435, 1.5708) = (5, 0.6435, \pi/2)$
21. $(3, 0.4636, 2.3005)$
23. $(2, 0.7854, 0.7854) = (2, \frac{\pi}{4}, \frac{\pi}{4})$ **25.** The graph is a line that makes an angle of $\frac{\pi}{4}$ with the z-axis and whose image on the xy-plane make an angle of $\frac{\pi}{4}$ with the x-axis.

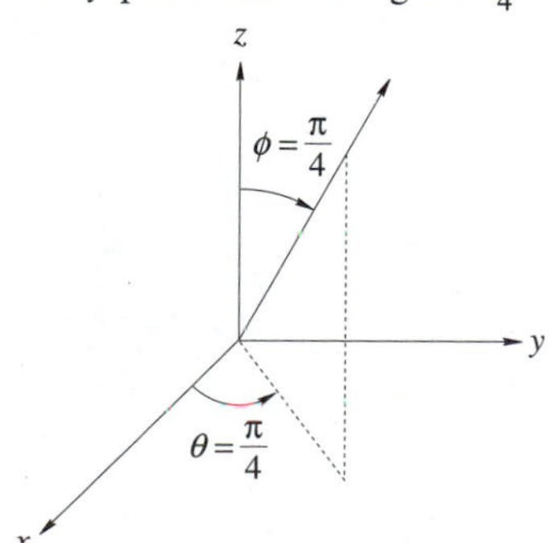

27. The region between the two spheres $\rho = 3$ and $\rho = 5$.

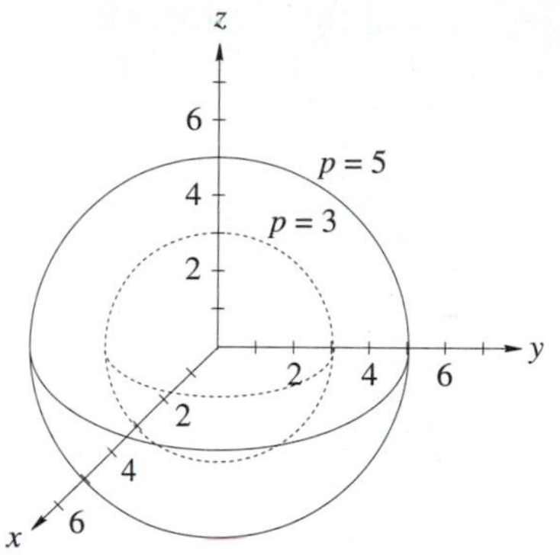

29. A cone shaped figure with spherical base.

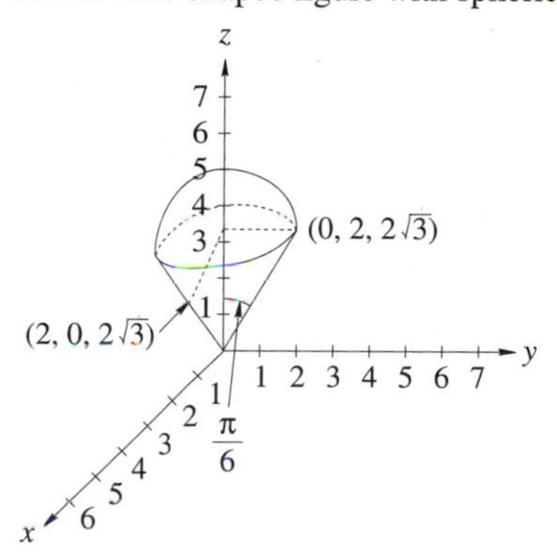

Exercise Set 11.7

1. $\bar{x} = 1.75$, $\bar{y} = 3.25$ **3.** $\bar{x} = \frac{32}{7} \approx 4.5714$, $\bar{y} = 0.8$
5. $\bar{x} = \frac{253}{195} \approx 2.6632$, $\bar{y} = \frac{195}{152} \approx 1.2829$ **7.** $\bar{x} = 1.25$, $\bar{y} = 3.4$ **9.** $\bar{x} = \frac{10}{21} \approx 0.4762$, $\bar{y} = \frac{5}{12} \approx 0.4167$
11. $\bar{x} = -0.5$, $\bar{y} = 4.4$ **13.** $r_y = \sqrt{2/5} \approx 0.6325$; $I_y = \dfrac{8\sqrt{2}}{15} \approx 0.7542$ **15.** $r_x = \sqrt{25/3} \approx 2.8868$; $I_x = 125$
17. $r_x = 1.9124$; $I_x = \frac{4096}{105}$ **19.** $r_y = 1.6928$; $I_y = 4.7$
21. **(a)** $\dfrac{296\sqrt{2}}{105}\rho$, **(b)** $\sqrt{74/35} \approx 1.45$
23. **(a)** $\dfrac{9472}{105}\rho \approx 90.21\rho$, **(b)** $\sqrt{\frac{296}{35}} \approx 2.91$

Review Exercises

1. A plane whose intercepts are $(6,0,0)$, $(0,2,0)$, and $(0,0,3)$

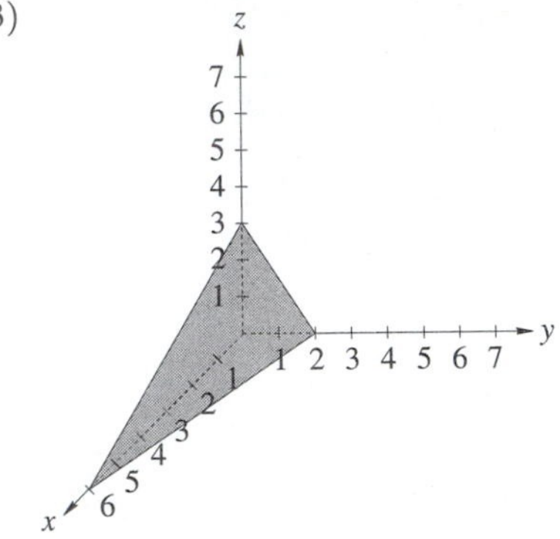

3. An elliptical cylinder whose axis is the x-axis

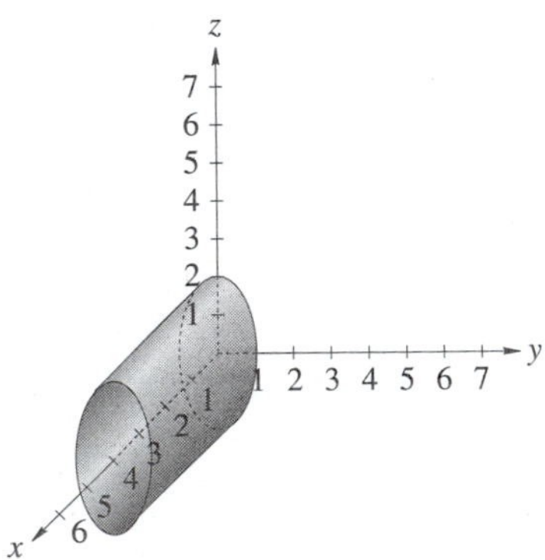

5. A parabola whose axis is the y-axis and has the trace in the xy-plane of $y = 4x^2$.

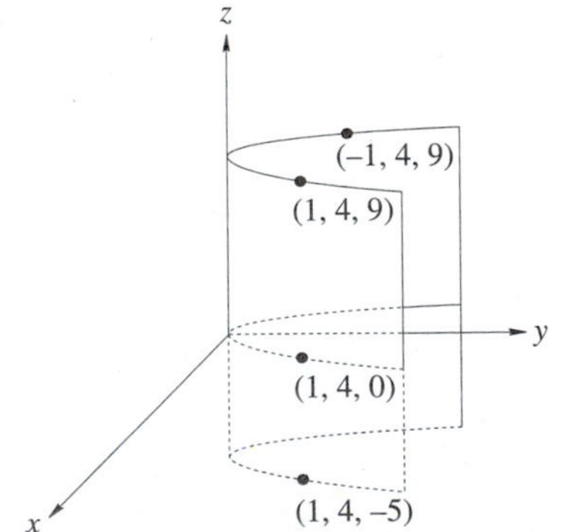

7. Hyperboloid of one-sheet. The trace in th xy-plane is the ellipse $\dfrac{x^2}{16} + \dfrac{y^2}{4} = 1$. The trace in the yz-plane is the hyperbola $\dfrac{y^2}{4} - \dfrac{z^2}{16} = 1$ and the trace in the xz-plane is the hyperbola $\dfrac{x^2}{16} - \dfrac{z^2}{16} = 1$.

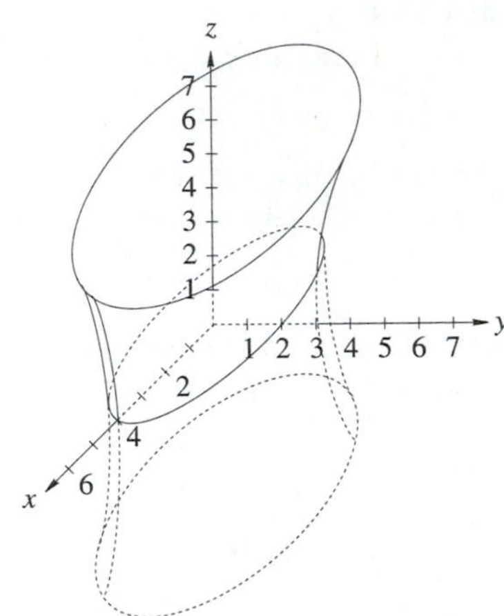

9. $\dfrac{\partial z}{\partial x} = 6x + 6y^3$, $\dfrac{\partial z}{\partial y} = 18xy^2$, $\dfrac{\partial^2 z}{\partial x^2} = 6$, $\dfrac{\partial^2 z}{\partial y^2} = 36xy$, $\dfrac{\partial^2 z}{\partial x\,\partial y} 18y^2$ **11.** $\dfrac{\partial z}{\partial x} = e^x \cos y - e^y \cos x$, $\dfrac{\partial z}{\partial y} = -e^x \sin y - e^y \sin x$, $\dfrac{\partial^2 z}{\partial x^2} = e^x \cos y + e^y \sin x$, $\dfrac{\partial^2}{\partial y^2} = -e^x \cos y - e^y \sin x, = -e^x \sin y - e^y \cos x$ **13.** 16 **15.** $\frac{4}{15}(5^{5/2} - 33) \approx 6.1071$ **17.** 148.5 **19.** rectangular: $(3.4641, 2, 2)$, spherical: $(4.4721, \pi/6, 1.1071)$ **21.** rectangular: $(-\sqrt{2}, \sqrt{2}, 2\sqrt{3})$, cylindrical: $(2, 3\pi/4, 3.4641)$ **23.** $\frac{128}{15} \approx 8.53$ **25.** $l = \omega = \sqrt[3]{600} \approx 8.4343$ in, $h = \frac{1}{2}\sqrt[3]{600} \approx 4.2172$ in **27.** 0.19Ω **29.** $R - 100\,\text{cm}^3/\text{min}$ **31.** $\bar{x} = \frac{248}{72} \approx 3.4444$, $\bar{y} = \frac{744}{72} = \frac{31}{3} \approx 10.3333$ **33.** $\bar{x} = \bar{y} = \frac{16}{35} \approx 0.4571$ **35.** $I_x = 11,250$, $r_x \approx \sqrt{150} \approx 12.2474$ **37.** $I_x = \frac{1}{9}$, $r_x = \sqrt{5/27} \approx 0.4303$ **39. (a)** $\dfrac{1{,}054{,}720}{3}\rho$, **(b)** $\sqrt{\dfrac{3{,}296}{3}} \approx 33.15$ **41.** 16

Chapter 11 Test

1. A plane whose intercepts are $(4,0,0)$, $(0,-8,0)$, and $(0,0,2)$

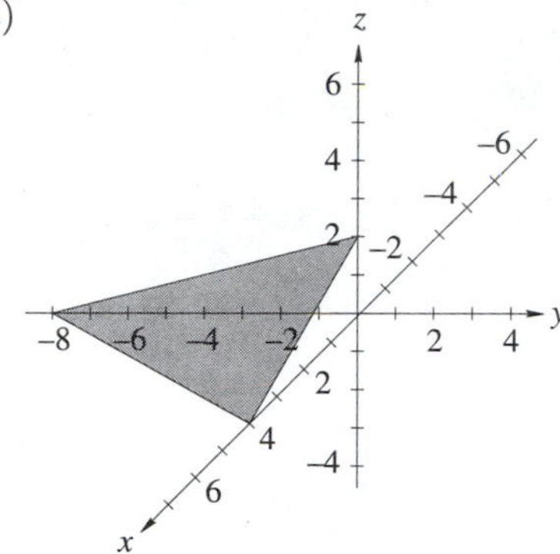

3. Hyperboloid of one-sheet. The trace in the xy-plane is the ellipse $\frac{x^2}{4}+\frac{y^2}{9}=1$. The trace in the yz-plane is the hyperbola $\frac{y^2}{9}-\frac{z^2}{4}=1$ and the trace in the xz-plane is the hyperbola

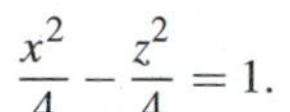

$\frac{x^2}{4}-\frac{z^2}{4}=1.$

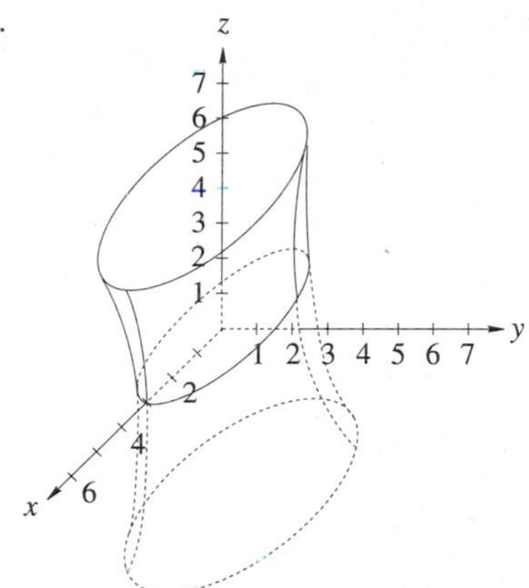

5. $\frac{\partial z}{\partial x}=12x^2-10xy^4$; $\frac{\partial z}{\partial y}=-20x^2y^3$
7. $8-4\sqrt{2}\approx 2.3431$ **9.** Rectangular coordinates: $\left(-\frac{3\sqrt{3}}{2},\frac{3}{2},8\right)$; Spherical coordinates: $\left(\sqrt{73},\frac{5\pi}{6},0.3588\right)$
11. 4π **13.** 16

ANSWERS FOR CHAPTER 12

Exercise Set 12.1

1. $\frac{1}{2};\frac{1}{3};\frac{1}{4};\frac{1}{5};\frac{1}{6};\frac{1}{7}$ **3.** $4;9;16;25;36;49$
5. $-1;2;-3;4;-5;6$ **7.** $\frac{1}{3};\frac{3}{5};\frac{5}{7};\frac{7}{9};\frac{9}{11};\frac{11}{13}$ **9.** $1;\frac{2}{3};\frac{4}{9};\frac{8}{27};\frac{16}{81};\frac{32}{243}$ **11.** $0;\frac{1}{9};\frac{1}{4};\frac{9}{25};\frac{4}{9};\frac{25}{49}$
13. $1;2;6;24;120;720$ **15.** $5;8;11;14;17;20$
17. $2;2;4;64;16{,}777{,}216;1.329228\times 10^{36}$
19. $1;2;9;262{,}144;5^{262{,}144};6^{(5^{262{,}144})}$ **21.** $1;\frac{1}{2};\frac{1}{2};\frac{1}{4};\frac{1}{8};\frac{1}{32}$ **23.** $0;2;2;0;-2;-2$ **25.** 18.75 ft or 18 ft 9 in
27. (a) \$30,400; (b) \$35,900

Exercise Set 12.2

1. $\frac{3}{2}+\frac{3}{4}+\frac{3}{8}+\frac{3}{16}$ **3.** $1-\frac{3}{2}+3-\frac{27}{4}$ **5.** 0
7. 91 **9.** 225 **11.** $\frac{1{,}593{,}342}{67{,}925}\approx 23.457$
13. Arithmetic; $S_{10}=165$ **15.** Geometric; $S_8=255$
17. Arithmetic; $S_{12}=192$ **19.** Arithmetic; $S_{20}=57.5$
21. Geometric; $S_{16}=0.5625$ **23.** Arithmetic; $S_{14}=114.8$
25. Geometric; $S_{15}=17920.253$ **27.** Geometric; $S_8=9840$ **29.** 392.32m when it hits the fourth time, 634.10m when it hits the tenth time. **31.** 104 ft
33. \$2540.47;\$3.99 **35.** (a) \$3,662,000; (b) \$7,682,000

Exercise Set 12.3

1. $x-\frac{x^3}{3!}+\frac{x^5}{5!}-\frac{x^7}{7!}$ **3.** $1+\frac{x^x}{2!}+\frac{x^4}{4!}+\frac{x^6}{6!}$
5. $1+3x+\frac{9}{2}x^2+\frac{27}{3!}x^3=1+3x+\frac{9}{2}x^2+\frac{9}{2}x^3$
7. $x^2-\frac{1}{2}x^4+\frac{1}{3}x^6-\frac{1}{4}x^8$
9. $1-\frac{12}{4!}x^4+\frac{1680}{8!}x^8-\frac{665280}{12!}=1-\frac{1}{2}x^4+\frac{1}{4!}x^8-\frac{1}{6!}x^{12}$
11. $x+x^2+\frac{1}{3}x^3-\frac{4}{5!}x^5$
13. $\frac{2}{2!}x^2-\frac{24}{4!}x^4+\frac{360}{6!}x^6-\frac{6720}{8!}x^8=x^2(1-x^2+\frac{1}{2}x^4-\frac{1}{6}x^6)$
15. $\frac{6}{2!}x^2-\frac{108}{4!}x^4+\frac{1458}{6!}x^6-\frac{17496}{8!}x^8=3x^2-\frac{9}{2}x^4+\frac{81}{40}x^5-\frac{243}{560}x^8$ **17.** $10\frac{4}{15}\approx 10.67\,\text{cm}^2$
19. $E(x)=1+\frac{1}{2}x^2+\frac{5}{24}x^4$

Exercise Set 12.4

1. $1+3x+\frac{9}{2}x^2+\frac{9}{2}x^3$ **3.** $1-\frac{x^2}{8}+\frac{x^4}{384}-\frac{x^6}{46{,}080}$
5. $1-\frac{x^6}{2!}+\frac{x^{12}}{4!}-\frac{x^{18}}{6!}$
7. $2x^2-\frac{8x^6}{3!}+\frac{32x^{10}}{5!}-\frac{128x^{14}}{7!}$
13. $\frac{1}{3}-\frac{1}{42}+\frac{1}{1320}\approx 0.3103$
15. $\frac{2}{3}(0.2)^{3/2}-\frac{2}{5\cdot 3!}(0.2)^{5/2}+\frac{2}{7\cdot 5!}(0.2)^{7/2}\approx 0.05844$
17. $1-x+\frac{2x^3}{3!}-\frac{4x^4}{4!}+\frac{x^5}{5!}+\cdots$
19. $x^2-x^4+\frac{x^6}{2!}-\frac{x^8}{3!}+\frac{x^{10}}{4!}-\frac{x^{12}}{5!}+\cdots$

21. $3.4835+3.5868j$ **23.** $6e^{4\pi j/3}$ **25.** 1.4519
27. $\frac{1}{3}(0.2)^3+\frac{1}{4}(0.2)^4+\frac{1}{5}(0.2)^5 \approx 0.0031$ **29.** 14.9251
31. **(a)** $t^2-\frac{t^6}{3!}+\frac{t^{10}}{5!}-\frac{t^{14}}{7!}+\frac{t^{18}}{9!}-\cdots$; **(b)** 0.0000027
33. **(a)** $P=P_0\left(2-3t+\frac{5}{2}t^2-\frac{3}{2}t^3\right)$,
(b) $\frac{dP}{dt}=P_0\left(-3+5t-\frac{9}{2}t^2\right)$, **(c)** $-4.82P_0$ ppm/week

Exercise Set 12.5

1. 0.7408 **3.** 0.9950 **5.** 0.0872 **7.** -0.0305
9. -0.6931 **11.** 0.0619 **13.** -0.0813 **15.** 0.2105
17. **(b)** For an accuracy of 0.0001, $x<\sqrt[3]{0.0006}\approx 0.0843$
19. $0.0021\,\mu\text{C}$ **21.** **(a)** $V=0.75e^{0.75t}=$
$0.75\left(1+0.75+\frac{(0.75)^2}{2!}+\frac{(0.75)^3}{3!}+\cdots\right)$; **(b)** 1.05105
23. **(a)** $e^{-t^2}\approx 1-t^2+\frac{1}{2}t^4-\frac{1}{6}t^6$
(b) $V(x)=V_0\int_0^x e^{-t^2}\,dt\approx V_0\left(x-\frac{1}{3}x^3+\frac{1}{10}x^5-\frac{1}{42}x^7\right)$
(c) $0.7722V_0$ **(d)** $0.0957V_0$

Exercise Set 12.6

1. $e^x=e^2(x-1+\frac{(x-2)^2}{2!}+\frac{(x-2)^3}{3!}+\frac{(x-2)^4}{4!}+\cdots)$
3. $\sin x=$
$\frac{\sqrt{2}}{2}[1+(x-\pi/4)-\frac{1}{2!}(x-\pi/4)^2-\frac{1}{3!}(x-\pi/4)^3+\cdots]$
5. $\frac{1}{x}=\frac{1}{2}(1-\frac{1}{2}(x-2)+\frac{1}{2\cdot 2!}(x-2)^2-\frac{1}{8}(x-2)^3+\cdots)$
7. $(x+1)^{-2}=1-2x+\frac{6x^2}{2!}-\frac{24x^3}{3!}+\cdots=$
$1-2x+3x^2-4x^3+\cdots$
9. $e^{-x}=e^{-1}[1-(x-1)+\frac{1}{2!}(x-1)^2-\frac{1}{3!}(x-1)^3+\cdots]$
11. $\sqrt[3]{x}=$
$2+\frac{1}{12}(x-8)-\frac{1}{9\cdot 16\cdot 2!}(x-8)^2+\frac{10}{27\cdot 2^8\cdot 3!}(x-8)^3+\cdots$
13. $e^2(1.1052)\approx 8.1662$ **15.** 0.6820 **17.** 0.5551
19. 0.7972 **21.** 0.4493 **23.** 1.9832 **25.** 0.14715

Exercise Set 12.7

1. $f(x)=\frac{-8}{\pi}(\sin x+\frac{1}{3}\sin 3x+\frac{1}{5}\sin 5x+\frac{1}{7}\sin 7x+\cdots)=$
$\frac{-8}{\pi}\sum_{k=1}^{\infty}\frac{1}{2k-1}\sin(2k-1)x$

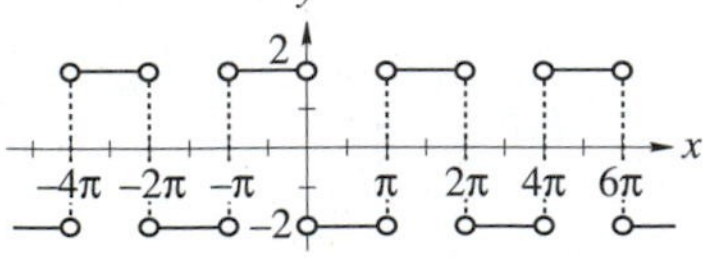

3. $h(x)=1-\frac{2}{\pi}\sin\frac{\pi x}{3}+\frac{1}{\pi}\sin\frac{2\pi x}{3}-\frac{2}{3\pi}\sin\frac{3\pi x}{3}+$
$\frac{2}{4\pi}\sin\frac{4\pi x}{3}-\cdots=1+\sum_{k=1}^{\infty}(-1)^k\frac{2}{k\pi}\sin\frac{k\pi x}{3}$

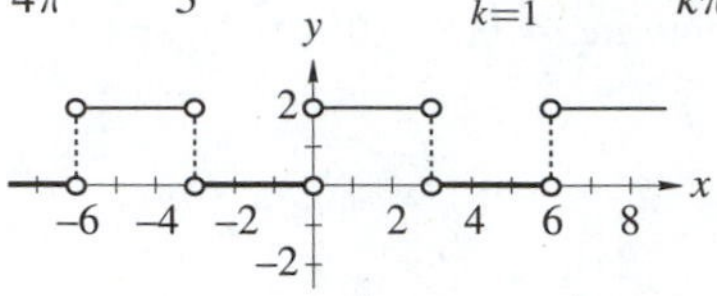

5. $f(x)=\frac{\pi}{2}-\frac{4}{\pi}\cos x-\frac{4}{9\pi}\cos 3x-\frac{4}{25\pi}\cos 5x-\cdots=$
$\frac{\pi}{2}-\frac{4}{\pi}\sum_{k=1}^{\infty}\frac{1}{(2k-1)^2}\cos(2k-1)x$

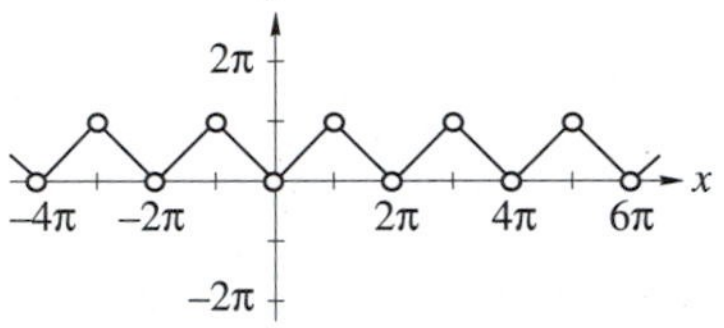

7. $h(x)=2\sin x-\sin 2x+\frac{2}{3}\sin 3x-\frac{1}{2}\sin 4x+\cdots=$
$2\sum_{k=1}^{\infty}\frac{(-1)^{k+1}}{k}\sin kx$

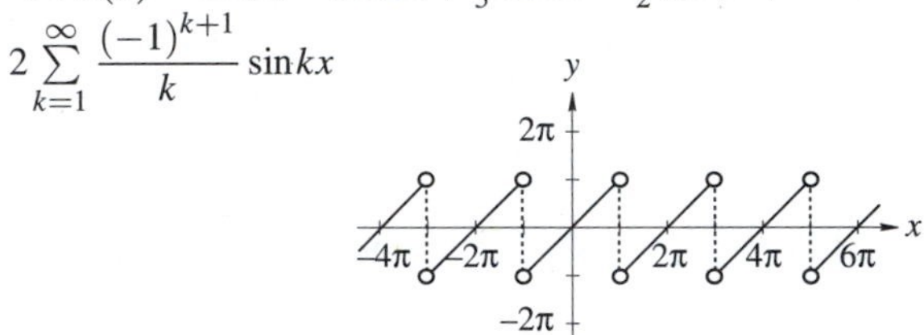

9. $f(x)=\frac{3}{\pi}(\sin x+\frac{1}{2}\sin 2x+\frac{1}{4}\sin 4x+\frac{1}{5}\sin 5x+\cdots)=$
$\frac{3}{\pi}\sum_{k=1}^{\infty}(\frac{\sin(3k-2)x}{3k-2}+\frac{\sin(3k-1)x}{3k-1})$

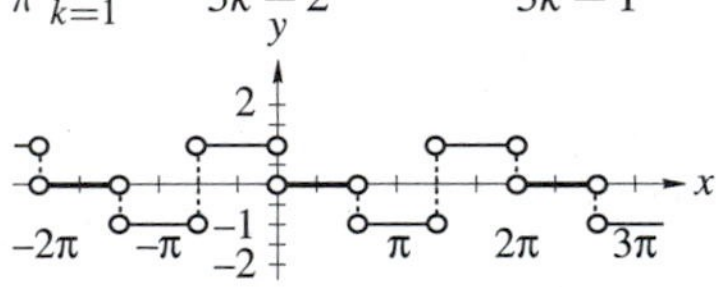

11. $h(x)=\frac{1}{\pi}+\frac{1}{2}\sin x-\frac{2}{\pi}(\frac{1}{3}\cos 2x+\frac{1}{15}\cos 4x+$
$\frac{1}{35}\cos 6x+\cdots)=\frac{1}{\pi}+\frac{1}{2}\sin x-\frac{2}{\pi}\sum_{k=1}^{\infty}(\frac{1}{4k^2-1}\cos 2kx)$

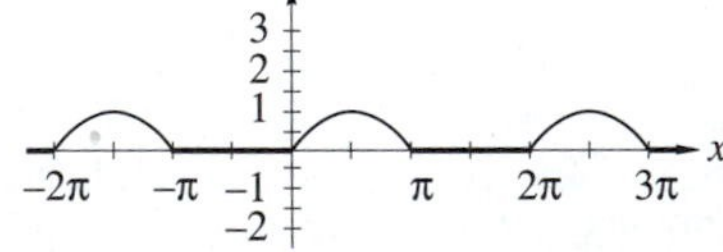

Review Exercises

1. $\frac{x}{2!}-\frac{x^3}{4!}+\frac{x^5}{6!}-\frac{x^7}{8!}+\frac{x^9}{10!}-\frac{x^{11}}{12!}+\cdots$
3. $x-\frac{5x^3}{3!}+\frac{16x^5}{5!}-\frac{64x^7}{7!}+\frac{247x^9}{9!}$ **5.** $1+\frac{3}{4}(x'+1)-\frac{s(x'+1)^2}{16\cdot 2!}+\frac{15(x'+1)^3}{64\cdot 3!}-\frac{135(x'+1)^4}{256\cdot 4!}+\cdots$

7. $x^2 - \frac{x^6}{3!} + \frac{x^{10}}{5!} - \frac{x^{14}}{7!} + \frac{x^{18}}{9!} - \cdots$ **9.** 0.6015 **11.** 0.7788 **13.** $\frac{\sqrt{3}}{2} - \frac{1}{2}(x - \frac{\pi}{6}) - \frac{\sqrt{3}}{2}\frac{(x-\pi/6)^2}{2!} + \frac{1}{2}\frac{(x-\pi/6)^3}{3!} + \frac{\sqrt{3}}{2}\frac{(x-\pi/6)^4}{4!} - \cdots$ **15.** $3 + \frac{1}{6}(x-9) - \frac{1}{4\cdot 3^3}\frac{(x-9)^2}{2!} + \frac{3}{8\cdot 3^5}\frac{(x-9)^3}{3!} - \frac{15}{16\cdot 3^7}\frac{(x-9)^4}{4!} + \cdots$
17. 0.8688 **19.** 3.0137
21. $\frac{1}{2} - \frac{2}{\pi}(\sin x + \frac{1}{3}\sin 3x + \frac{1}{5}\sin 5x + \cdots) = \frac{1}{2} - \frac{2}{\pi}\sum_{k=1}^{\infty}\frac{1}{2k-1}\sin(2k-1)x$

23. $\frac{\pi^2}{3} - 4\cos x + \cos 2x$-

$\frac{4}{9}\cos 3x + \frac{1}{4}\cos 4x - \frac{4}{25}\cos 5x + \cdots = \frac{\pi^2}{3} + 4\sum_{k=1}^{\infty}\frac{(-1)^k}{k^2}\cos kx$

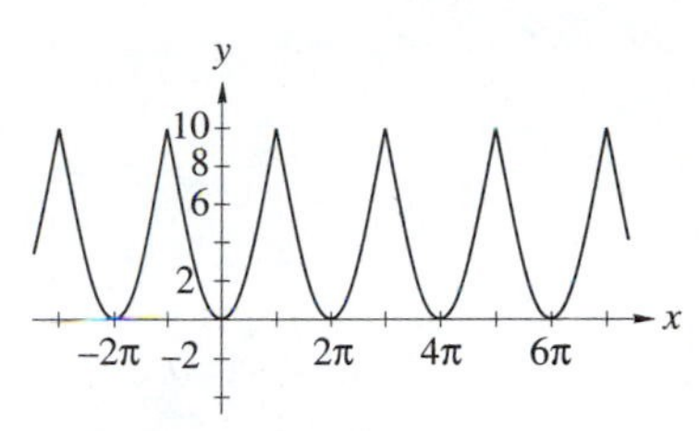

Chapter 12 Test

1. $\sin x^2 = x^2 - \frac{x^6}{3!} + \frac{x^{10}}{5!} + \cdots$
3. $\frac{1}{x} = \frac{1}{2} - \frac{x-2}{4} + \frac{(x-2)^2}{16} + \cdots$
5. $\sqrt[3]{x} = -1 + \frac{x+1}{3} + \frac{(x+1)^2}{9} + \frac{5(x+1)^3}{81} + \cdots$
7. $f(x) = \pi - 2\left(\sin x + \frac{\sin 2x}{2} + \frac{\sin 3x}{3}\cdots\right)$

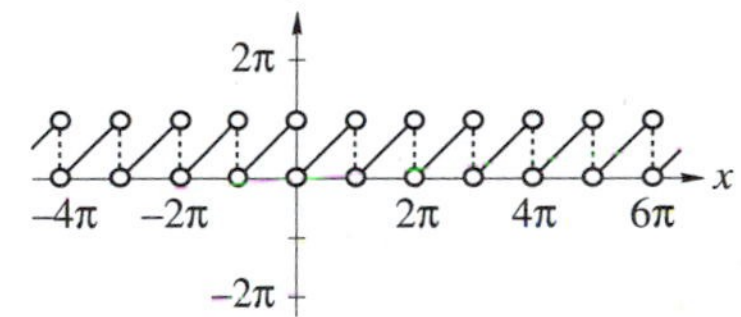

9. 18.1818

ANSWERS FOR CHAPTER 13

Exercise Set 13.1

17. $y = \frac{1}{2}x^4 - 4x + 6\frac{1}{2}$ **19.** $y = \tan x - 7$ **21.** **(a)** about $t = 3.178$ s;**(b)** 151.474 ft **23.** **(a)** 3 s;**(b)** $5\frac{1}{3}$ mph
25. **(a)** $t = 6$ s;**(b)** 128 ft/s downward
27. **(a)** $p(t) = -75t + 500$; **(b)** $t = 6\frac{2}{3}$ h **29.** 25 h
31. $L\left(-\frac{kR}{L}e^{-Rt/L}\right) + R\left(\frac{V}{R} + ke^{-Rt/L}\right) = V$

Exercise Set 13.2

1. $\frac{1}{2}y^2 - x^2 = C$ **3.** $\frac{5}{2}x^2 + y^3 = C$
5. $5x + 3\ln y = C$ **7.** $-4x^{-1} + y^{-3} = C$
9. $\frac{1}{4}x^4 + \ln\sqrt{y^2+4} = C$ **11.** $\ln\sqrt{y^2+4} = -e^{-x^2} + C$
13. $\frac{2}{3}e^{3x} + \ln y = C$ **15.** $\left(y^2 - 4\right)\sin x = C$
17. $\frac{1}{2}y^2 + 3\ln y = \frac{1}{3}x^3 - x + C$ **19.** $-\frac{1}{y} + \frac{1}{3}e^{\sin 3x} = C$
21. $-\frac{1}{y} + \frac{1}{4}x^4 = 5$ **23.** $\frac{3+y^2}{1+x^2} = 6$
25. $\ln y = x^3 - 2x + 1$
27. **(a)** $T(t) = 325 - 315e^{-0.231112t}$, **(b)** 2.42 hr ≈ 2 hr 25 min
29. 9:24 A.M. **31.** **(a)** If i represents the number of people who are infected, then $\frac{di}{dt} = ki(500{,}000 - i)$, **(b)** $i(t) = \frac{500{,}000}{1 + Ce^{-Bt}}$, **(c)** $i(t) = \frac{500{,}000}{1 + 4{,}999e^{-0.201620t}}$, **(d)** 5,578 people

Exercise Set 13.3

1. No, $\frac{\partial(x^2 - y)}{\partial y} = -1$, $\frac{\partial x}{\partial x} = 1$ **3.** Yes, $\frac{\partial(x^2+y^2)}{\partial y} = 2y = \frac{\partial 2xy}{\partial x}$ **5.** Yes, $\frac{\partial(x + y\cos x)}{\partial y} = \cos x = \frac{\partial \sin x}{\partial x}$ **7.** Yes, $\frac{\partial(2xy + x)}{\partial y} = 2x = \frac{\partial(y + x^2)}{\partial x}$ **9.** No, $\frac{\partial(x-y)}{\partial y} = -1$, $\frac{\partial(y+x)}{\partial x} = 1$ **11.** $\ln\frac{y}{x} = C_1$ or $\frac{y}{x} = C_2$ where $C_2 = e^{C_1}$
13. $\frac{x}{y} - \frac{5}{3}y^3 = C$ **15.** $\arctan\frac{y}{x} - x = C$
17. $\ln xy - x^3 = C$ **19.** $\ln xy = C$ **21.** $x^2y - x^3 = C$
23. $-x^3 + \arctan\frac{y}{x} = C$ **25.** $y = \ln|\sin(x^2+y^2)| + C$

27. $x^2y^2 = e^{x^3} + 9 - e$ **29.** $\arctan\frac{y}{x} = x - \frac{\pi}{12}$

Exercise Set 13.4

1. $ye^{x^2} = 3e^{x^2} + C$ **3.** $ye^{-2x} = -2e^{-2x} + C$
5. $yx = \frac{1}{4}x^4 + C$ **7.** $ye^{-6x} = \frac{1}{3}e^{6x}(-6x-1) + C$
9. $yx = e^x(x-1) + C$ **11.** $ye^{1/x} = -e^{1/x} + C$
13. $ye^{-2x} = \frac{e^{-2x}[-2\sin 2x - 2\cos 2x]}{8} + C = -\frac{1}{4}e^{-2x}(\sin 2x + \cos 2x) + C$ **15.** $yx^{-1/4} = \frac{8}{3}x^{3/4} + C$
17. $ye^{-\tan x} = -e^{-\tan x} + C$ **19.** $\frac{yx}{e^4x} = \frac{4}{3}x^3 + C$
21. $yx^2 = \frac{1}{4}x^4 + 8$ **23.** $yx^{-2} = -\frac{1}{3}\cos 3x + \frac{2}{3}$
25. **(a)** $w(t) = 120 + 40e^{-0.005t}$, **(b)** about 154.4 lb, **(c)** about 58 days, **(d)** 120 lb, **(e)** this person's weight will approach 120 lb if this diet is maintained for a long time.

Exercise Set 13.5

1. $N = 100e^{-0.0192t}$, t in days **3.** 152.0031 years
5. 1387 **7.** 50,208.4 years
9. **(a)** $T = 30 - 20e^{-0.02877t}$, t in minutes; **(b)** 31.8508 min; **(c)** 26.4404°C **11.** 101.59697 min = 1 h 21 min 35.82 s
13. $I = -10e^{-5t} + 10$ **15.** $I = -0.1e^{-50t} + 0.1$

Exercise Set 13.6

1. $y = kx$

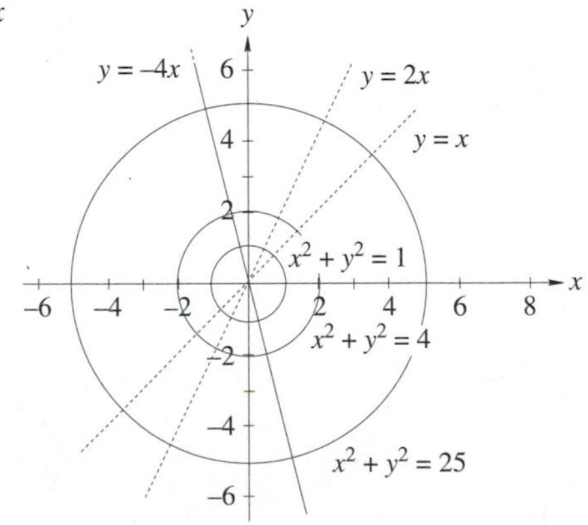

3. $y^2 + 2x = k$

5. $y^2 = kx$
7. $v(t) = 490\left(1 - e^{-t/50}\right)$ m/s
9. **(a)** $v(t) = 490 - 489e^{-t/50}$ m/s; **(b)** 89.64 m/s; **(c)** 490 m/s
11. 13.7203 lb; $Q = 25e^{-t/25}$
13. **(a)** $Q(t) = 24 - 22e^{-t/2}$; **(b)** 23.5971 lb
15. **(a)** $Q(t) = 400 - 390e^{-t/200}$, **(b)** about 15.806 kg

Review Exercises

1. $y = Ce^x - (x+1)$ **3.** $x^4 + 4y^{-1} = C$
5. $e^x - y^2 = C$ **7.** $\ln|xy| - y^2x = C$ **9.** $\frac{x-2}{y} = C$
11. $\ln|xy| = C$ **13.** $y = -\frac{1}{4} + Ce^{2x^2}$
15. $y = \frac{1}{2}(\sin x - \cos x) + Ce^x$ **17.** $\sin x + \frac{1}{2}y^2 = 2$
19. $yx^2 = \frac{1}{4}x^4 - \frac{1}{4}$ **21.** **(a)** $Q(t) = Ce^{-0.0527t}$ where $C = Q(0)$; **(b)** 13.1576 years ≈ 13 years 57 days 12 hours 48.5 min **23.** **(a)** 54.65 min; **(b)** 52.41°F
25. $q(t) = \frac{1}{89}(100\cos 2t + 160\sin 2t - 100e^{-5t/4})$
27. **(a)** $0.7 + 1.3e^{-t/20}$ ft³; **(b)** 0.7647 ft³; **(c)** 0.7032 ft³
29. **(a)** $50 - 40e^{-t/4000}4$ m³; **(b)** in about 1151 days or 3 years 56 days

Chapter 13 Test

1. Differentiating $y = 4e^{2x} + 2e^{-3x}$ we get $\frac{dy}{dx} = 8e^{2x} - 6e^{-3x}$ and $\frac{d^2y}{dx^2} = 16e^{2x} + 18e^{-3x}$. Substituting these in the given differential equation, we obtain $\left(16e^{2x} + 18e^{-3x}\right) + \left(8e^{2x} - 6e^{-3x}\right) - 6\left(4e^{2x} + 2e^{-3x}\right) = (16 + 8 - 24)e^{2x} + (18 - 6 - 12)e^{-3x} = 0$.
3. $\frac{1}{4}y^4 = x^5 + C$ **5.** $\arctan x + \frac{1}{2}(1 + y^2) = C$
7. $-\frac{y}{x} + \frac{1}{5}x^5 = C$ **9.** $y^2 - x = k$ **11.** ≈ 170.70 ft/s

ANSWERS FOR CHAPTER 14

Exercise Set 14.1

1. $y = c_1e^{2x} + c_2e^{5x}$ **3.** $y = c_1e^{2x} + c_2e^{6x}$
5. $y = c_1e^{\frac{1}{2}x} + c_2e^{-5x}$ **7.** $y = c_1e^{-\frac{1}{4}x} + c_2e^{2x}$
9. $y = c_1e^{x} + c_2e^{4x}$ **11.** $y = c_1e^{x} + c_2e^{-x}$
13. $y = c_1e^{\sqrt{7}x} + c_2e^{-\sqrt{7}x}$ **15.** $y = c_1e^{x/2} + c_2e^{-x}$
17. $y = c_1e^{x} + c_2e^{2x}c_3e^{3x}$
19. $y = c_1e^{-x} + c_2e^{-2x} + c_3e^{-3x}$ **21.** $y = 2e^{3x}$
23. $y = e^{2x} - e^{-4x}$ **25.** $y = 3e^{3x} + 2e^{-5x}$
27. $y = 5e^{-x} + 2e^{2x} - 4e^{-2x}$

Exercise Set 14.2

1. $y = (c_1 + c_2x)e^{-x}$ **3.** $y = (c_1 + c_2x)e^{2x}$
5. $y = (c_1 + c_2x)e^{x/3}$ **7.** $y = (c_1 + c_2x)e^{-3x/2}$
9. $y = e^{-2x}(c_1\cos x + c_2\sin x)$ **11.** $y = c_1 + c_2e^{-9x}$
13. $y = c_1e^{x} + e^{-x/2}(c_2\cos\frac{\sqrt{3}}{2}x + c_3\sin\frac{\sqrt{3}}{2}x)$
15. $y = c_1e^{-2x} + e^{x}(c_2\cos\sqrt{3}x + c_3\sin\sqrt{3}x)$
17. $y = c_1e^{x} + c_2e^{2x} + c_3e^{3x}$
19. $y = (c_1 + c_2x + c_3x^2 + c_4x^3)e^{-2x}$
21. $y = (c_1 + c_2x)e^{x} + (c_3 + c_4x + c_5x^2)e^{-2x}$
23. $y = (c_1 + c_2x)e^{3x} + c_3e^{(3+3\sqrt{2})x} + c_4e^{(3-3\sqrt{2})x}$
25. $y = e^{x/3}(c_1\cos\frac{\sqrt{3}}{3}x + c_2\sin\frac{\sqrt{3}}{2}x)$
27. $y = e^{x}(4\cos 2x + \frac{3}{2}\sin 2x)$ **29.** $y = (1 - x)e^{3/2x}$
31. $y = e^{-x} - 2xe^{-x}$ **33.** $y = e^{3x}(4\cos 5x - 3\sin 5x)$

Exercise Set 14.3

1. $y = (c_1 + c_2x)e^{5x} + 4/25$
3. $y = (c_1 + c_2x)e^{5x} + \frac{1}{4}e^{3x}$
5. $y = (c_1 + c_2x)e^{5x} + \frac{2}{5} + \frac{1}{16}e^{x}$
7. $y = (c_1 + c_2x)e^{5x} + \frac{21}{29}\sin 2x + \frac{20}{29}\cos 2x$
9. $y = c_1e^{2x} + c_2e^{-2x} - (\frac{1}{3}x^2 + \frac{4}{9}x + \frac{14}{27})e^{x} + \frac{3}{4}x$
11. $y = c_1\cos 3x + c_2\sin 3x + \frac{1}{4}\sin x + \frac{1}{2}\cos x$
13. $y = c_1\cos x + c_2\sin x + \frac{1}{5}e^{3x}$
15. $y = c_1e^{2x} + c_2e^{-2x} - 2x^2 - 1$
17. $y = c_1\cos x + c_2\sin x - \frac{1}{3}\cos 2x + 4$
19. $y = e^{-x/3}(c_1\cos\frac{\sqrt{2}}{3}x + c^2\sin\frac{\sqrt{2}}{3}x) + 6x^2 - 22x - 12$
21. $y = (2 + 9x)e^{x} + x^2 + 4x + 5$
23. $y = e^{2x} + \frac{1}{20}e^{-x} - \frac{3}{20}\sin 2x + \frac{1}{20}\cos 2x$
25. $y = -10e^{-x} + 5xe^{-x} + 3x^2 - 10x + 13$
27. $y = \cos x + \frac{2}{3}\cos 3x + \sin 3x$

Exercise Set 14.4

1. 4 lb/in. = 48 lb/ft **3.** 245 N/m
5. $x = \frac{-4}{\sqrt{75.2}}e^{1.6t}\sin\sqrt{75.2}t \approx -0.4613e^{1.6t}\sin 8.6718t$
7. $x = \frac{2}{\sqrt{33}}e^{4t}\sin\sqrt{33}t \approx 0.3482e^{4t}\sin 5.7446t$
9. $q(t) = e^{-30t}(\frac{1}{39}\cos 10t + \frac{9}{13}\sin 10t) - \frac{1}{26}\sin 100t - \frac{1}{39}\cos 100t$
11. **(a)** $q(t) = \frac{18}{11}\cos 50t - \frac{18}{11}\cos 60t$ **(b)** $\frac{1080}{11}\sin 60t$
13. $g(t) = e^{-\alpha t}\left(c_1\cos\sqrt{\omega_0{}^2 - \alpha^2}\,t + c_2\sin\sqrt{\omega_0{}^2 - \alpha^2}\,t\right)$
15. $q = e^{-2.1t}(c_1\cos 5.4t + c_2\sin 5.4t) + 0.375$

Review Exercises

1. $y = c_1 + c_2x$ **3.** $y = c_1e^{2x} + c_2e^{3x}$
5. $y = e^{3x}(c_1\cos 4x + c_2\sin 4x)$
7. $y = c_1e^{2x} + e^{-x}(c_1\cos\sqrt{3}x + c_2\sin\sqrt{3}x)$
9. $y = c_1e^{x} + c_2e^{-4x} - \frac{9}{4}x^2 - \frac{27}{8}x - \frac{117}{32}$
11. $y = c_1e^{-3x} + c_2e^{-4x} - \frac{13}{1394}\cos 5x + \frac{35}{1394}\sin 5x$
13. $y = (c_1 + c_2x)e^{-2x} + x - 1 + \frac{1}{16}e^{2x}$
15. $x = \frac{4}{\sqrt{759}}e^{-5t/4}\sin\frac{\sqrt{759}}{4}t$

Chapter 14 Test

1. $5m^2 + 2m - 3 = 0$ **3.** $y = c_1e^{-10x} + c_2e^{2x}$
5. $y = e^{-3x}(c_1\cos 2x + c_2\sin 2x)$
7. $y = c_1e^{-7x} + c_2e^{x} + \frac{2}{3}e^{2x}$
9. $y = c_1e^{-7x} + c_2e^{x} - 8x$ **11.** The charge is given by $q(t) = e^{-30t}(c_1\cos 20t + c_2\sin 20t) + 0.02\sin 10t + \frac{1}{75}\cos 10t$; the steady-state current is $i = 0.2\cos 10t - \frac{10}{75}\sin 10t$

ANSWERS FOR CHAPTER 15

Exercise Set 15.1

3.

x	y		Correct solution
	$h = 0,\ 1$	$h = 0.05$	$y = -\frac{1}{2}e^{x} + \frac{1}{2}e^{3x}$
0	0	0	0
0.5	–	0.5	0.5528
0.10	0.1	0.1101	0.12234
0.15	–	0.1818	0.2032
0.20	0.2405	0.2672	0.3004
0.25	–	0.3683	0.4165
0.30	0.4348	0.4878	0.5549
0.35	–	0.6285	0.7193
0.40	0.7002	0.7937	0.9141
0.45	–	0.9873	1.1446
0.50	1.0595	1.2138	1.4165
0.55	–	1.4784	1.7369
0.60	1.5422	1.7868	2.1138
0.65	–	2.1459	2.5566
0.70	2.1871	2.5636	3.0762
0.75	–	3.0488	3.6854
0.80	3.0046	3.6120	4.3988
0.85	–	4.2650	5.2337
0.90	4.1805	5.0218	6.2101
0.95	–	5.8980	7.3510
1.00	5.6807	6.9120	8.6836

5.

x	y	Correct solution
	$h = 0.05$	$y = \frac{1}{53}(55e^{7x} - 2\cos 2x - 7\sin 2x)$
0	1	1
0.05	1.35	1.4219
0.10	1.8275	2.0265
0.15	2.4770	2.8904
0.20	3.3588	4.1220
0.25	4.5538	5.8753
0.30	6.1717	8.3686
0.35	8.3600	11.9117
0.40	11.3182	16.9442
0.45	15.3154	24.0898
0.50	20.7149	34.2336

7.

x	y
0	1
0.1	1.1
0.2	1.222
0.3	1.3753
0.4	1.5735
0.5	1.8371
0.6	2.1995
0.7	2.1793
0.8	3.5078
0.9	4.8023
1.0	7.1895

9.

x	y
0	0
0.1	0.1
0.2	0.2010
0.3	0.3051
0.4	0.4147
0.5	0.5327
0.6	0.6633
0.7	0.8121
0.8	0.9887
0.9	1.2092
1.0	1.5062

11.

x	y
$\pi/2 \approx 1.57$	1
1.62	1.0270
1.67	1.0507
1.72	1.0711
1.77	1.0882
1.82	1.1022
1.87	1.1130
1.92	1.1209
1.97	1.1259
2.02	1.1281
2.07	1.1277
2.12	1.1249
2.17	1.1197
2.22	1.1123
2.27	1.1028
2.32	1.0913
2.37	1.0779

15. $T(1) \approx 82.694$, **(b)** $T(2) \approx 76.446$

Exercise Set 15.2

1. $y = \frac{1}{2}x^2 + 2x + \frac{3}{2}$, $y(2) = 7.5$ the exact value
3. $y = \frac{1}{3}x^6 + x^4 + x^2 + 1$, $y(1) = 3\frac{1}{3}$
5. $y = \frac{1}{63}x^7 + \frac{2}{15}x^5 + \frac{1}{6}x^4 + \frac{2}{3}x^3 + x^2 + x + 1$
7. $y = \frac{1}{8}x^4 + \frac{4}{3}x^3 + \frac{1}{2}x^2 + 4x + 1$
9. $y = -\frac{1}{2}\cos 2x - \frac{1}{4}\sin 2x + \frac{1}{2}x^2 + \dfrac{1-\pi}{2}x + \dfrac{4-2\pi-\pi^2}{8}$
11. $3(1-x) + \dfrac{x^2}{2!} - \dfrac{x^3}{3!} + \dfrac{x^4}{4!} - \dfrac{x^5}{5!} + x - 1 =$ $2 - 2x + \frac{3}{2}x^2 - \dfrac{x^3}{2} + \dfrac{x^4}{8} - \dfrac{x^5}{40}$. The first four terms are the same. **13.** $I(t) \approx 2\sin t + 3\cos t + 2t^2 - 2t - 3$ is the third approximation $I(0.5) \approx 0.0916$.

Exercise Set 15.3

1. $\dfrac{n}{s}$ **3.** $\dfrac{s}{s^2+a^2}$ **5.** $\dfrac{3!}{s^4}$ **7.** $\dfrac{6}{s^2+36}$
9. $\dfrac{3!}{(s-5)^4}$ **11.** $\dfrac{8s^2}{(s^2+16)^2}$ **13.** $\dfrac{5}{s^2+25} + \dfrac{s}{s^2+9}$
15. $\dfrac{30s}{(s^2+25)^2} - \dfrac{s}{s^2+16} + \dfrac{1}{s} = \dfrac{30s}{(s^2+25)^2} + \dfrac{16}{s(s^2+16)}$
17. $\dfrac{s+2}{(s+2)^2+9} - \dfrac{2}{(s-5)^2+1}$ **19.** $\dfrac{4}{s} + \dfrac{3}{s^2} + \dfrac{2}{s-1}$
21. $\dfrac{s^2+18}{s(s^2+36)}$

Exercise Set 15.4

1. $4e^{5t}$ **3.** $1 - \cos 4t$ **5.** $3\cos 3t$
7. $-2e^{-2t} + 7e^{-7t}$ **9.** $\frac{1}{5}t\sin 5t + \frac{3}{5}(\sin 5t + 5t\cos 5t)$
11. $(s^2+4s)L(f) - (s-4)$
13. $(3s^2-3s+1)L(f) - (2s+3)$
15. $(s^2+6s-1)L(f) - (s+5)$

Exercise Set 15.5

1. $\dfrac{1}{x} + \dfrac{-1}{x+1}$ **3.** $\dfrac{-4}{x} + \dfrac{2}{x-1} + \dfrac{2}{x+1}$
5. $\dfrac{-9/2}{x-3} + \dfrac{15/2}{x-5}$ **7.** $\dfrac{1}{x+1} + \dfrac{-2}{(x+1)^2}$
9. $\dfrac{1}{x^2+1} + \dfrac{x}{x^2+2}$ **11.** $1 - e^{-t}$ **13.** $-4 + 2e^t + 2e^{-t}$
15. $-\frac{9}{2}e^{3t} + \frac{15}{2}e^{5t}$ **17.** $e^{-t} - 2te^{-t}$ **19.** $\sin t + \cos\sqrt{2}t$
21. Use $L^{-1}\left[\dfrac{-2}{s} + \dfrac{-2}{s^2} + \dfrac{2}{s-1}\right] = -2 - 2t + 2e^t$
23. $L^{-1}\left[\dfrac{1/25}{s+1} + \dfrac{-s/25 + 4/25}{s^2+9}\right] =$ $\frac{1}{25}e^{-4t} - \frac{1}{25}\cos 3t + \frac{4}{25}\sin 3t$
25. $L^{-1}\left[\dfrac{-1/2}{s-1} + \dfrac{1/2}{(s-1)^2} + \dfrac{1/2s}{s^2+1}\right] =$ $-\frac{1}{2}e^t + \frac{1}{2}te^t + \frac{1}{2}\cos t$

Exercise Set 15.6

1. $y = -1 + e^t$ **3.** $y = te^{-6t} + 2e^{-6t}$
5. $y = \frac{301}{100}e^{-5t} + (\frac{1}{10}t + \frac{1}{100})e^{5t}$
7. $y = 2\cos 2t + \frac{3}{2}\sin 2t$
9. $y = \frac{1}{5}e^t - \frac{1}{5}(\cos 2t + \frac{1}{2}\sin 2t)$
11. $y = \frac{2}{27} + \frac{1}{9}t - \frac{2}{27}e^{3t} + \frac{10}{9}te^{3t}$
13. $y = e^{-3t}(\cos 2t + \frac{1}{2}\sin 2t)$
15. $y = \frac{1}{10}e^{-2t} + \frac{1}{10}e^{2t} - \frac{3}{5}\cos t$
17. $y = e^{-t}(\frac{2}{5}\cos 2t + \frac{13}{10}\sin 2t) + \frac{3}{5}e^{-2t}$
19. $y = -\frac{13}{50}e^{-3t} + \frac{5}{2}e^t - \frac{6}{25}e^{2t} + \frac{1}{5}te^{2t}$
21. $y = 2\cos t + \sin t$ **23.** $q(t) = 15e^{-10t}\sin 10t$, $i(t) = 150e^{-10t}(\cos 10t - \sin 10t)$, steady state current is 0 A
25. $q(t) = \frac{3}{25} - \frac{3}{20}e^{-10t} + \frac{3}{100}e^{-50t} i(t) = \frac{3}{2}(e^{-10t} - e^{-50t})$
27. $x \approx Ae^{(-64-2\sqrt{1022})t} + Be^{(-64+2\sqrt{1022})t}$ where $A = 1 - \dfrac{32}{\sqrt{1022}}$ and $B = 1 + \dfrac{32}{\sqrt{1022}}$ **29.** $y = \frac{1}{8}te^{-t}\sin 4t$

Review Exercises

1.

x	y, $h=0.1$	y, $h=0.05$
0.05	–	1.8
0.10	1.6	1.6299
0.15	–	1.4838
0.20	1.3184	1.3572
0.25	–	1.2467
0.30	1.1098	1.1495
0.35	–	1.0636
0.40	0.9497	0.9873
0.45	–	0.9190
0.50	0.8234	0.8578
0.55	–	0.8026
0.60	0.7217	0.7527
0.65	–	0.7074
0.70	0.6384	0.6661
0.75	–	0.6284
0.80	0.5691	0.5938
0.85	–	0.5621
0.90	0.5108	0.5329
0.95	–	0.5059
1.00	0.4612	0.4809

3.

x	y, $h=0.05$
0.05	59.9
0.10	59.8
0.15	59.6990
0.20	59.5961
0.25	59.4905
0.30	59.3812
0.35	59.2678
0.40	59.1496
0.45	59.0263
0.50	58.8975
0.55	58.7630
0.60	58.6228
0.65	58.4767
0.70	58.3249
0.75	58.1675
0.80	58.0047
0.85	57.8366
0.90	57.6635
0.95	57.4857
1.00	57.3034

5. $y=\frac{1}{8}x^4+\frac{4}{3}x^3+\frac{1}{2}x^2+4x+1$
7. $y=\frac{2}{9}(x-9)^3+\frac{2}{9}(x-9)^{3/2}+\frac{5}{9}$ **9.** $y=-4+5e^{t/2}$
11. $y=\frac{2}{3}e^{-2t}+\frac{1}{3}e^t$ **13.** $y=e^t$
15. $y=e^{-t}\cos 2t+\frac{1}{2}e^{-t}\sin 2t$
17. $y=\frac{3}{5}e^{-2t}+\frac{2}{5}e^{-t}\cos 2t+\frac{3}{10}e^{-t}\sin 2t$
19. $y=t-2t^2-\sin t$ **21.** $q(t)=\frac{10}{39201}(20\cos t+199\sin t-20e^{-10t}\cos 10t+3880.2e^{-10t}\sin 10t)$; $i(t)=\frac{10}{39201}(-20\sin t+199\cos t+39002e^{-10t}\cos 10t-4080.2e^{-10t}\sin 10t)$; steady state current $=\frac{10}{39201}(199\cos t-20\sin t)$

Chapter 15 Test

1. $\dfrac{5}{s+3}$ **3.** $\dfrac{s-5}{(s-5)^2+4^2}$
5. $L^{-1}(F)=e^{-6t}\sin\sqrt{7}t$
7. $L^{-1}(F)=0.5e^{-3t}-0.5\cos t+1.5\sin t$
9. $y=\frac{1}{6}e^{-2t}+\frac{5}{6}\cos\sqrt{2}t+\frac{1}{3\sqrt{2}}\sin\sqrt{2}t$
11. $q=L^{-1}(Q)=-49.925+49.985e^{-40t}-0.06\cos 20t-0.03\sin 20t$

ANSWERS FOR APPENDIX A

Exercise Set A.2

1. 73 **3.** 46 **5.** −149 **7.** 315 **9.** 3.6 **11.** 8.61 **13.** −23.602 **15.** −42.194 **17.** −3.34 **19.** 99.205 **21.** 200 **23.** −15 **25.** 0.25 **27.** 0.2222222 **29.** 0.0046447

Exercise Set A.3

1. 49 **3.** 625 **5.** 9.61 **7.** 339.6649 **9.** 6 **11.** 36 **13.** 3.2 **15.** 12.31 **17.** 729 **19.** 4084101 **21.** 103.823 **23.** 1.917875825 **25.** 9 **27.** 13 **29.** 2.3 **31.** 1.2693049 **33.** 0.0019531 **35.** 1.4922764 **37.** 29.090604 **39.** 0.000038 **41.** 2.7495843 **43.** 1.200937 **45.** 6.2114 **47.** 8.92 − 21 **49.** 8.437112 **51.** 2.71 − 12 **53.** 2.510438 **55.** 1.45666 − 02 **57.** 1.8658009 − 02 **59.** 2.19433226 **61.** 3.1493289 **63.** −0.8203622 **65.** −16.286738 **67.** 804.94849

Exercise Set A.4

Review Exercises

1. −42 **3.** −2025 **5.** 34.528 **7.** 3.309 **9.** 0.8 **11.** −0.0466853 **13.** 1225 **15.** −1024 **17.** 2.6045171 **19.** 5.0840476 **21.** 4.31 23 **23.** 3.47 31 **25.** 9.18×10^{41}

ANSWERS FOR APPENDIX B

1. 0.000 347 kg **3.** 79 200 W **5.** 23.85 mΩ **7.** 97.2 cm **9.** 8.35 m^2 **11.** 0.004 35 cm^3 **13.** 1.4 cm **15.** 92.6 t **17.** 162 km/h **19.** 14 kg **21.** 10.5 Ω **23.** 9.555 J **25.** 13.123 2 **27.** 52.92 **29.** 65.247 **31.** 4.258 8 **33.** 2.630 88 **35.** 97 **37.** 9.475 L **39.** 36 567 000 L

Index of Applications

This index categorizes the application problems in this book alphabetically. After each topic, the index lists the text page number, followed by the problem number in parentheses. For example, if you are looking for a problem on accounting, you would turn to page 41 and look for problem number 66.

Index

Properties of Exponents

1. $b^m b^n = b^{m+n}$
2. $(b^m)^n = b^{mn}$
3. $(ab)^n = a^n b^n$
4. $\left(\frac{a}{b}\right)^m = \frac{a^m}{b^m}, b \neq 0$
5. $\frac{b^m}{b^n} = b^{m-n}, b \neq 0$
6. $b^0 = 1, b \neq 0$
7. $b^{-n} = \frac{1}{b^n}, b \neq 0$

Properties of Radicals

1. $\sqrt[n]{ab} = \sqrt[n]{a}\sqrt[n]{b}$
2. $\sqrt[n]{\frac{a}{b}} = \frac{\sqrt[n]{a}}{\sqrt[n]{b}}$
3. $(\sqrt[n]{b})^n = b^{n/n} = b$
4. $\sqrt[n]{b} = b^{1/n}$
5. $\sqrt[m]{\sqrt[n]{b}} = \sqrt[mn]{b}$

Properties of Logarithms

1. $\log_b xy = \log_b x + \log_b y$
2. $\log_b \frac{x}{y} = \log_b x - \log_b y$
3. $\log_b x^p = p \log_b x$
4. $\log_b 1 = 0$
5. $\log_b b = 1$
6. $\log_b b^n = n$

Quadratic Formula

If $ax^2 + bx + c = 0$ and $a \neq 0$, then $x = \frac{-b \pm \sqrt{b^2 - 4ac}}{2a}$

Binomial Formula

$$(x+y)^n = x^n + \binom{n}{1}x^{n-1}y + \binom{n}{2}x^{n-2}y^2 + \cdots + \binom{n}{n-1}xy^{n-1} + y^n$$

Special Products and Factors

$a(x+y) = ax + ay$

$(x+y)(x-y) = x^2 - y^2$

$(x \pm y)^2 = x^2 \pm 2xy + y^2$

$(x \pm y)^3 = x^3 \pm 3x^2y + 3xy^2 \pm y^3$

$x^3 + y^3 = (x+y)(x^2 - xy + y^2)$

$x^3 - y^3 = (x-y)(x^2 + xy + y^2)$

Basic Operations on Complex Numbers

$(a+bj) \pm (c+dj) = (a \pm c) + (b \pm d)j$

$(a+bj)(c+dj) = (ac - bd) + (ad + bc)j$

$\frac{(a+bj)}{(c+dj)} = \frac{(ac+bd)}{c^2+d^2} + \frac{(bc-ad)}{c^2+d^2}j$

$|a+bj| = \sqrt{a^2+b^2}$

Properties of Inequalities

If a, b, and c are real numbers, and

If $a < b$, then $a + c < b + c$

If $a < b$ and $c > 0$, then $ac < bc$

If $a < b$ and $c < 0$, then $ac > bc$

If $a < b$ and $n > 0$, then $a^n < b^n$ and $\sqrt[n]{a} < \sqrt[n]{b}$

Properties of Absolute Value

If x and a are real numbers, $a > 0$, $|x| = \begin{cases} -x \text{ if } x < 0 \\ x \text{ if } x \geq 0 \end{cases}$

If $|x| < a$, then $-a < x < a$

If $|x| > a$, then $x < -a$ or $x > a$